INTRODUCTION TO DIGITAL LOGIC DESIGN

John P. Hayes
University of Michigan

ADDISON-WESLEY PUBLISHING COMPANY
Reading, Massachusetts • Menlo Park, California • New York
Don Mills, Ontario • Wokingham, England • Amsterdam • Bonn
Sydney • Singapore • Tokyo • Madrid • San Juan • Milan • Paris

Michael Payne *Executive Editor*
Jim Rigney *Senior Production Supervisor*
Sally Steele / Thompson Steele Book Production *Text Design, Art Development, Production Coordination, and Layout*
Joseph Vetere *Technical Art Consultant*
Anita M. Devine *Editorial Assistant*
Scientific Illustrators *Technical Illustration*
Mark Frigon *Cover Designer*
Peter Blaiwas *Cover Art Director*
Sarah McCracken *Prepress Services Manager*
Roy Logan *Senior Manufacturing Manager*

Library of Congress Cataloging-in-Publication Data

Hayes, John P. (John Patrick). 1944
Introduction to digital logic design / John Hayes.
p. cm.
Includes bibliographical references and index.
ISBN 0-201-15461-7
1. Logic design. 2. Logic, Symbolic and mathematical.
3. Electronic digital computers – Circuits. I. Title.
621.39'5 – dc20 93-21996
CIP

Photo Credits

Fig. 1.1a, Old Sturbridge Village; Fig. 1.1b, Blair Seitz/Photo Researchers
Fig. 1.2, Jeremy Burgess/SPL/Photo Researchers,
Fig. 1.14, photo courtesy of Omnibyte Corporation
Fig. 1.15, photo courtesy of Motorola
Fig. 1.23, photo courtesy of Hewlett Packard
Fig. 4.15, reprinted by permission of Texas Instruments
Fig. 5.43, copyright 1981 by International Business Machines Corporation, reprinted with permission.
Figs. 5.47, 5.54, 7.23, 8.4, 8.18, 8.23a, 8.25, 8.68, and Table 6.2, reprinted by permission of Texas Instruments

Reprinted with corrections May, 1994

8 9 10 -DOC- 00 99 98

PREFACE

INTRODUCTORY REMARKS

After teaching in the field of electrical engineering and computer science for twenty years, I am well aware of the challenges that confront instructors and students in an introductory logic design course. Thus, from the very beginning, I designed this text to be up-to-date, comprehensive, and pragmatic in its approach. It aims to provide balanced coverage of concepts, techniques, and practices; and to do so in a way that is, at the same time, both scholarly and supportive of the needs of the typical student reader. I wanted to give students a clear picture of fundamental concepts, effective problem-solving techniques, and appropriate exposure to modern technologies, design techniques, and applications. Through the mutual support of textual material, worked examples, end-of-chapter problems, and other pedagogical features, I have attempted to show how theoretical ideas, physical devices, and design methodologies can all come together to form a successful digital system.

Audience for Introduction to Digital Logic Design

This work is intended as a college text for a first course in digital logic design at about the sophomore or junior level. No specific prerequisites are assumed, nor is any knowledge of electrical circuits or electronics required. As a mainstream introduction to logic design, this book covers the basics of switching theory and logic design necessary to analyze and design combinational and sequential logic circuits at switch, gate, and register (or register-transfer) levels, with the main emphasis on the gate level. It can be used as the text for a one- or two-term course (semesters or quarters); it is useful for self-study.

Building Understanding from the Bottom Up

Introduction to Digital Logic Design builds student understanding from the bottom up— starting with simple binary numbers and codes, moving through the switch, gate, and register levels, and concluding with an introduction to system architecture. Each chapter progresses through more advanced levels of design and abstraction, tying together concepts previously introduced. Students will find that the circuits they learned to construct in the early chapters are used as primitive design elements in the later chapters, and that this progression greatly aids their comprehension and retention of the material.

Systematic Development of the Material

The nine chapters in this book are grouped into three parts that present digital logic design in a sequence familiar to most course instructors. Chapters 1–3 start students off with a general introduction to the digital design process, binary numbers and arithmetic, elementary logic circuits in terms of both switches and gates, and Boolean algebra. The next four chapters examine in detail the theory and techniques of designing gate-level logic circuits under the two traditional headings, combinational and sequential circuits. Here, students learn to analyze and design circuits of small and medium complexity, such as those found in SSI and MSI integrated circuits (ICs). Chapter 8 presents logic design at the higher register level, its components include storage registers, arithmetic elements, and other circuits constructed from moderate numbers of gates. The register level is appropriate to design using MSI ICs as components, as well as to the internal design of large- or very large-scale ICs. Using LSI/VLSI components, we can specify the structure and behavior of a digital computer, a topic addressed in the final chapter.

Sound Coverage of Basics

This text is designed to give a thorough grounding in core fundamentals like number systems, combinational and sequential design, and register-level logic. In addition, there are numerous opportunities to expand traditional coverage with the inclusion of important topics like wired logic, heuristic design methods, programmable logic devices (PLDs), and testability considerations. These optional sections are presented as self-contained modules so they can be easily incorporated into the rest of the chapter material, but not interfere with the organizational flow of the chapter. We have arranged the material so that students see slowly increasing levels of design detail and abstraction. The building-block philosophy allows students to gain conceptual understanding gradually. This successful teaching approach is supplemented by careful management of detail, thorough explanations, and clear, evenly-paced prose.

- **Balance of Concepts, Techniques, and Practices** Our primary goal in writing this text was to achieve a careful balance among the important issues that beginning students must face: understanding of principles, acquisition of basic problem-solving skills, application of principles and skills in analysis and design, and exposure to real-world implementation issues. We arrange many of the chapters by first developing the basic theory, and then showing how that theory is used in design. We thus try to provide a deeper, more thorough discussion of fundamental material, in addition to accessible, self-contained treatment of principles underlying the more difficult design problems.

• **Emphasis on Design Thinking** Introduction to Digital Logic Design attempts to integrate practical design issues into every chapter. Chapter 1 serves as a broad introduction to the discipline of engineering design. From then on, chapter by chapter, students learn to consider such design issues as timing and clocking, fan-in and fan-out, reliability, and cost—topics that often receive inadequate consideration in other texts. We explore issues like how number systems and codes are implemented in real-world digital systems; the design of testable sequential circuits using scannable flip-flops; and the entire design process as implemented in a modern CAD (computer–aided design) framework. Students receive exposure to key analysis and design criteria, and are made aware of the decisions and trade-offs that must be made as a working system is being created.

• **Modern Topical Coverage** Because it is important to a student's success in the workplace to be exposed to state-of-the-art topics in digital logic design, we have tried to capture the flavor of recent technological developments. In addition to the fundamental logic design topics, we address, in an integrated manner, application-specific logic structures like PLDs; the impact and requirements of VLSI technology; testing issues and design for testability; and the theory needed to understand such important implementation technologies as tri-state logic and CMOS circuits. We have made a point to consistently integrate computer-aided design throughout the text, so that students will be aware of the role and value of CAD in the design process. We focus on what CAD can and cannot do, rather than on the specifics of any single CAD system.

• **Technology-Independent Approach** This is perhaps the first text to use a technology-independent approach known as switch-level modeling to explain the structure and behavior of elementary logic circuits. This innovative technique treats transistors, pull-up resistors, and the like, as purely logical devices from which abstract logical models can be easily derived and explained. With *Introduction to Digital Logic Design,* students who have no electronics background will be able to understand how logic gates are designed and operate. Thus, students gain exposure to sophisticated design ideas—how tri-state circuits and PLDs work, for example—within the rigorous and complete framework provided by the switch-level approach.

HIGHLIGHTS OF TOPICAL COVERAGE

• **Unique Introductory Chapter** In the first chapter, we introduce the discipline of engineering design, which then serves as a foundation for the later, detailed coverage of design theory and practice. This chapter examines the nature and role of digital systems; provides an early orientation to key design issues; and acquaints students with basic concepts such as logical modeling, computer-aided design methodology, and design trade-offs like cost, reliability, and testability. The aim of Chapter 1 is to motivate students by giving them an initial taste of the design process and a broad view of the role of digital systems in the real world.

• **Implementation of Number Systems and Codes** Chapter 2 presents a thorough treatment of binary numbers and codes and their role in logic design. We attempt to enliven this sometimes dull subject with examples ranging from secret writing to the familiar universal product (bar) code. We provide many illustrations that show how abstract number ideas are put to use in actual systems. Our coverage of this standard topic exemplifies the blend of practical implementation and theoretical concepts found throughout this book. This chapter also contains a section on floating-point arithmetic that goes beyond the fixed-point numbers found in most introductory texts.

• **Early and Integrated Treatment of ICs and PLDs** We introduce integrated circuits, from SSI to VLSI in Chapter 3, and programmable logic devices (PLDs) in Chapter 4. The relatively early appearance of ICs and PLDs in the text should facilitate introduction of these devices in the digital laboratory experiments that often accompany a logic design course. Furthermore, the discussion of ICs and PLDs teaches students to analyze the trade-offs associated with alternative implementation technologies.

• **Presentation of Fundamental Theory** Recent texts have tended to gloss over basic principles, perhaps in reaction to past overemphasis on theory. As a result, students often do not acquire the basic elements of Boolean algebra, minimization theory, or the timing models needed to understand logic circuits or the CAD tools used in their design. We address this problem by providing coverage of basic concepts that is comprehensive, clearly related to current design practice, and accessible to the average student.

• **Coverage of Testing and Testability** The issue of testing circuits and circuit systems for physical faults occurring during or after manufacture has become a critical (and often expensive) issue in industry. Testing problems can no longer be an afterthought in the design process. In the last three sections of Chapter 7, we study the basic techniques for modeling faults and deriving tests at the gate level. We also present some practical design methods such as scan design to enhance a circuit's testability.

• **Systematic Approach to Register-Level Design** Using a well-defined set of building blocks of the kind found in typical hardware or software design libraries—most of which the student has seen designed in earlier chapters—Chapter 8 attempts to place register-level logic design on a solid footing. It exposes students to systems that are partitioned data processing (datapath) and control units, and introduces the design of programmable control units. We also show how a relatively complex sequential system can be precisely specified and implemented using formal descriptive methods and a CAD-like framework based on the well-regarded algorithmic-state-machine (ASM) approach.

• **Coverage of the Architecture Level** Chapter 9 aims to complete the student's perspective of digital systems by providing an overview of the architecture of computers and computer-based microcontrollers. Although optional, this chapter covers the basics of computer organization and microprocessor interfacing in a self-contained manner. It can also serve as a useful bridge to more advanced courses in the computer systems curriculum.

FEATURES THAT SUPPORT TEACHING AND ENHANCE LEARNING

• **Modular and Accessible Format** Recognizing that the topics covered in digital logic design courses vary widely, the book has been designed to allow course instructors to follow many paths through it. Each chapter consists of three or four major parts, which in turn consist of three or four numbered sections. These sections are designed to be relatively independent and self-contained so that, in the case of the less central topics, they can be skipped, if desired. We also make liberal use of subheadings within the numbered sections, both to modularize the material and to make it easy for the student to find a particular topic quickly.

• **Clear and Consistent Interior Design** We employ an open, two-color interior design that helps students to organize and absorb information efficiently. We have carefully aligned illustrations and text in our layout to motivate the reader and to provide a consistent format from chapter to chapter. Color is used where it seems pedagogically effective: to highlight key text elements and to identify important information within the illustrations.

• **Over 600 Two-Color Illustrations** We have included over 600 high-quality illustrations and corresponding captions to depict the key conceptual ideas and systems that logic design students need to know. Again, we have not employed color gratuitously; within the figures, for instance, color is used to label input/output signals, and to highlight other crucial pieces of information.

• **Clearly-Stated Objectives** To orient the student to the subject at hand, each chapter of *Introduction to Digital Logic Design* opens with an abbreviated table of contents and a short, introductory paragraph that summarizes the goals and contents of that chapter. These objectives aim to provide a sense of context for students and to help them tie together information presented in previous chapters with the new concepts to be presented. Shorter (re)statements of objectives appear throughout the book.

• **Comprehensive, Worked Examples** We present some 80 (numbered) examples to illustrate key design techniques and applications. They serve as small, self-contained case studies that summarize and integrate key principles developed in the preceding text. These and other (unnumbered) examples sprinkled throughout the text demonstrate a variety of useful analysis and design procedures, and also encourage effective problem-solving behavior on the student reader's part.

• **Problem-Solving Procedures** To help students solve problems systematically, we present clearly marked, step-by-step procedures to highlight and summarize important design and analysis techniques. These procedures are written in ordinary language to maximize comprehension.

• **End-of-Chapter Problems** We have provided nearly 600 homework problems ranging from simple drill exercises to subtle, thought-provoking problems to enable teachers to better support the needs of a diverse student group. These class-tested problem sets balance simple and difficult problems, as well as design and analysis issues.

• **Complete Instructor's Manual** A solutions manual will be available to all adopters of this text for course use. This comprehensive supplement contains complete, worked-out solutions to every end-of-chapter problem. For convenience, we have also included the problem statements in this manual.

• **Summary Margin Notes** Margin notes, one or two per page, summarize key facts and terminology. They also serve as a concise and easily accessible running summary of the text.

• **Further Readings Sections** A short, annotated set of references, which mainly comprise textbooks or other reference materials suitable for use by undergraduates, appears at the conclusion of each chapter to encourage further study.

• **Bulleted End-of-Chapter Summaries** Each chapter ends with a detailed, bulleted summary of key concepts and procedures found in that chapter.

• **Quality Control** To make this text as useful as possible, it has been thoroughly tested in its intended market. In addition to testing various drafts in the classroom at several universities, we obtained over 100 reviews of the manuscript, illustrations, homework problems, and table of contents from approximately forty-five individual reviewers. The end-of-chapter problems have been class-tested for accuracy and effectiveness.

COURSE OPTIONS

This book has ample material for a one- or two-term course on digital logic design. The material has been organized in a modular fashion to allow course instructors considerable flexibility in the topics they choose to cover, as well as in depth of coverage. In a typical 15-week course, for example, the following sequence of topics may be followed:

Week 1: Introduction to digital design; Binary numbers (Chapter 1, Sections 2.1–2.4)

Week 2: Binary arithmetic and codes; Switch-level concepts (Sections 2.5–2.9 and 3.1–3.3)

Week 3: Gate-level concepts; Basic gates and circuits; Boolean algebra and its uses (Sections 3.4–3.5 and 3.6–3.8)

Week 4: Combinational design; PLDs; Two-level minimization (Sections 4.1–4.2 and 4.5–4.7)

Week 5: Combinational design: The K-Map method (Sections 5.1–5.4)

Week 6: Combinational design: The tabular method (Sections 5.5–5.6)

Week 7: Combinational design: Multiplexers, decoders, adders, subtracters (Sections 5.8–5.9)

Week 8: Sequential circuit concepts: Delays and latches; State behavior (Sections 6.1–6.4)

Week 9: Sequential design: Triggering methods; Flip-flops and their uses (Sections 6.5 and 6.7)

Week 10: Sequential design: Circuit analysis; Clocking issues; (Sections 7.1–7.3)

Week 11: Sequential design: Synthesis methods; Design examples; Testing (Sections 7.4–7.7)

Week 12: Register-level design: Characteristics and design components (Sections 8.1–8.3)

Week 13: Register-level design: Datapath and control unit design methods (Sections 8.4–8.8)

Week 14: Architecture-level design: Basic computer organization (Sections 9.1–9.3)

Week 15: Computer organization (continued); Microprocessor-based systems (Sections 9.6–9.7)

In a two-term course sequence, the treatment of various topics listed above can be expanded, and extra topics can be added such as: floating-point numbers (Section 2.12), wired logic (Section 4.3), ROMs and PLAs (Section 4.4), heuristic design methods (Section 5.7), asynchronous circuits (Sections 6.8–6.10), and test generation and design for testability (Sections 7.8–7.9). In courses that emphasize computer design, there is an extensive discussion of CPU architecture and assembly-language programming concepts (Sections 9.4–9.5).

ACKNOWLEDGMENTS

I would like to express my deep appreciation to Harold Stone for his incisive comments concerning early versions of the manuscript. I am grateful to Michael Driscoll for his diligent and thoughtful technical review of later versions of the manuscript, and for his participation in the refinement of the chapter summaries. Special thanks are due to Keith Hoover for his thorough technical reviews, and for generously sharing his experience and insights as a teacher of logic design.

I also wish to thank the many individuals who, as chapter reviewers or otherwise, significantly contributed to the quality of *Introduction to Digital Logic Design*. They are as follows: Hosame Abu-Amara, Texas A&M University; Pradip Bose, IBM T. J. Watson Research Center; David W. Bray, Clarkson University; Swapan Chakrabarti, University of Kansas; Ranjan Chakravarty, University of Texas–Arlington; Yi Cheng, California Polytechnic State University Pomona; Steve Class, University of South Florida; Chita R. Das, Pennsylvania State University; Fred Dautermann, Texas Tech University; Edward S. Davidson, University of Michigan; Michael Driscoll, Portland State University; William Freedman, Drexel University; Stephanie Goldberg, Buffalo State College; Dong S. Ha, Virginia Polytechnic Institute and State University; James O. Hamblen, Georgia Institute of Technology; Lee Hollaar, University of Utah; Keith Hoover, Rose Hulman Institute of Technology; Walid Hubbi, New Jersey Institute of Technology; Barry W. Johnson, University of Virginia; Larry Jones, Motorola; Israel Koren, University of Massachusetts–Amherst; James T. Leffew, University of South Florida; Hobart McWilliams, Montana State University; Tom Miller, North Carolina State University; Zainalabedin Navabi, University of Tehran; Robert T. Nethken, Louisiana State University; Donald P. Orofino, Worcester Polytechnic Institute; A. Yavuz Oruc, University of Maryland; Christos Papachristou, Case Western Reserve University; Andrew R. Pleszkun, University of Colorado; Roland Priemer, University of Illinois–Chicago; V.V. Bapeswara Rao, North Dakota State University; Diane T. Rover, Michigan State University; David Schimmel, Georgia Institute of Technology; Martha Sloan, Michigan Technological University; Terry Smay, Iowa State University; Paul D. Stigall, University of Missouri–Rolla; Harold Stone, IBM; James Svoboda, Clarkson University; H.C. Torng, Cornell University; Anh Tran, University of Massachusetts–Lowell; Meghanad D. Wagh, Lehigh University; Karan Watson, Texas A&M University; Hee Yong Youn, University of Texas–Arlington; and Charles Zukowski, Columbia University.

Finally, this work could not have been completed without the enthusiastic support—from proof reading to providing midnight snacks—of my wife Terrie, to whom the book is lovingly dedicated.

John P. Hayes
Ann Arbor, Michigan

CONTENTS

1 INTRODUCTION

Why Digital? 1

1.1 Digital vs. Analog Systems 1
1.2 Binary Systems 6

Digital Systems 13

1.3 Structure and Behavior 13
1.4 Design Levels 17
1.5 Combinational vs. Sequential Systems 24
1.6 Design Goals 30
1.7 Computer-Aided Design 36

Chapter 1 Summary 43
Further Readings 44
Chapter 1 Problems 45

2 BINARY NUMBERS AND CODES

Number Systems 51

2.1 Positional Numbers 51
2.2 Number Base and Word Size 53
2.3 Binary-to-Decimal Conversion 57
2.4 Decimal-to-Binary Conversion 61

Binary Arithmetic 66

2.5 Unsigned Numbers 67
2.6 Signed Numbers 77

Binary Codes 88

2.7 A Binary Potpourri 88
2.8 Base-2^k Numbers 94
2.9 Binary-Coded Decimal Form 96

2.10 Character Codes 101
2.11 Error-Detecting Codes 104
2.12 Floating-Point Numbers 107

Chapter 2 Summary 114
Further Readings 116
Chapter 2 Problems 116

3 LOGIC ELEMENTS

The Switch Level 125

3.1 Switches 125
3.2 Switching Circuits 131
3.3 Complementary Circuits 136

The Gate Level 142

3.4 Logic Gates 142
3.5 Gate-Level Circuits 152

Boolean Algebra 160

3.6 A Definition of Boolean Algebra 160
3.7 Fundamental Properties 171
3.8 Expressions, Functions, and Circuits 177

Chapter 3 Summary 186
Further Reading 187
Chapter 3 Problems 187

4 COMBINATIONAL LOGIC

Circuit Structure 199

4.1 Basic Structures 199
4.2 Complete Gate Sets 213

Technology Considerations 219

4.3 Wired Logic 219
4.4 MOS Circuits 232
4.5 IC Implementation 238

Programmable Logic Devices 246

4.6 Programmable Circuits 247

4.7 ROMS and PLAs 257

Chapter 4 Summary 267
Further Readings 269
Chapter 4 Problems 269

5 COMBINATIONAL DESIGN

Two-Level Design 279

5.1 Minimal Forms 279
5.2 The K-Map Method 289
5.3 Petrick's Method 300
5.4 Larger Problems 304

Computer-Aided Design 316

5.5 The Tabular (Quine-McCluskey) Method 316
5.6 Multiple-Output Functions 330
5.7 Heuristic Methods 342

Useful Circuits 353

5.8 Data Transfer Logic 353
5.9 Adders and Subtracters 366

Chapter 5 Summary 375
Further Readings 377
Chapter 5 Problems 377

6 SEQUENTIAL LOGIC

Delays and Latches 391

6.1 Signal Storage 391
6.2 Propagation Delays 397
6.3 Latches 405
6.4 Timing and State Behavior 415

Clocks and Flip-Flops 424

6.5 Basic Concepts 425
6.6 Flip-Flop Design 434
6.7 Flip-Flop Behavior 443

Asynchronous Circuits 453
6.8 Hazards 453
6.9 The Huffman Model 457
6.10 Some Design Issues 463

Chapter 6 Summary 478
Further Readings 480
Chapter 6 Problems 480

7 SEQUENTIAL DESIGN

Sequential Circuit Analysis 491
7.1 Sequential vs. Combinational Logic 491
7.2 State Behavior 498
7.3 Timing Control and Clocks 511

Sequential Circuit Synthesis 519
7.4 The Design Process 519
7.5 The Classical Method 528
7.6 State Minimization 549

Design and Testing 556
7.7 Faults and Tests 556
7.8 Test Generation 564
7.9 Design for Testability 576

Chapter 7 Summary 586
Further Readings 588
Chapter 7 Problems 588

8 REGISTER-LEVEL DESIGN

The Register Level 599
8.1 General Characteristics 599
8.2 Combinational Components 614
8.3 Sequential Components 626

Datapath and Control Design 636
8.4 Datapath Units 636
8.5 Control Units 652
8.6 Programmable Controllers 668

ASM Design Methodology 674

8.7 Algorithmic State Machines 674
8.8 Integrating Data and Control 681

Chapter 8 Summary 699
Further Readings 701
Chapter 8 Problems 702

9 SYSTEM ARCHITECTURE

The Architecture Level 715

9.1 Basic Architecture 715
9.2 CPU and Memory 721
9.3 Input/Output Operations 732

The Central Processing Unit 740

9.4 CPU Operation 740
9.5 Instruction Sets and Programming 754

Computer-Based Systems 776

9.6 Microcontrollers 776
9.7 Input/Output Interfacing 782

Chapter 9 Summary 789
Further Readings 90
Chapter 9 Problems 791

BIBLIOGRAPHY 799

INDEX 803

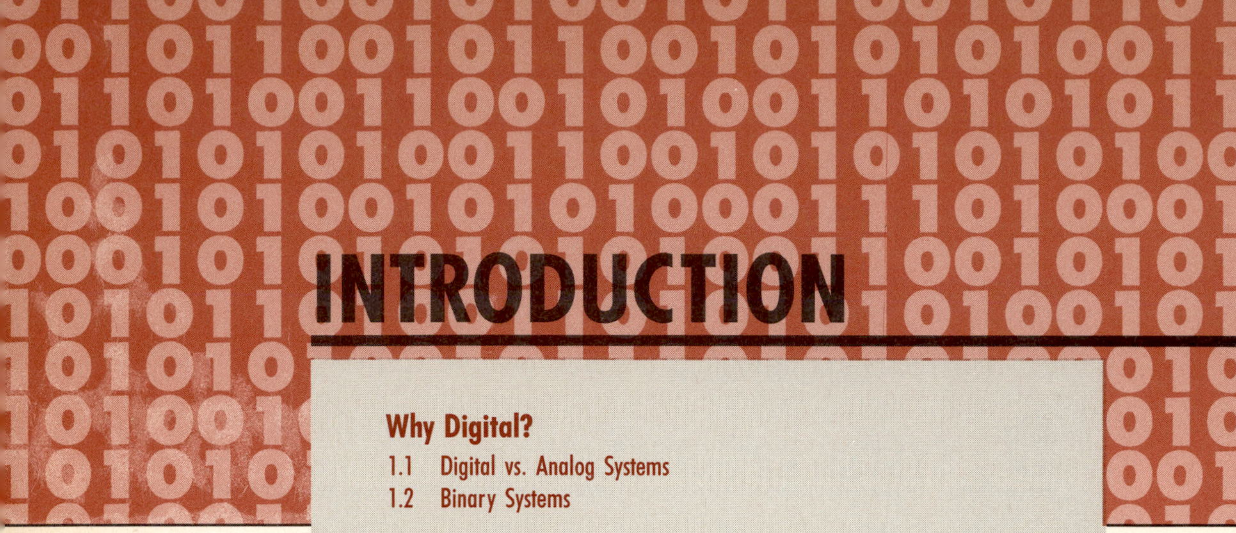

INTRODUCTION

Why Digital?

1.1 Digital vs. Analog Systems
1.2 Binary Systems

Digital Systems

1.3 Structure and Behavior
1.4 Design Levels
1.5 Combinational vs. Sequential Systems

The Design Process

1.6 Design Goals
1.7 Computer-Aided Design

Overview

Digital systems, from wristwatches to personal computers, have a pervasive influence on modern society. We begin by addressing the question, What are these systems and why are they used? We define analog and digital systems, including binary or two-valued systems, and discuss their role in information processing. As we shall see, digital design is a complex multilevel process in which computer-based tools play a key role.

1

Why Digital?

It is common to say that we live in an information-processing age dominated by computers and computer-based machines. Electronic machines assist our intellectual activities in ways that can be compared to mechanical machines such as the steam engine, introduced by the nineteenth-century industrial revolution to aid humankind's physical endeavors. The computer is a product of twentieth-century electronics technology, and its present form has been most heavily shaped by the development of integrated circuits during the last few decades. These tiny electronic devices serve as the building blocks not only for computers, but for a wide range of information-processing machines we call **digital systems**, which are designed to store, transform, and communicate information in digital form.

1.1 Digital vs. Analog Systems

We have two basic ways of representing information, which we term *analog* and *digital*. They are distinguished by the nature of the values (constants) that they allow information variables to assume.

Basic Ideas

Analog means continuous; digital means separated by gaps.

The values that an **analog** variable can take are not separated by gaps; instead, they form a smooth and seamless continuum. Thus, between any two values A and B on an analog measuring scale such as a ruler, we can find an intermediate value such as $C = (A + B)/2$. Between A and C lies $D = (A + C)/2$, between A and D lies $E = (A + D)/2$, and so on indefinitely. In fact, between *any* two points on the ruler there is an infinite number of other points, each defining a distance that represents, or serves as an "analog" for, some physical distance to be measured.

Digital information, on the other hand, is represented by a fixed number of noncontinuous or **discrete** symbols, called **digits**. The 10 decimal digits {0, 1, 2, 3, 4, 5, 6, 7, 8, 9} used to construct numbers are familiar examples. The word *digit* comes from the Latin *digitus*, meaning finger. The fact that we humans have 10 fingers probably explains our preference for decimal numbers. Computers, as we will see, prefer binary numbers that employ just two digit types, 0 and 1.[1]

A digital quantity such as a number is formed by a finite sequence of digits. This finiteness in both the digit types available and the sequence length implies that we cannot represent all the possible points on an analog scale such as a ruler by means of digital numbers. For example, three decimal digits allow us to represent 1.99 and 2.00, but nothing in between these two numbers. Thus, between these numbers on a digital scale there is a gap we cannot fill with only three digits. In fact, unless we use an infinite number of digits—a physical impossibility—we cannot represent all numbers that lie between any two points on a ruler or, indeed, any analog scale.

Physical quantities are mostly analog in nature.

We humans tend to prefer digital rather than analog quantities in our information-processing activities. Human languages, both written and spoken, are essentially digital, being composed of discrete elements (letters, words, and phonemes), which are combined into sequences of finite length (words, sentences, and books). Natural quantities such as time, distance, temperature, and weight tend to be analog—at least when viewed on a human scale. At the atomic level of quantum mechanics, these quantities assume a digital appearance again, as suggested by the word *quantum*, meaning a discrete amount.

Clocks

Information can be represented in either digital or analog form.

The contrast between analog and digital is illustrated by the familiar example of a clock, which is, of course, a time-measuring instrument. The traditional analog clock (Figure 1.1a) represents time on a continuous, and therefore analog, circular scale S. Distance along this scale is the analog of time duration. The exact time is specified by the point P on S to which the little hand (the hour hand) points. If S is marked only in hour units, an accurate time reading, say to the nearest minute, via the hour hand alone would require estimating the unmarked distance from P to the nearest hour mark on S. This error-prone estimation process is characteristic of analog devices and puts limits on the accuracy with which analog information can be converted to digital form.

The accuracy of the clock is improved by adding a minute hand and a minute scale, with one revolution corresponding to one hour instead of the 12 hours represented by a revolution of the hour hand. This effectively increases the length of the time scale by a factor of 12. Increasing the accuracy of an analog clock in this way is awkward and, beyond a certain limit, impractical.

[1]It has been observed that we might also use binary numbers if our hands had only two digits, for example, if they were "all thumbs."

(a)

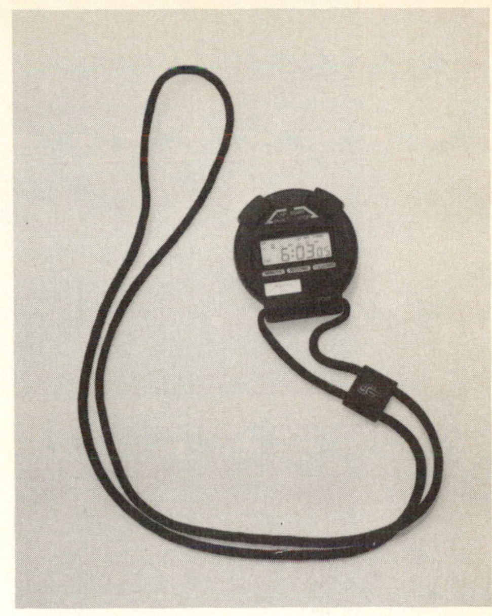

(b)

FIGURE 1.1 (a) A traditional analog clock and (b) a high-precision digital stopwatch.

By contrast, Figure 1.1b shows a digital clock face, which indicates hours and minutes directly in digital form. A digital clock's accuracy is relatively easy to increase simply by adding more digits to the output display.

Audio Recordings

Another common item existing in both analog and digital forms is the audio recording. The venerable long-playing record (LP) stores sounds by means of tiny, wavy grooves pressed into a circular track on a plastic-coated disk. The groove patterns are mechanical (spatial) analogs of the sound patterns they record (see Figure 1.2a). Both the groove and sound patterns are continuous quantities, so an LP is an analog storage device. A compact disk (CD), on the other hand, records sound in digital form. The original analog sound is sampled at discrete intervals, and the information content of each sample is converted into a fixed-length sequence of pits and lands (flat areas) on a plastic surface. These pit-land sequences are then pressed into a circular track in the CD, as depicted in Figure 1.2b. The digital data stored in the CD are read by a laser beam and converted into analog form for playback.

Digital formats are more accurate and reliable than analog.

As in the digital clock case, an advantage of the CD format is that accuracy—here, the fidelity with which the sound is recorded—can be made higher than that of its analog counterpart by using a sufficient number of digits for each sound sample. Consequently, CDs provide much better sound quality than LPs. A second advantage is that digital storage format of a CD is

(a) 

(b) 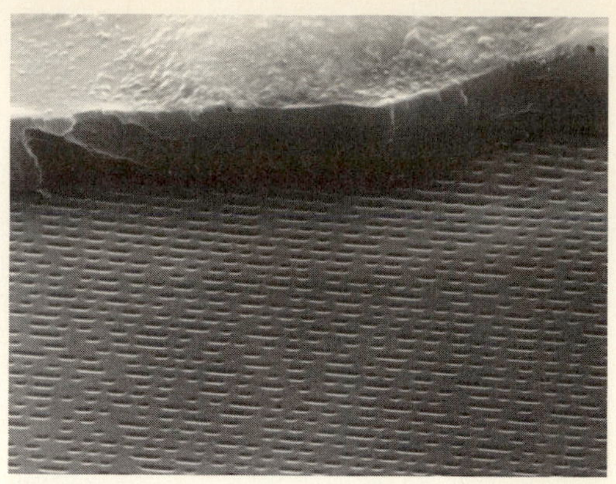

FIGURE 1.2 (a) Photomicrograph of a diamond stylus traveling through the grooves of a stereo LP record. (b) Photomicrograph of a compact disk cracked to show the musical layer.

easily processed by the computer-like circuits used to control the operation of a CD player. This allows a variety of value-enhancing functions to be added to a CD player that cannot be incorporated into an analog record player. An example is error correction to minimize the effects of scratches, dust particles, and the like. The data on a CD are encoded in such a way that by examining each digital sound sample before it is played back, certain errors—such as a land that is misread as a pit, or vice versa—due to minor flaws on the CD can be automatically identified and corrected. Similar flaws on an LP cannot be corrected; so, unlike a CD, an LP's sound quality deteriorates inexorably the more it is played.

Analog–Digital Conversion

In an analog world, digital systems need to convert data between digital and analog forms.

We see from the foregoing examples that even though an analog device can indicate an indefinitely large number of different values, converting these values to digital form is difficult, especially when a high level of accuracy is desired. When such accuracy is essential, it is often best to work exclusively with digital information.

Moreover, the power of a digital computer can be added at relatively low cost to enhance the usefulness and convenience of a digital system. In this way, the continuing improvements in the flexibility and "intelligence" of such mundane objects as television sets and cars can be largely attributed to their incorporation of digital information formats and digital control systems (including microcomputers). Some of these systems process both analog and digital information and are best classified as **mixed** or **hybrid systems**. However, the majority of present-day information-processing systems operate digitally, and rely on analog-to-digital and digital-to-analog conversion mechanisms when it is necessary to interact with an analog environment.

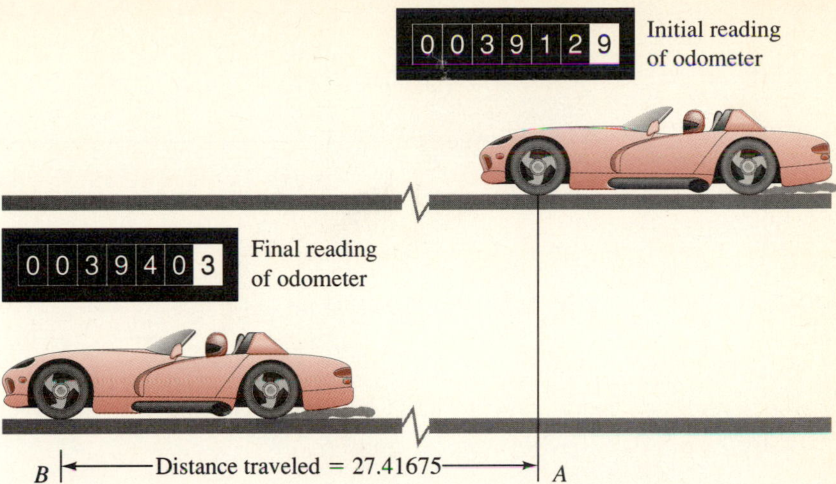

FIGURE 1.3 Measurement of analog distance by a car's digital odometer.

A familiar example of a device that converts information from analog to digital form is a car's odometer (see Figure 1.3). As the car moves forward, the small digit wheels forming the odometer rotate in (mostly) discrete spurts to indicate the distance travelled in kilometers or miles. The mechanism linking these wheels to the car's transmission system effectively converts the analog distance travelled into a digital number that is displayed to the driver. The accuracy of the distance measured by the odometer is determined by its rightmost or least significant digit, which is typically a tenth of the unit of distance being employed. In the situation depicted in the figure, the actual distance travelled is greater by 0.01675 (an insignificant amount) than the distance $3940.3 - 3912.9 = 27.4$ implied by the odometer readings.

Quantization

Many of the physical signals used to convey digital information are naturally analog rather than digital. For example, an electrical voltage v, which is the primary information-carrying medium in electronic circuits, is an analog quantity. Digital information is represented by a discrete subset $V_1, V_2, \ldots V_k$ of the analog values that v can assume. Each analog value of v is assigned to the discrete V_i that it is closest to, a process called **quantization** or **digitization**. Clearly, this process is inexact, so that quantization involves some unavoidable loss of accuracy.

Conversion between analog and digital forms is an inexact process.

Figure 1.4b shows how the continuously varying voltage of Figure 1.4a is quantized into four evenly spaced digital values: V_1, V_2, V_3, V_4. In representing these values, it is customary to draw vertical lines connecting adjacent digital values, as shown in Figure 1.4c; each of these lines represents an instantaneous change between two signal levels. The wavy diagrams like those of Figure 1.4 showing how digital or analog signals vary over time are called **timing diagrams** or **waveforms**.

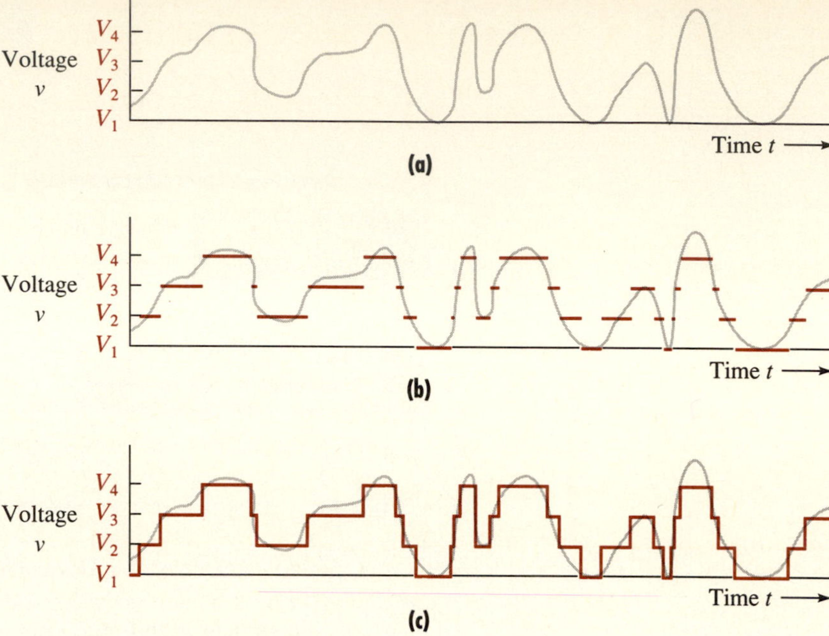

FIGURE 1.4 (**a**) Waveform of an analog voltage v. (**b**) Digital version of v after quantization into four levels. (**c**) Digital version with adjacent levels connected.

1.2 Binary Systems

For a number of reasons discussed below, it is desirable to restrict digital signals to only two distinct values or states. We call two-valued digital signals and the digital systems that process them **binary systems**. The storage format used by a compact disk (Figure 1.2b), for example, is inherently binary; the pits and lands are the two possible digital states. If we change the number of quantization levels k used in Figure 1.4 from four to two, we get the binary voltage waveform that appears in Figure 1.5. In this figure, only two digital voltages are recognized: a high voltage V_H and a low voltage V_L. The analog voltage waveforms used in practice to represent a binary information signal closely match the rectangular waveform shown by the solid lines in Figure 1.5. Another inherently binary quantity is the state of a switch, which can be either on or off. As we will see in later chapters, electronic switches play a key role in the processing of digital information.

Bits

Binary systems employ two binary digits (bits), denoted 0 and 1.

Because binary quantities are encountered in many different physical forms, it is convenient to have a common or abstract way of representing binary states. The standard approach is to use the digit symbols 0 and 1 to represent

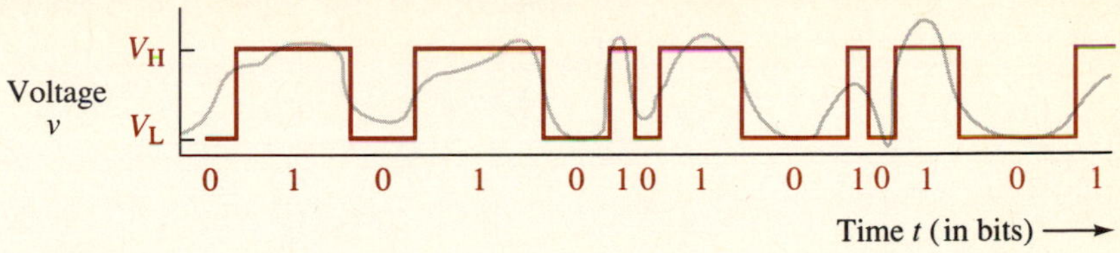

FIGURE 1.5 Binary waveform corresponding to Figure 1.4 and its bit (binary digit) representation.

the two possible values of a binary quantity at any point in time. These symbols are referred to as **bits**, a contraction of the term *binary digits*. If we use 0 and 1 to denote a pit and a land, respectively, then a CD sound sample from the CD in Figure 1.2b might be represented by the bit sequence

$$\ldots 100101110010101 \ldots \tag{1.1}$$

We could equally well use 1 to denote a land and 0 to denote a pit; this is simply a matter of convention. In a similar fashion, the binary voltage values V_L and V_H used in Figure 1.5 are conventionally represented by 0 and 1, respectively.

Binary Numbers

Binary data such as those in Equation (1.1) can be represented as a number within a binary number system. While the decimal number system ordinarily uses 10 distinct digits {0,1,2,3,4,5,6,7,8,9}, the binary system has only two, namely {0,1}. In a multidigit decimal integer such as 6709, each digit represents a different power of 10. Reading 6709 from right to left, we find 9 ones (one can be represented as 10^0), 0 tens (10^1), 7 hundreds (10^2), and 6 thousands (10^3). Hence, we can write

$$6709 = 6 \times 1000 + 7 \times 100 + 0 \times 10 + 9 \times 1$$

With binary numbers, we count by twos instead of tens.

In a multidigit binary number, on the other hand, each digit represents a power of two, so that we count with powers of two instead of powers of 10. For example, the 4-bit binary integer 1011 reads from right to left as 1 one (2^0), 1 two (2^1), 0 fours (2^2), and 1 eight (2^3). Consequently, we can express this binary number as

$$1011 = 1 \times 8 + 0 \times 4 + 1 \times 2 + 1 \times 1$$

which is, of course, eleven. Table 1.1 lists the first 16 integers, from 0 to 15, in both binary and decimal form. Observe that with four bits, only 16 numbers can be represented. In general, n bits allow us to specify up to 2^n distinct binary numbers.

TABLE 1.1 The first 16 integers in the binary and decimal number systems.

Number	*Binary*	*Decimal*	*Number*	*Binary*	*Decimal*
zero	0000	00	eight	1000	08
one	0001	01	nine	1001	09
two	0010	02	ten	1010	10
three	0011	03	eleven	1011	11
four	0100	04	twelve	1100	12
five	0101	05	thirteen	1101	13
six	0110	06	fourteen	1110	14
seven	0111	07	fifteen	1111	15

As the CD example of Figure 1.2b indicates, nonnumerical as well as numerical data can be encoded in binary form. Once this has been done, it becomes possible to treat the binary data in question as though they were numbers. In other words, numerical computations can be performed on the data to transform them in many different and useful ways. The data stored in a CD, for example, can be processed to improve sound quality during playback. This ability to process binary (or any digital) data using precisely defined numerical operations is a fundamental characteristic of digital systems and is a key to much of their power.

Why Binary?

From a human viewpoint, binary systems have several disadvantages. We are more familiar with decimal numbers, and therefore find them easier to use. As can be seen from Table 1.1, binary numbers tend to be longer (by a factor of $\log_2 10 \approx 3.322$) than their decimal counterparts. Thus, to human eyes, they are excessively long strings of digits with little variety in their symbols. However, binary systems have three significant advantages in the design of digital equipment.

1. Most information-processing systems are constructed from switches, which are binary devices.
2. The basic decision-making processes required of digital systems are binary.
3. Binary signals are more reliable than those formed by more than two quantization levels.

We now briefly explore each of these issues.

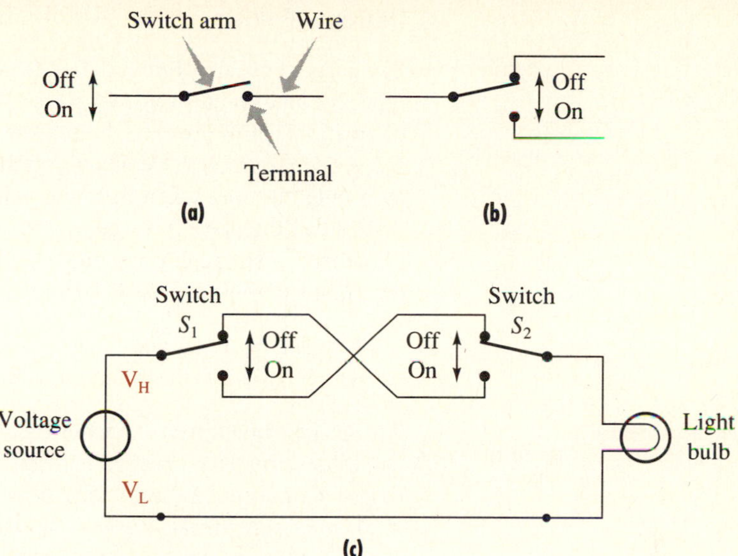

FIGURE 1.6 (a) A plain switch; (b) a transfer switch; (c) a switching circuit to control an electric light.

Switches

Switches are binary devices with two states: on (1) and off (0).

On-off switches are the basic building blocks of digital systems. They are inherently binary in that they have two natural states: on (closed) and off (open). Figures 1.6a and 1.6b show, in simplified form, two common types of manually operated switches used in household electric circuits. A digital system designed to control an electric light from two different places—for example, the top and bottom of a staircase—is depicted in Figure 1.6c. To turn on the light, exactly one of the switches S_1 or S_2 must be in the *on* state (turned down). So, there are two ways to turn on the light: in the first, S_1 is on and S_2 is off; in the second, S_1 is off and S_2 is on. In either situation, a closed electrical circuit is created, connecting the terminals or connection points of the voltage source to the terminals of the light bulb. The figure illustrates the case in which both switches are in the off state; hence, the light is turned off.

Besides the two switches, several of the quantities that appear in Figure 1.6 are binary. The light bulb is a binary device with two states: dark (off) and bright (on). When the light is turned off, the voltage measured at its upper terminal is V_L. When it is turned on, this voltage changes to V_H, due to the creation of an electrical path from the bulb to the V_H terminal of the voltage source. Hence, the switches effectively control the light bulb by transferring binary voltages to it. In this manner, the selective control of binary voltages and currents by (electronic) switches is the fundamental operating mode of electronic digital systems.

Decision Making

Logical decisions have the binary outcomes true (1) and false (0).

Digital systems are often required to perform tests, or make decisions, of this type: Is condition C_1 true? Or we might ask, Is condition C_2 false? Examples of such decisions are: Has button (switch) X been pushed? Has temperature t_{MAX} been reached? Decisions of this kind are inherently binary because their outcomes are taken from the value-pair {true, false}. The operation of a system can then be expressed by a set of statements of this form: If conditions C are true, then perform actions A. For instance, the function performed by the circuit of Figure 1.6c is specified by this conditional statement:

> If switch S_1 is on and switch S_2 is off, or else if switch S_2 is on and switch S_1 is off, then the light is bright; otherwise, the light is dark. (1.2)

Some common binary states are {true, false}, {on, off}, $\{V_H, V_L\}$, and $\{0, 1\}$.

All the key quantities encountered here in the role of device states or signal values, namely {true, false}, {on, off}, $\{V_H, V_L\}$, and {bright, dark}, are binary. Consequently, they all can be denoted abstractly by the bit symbols $\{1, 0\}$, and statements like (1.2) are seen to be inherently binary in nature.

Reliability

Binary signals are relatively insensitive to noise and related problems.

Finally, we show that binary systems have certain inherent reliability advantages over systems with more than two states. A nonbinary system employing, say, quantized voltage signals, has one or more intermediate values between the maximum (V_H) and minimum (V_L) allowed voltages. This reduces the distance separating adjacent signal values, making it easier for minor signal fluctuations, called **noise**, in the underlying analog voltage, which are unavoidable in practical systems, to cause a signal value to spill over into an adjacent value and be misinterpreted. Such errors are least likely in two-valued systems that have the largest separation between signal values. Compare, for example, the voltage waveforms of Figures 1.4c and 1.5. If the minimum and maximum voltages are the same in each figure, then the separation $s = V_H - V_L$ between adjacent values in the binary case reduces to $s/3$ when four values are used, or, more generally, to $s/(n-1)$ in an n-valued system.

In a nonbinary case like that of Figure 1.4c, which recognizes the four values $\{V_1, V_2, V_3, V_4\}$, a transition between nonadjacent values must pass through any intermediate values. For example, a transition from V_4 to V_2 must pass through a "wrong" value V_3, however briefly, due to the analog nature of electrical voltages. There is danger that this undesired but unavoidable intermediate value will be acted upon and lead to errors. Such errors can be minimized by passing through the intermediate values very quickly, but in a binary system there is no intermediate quantization level between the two recognized maximum (V_H) and minimum (V_L) values. Therefore, binary systems are not subject to errors due to intermediate values, and are the least sensitive to noise. This implies that binary systems are inherently less error-prone, and therefore are more reliable than nonbinary systems.

Logical Models

Digital design problems can be modeled using formal logic—hence logic design.

Statement (1.2), which describes a small switching circuit, has the format of sentences employed in ordinary logic to reason about things. It states certain premises (if C holds) and draws logical conclusions from them (then A follows). This suggests that logical reasoning is appropriate for describing digital systems. As with ordinary logic, it is useful to be able to reason from general principles rather than from specific instances. Consequently, we will formulate most of our design principles in general or abstract form rather than in terms of specific devices or states such as wall switches, light bulbs, or voltages. An abstract description of a digital system expressed in formal—that is, mathematically precise—logical statements is called the system's **logic design**, and is the central subject of this book.

In order to obtain a suitably general format for logical descriptions of binary systems, we denote all the binary quantities encountered above, namely {true, false}, {on, off}, $\{V_H, V_L\}$, and {bright, dark}, by the generic bit symbols $\{1, 0\}$. We further symbolize the connecting terms "and," "or," and "if . . . then," which refer to logical relationships, by the symbols $\wedge$, $\vee$, and $\Rightarrow$, respectively. (As we will see later, there are several standard sets of symbols for this purpose.) Statement (1.2) can now be rewritten in the following abstract terms:

$$\begin{aligned} ((S_1 = 1) \wedge (S_2 = 0)) \vee ((S_1 = 0) \wedge (S_2 = 1)) &\Rightarrow (L = 1) \\ ((S_1 = 0) \wedge (S_2 = 0)) \vee ((S_1 = 1) \wedge (S_2 = 1)) &\Rightarrow (L = 0) \end{aligned} \tag{1.3}$$

In these logical formulas, $\wedge$ is read as *and*, $\vee$ is read as *or*, and $\Rightarrow$ is read as *implies*. Comparing (1.2) and (1.3), we see that they say essentially the same things, but references to physical items such as switches, light bulbs, and the like in (1.2) have been replaced by abstract symbols in (1.3). A **logical** description like (1.3) may be contrasted with a **physical** description like (1.2).

Binary Adder

As another example of abstract binary operations, consider the task of adding two 1-bit binary numbers x and y. Despite its simplicity, this is an important operation in real digital circuits. Obviously, $0 + 0 = 0$, $0 + 1 = 1$, and $1 + 0 = 1$. The sum $1 + 1$ requires two bits to represent it, namely 10, the binary form of two. This can be expressed as follows: one plus one yields a sum bit $s = 0$ and a carry bit $c = 1$. If we ignore the carry bit and restrict the sum to the single bit s, then we obtain $1 + 1 = 0$. (We will see in the next chapter that this is a very useful special form of addition known as modulo-2 addition.) Thus, we obtain the following four rules for the 1-bit addition $x + y = s$.

$$\begin{aligned} 0 + 0 &= 0 \\ 0 + 1 &= 1 \\ 1 + 0 &= 1 \\ 1 + 1 &= 0 \end{aligned} \tag{1.4}$$

Numerical problems can always be converted to logical ones.

We can immediately reformulate the arithmetic rules in (1.4) as formal logical statements using the notation introduced above:

$$\begin{aligned} ((x = 1) \wedge (y = 0)) \vee ((x = 0) \wedge (y = 1)) &\Rightarrow (s = 1) \\ ((x = 0) \wedge (y = 0)) \vee ((x = 1) \wedge (y = 1)) &\Rightarrow (s = 0) \end{aligned} \tag{1.5}$$

Thus, we can interpret numerical operations, or indeed any type of binary operation, in purely logical terms.

At first sight, the light-control circuit of Figure 1.6c appears to have nothing to do with the addition of binary numbers. However, if we compare (1.3) with (1.5), we see that except for the names of the variables, the two descriptions are exactly the same! So, we can convert (1.5) to (1.3) by replacing x, y, and s with S_1, S_2, and L, respectively. Hence, the light-control circuit can be viewed as a primitive form of adder, in which the values of the input bits x and y are transferred manually into the states of the two switches (off is 0 and on is 1), and the result s is obtained by observing the light bulb (dark is 0 and bright is 1); see Figure 1.7. Although this light-bulb adder is hardly a useful device, it is nevertheless a rudimentary digital system that computes with binary numbers.

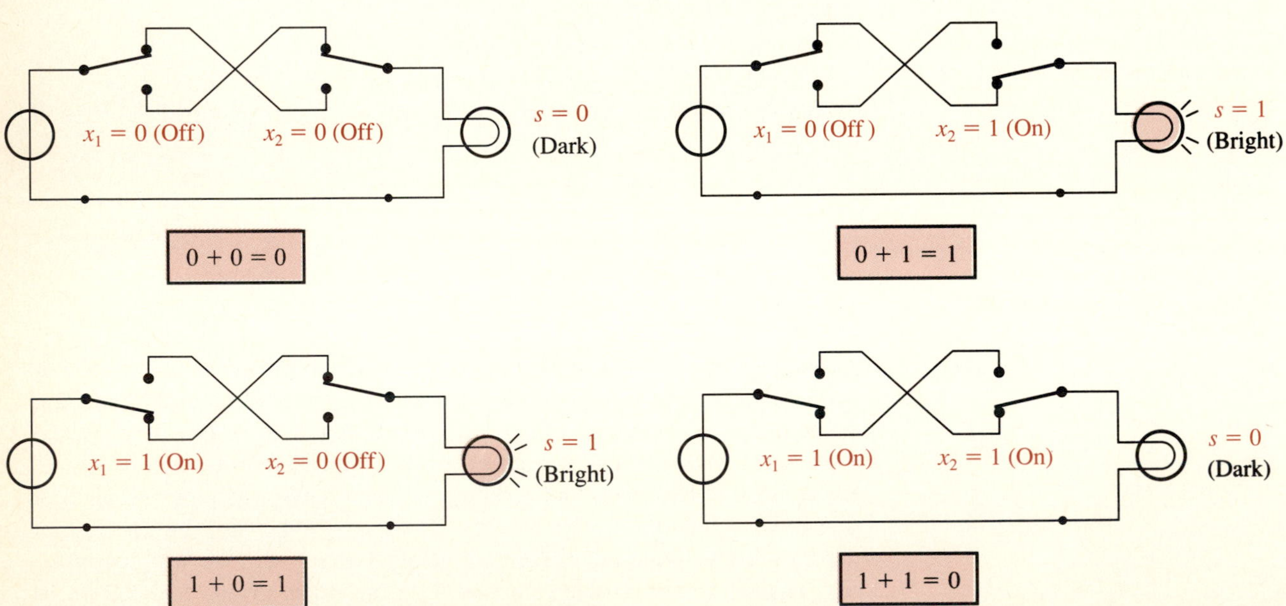

FIGURE 1.7 The light-control circuit as a (modulo-2) binary adder.

Digital Systems

Next we examine the notion of a system as a connected set of functional components, and we look at the fundamental distinction between a system's structure and behavior. We identify and discuss the various complexity levels of digital systems, as well as the major objectives of the system design process.

1.3 Structure and Behavior

The word **system** refers to an organized collection of objects that interact to form a unified whole. These objects are called the **components** of the system. A set of **connections** or **links** between the components enables them to interact with one another in ways that determine the overall function of the system. For example, a domestic audio system consists of "components" such as a CD player, a radio receiver or amplifier, a cassette tape player, and speakers; the connections between these audio components are electrical cables. The solar system is a quite different type of system composed of the sun and planets, and the forces such as gravity that connect them. A **digital system** is a system that processes information in digital or discrete form. Other terms used as synonyms for system in many contexts are *circuit* and *network*.

Structure

A system's structure indicates how it is put together.

In general, a system S has a **structure**, defined by its component set C and its intercomponent links L. We can equate the structure of system S to the pair (C, L); that is,

$$\text{Structure} = (\text{Components, Links})$$

Thus a system's structure indicates how it is assembled or put together, not what the system does.

Structure is described by block diagrams (schematics).

A convenient, general way to specify the structure of a system is a **block diagram**, in which a block or box represents each component, and lines represent the links. Figure 1.8 shows a block diagram for an audio system. Here the links denote cables between the components, and they carry electrical signals representing the sound being processed. Because these signals have specific directions, we can attach arrowheads to the links, indicating the direction of signal flow—for example, from the radio receiver to the speakers.

A block diagram like Figure 1.8 can, in principle, be used to represent any type of system, but specialized **schematic diagram** styles have been developed over the years in most scientific and engineering disciplines. For example, electrical engineers use diagrams like those of Figures 1.6 and 1.7

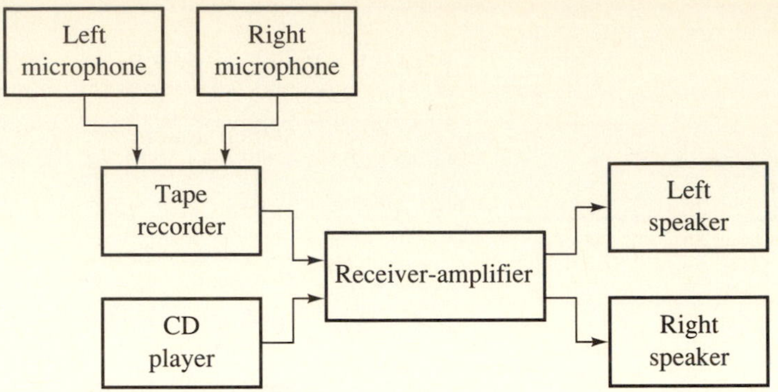

FIGURE 1.8 A block diagram defining the structure of an audio system.

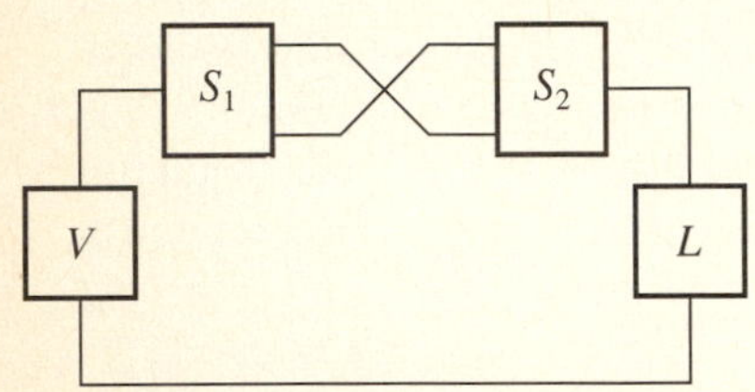

FIGURE 1.9 A block diagram defining the structure of the light control circuit shown in Figure 1.6.

to represent electrical systems.[2] In those figures, the generic block symbols are replaced by symbols with distinctive and standardized shapes that, to an experienced engineer, identify immediately what the blocks represent. A light bulb, for example, is depicted by a circle suggesting a bulb, and a loop suggesting the filament within the bulb. Figure 1.9 shows the circuit of Figure 1.6 redrawn in general block-diagram form. As we will see later, the components of digital systems are represented diagrammatically by both block symbols and distinctively shaped symbols.

Behavior

A system's behavior indicates what it does.

As stated above, the structure of a system defines its construction but not its behavior. Informally, a system's **behavior** specifies what it does or, slightly restated, the functions that the system performs. In the case of an information-processing system, we define behavior formally as the totality of the system's input/output relationships. This simply means that knowing the behavior, we can answer the question, What is the system's output response to every input applied to it? The system's behavior can therefore be summarized by

$$\text{Behavior} = (\text{Inputs}, \text{Outputs})$$

A common synonym for behavior in many contexts is (system) function.

[2]Electrical engineers generally employ the term *circuit* for a small or well-understood configuration, and reserve the term *system* for larger or more complex cases.

Formal Descriptions

Behavior is described by equations and truth tables.

Suppose we have a system S with n input distinct signals $X = x_1, x_2, \ldots, x_n$ and m output signals $Z = z_1, z_2, \ldots, z_m$. An input signal x_i might denote the state of a switch attached to some component, while an output signal z_j might denote the state of a display light on a panel or a character printed in a particular position by a printer. Input/output signals might also be voltage levels on wires. The system's behavior can be defined abstractly by an equation of the form

$$Z = F(X) \tag{1.6}$$

Here, F is a mathematical **function** or **mapping** from X to Z, which determines how each input value X is transformed to the corresponding output value Z.

Earlier in this chapter, in Statements (1.3), (1.4), and (1.5), we saw such equations being used to describe the behavior of binary systems. Logical equations of this type form a mathematical "language" for describing behavior; this language is more precise and succinct than ordinary language. We can, for instance, rewrite (1.2), which describes the behavior of the light-control circuit of Figure 1.6, in the following general form:

$$L = F(S_1, S_2) \tag{1.7}$$

Here F is a binary function defined to take the value 1 if $S_1 \neq S_2$; F is 0 if $S_1 = S_2$. Recall that 1 means a switch or light is turned on, while 0 means that it is turned off. By considering all combinations of input states, we can expand this equation into a complete description of the controller's behavior, as follows:

$$\begin{aligned} F(0,0) &= 0 \\ F(0,1) &= 1 \\ F(1,0) &= 1 \\ F(1,1) &= 0 \end{aligned} \tag{1.8}$$

Inputs		Output
S_1	S_2	L
0	0	0
0	1	1
1	0	1
1	1	0

(a)

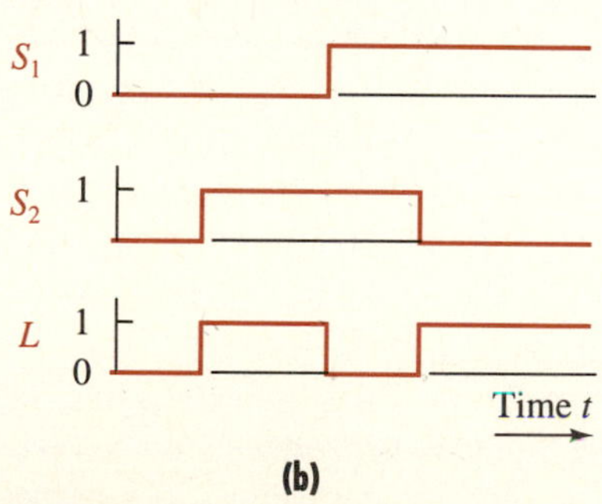

(b)

FIGURE 1.10 Behavioral descriptions of the light-control circuit: **(a)** a truth table; **(b)** a timing diagram.

We represent input/output behavior by various means, including mathematical equations, tables, timing diagrams, and descriptions in ordinary language or in special hardware description languages. Equations (1.6), (1.7), and (1.8) employ an abstract notation that is suitable for describing most types of behavior of interest to us.

A tabular representation of the behavior of the light controller appears in Figure 1.10a. It conveys precisely the same behavioral information as (1.8) by explicitly listing every combination of input values, along with the corresponding output values. Such an exhaustive listing of a system's function is called a **truth table**, because it shows when the system's output function value is 1 or 0, corresponding to the two "truth" values, true and false, assumed by statements in ordinary logic. Figure 1.10b gives the same information about circuit behavior in the form of a timing diagram that consists of a set of input/output waveforms.

Structure vs. Behavior

Structure and behavior are two complementary aspects of every system. They are independent in the sense that, given a purely behavioral description such as a truth table of some system S, we cannot determine S's structure, and vice versa. For example, Figures 1.6c and 1.9 show different structures that have the truth table of Figure 1.10a. Figure 1.9 is an abstract diagram that shows structure but not behavior. If we do not know the meaning of the various symbols V, S_1, and so on that appear in this block diagram, we cannot deduce the system's behavior. Hence, the block diagram is a purely structural description of the system in question.

Design is obtaining an efficient structure with a desired behavior.

Clearly, the behavior of a system is determined by the behavior of its components and the way they are interconnected. Hence, if we know the behavior of the various components appearing in a block diagram like Figure 1.9, we can deduce the overall system behavior. Usually, block diagrams identify the component types, and so define their function. Diagrams like Figure 1.6c contain enough information about the component's behavior to allow system behavior to be determined. Such diagrams describe explicitly a system's structure and implicitly its behavior.

Analysis and synthesis are important, complementary tasks.

The task of determining a system's behavior from a given, primarily structural, description is called **analysis**. The opposite and usually more difficult task of determining a system structure that has a given behavior is called **design** or **synthesis**. Although a given system has just one type of behavior, many different systems can be designed to exhibit the same behavior. System designers must determine a structure that not only has the desired behavior, but also meets additional design criteria, such as having low cost and high speed, using only components of a specified type, and so on.

Design Aids

When a system is complex—that is, has large numbers (hundreds or thousands) of components or links—ensuring that it exhibits the proper behavior is far from easy. Moreover, meeting secondary objectives such as low cost often involves complicated design trade-offs, the consequences of which are not easily foreseen. **Computer-aided design** or **CAD** tools play a central role in the successful analysis and synthesis of complex systems. CAD tools are typically computer programs that run on workstations, which are powerful desktop computers with high-quality graphic displays to facilitate the manipulation of design specifications such as schematic diagrams.

CAD tools are essential in the analysis and design of complex systems.

Computer-aided design programs provide facilities for managing the large amounts of data associated with the analysis or synthesis of complex digital systems. CAD programs can also implement the calculations associated with the more routine aspects of the design process. Finally, they can help determine whether a particular design has the desired behavior and meets other design criteria. Consequently, CAD tools play an important role in the design and analysis of digital systems. Note, however, that the more creative aspects of the design process cannot normally be automated; they continue to be carried out by human designers.

1.4 Design Levels

A system can be studied at various levels of abstraction.

Complex systems are modeled at various levels of detail or complexity, ranging from "high" levels that involve a small number of components performing relatively sophisticated functions, to "low" levels involving much larger numbers of simpler components. The design of the system is accomplished by a set of design steps carried out at several different levels of complexity.

Components as Systems

A fundamental characteristic of every system is that its components can also be regarded as systems or, as these components are often called, **subsystems**. For example, the CD player component in Figure 1.8 is itself a subsystem of some complexity whose function is to read the data stored on a compact disk and convert them into an analog electrical form suitable for transmission to a set of speakers. Within a CD player are various subsystems, including a small computer or microprocessor whose function is to control the CD player. This computer can, in turn, be viewed as a digital system, and so on. A component that has no recognized internal structure is termed **primitive**. Primitive devices are occasionally referred to as **black boxes**, because their structure (but not necessarily their behavior) is often unknown.

A system can therefore be examined at various levels of abstraction. Each such level, which we will refer to as a **design level**, is characterized by the set of components considered to be primitive at that particular level. At the so-called "high" or more abstract design levels, the system consists of relatively few components, each capable of performing quite complex functions. At lower or less abstract levels, the system is seen in terms of large numbers of primitives with relatively simple behavior. For example, the owner of an audio system need only understand the system at a high level, treating the CD player and the other system components as primitive devices. On the other hand, a person who designs or repairs CD players sees them at a low level, as complex electromechanical subsystems with large numbers of richly interconnected internal components.

Design Hierarchy

A nonprimitive component C_i at any design level of a system can be expanded or **refined** to a multicomponent subsystem S_j at the next lowest design level $i + 1$. Conversely, S_j can be contracted or **abstracted** to form C_i. This relationship among the levels of a system is called a **hierarchy**. As we will see, a hierarchical organization greatly facilitates the analysis, design, and maintenance of a large system.

Digital systems have three major levels: architecture, logic, and physical.

We can picture a system's hierarchy in the "tree" form illustrated in Figure 1.11 for the case of the home audio system. The higher design levels correspond to blocks at the top of the tree diagram; the lower design levels correspond to blocks at the bottom. Each block in a particular level represents a component or subsystem that is primitive at that level. The block is connected by lines to the set of blocks below it that represent its primitive components at the next level. Thus, as one moves from the top to the bottom of the tree, one is effectively looking inside the system at greater and greater levels of detail. Conversely, as one moves from bottom to top, less detail is seen; in other words, the system is viewed more abstractly.

Each design level in a hierarchy defines a system model that is **implemented** or **realized** by the next lowest level. For instance, a speaker component in Figure 1.11 is implemented by a subsystem composed of one or more tweeters, woofers, transformers, and other electrical components. The boundaries between the design levels in a system hierarchy, as well as the names of the levels, are not hard and fast. In general, the number and types of the design levels identified in a given system are subjective and largely a matter of descriptive or design convenience.

For many types of digital systems such as computers, however, we can single out three design levels that form a well-defined hierarchy.

1. The **architecture level,** also called the processor or system level, which is a high design level characterized by concern for overall system management

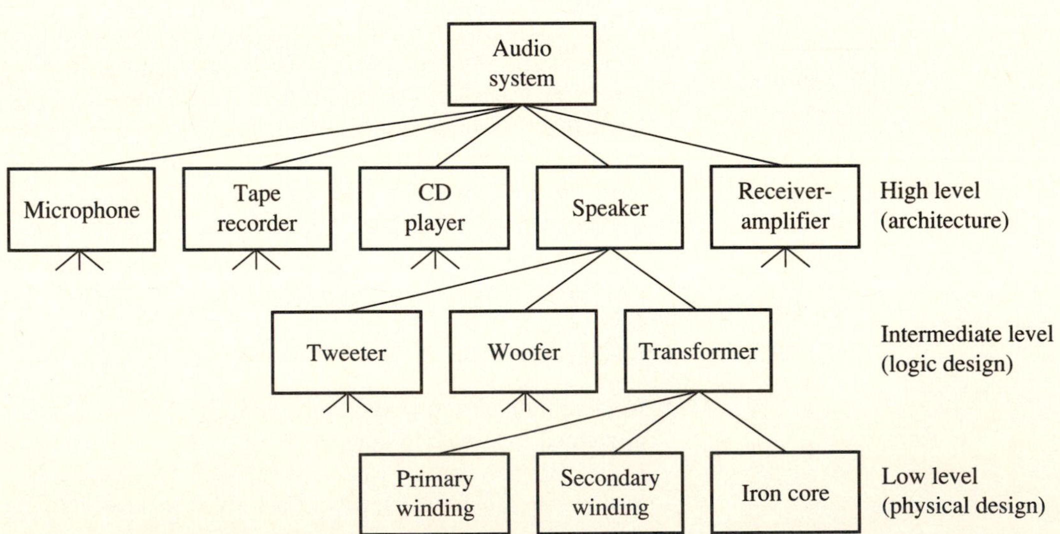

FIGURE 1.11 Hierarchical organization of an audio system.

2. The **logic level,** which is an intermediate design level concerned with the details of the system as seen by a technically sophisticated user
3. The **physical level,** which is a low design level concerned with the details needed to manufacture or assemble the system

Figure 1.12 illustrates these three design levels for the case of a typical digital computer system. The components (blocks, logic gates, and transistors) are the primitive devices at each level. For the architecture level, general block symbols are employed, whereas for the other two levels distinctively shaped symbols are used in accordance with standard practice. (The meanings of these special symbols will be explained in Chapters 3 and 4.)

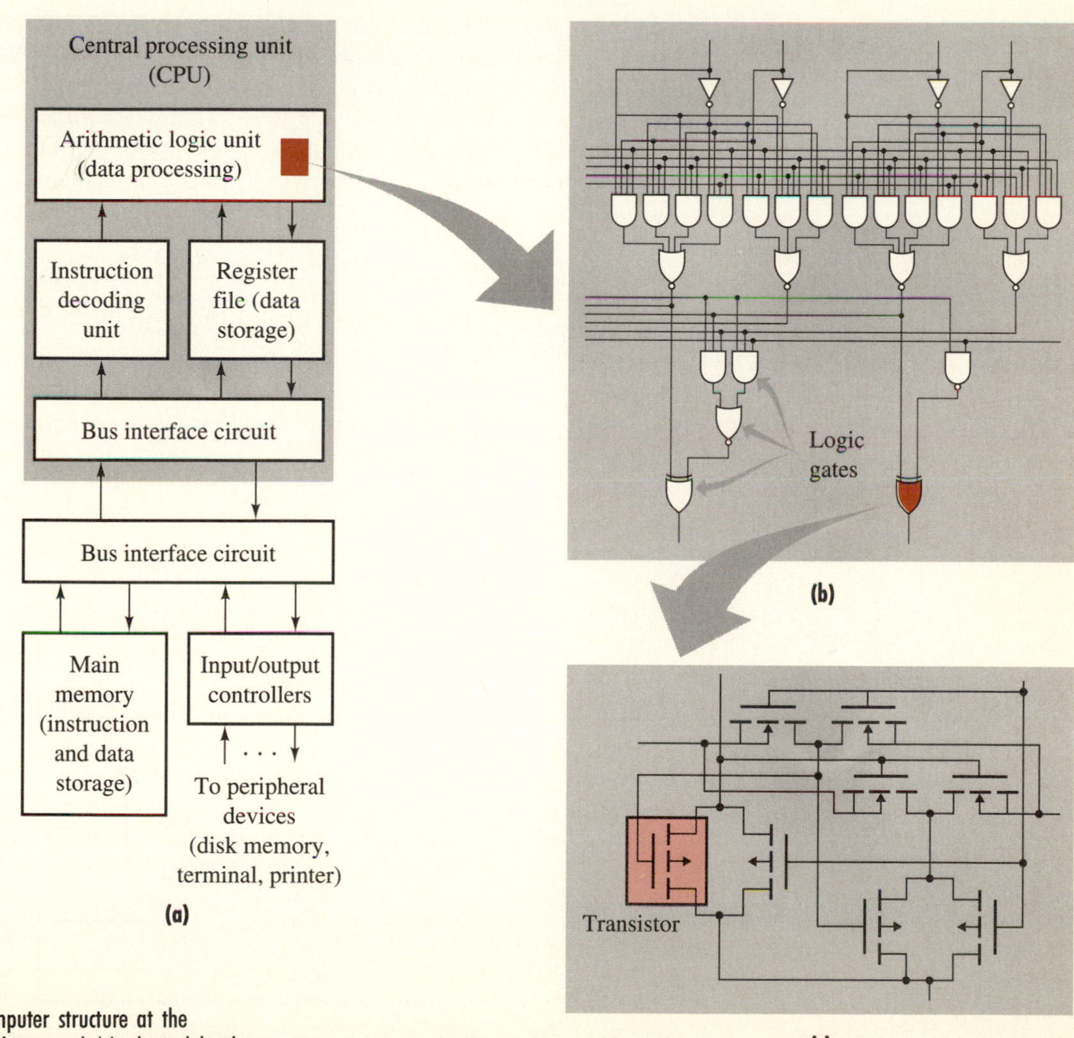

FIGURE 1.12 Computer structure at the (a) architecture, (b) logic, and (c) physical levels.

The Architecture Level

A system's architecture is typically its highest level of abstraction.

The **architecture** of a digital system identifies the highest-level components, their behavior, and their interconnections. In the case of a computer (see Figure 1.12a), these components include the central processing unit (CPU), the memory units used for instruction and data storage, and the components for attaching peripheral devices such as printers, keyboards, and video monitors to the computer (the input/output subsystem). Also included in the architecture are the set of instructions executed by the CPU, and certain internal details such as the CPU's local memory (the register file), its data-processing subsystem, and key features of its instruction-decoding mechanism.

Computers are often classified into families whose components all have roughly the same structure and behavior at the architecture level. Such families are said to have the same architecture, or to be **compatible** with one another. For example, "IBM PC compatibility" specifies a well-known architectural standard for personal computers.

The Logic Level

The logic level defines internal details of a digital circuit in a technology-independent fashion.

The internal structure of a CPU, a memory unit, or an input/output controller is defined at the logic level, a process called **logic design**. The task of the logic designer is to specify the subsystems needed to implement such tasks as "Compute the sum of two 32-bit binary numbers and store the result temporarily; if the sum exceeds 32 bits, then activate an appropriate 'overflow' signal." Figure 1.12b shows the logic design of a small portion of a CPU.

Two CPUs from different manufacturers may have entirely different internal structures or logic circuits. Because several distinct physical technologies exist for implementing the same logic circuit, this design level, like the architecture level above it, is to a large degree **technology independent** in that the designer is not concerned with such physical details such as electronic circuits or voltages. *The detailed design of digital systems at the logic level is the main focus of this book.* Later we will refine the logic level into three related and overlapping sublevels: the switch, gate, and register (or register-transfer) levels.

A typical logic circuit is composed of primitive components called **logic gates**, the behavior of which is defined entirely in terms of the binary quantities 0 and 1. Gates thus display "logical" behavior of the kind discussed earlier. They are used to compute certain simple functions involving a few binary variables. For example, the two-input, one-output gate shown shaded in Figure 1.12b performs a logical operation known as EXCLUSIVE-OR. The gate's output assumes the value 1 if and only if exactly one of the gate's inputs is 1; otherwise, the output value is 0. This input/output behavior is also defined by Figure 1.10.

By combining many gates in the manner illustrated in Figure 1.12b, logic circuits can be designed to perform addition and other operations on numerical and nonnumerical data of any desired size. Thousands of gates are needed to implement a typical CPU; a complete computer may contain many millions of gates. Therefore, a basic objective in logic design is to construct a logic circuit that performs a given computational task in a correct and cost-effective manner.

The Physical Level

The physical level defines the real devices used to manufacture a system.

Finally, the various subsystems of a logic design must be implemented by means of manufacturable physical devices. In most cases of interest in this book, these devices are **integrated circuits** or **ICs**, a term that refers to the fact that a large number of individual electronic components and their interconnections are placed in a single tiny chip of semiconductor material. Integrated circuits are available in various subtechnologies or IC "families." They all have electronic switches (transistors) as their fundamental components, but are distinguished by their cost, operating speed, size, power requirements, and other physical attributes.

Figure 1.12c shows an eight-transistor circuit belonging to the CMOS (complementary metal-oxide-semiconductor) IC family; this circuit constitutes a particular physical implementation of a two-input EXCLUSIVE-OR logic gate. Integrated circuits vary widely in their component density, which ranges from a few tens of transistors per IC (**small-scale integration**) to millions of transistors per IC (**very large-scale integration** or **VLSI**). Because of this wide range, a single IC of the appropriate density can serve as a primitive element at any design level.

An integrated circuit is manufactured on a small rectangular chip of semiconductor material such as silicon. The IC chip, in turn, is packaged in a small, hermetically sealed box, some examples of which appear in Figure 1.13. The IC is connected to other ICs via electrical terminals or pins on the outside of the box; the number of pins ranges from 10 or so to several

FIGURE 1.13 Representative IC packages: (a) a 28-pin dual in-line package (DIP); (b) a 40-pin surface-mount package; (c) a 325-pin grid array (PGA) package.

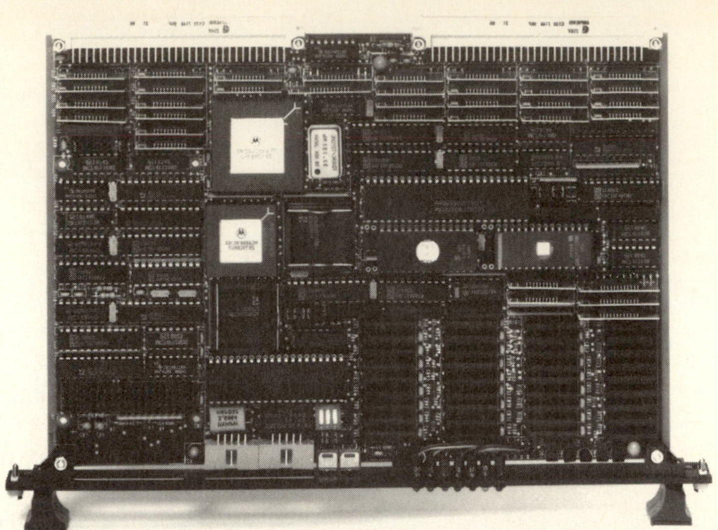

FIGURE 1.14 A printed-circuit board that forms a small computer.

hundred, depending on the package type. The ICs forming a computer, or one of its major subsystems such as the CPU or memory, are often mounted on a printed-circuit board, which forms both a support and an interconnection medium for the ICs. Figure 1.14 gives an example of a commercial "single-board computer" that implements a system like that of Figure 1.12a using about 100 ICs. The complexity of these ICs ranges from simple circuits containing a few tens of transistors, to CPU and memory ICs that contain in the range of 10^5 to 10^7 transistors. Figure 1.15 shows the Motorola 68040 microprocessor chip, introduced in 1990. This VLSI chip contains over a million transistors and constitutes a powerful CPU all by itself.

Hierarchical Design

Hierarchical design lowers costs and raises quality.

A multilevel, hierarchical approach to digital system design can reduce the cost of a system and improve its quality in a number of important ways:

1. The hierarchical approach to design makes a large system more manageable by *reducing complexity*. For example, a modern large-scale computer would be impossible to design or understand if it had to be handled entirely at the physical (or transistor) level. By grouping circuits of, say, 100 transistors into primitive logic-level components, a reduction of two orders of magnitude in component count (and, therefore, in design complexity) is achieved on moving from the physical to the logical level. Roughly speaking, a logic designer who creates a circuit containing 100 such logic components effectively produces

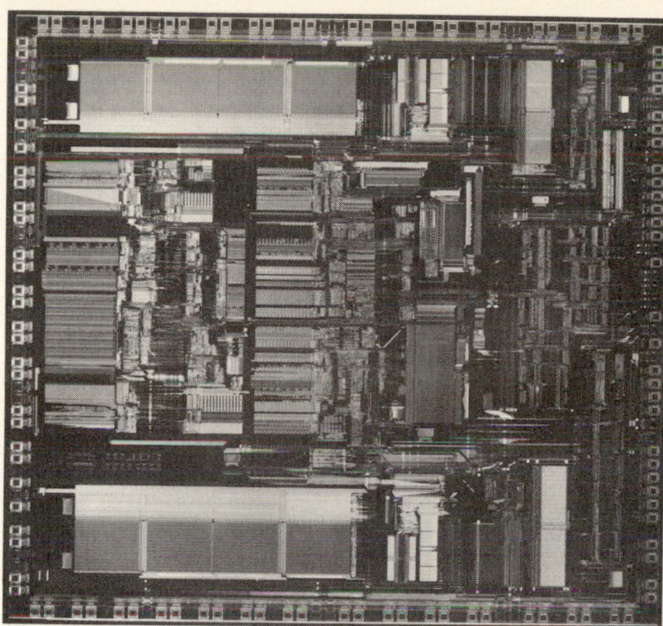

FIGURE 1.15 A photomicrograph of the Motorola 68040 microprocessor chip.

a 10,000-transistor design with one-hundredth of the effort required to produce the same design directly at the transistor level. A similar reduction in complexity occurs on moving from the logic to the architecture level.

2. The hierarchical approach introduces *division of labor* into the design process, which permits the components at different levels to be designed largely (although not entirely) independent of one another by specialists in that level, or in some instances by CAD synthesis programs. Therefore, a design team for a complex new system might contain system architects, logic designers, and IC designers, with each group concentrating on one design level. A logic designer needs to know little about IC design, and a system architect even less.
3. The subsystems or components specified at each level are self-contained entities, which can be designed, tested, and manufactured separately. This facilitates the use of *standard parts* that can often be mass produced at relatively low cost. Moreover, prior component designs can be stored in design libraries and reused, possibly with minor modifications, to reduce overall design time and cost.
4. Hierarchical designs are inherently more *robust* than nonhierarchical ones. Design errors, which are unavoidable in all nontrivial design projects, can be confined to a particular subsystem, and any necessary design corrections can be made within that subsystem, without compromising the entire design effort. This characteristic also makes the systems easier to maintain and repair after manufacture.

In summary, the organization of a digital system into a hierarchy of several design levels subdivides and shortens the design cycle, facilitates the use of standard components, and renders the resulting system more robust.

Top-Down vs. Bottom-Up Design

Hierarchical systems support two distinct styles of design known as bottom-up and top-down design. In the **bottom-up design** approach, the subsystems at or near the lowest design level of the hierarchy tree (Figure 1.11) are designed first. These designs are then passed as primitive components to the next highest level, where they are used to construct subsystems at that level, and so on. The architecture of the entire system is not specified until the top of the tree is reached. In **top-down design**, on the other hand, the hierarchy tree is traversed in the opposite direction, from top to bottom. In this case, the system architecture is specified at the highest level first. Then the components identified at that level are designed, and so on down the tree. In practice, design often proceeds simultaneously in several levels at once, making it difficult to classify the design process as either top-down or bottom-up.

Top-down design is generally preferred.

A predominantly top-down approach is generally preferred in practice, especially when the desired behavior of the components at the lower levels is easily defined, as is the case when standard parts or design libraries are used. Bottom-up design is most appropriate for a **full custom** or **custom design** project, in which the components at all design levels are specially designed for the project in question, and little use is made of prior designs or standard parts.

1.5 Combinational vs. Sequential Systems

Digital systems, especially when viewed at the logic level, are divided into two broad classes: those without memory, which are called **combinational** or, less frequently, **combinatorial**, and those with memory, which are called **sequential**. Next we examine this important distinction, which determines the organization of a large part of the material in this book.

Combinational Circuits

Combinational circuits act instantaneously; they have no memory.

Combinational digital systems, or **combinational circuits** as they are more often called, have a behavior that may be fully specified by a truth table or a set of equations that list for every input combination the corresponding output combination (hence the name "combinational"). Thus, the class of logic devices (modulo-2 adders) defined by Equation (1.8) and Figure 1.7 are combinational circuits. Such circuits map input data to output data; that is, they perform computations in a single step. In practice, there may be a slight delay before the output signals change in response to new input values.

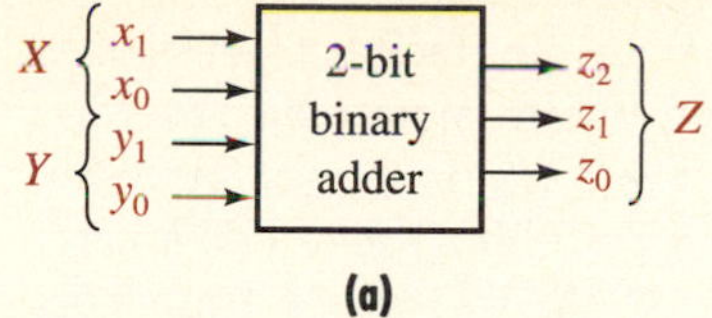

(a)

Inputs				Outputs		
x_1	x_0	y_1	y_0	z_2	z_1	z_0
0	0	0	0	0	0	0
0	0	0	1	0	0	1
0	0	1	0	0	1	0
0	0	1	1	0	1	1
0	1	0	0	0	0	1
0	1	0	1	0	1	0
0	1	1	0	0	1	1
0	1	1	1	1	0	0
1	0	0	0	0	1	0
1	0	0	1	0	1	1
1	0	1	0	1	0	0
1	0	1	1	1	0	1
1	1	0	0	0	1	1
1	1	0	1	1	0	0
1	1	1	0	1	0	1
1	1	1	1	1	1	0

(b)

FIGURE 1.16 (a) A 2-bit combinational adder; (b) its truth table.

This response time is usually very small—often only a nanosecond or less—so for most purposes we can regard the response of a combinational circuit as instantaneous. Therefore, we can also define a combinational circuit as one that, in the ideal case, has zero response time.

Figure 1.16 shows another type of adder, which is a combinational circuit that is a little more complex than our modulo-2 adder. This particular circuit is intended to perform ordinary addition on a pair of 2-bit binary numbers $X = x_1x_0$ and $Y = y_1y_0$, producing a 3-bit result $Z = z_2z_1z_0$, as can be verified by checking the adder's truth table (Figure 1.16b) against the list of binary numbers in Table 1.1. For example, when $X = 10$ and $Y = 11$, the adder produces the output $Z = 101$, which corresponds to the decimal addition $2 + 3 = 5$. When $X = Y = 01$, the adder produces the output $Z = 010$, corresponding to $1 + 1 = 2$ in decimal arithmetic.

In general, the behavior of any n-input combinational circuit employing binary signals can be described by a truth table with 2^n rows, one for every possible input combination. The 2-bit adder of Figure 1.16 has four input signals and so requires a 16-row truth table. An adder for two 16-bit numbers, which is a more practical size, would require a table containing an astronomical $2^{32} = 4{,}294{,}967{,}296$ rows! Consequently, we need more efficient methods than truth tables for expressing behavior. In later chapters, we will develop a number of special tools for this purpose, notably Boolean algebra and hardware description languages.

Sequential Circuits

Sequential circuits have memory and perform multistep operations.

Often, we break a complex computation into a sequence of simple steps, so instead of computing the result all at once, we compute a small part of it at a time. The same concept applies to digital systems, in which a circuit that performs a task in a sequence of steps is said to be "sequential." Such a circuit must remember partial results between steps, so being a sequential circuit and possessing memory are synonymous. The advantage of sequential operation is that simpler hardware is needed for each step. The disadvantage is a longer delay before the final result is produced. Thus, while a combinational implementation of some task is faster than a sequential one, it normally requires more hardware.

Sequential realizations are slower but cheaper than combinational realizations.

We can see this trade-off between hardware and speed in the case of binary addition. If we cast our pocket calculators aside and add two long (decimal) numbers using pencil and paper, we probably resort to a multistep method in which we add a pair of digits at a time, starting with the rightmost or least significant digits. If the sum computed in the current step exceeds one, we "carry one" to the next step. Thus, an n-digit addition becomes a sequence of n steps, in each of which we add digit x_i of the first number to the corresponding digit y_i of the second number. We also add to the sum a carry digit of 0 or 1 produced in the preceding step.

Another Adder

Adapting the above sequential addition method to binary numbers, we must add three bits x_I, y_i, and c_{old} in each step, where c_{old} is the last carry bit. A circuit implementing this step determines a sum bit z_i and a new carry bit c_i. For example, the addition step $x_i + y_i + c_{\text{old}} = 1 + 0 + 1$ produces $z_i = 0$ and a new carry bit $c_i = 1$. Such addition requires a three-input, two-output combinational circuit known as a "full" adder. (The modulo-2 adder considered earlier is a "half" adder because it has no provision for carries.) To retain the carry between steps, we need a 1-bit memory device that takes in c_i in the current step and returns it as c_{old} in the next step.

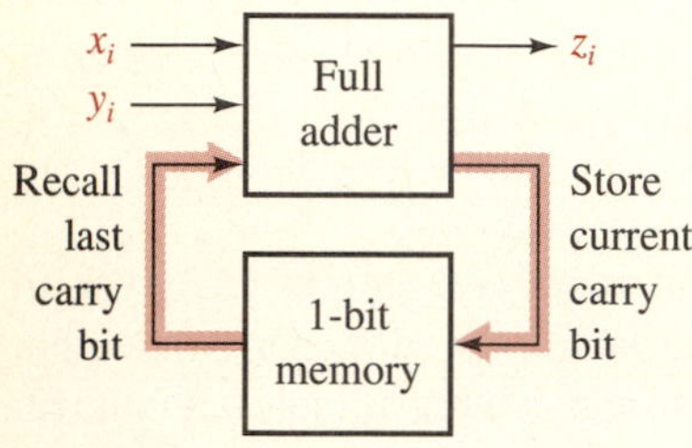

FIGURE 1.17 A sequential circuit that implements a binary adder.

Figure 1.17 shows a sequential adder circuit that implements the foregoing scheme. It is capable of adding any pair of n-bit numbers in $n + 1$ steps. This is illustrated in Figure 1.18, in which we perform the addition $2 + 3 = 5$ or, in binary, $10 + 11 = 101$. In the first step, shown in Figure 1.18a, we take in the two least significant bits $x_0 = 0$ and $y_0 = 1$. Because there was no previous step to generate a carry, we must initialize c_{old} to 0. Step 1 then produces the first digit $z_0 = 1$ of the sum, along with a carry output $c_0 = 0$. In step 2 (Figure 1.18b), the full adder computes $x_1 + y_1 + c_0 = 1 + 1 + 0$, which produces $z_1 = 0$ and $c_1 = 1$, so the memory now, for the first time, stores a 1. A final step, in which x_i and y_i are set to null values—that is, they become insignificant leading zeroes—produces the third and final digit $z_2 = 1$ of the sum. Thus, the answer is $z_2 z_1 z_0 = 101$, which, of course, is five once again.

Although the sequential adder of Figure 1.17 is not obviously a lot simpler than the combinational adder of Figure 1.16, the differences between the two designs become much more pronounced as n increases. For example, a 32-bit version of the combinational adder has a huge number of components, as suggested by the fact that it must have 64 input lines and 33 output lines. With no changes at all, our little sequential adder can do the same addition, taking roughly 32 times longer to do so.

Finally, we note that many design problems are inherently sequential because their input signals are not all available at the same time or their

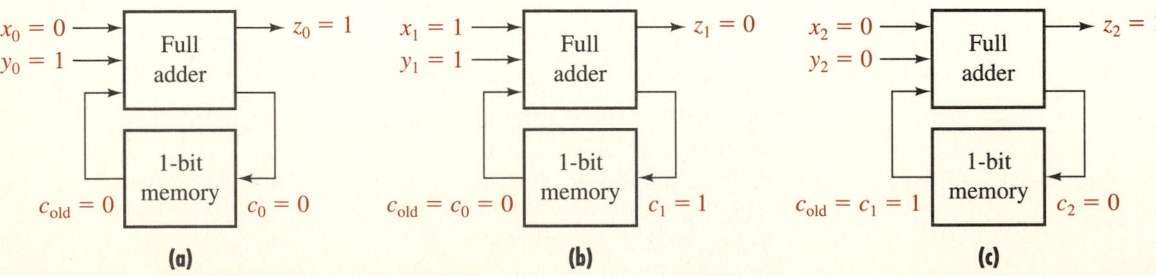

FIGURE 1.18 Using the sequential adder to compute $2 + 3 = 5$.

output actions are required in a certain sequence over time. In such cases, sequential circuits provide the only possible solutions to the problems in question.

Sequential Circuit Structure

Feedback and clocks are basic features of sequential circuits.

The sequential adder displays some features common to all sequential circuits. It is easily partitioned into two parts: a combinational part (the full adder) and a memory part (the carry store). There is feedback from the memory to the combinational circuit, so that past events recorded in the memory can affect current operation. No such feedback is required in combinational circuits.

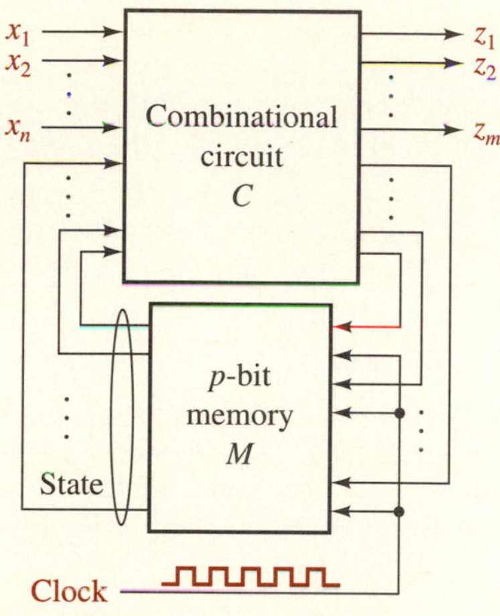

FIGURE 1.19 The structure of a general sequential circuit or finite-state machine.

A general sequential circuit has the structure depicted in Figure 1.19. The memory part M now can store a large but finite amount of information (p bits), and the combinational part C can also be quite complex. The information stored in M is referred to as the **state** of the circuit, so that another name for a sequential circuit is a **finite-state machine** or simply **state machine**. To prevent the various 1-bit memories that form M from getting out of step—so ensuring that the state information is stored and retrieved in an orderly and error-free fashion—a special timing signal known as a **clock signal** is distributed to all components of M. The clock signal has a waveform that consists of a perfectly regular sequence of 0s and 1s, and serves to synchronize the operation of the circuit, just as the conductor synchronizes the musicians in an orchestra.

As we remarked earlier, a large digital system can be decomposed hierarchically into subsystems, each of which may be either combinational or sequential. Therefore, the more complex systems take the form of sets of such circuits that interact with one another, perhaps in complicated ways. In such cases, the memory function is typically distributed throughout the system, instead of being centralized, as suggested by Figure 1.19.

Programmable Systems

Programs specify behavior in a systematic and flexible way.

Digital systems are often constructed to carry out long and complex sequences of actions, without human intervention. For example, switching on a CD player causes the following complex sequence of actions to take place automatically: the CD is gripped and begins to rotate; a laser beam is activated and pointed at a specific rotating track on the CD; the digital data on the track are read by the laser and decoded until the start of a piece of music is identified; the digitally encoded music is converted to analog form and errors are corrected, if necessary; and finally the analog signals are conveyed to speakers, from which they emerge as sound. Such action sequences implement an **algorithm** or **procedure**, a well-defined set of behavioral rules devised by the system's designers.

A convenient way to incorporate an algorithm into a digital system is to store it in the form of a program in the system's memory. A **program** is a

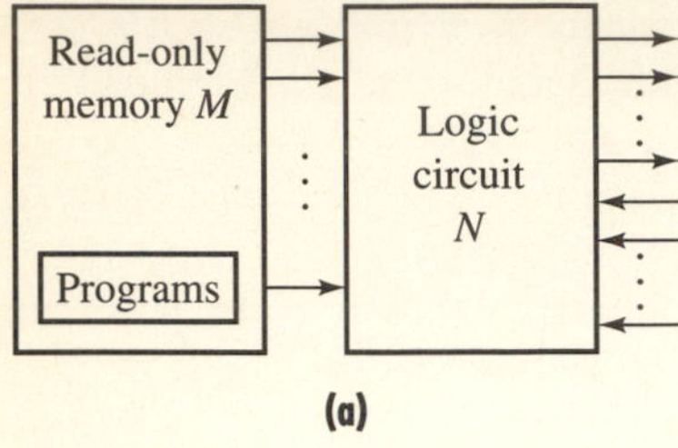

(a)

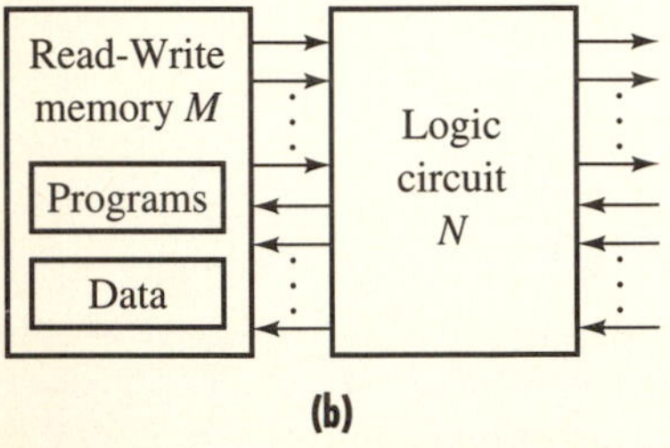

(b)

FIGURE 1.20 Programmable digital systems: (a) with read-only memory; (b) with read-write memory.

list of neatly organized instructions that control the behavior of the system, which is then said to be **programmable** (see Figure 1.20). The advantage of this design style is that programs can readily be changed, either during manufacture or later "in the field," so that the same basic design can be made to perform many different functions. We regard the memory unit M itself and the associated circuits N as the fixed **hardware** of the system; the program is the system's variable **software**. A very small example of software is the following three-instruction addition program, which mimics the behavior of the all-hardware sequential adder discussed above (Figure 1.18).

$$\begin{aligned} &C(-1) = 0 \\ &\text{FOR I} = 0 \text{ TO N} - 1 \text{ DO} \\ &C(I),Z(I) = X(I) + Y(I) + C(I - 1) \end{aligned} \qquad (1.9)$$

This symbolic format must be translated into a binary code for storing in M. To contrast a nonprogrammable system with a programmable one, the former may be referred to as **hardwired**.

A programmable system is sometimes built around a memory M of the "read-only" type, as in Figure 1.20a, which means that the contents of M cannot be changed without physically replacing M. Instructions are read from M and used to control the system's operation, but no new information can be written into M. This is contrasted with the situation shown in Figure 1.20b, where the memory is of the read-write variety, implying that its contents can be changed dynamically while the system is in operation. Hence, it is possible to store variable data along with programs in M, and the programs themselves can be changed from time to time, if desired.

Digital Computers

Computers are general-purpose, programmable systems.

The stored-program concept illustrated by Figure 1.20b leads to the most powerful of all programmable systems, the general-purpose digital computer. It has the structure shown in Figure 1.21, which is a high-level, architectural

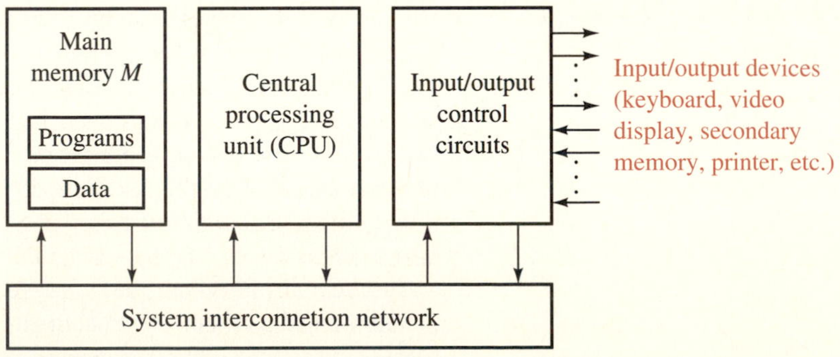

FIGURE 1.21 Overall structure of a general-purpose digital computer.

view of a computer. The principal components are the central processing unit (CPU), the read-write main memory M, and the input/output (I/O) control circuits. The main purpose of the CPU is to fetch and execute instructions that are stored in M. The data used by the instructions and the results they generate are also stored in M. The I/O control circuits provide a link between the computer and the outside world via keyboards, display screens, printers, and other I/O devices. Internal communication within the computer is organized around an interconnection network that takes the form of hundreds or thousands of wires and switches.

The CPU is the "brain" of the computer and contains the combinational and sequential circuits required to execute a set of a few dozen (in some cases a few hundred) different instruction types, from which the programs it executes are constructed. CPU instruction sets are designed to be general and flexible enough to allow all types of algorithms to be programmed. They include instructions of three main kinds:

1. Data-transfer instructions that move data unchanged between the CPU and M or between the I/O circuits and M
2. Data-processing instructions that implement arithmetic operations such as add, subtract, and multiply, as well as various nonnumerical or "logical" operations
3. Program control instructions that determine the order in which instructions are executed. The FOR instruction that appears in Statement (1.9) is of this type. Some control instructions perform tests on the results of previously executed instructions, and use the test outcomes to influence subsequent actions performed by the CPU.

Integrated circuit technology has made it possible to implement every major component of a computer in a single, low-cost IC chip. For example, most personal computers employ a single-chip CPU, referred to as a **microprocessor**. An entire computer of moderate program capacity and performance can also be placed on one chip. In the form known as **microcontrollers**, single-chip computers are employed to control an enormous range of useful devices, from microwave ovens to cars, by embedding application-specific programs in their memories. A typical microcontroller is equipped with a set of I/O connection points that can be reconfigured by software to match the characteristics of many I/O devices. The design of the hardware and software links between the microcontroller and the controlled system is termed **interfacing**. In a typical application of a microcontroller, much of the design effort goes into writing and debugging the interfacing programs executed by the microcontroller. We discuss such general programming issues briefly in the final chapter, but our main concern in this book is with nonprogrammable or application-specific programmable circuits.

Single-chip computers are widely used as embedded controllers.

The Design Process

We now examine the design process itself, focusing on the logic design level. The goals of digital system design and the difficulties encountered in achieving them are discussed, as well as the important role played by computer-aided design (CAD) tools.

1.6 Design Goals

Functional correctness is always the main design objective.

The primary design goal for every digital system is functional **correctness**—that is, the system must exhibit the proper behavior. Correctness implies that the specifications for the system are complete and free of errors, and that the final design meets those specifications precisely. Digital systems are quite intolerant of mistakes introduced in the design process. A small design error or **bug** that results in, say, a single bit of a binary number being computed incorrectly can have catastrophic consequences, and is likely to render the system unusable until the problem is fixed. Most systems are so complex in terms of the number of components and connections they contain that human error is an inevitable part of the design process. Therefore, comprehensive corrective or **debugging** procedures must be instituted to detect and eliminate such errors as soon as possible.

Cost

Minimizing cost is also a key objective.

Assuming correct behavior can be achieved, the next objective in most system design projects is to minimize the cost of the resulting system. This cost can be measured in various ways and is not always easy to determine. A basic cost measure for mass-produced systems is the manufacturing cost c_M per copy of the system, which, in the case of digital systems, is roughly determined by the number of physical components used. Examples of component count figures that are widely used as cost measures are the number of transistors or gates per IC and the number of ICs per printed-circuit board. So, reducing the number of components employed is an important way of reducing system cost. This goal can be achieved by replacing several ICs with a single, denser one, which explains the intensive efforts by IC manufacturers to obtain ever higher levels of integration.

System cost depends on the cost of design (c_D), manufacturing (c_M), and maintenance (c_S).

However, several other factors influence the total cost of a system. They include the cost c_D of the design effort itself, often classified as research and development (R&D) cost or nonrecurrent engineering (NRE) cost, and the support cost c_S for maintaining each copy of the system in good running condition during its working life. The net "lifetime" cost c_T of a system of which a total of N copies are manufactured is therefore approximated by this equation:

$$c_T = c_D/N + c_M + c_S \tag{1.10}$$

The relative importance of the factors appearing on the right-hand side of (1.10) in determining c_T depends on the type of system, the number of copies manufactured—that is, the size of the system's market—and its field support requirements.

The simplicity of (1.10) belies the difficulty of estimating the cost factors in question, especially early in the design cycle. The field support cost c_S tends to be the most difficult to measure, because it depends on the reliability of the individual components and their interconnections as well as the amount the customer is willing to pay for field support and servicing.

Graphics Board Example

Two examples illustrate the influence and range of the foregoing cost factors on design practice. First, consider a single-board graphics system intended as an add-on processor to enhance the user interface of a low-cost personal computer. The system is to be designed by a single logic designer working for a small company.

Manufacturing costs drop rapidly as production volume increases.

The cost of the target computer is $2,000, and the target price of the add-on board is to be no more than $500. To keep the costs low, we will use standard off-the-shelf ICs, which cost in the range of $1 to $10 each, depending on their complexity. Let us assume an average cost of $2 per IC for this project. The total cost of the 50 ICs needed for the graphics board is then $100, plus an additional $50 for the printed-circuit board, connectors, and so on, making $c_M = \$150$. If we allow a further $100 for support costs, and $100,000 for a design effort of around one designer-year, we obtain from (1.10) $c_T = 10^5/N + 250$. At a modest production level of $N = 1000$, this becomes $c_T = \$350$, which is within the overall cost goal, and also allows the manufacturer a moderate profit. Increasing production by a factor of 10, to 10,000, reduces the total cost per system to $260.

Microprocessor Example

As a second example, suppose that a large manufacturer of ICs wants to introduce a new high-performance microprocessor to serve as the single-chip CPU for the same family of personal computers. In this case, the R&D effort, which is likely to involve a great deal of custom design, can take a year or more and involve several dozen skilled design engineers.

The development cost c_D might be $100 million or so, a figure we shall assume. Initially, IC production costs tend to be high, but fall quickly as the manufacturer gains experience with the production problems posed by the design. Once efficient high-volume production levels have been reached, manufacturing costs normally drop into the range of $10 or less per IC. Support cost is difficult to measure for individual ICs, because, unlike printed-circuit boards, they cannot be repaired, and so must be completely replaced if they fail. If we ignore c_S in this case, the cost c_T of this microprocessor becomes $10^8/N + 10$, which is critically dependent on the production volume N. If $N = 10^8$, a representative figure for a successful microprocessor, then $c_T = \$11$, and the R&D costs have a very modest impact on the overall cost. On the other hand, if N reaches only 10,000 units, then c_T becomes \$1,010, or almost a hundred times more expensive, and consequently too costly for use in a \$2,000 computer.

Performance

Performance can make a key difference between two functionally equivalent systems.

The performance of a digital system refers mainly to its speed of operation. Two designs that perform the same function can differ greatly in performance. Thus, in addition to cost, a digital system is generally required to meet specific performance goals. Obviously, the performance of a system depends on that of its components and the way the components interact. Component performance is measured in various ways, depending on the design level. As we will see, the relationship between system and component performance is by no means simple; even measuring the performance of individual low-level components can be surprisingly difficult. Consequently, performance evaluation and control is an integral part of the design process.

Propagation Delay

A basic performance parameter that is useful at all design levels is **delay time** or **latency**; it is the time (in seconds) required for a primitive device to carry out a representative primitive operation. An equivalent parameter is operating **speed**, which is the inverse of delay; that is,

$$\text{Speed} = \frac{1}{\text{Delay}}$$

Speed is specified in operations per second (OPS). For example, the performance of a transistor or logic gate is measured by the delay between a signal change, say a 0 becoming a 1, on the device's input, and the resulting signal change appearing at the device's output. This is referred to as the **propagation delay** time t_{PD} and is typically around a nanosecond (1 ns or 10^{-9} s).

Fast systems tend to cost more.

Delay is inversely related to cost, because fast components almost always cost more than slow ones. It is also subject to physical limitations imposed by the circuit technologies available to implement a design. A fundamental limit in all technologies is the speed of light, which restricts signal propagation through space, or any physical medium, to approximately 300,000 km/s or 30 cm/ns. In general, it is desirable to make the delay of a system as small as the cost and technology constraints will allow.

Despite its apparent simplicity, propagation delay has its subtleties, of which the logic designer must be aware. For example, the propagation delay of a transistor is affected by its size and position within an IC, so that any particular copy of a transistor can vary by as much as 50 percent around the typical or "nominal" value of t_{PD}. As such components are connected, the uncertainty about the performance of the resulting circuit increases. For example, if two gates, each with $t_{PD} = 5 \pm 2.5$ ns, are connected in series (the output of one gate is connected to an input of the other), the resulting two-gate circuit has a propagation delay somewhere between 5 and 15 nanoseconds; that is, $t_{PD} = 10 \pm 5$ ns. As circuits get larger and we move to higher levels in the design hierarchy, uncertainty about performance increases. Special design steps, such as the use of clock signals to synchronize changes in different parts of the system, are usually taken to limit this uncertainty.

Other Performance Measures

Other useful performance measures are access time (memory) and bandwidth (communication links).

At higher design levels, performance measures that are directly related to a component or subsystem's function are employed. For example, the basic action performed by a memory unit is an "access" operation, in which the contents of some specified storage location are read from or written into. A common performance measure for a memory, therefore, is its **access time** t_A, defined as the delay between the presentation of an access request at the memory's input lines to the completion of the requested read or write operation. The performance of a communication link such as a bus is measured by the maximum number of units of information that it can transfer per second. This speed parameter, called **bandwidth**, is measured in bits per second (bits/s).

In the case of a CPU or other processor that executes a set of instructions, a useful performance parameter is the number of instructions executed per second, which is typically measured in millions of instructions per second or **MIPs**. Here, uncertainty similar to that of propagation delay exists, because execution time varies from instruction to instruction, and the same instruction can execute at different speeds with different data. Computer manufacturers like to cite the maximum achievable MIPS figures for their products, even though they are rarely, if ever, reached in practice. (It has been suggested that such peak MIPS figures are best regarded as performance levels that the manufacturer guarantees will never be exceeded.)

Other Design Goals

The design objectives we have discussed so far are correctness, low cost, and good performance, which apply to all types of digital systems. We turn now to some other design goals, the relative importance of which depends heavily on the type of system under consideration.

Compatibility

We often require a new design to be **compatible** with existing systems, which means that specific aspects of the structure and behavior must meet certain standards previously defined within a company or an industry. Design features subject to standards are the mechanical dimensions of IC packages and printed-circuit boards; the formats (the numbers of bits and their meaning) used for data and instructions; and the specifications (currents, voltages, delays, and the like) for the signals that appear at the interfaces where components meet. Such standards are essential if new systems are to work with old, and if systems designed by different manufacturers are to work together. Standardization also applies to the way in which a system is documented. For example, the symbols used for logic gates and transistors in Figure 1.12 conform to some widely used standards.

Power Consumption

Other design goals of varying importance are compatibility, low power, reliability, and testability.

All physical systems consume **power**, normally electrical power measured in watts (W) or kilowatts (1 kW = 10^3 W), much of which is eventually dissipated as heat into the system's surroundings. It is always desirable to keep a system's power consumption as low as its other design constraints permit. Larger systems are equipped with a **power supply** that provides them with electrical power at appropriate voltage and current levels. (Power is the product of current and voltage.) They may also have cooling fans to prevent excessive temperatures.

Power-speed trade-offs can strongly influence the choice of an IC family.

The major IC technologies are clearly distinguished by their power requirements. CMOS circuits have among the lowest power requirements (1 W or less per IC), and so are well suited to implementing VLSI circuits, in which the densely packed components can be easily damaged by excessive heat. Some IC families such as TTL (transistor-to-transistor logic) and ECL (emitter-coupled logic) consume more power, reducing their maximum component densities and requiring special attention to cooling. On the other hand, ECL and other high-power technologies have inherently lower propagation delays than does CMOS, and so yield the fastest systems. Power-speed trade-offs of this kind often determine the IC family to be employed in a particular design project.

Reliability

As we noted previously, the after-manufacture support cost c_S of a system forms a large part of the system's total lifetime cost. The ability of a digital system to operate for a long period of time without failing is a measure of its **reliability**; the higher the reliability, the lower c_S. Although modern ICs are very reliable, they are subject to failures from numerous sources, including manufacturing defects, mechanical damage due to vibration or mishandling, environmental interference due to power supply fluctuations or overheating, and various wear-and-tear phenomena such as electrical contact corrosion. Techniques to increase reliability include special packaging and shielding as well as adequate ventilation and cooling. In general, the smaller the number of physical components in a system, the higher the system's reliability. Thus, in addition to reducing manufacturing costs, higher levels of integration tend to improve reliability.

Testability

Designing for ease of testing can reduce design and support costs significantly.

A significant portion of c_D and c_S may be associated with testing a system to determine the presence and possibly the location of faults. **Testing** is accomplished by applying appropriate input signals (test patterns) and observing the resulting output responses; deviation from the normal or expected response indicates a fault. Because a critical fault may be due to the failure of a single wire or transistor, the complexity and therefore the cost of testing can be enormous, sometimes exceeding all other costs. Testing costs include the effort required to generate the test patterns and the equipment needed to apply the test patterns and evaluate the results. Testing is also complicated by the fact that access to an IC is very limited after the IC chip has been packaged; only a tiny fraction of the circuit's signal lines—namely, those with external connections on the package—can be probed directly by a tester. Testing costs can be reduced by incorporating into the design appropriate **design-for-testability** features that facilitate the generation and application of test patterns.

Design Trade-Offs

Design often requires compromises among competing goals.

Meeting all the foregoing design objectives simultaneously is difficult because they interact in complex ways that are poorly understood. Exact design techniques—that is, those with a well-understood underlying theory and design algorithms—can typically address only one or two of these goals at a time. Practical design often requires combining several design methods in ad hoc fashion in order to achieve a good compromise among many competing objectives. In addition, a new design must be constantly monitored during its development to determine if the relevant cost, performance, and other design parameters are within acceptable limits.

The design process involves several steps, indicated in flowchart form in Figure 1.22. First, a portion of the design—for instance, a subsystem at some level—is designed. Then it is evaluated to determine if the relevant design goals are met. If some current goal is not met, appropriate corrections or modifications are made to the design, and the result is reevaluated. Several iterations through the redesign-evaluation loop may be needed before a satisfactory result is obtained.

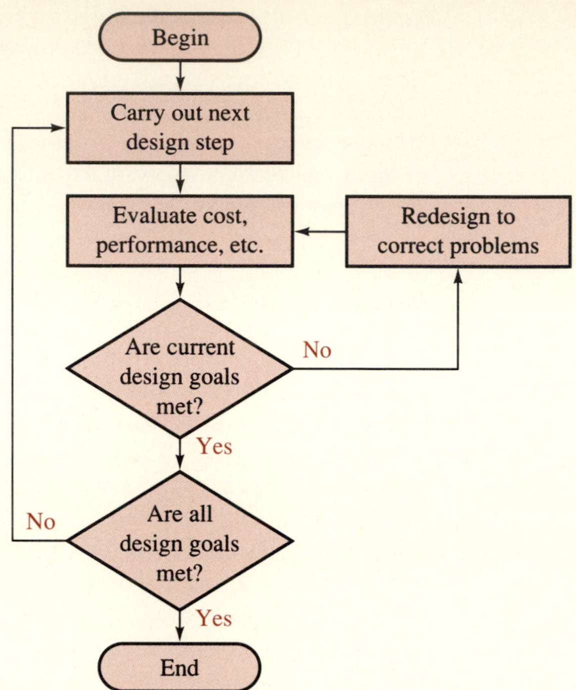

FIGURE 1.22 A flowchart of the digital system design process.

1.7 Computer-Aided Design

Digital system design is accomplished by a mixture of formal and informal methods of varying degrees of difficulty. It is greatly aided by programs that support computer-based representations of a design and automate the more tedious design and evaluation steps.

CAD Tools

Computer-aided design programs have three main roles in the digital system design process:

CAD tools serve as editors, simulators, and synthesizers.

1. To represent and document the target system in forms that can be processed efficiently both by humans and computers; this is the function of CAD programs called **editors**
2. To compute the behavior implied by a particular design or portion of a design, and to determine how well it meets performance and related design goals; this is the task of **simulator** programs
3. To automate the design process itself by deriving structures that implement all or part of some design step; such programs are termed **synthesizers**

These tasks increase in difficulty as we go from editing through simulation to automated synthesis. Editing is a relatively easy bookkeeping task; simulation requires extraction of implicit behavior from known but often complex designs; synthesis involves the difficult task of creating new designs. The synthesis steps that have been most successfully implemented are at the lower levels of the design hierarchy.

In general, CAD programs and the data they process are large, complex entities that require substantial computing resources. Until the 1980s, the high cost of computers and poor quality of their user interfaces caused CAD to play a minor role in system design. The advent of powerful, low-cost workstations and personal computers with good graphics interfaces changed this situation, however, making it feasible to equip individual designers with desktop computers capable of supporting a full range of CAD tools.

A modern **CAD workstation** for a professional designer is illustrated in Figure 1.23. It contains a powerful digital computer and a high-resolution color graphics screen on which schematic diagrams, signal waveforms, and the like can be displayed. The CAD algorithms are embedded in a suite of computer programs, often under the control of a management program known as a **CAD framework**. A designer interacts with the CAD system by means of a keyboard and a pointing device such as a mouse. Several workstations of this type can be linked via a computer network to allow a team of designers to share CAD software and cooperate closely on a common design task.

The design process can be summarized as the task of determining a system structure that achieves a desired behavior, subject to constraints on cost, performance, and so on. This implies a need for systematic ways to represent, manipulate, and modify—that is, edit—structural and behavioral information at all stages of design.

FIGURE 1.23 A Hewlett-Packard CAD workstation for digital system design.

Schematic Capture

For human designers, a convenient representation method for a system's structure is a schematic diagram, which depicts the components of the system by blocks or special graphic symbols and the interconnections by lines. As Figure 1.18 demonstrates, the style of schematic diagrams varies with the design level. Most CAD systems include an editing program for creating and modifying schematic diagrams on a computer screen, a process called **schematic capture**.

Circuits may be entered as diagrams (schematic capture).

Figure 1.24 shows a screen image presented by one such CAD tool, LogicWorks, which is intended for logic design using a personal computer such as the Apple Macintosh as the host workstation. This particular circuit editor is typical in that it allows logic diagrams to be "drawn" on a computer screen, using a mouse to manipulate an on-screen pointer (the arrowhead near the lower center of the figure). Predrawn symbols for the various components needed for logic diagrams are obtained from a pull-down menu labeled "Gates." For example, Figure 1.24 illustrates how a two-input EXCLUSIVE-OR gate is obtained by moving the pointer to the entry "XOR-2" on the Gates menu. Clicking the mouse button then causes a copy of a standard EXCLUSIVE-OR gate symbol (or a reasonable approximation) to appear. This gate symbol (shown highlighted in the figure) can then be moved anywhere on the screen by placing the pointer on it, pressing the mouse button, and dragging the symbol along with the pointer to its new location. Releasing the mouse button causes the gate to stay in its new position. Connections are drawn from one gate to another by a similar click-and-drag

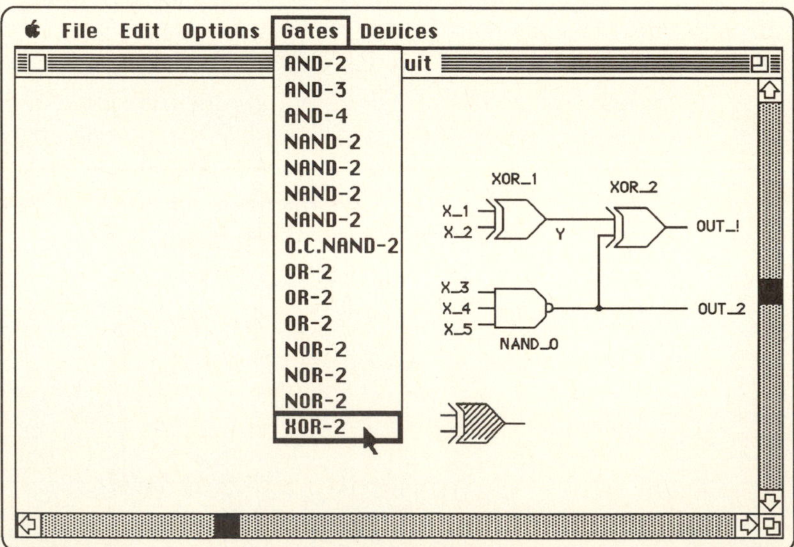

FIGURE 1.24 Logic-circuit editing using the LogicWorks program.

process in which the pointer serves as a "pencil." The pointer can also be changed to an "eraser" to delete unwanted components or connections. Making minor modifications is therefore quite easy.

A circuit editor provides more than just a passive picture of the system being designed; it also creates an internal computer model of the target system that can be used by simulators, synthesizers, and other CAD tools. In some cases, this model can serve as the input data to automated manufacturing equipment. Editors are also helpful in preparing the documentation, such as operating and maintenance manuals, needed by users of the new system.

Netlist Descriptions

Circuits can also be entered as netlists.

Another common way to represent a system's structure is simply to list its components and their interconnections. A straightforward tabular format for such a **netlist**, as it is usually termed, appears in Table 1.2, which is a netlist description of the circuit shown graphically in Figure 1.24. Distinguishing names are assigned to each component and connection present in the system. For each component we specify its type and provide a list of the connections to its input terminals (the input list) and to its output terminals (the output list). Therefore, to indicate that connection (wire) w_1 goes from an output of gate G_1 to an input of gate G_2, we place the name w_1 in both G_1's output list and G_2's input list. The name w_1 can also serve as the name of the (binary) signal that appears on the connection.

Schematic diagrams are represented by netlists on text-only equipment, such as "dumb" computer terminals that lack a graphics capability. The internal model of a digital system constructed by a circuit editor from a screen diagram often has a netlist format, because the latter is more easily stored and processed by a computer.

TABLE 1.2 A tabular description (netlist) of the logic circuit shown in Figure 1.24.

Name	*Type*	*Input list*	*Output list*
XOR_1	Two-input XOR gate	x_1, x_2	y
XOR_2	Two-input XOR gate	y, out_2	out_1
NAND_0	Three-input NAND gate	x_3, x_4, x_5	out_2

Other Descriptive Methods

We can represent the behavior of a digital system in various other ways, including signal waveforms, truth tables, and logic equations. It is a small step to go from a netlist format to a logic equation, if we can treat the component types as operators in an equation. For example, Table 1.2 corresponds to the following set of logic equations:

$$\begin{aligned} y &= \text{XOR}(x_1, x_2) \\ out_1 &= \text{XOR}(y, out_2) \\ out_2 &= \text{NAND}(x_3, x_4, x_5) \end{aligned} \tag{1.11}$$

HDLs permit designs to be represented at several complexity levels.

In addition to defining the circuit's structure, these equations also define its behavior, *provided* the behavior of the functions denoted by XOR and NAND in (1.11) are known. Equational descriptions of this type can be extended into **hardware description languages** (**HDLs**), which provide precise descriptions of structure and behavior at several hierarchical design levels. HDLs resemble programming languages in appearance, but their purpose is to specify hardware instead of algorithms. Two widely used HDLs are the languages known as Verilog and VHDL.

An important part of every design effort is its **documentation**, which is the preparation and production of design specifications, user guidebooks, repair manuals, and the like. CAD editors can help greatly in generating the system documents during the design process itself. That process cannot be called complete until all the necessary documentation has been prepared in machine-readable or hard-copy form.

Simulation

Simulators use computer models to check function and performance.

Logic designers make extensive use of CAD programs that can **simulate** a target design—that is, duplicate its input/output and internal behavior. A computer model M of the target design is constructed and, in combination with a simulation control program or **simulator**, is used to compute the outputs generated in response to representative inputs supplied by the designers. This behavioral information can be represented in various graphical or textual forms. The primary uses of a simulator are to check a design for functional correctness and to evaluate its performance. Simulators are key tools in determining whether the design goals have been met and whether redesign is necessary (see Figure 1.22).

Figure 1.25 shows how the simple logic circuit from Figure 1.24 is simulated by the LogicWorks program. A second window has been added to the screen image presented by this CAD tool to display signal waveforms. The waveforms for the input signals $x_1, x_2, \ldots, x_5$ have been "drawn" by the designer, using the host computer's mouse as a switch to turn the signals on or off at specific times on the horizontal time scale. The response signals y, out_1, out_2 have been computed and displayed by LogicWorks. Simulators also provide the option of displaying input/output data in tabular form. In any case, the simulated behavior is obtained in a format that can be compared to the desired behavior of the circuit being designed. Any discrepancies detected

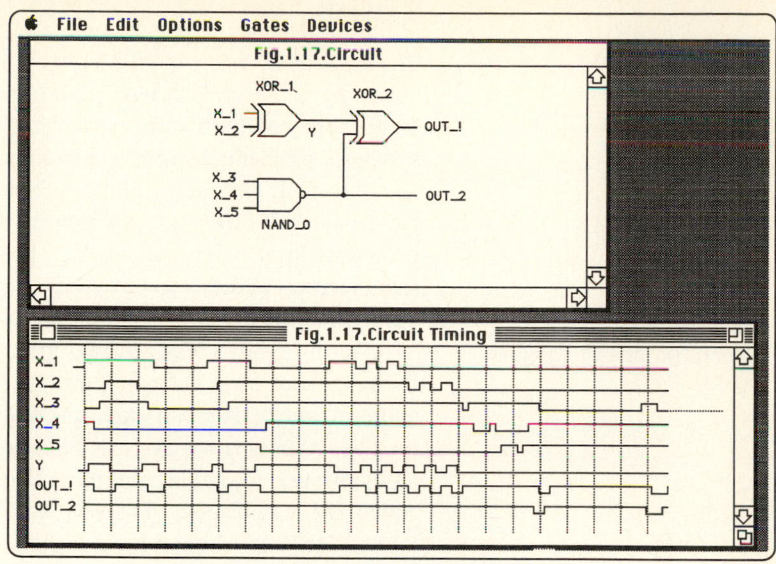

FIGURE 1.25 A logic circuit simulation using LogicWorks.

in the data produced by the simulator can be used to pinpoint and correct design errors. In addition to checking the correctness of a design, simulators are used to check performance (operating speed) and certain other design parameters. The advantage of a simulator is particularly apparent when developing a very complicated system.

Designers can also simulate systems by building prototypes on breadboards.

A new design can also be simulated by building a physical model or prototype of the proposed design. This may be done by employing off-the-shelf ICs to implement primitive components or nonprimitive subsystems, and by wiring the ICs together in an ad hoc assembly, referred to as a **breadboard**. It is generally more difficult and takes longer to fabricate, modify, and test a hardware prototype than a computer model generated by means of a CAD program. However, in some instances a prototype can more accurately represent important aspects of the final system's behavior and performance, because the computer-generated model provides a simplified picture of the real system. For example, the waveforms that appear in the computer simulation of Figure 1.25 are idealized versions of the actual noisy signal waveforms (see Figure 1.4). The latter can be observed by applying a suitable measuring instrument such as an oscilloscope to a breadboard or IC prototype. The fine details of signal timing are difficult to model and simulate accurately, so CAD models have not entirely replaced hardware prototyping. In practice, however, hardware prototypes are built only after a new design has passed extensive simulation tests using CAD tools.

Synthesis

The creation of a new system is in many ways the most rewarding part of the entire digital design process. Digital system design is so complex and varied that no one approach suffices, nor can the process be fully automated. Most synthesis problems require a judicious mix of design methods, both formal and informal, for their solution. The selection of the approaches reflects the designers' skill and experience, as well as the particular design task or design level being addressed. The methods employed range from small modifications of old, known systems to the generation of entirely new designs via sophisticated CAD synthesis programs.

Synthesis programs can use exact algorithms or inexact heuristics.

We distinguish between two main synthesis approaches: those based on formal, exact procedures (algorithms) and those based on informal, trial-and-error methods or on experience with prior designs. We refer to inexact methods as **heuristics**. As will become apparent, the logical design of digital systems has a core of algorithmic methods based on a branch of mathematics called Boolean algebra. There is also an extensive body of heuristic design knowledge with which a logic designer should be familiar.

CHAPTER 1 SUMMARY

Why Digital?

1.1 ■ We have two ways of representing information: **analog** and **digital**. These representations are distinguished by the nature of the values that they allow information variables to assume.

■ Digital information is represented by a fixed number of discrete symbols, called **digits**. An example is the 10 decimal digits we use to construct numbers.

■ In a system in which high accuracy is essential, digital is preferred over analog. A system that possesses both digital and analog qualities is called a **mixed** or **hybrid system.**

1.2 ■ Digital signals are restricted to discrete values or states. Two-valued digital signals and the digital systems that process them are called **binary**. A compact disk's storage format is an example.

■ Binary states are represented by 0 and 1 and are referred to as **bits**, a contraction of the term *binary digits*.

■ **On-off switches** are the basic building blocks of digital systems.

■ Unlike the decimal system, which uses 10 digits, binary numbers use only two distinct digits (0 and 1). In a multidigit binary number, each digit represents a power of two. Although a binary number system seems clumsy to us, it is important to understand it because most information-processing systems are constructed from switches, which are binary devices. In addition, the basic decision-making processes required of digital systems are binary, and binary signals are more reliable than those with more than two states.

Digital Systems

1.3 ■ A **system** is an organized collection of objects that interact to form a unified whole. These objects, called **components**, are able to interact with one another via connections or links. A good example of a system is a home stereo system.

■ A **digital system** is a system that processes information in digital form.

■ We specify the structure of systems on paper via **block diagrams,** in which each block or box represents a component of the system. Lines with or without arrows denote the communications links that join the components.

■ A system's **behavior** specifies what it does or the functions the system performs. We describe behavior by equations, tables, and timing diagrams.

■ A system's behavior is determined by the behavior of its components and the way they are interconnected. The process of determinining a system's behavior is called **analysis**. The opposite process—determining a system structure that has a given behavior—is called **design**.

■ **Computer-aided design** or **CAD tools** are crucial in successfully analyzing and designing complex systems. CAD tools are computer programs that run on powerful desktop computers called **workstations.**

1.4 ■ In every system, the components that make up that system can themselves be regarded as systems, or **subsystems**. Therefore, a system can be examined at various **design levels,** which are characterized by the set of components that are considered to be **primitive** (having no recognized internal structure) at that level. At the "high" end, a system has relatively few components, each capable of performing complex functions. At the "low" end, the system is seen in terms of large numbers of primitives with relatively simple behavior.

■ A primitive component at any design level can be **refined** to a multicomponent subsystem at the next lowest design level. Conversely, the subsystem can be **abstracted** to form the primitive component. This relationship among a system's levels is called a **hierarchy.** A hierarchical structure greatly facilitates the analysis, design, and maintenance of a large system.

- The **architecture** of a digital system identifies the highest-level components, their behavior, and their interconnections. In a computer, these components include the CPU, memory units, and peripheral devices.

- The internal structure of a CPU, memory unit, or input/output controller is defined at the **logic level** by a process called **logic design**. The detailed design of digital systems at the logic level is the main focus of this book. The various subsystems of a logic design must be implemented by means of manufacturable physical devices, such as **integrated circuits** or **ICs**. They are distinguished by their cost, operating speed, size, power requirements, and other physical attributes.

1.5 ■ Digital systems at the logic level are divided into two broad classes: those without memory (**combinational**) and those with memory (**sequential**). Combinational circuits act in a single step, whereas sequential circuits perform multistep operations.

- Functional specifications are often incorporated in digital systems in the form of **programs**, which are organized lists of words that control system behavior. Programmable logic devices can readily be changed during or after manufacture, so the same basic design can be made to perform many functions.

The Design Process

1.6 ■ The primary design goal for every digital system is functional **correctness.** The second major goal is minimizing **cost.** Another goal—good **performance**—is also important.

- Performance is measured in terms of **delay time** or **latency,** the time in seconds required for a primitive device to carry out representative operation. An equivalent parameter is operating **speed**, which is the inverse of delay.

1.7 ■ CAD programs have three main roles in the digital system design process: to represent and document the target system (**editors**); to compute the behavior implied by a design and determine how well it meets performance goals (**simulators**); and to automate the design process itself (**synthesizers**).

Further Readings

Thomas M. Fredericksen's book *Intuitive Digital Computer Basics* (McGraw-Hill, New York, 1988) is an elementary introduction to digital concepts for nonspecialists. A readable text on all aspects of the digital computer—hardware, software, and applications—can be found in *The Mystical Machine* (Addison-Wesley, Reading, Mass., 1986), by J. E. Savage, S. Magidson, and A. M. Stein. The *Handbook of Design Automation,* by S. Sapiro and R. J. Smith II (Prentice-Hall, Englewood Cliffs, N.J., 1986) gives a concise description of the role of CAD in the design process. The low-cost LogicWorks CAD system for the Apple Macintosh used for illustration purposes in this chapter is described in C. Dewhurst's *LogicWorks User's Guide* (Capilano Computing Systems, Ltd., New Westminster, B.C., 1991). A similar CAD program called MICRO-LOGIC II for the IBM PC and compatible computers can be found—both the program itself (on a floppy disk) and a user's manual—in *The Student Edition of MICRO-LOGIC II,* by L. D. Coraor (Addison-Wesley, Reading, Mass., 1989).

Chapter 1 Problems

Why Digital?

Digital vs. analog; binary systems; switching circuits; logical concepts.

1.1 Magnetic tape cassettes are used for both analog and digital audio recordings. Explain why a digital tape recording usually has better sound quality than an analog recording. What system characteristics would cause an analog tape to have better sound quality than a digital tape?

1.2 Discuss the advantages and disadvantages of replacing the traditional analog speedometer of a car with a digital version.

Digital information processing occurs in living systems.

1.3 Figure 1.26 (adapted from [Eccles, 1970]) shows a typical information-processing activity in a biological system—in this case, a cat's eye. The neural signals form a sequence of voltage pulses that control the angular movement of the cat's eye, as indicated. The eye motion is clearly analog, but the controlling nerve pulses are digital. Why do you think this is so? Are the neural signals binary?

1.4 The binary waveform depicted in Figure 1.5 is a rather crude approximation of the underlying analog waveform. Explain how the latter can be converted into a binary form that more accurately captures the original analog pattern.

1.5 Extend the list of binary numbers that appears in Table 1.1 by giving the binary forms of all integers from 16 to 40.

1.6 A very long binary sequence containing thousands of bits represents an integer N_2. Suppose that N_2 is converted into ordinary decimal form N_{10}. If N_2 contains n bits, estimate how many decimal digits are in N_{10}.

1.7 Suppose that numbers are to be represented using just the three digits 0, 1, 2; this is called the **ternary** number system. Following the style of Table 1.1, list the first 16 integers in ternary form.

1.8 Figure 1.27 shows a switching circuit containing three switches S_1, S_2, and S_3 to control an electric light. Switch S_3, of the "double-pole, double-throw" variety, consists of two separate single-pole, single-throw (transfer) switches that turn on and off simultaneously. By considering all possible settings of the switches, show that this circuit allows the electric light to be controlled independently from three different locations, such as the top, the bottom, and an intermediate landing of a staircase.

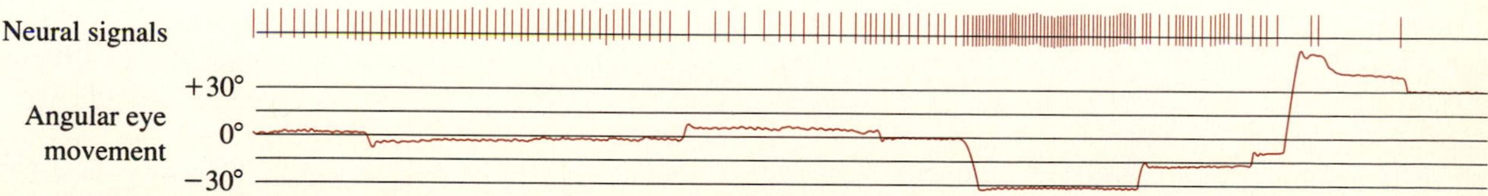

FIGURE 1.26 Sequence of nerve impulses to a cat's eye muscle and the angular eye movements they produce.

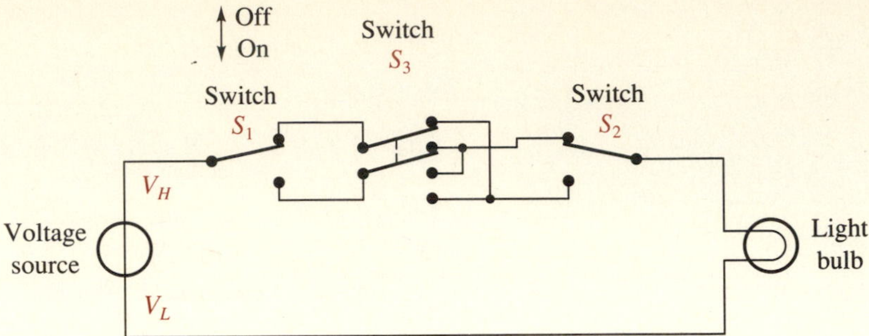

FIGURE 1.27 Another switching circuit to control an electric light.

1.9 How would you solve the preceding problem if an arbitrary number n of switches must be used instead of three?

1.10 Consider the switching circuit controlling two lights that appears in Figure 1.28. Derive a set of logical formulas in the style of Equation (1.3) that describe the circuit's function. Also derive a formula that indicates the conditions for both light bulbs to be turned on simultaneously.

1.11 As discussed in the text, the circuit of Figure 1.6c can be interpreted as an adder of single-bit numbers, in which the light bulb is an indicator of the output sum bit s, ignoring the carry bit c that results from the addition. Redesign this circuit to include a second light bulb that indicates the value of c. Assume that each switch S_i can be replaced by a pair of switches whose switch arms are mechanically linked together, but whose electrical terminals are independent, as in switch S_3 of Figure 1.27.

1.12 Express each of the following items in symbolic logical form using the style of Equations (1.3) and (1.5). **(a)** "If *ifs* and *ands* were pots and pans, there'd be no work for the tinker's hands." [Old Irish saying] **(b)** The arithmetic equation $x - y = z$, where x, y, and z are single-bit binary numbers, and $-$ denotes subtraction.

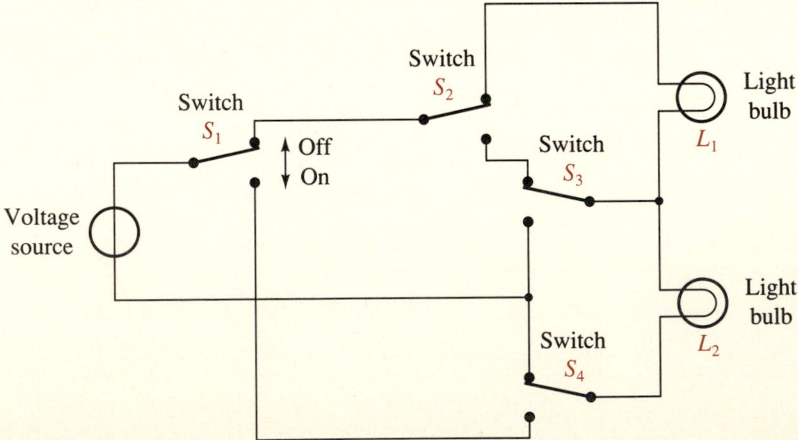

FIGURE 1.28 A switching circuit that controls two lights.

Digital Systems

Structure and behavior; design hierarchy; combinational and sequential circuits.

1.13 Give two ways of describing the structure of a small digital system that include no behavioral information. Give three ways of describing the behavior of a small digital system that include no structural information. Illustrate your answers with simple examples.

1.14 Identify each of the following items as either mainly a structural or a behavioral description: **(a)** an electrical wiring diagram; **(b)** a road map; **(c)** a computer program listing.

1.15 Briefly define each of the following terms: hierarchical system, integrated circuit, logic design, system architecture.

Complexity is closely related to level of description.

1.16 Discuss the significance of the following statement in the context of digital systems: "How complex or simple a structure is depends critically on the way we describe it" [Simon, 1981].

1.17 Give several advantages of hierarchy in digital systems. Identify and discuss one possible disadvantage of such a hierarchy.

1.18 Let S be a hierarchical system with three distinct design levels L_1 (highest), L_2, and L_3 (lowest). Suppose S contains 100 components of 10 different types taken from level L_1. Similarly, suppose each primitive component or subsystem S_i in L_1 and L_2 contains 100 components of 10 different types, which are unique to S_i and come from the level below it. Furthermore, suppose that a primitive at each level requires an average of 100 hours of design time. Estimate **(a)** the total time required to design S and **(b)** the total number of physical components included in S.

1.19 Human organizations often have a hierarchical structure to facilitate their management. For example, officers, noncommisioned officers, and troops define three levels of hierarchy in an army. Analyze this army hierarchy and show that it shares some of the characteristics of a digital system hierarchy.

1.20 Animals also have a multilevel hierarchical organization that can be expressed in terms of organs, cells, proteins, and similar well-defined biological components. Suggest how this hierarchy facilitates the evolution of complex living forms from simple ones.

1.21 Give an example of a human organization that is nonhierarchical in nature. What are some of the advantages and disadvantages of such an organization? Discuss how they might apply to a nonhierarchical digital system.

1.22 Briefly compare and contrast the top-down and bottom-up styles of hierarchical design. Illustrate your answer by referring to the design of a very large computer program such as an operating system for a new computer.

1.23 Construct a truth table for the light-control control circuit of Figure 1.28.

1.24 A digital lock for a bicycle has a knob with a pointer that rotates past a circular scale marked 0, 1, 2, . . . , 25. It has the following opening instructions: "Reset the lock by rotating the knob counterclockwise at least two full turns. Then turn the knob clockwise to 7, passing 0 once. Then turn the knob counterclockwise past 0 to 21, at which point the lock should open." Is this lock a combinational or sequential system?

1.25 Explain why all the functions of a pocket calculator could, in principle, be implemented by a combinational circuit. Explain why it would be totally impractical to do so.

The Design Process

Design goals; design costs; computer-aided design (CAD) tools

1.26 Explain why analog systems are more likely than digital systems to be tolerant of design errors. Illustrate your answer with an example of a common system that has both analog and digital implementations.

1.27 In some systems, manufacturing cost is of minor concern because of the overriding importance of other design objectives. Examples of such systems are the control computer for an unmanned space probe and a pacemaker for the human heart. Identify and briefly justify the probable main design goals in each case.

1.28 A company plans to design a single-board system that will employ 100 standard ICs and have a total manufacturing cost of $1,000 per system for a planned production run of 1,000 systems. Total R&D costs are estimated to be $50,000; the estimated manufacturer's support cost per system is expected to be $1,000 over the system's rather short one-year lifespan. **(a)** What must be the manufacturer's selling price to realize a net profit of 20 percent? **(b)** If the manufacturing cost per system can be reduced by $100 for every additional 1,000 systems produced down to a minimum cost of $500, how many must be produced in order for the manufacturer to be able to reduce the selling price by 10 percent without reducing the profit margin?

1.29 In Problem 1.28, the manufacturer can reduce manufacturing costs to $250 per system by designing a custom IC that will decrease the IC count from 100 to 50. This will increase R&D expenses to $500,000, but will also halve the manufacturer's support costs to $500 per system. What is the minimum number of systems that must be produced for the manufacturer to realize a net profit of 20 percent, if each system is sold for $1,000?

1.30 Cite two examples (not mentioned in the text) of common systems whose cost increases rapidly with operating speed or performance. Explain why this tends to be a universal property of systems, ranging from individual ICs to computers.

1.31 The performance of supercomputers is often measured in millions of floating-point operations per second (MFLOPS). It is found that computer A with an MFLOPS rating of f_{A} executes a particular computer program faster than computer B, whose MFLOPS rating of f_{B} is substantially greater than f_{A}. Cite some possible causes of this discrepancy.

1.32 Define each of the following terms in the CAD context: simulation, schematic capture, netlist, knowledge-based synthesis.

CAD systems decrease design time and increase design quality.

1.33 List some ways in which a good CAD system can reduce *time-to-market*—that is, the time span from initiating a new design to delivering a manufactured product to the first customer.

1.34 Derive a truth table for the three-gate circuit of Figure 1.25, using the waveforms given in the figure. What information do the waveforms convey that is not contained in the truth table?

1.35 A general algorithmic approach to most design problems is to conduct an exhaustive search among all possible designs that contain no more than N components. (It is usually easy to find the upper bound N on the size of the designs of interest.) Discuss why this approach can rarely be used in practice, illustrating your answer with a specific design problem of your choosing.

BINARY NUMBERS AND CODES

Number Systems

2.1 Positional Numbers
2.2 Number Base and Word Size
2.3 Binary-to-Decimal Conversion
2.4 Decimal-to-Binary Conversion

Binary Arithmetic

2.5 Unsigned Numbers
2.6 Signed Numbers

Binary Codes

2.7 A Binary Potpourri
2.8 Base-2^k Numbers
2.9 Binary-Coded Decimal Form
2.10 Character Codes
2.11 Error-Detecting Codes
2.12 Floating-Point Numbers

Overview

The Pythagorean school of ancient Greece had as its motto "All is number," a saying that might well apply to modern digital systems. The information processed by such systems is represented in various binary and essentially numerical forms with which the logic designer must be familiar. In this chapter, we examine the main number formats and number operations used in digital machines.

2

Number Systems

The methods we use to represent numbers have evolved in haphazard fashion since ancient times. Of more recent origin, binary numbers are better suited to information-processing machines than are the decimal numbers that humans prefer. In this section, we examine the binary and decimal number systems and their relationship to each other.

2.1 Positional Numbers

A number system is defined by its basic symbols, called **digits** or **numerals**, and the ways in which the digits can be combined to represent the full range of numbers we need. The number symbols with which we are most familiar today are known as **Arabic numerals** due to their development in medieval Arab culture. Arabic numerals consist of the 10 digits 1, 2, 3, 4, 5, 6, 7, 8, 9, and 0. They are derived from earlier forms, a few of which are illustrated in Figure 2.1. Note that a symbol for zero was not introduced until about 800 A.D.

The Decimal System

The decimal system uses positional notation: the weight given to each digit depends on its position.

The 10 Arabic numerals may be combined in various ways to represent any number, large or small. For example,

$$0.001 \qquad 1/1000 \qquad 10^{-3}$$

are three different ways of expressing one thousandth, using these Arabic digits and some auxiliary symbols. A fundamental way of constructing a number is to form a sequence or string of digits in which consecutive digits represent consecutive powers of 10. For instance, the four digits in the number 2979 represent, from left to right, thousands (2), hundreds (the first 9),

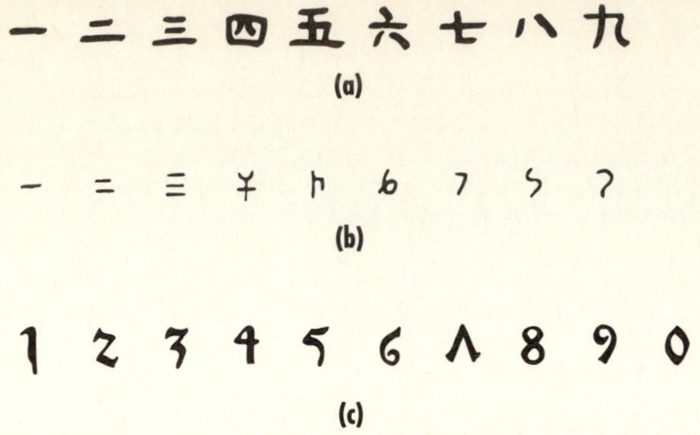

一 二 三 四 五 六 七 八 九

(a)

(b)

(c)

FIGURE 2.1 Some early forms of decimal digits: (**a**) Chinese (2000 B.C.); (**b**) Indian (300 B.C.); (**c**) Arabic (Germany, 15th century).

tens (7), and ones (the second 9). Hence, this four-digit number can be decomposed in the following way:

$$2979 = 2 \times 10^3 + 9 \times 10^2 + 7 \times 10^1 + 9 \times 10^0 \tag{2.1}$$

This method of representing numbers is called the **decimal system** (from the Latin *decem*, meaning *ten*). It is an example of a **positional notation** for numbers, in which each digit of a multidigit number has a fixed value (or **weight**) determined by its position. All the weights used are powers of the number 10 which, along with the number of available digit types, give the decimal system its name.

In the decimal system, fractions are denoted by sequences of digits whose weights are negative powers of 10. A number's whole (integer) and fraction parts, if any, are separated by the special . symbol, the **decimal point**. Thus, the decimal number 3.1416, which approximates pi (π), the ratio of a circle's circumference to its diameter, may be decomposed as follows:

$$3.1416 = 3 \times 10^0 + 1 \times 10^{-1} + 4 \times 10^{-2} + 1 \times 10^{-3} + 6 \times 10^{-4}$$

Other Number Systems

Positional numbers are more compact and easier to compute than nonpositional numbers.

Several number representations that are much older than our decimal system still find occasional use. An example is the "tally" method of counting, which has just a single digit or tally mark | that denotes one. The number n is thus represented by a sequence of n tally marks:

$$||||||| = 1 + 1 + 1 + 1 + 1 + 1 + 1 = 7 \tag{2.2}$$

The **Roman numerals** consist of a small set of letters that serve as numerical digits: I (one), V (five), X (ten), L (fifty), C (one hundred), D (five hundred), and M (one thousand). The number 2979 then takes the nine-digit form MMCMLXXIX, which has the interpretation

$$\text{MMCMLXXIX} = \text{M} + \text{M} - \text{C} + \text{M} + \text{L} + \text{X} + \text{X} - \text{I} + \text{X} \tag{2.3}$$

in Roman numerals. The value of a Roman number is the sum of its digits, except that a smaller digit preceding a larger one is subtracted from rather than added to the sum. In general, Roman numbers are much more compact than the corresponding tally numbers. However, Roman numbers are, in turn, less compact than decimal numbers. Furthermore, each digit position within a Roman number does not have a fixed weight, implying that positional notation is not used; compare Equations (2.1) and (2.3). These features make it far more difficult to perform calculations with Roman numbers than with decimal numbers. Pity the poor Roman schoolchild faced with sums such as LXXXIX + XCVIII = CLXXXVII.

Numbers as Codes

This chapter is mainly concerned with positional number systems employing a few useful bases such as two, powers of two, and 10. We will see that numbers serve as "codes" to denote all kinds of information, both numerical and nonnumerical. For example, the characters forming the text of a document are often encoded into number-like 8-bit words, while the sound elements recorded on an audio CD are 16-bit numbers.

FIGURE 2.2 Universal product code (UPC) symbol denoting a 10-digit decimal number.

Number codes sometimes take unusual forms designed for processing by machine. An example is the **universal product code** (**UPC**) used to place identification numbers on items for sale in stores. Each digit of a decimal number is represented in the UPC code by light and dark bars of specific thickness and spacing, a format designed for easy reading by optical scanning devices. (A detailed description of this code is given in Section 2.7.) Figure 2.2 shows an example of a 10-digit UPC number assigned to a specific product, in this case, a can of Campbell's Chicken Noodle Soup.

2.2 Number Base and Word Size

The radix or base of the binary system is $r = 2$; that of the decimal system is $r = 10$.

We can generalize the decimal number notation illustrated by Equation (2.1) to one in which the key quantity 10 is replaced by some other whole number r, called the **base** or **radix** of the number system. We then require r distinct digits rather than 10, and we must replace each decimal weight of the form 10^i with one of the form r^i. Thus, the three-digit base-r integer $x_2x_1x_0$ can be expressed as

$$x_2x_1x_0 = x_2 \times r^2 + x_1 \times r^1 + x_0 \times r^0$$

where the digits x_i are taken from an r-member set denoted by $\{0, 1, 2, \ldots, r-1\}$. A positional number notation of this type is called a **base-*r*** or ***r*-ary number system**. Some number systems of interest and their names are listed in Table 2.1.

By convention, the first r decimal digits, beginning with 0, serve as the digits for the r-ary systems with $r \leq 10$. The rarely used base-3 (ternary) system, for instance, employs the three digits $\{0, 1, 2\}$. When $r > 10$, the uppercase letters of the alphabet are appended to the decimal digits to provide

TABLE 2.1 Some number systems with their bases and digit sets.

System name	*Base r*	*Digits employed*
binary	2	0, 1
ternary	3	0, 1, 2
octal	8	0, 1, 2, 3, 4, 5, 6, 7
decimal	10	0, 1, 2, 3, 4, 5, 6, 7, 8, 9
hexadecimal	16	0, 1, 2, 3, 4, 5, 6, 7, 8, 9, A, B, C, D, E, F

the additional symbols needed. Thus, the very useful base-16 or **hexadecimal system** defined in Table 2.1 uses the six letters A, B, C, D, E, and F to represent the digits 10, 11, 12, 13, 14, and 15, respectively.

Which Base?

We attach the subscript r to an r-ary number N when we need to indicate its base.

Because the above number systems share the symbols used for digits, a given digit sequence can represent various numbers, depending on the base used. For example, the three-digit number 101 denotes five in the binary number system, because

$$101 = 1 \times 2^2 + 0 \times 2^1 + 1 \times 2^0$$

whereas 101 denotes one hundred and one in the decimal system. Whenever the base of an r-ary number N is unclear from the context, we will append a boldface subscript **r** to it thus: $N_{\mathbf{r}}$. Hence, we write

$$101_{\mathbf{2}} = 5_{\mathbf{10}}$$

and, as we will see shortly,

$$101_{\mathbf{10}} = 1100101_{\mathbf{2}}$$

Table 2.2 shows how the numbers 0 through 20 are represented in the base-r number systems defined in Table 2.1. Note that the larger the base, the fewer digits needed to represent a given number.

A number system can represent integers, fractions, or **mixed numbers**, which are numbers that have both an integer and a fraction part. In mixed decimal numbers, a decimal point separates the integer and fraction parts. In the case of general, r-ary numbers, this separator becomes the **radix point**. For example, the mixed ternary number $102.21_{\mathbf{3}}$ has the following interpretation:

$$102.21_{\mathbf{3}} = 1\times 3^2+0\times 3^1+2\times 3^0+2\times 3^{-1}+1\times 3^{-2} = 11.777\cdots_{\mathbf{10}} \quad (2.4)$$

The radix point in $102.21_{\mathbf{3}}$ is referred to as the *ternary point*.

Word Size

Digital systems process numbers with a fixed number of digits (the word size).

An important aspect of digital systems is that they can handle only numbers with a fixed number of digits n. Computers, for example, have a specified

TABLE 2.2 The numbers up to 20, represented in various bases.

Number	*Base 2*	*Base 3*	*Base 8*	*Base 10*	*Base 16*
zero	0	0	0	0	0
one	1	1	1	1	1
two	10	2	2	2	2
three	11	10	3	3	3
four	100	11	4	4	4
five	101	12	5	5	5
six	110	20	6	6	6
seven	111	21	7	7	7
eight	1000	22	10	8	8
nine	1001	100	11	9	9
ten	1010	101	12	10	A
eleven	1011	102	13	11	B
twelve	1100	110	14	12	C
thirteen	1101	111	15	13	D
fourteen	1110	112	16	14	E
fifteen	1111	120	17	15	F
sixteen	10000	121	20	16	10
seventeen	10001	122	21	17	11
eighteen	10010	200	22	18	12
nineteen	10011	201	23	19	13
twenty	10100	202	24	20	14

word size, which is the length (number of bits) of the binary numbers processed by the computer's internal instructions. Word sizes are typically powers of two, and range from 8 bits (a byte) in some microcomputers to 64 or more bits in large supercomputers.

Figure 2.3 shows two examples of binary numbers whose word size is eight. The first example (Figure 2.3a) denotes the integer 20, which, as you know from Table 2.2, is the 5-bit sequence 10100_2. To fit this number into an 8-bit format, it is extended to 8 bits by appending three 0s to its left end. These "leading zeroes" do not affect the numerical value of the word, and so are termed **insignificant**. (Digits whose removal changes a number's value are **significant**.) We regard 00010100 as the 8-bit word representing the number 20.

We can also use a fixed word size to represent fractions. Figure 2.3b gives an 8-bit representation of the fraction $\frac{29}{64}$, whose binary form is 0.011101_2. There are six significant digits, 011101, which are extended to 8 bits by appending two insignificant "trailing" zeroes, as shown in the figure. We conclude in this case that the 8-bit word representing $\frac{29}{64}$ is 01110100.

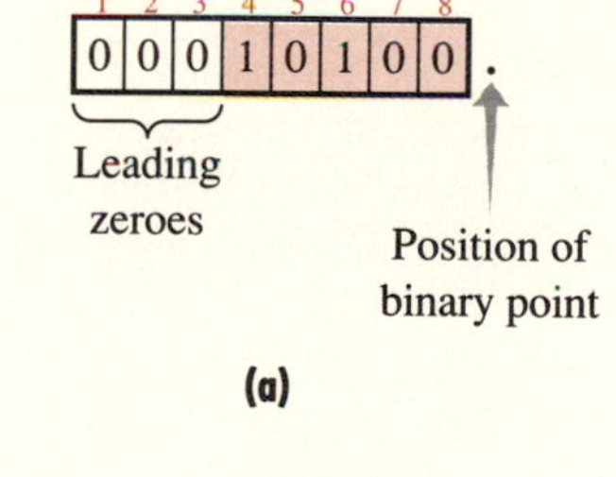

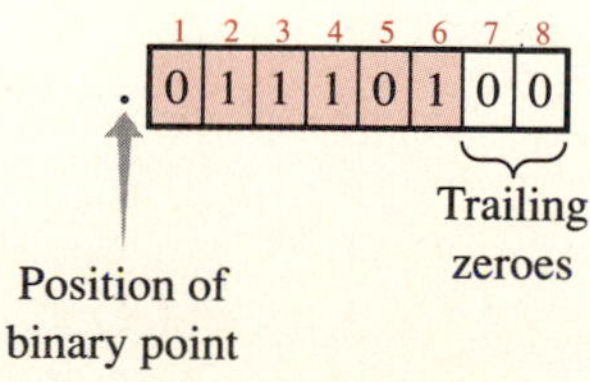

FIGURE 2.3 Two 8-bit binary numbers: (a) the integer 20; (b) the fraction $\frac{29}{64}$.

The Radix Point

Given a base-r number word N_r of fixed length, such as 01101100_2, we cannot determine its numerical value unless we know where the radix point is located. If 01101100 is an integer, there is an implicit binary point immediately on the right of the number word, which therefore denotes 1101100.0_2. If 01101100 is a fraction, the implicit binary point lies immediately to the left of the number, which is then interpreted as 0.0110111_2.

In general, a number word N_r has a fixed length n and a radix point assigned to some implicit but fixed position. If N_r is an integer, the radix point is immediately to the right of its rightmost digit. If N_r is a fraction, the radix point is immediately to the left of its leftmost digit. The radix point may also be assigned to some intermediate position to represent mixed numbers. If $r = 10$ and N_{10} represents dollars and cents, then a mixed format is often used, with the decimal point placed between the second and third digits from the right, as in 519.62_{10}. Figure 2.4 illustrates a word format of length 16 bits intended for mixed binary numbers, with a 10-bit integer part and a 6-bit fraction part. Observe that there is an implicit binary point between bit positions 10 and 11.

Once the position of a number's radix point has been agreed upon, it is not necessary to include it explicitly in the number representation. In the case of nonmixed numbers, it is sufficient to state whether the numbers under consideration are integers or fractions. The position of the radix point must be specified clearly for mixed numbers.

Precision

A fixed word length limits a number system's precision.

An r-ary number's **precision** or accuracy is the number of significant digits it contains. Suppose that such a number N contains $n + k$ significant digits, and we must make it fit within the n-digit word format used in a digital system. Clearly, we must remove k digits from N. The resulting n-digit number N^* differs from the "true" value N by some amount E, representing a numerical error. The value of E depends on r, the digits removed, and the position of the radix point. For example, to fit $N = 1101011.1101_2$ into an 8-bit word, we could remove its three rightmost or least significant bits, thereby reducing it to $N^* = 1101011.1_2$. The loss of three digits of precision means that N^* has an error of 0.0101_2 with respect to N. Although such errors appear small,

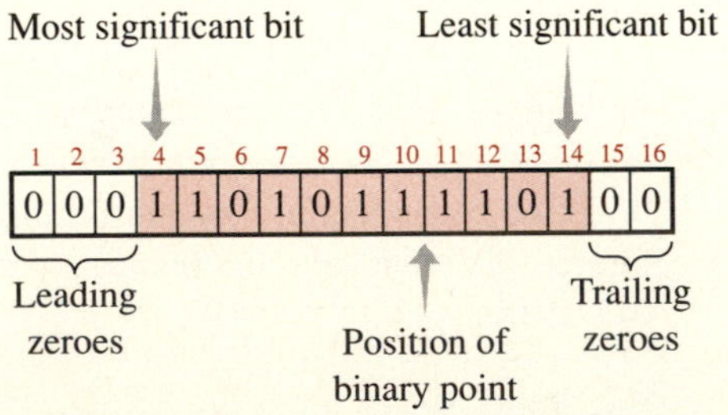

Most significant bit
Least significant bit
1 2 3 4 5 6 7 8 9 10 11 12 13 14 15 16
0 0 0 1 1 0 1 0 1 1 1 1 0 1 0 0
Leading zeroes
Position of binary point
Trailing zeroes

FIGURE 2.4 16-bit word representing the mixed binary number 1101011.1101_2.

they can accumulate over a long series of calculations to create big errors with serious consequences. This problem may be addressed by increasing the word size n or by using several words to represent a single number.

The precision with which a particular number can be represented is also affected by the number base we use. For example, the ternary number 102.21_3 appearing in Equation (2.4) corresponds to the decimal number $11.777\cdots_{10}$, which has a repeating fraction part of infinite length. Hence, we can represent this number exactly by a five-digit word if base 3 is used, but we cannot represent it exactly by a decimal number, no matter how long. In general, for every base r and word size n, there are many numbers that cannot be represented exactly by an n-digit r-ary word.

Base-r numbers that are of fixed length n and have an implicit radix point in a fixed position are referred to collectively as **fixed-point numbers.** Later we will see a different class of numbers called **floating-point numbers**, which allow extremely large and small numbers to be efficiently represented within an n-digit word format. We also postpone discussion of the representation of negative numbers, called **signed numbers.**

2.3 Binary-to-Decimal Conversion

As observed already, the binary number system is the primary one used in digital circuits. Because humans use the decimal system, it is important to know how to convert numbers from binary to decimal form, and vice versa. Indeed, hardware circuits or software programs to make such base conversions are found in almost every digital system (Figure 2.5).

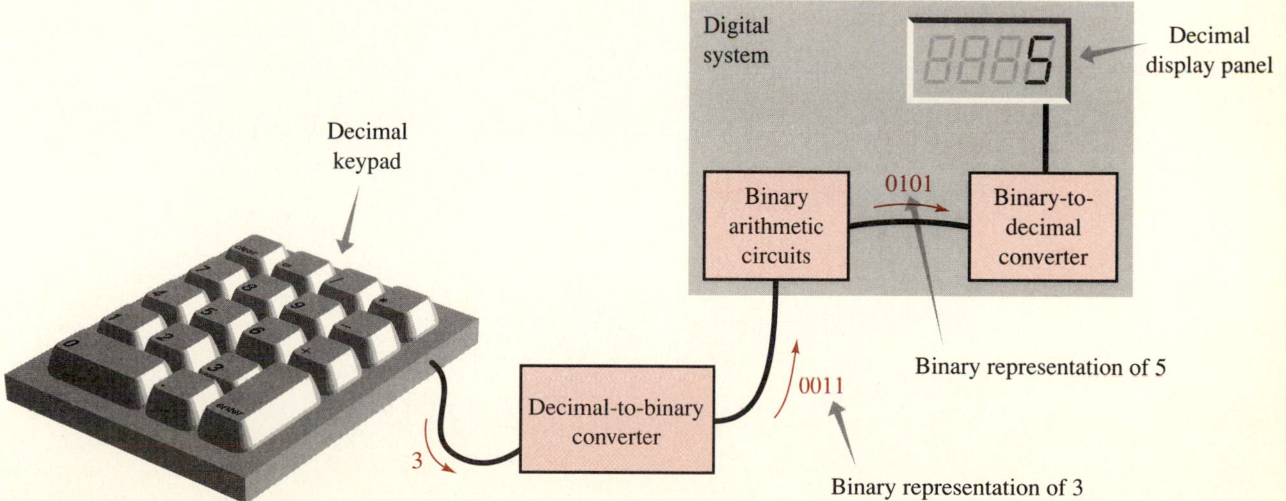

FIGURE 2.5 Conversion of numbers between binary and decimal forms.

Converting Integers

First we consider the task of converting to decimal form an unsigned binary integer N given in the general n-bit binary form

$$N_2 = b_{n-1}b_{n-2}\cdots b_1b_0 \tag{2.5}$$

where b_i is 0 or 1, and $0 \leq i \leq n-1$. Note that by numbering the bits from right to left starting with 0, we ensure that the weight of each bit b_i is 2^i for all i. Consequently, we can express N as

$$N = b_{n-1} \times 2^{n-1} + b_{n-2} \times 2^{n-2} + \cdots + b_1 \times 2^1 + b_0 \times 2^0$$

or, more concisely, as

$$N = \sum_{i=0}^{n-1} b_i \times 2^i \tag{2.6}$$

where b_i is defined by (2.5).

A possible binary-to-decimal conversion method is to expand the given binary number N_2 into a sum of powers of two along the lines of (2.6), and then evaluate this sum using ordinary decimal arithmetic. The result of the summation is the desired decimal form N_{10} of the given binary number. For example,

$$\begin{aligned} 11101 &= 1 \times 2^4 + 1 \times 2^3 + 1 \times 2^2 + 0 \times 2^1 + 1 \times 2^0 \qquad (2.7)\\ &= 16 + 8 + 4 + 0 + 1 \\ &= 29 \end{aligned}$$

Humans, who use decimal numbers, must communicate with machines, which use binary numbers.

Direct evaluation of this type is convenient for hand calculations. However, it is not well suited to machine implementation, because it requires a relatively complex (in terms of hardware or computing time) exponentiation operation to compute each power of two. Exponentiation can be avoided entirely by performing repeated multiplication by two. This is based on the observation that (2.6) can be rewritten as

$$\begin{aligned} N &= 2 \times \left(\sum_{i=1}^{n-1} (b_i \times 2^{i-1}) \right) + b_0 \\ &= 2 \times \left(2 \times \left(\sum_{i=2}^{n-1} (b_i \times 2^{i-2}) + b_1 \right) \right) + b_0 \\ &\qquad \cdots \\ &= 2 \times (2 \times \cdots (2 \times (2 \times b_{n-1} + b_{n-2}) \cdots) + b_1) + b_0 \end{aligned}$$

Hence, we can rewrite (2.7) as

$$\begin{aligned}11101 &= 2 \times (2 \times (2 \times (2 \times 1 + 1) + 1) + 0) + 1 \\ &= 2 \times (2 \times (2 \times 3 + 1) + 0) + 1 \\ &= 2 \times (2 \times 7 + 0) + 1 \\ &= 2 \times 14 + 1 \\ &= 29\end{aligned}$$

Base conversion is thus reduced to a sequence of n multiplications by two and n additions. The conversion process is summarized below as a semiformal procedure or algorithm named *BINDECi* (binary-to-decimal integer) for ease of reference.

Procedure

Procedure *BINDECi* for converting a binary integer to decimal format

1. Let $N_2 = b_{n-1}b_{n-2}\cdots b_0$ be the binary integer to be converted to the decimal form N_{10}. Set N_{10} to an initial value of zero.
2. Scan N_2 from left to right, and for each bit b_i in turn, compute $2 \times N_{10} + b_i$ and make this the new value of N_{10}. The final value of N_{10} obtained after n steps is the desired result.

Example 2.1

Converting a binary integer to decimal

Consider the problem of converting the 8-digit binary integer 01100101_2 to decimal form. With our particular subscript ordering, we have

$$N_2 = 01100101_2 = b_7b_6b_5b_4b_3b_2b_1b_0$$

Applying *BINDECi* to this number, we first set N_{10} to zero. The conversion then proceeds in the following eight steps:

$$\begin{aligned}
i &= 7 \quad & N_{10} &= 2 \times 0 + b_7 = 0 \\
i &= 6 & N_{10} &= 2 \times 0 + b_6 = 1 \\
i &= 5 & N_{10} &= 2 \times 1 + b_5 = 3 \\
i &= 4 & N_{10} &= 2 \times 3 + b_4 = 6 \\
i &= 3 & N_{10} &= 2 \times 6 + b_3 = 12 \\
i &= 2 & N_{10} &= 2 \times 12 + b_2 = 25 \\
i &= 1 & N_{10} &= 2 \times 25 + b_1 = 50 \\
i &= 0 & N_{10} &= 2 \times 50 + b_0 = 101
\end{aligned}$$

We conclude that $N_{10} = 101_{10}$. Note that all the above calculations are performed using ordinary decimal arithmetic, so that the result is automatically in the required base-10 form.

In contrast, direct evaluation of N_2 as a sum of powers of two proceeds as follows:

$$\begin{aligned}N_{10} &= 0 \times 2^7 + 1 \times 2^6 + 1 \times 2^5 + 0 \times 2^4 + 0 \times 2^3 + 1 \times 2^2 + 0 \times 2^1 + 1 \times 2^0 \\ &= 0 + 64 + 32 + 0 + 0 + 4 + 0 + 1 = 101\end{aligned}$$

Converting Fractions and Mixed Numbers

The base conversion procedure developed above can easily be modified to handle fractions or mixed binary numbers. If N is a fraction and $N_2 = b_{n-1}b_{n-2}\cdots b_1b_0$, then

$$\begin{aligned}N &= b_{n-1} \times 2^{-1} + b_{n-2} \times 2^{-2} + \cdots + b_1 \times 2^{1-n} + b_0 \times 2^{-n} \\ &= \sum_{i=0}^{n-1} b_i \times 2^{i-n}\end{aligned} \tag{2.8}$$

Mixed number N_2 has m fraction bits and $n - m$ integer bits.

Comparing (2.5) and (2.6) with (2.8), we see that each power of two in the summation is reduced by n, reflecting the shift of the binary point n places to the right in N_2. In general, if

$$N_2 = b_{n-1}b_{n-2}\cdots b_m.b_{m-1}\cdots b_1b_0 \tag{2.9}$$

↑ Binary point

and the binary point lies between bits b_m and b_{m-1}, then

$$N = \sum_{i=0}^{n-1} b_i \times 2^{i-m} \tag{2.10}$$

Suppose that we extract the factor 2^{-m} from each term of the summation in (2.10). We can then write

$$N = 2^{-m}\left(\sum_{i=0}^{n-1} b_i \times 2^i\right) = 2^{-m} \times N' \tag{2.11}$$

where N' is an integer, as can be seen from the values of the index i. The term 2^{-m} in (2.11) is a **scale factor** that has the effect of shifting the binary point of N' m places to the left to change it to mixed form. Similarly, 2^m is a scale factor that shifts the binary point m places to the right, and so will convert a mixed binary number to an integer.

We use the scaling ideas described above to construct the following procedure *BINDECm* for converting to decimal a mixed binary number N_2 of the form (2.9).

Procedure

Procedure *BINDECm* for converting a mixed binary number to decimal format

1. Multiply the given number N_2 by scale factor 2^m to change it into a binary integer N_2'.
2. Use the integer conversion procedure *BINDECi* to convert N_2' to the decimal form N_{10}'.
3. Finally, multiply N_{10}' by scale factor 2^{-m} to obtain the desired (mixed) decimal result N_{10}.

To illustrate *BINDECm*, suppose that $N_2 = 011001.01_2$, for which $m = 2$, is to be converted to decimal form N_{10}. To multiply N_2 by 2^m we shift the binary point $m = 2$ places to the right, producing the integer $N_2' = 01100101_2$. As demonstrated in Example 2.1, *BINDECi* converts this integer to $N_{10}' = 101_{10}$. Hence,

$$N_{10} = N_{10}' \times 2^{-2} = 101_{10} \times 0.25_{10} = 25.25_{10}$$

Note that *BINDECm* reduces to *BINDECi* when $m = 0$, and to a fraction conversion procedure when $m = n$.

2.4 Decimal-to-Binary Conversion

Next we consider the problem of converting a number N from base 10 to base 2. Initially we will assume that the numbers of interest are integers; we will consider the fraction and mixed cases later.

Converting Integers

We have seen that a binary number N_2 can be converted to decimal form by repeatedly multiplying it by two. As might be expected, the converse process, decimal-to-binary conversion, involves repeatedly dividing a given decimal integer N_{10} by two. The goal is to determine how many powers of two are contained in the number so that it can be expressed in the binary form

$$N_2 = b_{n-1}b_{n-2}\cdots b_2b_1b_0 = \sum_{i=0}^{n-1} b_i \times 2^i$$

Suppose we divide N_{10} by two to obtain a quotient Q_0 and a remainder R_0. If N_{10} is odd, the remainder $R_0 = 1$; if N_{10} is even, $R_0 = 0$. Hence, $R_0 = b_0$, the rightmost bit of N_2. Now divide the quotient Q_0 by two, producing a new quotient Q_1 and a remainder R_1. In this case, we must have $R_1 = b_1$. Division of Q_1 by two yields a remainder $R_2 = b_2$, and so on. Hence, a sequence of n divisions by two yields the n remainders $b_0, b_1, \ldots,$ which specify N_2. Each divide-by-two step cuts the current quotient approximately in half, so that it eventually reaches zero. Continued division beyond this point yields only $b_i = 0$; that is, it yields insignificant leading zeroes. A procedure named *DECBINi* (decimal-to-binary integer) that summarizes this method follows.

Procedure

Procedure *DECBINi* for converting a decimal integer to binary format

1. Let N_{10} be the decimal number to be converted to the binary format $N_2 = b_{n-1}b_{n-2}\cdots b_0$. Set an index (an iteration counter) i to an initial value of zero.
2. Divide N_{10} by two to obtain a remainder and a quotient. Set b_i to the remainder value and N_{10} to the quotient value.
3. If $N_{10} = 0$, halt. Otherwise, replace i with $i + 1$ and repeat step 2.

Example 2.2

Integer decimal-to-binary conversion

Let us apply *DECBINi* to the decimal integer $N_{10} = 249_{10}$. The integer quotient on division of N by two is denoted by $N \div 2$.

$$
\begin{array}{lll}
i = 0 & N_{10} = 249 \div 2 = 124 & b_0 = 1 \\
i = 1 & N_{10} = 124 \div 2 = 62 & b_1 = 0 \\
i = 2 & N_{10} = 62 \div 2 = 31 & b_2 = 0 \\
i = 3 & N_{10} = 31 \div 2 = 15 & b_3 = 1 \\
i = 4 & N_{10} = 15 \div 2 = 7 & b_4 = 1 \\
i = 5 & N_{10} = 7 \div 2 = 3 & b_5 = 1 \\
i = 6 & N_{10} = 3 \div 2 = 1 & b_6 = 1 \\
i = 7 & N_{10} = 1 \div 2 = 0 & b_7 = 1
\end{array}
$$

Because N_{10} is now zero, we halt with the answer $N_2 = b_7b_6b_5b_4b_3b_2b_1b_0 = 11111001_2$.

Converting Fractions and Mixed Numbers

We turn next to the task of converting a general decimal number of the form

$$N_{10} = d_{p-1}d_{p-2}\cdots d_m.d_{m-1}\cdots d_1d_0 \tag{2.12}$$

↑ Decimal point

to base 2. It might be expected that we could use the approach adopted in the binary-to-decimal conversion procedure *BINDECm*, where we would (1) convert N_{10} to an integer N'_{10} by multiplying it by the scale factor 10^m, (2) apply our integer decimal-to-binary procedure *DECBINi* to obtain N'_2, and (3) compute N_2 by multiplying N'_2 by 10^{-m}. The problem with this approach is that to get the final result N_2 in the proper binary form, we would have to use binary rather than ordinary decimal arithmetic for the final multiplication step $N_2^{\prime} \times 10^{-m}$. To avoid this major complication, we develop a different decimal-to-binary conversion method for mixed numbers.

The decimal number N_{10} in (2.12) can be separated into an integer part

$$N_{10}^{\mathrm{I}} = d_{p-1}d_{p-2}\cdots d_{m+1}d_m.0$$

and a fraction part

$$N_{10}^{\mathrm{F}} = 0.d_{m-1}d_{m-2}\cdots d_1 d_0 \tag{2.13}$$

as indicated by the superscript I or F. Conversion of N_{10}^{I} to N_2^{I} can be handled by *DECBINi*. To deal with the fraction, we introduce the new procedure *DECBINf*, which converts N_{10}^{F} to N_2^{F}. After conversion, the binary results are joined with a binary point to create the final answer $N_2 = N_2^{\mathrm{I}}.N_2^{\mathrm{F}}$.

Procedure

Procedure *DECBINf* for converting a decimal fraction to binary format

1. Let N_{10} be the decimal fraction to be converted to the binary number $N_2 = b_{-1}b_{-2}\cdots b_{-n+1}b_{-n}$. Set an index (an iteration counter) i to an initial value of one.
2. Multiply N_{10} by two. If the result is less than one, set b_{-i} to 0; otherwise, set b_{-i} to 1 and replace N_{10} with $N_{10} - 1$.
3. If $i = n$, halt. Otherwise, replace i with $i + 1$ and repeat step 2.

It is worthwhile for us to explore the implications of this procedure. Let $N_2^{\mathrm{F}} = 0.b_{-1}b_{-2}\cdots$ be the desired binary equivalent of the fraction N_{10}^{F}. *DECBINf* is based on the simple fact that multiplication of the fraction N_{10}^{F} by two using ordinary arithmetic yields a decimal number of the form $A = d_0.d_{-1}d_{-2}\ldots$ in which the ones digit d_0 is either 0 or 1. Now $d_0 = 1$ if and only if $N^{\mathrm{F}} \geq 0.5$, and $d_0 = 0$ if and only if $N^{\mathrm{F}} < 0.5$. Returning to N_2^{F}, we see that $b_{-1} = 1$ if and only if $N^{\mathrm{F}} \geq 0.5$, and $b_{-1} = 0$ if and only if $N^{\mathrm{F}} < 0.5$. Hence, $d_0 = b_{-1}$. In other words, the integer part of the result of multiplying the decimal fraction by two is the first bit of the binary fraction. For example, if $N_{10}^{\mathrm{F}} = 0.62705$, then multiplication by two yields $A = 1.25410$, implying that $b_{-1} = 1$ in this case. We can now take the fraction part of A and multiply it by two to obtain the next bit b_{-2} of N_2^{F}, and so on. In general, n multiplications by two are required to obtain the first n bits of a binary fraction.

Example 2.3

Decimal fraction conversion to binary format

To illustrate the conversion of fractions, let $N_{10} = 0.62705$, and suppose that an 8-bit result $N_2 = b_{-1}b_{-2}b_{-3}b_{-4}b_{-5}b_{-6}b_{-7}b_{-8}$ is desired. Application of *DECBINf* to N_{10} yields the following:

$$
\begin{array}{lll}
i = 1 & 0.62705 \times 2 = 1.25410 & b_{-1} = 1 \\
i = 2 & 0.25410 \times 2 = 0.50820 & b_{-2} = 0 \\
i = 3 & 0.50820 \times 2 = 1.01640 & b_{-3} = 1 \\
i = 4 & 0.01640 \times 2 = 0.03280 & b_{-4} = 0 \\
i = 5 & 0.03280 \times 2 = 0.06560 & b_{-5} = 0 \\
i = 6 & 0.06560 \times 2 = 0.13120 & b_{-6} = 0 \\
i = 7 & 0.13120 \times 2 = 0.26240 & b_{-7} = 0 \\
i = 8 & 0.26240 \times 2 = 0.52480 & b_{-8} = 0
\end{array}
$$

We conclude, therefore, that $N_2 = 0.10100000_2$.

Truncation and Rounding

Numbers with too many digits must be truncated or rounded.

The 8-bit binary fraction N_2 obtained in the last example is an inexact equivalent of N_{10}. The process of terminating number construction after a fixed number of bits n, or simply cutting off all but n digits of a number, is termed **truncation**. It can lead to significant inaccuracies in subsequent calculations that involve the truncated number. Had we continued the base conversion in Example 2.3 for several more steps, we would have obtained these results:

$$
\begin{array}{lll}
i = 9 & N_{10} = 1.0496 & b_{-9} = 1 \\
 & \cdots & \\
i = 14 & N_{10} = 1.5872 & b_{-14} = 1
\end{array}
$$

It follows that N is closer to 0.10100001_2 than to 0.10100000_2; hence, to make N_2 correct to eight binary places, the rightmost bit b_{-8} should be changed from 0 to 1. This can be accomplished by **rounding**, which replaces N with a number N^* that is as close to N as possible, within the specified constraints on number length or word size. Accurate rounding raises some subtle issues, which we only touch on briefly here.

Suppose that a binary fraction $N_2 = b_{-1}b_{-2}\cdots b_{-n}b_{-n-1}\cdots$ has been computed to at least $n + 1$ binary places. We can round off N_2 to n places using the following two-step procedure *ROUND*, which is easily implemented in hardware or software.

Procedure

Procedure *ROUND* for rounding a binary fraction to *n* places

1. Add to the given number $N_2 = b_{-1}b_{-2}\cdots b_{-n}b_{-n-1}$ the fraction $0.00\cdots01$, which has a single 1 in the bit position corresponding to b_{-n-1}.
2. Truncate the resulting sum to n bits.

This rounding technique sets N_2 to $b_{-1}b_{-2}\cdots b_{-n} + 2^{-n}$ if

$$0.00\cdots00b_{-n-1}b_{-n-2}\cdots \geq 2^{-n-1} \tag{2.14}$$

and to $b_{-1}b_{-2}\cdots b_{-n}$ otherwise. For example, if $N = 0.01111001_2$, then rounding to four places results in 0.1000_2, with an error (the rounded value of N minus the exact value) of $+0.00000111_2$. Rounding to five places yields 0.01111_2, with an error of -0.00000001_2.

Procedure Specification

Problem-solving procedures may be expressed formally by flowcharts or pseudocode.

Many of the computing techniques embodied in digital systems involve complex and intricate procedures or algorithms. The same is true of the computer-aided methods employed in their design. Such procedures often need to be defined in more precise and formal terms than those we have employed so far for such relatively simple procedures as *BINDECi* and *ROUND*.

Figure 2.6 presents examples of two of the most common formal specification methods, applied to the procedure *DECBINf* for decimal-to-binary conversion of fractions. The graphical description given in Figure 2.6a is an example of a flowchart. The description of Figure 2.6b is in **pseudocode**, a simplified program-like format written here in a language resembling Pascal. Pseudocode descriptions are characterized by the fact that minor details, such as the input/output operations needed for actual execution of the procedure, are omitted or written in abbreviated form. These details are enclosed in braces {...} to serve as explanatory comments.

The formal description methods illustrated by Figure 2.6 are intended to be largely self-explanatory. Execution of the specified procedure *DECBINf* corresponds to tracing a path through the flowchart or the pseudocode, performing the specified computations, and changing direction in accordance with the outcomes of the various decisions that must be made. In the flowchart, computational steps are represented by rectangular boxes, and decisions by diamond-shaped boxes. Both flowcharts and formal languages are used extensively in the design of digital systems. Flowcharts have a certain visual appeal to human designers, whereas language-based descriptions are more easily manipulated by computers.

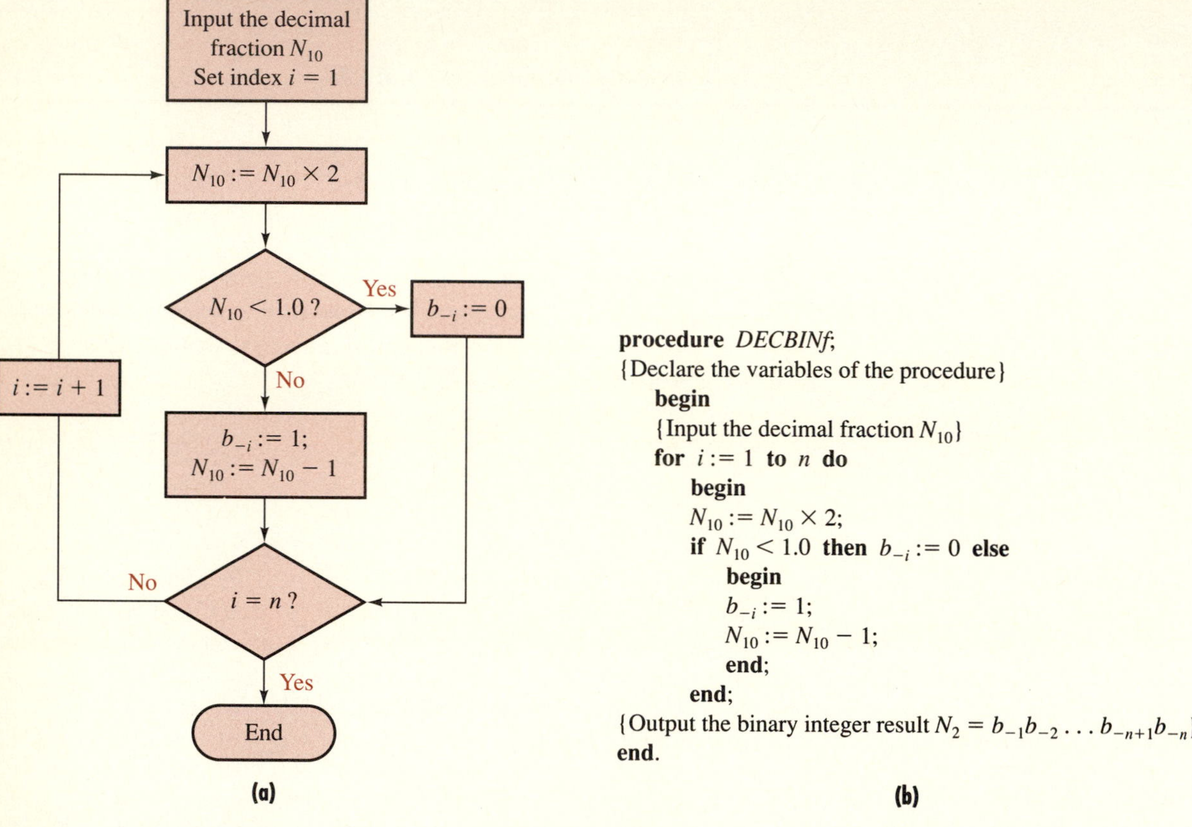


```
procedure DECBINf;
{Declare the variables of the procedure}
   begin
   {Input the decimal fraction N10}
   for i := 1 to n do
      begin
      N10 := N10 × 2;
      if N10 < 1.0 then b-i := 0 else
         begin
         b-i := 1;
         N10 := N10 - 1;
         end;
      end;
{Output the binary integer result N2 = b-1b-2 . . . b-n+1b-n}
end.
```

(b)

FIGURE 2.6 Two formal specifications of the procedure *DECBINf:* **(a)** a flowchart; **(b)** a formal language description (pseudocode).

Binary Arithmetic

First we examine the two basic operations, addition and subtraction, applied to unsigned binary numbers. We go on to consider the two most widely used codes, known as signed magnitude and two's complement, which represent negative as well as positive numbers—that is, signed numbers. We will see that each of these codes has advantages when implementing binary arithmetic by means of digital circuits.

2.5 Unsigned Numbers

The fundamental operations of arithmetic—addition, subtraction, multiplication, and division—can be performed in much the same way in all r-ary number systems. The main difference is the size r of the digit sets used; decimal operations employ the digit set {0,1,2,3,4,5,6,7,8,9}, while binary systems employ the set {0,1}. The reduction in the size of r from 10 to 2 generally increases the number of steps needed to complete a particular operation, but the individual steps are simplified by the fact that the digit choice is limited to two.

Addition

As in ordinary decimal arithmetic, the sum of two unsigned r-ary (mixed) numbers can be obtained by adding them digit by digit, starting at the right (least significant) end and moving to the left (most significant) end. Consider, for example, the addition of two 4-bit binary numbers

$$X = x_3x_2x_1x_0 = 1001 = \text{nine}$$

and

$$Y = y_3y_2y_1y_0 = 0011 = \text{three}$$

to form the sum

$$S = s_3s_2s_1s_0 = 1100 = \text{twelve} \tag{2.15}$$

We begin by adding $x_0 = 1$ to $y_0 = 1$. The result is 10 = two, which we interpret as a "sum" bit $s_0 = 0$ with a "carry" bit $c_0 = 1$. We then move left to add $x_1 = 0$ to $y_1 = 1$; we must, however, also include $c_0 = 1$ in the addition. Hence, we compute $x_1 + y_1 + c_0 = 0 + 1 + 1$, which is again 10 = two, implying that $s_1 = 0$ and $c_1 = 1$. A third addition step yields $x_2 + y_2 + c_1 = 0 + 0 + 1 = 1$, so this time the sum bit $s_2 = 1$, while the corresponding carry bit is $c_2 = 0$. A final addition, involving the leftmost bits yields $s_3 = 1$ and the desired result (2.15). Figure 2.7a gives a pictorial representation of this step-by-step addition procedure.

1 + 1 is not always 2; it's 10 in the binary system.

Now consider the general case, in which we want to add two r-ary numbers $X = \cdots x_{i+1}x_ix_{i-1}\cdots$ and $Y = \cdots y_{i+1}y_iy_{i-1}\cdots$. Each base-$r$ addition step $x_i + y_i$ yields a two-digit intermediate result c_is_i consisting of a sum digit s_i and a carry digit c_i. The sum digit can take any value in the set $\{0, 1, \ldots, r - 1\}$. The carry digit is always either 0 or 1, so it is correct to call it a bit, even in nonbinary systems. When adding x_i and y_i, we must also add the carry bit c_{i-1} due to earlier addition steps, giving it weight r. Thus, a complete or **full-add step**[1] is expressed as follows:

$$c_is_i = x_i + y_i + c_{i-1} \tag{2.16}$$

[1]The operation $c_is_i = x_i + y_i$ omitting the input carry bit is sometimes called a "half" addition.

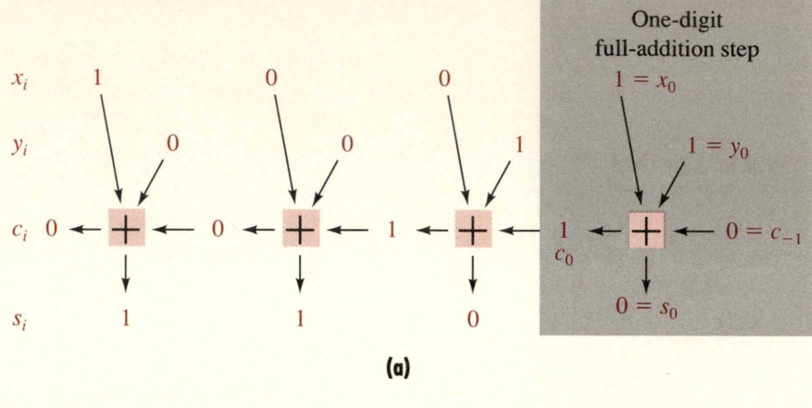

(a)

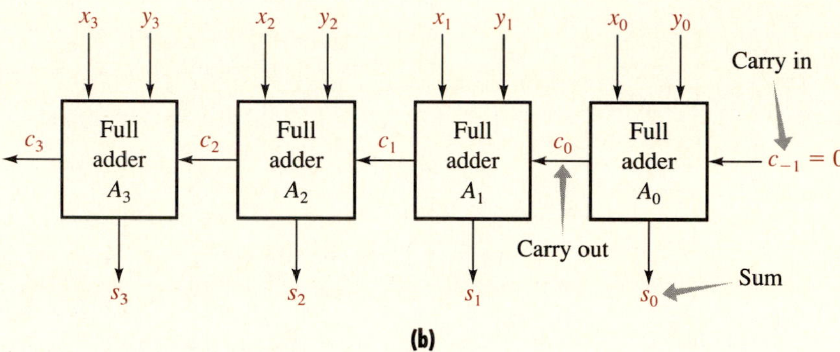

(b)

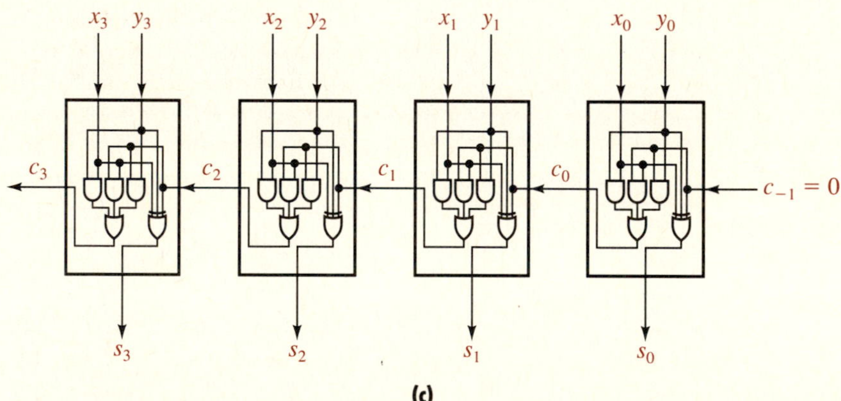

(c)

FIGURE 2.7 Addition of two 4-bit numbers $x_3x_2x_1x_0$ and $y_3y_2y_1y_0$: **(a)** the add procedure illustrated; **(b)** the block diagram of the adder; **(c)** the logic circuit of the adder.

Full addition computes the sum of a digit from X, the corresponding digit from Y, and a carry digit (bit); it is also referred to as **one-digit addition**.

Addition Procedure

In the binary system, (2.16) yields eight ways for the three input variables x_i, y_i, and c_{i-1} to determine the carry-sum pair $c_i s_i$:

These equations completely define 1-bit full addition.

$$\begin{aligned} c_i s_i \quad & x_i \;\; y_i \;\; c_{i-1} \\ 00 &= 0+0+0 \\ 01 &= 0+0+1 \\ 01 &= 0+1+0 \\ 10 &= 0+1+1 \\ 01 &= 1+0+0 \\ 10 &= 1+0+1 \\ 10 &= 1+1+0 \\ 11 &= 1+1+1 \end{aligned} \tag{2.17}$$

Equations (2.17) thus completely define full addition for the binary case.

To add two n-digit numbers, we repeat the full-add operation $c_i s_i = x_i + y_i + c_{i-1}$ n times, once for each of the n digit positions. We start on the right with $i = 0$, and proceed toward the left, digit by digit, until $i = n-1$. The final result is an n-digit sum $S_r = s_{n-1}s_{n-2}\cdots s_0$. Of the n carry values computed, we retain only the leftmost one $c_{out} = c_{n-1}$. The first (rightmost) addition step requires an input carry bit $c_{in} = c_{-1}$, which is set to 0. (The external carry bits c_{in} and c_{out} have several uses, which we will see later.) We may formalize the process of adding two n-digit numbers by a sequence of n full-addition steps in a procedure *ADDn*.

Procedure

Procedure *ADDn* to add two *n*-digit numbers and a carry bit c_{in}

1. Let $X = x_{n-1}x_{n-2}\cdots x_0$ and $Y = y_{n-1}y_{n-2}\cdots y_0$ be the n-digit numbers to be added. Set the external input carry bit $c_{in} = c_{-1}$ to 0, and set i to 0.
2. Perform the one-digit full addition $c_i s_i = x_i + y_i + c_{i-1}$.
3. If $i < n-1$, replace i with $i+1$ and repeat step 2; otherwise, halt with the $(n+1)$-digit result $c_{n-1}s_{n-1}s_{n-2}\cdots s_1 s_0$, where $c_{out} = c_{n-1}$.

Although we are mainly interested in binary addition, we can use *ADDn* with numbers of any base r. Consider, for instance, adding two numbers $X_5 = x_1x_0 = 32_5$ and $Y_5 = y_1y_0 = 34_5$ from the quinary or base-5 system. Applying *ADDn*, we first set $c_{-1} = 0$ and compute $x_0 + y_0 + c_{-1} = c_0 s_0 = 2+4+0 = 11$. Next we compute $x_1 + y_1 + c_0 = c_1 s_1 = 3+3+1 = 12$. Hence, the answer is $c_1 s_1 s_0 = 121_5$.

Procedure *ADDn* has the advantage of breaking down a multidigit sum into a sequence of very simple steps. This digit-by-digit approach is taught to elementary schoolchildren for adding decimal numbers—or was, before the advent of the ubiquitous pocket calculator. As we will see shortly, hardware implementations of *ADDn* are common in digital systems and share its simplicity.

Logic design involves mapping procedures such as *ADDn* into hardware.

Example 2.4

Adding two unsigned 8-bit binary numbers

Suppose we want to add two 8-bit numbers $X_2 = 10100001_2$ and $Y_2 = 00110011_2$. The first step of *ADDn* with $i = 0$ adds the rightmost bits $x_0 = 1$ and $y_0 = 1$ of X_2 and Y_2, as well as the input carry bit $c_{-1} = 0$ to produce

$$c_0 s_0 = 1 + 1 + 0 = 10$$

so $c_0 = 1$ and $s_0 = 0$. The next step yields

$$c_1 s_1 = x_1 + y_1 + c_0 = 0 + 1 + 1 = 10$$

and so on. The entire computation is conveniently laid out as follows for hand calculation:

	$X_2 =$	1	0	1	0	0	0	0	1
	$+Y_2 =$	0	0	1	1	0	0	1	1
Carries		0 0	1	0	0	0	1	1	
Sum	$S_2 =$	1	1	0	1	0	1	0	0

Note that although X_2, Y_2, and S_2 have been indexed as integers, the location of the binary point plays no role in addition, as long as it is assumed to be in the same position in each number—say, between bits j and $j - 1$. Thus *ADDn* applies to all types of mixed r-ary numbers.

Overflow

The final (leftmost) carry bit $c_{out} = c_7$ in the preceding example is 0 and so can be ignored. If, however, $c_7 \neq 0$, which would be the case if X_2 were changed to $X_2' = 11010001_2$, then the new sum $S_2' = X_2' + Y_2$ is a 9-bit number—namely, 100000100_2. In general, if all numbers are required to have a fixed length, in this case 8 bits, then we must conclude that the result is too long to fit in the available word size, a condition called **overflow**. If the word size is fixed at n and the numbers are integers, then all results should lie in the range defined by the smallest number $00 \cdots 0_2 =$ zero and the largest number $11 \cdots 1_2$ to avoid overflow. When $n = 8$, the largest representable integer is

$$11111111_2 = 2^8 - 1 = 255_{10}$$

Similarly, the largest representable 8-bit fraction is

$$11111111_2 = 1 - 2^{-8} = 0.99609375_{10}$$

Results of addition that lie outside the representable number range lose their most significant bit; that is, they overflow, a potentially serious error condition that can be detected by inspecting c_{out}.

Adder Circuit

The bit-by-bit addition scheme discussed above is the basis of a very useful circuit for adding n-bit numbers. As a foretaste of things to come, examine the overall structure of this adder, shown as a block diagram in Figure 2.7b for $n = 4$. It consists of n identical copies of a hardware component called a **full adder**, which performs the 1-bit full-add operation specified by (2.16) and (2.17). A full adder A_i has three binary input signals x_i, y_i, and c_{i-1}, and two binary output signals s_i and c_i. A set of n full adders $A_{n-1}A_{n-2}\ldots A_0$ are connected so that A_i supplies its output carry bit c_i to the full adder A_{i+1} on its left; it also supplies one bit s_i of the final result. Carry signals "ripple" through the circuit from right to left; for this reason the design of Figure 2.7b is known as a **ripple-carry adder**. Figure 2.7c provides another view of the ripple-carry adder at the logic level of abstraction, where internal components known as logic gates and their interconnections can be seen.

Observe from Figure 2.7 how the structure of the ripple-carry adder closely resembles that of the *ADDn* procedure from which it is derived. The fact that only one component type, namely a full adder, is used repeatedly simplifies the logic design process and reduces system cost. Furthermore, the way the full adders are laid out side by side, with interconnections confined to neighboring components, produces physical implementations that take up little space, a major requirement in VLSI circuits. Ripple-carry adders have one disadvantage, however: they are rather slow because of the time taken by the carry signals to travel through the circuit, a time that increases with n. As we will see in later chapters, there are many ways to do addition, and they involve making trade-offs between circuit size, operating speed, and other factors. The design of binary addition procedures and the corresponding logic circuits is a typical problem in digital logic design.

Subtraction

We can also perform subtraction in all number bases following the digit-by-digit approach used for addition. The result of subtracting y_i from x_i is a single d_i, the **difference digit,** if $y_i \leq x_i$. If $y_i > x_i$, then we must "borrow" a 1 from the next column on the left. We then replace x_i with $1x_i$, which is greater than y_i, and perform the subtraction $1x_i - y_i = d_i$. On moving left to the next column, the borrowed 1 must be subtracted along with y_{i+1} from x_{i+1} to restore the result to the correct value. In general, each one-digit subtraction step includes a **borrow digit** b_i, which has the value 0 or 1 and is specified by the following **full-subtraction** equation:

$$b_i d_i = x_i - y_i - b_{i-1} \tag{2.18}$$

When $r = 2$, Equation (2.18) defines eight ways for the three input variables x_i, y_i, and b_{i-1} to determine the borrow-difference pair $b_i d_i$:

This set of equations define 1-bit full subtraction.

$$\begin{aligned} b_i d_i \quad & x_i \quad y_i \quad b_{i-1} \\ 00 &= 0 - 0 - 0 \\ 11 &= 0 - 0 - 1 \\ 11 &= 0 - 1 - 0 \\ 10 &= 0 - 1 - 1 \\ 01 &= 1 - 0 - 0 \\ 00 &= 1 - 0 - 1 \\ 00 &= 1 - 1 - 0 \\ 11 &= 1 - 1 - 1 \end{aligned} \tag{2.19}$$

Subtraction involving multibit numbers can be performed by a bit-by-bit procedure *SUBn*, similar to *ADDn*, in which borrows replace carries and the addition step (2.16) is replaced by the corresponding subtraction step (2.18).

Example 2.5 **Subtracting two unsigned 8-bit binary numbers**

Consider subtraction of the 8-bit binary integer $Y_2 = 00110011_2$ from $X_2 = 10100001_2$. Following the bit-by-bit approach of *ADDn*, we set the external input borrow bit b_{-1} to 0. Then we subtract the rightmost bit x_0 of X_2 from the corresponding bit y_0 of Y_2. Thus:

$$b_0 d_0 = x_0 - y_0 - b_{-1} = 1 - 1 - 0 = 00$$

Moving left to the next bit position yields

$$b_1 d_1 = x_1 - y_1 - b_0 = 0 - 1 - 0 = 11$$

and so on. The eight steps to the solution are summarized below:

	$X_2 =$	1	0	1	0	0	0	0	1
	$-Y_2 =$	0	0	1	1	0	0	1	1
Borrows		0	1	1	1	1	1	1	0
Difference	$D_2 =$	0	1	1	0	1	1	1	0

An n-bit "ripple-borrow" subtracter circuit is constructed along the lines of Figure 2.7, with b_i replacing c_i, d_i replacing s_i, and a full-subtracter component S_i replacing each full adder A_i (Figure 2.8). The borrow signals ripple through the circuit from right to left in essentially the same manner as carries in the adder. The behavior of S_i is defined by (2.19), and thus differs from that of A_i, the behavior of which is specified by (2.17). Note that we cannot switch the input numbers X_2 and Y_2 to the subtracter without changing D_2. Moreover, if $X_2 < Y_2$, then $X_2 - Y_2$ is negative and so cannot be adequately represented in the unsigned number domain.

Circuits for addition and subtraction are very similar.

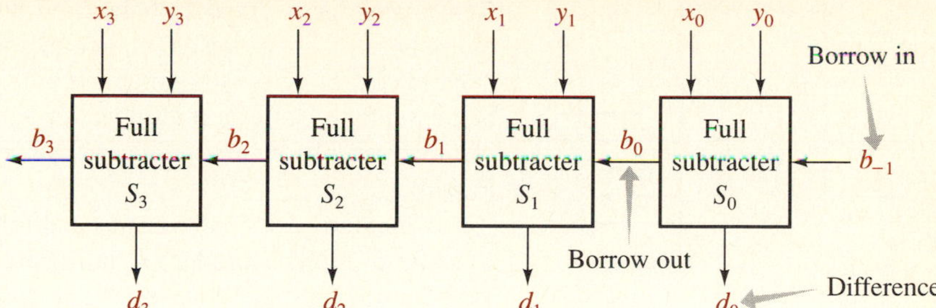

FIGURE 2.8 A block diagram of a 4-bit binary subtracter.

Multiplication and Division

Multiplication is harder than addition or subtraction; division is harder still.

The two most useful numerical operations after addition and subtraction are multiplication and division. The standard decimal techniques for these operations too can be adapted to handle binary numbers or, indeed, numbers of any base r. Again, we assume that the numbers are unsigned binary integers. The general multiplication problem is to find the **product** P formed by multiplying Y by X; that is,

$$P = X \times Y \tag{2.20}$$

If X and Y contain m and n digits, respectively, then P is an integer containing up to $m + n$ significant digits, and may have to be rounded off to fewer digits.

The division of Y by X yields a result Y/X, which may only be expressable as a mixed number, and may contain an arbitrary number of significant digits. For example, division of 32 by 9 yields $32/9 = 3.55555\ldots_{10}$. To ensure integer results, we compute a two-part result consisting of an integer Q, called the **quotient**, and a second integer R, called the **remainder**, which are related by the equation

$$Y = X \times Q + R \tag{2.21}$$

where $0 \leq R < X$. Using our previous example, we see that dividing 32 by 9 yields the integer quotient 3 and the remainder 5; that is, $32 = 3 \times 9 + 5$. Often, Q alone is taken to be the result because it represents the best integer approximation to Y/X; in this case, we write $Q = Y \div X$. Thus, although multiplication and division are complementary operations, as can be seen by comparing (2.20) and (2.21), division is by far the harder of the two processes.

Multiplier Circuit

Various multiplication circuits have been devised for unsigned numbers, most of which require lots of hardware. For example, it is possible to perform multiplication by repeated addition, noting that $X \times Y$ can be obtained by adding Y to itself X times. This implies that we could build a multiplier

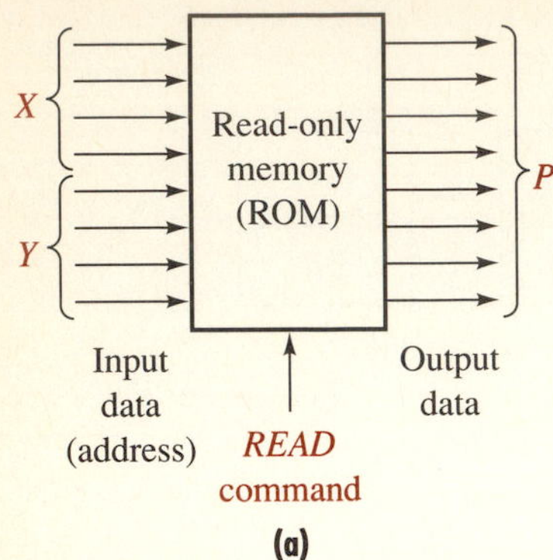

(a)

Input (address)		Output (stored data)
Y	X	P
0000	0000	00000000
0000	0001	00000000
0000	0010	00000000
⋮	⋮	⋮
1010	1101	10000010
1010	1110	10001100
⋮	⋮	⋮
1111	1101	11000011
1111	1110	11010010
1111	1111	11100001

(b)

FIGURE 2.9 (a) Read-only memory (ROM) implementing 4×4 multiplication; (b) an excerpt of its contents.

from many adder circuits of the kind we discussed earlier. An adder-based multiplier is a smart design choice in some situations. In other situations, it is more cost-effective to implement multiplication by using a memory device that essentially stores a multiplication table. The product $P = X \times Y$ is then obtained simply by retrieving it from the memory, a process called **table lookup**.

A memory unit configured to multiply two 4-bit unsigned numbers X and Y to form an 8-bit product P is illustrated by Figure 2.9. The information stored in the memory constitutes a 4×4 multiplication table. The input data sent to the memory consist of the 8-bit operand-pair XY, and the output is the 8-bit product P. We say that XY serves as the **address** of P, because the memory's control circuit uses XY to pinpoint the location where P is stored. To perform the multiplication 1010×1101, for example, XY is set to 10101101. The memory unit is then triggered by sending it a *READ* command that causes it to output the word stored at address 10101101, which in this case should be $P = 10000010$.

A memory IC provides a useful mechanism for implementing arithmetic and other operations. Such a memory is generally called a **read-only memory** or **ROM**, to contrast it with memory types whose contents can be changed during use. During manufacture, the contents of a ROM are fixed or "programmed" for some particular task, in this case, 4×4 multiplication. The same type of ROM can easily be programmed for many different tasks; we could reprogram the ROM of Figure 2.9a to perform division of 4-bit numbers, for instance. The main limitation of ROMs is that they grow in size exponentially with the number of address bits. The ROM for 4×4 multiplication stores $2^8 = 256$ 8-bit products, or a total of 2,048 bits. A 16×16 ROM-based multiplier circuit must store $32 \times 2^{32} = 2^{37}$, or approximately 1.374×10^{11} bits, which is impractical. Read-only memories are special cases of **programmable logic devices** or **PLDs**, which, as we will see later, play an important role in modern logic design.

Modular Arithmetic

Modular arithmetic confines all results to a fixed set of integers $\mathbf{Z}_m$.

An arithmetic operation performed on n-bit numbers may produce a result that is too long to be represented completely by n bits, a condition described earlier as overflow. For instance, the 8-bit addition

$$1111\ 1111 + 0000\ 0001 = 1\ 0000\ 0000 \qquad (2.22)$$

produces the 9-bit sum $N = 100000000_2 = 256_{10}$, which will not fit in an 8-bit word. If the overflow into the ninth bit position is ignored—which amounts to truncating the result of (2.22) to 8 bits—N becomes 00000000, that is, zero. This, of course, is not a good approximation to the true integer sum 256_{10}. It is, however, the correct sum within a special type of arithmetic called **modular arithmetic**, in which the results of all arithmetic operations are confined to some fixed set of m values such as $1, 2, \ldots, m-1, m$ or $0, 1, \ldots, m-2, m-1$. Modular arithmetic captures the finite word size

(a)

(b)

FIGURE 2.10 A 12-hour clock face that illustrates modulo-12 arithmetic.

of machine calculations and is, in fact, the fundamental type of arithmetic employed in digital systems.

A basic feature of modular arithmetic is that as a number N is incremented beyond a maximum representable value, it returns or "wraps around" to the smallest value. We are familiar with this concept in the measurement of time, where minutes and seconds are confined to the 60 integers $0, 1, 2, \ldots, 58, 59$, hours are confined to the 12 (decimal) integers $1, 2. \ldots, 12$, and so on. For example, if the current time is 10 o'clock on a 12-hour clock like that of Figure 2.10a, we know that the time three hours later will not be $10 + 3 = 13$ o'clock, but rather 1 o'clock, as depicted in Figure 2.10b. This is tantamount to saying that $10 + 3 = 1$ in the special modular arithmetic of 12-hour clocks.

Some Mathematical Details

To examine the general nature of modular arithmetic, let $\mathbf{Z}$ denote the infinite set of all possible (positive and negative) integers, and let $\mathbf{Z}_m$ denote the subset of $\mathbf{Z}$ consisting of the first m nonnegative integers $0, 1, \ldots, m-2, m-1$, where m is any positive integer. For example, $\mathbf{Z}_2 = \{0, 1\}$ and $\mathbf{Z}_8 = \{0, 1, 2, 3, 4, 5, 6, 7\}$. Now $\mathbf{Z}$ has the property that the result of applying any operation op from the set $\{+, -, \times, \div\}$ to members of $\mathbf{Z}$ is also a member of $\mathbf{Z}$; we say that $\mathbf{Z}$ is "closed" with respect to op. (Recall that we defined $N_1 \div N_2$ to be the integer quotient resulting from division of N_1 by N_2.) In other words, closure means that if N_1 and N_2 are in $\mathbf{Z}$, then so is $N = N_1 \; op \; N_2$.

Due to overflow, $\mathbf{Z}_m$ is not closed with respect to every choice of op. Thus, although 5 and 7 are both in $\mathbf{Z}_8$, their sum 12 is not. We can force closure, however, by redefining the result of

$$N = N_1 \; op \; N_2 \tag{2.23}$$

to be the remainder R after dividing N by the largest (positive or negative) integer k that satisfies

$$N = k \times m + R \tag{2.24}$$

where $0 \leq R < m$. For example, if $m = 8$, then $5 + 7$ yields the result 4 in $\mathbf{Z}_8$, because

$$5 + 7 = 12 = 1 \times 8 + 4$$

Let op_m denote the result of combining the original operation op over $\mathbf{Z}$ given by (2.23), with extraction of the remainder with respect to m defined by (2.24).

$$R = N_1 \; op_m \; N_2 \tag{2.25}$$

$\mathbf{Z}_m$ is clearly closed with respect to any choice of operation op_m from the set $\{+_m, -_m, \times_m, \div_m\}$. Equation (2.25) is usually written in the form

$$R = N_1 \; op \; N_2 \qquad (\text{mod } m) \tag{2.26}$$

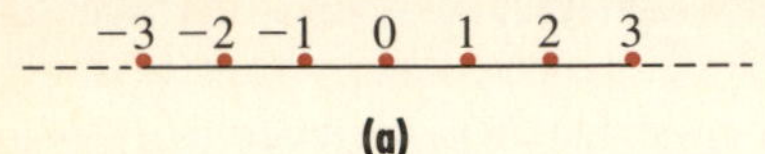

(a)

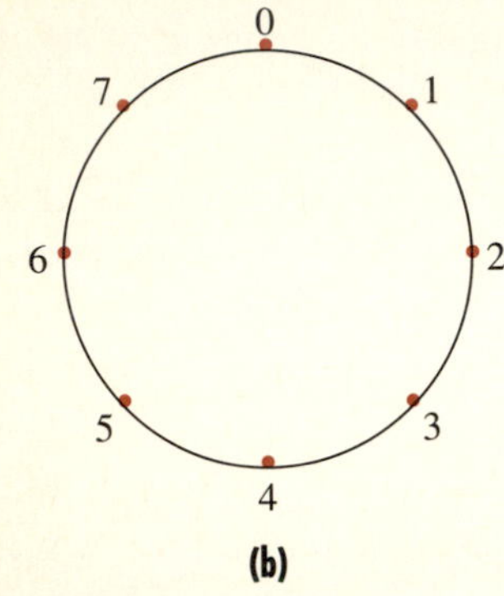

(b)

FIGURE 2.11 (a) The integers **Z** represented by points on an infinite line; (b) $\mathbf{Z}_m$ represented by points on a circle ($m = 8$).

and m is referred to as the **modulus**. If op denotes $+$, Equation (2.26) is read as "R equals N_1 plus N_2 modulo-m."

Suppose, for example, that $m = 4$, implying that $\mathbf{Z}_m = \mathbf{Z}_4 = \{0, 1, 2, 3\}$. This means that the following addition operations, among others, hold in $\mathbf{Z}_4$:

$$1 + 2 = 3 \quad (\text{mod } 4)$$
$$2 + 2 = 0 \quad (\text{mod } 4)$$
$$3 + 3 = 2 \quad (\text{mod } 4)$$

Other modulo-4 arithmetic operations such as those below behave in a similar fashion.

$$1 - 3 = 2 \quad (\text{mod } 4)$$
$$3 \times 3 = 1 \quad (\text{mod } 4)$$
$$3 \div 2 = 1 \quad (\text{mod } 4)$$

Modulo-2^n arithmetic results naturally from truncating all numbers to n bits.

We refer to an arithmetic system of this kind as a modular arithmetic with modulus m, or simply as **modulo-m arithmetic**. We now see that the "clock arithmetic" illustrated by Figure 2.10 is a kind of modulo-12 arithmetic. (It would conform exactly to our general definition of $\mathbf{Z}_{12}$ if we replaced 12 with 0 in the set of hours.) The relevance of modular arithmetics to digital design lies in the fact that if we perform the basic operations on n-bit unsigned integers and ignore overflow to obtain results that are n bits long, we have a modular arithmetic system with modulus $m = 2^n$, and numbers that are restricted to the range $0, 1, 2, \ldots, 2^n - 2, 2^n - 1$.

Extending the clock analogy, it may be useful to visualize ordinary integer arithmetic on **Z** as applying to discrete points lying on an infinite line, whereas modulo-m arithmetic on $\mathbf{Z}_m$ applies to m points lying on a circle C of circumference m units (Figure 2.11). Addition of X to Y in Z_m corresponds to a clockwise rotation of Y units from the zero point, followed by another clockwise rotation by X units. To subtract X, we perform a counterclockwise rotation of X units. When the result of an arithmetic operation in **Z** goes above $m - 1$, the corresponding operation in $\mathbf{Z}_m$ jumps to zero. Similarly, when a result in **Z** goes below zero, the corresponding operation in $\mathbf{Z}_m$ jumps to $m - 1$.

Example 2.6

Computing memory addresses

Memory devices such as the ROM of Figure 2.9 are among the most important components of digital systems. As we have seen, each storage location in a 2^n-word memory has associated with it a unique n-bit input word called its *address*. To access a stored item, either to read its contents or to change them, the memory control circuit must be supplied with the address of the item in question. Memory addresses are treated as unsigned binary integers in the modulo-2^n arithmetic $\mathbf{Z}_{2^n}$.

A fundamental addressing operation is to increment the current address A to form the next consecutive address $A + 1$. This requires modulo-2^n addition of the constant one. If the result exceeds the largest possible address value $2^n - 1$, the address wraps around from the high-address to the low-address end of the memory. For example, if $n = 16$, a sequence of increments will result in the following address stream:

$$\begin{array}{c} \ldots \\ 1111111111111101 \\ 1111111111111110 \\ 1111111111111111 \\ 0000000000000000 \\ 0000000000000001 \\ \ldots \end{array}$$

Occasionally, the current address A must be followed by a nonconsecutive address of the form $A' = A \pm B$, where B is a *branch address* relative to A. The new address A' is often computed by modulo-2^n addition or subtraction. For instance, if $n = 8$, $A = 10010000$, $B = 11110111$, and $A' = A + B$, the address-generating circuitry must perform the calculation

$$\begin{aligned} A' &= 10010000 + 11110111 \qquad (\text{mod } 2^8) \\ &= 10000111 \end{aligned}$$

The binary arithmetic operations encountered so far can be considered examples of modulo-2^n operations, provided all input and output operands are restricted to n bits, and carry or borrow bits generated from the leftmost (most significant) bit positions are ignored. As we will see in the next section, the same modular operations can be used with little or no modification for signed numbers, if the numbers are appropriately encoded.

2.6 Signed Numbers

We regard unsigned numbers as positive numbers with an implicit plus sign. Most numerical computations involve both positive and negative numbers, and therefore require an explicit representation of a number's sign. We next consider the two major ways of representing signed numbers: signed-magnitude and two's complement.

Signed-Magnitude Code

In the decimal system, the sign is indicated by placing the symbol + or − at the left end of the number, as in $+124_{10}$ and -3.1415926_{10}. This straightforward notation is called **signed-magnitude** or **SM representation,** because the part of the number following the sign is the magnitude or absolute value of the number, as shown in this example.

$$X = \underbrace{x_{n-1}}_{\text{Sign } s} \underbrace{x_{n-2}x_{n-3}\ldots x_1x_0}_{\text{Magnitude } |X|}$$

The leftmost bit of a signed-magnitude number is the sign, with 1 denoting minus.

Signed-magnitude code may be used with any base-r number. When $r = 2$, it is natural to represent the sign s by a single bit; the following **sign convention** is standard:

$$s = 0 \text{ denotes a positive number}$$
$$s = 1 \text{ denotes a negative number}$$

Figure 2.12 gives a sampling of 8-bit words forming SM binary numbers; the sign bit is colored. The process of obtaining $-N$ from N, called **negation**, is quite simple with SM numbers: just change the sign bit from 0 to 1 or from 1 to 0. We may position the implicit binary point anywhere to the right of the sign bit in order to obtain fraction or mixed formats.

When it is necessary to distinguish signed-magnitude numbers from unsigned numbers, we will append the boldface subscript **SM** to them thus:

$$11111100_{\mathbf{2SM}} = -124_{\mathbf{10}}$$

The same 8-bit word is interpreted differently as an unsigned binary number:

$$11111100_{\mathbf{2}} = 252_{\mathbf{10}}$$

Signed-magnitude binary numbers of length n can represent all integers in the range $+(2^{n-1} - 1)$ to $-(2^{n-1} - 1)$. For example, when $n = 3$, we can encode all the integers from $+3$ to -3 as follows:

$$\begin{array}{ll} +0 = 000_{\mathbf{2SM}} & -0 = 100_{\mathbf{2SM}} \\ +1 = 001_{\mathbf{2SM}} & -1 = 101_{\mathbf{2SM}} \\ +2 = 010_{\mathbf{2SM}} & -2 = 110_{\mathbf{2SM}} \\ +3 = 011_{\mathbf{2SM}} & -3 = 111_{\mathbf{2SM}} \end{array} \tag{2.27}$$

A signed-number system based on SM representation is referred to as a **signed-magnitude code**. Hence, (2.27) defines the complete SM code for

$+25_{10}$ = 0 0 0 1 1 0 0 1	-25_{10} = 1 0 0 1 1 0 0 1
$+124_{10}$ = 0 1 1 1 1 1 0 0	-124_{10} = 1 1 1 1 1 1 0 0
$+0_{10}$ = 0 0 0 0 0 0 0 0	-0_{10} = 1 0 0 0 0 0 0 0

FIGURE 2.12 Some 8-bit signed-magnitude binary numbers.

$n = 3$. Note that zero has two distinct representations, $000\ldots00$ (plus zero) and $100\ldots00$ (minus zero), which is a minor flaw of SM code.

Signed-Magnitude Operations

In digital circuit design, it is desirable that the operations on n-bit signed numbers resemble as much as possible the corresponding modulo-2^n operations on n-bit unsigned numbers, so that they can be implemented with simple circuits like the adder of Figure 2.7. The rules for addition and subtraction of SM numbers, however, are fairly complex and change with the numbers' signs and magnitudes, as demonstrated below.

Signed-magnitude addition and subtraction are relatively complex operations.

Consider the addition operation $Z = X + Y$ applied to n-bit SM numbers. If both input operands are positive, then an unsigned adder can be used, provided there is no overflow from the most significant magnitude bits into the sign bit z_{n-1} of the result. In other words, the correct result is given by modulo-2^n addition. If X and Y are both negative, then the result Z should also be negative. However, if the sign bits $x_{n-1} = 1$ and $y_{n-1} = 1$ are added like other (magnitude) bits, then the resulting sign bit z_{n-1} is 0 instead of 1, because $1 + 1 = 0$ (mod 2). To get the correct sign bit, therefore, z_{n-1} must be changed to 1 in a separate correction step.

Signed-magnitude addition is further complicated when X and Y are of opposite signs. As an illustration, let $X = 0100_{2SM} = +4_{10}$ and $Y = 1010_{2SM} = -2_{10}$. Let us compute the sum $X + Y = 2_{10}$. If we simply add X and Y as we do 4-bit unsigned numbers—that is, using modulo-16 addition—we obtain the erroneous result 1110, which denotes -6_{10} when interpreted as an SM number. What we need to do is first negate Y and then subtract (mod 16) the result $-Y = 0010_{2SM} = 2_{10}$ from X thus:

$$\begin{aligned} X + Y &= X - (-Y) \\ &= 0100 - 0010 \qquad \text{(mod 16)} \\ &= 0010 \end{aligned} \tag{2.28}$$

If, however, Y has larger magnitude than X, for example, if $Y = 1111_{2SM} = -7_{10}$, then (2.28) is no longer valid. The roles of X and Y must be switched as follows:

$$\begin{aligned} Y - X &= 1111 - 0010 \qquad \text{(mod 16)} \\ &= 1101 \end{aligned}$$

Hence, to perform SM addition and subtraction correctly in all cases, the signs and magnitudes of $X = x_{n-1}x_{n-2}\cdots x_0$ and $Y = y_{n-1}y_{n-2}\cdots y_0$ must be compared to determine the operation to be performed, the order of the operands, and any adjustments to the signs. The rules for the SM addition $X + Y = Z$ are summarized below in a procedure called *ADDsm*. It is assumed that no magnitude overflow occurs. Because $X - Y = X + (-Y)$, *ADDsm* also implicitly specifies SM subtraction.

Procedure

Procedure *ADDsm* for adding signed-magnitude numbers

1. $x_{n-1} = 0, y_{n-1} = 0$: Add the given numbers $X = x_{n-1}x_{n-2}\ldots x_0$ and $Y = y_{n-1}y_{n-2}\ldots y_0 \pmod{2^n}$ to form the result $Z = z_{n-1}z_{n-2}\ldots z_0$.
2. $x_{n-1} = 0$, $y_{n-1} = 1$: If $|X| < |Y|$, subtract X from Y (mod 2^n). If $|X| \geq |Y|$, set y_{n-1} to 0, and then subtract Y from X (mod 2^n).
3. $x_{n-1} = 1$, $y_{n-1} = 0$: If $|Y| < |X|$, subtract Y from X (mod 2^n). If $|Y| \geq |X|$, set x_{n-1} to 0, and then subtract X from Y (mod 2^n).
4. $x_{n-1} = 1$, $y_{n-1} = 1$: Add X and Y (mod 2^n), and set z_{n-1} to 1.

Two's-Complement Code

Given the difficulties associated with SM arithmetic, it is not surprising that most digital systems employ an alternative code, called **two's complement** or **2C.** This code has a simple procedure for addition and subtraction, and also has a unique representation for zero.

The 2C notation for positive numbers is exactly the same as the SM notation—namely, $X = x_{n-1}x_{n-2}\ldots x_1x_0$, where the leftmost bit $x_{n-1} = 0$ is the sign (plus) of X, and the remaining $n - 1$ bits denote the magnitude $|X|$ of X in the usual positional notation with base 2. Negative numbers are defined in a less direct way, which is best explained in terms of the negation process, often referred to as **two's complementation**, by which we obtain $-X$ from X. A basic two-step negation procedure *NEG2c* for 2C numbers is given below. We will see that this negation process, easily implemented by logic circuits, plays a central role in 2C arithmetic.

Procedure

Procedure *NEG2c* for negating a two's-complement number

1. Change the given number $X = x_{n-1}x_{n-2}\ldots x_1x_0$ to its ones'-complement form $\overline{X} = \overline{x}_{n-1}\overline{x}_{n-2}\ldots\overline{x}_1\overline{x}_0$, where $\overline{x}_i$ is 1 if $x_i = 0$ and $\overline{x}_i$ is 0 if $x_i = 1$ for $n - 1 \geq i \geq 0$.
2. Perform the n-bit addition $\overline{x}_{n-1}\overline{x}_{n-2}\ldots\overline{x}_1\overline{x}_0 + 00\ldots01 \pmod{2^n}$. The result is $-X$ in two's-complement (2C) code.

The first step of *NEG2c* converts all 0s to 1s and all 1s to 0s; the result $\overline{X}$ is called the **ones' complement** or simply the **complement** of X. (Note the different positions of the apostrophes in the terms two's complement and ones' complement.) The second step adds 1 to the least significant (rightmost) bit position of $\overline{X}$ using modulo-2^n addition. The result, called the two's complement of X, is the 2C representation of $-X$. We will indicate that X is a two's-complement number by adding the boldface subscript **2C** to it; the subscripts 2SM and 2C are not a standard usage, however. Figure 2.13 shows some representative positive and negative 8-bit 2C numbers.

$+25_{10} =$ 0 0 0 1 1 0 0 1 $-25_{10} =$ 1 1 1 0 0 1 1 1

$+124_{10} =$ 0 1 1 1 1 1 0 0 $-124_{10} =$ 1 0 0 0 0 1 0 0

$\pm 0_{10} =$ 0 0 0 0 0 0 0 0 $-1_{10} =$ 1 1 1 1 1 1 1 1

FIGURE 2.13 Some 8-bit two's-complement numbers.

Example 2.7

Negating an 8-bit 2C number

Let $X = 00001110_{2C}$ be an 8-bit positive number to be negated. The first (complementation) step of *NEG2c* yields $\overline{X} = 11110001$. The second step (adding 1 to the least significant bit position) produces $11110001 + 00000001 = 11110010 \pmod{2^8}$, yielding the final result $-X = 11110010_{2C}$. We may represent the entire process graphically as follows:

$$
\begin{array}{lrcccccccc}
 & X = & 0 & 0 & 0 & 0 & 1 & 1 & 1 & 0 \\
\text{Step 1}: & & \downarrow & \downarrow & \downarrow & \downarrow & \downarrow & \downarrow & \downarrow & \downarrow \\
 & \overline{X} = & 1 & 1 & 1 & 1 & 0 & 0 & 0 & 1 \\
\text{Step 2}: & & & & & & & & + & 1 \\
 & -X = & 1 & 1 & 1 & 1 & 0 & 0 & 1 & 0
\end{array}
$$

The leftmost bit of $-X$ is 1, and serves to indicate the sign. The remaining 7 bits clearly do not represent the number's magnitude in the usual way. If we now form the two's complement of $-X$, we first obtain $-\overline{X} = 00001101$. Addition of 00000001 produces 00001110, which, as expected, is the original number X.

Two's Complement vs. Ones' Complement

Ones' complement is not as useful as two's complement.

It might be asked, "Why not use the (ones') complement $\overline{X}$ of X rather than its two's complement to represent $-X$?" This is, indeed, a legitimate and occasionally used code for signed binary numbers; it is called the **ones'-complement** or **1C** code. It has, however, some disadvantages compared to 2C code. Zero has two distinct 1C representations, $00\ldots00_{1C}$ and $11\ldots11_{1C}$, whereas there is only one 2C representation. This causes problems in certain 1C arithmetic operations. Suppose, for example, that we add, say, 7 to the all-0 form of zero. This produces the expected result 7.

$$00000000 + 00000111 = 00000111 \pmod{2^8} \tag{2.29}$$

If, however, the all-1 form of zero is used, we obtain

$$11111111 + 00000111 = 00000110 \pmod{2^8} \tag{2.30}$$

that is, six! A one must be added to correct the result of this type of calculation. (There is an easy and systematic way to do this correction, however; see Problem 2.34.) The definition of the 2C code ensures that no similar correction step is needed when computing with 2C numbers.

Table 2.3 lists all the 4-bit 2C integers, with the equivalent SM, 1C, and decimal numbers shown for comparison. Note that the word 1000 is omitted from the 2C list. Although sometimes treated as the 2C representation of −8, this word, like the all-0 word representing zero, has some anomalous properties such as being its own two's complement, as can be verified by applying *NEG2c* to it.

Properties of Two's Complement

Two's-complement numbers have several properties that make them particularly attractive for use in digital circuits. Unlike SM code, 2C code has a unique representation of zero. If $X = 00000000_{2C}$ denoting (plus) zero, then $\overline{X} = 11\ldots11$ and

The 2C zero is unique.

$$11111111 + 00000001 = 00000000 \qquad (\text{mod } 2^8)$$

so plus and minus zero have the same 2C code—namely, the all-0 word. This makes it easier to detect a zero result and avoids certain arithmetic problems encountered by the other signed number codes.

All negative numbers have sign bit 1, while all nonnegative numbers have sign bit 0. This follows from the fact that step 1 of *NEG2c* always changes the sign of the X. Step 2 changes the sign of the number again if it generates a carry into the sign position; this only happens when X is zero.

TABLE 2.3 Various codes for 4-bit signed numbers.

Signed magnitude	*Two's complement*	*Ones' complement*	*Decimal*
0111	0111	0111	+7
0110	0110	0110	+6
0101	0101	0101	+5
0100	0100	0100	+4
0011	0011	0011	+3
0010	0010	0010	+2
0001	0001	0001	+1
0000	0000	0000	+0
1000	0000	1111	−0
1001	1111	1110	−1
1010	1110	1101	−2
1011	1101	1100	−3
1100	1100	1011	−4
1101	1011	1010	−5
1110	1010	1001	−6
1111	1001	1000	−7

The leftmost bit of a 2C number always denotes its sign.

Thus the sign of a 2C number is always given by its leftmost bit, using the standard sign convention, exactly as in the SM case.

A positive or negative 2C number $X = x_{n-1}x_{n-2}\ldots x_1x_0$ can be negated by applying procedure *NEG2c* to it. An alternative negation method is to subtract X from the $(n+1)$-bit binary number $100\ldots00$, as follows:

$$100\ldots00 - x_{n-1}x_{n-2}\ldots x_1x_0 = -X \qquad (\text{mod } 2^n) \tag{2.31}$$

The equivalence of the two negation methods follows immediately from the fact that $100\ldots00 = 11\ldots11 + 00\ldots01$ and $1 - x_i = \overline{x}_i$, so that the left-hand side of (2.31) may be rewritten as

$$\begin{aligned} &11\ldots11 - x_{n-1}x_{n-2}\ldots x_1x_0 + 00\ldots01 \\ &= \overline{x}_{n-1}\overline{x}_{n-2}\ldots\overline{x}_1\overline{x}_0 + 00\ldots01 \\ &= -X \qquad (\text{mod } 2^n) \end{aligned}$$

A 2C number may be negated by subtracting it from 2^n (mod 2^n).

With integer formats, $100\ldots00$ denotes 2^n; hence, we can rewrite (2.31) as

$$2^n - X_{2C} = (-X)_{2C} \qquad (\text{mod } 2^n) \tag{2.32}$$

which is the origin of the name "two's complement." For example, if $n = 5$ and $X_{2C} = 11011$, a negative 2C number, then $(-X)_{2C} = 100000 - 11011 = 00101$ (mod 32). *NEG2c* yields $(-X)_{2C} = 00100 + 00001 = 00101$ (mod 32). Applying (2.32) to $(-X)_{2C} = 00101$, a positive 2C number, gives $(-X)_{2C} = 100000 - 00100 = 00101$ (mod 32)—that is, the same result as *NEG2c*.

Operations of the form $X + Y$ and $X - Y$ on n-bit 2C integers can be performed directly using modulo-2^n addition and subtraction, respectively. This is obviously true when X and Y are positive. If an operand X is a negative 2C number of magnitude $|X|$, then (2.32) implies

$$X = 2^n - |X| = 0 - |X| = -|X| \qquad (\text{mod } 2^n) \tag{2.33}$$

2C numbers may be added and subtracted directly by modulo-2^n circuits.

Therefore, modulo-2^n arithmetic treats a negative 2C number X exactly like $-|X|$. The result is correct with respect to the integers $\mathbf{Z}$, provided its magnitude does not exceed $2^{n-1} - 1$; that is, provided no overflow occurs. Consequently, modulo-2^n adders and subtracters designed for n-bit unsigned numbers can also be used directly for 2C arithmetic on n-bit numbers. It is not necessary to change the order of the operands or to inspect and modify sign bits, as it is for SM code. The answer obtained here is the correct one in proper 2C format for *any* combination of positive and negative numbers (again provided there is no overflow).

Example 2.8

Adding and subtracting 2C numbers

We now illustrate 2C arithmetic for all combinations of positive and negative numbers. Consider the two 8-bit numbers $X = 01010011_{2C}$ and $Y = 00001011_{2C}$. The corresponding negative numbers are $-X = 10101101_{2C}$

and $-Y = 11110101_{2C}$. The reader can easily verify the following results, all of which use the basic (modulo-2^8) addition and subtraction procedures developed earlier for unsigned numbers.

$$\begin{aligned}
X + Y &= 01010011_{2C} + 00001011_{2C} = 01011110_{2C} \\
(-X) + Y &= 10101101_{2C} + 00001011_{2C} = 10111000_{2C} \\
X + (-Y) &= 01010011_{2C} + 11110101_{2C} = 01001000_{2C} \\
(-X) + (-Y) &= 10101101_{2C} + 11110101_{2C} = 10100010_{2C} \\
X - Y &= 01010011_{2C} - 00001011_{2C} = 01001000_{2C} \\
X - (-Y) &= 01010011_{2C} - 11110101_{2C} = 01011110_{2C} \\
(-X) - Y &= 10101101_{2C} - 00001011_{2C} = 10100010_{2C} \\
(-X) - (-Y) &= 10101101_{2C} - 11110101_{2C} = 10111000_{2C}
\end{aligned}$$

Again, the key point is that positive and negative numbers are handled uniformly without regard to their signs, and any output carry or borrow bits generated at the sign position are ignored. This uniformity greatly facilitates the design of arithmetic circuits.

The Two's-Complement Adder-Subtracter

A simple adder and complementer is all that is needed for 2C addition and subtraction.

Figure 2.14 outlines an arithmetic circuit for n-bit 2C numbers, showing its main components and connections. The adder at the heart of the unit is designed for unsigned binary numbers and so performs modulo-2^n addition. By activating a complementer—a relatively simple logic circuit—in one of its input data paths, the same adder is made to perform subtraction, exploiting the fact that $X - Y$ is the sum of X and the two's complement of Y. The complementer is controlled by an external control signal *OP_SELECT*, which selects the operation, addition or subtraction, that the arithmetic unit performs at any time. Two externally controlled components called data selectors (also known as multiplexers) determine which input operands to apply to the left and right inputs of the adder. With this arrangement, we can apply X or Y to the left inputs of the adder, and X, Y, $-X$, or $-Y$ to the right inputs, thus allowing $X+Y$, $X-Y$, $Y-X$, $2X$, and $2Y$ to be computed. The large arrows show Y and $-X$ being applied to the adder's left and right inputs, respectively; the adder is therefore set to compute $Y-X$. Finally, a circuit to detect overflow has been added to generate an overflow signal *OVF*. The internal details of this type of arithmetic unit will be covered in subsequent chapters.

A similar adder-subtracter circuit for SM numbers would be quite a bit more complicated due to the need for separate adders and subtracters, as well as circuits to compare the magnitudes of the input operands. However, it should be noted that SM code is better suited than 2C code for use in circuits that perform multiplication and division. To multiply X_{2SM} and Y_{2SM}, for

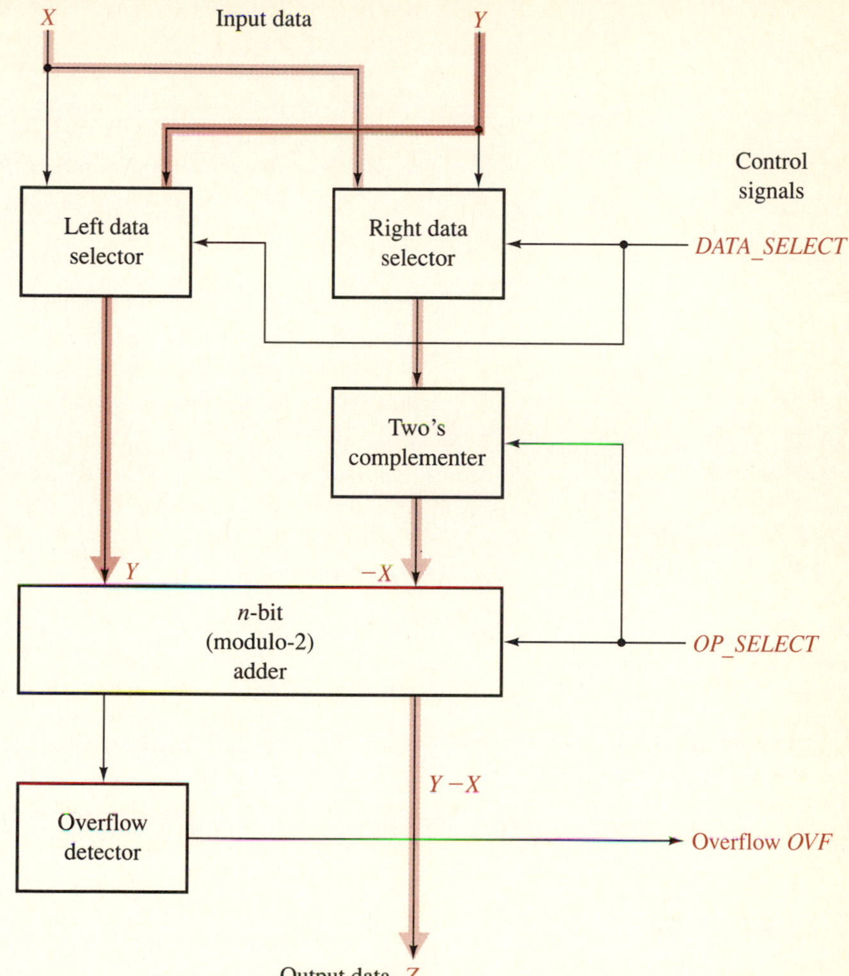

FIGURE 2.14 A block diagram of a circuit to add or subtract two's-complement numbers.

instance, we multiply their magnitudes as unsigned numbers to obtain the magnitude of the product P_{2SM}. The sign bit p_s of P_{2SM} is easily computed from this rule: $p_s = 0$ (positive) if and only if the sign bits of X_{2SM} and Y_{2SM} are the same. No similarly simple extension of unsigned multiplication exists for 2C numbers.

Radix-Complement Codes

The concepts underlying 2C code can be easily generalized to any radix r; the result is referred to as a **radix-complement** or **RC code.** The corresponding generalization of 1C code is called a **diminished radix-complement** or **DRC code.** The standard sign convention for these codes is that the leftmost digit is 0 for positive numbers and $r - 1$ for negative numbers, while $\overline{x}$ is defined for an r-ary digit x as $(r - 1) - x$.

If, for instance, $r = 10$, $n = 5$, and $X = 02914_{10}$, then the RC and DRC representations of $-X$, referred to as the **ten's complement** (10C) and **nines' complement** (9C) of X, respectively, are 97086_{10C} and 97085_{9C}. Letting Y be the negative 10C number 95010_{10C}, we compute the sum

$$\begin{aligned} X + Y &= 02914_{10C} + 95010_{10C} \\ &= 97924_{10C} \end{aligned}$$

using ordinary decimal addition and ignoring carries from the sign position. This, of course, is modulo-10^n arithmetic. Ten's-complement code is useful for decimal operations in digital circuits, where the decimal digits are encoded in a special binary format, which is discussed in Section 2.9.

Overflow

We saw earlier that if $Z = z_{n-1}z_{n-2}\ldots z_1z_0$ is an unsigned n-bit result of an arithmetic operation, then overflow can be detected by a carry or borrow signal $cb_{n-1} = 1$, generated along with the leftmost bit z_{n-1} of Z. This carry/borrow bit represents a significant magnitude bit of Z. In the case of n-bit signed numbers, the leftmost magnitude bit is z_{n-2} rather than z_{n-1}. The carry/borrow output cb_{n-2} from the second bit position from the left, which is also the carry/borrow input into the sign position, must be checked to detect overflow. It follows immediately that in SM addition or subtraction, overflow is indicated by the condition $cb_{n-2} = 1$; thus cb_{n-2} serves as the overflow signal *OVF*. In the corresponding 2C operations, however, the overflow conditions also depend on the input sign bits. These easily derived conditions are summarized in a procedure *OVFDET* to detect overflow.

Procedure

Procedure *OVFDET* for 2C overflow detection

1. If X and Y are both positive (that is, $x_{n-1} = y_{n-1} = 0$), then overflow occurs ($OVF = 1$) in $X + Y$ or $X - Y$ if the carry/borrow bit cb_{n-2} into the sign position is 1.
2. If X and Y are both negative (that is, $x_{n-1} = y_{n-1} = 1$), then overflow occurs if cb_{n-2} is 0.
3. If X and Y are of opposite signs (that is, $x_{n-1} \neq y_{n-1}$), then overflow cannot occur, because $|Z| \leq |X|$ or $|Z| \leq |Y|$.

To illustrate overflow detection via *OVFDET*, when $n = 4$, the result of adding $X = 0101_{2C}$ and $Y = 0100_{2C}$ (mod 16) is $Z = 1001_{2C}$. Overflow occurs because $c_2 = 1$, which also makes the sign bit $z_3 = 1$. Z represents -7_{10}, and the desired result, $+9_{10}$, is too long to represent with 4 bits. If we replace X and Y with their respective two's complements, we obtain $(-X) + (-Y) = 1011_{2C} + 1100_{2C} = 0111$, where now $b_2 = 0$ serves to indicate overflow. In this case, the result $Z = 0111$ denotes the 2C number $+7_{10}$ instead of the correct but unrepresentable sum -9_{10}.

Number Code Comparison

Finally, Table 2.4 presents a brief comparison of the various binary number codes we have encountered in terms of their ease of use in typical arithmetic circuits. All employ essentially the same representation of unsigned or positive numbers, and so can handle addition with the same adder circuitry. Subtraction brings the need to handle signed numbers, and here 2C code shines due to its ability to allow subtraction with a simple complement-and-add scheme. A similar but slightly more complicated approach works for 1C code. As we have seen, subtraction is relatively difficult with SM numbers, because it requires both an adder and a subtracter, as well as a control circuit to compare signs and magnitudes. In the domain of signed multiplication and division, however, SM code has the advantage. Multiplier circuits for both the 2C and 1C codes tend to be somewhat more complicated than their SM counterparts.

Of course, very general cost comparisons like those above must be reassessed for any particular design situation or implementation technology. For example, if ROM-based circuits are used, then the complexity of the design is measured by the capacity of the ROM, which, in turn, depends only on the size of the input and output data words. These parameters are independent of the way the numbers are encoded, so there is no significant difference in cost between, say, a 2C ROM-based adder-subtracter and an SM ROM-based adder-subtracter. In fact, ROMs are often the implementation technology of choice when unusual or exotic number codes are required.

TABLE 2.4 Comparison of the major number codes in terms of their ease of use in digital systems.

	Number code			
Operations	*Unsigned*	*SM*	*2C*	*1C*
Unsigned addition	Easy	Easy	Easy	Easy
Signed addition and subtraction	n/a	Hard	Easy	Fairly easy
Signed multiplication and division	n/a	Fairly hard	Hard	Hard

Binary Codes

The binary representation of information has a long and fascinating history. After a glance at the diversity of binary codes, we consider encoding numbers with bases that are a power of two. These codes often provide a convenient way of representing binary information. Later in this section we examine the encoding of decimal numbers into a binary format that preserves the base-10 positional notation of the decimal system. Finally, we discuss modern binary codes for nonnumerical data.

2.7 A Binary Potpourri

Although binary codes are central to current digital systems, their origin goes back several centuries [Heath, 1972]. A major motivation for such codes—one that continues undiminished to the present day—is to provide reliable and secure means for storing and transmitting information.

Secret Writing

Binary coding methods were devised as early as the 16th century.

It appears that the first person to devise a complete binary coding system was the English philosopher and statesman, Francis Bacon (1561–1626). His goal was a secret code or cipher for diplomatic messages. He encoded the 24 letters of the alphabet of his day using the two symbols *a* and *b*, as shown in Figure 2.15. Each letter clearly requires a five-symbol binary word to represent it, which corresponds to a 5-bit binary number in modern terms. Thus a message such as *Tace ut potes*,[2] meaning "Keep silence as best you can," would be encoded as

baaba aaaaa aaaba aabaa baabb baaba abbba abbab baaba aabaa baaab (2.34)

Bacon added a further twist by using a "biformed alphabet," in which messages are written using two slightly different letter styles or fonts, one

A	*B*	*C*	*D*	*E*	*F*	*G*	*H*
aaaaa	*aaaab*	*aaaba*	*aaabb*	*aabaa*	*aabab*	*aabba*	*aabbb*
I	*K*	*L*	*M*	*N*	*O*	*P*	*Q*
abaaa	*abaab*	*ababa*	*ababb*	*abbaa*	*abbab*	*abbba*	*abbbb*
R	*S*	*T*	*V*	*W*	*X*	*Y*	*Z*
baaaa	*baaab*	*baaba*	*baabb*	*babaa*	*babab*	*babba*	*babbb*

FIGURE 2.15 This binary code, devised by Francis Bacon, uses the two symbols *a* and *b*.

[2]The fact that like this one, Bacon's messages were frequently in Latin would, by itself, suffice to baffle most modern readers.

font representing a, the other b. The secret message could then be buried in an entirely different message whose two fonts convey the desired information. For example, choosing roman and italic styles to represent a and b, respectively, we could disguise *Tace ut potes* as follows:

*W*ha*t*ever you do *s*pea*k* an*d* be *hea*rd *w*he*ne*ver *an*d *w*her*e*ver *poss*ibl*e*

If the fonts of this message are translated into as and bs and grouped into fives, the secret contents are revealed in the form of (2.34).

A Measurement Problem

Consider the task of determining the angular position of a rapidly rotating shaft, a problem that is encountered in the design of hard-disk memories and in many types of industrial machine tools. Two subproblems must be solved: first, how do we measure the shaft's position without interfering with its motion? Second, how do we convert this measurement into binary form for processing by a digital system? Because angular position is an analog quantity, the second subproblem requires conversion of numbers from analog to digital.

Optical methods are useful for measuring position, velocity, etc.

The optical technique depicted in Figure 2.16 provides an elegant solution to the above problem. A special disk, called an **optical encoder**, through which a beam of light can be passed, is attached to the rotating shaft, and the light emerging from the disk is sensed and processed by electronic circuits. The encoder disk has translucent and opaque regions in patterns that vary with its position. We will assume that when the light beam to some sensor is blocked, the sensor produces a 1; when the light reaches the sensor, it generates a 0. Hence, the sensors produce a pattern of binary signals $X = x_3x_2x_1x_0$ whose value represents the current angular position of the disk or, more precisely, a 4-bit digital approximation of that angle.

Optical codes represent 0 and 1 by translucent (white) and opaque (dark) areas, respectively.

Figure 2.17 shows the patterns that appear on two optical encoder disks that contain four concentric tracks or channels. Each disk is divided into 16 angular segments of equal size (22.5 degrees), like a pie divided into 16 slices. Every segment has a unique pattern of translucent (0) and opaque (1) areas assigned to its four tracks. A 4-bit binary number is obtained by reading the four tracks via the light sensors. For example, when segment A in Figure 2.17a rotates to the sensors, the number read is 0000 because the A segment's tracks are all translucent. In segment B, however, the outer track is dark, so the segment is read as 0001. The next segment C is read as 0010, and so on. A complete rotation of the disk of Figure 2.17a produces the 16-number sequence $X = 0000, 0001, 0010, 00011, \ldots, 1111$. At that point, X returns to 0000, so the rotating disk embodies a form of modulo-16 arithmetic. By using n instead of four tracks, and 2^n instead of 16 segments, we can increase the precision to n bits.

When the boundary region between two segments passes the optical sensor, some or all of the 4 bits of X may change at slightly different

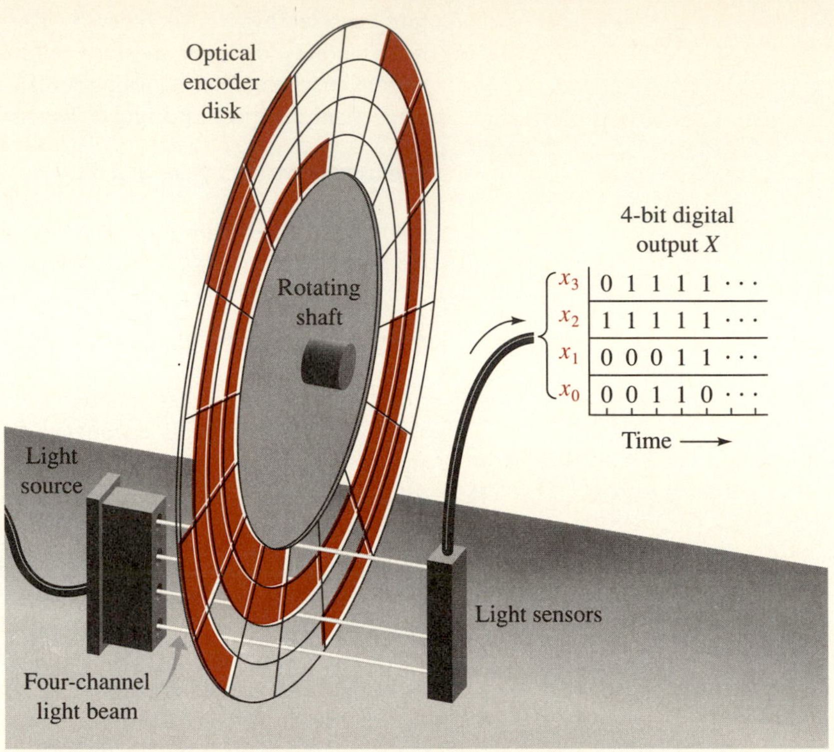

FIGURE 2.16 Optical measurement of the angular position of a rotating shaft.

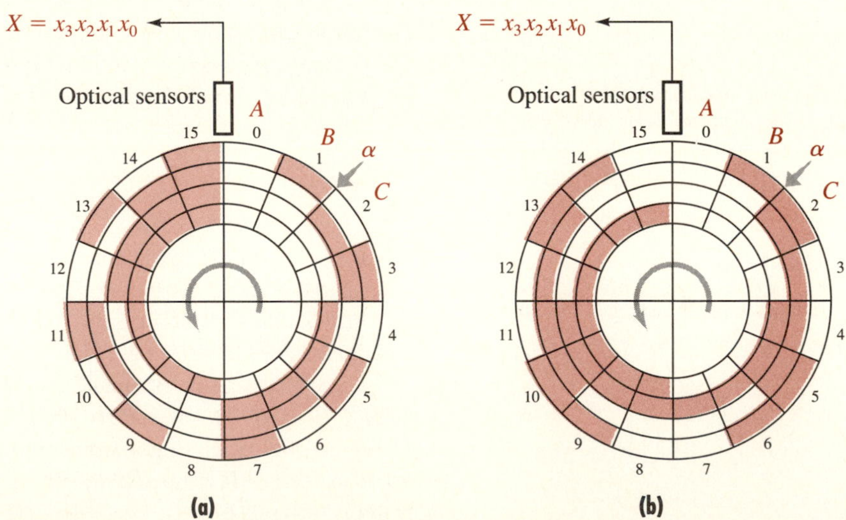

FIGURE 2.17 A rotary optical encoder disk that uses (a) ordinary binary code; (b) an alternative (Gray) code.

times, due to shaft wobble, variations in the response times of different light sensors, and other physical imperfections. Consider, for instance, the position α, marking the transition between segments B and C. When α rotates to the sensor, the value of X should change from 0001 to 0010. If the change in the outer track is perceived first, then the change in X follows the pattern

$$0001 \rightarrow 0000 \rightarrow 0010$$

producing a brief, spurious 0000 signal, which, if registered by the sensors, could lead to errors. On the other hand, if the next track changes first, we obtain the sequence

$$0001 \rightarrow 0011 \rightarrow 0010$$

containing a spurious 0011 signal. The worst case occurs in the transition from 1111 to 0000, when all four tracks should change together, and many possible spurious values may be produced.

Gray Codes

The problem of incorrect output signals during segment-to-segment transitions can be eliminated by using for the optical encoder disk a binary code in which consecutive segment numbers differ in exactly one bit. For example, the 2-bit code

$$\begin{aligned} 0 &= 00 \\ 1 &= 01 \\ 2 &= 11 \\ 3 &= 10 \end{aligned} \tag{2.35}$$

has this property. Observe that the positions of 10 and 11 in the number sequence are in not in the usual order. Another nonstandard 2-bit number code in which the adjacent numbers differ in one bit follows:

$$\begin{aligned} 0 &= 10 \\ 1 &= 11 \\ 2 &= 01 \\ 3 &= 00 \end{aligned} \tag{2.36}$$

Gray codes feature a single-digit change between successive numbers.

In general, an (unsigned) number code such as (2.35) and (2.36) in which only one digit changes when we move from i to $i+1$ is called a **Gray code,** after Frank Gray (1888–1969), a Bell Telephone Laboratories physicist who included it in a 1953 patent for improving telephone signal transmission. Similar coding schemes were developed in the nineteenth century for use in early telegraph equipment. The optical encoder disk shown in Figure 2.17b employs a 4-bit Gray code for its 16 segments. The number sequence $1, 2, 3, \ldots, 14, 15$ produced by reading the disk is now encoded as $0000, 00001, 0011, 0010, \ldots, 1001, 1000$. This code and that of (2.35) are examples of a class of binary codes known as **reflected Gray codes.**

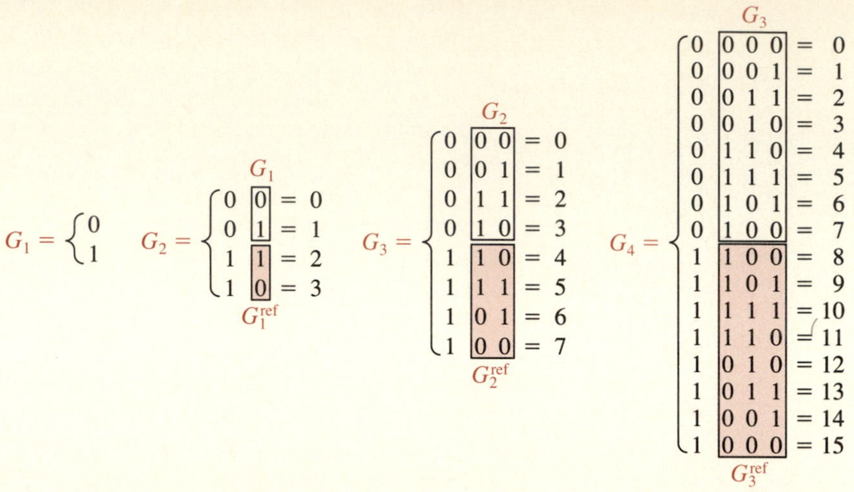
G1 = 0, 1
G2 =
G1
0 0 = 0
0 1 = 1
1 1 = 2
1 0 = 3
G1ref
G3 =
G2
0 0 0 = 0
0 0 1 = 1
0 1 1 = 2
0 1 0 = 3
1 1 0 = 4
1 1 1 = 5
1 0 1 = 6
1 0 0 = 7
G2ref
G4 =
G3
0 0 0 0 = 0
0 0 0 1 = 1
0 0 1 1 = 2
0 0 1 0 = 3
0 1 1 0 = 4
0 1 1 1 = 5
0 1 0 1 = 6
0 1 0 0 = 7
1 1 0 0 = 8
1 1 0 1 = 9
1 1 1 1 = 10
1 1 1 0 = 11
1 0 1 0 = 12
1 0 1 1 = 13
1 0 0 1 = 14
1 0 0 0 = 15
G3ref

FIGURE 2.18 Construction of the reflected binary Gray code G_n for $n \leq 4$.

Gray codes are useful in analog-to-digital conversion.

The four smallest reflected Gray codes, including the 4-bit code G_4 used in the encoder disk (Figure 2.17b), are illustrated in Figure 2.18. These examples suggest a general method for constructing such codes. To obtain the $n + 1$-bit reflected Gray code G_{n+1}, we start by taking a copy G_n of the next smallest code, and preceding every n-bit code word in G_n by 0. We denote the result, which gives the first half of the numbers in G_{n+1}, by $0G_n$. Taking another copy of G_n, we reverse the order of all its code words—that is, we *reflect* the code words—to obtain G_n^{ref}, which appears in color in Figure 2.18. We then precede G_n^{ref} by 1 to obtain the second half of G_{n+1}. We can summarize this code-construction process succinctly by the statement

$$G_{n+1} = \{0G_n, 1G_n^{\text{ref}}\} \qquad \text{where } G_1 = \{0, 1\}, \text{ and } n \geq 1$$

Finally, we observe that a reflected Gray code is an example of a useful nonpositional binary number system. It is easy to convert ordinary, positional binary code to Gray code, and vice versa; see Problem 2.40.

Universal Product Codes

UPC symbols can be read reliably by machines under quite adverse conditions.

Another interesting number code based on optical techniques is the universal product code, an example of which appeared in Figure 2.2. This is a bar code developed by the grocery industry in the 1970s to label products for sale in retail stores. Each code word is a UPC "symbol" that contains a decimal number uniquely identifying a product type. Figure 2.19a shows a UPC

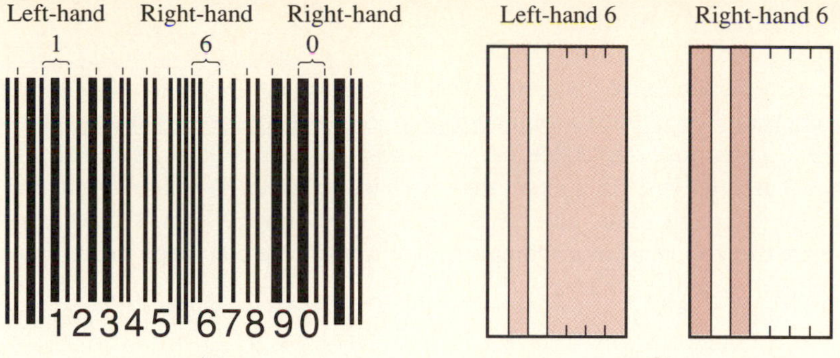

Decimal digit	Left-hand code	Right-hand code
0	0001101	1110010
1	0011001	1100110
2	0010011	1101100
3	0111101	1000010
4	0100011	0011100
5	0110001	0001110
6	0101111	1010000
7	0111011	1000100
8	0110111	1001000
9	0001011	1110100

(a) (b) (c)

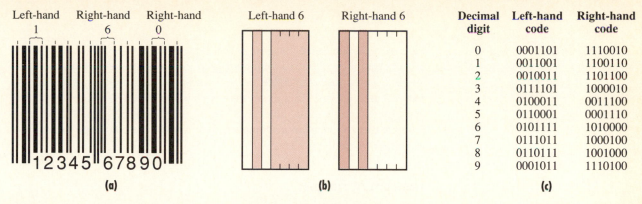

FIGURE 2.19 (a) A 10-digit UPC symbol; (b) representations of the digit 6; (c) digit codes.

symbol that encodes the number 1234567890. Like the disks of Figure 2.17, the UPC symbol is designed to be easily read by optical scanners attached to a digital system, such as a point-of-sale terminal in a store. This technique implies that the reading process should be insensitive to minor variations in printing quality and the size and orientation of the UPC symbol. Another requirement is that a UPC symbol must be difficult to tamper with, so that a thief well versed in binary codes cannot convert the label on, say, a can of caviar to that of a can of peas.

A decimal digit and a few special characters are represented in a UPC symbol by light (0) and dark (1) regions that form two dark parallel bars on a light background of width seven. Therefore, a UPC digit is composed of seven binary elements. The digit 6, for instance, takes the form 010111, consisting of dark bars of width one (1) and three (111), each of which has a light space (0) of width one on its left. A UPC digit has two forms, depending on whether it is used in the left half (first five digits) or the right half of a UPC symbol. The above example 0101111 is the left-hand form of 6; the right-hand version of 6 is 1010000 (Figure 2.19b). This left-right distinction facilitates the scanning and decoding process, during which a symbol can be scanned from either right to left or left to right. Certain other information is embedded in a UPC symbol such as sets of "guard" bars, which appear at both ends and in the middle of the symbol. The basic binary encoding scheme of the universal product code is summarized in Figure 2.19c.

Security and reliability are major goals of many coding methods.

A common concern of all the preceding binary coding methods is that the code be *secure* or *reliable* in some sense. Bacon's scheme was designed to be hard for uninitiated humans to decode. The Gray-coded optical disk works in situations where ordinary binary code causes errors. The UPC approach tolerates a wide range of irregularities in the reading process.

2.8 Base-2^k Numbers

Octal and hexadecimal codes provide a useful shorthand for binary data.

Earlier, we emphasized that conversion between binary and decimal systems is important for efficient communication between humans and machines. We also noted that conversion between those bases is tedious, especially in manual calculations. To address these issues, we often use base-2^k number systems, especially octal and hexadecimal, as shorthand (Figure 2.20). Not only are these systems more compact, they can also be converted to and from the binary system quite easily.

Introductory Example

A convenient relationship exists between binary and base-2^k numbers. To demonstrate this relationship, we consider converting the 9-bit unsigned binary number

$$N_2 = 100110101_2 \tag{2.37}$$

to octal form. By definition of the positional notation with $r = 2$, (2.37) can be expanded into powers of two as follows:

$$N = 1\times2^8+0\times2^7+0\times2^6+1\times2^5+1\times2^4+0\times2^3+1\times2^2+0\times2^1+1\times2^0 \tag{2.38}$$

We want to find N_8—that is, the octal representation of N, where $r = 8 = 2^3$. Therefore, we must express N in terms of powers of eight.

We can factor out powers of eight from consecutive groups of three terms in (2.38) in the following way.

$$N = (1\times2^2+0\times2^1+0\times2^0)\times8^2+(1\times2^2+1\times2^1+0\times2^0)\times8^1 \\ +(1\times2^2+0\times2^1+1\times2^0)\times8^0 \tag{2.39}$$

The three expressions in parentheses in (2.39) can be written as 3-bit binary numbers using the original bit patterns from (2.37).

$$N = 100_2\times8^2+110_2\times8^1+101_2\times8^0 \tag{2.40}$$

(a)			(b)		
1111010000000000	0010000001111000	0000000100000000	F400	2078	0100
1111010000000100	0001000000111000	0000000001100100	F404	1038	0064
1111010000001000	0010001000111100	0000000000000000	F408	223C	0000
1111010000001100	0000000000000000	1110001001010000	F40C	0000	E250
1111010000010000	0101001001001000	0101001101000000	F410	5248	5340
1111010000010100	0110011011111000	0011000111000001	F414	66F8	31C1
1111010000011100	0000000011111100	1111111111111111	F41C	00FC	FFFF

FIGURE 2.20 Display of a region of a computer's program memory (a) in binary and (b) in hexadecimal format.

These 3-bit numbers denote integers in the range zero to seven, and so can be represented by the octal digit set. Hence, we can rewrite (2.40) as

$$N = 4 \times 8^2 + 6 \times 8^1 + 5 \times 8^0$$

which immediately yields the desired octal result $N_8 = 465_8$.

Conversion Method

The preceding analysis leads to the conclusion that we can convert from base 2 to base 8 simply by mapping groups of three bits from N_2 directly into the corresponding octal digits.

Binary-to-octal conversion

$$N_2 = \underbrace{100}_{4}\,\underbrace{110}_{6}\,\underbrace{101}_{5}$$

It is obvious that reversing this procedure enables us to convert from octal to binary by replacing each octal digit with the corresponding 3-bit binary number. For example, given the octal integer $N_8' = 107251_8$, we make the transformation

Octal-to-binary conversion

$$N_8' = \underbrace{1}_{001}\,\underbrace{0}_{000}\,\underbrace{7}_{111}\,\underbrace{2}_{010}\,\underbrace{5}_{101}\,\underbrace{1}_{001}$$

yielding

$$N_2' = 001000111010101001_2 \tag{2.41}$$

In short, binary-octal conversions are most efficiently performed by scanning the given number and replacing each octal digit with its 3-bit binary equivalent, or vice versa.

Binary-hexadecimal conversions involve the similar process of replacing each hexadecimal digit with its 4-bit binary equivalent, or vice versa. For instance, we convert N_2', defined by (2.41), into hexadecimal form as follows (ignoring the two leading 0s):

Binary-to-hexadecimal conversion

$$N_2' = \underbrace{1000}_{8}\,\underbrace{1110}_{E}\,\underbrace{1010}_{A}\,\underbrace{1001}_{9}$$

implying that $N_{16}' = 8EA9_{16}$. Note that N_{16}' contains one fourth the number of digits in N_2'. This reduction in digit count, coupled with a loose resemblance to the decimal system, accounts for the widespread use of the hexadecimal system, and to a lesser extent the octal system, as a shorthand notation in the description of numbers, computer instruction codes, and other complex binary entities.

The foregoing conversion process applies to any base-2^k system. The key idea is to interchange base-2^k digits and their k-bit binary equivalents. This digit-by-digit replacement procedure involves no arithmetic operations such as the multiplication or division by two needed in binary-decimal conversion. The grouping of the binary digits in N_2 is such that the binary point always lies between two adjacent k-bit groups, so the 2^k-ary point is always in the corresponding position in the base-2^k number. Leading or trailing 0s may be appended to or deleted from the binary number as needed to make the number of bits an exact multiple of k.

Example 2.9

Hexadecimal-to-octal conversion

Suppose the eight-digit hexadecimal number N_{16} = A1F5213.C$_{16}$ is to be converted to base 8. We perform the conversion in two stages. First, we convert N_{16} to its binary equivalent N_2; then we convert N_2 to N_8. The first step is given by

$$N_{16} = \underbrace{\text{A}}_{1010}\;\underbrace{1}_{0001}\;\underbrace{\text{F}}_{1111}\;\underbrace{5}_{0101}\;\underbrace{2}_{0010}\;\underbrace{1}_{0001}\;\underbrace{3}_{0011}\;.\;\underbrace{\text{C}}_{1100} \tag{2.42}$$

resulting in the 32-bit number

$$N_2 = 1010000111110101001000010011.1100_2$$

For the second base conversion step, the length of the integer part of N_2 is increased from 28 to 30, an exact multiple of $k = 3$, by adding two leading 0s. The length of the fraction part of N_2 is made a multiple of three by deleting a trailing 0. (If the least significant bit were 1, we would add two trailing 0s instead.) The result is then converted to octal

$$N_2 = \underbrace{001}_{1}\,\underbrace{010}_{2}\,\underbrace{000}_{0}\,\underbrace{111}_{7}\,\underbrace{110}_{6}\,\underbrace{101}_{5}\,\underbrace{001}_{1}\,\underbrace{000}_{0}\,\underbrace{010}_{2}\,\underbrace{011}_{3}\,.\,\underbrace{110}_{6}$$

yielding $N_8 = 1207651023.6_8$.

2.9 Binary-Coded Decimal Form

Our discussion so far has assumed that decimal numbers are translated into some base-2 form for processing by digital circuits. An alternative approach that we examine next is to encode the decimal digits into binary form, but maintain the base-10 positional notation in which all digits are weighted by powers of 10. Such numbers are called **binary-coded decimal numbers** or, if the context is clear, simply *decimal* numbers. Following custom, we usually restrict the term *binary-coded decimal,* which is abbreviated **BCD**, to refer to the most widely used code of this sort.

BCD Code

BCD is simply the decimal system, with each digit encoded in binary.

An unsigned decimal number $N_{10} = d_{n-1}d_{n-2}\ldots d_1 d_0$ is converted into the standard BCD form by mapping each digit d_i separately into a 4-bit binary number $B_i = b_{i3}b_{i2}b_{i1}b_{i0}$, where $(B_i)_2 = (d_i)_{10}$. Thus, a 9 in N_{10} is mapped into 1001, an 8 into 1000, a 7 into 0111, and so on. For example, if $N_{10} = 7109_{10}$, then the decimal-to-BCD conversion process takes the form

Ordinary-decimal-to-BCD conversion

$$N_{10} = \underbrace{7}_{0111}\ \underbrace{1}_{0001}\ \underbrace{0}_{0000}\ \underbrace{9}_{1001} \tag{2.43}$$

leading to $N_{\underline{10}} = 0111000100001001_{\underline{10}}$, where the underlined subscript **$\underline{10}$** is our notation for binary-coded decimal. This conversion process is, in fact, the same as that used for changing a hexadecimal number to binary; compare (2.43) to (2.42). In this case, there are 10 digits instead of 16, so only 10 of the 16 possible 4-bit binary numbers are needed. Also, each 4-bit group must be assigned weight 10 rather than 16. For example, from (2.43) we get

$$N = 0111_2 \times 10^3 + 0001_2 \times 10^2 + 0000_2 \times 10^1 + 1001_2 \times 10^0$$

assuming N is an integer. The weight of an individual bit in $N_{\underline{10}}$ is of the form $j \times 10^i$, where j is 8, 4, 2, or 1. Standard BCD is therefore sometimes called **8421 decimal code.**

Conversion from BCD to ordinary decimal form is achieved by replacing 4-bit groups with the equivalent decimal digit. For instance,

BCD-to-ordinary-decimal conversion

$$N'_{\underline{10}} = \underbrace{0011}_{2}\ \underbrace{1000}_{8}\ \underbrace{0100}_{4}\ \underbrace{1001}_{9}\ \underbrace{0000}_{0}\ \underbrace{0101}_{5}$$

implying that $N'_{10} = 284905_{10}$. Conversion between binary (base 2) and BCD requires the decimal-binary conversion procedures covered in Sections 2.3 and 2.4, in addition to the decimal digit-encoding procedure discussed above.

BCD vs. Binary Form

Not all the possible binary patterns correspond to BCD numbers. In particular, the six 4-bit patterns {1010, 1011, 1100, 1101, 1110, 1111} are not needed to represent decimal digits, and therefore cannot appear in the digit positions of a BCD number. These six digit patterns appear in BCD numbers only as the result of an error, such as failure of a hardware component or a programming mistake.

BCD numbers are longer than binary, but are easier to convert to ordinary decimal.

The fact that some n-bit patterns are unused implies that a larger n is needed to represent a given range of numbers if BCD code is employed in place of binary. For example, to represent all integers from zero to a million requires 24 bits of BCD code, but only 20 bits of binary code. (Note that $2^{20} > 10^6$.) Therefore, BCD numbers have slightly greater storage requirements than binary numbers. Also, as we will see shortly, arithmetic operations on BCD numbers are more complicated than their binary counterparts. The main advantage of BCD code is that it eliminates most of the need for time-consuming base-10–to–base-2 and base-2–to–base-10 conversions. Digital computers are designed primarily to process binary (base-2) numbers, but many have features to support BCD operations.

BCD Adder Circuit

We now examine briefly the structure of a typical adder circuit designed to handle BCD code. This discussion will highlight the relationship between the binary and BCD codes and show how a modulo-16 circuit can be made to perform modulo-10 operations.

Although BCD code employs base 10 in place of base 2, its binary encoding scheme facilitates the use of binary arithmetic circuits, suitably modified to perform BCD arithmetic. Figure 2.21 illustrates this fact for the case of a BCD full-adder circuit, which is designed to add two BCD digits X and Y and an input carry digit c_{in}, to produce a sum digit S and an output carry digit c_{out}.

$$c_{\text{out}}S = X + Y + c_{\text{in}} \tag{2.44}$$

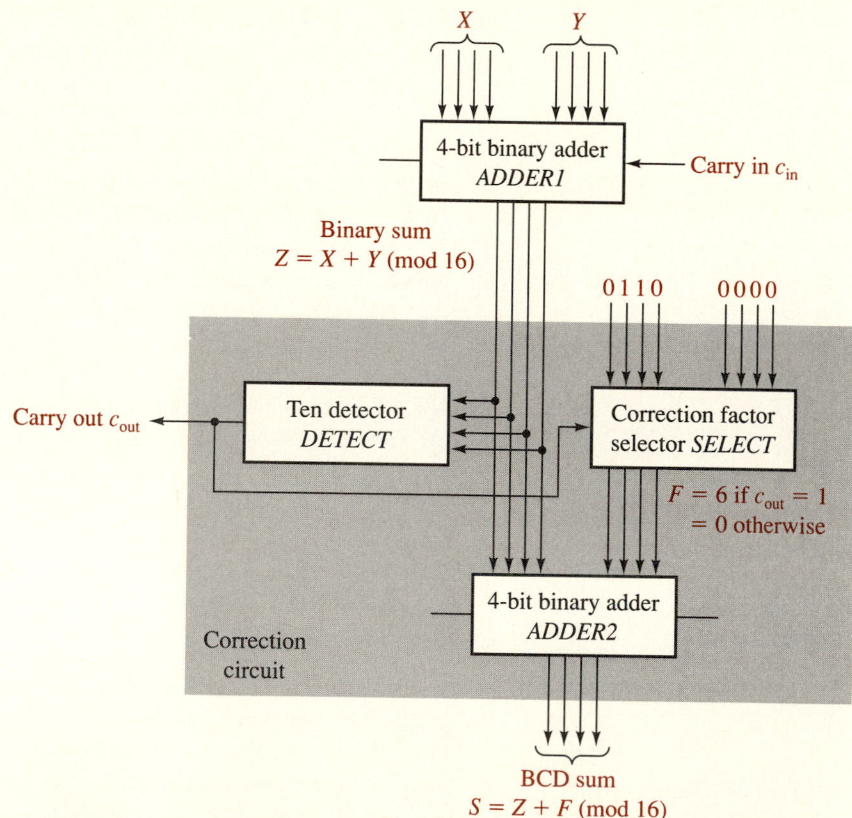

FIGURE 2.21 A one-digit (full) adder circuit for BCD numbers, based on binary addition with correction.

Here X, Y, and S are 4-bit words that can be treated as 4-bit binary numbers. The carries c_{out} and c_{in} are single bits; they can take only the values 0 and 1.

BCD addition can be realized by binary addition, with correction.

The resemblance of one-digit BCD addition to 4-bit binary addition suggests that we can use the binary adder circuit of Figure 2.7 to implement (2.44). This adder is designed to compute the sum of two 4-bit unsigned numbers using modulo-16 arithmetic. The top component *ADDER1* in Figure 2.21 is a copy of this 4-bit adder. It generates the 4-bit sum $Z = X + Y$ (mod 16), which is the desired BCD sum S if $Z \leq 9$ because, in that case, the binary and BCD sums coincide. If $S \geq 10$, we want to make $S = Z - 10$ and generate a carry signal of 1. For example, if $Z = 1110_2 = 14_{10}$, we must set $S = 0100_2 = 4_{10}$ and $c_{out} = 1$.

We conclude that the BCD and binary sums S and Z are related as follows:

$$\begin{aligned} S &= Z && \text{if } Z \leq 9 \\ &= Z - 10 \quad \text{(mod 16)} && \text{if } Z \geq 10 \end{aligned} \tag{2.45}$$

Equation (2.45) is unchanged if we add 16 to the right-hand side, yielding $S = Z + 6$. Hence, the relation between S and Z can be rewritten as

$$\begin{aligned} S &= Z + 0 \quad \text{(mod 16)} && \text{if } S \leq 9 \\ &= Z + 6 \quad \text{(mod 16)} && \text{if } S \geq 10 \end{aligned}$$

The result Z of binary addition can therefore be "corrected" to produce the desired BCD sum S by adding zero or six to Z. This correction step is implemented by the second 4-bit binary adder *ADDER2* in Figure 2.21. An additional component named *DETECT* is needed to detect when Z is 10 or more. The output signal of *DETECT* serves two purposes: it acts as the BCD carry signal c_{out} and it controls the data selector component *SELECT* that applies the correction factor $F = 0$ or 6 to *ADDER2*. Clearly, n copies of this one-digit BCD adder can be connected in the ripple-carry style of Figure 2.7b to form an n-digit BCD adder.

BCD Subtraction

As with binary numbers, the most convenient code for signed BCD numbers is radix-complement, which in this case is binary-coded ten's complement, and we indicate this by the subscript $\underline{\mathbf{10C}}$. We saw in Section 2.6 that a ten's-complement number is negated by replacing each digit d_i with $\overline{d}_i = 9 - d_i$, and by adding 00...01 to the resulting number. That is precisely what is done in the BCD case, recognizing that both d_i and $\overline{d}_i$ are now 4-bit words and that the addition of 00...01 is BCD addition modulo 10^n. As before, the zero digit $0000_{\underline{10C}}$ denotes the plus sign, while its ten's-complement negation, $1001_{\underline{10C}}$—that is, nine—denotes the minus sign.

Example 2.10 **Adding and subtracting three-digit BCD numbers**

Let $X = 050_{10C} = 000001010000_{\underline{10C}}$ and $Y = 037_{10C} = 000000110111_{\underline{10C}}$ be two three-digit positive decimal numbers in ten's-complement code. To negate X in its BCD ten's-complement format, we first form the nines' (diminished radix) complement $-X_{\underline{9C}}$ by replacing each 4-bit digit-group $b_{i3}b_{i2}b_{i1}b_{i0}$ in $X_{\underline{10C}}$ with $1001 - b_{i3}b_{i2}b_{i1}b_{i0}$ (mod 16) thus:

$$-X_{\underline{9C}} = 100101001001_{\underline{9C}}$$

which is equivalent to 949_{9C}. The ten's (radix) complement is generated by adding one to $-X_{\underline{9C}}$ via the modulo-10^3 BCD addition step

$$\begin{aligned} -X_{\underline{10C}} &= 100101001001_{\underline{10}} + 000000000001_{\underline{10}} \\ &= 100101010000_{\underline{10}} \qquad (\text{mod } 10^3) \end{aligned} \tag{2.46}$$

so that

$$-X_{\underline{10C}} = 100101010000_{\underline{10C}}$$

or, equivalently, $-X_{10C} = 950_{10C}$. Note how carries are handled in (2.46). In the same way, $-Y$ can be shown to be $100101100011_{\underline{10C}} = 963_{10C}$.

Now we illustrate BCD ten's-complement addition and subtraction with various combinations of X and Y. Because the word length (including the sign digit) is three, the arithmetic operations are modulo-10^3. This means all digits, including the signs, are treated in the same way, and any carries or borrows generated from the sign position are ignored. As in two's-complement arithmetic, the results are correct for all combinations of positive and negative numbers, provided no overflow occurs. Overflow can be detected by a suitable modification of the methods described for the two's-complement case:

$$\begin{aligned} X + (-X) &= 000001010000_{\underline{10C}} + 100101010000_{\underline{10C}} = 000000000000_{\underline{10C}} \\ X + Y &= 000001010000_{\underline{10C}} + 000000110111_{\underline{10C}} = 000010000111_{\underline{10C}} \\ X - Y &= 000001010000_{\underline{10C}} - 000000110111_{\underline{10C}} = 000000010011_{\underline{10C}} \\ X + (-Y) &= 000001010000_{\underline{10C}} + 100101100011_{\underline{10C}} = 000000010011_{\underline{10C}} \\ (-X) + (-Y) &= 100101010000_{\underline{10C}} + 100101100011_{\underline{10C}} = 100100010011_{\underline{10C}} \end{aligned}$$

Other Decimal Codes

Several alternative binary encoding schemes for decimal numbers have been proposed, although the BCD code is by far the most widely used. Two other examples of interesting binary codes for the decimal digits, the excess-3 and 2-out-of-5 codes, appear in Table 2.5. UPC symbols (Figure 2.19) also embody a 7-bit decimal code.

The **excess-3 code** or **E3 code** is so called because each E3 digit is obtained from the corresponding BCD digit by adding three (mod 16). For

TABLE 2.5 Three binary codes for the 10 decimal digits.

Digit	*BCD*	*Excess-3*	*2-out-of-5*
0	0000	0011	01100
1	0001	0100	10001
2	0010	0101	10010
3	0011	0110	00011
4	0100	0111	10100
5	0101	1000	00101
6	0110	1001	00110
7	0111	1010	11000
8	1000	1011	01001
9	1001	1100	01010

example, the E3 code for seven is given by 0111 + 0011 = 1010 (mod 16), so that $7_{10} = 1010_{E3}$. The main advantage of E3 over BCD is that the correct carry bit is generated automatically when two E3 digits are added using a 4-bit binary adder. This is a consequence of the fact that

Excess-3 code simplifies some aspects of addition and subtraction.

$$X_{E3} + Y_{E3} = \underline{X_{10}} + 3 + \underline{Y_{10}} + 3$$

which includes the +6 correction term needed for carry generation in BCD addition. E3 code also has certain properties that facilitate the implementation of subtraction; see Problem 2.55.

Two-out-of-5 code supports detection of single-bit errors.

The remaining decimal code specified in Table 2.5 is the **2-out-of-5 (2/5)** code, also known as the **biquinary code.** It is characterized by the fact that each digit consists of five bits, exactly two of which are 1. This code has the clear disadvantage of taking up 25 percent more storage space than the BCD or E3 codes. It is also less convenient than BCD or E3 for implementing arithmetic operations. The special advantage of 2/5 code is that it is an **error-detecting code** in the sense that any erroneous change in a single bit of a 2/5 digit (a 0 becoming a 1, or a 1 becoming a 0) is easily detected by a checking circuit, which periodically verifies that every decimal digit contains exactly two 1s. Of course, an error in which a 1 changes to a 0 and a 0 changes to 1 will not be detected, but such double-bit errors are much less likely than errors affecting only a single bit. As we will see later, binary numbers, BCD numbers, and, indeed, all forms of digital information can be endowed with similar error-detecting properties by appending extra "check" bits to them.

2.10 Character Codes

We know from experience that computers and other digital systems process a great deal of nonnumerical information. The preparation of letters, reports, and other documents, for instance, is a major application of computers. It is convenient to assign a short, fixed-length binary word or **character** to represent the smallest unit of information of interest in such applications. This character can be a numerical digit, a letter of the alphabet, or a special symbol such as , ! or +. These characters, collectively referred to as **alphanumeric data,** constitute the character sets used by digital input/output devices such as keyboards and printers. Character codes also specify nonprinting control functions such as backspace, line feed, delete character, and the like.

Character codes such as ASCII are based on an 8-bit word.

A suitable character size for many purposes is 8 bits or a byte. One bit of the character word may be used as a check bit for error-detection purposes, while the remaining 7 bits define the character code. Seven bits allow $2^7 = 128$ combinations, which suffice to encode all 26 letters of the English alphabet in both lower and uppercase forms, the 10 decimal digits, and the more common special characters such as comma and exclamation mark.

ASCII and EBCDIC

ASCII code uses seven data bits plus an optional parity bit.

One of the more widely used character codes, the **ASCII** (American Standard Code for Information Interchange) **code**, is defined in Table 2.6. ASCII (which is pronounced "as-key") code words are extended to 8 bits by appending an extra bit p at the left end of the basic 7-bit character code. (If p is not used for error detection, it is simply set to 0.) For example, the 6-character string "I am", including the space and quotation marks, is encoded in ASCII by the following six-word sequence:

0100010 1001001 0100000 1100001 1101101 0100010

Another character code, developed by IBM, is EBCDIC, which stands for Extended Binary-Coded Decimal Interchange Code, and is pronounced "ebb-see-dick." EBCDIC employs all 8 bits of a byte, and so encodes a 256-character set, which is broadly similar to the ASCII set. For example, the following EBCDIC code words are used for the 10 decimal digits:

$$\begin{array}{lllll} 0 = 11110000 & 1 = 11110001 & 2 = 11110010 & 3 = 11110011 & 4 = 11110100 \\ 5 = 11110101 & 6 = 11110110 & 7 = 11110111 & 8 = 11111000 & 9 = 11111001 \end{array} \tag{2.47}$$

To accommodate much larger character sets such as those of Chinese and other languages, a 16-bit character code known as Unicode has recently been developed as an international standard.

Number Representation

ASCII, EBCDIC, and similar character codes provide another way of representing decimal data: encoding each digit in an 8-bit word. We see from Table 2.6 that ASCII digits consist of a 4-bit field using the BCD digit

TABLE 2.6 The 7-bit ASCII character code.

Code	*Character*	*Code*	*Char.*	*Code*	*Char.*	*Code*	*Char.*
0000000	Null	0100000	Space	1000000	@	1100000	‘
0000001	Start heading	0100001	!	1000001	A	1100001	a
0000010	Start text	0100010	”	1000010	B	1100010	b
0000011	End text	0100011	#	1000011	C	1100011	c
0000100	End transmission	0100100	$	1000100	D	1100100	d
0000101	Enquire	0100101	%	1000101	E	1100101	e
0000110	Acknowledge	0100110	&	1000110	F	1100110	f
0000111	Bell	0100111	’	1000111	G	1100111	g
0001000	Backspace	0101000	(	1001000	H	1101000	h
0001001	Horizontal tab	0101001	)	1001001	I	1101001	i
0001010	Line feed	0101010	*	1001010	J	1101010	j
0001011	Vertical tab	0101011	+	1001011	K	1101011	k
0001100	Form feed	0101100	,	1001100	L	1101100	l
0001101	Carriage return	0101101	−	1001101	M	1101101	m
0001110	Shift out	0101110	.	1001110	N	1101110	n
0001111	Shift in	0101111	/	1001111	O	1101111	o
0010000	Data link escape	0110000	0	1010000	P	1110000	p
0010001	Device control 1	0110001	1	1010001	Q	1110001	q
0010010	Device control 2	0110010	2	1010010	R	1110010	r
0010011	Device control 3	0110011	3	1010011	S	1110011	s
0010100	Device control 4	0110100	4	1010100	T	1110100	t
0010101	Negative ack.	0110101	5	1010101	U	1110101	u
0010110	Idle	0110110	6	1010110	V	1110110	v
0010111	End block	0110111	7	1010111	W	1110111	w
0011000	Cancel data	0111000	8	1011000	X	1111000	x
0011001	End of medium	0111001	9	1011001	Y	1111001	y
0011010	Substitute	0111010	:	1011010	Z	1111010	z
0011011	Escape	0111011	;	1011011	[	1111011	{
0011100	File separator	0111100	<	1011100	\	1111100	|
0011101	Group separator	0111101	=	1011101	]	1111101	}
0011110	Record separator	0111110	>	1011110	∧	1111110	~
0011111	Unit separator	0111111	?	1011111	_	1111111	Delete

code, which is preceded by the fixed 3-bit pattern 011 (and possibly an additional parity bit). The 011 prefix, which is sometimes called the number's **zone field**, has no numerical significance; it merely serves to distinguish the decimal digits from the other, nonnumerical ASCII characters. As (2.47) indicates, EBCDIC has 1111 as its numerical zone field.

Before ASCII- or EBCDIC-encoded numbers are processed by arithmetic circuits, their zone fields are removed, thus reducing the numbers to ordinary BCD form, in which two digits rather than one are stored in a single byte. In this context, the two-BCD-digits-per-byte format is referred to as **packed decimal format;** the one-digit-per-byte format of the character codes is called **unpacked decimal format.** Computers sometimes have PACK and UNPACK instructions for converting binary-encoded decimal numbers between the packed and unpacked formats.

2.11 Error-Detecting Codes

Digital systems are subject to many sources of errors.

All digital systems occasionally experience errors; that is, they generate incorrect information. Common sources of errors are manufacturing flaws in the system's internal circuits, power supply surges during operation, and electromagnetic interference (EMI) from nearby equipment that is inadequately shielded. Internal signals that are physically close can also interfere with one another, a phenomenon called crosstalk. Less frequent but more obvious and more permanent errors can result from components failing due to age, overheating, mishandling, and the like.

Figure 2.22 illustrates the EMI phenomenon, in which interference radiating from a large electric motor changes an electrical signal that represents a transmitted 0 so that it is received as a 1. In this particular case, the transmitted word is the ASCII control character "delete data," but the received word is the digit 8. It is not difficult to imagine a single error of this sort having catastrophic consequences.

The 2/5 decimal code defined in Table 2.5 is error-detecting in the sense that a single error that consists of a 0 changing to a 1, or vice versa, is easily detected. This is a consequence of the fact that such an error always changes a valid 2/5 bit pattern or **code word** into an invalid bit pattern or **noncode word**. We now present a way of endowing other number types—indeed, any binary data words—with easily checked properties that can be used for single-error detection.

Parity

As observed above, the number of 1s, which we will denote by $W(X)$, in an n-bit binary word $X = x_{n-1}x_{n-2}\ldots x_1x_0$ can form the basis for error detection. We compute $W(X)$ by summing the n bits of X, a process that can be simplified by making the summation modulo some power of two. The simplest case has modulus two, when the summation becomes

$$P(X) = x_{n-1} + x_{n-2} + \cdots + x_1 + x_0 \qquad \text{(mod 2)}$$

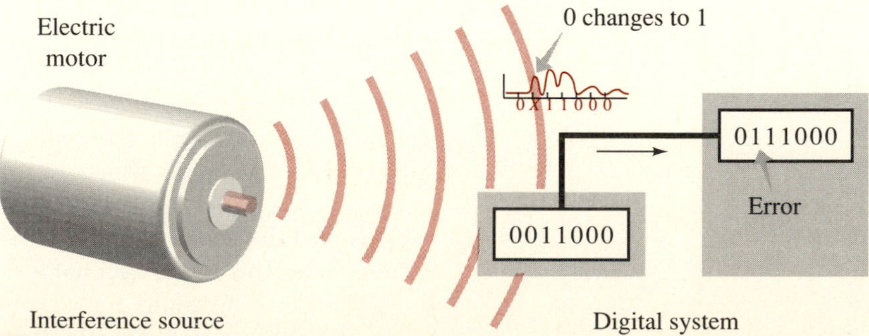

FIGURE 2.22 A single-bit error caused by electromagnetic interference (EMI).

Here, $P(X)$ is called the **parity function**[3] because the two possible values of $P(X)$ indicate the parity of $W(X)$—that is, whether $W(X)$ is even or odd. $P(X)$ is also referred to as the parity of X. If $W(X)$ is even, then $P(X) = 0$ and X is said to have **even parity**, whereas if $W(X)$ is odd, $P(X) = 1$ and X has **odd parity**. For example,

$$P(01101001) = 0 + 1 + 1 + 0 + 1 + 0 + 0 + 1 = 0 \quad (\text{mod } 2) \quad (2.48)$$

whereas

$$P(11001101) = 1 + 1 + 0 + 0 + 1 + 1 + 0 + 1 = 1 \quad (\text{mod } 2) \quad (2.49)$$

so the 8-bit words 01101001 and 11001101 have even and odd parity, respectively. All digits in the 2/5 code have even parity, whereas the parity varies from digit to digit in the BCD and E3 codes (Table 2.5).

Parity Check Bit

A parity check determines if the number of 1s in a word is even or odd.

It is obvious that an error that alters just one bit of X changes X's parity. Hence, we can detect such single-bit errors by periodically checking the parity of X. It is convenient to have a fixed parity, one or zero, associated with all error-free patterns of X—that is, with all valid code words. To this end, an extra bit p, called a **parity (check) bit**, is appended to X in order to fix the parity of the extended word Xp. The set of 2^n possible n-bit words Xp are considered to define a **parity code.** In the case of an **even parity code**, p is chosen to make the parity of Xp even; for an **odd parity code**, p is chosen to give Xp odd parity. For example, suppose that even parity is to be used. In that case, appending a parity bit p to the 8-bit word in (2.48) results in

$$P(01101001p) = 0 + 1 + 1 + 0 + 1 + 0 + 0 + 1 + p = p \quad (\text{mod } 2)$$

so p must be 0 to make the parity of Xp even. In the case of (2.49),

$$P(11001101p) = 1 + 1 + 0 + 0 + 1 + 1 + 0 + 1 + p = 1 + p \quad (\text{mod } 2)$$

so $p = 1$ gives Xp even parity.

In general, if $X = x_{n-1}x_{n-2}\ldots x_1x_0$, then an even parity code requires

$$p = x_{n-1} + x_{n-2} + \cdots + x_1 + x_0 \quad (\text{mod } 2) \quad (2.50)$$

while an odd parity code requires

$$p = x_{n-1} + x_{n-2} + \cdots + x_1 + x_0 + 1 \quad (\text{mod } 2)$$

If a parity bit is appended to the BCD digits, the resulting even parity code consists of the following 10 code words:

0 = 00000 1 = 00011 2 = 00101 3 = 00110 4 = 01001
5 = 01010 6 = 01100 7 = 01111 8 = 10001 9 = 10010

[3]The parity function $P(X)$ is also known as EXCLUSIVE-OR. This important function and its implementation are examined further in Chapter 3.

The remaining $2^5 - 10 = 22$ 5-bit words are the corresponding noncode words, and occur only as a result of errors.

Error-Detecting Circuits

Parity check circuits are mainly used in communication paths and memories.

We will assume that all valid code words Xp have even parity, with the parity check bit p defined by Equation (2.50). An error that affects any bit of Xp, including the parity bit p itself, makes the parity of Xp odd. This type of error is detected by computing the parity of Xp thus:

$$e = x_{n-1} + x_{n-2} + \cdots + x_1 + x_0 + p \quad \text{(mod 2)} \qquad (2.51)$$

The operation defined by (2.51) is called **parity checking**, and e is called a **parity error flag**. If $e = 0$, then no single-bit error has occurred, and Xp can be presumed to be correct. If $e \neq 0$, then some bit in Xp is erroneous, although we cannot tell which. Note that we cannot detect multiple-bit errors, unless those errors alter the parity of Xp and hence affect e.

Parity checks of this type are commonly used in digital systems to improve their reliability. Typical parts of a system that are guarded by parity checks are relatively long communication paths, such as the buses (cables) linking major units of the system, and memories that store information for extended periods of time. Buses and memories are particularly prone to errors due to environmental factors such as EMI (Figure 2.22).

A parity code for single-error detection is implemented in the manner illustrated in Figure 2.23. The n-bit parity function, whose behavior is specified by (2.50), is used to generate a parity bit p for the word X to be guarded. This **parity generator** may also be viewed as a modulo-2 summation circuit for n bits. The parity bit p is generated immediately before the data enter the error-prone region R of the system—for example, when X is written into a memory unit. p is appended to X, creating the code word Xp, which is the guarded version of the data item in question. When Xp emerges from R—for example, when the memory unit storing Xp is read—its parity e is computed by a circuit that implements the $(n + 1)$-bit parity function e defined by (2.51); this circuit is referred to as a **parity checker**. The parity checker is the error-detection circuit, and its output signal e serves as the parity error flag.

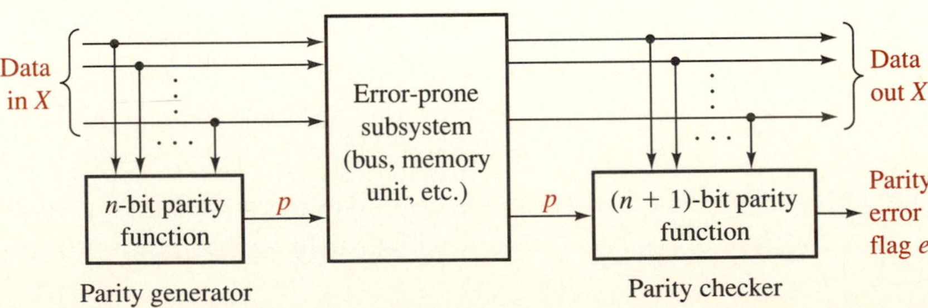

FIGURE 2.23 Circuits for single-bit error detection via parity checking.

The actions taken in response to detection of a parity error are determined by a system controller and range from retrying the failed operation to see if the error condition is temporary, to abandoning the operation entirely. By assigning several parity check bits to the same word, it is possible to not only detect a faulty bit, but also to pinpoint its location. The faulty 0 or 1 value can then be automatically changed back to 1 or 0, respectively, a process called **error correction** [Hamming, 1986]. Single-error correction circuitry is often built into the main memories of computers.

2.12 Floating-Point Numbers*

All the numbers we have encountered so far fall into the category of base-r fixed-point numbers, which are characterized by a radix point in a fixed position within each n-bit number word. Suppose that $r = 2$ and $n = 32$. Then, if the common unsigned integer format is employed, a fixed-point number X is confined to the range

$$0 \leq X \leq 2^{32} - 1 = 4{,}294{,}967{,}295 \approx 4.29 \times 10^9$$

Fixed-point numbers often provide insufficient range.

Although this representation is adequate for many purposes, it cannot represent the wide range of numbers encountered in scientific calculations, such as the mass of an electron in kilograms (1.67×10^{-27} kg) or the distance in Ångstroms to the nearest star outside the solar system (4.07×10^{26} Å).

Multiple Precision

One way to increase the range of representable numbers is to have k n-bit fixed-point numbers serve as a single kn-bit fixed-point number; this is referred to as a **multiple-precision number system.** For instance, if $n = 32$, two number words can be concatenated or chained to form a double-precision 64-bit number N. One word contains the more significant half of N, as well as N's sign bit; the other word contains the less significant half of N. Generally speaking, the processing of double-precision numbers takes twice as long, or else requires twice as many circuits, as the corresponding single-precision operation.

Suppose that two 32-bit words are used to represent a 64-bit signed number. Treated as a fraction, this format can represent numbers as small as $2^{-63} \approx 1.08 \times 10^{-19}$, which, however, is still insufficient to represent the electron mass figure. Moreover, a fraction format cannot represent numbers larger than one. A 64-bit integer format accommodates integers with magnitudes up to $2^{63} - 1 \approx 9.22 \times 10^{18}$; again, this cannot represent the astronomical distance cited above. Thus, multiple-precision arithmetic is costly and may still have inadequate range.

*This section may be omitted without loss of continuity.

Floating-Point Numbers

To handle very large and very small numbers of the type just considered, we normally switch to the so-called **scientific notation,** in which a number has the format $N = M \times 10^E$, where M and E are fixed point numbers called, respectively, the **mantissa** and the **exponent**. E is an integer that specifies the number of zeroes to be appended to M to obtain a fixed-point version of N. A positive value of E can specify an extremely large number; a negative E can specify an extremely small number. The version of scientific notation used in digital hardware is called *floating-point*, reflecting the fact that there is no longer a fixed position for the number's decimal point. In general, a **floating-point number** N is a word consisting of a pair of fixed-point binary-encoded numbers E and M, which are evaluated as

A binary floating-point number with exponent E and mantissa M denotes $M \times 2^E$.

$$N = M \times B^E$$

B is called the **base** of the floating-point number system, and is usually 2 or 10. M and E are signed numbers, which are binary if $B = 2$ or (binary-encoded) decimal if $B = 10$. Because B is fixed, it need not be included explicitly in the floating-point number format.

A representative 32-bit floating-point binary ($B = 2$) format appears in Figure 2.24. The sign s of the mantissa M is also the sign of the entire number N and is assigned the leftmost bit position to maintain compatibility with the sign convention for fixed-point binary numbers. Eight bits are used for the exponent E, with the remaining 23 bits used for the magnitude part of the mantissa. E represents a signed binary integer, so that exponent values here can range from $-2^7 = -128$ to $+2^7 = +128$; we denote this range by ± 128. (The precise exponent range depends on number-coding details.) The mantissa M is often treated as a fraction, so that mantissa values fall into the approximate range ± 1. With these assumptions, the representable numbers lie in the range $\pm 2^{-128}$ approximately—that is, $\pm 3.4 \times 10^{\pm 38}$. This particular 32-bit floating-point format can represent numbers whose fixed-point equivalent would require about 128 bits. Note, however, that the number of significant bits, called the *precision* of the representation (Section 2.3), is determined by the mantissa M and is less than that of the 32-bit fixed-point format, in which all nonsign bits can be significant. Thus, the floating-point format has less precision, but thanks to the exponent, it has a greatly increased

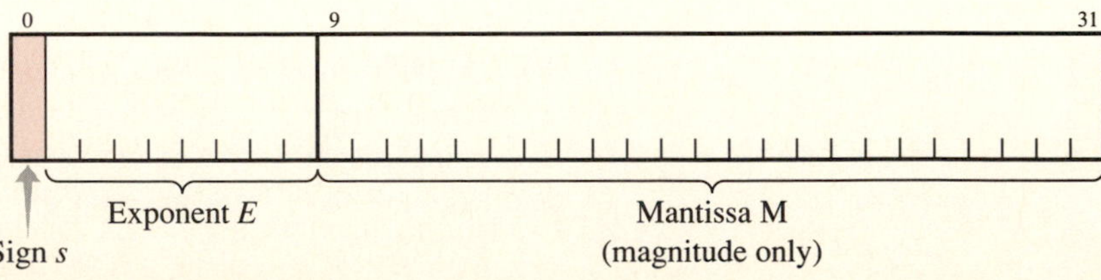

FIGURE 2.24 The format of a typical 32-bit floating-point number.

range of magnitudes. As a result, floating-point arithmetic is much less likely than fixed-point to result in overflow.

Besides the loss of precision noted above, floating-point numbers require considerably more hardware for processing than do fixed-point numbers. The analysis of floating-point arithmetic is also quite difficult. Consequently, circuits that perform arithmetic on floating-point numbers are not found in all types of digital computers. These circuits are also implemented in VLSI circuits called **floating-point processors**, which are intended to be coupled with central processing units that lack their own floating-point hardware. The Motorola 68040 microprocessor chip (Figure 1.15) is an example of a CPU that does contain a built-in floating-point processor.

Normalization

Normalization ensures a unique floating-point representation of each number.

We remarked earlier that some fixed-point number codes have multiple representations for zero, leading to minor implementation difficulties. This problem also appears in floating-point codes, except now *every* number has many different representations. For example, 0.001×10^3, 1.0×10^0, and $100000000.0 \times 10^{-8}$ all denote the number one. To reduce this profusion of equivalent forms, a specific one called the **normalized form** is defined to give every number a unique floating-point representation. Normalization fixes the position of the radix point of M relative to its most significant nonzero digit. Two typical normalization rules are (1) the most significant digit of M should immediately precede the radix point, and (2) the most significant digit should immediately follow the radix point. In the former case, the number one acquires the unique normalized representation 1.0×10^0. An unnormalized number is normalized by shifting the radix point of the mantissa and making compensating adjustments to the exponent. Thus, to normalize 5716.9001×10^5 to 5.7169001×10^8 according to rule (1), the mantissa's decimal point is shifted three positions to the left and the exponent is incremented by three.

Biasing

The representation of zero continues to pose problems. If the mantissa M is zero, then the exponent E can have any value, because $N = 0.0 \times B^E = 0$. However, if M is only approximately zero—for example, due to loss of precision in the computation of M—then to ensure that $M \times B^E$ is a good approximation to zero, E should be a large negative number. Therefore, it is usual to assign the exponent value $-E_{\max}$ to the representation of zero, where $E_{\max}$ is the largest available exponent magnitude. It is also desirable to assign the all-0 bit pattern to zero so that the fixed-point and floating-point representations of zero are identical.

To meet these two goals, it is necessary to encode $-E_{\max}$ into the all-0 pattern $00\ldots0$. This can be done by using an **excess-$E_{\max}$** code for the exponent, in which the numerical value of the exponent is obtained by subtracting E_{max} from the exponent field E of the floating-point number; compare this to the definition of excess-3 code (Section 2.9). When defined in this way, the exponent is said to have **bias** E_{max}. (A biased exponent is also called the **characteristic** of a floating-point number.) Exponent biasing thus allows the all-0 pattern to be used as the normalized form for floating-point zero, and ensures that imprecision in mantissa computations has minimal effect on the overall accuracy of the number system.

Biasing the exponent improves accuracy with very small numbers.

Example 2.11

IEEE 754 standard floating-point number format

Many distinct and incompatible floating-point formats have been introduced by computer manufacturers over the years. Since 1985, however, a standard format for binary floating-point numbers, sponsored by the Institute of Electrical and Electronics Engineers (IEEE) and known as the IEEE 754 standard [IEEE, 1985], has been available and widely adopted. The 32-bit version of the IEEE 754 format has the structure depicted in Figure 2.24, with a sign bit s, an 8-bit exponent field E, and a 23-bit mantissa field M.

The mantissa is a signed-magnitude binary number, with s serving as the sign bit. The choice of SM rather than 2C code here reflects, in part, the frequent use of multiplication and division in floating-point computations. The magnitude part of the mantissa is normalized to the form $1.bbb\ldots$, which places the binary point immediately to the right of the most significant nonzero mantissa bit. Because the mantissa is a binary number, the leading nonzero bit is always 1. In order to increase the number of mantissa bits, this leading 1 is not explicitly included in the floating-point format; it is therefore referred to as a "hidden" bit. The mantissa field M stores only 23 bits that follow the binary point. Hence, the numerical value of the complete mantissa, which is called the **significand** in the IEEE standard, is $(-1)^s \times 1.M$. Floating-point arithmetic circuits designed to process the IEEE 754 format must recognize this situation and insert or remove the hidden 1 bit, as appropriate. The complete signed-magnitude significand used in arithmetic calculations effectively has 25 bits, but is reduced to 24 bits for storage purposes.

The IEEE 754 standard has many features to minimize numerical errors.

The 8-bit exponent field E used in the IEEE 754 standard denotes a signed binary integer, and can represent exponent values from -127 to $+128$. (The asymmetry is due to the fact that $+0$ and -0 have a single representation.) For the reasons discussed above, the exponent is biased by 127, and so may be considered to be in excess-127 code. If E is considered to be an unsigned integer ranging from 0 to 255, the actual numerical value of the exponent is computed as $E - 127$. Thus, $E = 00000000$ denotes the exponent -127, $E = 00000001$ denotes -126, and so on. The pattern $E = 01111111$, which is the binary form of 127, denotes the zero exponent. The number one, for example, is encoded as

$$0 \;\; 01111111 \;\; 00000000000000000000000$$

which represents $+1.0\ldots0\times2^0$. In general, provided that $0 < E < 255$, the value of a floating-point number in this format can be expressed succinctly as

$$N = (-1)^s 2^{E-127} 1.M \tag{2.52}$$

Therefore, the representable numbers have magnitudes from $2^{-126} \approx 1.18 \times 10^{-38}$ to $2^{127}(2 - 2^{-23}) \approx 2^{128} \approx 3.40 \times 10^{38}$.

The representation of floating-point zero has all-0 E and M fields, and is an exception to (2.52). The IEEE 754 standard also recognizes several other exceptional cases that address various subtle aspects of floating-point arithmetic. For example, the largest exponent field $E = 11111111$, corresponding to 255, is reserved to denote "infinite" numbers that result from some types of overflow, as well as certain invalid bit combinations considered to be in a special "not a number" category, denoted NaN.

We conclude with a few more examples of floating-point numbers in the IEEE 754 format:

$$\begin{aligned}
1\ 00000000\ 00000000000000000000000 &= -0 \\
1\ 01111111\ 00000000000000000000000 &= -1 \\
0\ 01111111\ 11111111111111111111111 &= 2 - 2^{-23} \\
0\ 10000001\ 11100000000000000000000 &= +7.5 \\
0\ 11111111\ 10111100100011011111001 &= \text{NaN}
\end{aligned}$$

Addition

We now look at how floating-point numbers are added. Consider first an example of two numbers in ordinary decimal form (scientific notation): $N_1 = 9.9985 \times 10^9$ and $N_2 = 2.3128 \times 10^6$. Because their exponents are different, we cannot directly add N_1 and N_2. Instead, we modify one of the numbers to make its exponent equal to that of the other. It is convenient to take the smaller exponent, in this case, 10^6, and increase it to equal the larger exponent 10^9. This requires adding three to N_2's exponent, which is equivalent to multiplication of N_2 by 10^3. To keep the value of N_2 the same, a compensating reduction must be made to the mantissa 2.3128 by dividing it by 10^3. The division is implemented by right-shifting 2.3128 by three decimal places to produce 0.0023128.

In its new form, 0.0023128×10^9, N_2 can now be added directly to N_1, yielding $N_3 = 10.0008128 \times 10^9$. Note that the sum N_3 has the same exponent as N_1 and N_2, and that only the mantissas actually have to be added. Note further that the integer part of the result has two significant digits, making it unnormalized, if normalized numbers are required to have their most significant digit immediately to the left of the decimal point. Thus, another step to normalize the result requires right-shifting the mantissa one place (division by 10) and adding one to the exponent (multiplication by 10), so that N_3 becomes $1.00008128 \times 10^{10}$. The floating-point addition process $N_3 = N_1 + N_2$ illustrated above is summarized in the following procedure *FADD*.

Procedure

Procedure *FADD* to add two floating-point numbers

1. Compare the exponents of $N_1 = E_1M_1$ and $N_2 = E_2M_2$ and identify the smaller one, say E_2. This comparison can be implemented by special hardware (a comparator) or by a trial subtraction of the form $E_1 - E_2$.
2. Equalize the exponents of N_1 and N_2 by right-shifting $E_1 - E_2$ places, the mantissa M_2 of the number with the smaller exponent.
3. Add the mantissas to obtain M_3.
4. If the result is not normalized, then right-shift M_3 one place and add one to the exponent E_1 to obtain the exponent E_3 of the result $N_3 = E_3M_3$.

The four-step procedure *FADD* applies to binary as well as decimal floating-point numbers; it can also be easily extended to perform subtraction as well as addition. Note that about three fixed-point exponent or mantissa add/subtract operations are required by *FADD*, along with various shift operations. Further complications must be introduced to round off the result's mantissa, to handle overflow conditions, and to account for exponent biasing.

Example 2.12 **Adding two floating-point numbers**

Suppose that we are to add the two numbers

$$N_1 = 0 \;\; 10000001 \;\; 11100000000000000000000 = +7.5$$
$$N_2 = 0 \;\; 01111111 \;\; 10000000000000000000000 = +1.5$$

which use the IEEE 754 format. Subtraction to compare the exponents (step 1 of *FADD*) yields

$$\begin{aligned} E_1 - E_2 &= 10000001 - 01111111 \\ &= 00000010 \end{aligned}$$

This can be implemented by ordinary binary SM subtraction rather than excess-127 subtraction, because the bias values included in E_1 and E_2 cancel. We conclude that N_2's exponent is smaller by two than that of N_2, so in step 2, we right-shift $1.M_2$, including the hidden bit, two places to obtain 0.01100000000000000000000. The result is now added to $1.M_1$ thus (step 3):

$$1.11100\ldots00 + 0.01100\ldots00 = 10.01000\ldots00 \qquad (2.53)$$

In general, step 3 requires inspection of the sign bits of N_1 and N_2 to determine the proper operation to perform on the mantissas. To normalize the result, the unnormalized number appearing on the right-hand side of (2.53) is right-shifted and rounded off, yielding

$$M_3 = 00100000000000000000000$$

after the hidden bit (the leading 1) is removed. The exponent $E_3 = E_1$ is adjusted by adding one, and becomes $E_3 = 10000010$. The final result in the proper format for storage is therefore

$$0 \quad 10000010 \quad 00100000000000000000000 = +9.0$$

In conclusion, we mention that the multiplication and division operations on floating-point numbers are somewhat simpler than addition and subtraction, because they require no preliminary exponent equalization. Multiplication, for instance, is accomplished by adding the two exponents and multiplying the two mantissas of the given numbers, with a final normalization step, as before.

CHAPTER 2 SUMMARY

Number Systems

2.1 ■ A number system is defined by its basic symbols, called **digits** or **numerals**, and the ways those digits can be combined to represent the full range of numbers. The decimal system is made up of the 10 Arabic numerals 0, 1, 2, 3, 4, 5, 6, 7, 8, 9.

■ The decimal system employs a **positional notation,** in which each digit of a multidigit number has a fixed value or **weight**, determined by its position. Positional numbers are widely used because they are compact and easy to compute compared to nonpositional numbers.

2.2 ■ The **radix** or **base** of the binary system is $r = 2$, and the decimal system has a radix of $r = 10$. A positional number notation of this type is called a **base-*r*** or ***r*-ary** number system.

■ The binary system employs the two digits 0 and 1, which are referred to as **bits** (binary digits).

■ Digital systems process numbers with a fixed number of digits, called the **word size.** A fixed word length limits a number system's **precision**—the number of significant digits it contains.

2.3 ■ Because humans use the decimal system and digital circuits use the binary system, it is important to know how to convert numbers from one system to the other.

■ Conversion of binary integers to decimal can be accomplished by a series of multiplications by two and addition. Fractional binary numbers are **scaled** to integer form for conversion to decimal.

2.4 ■ Conversion of decimal integers to binary can be accomplished by a series of divisions by two from which the remainders form the equivalent binary number. Fractional decimal numbers can be converted to binary by repeated multiplication of the fraction by two.

■ The reduction in the size of r from 10 to 2 generally increases the number of steps needed to complete an arithmetic operation, but the individual steps are simplified by the fact that the digit choice is limited to two.

■ The loss of precision that results from using a fixed word size in digital systems is minimized by **rounding** procedures.

■ **Flowcharts** and **pseudocode** are two common methods for expressing problem-solving procedures.

Binary Arithmetic

2.5 ■ The fundamental arithmetic operations (addition, subtraction, multiplication, and division) can be performed in much the same manner in all r-ary number systems. For example, the sum of two binary integers is obtained by adding them digit by digit, moving from right to left.

■ Addition of two binary digits produces a 2-bit result consisting of a **sum** bit and a **carry** bit. Digital adder circuits are typically constructed so that each component of the circuit supplies an output carry bit to the component on its left and also contributes a sum bit to the final result.

■ Subtraction of two binary digits also produces two result bits: a **difference** bit and a **borrow** bit. Circuits for subtraction are very similar to adder circuits.

■ **Overflow** occurs when operations produce a result that contains more digits than are specified by a system's word size.

■ In **modular** or **modulo-*m* arithmetic**, the results of all arithmetic operations are confined to some fixed set of m values. A feature of modular arithmetic is that as a number is incremented beyond a maximum representable value, it "wraps around" to the smallest value. Modular arithmetic is the fundamental type of arithmetic employed in digital systems.

2.6 ■ In the decimal system, we sign numbers by placing the symbol + or − at the left end of the number in a notation called **signed-magnitude (SM)** notation. In the binary system, the sign is represented by a single bit: 0 for a positive number, 1 for a negative number.

■ Signed-magnitude addition and subtraction are relatively complex. For example, in order to add two SM numbers, we must compare the numbers to determine the operation to be performed, the order of the operands, and whether adjustments must be made to the signs.

■ **Two's-complement (2C)** code uses the same notation for positive numbers as SM notation. Negative numbers in 2C are defined by changing all the 0 digits in a given positive number X to 1 and vice versa to produce the **ones' complement** of X, and then adding one to the rightmost digit. 2C numbers are easy to add and subtract, making 2C code attractive for use in digital systems.

Binary Codes

2.7 ■ Many different binary codes have been designed for a variety of applications. Two major design goals of such codes are security and reliability.

2.8 ■ Base-2^k numbers, especially **octal** ($2^k = 8$) and **hexadecimal** ($2^k = 16$) numbers, are used as shorthand for binary data. Not only are these systems compact, but they can also be converted to and from the binary system easily.

2.9 ■ Decimal numbers can be converted to a 0/1 format by mapping each decimal digit into suitable binary code to form a binary-encoded number whose base is ten rather than two. In the popular **binary-coded decimal (BCD)** version of this approach, each decimal digit is replaced by the equivalent 4-bit binary number.

2.10 ■ **Character codes,** such as ASCII and EBCDIC, are based on a binary word, called a **character**, which represents the smallest unit of information in many applications. Character codes use seven or eight bits to encode a full set of keyboard symbols and functions.

2.11 ■ Because digital systems are subject to many sources of errors, error-detecting or error-correcting codes are widely used. These codes typically append one or more **parity check bits** to the information being guarded.

2.12 ■ The range of the representable numbers in a digital system can be greatly expanded beyond the range of **fixed-point numbers** by the use of **floating-point numbers**. Floating-point numbers consist of a pair of fixed-point numbers, the **mantissa** and the **exponent**.

■ Compared to the fixed-point format, floating point formats are less precise, require more hardware to process, and have far more complex arithmetic operations.

Further Readings

A readable introductory text covering numbers and their use in digital circuits is N. R. Scott's *Computer Number Systems and Arithmetic* (Prentice-Hall, Englewood Cliffs, N.J., 1985). A more advanced treatment of the same issues appears in S. Waser and M. J. Flynn, *Introduction to Arithmetic for Digital Systems Designers* (Holt, Rinehart and Winston, New York, 1982). Volume 2 (*Seminumerical Algorithms*) of D. E. Knuth's classic work *The Art of Computer Programming* (Addison-Wesley, Reading, Mass., 1969) has a deep and fascinating treatment of many relevant aspects of fixed-point and floating-point numbers, as well as their arithmetic. An introduction to error codes by a pioneer of this field can be found in R. W. Hamming's *Coding and Information Theory,* 2nd ed. (Prentice-Hall, Englewood Cliffs, N.J., 1986).

Chapter 2 Problems

Number Systems

Positional and nonpositional numbers; number bases and number words; binary-decimal conversion; other base conversion procedures; rounding procedures.

2.1 Show that the tally system illustrated by Equation (2.2) is a type of base-r positional number system. What is r in this case? Explain why a tally number system is unsuitable for use in digital equipment.

2.2 Give an example of a nonpositional number system in common use. What are its disadvantages compared to positional systems?

Number bases still used occasionally are the duodecimal (base 12), the vigesimal (base 20), and the sexagesimal (base 60).

2.3 Twelve has been advocated as a number base for everyday use. Indeed, fragments of a base-12 (duodecimal) system can be seen in such counting terms as a dozen and a gross (12×12). Specify the base-12 representations of the numbers from 20 to 40, as in Table 2.2.

2.4 Another number base with remnants in ordinary speech is 20 or a score. It occurs in the archaic English expression "four score," which has the exact counterpart *quatre-vingts* in the modern French word for 80. Define a suitable set of digit symbols for a base-20 system and use them to represent the numbers from 40 to 50.

2.5 Determine the base-4 and base-5 representations of the following decimal numbers: 10, 21, 50, 67, 100.

2.6 Convert each of the following numbers to ordinary decimal form: 24_5, 306_7, 147_8, ABC_{15}, 14_{64}.

2.7 Convert the following unsigned binary integers to decimal form: 101011_2, 1100110_2, 1010101010101010_2.

2.8 Convert the following fractional and mixed unsigned binary numbers to decimal, with fractions computed to four decimal places: 0.1111_2, 0.11010011_2, 001110.11111111_2.

2.9 Suppose an n-bit word is used to store an unsigned binary number. Give in decimal form the largest integer that can be stored when $n = 8, 16$, and 32. What is the smallest nonzero fraction that can be stored when $n = 8, 16$, and 32?

2.10 Prove that no finite binary fraction has a nonterminating decimal equivalent. (*Hint:* Observe that if F_2 is a finite nonzero binary fraction, then F_{10} must have 5 as its least significant digit.) Prove by means of a simple counterexample that the converse is false.

2.11 Convert the following decimal integers to binary form: 666_{10}, 101011_{10}, 239184356240_{10}.

2.12 Convert the following fractional and mixed decimal numbers to binary form, with fractions rounded off to four binary places: 0.1111, 947.613, 27094.6543.

2.13 With *BINDECi* as a model, design a procedure *rDECi* to convert integers from any integer base r to base 10. Test your procedure's correctness by applying it to the following cases: 120120120_3, 250616_7, $1234567890ABCDEF_{16}$.

2.14 Extend procedure *DECBINf* for converting fractions from decimal to binary to include steps to round off the result to n binary places.

2.15 The procedure *ROUND* is sometimes called **biased rounding** because it always rounds up whenever

$$0.00\cdots00b_{-n-1}b_{-n-2}\cdots \geq 2^{-n-1} \tag{2.54}$$

Unbiased rounding is more accurate than the biased approach used in *ROUND*.

is an equation rather than an inequality. In that case, an equally accurate result is obtained if N_2 is rounded down by skipping step 1 (addition) of *ROUND*. An **unbiased rounding** scheme can improve the accuracy of a sequence of computations by rounding up about as often as it rounds down whenever (2.54) is an equation. Devise an unbiased rounding procedure *ubROUND* for N_2, preferably one that does not involve counting the rounding up/down frequency. (*Hint:* Consider basing the up/down decision on some property of the truncated N_2.)

2.16 Construct a procedure *ROMDECi* in either the flowchart or the pseudocode style of Figure 2.6 for converting an integer expressed in Roman numerals to ordinary decimal form.

2.17 1300300013_r is the r-ary representation of some number N, which is known to be a perfect square. The base r is less than or equal to 10. Find all the values that r can assume. (*Hint:* Consider the possible values that the rightmost digit of $\sqrt{N}$ can assume.)

Binary Arithmetic

Unsigned numbers; unsigned addition and subtraction; modular arithmetic; signed numbers; signed-magnitude, two's-complement, and ones'-complement codes; overflow detection.

2.18 Suppose that $X = 010101_2$. Determine the sum $X + Y$ computed by a 6-bit adder for each of the following values of Y: 010101_2, 101011_2, 101111_2.

2.19 Draw a block diagram for a 2-bit ripple-carry adder. The adder's inputs are a_1a_0, b_1b_0, and c_{in}, while its outputs are s_1s_0 and c_{out}. Also construct a table listing the value of c_{out} for every combination of input values.

It's worth noting that $x + y$ (mod 2) is equal to $x - y$ (mod 2).

2.20 Let X = 00110101 and Y = 10110101 be unsigned numbers processed by a modulo-2^8 adder-subtracter. Determine the values computed for $X+Y$, $X-Y$, and $Y-X$.

2.21 **(a)** In the ripple-borrow subtraction circuit shown in Figure 2.8, a borrow signal b_i is sent from full-subtracter stage S_i *to* the stage S_{i+1} on its left. In manual subtraction, however, we say that S_i borrows a 1 *from* S_{i+1}, which suggests that the direction of the subtracter's borrow signal should be reversed. Explain this apparent discrepancy. **(b)** Comparing the behavior of the full adder defined by Equation (2.17) and the full subtracter defined by (2.19), we see that for each combination of the three input variables, the sum bit s_i and the difference bit d_i are the same, which seems to imply that the sum $S = X + Y$ computed by an adder is the same as the difference $D = X - Y$ computed by a subtracter. Explain this apparent contradiction.

2.22 Do the following calculations on decimal numbers, using the indicated modular arithmetic: 3 + 15 (mod 6); 3 − 15 (mod 6); 7 + 16 + 24 (mod 8); 12 × 3 − 6 (mod 11); [17 + 22 (mod 15)] ÷ [11 × 9 (mod 6)] (mod 7).

2.23 Evaluate each of the following expressions involving binary numbers: 101011−1101 (mod 16); 101011−1101 (mod 13); 1011×1101 (mod 16); 10110011+11000101 (mod 2^6); 10110011 + 11000101 (mod 2^8).

2.24 Two integers X and Y are **congruent** modulo m, denoted $X \equiv Y$ (mod m), if $X - Y$ is exactly divisible by m. Thus, every integer in $\mathbf{Z}$ is congruent (mod m) to a unique integer in $\mathbf{Z}_m = \{0, 1, \ldots, m-1\}$. Let $(x_{n-1}x_{n-2}\ldots x_1x_0)_{10}$ be the usual decimal representation of an unsigned integer X. Prove that

$$X \equiv x_{n-1} + x_{n-2} + \cdots + x_1 + x_0 \qquad (\text{mod } 9)$$

This result provides an easy way to determine if a given number is divisible by nine. What is it?

2.25 Give the correct 8-bit representation in the SM and 2C codes for each of the following decimal integers: +120, −120, −1, zero.

You can't identify a number word's code merely by inspection.

2.26 A register contains the 12-bit word N = 010110010111. What is N if it represents **(a)** an unsigned binary number; **(b)** a signed-magnitude binary number; and **(c)** a number in two's-complement code? What is N in each of these three cases if its leftmost bit is changed from 0 to 1?

2.27 The 10-bit registers A, B, and C contain the signed numbers 0001001010, 1001100011, and 110110011, respectively. Compute the quantities $A+B$, $A-B$, $B+C$, $B-C$, and $C-B$, assuming all numbers are in SM code.

2.28 Construct a set of rules for subtraction of n-bit SM numbers, analogous to those for SM addition given in Figure 2.12.

2.29 Explain why 2C code is generally preferred to the SM and 1C codes for representing signed binary numbers in digital systems.

2.30 Repeat Problem 2.27, this time assuming all numbers are in 2C code.

2.31 Convert the following decimal numbers to 16-bit two's-complement code and perform the indicated subtractions using a standard (unsigned) 16-bit adder and negation circuit. Show all carry/borrow bits in the style of Example 2.4: **(a)** 1748 − 5910; **(b)** 5910 − 2749; **(c)** (−2749) − 5910; **(d)** (−2749) − (−5910).

2.32 The bit pattern 1000 is not included among the numbers constituting the 4-bit 2C code of Table 2.3. It is often argued that 1000 is a legitimate representation of −8 and therefore should be included in this role in the 2C code. Analyze the validity of this claim by considering the results of the various arithmetic operations applied to 1000.

2.33 Here is yet another procedure *SCANEG2c* for negating a 2C number (two's complementation). The given 2C number X is scanned once from right to left, and certain bits are changed as indicated. Analyze this technique and demonstrate that it is equivalent to the original *NEG2c* procedure.

Procedure

Procedure *SCANEG2c* to negate a 2C number

1. Input the 2C number $X_{2C} = x_{n-1}x_{n-2}\cdots x_1x_0$. Set a control flag f to an initial value of 0.
2. Scan X_{2C} from right to left—that is, from $i = 0$ to $n-1$. If $f \neq 0$, then replace x_i with $\overline{x}_i$; otherwise, if $x_i = 1$, then set f to 1. When the scan is complete, X_{2C} has been changed to its two's complement.

2.34 The ones'-complement (1C) addition examples in Equations (2.29) and (2.30) will both be correct if the carry output bit c_7 generated from the sign (leftmost) bit position in the modulo-2^8 addition step $X + Y = Z$ is added to the rightmost bit of Z in a second modulo-2^8 addition step $Z + c_7$. Show that this "end-around carry" scheme gives the proper 1C result for all combinations of positive and negative 1C operands.

2.35 Use the radix-complement code with the appropriate radix r to represent each of the following signed numbers, assuming that each number must contain exactly six digits, some of which may be insignificant: $+100_{10}$, -100_{10}, -426_{10}, -426_9, $-39C_{16}$.

2.36 Let $N = 1001111010$ denote a signed binary integer. Find the smallest number M that must be added to N to produce overflow, assuming that the numbers are in **(a)** SM code; **(b)** 2C code. Also find the smallest numbers that must be subtracted from N to produce overflow for the same two codes.

Here's another useful formula for overflow detection.

2.37 Consider the problem of overflow detection in an n-bit adder for 2C numbers. Let c_{n-2} and c_{n-1} be the carry input and output bits, respectively, involved in adding the sign bits. Show that the overflow detection flag OVF can be defined by the equation

$$OVF = c_{n-2} + c_{n-1} \pmod 2 \qquad (2.55)$$

2.38 **(a)** Prove that the 2/5 code defined in Table 2.5 is not a positional or weighted number code in the usual sense. **(b)** It is, however, possible to assign integer weights

w_1, w_2, w_3, w_4, w_5 to the five bits of the 2/5 code word representing the digit N so that

$$w_1 + w_2 + w_3 + w_4 + w_5 = N \quad (\text{mod } p) \qquad (2.56)$$

where p is also an integer. Prove this claim by determining a set of values for w_i and p that satisfy (2.56).

Binary Codes

Gray codes; base-2^k numbers; binary-octal-hexadecimal conversion; decimal codes; BCD code; character codes; error detection; parity checking; floating-point numbers; IEEE 754 standard.

2.39 **(a)** What are the advantages and disadvantages of Gray codes over ordinary r-ary number codes? **(b)** Construct a special-purpose binary Gray code for counting modulo 6. Zero should be represented by 000, and incrementing the code for six by one should return the count to 000.

Binary-to-Gray code conversion is easy.

2.40 Let $N_{\mathbf{RG}} = g_{n-1}g_{n-2}\ldots g_1g_0$ be the code word representing a number N in the n-bit reflected Gray code G_n, and let $N_2 = b_{n-1}b_{n-2}\ldots b_1b_0$ be the ordinary binary version of the same number N. **(a)** Show that $N_{\mathbf{RG}}$ can be easily derived from N_2 by means of the formula

$$g_i = b_i + b_{i+1} \quad (\text{mod } 2) \qquad 0 \leq i < n-1$$

$$g_{n-1} = b_{n-1}$$

Use this formula to determine the code word in G_8 that denotes 97_{10}. **(b)** Obtain an analogous formula for deriving N_2 from $N_{\mathbf{RG}}$.

2.41 A binary code that was once of great importance is the Morse code developed by Samuel F. B. Morse (1791–1872), the American inventor of the telegraph (an early digital communication system). It uses dots and dashes to encode each letter of the alphabet, as shown in Figure 2.25. A feature of this code is that the shortest code words are assigned to the most frequently used letters. **(a)** It is obviously easy to map dots and dashes into 0s and 1s. However, the Morse code in this form is considered unsuitable for use in modern digital equipment. Explain why this is so. **(b)** For applications like long-distance radio communication that still occasionally employ the Morse code, it is mapped into an m-out-of-n binary code. What are suitable values of m and n for such a code, assuming only the letters in Figure 2.25 need to be encoded?

A·–	E·	I··	M––	Q––·–	U··–	Y–·––
B–···	F··–·	J·–––	N–·	R·–·	V···–	Z––··
C–·–·	G––·	K–·–	O–––	S···	W·––	
D–··	H····	L·–··	P·––·	T–	X–··–	

FIGURE 2.25 The international Morse code.

2.42 Convert each of the following binary numbers to bases 4, 8, and 16: 1010010_2, 110100.110100_2, 111111111111_2.

2.43 Convert the following numbers to binary: 1010010_4, 6210537_8, A15D.01CC_{16}.

2.44 Convert 4675542_8 to hexadecimal. Convert 1F08B725_{16} to octal.

2.45 Determine the unknown X_r in each of the following equations. Assume all numbers are unsigned integers and give your answers in r-ary form, where r is the subscript of X_r.

$$1010010_2 \times X_2 = 1102_4$$

$$12301230_8 + \text{FF89AB}_{16} = X_4 \qquad (\text{mod } 2^{24})$$

2.46 Convert 110100011011_2, 7869002_{10}, and 749C0F_{16} to BCD form.

2.47 Convert the following BCD numbers to binary and (ordinary) decimal form: $010100011001_{\underline{10}}$ and $0100010010101.10000010_{\underline{10}}$.

2.48 Construct a flowchart or pseudocode procedure to convert a mixed BCD number to ordinary decimal form. Include a feature to detect the presence of invalid bit combinations.

2.49 For each of the following ten's-complement numbers X, determine X and $-X$ in 10C BCD code: 910107_{10C}, 06612897_{10C}, 9.99999_{10C}.

2.50 Let $X = 916201_{10C}$ and $Y = 097782_{10C}$. Using six-digit BCD addition or subtraction, compute each of the following: $X+Y$, $X-Y$, $Y-X$, $X/10$.

2.51 Describe the conditions for overflow to occur in the addition or subtraction of n-digit ten's-complement numbers, both using ordinary decimal code and BCD code. As in two's-complement code, overflow may be detected by observing the input and output carry signals appearing at the sign position.

2.52 A given set of numbers require more storage space if they are encoded in BCD (base-10) format rather than binary (base-2) format. Thus, if a large set of long binary numbers occupy N bits of a computer's memory, the same set of numbers will occupy storage space cN if they are converted from binary to BCD code. Determine the "coding inefficiency" factor c, showing your calculations.

2.53 It has been proposed to reduce the "wasted" storage space of the BCD code by encoding pairs of decimal digits rather than single decimal digits into m-bit words. Thus, all digit-pairs from 00_{10} to 99_{10} would be represented by their k-bit binary equivalents. What should k be? What would the coding inefficiency factor c, defined in Problem 2.52, become in this case? Comment briefly on the disadvantages of this encoding scheme.

2.54 Construct a diagram for a four-digit BCD adder using the circuit of Figure 2.14 as a basic component, possibly with minor modifications.

2.55 Convert the following signed numbers to E3 and 2/5 code: 911450_{10C}, 111011110000_{2C}.

2.56 **(a)** Devise a procedure for adding two excess-3 digits analogous to that for BCD digits given by (2.45). **(b)** Perform the following E3 calculations showing all steps involved and all carries generated: $1010_{E3} + 0111_{E3}$ and $0011110001001000_{E3} + 0110001111001011_{E3}$.

Self-complementarity simplifies the implementation of decimal codes.

2.57 A decimal code is **self-complementing** if the nines'-complement of a digit $d_i = c_{i3}c_{i2}c_{i1}c_{i0}$ in the code is the same as $\overline{d}_i = \overline{c}_{i3}\overline{c}_{i2}\overline{c}_{i1}\overline{c}_{i0}$. For example, in the case of $1010_{E3} = 7$, we have $\overline{1}0\overline{1}\overline{0} = 0101$, while $(1100 - 1010) + 0011 = 1010 + 0011 = 0101 \pmod{16}$. The advantage of being self-complementing is that very simple hardware is needed to compute $\overline{d}_i$ compared to that required for $9 - d_i$. Prove that E3 code is self-complementing, but BCD code is not.

2.58 BCD is an example of a decimal code that is weighted but not self-complementing. (The latter is defined in the preceding problem.) Excess-3 is a nonweighted decimal code that is self-complementing. Show that the following decimal code is both weighted and self-complementing, and determine the weight of each bit.

0 = 0000   1 = 0001   2 = 0011   3 = 0100   4 = 1000

5 = 0111   6 = 1011   7 = 1100   8 = 1110   9 = 1111

2.59 **(a)** Decode the following ASCII character sequence:

0100100 1100101 0111101 1101101 1100011 1011110 0110010 0100100

(b) Encode the following digit sequence into both unpacked ASCII form and packed form: 91263380.

2.60 Write a brief note comparing and contrasting the 2/5 decimal code and the error-detecting code obtained by adding a single parity bit to BCD code.

2.61 Suppose that a single parity bit p is appended to the left end of every ASCII character. List the resulting 8-bit representations of the 10 decimal integers, assuming even parity is used.

2.62 How many distinct double-bit errors can occur in a 7-bit ASCII character? How many of these errors change the character's parity?

2.63 Odd-parity checking is used with 8-bit numbers that are sent to and from the central processor of a certain computer. The parity bit is appended to the right end of each number. Suppose that the transmitted number 110100101 is received as one of the following: 110100100; 110100101; 100100101; 100100100. In each of the four cases, answer the following questions. **(a)** Was there an error in transmission? **(b)** Does the parity-check system detect an error? **(c)** If there were no error-detecting circuits, would the processor have operated on incorrect information?

2.64 Write a note comparing and contrasting fixed-point and floating-point binary numbers from the viewpoints of precision, range, and hardware complexity.

2.65 What is exponent biasing and why is it used? What difficulties is biasing likely to cause when implementing floating-point arithmetic operations?

2.66 What is floating-point normalization and why is it used? Some computers allow a programmer to retain floating-point results in unnormalized form by overriding the automatic normalization step at the end of an operation such as addition of two floating-point numbers. Suggest a possible reason for this.

2.67 Using the IEEE 754 floating-point standard for 32-bit numbers, determine the correct floating-point representation for $+2.5$, -2.5, and 10^6.

2.68 Convert to ordinary decimal form the following numbers that employ the IEEE 754 standard:

$$N_1 = 1 \;\; 00101110 \;\; 00000000000000000000000$$

$$N_2 = 0 \;\; 10001111 \;\; 00000000000000000000001$$

2.69 Calculate the quantities $N_1 + N_1$, and $N_1 + N_2$ for the floating-point numbers defined in Problem 2.68. Do your calculations using binary arithmetic and show all intermediate results. The final results should be in standard IEEE 754 form.

64-bit number words are often used in large-scale scientific computations.

2.70 There is a 64-bit version of the IEEE 754 floating-point format that employs a sign bit s, an 11-bit exponent E, and a 52-bit mantissa M. Normal numbers are given by the formula $N = (-1)^s 2^{E-1023}(1.M)$, where $0 < E < 2047$. Determine the largest and smallest positive (nonzero) numbers that are representable in this format, giving your answers as 16-digit hexadecimal strings.

2.71 The IBM System/370 series of computers uses a floating-point number format in which the base $B = 16$. Analyze the advantages and disadvantages of this format compared to those formats with $B = 2$, such as the IEEE 754 format.

2.72 **(a)** Explain in general terms when and how overflow is detected during the addition of floating-point numbers. **(b)** The exceptional condition that a number is too small to be represented in a particular floating-point arithmetic format is termed **underflow**. Using the IEEE 754 format as an example, construct a specific instance of a floating-point subtraction operation that results in underflow.

LOGIC ELEMENTS

The Switch Level

3.1 Switches

3.2 Switching Circuits

3.3 Complementary Circuits

The Gate Level

3.4 Logic Gates

3.5 Gate-Level Circuits

Boolean Algebra

3.6 A Definition of Boolean Algebra

3.7 Fundamental Properties

3.8 Expressions, Functions, and Circuits

Overview

Logic circuits are designed to perform useful operations on binary quantities, whose interpretation ranges from true-or-false decisions in logical reasoning to high- or low-voltage signals in electronic circuits. This chapter introduces the elements of logic design. By "elements" we mean both the basic components, switches and gates, that are used to construct digital circuits, and the theoretical principles underlying circuit structure and behavior, which are firmly grounded in a branch of mathematics called Boolean algebra.

3

The Switch Level

Switches that control the flow of an electric current or a gas supply are familiar to us in everyday life. A switch has two basic states: open and closed. In one state, the controlled item is permitted to flow through the switch; in the other state, the flow is stopped. If we equate the states of a switch with the binary values 0 and 1, we see that a switch can control the flow of binary information.

3.1 Switches

In this section, we discuss some common situations that demonstrate the basic properties of switches and switching circuits. We then introduce the concept of an abstract or ideal switch, and show how several such switches can be connected to form an important class of circuits known as *logic gates*.

An Electric Circuit

An electric circuit may consist of a power supply, a set of switches, and a load.

The most familiar example of a switch is the device that controls the flow of current through an electric circuit. Consider the following task faced by a homeowner who is installing a new electric light in a room. A light fixture L is to be connected to the electric power supply PS and to a manual switch S_2, so that turning S_2 on and off turns the light on and off. In addition, a circuit breaker S_1 is to be installed near the entry point of the power supply, so that turning S_1 off will turn L off, independently of S_2. A little thought shows that L, PS, S_1, and S_2 should be connected in a loop or "circuit," as illustrated by Figure 3.1. The connections are made by wires through which an electric current can flow.

The behavior of this simple electric circuit is as follows. When at least one of the two switches S_1 or S_2 is in the off or open state, the circuit is said

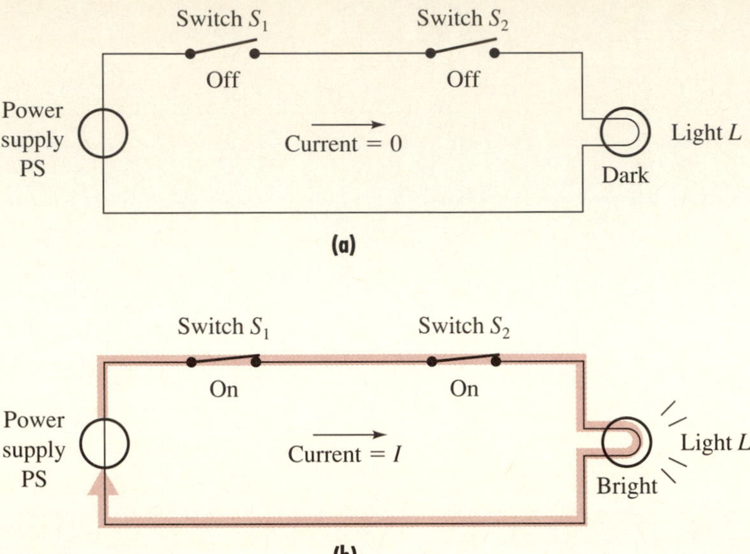

FIGURE 3.1 An electric circuit that controls a light; (**a**) switches turned off; (**b**) switches turned on.

to be open, so no current can flow through it and L is dark (Figure 3.1a). When both switches are turned on or closed, the circuit forms a closed path around which an electric current can pass. This current flows from the power supply PS through the switches S_1 and S_2 to the light L, and then back to PS. This process causes L's state to change from dark to bright, as indicated in Figure 3.1b. Therefore, closing the two switches causes the current flow to jump from zero to some value $I > 0$, which, in turn, causes L to emit light energy. Opening either switch changes the current from I back to zero. L could be replaced by any "load" device that consumes energy from the electric current passing through it.

An Hydraulic Circuit

A useful analogy can be made between hydraulic and electric circuits.

Figure 3.2 depicts another familiar type of circuit—a piece of domestic plumbing—that is controlled by devices that can be modeled as switches. This time the item controlled is water rather than electricity, so the circuit is hydraulic rather than electric. However, again we have an input source, the water supply PS, and an output load L, which happens to be a bathroom showerhead. The source and load are again linked by a pair of switches designed to turn a flow on and off. These are, of course, usually called faucets, taps, or (water) valves, rather than switches. The connecting lines are now

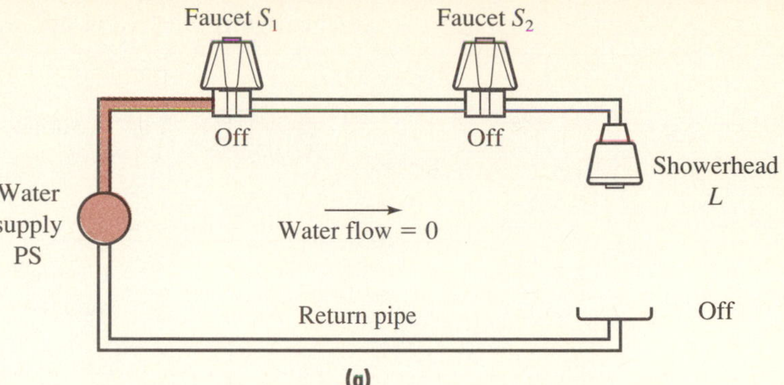

(a)

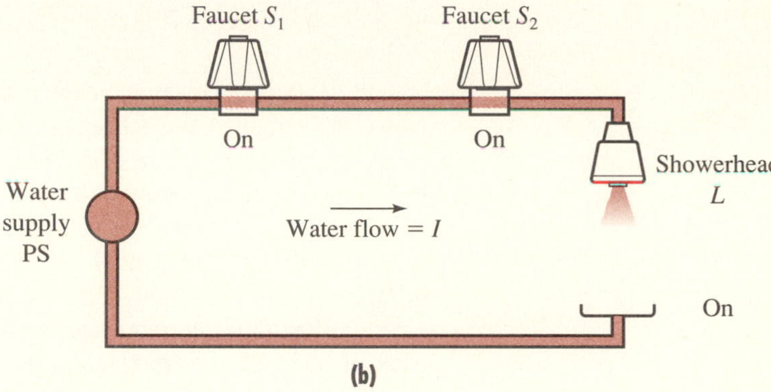

(b)

FIGURE 3.2 A water circuit that controls a shower: (**a**) faucets turned off; (**b**) faucets turned on.

water pipes instead of electric wires, and the flow of electrons is replaced by the flow of H_2O. As in the electric circuit, a return path sends the water from the shower back to the water source in ecologically correct fashion.

Physical Switches

All switches have binary input/output signals and two states: open and closed.

Using switches of various kinds, we can control the more intangible or abstract entity we call information. These switches are most often electrical in nature, but, as we will see in a moment, there are many different types of electrical switches. Therefore, consider a general or abstract notion of a **switch** S as a component that uses one binary variable x to control the state of another binary variable z. The first variable x is called the **input** or **control variable**; z is called the **output** or **data variable**.

The examples of Figure 3.1 and 3.2 illustrate switches that control the flow of electrons and water molecules, respectively. Two more examples of electrical switches are sketched in Figure 3.3. The switch of Figure 3.3a

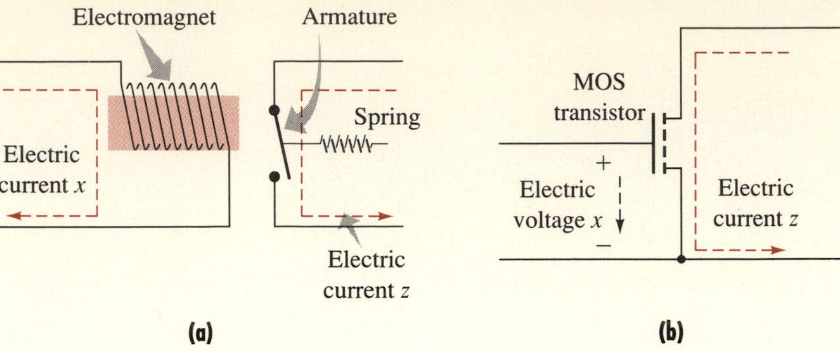

FIGURE 3.3 Two more examples of electrical switches: (**a**) an electromechanical relay; (**b**) an MOS transistor.

is called an electromechanical **relay** and is used mainly in industrial control circuits. It contains an electromagnet that consists of a wire coil wrapped around a magnetizable iron core. The relay's input variable x is an electric current applied to the coil. When activated by a nonzero electric current $x = I$, the iron core becomes a magnet and pulls in an iron lever (the armature), thereby closing an electric switch or contact that controls an output current z. When the input current supply is cut off—that is, when $x = 0$—no current flows through the electromagnet, and the iron core of the electromagnet becomes demagnetized, causing the armature to be released and pulled into the open position by a spring.

MOS transistors are among the most important switches in digital systems.

The switch shown in symbolic form in Figure 3.3b is another electrical or, more precisely, electronic[1] switch called a **transistor**. This particular transistor is of a type known as a **metal-oxide semiconductor—MOS** for short—and is one of the principal components of modern digital circuits. Its controlling input variable x is an electric voltage that can assume two values $\{V_L, V_H\}$, where the subscripts L and H denote low and high, respectively. We can take the transistor's output variable z to be an electric current that takes the binary values $\{I_L, I_H\}$. However, it is more practical when dealing with MOS transistor circuits to treat all input/output variables x and z as binary voltages that assume the same set of values $\{V_L, V_H\}$. Note that voltage is the electrical "pressure" that drives an electric current. In a certain sense, voltage and current are inseparable parameters in electric circuits: Without a voltage difference, no electric current can flow; if a current flows, some voltage difference is needed to drive it. The quantity analogous to voltage in an hydraulic circuit like that of Figure 3.2 is water pressure.

Table 3.1 summarizes some key characteristics of the four types of switches we have just discussed.

[1]Electronic circuits are distinguished from other electric circuits by their use of very small currents and the fact that they can be switched on and off at very high frequencies.

TABLE 3.1 Characteristics of some technologies involving switching concepts.

Physical technology	*Switches*	*Pressure parameter*	*Current parameter*
Domestic electrical wiring	Manual switches	Voltage (volts)	Electric current (amperes)
Industrial electrical wiring	Relays	Voltage (volts)	Electric current (amperes)
Electronic circuits	Transistors	Voltage (volts)	Electric current (microamperes)
Domestic plumbing	Faucets	Water pressure (pascals)	Water flow (liters/second)

Technology Independence

The physical switches cited in Table 3.1 have certain basic features in common. All involve opening and closing a transmission path, and all have two stable configurations or states, which correspond to the fully open and fully closed conditions, respectively. We next define a general switch model that attempts to capture these common characteristics while ignoring such minor details as the nature of the transmission medium or the particular switching mechanism used. Our goal is to identify an **ideal switch** that is a simplified model of a broad class of real switches. Because it ignores many physical details, the ideal switch can also be said to be **technology independent**.

Technology independence makes it easy to handle new kinds of switches and circuits.

Technology independence greatly simplifies the analysis and design of switching circuits, because so many different physical techniques are available for building switches. For example, since its first appearance in the 1930s, the modern digital computer has passed through several eras or "generations" characterized by drastic changes in switch technology. The earliest computers employed relays of the kind depicted in Figure 3.3a. Subsequently, relays were replaced by electronic switches composed of vacuum tubes (electronic valves). Around 1960 the transistor became the dominant switch type in digital equipment, a role it still retains. Transistor switches also come in several varieties besides the MOS type of Figure 3.3b.

Digital systems of specialized kinds are occasionally built from nonelectrical technologies such as mechanical, hydraulic, or pneumatic devices, which replace the motion of electrons by those of solids, liquids, and gases, respectively. An important emerging technology for digital systems employs light as the transmission medium and replaces electronic switches and circuits with their photonic equivalents.

Ideal Switches

Ideal switches use 0 and 1 signals only, and are technology independent.

A useful first step in defining an ideal switch is to introduce standard names for the two values that the switch's input and output signals x and z can assume. By convention, the binary digits (bits) {0, 1} are used for this purpose

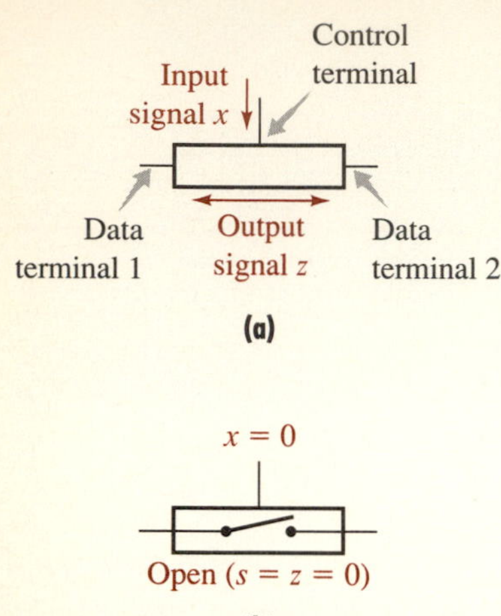

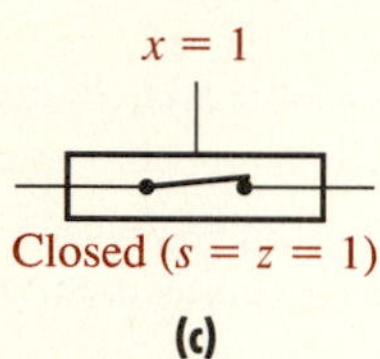

FIGURE 3.4 An ideal switch: (a) its graphic symbol; (b) the open or off state; (c) the closed or on state.

where 0 denotes an inactive or off signal, and 1 denotes an active or on signal. We therefore treat {0, 1} as abstract *information* signals, and use them to replace any of the signal-value pairs {off, on}, {open, closed}, {V_L, V_H}, and so on, which we encountered earlier. Besides its conciseness, the pair {0, 1} has the advantage of being the standard digit set used for binary numbers, as discussed in Chapter 2.

We also need a general symbolism for an ideal switch and its input/output connections. Figure 3.4a shows the graphic symbol we will use for a switch. It reduces the many possible physical switch mechanisms to an oblong "black box" with three connection points called **terminals**, to which external signals may be applied or from which internal signals may be drawn. The lines attached to the terminals can represent wires, pipes, or other transmission media appropriate to the switch technology at hand. We have a single input terminal that allows a signal to enter the switch and change its state. (Occasionally, it is appropriate to add a fourth terminal to allow a control signal to leave the switch.)

A terminal is designated an **input terminal** if it only allows signals to enter the switch, or an **output terminal** if it only allows signals to leave the switch. The control variable x is applied to an input terminal of the switch. A terminal that can function both as an input and an output is said to be **bidirectional**. The direction of signal flow associated with a terminal can be indicated by an arrow, as in Figure 3.4a; a double-headed arrow is used for bidirectional signals. Often the direction of a signal or terminal will be clear from the context and need not be explicitly specified. The data terminals of our ideal switch will be assumed to be bidirectional, so that the output quantity z can "flow" in via one data terminal and out via the other.

The effect of the switch on the output z is determined by the switch's state s. We see from Figures 3.4b and 3.4c that our ideal switch has two states: $s = 0$, which denotes open or off and implies that there is no closed path through the switch connecting the z terminals; and $s = 1$, which denotes the closed or on state, where the z terminals are connected via a path through the switch. When $s = 0$, the output variable z cannot pass through the switch, so we have $z = 0$. When $s = 1$, the z variable can pass through the switch, so $z = 1$. Thus we can equate the switch state s and the value of the data variable z.

We should emphasize that the input/output signals x and z, as well as the switch's state s, are all abstract binary quantities denoting information. Consequently, we have considerable freedom to define the manner in which 0 and 1 signals "flow" through a circuit of ideal switches. Although our definition is consistent with the behavior of the physical devices and circuits we want to represent, we should not necessarily equate the flow of 0s and 1s to, say, the flow of electrons or other information carriers in the underlying physical circuit. Often the amount of current actually flowing in the circuit is negligible, but there is a significant change in voltage (pressure) levels that serves to convey the relevant information.

Positive and Negative Switches

A positive switch is closed by $x = 1$; a negative switch is closed by $x = 0$.

The switch model S considered so far has the property that $x = 1$ turns it on, while $x = 0$ turns it off; this is called a **positive switch**. It is very useful to define another type of ideal switch $\bar{S}$ in which the control signal levels are exchanged, so that $x = 0$ turns the switch on, and $x = 1$ turns it off; this is a **negative switch**.

Figure 3.5 defines these two switch types and introduces a new graphic symbol for a negative switch. This is the same switch symbol as before; it has a small circle called the **negation symbol** placed at the control terminal of the switch. The inversion symbol can be thought of as converting an external 0 or 1 arriving at the control input of the switch into a 1 or 0, respectively, as it enters the switch. The conversion process $0 \leftarrow 1$ and $1 \leftarrow 0$ is variously known as **inversion** and **complementation**, and is of fundamental importance, as we saw in Chapter 2. The circle symbol for inversion employed in Figure 3.5b is a standard one used throughout logic design.

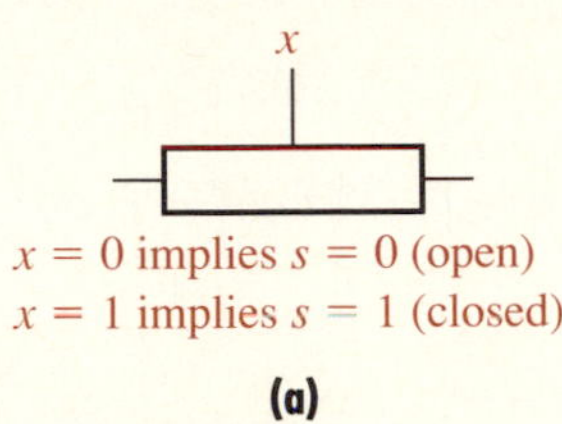

(a)

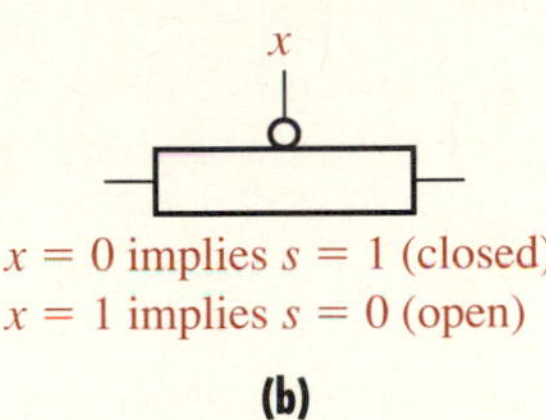

(b)

FIGURE 3.5 Two basic ideal switch types: (**a**) positive; (**b**) negative.

Positive and negative switches are found in most switch technologies. The "normally open" relay of Figure 3.3a is an instance of a positive switch. The corresponding negative switch is a "normally closed" relay, which is designed so that the spring pulls the armature switch closed, making $z = 1$ when the electromagnet is inactive—that is, when the controlling input current $x = 0$. Activating the electromagnet ($x = 1$) pulls the switch open; because this opens the output circuit, z changes from 1 to 0. As we will see shortly, transistors also consist of positive and negative varieties.

3.2 Switching Circuits

Switches are components or building blocks that can be interconnected by lines to make **switching circuits** or **networks**. A **line** L serves to transmit a binary signal applied to L to all lines and terminals to which L is connected. In other words, a line is a perfect conductor of binary signals. This section shows various ways in which ideal switches are used to construct logic gates and other relatively simple switching circuits.

AND, OR, and NOT Circuits

An AND circuit generates a 1 if and only if all its inputs are 1.

We begin by considering three of the most fundamental connection methods for switches. The first of these is illustrated by Figure 3.6a. This is an idealized model of the electrical and hydraulic circuits we saw in Figures 3.1 and 3.2. It contains two switches S_1 and S_2, which are said to be connected in **series**. A data terminal of one switch is connected to a data terminal of the other, while the remaining two data terminals are connected to a power supply and an output load. We will now view the power supply as an abstract supplier of 0s and 1s, and the load as an abstract consumer of 0s and 1s. The figure shows a complete circuit, with a return path connecting the load back

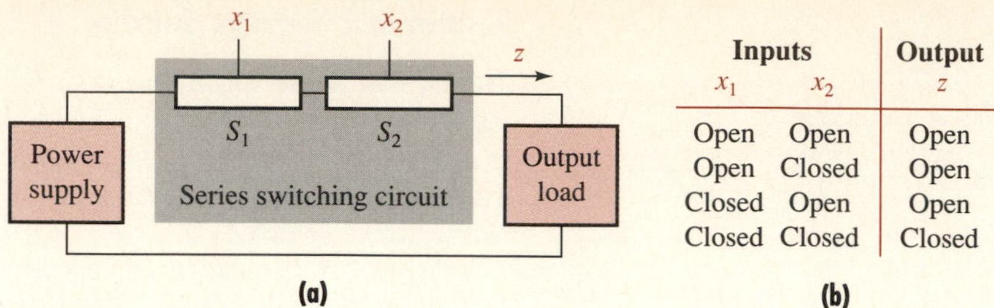

Inputs		Output
x_1	x_2	z
Open	Open	Open
Open	Closed	Open
Closed	Open	Open
Closed	Closed	Closed

(b)

FIGURE 3.6 (a) A switching circuit that implements the AND function; (b) its behavior in terms of open-closed states.

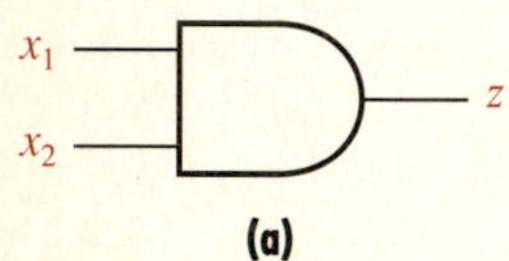

(a)

Inputs		Output
x_1	x_2	z
0	0	0
0	1	0
1	0	0
1	1	1

(b)

FIGURE 3.7 (a) A gate representation of the AND circuit of Figure 3.6; (b) its truth table.

to the power supply. Clearly, this circuit forms a closed path represented by $z = 1$ if and only if the control inputs x_1 *and* x_2 are both 1, causing S_1 *and* S_2 to be turned on. Taking a cue from the foregoing emphasis on the word *and*, the circuit of Figure 3.6 is said to implement or realize the **AND function.**

The AND circuit's behavior is spelled out in terms of open and closed states in Figure 3.6b. The two switches allow four possible open-closed combinations that together define the inputs to the circuit. If either switch is open, then the circuit is open and $z = 0$. Only when the first *and* second switches are closed is the entire circuit closed, making $z = 1$.

When dealing with larger circuits, we will want to focus mainly on the circuit's behavior in terms of 0s and 1s; this is referred to as its **logical behavior**. Consequently, we will represent the AND circuit and its behavior in the more abstract forms depicted in Figure 3.7, in which the AND circuit has been reduced to a single idealized component called a **logic gate**. In this simplified form, the power supply, load device, and return path are all ignored. The circuit is seen as having two inputs x_1, x_2 and an output z, all of which assume only the binary values 0 and 1. Because of its importance, the AND gate has the standard graphic symbol shown in the figure. Figure 3.7b restates the table of Figure 3.6b, with 0 and 1 corresponding to open and closed, respectively. A complete tabular description of a circuit's function in terms of 0s and 1s, such as in Figure 3.7b, is a **truth table**.

An OR circuit generates a 1 if and only if at least one of its inputs is 1.

A second basic way to connect two switches is shown in Figure 3.8a. In this **parallel circuit**, both the data terminals of one switch are connected to the corresponding terminals of the other switch. If either the first *or* the second switch (or both) are closed, then the entire circuit is closed and $z = 1$. The stress on the word *or* here leads to the name of the function realized by this circuit: the **OR function**. As before, the OR circuit is represented more abstractly at the gate level, using its own symbol; see Figure 3.8b. Again, we can represent the circuit behavior in terms of switch states (Figure 3.8c) or 0s and 1s (Figure 3.8d).

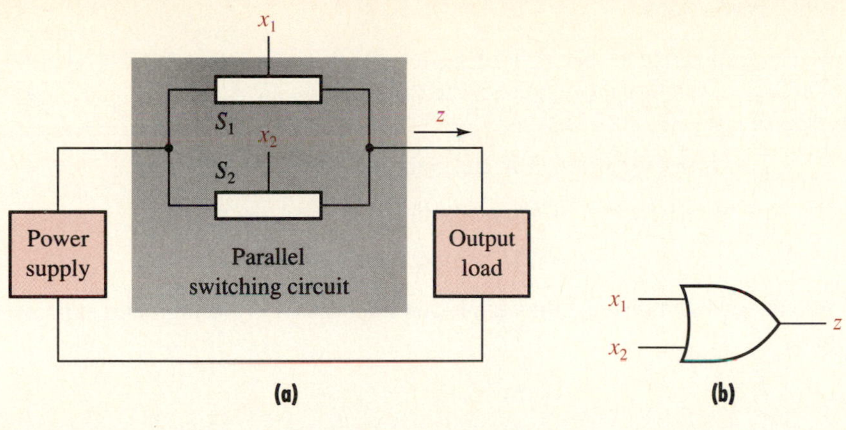

Inputs		Output
x_1	x_2	z
Open	Open	Open
Open	Closed	Closed
Closed	Open	Closed
Closed	Closed	Closed

(c)

Inputs		Output
x_1	x_2	z
0	0	0
0	1	1
1	0	1
1	1	1

(d)

FIGURE 3.8 (**a**) A switching circuit that implements the OR function; (**b**) its gate representation; (**c**) its behavior in terms of open-closed states; (**d**) its truth table.

A NOT circuit generates a 1 if and only if its single input is 0.

A third type of switching circuit implements the **NOT function**, also known as **inversion** or **complementation**. This simple one-switch circuit, called an **inverter** or NOT gate, can be implemented as shown in Figure 3.9a. It has a single input variable x connected to the control input of a negative switch, and a single output z. Its behavior is such that when $x = 0$, the switch—and therefore the entire circuit—is closed, making $z = 1$; when $x = 1$, the circuit is open and $z = 0$. Thus, we can say that z is *not* x. The inverter has its own standard graphic symbol, shown in Figure 3.9b. Note the inversion circle that appears on the inverter's output terminal has the same meaning as the circle on the negative switch's control terminal. Figure 3.9c is the circuit's truth table.

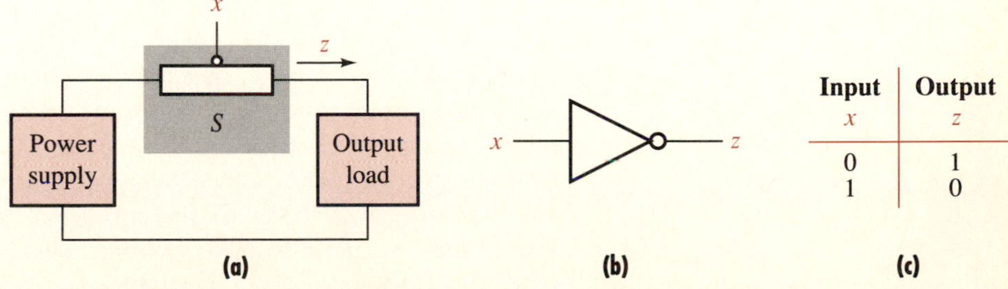

Input	Output
x	z
0	1
1	0

FIGURE 3.9 (**a**) A switching circuit that implements the NOT function; (**b**) its gate representation; (**c**) its truth table.

Logic Expressions

We can summarize the behavior of the foregoing AND, OR, and NOT circuits as follows, denoting the outputs of the three circuits by z_{AND}, z_{OR}, and z_{NOT}, respectively.

$$z_{\text{AND}}(x_1, x_2) = 1 \text{ if and only if } x_1 = 1 \textit{ and } x_2 = 1$$

$$z_{\text{OR}}(x_1, x_2) = 1 \text{ if and only if } x_1 = 1 \textit{ or } x_2 = 1$$

$$z_{\text{NOT}}(x) = 1 \text{ if and only if } x \neq 1$$

A logic expression is a concise way to describe a switching circuit.

We can rewrite the foregoing definitions more concisely using **logic expressions**:

$$\begin{aligned} z_{\text{AND}}(x_1, x_2) &= x_1 \cdot x_2 \\ z_{\text{OR}}(x_1, x_2) &= x_1 + x_2 \\ z_{\text{NOT}}(x) &= \overline{x} \end{aligned} \tag{3.1}$$

Logic expressions like these pervade logic design; they will be defined more precisely later in this chapter. They are similar to the expressions used in ordinary algebra, but apply to binary or logical entities rather than to numbers. Here, $x_1 \cdot x_2$, which is also denoted by $x_1 x_2$, is read as "x_1 and x_2," $x_1 + x_2$ is read as "x_1 or x_2," and $\overline{x}$ is read as "not x." The operations $\cdot$ and $+$ in (3.1) are to be viewed as logical operations, and should not be confused with the (binary) arithmetic operations, times and plus, that employ the same symbols. Later, when we explore in detail the relationship between logical and numerical expressions, we will see that logical operations have their own algebra, which is called Boolean algebra.

General Circuits

More complex switching circuits in the foregoing style can be obtained by combining the series and parallel connections in different ways. We illustrate this point with a small design exercise.

Example 3.1

Switching circuit for a burglar alarm

Suppose that a burglar alarm system is to be installed in a two-room apartment. The alarm receives input 0/1 signals from a set of switches hidden about the apartment. A master disarming switch x_M turns the alarm off, independently of all the other inputs. One room contains three normally closed switches $x_{1,1}$, $x_{1,2}$, and $x_{1,3}$ embedded in doors and windows; if any of these switches opens, thereby generating a 0 signal, the alarm should sound. The second room contains two normally open switches $x_{2,1}$ and $x_{2,2}$ of the same type; they produce 1 signals to indicate a possible intruder. Finally, there is a manual switch x_1 concealed in the first room; it can be turned off to disable the three door-and-window switches of that room only. The problem is to

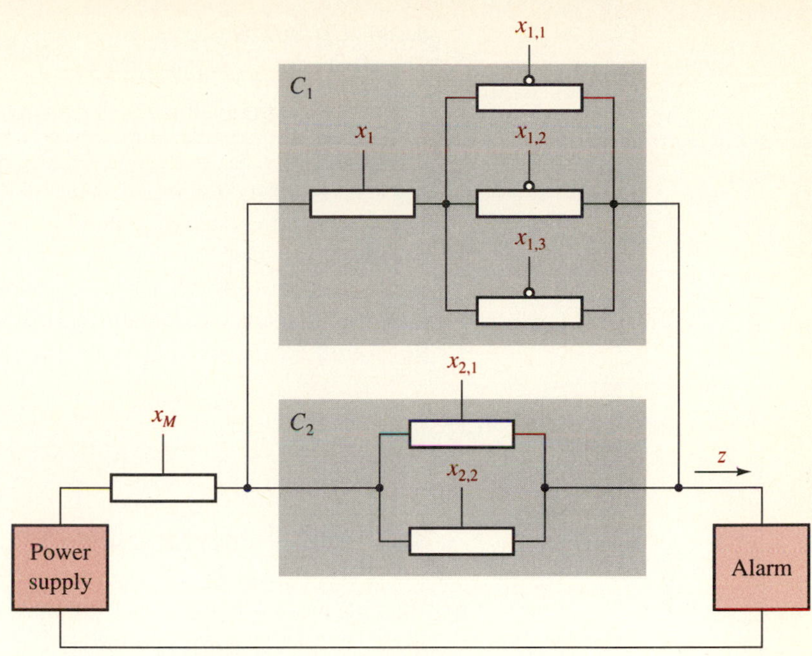

FIGURE 3.10 A switching circuit for the burglar alarm.

design a switching circuit of positive and negative ideal switches to meet the foregoing specifications.

It's often helpful to decompose a large circuit into several smaller ones.

We can approach the problem by breaking it into several independent parts: a circuit C_1 for room 1, a circuit C_2 for room 2, and a "main" circuit that combines C_1 and C_2 with the master switch x_M (Figure 3.10). We will denote the alarm bell by z so that setting $z = 1$ sounds the alarm. We also denote by z_1 and z_2 the warning signals generated by C_1 and C_2, respectively. The behavior of the alarm system is then defined by the following statement:

$$z = 1 \text{ if and only if } x_M = 1 \textit{ and } (z_1 = 1 \textit{ or } z_2 = 1) \tag{3.2}$$

which can be rewritten as a composite logic expression

$$z = x_M \cdot (z_1 + z_2)$$

Thus, the alarm circuit should have the overall structure of Figure 3.10, with C_1 and C_2 in parallel, and with the combined C_1–C_2 circuit in series with switch x_M.

Turning to the subcircuit C_2, we can easily see that it should satisfy the following requirement:

$$z_2 = 1 \text{ if and only if } x_{2,1} = 1 \textit{ or } x_{2,2} = 1$$

which corresponds to the logic expression $z_2 = x_{2,1} + x_{2,2}$. Hence, we can design C_2 with two positive switches $x_{2,1}$ and $x_{2,2}$ connected in parallel. The circuit for C_1 is specified by

$$z_1 = 1 \text{ if and only if } x_1 = 1 \textit{ and } (x_{1,1} = 0 \textit{ or } x_{1,2} = 0 \textit{ or } x_{1,3} = 0)$$

or, equivalently, $z_1 = x_1 \cdot (\overline{x}_{1,1} + \overline{x}_{1,2} + \overline{x}_{1,3})$. This subcircuit requires three negative switches connected in parallel, with a fourth switch for x_1 in series with them collectively. Putting the pieces together leads directly to the complete circuit of Figure 3.10. For obvious reasons, this type of switching circuit is described as a **series-parallel circuit**. If we combine the expressions for its parts, we obtain the following logic expression that describes the overall control circuit for the burglar alarm:

$$z = x_M \cdot [x_1 \cdot (\overline{x}_{1,1} + \overline{x}_{1,2} + \overline{x}_{1,3}) + (x_{2,1} + x_{2,2})]$$

3.3 Complementary Circuits*

We turn now to a somewhat different class of switching circuits that form the basis for the very important technology known as complementary MOS, abbreviated CMOS and pronounced "see-moss." We also introduce the notion of signal amplification (buffering).

Current vs. Complementary Circuits

The ideal switching circuits examined in the preceding section model physical circuits composed of manual electrical switches, electromagnetic relays, hydraulic switches, and a few other technologies. As suggested by Figure 3.11a, the various switches in the circuit C act like a single composite switch that opens and closes a path for a "current" signal z, such as an electric current, in response to a set of control inputs $x_1, x_2, \ldots, x_n$. We will therefore refer to such circuits as **current-switching circuits**.

Complementary networks switch voltages instead of currents.

A second category of switching circuit, called a **complementary switching circuit**, in which all input/output signals are of the pressure type is depicted in Figure 3.11b. In the case of electronic circuits, this means that 0 and 1 correspond to voltage levels. We now regard the power supply PS as a source of pressure-type 0 and 1 signals, which are available at its two output terminals. The circuit's output signal z is made a 0 or a 1 by creating a closed path from z to either the 0 terminal or the 1 terminal of PS.

As the figure suggests, we can obtain any desired logical behavior with a pair of current-switching circuits C and $\overline{C}$ that connect the output terminal z to the 1 and 0 terminals of PS, respectively. C is the same circuit that appears in the current-switching implementation of z. It is now called a **pull-up circuit** because it connects z to the source of 1 and so "pulls z up"

*This section can be omitted without loss of continuity.

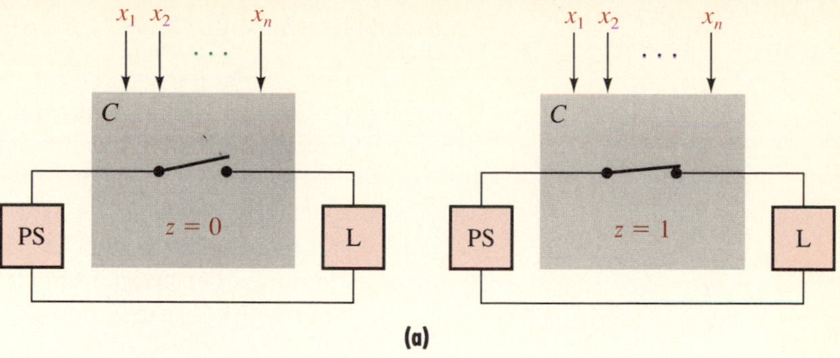

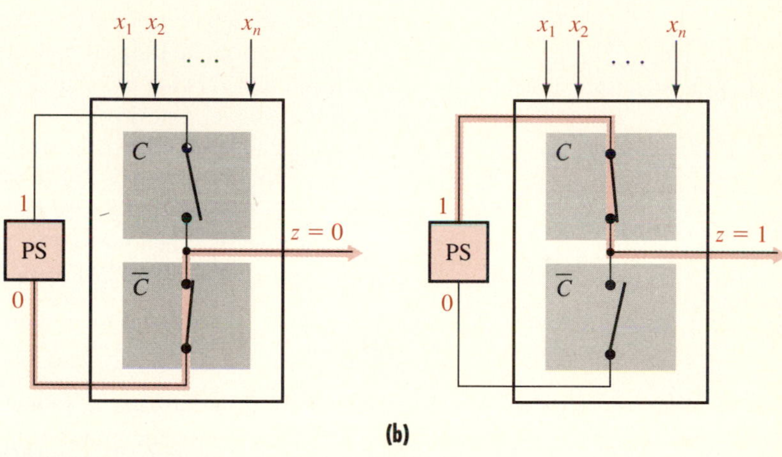

FIGURE 3.11 The general structure of (**a**) a current-switching circuit, and (**b**) a complementary switching circuit.

from 0 to 1 when appropriate control signals are applied. Similarly, $\overline{C}$ is called a **pull-down circuit** because it connects z to the 0 source for certain input combinations.

Our complementary switching circuits closely model an important class of CMOS circuits.

The circuit of Figure 3.11b must be designed so that every combination of the control variables $x_1, x_2, \ldots, x_n$ closes at least one path through C and opens all paths through $\overline{C}$, or vice versa. In other words,

$$C \text{ is closed if and only if } \overline{C} \text{ is open} \tag{3.3}$$

or, equivalently,

$$C \text{ is open if and only if } \overline{C} \text{ is closed} \tag{3.4}$$

Consequently, z is always connected by a closed path to one and only one of the terminals of PS. A switching circuit designed in the style of Figure 3.11b that satisfies (3.3) and (3.4) is called a complementary switching circuit. The complementary design style is used extensively in CMOS transistor circuits. The circuits composed of ideal switches that we discuss in this section closely model physical circuits implemented via CMOS technology.

Complementary Inverters

We now present an implementation of the inverter or NOT gate as a complementary switching circuit; a current-switching version appears in Figure 3.9. The new inverter uses two ideal switches, a negative switch S_1 that forms the pull-up circuit C, and a positive switch S_2 that forms the pull-down circuit $\overline{C}$, as depicted in Figure 3.12a. The single input variable x is connected to the control inputs of both switches. Note that C is the switching circuit that appears in our earlier inverter circuit (Figure 3.9a).

We can verify that the circuit shown in Figure 3.12 acts as an inverter by considering the effects of the possible input combinations, namely, $x = 0$ and $x = 1$. In the first case (Figure 3.12b), $x = 0$ turns S_2 off and turns S_1 on. This closes a path from the power supply's 1-terminal to z. As indicated by the large arrow, a 1-signal can then "flow" through S_1, making $z = 1$. (Note again that this *logical* signal flow is distinct from any *physical* signal flow—for example, a positive or negative electric current—in the underlying physical technology.) Changing x to 1, as shown in Figure 3.12c, reverses the states of the switches; S_1 turns off and S_2 turns on. The path from the 1 source to z is broken, and a new path is formed via S_2 from the 0 source to z, making $z = 0$. When viewed as a gate, this circuit is represented by the graphic symbol introduced in Figure 3.9b.

Buffering

Complementary circuits buffer or amplify the signals they process.

Examination of the inverter in Figure 3.12 shows that the output signal z is always obtained from the power supply PS and not from the input variable x. This permits a "weak" x—that is, one derived from a low-powered signal source—to control a higher powered or "strong" signal z derived from PS. Consequently, the inverter performs the important secondary function of signal **buffering** or **amplification**, which means it boosts the energy level of the signals it processes. Because all physical signals are weakened by passage through physical switches and connections, periodic amplification of signals is essential for reliable operation.

A buffer is a gate designed solely for amplifying signals.

A useful switching circuit closely related to an inverter is a **buffer**, which has the sole function of signal amplification. A buffer's output z is 0 or 1 whenever its input x is 0 or 1, respectively. Hence, if signal strengths are ignored, the function of a buffer can be written as $z = x$, which is the **identity function**. We can construct a complementary buffer by exchanging the switches S_1 and S_2 in the inverter circuit of Figure 3.12; the result appears in Figure 3.13a. Again, z is always a strong signal obtained from PS, but here $x = 0$ creates a path from the 0 signal of PS to z, and $x = 1$ creates a path from the 1 signal of PS to z. (Not all switch technologies allow amplifiers to be constructed in quite the simple way shown in Figure 3.13.)

Observe that the triangle-shaped buffer symbol given in Figure 3.13b also appears in the symbol for an inverter (Figure 3.9b). The inverter symbol can now be seen as a composite of two more basic elements: a buffer symbol (the triangle part) and an inversion symbol (the small circle).

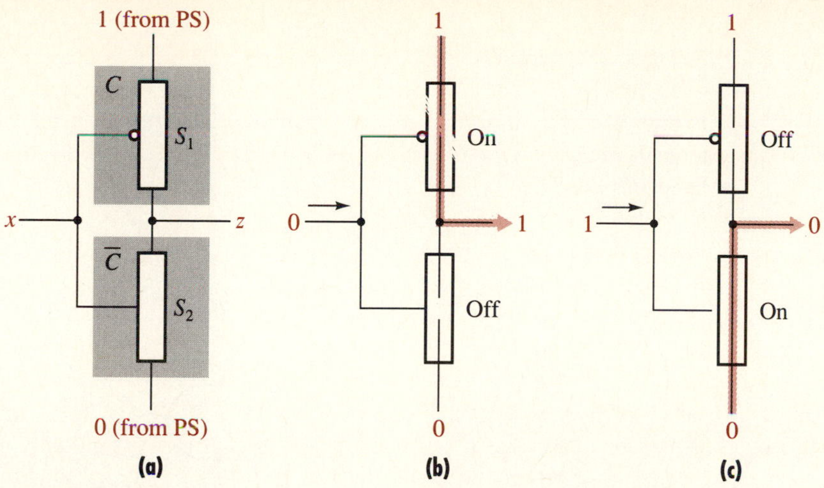

FIGURE 3.12 (a) A complementary inverter (NOT gate); (b) its behavior with $x = 0$; (c) its behavior with $x = 1$.

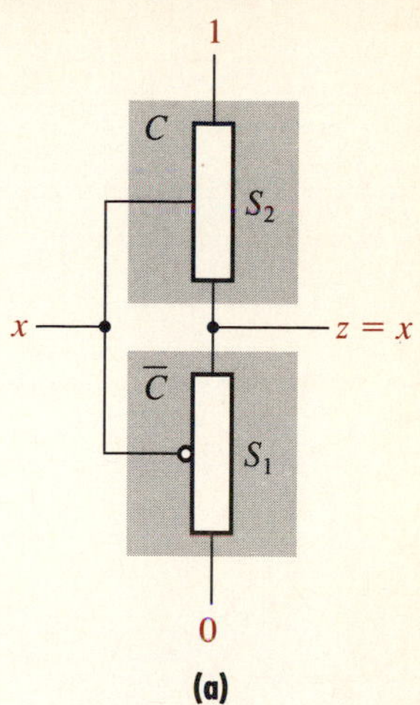

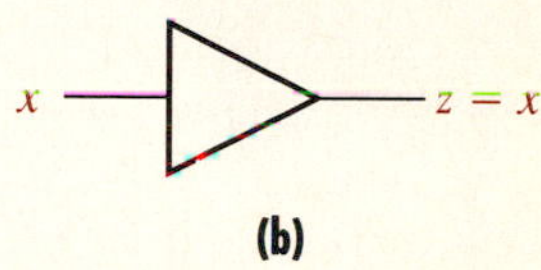

FIGURE 3.13 A complementary buffer: (a) its switch-level circuit; (b) its gate symbol.

NAND Gates

The NAND function is equivalent to AND followed by NOT.

The complementary technique can be used to realize the AND and OR gates described earlier; we leave this as an easy exercise. Instead, we describe next a complementary switching circuit that forms a new type of gate. The output signal z of a **NAND gate** is 0 if and only if all the input signals of the gate are 1. Hence, a 0 applied to any input line changes the output to 1. The name NAND is a contraction of "not-and," and follows from the rather awkward characterization of the circuit as *not* producing the output 1 whenever $x_1 = 1$ *and* $x_2 = 1$. We can denote this by the logic equation

$$z_{\text{NAND}} = \overline{x_1 \cdot x_2} \tag{3.5}$$

As we will see later, NAND is especially important because it is a **universal gate** in the sense that any desired logic function can be implemented by a circuit composed entirely of NAND gates.

Figure 3.14a shows a complementary switching circuit that realizes the two-input NAND function. It comprises four switches, two positive and two negative. The positive switches are in series and form the pull-down circuit $\overline{C}$. The negative switches are in parallel and form the pull-up circuit C. If both input variables x_1 and x_2 are set to 1, a path is created from the 0-terminal of the power supply PS through $\overline{C}$ to the output line z, making $z = 0$. At the

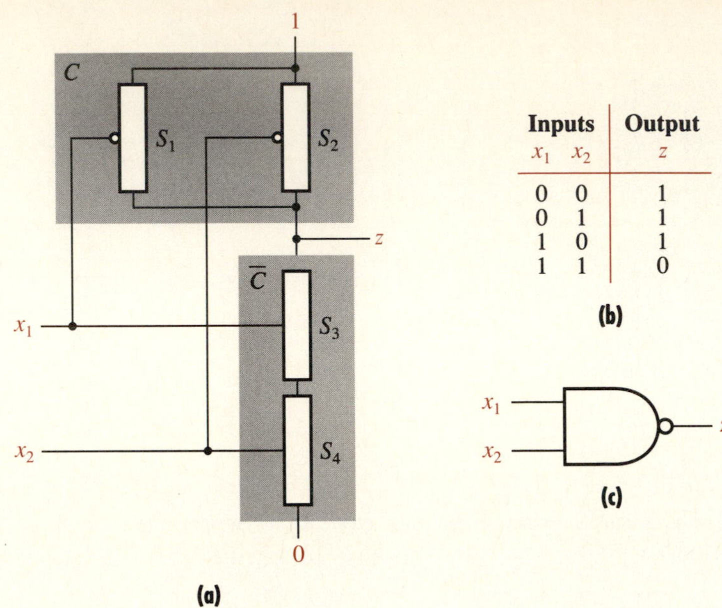

FIGURE 3.14 (a) A complementary NAND circuit; (b) its truth table; (c) its gate symbol.

same time, the negative switches S_1 and S_2 are turned off, thereby opening C and isolating z from the 1-terminal of PS. If either x_1 or x_2 becomes 0, the path through $\overline{C}$ is opened, and a closed path is created through C via S_1 or S_2, making $z = 1$. Thus, the desired NAND behavior summarized in the truth table of Figure 3.14b is obtained. A standard symbol for representing a NAND gate appears in Figure 3.14c. Note that it combines the AND (Figure 3.7a) and inversion symbols in a manner consistent with Equation (3.5).

Example 3.2

A binary adder circuit

The complementary design method can be applied to the implementation of any logic function. To illustrate this point, we will design a complementary switching circuit to perform a simple operation in binary arithmetic—namely, computing the sum (mod 2) z of two bits x_1 and x_2. Ignoring carries, we require only the following behavior:

$z = 1$ if and only if $(x_1 = 1$ *and* $x_2 = 0)$ *or* $(x_1 = 0$ *and* $x_2 = 1)$

That is,

$$z = (x_1 \cdot \overline{x}_2) + (\overline{x}_1 \cdot x_2) \tag{3.6}$$

The structure of Equation (3.6) suggests that we need a pull-up containing four switches, two positive and two negative. A positive switch controlled by x_1 should be connected in series with a negative switch controlled by x_2,

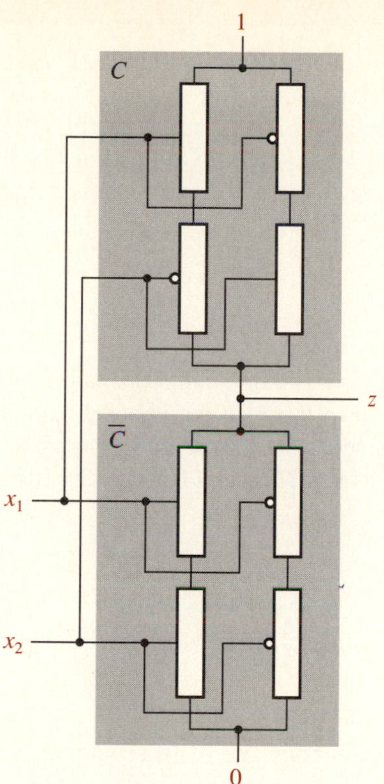

FIGURE 3.15 A complementary adder circuit, which is also an EXCLUSIVE-OR gate.

as required by the term $x_1 \cdot \overline{x}_2$ in (3.6). A similar pair of series-connected switches is needed to implement the second term $\overline{x}_1 \cdot x_2$. Finally, the OR operation + of (3.6) suggests that the two series subcircuits should be connected in parallel. The resulting series-parallel circuit appears as C in Figure 3.15.

To obtain the pulldown circuit $\overline{C}$ for the adder, observe that its function can be described by

$$z = 0 \text{ if and only if } (x_1 = 0 \textit{ and } x_2 = 0) \textit{ or } (x_1 = 1 \textit{ and } x_2 = 1)$$

That is,

$$\overline{z} = (\overline{x}_1 \cdot \overline{x}_2) + (x_1 \cdot x_2) \tag{3.7}$$

Clearly, Equation (3.7) has the same form as (3.6) except for the positions of the inversion symbols (overbars). Hence, $\overline{C}$ is also a four-switch series-parallel circuit (Figure 3.15).

We have derived the adder circuit in a rather informal manner. Verify its correctness by checking that paths through C and $\overline{C}$ are opened and closed correctly for each of the four possible combinations of the input signals x_1 and x_2.

EXCLUSIVE-OR Gates

Another useful type of gate is known as **EXCLUSIVE-OR**, or **XOR** for short. With two inputs x_1 and x_2, its specification calls for the output z to be 1 if and only if either x_1 or x_2, but not both, is 1. The exclusion of $x_1 = x_2 = 1$ from producing $z = 1$ is responsible for the gate's name, and serves to differentiate EXCLUSIVE-OR from the (inclusive) OR function introduced earlier.

It is easy to see that the adder of the preceding example has exactly the same specifications as EXCLUSIVE-OR. Hence, the complementary circuit of Figure 3.15 also happens to realize a two-input EXCLUSIVE-OR circuit. When viewed at the gate level, EXCLUSIVE-OR is represented by the special symbol given in Figure 3.16a. Note the slight difference from the OR symbol of Figure 3.8b: the EXCLUSIVE-OR gate has an extra curved line on its input side.

An EXCLUSIVE-OR circuit generates a 1 if exactly one of its two inputs is 1.

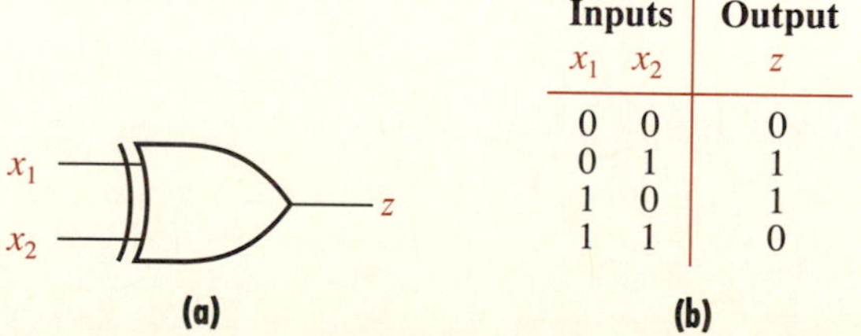

Inputs x_1	x_2	Output z
0	0	0
0	1	1
1	0	1
1	1	0

FIGURE 3.16 (a) An EXCLUSIVE-OR gate; (b) its truth table.

The Gate Level

We switch now to a higher level of abstraction in our study of digital circuits. The various switching circuits called logic gates in the preceding sections are treated here as primitive or indivisible design components. We define the main gate types in their most general form. We also introduce combinational logic circuits constructed by connecting gates together, and discuss some practical constraints affecting the design of such circuits.

3.4 Logic Gates

Logic gates implement functions such as AND, OR, and NOT.

Logic gates are switching circuits that perform certain simple operations on binary signals. These operations are chosen to facilitate the implementation of useful operations; that is, they are intended to be good general-purpose design components. They are also chosen for their ease of implementation in the relevant physical technologies.

A General Definition

In general, a **logic gate** G is a switching circuit characterized by the following parameters:

1. A set of n binary input signals or variables $X = (x_1, x_2, \ldots, x_n)$, where each x_i is 0 or 1, and n ranges from one to 10 or so
2. A single binary output signal z
3. A function f from X to z that may be expressed as

$$z = f(X) = f(x_1, x_2, \ldots, x_n)$$

and is one of about six basic types

This function defines the value of z for each of the 2^n possible combinations of binary values that the n input variables X can assume. Frequently, no distinction is made between the line z and the function f, and both are represented interchangeably by z or f.

In circuit diagrams, a gate G can be denoted by a box symbol of the kind appearing in Figure 3.17a. The most important gates also have a special graphic symbol, the distinctive shape of which identifies the gate's function f. Because of their relative simplicity and more widespread use, distinctive-shape symbols are used throughout this book. Both the box and distinctive-shape symbols that we use for gates and other logic elements correspond to widely used international standards.[2]

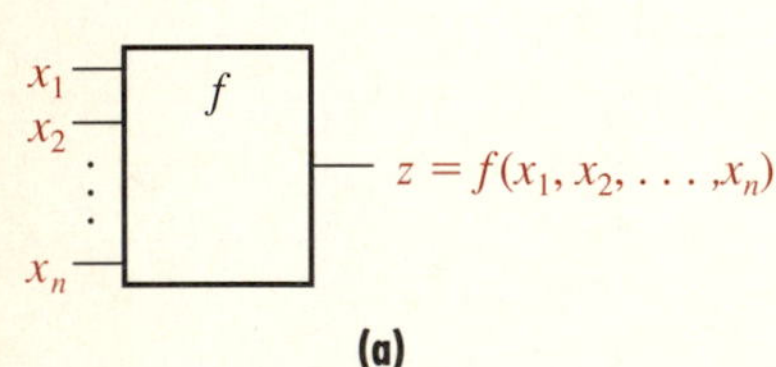

(a)

Inputs				Output
x_1	x_2 ⋯	x_{n-1}	x_n	$z = f$
0	0	0	0	a_0
0	0	0	1	a_1
0	0	1	0	a_2
0	0	1	1	a_3
	...			
1	1	1	0	a_{2^n-2}
1	1	1	1	a_{2^n-1}

(b)

FIGURE 3.17 (a) The graphic symbol for an n-input logic gate that realizes a general function f; (b) its truth table.

[2]The box symbols follow a standard defined by the International Electrotechnical Commission (IEC) and adopted by most national standards organizations [Kampel, 1985]. The equivalent standard in the United States is the IEEE Standard 91-1984: *Standard Graphic Symbols for Logic Functions* [IEEE, 1984]. The older distinctive-shape symbols remain popular and they are recognized (but not encouraged) by the foregoing IEEE standard.

The gate function f can be defined by a truth table, which is simply a complete listing of the gate's input/output signal combinations. In the general example depicted in Figure 3.17b, each of the 2^n a_i values in the output column represents 0 or 1; the pattern of 0s and 1s in this column determines f. Gate behavior may also be specified by an informal narrative description or by a more formal mathematical one such as an algebraic expression.

Single-Input Gates

Figure 3.18 shows the two possible gate functions for which $n = 1$; that is, $X = x_1 = x$. The first of these single-input gates is a **buffer**, whose output z is 1 whenever the input x is 1; z is 0 whenever x is 0. Therefore, a buffer realizes the identity function

$$z = x$$

which is indicated by a 1 (one) function designator in the buffer box symbol. The distinctive-shape symbol for a buffer is a triangle.

The second single-input gate type is an **inverter** or **NOT gate**, whose function is written z = NOT(x) or, more compactly, as

$$z_{\text{NOT}} = \overline{x} \tag{3.8}$$

An inverter converts a 0 to 1 and a 1 to 0, an operation known variously as inversion, complementation, or the NOT function. NOT is denoted by an overbar in functional expressions and by a small circle, the inversion symbol, in circuit diagrams. This circle is seen as converting a 1 (0) signal that appears at its input side to a 0 (1) signal on its output side, and can be added to any terminal of any gate symbol. (Some authors prefer to use a prime rather than an overbar for inversion, and so replace (3.8) with $z_{\text{NOT}} = x'$.) Figure 3.19 depicts two variant buffer and inverter symbols that find occasional use.

Switch-level implementations of buffer and NOT gates were presented earlier in this chapter. As noted then, a buffer's role is to amplify or increase the strength of binary signals. Physical NOT gates (and, in fact, most gate types) also perform amplification, as suggested by the inclusion of the

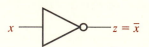

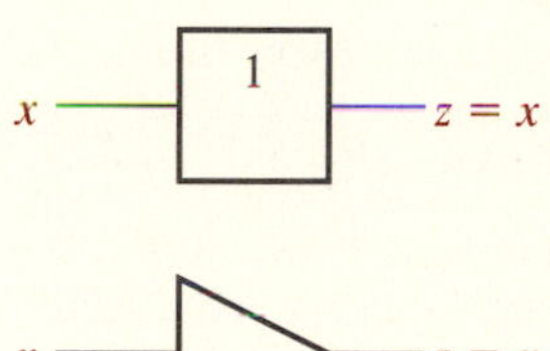

Input x	Output z
0	0
1	1

(a)

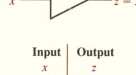

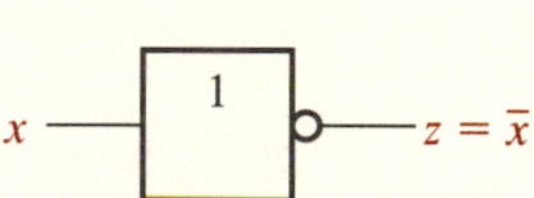

x — inverter — $z = \overline{x}$

Input x	Output z
0	1
1	0

(b)

FIGURE 3.18 One-input gates: (a) a buffer or amplifier; (b) an inverter or NOT gate.

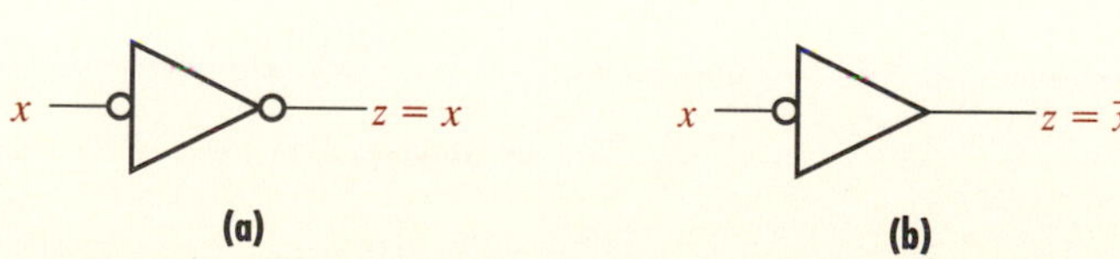

FIGURE 3.19 Alternative gate symbols: (a) a buffer; (b) an inverter.

Most gates have both box-shaped and distinctive-shape graphic symbols.

amplifier symbol within the corresponding (box or distinctive-shape) NOT symbol. The manner in which signal amplification is implemented is, like the underlying switching circuits, invisible at the gate level.

AND and OR Gates

Two important multi-input gates are specified in Figure 3.20. An n-input **AND gate** is characterized by the fact the output z is 1 if and only if x_1 is 1 *and* x_2 is 1 *and* ... *and* x_n is 1. In other words, z is 0 if at least one input is 0; it is 1 only if all inputs are 1. The AND operation can be interpreted as the multiplication of a set of 1-bit numbers; a 0 among the input values makes the result (product) 0; the product is 1 if and only if all the inputs are 1. For this reason, the AND function is written as a product expression

$$z_{\text{AND}}(x_1, x_2, \ldots, x_n) = x_1 \cdot x_2 \cdots x_n$$

where the $\cdot$ symbol between the variables is read as *and* rather than *times*. Alternative AND symbols in common use are $\wedge$ and &; the latter is the AND designator in the standard box symbol for an AND gate (Figure 3.20a). As with multiplication, the $\cdot$ symbol is usually omitted from AND expressions,

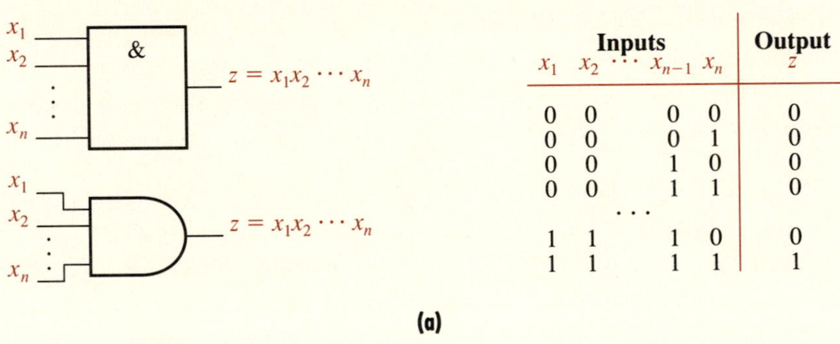

Inputs x_1	x_2	$\cdots$	x_{n-1}	x_n	Output z
0	0		0	0	0
0	0		0	1	0
0	0		1	0	0
0	0		1	1	0
		$\cdots$			
1	1		1	0	0
1	1		1	1	1

(a)

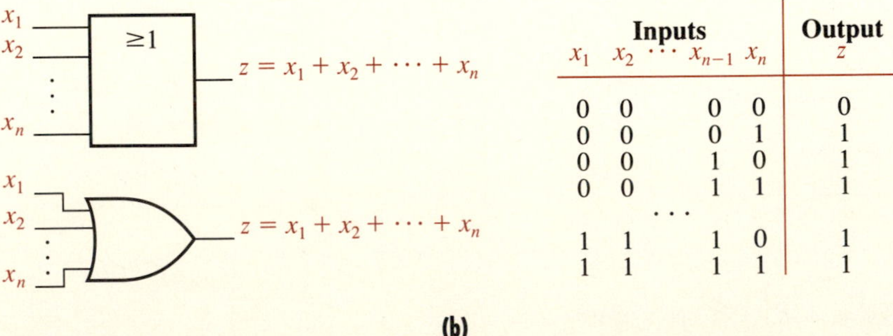

Inputs x_1	x_2	$\cdots$	x_{n-1}	x_n	Output z
0	0		0	0	0
0	0		0	1	1
0	0		1	0	1
0	0		1	1	1
		$\cdots$			
1	1		1	0	1
1	1		1	1	1

(b)

FIGURE 3.20 The definition of two basic n-gate types: **(a)** AND; **(b)** OR.

so that $x_1 \cdot x_2 \cdot x_3$ reduces to $x_1x_2x_3$, the form that we will generally use from now on. The distinctive-shape symbol for an AND gate is an elongated box, one side of which is changed to a semicircle. The center of the circular side is the output terminal, and the input terminals are distributed along the opposite (straight) side, as indicated in the figure.

An AND gate produces 1 if and only if all its inputs are 1; an OR gate produces 0 if and only if all its inputs are 0.

The OR operation takes its name from the fact that the output z is 1 if and only if x_1 is 1 *or* x_2 is 1 *or* ... *or* x_n is 1. In other words, the output z of an **OR gate** is 1 if and only if the number of 1s applied to the input lines is one or greater. This interpretation leads to the use of the symbol ≥ 1 in the OR box symbol of Figure 3.20b. By a somewhat weak analogy with numerical addition, the OR function is usually written as a sum expression:

$$z_{\mathrm{OR}}(x_1, x_2, \ldots, x_n) = x_1 + x_2 + \cdots + x_n$$

Thus, + denotes OR in this context, and is read as *or* rather than *plus*; an alternative OR symbol is $\bigvee$. The distinctive-shape symbol for an OR gate resembles the corresponding AND symbol, but the input side is curved and the output side comes to a point (Figure 3.20b).

NAND and NOR Gates

A NAND gate produces 0 if and only if all its inputs are 1; a NOR gate produces 1 if and only if all its inputs are 0.

At this point we have defined three operations AND, OR, and NOT, which we will denote in logic expressions by juxtaposition, plus, and overbar, respectively. As we will see, these operations and the corresponding gates constitute a complete set of building blocks for realizing all relevant functions.

Several other gate operations are also very useful and can be more easily realized using electronic circuits. Of these, the NAND and NOR operations defined in Figure 3.21 are the most important. We encountered the NAND function earlier in this chapter; a complementary switching circuit implementing a two-input NAND gate appears in Figure 3.14. As their names suggest, NAND and NOR are formed by appending NOT to AND and OR, respectively. Their gate symbols are therefore formed by appending the graphic inversion symbol (a small circle) to the corresponding AND or OR symbols, as shown in the figure. Similarly, NAND and NOR are denoted in logic expressions by placing a bar over the corresponding product (AND) or sum (OR) expressions, as follows:

$$z_{\mathrm{NAND}}(x_1, x_2, \ldots, x_n) = \overline{x_1x_2\cdots x_n}$$

$$z_{\mathrm{NOR}}(x_1, x_2, \ldots, x_n) = \overline{x_1 + x_2 + \cdots + x_n}$$

All the gate types introduced so far have this property: There is exactly one input combination that produces a 0 (in the case of OR and NAND) or a 1 (in the case of AND and NOR) on the output line. The remaining $2^n - 1$ input combinations produce the complementary output value.

NAND and NOR gates are often easier to manufacture than AND and OR gates.

Like the one-input buffer and inverter of Figure 3.18, AND, OR, NAND, and NOR gates can be modified by the addition of inversion symbols or, equivalently, complete inverters, to some or all of their input lines. Because

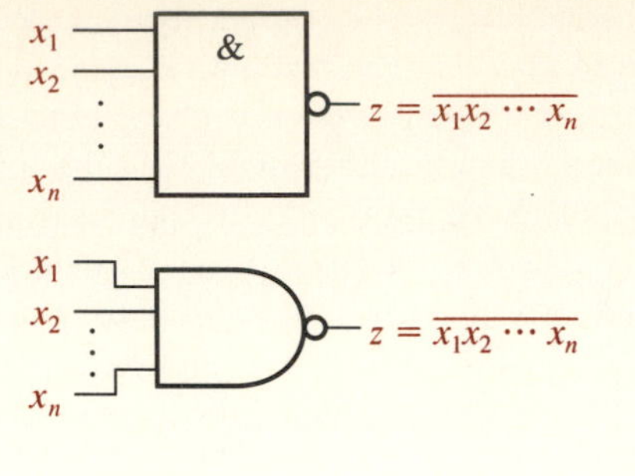

Inputs				Output
x_1	x_2	$\cdots$ x_{n-1}	x_n	z
0	0	0	0	1
0	0	0	1	1
0	0	1	0	1
0	0	1	1	1
		$\cdots$		
1	1	1	0	1
1	1	1	1	0

(a)

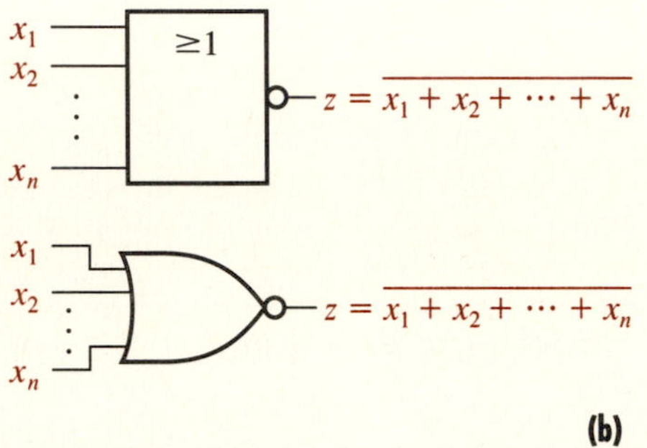

Inputs				Output
x_1	x_2	$\cdots$ x_{n-1}	x_n	z
0	0	0	0	1
0	0	0	1	0
0	0	1	0	0
0	0	1	1	0
		$\cdots$		
1	1	1	0	0
1	1	1	1	0

(b)

FIGURE 3.21 Two additional n-input gate types: (a) NAND; (b) NOR.

physical NAND and NOR gates are easy to manufacture, it is common to realize AND and OR gates by appending an inversion to NAND and NOR gates, respectively, rather than the other way around.

Example 3.3

Gates with inverted inputs

Suppose that we require a three-input logic element that produces the function $z(x_1, x_2, x_3)$ with the following behavior: $z = 0$ if $x_1 = 1$, $x_2 = 0$, and $x_3 = 0$; otherwise, $z = 1$. The fact that z has a unique input combination—namely, $(x_1, x_2, x_3) = (1, 0, 0)$—that makes z assume one of its two values (0) suggests that this function can be realized by a variant of one of the foregoing gates. If we change the specification of z by inverting x_1, we obtain a function z^* that is 0 if and only if $(x_1, x_2, x_3) = (0, 0, 0)$. Comparing this statement to Figure 3.20b, we see that z^* is the three-input OR function. Hence, the desired function z can be realized by a three-input OR gate with its x_1 input inverted, as shown in Figure 3.22a. It follows that z can be represented by the sum expression

$$z(x_1, x_2, x_3) = \overline{x}_1 + x_2 + x_3 \tag{3.9}$$

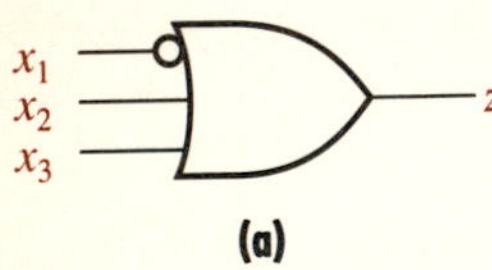

(a)

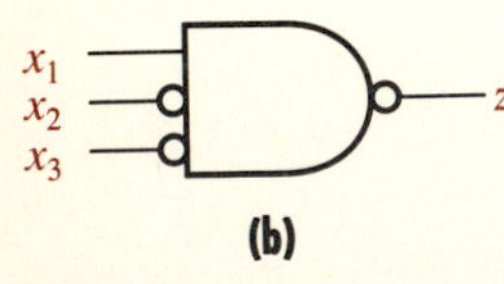

(b)

FIGURE 3.22 Single-gate implementations of $z = \overline{x}_1 + x_2 + x_3$ based on: (a) OR; (b) NAND.

Alternatively, we can negate x_2 and x_3 in the definition of z, yielding the function z'' that is 0 if and only if $(x_1, x_2, x_3) = (1, 1, 1)$. Because z'' is the three-input NAND function, the original function z can now be implemented by a three-input NAND gate that has its x_2 and x_3 inputs inverted. z can also be expressed in the NAND-like form

$$z(x_1, x_2, x_3) = \overline{x_1\overline{x}_2\overline{x}_3} \tag{3.10}$$

Obviously, Equations (3.9) and (3.10) must be equivalent, a fact that is not apparent from the equations alone. Boolean algebra, introduced later in this chapter, provides the necessary mathematical tools for identifying this sort of equivalence.

EXCLUSIVE-OR and -NOR Gates

An XOR gate produces 1 if and only if an odd number of its inputs are 1.

Only a few other logic functions are designated as gate functions. Perhaps the most important of these is XOR (EXCLUSIVE-OR). The basic two-input XOR function was introduced earlier (Figure 3.16) when it was observed to have the defining property that $z_{\text{XOR}}(x_1, x_2)$ is 1 if and only if $x_1 = 1$ *or* $x_2 = 1$, *excluding* the case where $x_1 = x_2 = 1$. It differs from the two-input *inclusive* OR function $z_{\text{OR}}(x_1, x_2) = x_1 + x_2$ only in that for the latter, OR(1, 1) = 1 (that is, the input combination $x_1 = x_2 = 1$ is included among those that make the output 1).

The generalization of XOR to n input variables is most easily specified in terms of the parity of the number of 1s among the n input variables:

$$\begin{aligned} z_{\text{XOR}}(x_1, x_2, \ldots, x_n) &= 1 \text{ if an odd number of inputs are 1} \\ &= 0 \text{ otherwise} \end{aligned} \tag{3.11}$$

For this reason, XOR is also known as the **odd-parity function**, and is the basis of the error-handling circuits discussed in Section 2.11. This versatile function can also be interpreted as (numerical) summation modulo 2. Consequently, another definition of XOR equivalent to Equation (3.11) is

$$z_{\text{XOR}}(x_1, x_2, \ldots, x_n) = x_1 + x_2 + \ldots + x_n \pmod 2$$

EXCLUSIVE-OR is important enough to merit its own circuit and operator symbols; see Figure 3.23. The distinctive-shape graphic symbol for an

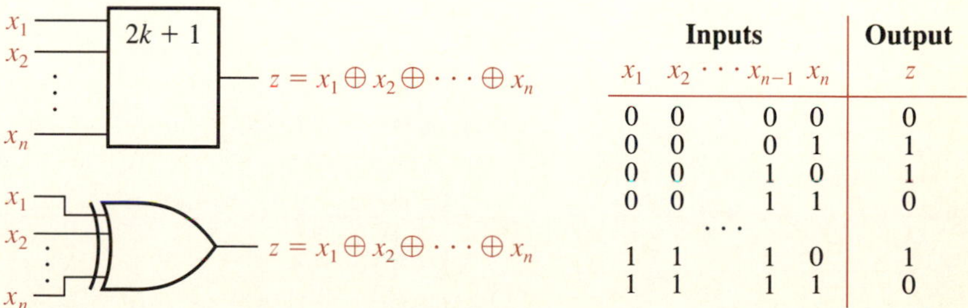

Inputs				Output
x_1	$x_2 \cdots$	x_{n-1}	x_n	z
0	0	0	0	0
0	0	0	1	1
0	0	1	0	1
0	0	1	1	0
	. . .			
1	1	1	0	1
1	1	1	1	0

FIGURE 3.23 Definition of an n-input EXCLUSIVE-OR gate (n assumed to be even in the truth table).

XOR gate is the OR symbol with an additional curved line on the input side. The use of the generic odd number $2k + 1$ as the function designator in the standard box symbol reflects the fact that the output is 1 if and only if $2k + 1$ inputs are 1, for $k = 0, 1, 2, \ldots$. In logic expressions, the XOR operator is $\oplus$, which is read as *exclusive-or, ring-sum,* or *sum modulo 2*. Thus, we can write

$$z_{\mathrm{XOR}}(x_1, x_2, \ldots, x_n) = x_1 \oplus x_2 \oplus \cdots \oplus x_n$$

An XNOR gate produces 1 if and only if an even number of its inputs are 1.

Another useful logic element, the **EXCLUSIVE-NOR** or **XNOR gate**, is obtained by inverting the output of XOR:

$$z_{\mathrm{XNOR}}(x_1, x_2, \ldots, x_n) = \overline{x_1 \oplus x_2 \oplus \cdots \oplus x_n}$$

EXCLUSIVE-NOR is also known as the **even-parity function** for obvious reasons.

Logic Conventions

In real digital circuits, two technology-dependent physical signal values represent the logic values 0 and 1. We can indicate these physical signals in a technology-independent way by H (for high) and L (for low); common examples are high and low voltages in the electrical domain. Logical 0 is naturally equated with the low value L, while logical 1 is equated with the high value H. This convention

$$\mathrm{L} = \text{logical } 0 \quad \mathrm{H} = \text{logical } 1$$

Positive logic maps L (low) onto 0, and H (high) onto 1.

is termed **positive logic**. The opposite convention, H = logical 0 and L = logical 1, is called **negative logic**, but is rarely used.

If a gate G is specified in terms of the physical signals {L,H}, then the gate's logical function will change as we switch the logic convention from positive to negative, or vice versa. This concept is illustrated in Figure 3.24a, where gate G_1 is an AND gate under the positive logic convention, but an OR gate under the negative convention. Similarly, G_2 in Figure 3.24b is a NAND gate with the positive convention, but a NOR gate with the negative convention.

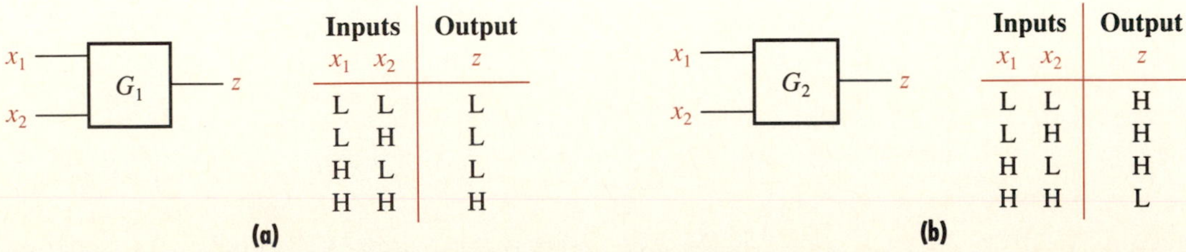

(a)

Inputs x_1	Inputs x_2	Output z
L	L	L
L	H	L
H	L	L
H	H	H

(b)

Inputs x_1	Inputs x_2	Output z
L	L	H
L	H	H
H	L	H
H	H	L

FIGURE 3.24 Two gates: (**a**) AND (positive logic) or OR (negative logic); (**b**) NAND (positive logic) or NOR (negative logic).

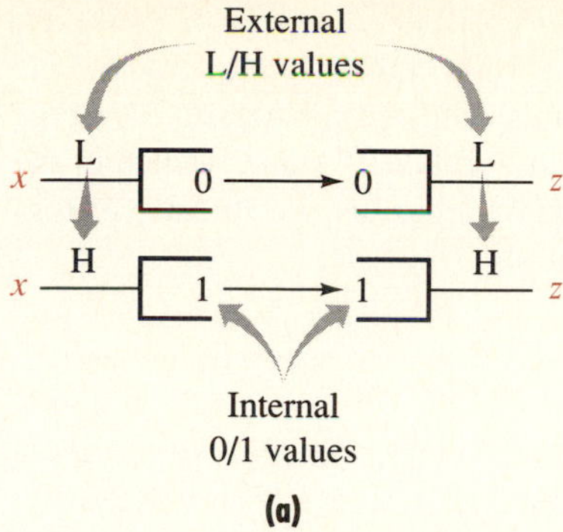

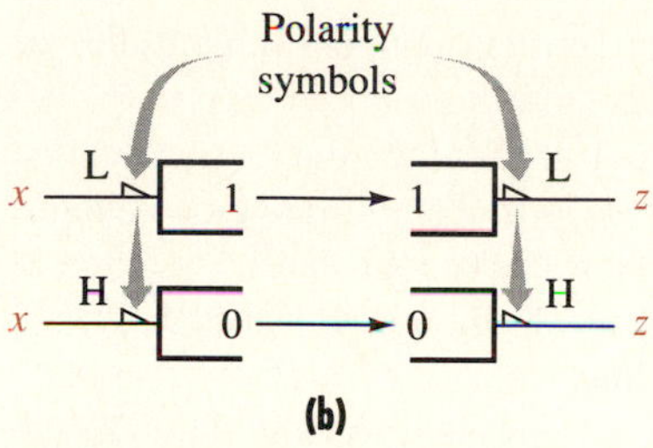

FIGURE 3.25 Direct-polarity indication: **(a)** with no polarity symbol; **(b)** with a polarity symbol.

Polarity Indication

Sometimes the positive and negative conventions are combined together in so-called **mixed logic circuits**. Such circuits can be quite confusing and are, perhaps, best avoided entirely. However, they can be converted to all-positive or all-negative circuits for analysis purposes by the judicious insertion of inversion symbols. A better approach is to use **direct-polarity indication**, as provided for in the IEC 617 (IEEE 91) standard for graphic symbols. This standard makes it possible to indicate independently for every signal the logical (0/1) interpretation of its L and H signals.

The direct-polarity indication principle is demonstrated in Figure 3.25. Every signal x is assumed to have L and H values. If x is connected to an input of a logic gate G or other 0/1 component, as in Figure 3.25a, then the internal value of x seen by G is assumed to be 0 if x = L and 1 if x = H. Similarly, an internal 0 or 1 appears externally as z = L or z = H, respectively. In other words, when no special symbols are used, the default convention is positive logic. If, however, the internal value of x seen by G is supposed to be 1 when x = L and 0 when x = H, then a special **polarity symbol** is placed at the point of entry of x, as shown in Figure 3.25b. This symbol takes the form of a half-arrowhead that points in the direction of signal flow. It can be thought of as converting an H or an L that is external to the logic component into a 0 or a 1, respectively, that is internal to the component. A polarity symbol can be placed on an output line also, with the same interpretation.

Figure 3.26 illustrates direct-polarity indication. The two gates shown have exactly the same effect on L/H signals, as specified in Figure 3.26c. Suppose, for instance, that two Ls are applied to the AND gate of Figure 3.26a. x_1 = L is converted into an internal 0, whereas the polarity symbol means that x_2 = L is converted into an internal 1. Hence, the AND gate receives the logic input 01, which causes it to generate a 0 at its output. This internal 0 appears on line z as an L, in accordance with the first row of the table in Figure 3.26c.

The logical interpretation of every H/L signal can be specified with polarity symbols.

Finally, note the distinction between the inversion symbol (small circle) and the polarity symbol (half-arrowhead). The inversion symbol converts an external 0 to an internal 1, and vice versa. It does not affect L and H

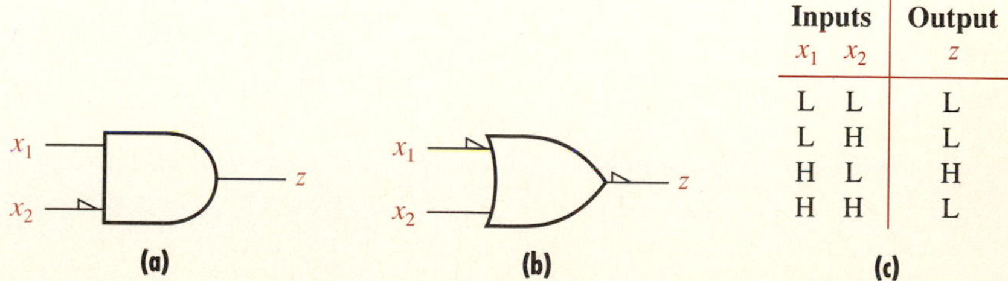

Inputs x_1	x_2	Output z
L	L	L
L	H	L
H	L	H
H	H	L

FIGURE 3.26 **(a)** An AND gate and **(b)** an OR gate with direct polarity indication; **(c)** the L/H table for both gates.

signals, so inversion and polarity symbols are not interchangeable. However, it is easy to see that if we use only positive logic and replace all Ls and Hs with 0s and 1s, respectively, then we can also replace all polarity symbols with inversion symbols, without affecting any of the 0/1 signal values in the circuit. Consequently, if we use only positive logic, as we intend to from here on, we have no need for polarity symbols.

Voltage Conventions

Because of the importance of electronic digital circuits, we will briefly discuss the case in which L and H denote a low voltage V_L and a high voltage V_H, respectively. V_L is often referred to as **ground**, denoted by GND. Depending on the circuit technology used, V_H can be written as PWR (power), V_{CC} (in TTL-style circuits), or V_{DD} (in MOS-style circuits). Typical magnitudes of these voltages are $V_L = 0$ volts and $V_H = 5$ volts.

Electronic circuits are often TTL compatible, with $V_H = 2$ to 5 volts and $V_L = 0$ to 0.8 volts.

Voltage is an analog quantity and is subject to slight fluctuations called **noise**. Hence, it is necessary to assign V_L and V_H to a small range of voltages near their nominal values. For example, many ICs recognize any voltage in the range 0 to 0.8 volts as V_L, and any voltage in the range 2.0 to 5.0 volts as V_H.

A widely used voltage standard for digital circuits is specified by Figure 3.27. It was originally defined for certain IC families that employ what is known as **transistor-transistor logic (TTL)**. Despite its name, this **TTL voltage standard** is met by many non-TTL IC families; such circuits are then said to be **TTL compatible**. The TTL standard requires a 5-volt (designated by V) power supply, the negative and positive terminals of which define the

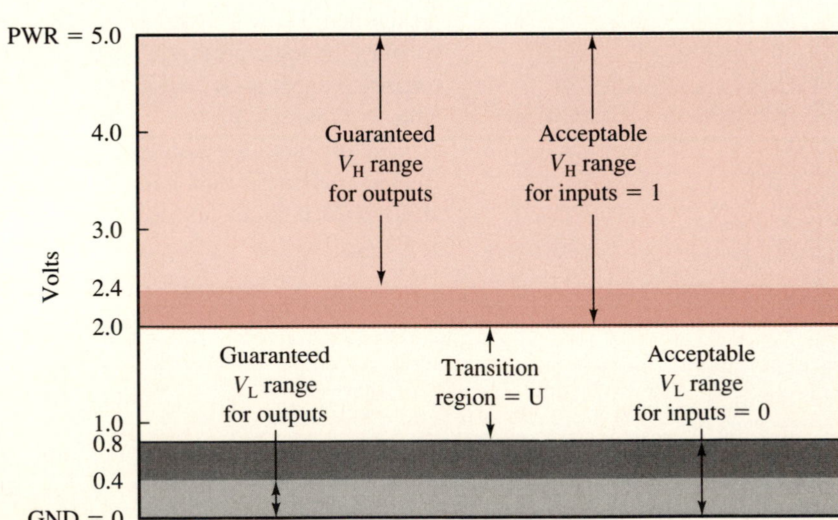

FIGURE 3.27 The TTL standard voltage ranges for logic signals.

power (PWR) and ground (GND) voltage levels, respectively. Circuits that meet this standard accept input voltages in the range 0.0 to 0.8 V as logical 0, and input voltages in the range 2.0 to 5.0 V as logical 1. Voltage signals in the intermediate transition range (0.8 to 2.0 V) assume an "unknown" or indeterminate digital value.

To allow for the impact of noise on logic signals, the output signals generated by TTL-compatible circuits are required to meet tighter constraints for the 0 and 1 values: 1 signals may not go below 2.4 V, and 0 signals may not go above 0.4 V. Hence, a signal in transit between two ICs can deviate from its original voltage level without misinterpretation by as much as ±0.4 V due to noise; the noisy output signal still lies within the acceptable input voltage range. The two ±0.4 V safety bands, or **noise margins** as they are called, decrease susceptibility to noise-induced errors and thus improve overall reliability.

Summary

Figure 3.28 summarizes the six gates introduced above for $n = 3$ inputs, assuming, as we will henceforth, that positive logic is used. The NAND, NOR, and XNOR gates are considered to be **inverting gates;** AND, OR, and

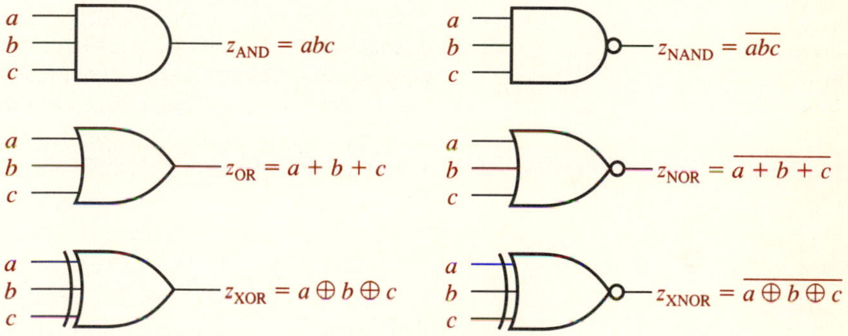

(a)

Inputs			Outputs					
a	b	c	z_{AND}	z_{NAND}	z_{OR}	z_{NOR}	z_{XOR}	z_{XNOR}
0	0	0	0	1	0	1	0	1
0	0	1	0	1	1	0	1	0
0	1	0	0	1	1	0	1	0
0	1	1	0	1	1	0	0	1
1	0	0	0	1	1	0	1	0
1	0	1	0	1	1	0	0	1
1	1	0	0	1	1	0	0	1
1	1	1	1	0	1	0	1	0

(b)

FIGURE 3.28 (a) The six basic three-input logic gates; (b) their truth tables.

XOR are **noninverting gates**. All the inverting gates reduce to a simple inverter when n is reduced to 1. The noninverting gates reduce to buffers when $n = 1$. Inversions may be inserted into the input/output lines of any gate if allowed by the implementation technology.

3.5 Gate-Level Circuits

Next we discuss some fundamental properties of circuits obtained by connecting two or more gates. If their time-dependent behavior is ignored, such circuits (and the underlying switch-level circuits) are known as *combinational logic circuits*. Every binary (logic) function can be realized by a combinational circuit, usually in many different ways. Here we present some examples of combinational circuits and consider the major constraints on their design.

Combinational Circuits

Ideal combinational circuits respond instantly to input changes.

An n-input, m-output **combinational circuit** N is a multigate circuit that realizes a set of $m \geq 1$ functions:

$$F = (f_1(X), f_2(X), \ldots, f_m(X)) \tag{3.12}$$

where $X = (x_1, x_2, \ldots, x_n)$ and each f_i is a binary function of the form f_i: $X \rightarrow z_i$, for $1 \leq i \leq m$. A gate is an example of an n-input, single-output combinational circuit. The function F in Equation (3.12) is a complex or multi-output function that defines the behavior of N, while its internal gates and their interconnections define N's structure. In general, the same function F can be realized by many different circuits. Realizing a desired single- or multi-output function with a cost-effective, gate-level combinational circuit is a central issue in logic design.

The output signals $F(X)$ of a combinational circuit N at any time t are completely determined by the combination of values assumed by the input signals X at that time; hence the term *combinational*. Any changes in these inputs—say, from X_i to X_j—are assumed to change the circuit's output signals from $F(X_i)$ to $F(X_j)$.

In real circuits, there is a small delay—which we often ignore—before a response appears.

Logic gates and the combinational circuits they form are idealized in the sense that all signal changes are assumed to take place instantaneously. In physical circuits, however, signals cannot be transmitted or altered instantaneously, so there is a nonzero delay between a circuit's input change and the output change it produces. This switching delay or response time is quite small in electronic circuits, typically of the order of a nanosecond (10^{-9} s) per gate. Consequently, when addressing the functional or logical behavior of a circuit, we usually assume that all lines and gates have zero delay; in effect, we eliminate time from consideration. Combinational logic circuits are therefore equated with delay-free circuits in which all signal changes can occur instantaneously. Shortly, we will consider the impact of the nonzero delays of real devices and interconnections.

Example 3.4

Gate-level design for a full adder

Figure 3.29 shows an example of a three-input, two-output multigate circuit that generates a function $F(X) = (f_1(x_1, x_2, x_3), f_2(x_1, x_2, x_3))$. This useful circuit consists of three AND gates, an OR gate, and an EXCLUSIVE-OR gate. It is called a **full adder** and is designed to compute the numerical sum of the three input bits in unsigned binary code. (See also Section 2.5, in which the full adder is discussed in the context of binary arithmetic.) For example, when all three inputs are 1, the output $f_1 f_2 = 11$, which is the binary representation of the number three. Figure 3.29b defines the full adder's behavior F by a truth table.

We can also construct logic expressions for F using the operators introduced earlier. For example, the five gates used in the circuit N of Figure 3.29a define these five equations:

$$\begin{aligned} f_1 &= y_1 + y_2 + y_3 \\ y_1 &= x_1 x_2 \\ y_2 &= x_2 x_3 \\ y_3 &= x_1 x_3 \\ f_2 &= x_1 \oplus x_2 \oplus x_3 \end{aligned} \tag{3.13}$$

Again, note that the + symbol in Equation (3.13) is the logical OR operation, and should not be confused with the numerical add operation performed by the entire circuit. When we need to include the latter in an arithmetic expression, we will use a boldface **+** to distinguish it thus:

$$f_1 f_2 = x_1 \boldsymbol{+} x_2 \boldsymbol{+} x_3$$

We refer to $X = (x_1, x_2, x_3)$ and $F = (f_1, f_2)$ as the **primary** input and output lines or signals, respectively, of N. The y_i's used in (3.13) constitute N's **internal** lines or signals. Unlike the primary input/output lines,

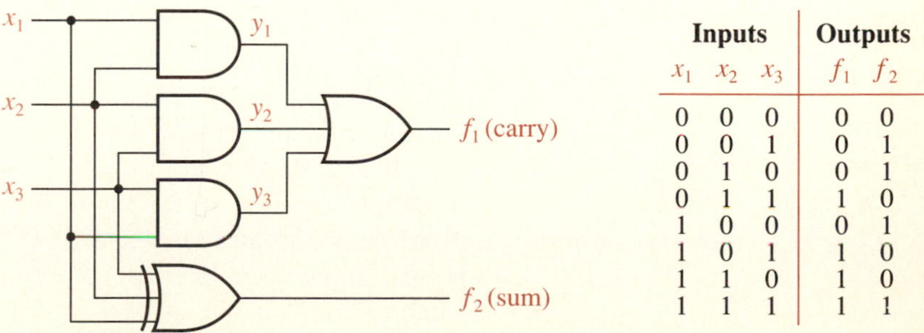

Inputs			Outputs	
x_1	x_2	x_3	f_1	f_2
0	0	0	0	0
0	0	1	0	1
0	1	0	0	1
0	1	1	1	0
1	0	0	0	1
1	0	1	1	0
1	1	0	1	0
1	1	1	1	1

FIGURE 3.29 (**a**) A three-input, two-output combinational circuit (a full adder); (**b**) its truth table.

the internal lines are not directly accessible from outside the circuit. Substituting for y_i in the expression for f_1 in (3.13) yields the following two logic equations, which define F in terms of the primary inputs and outputs:

$$\begin{aligned} f_1 &= x_1x_2 + x_2x_3 + x_1x_3 \\ f_2 &= x_1 \oplus x_2 \oplus x_3 \end{aligned} \tag{3.14}$$

Logic equations like (3.14) are often used as the initial specification of the desired behavior of a combinational circuit.

Design Goals

A given logic function can be implemented in many ways using various gate types; the designer must choose the implementation that best meets certain design goals. Figure 3.30 shows two possible designs for the full adder, which are functionally equivalent to the circuit of Figure 3.29. The first circuit (Figure 3.30a) uses NAND and NOT gates in place of the original three gates types, AND, OR, and EXCLUSIVE-OR. In fact, if the inverters are taken to be single-input NAND gates, as indeed they are, this is an all-NAND circuit. Although the original five-gate design seems the simpler of the two, circuits in the style of Figure 3.30a are often preferred, because in most IC technologies NANDs are more easily implemented than AND, OR, and EXCLUSIVE-OR gates. Other IC technologies favor NOR gates.

Minimizing the total number of gates is a major design goal.

A central problem in logic design is to obtain a combinational circuit that realizes a given set of functions, uses a given set of gate types, and has the lowest possible cost. Circuit cost is conveniently measured in a technology-independent fashion by the total number of gates used. Clearly, the original full-adder circuit of Figure 3.29 with its five gates is cheaper than the 12-gate design of Figure 3.30a, if there are no constraints on the permissible gate types. If only NAND gates are allowed, the question then arises: Is the circuit of Figure 3.30a the cheapest all-NAND design? The answer to this difficult question is no, as demonstrated by the circuit in Figure 3.30b, which is an all-NAND adder of complex structure that uses only eight gates. Finding minimum-cost or low-cost circuits of this sort is a major topic in logic design that we will examine in depth in the next few chapters.

Our discussion so far has assumed that logic gates and circuits can be of arbitrary size and can respond instantaneously to signal changes on their primary input lines. Physical constraints limit both the size and speed of real combinational circuits in important ways, of which the logic designer must be aware. Again, we address these issues here in a technology-independent fashion.

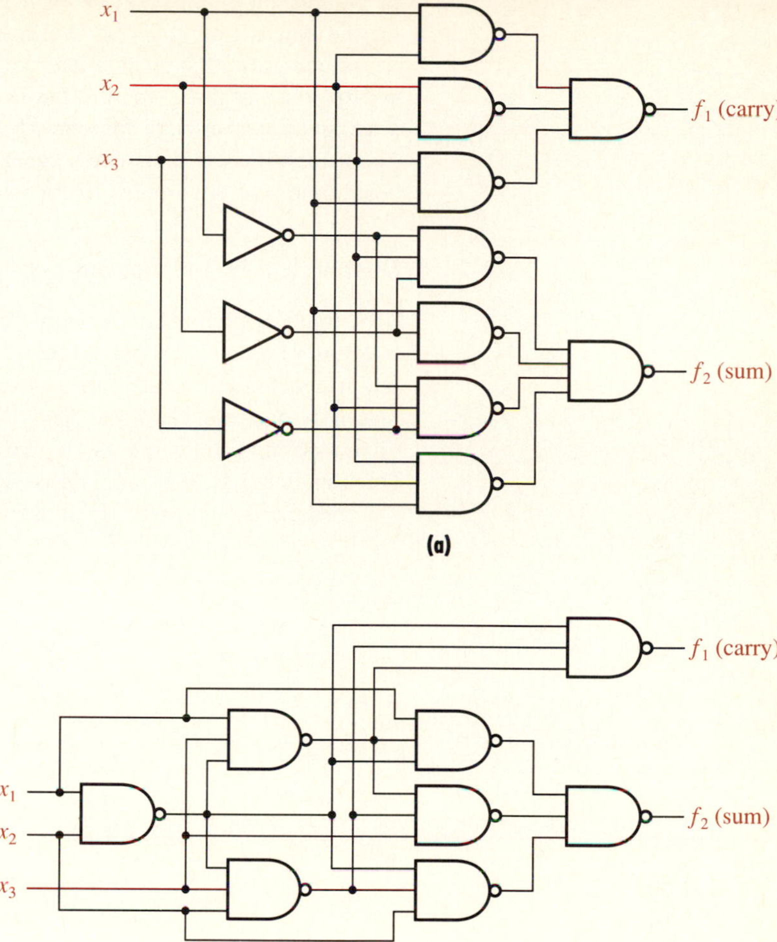

FIGURE 3.30 Two more implementations of the full adder.

Fan-in and Fan-out

All gates have limits on their fan-in and fan-out.

The transfer and processing of information require the expenditure of energy, albeit in very small amounts, and a physical device can absorb or produce only a certain amount of energy without failing. This places an upper bound on the number of input sources that may supply signals to a gate G, and another bound on the maximum number of output devices that may be supplied with signals by G's output line. The number of input lines of G is termed its **fan-in**; the number of lines, which are typically inputs to other gates, that are connected to G's output line is termed its **fan-out**. We may also associate a fan-out parameter with each primary input line x_i. In the all-NAND circuit of Figure 3.30b, for instance, gate fan-in is either two or three, and gate fan-out ranges from one to five. Each primary input has fan-out three.

In the usual design situation, we are given maximum permissible values $\phi_{in}^{max}(G)$ and $\phi_{out}^{max}(G)$ for the fan-in and fan-out of each gate type G, and a maximum permissible fan-out $\phi_{out}^{max}(x_i)$ for each primary input signal x_i. The values of these parameters depend on the particular physical technology being used, but are around 10 for many electronic circuit types.

Meeting Fan-in and Fan-out Constraints

Sometimes a preliminary design will contain a gate G that exceeds the maximum fan-in or fan-out values allowed by the relevant technology. These fan-in and fan-out limits can be overcome in various ways. G can be replaced by a gate of the same type that has more inputs or has higher output drive capability. If such a gate is not available, then G can be replaced by a functionally equivalent multigate circuit N that meets the current fan-in/fan-out constraints. Figure 3.31a illustrates this replacement of G for an AND

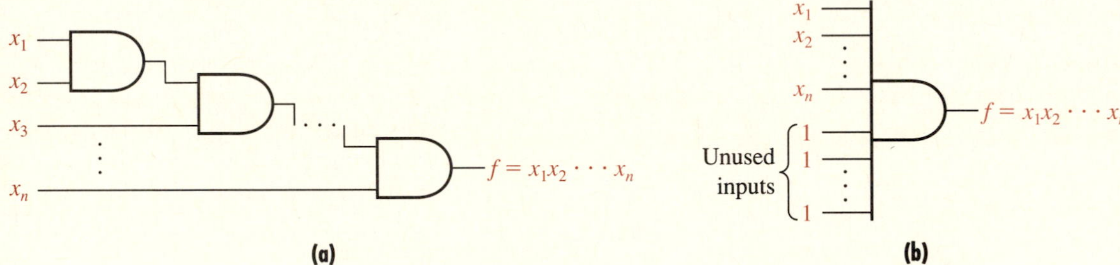

FIGURE 3.31 (a) Increasing and (b) decreasing the fan-in of an AND gate.

gate. If the rightmost gate only is made a NAND, the result is an n-input NAND circuit composed of gates with fan-in of two. The same approach can be applied to all the other standard gate types. To decrease fan-in, all that is required is to apply appropriate constant values, 1 in the AND and NAND cases, and 0 in the OR and NOR cases, to the unused gate inputs. Figure 3.31b illustrates this simple process for an AND gate.

To increase the fan-out of G, one or more buffers can be inserted in the lines connected to G's output line. Figure 3.32 shows how the fan-out of a NOR gate is increased from four to 16 in this fashion, using four buffers. The NOR gate supplies its output signal only to the four buffers, each of which, in turn, supplies four copies of this signal to other gates. Sometimes a buffer is replaced by NOT gates when advantage can be taken of the inversion it produces. Note that because most practical gates have $\phi_{out}(G) > 1$, essentially any functionally acceptable gate type can be inserted into a line to increase its effective fan-out.

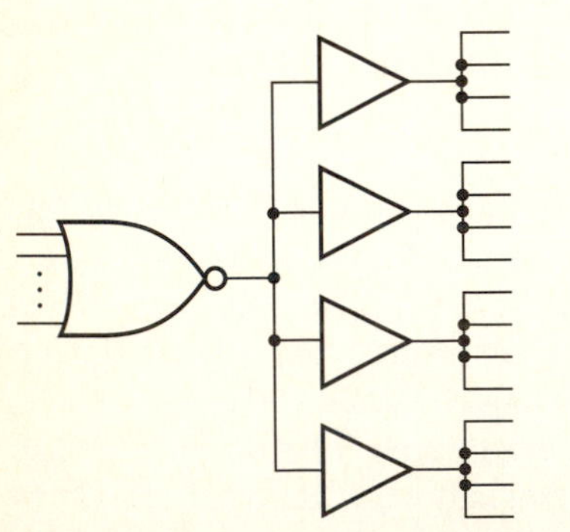

FIGURE 3.32 Using buffers to increase the fan-out of a NOR gate from four to 16.

Example 3.5 **Modifying a circuit to meet fan-in and fan-out constraints**

Suppose that the five-gate full adder of Figure 3.29 must be built from AND, OR, and EXCLUSIVE-OR gates that have unusually strict limits on fan-in and fan-out. Specifically, we will assume that all available gates have a maximum fan-in and fan-out of two, and that the maximum fan-out from any primary input line is also two.

We see from the circuit of Figure 3.29 that its three AND gates satisfy these constraints, but the OR and EXCLUSIVE-OR gates have fan-in of three and violate the fan-in constraint. We will therefore replace each of the latter gates by a pair of two-input gates of the same type, connected as shown in Figure 3.33.

It remains to satisfy the fan-out limit imposed on the primary input lines. Observe that each x_i fans out to three gates. Assuming we may use only the given two-input gate types, we will make one of them serve as a buffer B. As Figure 3.33 shows, we have opted to use a two-input AND gate as B, with one of its inputs set permanently to the 1 value. We might equally well choose an OR or XOR gate, with one input line set to 0. By inserting a copy of B in the path from x_i, we can limit x_i to supplying one original gate and B itself. The other two gates originally connected to x_i are now connected to the output of B.

Meeting fan-in and fan-out constraints may increase gate number and propagation delay.

The resulting circuit (Figure 3.33) realizes the original two functions, but satisfies all the given fan-in/fan-out constraints. To meet these constraints, we have doubled the number of gates and also increased the maximum path length from the primary inputs to the primary outputs, which, as we will see next, adversely affects the circuit's performance.

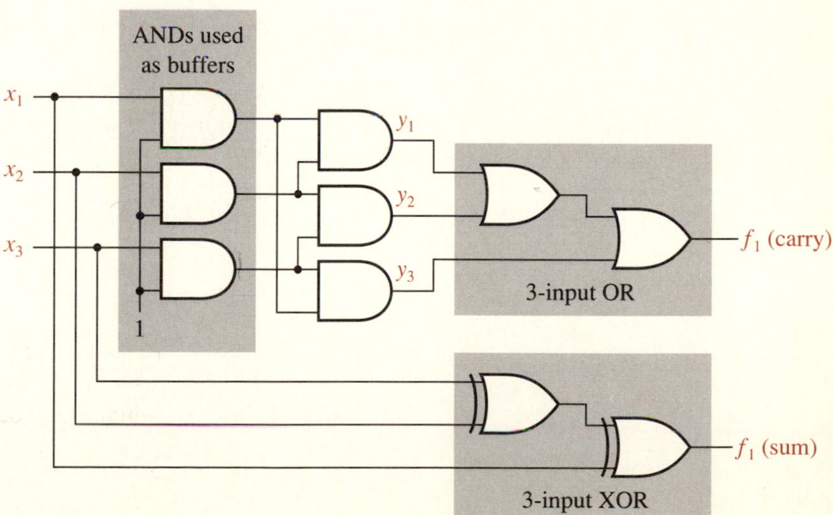

FIGURE 3.33 The full adder of Figure 3.30 after redesign to meet fan-in and fan-out limits of two.

Propagation Delay

All physical gates and lines have nonzero propagation delays.

Physical signals travel at a finite rate bounded above by the speed of light, which is approximately 3×10^8 m/s or 30 cm/ns. Various physical phenomena, such as resistance and capacitance in the case of electronic circuits, further substantially reduce the speed at which signals can be transferred through physical circuits. These quantities vary widely from place to place, even within a single gate, and they are difficult to control during circuit manufacture. Hence, the propagation delay cannot be measured or predicted with much precision.

A physical logic gate, therefore, imposes various technology-dependent delays on the signals passing through the gate and its input/output lines. We will assume that all such delays can be lumped into a single average or "nominal" delay parameter t_{pd}, referred to as the gate's **propagation delay**. Typical values of t_{pd} for electronic circuits are in the range 10^{-10} to 10^{-8} s, that is, 0.1 to 10 ns.

The effect of propagation delay is illustrated by the **timing diagram** of Figure 3.34, which shows by means of waveforms the changes in the output signal of a gate, in this case a two-input NAND gate, as the inputs change over time. For the idealized zero-delay gate of the type we have been implicitly considering so far, $t_{pd} = 0$, and the output responds instantly to input changes at any time t, as depicted in Figure 3.34a. This behavior is defined by the following sequential or time-dependent NAND equation

$$z(t) = \overline{x_1(t)x_2(t)} \tag{3.15}$$

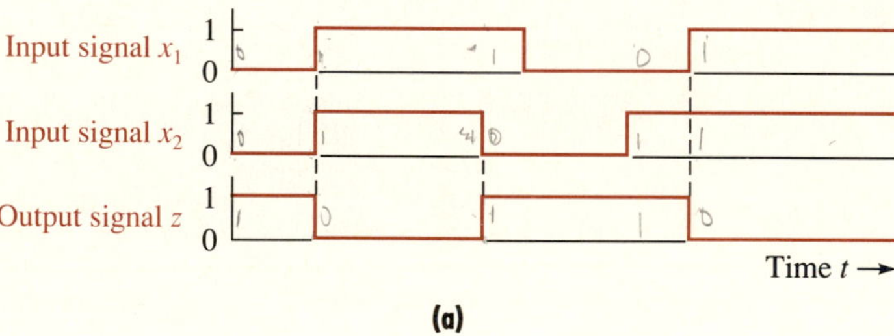

(a)

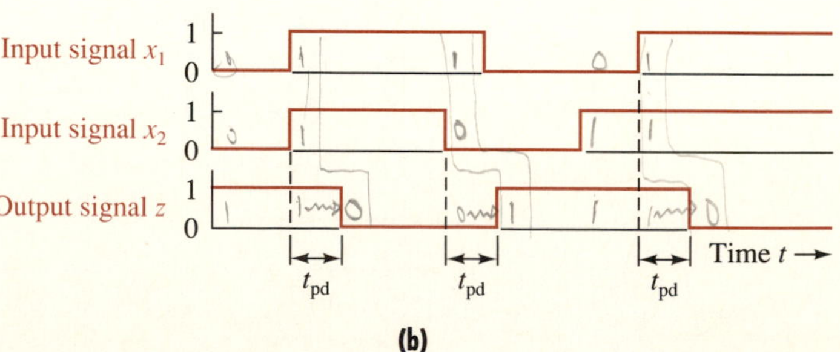

(b)

FIGURE 3.34 Timing diagrams of NAND behavior: (a) a zero-delay gate; (b) a gate with (nominal) delay t_{pd}.

$x(t) \rightarrow$ [t_{pd}] $\rightarrow z(t) = x(t - t_{pd})$

(a)

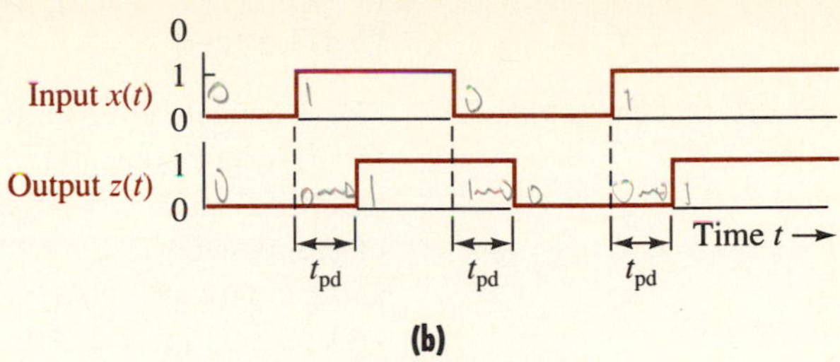

FIGURE 3.35 (a) A delay element and (b) its behavior.

A more realistic delay assumption is made in Figure 3.34b, where $t_{pd} > 0$, and the output changes lag the input changes that produce them by t_{pd}. The corresponding NAND equation in this case is

$$z(t) = \overline{x_1(t - t_{pd})x_2(t - t_{pd})} \tag{3.16}$$

The waveform for z in Figure 3.34b is the same as that of Figure 3.34a, but is shifted to the right by t_{pd}.

Delay Elements

We can represent dynamic behavior like that of Equation (3.16) by introducing a logic device called a **delay element** D (Figure 3.35a). Its effect is to delay an input signal $x(t)$ by an amount t_{pd}, so that the output of D is specified by

$$z(t) = x(t - t_{pd})$$

as shown in Figure 3.35b.

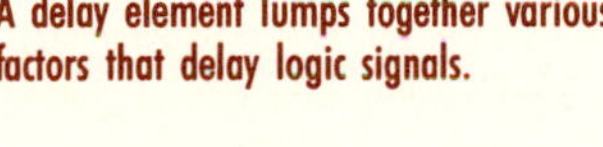

A delay element lumps together various factors that delay logic signals.

We can now model the NAND behavior of Figure 3.34 by means of the circuit of Figure 3.36, where we have inserted a delay element D into the output line of a (zero-delay) NAND gate. We should emphasize that D does not correspond to a physical logic device. Rather, it lumps together the varied and complex physical delays associated with the gate. It is therefore more realistic to place the delay element inside the logic gate, and simply use the usual gate symbol with an added label that indicates its delay, when necessary (Figure 3.36b). We will generally follow the latter approach.

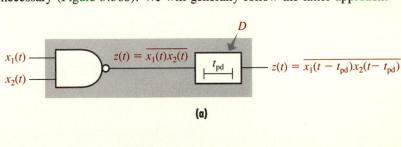

$x_1(t)$, $x_2(t)$ → NAND → $z(t) = \overline{x_1(t)x_2(t)}$ → D [t_{pd}] → $z(t) = \overline{x_1(t - t_{pd})x_2(t - t_{pd})}$

(a)

$x_1(t)$, $x_2(t)$ → NAND [t_{pd}] → $z(t) = \overline{x_1(t - t_{pd})x_2(t - t_{pd})}$

(b)

FIGURE 3.36 NAND gate models with (nominal) delay t_{pd}.

Circuit Speed

Circuit delay increases with depth (number of levels).

The practical significance of gate propagation delays is that they determine the maximum operating speed of a logic circuit. Let $d(C)$ be the maximum number of gates lying along any path from a primary input to a primary output of a circuit C; $d(C)$ is called the **depth** or the **number of levels** of C. The adder circuit of Figure 3.29 has depth two, for instance, whereas the modified version of Figure 3.33b has depth four. (Some paths in C may be shorter than $d(C)$.) If all gates in C are characterized by a common propagation delay t_{pd}, then the time required to propagate a signal through C is, at most, $d(C)t_{\text{pd}}$. If $d(C)t_{\text{pd}}$ is taken as the time C requires to perform one operation or computation step, then C can perform, at most, $1/[d(C)t_{\text{pd}}]$ operations per second.

By definition, ideal combinational logic elements act instantaneously; they do not contain delays. Delay elements and any logic circuit containing delays are termed **sequential** rather than combinational. Therefore, the consideration of delays, operating speed, and related time-dependent issues is properly a part of sequential theory, to which we will turn in Chapter 6. However, as we will see long before that, minimizing the number of logic levels in order to maximize speed is considered an important goal in combinational circuit design.

Boolean Algebra

This section examines *Boolean algebra*, which provides the theoretical basis for logic design. It discusses the properties of Boolean algebra that are useful in logic design at the switch and gate levels.

3.6 A Definition of Boolean Algebra

We have seen in this chapter that we can employ mathematical operations similar to those of high-school algebra to describe the structure and behavior of logic circuits at the switch and gate levels. Figure 3.37 shows how we have used the product, sum, and overbar operations x_1x_2, $x_1 + x_2$, and $\overline{x}$, respectively, to represent some basic circuits and components. Our goal now is to explain the algebra underlying these operations and place it on a firm mathematical basis suitable for analyzing and synthesizing general logic circuits.

Origins

Boolean algebra was developed in the mid-1800s as a mathematical theory of logical reasoning.

Boolean algebra is named after the self-taught English mathematician George Boole (1815–1864), whose influential 1854 book *The Laws of Thought* showed for the first time how to reduce human reasoning (logic) to a symbolic form resembling ordinary numerical algebra. Boole used the numbers 1 and 0 to symbolize the logic values true and false, respectively, and a binary variable symbol such as x to represent a true-or-false statement, which is

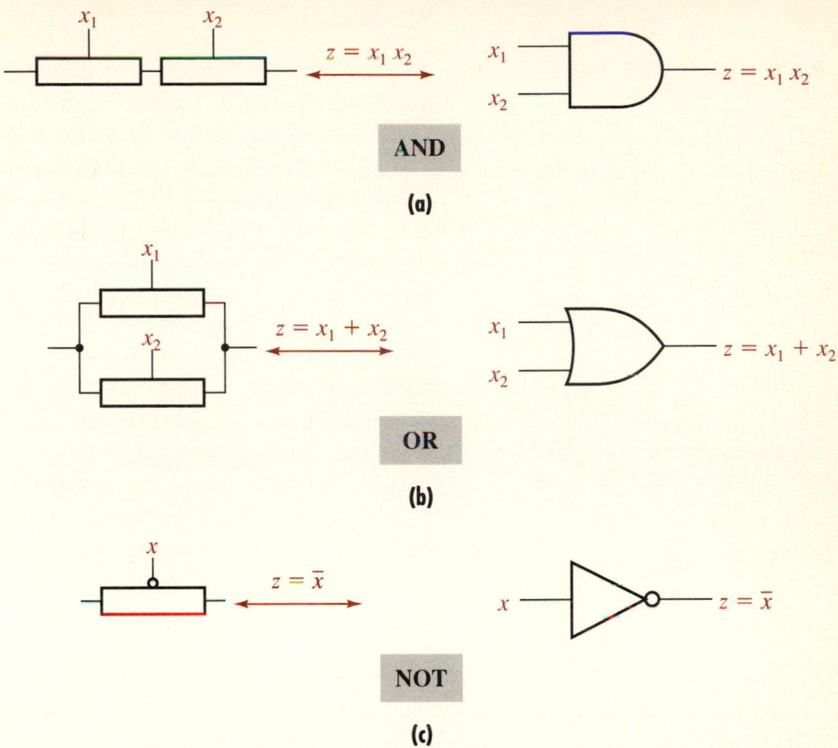

FIGURE 3.37 Logic circuits corresponding to the (a) AND, (b) OR, and (c) NOT operations.

called a **proposition** in this context. Examples of propositions are "*A* is *B*" and "*A* is not *B*." If x symbolizes a proposition such as "*A* is *B*," then $x = 1$ indicates that the proposition is true, while $x = 0$ means it is false.

Under the name **propositional calculus**, Boole's algebra became one of the foundation stones of the mathematical theory of logic. In the late 1930s, mainly as a result of the work of Claude E. Shannon, an M.I.T. graduate student in electrical engineering, the same algebra was seen to provide the theory needed for the analysis and design of digital circuits built from electromechanical relays [Shannon, 1938]. The two natural states of a relay switch, on and off, are the analogs of the true and false values of propositional calculus, and can similarly be represented abstractly by 1 and 0. Such switches are also combined into circuits that are the counterpart of propositions in mathematical logic.

Relays were soon superseded in computers by much faster switches and gates built from electronic vacuum tubes. In the 1950s, another electronic switch, the transistor, began replacing the vacuum tube. In various highly integrated forms, transistors continue to be the fundamental switching devices of present-day electronic circuits. Because it addresses the structure and behavior of circuits in an abstract, technology-independent form, Boolean algebra has retained its position as the relevant theory through all these profound changes in the physical technologies used to realize digital systems.

Algebra

High-school algebra involves the real numbers R and the operations of $\{+, -, \times, /\}$ on R.

An **algebra** $\mathscr{A}$ is a mathematical theory involving a set of **elements** K and a set of **operations** Φ that act on the members of K. The algebra $\mathscr{R}$ fondly remembered from high school has $K = R$, the infinite set of real numbers. For its basic operation set Φ, ordinary (high-school) algebra employs the four arithmetic operations addition $(+)$, subtraction $(-)$, multiplication $(\times, \cdot$, or simply juxtaposition), and division $(/)$ defined on R. Another numerical algebra we discussed in Chapter 2 is the m-element algebra formed by the integers Z_m with the arithmetic operations modulo m.

Boolean algebra involves a set K and the operations $\{and, or, not\}$ on K.

Among Boolean algebras, infinite ones exist, as well as finite Boolean algebras containing 2^n distinct elements for all $n \geq 1$. The smallest Boolean algebra contains just two elements corresponding to the generic 0 and 1 states and is, as we will see, particularly interesting. However, we will also encounter large Boolean algebras—for example, that formed by all combinational functions of up to n binary variables, which contain 2^{2^n} elements.

In general, the elements K form the constants of the algebra, and unspecified members of K are denoted by the variable symbols x, y, z, etc. The operations in Φ define functions of the form

$$z = f(x_1, x_2, \ldots, x_m) \tag{3.17}$$

where f maps each combination of m distinct values from K into some z that is also in K. Often, the function f has a special **operator** symbol such as $+$, $\times$, or $*$, which allows Equation (3.17) to be written in alternative ways, such as the familiar form

$$z = x_1 * x_2 * \cdots * x_m$$

Axioms (Postulates)

Axioms define an algebra's basic properties, from which all other properties can be deduced.

The manner in which algebraic operations can be used is governed by certain basic rules, which also form part of the algebra and are referred to as **axioms** or **postulates**. For example, number algebra $\mathscr{R}$ has, among others, the following axioms:

1. If a and b are members of K, then $a + b$, $a - b$, $a \cdot b$, and a/b are also members of K. This axiom is called the **closure property** of $\mathscr{R}$.
2. For all a and b in K, $a + b = b + a$ and $a \cdot b = b \cdot a$. This axiom, which says that the order of the operands is unimportant in the case of addition and multiplication, is called **commutativity**.
3. For all a, b, and c in K, $a \cdot (b + c) = a \cdot b + a \cdot c$. This axiom, which relates addition and multiplication, is called the **distributivity** of multiplication over addition.
4. K contains the special element 0 (zero), which is such that for any a in K, $a + 0 = a$ and $a \cdot 0 = 0$. Note that these properties of zero are not shared by the other, nonzero elements of the algebra.

With a little experience, these axioms appear obvious, and we apply them unconsciously when solving problems in algebra. However, the axioms of $\mathcal{R}$ are not without their subtleties. For example, the closure of a/b does not apply to the case in which $b = 0$, because division by zero is considered undefined. The commutativity property of addition and multiplication does not extend to subtraction and division because, in general, $a - b \neq b - a$ and $a/b \neq b/a$. As we will see, the axioms of Boolean algebra differ from those of ordinary algebra in more profound ways. It is therefore necessary to spell out these axioms completely and precisely.

Huntington's Postulates

Huntington's postulates provide a complete and concise definition of Boolean algebra.

Various sets of axioms have been proposed for Boolean algebra. The one we use here is the first of several sets published by the American mathematician Edward V. Huntington (1874–1952) in 1904 [Huntington, 1904, 1933] and referred to as **Huntington's postulates**. They employ three operations *and*, *or*, and *not*, which we denote by the special symbols juxtaposition (or, occasionally, by $\cdot$), $+$ and $\bar{\ }$, respectively. As might be expected, the three operations of Boolean algebra correspond exactly to the AND, OR, and NOT gate operations of Figure 3.37.

Table 3.2 presents the 11 Huntington's postulates. The first eight axioms are in matching a–b pairs, with the roles of *and* and *or* switched in each pair. We will discuss the significance of this pairing later. It can be shown

TABLE 3.2 Huntington's postulates (axioms) for Boolean algebra [Huntington, 1904].

Axiom number	*Statement of axiom*	*Name*
1a 1b	If a and b are in K, then $a + b$ is in K If a and b are in K, then ab is in K	Closure
2a 2b	There is an element 0 in K such that $a + 0 = a$ There is an element 1 in K such that $a1 = a$	Zero axiom Unit axiom
3a 3b	For all a and b in K, $a + b = b + a$ For all a and b in K, $ab = ba$	Commutativity
4a 4b	For all a, b, and c in K, $a + bc = (a + b)(a + c)$ For all a, b, and c in K, $a(b + c) = ab + ac$	Distributivity
5a–b	For each a in K, there is an inverse or complement element $\bar{a}$ in K such that: 5a): $a + \bar{a} = 1$ and 5b): $a\bar{a} = 0$	Inverse axioms
6	There are at least two distinct elements in K	

that Huntington's postulates are **consistent**, meaning that they cannot lead to any contradictions. They are also **independent** in the sense that no one of the axioms can be deduced from the others; see Problem 3.53. In this way, Huntington's postulates provide a concise and complete starting point from which everything about any Boolean algebra $\mathcal{B}$ can be deduced. These axioms have also been found to be among the most useful when working with $\mathcal{B}$.

Boolean Algebra vs. Number Algebra

Boolean *and* and *or* obey some, but not all, of the laws of multiplication and addition for numbers.

A glance at Table 3.2 shows that most of the axioms of $\mathcal{B}$—specifically, axioms 1a–3b, 4b, and 6—are also valid for the number algebra $\mathcal{R}$, with *and* and *or* corresponding to multiplication and addition, respectively. Indeed, this correspondence is the justification for using the numerical operation symbols $+$ and $\cdot$ in Boolean algebra. As axioms 2a–b indicate, the special elements 0 and 1 of $\mathcal{B}$ behave somewhat like the numbers 0 and 1 of $\mathcal{R}$; hence, the use of a common notation here, too.

The analogy between the Boolean and number algebras is imperfect, however, because the definition of $+$ yields $1 + 1 = 1$, which is true in $\mathcal{B}$ but not in $\mathcal{R}$. The Boolean inverse or complement operation *not* has no close counterpart in $\mathcal{R}$, so axiom 5 is unique to Boolean algebra. Among the other axioms, the first distributivity axiom 4a, $a + bc = (a + b)(a + c)$, also does not hold in $\mathcal{R}$; for instance, $2 + 2 \cdot 3 = 8 \neq (2 + 2) \cdot (2 + 3) = 20$. On the other hand, the second distributivity axiom 4b is perfectly valid in $\mathcal{R}$. Axiom pairs 3 and 4 are also known as the **commutative** and **distributive laws**, respectively.

Switch-Level Interpretation

We can easily interpret Huntington's postulates in terms of current-switching circuits. Here K is the pair $\{0, 1\}$ or, equivalently, {open, closed}. The variables x and $\overline{x}$ correspond to positive and negative switches, respectively, with x applied to their control inputs. We equate *and* and *or* to the series and parallel connections, respectively. It is not hard to see that every series-parallel switching circuit C_i corresponds to an *and-or-not* expression E_i that mirrors its structure. E_i is called a **Boolean expression** and defines a binary (Boolean) function $z(x_1, x_2, \ldots, x_n)$, which C_i is said to **realize** or **implement**.

The laws of Boolean algebra define pairs of equivalent logic circuits.

Two switching circuits with the same input variables applied to them are equivalent if and only if each realizes the same Boolean function z. This means that any input combination $x_1, x_2, \ldots, x_n$ that creates a closed path in one circuit also creates a closed path in the other. Therefore, if expressions E_1 and E_2 are equivalent, we would expect the corresponding switching circuits to be equivalent. Consequently, a postulate of the form $E_1 = E_2$ corresponds to a pair of equivalent circuits.

Figure 3.38a gives a pair of equivalent switching circuits that model the zero postulate (axiom 2a) of Boolean algebra. Here $a + 0$ denotes a

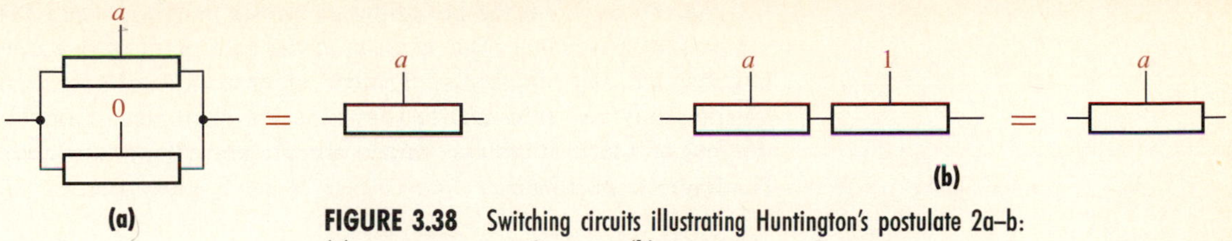

(a)

(b)

FIGURE 3.38 Switching circuits illustrating Huntington's postulate 2a–b: (a) zero axiom $a + 0 = a$; (b) unit axiom $a \cdot 1 = a$.

switch with a applied to it that is connected in parallel with a switch with 0 applied. Because the second switch is kept permanently open, it can be removed, yielding the single switch a that is equivalent to the original circuit. This circuit equivalence is obviously expressed by $a + 0 = a$. Similarly, we see that a permanently closed switch can be placed in series with another switch without affecting the circuit's function; this models the unit axiom $a1 = a$ (Figure 3.38b).

The remainder of Huntington's postulates (except axioms 1 and 6, which do not assert any equivalences) may be interpreted along the same lines. Note that the commutative laws (axioms 3a–b) assert the fact that in a series or parallel connection, the order of the switches does not affect the function realized.

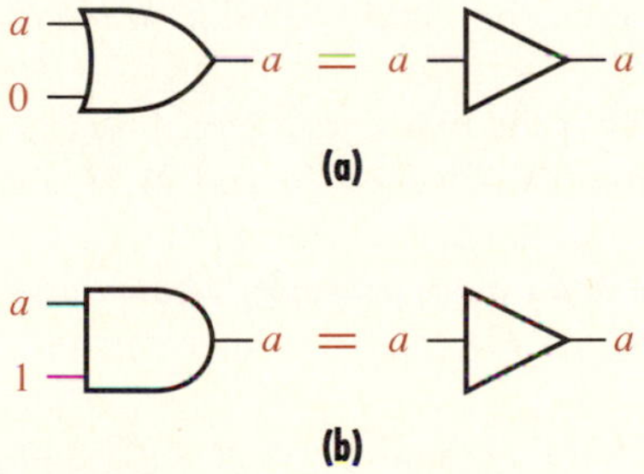

(a)

(b)

FIGURE 3.39 Gate-level circuits illustrating Huntington's postulates 2a–b: (a) zero axiom $a + 0 = a$; (b) unit axiom $a \cdot 1 = a$.

Gate-Level Interpretation

We can also interpret Huntington's postulates in terms of gate-level circuits. Again $K = \{0, 1\}$, but in this case, *and*, *or*, and *not* correspond to the gates of the same names. Figure 3.39 shows circuit pairs that represent the zero and unit axioms; compare it to Figure 3.38. We can regard these axioms as asserting here that the fan-in of a two-input AND or OR gate can be reduced to one by applying a 1 or 0, respectively, to an input of the gate, a fact we noted earlier. Equivalent circuits corresponding to the distributive laws 4a–b are given in Figure 3.40.

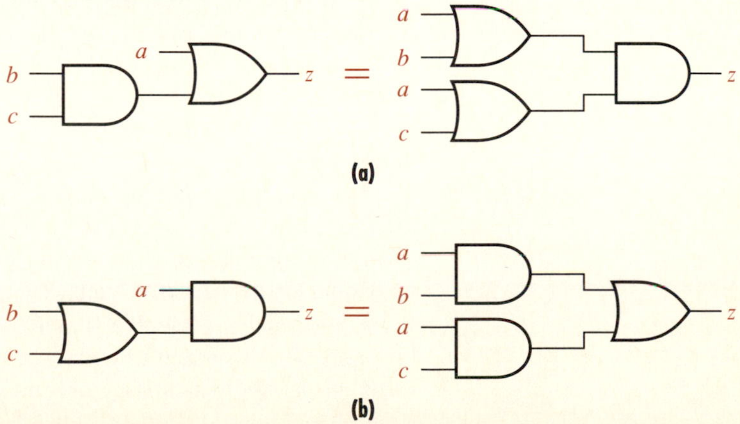

(a)

(b)

FIGURE 3.40 Gate-level circuits that illustrate the distributivity postulates 4a–b.

We can see from the preceding examples that Boolean algebra provides a means of describing logic circuits at the gate and the switch levels. From now on, we will mainly use Boolean algebra as a tool for analyzing gate-level circuits. We turn next to some specific examples of Boolean algebras, the first of which is the two-valued algebra underlying our interpretations of Huntington's postulates.

Switching Algebra $\mathcal{B}_2$

Inputs		Outputs		
a	b	ab	$a+b$	$\bar{a}$
0	0	0	0	1
0	1	0	1	1
1	0	0	1	0
1	1	1	1	0

FIGURE 3.41 A truth table defining *and, or,* and *not* for the two-element Boolean algebra $\mathcal{B}_2$ (switching algebra).

A Boolean algebra is defined by its element set K and its operation set $\Phi = \{and, or, not\}$. Recall that the names of the operations come from the following definitions: ab is 1 if and only if a is 1 *and* b is 1; $a + b$ is 1 if and only if a is 1 *or* b is 1 (or both); and $\bar{a}$ is 1 if and only if a is *not* 1. The operations in Φ can also be defined conveniently by means of function or truth tables. Figure 3.41 defines the Boolean algebra $\mathcal{B}_2$ that contains only two elements. Huntington's postulates 2a–b require that the two elements of $\mathcal{B}_2$ must also serve as the special 0 and 1 elements; hence, we can write $K = \{0, 1\}$. By axiom 6, no Boolean algebra can contain fewer elements, so $\mathcal{B}_2$ represents the smallest possible Boolean algebra, and is often referred to as **switching algebra**.

The 0 and 1 elements of $\mathcal{B}_2$ correspond to the two signal values used in binary logic circuits. The Boolean operations *and* and *or* defined in Figure 3.41 correspond, respectively, to series and parallel connections (switch-level interpretation) or AND and OR gates (gate-level interpretation). The inverse element $\bar{a}$ is associated with a negative switch (switch-level interpretation) or an inverter (gate-level interpretation).

The smallest Boolean algebra $\mathcal{B}_2$ has just the two elements 0 and 1.

It is easy to check that $\mathcal{B}_2$ satisfies all of Huntington's postulates. Closure follows from the fact that all entries in the output columns of Figure 3.41 are taken from $K = \{0, 1\}$. The zero and unit axioms are satisfied, because by letting a assume the only possible values 0 and 1, and applying Figure 3.41 to the equations $a + 0 = a$ and $a \cdot 1 = a$, we obtain:

$$0 + 0 = 0$$

$$1 + 0 = 1$$

$$0 \cdot 1 = 0$$

$$1 \cdot 1 = 1$$

These equations prove exhaustively that Huntington's postulates 2a–b hold. The commutative and distributive laws 3a–4b can be similarly verified. With 0 and 1 serving as each other's inverses, axiom 5 is satisfied as, obviously, is axiom 6. This approach to proving that $\mathcal{B}_2$ is a Boolean algebra by enumerating all possibilities quickly becomes impractical as the size of the algebra increases.

Example 3.6

Deciding if the three-valued algebra $\mathscr{E}_3$ is Boolean

Most simulation programs intended to assist the logic designer employ a third logic value X, which is assigned to logic signals whose present value is 0 or 1, but which of these applies is unknown. For example, after a circuit is initially switched on, certain signals may power up to an unknown state, which is conveniently denoted by X. We can easily extend the usual Boolean operator set $\{and, or, not\}$ to handle X, as shown in Figure 3.42. This table reduces to the two-valued Boolean algebra $\mathscr{B}_2$ when all rows containing X are removed; compare it to Figure 3.41. Moreover, the three Boolean operations act on X in a very natural fashion. If X is an input to an AND gate, then the output z is X—that is, unknown—unless another input to the gate is 0, in which case the 0 overrides the X and makes $z = 0$. Similarly, the output of an OR gate with an X input is X, unless an overriding 1 is applied to the gate. Finally, if the input to an inverter is unknown, the output is also unknown, suggesting the assignment $\overline{X} = X$.

Thus, the set $K = \{0, 1, X\}$ with the operator set $\Phi = \{and, or, not\}$ forms a well-defined algebra which we will denote by $\mathscr{E}_3$. But is it a Boolean algebra? To answer this, we will determine if $\mathscr{E}_3$ satisfies all of Huntington's postulates. Clearly, axioms 1 (closure) and 6 ($|K| \geq 2$) are satisfied by $\mathscr{E}_3$. The zero and one axioms are also satisfied, with 0 and 1 retaining the same roles they have in $\mathscr{B}_2$. This follows from the fact that $X + 0 = X$ and $X \cdot 1 = X$. A check of the various possibilities indicates that distributivity (axioms 4a–b) is satisfied. For example, Figure 3.42 implies that

$$X + X \cdot 1 = X + X = X$$
$$(X + X) \cdot (X + 1) = X \cdot 1 = X$$

so axiom 4a is satisfied for $(a, b, c) = (X, X, 1)$. All other assignments of values from $\mathscr{E}_3$ to axiom 4 are similarly satisfied.

Inputs		Outputs		
a	b	ab	$a+b$	$\overline{a}$
0	0	0	0	1
0	1	0	1	1
1	0	0	1	0
1	1	1	1	0
0	X	0	X	1
1	X	X	1	0
X	0	0	X	X
X	1	X	1	X
X	X	X	X	X

FIGURE 3.42 A truth table defining *and, or,* and *not* for the three-element algebra $\mathscr{E}_3$ containing 0,1, and X (unknown).

It remains to see if the inverse axioms 5a–b hold for $\mathscr{E}_3$. The requirement (axiom 5a) that $a + \overline{a} = 1$ implies that $X + \overline{X} = 1$, and the last row of Figure 3.42 asserts that $X + \overline{X} = X + X = X$; hence, we conclude that $X = 1$. Similarly, axiom 5b and the last row of Figure 3.42 imply that $X = 0$. We must therefore conclude that $X = 0 = 1$, so that $\mathscr{E}_3$ contains just one element, in violation of axiom 6. Hence, if the elements of $\mathscr{E}_3$ are all distinct, they cannot satisfy the inverse axioms of Boolean algebra, so we conclude that $\mathscr{E}_3$ is almost, but not quite, a Boolean algebra. Failure to satisfy the inverse axioms is common in nonBoolean algebras.

Other Boolean Algebras

Boolean algebras are found in various domains.

Boolean algebra has been found useful in a number of different areas, each of which has developed its own terminology and notation. Table 3.3 identifies some important examples and lists their elements and operation sets. We

TABLE 3.3 Some examples of Boolean algebras.

Domain	*Elements*	*Operations*
Propositional calculus	K = all true-false statements (of up to n variables) 0 = false statement (contradiction) 1 = true statement (tautology)	*and* = and *or* = or *not* = not
Set theory	K = all 2^n subsets of n objects 0 = the empty set $\varnothing$ containing no members 1 = the universal set U consisting of all n objects	*and* = intersection $\cap$ *or* = union $\cup$ *not* = complement $\bar{\ }$
Logic circuits	K = all 2^{2^n} binary (logic) functions $f(X)$ of up to n variables 0 = the constant function $f_Z(X) = 0$ for all X 1 = the constant function $f_U(X) = 1$ for all X	*and* = AND function $\cdot$ *or* = OR function + *not* = NOT function $\bar{\ }$

remarked earlier that Boolean algebra underlies propositional calculus. In this case, the elements of interest are propositions (true-or-false statements) in ordinary language. The basic operations of propositional calculus are the language connectives and, or, and not, which are used to combine simple propositions into more complex ones. The Boolean algebra defined by the n-variable logic functions will be examined later; it is the central Boolean algebra in logic design. The two-element switching algebra $\mathcal{B}_2$ is this algebra for the special case of the 0-variable—that is, constant—logic functions. Set theory provides another interesting class of Boolean algebras, which we consider next.

Set Algebras

Every set of objects has associated with it a collection of subsets, as well as certain operations that combine the subsets to form new ones. These subsets form an easily understood type of Boolean algebra, referred to as **set algebra**. Because the notion of a set is so general, it turns out that *every* Boolean algebra can be viewed as a set algebra by appropriately interpreting it. This gives us some useful new insights into the nature of Boolean algebra.

Set algebras employ intersection and union for *and* and *or*, respectively.

First, we need to formalize sets and their operations. A finite **set** s is simply a collection of n objects that constitute the members of s. An example is the set of integers between 1.7 and 8.9, which has seven members, namely $\{2, 3, 4, 5, 6, 7, 8\}$. The statement "x is a member of s" will be abbreviated to $x \in s$. The **intersection** of two sets s_1 and s_2, which we denote by $s_1 \cap s_2$, is the set of objects, each of which is included in both s_1 *and* s_2. The **union** $s_1 \cup s_2$ of two sets s_1 and s_2 is the set of objects formed by combining s_1 and s_2 into a single set. In other words, an object in $s_1 \cup s_2$ appears in s_1 *or* s_2. For example, if $x = \{1, 4, 5, 6\}$ and $y = \{1, 2, 4, 7\}$ are two four-member sets of integers, then $x \cap y = \{1, 4\}$ and $x \cup y = \{1, 2, 4, 5, 6, 7\}$. Note the convention of enclosing the members of a set in braces. Finally, the **complement** $\bar{s}$ of s is the set of all nonmembers of s.

The special zero element 0 required by Huntington's postulate 2a is identified with the empty set $\{\}$, which has no members and is usually denoted by the symbol $\varnothing$. This is a perfectly meaningful concept; the set of integers between 1.1 and 1.9 and the set of living dodo birds are examples of $\varnothing$. The special unit element 1 of axiom 2b is the "universal" set U, which contains all the objects in the domain of interest.

With these definitions, it is straightforward to show that the set K of all possible subsets of U with the operations $\cap$, $\cup$, and $^-$ as defined above satisfies Huntington's postulates. It is therefore a Boolean algebra $\mathcal{S}_{2^n}$, which we call a set algebra. For example, axiom 5 asserts the following: $s \cup \bar{s} = U$ (that is, the union of a set and its complement is the universal set); and $s \cap \bar{s} = \varnothing$ (that is, the intersection of a set and its complement is the empty set). Note that an n-member set such as $Z_n = \{0, 1, 2, \ldots, n-1\}$ has exactly 2^n subsets. This follows from the fact that every subset of Z_n may or may not contain element 0, implying a two-way choice; it may or may not contain element 1, implying another two-way choice; it may or may not contain element 2, implying yet another two-way choice, and so on, for a total of n two-way choices. Hence, the total number of ways of choosing a subset from Z_n is

$$\underbrace{2 \times 2 \times \cdots \times 2}_{n \text{ times}} = 2^n$$

For instance, $Z_2 = \{0, 1\}$ has the following four subsets: $\varnothing$, $\{0\}$, $\{1\}$, $\{0, 1\} = U$, and defines the four-element set algebra $\mathcal{S}_4$. The subsets of $Z_3 = \{0, 1, 2\}$ define the eight-element set algebra $\mathcal{S}_8$; the members of K in this case are $\varnothing$, $\{0\}$, $\{1\}$, $\{2\}$, $\{0, 1\}$, $\{0, 2\}$, $\{1, 2\}$, $\{0, 1, 2\} = U$.

Example 3.7

Representing a Boolean algebra with a Venn diagram

We can study set algebras and other Boolean algebras graphically by means of Venn diagrams, named after their inventor, the English logician John Venn (1834–1923). The idea is to represent the universal set U with points on a plane, as in Figure 3.43a. We can then identify a subset a, whose value is an element of a set algebra S_n, by enclosing the corresponding points with a line such as a circle. It is usual to say that a is the enclosed area, and not explicitly specify the points of U that a encloses. The part of U that lies outside a represents $\bar{a}$, as indicated in Figure 3.43b. Now suppose that we draw two overlapping circles representing a and b. The common area intersected by both circles corresponds to $a \cap b = ab$ (Figure 3.43c), and the combined area of the two circles corresponds to $a \cup b = a + b$ (Figure 3.43d).

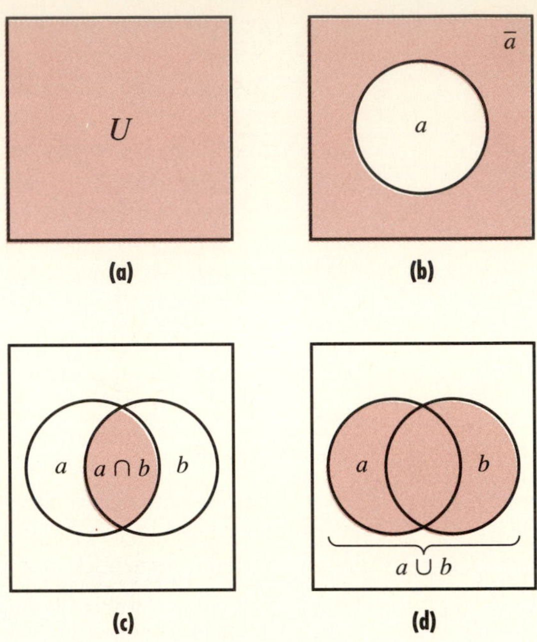

FIGURE 3.43 Venn diagrams in which the colored areas represent: (**a**) the universal set (unit element) U; (**b**) $\bar{a}$; (**c**) $a \cap b$; (**d**) $a \cup b$.

Venn diagrams demonstrate Boolean relationships in graphical form.

We can use Venn diagrams to illustrate Huntington's postulates and the other laws of Boolean algebra. Figure 3.44 demonstrates the distributive law 4a, namely, $a + bc = (a + b)(a + c)$, which in terms of set algebra is $a \cup (b \cap c) = (a \cup b) \cap (a \cup c)$. The top right Venn diagram shows the colored regions corresponding to $a \cup (b \cap c)$, while the bottom right Venn diagram shows $(a \cup b) \cap (a \cup c)$. The distributive law follows from the fact that their two colored regions are the same.

Venn diagrams can be used in this way to demonstrate all of Huntington's postulates, as well as other laws of Boolean algebra. With more than three variables, it is difficult to show pictorially the overlap among the corresponding regions (which can no longer all be circles); see Problem 3.63. Chapter 5 introduces a related type of diagram (a Karnaugh map) that is more useful for representing and analyzing Boolean functions.

We noted in passing that an n-member set has 2^n subsets, which implies that every set algebra has exactly 2^n elements. Because every Boolean algebra $\mathcal{B}$ is equivalent to some set algebra, it follows that the number $|K|$ of elements in $\mathcal{B}$ must be an exact power of two. Hence, the only possible values of $|K|$ are 2, 4, 8, 16, and so on. This fact immediately means that the three-element algebra $\mathcal{E}_3$ defined by Figure 3.42 cannot be Boolean.

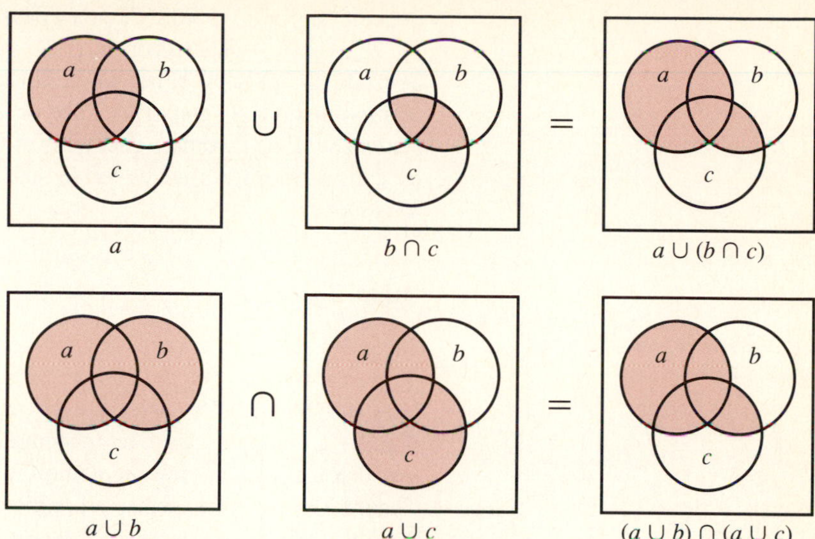

FIGURE 3.44 Venn diagrams that demonstrate Huntington's distributivity postulate 4b: $a + bc = (a + b)(a + c)$ or, equivalently, $a \cup (b \cap c) = (a \cup b) \cap (a \cup c)$.

3.7 Fundamental Properties

We derive further useful properties of Boolean algebra in this section. The main results are presented as theorems derived directly or indirectly from the defining axioms (Huntington's postulates). Along with the axioms, these results provide the rules needed to simplify logic circuits, change the gate types used, and so on. They also enable the logic designer to analyze new and often complex aspects of circuit structure and behavior that are part of the designer's everyday work.

Theorems and Proofs

Theorems are true statements that can be deduced from the axioms.

A true statement or proposition about an algebra constitutes a **theorem** of the algebra. Many theorems of Boolean algebra take the form $E_1 = E_2$, in which E_1 and E_2 are expressions composed of variables, constants from K, and operators from Φ. Because a Boolean expression E_i can be mapped into a logic circuit C_i, the theorems of Boolean algebra can give us useful insights into the design of logic circuits.

The axioms of Boolean algebra are the "self-evident" theorems that are assumed to hold without further justification. Other theorems of the algebra must be derived from the axioms by some proof process. We can often justify a theorem informally by demonstrating its validity for a particular type of circuit or, when applicable, by using truth tables or Venn diagrams. However, to cover all the cases we might encounter, we need to be able to construct general, algebraic proofs. Such a **proof** takes the form of a sequence of steps $T_1, T_2, \ldots, T_p$, where each step T_i specifies a proposition whose validity depends on applying either an axiom or a previously proven theorem to the relation asserted in step T_{i-1}. The final step T_p is, or leads directly to, the statement of the theorem. Theorems of the form $E_1 = E_2$ can often be proven by constructing truth tables for E_1 and E_2, and verifying that the two tables are identical.

Useful Theorems

We often use theorems to simplify logic expressions or circuits.

Table 3.4 lists the more useful theorems of Boolean algebra. For reference purposes, we continue the numbering scheme begun with Huntington's postulates in Table 3.2. The first theorem pair 7a–b states that the special elements 0 and 1 are unique. The **idempotence** ("same strength") **theorems** 8a–b assert that *and*ing or *or*ing an element to itself does not change that element. Theorems 9a–b give further properties of 0 and 1. Theorems 10a–b, called the **absorption laws**, state that certain terms involving a and b may be combined with a, leaving a unchanged; intuitively speaking, a "absorbs" the term containing b. Note that theorems 8a–10b show how a Boolean expression can be replaced with a simpler one by dropping unnecessary terms. As we will see later, properties like these are basic to the simplification of logic circuits.

De Morgan's laws are used to convert between *and* and *or* forms.

The uniqueness of the inverse or complement of every element is asserted by theorem 11. Theorem pair 12a–b, known as the **associative laws**, states that the order in which two copies of the *and* or *or* operation are applied do not affect the final outcome. This implies that the parentheses in expressions like $(a + b) + c$ and $a(bc)$ are superfluous, so we can write $a + b + c$ and abc without ambiguity. The next theorem pair, **De Morgan's laws**, shows how *and* expressions can be converted to *or* expressions and vice versa by appropriate application of *not* operators. These very useful theorems have the distinction of being named after a person, the English mathematician Augustus De Morgan (1806–1871). The final theorem in Table 3.4, which is known as **involution**, asserts that two successive inversions cancel each other.

A few of the foregoing theorems have natural extensions to larger numbers of variables. For example, De Morgan's laws can be extended from the two-variable form of theorem 13 to n variables, for any $n \geq 2$.

$$\overline{x_1 + x_2 + \cdots + x_n} = \overline{x}_1\overline{x}_2\cdots\overline{x}_n$$

$$\overline{x_1x_2\cdots x_n} = \overline{x}_1 + \overline{x}_2 + \cdots + \overline{x}_n$$

Informal Proofs

Many of the theorems of Table 3.4 have easy interpretations in terms of logic circuits or set theory, and so require little justification. For example, theorem 8 (idempotence) implies that two or more applications of the same signal a

TABLE 3.4 Some useful theorems of Boolean algebra.

Theorem number	*Statement of theorem*	*Name*
7a 7b	The zero element 0 is unique The unit element 1 is unique	Uniqueness of 0 and 1
8a 8b	For each a in K, $a + a = a$ For each a in K, $aa = a$	Idempotence
9a 9b	For each a in K, $a + 1 = 1$ For each a in K, $a0 = 0$	Unit property Zero property
10a 10b	For all a and b in K, $a + ab = a$ For all a and b in K, $a(a + b) = a$	Absorption
11	For each a in K, the inverse $\bar{a}$ is unique	Uniqueness of inverse
12a 12b	For all a, b, and c in K, $(a + b) + c = a + (b + c)$ For all a, b, and c in K, $(ab)c = a(bc)$	Associativity
13a 13b	For all a and b in K, $\overline{a + b} = \bar{a}\bar{b}$ For all a and b in K, $\overline{ab} = \bar{a} + \bar{b}$	De Morgan's laws
14	For each a in K, $\bar{\bar{a}} = a$	Involution

to the inputs of an AND or OR gate do not change the gate's output function z. This case is proven in Figure 3.45, from which we can extrapolate the same result for other Boolean algebras.

Theorems can often be proved informally by constructing circuits or truth tables.

A less obvious result is theorem 10 (absorption), which we also prove for logic circuits in Figure 3.46. Here we construct two combinational circuits that match the expressions appearing in the theorem. Then we construct truth

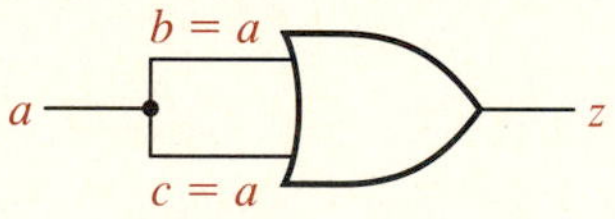

a	b	c	$z = b + c$
0	0	0	0
1	1	1	1

(a)

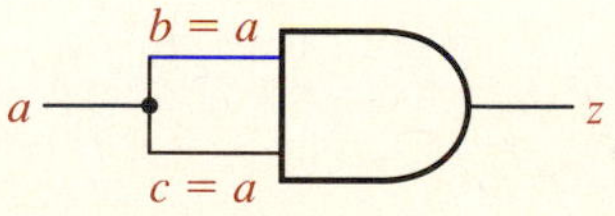

a	b	c	$z = bc$
0	0	0	0
1	1	1	1

(b)

FIGURE 3.45 Demonstration of the idempotence theorem: **(a)** $a + a = a$; **(b)** $a \cdot a = a$.

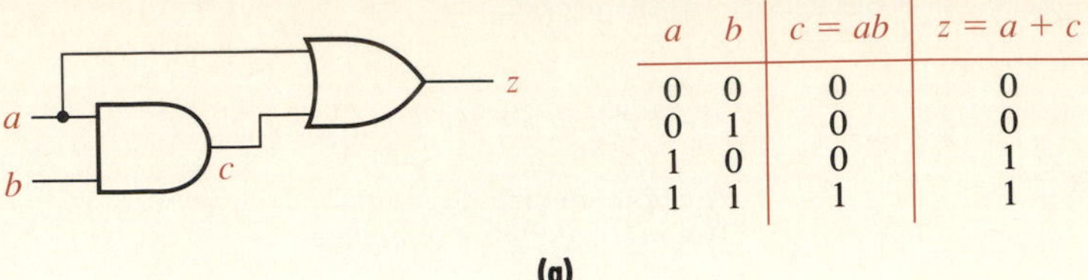

a	b	$c = ab$	$z = a + c$
0	0	0	0
0	1	0	0
1	0	0	1
1	1	1	1

(a)

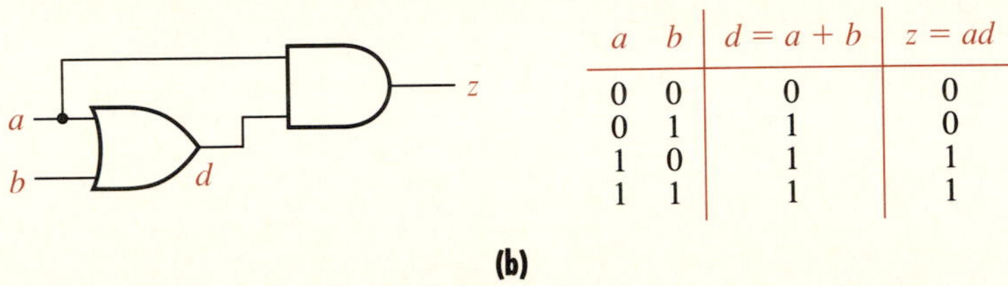

a	b	$d = a + b$	$z = ad$
0	0	0	0
0	1	1	0
1	0	1	1
1	1	1	1

(b)

FIGURE 3.46 Demonstration of the absorption theorem: (a) $a + ab = a$; (b) $a \cdot (a + b) = a$.

tables for each circuit, using the definition of AND and OR gates. It can be seen from the figure that the output column z is the same as the a column in each case, implying that the circuits realize the function a. Hence, the corresponding expressions $a + ab$ (Figure 3.46a) and $a(a + b)$ (Figure 3.46b) are both equivalent to a, as asserted by the absorption theorem.

Finally, we use the combinational circuits of Figure 3.47 to validate the first De Morgan law (theorem 13a), which states

$$\overline{a + b} = \bar{a}\bar{b} \tag{3.18}$$

Figure 3.47 gives two circuits that directly correspond to the two sides of Equation (3.18). We then derive truth tables for each circuit, a relatively trivial task. It can be seen that the two functions z_1 and z_2 realized by these circuits have the same truth tables, implying the validity of (3.18).

Formal Proofs

We now present formal proofs of some representative theorems of Boolean algebra, using general algebraic techniques that can readily be adapted to new problems. First we consider theorems 7a–b, which state that 0 and 1 are unique. Unlike statements of the form $E_1 = E_2$, theorems on existence or uniqueness are not easily proven by constructing truth tables or by similar approaches.

Theorem 7a The 0 element of a Boolean algebra is unique.

a b c z_1

a	b	$c = a + b$	$z_1 = \overline{c}$
0	0	0	1
0	1	1	0
1	0	1	0
1	1	1	0

(a)

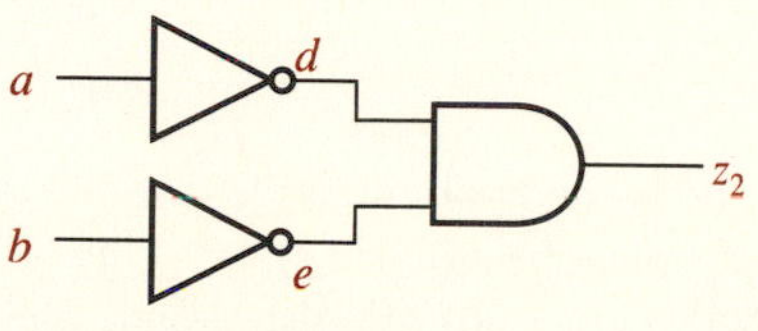

a	b	$d = \overline{a}$	$e = \overline{b}$	$z_2 = de$
0	0	1	1	1
0	1	1	0	0
1	0	0	1	0
1	1	0	0	0

(b)

FIGURE 3.47 Demonstration of De Morgan's law $\overline{a + b} = \overline{a}\,\overline{b}$.

Proof

Proof by contradiction: Assume the theorem is false and deduce a contradiction.

We establish this theorem using the method of proof by contradiction (also known by its Latin name *reductio ad absurdum*): The theorem is assumed false, and this assumption is shown to lead to a contradiction, thus proving the theorem. Suppose that there are two distinct zero elements denoted 0_1 and 0_2. Then, by Huntington's postulate 2a,

$$a + 0_1 = a \tag{3.19}$$

$$a + 0_2 = a \tag{3.20}$$

Now, setting a to 0_2 in Equation (3.19) and to 0_1 in Equation (3.20), we get

$$0_2 + 0_1 = 0_2 \tag{3.21}$$

$$0_1 + 0_2 = 0_1 \tag{3.22}$$

By the first commutative law (axiom 3a), the left-hand sides of (3.21) and (3.22) must be equal thus: $0_2 + 0_1 = 0_1 + 0_2$. Hence, the right-hand sides of Equations (3.21) and (3.22) are also equal, yielding $0_1 = 0_2$. This clearly contradicts the initial assumption that 0_1 and 0_2 are distinct; therefore, there can only be one zero element in a Boolean algebra.

Our proof of the second part of theorem pair 7a–b closely matches that given above.

Theorem 7b The 1 element of a Boolean algebra is unique.

Proof

Suppose, by way of contradiction, that there are two distinct unit elements denoted 1_1 and 1_2. Then, by Huntington's postulate 2b,

$$a1_1 = a \tag{3.23}$$

$$a1_2 = a \tag{3.24}$$

Now, setting a to 1_2 in Equation (3.23) and to 1_1 in (3.24), we get

$$1_2 1_1 = 1_2 \tag{3.25}$$

$$1_1 1_2 = 1_1 \tag{3.26}$$

By the second commutative law (axiom 3b), the left-hand sides of (3.25) and (3.26) must be equal: $1_2 1_1 = 1_1 1_2$. Hence, the right-hand sides of Equations (3.25) and (3.26) are also equal, yielding $1_1 = 1_2$. This clearly contradicts the original assumption that 1_1 and 1_2 are distinct. Hence, there can only be one unit element in a Boolean algebra.

Next, we prove formally the first of the two idempotence theorems, which are illustrated by Figure 3.45.

Theorem 8a

For each a in K, $a + a = a$.

Proof

Proof by algebraic manipulation: Convert one expression of $E_1 = E_2$ into the other.

This proof employs algebraic manipulation to convert the left-hand side of the given relation $a+a = a$ into the right-hand side. Each step of the conversion must be justified by some axiom or a previously proven theorem. We begin by applying Huntington's postulate 2b to the left-hand side of $a + a = a$ to obtain $a + a = (a + a)1$. We then use axiom 5a to replace the 1 on the right-hand side of Equation (3.26) by $a + \overline{a}$, as follows:

$$a + a = (a + a)(a + \overline{a})$$

Invoking the distributivity axiom 4a, we get

$$a + a = a + a\overline{a}$$

Using axiom 5b,

$$a + a = a + 0$$

Finally, applying the zero axiom $a + 0 = a$, we get $a + a = a$, which is the desired result.

The proof technique of algebraic manipulation is very versatile, and is applicable to many problems in Boolean algebra and logic design.

Example 3.8 **Proving the equivalence of two Boolean expressions**

Suppose we have to prove that the following expressions are equivalent:

$$E_1 = \bar{a}\bar{b}h + \bar{c}fgh + \bar{d}fgh + \bar{e}fgh$$

$$E_2 = \overline{(a+b)(cde+\overline{fg})}h$$

They involve eight variables, and so correspond to very large truth tables containing $2^8 = 256$ rows. We will prove equivalence by converting one expression into the other using the laws of Boolean algebra. Comparing the two expressions, we see that E_1 has a "sum-of-products" form that appears simpler than E_2, which contains large inverted subexpressions. We will therefore try to convert the more complicated expression E_2 into the simpler one.

E_2 has the form $\overline{E}h$, so we first break $\overline{E}$ into smaller pieces. De Morgan's laws are particularly useful for this. Applying theorem 13b to E_2 yields

$$E_2 = \overline{(a+b)(cde+\overline{fg})}h = (\overline{(a+b)} + \overline{(cde+\overline{fg})})h \tag{3.27}$$

We now apply De Morgan's other law (theorem 13a) to the two larger complemented subexpressions on the right-hand side of Equation (3.27).

$$E_2 = (\bar{a}\bar{b} + (\overline{cde})\overline{\overline{fg}})h$$

Involution (theorem 14) allows us to replace $\overline{\overline{fg}}$ with fg, and another application of De Morgan's laws breaks up $\overline{cde}$.

$$E_2 = (\bar{a}\bar{b} + (\bar{c}+\bar{d}+\bar{e})fg)h \tag{3.28}$$

At this point, two applications of the distributive law (axiom 4b) will convert Equation (3.28) to sum-of-products form.

$$\begin{aligned} E_2 &= (\bar{a}\bar{b} + \bar{c}fg + \bar{d}fg + \bar{e}fg)h \\ &= \bar{a}\bar{b}h + \bar{c}fgh + \bar{d}fgh + \bar{e}fgh \end{aligned}$$

E_2 has been changed to E_1, and the equivalence is proven.

3.8 Expressions, Functions, and Circuits

Boolean expressions are of central importance in logic design because they provide an efficient way to describe the function and, to a certain extent, the structure of logic circuits. We consider next the relationship between Boolean expressions and functions, and how they, in turn, relate to circuits.

Boolean Functions

A function $z(X) = z(x_1, x_2, \ldots, x_n)$ whose input values $x_1, x_2, \ldots, x_n$ and output value z are all taken from a Boolean algebra is a **Boolean function**. The particular class of Boolean functions whose input/output values come from the switching algebra $\mathcal{B}_2$ are the **switching** or **logic functions**. $\mathcal{B}_2$ is the two-element Boolean algebra for which $K = \{0, 1\}$ and $\Phi = \{+, \cdot, \neg\}$, as specified in Figure 3.42. Unless otherwise stated, we will be interested in switching functions only and will therefore use the terms Boolean function, logic function, and switching function interchangeably.

Boolean functions map n-bit binary words onto $\{0, 1\}$.

The Boolean function $z(x_1, x_2, \ldots, x_n)$ can be defined by a 2^n-row truth table that lists all 2^n combinations of values of $X = (x_1, x_2, \ldots, x_n)$, along with the corresponding function or output value $z(X)$. We have also seen functions defined by means of Boolean expressions of various kinds. These Boolean expressions resemble those of number algebra and are formed in the same general way. For example, the expression $(a + b)c + (a + b)\overline{d}$ specifies the four-variable Boolean function $z(a, b, c, d)$ defined by Figure 3.48. We can therefore write

$$z = (a + b)c + (a + b)\overline{d} \tag{3.29}$$

A comparison between Equation (3.29) and Figure 3.48 suggests that, as descriptions of Boolean functions, Boolean expressions are usually much more concise than truth tables.

a	b	c	d	$z(a, b, c, d)$
0	0	0	0	0
0	0	0	1	0
0	0	1	0	0
0	0	1	1	0
0	1	0	0	1
0	1	0	1	0
0	1	1	0	1
0	1	1	1	1
1	0	0	0	1
1	0	0	1	0
1	0	1	0	1
1	0	1	1	1
1	1	0	0	1
1	1	0	1	0
1	1	1	0	1
1	1	1	1	1

FIGURE 3.48 A truth table for a four-variable Boolean function $z(a, b, c, d)$.

Expression Evaluation

The truth table of the function defined by a Boolean expression can be determined by evaluating the expression for every possible input combination, using the operator definitions and the laws of Boolean algebra. We illustrate this concept for the expression E of Equation (3.29), where the corresponding truth table of Figure 3.48 can be obtained by evaluating E for each of the 16 possible combinations of the four input variables a, b, c, d. For example, to determine $z(0, 1, 1, 0)$, we set a, b, c, and d to 0, 1, 1, and 0, respectively, in (3.29) and evaluate E thus:

$$\begin{aligned} z(0, 1, 1, 0) &= (0 + 1)1 + (0 + 1)1 \\ &= (0 + 1)1 \quad \text{(Idempotence 8a)} \\ &= (1)1 \qquad\quad \text{(Definition of } \textit{or}\text{)} \\ &= 1 \qquad\qquad\;\, \text{(Definition of } \textit{and}\text{)} \end{aligned}$$

Boolean expressions provide compact function definitions.

It is apparent that a particular function can be specified by many—in fact, arbitrarily many—different Boolean expressions. It is easily shown, either by determining the corresponding truth table or by algebraic manipulation, that

$$z = ac + bc + a\overline{d} + b\overline{d} \tag{3.30}$$

specifies the same function z as (3.29). The expressions appearing in (3.29) and (3.30) are said to be **equivalent Boolean expressions.** Thus, every Boolean expression defines a unique Boolean function, whereas a Boolean function can be defined by many different, but equivalent, Boolean expressions.

Well-Formed Expressions

Well-formed (WF) Boolean expressions follow precise construction rules.

Boolean expressions are formed according to certain rules, which we now spell out. As we will see, these rules also define well-formed—that is, properly designed—logic circuits. Let $\mathcal{B}$ be a Boolean algebra with element set K and operator set $\Phi = \{+, \cdot, \bar{}\}$. The simplest valid or **well-formed (WF) Boolean expression** on $\mathcal{B}$ is a single variable or constant symbol, where a constant is any member of K. Thus 0, a, and x_i are all examples of WF Boolean expressions. We can construct general WF expressions by repeated application of the following recursive rule: If $E_1, E_2, \ldots, E_k$ are WF Boolean expressions, then so are all expressions of the form $(E_1 + E_2 + \cdots + E_k)$, $(E_1 E_2 \cdots E_k)$, and $(\overline{E}_1)$. Hence, $(\overline{x}_i)$, $(a + b + c)$, and $((a + b)(\overline{c}))$ are WF Boolean expressions. Examples of algebraic expressions that are *not* WF Boolean expressions are these: $)abc)$, $a + (+b)$, and $(a \div b)$. We refer to WF expressions of the foregoing kind as **fully parenthesized WF** (or **FPWF**) **expressions**, because they use parentheses to define the scope of each *and*, *or*, and *not* operation.

Operator precedence rules allow parentheses to be deleted.

A WF expression can be simplified by deleting some of its parentheses without changing the Boolean function being described. Following custom, we can (optionally) delete the parentheses around a single complemented variable, writing $(\overline{x})$ as $\overline{x}$. We can also delete the outermost parentheses around entire expressions, so that $((a + b)(\overline{c}))$ can be reduced to $(a + b)\overline{c}$. We also take advantage of the associative laws of Boolean algebra to drop the parentheses from all-*or* or all-*and* expressions. For example, we can write $(a + b) + (c + d)$ unambiguously as $a + b + c + d$, and $((ab)c)d$ as $abcd$.

In numerical expressions such as $a + b \times c$, we perform multiplication before addition. In other words, this expression is interpreted as $a + (b \times c)$ rather than as $(a + b) \times c$. This rule for ordering the operators $+$ and $\times$ allows parentheses to be dropped without causing ambiguity. We have similar **precedence rules** for the operators in Boolean (sub)expressions: *not* is evaluated first, then *and*, and finally *or*. Thus, $a + b\overline{c}$ is unambiguously evaluated as $(a + (b(\overline{c})))$ and not as $((a + b)(\overline{c}))$. The latter expression can be simplified to $(a + b)\overline{c}$, but the parentheses around $a + b$ must be retained to avoid misinterpretation as $a + b\overline{c}$.

We regard FPWF expressions that are simplified according to the foregoing rules as well formed, but not fully parenthesized. In general, if any doubt exists when manipulating Boolean expressions, retain parentheses around doubtful (sub)expressions. In most cases, we want to keep the number of parentheses in Boolean expressions to a minimum. However, in the following discussion and in a few other instances, the fully parenthesized form is more desirable.

Duality

There is an obvious symmetry between the *and* and *or* operators of Boolean algebra, and between the 0 and 1 elements. We now develop a precise statement of this symmetry.

Dual expression E^d is formed from E by replacing $+$, $\cdot$, 0, and 1 with $\cdot$, $+$, 1, and 0, respectively.

Let E be any FPWF Boolean expression involving only variables, the operators *and*, *or*, and *not*, and the constants 0 or 1. Then the expression E^d obtained from E by replacing all occurrences of *and*, *or*, 0, and 1 with *or*, *and*, 1, and 0, respectively, is called the **dual** of E. For example, if

$$E = ((x_1 + x_2)(\overline{x}_3)) \tag{3.31}$$

then its dual is

$$E^d = ((x_1 x_2) + (\overline{x}_3)) \tag{3.32}$$

Note that overbars and parentheses are not affected by this **dualization** process. Note also that the dual of E^d is the original expression E; that is,

$$(E^d)^d = E$$

A more complex illustration follows.

$$z = (\overline{(((a + b)c) + ((\overline{d})(a + 0)))})$$

$$z^d = (\overline{(((ab) + c)((\overline{d}) + (a1)))})$$

Although the number of parentheses here may seem excessive, dualization of WF expressions that are not fully parenthesized can lead to errors. For instance, we can simplify (3.32) to the parenthesis-free form $x_1 x_2 + \overline{x}_3$ by taking advantage of the precedence of *and* over *or*. However, applying the dualization process to this simplified expression yields $x_1 + x_2\overline{x}_3$, which corresponds to $x_1 + (x_2\overline{x}_3)$ and not to $(x_1 + x_2)\overline{x}_3$. The latter is the correct dual expression, according to Equation (3.31).

In the above examples, $E^d \neq E$. The relation $E^d = E$ holds in such cases as $\overline{a}^d = \overline{a}$ and $(ab+bc+ac)^d = (a+b)(b+c)(a+c) = ab+bc+ac$, as the reader can readily verify. Boolean expressions for which $E^d = E$ (that is, E and E^d define the same function) are said to be **self-dual**.

The Principle of Duality

Theorem $E_1 = E_2$ holds if and only if the dual $E_1^d = E_2^d$ holds.

We now state an important Boolean algebra "metatheorem"—that is, a theorem about theorems—known as the **Principle of Duality**, which simplifies the proof of many results. It also explains the a–b dichotomy in our naming of the axioms and theorems of Boolean algebra:

> If $E_1 = E_2$ is a valid equation between Boolean expressions that involve only variables, the operators *and*, *or*, and *not*, and the constants 0 and 1, then $E_1^d = E_2^d$ is also a valid equation of Boolean algebra.

Thus, if we know that $E_1 = E_2$ is a theorem, then we can immediately conclude that $E_1^d = E_2^d$ is also a theorem (the dual theorem). On examining the equations appearing in the axioms and theorems presented in Tables 3.2 and 3.4, we see that they occur as dual pairs, except in the case of theorem 14 (involution), which is self-dual, and axiom 6 and theorem 11, to which duality does not apply.

The proof of the Principle of Duality itself is quite simple. Note first that every valid Boolean equation $E_1 = E_2$ is, by definition, a provable theorem of Boolean algebra. Hence, we can always construct—at least in principle—a proof $T = T_1, T_2, \ldots, T_p$ of this equation, in which every step of T involves one of Huntington's postulates. Now suppose that we modify the proof T by replacing every appearance of an axiom P by its dual P^d. In most cases, this will mean replacing the a-form of the axiom with the b-form, or vice versa. The result of this dualization of the proof T is the sequence $T_1^d, T_2^d, \ldots, T_p^d$. This must be a proof T^d of $E_1^d = E_2^d$, which therefore is a theorem of Boolean algebra, as asserted by the Principle of Duality.

Dual Theorems

The proofs of theorems 7a and 7b (uniqueness of the 0 and 1 elements) given above can now be seen to be dual, in that an axiom used in a step of one proof is the dual of the axiom used in the corresponding step of the other proof. The proof for theorem 8a (idempotence of *or*) is summarized below, with the axioms used in each step indicated on the right. Because it is easier to remember names than numbers, we cite theorems and axioms by their names as well. Note, too, that some people's theorems are other people's axioms, and vice versa, so the numbers assigned to them in this book have no special significance. To facilitate dualization, we use FPWF expressions.

$$\begin{aligned}
(a + a) &= ((a + a)1) && \text{(Unit property 2b)}\\
&= ((a + a)(a + \overline{a})) && \text{(Inverse property 5a)}\\
&= (a + (a\overline{a})) && \text{(Distributivity 4a)}\\
&= (a + 0) && \text{(Inverse property 5b)}\\
&= (a) && \text{(Zero property 2a)}
\end{aligned}$$

Proof by duality: Dualizing one theorem produces another (the dual theorem).

Now, if we mechanically dualize the foregoing five-step proof, we immediately obtain the following proof of Theorem 8b (idempotence of *and*).

$$\begin{aligned}
(aa) &= ((aa) + 0) && \text{(Zero property 2a)}\\
&= ((aa) + (a\overline{a})) && \text{(Inverse property 5b)}\\
&= (a(a + \overline{a})) && \text{(Distributivity 4b)}\\
&= (a1) && \text{(Inverse property 5a)}\\
&= (a) && \text{(Unit property 2b)}
\end{aligned}$$

Henceforth, we simply invoke the Principle of Duality to prove any statement about Boolean algebra whose dual is known to be true.

Expression-Circuit Correspondence

There is a fairly close correspondence between gate-level logic circuits and WF Boolean expressions. Consider the expression $E = E_1 + E_2$, in which the (sub)expressions E, E_1, and E_2 specify the functions z, z_1, and z_2,

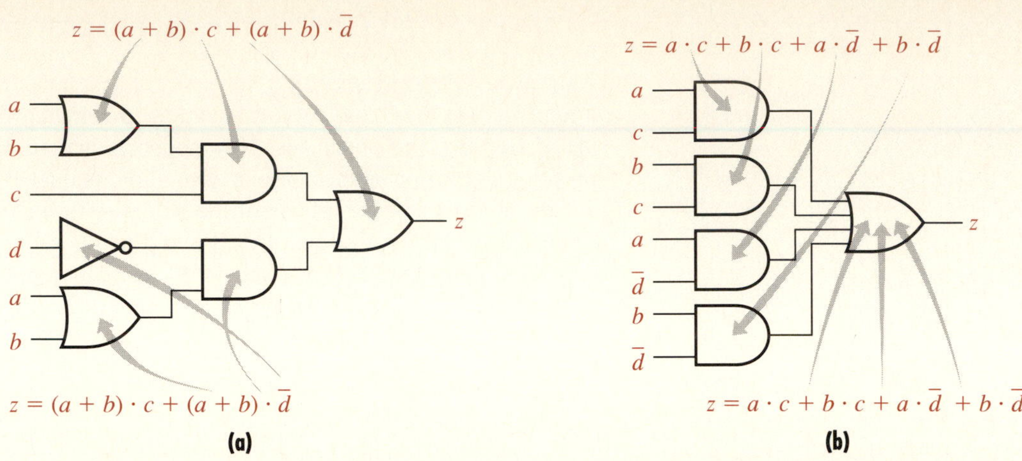

FIGURE 3.49 Expression-circuit correspondence for two realizations of the Boolean function z.

respectively. The circuit corresponding to E contains two (possibly trivial) subcircuits N_1 and N_2, which implement z_1 and z_2, respectively, and an OR gate fed by the outputs of N_1 and N_2. The output of this OR gate is z. The expression $E = E_1E_2$ denotes a similar circuit, with an AND gate replacing the OR gate, while the circuit obtained by inserting a NOT gate in the output line of N_1 corresponds to the expression $E = \overline{E}_1$.

Figure 3.49a shows a circuit corresponding in the above way to the Boolean expression for z that appears in Equation (3.29). Note how each operator in the Boolean expression maps to a distinct gate. In the case of Figure 3.49b, which corresponds to the Boolean expression in (3.30), all the + operators represent a single OR gate. Also, no NOT gates correspond to the overbars in the expression for z; instead, the primary input signals appear in both true and complemented form. This implies that the variables in question are complemented elsewhere, a common situation in practice.

There is a useful but imprecise correspondence between Boolean expressions and logic circuits.

The correspondence between a Boolean expression and a circuit is sometimes much less precise than it is in the foregoing instances. Figure 3.50 shows another realization of z that corresponds approximately to Equation (3.29). The two occurrences of the subexpression $(a + b)$ in (3.29) are represented by two separate OR gates in Figure 3.49a, while Figure 3.50 represents them by a single, shared OR gate whose output fans out to two destinations. As this example shows, the ordinary notation for WF Boolean expressions cannot distinguish between multiple copies of the same subcircuit and a single, shared copy of that subcircuit.

Thus, we mainly view Boolean expressions as defining the function or behavior of a logic circuit. We use only certain restricted types of Boolean expressions to represent circuit structure, and then only under precisely stated conditions.

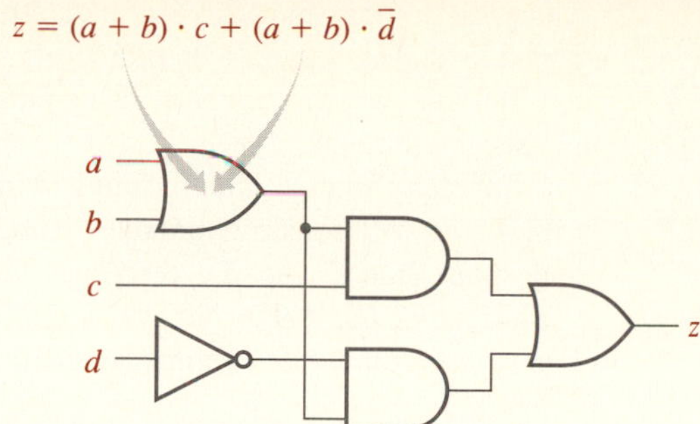

FIGURE 3.50 Expression-circuit correspondence for another realization of z.

Design Verification

Proving the equivalence of logic expressions and circuits is central to verifying that a new design is correct.

The initial design specification of a logic circuit of low to medium complexity often takes the form of a Boolean expression E or a set of such expressions. The structure of the final circuit N, however, may bear little outward resemblance to E, even though both realize the same function z. For example, E may be in a simple, easily manipulated AND-OR form, whereas N may be composed entirely of NAND gates that meet technology-specific fan-in and fan-out constraints.

The question then naturally arises: Have any errors been introduced in the design process used to derive N from E? One way to answer this question is to derive truth tables independently from N and E and compare the tables bit by bit. This can be done manually or by using a simulator program, although few if any CAD programs can handle general Boolean expressions. An alternative approach, illustrated below, is to derive directly from N a second Boolean expression E_N that represents its function. The design verification task is then to determine if $E_N = E$. This is tantamount to proving that $E_N = E$ is a theorem of Boolean algebra. Of course, if $E_N \neq E$ due to a design error, then we must show that a discrepancy exists between the two expressions and try to identify its cause.

Example 3.9 **Verifying the correctness of a logic circuit**

Figure 3.51 presents a simple circuit N whose purpose is to compare the values of two bits A_i and B_i. The output function z_i of N is supposed to be 1 when $A_i = B_i$; it is supposed to be 0 otherwise. Hence, we can define the desired behavior of N by the Boolean expression E in

$$E = A_iB_i + \overline{A}_i\overline{B}_i \qquad (3.33)$$

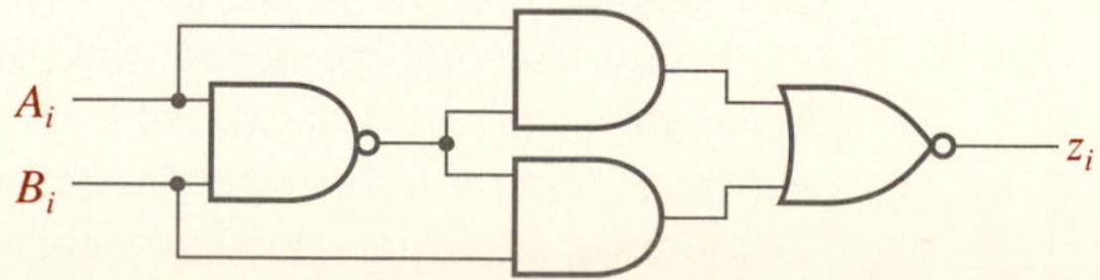

FIGURE 3.51 Logic circuit N for a 1-bit comparator.

The 1-bit comparator N is useful as a building block for circuits designed to compare the magnitudes of two n-bit binary numbers—for instance, in the 74LS85 IC, which compares 4-bit numbers and contains four copies of N [Texas Instruments, 1988].

To verify that N actually implements the function specified by Equation (3.33), we will construct another expression E_N that has the structure of N, and show that E and E_N represent the same Boolean function. From the definitions of the AND, NAND, and NOR gates (Section 3.4), the circuit of Figure 3.51 corresponds directly to the following Boolean expression:

$$E_N = \overline{A_i\overline{A_iB_i} + \overline{A_iB_i}B_i} \tag{3.34}$$

We proceed to show that Equations (3.33) and (3.34) are equivalent by applying the laws of Boolean algebra to (3.34), omitting only a few obvious steps.

$$\begin{aligned}
E_N &= \overline{A_i(\overline{A}_i + \overline{B}_i) + (\overline{A}_i + \overline{B}_i)B_i} && \text{(De Morgan 13b)}\\
&= \overline{A_i\overline{A}_i + A_i\overline{B}_i + \overline{A}_iB_i + \overline{B}_iB_i} && \text{(Distributivity 4b)}\\
&= \overline{A_i\overline{B}_i + \overline{A}_iB_i} && \text{(Inverse axiom 5b)}\\
&= (\overline{A_i\overline{B}_i})(\overline{\overline{A}_iB_i}) && \text{(De Morgan 13a)}\\
&= (\overline{A}_i + B_i)(A_i + \overline{B}_i) && \text{(De Morgan 13b)}\\
&= \overline{A}_iA_i + A_iB_i + \overline{A}_i\overline{B}_i + B_i\overline{B}_i && \text{(Distributivity 4b)}\\
&= A_iB_i + \overline{A}_i\overline{B}_i && \text{(Inverse axiom 5b)}
\end{aligned}$$

Hence, $z_i = E = E_N$, verifying that N does, indeed, perform the specified 1-bit comparison, and so has been designed correctly.

Function Algebra

Consider the set F_n of all Boolean functions of n or fewer variables. Each one can be defined by its own unique 2^n-row truth table. There are 2^{2^n} distinct ways to assign 0s and 1s to the output column of a 2^n-row truth table; therefore, F_n has 2^{2^n} members $\{z_0, z_1, z_2, \ldots, z_{2^{2^n}-1}\}$. We can define the operations $\Phi = \{and, or, not\}$ on F_n for arbitrary functions $z_1(X)$ and $z_2(X)$ as follows, employing the usual operator symbolism:

$$\begin{aligned}
z_1(X)z_2(X) &= 1 \text{ if and only if } z_1(X) = 1 \text{ and } z_2(X) = 1\\
z_1(X) + z_2(X) &= 1 \text{ if and only if } z_1(X) = 1 \text{ or } z_2(X) = 1\\
\overline{z}_1(X) &= 1 \text{ if and only if } z_1(X) = 0
\end{aligned}$$

It is a straightforward task to verify that F_n and Φ meet all the axioms of Boolean algebra; we denote this **Boolean function algebra** by $\mathcal{F}_n$. The zero element of $\mathcal{F}_n$ is the (zero) function z_Z such that $z_Z(X) = 0$ for all X, while the unit element is the (unit) function z_U such that $z_U(X) = 1$ for all X. Therefore, the functions represented by Boolean expressions and realized by logic circuits are all members of $\mathcal{F}_n$ for some sufficiently large n. This Boolean algebra is therefore the fundamental algebra of logic design.

When $n = 0$, $\mathcal{F}_n$ reduces to the two-element switching algebra $\mathcal{F}_0 = \mathcal{B}_2$, the only members of which are the "zero-variable" functions or constants $z_Z = 0$ and $z_U = 1$. The elements of $\mathcal{F}_1$ are again the constants 0 and 1, and the two possible one-variable functions $z_1(x_1) = x_1$ and $z_2(x_1) = \overline{x}_1$. Moving to $n = 2$, the number of functions jumps to 16, all of which are listed in Figure 3.52.

There are $2^{2^2} = 16$ Boolean functions of up to two variables.

x_1	x_2	z_0	z_1	z_2	z_3	z_4	z_5	z_6	z_7	z_8	z_9	z_{10}	z_{11}	z_{12}	z_{13}	z_{14}	z_{15}
0	0	0	0	0	0	0	0	0	0	1	1	1	1	1	1	1	1
0	1	0	0	0	0	1	1	1	1	0	0	0	0	1	1	1	1
1	0	0	0	1	1	0	0	1	1	0	0	1	1	0	0	1	1
1	1	0	1	0	1	0	1	0	1	0	1	0	1	0	1	0	1
		↑ 0	↑ *and*		↑ x_1		↑ x_2	↑ *xor*	↑ *or*	↑ *nor*	↑ *xnor*	↑ $\overline{x}_2$		↑ $\overline{x}_1$		↑ *nand*	↑ 1

FIGURE 3.52 A truth table for all 16 Boolean functions in $\mathcal{F}_2$.

Many of the Boolean functions in $\mathcal{F}_2$ are familiar gate functions and have the names indicated at the bottom of the figure. Observe that $\mathcal{F}_2$ contains the two constant functions and four functions of exactly one variable, for example, $z_{10}(x_1, x_2) = \overline{x}_2$. Functions of this type that are actually determined by fewer than n variables are called **degenerate** n-variable functions. Hence, $\mathcal{F}_2$ contains six degenerate functions. The remaining 10 functions in Figure 3.52 are nondegenerate two-variable functions.

Every signal in a WF logic circuit represents a Boolean function.

As we have seen, an n-input gate-level logic circuit composed of AND, OR, and NOT gates provides a direct implementation of a set of n-variable Boolean functions. In fact, every line in the circuit realizes some function from the Boolean algebra $\mathcal{F}_n$. Figure 3.53 illustrates this concept for a small five-variable circuit. Each gate combines the Boolean functions that appear on its input lines to create another Boolean function that it produces on its output line. Thus, the circuit's signals can be equated to Boolean functions that "flow" through the circuit from the primary inputs to the primary outputs and are transformed into new functions in the process.

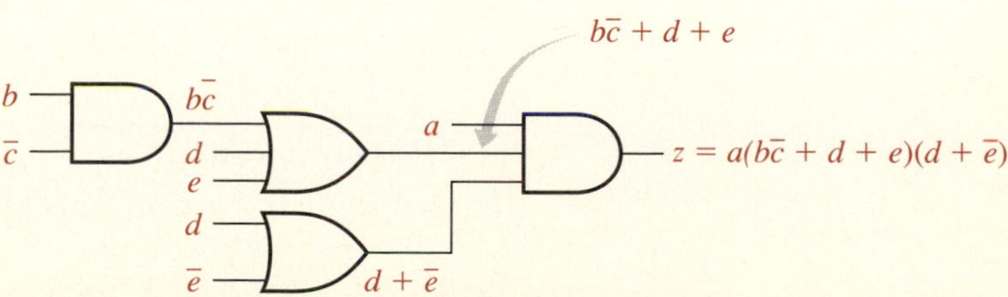

FIGURE 3.53 Examples of functions that are realized in a logic circuit.

CHAPTER 3 SUMMARY

The Switch Level

3.1 ■ **Switches** control the flow of electric currents or gas supplies, and have two basic states: open and closed. When the switches are closed so that a current can flow through them, they form a **circuit**, which consists of a power supply, a set of switches, and a load. Using switches of various kinds, we can control the more intangible entity we call information.

■ An **ideal switch** is a general model that attempts to capture the common characteristics of switches while ignoring minor details. Ideal switches use 0 and 1 signals only. Because ideal switches ignore many physical details of real switches, they are said to be **technology independent.** Technology independence greatly simplifies the analysis and design of switching circuits.

3.2 ■ Switches are building blocks that can be interconnected by lines to make **switching circuits**, also known as **networks.** Some of the simplest switching circuits are AND, OR, and NOT circuits. These circuits can be put together in different ways to form more complex circuits.

3.3 ■ A **current-switching circuit** contains various switches that act like a single composite switch that opens and closes a path for a "current" signal in response to a set of control inputs. In a **complementary switching circuit,** voltages, not currents, are switched via **pull-up** and **pull-down** subcircuits. This type of circuit is used in **complementary MOS** or **CMOS** technology.

The Gate Level

3.4 ■ **Logic gates** are small switching circuits that perform certain simple operations on binary signals. They implement useful operations in circuit design, such as AND, OR, and NOT.

3.5 ■ **Combinational circuits** are multigate circuits that perform logical operations and respond instantly to input changes.

■ A central problem in logic design is to obtain a combinational circuit that realizes a given set of functions, uses a given set of gate types, and has the lowest possible cost.

■ Physical constraints limit both the size and speed of combinational circuits in important ways. The transfer and processing of information require the expenditure of energy; a physical device can absorb or produce only a certain amount of energy without failing. This places an upper bound on the number of input sources that may supply signals to a gate (**fan-in**), and another bound on the maximum number of output devices that may be supplied with signals (**fan-out**).

Boolean Algebra

3.6 ■ **Boolean algebra** provides the theoretical basis for logic design. Boolean algebra was developed in the mid-1800s as a mathematical theory of logical reasoning. Most of the laws of Boolean algebra define pairs of equivalent logic expressions.

■ **Axioms** define an algebra's basic properties, from which all other properties can be deduced. **Huntington's postulates** provide a complete and concise set of axioms for Boolean algebra.

3.7 ■ **Theorems** are true statements that can be deduced from axioms. Theorems can often be proved informally by constructing circuits or truth tables.

3.8 ■ Boolean expressions can represent the function and, to a limited extent, the structure of logic circuits. The axioms and theorems of Boolean algebra are extremely useful for simplifying Boolean expressions and the corresponding circuits.

Further Readings

Under the name "contact networks," current-switching circuits were the main subject of early logic design, with electromechanical relays serving as the implementation technology; see, for example, the classic text by S. H. Caldwell, *Switching Circuits and Logical Design* (John Wiley and Sons, New York, 1958). Complementary switching circuits (which is our term) are nonclassical switch-level circuits that model CMOS transistor circuits. The latter are discussed briefly in the switch-level context in N. Weste and K. Eshragian's *Principles of CMOS VLSI Design* (Addison-Wesley, Reading, Mass., 1985). Gate-level circuit concepts, with electronic circuits as the target technology, are covered in most modern textbooks on switching theory and logic design; for example, K. J. Breeding's *Digital Design Fundamentals,* 2nd ed. (Prentice-Hall, Englewood Cliffs, N.J., 1992). An interesting contrast is provided by E. C. Fitch and J. B. Surjaatmadja's *Introduction to Fluid Logic* (Hemisphere Publishing Corp., Washington, D.C., 1978), which is a logic design text aimed at digital systems operated by air, water, oil, and other fluids. A broad and insightful treatment of Boolean algebra and its applications can be found in F. M. Brown's *Boolean Reasoning* (Kluwer Academic Publishers, Boston, 1990).

Chapter 3 Problems

The Switch Level

Switches; current-switching circuits; complementary switching circuits; analysis and design of logic gates and all-switch circuits.

3.1 **(a)** Identify three different types of nonelectrical switches and the signal media they control. **(b)** Discuss the advantages and disadvantages of using ideal switches to model the structure and behavior of physical switches.

3.2 Determine a truth table and a logic expression for the series-parallel current-switching circuit of Figure 3.54.

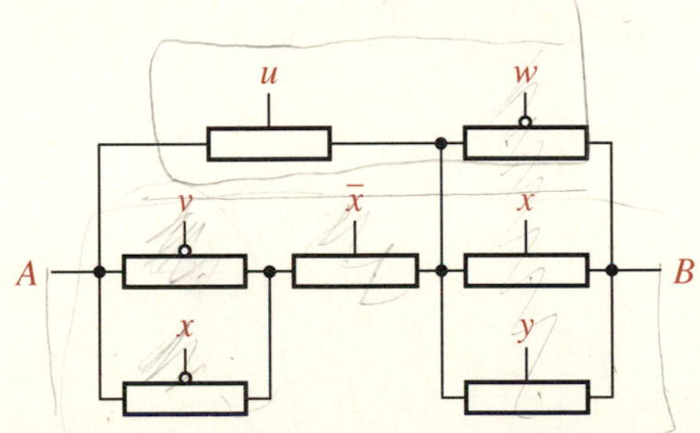

FIGURE 3.54 Example of a series-parallel current-switching circuit.

3.3 Determine a truth table and a logic expression for the circuit of Figure 3.54 if all positive switches are changed to negative, and vice versa.

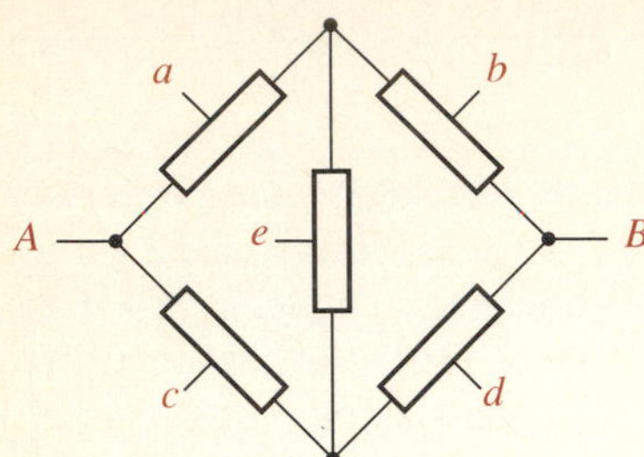

FIGURE 3.55 A bridge circuit, which is not series-parallel.

3.4 Not all current-switching circuits are of the series-parallel type. Figure 3.55 shows an example of a **bridge circuit,** which is nonseries-parallel. **(a)** Determine a truth table for this circuit. **(b)** Using as few switches as possible, construct a series-parallel circuit that is functionally equivalent to the bridge circuit.

3.5 Construct a current-switching circuit and a truth table for **(a)** the two-input NAND function $z_{\text{NAND}}(x_1, x_2)$, and **(b)** the three-input NAND function $z_{\text{NAND}}(x_1, x_2, x_3)$.

3.6 Construct a current-switching circuit and a truth table for **(a)** the three-input OR function $z_{\text{OR}}(x_1, x_2, x_3)$, and **(b)** the three-input NOR $z_{\text{NOR}}(x_1, x_2, x_3)$.

3.7 Determine a truth table for the logic function realized by the complementary switching circuit of Figure 3.56.

3.8 Suppose that all positive switches are changed to negative, and vice versa, in the circuit of Figure 3.56. Is the result a complementary switching circuit? What function does it implement?

3.9 Design current-switching circuits to implement each of the following Boolean functions: **(a)** $z_1 = (a + b)\bar{c}$; **(b)** $z_2 = (a + b)(c + d)e$; **(c)** $z_3 = (ab + c)d$; **(d)** $z_4 = (((a + b)c + d)e + f)g$.

3.10 Design complementary switching circuits that realize **(a)** the two-input AND function $z_{\text{AND}}(x_1, x_2)$, and **(b)** the three-input AND function $z_{\text{AND}}(x_1, x_2, x_3)$.

3.11 Construct a complementary switching circuit and a truth table for **(a)** the three-input OR function $z_{\text{OR}}(x_1, x_2, x_3)$, and **(b)** the three-input NOR $z_{\text{NOR}}(x_1, x_2, x_3)$.

3.12 Figure 3.57 shows a complementary circuit for which the pull-down part $\overline{C}$ is specified, but the pull-up part C is not. Design a suitable circuit for C and determine the function z realized by the entire circuit.

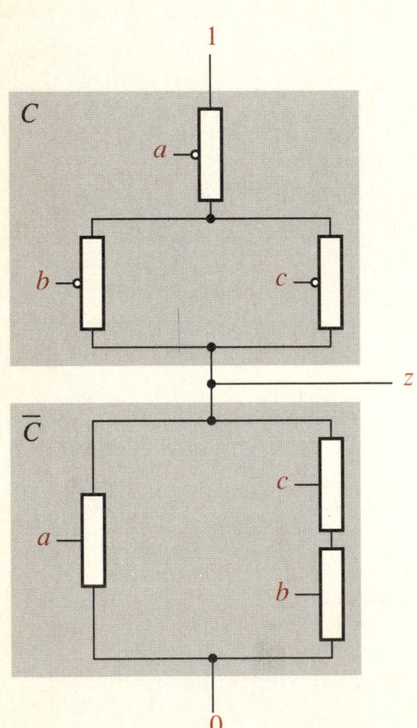

FIGURE 3.56 Example of a complementary switching circuit.

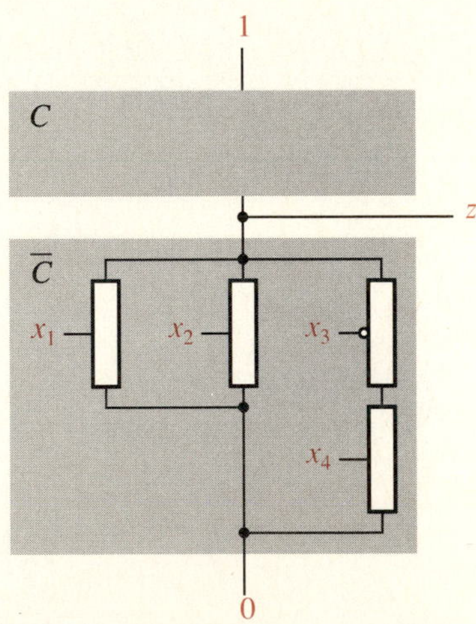

FIGURE 3.57 A partially specified complementary switching circuit.

Inputs x_1	x_2	Output z
0	0	0
0	1	0
1	0	1
1	1	0

FIGURE 3.58 Example of a two-input truth table.

Inputs x_1	x_2	Output z
L	L	H
L	H	L
H	L	L
H	H	L

FIGURE 3.59 Example of a two-input L/ H table.

3.13 Construct a complementary switching circuit to implement the truth table in Figure 3.58.

3.14 Design a complementary switching circuit that implements the three-variable function $z(x_1, x_2, x_3)$ with the following specifications: $z = 0$ if and only if at least two of the input variables x_1, x_2, x_3 assume the 1 value; $z = 1$ otherwise.

3.15 Construct complementary switching circuits that realize the following Boolean expressions: **(a)** $x_1 + x_1x_2x_3$; **(b)** $(x_1 + x_2)(x_3 + x_4 + x_5)$; **(c)** $(x_1 + x_2 + x_3)(\overline{x}_1 + \overline{x}_2 + \overline{x}_3)$; **(d)** $\overline{x_1x_2 + \overline{x}_1\overline{x}_2}$.

3.16 Design complementary switching circuits that implement the four functions defined in Problem 3.9.

3.17 Define the positive and negative logic conventions. Explain why the logic values 0 and 1 are often assigned to two continuous ranges of physical values rather than to two fixed physical values.

3.18 The table of Figure 3.59 defines the behavior of a particular physical device D in terms of H (high) and L (low) values. What is D's logical behavior if **(a)** the positive logic convention is used, and **(b)** negative logic is used?

3.19 Suppose that an electronic circuit C realizes the logic function $f(x_1, x_2, \ldots, x_n)$ when the positive logic convention is used. Devise a general rule for determining the function $g(x_1, x_2, \ldots, x_n)$ realized by the same circuit C when the negative logic convention is used. Under what circumstances, if any, is it possible for f and g to be exactly the same logic functions?

3.20 Determine L/H tables and truth tables (assuming the positive logic convention) for the two gates with direct polarity indicators that appear in Figure 3.60.

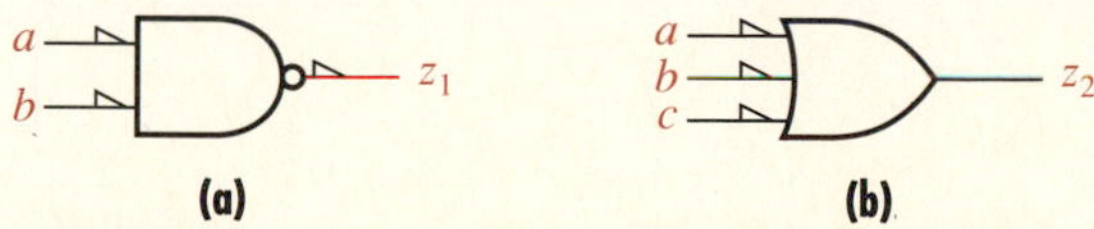

FIGURE 3.60 Gate examples with direct polarity indication.

Not all useful switching circuits are of the current-switching or complementary types.

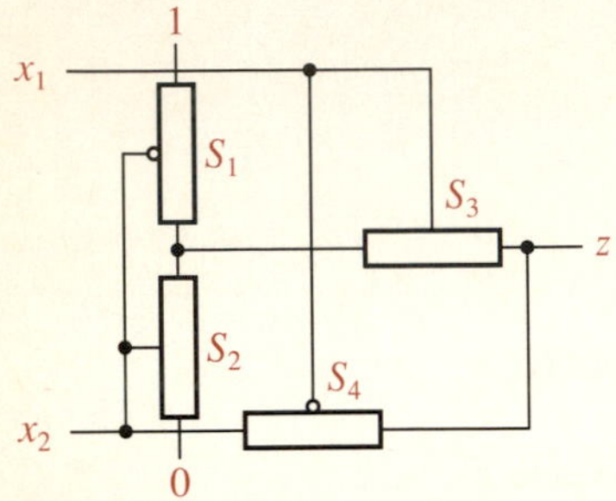

FIGURE 3.61 Another implementation of an EXCLUSIVE-OR gate.

3.21 Figure 3.61 shows a circuit that is almost, but not quite, a complementary switching circuit. Prove that it realizes the EXCLUSIVE-OR function $z = x_1 \oplus x_2$ and explain why it is not complementary.

3.22 In the EXCLUSIVE-OR circuit of Figure 3.61, the signal supplied to the output z by the switch S_3 comes directly from the power supply PS, while that supplied by S_4 comes from the input line x_2. Because the latter is not restored (amplified), this circuit is referred to as a **nonrestoring gate.** Describe how to modify the circuit to make it into a **restoring gate**, in which z is always driven by PS.

3.23 Without increasing the number of switches used, redesign the circuit of Figure 3.61 so that the output is inverted—that is, z becomes $\bar{z}$. The resulting circuit implements an EXCLUSIVE-NOR gate.

The Gate Level

Gate types and symbols; positive and negative logic; combinational circuits; truth tables; fan-in and fan-out constraints; propagation delay.

3.24 Give the box symbol (IEC 617 standard), distinctive-shape symbol, and truth table for each of the following: **(a)** a three-input OR gate; **(b)** a three-input NAND gate; **(c)** a four-input EXCLUSIVE-OR gate.

3.25 Construct a single-gate implementation of the six-input function $z(x_1, x_2, x_3, x_4, x_5, x_6)$ defined as follows: $z = 1$ if and only if $x_1 = x_2 = x_3 = x_4 = 0$ and $x_5 = x_6 = 1$. Your design should contain as few inversion symbols as possible.

3.26 Consider a single n-input EXCLUSIVE-OR gate G with output z. Prove that the output function z' that results from the inversion of some input x_i of G is exactly the same as that obtained by inverting some other input $x_j, j \neq i$. What happens if both x_i and x_j are simultaneously inverted?

3.27 Figure 3.62 shows the box symbol—it has no distinctive-shape symbol—for an uncommon n-input component called the **IDENTITY gate**. It produces the output 1 if and only if all its input signals are identical, either all 1 or all 0. Show that the IDENTITY element can be realized by a single common gate if $n = 2$ and by a three-gate circuit if $n \geq 3$.

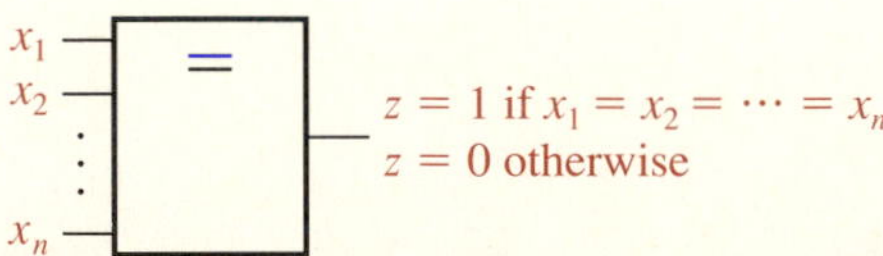

FIGURE 3.62 Graphic symbol for the IDENTITY gate.

3.28 A four-input function $z(x_1, x_2, x_3, x_4)$ has the value 1 if and only if either of the following two conditions holds: **(a)** all the inputs have the same value; **(b)** half the inputs are 0s and half are 1s. Determine whether or not z has a single-gate realization using a standard gate type (with input/output inversions as needed).

3.29 Define each of the following terms: fan-in, fan-out, nominal gate delay, circuit depth.

3.30 Using two-input AND and NAND gates as the only building blocks, design multigate circuits equivalent to each of the following gates: **(a)** a four-input AND gate; **(b)** a 20-input AND gate; and **(c)** a four-input NAND gate.

3.31 With two-input OR and NOR gates as the sole components, design circuits equivalent to each of the following: **(a)** a five-input OR gate; **(b)** a 15-input OR gate; and **(c)** a four-input NAND gate.

3.32 Obtain a truth table for the four-variable function $f(a, b, c, d)$ realized by the circuit of Figure 3.63.

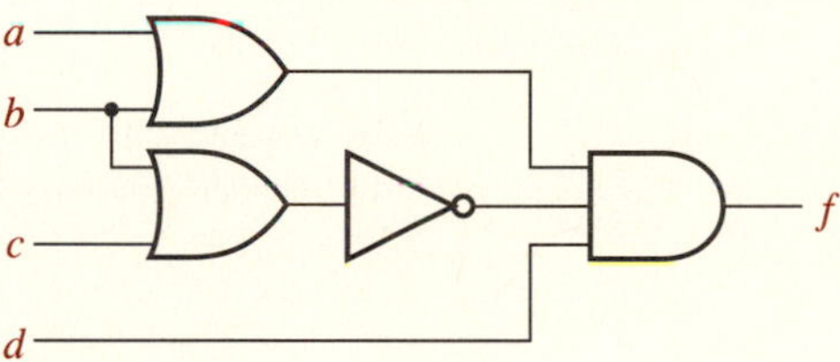

FIGURE 3.63 A single-output logic circuit.

3.33 Construct truth tables for each of the following logic functions: **(a)** $z_1 = ab + \bar{c}$; **(b)** $z_2 = \overline{(a\bar{b} + ac)}d$; **(c)** $z_3 = ab + ac + ad$; **(d)** $z_4 = ab \oplus cd$.

3.34 Write down a Boolean expression for the function f produced by the circuit of Figure 3.63.

3.35 Obtain Boolean expressions—no simplification is necessary—for the two functions z_1 and z_2 realized by the circuit of Figure 3.64.

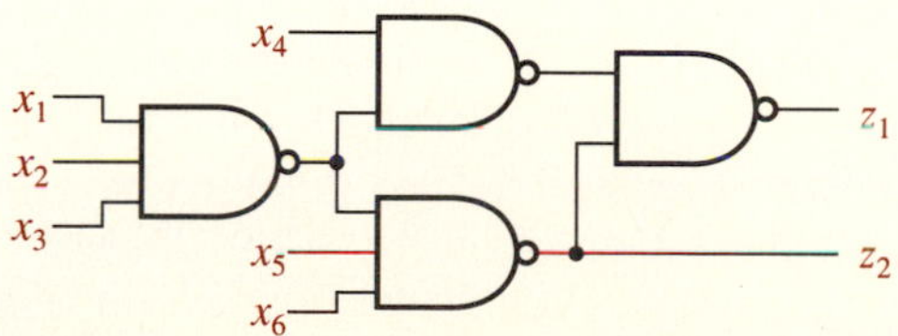

FIGURE 3.64 A two-output logic circuit.

3.36 Construct Boolean expressions and truth tables for each of the three functions f_1, f_2, and f_3 specified in Figure 3.65.

3.37 Redraw the circuit of Figure 3.65 with each gate G replaced by the gate $\overline{G}$ that realizes the complement of G. In other words, change the AND gate to a NAND gate, the inverters to noninverting buffers, and so on. Construct a truth table defining the three functions implemented by the modified circuit.

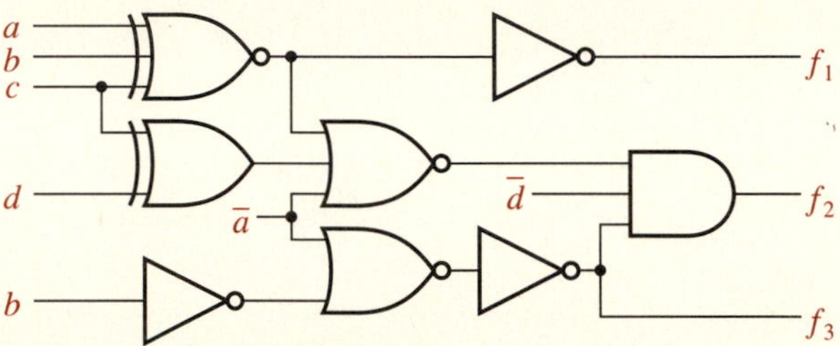

FIGURE 3.65 A three-output logic circuit.

3.38 An alternative implementation of the full adder circuit of Figure 3.29 appears in Figure 3.66. Identify the gate types that must be used for G_1 and G_2.

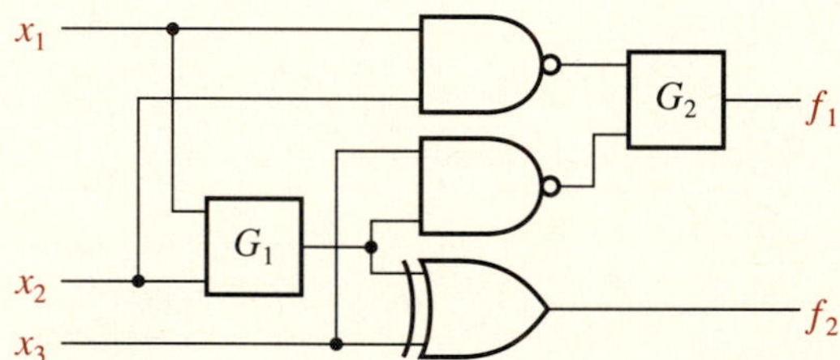

FIGURE 3.66 An alternative realization of a full adder.

3.39 The all-NAND full-adder circuit of Figure 3.30a is to be redesigned using NAND gates whose fan-in is either two or three, and whose maximum fan-out is 10. Inverters of maximum fan-out two are available, and the maximum fan-out of the primary input lines is also two. Redesign the adder so that all these fan-in/fan-out constraints are met.

3.40 Consider the adder redesign task described in Problem 3.39. Suppose now that the only NAND gates available have fan-in two. In addition, however, AND gates of fan-in two and three, and maximum fan-out five, are available as components. Redesign the full adder to meet these revised constraints.

3.41 Suppose you are to use any standard gate types to design a 300-input NAND gate whose fan-in is at most three and whose propagation delay is 1 ns. What is the minimum number of gates needed for this huge circuit, and what is its minimum propagation delay?

3.42 The N-input OR function is to be realized by a circuit C composed solely of n-input OR gates, where $n \leq N$. Describe how to design C so that the minimum number of gates $c(C)$ and the minimum circuit depth $d(C)$ are achieved. Give general expressions for $c(C)$ and $d(C)$ as functions of n and N.

3.43 Repeat Problem 3.40, with a new design objective: to minimize the adder's worst-case propagation delay, while meeting all the stated fan-in/fan-out constraints. Treat minimizing the number of gates as a secondary design objective.

Boolean Algebra

Boolean operators; Huntington's postulates; logic circuit interpretation; Boolean and nonBoolean algebras; theorems and proofs; expressions and functions; duality.

3.44 Define each of the following axioms for the Boolean algebra of sets using the standard set operators $\cap$, $\cup$, and $\overline{}$: commutativity, distributivity, and De Morgan's laws.

3.45 A useful operation in set algebra is the **difference operation**, denoted by $\sim$. The difference $s_1 \sim s_2$ is defined as the set of members of s_1 that are not also members of s_2. Show how the complement operator $\overline{}$ for sets can be defined in terms of $\sim$.

3.46 Construct truth tables and all-switch logic circuits that demonstrate the validity of the following Huntington's postulates: distributivity and the zero/unit axioms.

3.47 Construct truth tables and gate-level logic circuits that demonstrate the validity of the following Huntington's postulates: commutativity and the inverse axioms.

3.48 Identify three of Huntington's postulates that do not hold for ordinary (real number) algebra, illustrating your answer with specific examples.

3.49 Determine whether the *nor* operator $\downarrow$ defined by $a \downarrow b = \overline{a + b}$ is associative, commutative, or distributive over *or*.

3.50 Prove that the commutative and associative laws hold for the *exclusive-or* operator $\oplus$ defined by $a \oplus b = \overline{a}b + a\overline{b}$. Does *exclusive-or* distribute over *and* and vice versa?

3.51 We can extend any Boolean operation $\phi \in \Phi$ defined for $K_2 = \{0, 1\}$ to the set K_2^n of all n-tuples or "vectors" formed from K_2, as follows: Let $A = (a_1, a_2, \ldots, a_n)$ and $B = (b_1, b_2, \ldots, b_n)$ be any two elements of K_2^n. Then,

$$\phi(A, B) = (\phi(a_1, b_1), \phi(a_2, b_2), \ldots, \phi(a_n, b_n))$$

Thus, ϕ is applied bitwise to corresponding bits of A and B. It is not difficult to show that K_2^n and the bitwise versions of any functionally complete set of operations form a Boolean algebra. Construct tables defining the extended *and*, *or*, *not*, and *exclusive-or* operations for the four-element Boolean algebra whose element set is K_2^2.

What mainly distinguishes Boolean algebras from modular integer algebras is the idempotence property.

3.52 The integers Z along with the usual add (+) and multiply (×) operations form an algebra $\mathfrak{Z}$ known as a **ring**. The defining axioms for the ring of integers are listed below.

1. The operations + and × are closed, associative, and commutative.
2. Multiply distributes over add; that is, $a \times (b + c) = (a \times b) + (a \times c)$.
3. Z contains two distinct elements 0 and 1 such that $a + 0 = a$, and $a \times 1 = a$.
4. Each element a has an "additive inverse" $-a$ such that $a + (-a) = 0$.

If this ring also obeys the idempotence law for ×, that is, $a \times a = a$, it is called a **Boolean ring**. Interpreting the ring operations + and × as *exclusive-or* and *and*, respectively, show that the foregoing Boolean ring is also a Boolean algebra. Thus, the axioms of a Boolean ring provide an alternative to Huntington's postulates for defining a Boolean algebra.

3.53 To demonstrate the independence of any particular axiom A_i in Huntington's postulates, we need only exhibit an algebra (K_i, Φ_i) that satisfies all the axioms except A_i. For most of the axioms, we can take $K_i = \{0, 1\}$ and define + and · in an appropriate, nonstandard way for each case. **(a)** Prove that the two nonstandard algebras defined in Figure 3.67 satisfy all Huntington's postulates except the specified ones, thus proving that axioms 2a–b are independent. **(b)** Use this general technique to prove the independence of axiom 6.

Axiom	0 + 0	0 + 1	1 + 0	1 + 1	0 · 0	0 · 1	1 · 0	1 · 1
2a	0	0	0	0	0	0	0	1
2b	0	1	1	1	1	1	1	1

FIGURE 3.67 Proof of independence of Huntington's postulates 2a–b.

Here's an interesting Boolean algebra based on arithmetic operations.

3.54 A Boolean algebra $\mathcal{B}$ based on arithmetic operations is defined as follows: K is the eight-member set of integers $\{1, 2, 3, 5, 6, 10, 15, 30\}$. The element $a + b$ is the least common multiple (*lcm*) of a and b; ab is the greatest common divisor (*gcd*) of a and b. The inverse $\bar{a}$ of a is $30/a$. For example, if $a = 6$ and $b = 10$, then $a + b = 30$, $ab = 2$, and $\bar{a} = 5$. **(a)** What are the unit and zero elements of $\mathcal{B}$? **(b)** Prove that $\mathcal{B}$ is closed with respect to the *lcm* and *gcd* operations.

3.55 Assuming the usual properties of the number algebra $\mathcal{R}$ such as the commutativity of addition and multiplication, prove that $\mathcal{B}$ as defined in Problem 3.54 is a Boolean algebra by demonstrating that it satisfies all of Huntington's postulates.

AND	0	1	R	F	X
0	0	0	0	0	0
1	0	1	R	F	X
R	0	R	R	X	X
F	0	F	X	F	X
X	0	X	X	X	X

x	$\bar{x}$
0	1
1	0
R	F
F	R
X	X

FIGURE 3.68 Definition of *and* and *or* for a five-element algebra $\mathcal{A}_5$.

3.56 Logic circuit simulation programs employ extra logic values in addition to $\{0, 1\}$ to capture dynamic behavior within a purely logical framework. Several use the five-valued algebra $\mathcal{A}_5$ with elements $\{0, 1, \text{R}, \text{F}, \text{X}\}$, for which the AND and NOT operations are defined as in Figure 3.68. Here, 0 and 1 are interpreted as static logic signals in the usual sense, R is interpreted as a rising (0-to-1) signal, F is a falling (1-to-0) signal, and X is an unknown signal that could be 0,1, R, or F. **(a)** Construct a suitable truth table for the OR operation on $\mathcal{A}_5$. **(b)** Prove that $\mathcal{A}_5$ cannot be a Boolean algebra by showing how it violates one of Huntington's postulates.

3.57 Consider the five-element algebra $\mathcal{A}_5$ defined in Problem 3.56. The two-element Boolean algebra $\mathcal{B}_2$ is a subalgebra of $\mathcal{A}_5$ in that if we delete the elements R, F, and X from the AND, OR, and NOT definitions for $\mathcal{A}_5$, we are left with a definition of $\mathcal{B}_2$. Suppose we delete X alone from $\mathcal{A}_5$. Is the four-element result $\mathcal{A}_4$ a Boolean algebra?

3.58 Consider the four-element algebra $\mathcal{B}_4$ defined in Figure 3.69.[3] **(a)** Prove that U and Z are the unit (1) and zero (0) elements, respectively, of $\mathcal{B}_4$. **(b)** Prove that $\mathcal{B}_4$ is a Boolean algebra by verifying that it satisfies all Huntington's postulates.

a	b	ab	$a+b$	$\bar{a}$
L	L	L	L	H
L	H	Z	U	H
L	Z	Z	L	H
L	U	L	U	H
H	L	Z	U	L
H	H	H	H	L
H	Z	Z	H	L
H	U	H	U	L
Z	L	Z	L	U
Z	H	Z	H	U
Z	Z	Z	Z	U
Z	U	Z	U	U
U	L	L	U	Z
U	H	H	U	Z
U	Z	Z	U	Z
U	U	U	U	Z

FIGURE 3.69 Function table defining *and*, *or*, and *not* for the four-element Boolean algebra $\mathcal{B}_4$ with $K = \{\text{L}, \text{H}, \text{U}, \text{Z}\}$.

[3]This algebra is of interest in switch-level circuits in which L and H correspond to the high- and low-voltage levels, Z corresponds to an open circuit (the so called "high-impedance state"), and U corresponds to a short circuit. We will revisit these logic values in Chapter 4.

a	b	ab	$a+b$	$\bar{a}$
Z	Z	Z	Z	U
Z	U	Z	U	U
U	Z	Z	U	Z
U	U	U	U	Z

FIGURE 3.70 Truth table for the Boolean algebra $\mathcal{B}_4$ obtained from Figure 3.69.

3.59 Figure 3.69 is called a **function table** for *and, or,* and *not*, because it lists the values of these functions for all 16 combinations of a and b. We can reduce it to a four-row truth table by deleting all rows except the four in which a and b assume only the unit (U) and zero (Z) values; see Figure 3.70. It can be shown that any Boolean function is completely defined by its truth table, so the other 12 rows of Figure 3.69 are not really needed. Demonstrate this by deducing the complete function table for $\bar{a}$ from the truth table.

3.60 Demonstrate that the set algebra $\mathcal{S}_4$ with element set $\{\varnothing, \{0\}, \{1\}, \{0, 1\}\}$ is isomorphic to the Boolean algebra $\mathcal{B}_4$ of Figure 3.69 by defining a suitable mapping ϕ from $\mathcal{S}_4$ to $\mathcal{B}_4$.[4]

3.61 Construct truth tables and gate-level logic circuits that demonstrate the validity of theorem 12a–b, the associative laws of Boolean algebra.

3.62 Use Venn diagrams to demonstrate De Morgan's laws for the two- and three-variable cases.

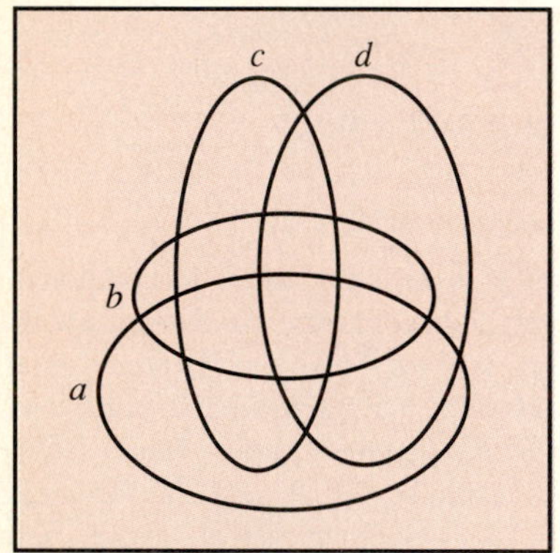

FIGURE 3.71 Venn diagram for four-variable problems.

3.63 Figure 3.71 shows a Venn diagram for four variables a, b, c, d. Use it to prove the following theorem:

$$(abd + \bar{a}b + \bar{b}d + \bar{c})(c + ab + bd) = b(a + c)(\bar{a} + \bar{c}) + d(b + c)$$

3.64 Give a formal algebraic proof of theorem 9a–b (properties of 0 and 1) using only Huntington's postulates in your proof. Cite the axioms used by name and number, as in the text.

3.65 Prove theorem 10 (absorption) using only Huntington's postulates and theorems 7–9. Cite the postulates and theorems used by name and number.

3.66 Use truth tables to show the validity of the following Boolean equations: **(a)** $a + abc = a$; **(b)** $\bar{a}b + a\bar{b} = \overline{ab + \bar{a}\bar{b}}$; **(c)** $(ab + ac)(ab + ad) = a(b + cd)$; **(d)** $\overline{\overline{ab} + \overline{bc}} = abc$.

3.67 Use the laws of Boolean algebra, that is, prove algebraically, that the four Boolean equations defined in Problem 3.66 are valid. Cite the laws you use in your proofs.

3.68 Prove algebraically that the following relations hold for every Boolean algebra. **(a)** $(ab + c + d)(\bar{c} + d)(\bar{c} + d + e) = ab\bar{c} + d$. **(b)** If $a\bar{b} + \bar{a}b = a\bar{c} + \bar{a}c$, then $b = c$.

[4]To be an *isomorphism* from algebra $\mathcal{A}$ to algebra $\mathcal{A}'$, ϕ should map every element a of $\mathcal{A}$ onto an element $\phi(\mathbf{a})$ of $\mathcal{A}'$ so that there is a one-to-one correspondence between the two element sets. Similarly, ϕ should define a one-to-one correspondence between the operators of $\mathcal{A}$ and those of $\mathcal{A}'$. Finally, if $a * b = c$ is the result of applying an operation $*$ of $\mathcal{A}$, then we should have $\phi(\mathbf{a})\phi(*)\phi(b) = \phi(c)$ in $\mathcal{A}'$.

3.69 Use algebraic techniques to determine whether or not the following Boolean equation is valid:

$$(abd + \overline{a}b + \overline{b}d + \overline{c})(c + ab + bd) \stackrel{?}{=} b(a + c)(\overline{a} + \overline{c}) + d(b + c)$$

3.70 Construct logic circuits that implement the following Boolean expressions directly: **(a)** $f(A, B, C, D) = A\overline{B} + A\overline{C} + \overline{A}D$; **(b)** $z(a, b, c, d) = \overline{ab \oplus acd}$; **(c)** $Z(p, q, r, s) = ((pq + r)s + \overline{p})\overline{r}$.

3.71 Use De Morgan's laws to convert the expression $((pq + r)s + \overline{p})\overline{r}$ into an equivalent expression that contains the *and* and *not* operators only. Then construct a circuit composed of AND and NOT gates that implements the new expression.

3.72 Determine the dual of each of the following Boolean expressions, introducing parentheses as needed: **(a)** $\overline{x_1 + x_2x_3x_4}$; **(b)** $(a + b)(c + de)f(gh + \overline{a})$.

Here we define dual function in a way that is independent of how the function is expressed.

3.73 (a) The **dual** of a Boolean function $f(X) = f(x_1, x_2, \ldots, x_n)$ can be defined as $f^d(X) = \overline{f}(\overline{x}_1, \overline{x}_2, \ldots, \overline{x}_n)$, which is obtained by complementing each input variable x_i of f and by complementing the (output) function f. Prove that the rules we gave in the text for forming the dual of WF Boolean expressions involving *and*, *or*, *not*, 0, and 1 constitute a special case of this more general definition.

3.74 **(a)** Recall that a Boolean function is self-dual if $f(X) = f^d(X)$—that is, if the function and its dual are the same. Determine which of the 16 distinct Boolean functions of up to two variables tabulated in Figure 3.52 are self-dual. **(b)** How many of the 2^{2^n} functions of n or fewer variables are self-dual?

3.75 Determine which of the following Boolean expressions are self-dual: **(a)** a; **(b)** ab; **(c)** $ab + \overline{a}\overline{b}$; **(d)** $a\overline{b} + \overline{a}\overline{c} + \overline{b}\overline{c}$.

COMBINATIONAL LOGIC

Circuit Structure

4.1 Basic Structures

4.2 Complete Gate Sets

Technology Considerations

4.3 Wired Logic

4.4 MOS Circuits

4.5 IC Implementation

Programmable Logic Devices

4.6 Programmable Circuits

4.7 ROMs and PLAs

Overview

This chapter examines the structure of gate-level combinational circuits that realize arbitrary Boolean functions. It is concerned not only with implementing the specified functions correctly but also with doing so efficiently. First, we discuss some fundamental structural properties of gate-level circuits. Then we consider aspects of the switch level that can represent a wider variety of circuit technologies than the all-switch models we have seen so far. A particularly important switch-level structure, the programmable logic device or PLD, is examined.

4

Circuit Structure

We begin by demonstrating that all Boolean functions can be realized by circuits of AND, OR, and NOT gates that have a standard or "canonical" structure. From these structures we identify the classes of gates that are functionally complete, meaning that they alone suffice to realize any possible function.

4.1 Basic Structures

A combinational switching function $z(x_1, x_2, \ldots, x_n)$, also referred to as a logic or Boolean function, maps the 2^n combinations of the n binary input variables $x_1, x_2, \ldots, x_n$ into the binary output variable z. An n-input, single-output combinational circuit N_z is said to realize or implement z if it performs the same mapping as z on its input/output signals. By "combinational" we mean that the output changes instantly in response to input changes; in other words, the circuit has no memory. Although physical gates and their interconnections always have some memory in the form of nonzero propagation delays, such delays are very small. They are factored into combinational logic design mainly to the extent that we recognize that the propagation delay, and therefore the operating speed, of a circuit is determined by its maximum depth.

Well-Formed (WF) Circuits

Well-formed circuits follow a small set of interconnection rules.

Combinational circuits are generally constructed from the standard gate types introduced in Chapter 3. These gates can be interconnected according to certain rules that define **well-formed** or **WF combinational circuits**. The rules, which are fairly obvious, are stated below and illustrated in Figure 4.1.

1. A single line (wire) or gate with constant or variable (binary) primary input signals is WF.

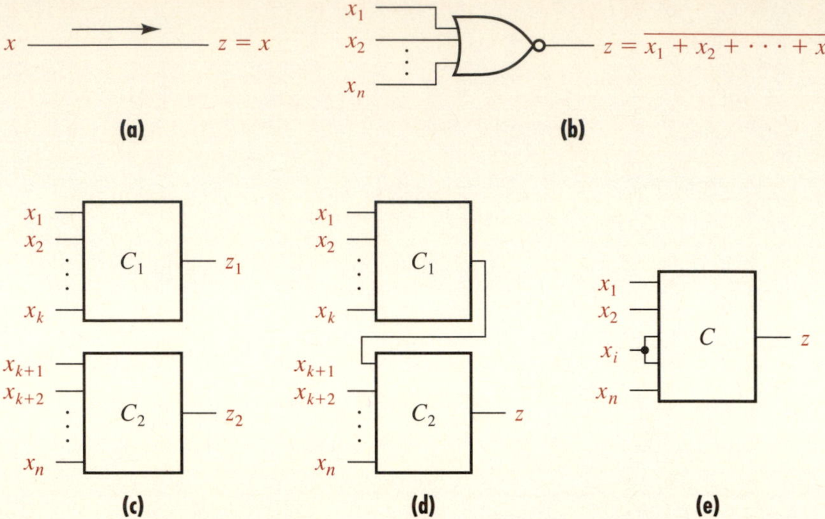

FIGURE 4.1 Well-formed structures: (**a**) a line (wire): (**b**) a single (NOR) gate; (**c**) two juxtaposed WF circuits; (**d**) an output-to-input connection; (**e**) an input-to-input connection.

2. A circuit formed by the juxtaposition of two disjoint WF circuits C_1 and C_2—that is, by placing C_1 and C_2 side by side—is WF.
3. Let C_1 and C_2 be disjoint (disconnected) WF circuits. A circuit obtained by connecting an output line of C_1 to an input line of C_2 or to a new primary output is WF.
4. If C is WF, the circuit obtained by connecting two primary inputs of C to form a single primary input is WF.

Well-formed circuits are free of inconsistencies.

Any circuit obtained by repeated application of the above rules in any order is said to be well formed. A circuit meeting the WF definition is free of logical contradictions, and is said to be a **well-behaved circuit**.

A representative WF combinational circuit appears in Figure 4.2. It has four primary inputs a, b, c, d and three primary outputs z_1, z_2, z_3. The func-

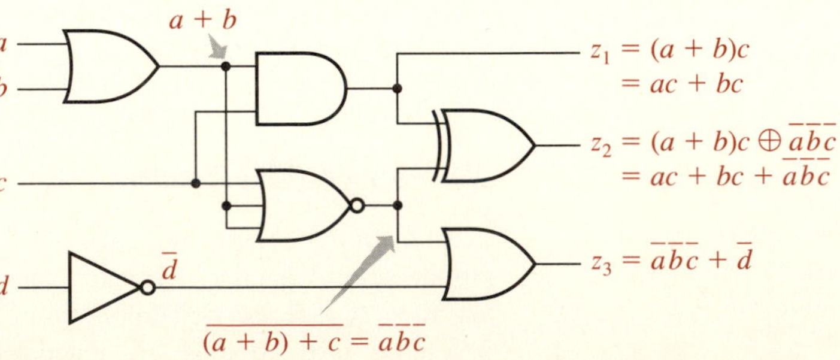

FIGURE 4.2 A well-formed combinational circuit and the functions it realizes.

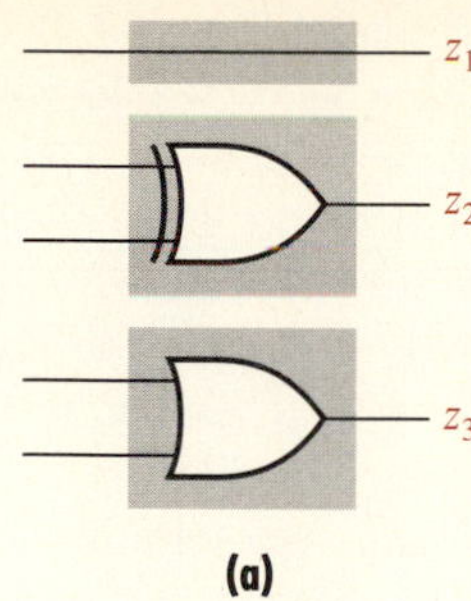

(a)

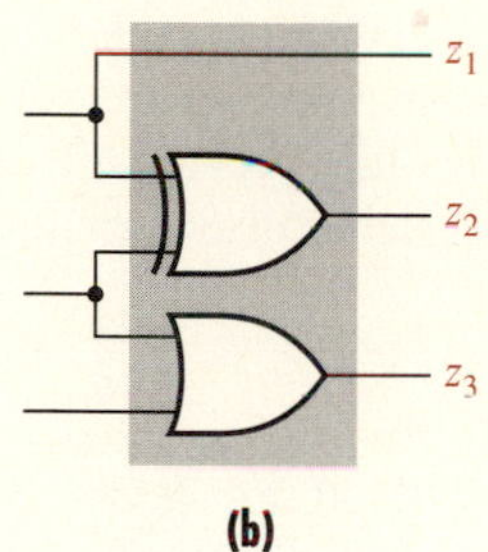

(b)

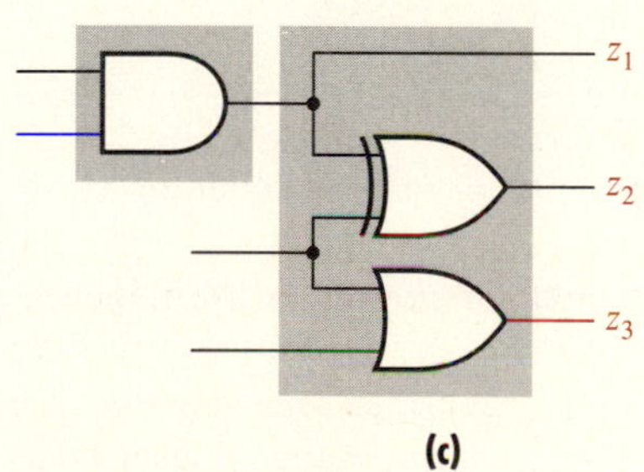

(c)

FIGURE 4.3 Some steps in constructing the well-formed circuit of Figure 4.2.

Well-formed circuits may not have two outpoints tied together; they also may not contain feedback.

tions it computes may be deduced from the definitions of the component gates, and are as indicated on Figure 4.2. The fact that the circuit is well formed according to the above definition can be seen in Figure 4.3. The EXCLUSIVE-OR gate, the OR gate, and the line feeding z_1, z_2, and z_3, respectively, form three individual WF circuits by rule 1. The circuit of Figure 4.3a is obtained by juxtaposing these three circuits according to rule 2. Applying rule 4 twice produces the WF circuit shown in Figure 4.3b. The AND gate can now be connected to this circuit according to rule 3 (Figure 4.3c). Continuing in this fashion, we can wire the entire circuit of Figure 4.2 according to the WF circuit rules.

Non-WF Circuits

We can, perhaps, best appreciate the significance of well-formedness by considering some circuits that are *not* WF. The outputs of the two gates in Figure 4.4a are connected or tied together to form a structure that is implicitly excluded from the definition of a WF circuit. This circuit is not well behaved because it is possible for the two gates to produce conflicting signals on their common output line z. In this example, the OR gate is producing 1 while the AND gate outputs 0; this implies that z is both 0 and 1 at the same time—a logical contradiction. If physical gates are connected in the manner of Figure 4.4a, then z is likely to assume an indeterminate value. Such indeterminate signals must be avoided; their presence constitutes a serious design error. Inconsistent or indeterminate conditions cannot occur if the rules for WF circuit construction are followed.

In Figure 4.4b, we illustrate a different structural problem that leads to indeterminate signals. In this case, an output of a WF circuit is connected back to one of its inputs, a configuration called **feedback**. Again, the circuit is not well behaved, because the signals y and z may be indeterminate. Suppose that $y = 0$ with $a = 0$ and $b = 1$ as shown. The AND gate requires $z = 0$. The NOR gate has 0s on both of its inputs, which implies that y should be 1. But if $y = 1$, then $z = 1$. Thus, we again have signals that are inconsistent with the logical functions realized by the circuit. One might ask: What would

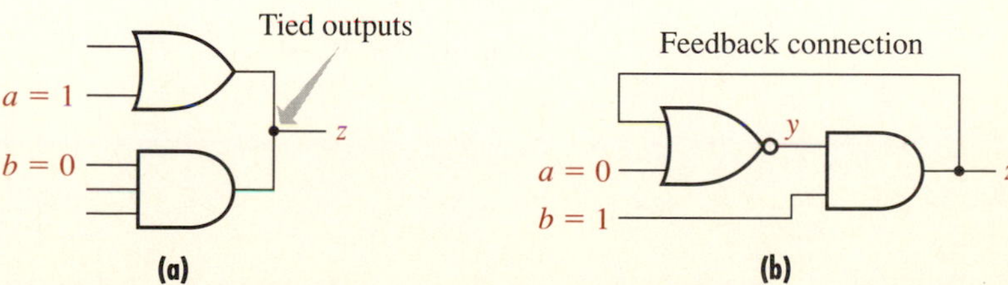

FIGURE 4.4 Two combinational circuits that are not well formed.

happen if two physical gates were configured as in Figure 4.4b? The answer is that the nonzero propagation delays of such gates would come into play to "resolve" the contradiction. In particular, if y becomes 1 at time t—that is, $y(t) = 1$—then z does not become 1 until time $t + t_{\text{pd}}$, where t_{pd} is the gate propagation delay. Thus, we can have $y(t) = 1$ and $z(t) = 0$ without contradiction. However, after z changes to 1, it forces y to change to 0 at $t + 2t_{\text{pd}}$, which then forces z back to 0 at $t + 3t_{\text{pd}}$, and so on. Thus, the y and z values change repeatedly, a process called **oscillation**.

To avoid the possibility of such oscillations, our definition of WF combinational circuits given above does not permit feedback. Consequently, WF circuits are sometimes called **feedforward circuits.** Surprisingly, consistent or well-behaved circuits exist that contain feedback (see Problem 4.3), but such circuits are usually avoided. In Chapter 6 we will revisit circuits that contain feedback, taking propagation delays into account. Because of their nonzero delays, such circuits are sequential rather than combinational.

Minterms

Next we consider some WF circuit structures in which we impose further constraints on the gate types used and the manner in which the gates are connected. Our goal is to identify structures that are **canonical** or universal in the sense that every function can be realized using these particular structures. As we will see, the canonical circuits can be represented by Boolean expressions that exhibit the same canonical forms.

A minterm is a function that can be expressed as a product of every x_i or $\overline{x}_i$.

Let $m_i(x_1, x_2, \ldots, x_n) = m_i(X)$ denote the n-variable function whose truth table has 1 in row i and 0 everywhere else, where $0 \leq i \leq 2^n - 1$. The function m_i is called the ith **minterm function,** or simply the ith minterm. It is obvious that there are 2^n such functions for any n. Figure 4.5a lists all eight minterms of three variables. It is also clear that each minterm can be realized by a single n-input AND gate with an appropriate subset of its inputs inverted; this is illustrated for the three-variable case by Figure 4.5b.

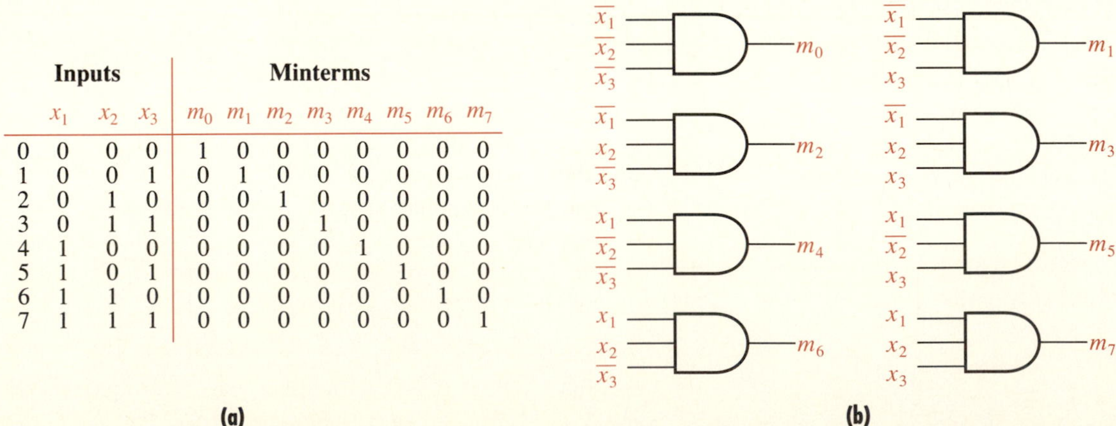

	Inputs			**Minterms**							
	x_1	x_2	x_3	m_0	m_1	m_2	m_3	m_4	m_5	m_6	m_7
0	0	0	0	1	0	0	0	0	0	0	0
1	0	0	1	0	1	0	0	0	0	0	0
2	0	1	0	0	0	1	0	0	0	0	0
3	0	1	1	0	0	0	1	0	0	0	0
4	1	0	0	0	0	0	0	1	0	0	0
5	1	0	1	0	0	0	0	0	1	0	0
6	1	1	0	0	0	0	0	0	0	1	0
7	1	1	1	0	0	0	0	0	0	0	1

FIGURE 4.5 (a) All three-input minterms; (b) their AND-gate implementations.

In general, if index i has the binary representation $(b_{i,1}b_{i,2}\cdots b_{i,n})_2$, then the AND gate that realizes m_i has input x_j inverted if and only if bit $b_{i,j} = 0$. For instance, if $i = 12_{10} = 1100_2$, then $m_{12}(x_1, x_2, x_3, x_4)$ is realized by an AND gate with inputs x_3 and x_4 inverted; hence, we can write

$$m_{12}(x_1, x_2, x_3, x_4) = x_1x_2\overline{x}_3\overline{x}_4 \tag{4.1}$$

For any $i = (b_{i,1}b_{i,2}\cdots b_{i,n})_2$, m_i can be expressed in the product form

$$m_i(X) = \dot{x}_1\dot{x}_2\ldots\dot{x}_n$$

where $\dot{x}_j = \overline{x}_j$ if $b_{i,j} = 0$, and $\dot{x}_j = x_j$ if $b_{i,j} = 1$. It is convenient to be able to refer to a variable x or its complement $\overline{x}$ by the single name **literal**. Thus, a minterm is an n-variable Boolean function that can be written as the product of n distinct literals, as in Equation (4.1), where $n = 4$.

The Sum-of-Minterms Form

A Boolean function has a minterm for every 1 in its truth table.

Now consider a general n-variable function $f(X)$ whose truth table has 1s in rows $a, b, \ldots, k$ and 0s everywhere else. It is apparent from the definitions of minterm and the OR function that $f(X)$ can be written in the following **sum-of-minterms** form

$$f(X) = m_a(X) + m_b(X) + \cdots + m_k(X) \tag{4.2}$$

where, as usual, + denotes the OR operation. If input combination X_i, corresponding to truth table row i, makes $f(X_i) = 1$, then m_i appears in Equation (4.2), which therefore becomes

$$\begin{aligned} f(X_i) &= m_a(X_i) + m_b(X_i) + \cdots + m_i(X_i) + \cdots + m_k(X_i) \\ &= 0 + 0 + \cdots + 1 + \cdots + 0 \\ &= 1 \end{aligned}$$

Conversely, if $f(X_i) = 0$, then m_i does not appear in (4.2), which therefore becomes

$$\begin{aligned} f(X_i) &= 0 + 0 + \cdots + 0 + \cdots + 0 \\ &= 0 \end{aligned}$$

The particular minterms appearing in (4.2) are referred to as the **minterms of** f, and they uniquely define f.

Inputs				Outputs		
a	b	c	d	z_1	z_2	z_3
0	0	0	0	0	1	1
0	0	0	1	0	1	1
0	0	1	0	0	0	1
0	0	1	1	0	0	0
0	1	0	0	0	0	1
0	1	0	1	0	0	0
0	1	1	0	1	1	1
0	1	1	1	1	1	0
1	0	0	0	0	0	1
1	0	0	1	0	0	0
1	0	1	0	1	1	1
1	0	1	1	1	1	0
1	1	0	0	0	0	1
1	1	0	1	0	0	0
1	1	1	0	1	1	1
1	1	1	1	1	1	0

FIGURE 4.6 The truth table for the circuit of Figure 4.2.

Figure 4.6 is a truth table for the circuit of Figure 4.2. From the number of 1s in the output columns, we see that z_1, z_2, and z_3 have six, eight, and nine minterms, respectively. The sum-of-minterms expressions for the three functions can be written directly from this table:

$$\begin{aligned} z_1 &= \overline{a}bc\overline{d} + \overline{a}bcd + a\overline{b}c\overline{d} + a\overline{b}cd + abc\overline{d} + abcd \\ z_2 &= \overline{a}\overline{b}\overline{c}\overline{d} + \overline{a}\overline{b}\overline{c}d + \overline{a}bc\overline{d} + \overline{a}bcd + a\overline{b}c\overline{d} + a\overline{b}cd + abc\overline{d} + abcd \\ z_3 &= \overline{a}\overline{b}\overline{c}\overline{d} + \overline{a}\overline{b}\overline{c}d + \overline{a}\overline{b}c\overline{d} + \overline{a}b\overline{c}\overline{d} + \overline{a}bc\overline{d} + a\overline{b}\overline{c}\overline{d} + a\overline{b}c\overline{d} + \\ &\quad ab\overline{c}\overline{d} + abc\overline{d} \end{aligned}$$

Algebraic Construction

We can also algebraically derive a sum-of-minterms expression for a function using the properties of Boolean algebra. For example, the four-variable function z_2 is defined by Figure 4.2 as

$$z_2 = ac + bc + \bar{a}\bar{b}\bar{c} \tag{4.3}$$

We can introduce any missing variable x into a product term T of Equation (4.3) by ANDing the factor $(x + \bar{x})$ with T to change it to $T(x + \bar{x})$. In the case of (4.3), we have

$$z_2 = a(b + \bar{b})c(d + \bar{d}) + (a + \bar{a})bc(d + \bar{d}) + \bar{a}\bar{b}\bar{c}(d + \bar{d}) \tag{4.4}$$

This does not alter the equation because $x + \bar{x} = 1$ and $T \cdot 1 = T$. If we "multiply out" Equation (4.4) using the distributive laws and delete any duplicated terms produced, we obtain the same sum-of-minterms expression

$$z_2 = \bar{a}\bar{b}\bar{c}\bar{d} + \bar{a}\bar{b}\bar{c}d + \bar{a}bc\bar{d} + \bar{a}bcd + a\bar{b}c\bar{d} + a\bar{b}cd + abc\bar{d} + abcd \tag{4.5}$$

as before.

An all-minterm equation such as (4.5) is a standard way of representing any logic function z, and is called the **canonical sum-of-minterms form** or **canonical disjunctive form** (*disjunction* is another term for OR). This type of Boolean expression is simply an OR-sum of all z's minterms, with each minterm expressed as an AND-product term containing a literal for every primary input variable. The order in which the product terms are written, and the order of the literals within each product term, are insignificant—a consequence of the commutative laws of Boolean algebra.

Expression-Circuit Correspondence

We can immediately translate the canonical sum-of-minterms expression for z into an equally canonical combinational circuit N_z that realizes z. Each of the k minterms of $z(x_1, x_2, \ldots, x_n)$ is implemented by an n-input AND gate as in Figure 4.5, and the outputs of all the AND gates are fed to a single k-input OR gate G, whose output is the desired function z. Because most of the inputs to the ANDs are required in both true and complemented form, it is convenient to generate each complemented variable $\bar{x}_i$ once via a single inverter and feed $\bar{x}_i$ from this inverter to any AND gates that require it.

In general, if f has k minterms, we can realize it by a sum-of-minterms circuit that consists of k AND gates, one k-input OR gate, and up to n inverters or NOT gates. (Note that we are ignoring fan-in and fan-out constraints here.) Figure 4.7 shows the canonical sum-of-minterms implementation of the four-variable function z_2 defined by Equation (4.5). Because z_2 has eight

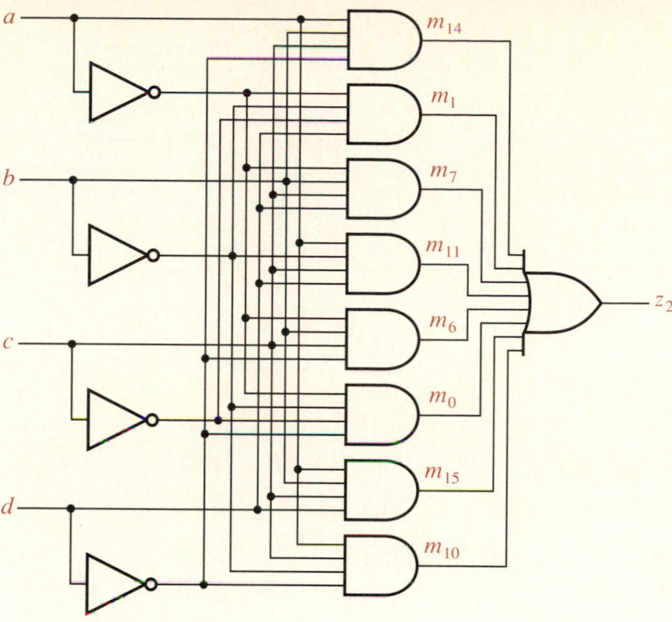

FIGURE 4.7 Canonical sum-of-minterms implementation of the four-variable function z_2.

minterms, we need eight four-input AND gates and one eight-input OR gate. Four inverters are used to generate the complements of the primary input variables.

Example 4.1

Sum-of-minterms form of $x_1 \oplus x_2 \oplus x_3$

Suppose we want to obtain a canonical sum-of-minterms realization of the three-variable EXCLUSIVE-OR function

$$z_{XOR}(x_1, x_2, x_3) = x_1 \oplus x_2 \oplus x_3$$

By definition, $f = 1$ for every input combination with an odd number of 1s. Clearly, half the eight possible input combinations—namely, 001, 010, 100, and 111—have this property. Hence, z_{XOR} has the four minterms m_1, m_2, m_4, and m_7, and can be expressed in canonical sum-of-minterms form as follows:

$$\begin{aligned} z_{XOR} &= m_1 + m_2 + m_4 + m_7 \\ &= \overline{x}_1\overline{x}_2x_3 + \overline{x}_1x_2\overline{x}_3 + x_1\overline{x}_2\overline{x}_3 + x_1x_2x_3 \end{aligned} \tag{4.6}$$

Observe that each product term in Equation (4.6) contains an odd number of noninverted variables. The canonical sum-of-minterms circuit corresponding to (4.6) appears in Figure 4.8.

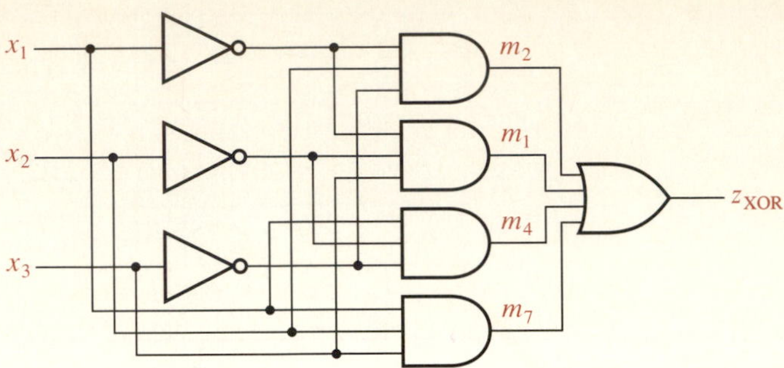

FIGURE 4.8 Canonical sum-of-minterms implementation of the three-variable EXCLUSIVE-OR function.

Maxterms

A maxterm is a function that can be expressed as a sum of every x_i or $\overline{x}_i$.

By forming the duals of the foregoing concepts, we obtain a second class of canonical expressions. The ith **maxterm function** of n variables, denoted M_i, is a Boolean function whose truth table has value 0 in the ith row of the output column and 1s everywhere else. A maxterm can be expressed in the following OR or sum form:

$$M_i(x_1, x_2, \ldots, x_n) = \dot{x}_1 + \dot{x}_2 + \ldots + \dot{x}_n$$

where literal $\dot{x}_j = x_j$ ($\overline{x}_j$) if and only if the jth bit from the left in the binary representation of the number i is 0 (1). For example, $M_3(x_1, x_2, x_3, x_4) = x_1 + x_2 + \overline{x}_3 + \overline{x}_4$ and $M_{14}(x_1, x_2, x_3, x_4) = \overline{x}_1 + \overline{x}_2 + \overline{x}_3 + x_4$.

Note that the pattern of inversions in the sum term that represents M_i is the complement of the inversion pattern in the product term that represents the minterm m_i. For example, $M_{11}(x_1, x_2, x_3, x_4) = \overline{x}_1 + x_2 + \overline{x}_3 + \overline{x}_4$, whereas $m_{11}(x_1, x_2, x_3, x_4) = x_1\overline{x}_2x_3x_4$.

The Product-of-Maxterms Form

A function is uniquely defined by a decimal-coded list of its minterms or maxterms.

Every function $z_i(X)$ has a unique set of maxterms $M_{i_1}, M_{i_2}, \ldots, M_{i_p}$ that corresponds to the set of 0s appearing in the output column of its truth table. The canonical **product-of-maxterms form** is then the AND function or logical product of these maxterms

$$z_i(X) = M_{i_1} \cdot M_{i_2} \cdot \ldots \cdot M_{i_p}$$

Figure 4.9 lists the minterms and maxterms of the three-variable function $z = \overline{a} + bc$, which has the following canonical product-of-maxterms expression:

$$\begin{aligned} z &= M_4M_5M_6 \\ &= (\overline{a} + b + c)(\overline{a} + b + \overline{c})(\overline{a} + \overline{b} + c) \end{aligned} \qquad (4.7)$$

a	b	c	z	m_0	m_1	m_2	m_3	M_4	M_5	M_6	m_7
0	0	0	1	1	0	0	0	1	1	1	0
0	0	1	1	0	1	0	0	1	1	1	0
0	1	0	1	0	0	1	0	1	1	1	0
0	1	1	1	0	0	0	1	1	1	1	0
1	0	0	0	0	0	0	0	0	1	1	0
1	0	1	0	0	0	0	0	1	0	1	0
1	1	0	0	0	0	0	0	1	1	0	0
1	1	1	1	0	0	0	0	1	1	1	1

FIGURE 4.9 Minterms and maxterms of the function $z = \bar{a} + bc$.

The corresponding canonical sum-of-minterms expression for z is

$$\begin{aligned} z &= m_0 + m_1 + m_2 + m_3 + m_7 \\ &= \bar{a}\bar{b}\bar{c} + \bar{a}\bar{b}c + \bar{a}b\bar{c} + \bar{a}bc + abc \end{aligned} \tag{4.8}$$

Logic circuits that realize the above canonical forms appear in Figure 4.10. Here we assume that the primary inputs are available in true and complemented form, so the circuits both have depth two.

Sum-of-minterms and product-of-maxterms expressions are two functionally equivalent descriptions of the same function. This equivalence can be demonstrated by deriving a truth table from each expression, or by transforming algebraically one expression into the other.

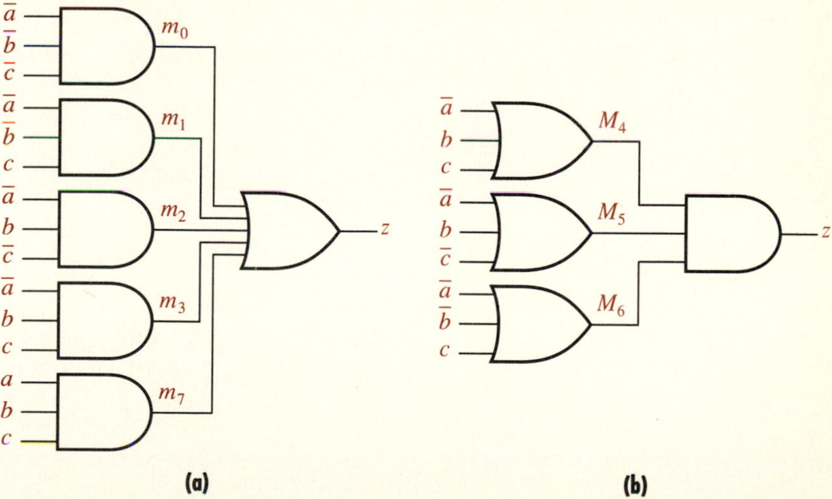

FIGURE 4.10 Circuits that implement the canonical (a) sum-of-minterms and (b) product-of-maxterms forms of $z = \bar{a} + bc$.

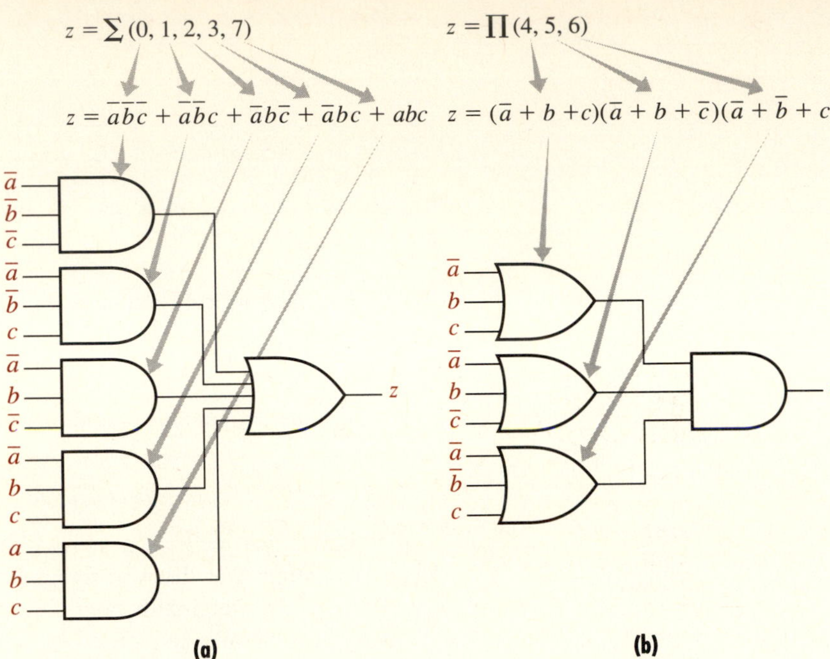

FIGURE 4.11 Expression-circuit correspondence for (**a**) the sum-of-minterms and (**b**) the product-of-maxterms forms of $z = \bar{a} + bc$.

Decimal Notation

A Boolean function can be defined by listing the indexes of all its minterms or maxterms.

The canonical expressions lead to a convenient shorthand notation for defining Boolean functions. In this **decimal notation**, only the index (subscript) i of m_i or M_i is retained, and it is written as an ordinary decimal number. A sum-of-minterms expression is then written as a list of the minterm indexes preceded by the symbol $\sum$ to indicate sum of products. Similarly, a product-of-maxterms expression is written as a list of maxterm indexes preceded by the symbol $\prod$ to indicate product of indexes. For example, Equations (4.7) and (4.8) can be rewritten as

$$z = \prod(4, 5, 6) = \sum(0, 1, 2, 3, 7)$$

Figure 4.11 summarizes our notation for canonical expressions and the corresponding circuits. Note again that in translating minterms into binary code, x_i and $\bar{x}_i$ map into 1 and 0, respectively; this relation is reversed for maxterms.

Two-Level Forms

The foregoing canonical expressions and the corresponding circuit realizations are considered to be **two-level forms**. In the circuit case, this means that the circuit depth is two, provided that, as in Figure 4.11, input inverters

Every Boolean function has a one- or two-level realization.

are omitted or not counted. The fact that every Boolean function can be expressed in the canonical forms means that every function has a two-level realization constructed from AND and OR gates.

Except for the few cases that can be realized in one level—namely, the basic gate functions—a two-level realization is the best possible design in terms of delay or operating speed. However, the canonical circuits are far from the best in terms of the number of gates they contain. If an n-variable function z has k minterms, it has $2^n - k$ maxterms, so the number of gates in a canonical implementation of z can approach 2^n, possibly a very large figure. This also entails fan-in and fan-out requirements that may exceed technology-imposed limits. Consequently, simpler two-level forms are usually sought for practical designs.

Figure 4.12 depicts alternative two-level realizations of $z = \overline{a} + bc$ that are functionally equivalent to the canonical circuits of Figure 4.11 but contain fewer gates. In fact, it can be shown that these circuits contain the fewest possible gates among all two-level realizations of the given function. A major issue in logic design, which we will study in Chapter 5, is how to obtain two-level designs of this type with the lowest possible hardware cost.

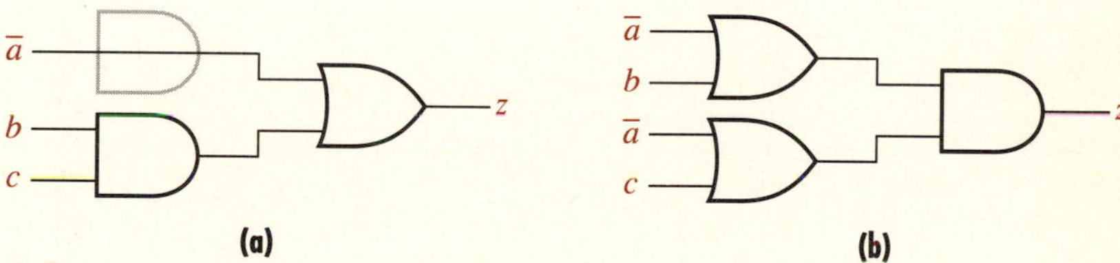

FIGURE 4.12 Minimum-cost two-level designs that are functionally equivalent to the canonical circuits of Figure 4.11.

The Sum-of-Products Form

A Boolean function has many possible two-level implementations that contain widely varying numbers of gates. Consider the following four-variable function expressed in canonical sum-of-minterms form:

$$z_3(a, b, c, d) = \overline{a}\overline{b}\overline{c}\overline{d} + \overline{a}\overline{b}\overline{c}d + \overline{a}\overline{b}c\overline{d} + \overline{a}b\overline{c}\overline{d} + \overline{a}bc\overline{d} + a\overline{b}\overline{c}\overline{d} + a\overline{b}c\overline{d} + ab\overline{c}\overline{d} + abc\overline{d} \quad (4.9)$$

or, more concisely, as $z_3 = \sum(0, 1, 2, 4, 6, 8, 10, 12, 14)$. The canonical expression (4.9) consists of nine product terms, each of which contains four literals. Figure 4.13a shows a canonical sum-of-minterms circuit that implements z_3. This circuit contains nine four-input AND gates, each of which generates one of the minterms listed in (4.9), while the nine-input OR gate generates the sum of the nine minterms.

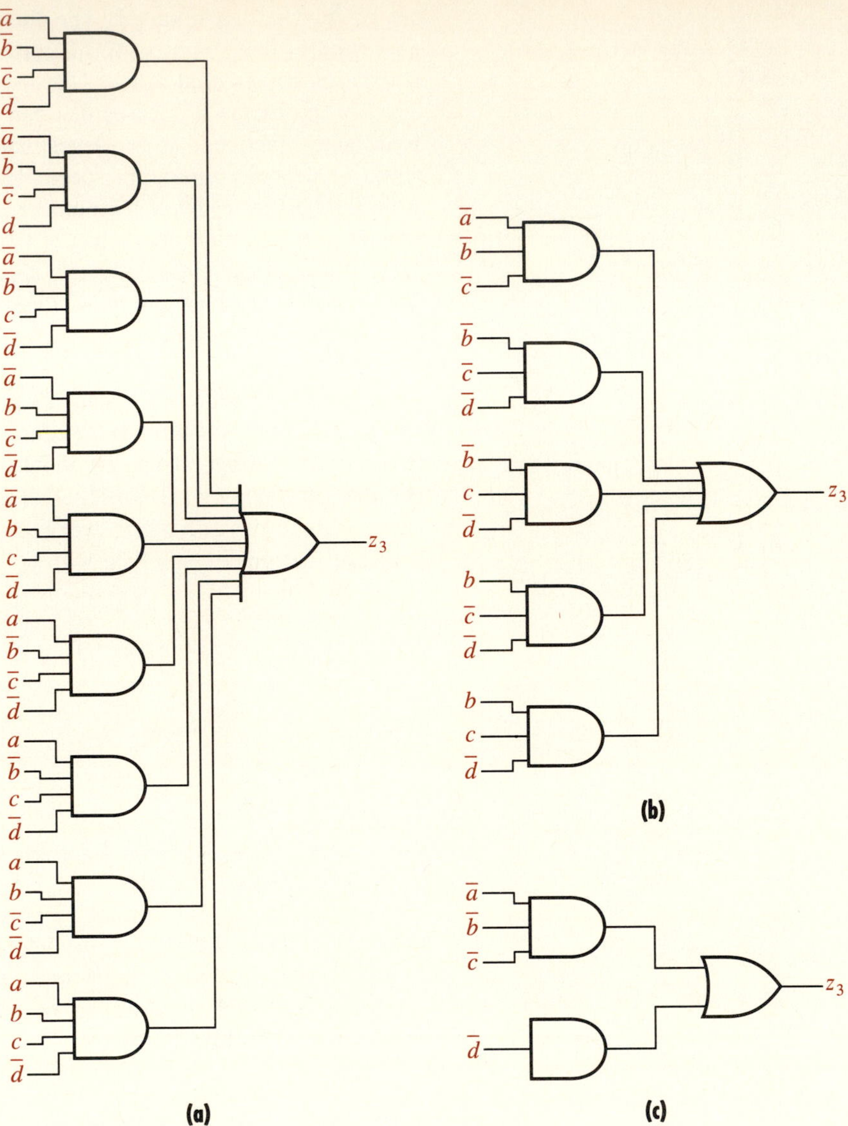

FIGURE 4.13 Three AND-OR (sum-of-products) realizations of a four-variable function z_3.

Two simpler two-level realizations of z_3 appear in Figure 4.13b and Figure 4.13c. These circuits correspond to the Boolean expressions

$$z_3(a, b, c, d) = \bar{a}\bar{b}\bar{c} + \bar{b}\bar{c}\bar{d} + \bar{b}c\bar{d} + b\bar{c}\bar{d} + bc\bar{d} \tag{4.10}$$

and

$$z_3(a, b, c, d) = \bar{a}\bar{b}\bar{c} + \bar{d} \tag{4.11}$$

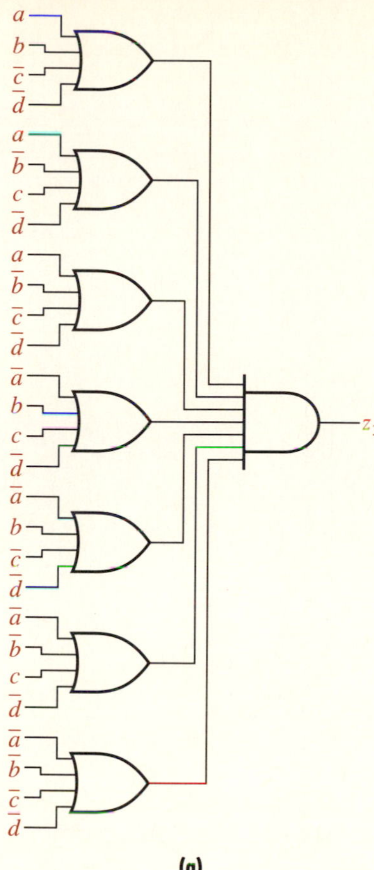

(a)

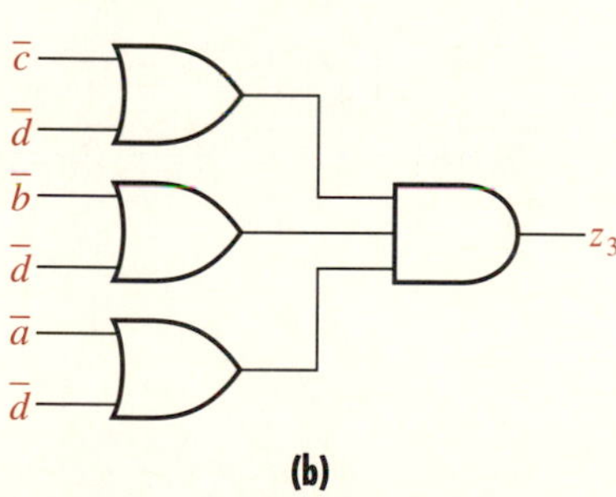

(b)

FIGURE 4.14 Two OR-AND (product-of-sums) realizations of z_3.

respectively. We refer to all the expressions for z_3 in Equations (4.9), (4.10), and (4.11) as **sum-of-products (SOP) expressions** for the function in question, and the corresponding circuits in Figure 4.13 as SOP or AND-OR circuits.

The Product-of-Sums Form

Figure 4.14a shows a two-level OR-AND circuit for the same function z_3, which implements the canonical product-of-maxterms expression

$$\begin{aligned} z_3(a, b, c, d) = &(a + b + \bar{c} + \bar{d})(a + \bar{b} + c + \bar{d})(a + \bar{b} + \bar{c} + \bar{d}) \\ &(\bar{a} + b + c + \bar{d})(\bar{a} + b + \bar{c} + \bar{d})(\bar{a} + \bar{b} + c + \bar{d}) \\ &(\bar{a} + \bar{b} + \bar{c} + \bar{d}) \end{aligned} \tag{4.12}$$

that is, $z_3 = \prod(3, 5, 7, 9, 11, 13, 15)$. Figure 4.14b implements the simpler but functionally equivalent expression

$$z_3(a, b, c, d) = (\bar{c} + \bar{d})(\bar{b} + \bar{d})(\bar{a} + \bar{d}) \tag{4.13}$$

We refer to the expressions for z_3 in Equations (4.12) and (4.13) as **product-of-sums (POS) expressions**, and the corresponding circuits in Figure 4.14 as POS or OR-AND circuits.

Summary

In general, an SOP circuit corresponds to an SOP expression for an n-variable function z of the following general form:

$$z = \sum_{i=1}^{s} \dot{x}_{i1}\dot{x}_{i2} \ldots \dot{x}_{in_i}$$

where $\dot{x}_{ij}$ denotes either of the literals x_{ij} or $\overline{x}_{ij}$, and $\dot{x}_{i1}\dot{x}_{i2} \ldots \dot{x}_{in_i}$ denotes a product (AND) term of up to n literals. A POS implementation of z corresponds to a POS expression of the form

$$z = \prod_{i=1}^{p}(\dot{x}_{i1} + \dot{x}_{i2} + \ldots + \dot{x}_{im_i})$$

where $\dot{x}_{i1} + \dot{x}_{i2} + \ldots + \dot{x}_{im_i}$ is a sum (OR) term of up to n literals. The canonical forms are special cases in which the component terms all have exactly n literals.

Two-level designs are widely used in practice, either by themselves or to implement major components of other circuits. Figure 4.15 shows the logic circuit of the 74X150, a commercial integrated circuit that is basically a minimal two-level AND-OR circuit, with buffers and inverters added to some of its input lines. Note that the output OR gate has large fan-in (16). Some input lines such as STROBE and the "select" lines A,B,C,D have large fan-out, which accounts for the presence of the input buffers. Large

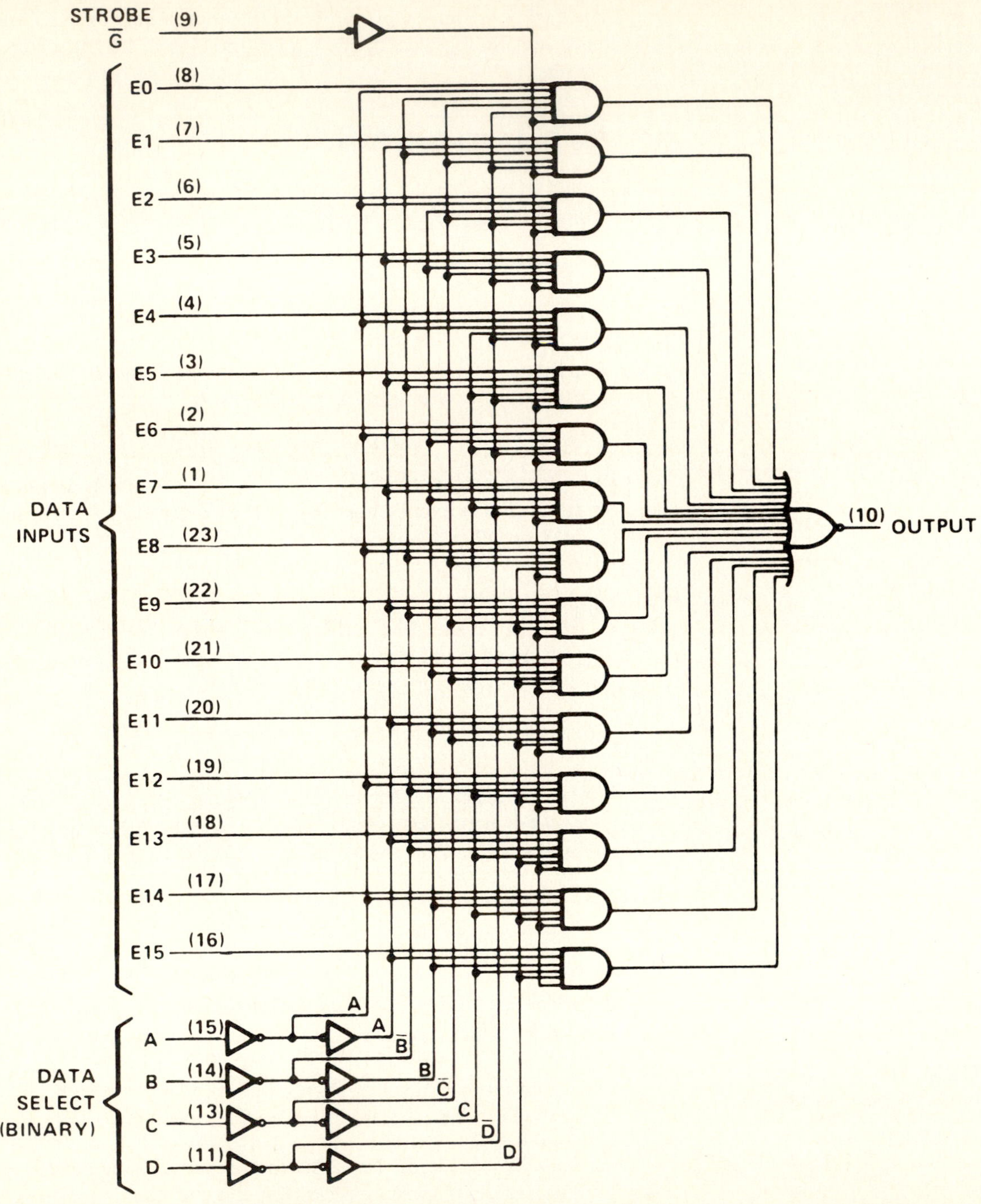

FIGURE 4.15 An example of a two-level AND-OR logic circuit, the 74X150 multiplexer IC. [Courtesy of Texas Instruments]

fan-in and fan-out requirements of this kind are typical of two-level logic. The 74X150 realizes an important type of circuit called a *multiplexer,* which we will study in Chapter 5.

4.2 Complete Gate Sets

We have seen that any Boolean function can be implemented by various types of two-level AND-OR or OR-AND circuits, in the manner illustrated by Figures 4.12, 4.13, and 4.14. In practice, however, ANDs and ORs may not be the most efficient gate types to use. For example, most transistor technologies favor NAND or NOR gates because these gates can be manufactured using the fewest transistor switches. In such cases, AND and OR gates are normally implemented by adding output inverters to NAND and NOR gates, respectively, so that the former are more expensive, slower, and no longer truly "primitive" gates. EXCLUSIVE-OR and EXCLUSIVE-NOR tend to require even more complex implementations at the switch level, as we have seen in Chapter 3. Thus, while AND, OR, and NOT are natural Boolean operations for analysis purposes, NAND and NOR are more often used in actual design.

We therefore ask the question: What gate types do we need to realize arbitrary logic functions? Because it is also useful to be able to convert circuits and subcircuits from one gate type to another, we will also consider the relationships between the various gate types.

AND, OR, and NOT Gates

A gate set is functionally complete if it alone can be used to realize any Boolean function.

The fact that every Boolean function f has a canonical sum-of-minterms form means that the set of gate types {AND, OR, NOT} is **functionally** or **logically complete** in the sense that if we have enough gates of these types, we can design a combinational circuit to realize f. No restrictions need be placed on the numbers of input lines to the AND and OR gates; obviously, each should have at least two inputs. If we have, say, AND gates with just two inputs per gate, then $n - 1$ of these gates can be connected in various ways to produce the n-input AND function for any $n \geq 2$, as discussed in Section 3.5. The resulting circuit is functionally equivalent to a single n-input AND gate, although the multigate circuit has undesirable properties such as higher cost. An N-input AND gate with $N > n$ input lines can be made to realize the n-input AND function by permanently setting $N - n$ of its input lines to the constant 1 value. The OR function can also be realized with an arbitrary number of inputs in essentially the same way.

{AND, NOT} and {OR, NOT} are functionally complete.

Knowing that the set {AND, OR, NOT} is complete, we can prove that any other gate set S is functionally complete by showing that each of the gate types AND, OR, and NOT can be realized by equivalent circuits that are

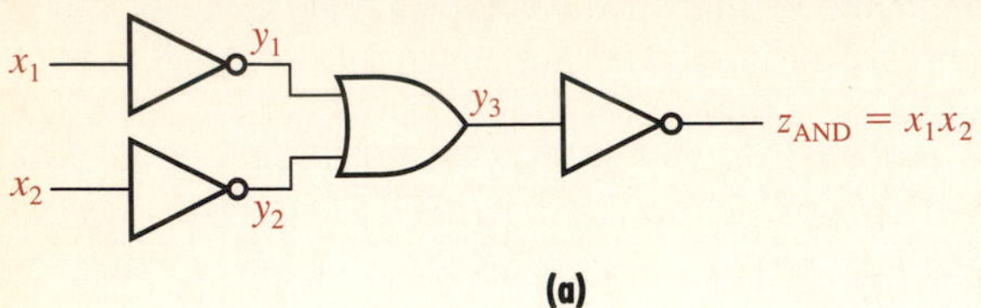

(a)

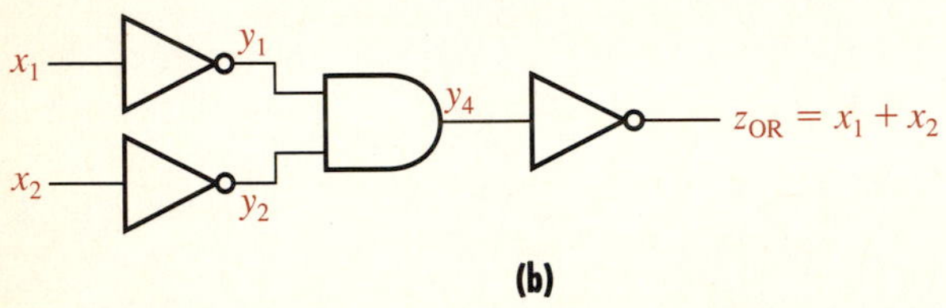

(b)

Inputs		Secondary and primary outputs					
x_1	x_2	y_1	y_2	y_3	y_4	z_{AND}	z_{OR}
0	0	1	1	1	1	0	0
0	1	1	0	1	0	0	1
1	0	0	1	1	0	0	1
1	1	0	0	0	0	1	1

(c)

FIGURE 4.16 **(a)** Implementing AND using {OR, NOT}; **(b)** implementing OR using {AND, NOT}; **(c)** a truth table for the functions generated in (a) and (b).

composed solely of gates from S. First, note that the subsets {AND, NOT} and {OR, NOT} of {AND, OR, NOT} are each complete. This follows from the fact that AND can be realized by a circuit composed of ORs and NOTs, while OR can be realized by a circuit composed of ANDs and NOTs, as illustrated in Figure 4.16. The relationships involved here are readily verified from the truth table in Figure 4.16c. They can also be expressed as

$$x_1x_2 = \overline{\overline{x}_1 + \overline{x}_2}$$
$$x_1 + x_2 = \overline{\overline{x}_1\overline{x}_2} \tag{4.14}$$

which are basically restatements of De Morgan's laws.

{AND} and {OR} are not functionally complete.

It is easily proven that the one-member sets {AND} and {OR} are not functionally complete. Consider any well-formed one-output all-AND circuit N, and select any primary input x of N (see Figure 4.17). This primary input is linked to the primary output z of N by one or more paths that contain AND gates only. Suppose that a 0 is applied to x. This causes the outputs of all gates $\{G_{1,i}\}$ connected to x to become 0; hence, the outputs of all gates $\{G_{2,i}\}$ that have inputs connected to outputs of $\{G_{1,i}\}$ become 0, and so on, until

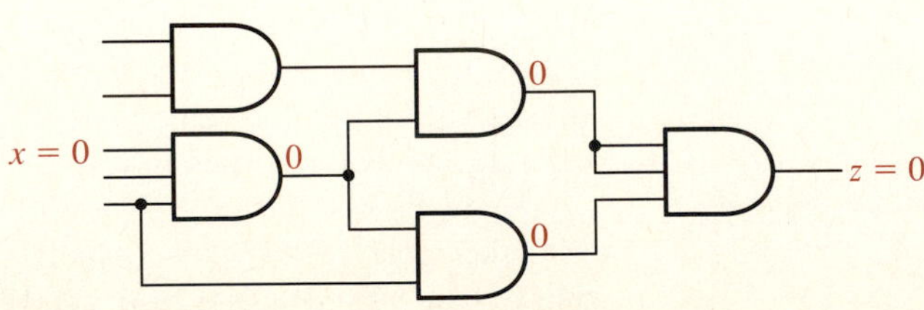

FIGURE 4.17 An example of an all-AND circuit.

eventually the primary output z of N becomes 0. In a WF circuit, there is at least one path from every primary input to the primary output; hence, z can be 1 only if all primary input signals are 1. We must therefore conclude that N can realize only the AND function. Thus, with AND gates as the sole building blocks, only AND functions can be implemented, so {AND} must be functionally incomplete. A similar argument shows that all-OR circuits can only implement OR functions; consequently, {OR} is also incomplete.

NAND and NOR Gates

{NAND} and {NOR} are functionally complete.

A surprising result with important practical consequences is that the one-member sets {NAND} and {NOR} are functionally complete. In fact, they are the *only* single gates with this property among the standard gate types. Figure 4.18 shows how all-NAND circuits can implement the AND, OR, and NOT functions, thus proving the functional completeness of {NAND}. Note that reducing the number of primary inputs of a NAND gate to one by applying 1 to the unwanted input lines produces a NOT circuit. (We assume that constant signals 0 and 1 are freely available for this purpose.) By connecting such a NOT circuit to the output of a NAND gate G, the inversion on the output of G is canceled, yielding the AND function. OR is realized by a combination of NAND and NOT circuits in the manner of Figure 4.18c.

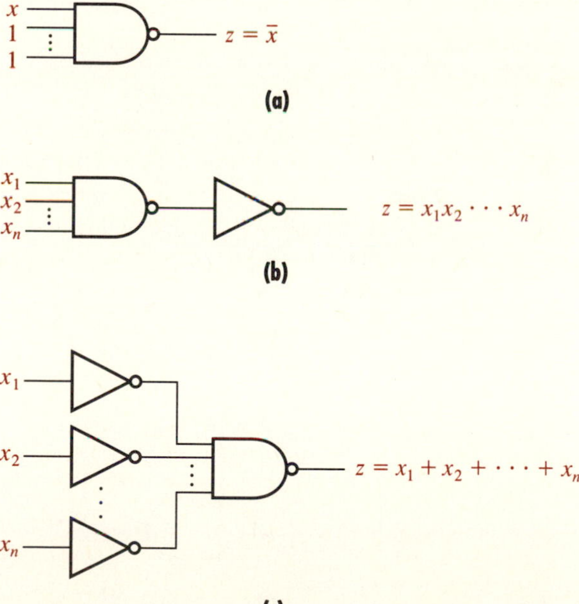

FIGURE 4.18 All-NAND implementations of (**a**) NOT, (**b**) AND, and (**c**) OR, which demonstrate that NAND is functionally complete.

An inverter is also a one-input NAND or NOR gate.

The proof that {NOR} is also functionally complete is very similar, and is left as an exercise. It is convenient to regard NOT as simply the special case of the n-input NAND and NOR functions, where $n = 1$. Thus, we include inverters or negation symbols with impunity in circuits that we consider to be all-NAND or all-NOR circuits.

Two-Level NAND or NOR Circuits

The gate correspondence given by Figure 4.18 makes it easy to convert a circuit composed of AND, OR, or NOT gates into an all-NAND circuit with the same behavior. Figure 4.19b shows the result of replacing all gates in the two-level AND-OR circuit of Figure 4.19a with the equivalent NAND circuits, according to Figure 4.18. This produces a pair of inverters in series between the first and last levels of the circuit. It follows from the involution law ($\bar{\bar{a}} = a$) that we can remove each cascaded inverter-pair, as shown in Figure 4.19c. The resulting NAND circuit has the same behavior *and the*

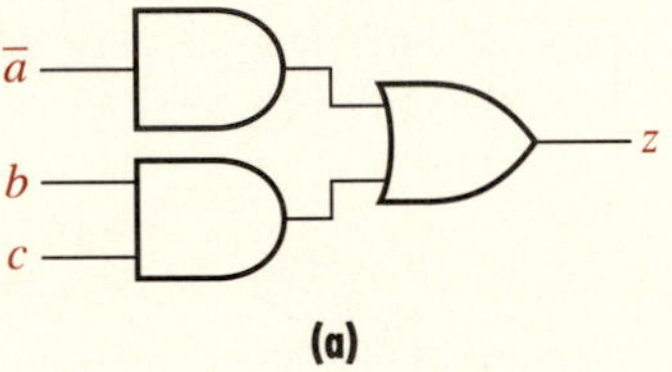

(a)

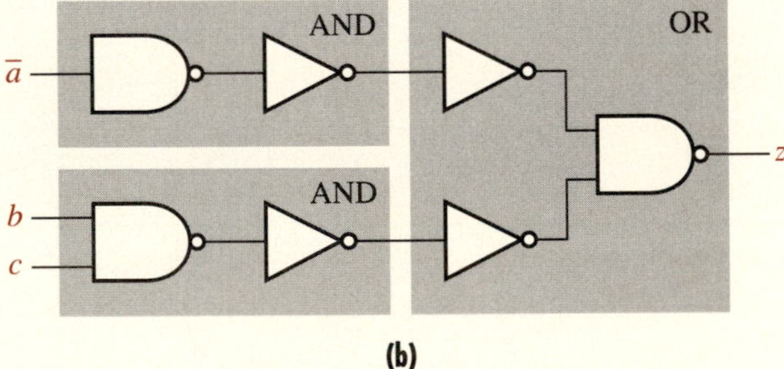

(b)

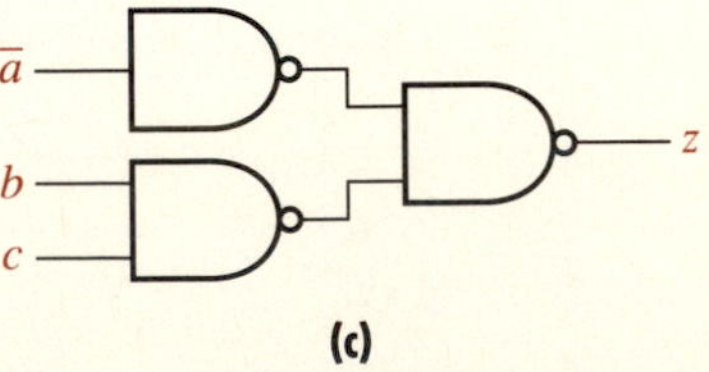

(c)

FIGURE 4.19 (a) A two-level AND-OR circuit; (b) an all-NAND equivalent circuit; (c) a NAND-NAND circuit obtained by removing the internal inverters.

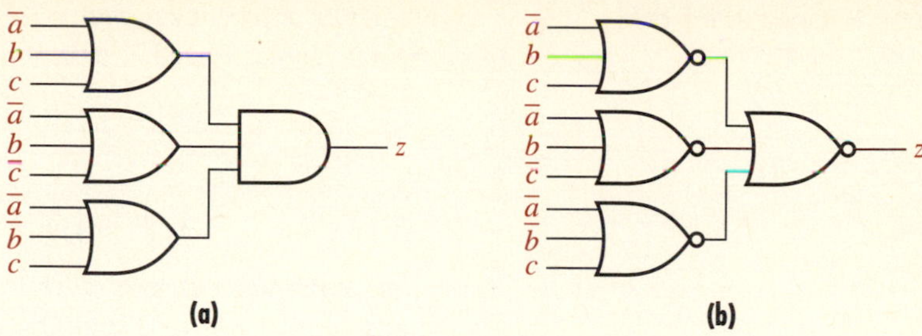

FIGURE 4.20 (**a**) A two-level OR-AND circuit; (**b**) the equivalent NOR-NOR circuit.

same structure as the original {AND, OR, NOT} circuit. In general, every gate in any two-level AND-OR circuit (AND gates followed by an output OR gate) can be replaced by a NAND gate of the same size (fan-in), without altering the output function of the overall circuit. We consider such circuits to be of the SOP type and refer to them as **NAND-NAND circuits**.

If we substitute equivalent NAND circuits for the gates in an OR-AND or POS circuit, inverters are introduced into the primary input/output lines, which do not cancel out. Consequently, the number of gates and the number of levels are increased, making this an unattractive option. However, we *can* transform an OR-AND circuit efficiently into an all-NOR equivalent. We can replace each AND and OR gate in an OR-AND circuit with a NOR gate, without changing the circuit's output. This easily verified fact is a consequence of the duality between NAND and NOR expressions. The result is a type of POS circuit called a **NOR-NOR circuit**. An example appears in Figure 4.20, which shows the canonical product-of-maxterms circuit of Figure 4.10b and its all-NOR equivalent.

SOP and POS forms have direct all-NAND and all-NOR realizations.

Thus, we can directly convert SOP and POS expressions into NAND-NAND and NOR-NOR circuits, respectively. The all-NAND or all-NOR circuits are comparable in gate cost to realizations that use AND and OR gates, but have the advantage of using a single component type.

NAND and NOR Operators

In view of the advantages of NAND and NOR circuits cited above, it might be asked, Why not introduce operator symbols for NAND and NOR and do our analysis in terms of these operators rather than in terms of $+$, $\cdot$, and $\bar{\ }$? Indeed, there are standard operator symbols available for this purpose: $|$ for NAND and $\downarrow$ for NOR. Hence, following the construction of Figure 4.18, we can replace the AND, OR, and NOT operators with $|$, as follows:

$$a \cdot b = (a \mid b) \mid (a \mid b)$$
$$a + b = (a \mid a) \mid (b \mid b)$$
$$\bar{a} = a \mid a = a \mid 1$$

The NAND and NOR operators lack useful properties such as associativity.

Despite its apparent simplicity, the | operator is very inconvenient to use. In particular, unlike + and ·, it is not associative, because

$$(a \mid b) \mid c = \overline{(\overline{ab})c} = ab + \overline{c}$$

whereas

$$a \mid (b \mid c) = \overline{a(\overline{bc})} = \overline{a} + bc$$

Hence, an expression such as $a \mid b \mid c$ is not well formed and is inherently ambiguous, unlike its AND and OR counterparts abc and $a + b + c$. An analysis of the ↓ operator for NOR leads to similar conclusions. Consequently, we will not use | or ↓, and will continue to represent NAND and NOR indirectly by means of AND, OR, and NOT.

EXCLUSIVE-OR and -NOR Gates

In conclusion, we consider briefly the completeness question for EXCLUSIVE-OR. Note that if one input of a two-input EXCLUSIVE-OR gate is set to 1, the gate becomes an inverter with respect to the other input. This can be expressed formally by

$$z = x \oplus 1 = \overline{x}$$

{EXCLUSIVE-OR, NOT} is not functionally complete.

Thus, the gate-set {NOT, EXCLUSIVE-OR} realizes the same set of functions as {EXCLUSIVE-OR}; that is, the addition of NOT gates does not increase the computational power of EXCLUSIVE-OR circuits. It can be shown that all-EXCLUSIVE-OR circuits can realize only EXCLUSIVE-OR functions in which some inputs or the output may be inverted (see Problem 4.20). We conclude from this that {EXCLUSIVE-OR} and related gate sets such as {NOT, EXCLUSIVE-OR} and {EXCLUSIVE-NOR} are not functionally complete. Although the EXCLUSIVE-OR operator is not functionally complete by itself, it becomes so when combined with either AND or OR, as the following easily verified results show:

$$a \cdot b = (a \oplus b) \oplus (a + b)$$
$$a + b = (a \oplus b) \oplus (ab)$$
$$\overline{a} = a \oplus 1$$

EXCLUSIVE-ORs and -NORs are found mainly in special types of logic circuits, especially for error detection and correction, or else they are used in conjunction with other gate types. The EXCLUSIVE-OR operator ⊕ is quite useful, and we will employ it from time to time as an adjunct to the AND, OR, and NOT operators. As we saw in Chapter 3, ⊕ is also interpreted as the odd parity operation and addition modulo 2.

Example 4.2

Deciding if a given component is functionally complete

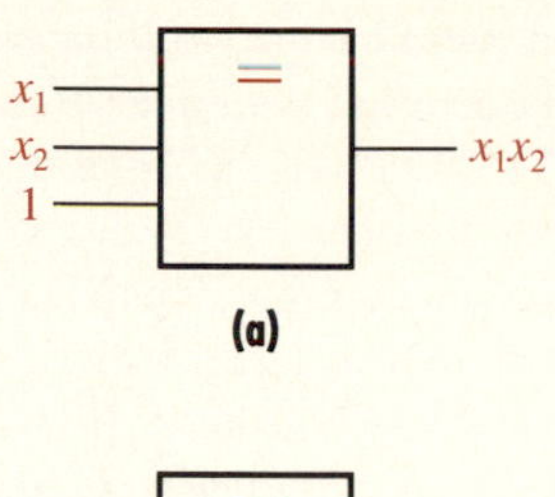

(a)

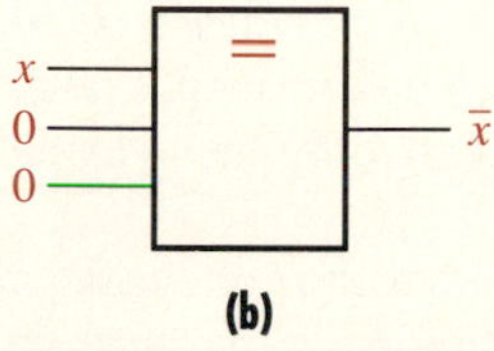

(b)

FIGURE 4.21 Use of an IDENTITY gate to realize (a) AND and (b) NOT.

Suppose we are asked to determine if a particular component type is functionally complete. A case in point is a somewhat uncommon element called an **IDENTITY gate**, which produces the output $z(x_1, x_2, \ldots, x_n) = 1$ if and only if $x_1 = x_2 = \ldots = x_n$. In other words, an IDENTITY gate indicates whether all its input signals are identical.

We can generally solve problems of this kind by showing how to implement {AND, NOT}, {NAND}, or some other functionally complete set of operations from the given element, or by showing this to be impossible. In the IDENTITY case, if the maximum fan-in k of the available gates is two, then IDENTITY = EXCLUSIVE-NOR, and from the preceding discussion, {IDENTITY} is functionally incomplete. If, however, $k \geq 3$, then an IDENTITY element can be made realize both the AND and NOT functions, as demonstrated in Figure 4.21. In the NOT case, for instance, if $x = 1$, the input to the IDENTITY gate is 100, making $z = 0$. However, changing x to 0 makes the inputs identical, so z becomes 1, and we have implemented the NOT function. We know that {AND, NOT} is functionally complete; therefore, IDENTITY is functionally complete, provided gates of fan-in three or more are available.

Technology Considerations

In the next few sections, we examine technology-related issues that we need in order to understand some key physical circuit types used to implement combinational circuits, especially two-level circuits. First, we revisit the switch level and introduce the notions of wired logic and logical strength. This discussion will enable us to deal with logic circuits that are based on MOS transistors, as well circuits that exploit the signal value Z known as the high-impedance state. Then we look at some representative IC technologies, including standard, off-the-shelf ICs.

4.3 Wired Logic

The various complementary switching circuits discussed in Section 3.3 are carefully designed so that the output terminal z of a logic element has 0 or 1, but not both, applied to it at all times. Two questions naturally arise: What happens in a switching circuit if neither 0 nor 1 is applied to z? What happens if both 0 and 1 are applied simultaneously to z? These possibilities give rise to two new "logic" values Z and U, as well a new class of logic circuits.

Electrical Interpretation of Z and U

The nature of the Z and U values can, perhaps, be most easily visualized from the simple, complementary-style electric circuit of Figure 4.22. Its power supply consists of a pair of batteries connected via two switches to a binary voltmeter that serves to output a voltage z. Normal operation with one switch open and the other closed is indicated in Figures 4.22a and 4.22b. Note that a small current i flows through the voltmeter to move it either left (1) or right (0) from its neutral, vertical position. The amount of this current depends on the circuit's impedance or resistance to current flow, to which the batteries, the switches, the meter, and the interconnecting wires all contribute. If both switches are open as in Figure 4.22c, the meter is said to be **open-circuited** or **floating**. The meter receives no energy from the power supply, and z assumes a neutral or inactive state. This state is indicated by the symbol Z and is generally called the **high-impedance state.**

Z results from an open circuit to the power source PS; U results from a short circuit to PS.

In the final configuration of Figure 4.22d, both switches are closed and a **short circuit** is said to be present. A path of negligible impedance now exists between the 0 and 1 terminals of the power supply. This path will

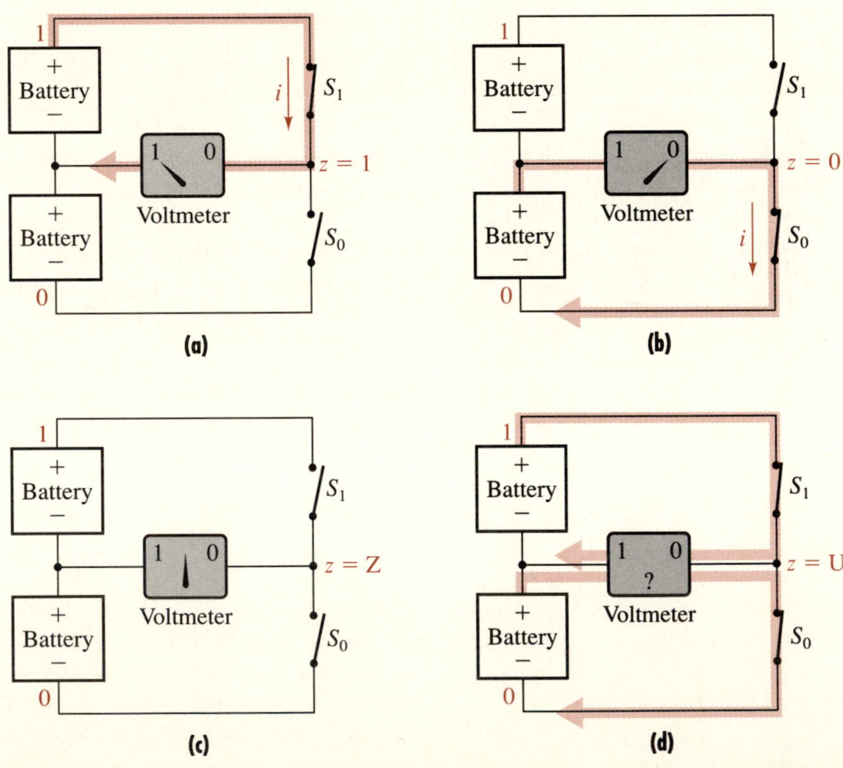

FIGURE 4.22 Electrical demonstration of the four signal states **(a)** 1, **(b)** 0, **(c)** Z, and **(d)** U.

cause a very large (short-circuit) current to flow, as suggested in the figure; this current is likely to damage the circuit, not to mention deplete the batteries. The voltmeter reading is now unpredictable; we denote it by U for **unknown** or **uncertain state.** States analogous to Z and U are found in other switching technologies. Henceforth, we deal with them all in general, technology-independent terms.

Z and U as Logic Values

In a normal logic circuit, all signals external to the circuit are required to assume values from the two-valued set {0, 1}. The high-impedance state Z may occur on internal lines that happen to be disconnected from any 0 or 1 source by the present input signal combination; this is a perfectly normal inactive condition. The unknown signal U, on the other hand, is not a normal signal and should never occur anywhere in a circuit. Thus, generation of a U signal by a logic circuit, which requires a non-WF circuit structure such as those of Figure 4.4, usually indicates a design error.

Z appears on logic lines that are disconnected from all signal sources.

There is a common situation in which Z and U occur naturally—namely, during the brief periods when a signal z is in transition between the 0 and 1 levels. When the logic values are implemented by analog physical quantities, as is usually the case, a signal changing from, say, 0 to 1 must pass through intermediate, transition values that we can denote collectively by z = U. If, on the other hand, z is first disconnected from 0 and then connected to 1, there may be a brief period during which z is connected to neither 0 nor 1, in which case we could write z = Z. The effect of such temporary Z and U states can be minimized by ensuring that all 0/1 changes take place very rapidly, and by employing certain design techniques (to be discussed later) that mask the effects of transient states that appear during signal changes. Thus, proper design effectively eliminates Z and U conditions in this case.

Z and U can represent fault signals and transient normal signals.

Permanent Z and U values can appear in a circuit as the result of certain types of physical faults. Suppose, for instance, that switch S_1 in Figure 4.23 is "stuck closed," a not uncommon fault condition in which the switch's data path is permanently closed, independent of the level of the control variable x. Then the input condition $x = 1$ will also close S_2 and connect z to both 0 and 1, a condition represented by z = U. As noted above, such a U condition, if sustained, is likely to damage the circuit. Similarly, a "stuck open" fault that affects S_1 will result in z = Z when $x = 0$.

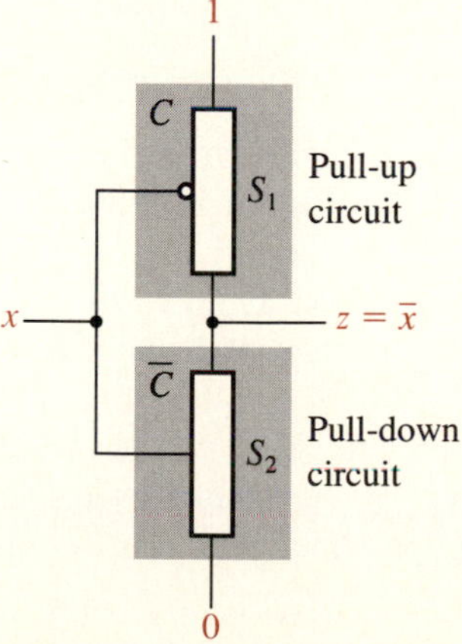

FIGURE 4.23 A complementary inverter circuit.

In some important classes of logic circuits such as "tri-state" and "open-collector" circuits, the normal output signal levels are taken from the three-valued set {0, 1, Z}. In these circuits, interconnecting wires are seen as performing basic AND or OR operations, and so they implement what is often called **wired logic**. Such wired logic functions are, in fact, performed by two-valued switching circuits as well, and provide a useful, technology-independent way of explaining their behavior. To understand this, we need to delve into some recently developed nonclassical aspects of switching theory [Hayes, 1987].

The Connection Function

We can regard a line or, more generally, a set of connected lines L as implementing a function of the form

$$z = \#(x_1, x_2, \ldots, x_n) \tag{4.15}$$

The connection function # defines the interaction of signals taken from {0, 1, U, Z}.

where $x_1, x_2, \ldots, x_n$ denotes the set of all n signals from the set {0, 1, U, Z} applied to L. The function implemented by L and denoted by the sharp symbol # in Equation (4.15) is called the **connection function**. The level or state of z is the net result of the interactions among the signals applied to L; it is the signal that L applies to any new device attached to it. If all the x_i signals had the same level 0 or 1, we would expect

$$\begin{aligned} z &= \#(0, 0, \ldots, 0) = 0 \\ z &= \#(1, 1, \ldots, 1) = 1 \end{aligned} \tag{4.16}$$

Thus, # is the identity function, and L's role is simply that of distributing the constant 0 or 1 level among the various terminals connected by L. The direction of signal flow through L is from the **source** or **driving terminals** to the **sink** or **driven terminals**.

If some x_i applied to L was changed to the high-impedance state Z, we would not expect the output signal z defined by Equation (4.16) to change. The Z signal is thus weaker than either 0 or 1 in the sense that it is always overridden by 0 and 1, as the following equations indicate:

$$\begin{aligned} z &= \#(0, 0, \ldots, 0, \text{Z}, \text{Z}, \ldots, \text{Z}) = 0 \\ z &= \#(1, 1, \ldots, 1, \text{Z}, \text{Z}, \ldots, \text{Z}) = 1 \end{aligned}$$

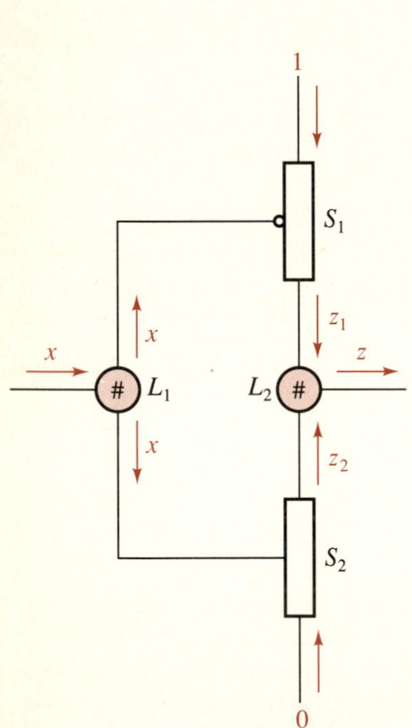

FIGURE 4.24 Connection operations in a complementary inverter circuit.

Internal Operation of an Inverter

Figure 4.24 illustrates the preceding concepts for the complementary inverter circuit of Figure 4.23. The main input line L_1 of the inverter is driven by the external (source) signal x. The control terminals of the switches S_1 and S_2 are sinks of L_1, and are driven by x in the obvious fashion. Assuming that the values of x are confined to the set {0, 1}, the state of L_1 is simply $\#(x) = x$ and is also confined to {0, 1}.

The behavior of the inverter's output line L_2 is slightly more complicated. It has two driving signals z_1 and z_2, which are derived from S_1 and S_2, respectively; the output line z is L_2's sole sink. Consequently, the behavior of L_2 can be expressed as

$$z = \#(z_1, z_2) \tag{4.17}$$

Unlike x, z_1 and z_2 can assume the level Z in addition to 0 and 1. This

TABLE 4.1 Interpretation of the four logic values {0, 1, U, Z}.

Signal value	*Name*	*Meaning*	*Connection behavior*	*Occurrence*
0	Logical 0	Normal function level	$z = \#(0, 0, \ldots, 0) = 0$	In all logic circuits
1	Logical 1	Normal function level	$z = \#(1, 1, \ldots, 1) = 1$	In all logic circuits
Z	High-impedance state	Open-circuit or floating state	$z = \#(\mathrm{Z}, \mathrm{Z}, \ldots, \mathrm{Z}) = \mathrm{Z}$ $z = \#(0, 0, \ldots, 0, \mathrm{Z}, \mathrm{Z}, \ldots, \mathrm{Z}) = 0$ $z = \#(1, 1, \ldots, 1, \mathrm{Z}, \mathrm{Z}, \ldots, \mathrm{Z}) = 1$	In internal lines, tri-state circuits, and faulty circuits
U	Unknown state	Short-circuit or conflict state	$z = \#(0, 1, \ldots) = \mathrm{U}$ $z = \#(\mathrm{U}, \ldots) = \mathrm{U}$	Forbidden state due to design errors or physical faults

happens when the corresponding switch is open. For example, when S_1 is open and S_2 is closed, a situation caused by $x = 1$ (Figure 4.24), then Equation (4.17) becomes

$$z = \#(\mathrm{Z}, 0) = 0$$

When x becomes 0, (4.17) changes to

$$z = \#(1, \mathrm{Z}) = 1$$

Thus, we see that z assumes the values 0 and 1 only in accordance with the NOT function $z = \bar{x}$.

The U value cannot appear anywhere in the inverter circuit under normal conditions. However, if switch S_1 suffers from the "stuck-closed" fault condition mentioned above, then $x = 1$ results in

$$z = \#(1, 0) = \mathrm{U}$$

Moreover, any line that is then driven by U assumes the U value itself, thus implying that U overrides all other signal values, making it the "strongest" member of our extended set of logic values.

Hence, we see that the connection function # defines a kind of logical operation on {0, 1, U, Z} that allows stronger signals to override weaker ones when applied to the same line or interconnection. There is a hierarchy of strength levels indicated by Figure 4.25, where U is the strongest, Z is the weakest, and 0 and 1 are intermediate signals of equal strength. Table 4.1 summarizes the properties of Z and U derived from the foregoing considerations.

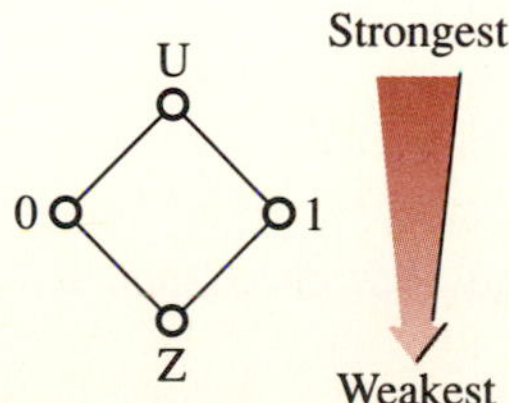

FIGURE 4.25 Strength relationship among the logic values {0, 1, U, Z}.

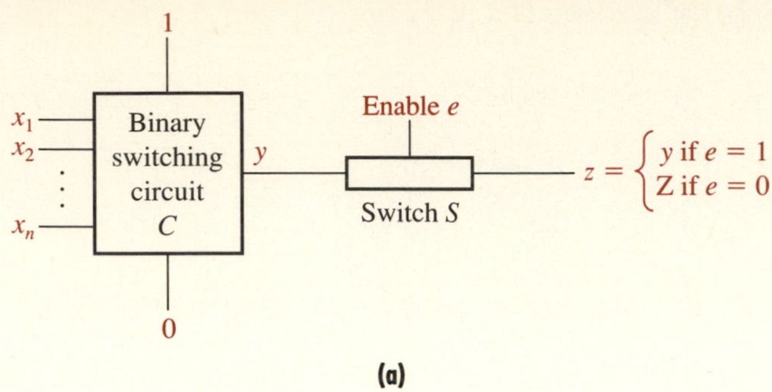

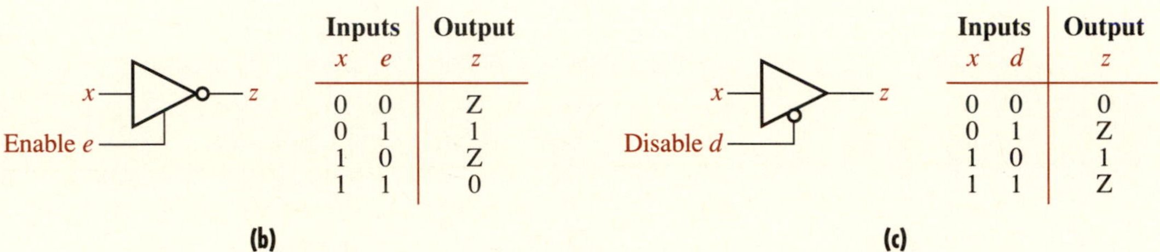

Inputs x	Inputs e	Output z
0	0	Z
0	1	1
1	0	Z
1	1	0

Inputs x	Inputs d	Output z
0	0	0
0	1	Z
1	0	1
1	1	Z

FIGURE 4.26 (a) A general tri-state circuit C; (b) a tri-state inverter with an enable line; (c) a tri-state buffer with a disable line.

Tri-State Logic

When disabled, a tri-state circuit outputs a Z.

Tri-state circuits have the high-impedance state Z as a normal output value, in addition to the usual 0 and 1 values. In principle, we can convert any logic circuit C to tri-state form simply by inserting a switch S in its output line z, as indicated in Figure 4.26a. (Physical tri-state circuits may require more complex, technology-dependent implementations.) The control input signal e of S is called an **enable signal** if $e = 1$ makes $z = y$ (as in Figure 4.26a), and $e = 0$ makes $z = \text{Z}$. When $e = 1$, the output z is said to be **enabled**, and assumes its normal 0-1 levels determined by C. When $e = 0$, z is **disabled**, and is unaffected by C. If the roles of $e = 0$ and $e = 1$ are reversed—for example, by replacing S in Figure 4.26a with a negative switch—e is referred to as a **disable control line**. Figures 4.26b and 4.26c define tri-state versions of the inverter and noninverting buffer circuits introduced earlier. Note the addition of the extra enable or disable input line in each case. The active (1) level of the disable line d in the buffer of Figure 4.26c sets the output z to the high-impedance state Z; the inactive (0) state enables z.

Tri-state logic is employed mainly to transfer data from one part of a system to another. A device that allows one and only one of n source devices $S_1, S_2, \ldots, S_n$ to be logically connected to a common destination device D at any time is called an n-way **multiplexer**. We now examine a tri-state circuit that implements a small multiplexer.

Example 4.3

A tri-state four-way multiplexer

Suppose we want to create a controlled path from four sources S_0, S_1, S_2, and S_3 of 1-bit data items to a common destination device D. The particular source to use at any time is specified by a 2-bit address $A = a_1a_0$. If $A = 00$, S_0 should be linked to D; if $A = 01$, S_1 should be linked to D, and so on. Meeting these specifications requires a four-way multiplexer.

A tri-state realization of the multiplexer appears in Figure 4.27. D is attached to a shared "bus" line L, which is driven by four tri-state buffers, $B_0 : B_3$, one from each source. To transfer the data bit x from some source—say, S_1—to D, we must enable buffer B_1 and disable the other three. To do this requires some logic to control the enable lines of the four buffers. This logic takes the form of an **address decoder** composed of four AND gates, each of which generates a different minterm defined on a_0, a_1. For example, S_1's buffer B_1 has minterm $m_1(a_1, a_0)$, which denotes address 1, applied to its enable line. Consequently, when $m_1(a_1, a_0) = 1$, specifying S_1 as the data source, B_1 alone is enabled; the data bit x, whose value is 0 or 1, produced by S_1 passes through B_1 onto the bus L. At the same time, the other three buffers drive their output lines to the high-impedance state Z. Consequently, L assumes the value $\#(Z, x, Z, Z) = x$, which it passes on to D. Note that the address-decoding logic ensures that only one buffer is enabled at a time. Enabling two buffers simultaneously would lead to the possibility of $\#(0, 1, Z, Z) = U$ being transmitted to D, an obviously unacceptable situation.

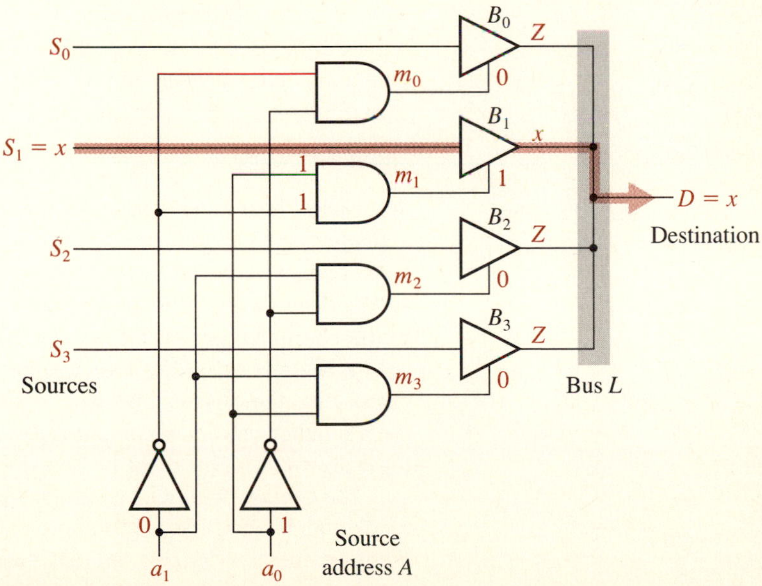

FIGURE 4.27 A four-way multiplexer designed with tri-state logic.

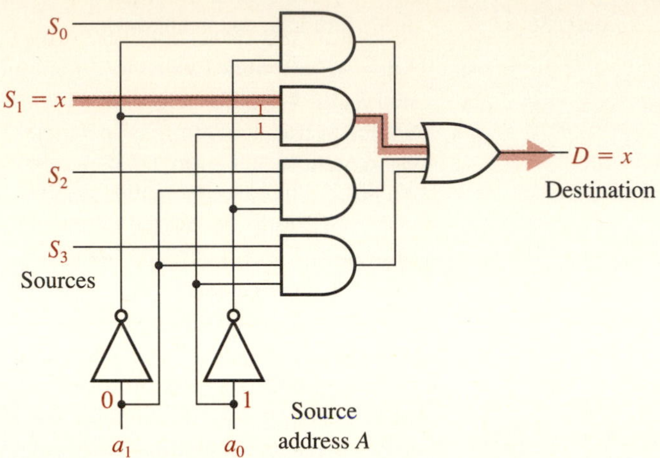

FIGURE 4.28 A four-way multiplexer designed with ordinary (two-state) logic.

In contrast, Figure 4.28 shows another four-way multiplexer, this time constructed from ordinary (all-binary) gates. This two-level logic circuit is similar to the preceding one, except that a four-input OR gate replaces the bus line L. Its behavior may be described by the SOP equation

$$D = S_0\bar{a}_1\bar{a}_0 + S_1\bar{a}_1a_0 + S_2a_1\bar{a}_0 + S_3a_1a_0$$

Again, the destination signal D assumes the value x of the source S_1 in the figure, whose binary address is placed on the address lines $A = a_1a_0$. Because line L in Figure 4.27 is being used in a manner similar to the OR gate, it is sometimes referred to as a **wired OR gate**. Depending on physical implementation details, wired logic of this type can lead to a reduction in the number of gates needed and other cost savings.

Signal Strength

Logical strength is a measure of a signal's ability to override other signals.

We usually treat signals in logic circuits as binary variables characterized by a single "value" parameter that can be 0 or 1. These basic values are occasionally supplemented by a few other values such as Z and U. At the switch level, it can be useful to append a second abstract parameter called **strength** to the possible signal values, so that every signal can be characterized by a pair of discrete quantities indicated by (*value, strength*). As we will see, this characterization makes it possible to analyze in technology-independent fashion a number of very important types of logic circuits such as programmable logic arrays, which cannot be described completely at the gate level.

The (*value, strength*) parameter pair typically has "pressure" and "current" components, the interpretations of which depend on the physical technology being used; for example, (*voltage, current*) in electrical switching circuits, and (*pressure, flow*) in hydraulic circuits. In analog electric circuit theory, signals are normally characterized in this way, but there the pair-values are continuous rather than discrete. At the switch level, the value parameter in (*value, strength*) can be any member of the set {0, 1, Z, U}, introduced in the preceding section. The number of strength values recognized is not fixed, but can be varied with the level of accuracy desired. Here we restrict our attention to the two or three strength levels needed to characterize MOS transistor circuits. It should be emphasized that the functional information will continue to be carried by the two logic values 0 and 1. Like the high-impedance value Z, strength serves in an auxiliary role that facilitates certain styles of circuit design.

Logical strength reflects the ability of a physical signal to supply energy to, or **drive**, signal lines and devices. The power supply PS, which is normally the largest energy source available in a logic circuit, is then viewed as supplying the strong signals of the form (0,STRONG) and (1,STRONG), which we will continue to denote by 0 and 1, respectively. Weak signals of the form (0,WEAK) and (1,WEAK) are formed by weakening or "attenuating" the strong signals; for brevity, we will write $\tilde{0}$ and $\tilde{1}$ for (0,WEAK) and (1,WEAK), respectively. Because the high-impedance state defines the weakest possible signal source—namely, no signal at all—we can write Z = (Z,STRONG) = (Z,WEAK) = $\tilde{Z}$. The weak U value $\tilde{U}$ = (U,WEAK) may be defined analogous to U = (U,STRONG) by the relation $\tilde{U} = \#(\tilde{0},\tilde{1})$, where # is the connection function introduced earlier in this section. In other words, a weak U is the result of a conflict between a weak 0 and a weak 1.

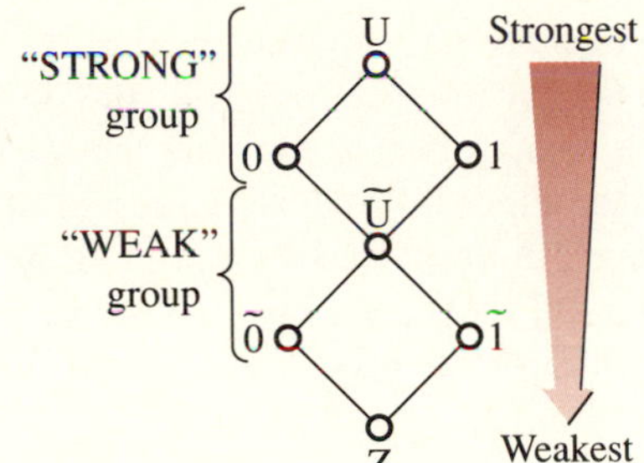

FIGURE 4.29 Strength relationship among the logic values {0, 1, U, Z, $\tilde{0}$, $\tilde{1}$, $\tilde{U}$}.

We thus arrive at a set of seven distinct value-strength pairs {0, 1, U, Z, $\tilde{0}$, $\tilde{1}$, $\tilde{U}$}, which we will treat as our set of possible signal values. The strength relationship among these signal values is depicted graphically in Figure 4.29. This is a refinement of Figure 4.24, in which the original {0, 1, U} values have been split into a strong group denoted by {0, 1, U} and a weak group denoted by {$\tilde{0}$, $\tilde{1}$, $\tilde{U}$}.

x z

(a)

Input x	Output z
0, $\tilde{0}$	$\tilde{0}$
1, $\tilde{1}$	$\tilde{1}$
U, $\tilde{U}$	$\tilde{U}$
Z	Z

(b)

FIGURE 4.30 (a) An attenuator symbol; (b) its function table.

Attenuators

As might be expected, buffers and other amplifying circuits increase the strength of the signals they process. In logical terms, they convert $\tilde{x}$ to x. A primitive component that converts a strong signal to the corresponding weak one is termed an **attenuator**. An attenuator is an idealized model of an energy-dissipating device such as an electrical resistor or a mechanical dashpot (a shock absorber). Figure 4.30a shows the symbol we will use for an attenuator; Figure 4.30b is a function table defining its behavior. In general, the attenuator converts x or $\tilde{x}$ to $\tilde{x}$. A weak signal is unchanged by the attenuator because we have no weaker strength value to assign to it. In this respect, an attenuator differs from physical devices such as resistors, which weaken signals over a continuous range of values. The arrow directions in Figure 4.30a can be reversed and x and z interchanged, because the attenuator is symmetric and bidirectional.

Attenuators convert strong signals to weak.

The usefulness of the logical strength concept derives from the fact that when a strong signal (v,STRONG) and a weak one (u,WEAK) are applied to a line L, the strong always overrides the weak; that is,

$$\#[(v, \text{STRONG}), (u, \text{WEAK})] = (v, \text{STRONG}) \qquad (4.18)$$

For example, $\#(\tilde{0}, 0) = 0$, $\#(1, \tilde{0}) = 1$, and $\#(\tilde{\text{U}}, 1, 0) = \text{U}$. By serving as sources of weak 1s and 0s, attenuators allow the operation defined in Equation (4.18) to implement a logic function z.

Attenuators as Pull-Up/Pull-Down Devices

Figure 4.31a depicts an all-switch complementary circuit structure that can be used to realize any Boolean function, as we saw in Section 3.3. It consists of two subcircuits: a pull-up circuit C that is responsible for applying 1 to the output line z, and a pull-down circuit $\overline{C}$ that applies 0 to z. C is designed so that every input combination X making $z(X) = 1$ closes a path from the 1-terminal of the power supply PS to z. At the same time, $\overline{C}$ performs the complementary task of opening all paths from the 0-terminal of PS to

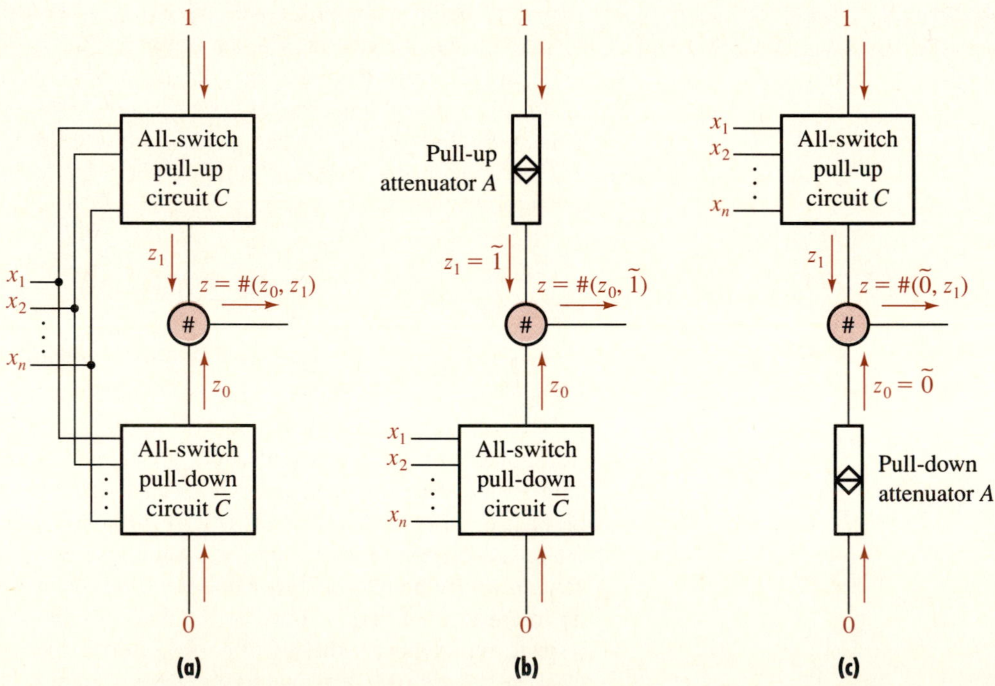

FIGURE 4.31 (a) A complementary switching circuit; (b) replacement of the pull-up circuit by an attenuator A; (c) replacement of the pull-down circuit by A.

the output. This makes $z = \#(1, \mathrm{Z}) = 1$, and C is seen as pulling z up to the 1 value. For each input combination that makes $z = 0$, the roles of C and $\overline{C}$ are reversed, so that $z = \#(\mathrm{Z}, 0) = 0$, and $\overline{C}$ pulls z down to 0.

A pull-up or pull-down circuit can be replaced by an attenuator.

We can use an attenuator to replace either C or $\overline{C}$. Figure 4.31b shows an attenuator A in its more common role as a pull-up device to replace C. In this configuration, A permanently applies a weak $\tilde{1}$ signal to the output line z. If the path from z to the (strong) 0-terminal of PS through the pull-down circuit $\overline{C}$ is open, then $z = \#(\tilde{1}, \mathrm{Z}) = \tilde{1}$; that is, A pulls the output up to the $\tilde{1}$ value. If $\overline{C}$ connects z to 0, then z assumes the value $z = \#(\tilde{1}, 0) = 0$. In the circuit of Figure 4.31c, the attenuator serves as a pull-down device. A applies $\tilde{0}$ to z, which becomes the value of z, unless C connects z to the 1-terminal of PS.

The advantage of the attenuator-based design is that we replace a multi-switch subcircuit with a single attenuator whose complexity is comparable to a single switch. A disadvantage is that the 0 and 1 output signals that appear on z have differing strengths, resulting in an asymmetry in z's ability to drive other devices. Moreover, physical logic circuits tend to have higher power consumption when attenuators are used. These factors do not affect the circuit's information-processing function, which, as noted earlier, depends entirely on the 0 and 1 levels of the output signals generated. As discussed further in Section 4.4, the circuits of Figures 4.31a, 4.31b, and 4.31c model CMOS, *n*MOS, and *p*MOS transistor circuits, respectively. In the latter two cases, a specially configured transistor acts as an attenuator; the remaining transistors act as switches.

The property of signal attenuation is inherent in all physical devices to some degree. Thus, although ideal lines and switches do not attenuate the signals they handle, the corresponding physical devices—for example, electrical wires and transistors—do so. The extent to which these effects need to be taken into account in logic design varies with the circuit technology being used.

Attenuator-Based Gates

Figure 4.32 shows alternative implementations of some basic gate types, using attenuators as well as switches. Figure 4.32c depicts a two-input NAND gate. Comparing this design to the all-switch complementary NAND circuit of Figure 3.14, we see that the two negative switches that form the pull-up circuit have been replaced by a single attenuator A. If either of the positive switches is in the off state, then $z_0 = \mathrm{Z}$, and the attenuator pulls the output z up to the weak 1 level, because $z = \#(z_1, z_0) = \#(\tilde{1}, \mathrm{Z}) = \tilde{1}$. When both positive switches are switched on, corresponding to $x_1 = x_2 = 1$, we obtain $z = \#(\tilde{1}, 0) = 0$. Thus, ignoring signal strengths, we obtain the NAND function $\overline{x_1 x_2}$. The behavior of the AND and OR gates may be analyzed in a similar manner.

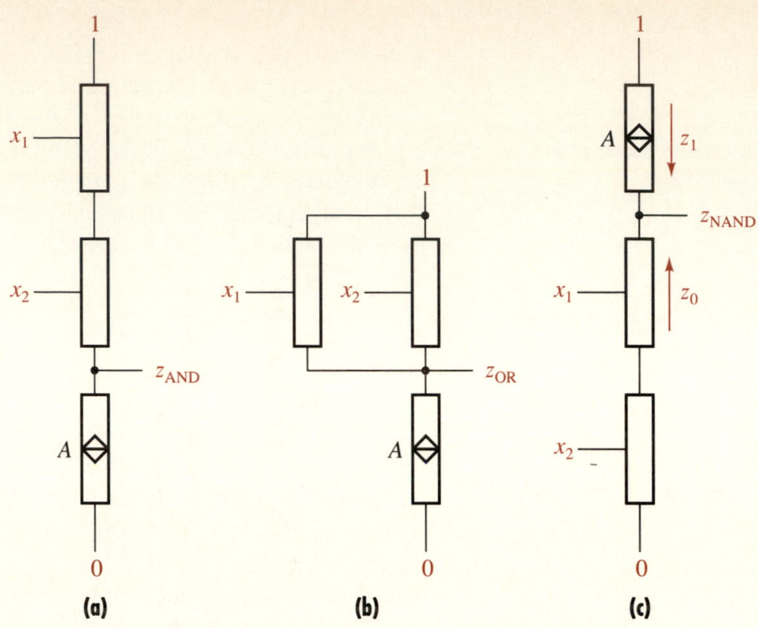

FIGURE 4.32 Attenuator-based designs for (a) AND, (b) OR, and (c) NAND gates.

Open-Circuit Logic

OC gates share an external attenuator to form two-level logic circuits.

There are several specialized logic design styles that implement multilevel circuits using attenuators and wired logic. A technique of this type makes use of logic gates called **open-circuit** or **OC gates**, which have attenuator-based designs like those of Figure 4.32, but omit the attenuator. As a result, the output circuits of OC gates are "open" or incomplete; they must be completed by connecting the gate's output to a suitable attenuator. However, several OC gates can have their outputs wired to the same attenuator, resulting in a low-cost implementation of a two-level circuit. The major forms of OC gates are open-collector and open-emitter gates (found in some bipolar transistor circuit families) and open-source and open-drain gates (found in some MOS circuit families). We refer to circuits constructed from OC gates as **OC logic circuits**; the terms wired, dot, and implied logic are also used.

When the output lines $y_1, y_2, \ldots, y_k$ of a set of matching OC gates are connected to an attenuator, the result that appears on the common output connection point P is either $z = y_1 y_2 \ldots y_k$ or $z = y_1 + y_2 + \ldots + y_k$. Thus, P is often termed a **wired AND** or a **wired OR gate** because no physical output gate is present in the usual sense; the logic functions are performed by the connection operator # implemented by P. Open-circuit AND gates are designed so that connecting their outputs forms a wired OR gate; consequently, a set of k OC AND gates can implement a k-term SOP expression. Similarly, a set of k OC OR gates can implement a k-term POS expression. The usefulness of such wired logic circuits in practice is limited

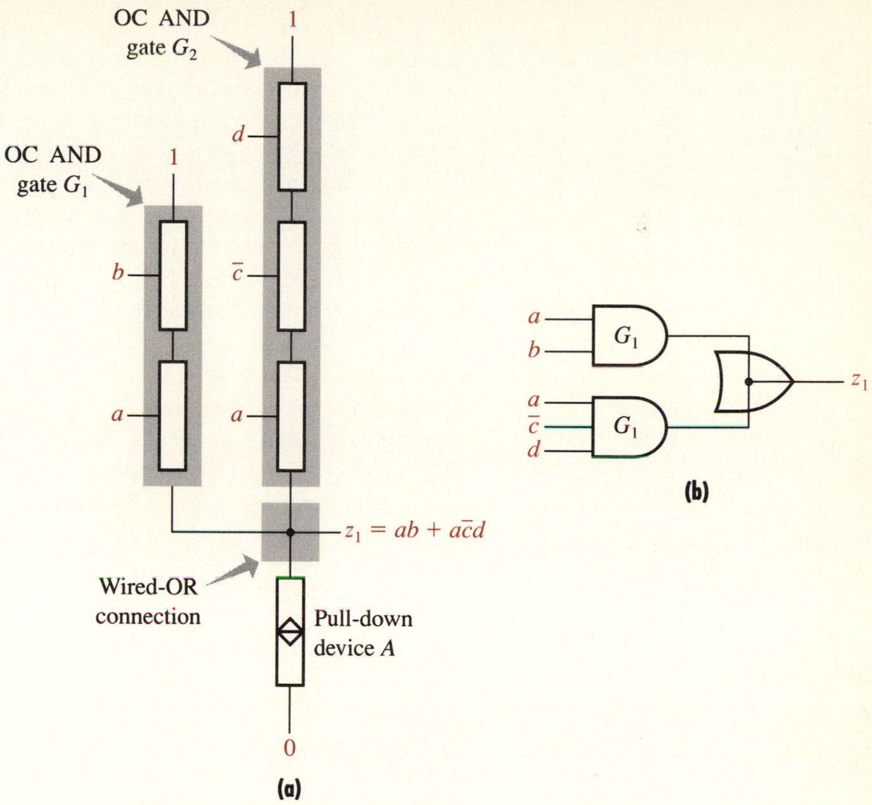

FIGURE 4.33 A two-level circuit constructed from OC gates: (a) a switch-level model; (b) a gate-level model.

by the fact that they tend to be slower and more sensitive to noise than conventional logic circuits.

A two-level logic circuit that employs OC logic to realize the function $z_1 = ab + a\bar{c}d$ is illustrated in Figure 4.33. It contains a pair of OC AND gates (compare them to the AND gate of Figure 4.32a) and an output circuit composed of a wired connection and a pull-down attenuator A. A switch-level model of this circuit appears in Figure 4.33a. The OC AND gates together form a pull-up network that creates a closed path from the (strong) 1 source to the output z_1 if $ab = 1$ or $a\bar{c}d = 1$. If either of these conditions is met, z_1 is pulled up to 1; otherwise, the pull-down device holds z_1 at the weak 0 value $\tilde{0}$. The value of z_1 is the # function of the three signals applied to it by the two OC AND gates and the attenuator, and is clearly equivalent to the desired OR function. A common gate-level representation of this type of circuit is given in Figure 4.33b.

A related but much more important technology for implementing two-level logic is the programmable logic device or PLD, which we discuss in detail in the last part of this chapter.

4.4 MOS Circuits*

One of the most important physical technologies for digital circuit manufacture is **MOS** (metal-oxide-semiconductor) **IC technology**. It is an electronic technology that employs two related types of transistors, n-channel or nMOS transistors and p-channel or pMOS transistors, as the principal switch-level components. MOS transistors use a single type of electric charge carrier for information transfer: n-channel transistors use electrons, whose electric charge or polarity is negative (n), while p-channel transistors use holes (a hole is the absence of an electron) whose polarity is positive (p). MOS transistors are contrasted with a very different class of transistors called **bipolar transistors**, in which positive and negative charge carriers play equal roles. (MOS transistors are occasionally called *unipolar*.) Bipolar transistors are the basis for such digital circuit families as TTL (transistor-transistor logic) and ECL (emitter-coupled logic) circuits. MOS transistors more closely approximate ideal switches than do bipolar transistors; consequently, MOS rather than bipolar circuits are more easily studied using the switch-level approach developed in this book.

MOS transistors and circuits are contrasted with bipolar transistors and circuits.

MOS Switches

An MOS transistor fabricated as part of a silicon integrated circuit is a three-terminal electronic device composed of several types of materials, including metal (usually aluminum) connectors, oxide (silicon dioxide) insulating layers, and several semiconductor regions. These regions are characterized as either p-type or n-type, depending on the polarity of the electrical charge carriers they employ. Cross-sections of the two basic types of MOS transistors are shown in greatly simplified form in Figure 4.34; these are also the standard graphic symbols used in electronic circuit diagrams. The dimensions of a typical transistor are measured in micrometers (microns), which is why millions can be placed on a single IC chip.

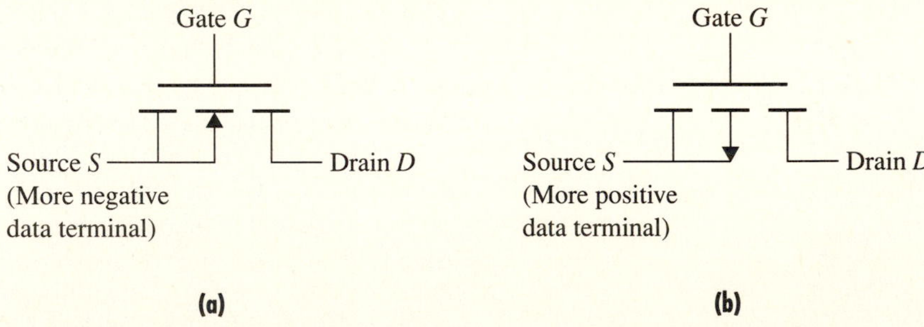

FIGURE 4.34 Graphic symbols for MOS transistors: **(a)** n-channel; **(b)** p-channel.

*This section can be omitted without loss of continuity.

In MOS circuits, the logical 0 signal level normally corresponds to a low-voltage range V_L; the logical 1 level corresponds to a high-voltage range V_H. Many digital ICs follow the TTL standard (see Figure 3.27), which recognizes any voltage in the range 0 to 0.8 volts as V_L, and any voltage in the range 2.0 to 5.0 volts as V_H.

The source S and drain D terminals of an MOS transistor T form the endpoints of the main data path through the transistor. The remaining terminal, called the gate,[1] serves as a control input to regulate the flow of electric current through the source-drain data path. The magnitude of the source-drain current of T is a function of the voltage level applied between the S and G terminals. In the case of an n-channel transistor, a high voltage on G allows a large current to flow between S and D, whereas a low voltage on G reduces this current to a low, near-zero level. In a p-channel transistor, the polarity of the control voltage is reversed: a low voltage on G allows a high source-drain current to flow, and a high voltage on G produces a low source-drain current. Thus, n-channel and p-channel MOS transistors correspond to positive and negative switches, respectively.

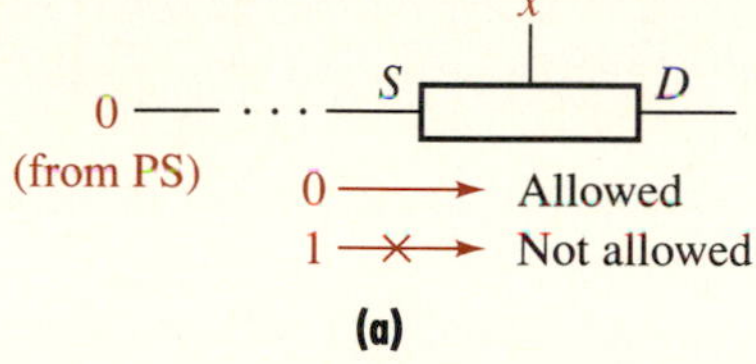

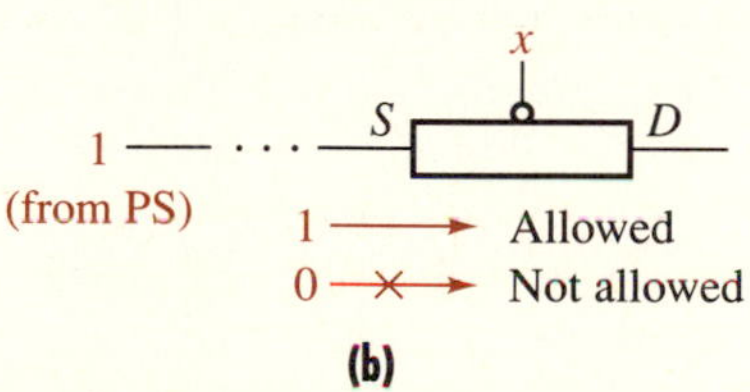

FIGURE 4.35 MOS transistor switch models: **(a)** n-channel; **(b)** p-channel.

An MOS transistor switch can transmit a 0 or a 1, but not both.

The source and drain are physically very similar. The data terminal referred to as the source is the one with the more negative applied voltage in the case of an n-channel transistor; it is the data terminal with the more positive voltage in the p-channel case. The voltage differences that determine the source and the drain make an MOS transistor asymmetric in its ability to conduct electrical (and hence logical) signals. This asymmetric structure reflects the fact that the control and data voltage signals of a transistor are *not* independent, as they are implicitly assumed to be in an ideal switch. Consequently, an MOS transistor's ability to transmit signals in both directions (from source to drain and from drain to source) is restricted. In particular, an n-channel transistor can pass an electric current from its source to its drain, but not vice versa. This is equivalent to saying that the transistor can pass a 0 but not a 1 between its source and its drain terminals. A p-channel transistor is similarly unidirectional with respect to current flow and can transmit a 1 but not a 0 through its source-drain path.

The foregoing constraints, summarized in Figure 4.35, suggest that an MOS transistor should be thought of as a nearly ideal switch that can transmit a 0 (in the n-channel case) or a 1 (in the p-channel case), but not both.[2] Thus, p-channel switches are normally used to control paths in the pull-up part of a circuit; n-channel switches control paths in the pull-down part.

[1]The term *gate* used here for a particular terminal of an MOS transistor should not be confused with the same term used for certain basic digital circuits (logic gates).

[2]In certain circuit configurations, an MOS transistor T can take advantage of the fact that V_L and V_H correspond to voltage ranges rather than single values to transmit both 0 and 1 satisfactorily. In such cases, T is referred to as a *pass transistor*.

CMOS Logic

CMOS logic employs all-switch pull-up and pull-down circuits.

If we reexamine the complementary switching circuits presented in Section 3.3, we find that several of them meet the signal-flow constraints implied by Figure 4.35, and so represent valid MOS circuits, if each transistor is modeled by the appropriate positive or negative switch. Figure 4.36a shows the two-switch inverter discussed in Section 4.3 (Figure 4.23) and an equivalent MOS realization. Because this type of circuit pairs n- and p-channel transistors in a complementary fashion, it is known as **complementary MOS** or **CMOS**. The circuit of Figure 4.36b is therefore referred to as a **CMOS inverter**. CMOS logic circuits have a number of attractive properties, including ease of design and extremely low power consumption.

A circuit that fails to meet the directionality constraints of Figure 4.35 is the two-switch complementary buffer of Figure 3.13. This circuit requires the positive switch S_2 to transmit a 1 from the power supply PS to the output z in violation of the rules of Figure 4.35. A CMOS buffer can be obtained by cascading two CMOS inverters of the kind that appears in Figure 4.36a, so that the two inversions cancel each other.

Figure 4.37 shows a four-transistor CMOS NAND circuit that is precisely modeled by the four-switch complementary circuit of Figure 3.14. The two p-channel transistors T_1 and T_2 form a pull-up circuit in which each transistor can supply a 1 but not a 0 to the output line z_{NAND}. The n-channel transistors T_3 and T_4 form the pull-down circuit and are required to supply 0 but not 1 to z_{NAND}.

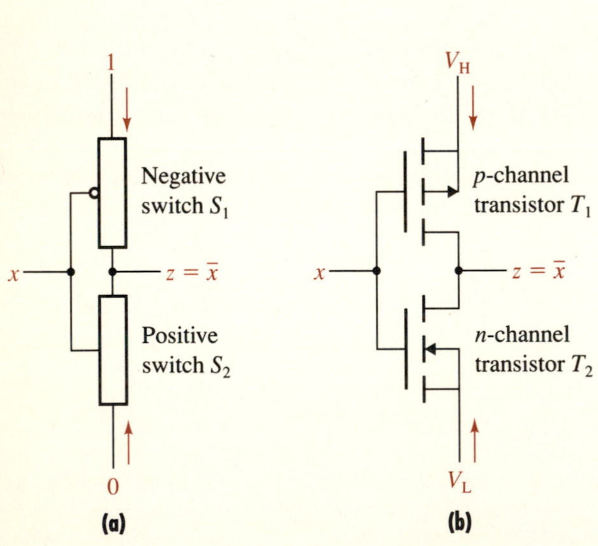

FIGURE 4.36 (a) A two-switch inverter model; (b) a CMOS inverter.

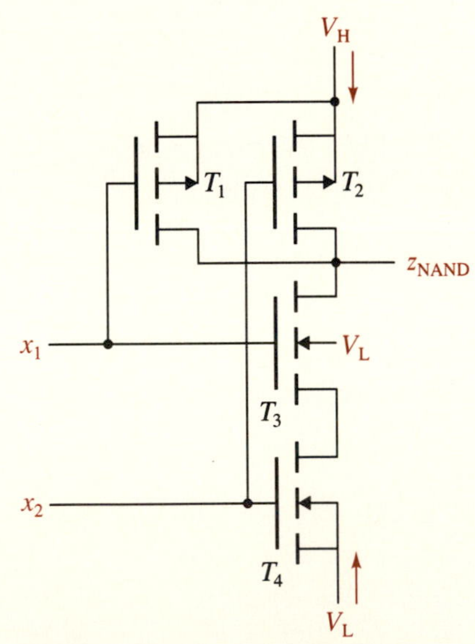

FIGURE 4.37 The CMOS NAND gate modeled by Figure 3.14.

Transmission Gates

Transmission gates are very close to ideal bidirectional switches.

CMOS technology allows a true bidirectional MOS switch to be constructed from a complementary pair of transistors, as depicted in Figure 4.38. This important device, called a **transmission gate** *TG*, can transmit a 0 and a 1 with equal facility. *TG*'s behavior is much closer to an ideal switch than that of a single MOS transistor. *TG* has two control signals x_p and x_n, which are applied to the p-channel transistor T_1 and the n-channel transistor T_2, respectively. The behavior of *TG* is as follows:

1. To turn *TG* on, set x_p to 0 and x_n to 1. Suppose that a 0 is applied to data terminal z for transmission to z'. (The transmission gate is quite symmetrical, so z and z' are completely interchangeable.) The n-channel transistor T_2 has the proper gate voltage applied to turn it on with z serving as the source (more negative) terminal, whereas the p-channel T_2 is turned off. When a 1 is applied to z, T_2 turns off, but T_1 turns on, thus providing the necessary data transmission path for the 1 signal.
2. To turn *TG* off, set x_p to 1 and x_n to 0. Both transistors are switched off, independent of the signals applied to z and z'.

A transmission gate provides complete bidirectional switching while meeting the constraints of Figure 4.35, because one of its two transistors can always provide a data transmission path for a variable signal applied to either of its data terminals. If a control variable is available in both normal (true) form x and complemented form $\overline{x}$, then we can apply x and $\overline{x}$ to *TG*'s x_n and x_p terminals, respectively, thereby allowing a single variable to control *TG*.

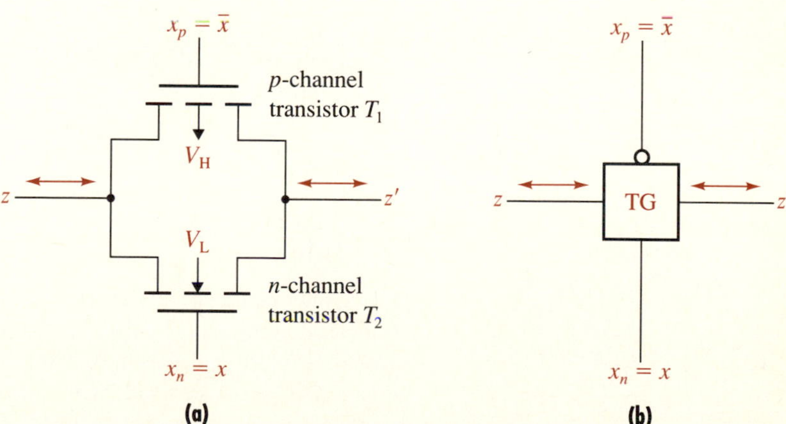

FIGURE 4.38 A CMOS transmission gate: (a) its transistor circuit; (b) its symbol.

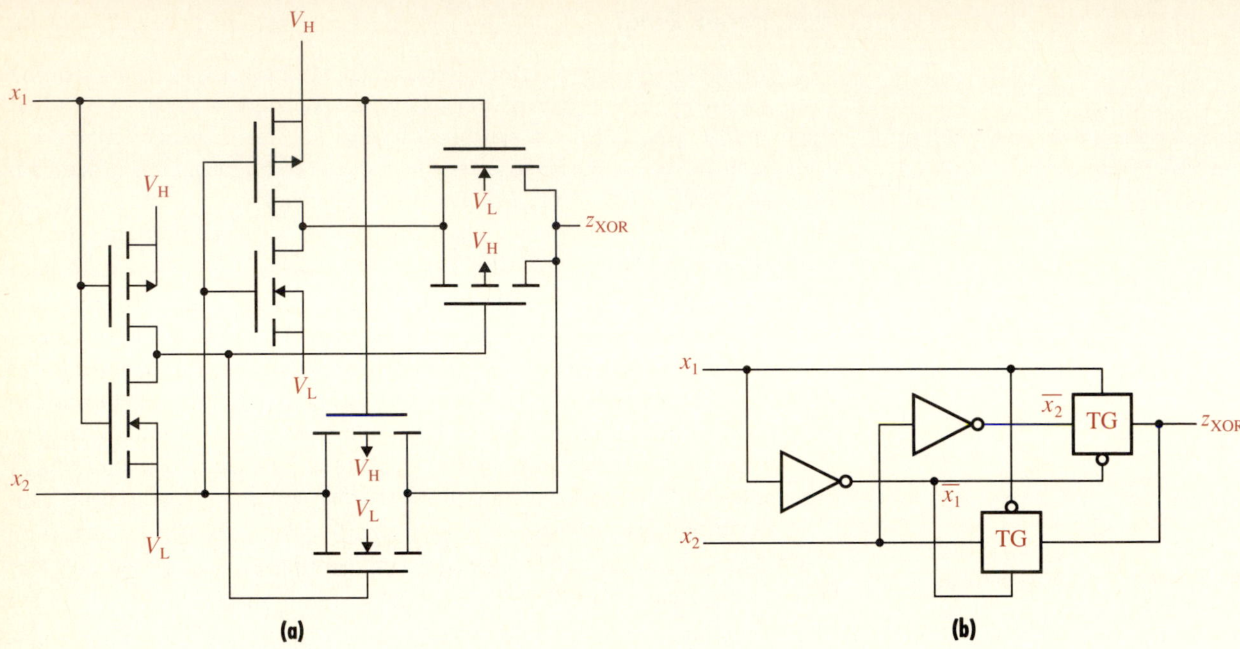

FIGURE 4.39 A CMOS EXCLUSIVE-OR gate: (a) its transistor circuit; (b) its logic circuit.

Example 4.4

A CMOS realization of an EXCLUSIVE-OR gate

Inverters and transmission gates are basic components for building more complex CMOS circuits. This concept is illustrated by Figure 4.39, which shows a CMOS EXCLUSIVE-OR gate derived from the noncomplementary all-switch design of Figure 3.61. Here, two standard CMOS inverters generate $\overline{x}_1$ and $\overline{x}_2$, which, along with x_1 and x_2, control a pair of transmission gates. If $x_1 = 1$, the upper transmission gate is opened, applying $\overline{x}_2$ to z_{XOR}, and the lower one is closed. Hence, $z_{XOR} = \#(\overline{x}_2, Z) = \overline{x}_2$. Hence, $z_{XOR} = 1$ if $x_1 = 1$ and $x_2 = 0$. Similarly, the lower transmission gate makes $z_{XOR} = 1$ when $x_2 = 1$ and $x_1 = 0$, so $z_{XOR} = x_1 \oplus x_2$.

*n*MOS Logic

*n*MOS circuits employ an attenuator pull-up circuit and an all-switch pull-down circuit.

CMOS integrated circuits require both *n*- and *p*-channel transistors to be fabricated in the same chip. It is somewhat simpler to build ICs from a single transistor type, in which case the attenuator-based approach illustrated by Figure 4.32 can be used, with a specially configured transistor serving as the attenuator. In this category, *n*MOS circuits composed solely of *n*-channel transistors have replaced *p*MOS circuits composed of *p*-channel transistors because they are faster. This speed is due to the fact that electrons have

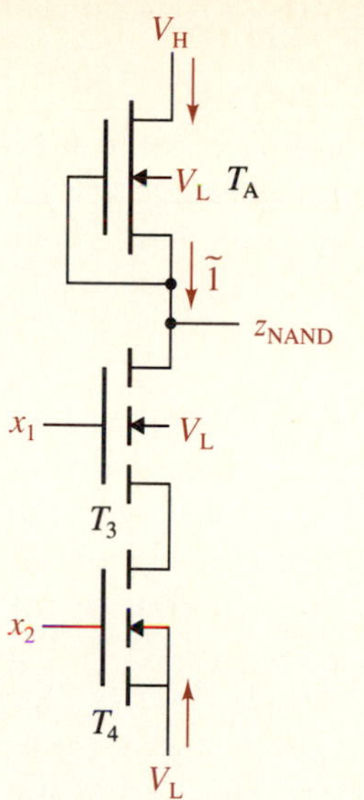

FIGURE 4.40 An nMOS NAND gate modeled by Figure 4.32c.

higher mobility, and therefore can travel faster through a circuit than holes can. CMOS and nMOS circuits are widely used in digital system design today, with CMOS preferred for its lower power consumption.

Figure 4.40 shows an nMOS version of the two-input NAND circuit of Figure 4.37, in which the pull-up circuit composed of transistors T_1 and T_2 is replaced with a special kind of nMOS transistor T_A, the physical design and source-gate connection of which cause it to behave like a logical attenuator. (Note the slight difference in the symbol for T_A; the dashed vertical line is replaced by a solid one, identifying it as a "depletion mode" MOS transistor in contrast with the more common "enhancement-mode" type.) This type of nMOS design contains fewer transistors, and therefore typically requires less chip area, than does the corresponding CMOS circuit. However, it also has the undesirable property of higher power consumption due to the attenuation property of the pull-up transistor T_A.

Figure 4.41 presents both CMOS and nMOS implementations of the three-input NOR function. Comparing these to the corresponding NAND circuits, we see that the transistors in the pull-up or pull-down circuits that are connected in series in the NAND circuits are connected in parallel in the NOR circuits, and vice versa.

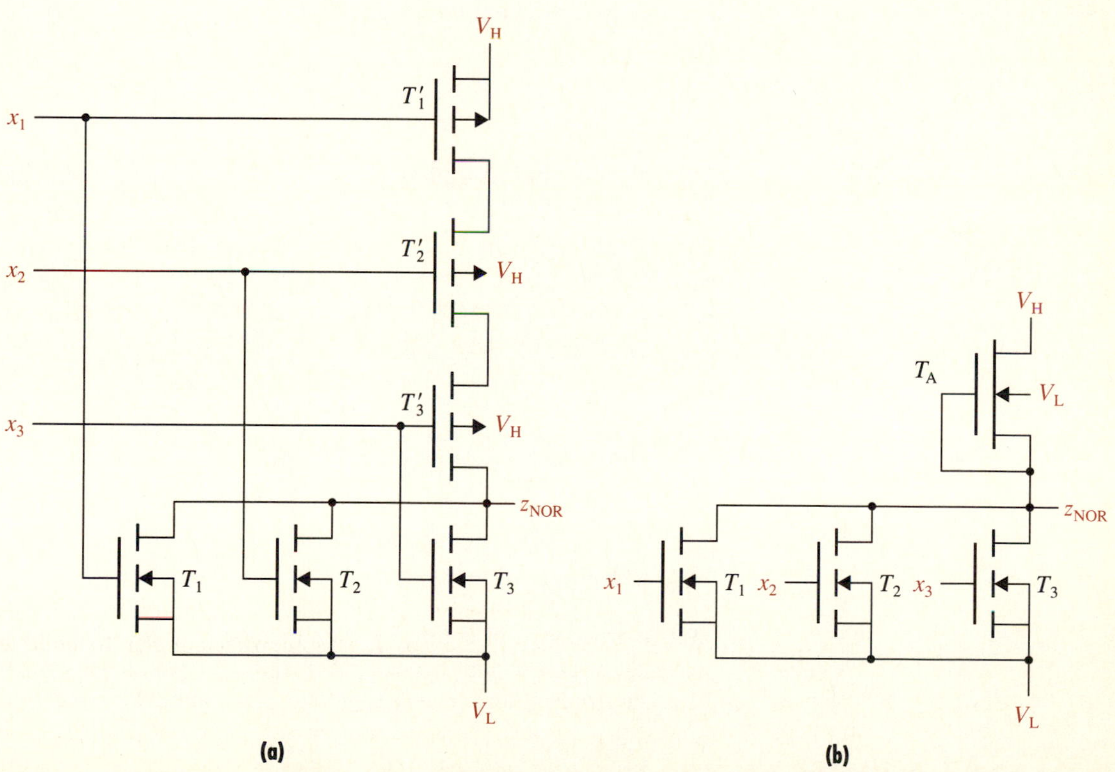

FIGURE 4.41 Three-input NOR gate implementations: (a) CMOS and (b) nMOS.

4.5 IC Implementation*

Integrated circuits at all levels of integration provide the primary technology for building digital systems. Next we discuss the implementation of gate-level designs in systems that employ small-scale and very large-scale ICs.

Small-Scale Integration

At the small-scale integration (SSI) level, a few gates—typically from 1 to 10—are packaged in a single IC, which is the basic physical design component. Two examples of small-scale ICs are illustrated in Figure 4.42. The first contains three three-input NAND gates and is housed in a 14-terminal package to accommodate the nine input and three output terminals of the NAND gates, along with power (PWR or V_H) and ground (GND or V_L) terminals for the electric power supply. The second example contains four two-input EXCLUSIVE-OR gates in the same type of 14-terminal IC package. In design at the SSI level, a logic circuit C is mapped into a set of ICs that contain all the gates (or their logical equivalents) specified in C. The ICs, whose number may range from one or two to 50 or more, are typically mounted on a printed-circuit board that provides mechanical support for the ICs, as well as the necessary electrical interconnections among them.

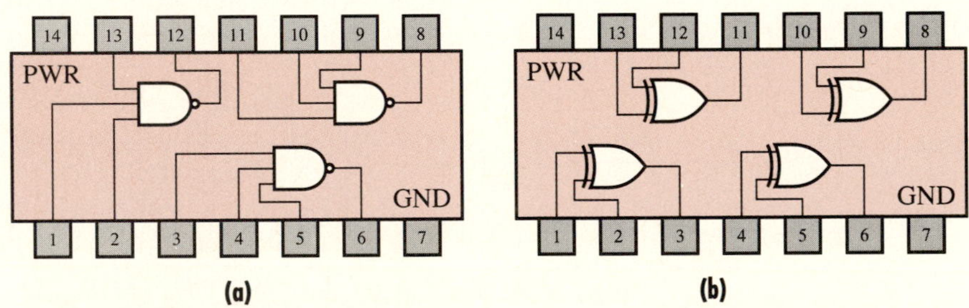

FIGURE 4.42 Two small-scale integrated circuits: (**a**) the 74X10 triple three-input NAND; (**b**) the 74X86 quadruple two-input EXCLUSIVE-OR.

The 74X Series

The ICs depicted in Figure 4.42 are members of the 74X series of SSI/MSI circuits, which contains hundreds of different gate-level components, each characterized by an industry-standard part number prefixed by 74. This IC

*This section can be omitted without loss of continuity.

TABLE 4.2 Key characteristics of some 74X-series IC families.

IC family	*Propagation delay (ns)*	*Power per gate (mW)*	*Max. fan-out within family*	*Max. fan-out to LS gates*
Standard TTL	10	10	10	40
Schottky TTL (S)	3	19	10	50
Low-power Schottky TTL (LS)	10	2	10	20
Advanced low-power Schottky TTL (ALS)	4	1	80	20
High-speed CMOS (HC)	8	$<10^{-5}$	> 500	10

nomenclature originated with Texas Instruments, Inc., in the 1970s and has since been adopted by all IC manufacturers. The X part of the name denotes a specific transistor circuit family, such as "standard" TTL, the first family in the series (X = blank); Schottky TTL (X = S), a faster TTL technology with greater power needs; low-power Schottky TTL (X = LS); and high-speed CMOS (X = HC). 74X is followed by two or three digits that uniquely identify the logical functions provided by each IC. Thus, the logic diagram of the IC designated 74X86 in Figure 4.42b represents the 7486 standard TTL version, the 74LS86 low-power Schottky version, and so on. It also represents the "mil-spec" (military specification) implementations of the same circuit, which replace 74X with 54X in the standard nomenclature and are required to have higher reliability than the ordinary or "commercial-grade" 74X parts.

The 74X series provides many basic logic functions in standard formats.

Table 4.2 summarizes a few important characteristics of the main 74X IC families. Their circuit technologies differ in such areas as propagation delay (speed), power consumption, and fan-out limits. New 74X families with improved physical characteristics are introduced from time to time. Note that all IC types that share the same 74X-series suffix number have the same *logical* structure. Also, they are often manufactured in the same types of packages, including dual-in-line packages (DIPs) and other package types. The 14-pin DIP configurations for the 74X10 and 74X86 are outlined in Figure 4.42. Detailed descriptions of each 74X family, including its electrical and mechanical (packaging) characteristics, as well as the 74X-series ICs available in that family, can be obtained from the family-specific (logic) **data books** published by IC manufacturers. In addition to the combinational SSI components discussed here, the 74X series contains a full range of sequential SSI components, as well as many medium-scale (MSI) ICs.

IC Family Selection

The IC family type—that is, the X chosen by a designer to realize a particular logic function—is determined by the trade-offs desired from among the cost, speed, power requirements, and other physical attributes of the available IC families. Once a family has been selected, it is desirable to use ICs from that family throughout a particular digital system, because all members of a family are designed to be physically compatible with one another. Different families can be mixed in the same system, but special care must be taken to ensure that they are connected in ways that are compatible in terms of physical signal levels, fan-out, and so on.

74X-series components are implemented in many circuit styles (TTL, CMOS, etc.).

As an example, we can infer from Table 4.2 that 74HC gates draw very little power from the circuits that drive them[3]; consequently, all 74X families can drive, or fan out to, a very large number (hundreds) of 74HC-series gates. On the other hand, a 74HC-series gate can only fan out to 10 74LS-series gates, as indicated in the rightmost column of the table. Some CMOS and TTL families employ incompatible voltage ranges to represent logical 0 and 1. Consequently, special electrical circuits are needed to interface such families to one another, and are described in the relevant manufacturers' data books. A small sampling of the hundreds of gate-level combinational logic functions available in the 74X series appears in Table 4.3.

TABLE 4.3 A sampling of small-scale combinational logic circuits in the 74X IC series.

IC part number	*Logic functions provided by the IC*	*Number of pins (DIP)*
74X00	Four two-input NAND gates	14
74X02	Four two-input NOR gates	14
74X04	Six inverters (NOT gates)	14
74X08	Four two-input AND gates	14
74X10	Three three-input NAND gates	14
74X20	Two four-input NAND gates	14
74X21	Two four-input AND gates	14
74X27	Three three-input NOR gates	14
74X30	One eight-input NAND gate	14
74X32	Four two-input OR gates	14
74X54	AND-NOR circuit with two two-input and two three-input ANDs whose outputs are connected to a four-input NOR	14
74X86	Four two-input EXCLUSIVE-OR gates	14
74X126	Four tri-state buffers	14
74X133	One 13-input NAND gate	16
74X134	One 12-input tri-state NAND gate	16
74X245	Eight tri-state bus transceivers with common enable logic	20
74X366	Six tri-state buffers with common enable from two-input NOR	16

[3]The power dissipation figure for a CMOS gate given in Table 4.2 is for static operation only. The power consumption of this and other CMOS families increases, and maximum allowable fan-out decreases, with operating frequency—that is, with the rate at which signals change value.

The Design Process

Design using SSI ICs involves obtaining a logic circuit C for the desired functions in a form that can make efficient use of the IC types available in the chosen family of components. Commercial IC families contain all the basic gate types discussed so far, usually in several versions, and so provide the functional completeness necessary to implement any set of logic functions. The task of assigning the gates of C to available IC types is termed **logic partitioning** or **technology mapping**, and usually has the goal of minimizing the number of ICs employed. Other design criteria that affect such parameters as propagation delay and fan-in/fan-out constraints must also be accounted for. Judicious transformations between gate types are often necessary to meet all the design objectives.

Example 4.5

SSI IC realization of a full adder

Suppose that we want to implement the full adder circuit defined earlier (Figure 3.29) and repeated in Figure 4.43a, using 74X-series ICs from the small catalog of parts in Table 4.3. The given circuit contains three two-

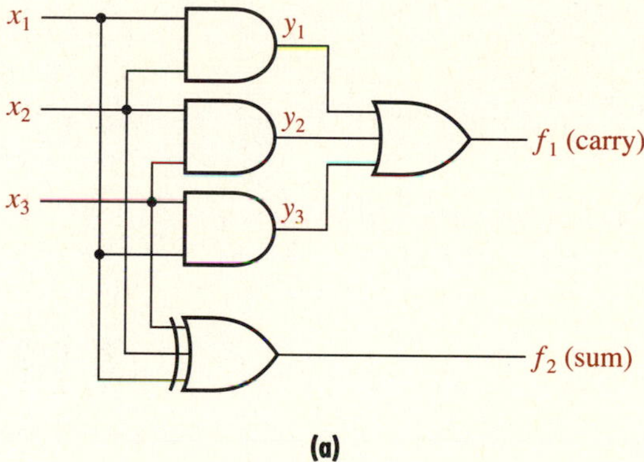

(a)

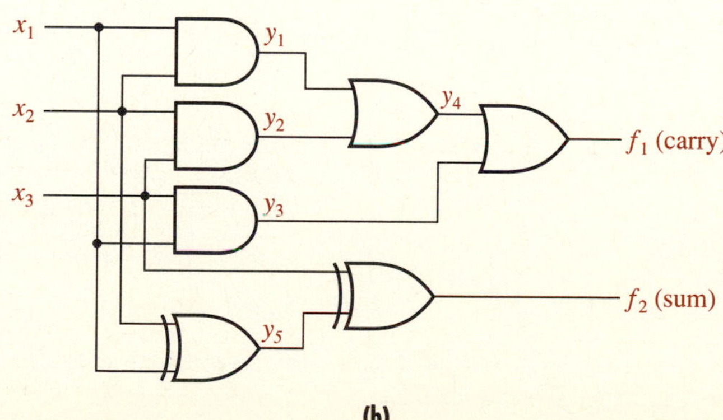

(b)

FIGURE 4.43 A full adder: (a) the initial design; (b) the modified design for SSI IC implementation.

input ANDs, one three-input OR, and one three-input EXCLUSIVE-OR gate. Our IC catalog includes the 74X08 quadruple two-input AND circuit, which can clearly take care of the AND gates. There are no three-input ORs or EXCLUSIVE-ORs available; however, we have two-input versions of these gates in the 74X32 and 74X86, respectively. We can use a pair of two-input gates in the obvious fashion to implement each of the required gates of fan-in three (see Figure 4.43b).

A logic circuit can often be mapped into standard ICs in several ways.

The resulting circuit is easily mapped into three ICs from Table 4.3: a 74X08, a 74X32, and a 74X86, which contain four OR, four AND, and four EXCLUSIVE-OR gates, respectively. Figure 4.44 shows all the connections

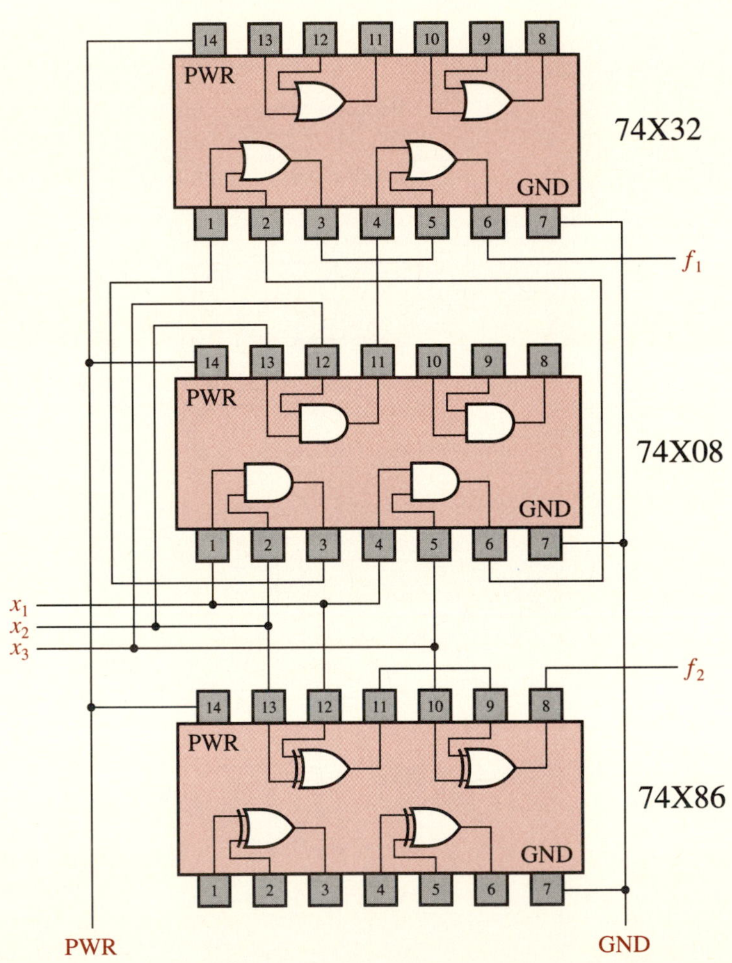

FIGURE 4.44 SSI implementation of the full adder of Figure 4.43 using three ICs from Table 4.3.

among the three ICs, assuming that standard DIP packages are used. Note that because the original three-input OR has been replaced by a two-level circuit composed of two OR gates, the circuit depth has been increased from two to three, an undesirable change from a speed viewpoint. One AND gate and two each of the available ORs and EXCLUSIVE-ORs are not used, so this is not a particularly efficient design in terms of IC utilization.

Several alternative designs that employ only two ICs are possible (see Problems 4.49 and 4.50). We can transform the subcircuit that realizes f_1 into an all-NAND circuit simply by replacing each AND and OR gate with the corresponding NAND gate. Partitioning of the resulting NAND circuit is simplified by the fact that we have many types of NAND ICs in the 74X catalog, including three-input NANDs. Thus, we could replace the 74X08 AND and the 74X32 OR ICs of Figure 4.44 with the 74X00 and the 74X10 NAND ICs, respectively, yielding a two-level instead of a three-level IC implementation.

We observe in passing that several adder circuits that are more complex and more useful than a full adder are included in the 74X series. The 74X83 4-bit adder, for instance, which is examined in Chapter 5, is a nine-input, five-output combinational circuit that computes the sum of two 4-bit binary numbers. Because of its complexity (it contains nearly 50 gates), the 74X83 is classed as an MSI circuit.

Very Large-Scale Integration

Chip area minimization is a major goal of VLSI design.

Integrated-circuit technology allows thousands of gates to be placed on a single IC. The creation of such circuits, loosely referred to as **LSI (large-scale integrated)** or, reflecting a certain inflation in language, **VLSI (very large-scale integrated)** design, requires the designer to be familiar with digital design at various complexity levels, including the switch and gate levels. The design process also depends heavily on the use of CAD tools to speed up the design stage and to manage the complex task of manipulating extremely large numbers of components. In logic design at the SSI level, a primary objective is to reduce overall hardware cost as measured by the total number of gates or ICs used. In VLSI design, however, the cost of a logic circuit is measured by the area it occupies on an IC chip. This area is roughly proportional to the number of gates used, so obtaining a circuit that uses a minimal or near-minimal number of gates continues to be an important design goal.

Standard Cells

VLSI designs can be implemented using standard predesigned cells.

Several special techniques exist to incorporate gate-level design techniques into the design of very complex ICs. The **standard-cell approach** employs gates and other predesigned components, collectively called **cells** in this context, as building blocks. These cells are stored in **standard-cell libraries** in CAD systems, and are grouped into families that correspond to particular circuit technologies—for example, 0.8-micron CMOS (CMOS circuits with a minimum physical feature size of 0.8 μm), 1-micron TTL, and so on. Therefore, a cell library loosely corresponds to an SSI/MSI data book. This correspondence is reinforced by the fact that some commercial cell libraries include cells that perform the same functions as 74X-series ICs.

A designer can call up a cell for, say, the 74X86 EXCLUSIVE-OR function, and display it on a CAD workstation in a logic-diagram style. The designer can also display the low-level physical or geometric layout of a circuit to position the cell on the chip or to address the problem of area minimization. Moreover, in this design environment, a cell can easily be combined with others to form new, larger cells, which can be added to the standard-cell library. Once a design based on standard cells has been completed, it can be converted into a special set of IC masks for manufacture.

Gate Arrays

VLSI designs can also be implemented using gate arrays.

A different approach to gate-level design of complex ICs employs a partially predefined IC called a **gate array** or a **sea of gates**, in which a large number of identical gates are laid out on a regular two-dimensional grid on a chip of fixed size and fixed circuit technology. Thus, the cell types in this case are restricted to single gates, and the physical positions of the gates on the target IC chip are fixed. Only the logical interconnections among the gates remain to be defined by the circuit designer. These interconnections can be specified in just one or two of the five to 10 steps typically needed in IC manufacture, thereby reducing the cost of the manufacturing process. Obviously, the gate types selected should be functionally complete; a good choice are NAND gates of small fan-in. Various CAD programs can assist in the design of a gate-array-based circuit, especially a routing program that can determine the physical positions of the interconnection wires in the routing channels to ensure that all required connections can be achieved within the given space constraints.

Example 4.6 Gate array realization of a full adder

Suppose we want a gate-array implementation of the full adder circuit (Figure 4.43a) that we have used in several examples. The gate-array IC is assumed to contain a large two-dimensional array of identical four-input NAND cells.

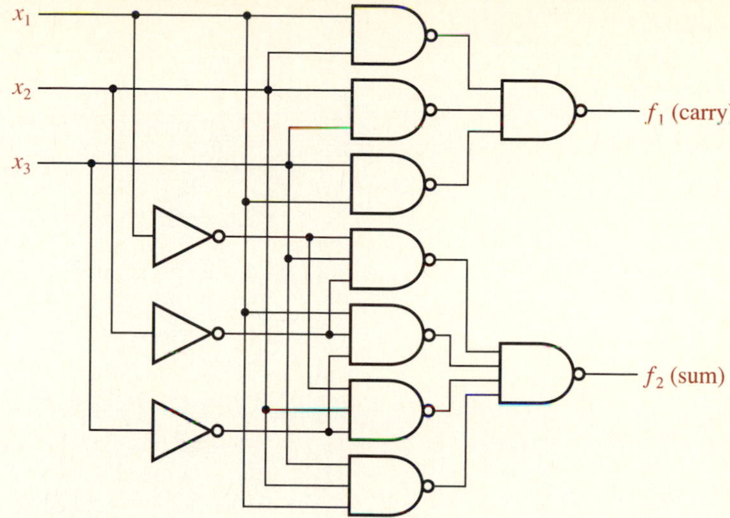

FIGURE 4.45 An all-NAND implementation of the full adder of Figure 4.43.

Our first step is to convert the original circuit, which utilizes three different gate types, to all-NAND form. The two-level AND-OR circuit that realizes f_1 is converted to NAND-NAND form simply by replacing each AND and OR gate with a NAND gate with the same fan-in. The EXCLUSIVE-OR function f_2 has the canonical sum-of-minterms expression

$$f_2 = = \overline{x}_1\overline{x}_2x_3 + \overline{x}_1x_2\overline{x}_3 + x_1\overline{x}_2\overline{x}_3 + x_1x_2x_3 \tag{4.19}$$

which is its simplest SOP form (see also Example 4.1). Again, we can map the SOP expression of Equation (4.19) directly into a two-level NAND-NAND circuit, as shown in Figure 4.45. The largest NAND gate has four inputs, so, unlike the SSI implementation of Figure 4.44, no redesign needs to be done to change gate fan-in.

Figure 4.46 shows the layout of the gate-array IC. The NAND gates are arranged in rows and columns separated by horizontal and vertical routing (wiring) channels of fixed width. The adder has a total of 12 gates and can be implemented by a small 6×2 subarray of NAND gates. The interconnections among the NAND gates follow those of Figure 4.45. Three of the 12 NAND gates are configured as inverters, and the remainder as two-, three-, or four-input NANDs. Each NAND cell must be connected to PWR and GND, and unused NAND inputs should be connected to PWR (logical 1); for simplicity these connections have been omitted from Figure 4.46.

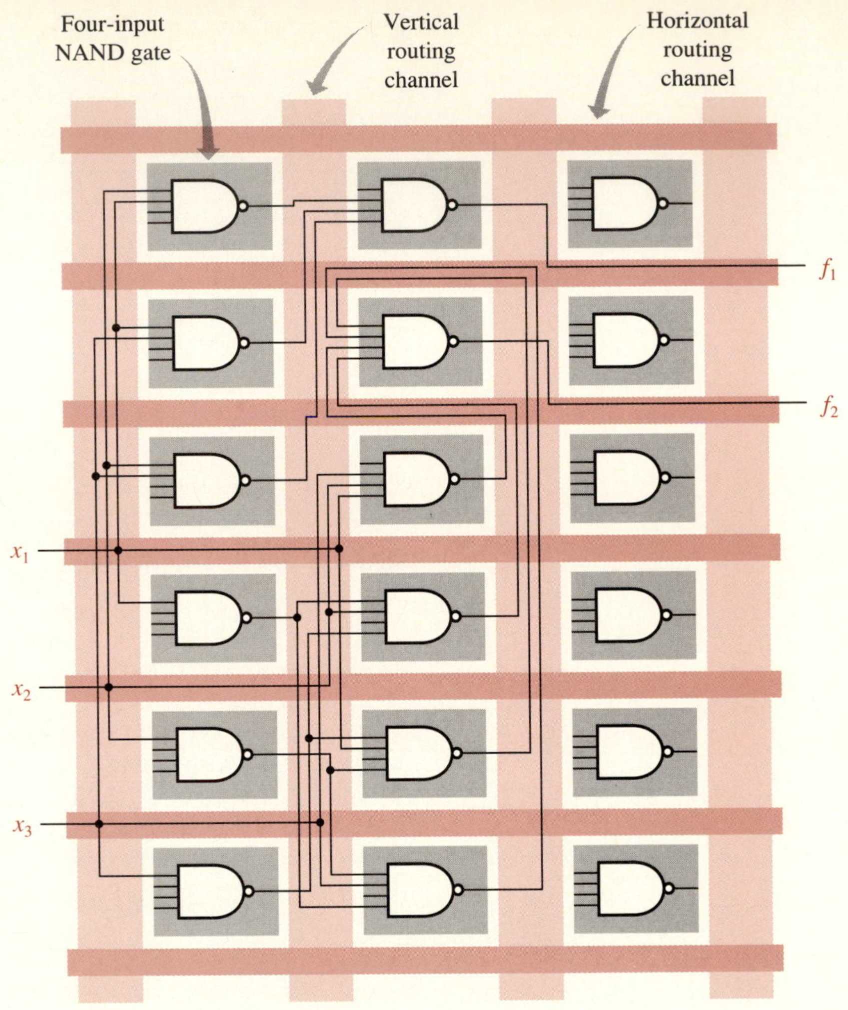

FIGURE 4.46 A gate-array implementation of the full adder.

Programmable Logic Devices

We turn next to an important and flexible class of combinational circuits that employ programmable components. Programmable logic circuits not only have the flexibility to be used for many different applications, they tend to be easier to design and manufacture.

4.6 Programmable Circuits

The term **programmable logic device** or **PLD** is broadly applied to ICs that have a fixed underlying set of components such as switches, gates, or more complex cells, but whose interconnections can be tailored in application-specific ways that make them general-purpose design components. The gate-array implementation of the adder in Figure 4.46 is an example of a PLD. The number of components available in PLDs can be very large, so that they span the range in complexity from medium-scale to very-large-scale integration. Many design techniques and IC technologies have been developed for PLDs; we consider several of the more important types here. PLDs constitute a key technology for implementing **application-specific integrated circuits** or **ASICs**.

The Programming Process

The application-specific interconnections of a PLD are produced in two main ways. The first requires one or two special steps in the IC chip manufacturing process, in which case the PLD is said to be **mask-** or **factory-programmed**. The second approach allows a designer or end user to perform the customizing process on the packaged IC by injecting electrical signals into it using a specialized piece of equipment called a **PLD programming unit**. These signals open or close switches to establish the desired connection patterns among the components. ICs customized in this second manner are called **field programmable**. In some instances, a field-programmable IC is also erasable, so that it can be repeatedly programmed—a convenience when a new design is under development.

A PLD has fixed components, but interconnections that are tailored to an application.

PLD programming units are relatively small and inexpensive devices that are easily attached as peripheral devices to CAD workstations or personal computers (see Figure 4.47). They constitute a simple "desktop factory"

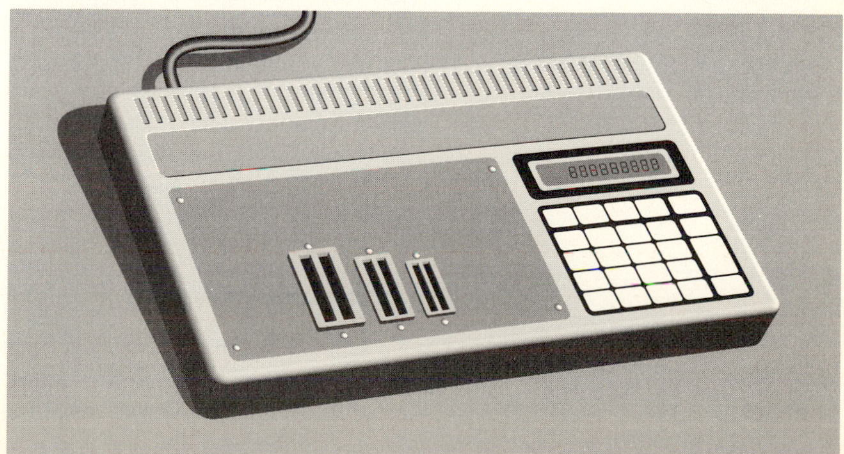

FIGURE 4.47 An example of a PLD programming unit.

for producing a self-contained and fully functional IC realization of a logic circuit. Field-programmable PLDs are widely used for implementing prototype designs and for low-volume manufacture of ASICs. Mask-programmable PLDs are preferred for high-volume ASIC manufacture.

Crosspoint Switches

PLDs can be programmed by blowing fuses or antifuses at crosspoints.

A programmable connection between two lines A and B in a PLD is logically equivalent to a switch that can be programmed to be closed (meaning that A and B are electrically and logically connected and carry the same signal) or open (meaning that A and B are disconnected). Figure 4.48a shows a switch model of such a programmable connection, which is often called a **crosspoint**. Programming can be viewed logically as permanently setting x to 0 (switch open) or 1 (switch closed). Because a typical PLD contains thousands of crosspoints, the shorthand notation of Figure 4.48b is commonly used, with an $\times$ denoting a programmable connection in place of the dot used for a normal permanent connection.

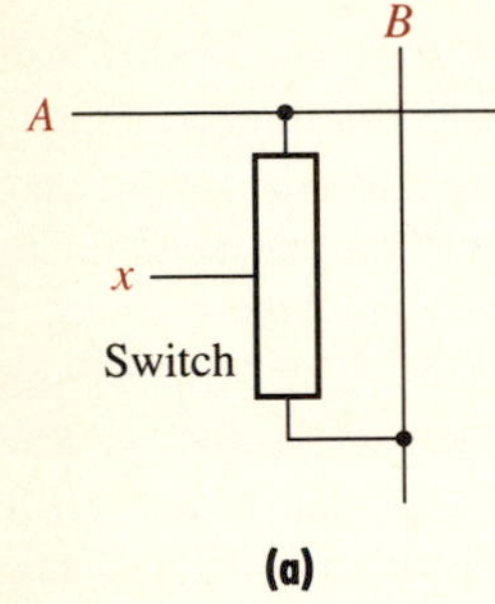

(a)

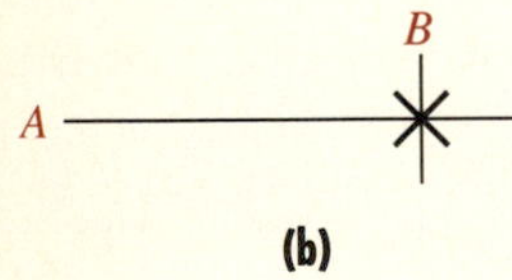

(b)

FIGURE 4.48 A programmable crosspoint of a PLD: **(a)** its switch representation; **(b)** its shorthand notation.

Various physical devices are used to implement crosspoint switches. In mask-programmed PLDs, parts of the wires at the crosspoint can be omitted, eliminating the need for a physical switch, a process called **discretionary wiring**. Alternatively, a physical switch can be present in the form of a **fuse** that normally connects A and B but is opened or "blown" by injecting a relatively large electrical current into the fuse. An alternative switch is an **antifuse**, which is normally an insulator but can be made to conduct an electric current by suitable programming. Once blown, the state of a fuse or antifuse is generally irreversible, and the resulting PLD is of the "read-only" variety.

Many PLDs, including most of the reprogrammable kind, employ transistors as crosspoint switches. The signal x that determines the switch state can take the form of an electrical charge package that can be altered by signals from the PLD programming unit. For example, there are field-programmable PLDs that are electrically programmable, but exposing the PLD to ultraviolet light resets all xs to logical 1. This effectively erases the PLD's old program, permitting it to be reprogrammed. Observe that in all cases, the programming of a PLD requires special signal conditions (extra-high voltages or currents, ultraviolet light, etc.) that are not present when the PLD is in normal use. Once placed in operation, a PLD has essentially the same electrical and logical characteristics as any other IC from the same circuit family.

Programmable Arrays

As discussed earlier, switches can be used to construct logic gates in various ways. PLDs employ arrays of switches to implement large gates efficiently—that is, with minimum area or a minimum number of connections. Figure 4.49

Output
Pull-up device (attenuator)
Inputs x_1 x_2 x_3 x_{n-1} x_n

(a)

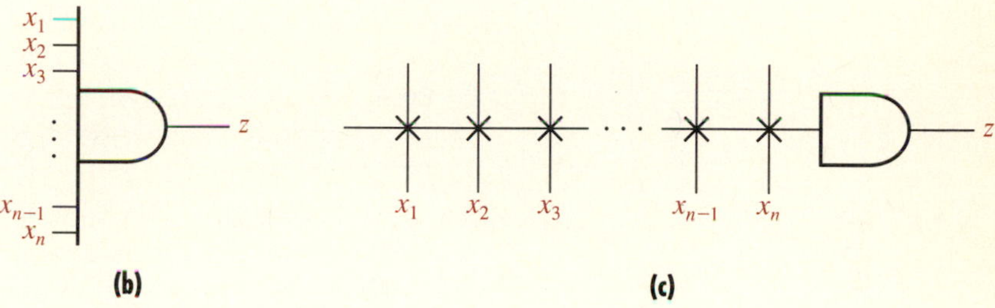

FIGURE 4.49 A programmable AND gate: (a) its switch-level model; (b) its gate-level model; (c) its shorthand notation.

shows a typical n-input AND gate implemented in this fashion by a row of switches. Here the switches are negative rather than positive, so that they are turned on by $x = 0$ and turned off by $x = 1$. The pull-up device represented by an attenuator can drive the output line z to a weak logical 1, provided all the switches are turned off. If any input x_i is 0, however, it turns on the ith switch, which then applies a strong 0 to the z line, pulling it down to the logical 0 level and thus realizing the AND function on z. The shorthand notation of Figure 4.49c again uses ×s to denote programmable crosspoints, but here they are interpreted as programmable AND inputs. OR gates can be constructed in similar fashion (Figure 4.50). In this case, the programmable crosspoints define the inputs of an OR gate.

A PLD realizes a Boolean function in two-level SOP or POS form.

By combining the gate structures of Figures 4.49 and 4.50, a variety of useful PLDs can be constructed from switches. They take the form of a large two-dimensional array of switches, part of which realizes a set of AND gates (the AND plane), the other of which realizes a set of OR gates (the OR plane). The AND and OR planes are connected to give two-level AND-OR or OR-AND—that is, sum-of-products or product-of sums—structures that can implement arbitrary Boolean functions. Depending on the organization and degree of programmability of the AND and OR planes, we obtain several different PLD types, including read-only memories (ROMs) and programmable logic arrays (PLAs), which we examine further in Section 4.7.

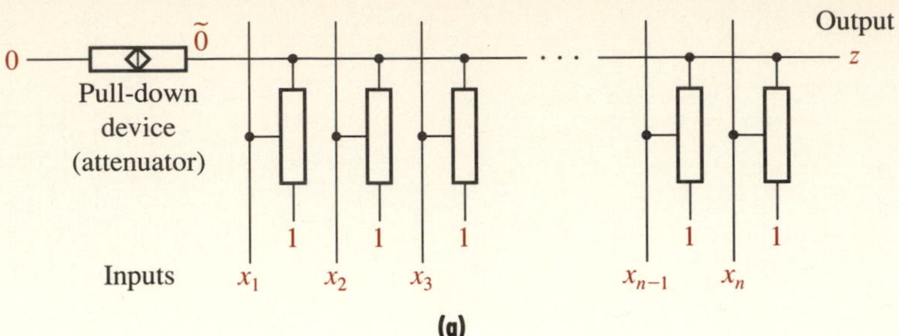

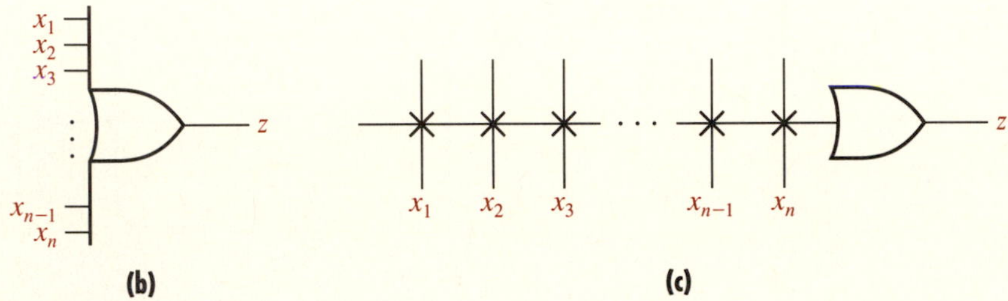

FIGURE 4.50 A programmable OR gate: (**a**) its switch-level model; (**b**) its gate-level model; (**c**) its shorthand notation.

A PLA Example

To illustrate these ideas, Figure 4.51 presents a small example of a PLA. This unprogrammed circuit consists of eight horizontal AND gates that form the AND plane, and two vertical OR gates that form the OR plane. Each AND gate has six programmable inputs taken from the three variables x_1, x_2, x_3 and their complements, and hence can be programmed to realize any product of one, two, or three literals. The eight rows of the PLA can be thought of as a realization of the product terms. The outputs of these rows are the inputs of the two OR gates. Again, the input connections to the OR gates are programmable, so that any subset of the row terms can be summed by either of the output columns. Thus, this particular PLD realizes a pair of Boolean functions z_1 and z_2 in AND-OR sum-of-products form. Because we can use the eight rows to realize all eight minterms of three variables, we can program the PLA to realize *any* two functions z_1 and z_2 in their canonical sum-of-minterms form.

Figure 4.51 shows the complete switch-level model of the PLA. A gate-level model of the same PLA appears in Figure 4.52, which clearly shows the circuit's two-level logical structure. However, it does not reflect the PLA's physical interconnection structure very well. We generally represent PLAs and other PLDs using the simplified representation of Figure 4.53, which

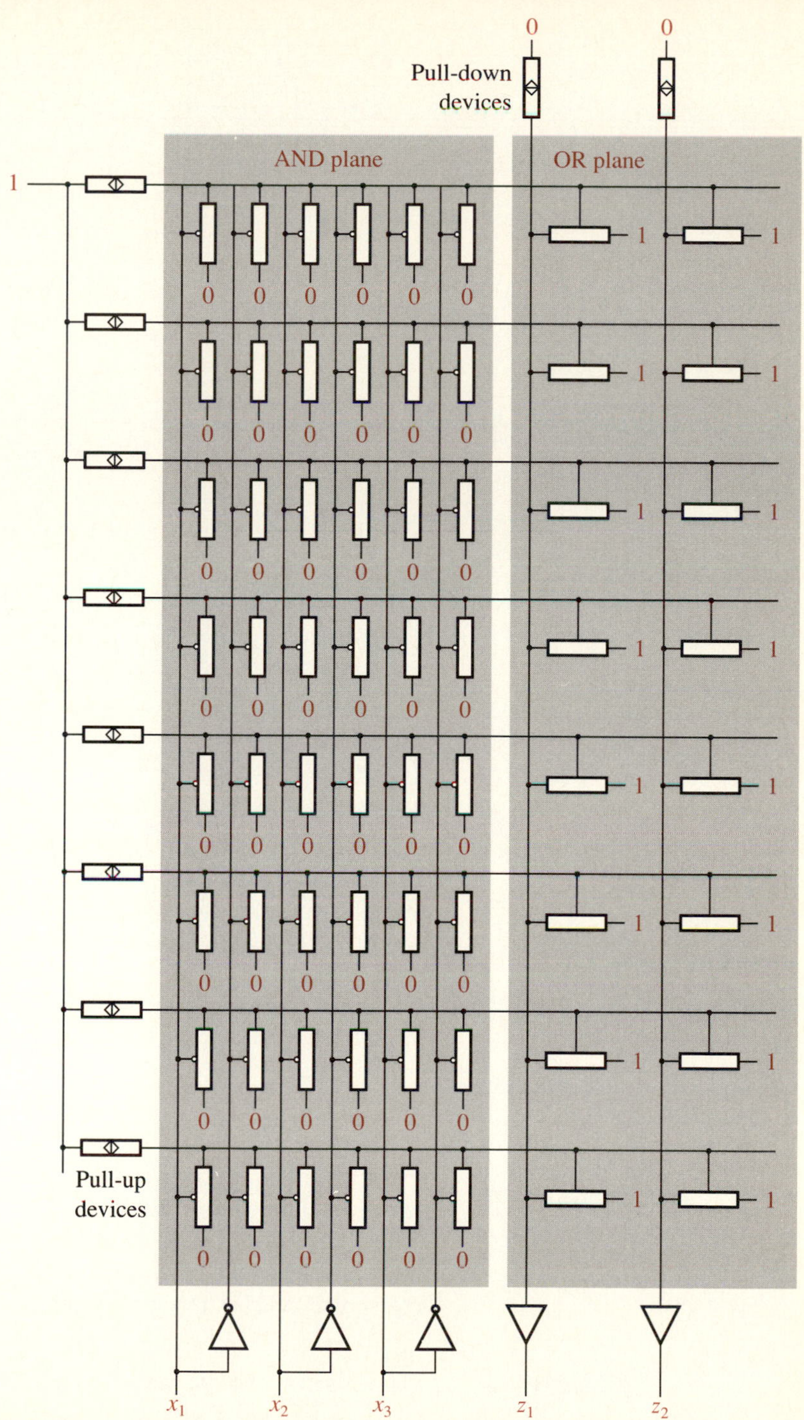

FIGURE 4.51 A switch-level model of a small unprogrammed programmable logic array (PLA).

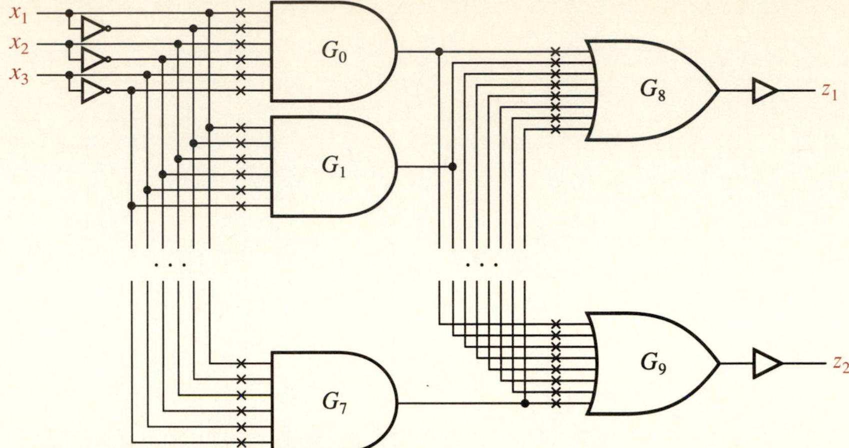

FIGURE 4.52 A gate-level model of the PLA.

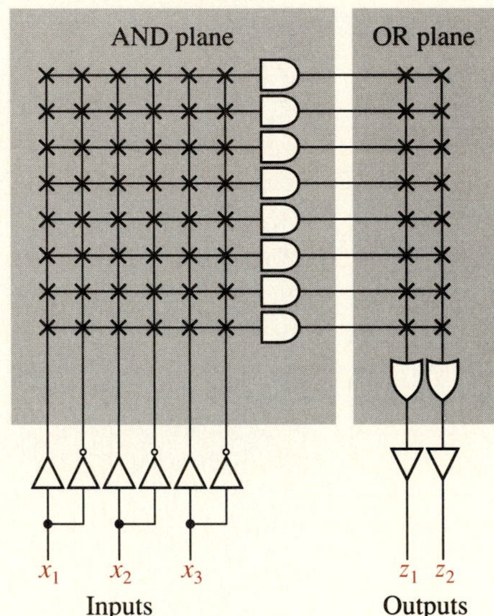

FIGURE 4.53 A simplified representation of the PLA: a PLA diagram.

gives both the physical and logical structures in a compact format. Here the rows and columns use the AND and OR formats of Figures 4.49c and 4.50c, respectively. This representation, which we call a **PLA diagram**, has the advantage that the pattern of ×s accurately represents the pattern of crosspoint switches available to program the PLA.

Example 4.7

PLA realization of a full adder

Suppose we want to use the PLA of Figures 4.51, 4.52, and 4.53 to realize our old friend, the full adder. We first need to obtain a two-level SOP realization of the adder; a suitable one appears in Figure 4.54, corresponding to

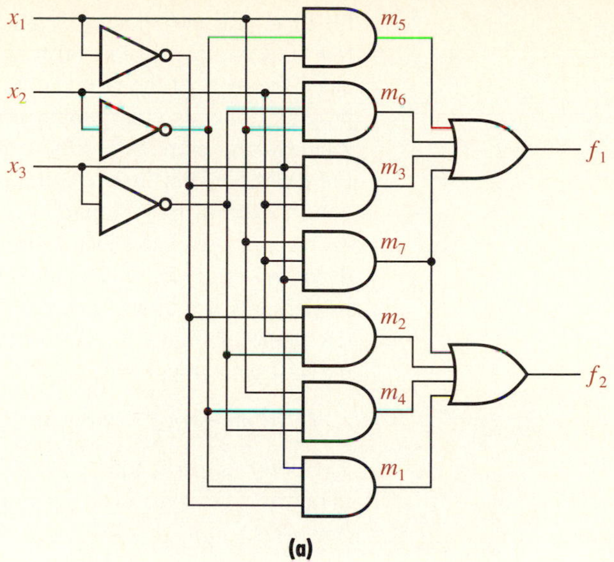

(a)

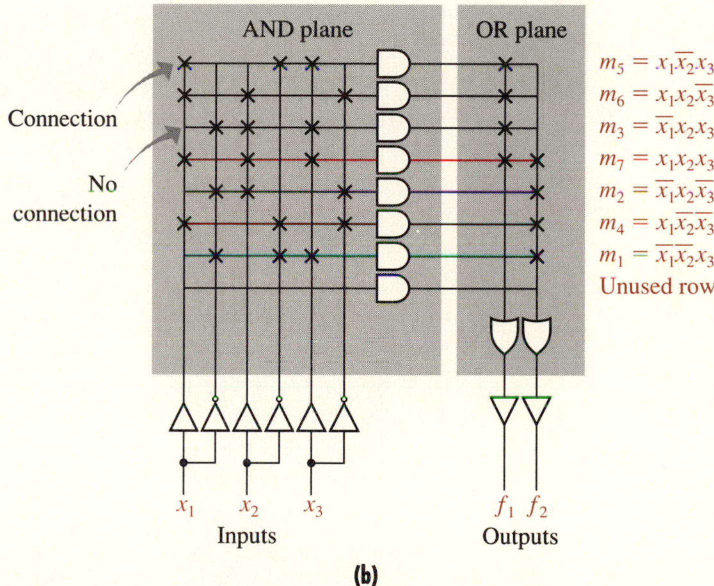

(b)

FIGURE 4.54 A sum-of-minterms realization of a full adder: **(a)** a gate-level model; **(b)** a PLA diagram.

the pair of canonical sum-of-minterms expressions

$$f_1 = x_1\overline{x}_2x_3 + x_1x_2\overline{x}_3 + \overline{x}_1x_2x_3 + x_1x_2x_3$$
$$f_2 = \overline{x}_1\overline{x}_2x_3 + \overline{x}_1x_2\overline{x}_3 + x_1\overline{x}_2\overline{x}_3 + x_1x_2x_3$$

Each expression contains four minterms, one of which is shared. We can map each of the seven distinct minterms (three-input AND gates) into a row of the PLA. Observe that each row is supplied with every input variable in both true and complemented form, one of which is connected to the row (indicated by an × at the crosspoint); the other is disconnected. By appropriately programming its crosspoints, the row can be made to generate any desired minterm or, more generally, any desired product of up to three variables. For example, the first row in the AND plane of Figure 4.54b has connections to the primary inputs x_1, $\overline{x}_2$, and x_3, and so generates the minterm $m_5 = x_1\overline{x}_2x_3$. The second row has connections to x_1, x_2, and $\overline{x}_3$, and produces $m_6 = x_1x_2\overline{x}_3$, and so on. Each of the first seven rows of the PLA implements a minterm $m_1 : m_7$; the last row is not required. Turning to the OR plane of Figure 4.54b, we see that the first column is connected to only the first four rows, and so computes the four-term (carry) function

$$f_1 = m_5 + m_6 + m_3 + m_7 = x_1\overline{x}_2x_3 + x_1x_2\overline{x}_3 + \overline{x}_1x_2x_3 + x_1x_2x_3$$

The other (sum) function f_2 is realized similarly as a logical sum of its four minterms. Observe how the minterm m_7 is shared between the two output functions.

Reprogramming

Suppose that, due to an "engineering change" affecting the specifications of the preceding full adder, we find it necessary to replace the sum output $f_2 = x_1 \oplus x_2 \oplus x_3$ with $f_2^* = x_1 \oplus x_2$. Converting the latter to sum-of-minterms form yields

$$\begin{aligned} f_2^* &= m_2 + m_3 + m_4 + m_5 \\ &= \overline{x}_1x_2\overline{x}_3 + \overline{x}_1x_2x_3 + x_1\overline{x}_2\overline{x}_3 + x_1\overline{x}_2x_3 \end{aligned}$$

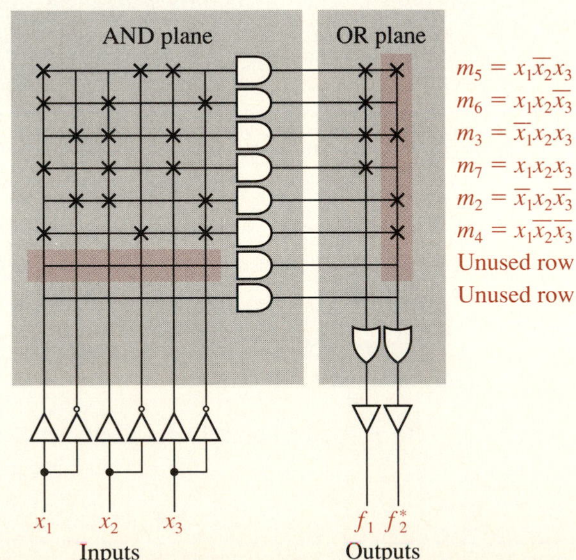

FIGURE 4.55 A reprogrammed PLA with f_2 replaced by f_2^*.

These minterms have already been generated for f_1 and f_2. All we need to do is drop the old minterm m_1 of f_2, which is no longer needed, and change the crosspoints in the f_2 output column to implement f_2^*. The resulting six-row PLA appears in Figure 4.55, with the modified portions colored.

It is apparent that the dimensions of the PLA limit the numbers and types of functions that can be realized. In general, we try to express the functions in the simplest two-level form to minimize the size of the PLA needed to implement them.

Computer-Aided Design

CAD tools can almost completely automate PLD design.

PLDs lend themselves very well to computer-aided design and manufacture. Indeed, once the logic of the target system has been specified, the process of converting it into a PLD can be almost entirely automated. The tools required are a computer, a PLD programming unit, and software for entering the PLD description into the host computer and generating a description of the PLD as a "program" that can be applied to the PLD programming unit to direct the PLD manufacturing process. The CAD software usually includes an optimization part that attempts to minimize the size of the PLD, as measured by layout area.

The main steps in the computer-aided design of a PLD are illustrated by Figure 4.56. The starting design can be created in the form of a logic (schematic) diagram, or else in textual form by means of a hardware description language (HDL). A general-purpose HDL such as Verilog or VHDL can be used, or any of a large number of HDLs designed just for PLD specification. Examples of the latter, developed by various PLD manufacturers and CAD tool suppliers, are PALASM (introduced by Advanced Micro Devices, Inc.), ABEL (Data I/O Corporation), and CUPL (Logical Devices, Inc.). These languages allow circuit function to be specified using Boolean equations, truth tables, or equivalent forms.

Figure 4.56 shows the input specification for the functions f_1 and f_2^* of the modified adder in two forms: a logic diagram in sum-of-minterms form and an HDL description loosely based on the VHDL language. The HDL input is built around a case statement that explicitly lists every input combination and the corresponding output combinations. It is thus essentially a truth table for the circuit in question. For example,

$$\begin{aligned}&\textbf{case } X \textbf{ is}\\&\qquad \textbf{when } 000 \;=>\; F <= 00;\end{aligned} \tag{4.20}$$

is a VHDL construction with the following meaning: if the input variable $X = x_1x_2x_3$ assumes the value 000, then the output function $F = f_1f_2^*$ should be 00. In Statement (4.20), => can be read as *implies* and <= as *takes the value*.

Whether the target circuit C in Figure 4.56 is entered as a logic diagram or a HDL description, the PLD design process starts with an SOP

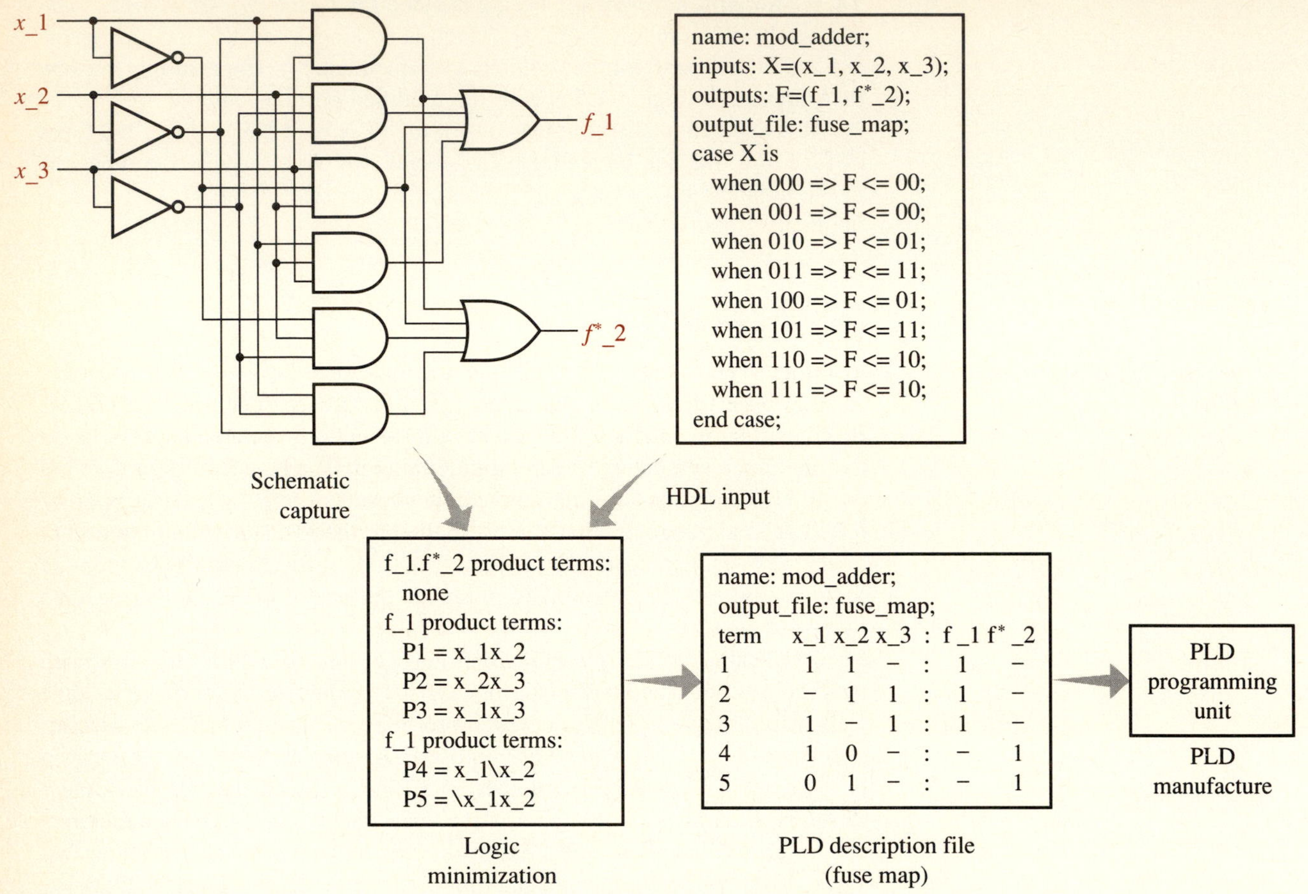

FIGURE 4.56 An overview of the PLD design process.

description of C. A typical optimization task, then, is to find the simplest SOP form for C, a task that is carried out by methods described in Chapter 5. In the present instance, it is easy to find a minimal SOP form that corresponds to the following Boolean equations:

$$f_1 = x_1x_2 + x_2x_3 + x_1x_3$$

$$f_2^* = \overline{x}_1x_2 + x_1\overline{x}_2$$

A PLA that implements these equations has only five rows, one less than the PLA of Figure 4.55. The PLA is described in Figure 4.56 by means of a text file, sometimes called a **fuse map**, which conveys essentially the same information as a PLA diagram. The fuse-map file can then be transferred to a PLD programming unit, where it can be used to "burn" the desired program into a suitable PLD chip.

4.7 ROMs and PLAs

Next we consider in some detail two of the major types of programmable logic devices, the *read-only memory (ROM)* and the *programmable logic array (PLA)*.

Read-Only Memories

A ROM realizes a function in canonical sum-of-minterms form.

Read-only memories or ROMs are among the oldest PLDs, but continue to have many important uses. A ROM is characterized by the fact that it generates all 2^n minterms of n variables in its AND plane. This structure permits each output column in the OR plane to produce any function of up to n variables if expressed in canonical sum-of-minterms form. Because the AND functions are predetermined as the minterms, the AND plane need not be programmable; all programming is therefore confined to the crosspoints in the OR plane. Figure 4.57 depicts an unprogrammed four-variable example, which has $2^4 = 16$ rows and four output columns. The ×s in each AND-

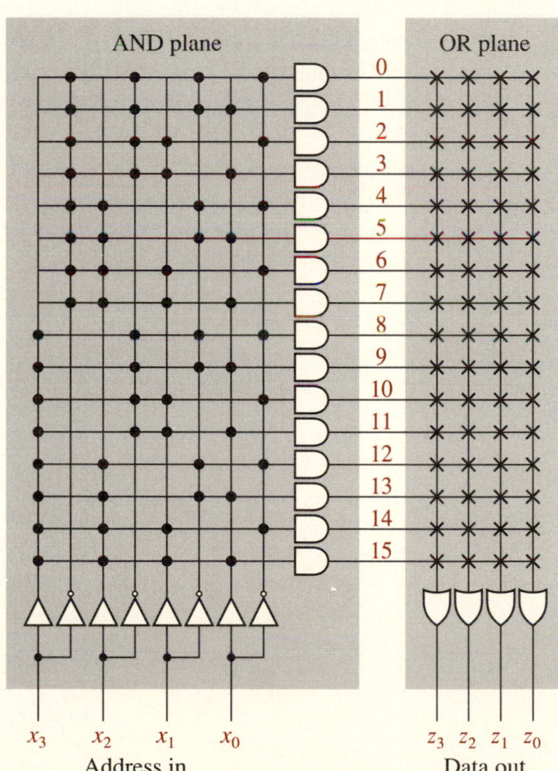

FIGURE 4.57 Internal structure of a 16 × 4-bit ROM.

plane row have been replaced by dots to indicate fixed connections, with the dot pattern in row i defining minterm m_i for $0 \leq i \leq 2^n - 1$. This PLD is thus able to realize any four of the $2^{2^4} = 65,536$ Boolean functions of four or fewer variables. A programmable logic element configured in this general fashion is referred to as a *ROM*. Field- or user-programmable ROMs are called **programmable ROMs** or **PROMs**.

Behavior

An n-input, m-output ROM, also called a $(2^n \times m)$-bit ROM, is a universal combinational component capable of implementing simultaneously any set of m logic functions that depend on up to n variables. It is a memory device in the sense that the OR plane stores 2^n data words that are addressed by the AND plane. In fact, the AND plane is functionally a 1-out-of 2^n decoder, and thus performs the address-decoding operation characteristic of all memories. A ROM is only a sequential circuit to the extent that there is significant propagation delay through it, which constitutes the ROM's **read access time**.

The operation of a ROM is as follows. In response to the n-bit input X representing the binary number i, the ROM generates a 1 on the output of the ith row of the AND plane, which is then applied to the OR plane. In the case of Figure 4.57, this 1-signal causes the output Z to assume the values $D_i = d_{i,3}d_{i,2}d_{i,1}d_{i,0}$, uniquely determined by the pattern of $\times$s (crosspoints in the on state) in the OR part of the ith row. For example, if all bits of D_i are programmed to be connected or on—that is, $D_i = \times\times\times\times$—then $Z = 1111$; if only the left two crosspoints are programmed to be on, then $D_i = \times\times--$ and $Z = 1100$.

A ROM stores a function's complete truth table.

Note that the pattern of crosspoints in a ROM, including the fixed connections in the AND plane, closely resembles a truth table. This can be seen from Figure 4.58, which shows an (8×2)-bit ROM realization of a full adder.

Inputs			Outputs	
x_i	y_i	c_{i-1}	s_i	c_i
0	0	0	0	0
0	0	1	1	0
0	1	0	1	0
0	1	1	0	1
1	0	0	1	0
1	0	1	0	1
1	1	0	0	1
1	1	1	1	1

(a)

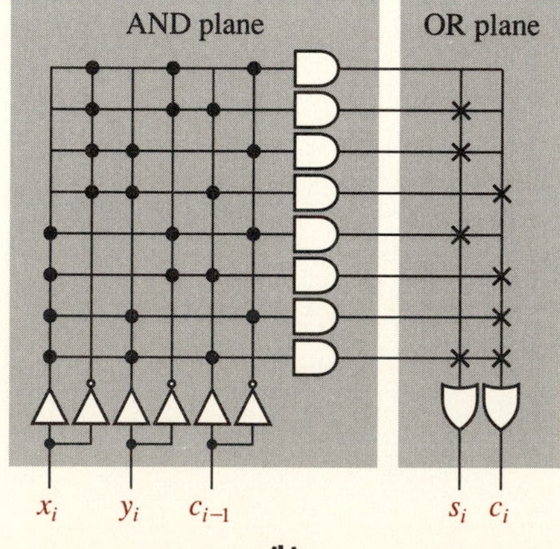

(b)

FIGURE 4.58 ROM implementation of a full adder: (a) its truth table; (b) its ROM diagram.

We can therefore regard a ROM as storing the complete truth table of a set of combinational functions. The process of reading stored information from a ROM is often called **table lookup**.

Applications

Read-only memories are very well suited to implementing a combinational circuit in which the relationship $Z(X)$ between the input X and output Z is complex and difficult to specify. This is the case when converting from one character code to another—for example, from the 7-bit ASCII code (Table 2.6) to the 8-bit EBCDIC. This particular code-conversion task would require a "mapping ROM" with seven inputs, eight outputs, and the ability to store $2^7 = 128$ 8-bit words.

ROMs are well suited to complex encoding and decoding applications.

Some of the more complicated mathematical functions are also suitable for ROM implementation. For example, a multiplier for two 4-bit unsigned numbers X and Y can be constructed from a ROM that stores the truth table of Figure 4.59. The ROM is addressed by an 8-bit word consisting of the

	Inputs								**Outputs**							
	x_3	x_2	x_1	x_0	y_3	y_2	y_1	y_0	z_7	z_6	z_5	z_4	z_3	z_2	z_1	z_0
0	0	0	0	0	0	0	0	0	0	0	0	0	0	0	0	0
1	0	0	0	0	0	0	0	1	0	0	0	0	0	0	0	0
2	0	0	0	0	0	0	1	0	0	0	0	0	0	0	0	0
3	0	0	0	0	0	0	1	1	0	0	0	0	0	0	0	0
4	0	0	0	0	0	1	0	0	0	0	0	0	0	0	0	0
5	0	0	0	0	0	1	0	1	0	0	0	0	0	0	0	0
6	0	0	0	0	0	1	1	0	0	0	0	0	0	0	0	0
7	0	0	0	0	0	1	1	1	0	0	0	0	0	0	0	0
8	0	0	0	0	1	0	0	0	0	0	0	0	0	0	0	0
9	0	0	0	0	1	0	0	1	0	0	0	0	0	0	0	0
10	0	0	0	0	1	0	1	0	0	0	0	0	0	0	0	0
11	0	0	0	0	1	0	1	1	0	0	0	0	0	0	0	0
12	0	0	0	0	1	1	0	0	0	0	0	0	0	0	0	0
13	0	0	0	0	1	1	0	1	0	0	0	0	0	0	0	0
14	0	0	0	0	1	1	1	0	0	0	0	0	0	0	0	0
15	0	0	0	0	1	1	1	1	0	0	0	0	0	0	0	0
16	0	0	0	1	0	0	0	0	0	0	0	0	0	0	0	0
17	0	0	0	1	0	0	0	1	0	0	0	0	0	0	0	1
⋮				. . .								. . .				
247	1	1	1	1	0	1	1	1	0	1	1	0	1	0	0	1
248	1	1	1	1	1	0	0	0	0	1	1	1	1	0	0	0
249	1	1	1	1	1	0	0	1	1	0	0	0	0	1	1	1
250	1	1	1	1	1	0	1	0	1	0	0	1	0	1	1	0
251	1	1	1	1	1	0	1	1	1	0	1	0	0	1	0	1
252	1	1	1	1	1	1	0	0	1	0	1	1	0	1	0	0
253	1	1	1	1	1	1	0	1	1	1	0	0	0	0	1	1
254	1	1	1	1	1	1	1	0	1	1	0	1	0	0	1	0
255	1	1	1	1	1	1	1	1	1	1	1	0	0	0	0	1

FIGURE 4.59 The truth table for a 4 × 4-bit multiplier suitable for ROM implementation.

pair X, Y; it outputs the 8-bit result $Z = X \times Y$. Thus, in each row of the ROM's OR plane, we store an 8-bit number that is the binary product of the first and second halves of that row's address; the ROM effectively stores a binary multiplication table. For example, the bottom row of Figure 4.59, which has address $11111111_2 = 255_{10}$, stores 11100001_2, which represents $1111_2 \times 1111_2$ or $15_{10} \times 15_{10} = 225_{10}$. The total number of crosspoints—a measure of the ROM's hardware complexity and physical size—is $24 \times 256 = 6{,}144$ crosspoints, which indicates an IC of moderate complexity.

Limitations

Despite the fact that they consist of only two logic levels, ROMs tend to be slow compared to ordinary logic circuits because of the extremely large fan-in and fan-out of their AND and OR gates. Another factor that limits the use of ROMs is the fact that their size approximately doubles with each new primary input. Thus, while the ROM that multiplies 4-bit numbers stores 256 8-bit words and has 6K crosspoints, a similar ROM multiplier for 8-bit numbers would have to store $2^{16} = 65{,}536 = 64\text{K}$ 16-bit words and have $2{,}097{,}152 = 2{,}048\text{K}$ crosspoints. An IC of this size is near the limits of the current VLSI range. A ROM multiplier for the 32-bit integers used in many general-purpose computers would have the crosspoint count of $96 \times 2^{32} = 393{,}216\text{M} \approx 4.123 \times 10^{11}$, which hardly seems practical for the foreseeable future.[4]

Programmable Logic Arrays

We turn next to some types of programmable devices that permit functions to be realized in a much more compact form than does a ROM. A ROM contains a row or, equivalently, an AND gate, for each of the 2^n potential minterms of the functions it realizes. In many cases, a particular minterm is not needed at all; in the ROM multiplication table of Figure 4.59, for example, none of the eight output functions has any minterms in rows 0 through 16. Furthermore, a row of a PLD can realize any product term (AND function) of its n input variables, not just an n-literal minterm. Hence, if the AND plane is programmable, we can replace a set of minterms that are covered by a common AND term P with a single row that realizes P. By merging

[4]Note, however, that most past predictions about ultimate technological limits on the capacity of memory ICs have proved far too conservative.

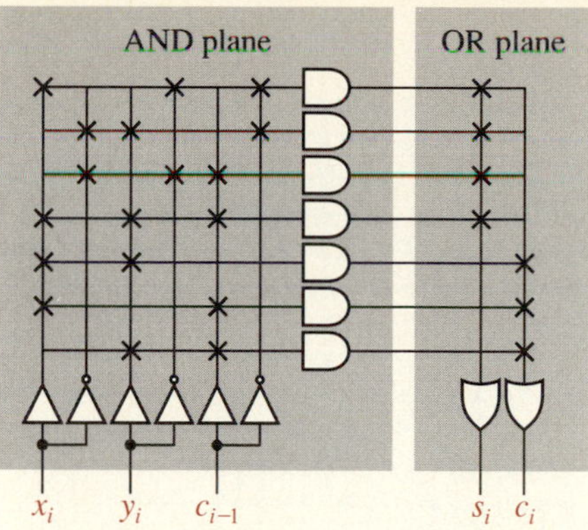

FIGURE 4.60 A PLA implementation of a full adder.

rows in this way, a PLD implementation is obtained that is smaller—often much smaller—than a ROM implementation of the same set of functions. A logic device with AND and OR planes that are both programmable is called a **programmable logic array** or **PLA**. Like ROMs, PLAs are widely used in both mask-programmable and field-programmable versions.

A PLA realizes a function in (minimal) sum-of-products form.

Figure 4.54 gives a PLA design for a full adder; a slightly different PLA implementation of the same adder function appears in Figure 4.60. Unlike the ROM design in Figure 4.58, the AND plane is programmed, and the number of rows has been reduced from eight to seven. Each row is now generating a general product term rather than a minterm, and the two output functions are being generated by the PLA in the optimal SOP form specified by

$$s_i = x_i\overline{y}_i\overline{c}_{i-1} + \overline{x}_iy_i\overline{c}_{i-1} + \overline{x}_i\overline{y}_ic_{i-1} + x_iy_ic_{i-1}$$
$$c_i = x_iy_i + x_ic_{i-1} + y_ic_{i-1}$$

As we noted already, it is desirable to obtain minimal SOP forms for PLA implementation. Depending on the underlying IC technology, the AND-OR logical structure of a PLA can be replaced by other two-level forms such as OR-AND or NOR-NOR.

PLA Extensions

To increase their flexibility, programmable logic arrays are often designed with additional logic circuits beyond the basic AND-OR planes. An interesting example is Signetics Corporation's PLS153 FPLA [Alford, 1989], the logical structure and layout of which appear in Figure 4.61. This PLA diagram has its AND and OR planes rotated with respect to the PLA diagrams we have

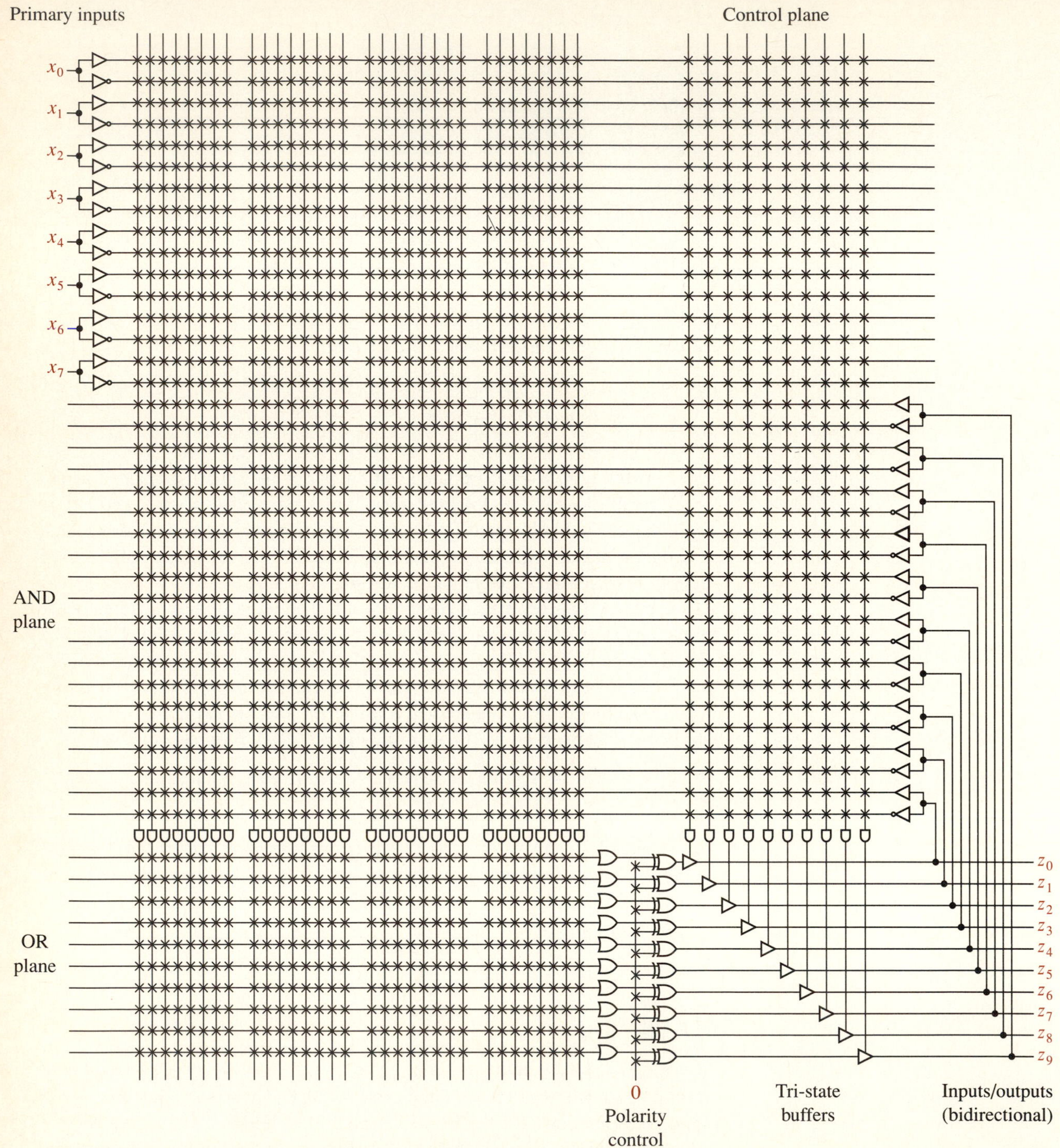

FIGURE 4.61 The Signetics PLS153 field-programmable PLA.

been using. The PLS153 comes in a small 20-pin IC package. As can be seen, there are eight primary input pins and 10 pins that have tri-state control and can be programmed to serve either as inputs or outputs. (Two additional pins are needed for power and ground.) Thus, the PLS153 can implement a single 17-variable function, two 16-variable functions, and so on.

Some PLAs can realize multilevel logic circuits.

The 10 bidirectional pins of the PLS153 are driven by tri-state buffers whose (vertical) enable inputs are derived from a special programmable "control plane," which produces AND functions of the PLA's input signals. When tri-state buffer i is disabled, pin z_i becomes an input pin, and any externally applied signal, along with its complement, enter the main AND plane in the usual fashion, but from the right rather than the left. When enabled, tri-state buffer i drives pin z_i as a primary output. However, in this case, because of the path from z_i back to the AND plane, z_i can also serve as a secondary input—that is, as an internal signal of a multilevel combinational design. Thus, the PLS153 is not restricted to two-level designs.

Another useful feature of the PLS153 is a polarity control in the output logic. Each signal z that emerges from the OR plane passes through a two-input EXCLUSIVE-OR gate, whose second input can be programmed to be 0 (its × connection is left intact) or 1 (the connection is removed). In the latter case, z changes to $z \oplus 1 = \bar{z}$. Suppose that a function z has a POS expression $E = S_1 S_2 \ldots S_k$ that is simpler than any SOP expression for z_i. Invoking De Morgan's laws, we can implement $\bar{z}$ in the SOP form $E' = \bar{S}_1 + \bar{S}_2 + \cdots + \bar{S}_k$, which has the same complexity as E. Complementing E' via the output parity control produces the original function z, because

$$z = S_1 S_2 \ldots S_k = \overline{\bar{S}_1 + \bar{S}_2 + \cdots + \bar{S}_k}$$

Example 4.8

PLA realization of a 9-bit parity function

To illustrate design with a commercial PLA IC, we use the PLS153 to design a common type of parity-checking circuit that computes the nine-variable EXCLUSIVE-OR or parity function

$$z_1 = x_0 \oplus x_1 \oplus x_2 \oplus x_3 \oplus x_4 \oplus x_5 \oplus x_6 \oplus x_7 \oplus x_8 \quad (4.21)$$

and its complement $z_0 = \bar{z}_1$. This problem requires all eight of the PLS153's fixed primary inputs for $x_0{:}x_7$ and one of the bidirectional pins for the ninth input x_8. Two of the latter are also used as primary output pins for z_0 and z_1.

An n-variable EXCLUSIVE-OR has a minimal SOP form containing 2^{n-1} product terms, each of which is also a minterm. A straightforward two-level PLA realization of z_0 and z_1 would have to generate all $2^{9-1} = 256$ nine-variable minterms, clearly an impossibility for the PLS153, which can generate at most 32 AND terms. We therefore turn to a multilevel design.

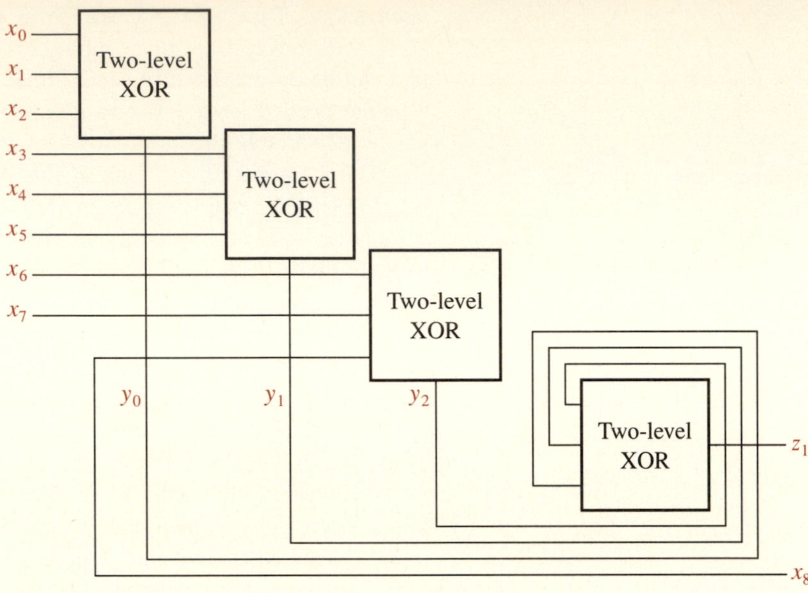

FIGURE 4.62 The global structure of the nine-input parity circuit.

Now Equation (4.21) reduces to the three-variable EXCLUSIVE-OR form

$$z_1 = y_0 \oplus y_1 \oplus y_2$$

where

$$y_0 = x_0 \oplus x_1 \oplus x_2 \qquad y_1 = x_3 \oplus x_4 \oplus x_5 \qquad y_2 = x_6 \oplus x_7 \oplus x_8$$

In this form, we can realize the parity circuit as outlined in Figure 4.62. The y_is are generated by three separate two-level circuits that implement the three-variable EXCLUSIVE-OR function. The y_is are then returned to the AND array, where they serve as inputs to a fourth EXCLUSIVE-OR circuit that computes the final output z_1. Note that this does *not* introduce feedback. The circuit of Figure 4.62 has a well-formed feed-forward structure, with the equivalent of four levels of AND and OR gates.

A three-variable EXCLUSIVE-OR has only four AND terms in its minimal SOP form. In fact, the minimal SOP form is also the canonical sum-of-minterms form given in Figure 4.8. Hence, z_1, y_0, y_1, and y_2 can be realized using 4×4 AND terms, or 16 columns in the AND plane of the PLS153 (Figure 4.63). Three rows of the OR plane compute the y_is and a fourth row forms z_1. A fifth row with the same OR-plane connections as z_1 but with 1 rather than 0 applied to its output polarity control gate computes $z_0 = \bar{z}_1$. The tri-state buffers for the primary and secondary outputs must

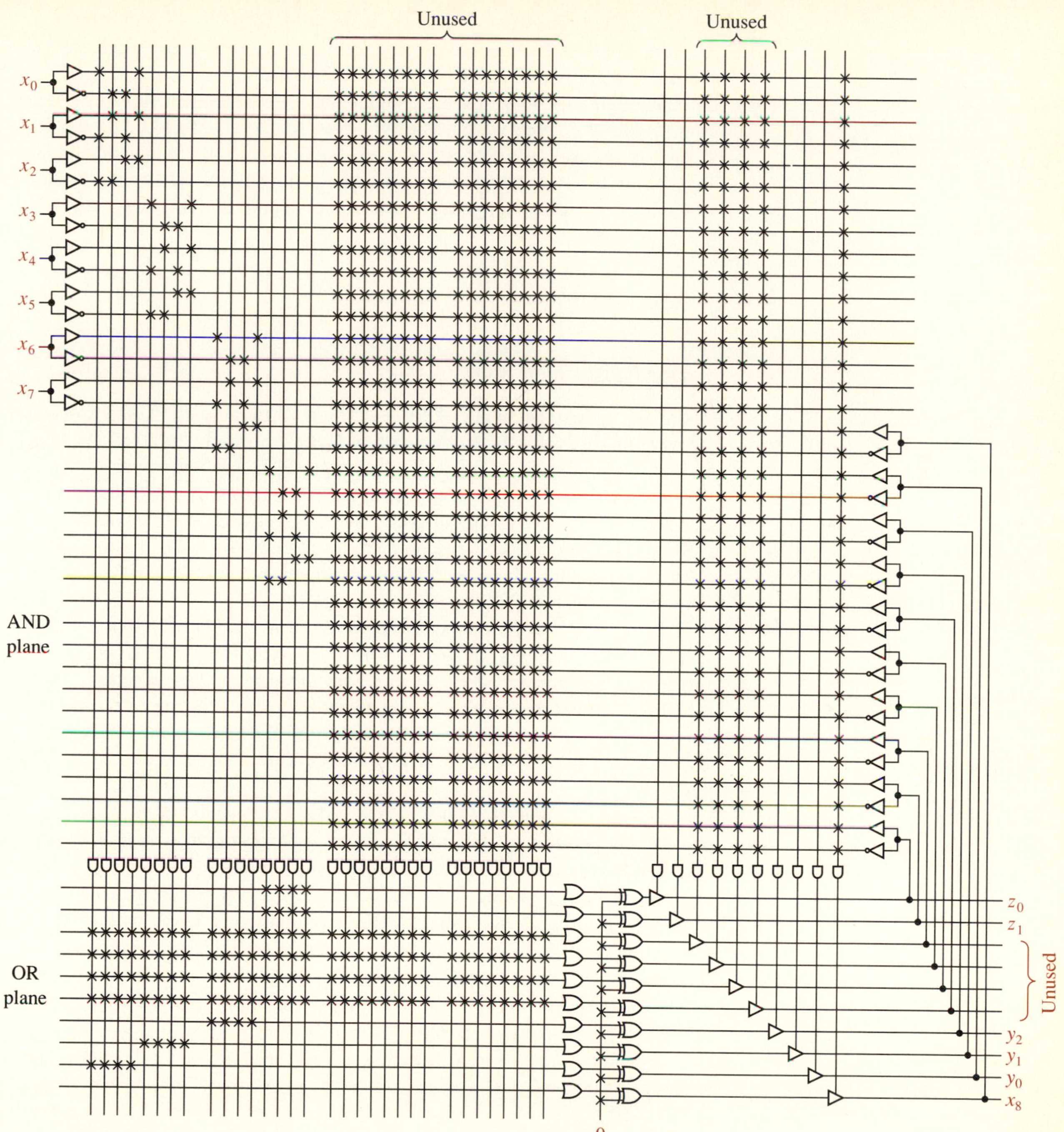

FIGURE 4.63 PLS153 implementation of the nine-input parity circuit.

be permanently enabled by disconnecting all crosspoints that link them to the control plane. In the case of primary input x_8, which is assigned to the bottom row of Figure 4.63, its tri-state buffer must be disabled by connecting its enable line to at least one input signal and its complement. It can be seen from Figure 4.63 that we have a four-level AND-OR-AND-OR design for the parity checker, and that it fits comfortably within the boundaries of the PLS153. In fact, four input/output pins along with half the AND plane remain unused and could be employed to implement additional functions.

Programmable Array Logic

A PAL is a PLA with a fixed set of OR connections.

There are many other types of PLD structures. Variants of PLAs known as **programmable array logic elements** or **PALs** are widely used and available in many logic families. The AND plane of a PAL is programmable, but its OR plane is not. The OR plane has fixed connections that link disjoint but fixed groups of AND rows, typically six to 10 at a time. For example, 72 AND rows might be ORed into nine groups of eight rows. The resulting PAL then has nine output lines, each of which implements an SOP expression that contains, at most, eight terms.

A PAL limits the number of product terms that any output function can have, and does not allow two outputs to share the same AND line. Its advantages are smaller size and therefore lower cost than a general PLA; the PAL can also be faster and somewhat easier to program. Bidirectional input/output lines, output polarity control, and feedback to the AND plane, as illustrated by the PLS153 (Figure 4.61), can also be incorporated into PALs.

We can see now that ROMs and PALs are special cases of PLAs in which only one of the AND or OR planes is programmable. A ROM generates a fixed set of AND terms—namely, all 2^n minterms—not all of which are needed in most applications. Assuming a function being implemented is expressed in optimal SOP form, a PAL generates only those product terms that are needed, but it has, in effect, OR gates of limited fan-in. Only in a general PLA in which both the AND and OR plane are programmable can the circuit be sized—at least in principle—to fit a given set of functions precisely.

Field-Programmable Gate Arrays

Another broad class of PLDs are the **field-programmable gate arrays** or **FPGAs**. These consist of two-dimensional arrays of register-level logic circuits called **cells**, which are separated by wiring channels, as shown in

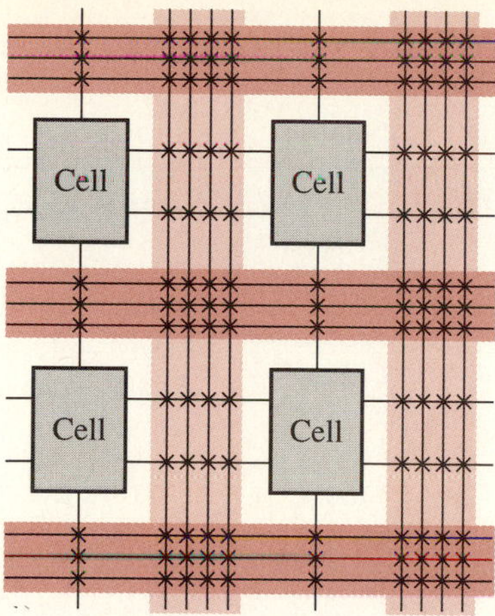

FIGURE 4.64 The layout of part of a field-programmable gate array (FPGA).

Figure 4.64. The wiring channels have programmable connections that allow the cells to be linked in essentially arbitrary ways. The cell types are rarely restricted to gates, despite the name "gate array." They are typically small multifunction combinational circuits capable of realizing many Boolean functions of a few variables, depending on how their input/output connections are programmed. A cell can also contain a few flip-flops. Like other field-programmable devices, FPGAs are primarily suited to implementing prototype designs or circuits that are to be manufactured in small quantities.

CHAPTER 4 SUMMARY

Circuit Structure

4.1 ■ Combinational switching functions map binary inputs to outputs in a memoryless fashion (i.e., the current output depends only on the current inputs). Combinational circuits implement combinational switching functions.

■ **Combinational circuits** are generally constructed from a few standard gate types, which can be interconnected according to certain rules that define **well-formed (WF) circuits**. A circuit that meets the WF definition is free of logical contradictions and is called a **well-behaved circuit.**

■ Useful in **canonical** or universal structures, a **minterm** is a function that can be expressed as a product of a literal (x_i or $\overline{x}_i$) for each input of the function. A minterm is 1 for exactly one pattern of input values. A Boolean function has a minterm for every 1 in its truth table. From this truth table we can derive its **sum-of-minterms form** as an OR sum of the minterms.

■ **Maxterms** are the duals of minterms. A maxterm is a function that can be expressed as a sum of every x_i or $\overline{x}_1$. The **product-of-maxterms form** is then the AND function or logical product of these maxterms.

- **Decimal notation** is a convenient shorthand for defining Boolean functions. A function is uniquely defined by a decimal-coded list of its minterms or maxterms.
- Every Boolean function has a **one-** or **two-level** realization that uses the standard gate types.

4.2 ■ A **gate set** is functionally complete if it alone can be used to realize any Boolean function. For instance, {AND, NOT} and {OR, NOT} are functionally complete, but {AND} and {OR} are not. {NAND} and {NOR}, on the other hand, are functionally complete. This means that we can convert a circuit composed of AND, OR, and NOT gates into an all-NAND or all-NOR circuit with the same behavior.

Technology Considerations

4.3 ■ We encountered two new logic values Z and U. Z results from an open circuit to the power source PS; U results from a simultaneous connection to the 1 and 0 terminals of PS. The state represented by Z is called the **high-impedance state**; U refers to the **unknown** or **uncertain state.** It does not occur in WF circuits except during switching.

- The **connection function** # defines the resulting signals when values from {0,1,U,Z} are combined.
- In **tri-state circuits**, the high-impedance state Z is a normal output value, in addition to 0 and 1. We can convert any logic circuit to tri-state form by inserting a switch in its output line. The control input signal of the switch is called an **enable signal**. When **disabled**, a tri-state circuit outputs a Z. Tri-state circuits allow several devices to share a signal path.
- Practical systems make use of signals of the same logic values but with differing strengths. Strong signals can overpower weak signals. This interaction can be modeled with a more complex connection function.
- A component such as an electrical resistor that converts a strong signal to a weak one is called an **attenuator**. Attenuators can be used as pull-up or pull-down circuits in logic gates.
- Several specialized logic design styles implement multilevel circuits using attenuators and wired logic. One technique is called **open-circuit (OC) logic**, for which the gates have attenuator-based designs but omit the attenuator. An external attenuator must be added and is shared among several circuits.

4.4 ■ **MOS technology,** which employs two related types of transistors (n-channel and p-channel), is one of the most important technologies for digital circuit manufacture. MOS transistors and circuits differ from **bipolar transistors**, in which positive and negative charge carriers play equal roles.

- **Complementary MOS (CMOS)** logic employs all-switch pull-up and pull-down circuits. A CMOS **transmission gate** is very close to an ideal bidirectional switch. CMOS logic is easy to design and uses less power than other logic families.

4.5 ■ Digital circuits can be implemented using off-the-shelf small-scale to very large-scale ICs. A different approach uses a partially predefined IC called a **gate array**, in which a large number of identical gates are laid out on a two-dimensional grid.

Programmable Logic Devices

4.6 ■ **Programmable logic devices (PLDs)** are ICs that have a fixed underlying set of components (switches, gates, or complex cells), but whose interconnections can be application-specific. Examples of PLDs are **read only memories (ROMs)** and **programmable logic arrays (PLAs).**

- Typically, PLDs consist of an AND plane, which generates products of inputs, and an OR plane, which generates sums of product terms. Field programmable PLDs are programmed by blowing fuses or antifuses to include inputs in product terms and product terms in sums.

- **Computer-aided design (CAD)** tools can almost completely automate PLD design.

4.7
- A ROM has a fixed AND plane and realizes a function in canonical sum-of-minterms form. A ROM stores a function's complete truth table and is well suited to complex and difficult to specify functions.
- A PLA uses programmable AND and OR planes to realize functions in (minimal) sum-of-products or product-of-sums form and uses less IC area than a ROM.
- Variants of PLAs known as **programmable array logic (PAL)** elements are widely used. A PAL is a PLA with a fixed set of OR connections. A **field-programmable gate array (FPGA)** is a two-dimensional PLD array of gates or more complex cells.

Further Readings

The classic paper defining well-formed and well-behaved combinational circuits at the gate level is A. W. Burks and J. B. Wright, "Theory of logical nets," *Proceedings of the IRE* (now called the *Proceedings of the IEEE*), vol. 41, pp. 1357–1365, October 1953. The concept of logical strength originated around 1980 and has yet to make its way into most textbooks. A short survey of the subject can be found in John P. Hayes's article, "An introduction to switch-level modeling," *IEEE Design and Test*, vol. 4, no. 4, pp. 18–25, August 1987. Switch-level design is also discussed in the context of CMOS circuit technology in the book by N. Weste and K. Eshragian, *Principles of CMOS VLSI Design* (Addison-Wesley, Reading, Mass., 1985). Roger C. Alford's *Programmable Logic Designer's Guide* (Howard W. Sams, Indianapolis, 1989) is a comprehensive introduction to PLDs and their design.

Chapter 4 Problems

Circuit Structure

Well-formed circuits; minterms and maxterms; canonical sum-of-minterms and product-of-maxterms forms; two-level SOP and POS forms; functional completeness; complete and incomplete gate sets.

Well-behaved combinational circuits that contain feedback exist.

4.1 Define each of the following terms: canonical form; functional completeness; maxterm; minterm; well-behaved circuit; well-formed circuit.

4.2 Construct examples of all-NOR circuits with at most three gates that are not well formed and that **(a)** contain feedback; **(b)** do not contain feedback.

4.3 Figure 4.65 shows an all-NAND circuit C that contains feedback. **(a)** Determine all the functions realized by C and show that they imply no contradiction. Hence, C is well behaved, despite not being well formed. **(b)** Well-behaved but not well-formed circuits like this one are not used in practice. Suggest some possible reasons for this.

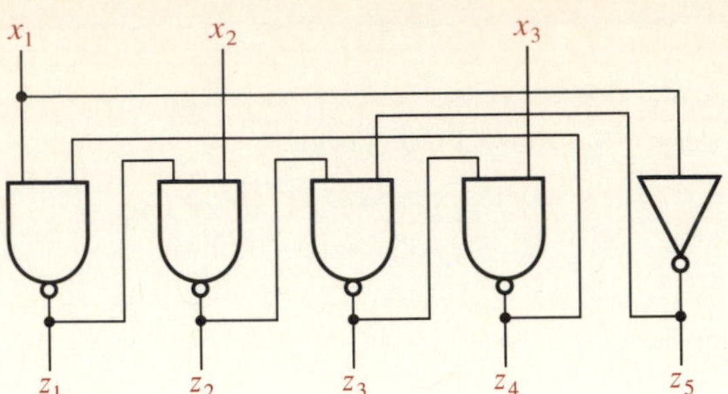

FIGURE 4.65 A combinational circuit that contains feedback.

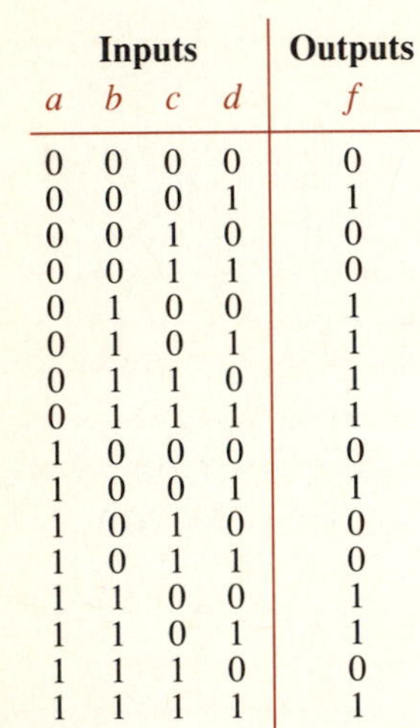

Inputs				Outputs
a	*b*	*c*	*d*	*f*
0	0	0	0	0
0	0	0	1	1
0	0	1	0	0
0	0	1	1	0
0	1	0	0	1
0	1	0	1	1
0	1	1	0	1
0	1	1	1	1
1	0	0	0	0
1	0	0	1	1
1	0	1	0	0
1	0	1	1	0
1	1	0	0	1
1	1	0	1	1
1	1	1	0	0
1	1	1	1	1

FIGURE 4.66 The truth table for a four-variable function $f(a, b, c, d)$.

4.4 Draw WF logic circuits that correspond directly to the following Boolean expressions: **(a)** $x_1\overline{x}_3 + \overline{x}_1 x_2 x_3$; **(b)** $(a + bc)(\overline{a}d + \overline{b}\,\overline{d}\,\overline{e})f$; **(c)** $(((AB + C) + DEF) + G(H + IJ))$.

4.5 Implement the three Boolean expressions that appear in Problem 4.4 using NAND gates only. Maintain as close a correspondence as you can between the structures of the expressions and the circuits.

4.6 Consider the truth table for the three-variable functions z_1, z_2, and z_3 appearing in Figure 4.6. **(a)** Construct a truth table for the function $z_4 = z_1 + z_2 z_3$. **(b)** List in standard form (sum or product expressions) all the minterms and maxterms of z_4.

4.7 Give in standard sum or product form Boolean expressions that define all the minterms and maxterms of the function $f(a, b, c, d)$ defined in Figure 4.66.

4.8 List in standard product form all the minterms for each of the following functions: **(a)** $x_1 \oplus x_2 \oplus x_3$; **(b)** $x_1 + \overline{x}_2 + \overline{x}_3$; **(c)** $x_1 x_2 + x_3 x_4$.

4.9 List in standard sum form all the maxterms for each of the following functions: **(a)** $a\overline{(b \oplus c \oplus d)}$; **(b)** $ab + c(\overline{a} + \overline{d})$; **(c)** $a + (b + (c + d + e))$.

4.10 Determine the truth or falsity of the following statements, proving your answer in each case. Assume $n > 1$. **(a)** Every n-variable function having exactly $2^n - 1$ minterms can be realized by a single AND or OR gate, possibly with input/output inversions. **(b)** No n-variable function having exactly $2^n - 1$ minterms can be realized by a single EXCLUSIVE-OR or EXCLUSIVE-NOR gate, even with input/output inversions.

4.11 **(a)** Express the following function in the usual (long) sum-of-products form.

$$z(a, b, c, d) = \sum(0, 1, 4, 5, 8, 10, 11, 12, 13, 15) \qquad (4.22)$$

(b) Construct a two-level logic circuit that corresponds to Equation (4.22).

4.12 Write down sum-of-minterms and product-of-maxterms expressions that specify the function $f(a, b, c, d)$ of Figure 4.66. Also give the shorthand decimal representation of these expressions.

4.13 **(a)** Express the function $z(a, b, c, d)$ given by Equation (4.22) of Problem 4.11 in product-of-sums form. **(b)** Using NAND or NOR gates only, construct a two-level logic circuit that corresponds to your POS expression.

4.14 Determine all the minterms and maxterms of the following Boolean functions: **(a)** $z_1 = (x_1 + \overline{x}_1 x_2 \overline{x}_3) \downarrow (x_1 \oplus \overline{x}_2)$; **(b)** $z_2 = ((x_1 \equiv x_2) \equiv x_3)$. The symbol $\equiv$ denotes the **equivalence** operator defined by $(a \equiv b) = 1$ if and only if $a = b$.

4.15 Convert each of the following four-variable Boolean expressions to their canonical sum-of-minterms and product-of-maxterms forms expressed in the shorthand decimal notation. **(a)** $((x_1 \oplus x_2) \oplus x_3) \oplus x_4$; **(b)** $((x_1 \mid x_1) \downarrow (x_2 + x_3 + x_1 x_4))\overline{x_2 x_3}$.

4.16 A two-level all-NAND implementation of a complex logic function $z(x_1, x_2, \ldots, x_n)$ has been fabricated in a semi-custom gate-array IC at considerable expense. As the IC is about to be installed in a system, it is discovered that all the gates it contains are actually NORs instead of NANDs. It is now too late to fix the IC itself, but it is possible to add some extra logic to the IC's input/output lines. The IC designer will be shot at dawn unless he comes up with a "quick fix" for this problem that will allow the IC in question to be used, despite its serious design flaw. What should he do to save his life?

4.17 Prove that the one-member gate set {NOR} is functionally complete.

4.18 Prove that the gate set {AND, OR} is not functionally complete.

4.19 Determine whether the gate set {AND, EXCLUSIVE-OR} is functionally complete. Prove your answer.

4.20 **(a)** Prove that if $z(x_1, x_2, \ldots, x_n)$ is realized by a circuit C composed entirely of two-input EXCLUSIVE-OR gates, then z is either an EXCLUSIVE-OR or EXCLUSIVE-NOR function. (*Hint:* Consider the effect on z of altering the value of a single primary input x_i.) **(b)** Use the result of part (a) to prove that {EXCLUSIVE-OR} and {EXCLUSIVE-NOR} are not functionally complete.

4.21 We define the DISAGREE function $z(x_1, x_2, \ldots, x_n)$ to be 1 if and only if $x_i \neq x_j$ for some i and j. Determine whether {DISAGREE} is functionally complete.

4.22 Show how to define the NAND operator $\mid$ in terms of the NOR operator $\downarrow$, and vice versa.

4.23 A manufacturer produces three types of ICs in identical packages distinguished only by their labels. The first contains a five-input NAND gate, the second contains a five-input NOR gate, and the third contains a five-input EXCLUSIVE-NOR gate. Due to a manufacturing problem, several thousand ICs of all three types are produced without their labels. Devise the simplest test that the manufacturer can apply to each of the unlabeled ICs to determine its type.

Technology Considerations

Wired logic; Z and U values; tri-state logic; logical strength; attenuator-based logic; MOS circuits; CMOS and NMOS logic; IC implementation; 74X series ICs; VLSI.

4.24 Define the high-impedance level Z and explain its role in tri-state circuits. Explain why Z cannot be replaced by U in this application.

4.25 Tri-state gates normally have their outputs tied together in the manner of Figure 4.4a. Explain why this is a WF structure for tri-state circuits but not for ordinary binary circuits.

4.26 Evaluate each of the following connection expressions: $\#(0, Z)$; $\#(0, 1, Z)$; $\#(0, 1, U)$; $\#(0, 0, Z, Z, Z)$; $\#(0, 0, 1, 1)$.

4.27 Give an all-switch switching circuit, graphic symbol, and function table for each of the following: **(a)** a tri-state two-input NOR gate; **(b)** a tri-state two-input EXCLUSIVE-OR gate.

4.28 The three signal values {0, 1, Z} of tri-state logic and their corresponding gate operations cannot form a Boolean algebra. Prove this by showing how they violate a specific member of Huntington's postulates.

The switch-level values {0, 1, Z, U} define a four-element Boolean algebra.

4.29 The four signal values {0, 1, Z, U} form a Boolean algebra $\mathcal{B}_4$, in which the connection operator # plays the role of *or*, and U and Z form the unit and zero elements, respectively. Construct tables defining the *and*, *or*, and *not* functions for $\mathcal{B}_4$. Suggest a possible physical interpretation for the *and* operator analogous to that of #.

4.30 Explain the following statement: "In a switch-level circuit, the key information-processing component is a connector." If this viewpoint is taken, what is the information-processing role of a circuit's switches?

4.31 Because a line L is inherently bidirectional, two independent signals z_1 and z_2 that take levels from {0, 1, Z, U} and simultaneously flow in opposite directions can be associated with L. If a signal-measuring device is attached to L, what level will it display?

4.32 We want to model the effects of some representative physical faults on the behavior of a switching circuit. For each of the following faults, determine its effect on the NAND circuit in Figure 4.67 by constructing a table in terms of {0, 1, Z, U}, defining the output function z with the fault in question present. **(a)** Switch S_1 "stuck on," implying that it is always in the on state, independent of the value of x_1. **(b)** Switch S_4 "stuck off," implying that it is always in the off state, independent of the value of x_2. **(c)** A "short-circuit" fault that permanently connects the input control terminals of switches S_3 and S_4.

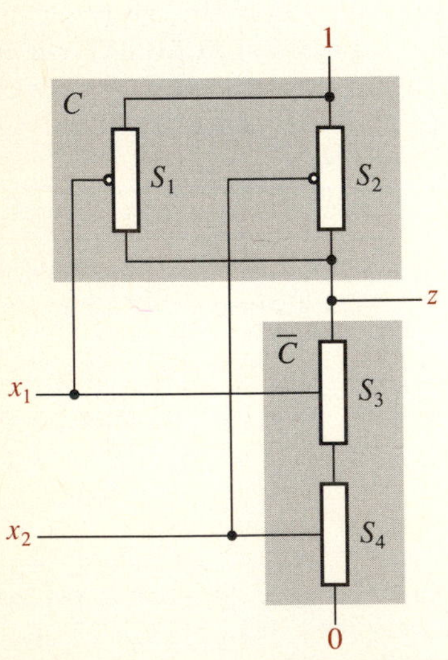

FIGURE 4.67 A complementary NAND circuit.

4.33 Define the terms logical strength and attenuator. Identify an electrical and a mechanical quantity that is modeled by each of these concepts.

4.34 Compare and contrast nMOS and CMOS technologies from the viewpoints of design complexity, circuit area, and power consumption.

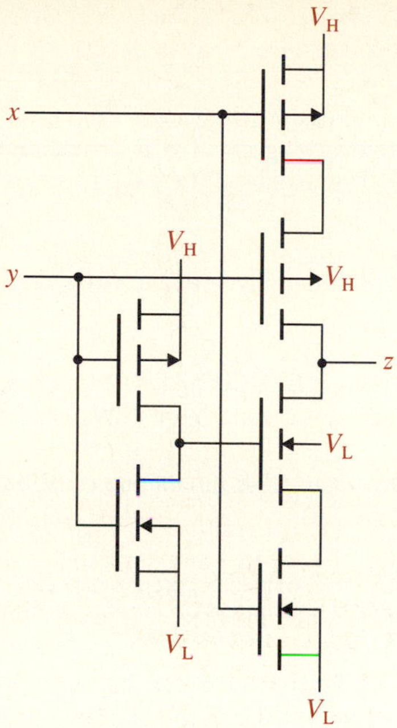

V_H
x
V_H
y
V_H
z
V_L
V_L
V_L

FIGURE 4.68 A CMOS circuit.

4.35 Describe how an attenuator can be used as a pull-down device that can drive a signal to the weak 0 level. Design a three-input NAND gate that employs an attenuator in this role. What physical technology does your circuit model?

4.36 Construct switch-level implementations of each of the following gates, using an attenuator as a pull-up device: **(a)** a three-input NOR gate; **(b)** a two-input EXCLUSIVE-OR gate.

4.37 Figure 4.68 shows a transistor circuit N_C for a CMOS cell that realizes a function $z(x, y)$. **(a)** Construct a switch-level model N_S for N_C. **(b)** Construct a function table for N_S, assuming that the inputs x and y can take only the values 0 and 1. **(c)** What useful function does this circuit perform when viewed at the gate level?

4.38 **(a)** What is a transmission gate? What are its advantages over a single-transistor switch? **(b)** Explain why the input combinations $(x_p, x_n) = (0, 0)$ and $(1, 1)$ are not used to operate the transmission gate *TG* of Figure 4.38.

4.39 A company proposes to manufacture the noninverting buffer of Figure 4.69a by realizing each switch with a transmission gate. Give a logic diagram for the resulting circuit and explain why it is impractical.

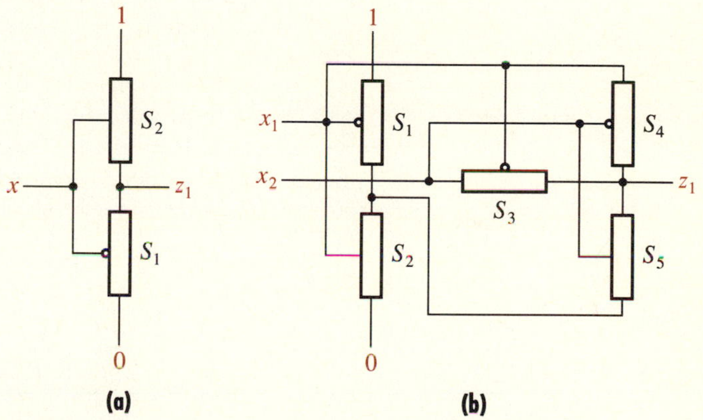

FIGURE 4.69 Two switching circuits: **(a)** a noninverting buffer; **(b)** an EXCLUSIVE-OR gate.

4.40 Suppose the noninverting buffer of Figure 4.69a is implemented by realizing each switch with an MOS transistor. Explain why the resulting two-transistor circuit is not a valid CMOS circuit. Give a valid CMOS design for such a buffer.

4.41 **(a)** Prove that the all-switch circuit of Figure 4.69b realizes the two-variable *exclusive-or* function. **(b)** Construct a CMOS transistor circuit that implements this design.

4.42 Determine the logic functions realized by the MOS circuits of Figures 4.36b, 4.40, and 4.41b if the negative logic convention is used, where V_L and V_H are interpreted as 1 and 0, respectively.

4.43 Design a CMOS implementation of a tri-state inverter with a disable line. Give a CMOS circuit diagram, a logic symbol, and a function table for your design.

4.44 Let N be a CMOS circuit with pull-up and pull-down parts C and $\overline{C}$, respectively. Let C and $\overline{C}$ both be series-parallel current-switching circuits. If C realizes the function $f(x_1, x_2, \ldots, x_n)$, show that $\overline{C}$ realizes the function $f^d(x_1, x_2, \ldots, x_n)$, which is the dual of f.

4.45 Design nMOS and pMOS implementations of **(a)** an inverter, and **(b)** a non-inverting amplifier.

4.46 Design an nMOS circuit that realizes the two-input EXCLUSIVE-OR function and compare its complexity to the CMOS circuit that appears in Figure 4.39a.

4.47 Construct a pMOS implementation of a three-input NOR circuit and describe its operation.

4.48 Write a brief essay comparing and contrasting MOS transistor circuits and electromechanical relay circuits from the viewpoints of cost, performance, and reliability.

4.49 Show that the OR gate in the full adder of Figure 4.43a can be replaced by an EXCLUSIVE-OR gate without altering the logic function being realized. Use this fact to implement the modified design, using just two ICs from the catalog of Table 4.3.

AND-OR-INVERT ICs realize small two-level logic circuits.

4.50 Figure 4.70 shows the 74X54 AND-NOR circuit, which is one of several general-purpose, two-level SSI logic circuits in the 74X series. (These circuits are usually referred to as *AND-OR-INVERT circuits* in logic data books.) Demonstrate how the 74X54 can be used with one other 74X-series IC to implement the full adder of Figure 4.43a.

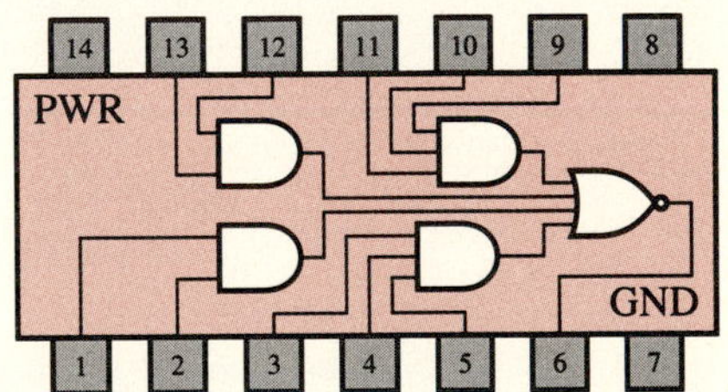

FIGURE 4.70 The 74X54 AND-OR-INVERT circuit.

4.51 Consider the all-NAND full adder given in Figure 4.71b. Partition this circuit into as few ICs as possible, using only NAND ICs from Table 4.3.

4.52 Consider the full adder in Figure 4.71a. Partition this circuit into as few ICs as possible, using only NAND and inverter ICs from Table 4.3.

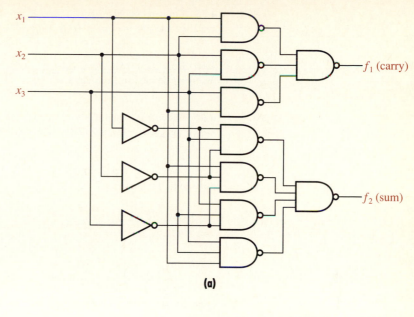

x_1
x_2
x_3
f_1 (carry)
f_2 (sum)

(a)

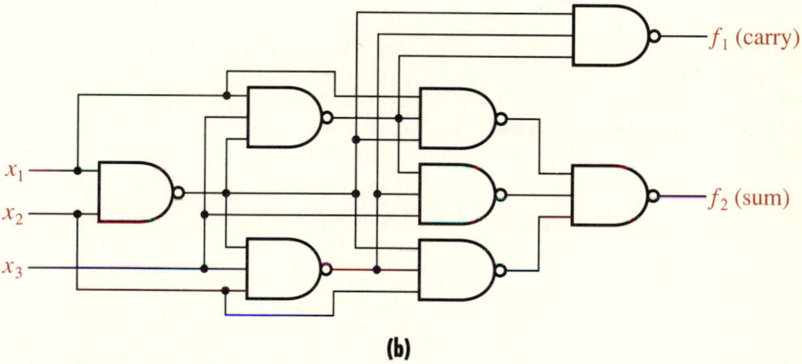

(b)

FIGURE 4.71 Two all-NAND implementations of the full adder.

4.53 Show how to implement the logic function of the 74X54 IC (Figure 4.70), using a NAND gate array of the kind that appears in Figure 4.46.

Programmable Logic Devices

Programmable circuits; read-only memories (ROMs); programmable logic arrays (PLAs); programmable array logic (PALs).

4.54 Write a brief essay comparing the three PLD types—ROMs, PLAs, and PALs—as implementation devices for large application-specific combinational circuits.

4.55 Convert the PLA circuit of Figure 4.51 into one that **(a)** uses only positive switches and attenuators, and **(b)** uses only negative switches and attenuators.

4.56 How many rows and columns are needed in a ROM that realizes the three functions defined in Problem 4.4?

4.57 Show the layout of a ROM that implements the addition of two unsigned 2-bit binary integers with input and output carries.

4.58 Show how to program the 16 × 4 ROM of Figure 4.57 to realize the following three logic functions:

$$z_0 = x_0 + \overline{x}_1 x_3 + x_1 \overline{x}_3$$

$$z_1 = \overline{x}_0 x_2 \overline{x}_3 + \overline{x}_1 x_3 + x_1 \overline{x}_3 + \overline{x}_2 \overline{x}_3$$

$$z_2 = x_3(x_0 \oplus x_1 \oplus x_2)$$

4.59 Program the ROM of Figure 4.57 to implement a multiplier of two 2-bit numbers X and Y to form a 4-bit product $P = X \times Y$.

Large-scale code conversion is an important application of ROMs.

4.60 **(a)** Program the small ROM of Figure 4.57 to make it into a code-conversion device that translates a decimal digit from BCD to excess-3 code. **(b)** How much smaller can the code converter be made by using a PLA instead of a ROM?

4.61 Japanese-language computers use a keyboard in which letters are entered in an alphabet-like phonetic script called *katakana*, which has approximately 40 different symbols. Most Japanese words are normally written in nonphonetic, *kanji* ideograms, which are of Chinese origin. Assume that 2,000 *kanji* characters and 25,000 Japanese words suffice for most purposes. Assume also that the average Japanese word can be written either as five *katakana* symbols or as two *kanji* symbols. Estimate the dimensions of a ROM needed to translate keyboard-entered *katakana* words into *kanji* words for display on a computer screen.

4.62 Repeat Problem 4.59, replacing the ROM with a PLA of minimal or near-minimal size.

4.63 Making as few changes as possible, redesign the full-adder PLA of Figure 4.60 so that the sum output is $\overline{s}_i$ instead of s_i.

4.64 Draw the PLA diagram of a PLA of minimum size that implements a full subtracter.

4.65 Modify the PLA implementation of the 9-bit parity checker in Figure 4.63 to accommodate an extra input bit. In other words, the PLA should compute the even and odd parity of a 10-bit word $x_0{:}x_9$.

4.66 Program the PLS153 PLA of Figure 4.61 to implement a 2-bit adder with carry-in and carry-out signals.

4.67 Consider the PLS153 PLA of Figure 4.61. What is the largest n such that any Boolean function of n variables can be implemented in two-level SOP form by the PLS153? Identify a specific $(n + 1)$-variable function that it cannot implement in this way.

COMBINATIONAL DESIGN

Two-Level Design

5.1 Minimal Forms
5.2 The K-Map Method
5.3 Petrick's Method
5.4 Larger Problems

Computer-Aided Design

5.5 The Tabular (Quine-McCluskey) Method
5.6 Multiple-Output Functions
5.7 Heuristic Methods

Useful Circuits

5.8 Data Transfer Logic
5.9 Adders and Subtracters

Overview

We turn to the problem of designing a gate-level logic circuit to realize an arbitrary Boolean function or set of functions. We will be concerned with not only implementing the given functions correctly, but also with doing so efficiently—in particular, with a minimum number of gates. We show how to obtain an efficient two-level (depth-two) implementation using two approaches: one that is suited to hand calculations and another intended for computer-aided design. We conclude the chapter by examining some classes of combinational circuits that are frequently encountered in practice.

5

Two-Level Design

Every logic function can be implemented by a two-level circuit that corresponds to a sum-of-products or product-of-sums Boolean expression. As we have seen, there is a profusion of techniques for implementing two-level logic: SSI circuits, open-circuit logic, ROMs, and PLAs, to name a few. We investigate some methods for constructing two-level designs that contain a minimum number of gates and connections, and so have the lowest implementation cost.

5.1 Minimal Forms

First we recall the basic properties of two-level expressions and circuits. Then we characterize minimal expressions and show how such expressions can be determined using the laws of Boolean algebra.

Canonical Forms

The canonical two-level forms are sum-of-minterms and product-of-maxterms expressions.

In Chapter 4 we saw that a typical Boolean function can be expressed in two-level form in many different ways. The most basic of these are the canonical sum-of-minterms and the product-of-maxterms forms. In the sum-of-minterms case, an n-variable function $z(X) = z(x_1, x_2, \ldots, x_n)$ can be expressed in the OR (sum) form

$$z(X) = m_a(X) + m_b(X) + \cdots + m_k(X) \tag{5.1}$$

Each subfunction $m_i(X)$ in Equation (5.1) corresponds to an input combination $i_{\mathbf{10}} = (b_{i,1}b_{i,2}\cdots b_{i,n})_{\mathbf{2}}$, for which $z(X) = 1$. We call $m_i(X)$ a minterm of $z(X)$, and express it in the AND (product) form

$$m_i(X) = \dot{x}_1\dot{x}_2\cdots\dot{x}_n$$

where the literal $\dot{x}_j = \overline{x}_j$ if $b_{i,j} = 0$, and $\dot{x}_j = x_j$ if $b_{i,j} = 1$. A product-of-maxterms expression has the general form

$$z(X) = M_p(X)M_q(X)\cdots M_t(X) \tag{5.2}$$

where $M_i(X)$ is a maxterm function that can be expressed as

$$M_i(X) = \dot{x}_1 + \dot{x}_2 + \cdots + \dot{x}_n$$

In this form, $\dot{x}_j = x_j$ if $b_{i,j} = 0$, and $\dot{x}_j = \overline{x}_j$ if $b_{i,j} = 1$. We also write Equations (5.1) and (5.2) in the condensed, decimal forms

$$z(X) = \sum(a, b, \ldots, k) = \prod(p, q, \ldots, t)$$

As an example, consider the four-variable function $z_1(a, b, c, d)$, which is defined as 1 if and only if at least two of its four inputs are 1. Thus, the input combinations that make $z_1 = 1$ are 0011, 0101, 0110, 0111, 1001, 1010, 1011, 1100, 1101, 1110, and 1111. These combinations, interpreted as 4-bit binary numbers, correspond to the 11 decimal numbers 3, 5, 6, 7, 9, 10, 11, 12, 13, 14, and 15, respectively. The other five input combinations that correspond to 0, 1, 2, 4, and 8 make $z_1 = 0$. Consequently, we have the sum-of-minterms form

$$\begin{aligned} z_1(a, b, c, d) &= \sum(3, 5, 6, 7, 9, 10, 11, 12, 13, 14, 15) \\ &= \overline{a}\,\overline{b}cd + \overline{a}b\overline{c}d + \overline{a}bc\overline{d} + \overline{a}bcd + a\overline{b}\,\overline{c}d + a\overline{b}c\overline{d} \\ &\quad + a\overline{b}cd + ab\overline{c}\,\overline{d} + ab\overline{c}d + abc\overline{d} + abcd \end{aligned} \tag{5.3}$$

and the product-of-maxterms form

$$\begin{aligned} z_1(a, b, c, d) &= \prod(0, 1, 2, 4, 8) \\ &= (a + b + c + d)(a + b + c + \overline{d})(a + b + \overline{c} + d) \\ &\quad (a + \overline{b} + c + d)(\overline{a} + b + c + d) \end{aligned} \tag{5.4}$$

We can map (5.3) into an AND-OR circuit containing 11 four-input AND gates that feed a single 11-input OR gate. Similarly, (5.4) maps into an OR-AND circuit that contains five four-input OR gates that feed a single five-input AND.

Minimal Two-level Forms

Minimal SOP (POS) expressions contain the fewest product (sum) terms.

It is easy to see that the canonical forms are not the simplest sum-of-products (SOP) or product-of-sums (POS) forms for the preceding example. Because $z_1(a, b, c, d)$ is defined as 1 if and only if any two of its inputs are 1, it follows that we can express z_1 as the sum of all two-variable product terms $x_i x_j$; in other words,

$$z_1(a, b, c, d) = ab + ac + ad + bc + bd + cd \tag{5.5}$$

Comparing Equation (5.3) with (5.5), we see that the latter requires five fewer gates in an AND-OR implementation (Figure 5.1a), and the AND gates are much smaller because their fan-in is reduced from four to two. As will be apparent later, Equation (5.5) is a minimal SOP expression for z_1 because it cannot be expressed as a sum of fewer than six product terms, and no six-product expression for z_1 has fewer literals—that is, AND-gate inputs—than (5.5).

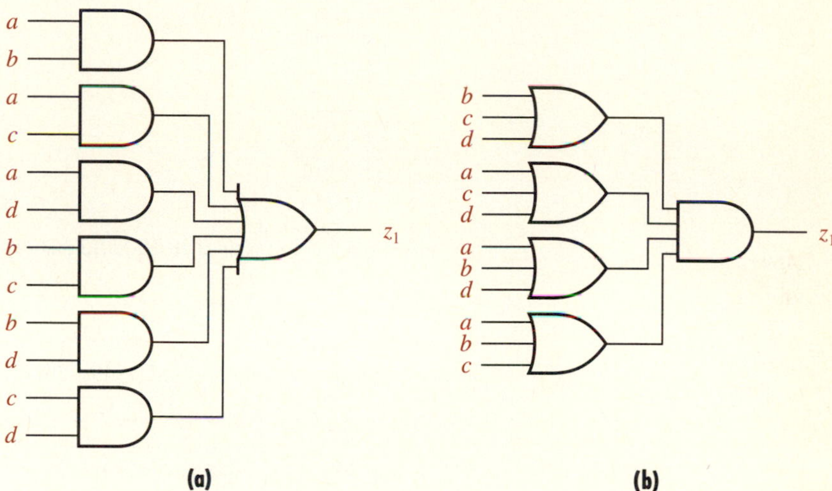

FIGURE 5.1 (a) Minimal SOP (AND-OR) and (b) POS (OR-AND) realizations of z_1.

Somewhat less obvious is the fact that we can express the same function z_1 in the following POS form:

$$z_1(a, b, c, d) = (b + c + d)(a + c + d)(a + b + d)(a + b + c) \quad (5.6)$$

[The equivalence of Equations (5.5) and (5.6) can be proven by "multiplying out" the latter into SOP form, using distributivity and other laws of Boolean algebra.] This expression, which can also be shown to be the simplest POS expression for z_1, is implemented by four three-input AND gates that feed a four-input OR gate (see Figure 5.1b). This POS design is clearly simpler than the SOP circuit of Figure 5.1a, and thus represents the best two-level realization of z_1.

The major goal of the next few sections is to develop systematic methods of obtaining minimal two-level implementations of Boolean functions. As discussed earlier, such circuits also have efficient realizations that use other gate types. For example, the AND-OR form of Figure 5.1a can be replaced by the NAND-NAND realization of Figure 5.2a. This follows from the fact that

$$ab + ac + ad + bc + bd + cd = \overline{(\overline{ab})(\overline{ac})(\overline{ad})(\overline{bc})(\overline{bd})(\overline{cd})}$$

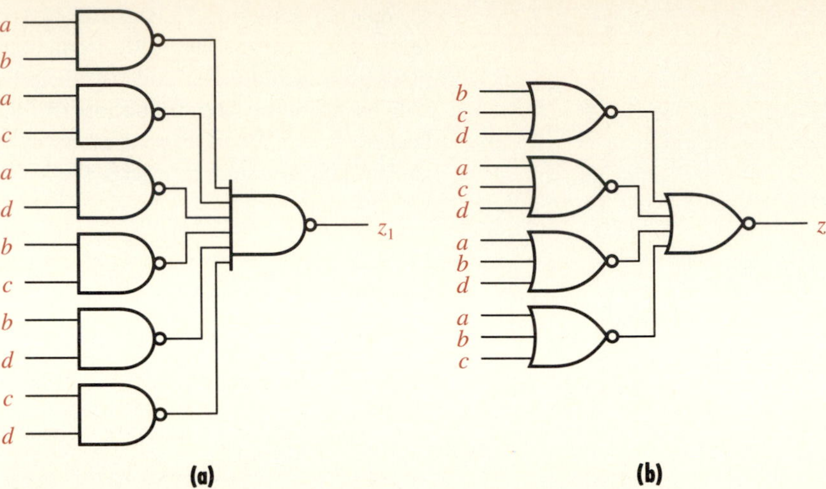

FIGURE 5.2 (a) Minimal NAND-NAND and (b) NOR-NOR realizations of z_1.

in accordance with De Morgan's law (theorem 13a). The dual version of this law (theorem 13b) applied to Equation (5.6) implies

$$(b + c + d)(a + c + d)(a + b + d)(a + b + c) = \overline{\overline{(b + c + d)} + \overline{(a + c + d)} + \overline{(a + b + d)} + \overline{(a + b + c)}}$$

leading to the NOR-NOR design of Figure 5.2b.

Algebraic Minimization

We minimize the number of literals as a secondary goal.

We can use the properties of Boolean algebra to simplify SOP or POS expressions by removing unnecessary literals or entire product or sum terms. Suppose two product terms p_1 and p_2 that appear in an SOP expression are identical except for one literal, which appears as x in one term and as $\overline{x}$ in the other. Two such terms are said to be **logically adjacent** in x. We can rewrite the sum $p_1 + p_2$ as $px + p\overline{x}$. Now

$$\begin{aligned} px + p\overline{x} &= p(x + \overline{x}) && \text{(Distributivity 4b)} \\ &= p(1) && \text{(Inverse axiom 5a)} \\ &= p && \text{(Unit axiom 2b)} \end{aligned}$$

Hence, the two product terms p_1 and p_2 that are adjacent in x can be replaced by a single term p, which is obtained by deleting the literal involving x from either of them. By the duality principle, in a POS expression we can replace the product $s_1 s_2$ of the sum terms $s_1 = s + x$ and $s_2 = s + \overline{x}$, which are also called adjacent in x, with the single, simpler term s. We summarize the foregoing argument in slightly more general form in the following **minimization theorem.**

Theorem 15

Let E be any Boolean expression and x a variable; then (a) $Ex + E\overline{x} = E$ and (b) $(E + x)(E + \overline{x}) = E$.

The key minimization step is replacing $Ex + E\overline{x}$ or $(E + x) \cdot (E + \overline{x})$ with E.

This theorem is the basis for most methods of obtaining two-level designs of minimum complexity; hence its name. By systematically identifying all logically adjacent terms associated with an SOP or POS expression for a Boolean function and replacing them with smaller terms, we can simplify the expression (and the corresponding two-level circuits). By considering a sufficiently large set of possible terms, we can find the simplest possible expression for the function.

Example 5.1

Simplifying a Boolean expression by algebraic manipulation

Consider the three-variable function with the canonical SOP form

$$z_2(a, b, c) = \sum(0, 1, 2, 3, 7) = \overline{a}\,\overline{b}\,\overline{c} + \overline{a}\,\overline{b}c + \overline{a}b\overline{c} + \overline{a}bc + abc \quad (5.7)$$

The first two minterms of Equation (5.7), $m_0 = \overline{a}\,\overline{b}\,\overline{c}$ and $m_1 = \overline{a}\,\overline{b}c$, are adjacent in c. Hence, by theorem 15a, we can replace them with the common subexpression $\overline{a}\,\overline{b}$. The third and fourth terms are also adjacent in c, and can be replaced with $\overline{a}b$. Now the fourth and fifth terms, representing m_3 and m_7, are adjacent in a. By the idempotence property of Boolean algebra, we can add another copy of the m_3 term to the expression and then combine it with m_7 according to theorem 15a, yielding the replacement term bc. At this point, Equation (5.7) has been simplified to the form

$$z_2(a, b, c) = \overline{a}\,\overline{b} + \overline{a}b + bc \quad (5.8)$$

We can apply theorem 15a once more to the first two terms in (5.8) that are adjacent in b. The result is

$$z_2(a, b, c) = \overline{a} + bc \quad (5.9)$$

which cannot be further simplified.

The foregoing algebraic simplification process for z_2 can be summarized thus:

$$z_2(a, b, c) = \underbrace{\underbrace{\overline{a}\,\overline{b}\,\overline{c} + \overline{a}\,\overline{b}c}_{\overline{a}\,\overline{b}} + \underbrace{\overline{a}b\overline{c} + \overbrace{\overline{a}bc}}_{\overline{a}b}}_{\overline{a}} \overbrace{+ abc}^{bc} \quad (5.10)$$

The same technique can be applied to the canonical POS expression for z_2 using the dual form of the minimization theorem:

$$z_2(a, b, c) = \underbrace{(\overline{a} + b + \overline{c})(\overline{a} + b + c)}_{(\overline{a} + b)} \overbrace{}^{(\overline{a} + c)}(\overline{a} + \overline{b} + c)$$

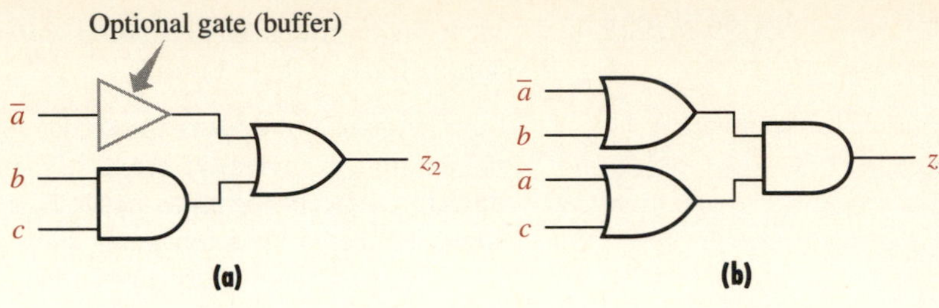

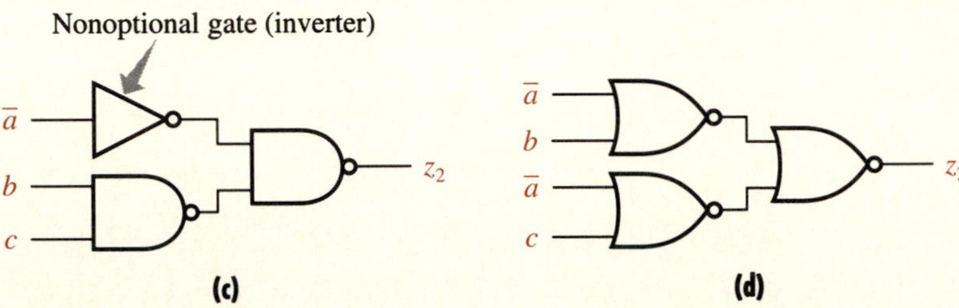

FIGURE 5.3 Minimal two-level realizations of z_2: (a) AND-OR, (b) OR-AND, (c) NAND-NAND, (d) NOR-NOR.

The result in this case is

$$z_2(a, b, c) = (\overline{a} + b)(\overline{a} + c) \tag{5.11}$$

Note the equivalence between (5.9) and (5.11), implied by distributivity.

The reduced, noncanonical expressions (5.9) and (5.11) specify the two-level circuits of Figures 5.3a and 5.3b, which are the simplest possible. No AND gate is actually needed for the $\overline{a}$ term that appears in (5.7), because a one-input AND gate—that is, a buffer—is functionally equivalent to a line or wire. Thus, the AND-OR design derived from the SOP expression for z_2 is simpler than the OR-AND design derived from the POS expression. In general, there is no way to tell in advance which of the two expression types, SOP or POS, leads to the simpler circuit. Figures 5.3c and 5.3d show all-NAND and all-NOR versions of these minimal circuits. Note that although the buffer can be omitted from the AND-OR design, the corresponding NAND-NAND circuit must contain a one-input NAND (an inverter) for the output to be functionally correct, assuming that both circuits use $\overline{a}$ as the primary input signal.

Terms that consist of one literal may require a gate in a two-level realization.

The Problem Statement

We are now in a position to specify precisely the main design problem of interest in this chapter:

Two-level minimization problem:

> Given a Boolean function $z(x_1, x_2, \ldots, x_n)$, determine an SOP or POS expression E for z that (1) contains the minimum number of product terms (in the SOP case) or sum terms (in the POS case), and (2) is such that no expression for z containing the minimum number of terms has fewer literals.

The expression E is called a **minimal SOP (POS) expression,** and a circuit N that corresponds directly to E is called a **minimal SOP (POS) circuit.**

Cost Implications

Minimizing the number of gates reduces component cost.

The primary goal of the minimization problem is to obtain a logic circuit that has, at most, two levels—that is, depth two—and contains the smallest possible number of gates. Although the actual cost of implementing a new two-level design depends on many intertwined factors, a design with the fewest gates generally requires the least hardware, and so is the least costly to implement. For example, if the target circuit N is implemented with SSI ICs, minimizing the number of gates will tend to minimize the number of ICs needed. We stress, however, that gate and IC minimization are *not* identical problems. If, for instance, a six-gate NAND circuit can be implemented using three 74X20s, each of which contains two four-input NAND gates, then reducing the gate count from six to five will not reduce the number of 74X20s needed.

Gate minimization is important in the design of ASICs and other VLSI circuits because it tends to reduce the necessary chip area, which is a key measure of the manufacturing cost of such circuits. In the case of a PLA, for example, eliminating a product term eliminates an entire row from the AND plane in an SOP realization. Term elimination also reduces the fan-in and, indirectly, the propagation delay of the output OR gate.

Minimizing the number of literals reduces connection cost.

The secondary minimization goal—minimizing the number of literals used—also aims at lowering manufacturing costs. For example,

$$z_2(a, b, c) = (\overline{a} + b)(\overline{a} + \overline{b} + c) \tag{5.12}$$

is functionally equivalent to Equation (5.11)—that is, $z_2(a, b, c) = (\overline{a} + b)(\overline{a} + c)$. Both expressions meet the primary goal of having the minimum number of sum terms: two. However, (5.11) is the more desirable (less costly) of the two because it contains fewer literals, and the corresponding circuits have gates with smaller net fan-in. Looking at it another way, minimizing the number of literals tends to minimize the total number of connections that appear in the target circuit. In the case of VLSI circuits, this also tends to reduce chip area.

Alternative Formulations

A possible alternative to the above minimization problem is to make the secondary objective the sole goal, and attempt to minimize the total number of literals that appear in an SOP or POS expression. This can, in some instances, lead to two-level designs that contain a nonminimal number of gates, but have fewer total gate inputs (net fan-in) than any minimum-gate design. The minimization techniques we will consider can be adapted to solve this problem, which is of interest in the case of implementation technologies for which the cost of the interconnections between gates is relatively high, as reflected, for example, in the chip area the connections occupy on an IC. It appears, however, that in most practical cases, changing the primary design goal from minimizing the number of gates to minimizing the number of lines has little impact on the overall cost of the resulting circuit, so we will not consider this alternative minimization problem.

Occasionally, we find the two-level minimization problem formulated without the secondary goal; that is, the problem is reduced to minimizing only the number of gates. As we have seen, the primary goal can be satisfied by designs that contain unnecessary literals that can add to implementation costs. This simplified and less realistic minimization problem has been called "sloppy" minimization [Muroga, 1979].

Fan-in/Fan-out Constraints

The minimization problem does not limit fan-in or fan-out.

A very important assumption is implicit in all the foregoing minimization problems—namely, *no limits are placed on fan-in or fan-out*. The output gate of the minimal circuit N—that is, the gate directly connected to the output line z—can have an arbitrarily large number of inputs, which is the number k of terms in the minimal expression E. Although we attempt to minimize k, it can, in the worst case, approach 2^n for n variables, and so can be excessively large, even for small values of n. Moreover, a given literal $\dot{x}$ may appear in up to k terms of E, implying that the external source of the $\dot{x}$ signal must have a maximum fan-out of at least k.

Practical logic elements have maximum fan-in and fan-out levels that are limited by technology considerations; see Section 3.5. As discussed there, fan-in and fan-out limits can be bypassed by the use of additional gates, but then the circuit no longer has two levels. Although the fan-in and fan-out of conventional gates is limited to 10 or so, these limits can be raised to as high as 50 using PLDs, possibly at the cost of increasing propagation delays.

Redundancy

Deleting redundant terms or literals alone does not ensure minimality.

It might be thought that our minimization goals could be achieved by simply deleting unnecessary or **redundant** literals or terms from any given expression for the target function. For example, the second $\overline{b}$ literal in (5.12) is redundant. On deleting this literal, we obtain the minimal POS form $z_2(a, b, c) = (\overline{a} + b)(\overline{a} + c)$.

That this approach will not work in general is illustrated by the following three SOP expressions, all of which represent the same five-variable function $z_3(a, b, c, d, e)$, as can readily be checked.

$$z_3 = ac\bar{e} + \bar{b}ce + \bar{a}\bar{c}d\bar{e} + c\bar{d}\bar{e} + \bar{a}bde + \bar{a}\bar{c}\bar{d}e + \bar{a}\bar{b}c\bar{d} \quad (5.13)$$

$$= ac\bar{e} + \bar{b}ce + \bar{a}\bar{c}d\bar{e} + bc\bar{e} + \bar{a}\bar{c}\bar{d}e + \bar{a}\bar{b}c\bar{d} \quad (5.14)$$

$$= ac\bar{e} + \bar{b}ce + \bar{a}\bar{c}d\bar{e} + bc\bar{e} + c\bar{d}\bar{e} + \bar{a}\bar{b}\bar{d}e \quad (5.15)$$

These expressions are all **irredundant;** deleting a term or literal from any of them changes the function specified. Moreover, they all have different costs, as measured by the number of terms and literals they contain. The first expression, (5.13), contains seven product terms and 25 literals. Because each of the second two expressions for z has only six terms, (5.13) cannot be minimal. The numbers of literals in (5.14) and (5.15) are 21 and 20, respectively; hence, (5.14) also cannot be minimal, despite the fact that it contains the minimum number of terms. It turns out that (5.15), with six terms and 20 literals, is, indeed, a minimal SOP expression for z_3. To distinguish among the possibilities represented by Equations (5.13), (5.14), and (5.15), we need to characterize the types of terms, known as **prime terms,** that can appear in minimal SOP and POS expressions.

Prime Implicants

Let $p(X)$ be a product term—that is, a Boolean function that can be written as a product of literals of X. If for the function $z(X)$, the relationship

$$\text{for all } A \text{ such that } p(A) = 1, z(A) = 1$$

holds, then p is called an **implicant** of z. This name comes from the fact that $p(A) = 1$ *implies* $z(A) = 1$. Another way to say this is that every minterm of p is also a minterm of z. Note that $z(B) = 1$ does not necessarily imply $p(B) = 1$; that is, z may have minterms not in p. Clearly, every product term that appears in an SOP expression for z is an implicant of z. For example, ab, bc, and $a\bar{b}c$ are all implicants of z_4, where

$$z_4 = ab + bc + a\bar{b}c \quad (5.16)$$

Other implicants of z_4 that do not appear in (5.16) are its remaining three minterms $\bar{a}bc$, $ab\bar{c}$, and abc, the nonminterm ac, and others.

A prime implicant is a product term that has no redundant literals.

An implicant p of z is called a **prime implicant** if any product term obtained by deleting a literal from p is not an implicant of z. For example, the implicant ab is a prime implicant of z_4 because neither of the terms a or b obtained by deleting a literal from ab implies z_4. To see this, consider the input combination $(a, b, c) = (1, 0, 0)$. Clearly, $a(1, 0, 0) = 1$, viewing a as a degenerate three-variable function, whereas $z_4(1, 0, 0) = 0$. Similarly, $b(0, 1, 0) = 1$, but $z_4(0, 1, 0) = 0$. The implicant $a\bar{b}c$ in (5.16) is not prime, however, because deleting $\bar{b}$ yields ac, which is an implicant of z_4—in fact, it is a prime implicant. Intuitively speaking, a prime implicant is an "indivisible" implicant in that it ceases to be an implicant if any of its literals are removed. A prime implicant can be loosely compared to a prime factor of an integer, which is an indivisible factor.

Prime Implicates

A prime implicate is a sum term with no redundant literals.

As might be expected, we have a dual concept that applies to sum terms. A sum term s that satisfies the relation

$$\text{for all } A \text{ such that } s(A) = 0, z(A) = 0$$

is called an **implicate** of z. Equivalently, s is an implicate of z if all its maxterms are maxterms of z. Clearly, the sum terms that appear in a POS expression for z are implicates of z, as are all its maxterms. If no sum term obtained by deleting a literal from implicate s is also an implicate of z, then s is a **prime implicate** of z.

For example, z_4 has the POS form

$$z_4 = (a + b)(a + \overline{b} + c)(\overline{a} + b + c) \tag{5.17}$$

in which only the sum term $a + b$ is a prime implicate. It is easily shown that $b + c$ and $a + c$ are prime implicates of z_4; hence, the "larger" implicates $a + \overline{b} + c$ and $\overline{a} + b + c$ that appear in (5.17) cannot be prime.

Theorem 16

Let E be a minimal SOP (POS) expression for a Boolean function z. Every term in E is a prime implicant (implicate) of z.

Proof

Minimal SOP (POS) expressions are composed exclusively of prime implicants (implicates).

Consider the case in which E is an SOP expression. Obviously, the product terms that appear in E are implicants of z; otherwise, E would not specify z. By way of contradiction, suppose that one of these product terms p_i is not prime. Then there is a literal $\dot{x}_j$ that can be deleted from p_i, changing it into a new implicant p_i' of z. Let E' be the SOP expression obtained from E by replacing p_i with p_i', and let E' define the function z'. Clearly, every minterm of z is also a minterm of z'. If z' contains any minterm not in z, it must be of the form $p\bar{\dot{x}}_h$, which is a minterm of p_i' but not of $p_i = p_i'\dot{x}_h$. Because p_i' is an implicant of z, all its minterms are also implicants of z. Hence, z' contains the same set of minterms as z, so $z = z'$ and $E = E'$. This means that E' is a simpler expression for z than E, which implies that the latter is nonminimal—a contradiction. Therefore, the assumption that p_i is not prime is false, so every term in E is a prime implicant. A dual argument applies to the POS case.

In order to obtain a minimal two-level implementation of an n-variable function z, we need to determine the smallest set of prime implicants (or implicates) whose sum (or product) specifies z. The algebraic simplification process we employed earlier, which made repeated use of theorem 16, can now be seen as an ad hoc method for deriving prime implicants or implicates from canonical expressions. We turn next to a more systematic way of obtaining prime terms and using them to construct minimal expressions and circuits.

5.2 The K-Map Method

Minimal two-level designs for Boolean functions of up to five or six variables can be obtained manually using a technique called the *Karnaugh-map method*. We will later extend this approach to handle functions of more variables, in which case the computer is an essential design tool.

Karnaugh Maps

A K-map is a graphical tool for two-level minimization.

A **Karnaugh map**—for short, a **K-map**—is a modified truth table intended to allow minimal SOP and POS expressions to be obtained by visual inspection. It was first described by Maurice Karnaugh of AT&T Bell Laboratories in the 1950s [Karnaugh, 1953]. The K-map is related to the Venn diagram and other graphic representations of Boolean functions.

Figure 5.4 illustrates the relationship between a truth table and a K-map for the three-variable Boolean function $z_2 = \overline{a} + bc$. Each row i of the truth table corresponds to a square region or **cell** i of the K-map. Each cell i is identified by binary row-column coordinates formed from the combination of the input variables—in this case, a, b, c—that represent the 3-bit binary number i. We will also label cells internally (in the top left corner) with a small decimal version of i. The cell at the left end of the top row of the K-map, for instance, has the coordinates $abc = 000_2 = 0_{10}$ and contains the entry 1, indicating that $z_2(0, 0, 0) = 1$. This cell, referred to as cell 0, corresponds to the first row (row 0) of the truth table. The rightmost cell in

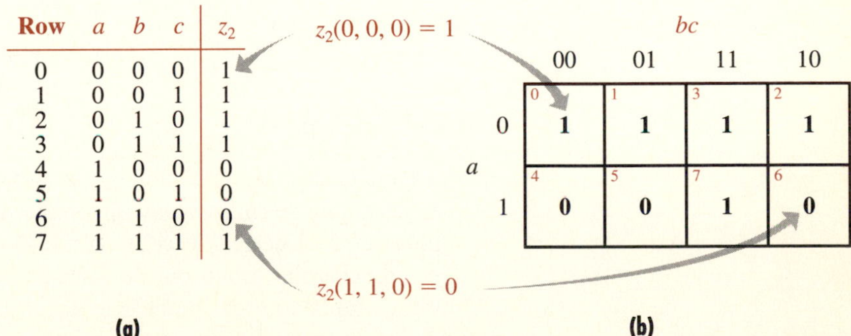

Row	a	b	c	z_2
0	0	0	0	1
1	0	0	1	1
2	0	1	0	1
3	0	1	1	1
4	1	0	0	0
5	1	0	1	0
6	1	1	0	0
7	1	1	1	1

(a)

(b)

FIGURE 5.4 Representations of the three-variable function $z_2(a, b, c)$: (a) a truth table; (b) a Karnaugh map (K-map).

the second row of the K-map (cell 6) has the label $abc = 110_2 = 6_{10}$ and indicates that $z_2(1, 1, 0) = 0$; it obviously corresponds to row 6 of the truth table.

Thus, the entry, 0 or 1, placed in the cell i of a K-map is the value of the function z that appears in row i of the truth table. A cell that contains a 1 is referred to as a **1-cell**; a cell that contains a 0 is a **0-cell**. Because the presence of a 1 in row i of a truth table implies that m_i is a minterm of z, every 1-cell of the K-map corresponds to a minterm of z. Similarly, every 0-cell corresponds to a maxterm of z.

Logical vs. Physical Adjacency

A K-map is a modified truth table that displays logical adjacencies.

The main idea behind a K-map is that it attempts to position the 0s and 1s so that logically adjacent cells are also physically adjacent. This makes it easy to recognize terms that can be combined into a single, simpler term. Recall that two minterms or maxterms are logically adjacent if they differ in a single literal. Hence, logically adjacent minterms or maxterms correspond to cells whose binary coordinates differ in one variable. For example, the 1-cells labeled 3 and 7 correspond to the minterms $m_3 = \bar{a}bc$ and $m_7 = abc$, which are adjacent in a. These two 1-cells are positioned one over the other in the third column of Figure 5.4b. The 0-cells that correspond to the adjacent maxterms $M_4 = \bar{a}+b+c$ and $M_5 = \bar{a}+b+\bar{c}$ are side by side in the bottom row of the K-map.

The cells at both ends of a row or column of a K-map are logically adjacent.

It is not immediately apparent from the K-map of Figure 5.4b that cells in the first and last columns are logically adjacent if they lie in the same row. For instance, the 1-cells for minterms $m_0 = \bar{a}\bar{b}\bar{c}$ and $m_2 = \bar{a}b\bar{c}$ at the ends of the top row are adjacent because they differ only in the b literal. When using a three-variable K-map, we view the first and last columns as physically adjacent, which would be the case if we wrapped this K-map around a cylinder. Because a K-map attempts to represent adjacencies like this that cannot be physically realized on a plane surface, some effort may be needed to spot all logical adjacencies implied by a K-map. This problem increases with the number of variables, as we will see.

Cell Grouping

In general, if a set of 2^k minterms or maxterms of z can be represented by a single product term (which will contain $n-k$ literals), then the corresponding group of 2^k 1-cells or 0-cells on the K-map for z will be clustered together in a rectangular group on the K-map. The largest groups correspond to the prime implicants or implicates of z. Figure 5.5 shows these cell groups for the example of Figure 5.4. In the SOP case illustrated by Figure 5.5a, there are two such groups: the four encircled 1-cells in the top row correspond

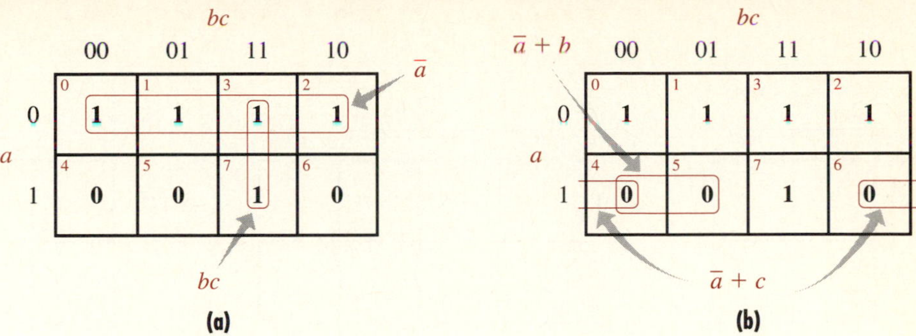

FIGURE 5.5 (a) Prime implicants and (b) prime implicates of z_2 marked on a K-map.

to the prime implicant $\bar{a}$; the second group of two encircled 1-cells in the third column of the figure correspond to the prime implicant bc. Compare the grouping of adjacent terms in Equation (5.16), which we repeat below.

$$z_2(a, b, c) = \underbrace{\underbrace{\bar a\bar b\bar c + \bar a\bar b c}_{\bar a\bar b} + \underbrace{\bar a b\bar c + \bar a b c}_{\bar a b}}_{\bar a} + abc \qquad \overbrace{\bar a bc + abc}^{bc}$$

The two groups of 0-cells that define the prime implicates of z are similarly encircled in Figure 5.5b. In design with K-maps, all prime terms are systematically identified and grouped in this fashion.

K-maps for up to Four Variables

The binary row or column labels of a K-map form a Gray code.

Figure 5.6 shows K-maps for two-, three-, and four-variable functions constructed using the method described above. Most two-variable functions can be realized by a single gate; hence, the two-variable K-map (Figure 5.6a) is not of much use. As observed above, the two columns at the end of the three-variable K-map are logically adjacent, but this cannot be conveniently shown on a flat surface. If the column labels are read from left to right, they do not follow the normal binary number sequence 00,01,10,11; instead, they follow the reflected Gray-code sequence 00,01,11,10. As discussed in Section 2.7, a Gray code has the property that successive numbers (code words) differ only in the value of one variable.

The four-variable K-map shown in Figure 5.6c is formed by doubling the three-variable map in the vertical direction and introducing a fourth variable. Note how the binary labels of the four rows are in the same Gray-code order as the labels of the four columns. In this case, the end rows as well as the end columns should be regarded as logically adjacent. Exact representation of this

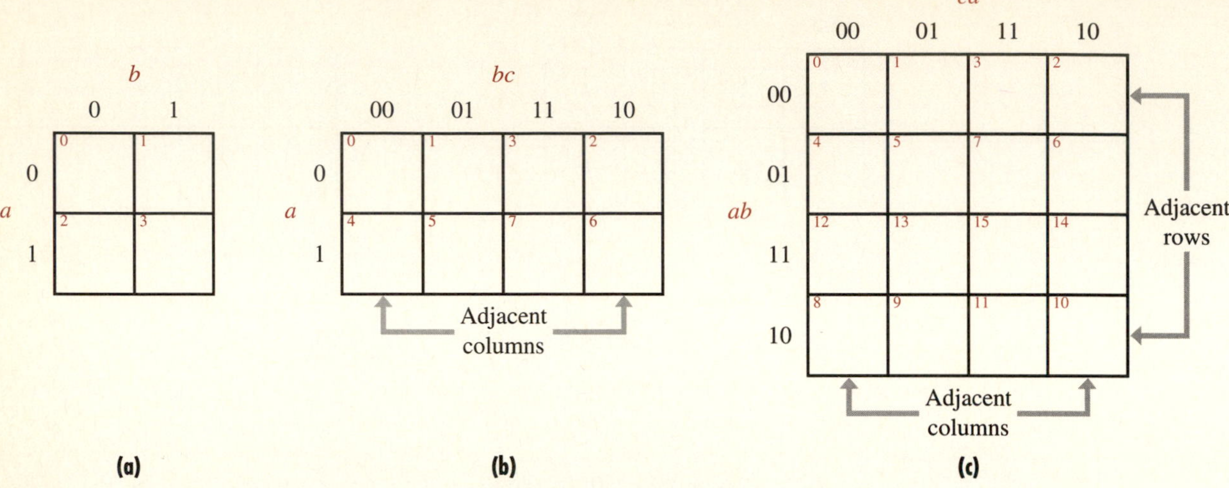

FIGURE 5.6 K-maps for Boolean functions of two, three, and four variables.

adjacency would require drawing the four-variable K-map on the surface of a *torus*—that is, a donut-shaped body. For example, cell 0 at the top left and cell 8 at the bottom left of the four-variable map of Figure 5.6c are logically adjacent, because the corresponding minterms $m_0 = \overline{a}\overline{b}\overline{c}\overline{d}$ and $m_8 = a\overline{b}\overline{c}\overline{d}$ are adjacent in a, as are the maxterms M_0 and M_8. We examine some larger K-maps later.

Minimization

Two-level minimization using the K-map approach is a two-step process that begins with a map M whose entries specify the target function $z(x_1, x_2, \ldots, x_n)$. The task of finding a minimal SOP form for z is summarized in the following procedure.

Procedure

K-map minimization procedure

1. Identify all prime implicants of z by encircling appropriate maximum-sized groups of 2^k 1-cells of M, where $1 \leq k \leq n$.
2. Select a minimum set of prime implicant groups that contain or cover all the 1-cells of z. Where several such sets exist, select one that has the lowest total connection (literal) cost.

The same approach applies to finding a minimal POS form, with prime implicates replacing prime implicants, and 0-cells replacing 1-cells. We next examine this minimization process in detail, focusing on SOP expressions; straightforward dual arguments apply to the POS case.

Prime Implicant Identification

A product term is represented on a K-map by a group of 2^k 1-cells, called a **product group**. The 1-cells are clustered together in such a way that the sum of the minterms they represent is functionally equivalent to a single product term containing $n - k$ literals. This equivalence can be demonstrated by repeated application of the relation

$$Ex + E\overline{x} = E$$

from theorem 15. A prime implicant is a product group that is not contained in any larger product group. Our goal here is to identify such **prime implicant groups** by inspecting the K-map.

Figure 5.7 shows representative sets of prime implicant groups of sizes one, two, four, and eight as they appear on the four-variable K-map. It can be seen that the 1-cell groups, which are encircled by lines, are rectangular in shape. In some cases, the groups involve 1-cells in end rows or columns, and the encircling line is broken into several pieces. To simplify the discussion, we denote the group of cells whose decimal labels are $i_1, i_2, \ldots, i_m$ by $P_{i_1,i_2,\ldots,i_m}$.

One-Cell Groups

Isolated minterms form one-cell prime implicants.

The easiest prime implicants to identify are those that correspond to a single 1-cell surrounded by 0-cells. In this case, the 1-cell is not adjacent to any other 1-cell on the K-map of the function in question. Figure 5.7a shows a function with four 1-cells of this type. Each of these 1-cells defines a prime implicant that is also a minterm of z, and so is expressed as a product of $n = 4$ literals, as shown in the figure. For example, the cell with decimal index 3 forms a one-cell group P_3, which specifies the prime implicant $\overline{a}\overline{b}cd$. Where there is no ambiguity, we equate a group with the prime implicant it represents; hence, in this case we write $P_3 = m_3 = \overline{a}\overline{b}cd$.

To determine if a 1-cell defines a prime implicant, all n of its neighbors in the horizontal and vertical (but not diagonal) directions must be inspected and be seen to be 0-cells. In the case of a 1-cell in an end row or column of the K-map, care should be taken not to overlook the logically adjacent one-cell neighbor(s) in other end rows or columns. Thus, cell 3 of Figure 5.7a has cells 1 and 2 as neighbors in the top row, and a more distant neighbor, cell 11, in the bottom row.

Two-Cell Groups

Figure 5.7b illustrates the various ways prime implicant groups can be formed from a pair of 1-cells. Each such pair is specified by a product of $n - 1 = 3$ literals. To ensure primality, a two-cell group P_{i_1,i_2} of 1-cells should not be adjacent to another two-cell group of 1-cells; that is, P_{i_1,i_2} should not form

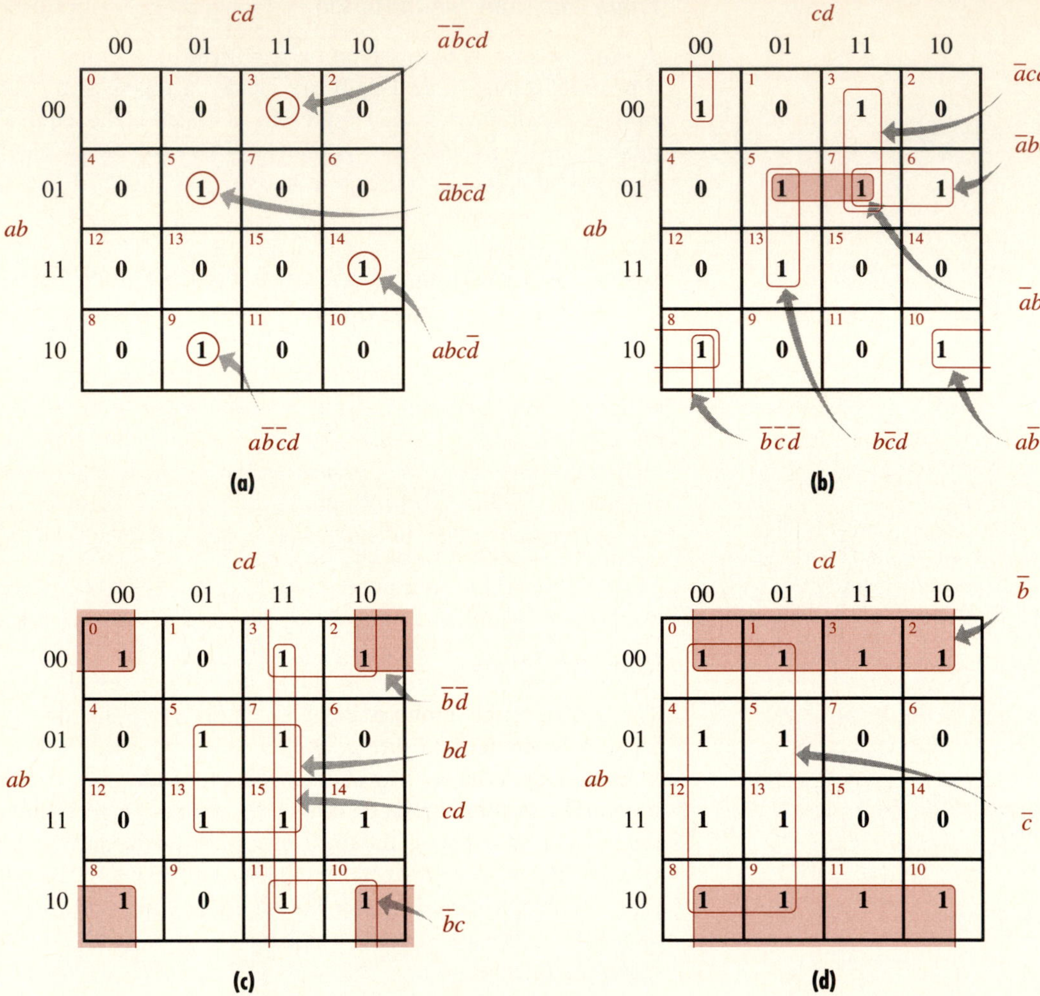

FIGURE 5.7 Prime implicant groups that involve **(a)** one, **(b)** two, **(c)** four, and **(d)** eight 1-cells.

part of a rectangular group of four 1-cells. Consider, for instance, the pair of 1-cells $P_{5,7}$ that is shown in color in Figure 5.7b. By theorem 15, this group corresponds to

$$\begin{aligned} m_5 + m_7 &= \bar{a}b\bar{c}d + \bar{a}bcd \\ &= \bar{a}bd \end{aligned}$$

That is, $P_{5,7} = \bar{a}bd$. If they contained 1s, the two-cell groups that would be physically adjacent to $P_{5,7}$ are $P_{1,3}$ (immediately above it), $P_{13,15}$ (immediately below), and $P_{4,6}$ (in the same row as $P_{5,7}$). Because cells 1, 4, and 15 are all 0-cells, none of the pairs of 1-cells in question is logically adjacent to

$P_{5,7}$, so $P_{5,7}$ denotes a prime implicant. Again, note how a prime implicant group can cross the boundaries of the K-map, as in the case of $P_{0,8} = \overline{b}\overline{c}\overline{d}$ and $P_{8,10} = a\overline{b}\overline{d}$.

Larger Groups

Prime implicants formed by groups of four 1-cells are illustrated in Figure 5.7c. The square group $P_{5,7,13,15}$ in the center of the K-map is formed from the two two-cell groups $P_{5,7}$ and $P_{13,15}$, which now represent nonprime implicants. $P_{5,7,13,15}$ defines the product term bd given by

$$\begin{aligned} m_5 + m_7 + m_{13} + m_{15} &= \overline{a}b\overline{c}d + \overline{a}bcd + ab\overline{c}d + abcd \\ &= \overline{a}bd + abd \\ &= bd \end{aligned}$$

Again, primality requires that $P_{5,7,13,15}$ not be embedded in a rectangular group of eight 1-cells. Note that if cell 15 contained a 0, as in Figure 5.7b, $P_{5,7,13,15}$ would not even be an implicant. The group $P_{3,7,11,15}$, which comprises all cells in the third column of the K-map, represents the prime implicant cd. A third group $P_{2,3,10,11}$ is formed from the right halves of the first and last row. Easily overlooked is the fourth four-cell group $P_{0,2,8,10}$, which comprises the four corner cells of the K-map and represents the prime implicant $\overline{b}\overline{d}$; this group is in color in Figure 5.7c.

Two examples of groups of eight 1-cells appear in Figure 5.7d; their interpretation is fairly obvious. Each corresponds to a prime implicant that can be denoted by a single literal. Finally, we observe that if all 16 cells of the K-map are 1-cells, the result is the constant function $z(a, b, c, d) = 1$, which can be viewed rather trivially as a prime implicant of zero variables.

The General Case

A group of 2^k 1-cells denoted by a product term and not in a larger product group is a prime implicant.

Every 2^k-cell group P that defines a prime implicant can be expressed as a product of $n - k$ literals; many examples appear in Figure 5.7. In general, if every cell in P has the fixed binary coordinate 0 (1) for the variable x, then $\overline{x}$ (x) appears in the product term for P. If two cells in P have different coordinates for x, then neither x nor $\overline{x}$ appears in the product term for P; they are effectively eliminated by the merger step $Ex + E\overline{x} = E$, defined by theorem 15. Thus, the coordinates of the cells that contribute to P consist of all 2^k possible combinations of values of the nonfixed variables, combined with the fixed coordinate pattern of the remaining variables. The colored four-cell prime implicant $P_{0,2,8,10} = \overline{b}\overline{d}$ in Figure 5.7c, for instance, contains the cells with coordinates 0000, 0010, 1000, and 1010. The coordinates b and d are both fixed at 0; the other coordinates a and c vary in all four possible ways. This observation provides a quick way to "read" the product expression of a prime implicant group directly from the K-map.

Prime Implicant Covers

The two-level SOP minimization problem is solved by a minimal prime implicant cover.

Once all prime implicants of the function z under consideration are known, the problem is to select a suitable subset $\mathcal{S}$ of these prime implicants whose OR-sum specifies z. In K-map terms, the cell groups that form $\mathcal{S}$ must enclose all the 1-cells or minterms of z, in which case $\mathcal{S}$ is said to form a **prime implicant cover** of z. If $\mathcal{S}$ is a cover with the minimum number of prime implicants, and also has the fewest literals among all such covers, then $\mathcal{S}$ is a **minimal cover** of z. It is clear that $\mathcal{S}$ then provides a solution to the two-level minimization problem, because if the members of $\mathcal{S}$ are expressed as product terms, the sum of these product terms is a minimal SOP expression for z.

Figure 5.7b illustrates the fact that, in general, not all prime implicants are needed to cover a given function. The circled groups of 1-cells represent all prime implicants of the function in question. If the colored group $P_{5,7} = \bar{a}bd$ is eliminated from the K-map, the remaining prime implicant groups suffice to cover z.

Essential Prime Implicants

An essential prime implicant is the sole cover of some minterm.

Certain prime implicants must appear in every minimal cover and are easy to identify on a K-map. In particular, a prime implicant is said to be an **essential prime implicant** if and only if it covers some 1-cell or minterm m_{i_j} that is not covered by any other prime implicant of z. Clearly, a minimal cover of z must contain all z's essential prime implicants. For example, the prime implicant $P_{6,7} = \bar{a}bc$ in Figure 5.7b is essential because it is the only prime implicant that covers $m_6 = \bar{a}bc\bar{d}$. Prime implicant $P_{5,7,13,15} = bd$ in Figure 5.7c is also essential; it alone covers $m_5 = \bar{a}b\bar{c}d$ and $m_{13} = ab\bar{c}d$.

The first step in finding a minimal cover is to identify all essential prime implicants. This is easily done by inspecting the K-map after all cell groups that define prime implicants have been marked. If some minterm (1-cell) is enclosed in or covered by exactly one prime implicant group P_A, then P_A is essential. P_A can safely be removed from further consideration, because we know that it will appear in all minimal SOP expressions for z. Furthermore, we need no longer be concerned about covering any of the minterms covered by P_A; they will always be covered by P_A in whatever final solution we obtain. Hence, the only minterms that now need to be considered are those *not* covered by essential prime implicants. In many (but not all) cases, a suitable set of inessential prime implicants to cover these remaining minterms can easily be obtained by inspection. Before we delve into this selection process, we present a complete and typical example of two-level minimization by means of K-maps.

Example 5.2 **Finding minimal two-level forms for a four-variable function**

Suppose we are asked to find minimal SOP and POS expressions for the four-variable function

$$z_1(a, b, c, d) = \sum(1, 3, 5, 10, 11, 12, 13, 14, 15) \qquad (5.18)$$

This decimal sum-of-minterms specification is easily transferred to a four-variable K-map, as shown in Figure 5.8a. The first task in the SOP minimization problem is to identify all the prime implicant groups, which is done in the same figure. Two obvious prime implicant groups of four cells cover six of the nine minterms. The remaining three minterms yield four different

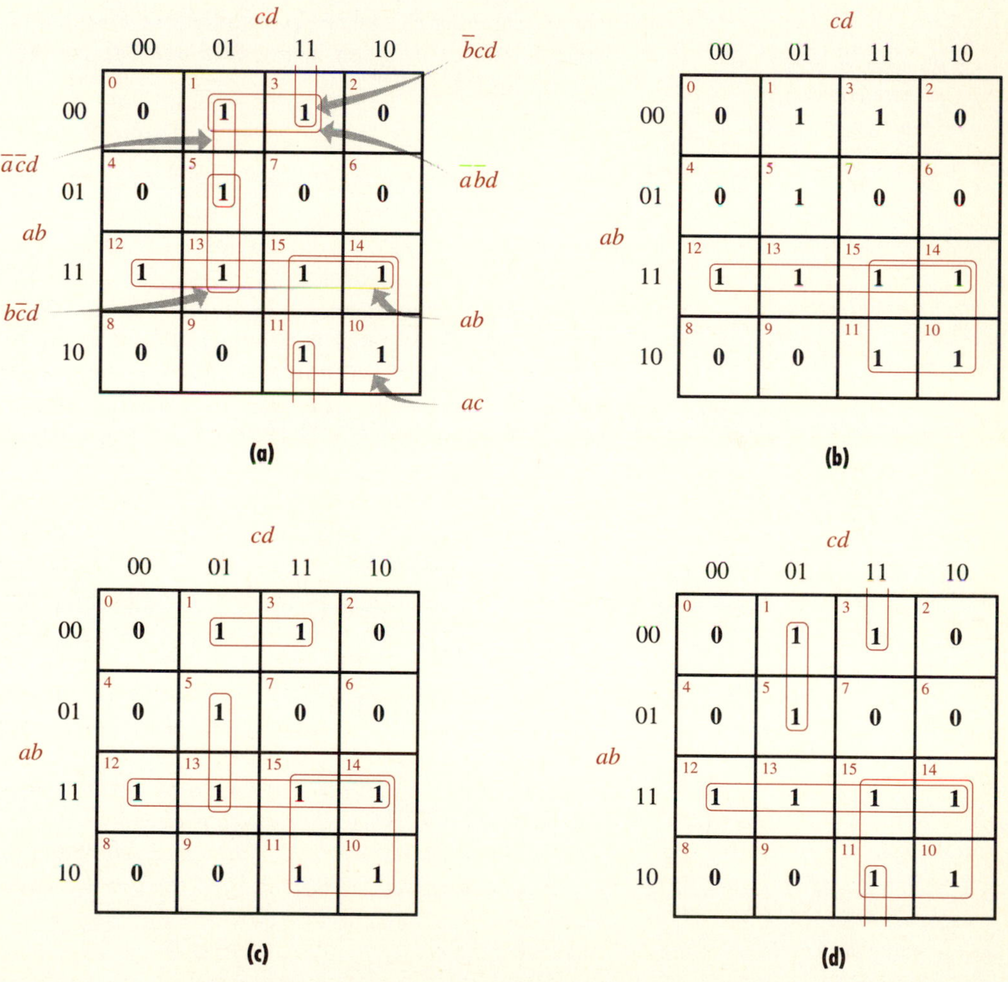

FIGURE 5.8 A four-variable function: (**a**) its prime implicants; (**b**) its essential prime implicants; (**c**) a minimal SOP cover; (**d**) another minimal SOP cover.

prime implicant groups, each of which contains two 1-cells. Thus, z_1 has a total of six prime implicants.

Inspection of all the 1-cells indicates that both four-cell groups are essential, because each covers a minterm not covered by any other prime implicant. $P_{12,13,14,15} = ab$ is the sole cover of m_{12}; only $P_{10,11,14,15} = ac$ covers m_{10}. The essential prime implicants are shown by themselves in Figure 5.8b. They and the six minterms they cover can be set aside, and we can turn our attention to covering the remaining three minterms m_1, m_3, and m_5 with the four remaining prime implicants. Because no single prime implicant covers all three, we need at least two more prime implicants. We can proceed to try in turn every possible pair taken from the four two-cell prime implicant groups; there are six possibilities. It turns out that three of the six pairs cover m_1, m_3, and m_5. For example, $P_{1,3} = \bar{a}\bar{b}d$ and $P_{5,13} = b\bar{c}d$ form one such pair. Combining $P_{1,3}$ and $P_{5,13}$ with the two essential prime implicants yields the minimal cover given in Figure 5.8c. The corresponding minimal SOP expression for z_1 is

A typical minimal cover contains both essential and inessential prime terms.

$$z_1 = ab + ac + \bar{a}\bar{b}d + b\bar{c}d \tag{5.19}$$

Note that every inessential prime implicant covers two 1-cells, so each is a product of the same number of literals—namely, three. Thus, Equation (5.19) contains the smallest possible number of literals (10) among all SOP covers consisting of four product terms. A second minimal SOP solution is given in Figure 5.8d; the reader can easily deduce the third and final one.

To obtain a minimal POS expression for z_1 we use the same grouping and selection procedures, this time applying them to the 0-cells or maxterms rather than the 1-cells. The result of grouping the maxterms to identify the prime implicates appears in Figure 5.9a. Here we are forming the largest

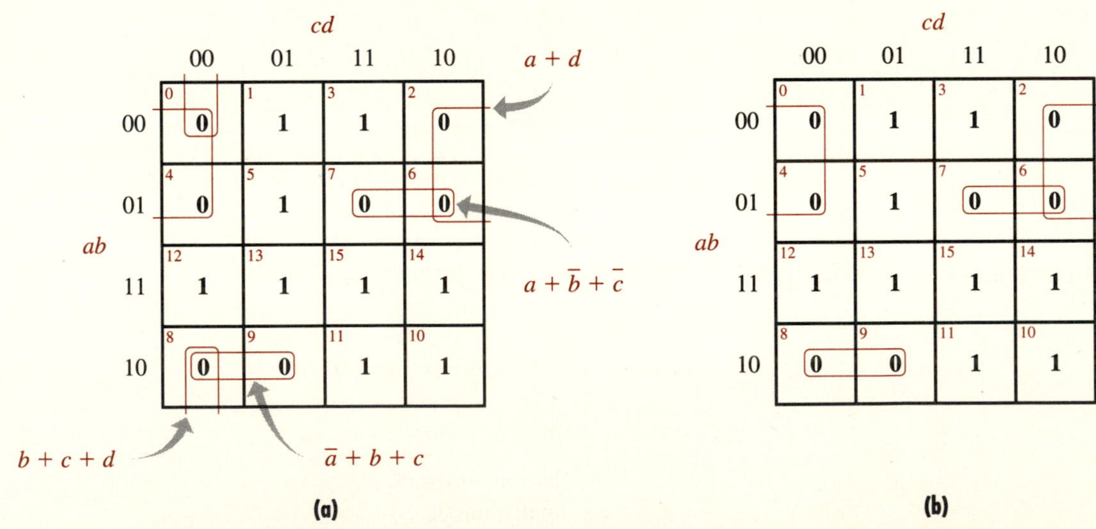

FIGURE 5.9 Function z_1: **(a)** its prime implicates; **(b)** its unique minimal POS cover.

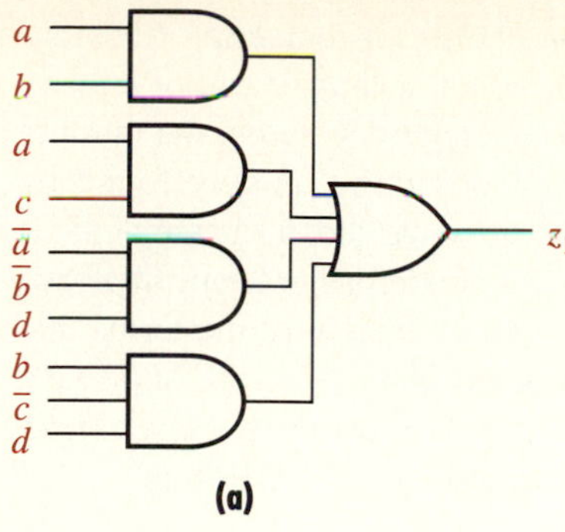

(a)

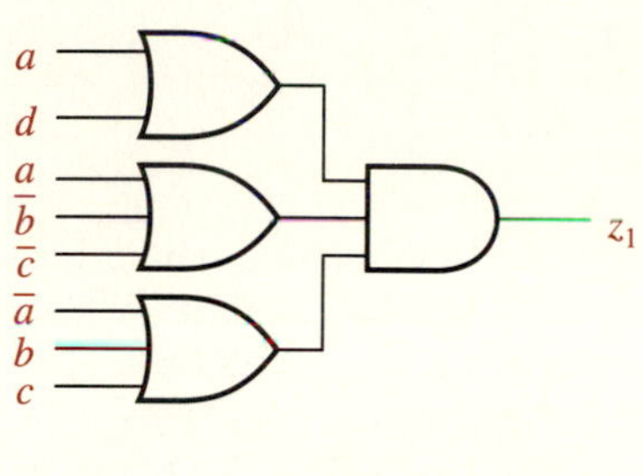

(b)

FIGURE 5.10 (a) Minimal SOP and (b) POS implementations of z_1.

2^k-member 0-cell groups that are represented by sum terms not contained in any larger sum groups. There is one four-cell prime implicate group and three two-cell prime implicate groups, so z_1 has a total of four prime implicates.

In expressing a prime implicate group S as a sum term, we choose the same literals that appear in the prime implicant group P that covers the same cells of the K-map. However, the covered cells now represent maxterms rather than minterms, so the literals that appear in true form in S are complemented in P, and vice versa. For example, z_1's only four-cell prime implicate corresponds to the sum term $S_{0,2,4,6} = a + d$. If the four cells in question were 1-cells, we would obtain the product term $P_{0,2,4,6} = \overline{a}\overline{d}$. Care must be taken when reading a sum expression for a prime implicate group from the binary coordinates on a K-map. A fixed x coordinate of 0 (1) results in x ($\overline{x}$) appearing in S.

Inspection of the seven maxterms of Figure 5.9a shows that three of the four prime implicates are essential; that is, they cover maxterms not covered by any other prime implicate. For instance, $S_{0,2,4,6} = a + d$ is the sole cover of the maxterms M_2 and M_4. The three essential prime implicates of z_1 are shown by themselves in Figure 5.9b; they must all appear in any minimal POS solution. In fact, because they cover all the maxterms of z_1, they constitute a unique minimal POS solution that corresponds to

$$z_1 = (a + d)(a + \overline{b} + \overline{c})(\overline{a} + b + c) \tag{5.20}$$

This is a particularly simple instance of the minimization problem, but one that is by no means unusual.

Figure 5.10 shows two-level AND-OR and OR-AND circuits that implement (5.19) and (5.20), respectively. In this particular case, the OR-AND circuit derived from the minimal POS expression is the simpler of the two.

Prime Implicant Selection

Absolutely inessential prime terms cannot be in any minimal cover.

The unused prime implicate $S_{0,8} = b + c + d$ that appears in Figure 5.9a has the property that all the maxterms it covers are also covered by the set of essential prime implicates. A prime implicate (and, dually, a prime implicant) with this property is said to be an **absolutely inessential prime term** because it makes no covering contribution and so cannot appear in any minimal cover. Absolutely inessential prime terms can be permanently discarded in the minimization process.

At this point, we have two rules for prime term selection when forming a minimal cover of a Boolean function:

1. Remove all essential prime terms and set them aside for inclusion in the final minimal cover.
2. Discard all absolutely inessential prime terms.

This leaves the problem of selecting from the set of remaining (inessential) prime terms a minimal cover for the minterms or maxterms not already covered by the essential terms. In most cases handled by K-maps, the number of potential covers is small and a minimal cover can be quickly found by exhaustive examination of all potential covers, as we did in Example 5.2. We begin with potential covers that consist of a single prime term, then two terms, then three terms, and so on, until we find a set of k prime terms that forms a complete cover. We then examine all such covers of size k and select one with the lowest literal cost. Combining the members of this cover with the essential prime terms produces a minimum-cost two-level design.

5.3 Petrick's Method

Next we describe a systematic, nonexhaustive way of finding minimum prime implicant covers in less obvious cases. The essence of this technique, which was developed by S. R. Petrick in the 1950s, is to express all the covering conditions that must be satisfied by the current prime implicants and minterms in the form of a product-of-sums expression $\mathbf{C}_{\text{POS}}$ that is easily derived from the K-map [Petrick, 1960]. $\mathbf{C}_{\text{POS}}$ is then converted to sum-of-products form $\mathbf{C}_{\text{SOP}}$ by a straightforward application of the laws of Boolean algebra. Each term in $\mathbf{C}_{\text{SOP}}$ that contains the minimum number of literals specifies a minimal prime implicant cover, and hence a solution to the two-level minimization problem.

Covering Propositions

Petrick's method represents all covering conditions with a POS proposition.

To illustrate the concepts involved, consider the problem that arose in Example 5.2: covering the three minterms $\mathcal{M} = \{m_1, m_3, m_5\}$ using a minimal subset of the four inessential prime implicants $\mathcal{P} = \{P_{1,3}, P_{1,5}, P_{5,13}, P_{3,11}\}$; see Figure 5.8a. To ensure that a minterm—say, m_1—in $\mathcal{M}$ is covered by $\mathcal{P}$, we must include in every cover of $\mathcal{M}$ by $\mathcal{P}$ at least one of the prime implicants that cover m_1—namely, $P_{1,3}$ *or* $P_{1,5}$. This type of condition must be met by m_1 *and* m_3 *and* m_5. Hence, all the ways that the minterms can be covered by the prime implicants are stated in the following **covering proposition**:

> A cover C of $\mathcal{M}$ by $\mathcal{P}$ must contain $P_{1,3}$ *or* $P_{1,5}$ to cover m_1, *and* it must contain $P_{1,3}$ *or* $P_{3,11}$ to cover m_3, *and* it must contain $P_{1,5}$ *or* $P_{5,13}$ to cover m_5.

This is a proposition **C** within the Boolean algebra of true-false propositions (propositional calculus), so we can manipulate it, using the laws of Boolean algebra. If we let the literal $\mathbf{P}_i$ denote the proposition "prime implicant P_i is in the cover," and denote *or* and *and* by + and juxtaposition, respectively, then the above covering proposition can be written in concise symbolic form, as follows:

$$\mathbf{C} = (\mathbf{P}_{1,3} + \mathbf{P}_{1,5})(\mathbf{P}_{1,3} + \mathbf{P}_{3,11})(\mathbf{P}_{1,5} + \mathbf{P}_{5,13}) \tag{5.21}$$

The covering proposition (5.21) is a Boolean expression in POS form. We convert it to SOP form by "multiplying it out"—that is, by repeated application of the distributive laws of Boolean algebra.

$$\begin{aligned}\mathbf{C} = {} & \mathbf{P}_{1,3}\mathbf{P}_{1,3}\mathbf{P}_{1,5} + \mathbf{P}_{1,3}\mathbf{P}_{1,3}\mathbf{P}_{5,13} + \mathbf{P}_{1,3}\mathbf{P}_{3,11}\mathbf{P}_{1,5} + \mathbf{P}_{1,3}\mathbf{P}_{3,11}\mathbf{P}_{5,13} \\ & + \mathbf{P}_{1,5}\mathbf{P}_{1,3}\mathbf{P}_{1,5} + \mathbf{P}_{1,5}\mathbf{P}_{1,3}\mathbf{P}_{5,13} + \mathbf{P}_{1,5}\mathbf{P}_{3,11}\mathbf{P}_{1,5} + \mathbf{P}_{1,5}\mathbf{P}_{3,11}\mathbf{P}_{5,13}\end{aligned}$$

The number of product terms generated by Petrick's method can be reduced by applying some other laws of Boolean algebra, provided no terms are eliminated that might lead to a minimal cover. For example, we can apply the idempotence laws $a + a = a$ and $aa = a$ to eliminate repeated product terms and repeated occurrences of $\mathbf{P}_i$ within product terms. Applying idempotence to Equation (5.21) yields

$$\begin{aligned}\mathbf{C} = {} & \mathbf{P}_{1,3}\mathbf{P}_{1,5} + \mathbf{P}_{1,3}\mathbf{P}_{5,13} + \mathbf{P}_{1,3}\mathbf{P}_{3,11}\mathbf{P}_{1,5} + \mathbf{P}_{1,3}\mathbf{P}_{3,11}\mathbf{P}_{5,13} \\ & + \mathbf{P}_{1,5}\mathbf{P}_{1,3}\mathbf{P}_{5,13} + \mathbf{P}_{1,5}\mathbf{P}_{3,11} + \mathbf{P}_{1,5}\mathbf{P}_{3,11}\mathbf{P}_{5,13}\end{aligned} \tag{5.22}$$

Each product term in (5.22) represents a complete cover of $\mathcal{M}$ by $\mathcal{P}$. For example, the first term $\mathbf{P}_{1,3}\mathbf{P}_{1,5}$ represents the cover that consists of the prime implicants $P_{1,3}$ and $P_{1,5}$. Those product terms that contain the fewest occurrences of $\mathbf{P}_i$ and whose occurrences of $\mathbf{P}_i$ correspond to P_{ij} with the fewest literals represent minimal SOP covers. In this case, three product terms contain only two $\mathbf{P}_i$: $\mathbf{P}_{1,3}\mathbf{P}_{1,5}$, $\mathbf{P}_{1,3}\mathbf{P}_{5,13}$, and $\mathbf{P}_{1,5}\mathbf{P}_{3,11}$. Each corresponds to a set of prime implicants that involve a total of four literals. Hence, there are three possible minimal covers of $\mathcal{M} = \{m_1, m_3, m_5\}$: $\{P_{1,3}, P_{1,5}\}$, $\{P_{1,3}, P_{5,13}\}$, and $\{P_{1,5}, P_{3,11}\}$.

Procedure

Petrick's method for finding a minimal cover

1. Construct the covering proposition $\mathbf{C}$ that consists of the product of $|\mathcal{M}|$ sum terms of the form $\mathbf{S} = \mathbf{P}_{i_1} + \mathbf{P}_{i_2} + \cdots + \mathbf{P}_{i_k}$, one for each $m_i \in \mathcal{M}$, where $\mathbf{S}$ includes a symbol $\mathbf{P}_{i_j}$ for every member P_{i_j} of $\mathcal{P}$ that covers m_i.
2. Convert $\mathbf{C}$ to sum-of-products form by repeatedly applying the distributive laws, and eliminate redundant literals and terms by repeatedly using idempotence and certain other laws of Boolean algebra.
3. Examine all product terms in $\mathbf{C}$ that contain the fewest $\mathbf{P}_{i_j}$ symbols. Select one of these terms $\mathbf{P}_{i_1}\mathbf{P}_{i_2}\ldots\mathbf{P}_{i_k}$ for which the corresponding set C of prime terms $P_{i_1}, P_{i_2}, \ldots, P_{i_k}$ contains the smallest total number of x_k literals; C is the desired minimal cover.

The Petrick method for finding a minimal SOP cover is summarized in the preceding procedure. Here $\mathcal{M}$ is a set of minterms of the function z under consideration, and $\mathcal{P}$ is the set of prime implicants of z that are required to cover $\mathcal{M}$.

It should be noted that the method can be applied with no change whatsoever to the problem of finding a minimal POS cover. All that is required is to equate $\mathcal{M}$ with a set of maxterms and $\mathcal{P}$ with a set of prime implicates of z. In fact, our procedure presents Petrick's method in a general form that is applicable to a variety of covering problems in various disciplines, depending on the interpretation given to $\mathcal{M}$ and $\mathcal{P}$.

This elegant procedure is more complex from a computational viewpoint than first meets the eye, because the (logical) multiplication process in step 2 can produce an enormous number of terms. For example, if each sum term **S** in the POS form of **C** contains r literals, and there are s sum terms in **C**, the multiplication process will generate up to r^s product terms. If $r = s = 4$, for example, $r^s = 4^4 = 256$, but if $r = s = 8$, then $r^s = 8^8 = 16{,}777{,}216$. In design problems of up to four or so variables, r and s are quite small, so Petrick's method is practical for the relatively rare cases in which a minimal cover cannot be determined from a K-map by inspection.

Computation Shortcuts

As remarked already, the number of product terms generated by Petrick's method can be greatly reduced by judicious application of the laws of Boolean algebra. Specifically, we can apply the idempotence laws $a + a = a$ and $aa = a$ to delete repeated terms and literals, and the absorption law $a + ab = a$ to eliminate a term of the form ab, provided a is retained. For example, applying absorption to Equation (5.22) eliminates four of its seven product terms. This type of simplification can be done at any stage in the conversion process from POS to SOP form.

Covering propositions can be simplified via various laws of Boolean algebra.

Covering propositions can also be simplified by means of the distributive law $(a+b)(a+c) = a+bc$, which effectively replaces two sum terms with one. Applying this law to the first two sum terms of (5.21), we get

$$\mathbf{C} = (\mathbf{P}_{1,3} + \mathbf{P}_{1,5}\mathbf{P}_{3,11})(\mathbf{P}_{1,5} + \mathbf{P}_{5,13})$$

so that on multiplying out to SOP form we obtain four rather than eight product terms. The following generalization of the basic distributive law is very useful:

$$(a_1 + a_2 + \cdots + a_k + b_1 + b_2 + \cdots + b_p)(a_1 + a_2 + \cdots + a_k + c_1 + c_2 + \cdots + c_q)$$
$$= a_1 + a_2 + \cdots + a_k + b_1c_1 + b_2c_1 + \cdots + b_pc_1 + b_2c_1 + \cdots + b_pc_q \tag{5.23}$$

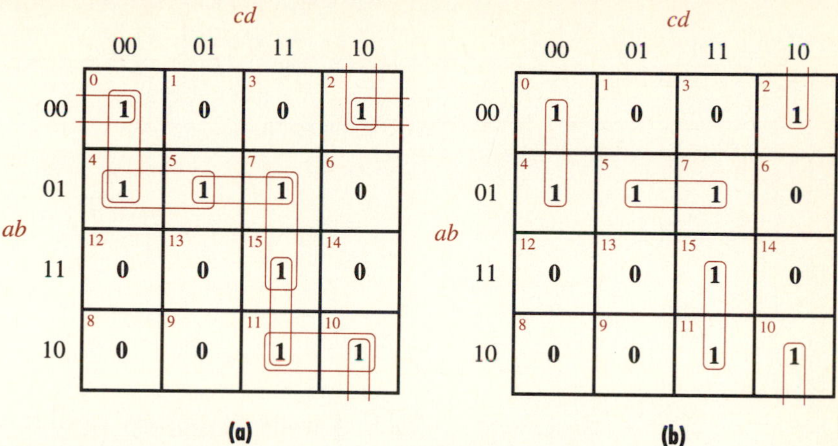

FIGURE 5.11 (a) Prime implicants and (b) a minimal SOP cover of z_2.

Example 5.3

Using Petrick's method to find a minimal prime implicant cover

We illustrate Petrick's method for a difficult type of Boolean function

$$z_2(a, b, c, d) = \sum(0, 2, 4, 5, 7, 10, 11, 15) \tag{5.24}$$

all of whose prime implicants are inessential and form a closed sequence of overlapping cell groups—a so-called "cycle"—on the K-map (Figure 5.11a). There are eight easily derived prime implicants, each covering two minterms. Hence, a minimal SOP cover must contain at least four prime implicants to cover the eight minterms. Each minterm is covered by two prime implicants, so none of the latter are essential.

Applying Petrick's method to the problem of covering the minterms of z_2 by its prime implicants yields the covering proposition

$$\begin{aligned}\mathbf{C} = {} & (\mathbf{P}_{0,2} + \mathbf{P}_{0,4})(\mathbf{P}_{0,4} + \mathbf{P}_{4,5})(\mathbf{P}_{4,5} + \mathbf{P}_{5,7})(\mathbf{P}_{5,7} + \mathbf{P}_{7,15}) \\ & (\mathbf{P}_{7,15} + \mathbf{P}_{11,15})(\mathbf{P}_{11,15} + \mathbf{P}_{10,11})(\mathbf{P}_{10,11} + \mathbf{P}_{2,10})(\mathbf{P}_{2,10} + \mathbf{P}_{0,2})\end{aligned} \tag{5.25}$$

If we simply multiplied out this POS expression, retaining all terms, we would obtain $2^8 = 256$ terms, each containing eight literals. Instead, observing that successive pairs of sum terms in (5.25) contain a common literal, we apply the distributive law $(a + b)(a + c) = a + bc$, as discussed above, to pairs of sum-terms that contain a common literal $\mathbf{P}_i$.

$$\begin{aligned}\mathbf{C} = {} & (\mathbf{P}_{0,4} + \mathbf{P}_{0,2}\mathbf{P}_{4,5})(\mathbf{P}_{5,7} + \mathbf{P}_{4,5}\mathbf{P}_{7,15})(\mathbf{P}_{11,15} + \mathbf{P}_{7,15}\mathbf{P}_{10,11}) \\ & (\mathbf{P}_{2,10} + \mathbf{P}_{10,11}\mathbf{P}_{0,2})\end{aligned}$$

A further multiplication produces

$$\begin{aligned}\mathbf{C} = {} & \mathbf{P}_{0,4}\mathbf{P}_{5,7}\mathbf{P}_{11,15}\mathbf{P}_{2,10} + \mathbf{P}_{0,2}\mathbf{P}_{4,5}\mathbf{P}_{7,15}\mathbf{P}_{10,11} \\ & + \text{additional terms containing five or more literals}\end{aligned} \tag{5.26}$$

We conclude that there are only two covers that contain four prime implicants, as specified in (5.26). Moreover, both involve the same number of literals (eight); hence, both covers are minimal. Figure 5.11b shows the minimal cover that corresponds to $\mathbf{P}_{0,4}\mathbf{P}_{5,7}\mathbf{P}_{11,15}\mathbf{P}_{2,10}$ and yields the minimal SOP expression

$$\begin{aligned} z_2 &= P_{0,4} + P_{5,7} + P_{11,15} + P_{2,10} \\ &= \bar{a}\bar{c}\bar{d} + \bar{a}bd + acd + \bar{b}c\bar{d} \end{aligned}$$

5.4 Larger Problems

The K-map method is limited to about five or six variables.

In principle, Karnaugh maps can be constructed for any number of variables n. In general, an n-variable K-map is formed from two copies of an $(n-1)$-variable K-map, with the copies positioned and labeled in a way that maximizes the correspondence between logical and physical adjacency. However, as should be apparent from the discussion so far, it becomes increasingly difficult to identify these adjacencies as n increases. Moreover, beyond five or six variables, the number of prime implicants or implicates can become unmanageable.

A Five-Variable K-Map

The K-map we will use for representing a five-variable Boolean function $z(a, b, c, d, e)$ is shown in Figure 5.12a. It consists of two copies of the four-variable map, one of which is reflected or flipped horizontally. This gives a total of eight columns, the binary labels of which form a 3-bit reflected Gray code. Now, for the first time, we have internal columns that are logically, but not physically, adjacent, which significantly complicates the process of identifying the prime implicants. An alternative form of five-variable K-map preferred by some authors appears in Figure 5.12b. It consists of two four-variable maps: one for $e = 0$, another for $e = 1$. These maps must be conceptually superimposed, one on top of the other, in order to display the adjacencies between the columns.

As an example of a five-variable function, consider

$$z_3(a, b, c, d, e) = \sum(0, 2, 4, 6, 10, 11, 14, 15, 16, 18, 20, 22, 25, 27, 31)$$

Figure 5.13 shows that it has four prime implicants, all of which are essential. The four-cell prime implicant $P_{11,15,27,31}$, indicated by the light color, is split between the third and sixth columns. Observe that in their 5-bit binary forms, the cell labels of this prime implicant are 01011, 01111, 11011, and 11111, which differ only in the a and c positions. Because $b = d = e = 1$ in each

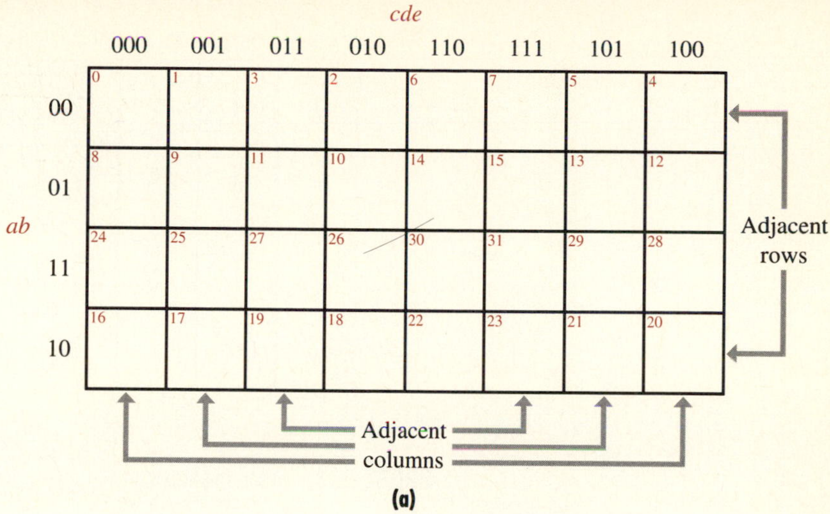

(a)

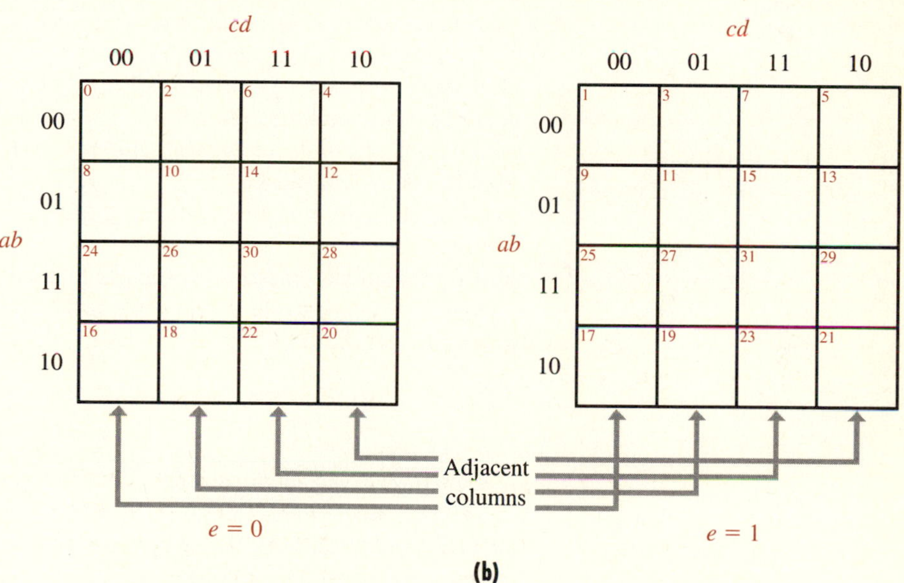

(b)

FIGURE 5.12 Five-variable K-maps: (a) our form; (b) an alternative form.

case, we conclude that $P_{11,15,27,31} = bd\,e$. There is also an eight-cell prime implicant $P_{0,2,4,6,16,18,20,22} = \overline{b}\overline{e}$, indicated by the 1-cell grouping in dark color, which is broken into six separate pieces.

Example 5.4

Designing the control logic for a patient monitor

Suppose that a monitoring system is attached to a bedridden patient in a hospital. A sensor S_1 determines when the patient is awake ($S_1 = 0$) or

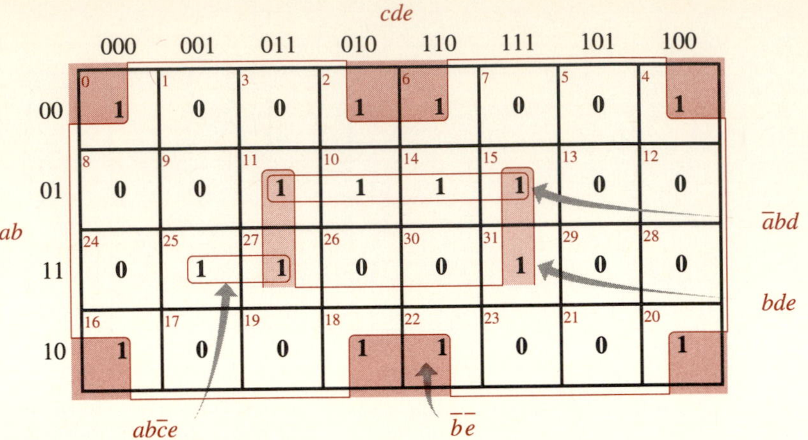

FIGURE 5.13 Prime implicants of the five-variable function f.

asleep ($S_1 = 1$). A second sensor measures certain vital signs, including pulse and blood pressure, and issues a status signal S_2 in the form of a 4-bit binary number that indicates the patient's condition. When all is normal and the patient is awake, S_2 should fall in the range 6 to 12; when the patient is asleep, the normal range for S_2 changes to between 2 and 8. Design a minimum-cost two-level logic circuit that will sound an alarm if S_2 falls outside the normal ranges.

The problem is to find a minimal SOP or POS form of a five-variable function $f_1(a, b, c, d, e)$, where $S_1 = a$ and $S_2 = b, c, d, e$. Let $f_1 = 1$ activate the alarm. Encoding $abcde$ into a decimal number X, we see that $f_1 = 1$ if X falls outside the ranges 6 to 12 and 18 to 24. Hence, f_1 has the following specification:

$$f_1 = \sum(0, 1, 2, 3, 4, 5, 13, 14, 15, 16, 17, 25, 26, 27, 28, 29, 30, 31) \tag{5.27}$$

A K-map that corresponds to (5.27) appears in Figure 5.14.

First, we determine a minimal SOP expression for f. Figure 5.14a shows all prime implicants for f, which are 10 in number. Note, in particular, the two two-part prime implicant groups that are colored. Six of the prime implicants are essential, because each covers a 1-cell not covered by any other prime implicant. The essential prime implicant groups are shown together in Figure 5.14b. Only two 1-cells, m_{13} and m_{25}, are not covered by the essential terms. Now m_{13} is covered by two prime implicants, bce and $\bar{a}c\bar{d}e$. Because bce has fewer literals, it is selected for inclusion in the final solution. Similarly, the prime implicant abe is chosen to cover m_{25}. Hence, we have a unique minimal SOP solution

$$f_1 = \bar{a}\bar{b}\bar{c} + bcd + \bar{a}\bar{b}\bar{d} + bce + abc + abe + abd + \bar{b}\bar{c}\bar{d} \tag{5.28}$$

which contains eight prime implicants.

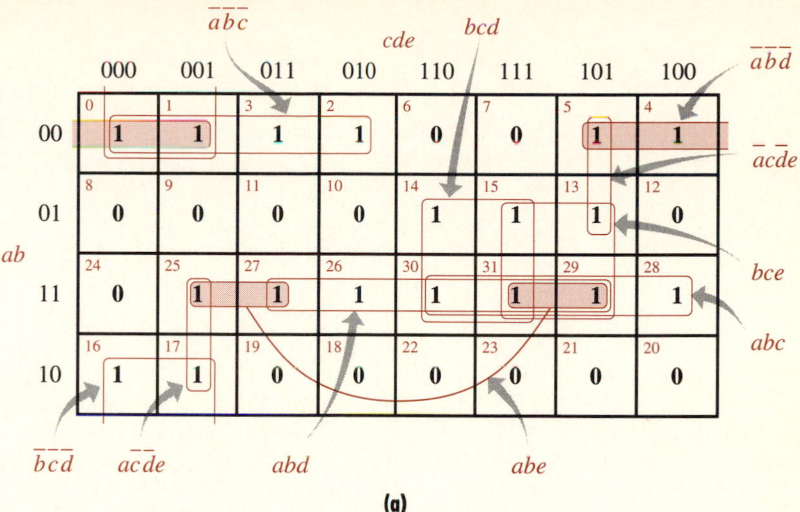

(a)

ab \ cde	000	001	011	010	110	111	101	100
00	(0) 1	(1) 1	(3) 1	(2) 1	(6) 0	(7) 0	(5) 1	(4) 1
01	(8) 0	(9) 0	(11) 0	(10) 0	(14) 1	(15) 1	(13) 1	(12) 0
11	(24) 0	(25) 1	(27) 1	(26) 1	(30) 1	(31) 1	(29) 1	(28) 1
10	(16) 1	(17) 1	(19) 0	(18) 0	(22) 0	(23) 0	(21) 0	(20) 0

(b)

FIGURE 5.14 (a) All prime implicants and (b) all essential prime implicants of the five-variable function f_1.

The prime implicates of f_1 appear in the five-variable K-map of Figure 5.15. All six are essential, implying the following minimal POS form for f_1:

$$f_1 = (b+\bar{c}+\bar{d})(a+\bar{b}+d+e)(\bar{a}+b+\bar{c})(\bar{a}+b+\bar{d})(a+\bar{b}+c)(\bar{b}+c+d+e) \quad (5.29)$$

Comparing Equations (5.28) and (5.29), we see that the minimal SOP form has eight terms and 24 literals, whereas the minimal POS form has only six terms and 20 literals. We therefore conclude that the POS form is optimal.

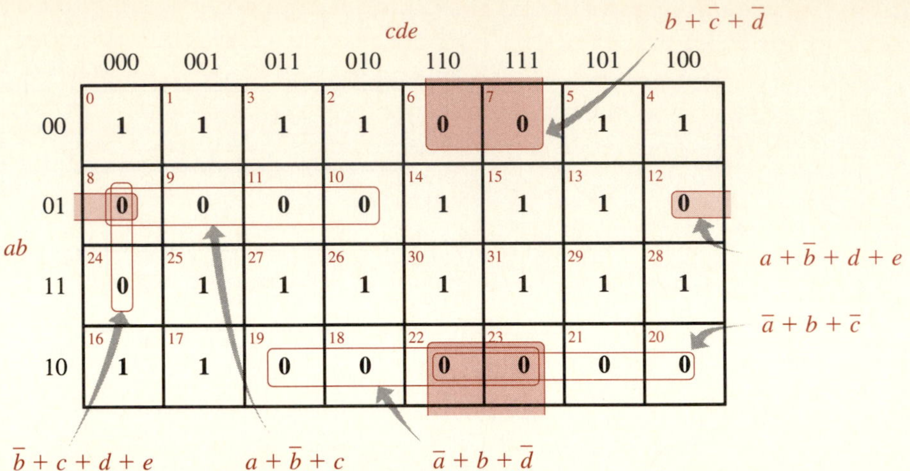

FIGURE 5.15 The prime implicates of f_1, all of which are essential.

Incompletely Specified Functions

A don't-care X is an output value that may be 0 or 1.

It is not unusual for a logic function $z(x_1, x_2, \ldots, x_n)$ to be defined so that certain output values are unrestricted or unspecified; such functions are said to be **incompletely specified functions**. An unspecified output value of z is traditionally called a **don't-care value**; the circuit user does not "care" if z is 0 or 1 when the corresponding input combination is present. Hence, if z is defined with k don't-cares, any of the 2^k assignments of 0s and 1s to these output values will result in a completely specified function that satisfies z's specifications. Don't-cares are typically associated with input combinations that are never applied to the circuit in question under normal operating conditions. Consequently, the corresponding function values never actually appear at the circuit's outputs.

Consider the design of a circuit C that determines if a BCD digit X is odd. This BCD parity-check circuit implements a four-variable function $z_4(X) = z_4(x_1, x_2, x_3, x_4)$, whose four input bits represent any one of the 10 decimal integers. We require $z_4(X) = 1$ whenever X denotes one of the odd digits 1, 3, 5, 7, or 9; $z_4(X) = 0$ whenever X denotes 0, 2, 4, 6, or 8. The six input combinations that represent the numbers 10 through 15 are unused, so the corresponding six output values are don't-cares. We use the symbol X to denote a don't-care value; the symbol d is also frequently used for the same purpose. Figure 5.16a shows the K-map specification for z_4, with an X entered in each don't-care position, which we refer to as an **X-cell**. By an

x_1x_2 \ x_3x_4	00	01	11	10
00	(0) **0**	(1) **1**	(3) **1**	(2) **0**
01	(4) **0**	(5) **1**	(7) **1**	(6) **0**
11	(12) **X**	(13) **X**	(15) **X**	(14) **X**
10	(8) **0**	(9) **1**	(11) **X**	(10) **X**

(a)

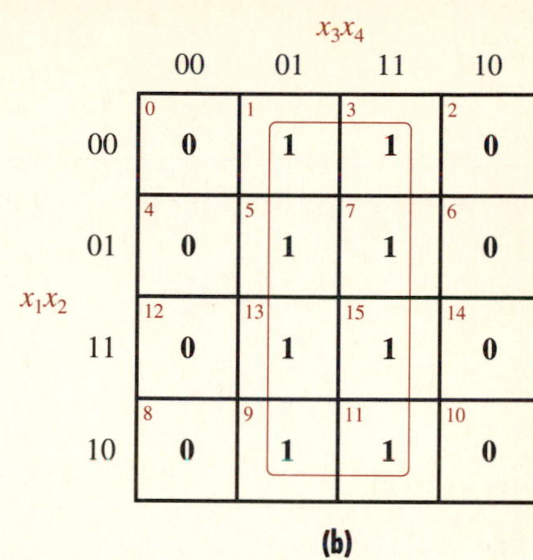

(b)

FIGURE 5.16 (a) Prime implicants of the incompletely specified function z_4; (b) a minimal SOP cover.

obvious extension of our earlier decimal shorthand for specifying functions, we can express z_4 as follows:

$$z_4(x_1, x_2, x_3, x_4) = \sum(1, 3, 5, 7, 9) + \Delta(10, 11, 12, 13, 14, 15)$$

or

$$z_4(x_1, x_2, x_3, x_4) = \prod(0, 2, 4, 6, 8) \cdot \Delta(10, 11, 12, 13, 14, 15)$$

where the large delta (for *don't-care*) identifies the X-cells.

Minimization with Don't-Cares

The logic designer can take advantage of the presence of don't-cares to reduce the number of gates in a two-level implementation of z, as we show next. In particular, he or she can set some don't-care value to 1 and others to 0 in a way that reduces the number of prime implicant (or implicate) groups and maximizes their size or, equivalently, minimizes the number of literals needed to specify them. Suppose, for example, that a 1-cell m_i of z is adjacent to $n-1$ 0-cells and to a single X-cell. If we change the latter's contents from X to 0, then m_i becomes an isolated 1-cell, and therefore denotes an n-literal prime implicant of z. If, however, we set X to 1, we obtain a two-cell group $P_{i,j}$ that can be represented by $n-1$ literals. Clearly, $P_{i,j}$ is a lower-cost alternative to m_i in any cover for z.

Rules for handling don't-cares during minimization

With this idea, we can extend our two-level SOP minimization procedure to incompletely specified Boolean functions by modifying the rules for identifying and selecting prime implicants, as follows:

1. When identifying prime implicants, treat all Xs like 1s; in other words, use the don't-cares to form the largest possible prime implicant groups.

2. When selecting prime implicants to include in a minimum cover, do not attempt to cover the Xs; in other words, don't-care values that are not included in groups needed to cover the original 1s are set to 0.

To find a minimal POS expression, 0s and 1s are interchanged in the above rules, and prime implicates replace prime implicants.

Applying the foregoing rules to construct an SOP expression for z_4 produces the three prime implicant groups depicted in Figure 5.16a. The eight-cell group contains the five minterms of z_4, along with three don't-care cells. This group, which corresponds to the prime implicant x_4, is clearly essential. The two four-cell groups are absolutely inessential and can therefore be eliminated. Hence, we obtain the very simple realization $z_4^* = x_4$, shown in Figure 5.16b. Observe how the Xs that contribute to the prime implicant x_4 have been set to 1, while the remaining Xs that need not be covered have been set to 0. Note that z_4^* is not z_4; it is one of many possible completely specified functions that are compatible with the definition of z_4.

Example 5.5

Minimizing an incompletely specified five-variable function

Suppose we want minimal SOP and POS expressions for the five-variable Boolean function

$$z_5(a, b, c, d, e) = \sum(2, 7, 12, 13, 18, 20, 24, 25, 27) + \Delta(0, 3, 8, 10, 16, 22, 26, 28, 29, 30)$$

Figure 5.17a shows the K-map for z_5. Treating all X-cells as 1s yields the eight prime implicant groups shown in Figure 5.17b; some care is needed

ab \ cde	000	001	011	010	110	111	101	100
00	0: X	1: 0	3: X	2: 1	6: 0	7: 1	5: 0	4: 0
01	8: X	9: 0	11: 0	10: X	14: 0	15: 0	13: 1	12: 1
11	24: 1	25: 1	27: 1	26: X	30: X	31: 0	29: X	28: X
10	16: X	17: 0	19: 0	18: 1	22: X	23: 0	21: 0	20: 1

(a)

FIGURE 5.17 (a) A K-map of the function z_5; (b) prime implicants; (c) essential prime implicants; (d) a minimal SOP cover.

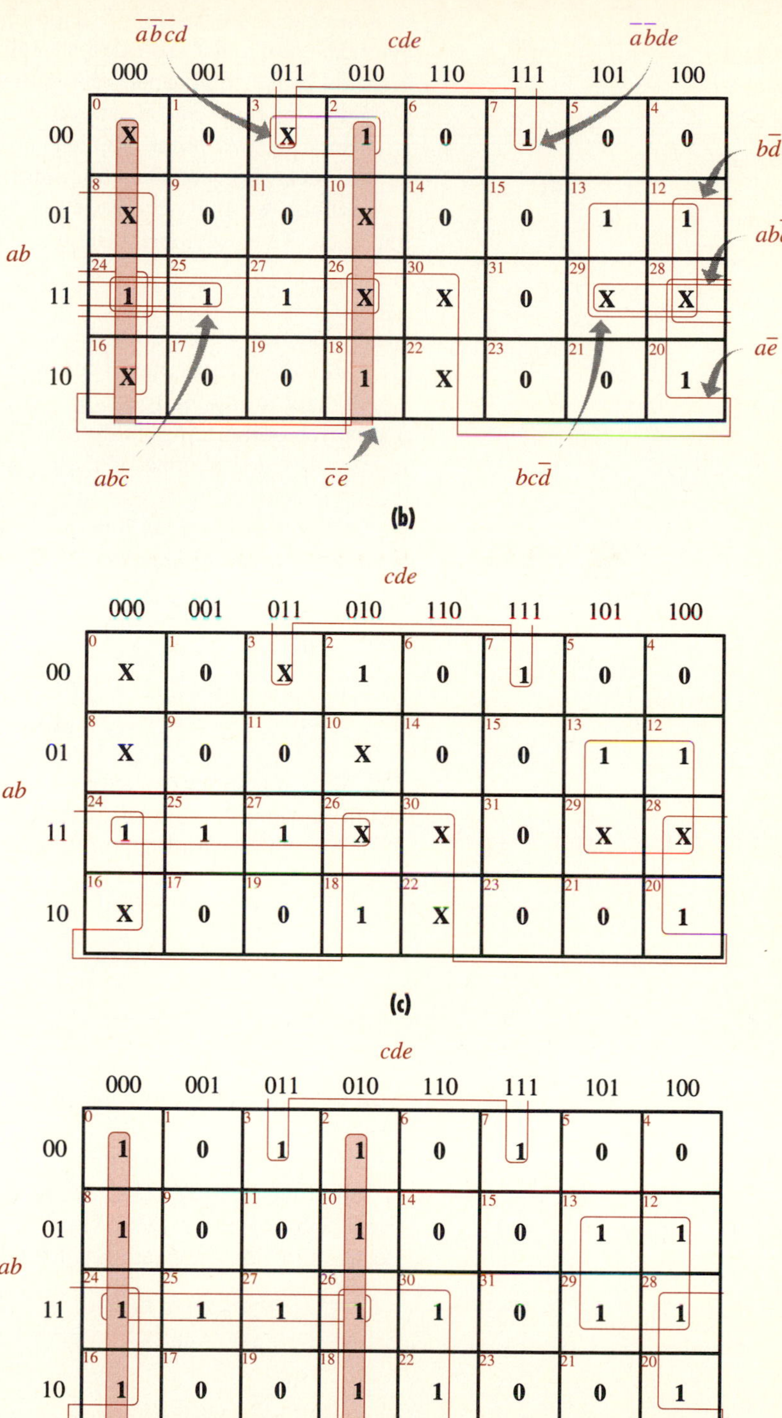
$\bar{a}\bar{b}\bar{c}d$
cde
$\bar{a}\bar{b}de$
$b\bar{d}\bar{e}$
$ab\bar{d}$
$a\bar{e}$
$ab\bar{c}$
$\bar{c}\bar{e}$
$bc\bar{d}$
ab

(b)

(c)

(d)

to ensure that no possible cell grouping is overlooked. Next, we identify the prime terms that are essential with respect to the 1-cells. There are four essential prime implicants, represented by the cell groups of Figure 5.17c. Only the minterm m_2 is not covered by an essential prime implicant. It is covered by two prime implicants: $P_{2,3} = \bar{a}\bar{b}\bar{c}d$ and $P_{0,2,8,10,16,18,24,26} = \bar{c}\bar{e}$. To meet the secondary minimization criterion of using as few literals as possible, we must choose $P_{0,2,8,10,16,18,24,26}$ as the fifth and final prime implicant. Thus, we obtain the unique minimal SOP expression

$$z_5^*(a, b, c, d, e) = a\bar{e} + ab\bar{c} + bc\bar{d} + \bar{a}\bar{b}de + \bar{c}\bar{e} \tag{5.30}$$

which contains 14 literals and corresponds to Figure 5.17d.

Determination of the prime implicates of z_5 requires grouping 0- and X-cells and results in 16 prime implicates. For clarity, this relatively large number of prime terms is divided between two K-maps (Figure 5.18a). It is easy to see that none of these terms are essential; every one of the 13 0-cells is covered by at least two prime implicates. Application of Petrick's method to selecting a minimal prime implicate cover results in the following formidable covering proposition:

$$\begin{aligned}\mathbf{C} = {} & (\mathbf{P}_A + \mathbf{P}_F + \mathbf{P}_H + \mathbf{P}_I + \mathbf{P}_J + \mathbf{P}_N)(\mathbf{P}_A + \mathbf{P}_E)(\mathbf{P}_A + \mathbf{P}_J)(\mathbf{P}_E + \mathbf{P}_K)\\ & (\mathbf{P}_B + \mathbf{P}_H + \mathbf{P}_N)(\mathbf{P}_B + \mathbf{P}_N + \mathbf{P}_O)(\mathbf{P}_D + \mathbf{P}_K + \mathbf{P}_L + \mathbf{P}_O)(\mathbf{P}_L + \mathbf{P}_O)\\ & (\mathbf{P}_F + \mathbf{P}_G + \mathbf{P}_I + \mathbf{P}_J)(\mathbf{P}_F + \mathbf{P}_G)(\mathbf{P}_G + \mathbf{P}_J + \mathbf{P}_P)\\ & (\mathbf{P}_G + \mathbf{P}_M + \mathbf{P}_P)(\mathbf{P}_C + \mathbf{P}_L + \mathbf{P}_M + \mathbf{P}_P)\end{aligned} \tag{5.31}$$

We can simplify this expression by applying the generalized distributive law of Equation (5.23). This law also allows us to delete the three sum terms of (5.31), all of whose literals appear in some other sum term:

$$\begin{aligned}\mathbf{C} = {} & (\mathbf{P}_A + \mathbf{P}_E\mathbf{P}_F + \mathbf{P}_E\mathbf{P}_H + \mathbf{P}_E\mathbf{P}_I + \mathbf{P}_E\mathbf{P}_J + \mathbf{P}_E\mathbf{P}_N)(\mathbf{P}_B + \mathbf{P}_N + \mathbf{P}_H\mathbf{P}_O)\\ & (\mathbf{P}_K + \mathbf{P}_E\mathbf{P}_D + \mathbf{P}_E\mathbf{P}_L + \mathbf{P}_E\mathbf{P}_O)(\mathbf{P}_G + \mathbf{P}_J + \mathbf{P}_F\mathbf{P}_M + \mathbf{P}_I\mathbf{P}_P)\\ & (\mathbf{P}_M + \mathbf{P}_P + \mathbf{P}_C\mathbf{P}_G + \mathbf{P}_G\mathbf{P}_L)\end{aligned} \tag{5.32}$$

As the reader can verify, converting this five-term POS expression to full SOP form is very tedious, even with the simplification provided by Boolean algebra. We can easily see, however, that there will be no four-literal product terms in the final expression for **C**, but numerous five-literal terms; the latter must therefore include the minimal covers. The minimum number of literals in any prime implicate of z_5 is three, so the minimum literal cost of a five-term POS cover for z_5 is 15. Many such solutions can be derived from Equation (5.32); the example in Figure 5.18b corresponds to

$$\begin{aligned}z_5^{**} &= P_A P_B P_G P_K P_L\\ &= (a + b + d)(a + \bar{b} + c)(\bar{a} + b + \bar{e})(\bar{c} + \bar{d} + e)(\bar{b} + \bar{c} + \bar{d})\end{aligned} \tag{5.33}$$

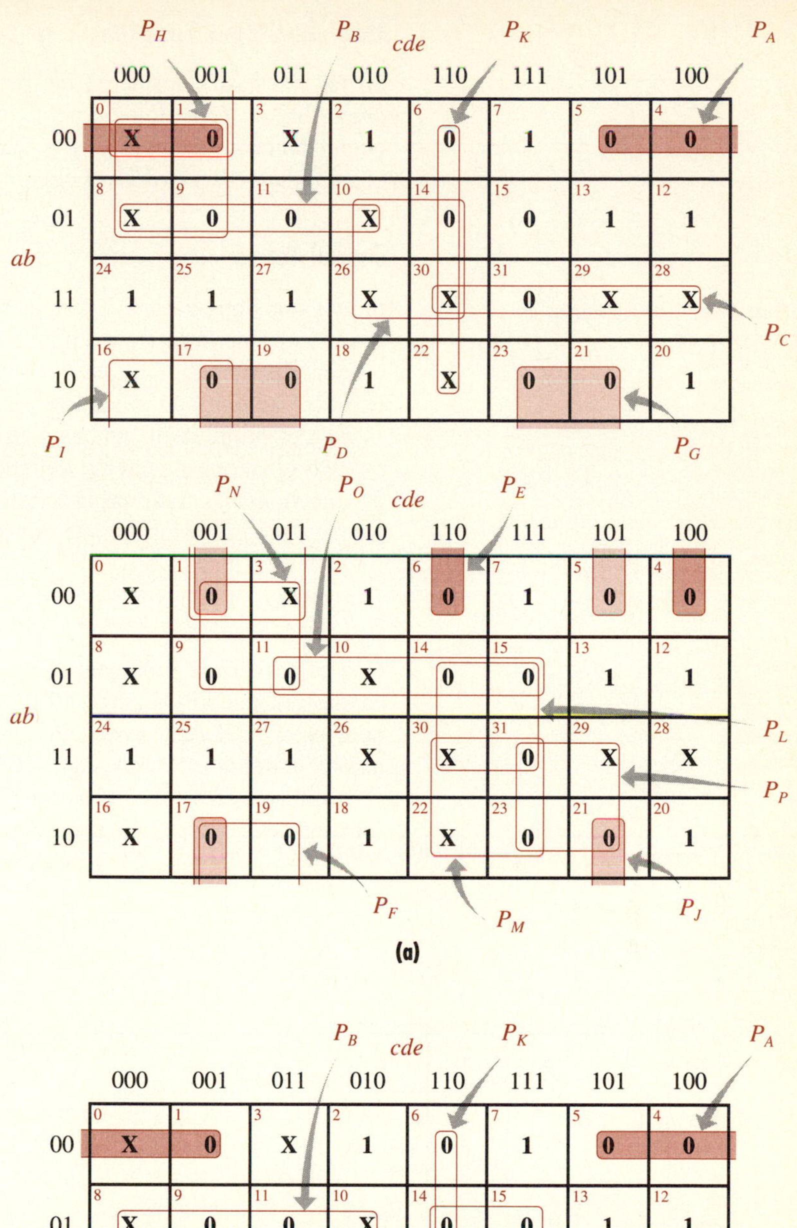

FIGURE 5.18 (a) Prime implicates of z_5; (b) a minimal POS cover.

Multiple-Output Functions

So far, we have considered the design of logic circuits with only a single output line z. Most logic circuits have multiple outputs and realize a set of output functions $Z = (z_1, z_2, \ldots, z_m)$. We now briefly consider the design of minimum-cost, two-level circuits of this type, using K-maps.

Shared Terms

Several outputs may share prime or nonprime terms.

If two functions z_i and z_j in Z have minimal expressions that contain a common prime term P, it suffices to use one gate G_P to generate P and to connect the output of G_P to the output gates used by z_i and z_j. Thus, lines between the first and second levels of the circuit may fan out to permit the sharing of prime terms among two or more outputs. Consider, for instance, the two functions z_1 and z_2, defined by the K-maps of Figure 5.19. Each of the functions has three prime implicants, all of which are essential. Hence, we have the following unique minimal SOP expressions for these functions:

$$\begin{aligned} z_1 &= ac + \overline{a}b\overline{c} + \overline{a}\overline{c}d \\ z_2 &= ac + \overline{a}\overline{c}\overline{d} + \overline{a}\overline{b}\overline{c} \end{aligned} \tag{5.34}$$

One of the prime implicants, $P_A = ac$, is common to z_1 and z_2 and so can be generated by a single AND gate. This sharing effectively reduces the number of AND gates needed to realize Equations (5.34) from six to five, and the number of literals from 16 to 14.

A shared term need not necessarily be a prime term for each individual function, as Figure 5.19 illustrates. We can use the minterm $P^* = m_1 =$

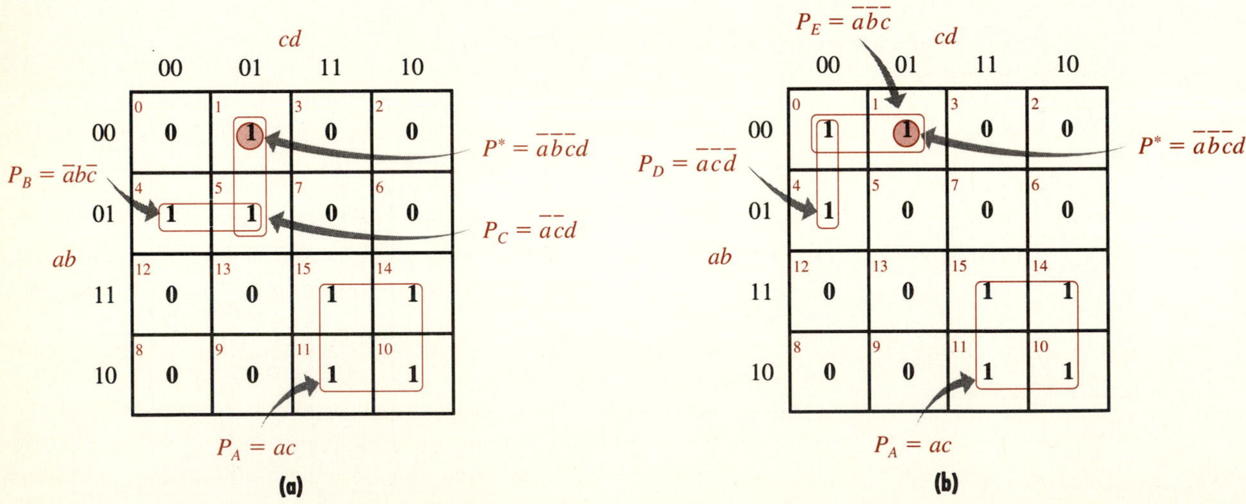

FIGURE 5.19 K-maps for the functions (a) z_1 and (b) z_2.

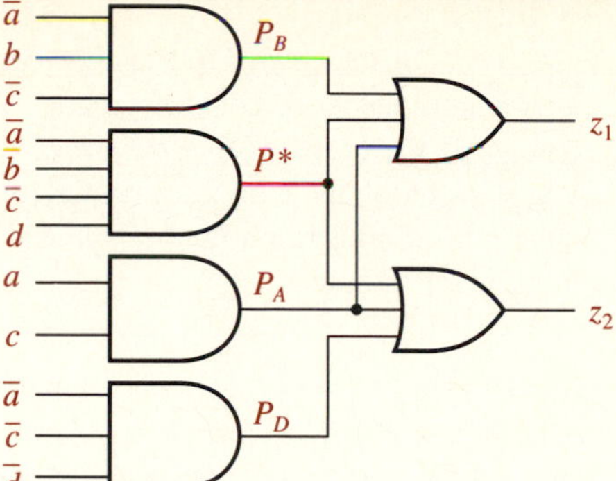

FIGURE 5.20 A minimal AND-OR implementation of the functions of Fig. 5.19.

$\bar{a}\bar{b}\bar{c}d$ (colored in the figure) to replace the prime implicant $P_C = \bar{a}\bar{c}d$ of z_1 and the prime implicant $P_E = \bar{a}\bar{b}\bar{c}$ of z_2. The resulting SOP expressions

$$\begin{aligned} z_1 &= ac + \bar{a}b\bar{c} + \bar{a}\bar{b}\bar{c}d \\ z_2 &= ac + \bar{a}\bar{c}\bar{d} + \bar{a}\bar{b}\bar{c}d \end{aligned} \tag{5.35}$$

are nonminimal as far as each individual output function is concerned. However, the nonprime term $P^* = \bar{a}\bar{b}\bar{c}d$ can be shared, so that (5.35) has the six-gate, 12-literal realization depicted in Figure 5.20. This is clearly better than any two-level design in which there is no sharing of terms, or in which sharing is limited to the prime implicants of the individual output functions.

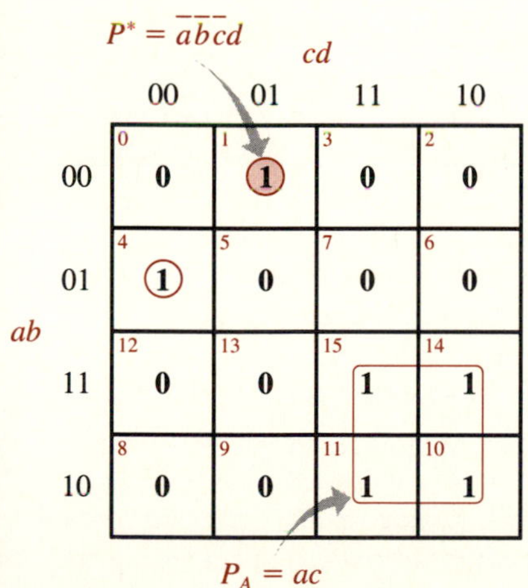

FIGURE 5.21 Prime implicants of the product function z_1z_2 from Equations (5.34).

Multiple-Output Minimization

We can easily generalize the two-level minimization problem from one to m functions. Suppose we want a minimal SOP implementation of $Z = z_1, z_2, \ldots, z_m$. We will now need up to m output OR gates instead of one. Our primary goal continues to be to minimize the total number of gates, while the secondary goal is to minimize the total number of connections to the input AND gates and the connections to the output OR gates. The latter requirement is to ensure that we include no unnecessary connections between the two levels.

The example of Figures 5.19, 5.20, and 5.21 suggest that to minimize the m functions $Z = z_1, z_2, \ldots, z_m$, we must consider as potential AND terms the prime implicants of all the product functions that can formed from the m given functions. The multiple-output covering problem is greatly complicated by the large number of such terms that must be considered. It is further complicated by the need to account for the number of input connections used in the OR gates. We will return to this problem again in Section 5.6.

Computer-Aided Design

We next consider ways to address difficult two-level minimization problems that cannot easily be solved by hand. We present a systematic design method, known as the *tabular* or *Quine-McCluskey method*, which is suitable for solving such problems with the aid of a computer. We also discuss approximate or heuristic minimization methods and the problem of designing minimum-cost multilevel circuits.

5.5 The Tabular (Quine-McCluskey) Method

The covering problem encountered in Example 5.5 demonstrates the difficulty of exact minimization by purely manual techniques, even for relatively small problems. The amount of computation involved can be very great, and the process of visual identification and selection of prime terms via K-maps is inherently error-prone.

Logic Simulation Programs

CAD programs are essential for solving very large design problems.

The most basic requirement of any new design is that it be functionally correct; that is, the design must meet the problem specifications. To verify functional correctness, it is usually desirable to employ a logic simulation program to create and test a computer model of the proposed design. Such a simulation is illustrated in Figure 5.22, in which the DesignWorks simulator is employed to verify the functional correctness of the two-level AND-OR circuit defined by (5.30) in Example 5.5. Figure 5.22a shows the circuit model as drawn on a computer screen and entered into the simulator. (The menu-driven design entry process for this kind of simulator is illustrated by Figure 1.24.) A switch and a hexadecimal keyboard, both of which can be altered with mouse clicks, provide the test input patterns; probes on lines of interest indicate their current signal values.

The timing chart Figure 5.22b records a complete sequence of user-entered input patterns for lines $a{:}e$, and the corresponding computer-generated output patterns on lines y and z_5 or any other designated lines. All 32 possible 5-bit input patterns are applied here, so the primary output waveform can be mapped into the entire contents of the K-map (Figure 5.17d), which was the starting point in the design process. At time 98, for instance, the input pattern changes from $abcde = 10111_2 = 23_{10}$ to $11000_2 = 24_{10}$, and the output z_5 changes at time 101 (each gate has a default propagation delay of one time unit) from 0 to 1, as specified by the K-map. For larger circuits, the interactive (and therefore slow) on-screen switches and keys can be replaced with programmed commands that automatically generate and apply the input sequences used to check the simulated circuit.

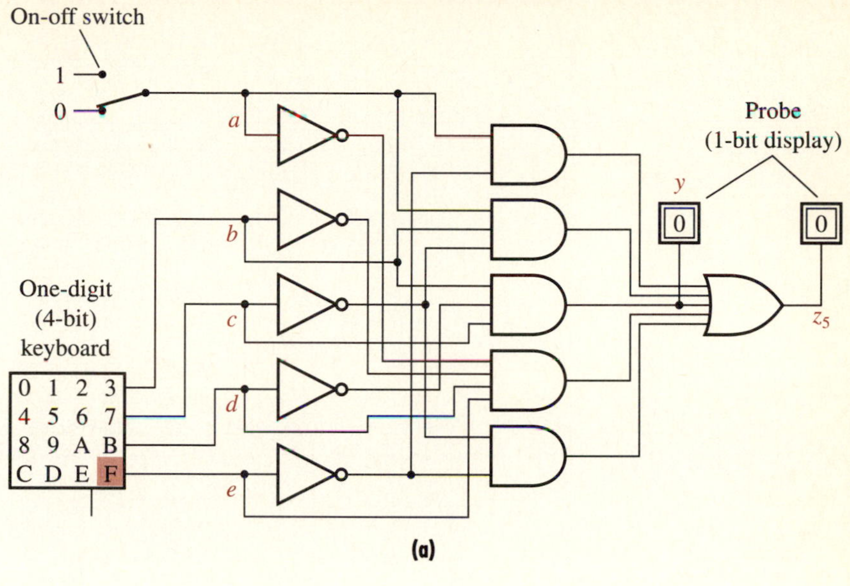

(a)

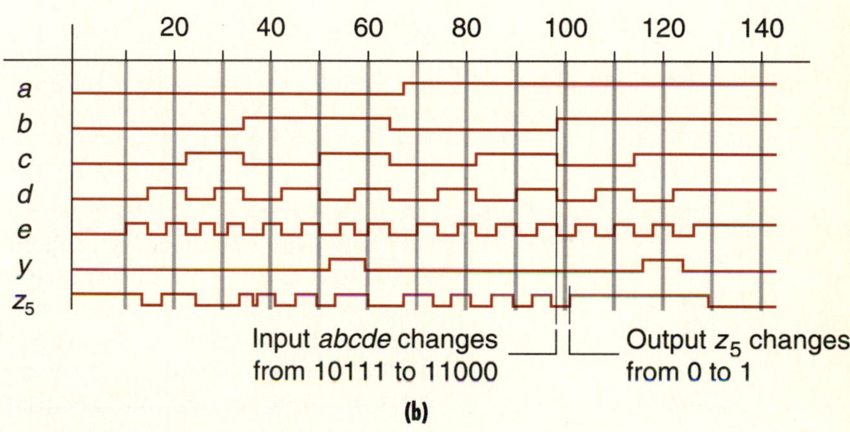

(b)

FIGURE 5.22 Design verification using the DesignWorks simulator: (a) a circuit model; (b) simulated waveforms.

Logic Minimization Programs

Minimization programs are exact in solving many problems of up to 20 variables.

CAD programs have also been written to perform two-level minimization. They can provide exact solutions automatically for small and medium-sized problems (up to about 20 variables or so) and approximate (near-minimal) solutions for much larger problems. One such program for exact minimization, McBOOLE, was developed at McGill University in the mid-1980s [Dagenais et al., 1986]. A widely used program for both exact and approximate minimization is ESPRESSO, which is based on work at IBM Watson Research Center and the University of California, Berkeley [Brayton et al., 1984].

/* INPUT FILE	/* OUTPUT FILE
*/	McBOOLE: 1K CPU time (user sys total) 0.00 0.00 0.00
00000 x	19 nodes, var: in 5 out 1 literal: in 95 out 9
00010 1	
00011 x	McBOOLE: 1K CPU time (user sys total) 0.00 0.00 0.00
00111 1	The function has 8 prime implicants
01000 x	
01010 x	McBOOLE: 1K CPU time (user sys total) 0.00 0.00 0.00
01100 1	The function has 4 essential PI
01101 1	
10000 x	McBOOLE: 1K CPU time (user sys total) 0.00 0.00 0.00
10010 1	The solution contains 5 nodes
10100 1	
10110 x	McBOOLE: 1K CPU time (user sys total) 0.00 0.00 0.00
11000 1	Final solution, literal: 14 input 5 output
11001 1	
11010 x	*/
11011 1	00x11 1
11100 x	xx0x0 1
11101 x	x110x 1
11110 x	1xxx0 1
/*	110xx 1
	/*

FIGURE 5.23 Computer solution of Example 5.5, using the McBOOLE minimization program.

McBOOLE was applied to the preceding example. On a small computer, it produced in a fraction of a second the result we so laboriously obtained using K-maps. The input data given to McBOOLE for this small problem appear as the left column of Figure 5.23. This is a truth-table description of the five-variable function z_5 of Example 5.5, which lists all 1 and don't-care (X) rows explicitly; the rows not listed are those for which $z_5 = 0$. The output data produced by McBOOLE appear as the right column of Figure 5.23. The prime implicants in the solution obtained by McBOOLE are listed in a condensed truth-table format at the end, with 1s denoting true literals, 0s denoting complemented literals, and Xs indicating missing variables. For example, in the last row, 110XX indicates that $ab\bar{c}$ is a term in the minimal SOP expression for z_5.

We turn next to a two-level minimization technique that is suitable for implementation in CAD programs—it is used in McBOOLE, for example—and for manual calculations that involve more variables than can be handled by the K-map method. This approach, developed in the 1950s, is known as the **tabular** or **Quine-McCluskey method**, after its inventors, Willard V. Quine of Harvard University and Edward J. McCluskey of AT&T Bell Laboratories.

A General Approach

The tabular method uses binary notation for terms, and tables to obtain minimal covers.

The essence of the tabular method is to represent the terms of interest by tables of number-like binary words, which are processed systematically to derive all the prime terms from the minterms (in the SOP case) or maxterms (in the POS case). The visual grouping of cells on the K-map is replaced by a systematic comparison and merging of the binary words.

Once the prime terms have been obtained, a table is also employed to represent the covering relations between prime implicants (or implicates) and the minterms (or maxterms). A minimum cover is then derived from this table. The processing of the binary tables involved in the tabular method is fairly easy to program on a computer, unlike the graphical techniques inherent in the K-map approach. The following discussion focuses on SOP minimization; however, the results easily extend to the POS case by the principle of duality.

Code Words

In the tabular approach, a product term $\dot{x}_1\dot{x}_2\ldots\dot{x}_k$ of $k \leq n$ literals, where n is the number of input variables of the function under consideration, is represented by words of length n in which 1 in position i denotes the uncomplemented literal x_i, 0 denotes a complemented literal $\overline{x}_i$, and a dash (-) indicates that neither x_i nor $\overline{x}_i$ is present. We refer to these words as binary[1] **code words.** For example, with $n = 6$,

010011	denotes $\overline{x}_1 x_2 \overline{x}_3 \overline{x}_4 x_5 x_6$
1-1-01	denotes $x_1 x_3 \overline{x}_5 x_6$
-----0	denotes $\overline{x}_6$

The product term $x_1\overline{x}_2 x_3\overline{x}_6$ is denoted by the code word 101--0.

As before, two product terms $P_1 = Px$ and $P_2 = P\overline{x}$ are adjacent if they differ in only a single variable. By the minimization theorem 15, $Px + P\overline{x} = P$, so we can replace both P_1 and P_2 with a common covering term P that contains one less literal. Two code words S_1 and S_2 are adjacent if a 0 appears in some position i of S_1, a 1 appears in the same position i of S_2, and S_1 and S_2 agree completely in all the remaining $n - 1$ positions. An adjacent pair S_1 and S_2 can be replaced by the code word S obtained by taking, say, S_1, and inserting a dash (-) in place of 0 in the ith position. For example,

$$001100 + 001110 = 0011\text{-}0$$

$$11\text{-}1\text{-}1 + 11\text{-}1\text{-}0 = 11\text{-}1\text{--}$$

$$1101\text{--} + 1111\text{--} = 11\text{-}1\text{--}$$

[1] Strictly speaking, the code words involve three rather than two symbols due to the presence of dashes. However, it is usual to consider these words to be two-valued rather than three-valued.

where + denotes *or*. The number of 0s and 1s in a code word S_i of the foregoing sort is the same as the number of literals in the corresponding product term P_i; we call this number the **literal cost.**

Prime Implicant Identification

We will explain the tabular method by applying it to a small problem (Example 5.2) that we solved earlier using the K-map method. The objective is a minimal SOP form for the four-variable function

$$z_1(a, b, c, d) = \sum(1, 3, 5, 10, 11, 12, 13, 14, 15)$$

First, we list all nine minterms of z_1 in the code-word notation, where $m_1 = 0001$, $m_3 = 0011$, and so on. It is convenient to group together those minterms that have the same number of 1s; the result is list 1 of Figure 5.24a,

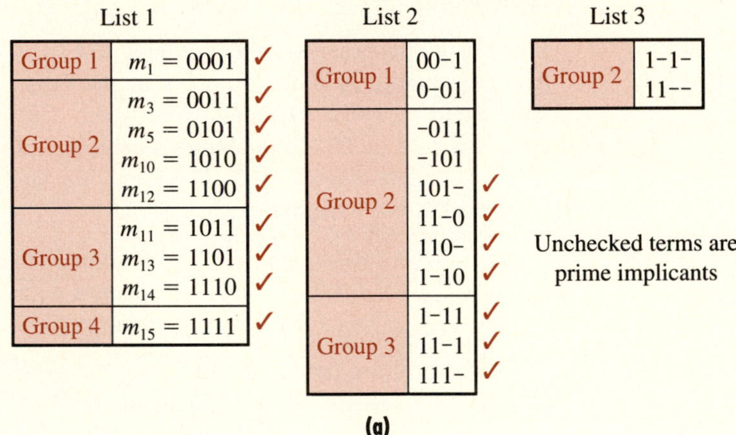

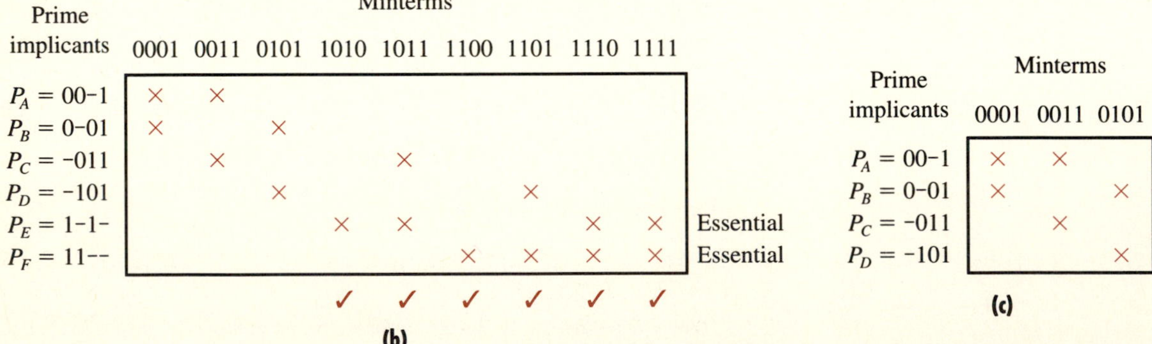

FIGURE 5.24 The tabular minimization method: (a) prime implicant identification; (b) a cover table; (c) a reduced cover table.

in which group i contains all the minterms with exactly i 1s and, by default, $4 - i$ 0s.

Next we systematically compare all members of group i with all members of group $i + 1$ in list 1, and identify all pairs of minterms that are adjacent. Thus, we first compare $m_1 = 0001$ (the sole member of group 1) with $m_3 =$ 0011 (the first member of group 2). These code words differ only in position c, the third from the left; hence, we can cover the two minterms with 00-1, which becomes the first entry in list 2. In this instance, we are using the fact that

$$\bar{a}\bar{b}\bar{c}d + \bar{a}\bar{b}cd = \bar{a}\bar{b}d$$

or, equivalently,

$$0001 + 0011 = 00\text{-}1$$

Now we check off m_1 and m_3 to indicate that they are covered by a product term and so cannot be prime implicants. We next compare $m_1 = 0001$ with the second entry $m_5 = 0101$ in group 2. These minterms also are adjacent, and result in the second entry 0-01 in list 2. The minterm $m_5 = 0101$ is then checked off. On comparing m_1 to the two remaining members of group 2, we find no more adjacent terms.

Each code word with k 1s is compared to all code words with $k + 1$ 1s and matching dashes.

At this point, group 1 of list 2 is complete. We proceed to compare the four members of group 2 with the three members of group 3 in list 1. The adjacencies identified result in the six distinct terms that form group 2 of list 2. A final round of comparisons produces the last group in list 2. Because every minterm has been found to be adjacent to at least one other minterm, all entries in list 1 have been checked off. Now we move to list 2, and again compare each group i to the neighboring group $i + 1$. Every adjacent pair of code words results in an entry in a third list. Note that 11-- can be produced in several ways; for example, by combining 111- with 110- or by combining 11-1 with 11-0. In such cases, only one copy of each repeated term is retained. Notice also that for two code words to be adjacent, all their dashes must appear in exactly the same positions.

The comparison process must eventually terminate, because each list contains entries with fewer 1s than the preceding list. At the end, the unchecked entries in all the lists define the desired prime implicants. In the present example, six prime implicants have been identified: four in list 2 and two in list 3.

Prime Implicant Selection

The next phase of tabular minimization is to determine a minimum set of prime implicants that cover the original function z. This is done with the aid of the table of Figure 5.24b, which is known as a **cover table** or a **prime implicant table.** It has a row for every prime implicant and a column for every minterm of z. An $\times$ is entered at the intersection of row i and column

j if prime implicant P_i of row i covers, or is implied by, minterm m_j of row j. The desired minimum cover is obtained by systematically processing the cover table. A set of rows that together have at least one × in every column of the cover table is referred to as a **row cover** or a **prime implicant cover.**

The minimization problem may now be restated in the following tabular form: Find a cover of the given function z that (1) contains the minimum number of rows and (2) is such that no other minimum-row cover has fewer 0s and 1s in its row labels. Clearly, a solution to this problem constitutes a minimal prime implicant cover. We can easily map this minimal cover into a minimal SOP Boolean expression or a minimal two-level logic circuit for z.

Essential Prime Implicants

As in the K-map approach, the first step in constructing a minimum cover is to identify any prime implicant that is the sole cover of some minterm. Such a prime implicant is essential because it must be a member of every cover, including every minimal cover. The essential prime implicants are easily identified by scanning the cover table for columns containing exactly one ×. (All columns should initially contain at least one ×.) If such a column m_j is found, and its sole × is in row P_i, then P_i is an essential prime implicant. If P_i is omitted from a solution, then minterm m_j will not be covered. In Figure 5.24b, we find that the last two prime implicants, P_E and P_F, are essential. P_E = 1-1- is the only row covering m_{10} = 1010, while P_F = 11-- is the only row covering m_{12} = 1100.

Essential rows and the columns they cover are removed first from a cover table.

Because the essential prime implicants appear in every complete cover, the columns they cover will be covered in every possible solution. Therefore, we can check off these columns as covered; this applies to the six rightmost columns in Figure 5.24b, which have check marks at the bottom. It is useful at this point to delete the rows that correspond to essential prime implicants and the columns they cover from the cover table. The essential rows are set aside for inclusion in the final minimal cover. The result of this deletion is a **reduced cover table** of the kind illustrated by Figure 5.24c. It remains to obtain a minimum cover for this smaller table. Clearly, one row cannot cover the last three minterms. There are several two-row covers, all involving a total of six 0s and 1s in their labels. For the moment, we arbitrarily take P_A and P_B as the minimum cover for the reduced table. Combining these with the two essential prime implicants P_E and P_F that we placed aside, we obtain the final solution

$$\{P_A, P_B, P_E, P_F\} = \{00\text{-}1, 0\text{-}01, 1\text{-}1\text{-}, 11\text{-}\text{-}\}$$

This maps directly into the optimal SOP expression

$$z_1 = \overline{a}\overline{b}d + \overline{a}\overline{c}d + ac + ab$$

obtained in Example 5.2 via a K-map.

Row Dominance

We can reduce a cover table further by exploiting some relationships among rows and columns. Consider again the reduced cover table of Figure 5.24c. The top row P_A = 00-1 covers the first two columns; the third row P_C = 0-01 covers the second column only. Moreover, both row labels contain a total of three 0s and 1s, so the cost of each row is the same. We can delete row P_C from the table, because any contribution it could make to a minimum cover is made at the same cost by P_A, and P_A has the added advantage of covering one more column. We find the same relationship between rows P_B and P_D. Thus, we can delete rows P_C and P_E from the table, leaving only rows P_A and P_B. These are now essential with respect to the remainder of the table, and so yield the final cover $\{P_A, P_B\}$. Observe that this is not a unique minimal cover, however; alternative solutions are $\{P_A, P_D\}$ and $\{P_B, P_C\}$.

The relationship between P_A and P_C discussed above is termed **row dominance** and is defined as follows. Row P_i dominates row P_j, denoted $P_i \supseteq P_j$, if row P_i contains an $\times$ in every column in which row P_j contains an $\times$. In general, P_i contains more occurrences of $\times$ than does P_j; in the special case where they contain the same number of occurrences of $\times$ (which must, of course, be in the same columns), we write $P_i = P_j$. Figure 5.25 illustrates row dominance in the general case. Row dominance is useful because we can delete a dominated row P_j from a cover table, *provided* its literal cost is greater than or equal to that of the dominating row P_i.

Column Dominance

Dominating columns and lower-cost dominated rows may be deleted from cover tables.

We can also find dominance relations among the columns of a cover table. This property, called **column dominance**, is illustrated by Figure 5.25. We say that column m_i dominates column m_j, denoted $m_i \supseteq m_j$, if column m_i contains an $\times$ in every row in which column m_j contains an $\times$. In this case,

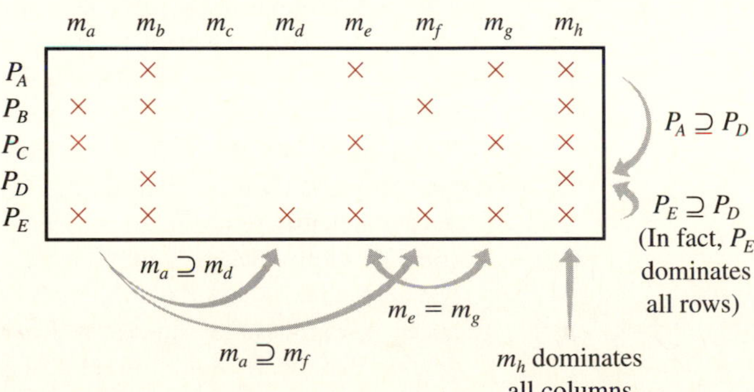

	m_a	m_b	m_c	m_d	m_e	m_f	m_g	m_h
P_A		×			×		×	×
P_B	×	×				×		×
P_C	×				×		×	×
P_D		×						×
P_E	×	×		×	×	×	×	×

FIGURE 5.25 A cover table that illustrates row and column dominance.

the dominat*ing* column m_i rather than the dominat*ed* column m_j is deleted from the cover table. This follows from the fact that if we retain m_j, the rows in any solutions that cover m_j must also cover m_i. Unlike the rows, there is no literal cost to be taken into account when comparing columns.

We summarize the cover table reduction possibilities provided by row and column dominance as follows:

1. If column m_i dominates column m_j, the dominating column m_i may be deleted from the cover table.
2. If row P_i dominates row P_j, the dominated row P_j may be deleted from the cover table, but only if P_j's literal cost equals or exceeds that of P_i.

If we delete m_i because $m_i \supseteq m_j$, we must retain m_j in the cover table, unless m_j itself dominates some other column m_k that is retained. In general, if we have $m_a \supseteq m_b \supseteq \cdots \supseteq m_k$, we must retain only m_k. A similar situation holds for a chain of row dominance relations $P_a \supseteq P_b \supseteq \cdots \supseteq P_k$; only P_a must be retained. Note here that the literal costs of the rows are not necessarily in the same order as their dominance relations, so it may not be possible to delete some or, indeed, any of the rows that satisfy $P_a \supseteq P_b \supseteq \cdots \supseteq P_k$.

The Branching Method

The branching method systematically generates trial covers until it finds a minimal one.

The exploitation of essential rows and dominance relations reduces the number of columns that remain to be covered. If all the columns can be handled in this way—that is, if the cover table can eventually be reduced to zero, as in the example of Figure 5.24—we have solved the original covering problem. The foregoing table reduction procedures by themselves simplify but do not always solve the covering problem. After we have made all possible simplifications, we may still be left with a relatively large prime implicant table to cover. In such cases, we can resort to Petrick's method, which, although not especially efficient, is amenable to computer solution [Cutler and Muroga, 1987].

An alternative to Petrick's method often used in logic minimization programs—for example, in McBOOLE—is known as the **branching** or **branch-and-bound method**. An arbitrary row P_1 is removed from the current cover table and included in a tentative solution. The columns covered by P_1 are also removed. Then the reduced table is examined for further simplification, such as newly created essential rows or dominance relations that might lead to a final solution S_1 of cost c_1. The process is repeated for all k_1 other possible choices of P_1. If a set of complete solutions (covers) results, the one with the lowest cost provides the final result.

If the selection of one prime implicant in the above manner does not lead to an overall solution of guaranteed minimum cost—that is, if we still have irreducible cover tables—we select a second P_2 from the k_2 available choices and investigate possible solutions with P_1 and P_2 as required members. Observe that the number of possible cases to consider is now k_1k_2. This process may continue through several selection steps. Eventually, we will run out of possible choices and will terminate with an optimal solution. Thus, the branching method generates many possible solutions by a searching process that can grow exponentially with the size of the function under consideration. Optimality is guaranteed only if a complete set of branchings is made.

An Application

Let us apply the branching method to the cover table of Figure 5.26a, which is an example of a cyclic table that has no dominating rows or columns. First, we select P_A for inclusion in a tentative solution; that is, we "branch on P_A." We therefore delete row P_A and the two columns m_a and m_b covered by P_A to obtain the five-row reduced table shown in Figure 5.26b. Now $P_C \supseteq P_B$, and both have literal cost 3, so the dominated row P_B is deleted. Also, $P_E \supseteq P_F$, and P_F has higher literal cost (3 instead of 2); hence, row P_F is deleted, yielding the three-row reduced table of Figure 5.26b. At this point, rows P_C and P_E have become essential, and between them they cover all four remaining columns. Hence, we have found a three-row cover

$$\{P_A, P_C, P_E\} = \{\text{-0-0}, \text{010-}, \text{--11}\} \tag{5.36}$$

for the original table, which has literal cost $2+3+2 = 7$. This is the lowest cost solution when P_A is forced into the cover. It will be a minimum-cost solution of the original problem *only if* there is no cover of fewer than three rows and no other three-row cover of lower literal cost.

We must still examine the five other solutions obtained by putting each of P_B, P_C, P_D, P_E, and P_F, in turn, into a tentative solution. The calculations required for branching on P_B appear in Figure 5.26c. Again, we start with a five-row table obtained by removing the row on which we are branching. We see that as a result of removing P_B, we have $P_D \supseteq P_C$ and $P_F \supseteq P_A$. We can delete P_C but not P_A, because the latter has lower cost than the row dominating it. The resulting four-row table leads directly to the unique minimal cover

$$\{P_B, P_D, P_F\} = \{\text{0-00}, \text{01-1}, \text{001-}\} \tag{5.37}$$

This solution also has three rows, but its literal cost is $3+3+3 = 9$, which is clearly worse than the previous cover (5.36). It can be shown that solutions obtained by branching on the other four prime implicants always lead to three-term solutions with the same literal cost as either (5.36) or (5.37). We conclude that (5.36) defines a minimum-cost cover.

	m_a	m_b	m_c	m_d	m_e	m_f
P_A = p0–0	×	×				
P_B = 0–00	×			×		
P_C = 010–				×	×	
P_D = 01–1					×	×
P_E = ––11			×			×
P_F = 001–		×	×			

(a)

	m_c	m_d	m_e	m_f	
P_B = 0–00		×			Dominated by P_C
P_C = 010–		×	×		
P_D = 01–1			×	×	
P_E = ––11	×			×	
P_F = 001–	×				Dominated by P_E

	m_c	m_d	m_e	m_f	
P_C = 010–		×	×		Essential
P_D = 01–1			×	×	
P_E = ––11	×			×	Essential

(b)

	m_b	m_c	m_e	m_f	
P_A = –0–0	×				
P_C = 010–			×		Dominated by P_D
P_D = 01–1			×	×	
P_E = ––11		×		×	
P_F = 001–	×	×			

	m_b	m_c	m_e	m_f	
P_A = –0–0	×				
P_D = 01–1			×	×	Essential
P_E = ––11		×		×	
P_F = 001–	×	×			

(c)

FIGURE 5.26 (a) A cyclic cover table: (b) branching on P_A; (c) branching on P_B.

A Summary of the Tabular Method

The tabular or Quine-McCluskey method for SOP minimization is summarized in the procedure *TABMIN* below.

Procedure

Tabular method *TABMIN* for finding a minimal SOP form

1. Phase 1 (prime implicant identification): Form a list (table) $L_i = L_1$ of all minterms of z represented by binary code words. Divide L_i into groups $G_0, G_1, \ldots, G_m$, where G_h contains all minterms with exactly h 1s in their binary representations, $0 \leq h \leq m$.
2. Scan L_i and compare each entry E' of G_h with each entry E'' of G_{h+1}, for all h such that $0 \leq h < m$. If E' and E'' are logically adjacent and $E = E' + E''$, check off both entries and add E to group G_h of the new list L_{i+1}, if it is not already in L_{i+1}.

3. If L_{i+1} is not empty, change i to $i+1$ and repeat step 2. If L_{i+1} is empty, the unchecked entries in the L_i lists are the prime implicants of z; proceed to step 4.
4. Phase 2 (prime implicant selection): Set up a cover table T whose rows are all the prime implicants and whose columns are all the minterms. Enter an $\times$ at the intersection of row i and column j if the ith prime implicant P_i covers the jth minterm m_j.
5. Identify all the essential rows of T and add to the solution set S. Reduce T by removing the essential rows and all columns covered by these rows. Reduce T by removing dominating columns, and rows that are dominated by rows of less or equal literal cost.
6. Continue applying step 5 until no further reduction of T is possible. If T becomes empty, then terminate with S as the solution; if T is nonempty, then proceed to step 7.
7. Use the branching method or Petrick's method to select a minimum-cost subset of the remaining rows to add to S.

The modifications to this procedure for POS minimization or the minimization of incompletely specified functions are the same as those presented earlier for the K-map approach. To handle POS minimization, we substitute maxterms for minterms and prime implicates for prime implicants. The same code word notation is used, but with the code words interpreted as sums rather than products, and with 0 corresponding to x_i and 1 corresponding to $\overline{x}_i$. To handle incompletely specified functions, we treat don't-care entries like minterms (SOP case) or maxterms (POS case) in phase 1 (generation of prime terms); they are ignored in phase 2 (the covering problem). The following example illustrates all these issues.

Example 5.6

Minimizing a five-variable function by the tabular method

Consider the task of finding a minimal POS expression for the five-variable logic function

$$\begin{aligned} z_5(a,b,c,d,e) = &\prod(1,4,5,6,9,11,14,15,17,19,21,23,31) \\ &\cdot \Delta(0,3,8,10,16,22,26,28,29,30) \end{aligned} \tag{5.38}$$

which we tackled already by the K-map approach in Example 5.5. To calculate the prime implicates, we treat all the don't-cares of (5.38) as if they were maxterms, resulting in a total of 23 provisional maxterms, which are used to create list 1 in Figure 5.27. Systematic comparison of terms via the tabular method—a time-consuming task without computer assistance, as the reader can verify—results in the two additional lists given in the figure. There are 16 unchecked entries which define all the possible prime implicates.

List 1

Group 0	$M_0 = 00000$	✓
Group 1	$M_1 = 00001$	✓
	$M_4 = 00100$	✓
	$M_8 = 01000$	✓
	$M_{16} = 10000$	✓
Group 2	$M_3 = 00011$	✓
	$M_5 = 00101$	✓
	$M_6 = 00110$	✓
	$M_9 = 01001$	✓
	$M_{10} = 01010$	✓
	$M_{17} = 10001$	✓
Group 3	$M_{11} = 01011$	✓
	$M_{14} = 01110$	✓
	$M_{19} = 10011$	✓
	$M_{21} = 10101$	✓
	$M_{22} = 10110$	✓
	$M_{26} = 11010$	✓
	$M_{28} = 11100$	✓
Group 4	$M_{15} = 01111$	✓
	$M_{23} = 10111$	✓
	$M_{29} = 11101$	✓
	$M_{30} = 11110$	✓
Group 5	$M_{31} = 11111$	✓

List 2

Group 0	0000-	✓
	00-00	✓
	0-000	✓
	-0000	✓
Group 1	000-1	✓
	00-01	✓
	0-001	✓
	-0001	✓
	0010-	✓
	001-0	
	010-0	✓
	0100-	✓
	1000-	✓
Group 2	0-011	✓
	-0011	✓
	-0101	✓
	0-110	✓
	-0110	✓
	0101-	✓
	010-1	✓
	01-10	✓
	-1010	✓
	100-1	✓
	10-01	✓
Group 3	01-11	✓
	0111-	✓
	-1110	✓
	10-11	✓
	101-1	✓
	1011-	✓
	11-10	✓
	1-110	✓
	1110-	✓
	111-0	✓
Group 4	-1111	✓
	1-111	✓
	111-1	✓
	1111-	✓

List 3

Group 0	00-0-
	-000-
	0-00-
Group 1	0-0-1
	-00-1
	-0-01
	010--
Group 2	--11-
	01-1-
	-1-10
	10--1
Group 3	-111-
	1-1-1
	111--
	1-11-

FIGURE 5.27 The tabular computation of prime implicates for the five-variable function z_5.

We now proceed to phase 2 and set up the covering table of Figure 5.28a. Here we drop the don't-cares and attempt to cover only the 13 maxterms specified in Equation (5.38). The prime implicates have the same names used in the earlier K-map computation (Figure 5.18). There are no essential

Prime implicates	00001	00100	00101	00110	01001	01011	01110	01111	10001	10011	10101	10111	11111
P_A = 00-0-	×	×	×										
P_B = 010--					×	×							
P_C = 111--													×
P_D = -1-10							×						
P_E = 001-0		×		×									
P_F = -00-1	×								×	×			
P_G = 10--1									×	×	×	×	
P_H = 0-00-	×				×								
P_I = -000-	×								×				
P_J = -0-01	×		×						×		×		
P_K = --110				×			×						
P_L = -111-							×	×					×
P_M = 1-11-												×	×
P_N = 0-0-1	×				×	×							
P_O = 01-1-						×	×	×					
P_P = 1-1-1											×	×	×

Maxterms

(a)

	00100	00101	00110	01001	01011	01111	10011	10101	10111	11111	
P_A = 00-0-	×	×									
P_E = 001-0	×		×								
P_F = -00-1							×				
P_G = 10--1							×	×	×		
P_J = -0-01		×						×			
P_K = --110			×								
P_L = -111-						×				×	
P_N = 0-0-1				×	×						E (essential)
P_O = 01-1-					×	×					
P_P = 1-1-1								×	×	×	

(b)

	00100	00101	00110	01111	10011	10111	11111	
P_A = 00-0-	×	×						
P_E = 001-0	×		×					
P_G = 10--1					×	×		E
P_J = -0-01		×						
P_K = --110			×					
P_L = -111-				×			×	
P_O = 01-1-				×				
P_P = 1-1-1						×	×	

(c)

	00100	00101	00110	01111	11111	
P_A = 00-0-	×	×				E
P_E = 001-0	×		×			
P_K = --110			×			
P_L = -111-				×	×	E
P_P = 1-1-1					×	

(d)

FIGURE 5.28 (a) Initial and (b), (c), (d) reduced covering tables for z_5.

prime implicates because all columns contain at least two occurrences of $\times$. There is, however, a considerable amount of row and column dominance that we can exploit. The row dominance relations include $P_N \supseteq P_B$, $P_P \supseteq P_C$, $P_O \supseteq P_D$, $P_N \supseteq P_H$, $P_F \supseteq P_I$, and $P_P \supseteq P_M$. The six dominated rows all have the same number of literals (three) as the rows that dominate them; therefore, all dominated rows are deleted. Among the columns, we find the following dominance relations: $M_1 \supseteq M_5$, $M_{14} \supseteq M_{15}$, and $M_{17} \supseteq M_{21}$, so we can delete the three dominating columns M_1, M_{14}, and M_{17}.

The resulting reduced cover table appears in Figure 5.28b. The prime implicate P_N has now become essential, and some new dominance relations have appeared. $P_E \supseteq P_K$, but P_K has lower literal cost; $P_G \supseteq P_F$, and both have the same cost; and $M_{21} \supseteq M_{23}$. We therefore remove rows P_N and P_F (but not P_K), the dominating column M_{21}, and the two columns M_9 and M_{11} covered by P_N to obtain the cover table of Figure 5.28c. Now we find that P_G has become essential, while $P_A \supseteq P_J$ and $P_L \supseteq P_O$, and all these rows have equal cost.

Removal of the essential and dominated rows and the two columns covered by P_G produces the much shrunken table of Figure 5.28d. Now P_A and P_L are seen to be essential, and between them they cover all remaining columns except M_6. The latter is covered by P_E and P_K, which have costs 4 and 3, respectively, so we must choose P_K. At this point all columns are covered, and we have a minimum-cost cover that consists of P_K and the four prime implicates identified as essential at various stages: P_A, P_G, P_L, and P_N. The corresponding POS expression is

$$
\begin{aligned}
z_5^* &= P_A P_G P_K P_L P_N \\
&= (a + b + d)(\bar{a} + b + \bar{e})(\bar{c} + \bar{d} + e)(\bar{b} + \bar{c} + \bar{d})(a + c + \bar{e})
\end{aligned}
$$

■

5.6 Multiple-Output Functions*

Most logic circuits have two or more outputs.

The tabular approach to two-level logic minimization can be extended to a set of Boolean functions—that is, to a multiple-output Boolean function of the form $Z = z_1, z_2, \ldots, z_m$—by considering how product or sum terms can be shared among the component functions of Z.

Multiple-Output Circuits

As in the single-output minimization problem, our aim here is to design a two-level AND-OR circuit, or an equivalent sum-of-products form such as a NAND-NAND circuit, that has minimum cost. (The dual problem of finding a minimal POS-style circuit will not be considered explicitly.) The primary

*This section can be omitted without loss of continuity.

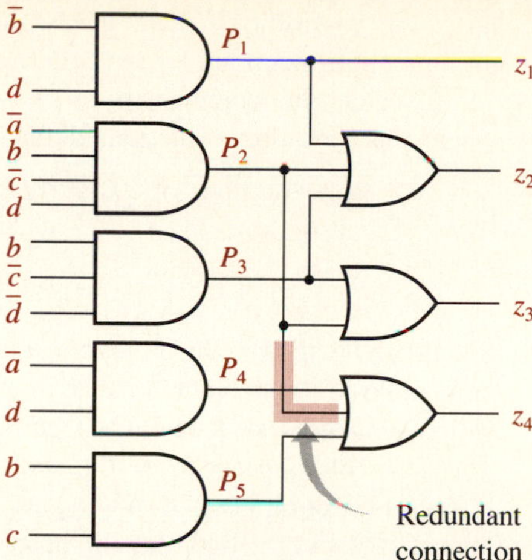

FIGURE 5.29 An example of a four-output, two-level AND-OR circuit.

cost goal is to minimize the number of gates used; the secondary goal is to minimize the total number of connections employed, as measured by the total number of inputs to all the gates, including those in the second level of the circuit.

In the example circuit C of Figure 5.29, where $m = 4$, there are three OR gates and five AND gates; the latter generate the five product terms $P_1{:}P_5$. This circuit corresponds to the following set of $m = 4$ SOP equations:

$$\begin{aligned} z_1 &= P_1 = \bar{b}d \\ z_2 &= P_1 + P_2 + P_3 = \bar{b}d + \bar{a}b\bar{c}d + b\bar{c}\bar{d} \\ z_3 &= P_2 + P_3 = \bar{a}b\bar{c}d + b\bar{c}\bar{d} \\ z_4 &= P_2 + P_4 + P_5 = \bar{a}b\bar{c}d + \bar{a}d + bc \end{aligned} \quad (5.39)$$

where each distinct P_i term is realized by a single AND gate. The gate cost of C is eight. The AND gates have a total of 13 inputs (literals); the OR gates have a total of eight inputs. Hence, the connection cost of this circuit is taken to be $8 + 13 = 21$. Observe that we cannot simply count the terms and literals in Equations (5.39) in order to measure the connection cost of C.

Subtle tradeoffs exist among the gates and connections of the two levels.

Suppose that a two-level AND-OR circuit C that realizes Z has g_{AND} AND gates and g_{OR} OR gates, and that its gate-input counts are c_{AND} and c_{OR} for the AND and OR gates, respectively. Then to say that C is minimal means first that there is no SOP circuit that implements Z that has fewer than $g_{\text{AND}} + g_{\text{OR}}$ gates, and second that among all SOP realizations of Z with $g_{\text{AND}} + g_{\text{OR}}$ gates, there is none with fewer than $c_{\text{AND}} + c_{\text{OR}}$ gate inputs. In the single-output case, we have only one output OR gate, so $g_{\text{OR}} = 1$ and $c_{\text{OR}} = g_{\text{AND}}$. For an m-output circuit with $m > 1$, however, we can have up to m OR gates, each with a different number of inputs. This

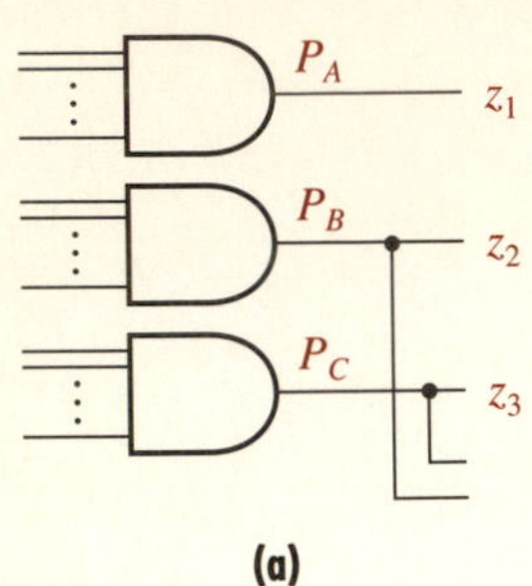

(a)

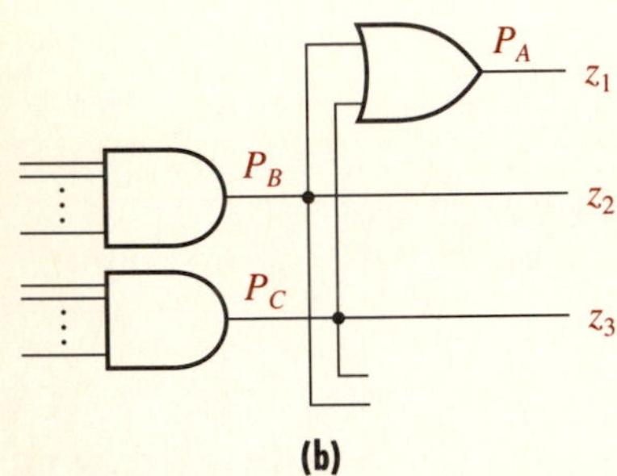

(b)

FIGURE 5.30 Two implementations of a three-output function.

raises the possibility of trading AND gates for OR gates, which makes strict minimization much harder with multiple outputs.

Consider the fragment of a two-level circuit depicted in Figure 5.30, where there are three functions defined by

$$z_1 = P_A = P_B + P_C$$
$$z_2 = P_B$$
$$z_3 = P_C$$

Should we realize z_1 by P_A alone as shown in Figure 5.30a, which requires only an AND gate, or realize it in the form $P_B + P_C$ as in Figure 5.30b, which dispenses with the AND gate but introduces a two-input OR gate? The gate cost appears to be the same in each case, but there is a trade-off in the connection cost. If AND gate P_A in Figure 5.30a has three or more inputs, replacing it with the OR gate *may* reduce the overall connection cost of the circuit. If this AND gate has only one input—that is, if $P_A = \dot{x}_i$—then the AND could be deleted, yielding a circuit that has lower connection cost *and* lower gate cost than the circuit of Figure 5.30b. The cost calculation is further complicated by the presence of other output functions that can use any of P_A, P_B, and P_C.

Problem Statement

We will therefore simplify the overall problem by requiring every output line z_i to be connected to an OR gate, and every input literal $\dot{x}_j$ to be connected to an AND gate. This fixes the number of OR gates at m for an m-output circuit and reduces the gate minimization part to minimizing the number of AND gates, just as it is in single-output minimization. It may, however, mean including some one-input AND or OR gates—that is, noninverting buffer gates—in the design. In the example of Figure 5.30, we must add three OR gates to Figure 5.30a and two to Figure 5.30b to meet the foregoing assumption. However, we can then say with certainty that the first circuit has higher gate cost.

It should also be noted that some physical technologies, such as many kinds of PLAs, require the above assumption to be met. We can also remove any superfluous single-input gates in a final, post-processing step, if desired. Our formal statement of the SOP form of the two-level minimization problem follows:

Multiple-output, two-level minimization problem.

> Given a set of m Boolean functions $Z = z_1, z_2, \ldots, z_m$, determine a two-level AND-OR circuit that implements Z and is such that (1) each primary input (output) signal is connected to an AND (OR) gate; (2) the circuit has the minimum number of AND gates; and (3) no circuit that satisfies the preceding conditions has a smaller total number of gate inputs.

As we will see, the secondary, connection-minimization condition is still difficult to satisfy exactly.

Multiple-Output Prime Implicants

If a term P_i is shared by a set of k functions $z_{i1}, z_{i2}, \ldots, z_{ik}$ in a minimum set of SOP expressions such as Equations (5.39), P_i must be a prime implicant of the k-fold product function $z_{i1}z_{i2}\cdots z_{ik}$. For example, $P_2 = \bar{a}b\bar{c}d$ appears in the last three equations of Equation (5.39); hence, P_2 is a prime implicant of

$$z_2z_3z_4 = (\bar{b}d + \bar{a}b\bar{c}d + b\bar{c}\bar{d})(\bar{a}b\bar{c} + b\bar{c}\bar{d})(\bar{a}b\bar{c}d + \bar{a}c + bc)$$

No literal can be deleted from P_2. Otherwise, we could remove the corresponding input line from the four-input AND gate of C that produces P_2, without affecting any of the three functions, z_2, z_3, or z_4. It is easily shown that P_2 is not a prime implicant of any of these individual functions. For instance, $z_3 = \bar{a}b\bar{c} + b\bar{c}\bar{d}$, implying that the last literal d of $P_2 = \bar{a}b\bar{c}d$ is redundant as far as z_3 alone is concerned.

A multiple-output prime implicant is a prime implicant of a product of output functions.

A product term that is a prime implicant of the k-fold product function $z_{i1}z_{i2}\cdots z_{ik}$, where each z_{ij} is taken from the set $Z = \{z_1, z_2, \ldots, z_m\}$ is called a **multiple-output prime implicant** of Z. If a multiple-output prime implicant P_i is a prime implicant of some product of k z_{ij}'s, but not of $k+1$ z_{ij}'s, P_i is called a **paramount prime implicant** of Z [Muroga, 1979]. In the foregoing example, $P_2 = \bar{a}b\bar{c}d$ is a paramount prime implicant because it is a prime implicant of $z_2z_3z_4$ but not of $z_1z_2z_3z_4 = 0$. In fact, each of the occurrences of P_i in Equation (5.39) is a paramount prime implicant.

Minimal forms contain only paramount multiple-output prime implicants.

It is clearly sufficient to restrict attention to paramount prime implicants in order to obtain a minimal two-level SOP design with the minimum number of gates. It is also the case, but is less obvious, that there is a minimum-gate circuit with a minimum connection cost among the circuits that generate only paramount prime implicants. For example, the circuit of Figure 5.29 has a minimum number of gates, every AND gate generates a paramount prime implicant, and the OR gate for z_i has a connection to each paramount prime implicant it covers. Not all such connections from AND to OR are necessary. The colored connection is redundant and can be removed because $P_2 = \bar{a}b\bar{c}d$ is covered by $P_4 = \bar{a}d$, and so is not needed by z_4.

Closely following the single-output procedure *TABMIN*, we solve the multiple-output minimization problem in three main steps. First we generate a complete set of (paramount) prime implicants, then we select subsets of these prime implicants that meet the primary objective, and finally we determine a prime implicant cover that also meets the secondary cost objective. The prime implicants are computed using lists that represent each term $P_j = \dot{x}_{j1}\dot{x}_{j2}\cdots\dot{x}_{jh}$ in the binary code-word form, and also indicate which product functions $z_{i1}z_{i2}\cdots z_{ik}$ have P_j as a prime implicant. Prime implicant selection is done using a tabular method (or its programmed equivalent) guided by the cost criteria.

Prime Implicant Identification

We add tags to code words that represent terms of multiple-output functions.

Previously, we represented a product term P_i of a single Boolean function $z_1(x_1, x_2, \ldots, x_n)$ by a code word of length n composed of the symbols 0, 1, and a dash (-), which denote, respectively, the presence of $\overline{x}_j$, the presence of x_j, and the presence of neither in the product expression $\dot{x}_{j1}\dot{x}_{j2}\cdots\dot{x}_{jh}$ for P_j. For example, $P_1(a, b, c, d) = \overline{b}d$ has the code word -0-1. Suppose there are m output functions $z_1, z_2, \ldots, z_m$, any subset of which can be implied by P_i. We append to the code word for P_i an m-bit binary string $t_1t_2 \ldots t_m$, called a **tag,** specified thus: $t_j = 1$ if P_i implies z_j; otherwise, $t_j = 0$. Hence, to indicate that $P_1 =$ -0-1 implies z_1 and z_2, but not z_3 and z_4 (which is the case in Figure 5.29), we assign the tag 1100 to P_1 and write code word and tag together, separated by a colon:

$$P_1 = \text{-0-1: 1100}$$

The tabular identification of paramount prime implicants follows the style developed earlier for $m = 1$. We first list the code word for each minterm contained in any of the given functions, indicating in its tag the functions to which the minterm belongs. Then beginning with the minterms, we systematically identify pairs of code words p_i, p_j that are logically adjacent and construct the code word that represents $p_{ij} = p_i + p_j$. A tag for p_{ij} is computed as the bitwise logical product of the tags for p_i and p_j because it should contain a 1 in a particular position if and only if the tags for p_i and p_j both have 1s in that position. If the tag for p_{ij} is not all 0s, meaning that it contributes to at least one output function, the code word and tag for p_{ij} are entered into the table. For example, if we have $p_a =$ 101-0: 110 and $p_b =$ 100-0: 111, then 10--0: 110 denoting $p_{ab} = p_a + p_b$ becomes a new entry of the table. On the other hand, if we have $p_c =$ --00: 0100 and $p_d =$ --01: 1001, then $p_{cd} =$ --0-: 0000, but p_{cd} is discarded because of its 0000 tag.

Whether a merged code word is checked off depends on the tags.

A new entry p_{ij} with a nonzero tag constructed in the above manner may or may not serve as a replacement for p_i or p_j. If p_i and p_{ij} have identical tags—that is, if they imply precisely the same set of functions—p_i can be checked off, and so will not be considered as a possible member of a minimum cover. In this case, p_{ij} has one less literal than p_i and contributes to precisely the same set of output functions as p_i. However, if p_i and p_{ij} have different tags, it is possible for both p_i and p_{ij}, or terms derived from them, to appear together in a minimal solution because there is at least one output function that is implied by p_i but not by p_{ij}. Thus, in the foregoing example, with $p_a =$ 101-0: 110, $p_b =$ 100-0: 111, and $p_{ab} =$ 10--0: 110, we would check off p_a but not p_b.

Figure 5.31 shows the application of the foregoing procedure to the four four-variable functions defined by Equations (5.39). The first list contains the minterms of all the functions grouped, as before, by the number of 1s they contain. The 4-bit tag that follows each code word specifies which of

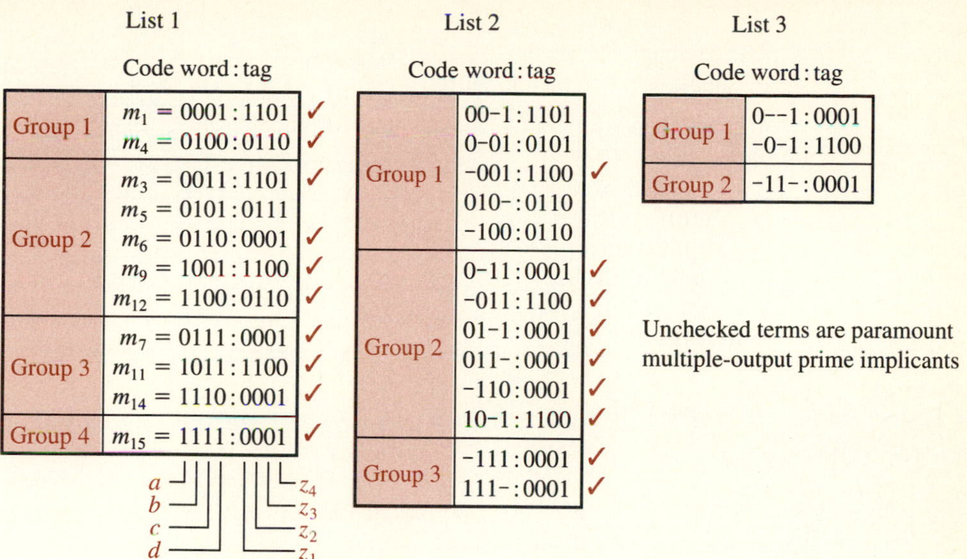

FIGURE 5.31 The tabular computation of the paramount prime implicants for a set of four four-variable functions.

the four functions $Z = z_1, z_2, z_3, z_4$ contain the minterm. The second list is formed by comparing elements of adjacent groups in list 1. For instance, the first item of list 2 is formed by merging the minterms $m_1 = 0001$ and $m_3 = 0011$ to produce the new code word 00-1. Because the two minterms have the common tag 1101, this tag is also assigned to the new code word, and both m_1 and m_3 are checked off. The second entry of list 2 comes from minterms m_1 and m_5; its tag 0101 is the product of the tags 1101 and 0111 of m_1 and m_5, respectively. Because the new tag differs from both the old ones, neither is checked off as a result of this operation, although m_1 retains its previous check mark. Continuing until no more mergers are possible yields the set of lists in Figure 5.31. At the end, eight unchecked entries represent all the paramount prime implicants of Z.

Prime Implicant Selection

The next step is to construct a prime implicant table to guide the selection of an optimal prime implicant cover of all the component functions of Z. Such a table for our four-variable example appears in Figure 5.32a. It is necessary to consider all the minterms of each output function z_i separately, and it is convenient to group the minterm columns by output function as shown. A column labeled j is included in column-group z_i if m_j is a minterm of z_i.

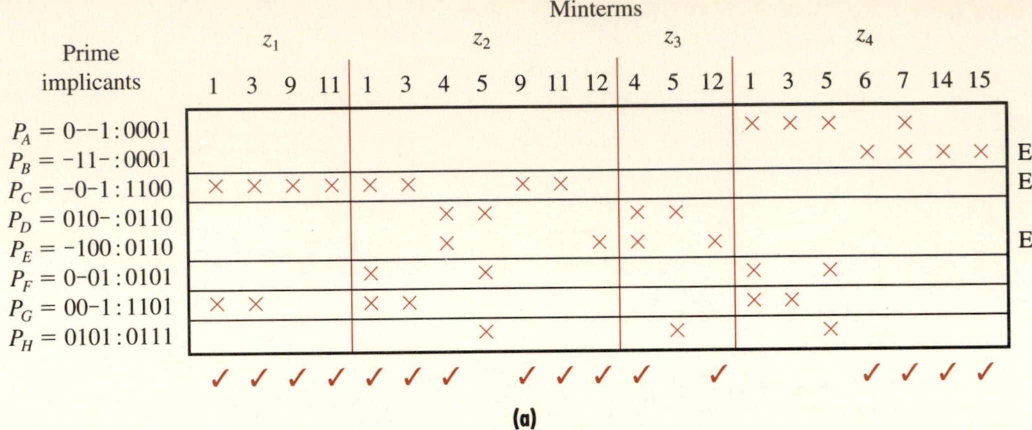

Prime implicants	z_1				z_2							z_3			z_4							
	1	3	9	11	1	3	4	5	9	11	12	4	5	12	1	3	5	6	7	14	15	
P_A = 0--1:0001															×	×	×		×			
P_B = -11-:0001																		×	×	×	×	E
P_C = -0-1:1100	×	×	×	×	×	×			×	×												E
P_D = 010-:0110							×	×				×	×									
P_E = -100:0110							×				×	×		×								E
P_F = 0-01:0101					×			×							×		×					
P_G = 00-1:1101	×	×			×	×									×	×						
P_H = 0101:0111								×					×				×					
	✓	✓	✓	✓	✓	✓	✓		✓	✓	✓	✓		✓				✓	✓	✓	✓	

(a)

Prime implicants	z_2	z_3	z_4		
	5	5	1	3	5
P_A = 0--1:0001			×	×	×
P_D = 010-:0110	×	×			
P_F = 0-01:0101	×		×		×
P_G = 00-1:1101			×	×	
P_H = 0100:0111	×	×			×
	D		D		

(b)

Prime implicants	z_3	z_4	
	5	3	5
P_A = 0--1:0001		×	×
P_D = 010-:0110	×		
P_F = 0-01:0101			×
P_G = 00-1:1101		×	
P_H = 0101:0111	×		×

(c)

FIGURE 5.32 Prime implicant tables for the four-output example: (a) the initial table; (b) the reduced table after selecting essential prime implicants; (c) the reduced table after deleting dominating columns.

The rows correspond to paramount prime implicants and are labeled by the corresponding code words and tags. It is also convenient to group together all prime implicants with the same tag, as shown in the figure. If P_k covers m_j, an × is entered in the j column of z_i if P_k's tag has a 1 in the z_i position. Thus, in the case of P_C = -0-1: 1100, which defines the third row of Figure 5.32a, P_C covers m_3 = 0011 and an × is entered into column 3 of groups z_1 and z_2. There is no × under column 3 of group z_4 for this particular row because P_C's tag 1100 has a 0 in the rightmost position that corresponds to z_4.

Essential prime implicants appear in every solution.

We can attempt to reduce the multiple-output prime implicant table, using the approach developed for the single-output case, but some changes are necessary. Essential rows and row and column dominance are defined exactly as before. And, as before, our first step is to identify and select all essential rows; these are marked by an E in Figure 5.32a. Column 11 of group z_1 and column 9 of group z_2 are two of several columns whose only $\times$ is in row P_C; hence, P_C is an essential prime implicant with respect to both z_1 and z_2. Note that, in general, an essential multiple-output prime implicant P need not be essential to *all* the output functions it covers; hence, we cannot say right away to how many OR gates the AND gate that realizes P will eventually be connected. The contribution to the connection cost made by a selected prime implicant, whether essential or nonessential, cannot usually be determined until a complete solution has been constructed. This is one reason why we postpone consideration of connection costs until the final phase of our minimization procedure. Figure 5.32a has three essential rows, P_B, P_C, and P_E, so the corresponding paramount prime implicants will appear in all minimal solutions to this problem.

We do not use row dominance to reduce multiple-output prime implicant tables.

Next we delete the three essential rows and all the columns that have an $\times$ in any of these rows, producing the much-reduced table of Figure 5.32b. We may use column dominance as before to eliminate any dominating columns. Recall that column m_i dominates column m_j if m_i has an $\times$ in every column in which m_j has an $\times$. In the example, column 5 of z_2 dominates column 5 of z_3, so the former can be deleted. It is easy to see that in this table there are two dominating columns, marked by Ds. Note that no gate or connection cost is associated with column dominance. This is *not* the case with row dominance where, in the single-output case, we could delete a dominated row if it had a literal or AND-connection cost that was not less than that of the dominating row. In multiple-output minimization, the connection cost of a row depends on OR as well as AND connections, and cannot be so easily measured while selecting prime implicants. Hence, we simply do not allow row dominance in our multiple-output minimization algorithm.

On deleting the two dominating columns from Figure 5.32b, we obtain the final, irreducible five-row, three-column table of Figure 5.32c. For such tables, we use Petrick's method or the branching method to determine *all* possible minimal sets of prime implicant covers. For the running example, the covering proposition is

$$\begin{aligned} \mathbf{C} &= (\mathbf{P}_D + \mathbf{P}_H)(\mathbf{P}_A + \mathbf{P}_G)(\mathbf{P}_A + \mathbf{P}_F + \mathbf{P}_H) \\ &= \mathbf{P}_A\mathbf{P}_D + \mathbf{P}_A\mathbf{P}_H + \mathbf{P}_G\mathbf{P}_H + \text{terms containing three literals} \end{aligned}$$

Hence, we have three minimal choices that, on adding the three essential prime implicants, yield three potential solutions

$$\{P_A, P_B, P_C, P_D, P_E\} \tag{5.40}$$

$$\{P_A, P_B, P_C, P_E, P_H\} \tag{5.41}$$

$$\{P_B, P_C, P_E, P_G, P_H\} \tag{5.42}$$

with five AND gates, the minimum possible number. Each set of five AND gates must connect to the four OR gates that every allowed realization of this four-output function must have.

Output Connection Minimization

It remains to determine what connections are needed from the five AND gates to the four OR gates in the three minimum-gate solutions (5.40), (5.41), and (5.42). In other words, we need a minimum prime implicant cover for each z_i, choosing the prime implicants from only the five in the current minimum-gate solution. Because we have already minimized the number of prime implicants, the possible choices are usually few in number. However, in the worst case, we might have to resort to up to m applications of Petrick's method to minimize the number of output connections.

Consider the solution to our running example given by (5.40). From the prime implicant table of Figure 5.32a, we find a cover for each output function in turn, using only the five rows P_A, P_B, P_C, P_D, and P_E. It is immediately seen that z_1 is covered by one essential prime implicant—namely, P_C. The next function z_2 has three essential prime implicants, P_C, P_D, and P_E, which collectively cover it. Similarly, the remaining two output functions z_3 and z_4 are covered by pairs of prime implicants that are essential with respect to each function. The resulting circuit appears in Figure 5.33a and has connection cost $c_{\text{AND}} + c_{\text{OR}} = 12 + 8 = 20$. For (5.41), we also find a unique solution given by the four sets of essential covers for each individual function (Figure 5.33b). In this case, z_4 is implied by one inessential term—namely, P_H. Hence, a connection from AND gate P_H to output gate z_4 will not alter the output function but is clearly redundant. The connection cost of this circuit is $13 + 8 = 21$. This last case is similar, and is left as an exercise. The corresponding circuit appears in Figure 5.33c and has connection cost $14 + 9 = 23$. We conclude that the design of Figure 5.33a is optimal, with gate cost 9 and connection cost 20.

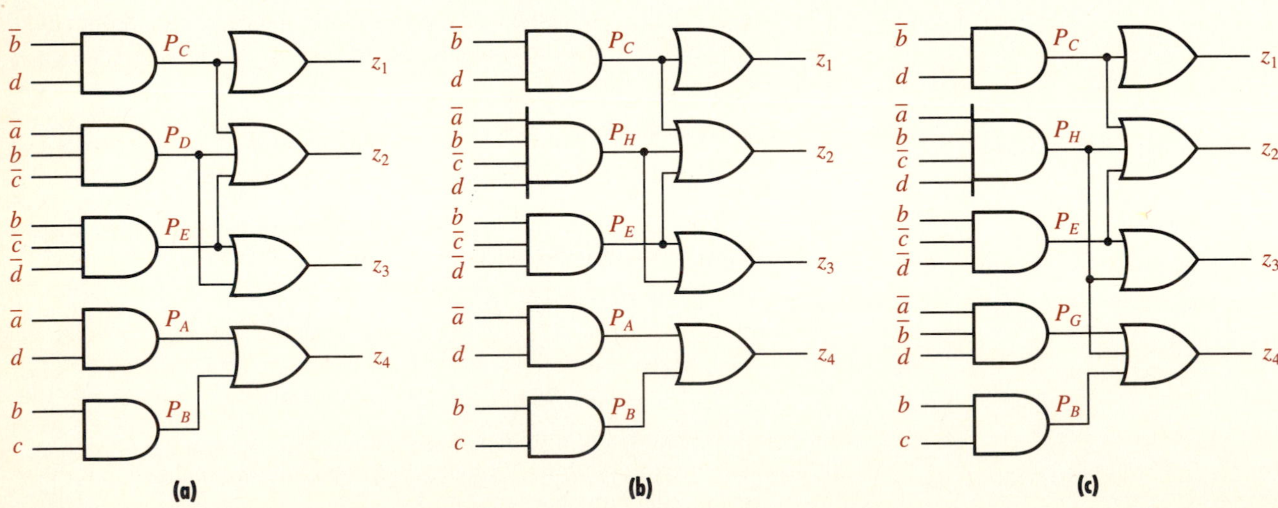

FIGURE 5.33 Three minimum-gate designs with differing connection costs.

Procedure

Procedure *MULTIMIN* for two-level, multiple-output minimization

1. Phase 1 (paramount prime implicant identification): Form a list $L_i = L_1$ of all minterms of $Z = z_1, z_2, \ldots, z_m$ represented by binary code words $E_{h,j}$ with tags $T_{h,j}$, indicating which functions contain each minterm. Divide L_i into groups $G_0, G_1, \ldots, G_p$, where G_h contains all minterms with exactly h 1s in their code words, $0 \leq h \leq p$.
2. Scan L_i and compare each code word $E_{h,j}$ of G_h with each code word $E_{h+1,k}$ of G_{h+1} for all h. If the code words are logically adjacent with $E = E_{h,j} + E_{h+1,k}$, compute the bitwise product $T = T_{h,j} \cdot T_{h+1,k}$ of their tags. Add $E{:}T$ to group G_h of the (new) list L_{i+1} if T is nonzero and $E{:}T$ is not already in L_{i+1}. Check off $E_{h,j}$ if its tag is T, and also $E_{h+1,k}$ if its tag is T.
3. If L_{i+1} is not empty, change i to $i + 1$ and repeat step 2. If L_{i+1} is empty, the unchecked entries in lists $L_1, L_2, \ldots, L_i$ are the paramount prime implicants of Z; proceed to step 4.
4. Phase 2 (gate minimization): Set up a cover table CT in which the rows are all the paramount prime implicants and the columns are all the minterms of each function listed separately. Enter an $\times$ at the intersection of row i and column j if the ith prime implicant P_i covers the jth minterm m_j.
5. Identify all the essential rows of CT and select them for inclusion in the potential solutions set $\mathcal{S} = \{S_1, S_2, \ldots, S_q\}$. Delete from CT all columns covered by these rows and the essential rows themselves. Delete any rows that contain no occurrences of $\times$. Reduce T by removing all columns that dominate a retained column. (Row dominance should not, in general, be used.)
6. Continue applying step 5 until no further reduction of CT is possible. If T is empty, proceed to step 9, with $\mathcal{S}$ as the minimum-gate solution set. If CT is nonempty, go to step 7.
7. Use Petrick's method or the branching method to select a minimum subset of the remaining rows to add to each set in $\mathcal{S}$.
8. Phase 3 (output connection minimization): For each set S_r in $\mathcal{S}$, determine the essential set $\mathcal{P}_{r,s} = P_{r,s,t}$ of prime implicants in $S_{r,s}$ needed to cover each individual output function z_s of Z.
9. If $z_s \neq \sum P_{r,s,t}$, determine an additional minimal set of prime implicants to add to $\mathcal{P}_{r,s}$ to obtain a minimum-term cover, by using row and column dominance, Petrick's method, and so on.
10. Compute the combined connection costs $c = c_{\mathrm{AND}} + c_{\mathrm{OR}}$ of $\sum_{s=1}^{m} \mathcal{P}_{r,s}$ for $1 \leq r \leq q$. Any $\mathcal{P}_{r,s}$ that has minimum c serves as the final solution.

The foregoing steps pull together all the preceding threads into a complete and fairly formidable-looking minimization procedure *MULTIMIN* for multiple-output functions. Some of these steps can be bypassed or simplified if we settle for a design that has nearly but not exactly the minimum cost. For example, we can dispense with steps 8 through 10 (the entire output-connection minimization phase), and settle for one of the minimum-gate designs produced in step 7 that includes all the prime implicants that imply each z_i, rather than a minimum set of such prime implicants.

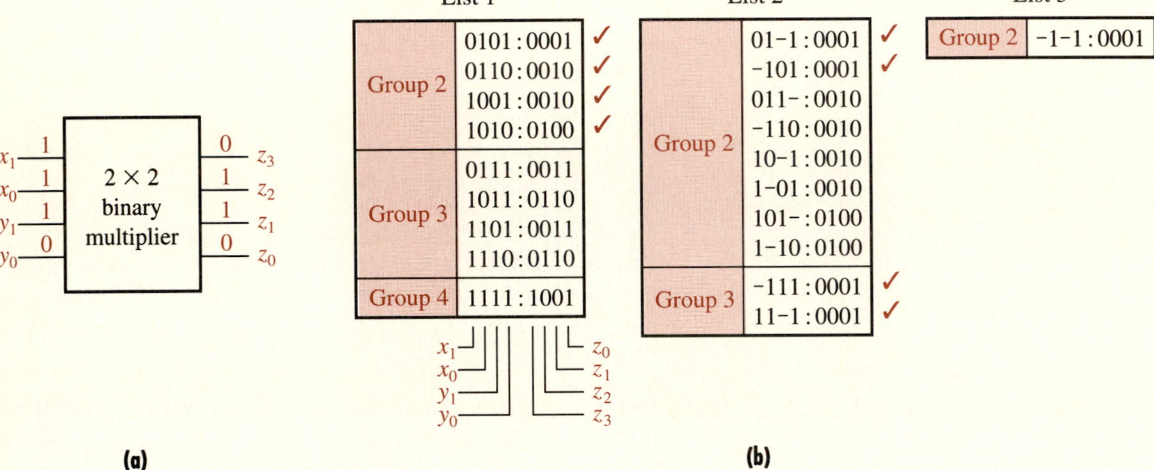

FIGURE 5.34 (a) A 2 × 2 multiplier; (b) computation of its multiple-output prime implicants.

Example 5.7

Design of a 2 × 2 multiplier

Consider the design of a four-input, four-output circuit that performs the useful function of multiplying two 2-bit unsigned binary numbers $X = x_1x_0$ and $Y = y_1y_0$ to produce a 4-bit product $Z = z_3z_2z_1z_0 = X \times Y$. The minterm list is shown in Figure 5.34b. It is easy to see that our code word–tag representation of the minterms amounts to the set of all nonzero entries in the multiplier's truth table. For example, the last minterm in group 2 is m_{10} = 1010, which corresponds to 2 × 2, while its tag is 0100 = 4. The process of constructing the paramount prime implicant via steps 1 through 3 of *MULTIMIN* is easy. It results in the 12 paramount prime implicants represented by unchecked terms in Figure 5.34b.

Next we construct the covering table of Figure 5.35a. It contains two essential prime implicants, P_G and P_H, which cover all columns of z_3 and z_0. Removing the two essential rows and the five minterm columns they cover produces the reduced table of Figure 5.35b. Unfortunately, there is no column dominance present in this table, so we switch to Petrick's method to

Prime implicants	Minterms z_3	z_2			z_1						z_0				
	1111	1010	1011	1110	0110	1001	0111	1011	1101	1110	0101	0111	1101	1111	
P_A = 101–:0100		×	×												
P_B = 1–10:0100		×		×											
P_C = 011–:0010					×		×								
P_D = –110:0010					×					×					
P_E = 10–1:0010						×		×							
P_F = 1–01:0010						×			×						
P_G = –1–1:0001											×	×	×	×	E
P_H = 1111:1001	×													×	E
P_I = 1011:0110			×					×							
P_J = 1110:0110				×						×					
P_K = 0111:0011							×					×			
P_L = 1101:0011									×				×		

(a)

Prime implicants	Minterms z_2			z_1					
	1010	1011	1110	0110	1001	0111	1011	1101	1110
P_A = 101–:0100	×	×							
P_B = 1–10:0100	×		×						
P_C = 011–:0010				×		×			
P_D = –110:0010				×					×
P_E = 10–1:0010					×		×		
P_F = 1–01:0010					×			×	
P_I = 1011:0110		×					×		
P_J = 1110:0110			×						×
P_K = 0111:0011						×			
P_L = 1101:0011								×	

(b)

FIGURE 5.35 (a) A covering table for the multiplier; (b) a reduced table.

obtain a minimal cover.

$$\begin{aligned}\mathbf{C} &= (\mathbf{P}_A + \mathbf{P}_B)(\mathbf{P}_A + \mathbf{P}_I)(\mathbf{P}_C + \mathbf{P}_D)(\mathbf{P}_C + \mathbf{P}_K)(\mathbf{P}_E + \mathbf{P}_F)(\mathbf{P}_F + \mathbf{P}_L)\\ &\quad (\mathbf{P}_B + \mathbf{P}_J)(\mathbf{P}_D + \mathbf{P}_J)(\mathbf{P}_E + \mathbf{P}_I)\\ &= (\mathbf{P}_A + \mathbf{P}_B\mathbf{P}_I)(\mathbf{P}_C + \mathbf{P}_D\mathbf{P}_K)(\mathbf{P}_F + \mathbf{P}_E\mathbf{P}_L)(\mathbf{P}_J + \mathbf{P}_B\mathbf{P}_D)(\mathbf{P}_E + \mathbf{P}_I)\\ &= (\mathbf{P}_A\mathbf{P}_C + \mathbf{P}_B\mathbf{P}_C\mathbf{P}_I + \mathbf{P}_A\mathbf{P}_D\mathbf{P}_K + \mathbf{P}_B\mathbf{P}_D\mathbf{P}_I\mathbf{P}_K)(\mathbf{P}_F\mathbf{P}_J + \mathbf{P}_E\mathbf{P}_J\mathbf{P}_L\\ &\qquad + \mathbf{P}_B\mathbf{P}_D\mathbf{P}_F + \mathbf{P}_B\mathbf{P}_D\mathbf{P}_E\mathbf{P}_L)(\mathbf{P}_E + \mathbf{P}_I)\\ &= \mathbf{P}_A\mathbf{P}_C\mathbf{P}_E\mathbf{P}_F\mathbf{P}_J + \mathbf{P}_A\mathbf{P}_C\mathbf{P}_E\mathbf{P}_L\mathbf{P}_J + \mathbf{P}_A\mathbf{P}_C\mathbf{P}_F\mathbf{P}_I\mathbf{P}_J + \mathbf{P}_B\mathbf{P}_C\mathbf{P}_F\mathbf{P}_I\mathbf{P}_J\\ &\quad + \mathbf{P}_B\mathbf{P}_C\mathbf{P}_D\mathbf{P}_F\mathbf{P}_I + \mathbf{P}_B\mathbf{P}_D\mathbf{P}_F\mathbf{P}_I\mathbf{P}_K + 18 \text{ terms of size 6 or greater}\end{aligned}$$

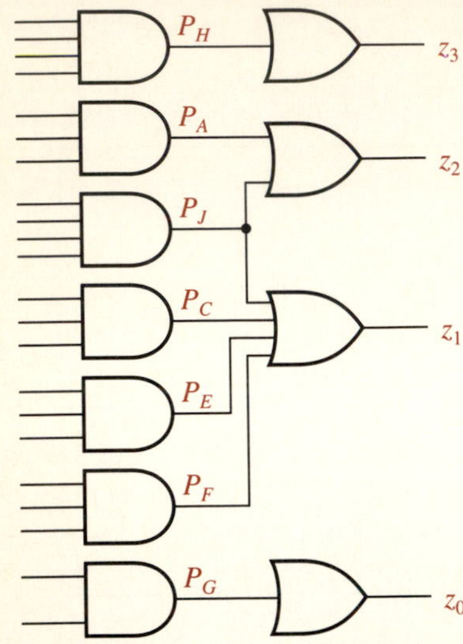

FIGURE 5.36 An optimal design for the multiplier with gate cost 11 and connection cost 30.

Thus, there are six five-term minimal covers for the reduced cover table. Restoring the two essential terms yields the following six solutions with the minimum number of seven AND gates:

$$\begin{aligned}
&\{P_A, P_C, P_E, P_F, P_G, P_H, P_J\}\\
&\{P_A, P_C, P_E, P_G, P_H, P_L, P_J\}\\
&\{P_A, P_C, P_F, P_G, P_H, P_I, P_J\}\\
&\{P_B, P_C, P_F, P_G, P_H, P_I, P_J\}\\
&\{P_B, P_C, P_D, P_F, P_G, P_H, P_I\}\\
&\{P_B, P_D, P_F, P_G, P_H, P_I, P_K\}
\end{aligned} \tag{5.43}$$

A check shows that two of these solutions have a connection cost of 30; the others have cost 31. Consider the first of the solutions in (5.43), which consists of the seven prime implicants P_A, P_C, P_E, P_F, P_G, P_H, and P_J. From Figure 5.35 we see that outputs z_3 and z_0 have the single-term covers P_H and P_G, respectively. In the case of z_2, P_A and P_J are essential, and together cover this output. The remaining four prime implicants P_C, P_E, P_F, and P_J are essential with respect to z_1, and collectively cover it. Hence, we have no choice but to realize the four output functions in the form

$$\begin{aligned}
z_3 &= P_H\\
z_2 &= P_A + P_J\\
z_1 &= P_E + P_F + P_G + P_J\\
z_0 &= P_G
\end{aligned}$$

which corresponds to the AND-OR design of Figure 5.36, whose connection cost is 30. If the two single-output OR gates are removed, the gate and connection costs reduce to 9 and 28, respectively.

5.7 Heuristic Methods

Heuristic methods try to find acceptable solutions quickly.

For large two-level minimization problems that involve, say, functions of 20 or more variables, even a powerful minimization program like McBOOLE is too slow to find an optimal solution in a reasonable amount of time. If we are willing to settle for a functionally correct solution of reasonably low but not necessarily minimal cost, we can often find a solution quickly, using ad hoc or intuitive methods, termed **heuristic procedures** or simply **heuristics.** These methods underlie most types of CAD programs used in digital design. Heuristics generally give acceptable solutions and sometimes even optimal ones, but there is no guarantee that they will do so in any particular case.

"Minimal" designs are not necessarily the best.

It is also worth noting that exact minimization does not necessarily lead

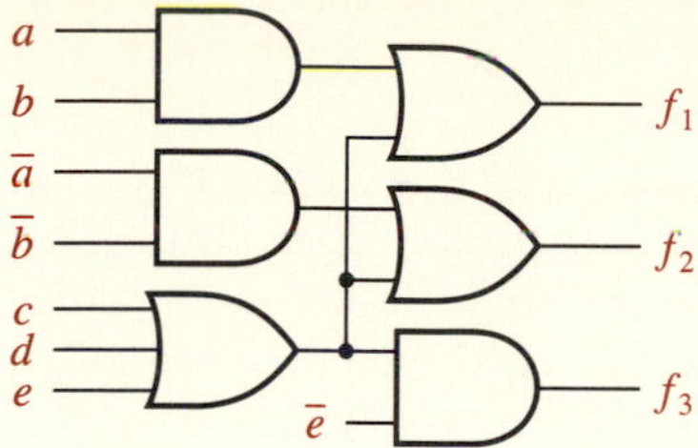

FIGURE 5.37 A two-level circuit of lower cost than any SOP or POS design.

to the best or most practical two-level design. For example, if the design is to be implemented by integrated circuits, we will probably be more interested in minimizing the number of ICs used rather than the number of gates or connections. It is possible that a minimal two-level design will use a nonminimal number of ICs and, conversely, a design that employs the fewest ICs may use a nonminimum number of gates.

The assumptions underlying our minimization problem also limit the solution possibilities. For example, *MULTIMIN* and its dual version can generate only two-level SOP or POS designs, with one gate type in each level. What if we could mix AND and OR gates in the same level? Then we would have the possibility of a further reduction in cost. Figure 5.37 gives an example of a multiple-output, two-level circuit that has AND and OR gates in both its levels, and thus cannot be called SOP or POS. It can be shown easily that this design employs fewer gates (our primary objective) and gate connections (our secondary objective) than any SOP or POS implementation of the same three functions (Problem 5.40).

A Covering Heuristic

To illustrate the heuristic approach, we present a straightforward method we call *GREEDYCOV* that solves the covering problem encountered in two-level logic minimization. Suppose we must obtain a minimum-cost cover for a set of minterms or maxterms $\mathcal{M}$ using prime terms chosen from a set $\mathcal{P}$. The *GREEDYCOV* procedure to solve this problem may be summarized as follows: Identify the essential and absolutely inessential members of $\mathcal{P}$; discard the latter and retain the former in the solution set $\mathcal{P}_S$. While $\mathcal{P}$ is nonempty, select the next element P_i of $\mathcal{P}$ that covers the most members of $\mathcal{M}$ not yet covered by $\mathcal{P}_S$ and that has the fewest literals. Remove P_i from $\mathcal{P}$ and add to $\mathcal{P}_S$. Continue until $\mathcal{M}$ is completely covered; the resulting $\mathcal{P}_S$ is the solution. A more formal description of *GREEDYCOV* follows.

Procedure

Procedure *GREEDYCOV* to find a minimal cover of $\mathcal{M}$ by $\mathcal{P}$

1. Initialize $\mathcal{P}_S$ to be the empty set.
2. Identify the set $\mathcal{P}_E$ of essential terms in $\mathcal{P}$. Add $\mathcal{P}_E$ to $\mathcal{P}_S$. Remove $\mathcal{P}_E$ from $\mathcal{P}$, and remove the elements covered by $\mathcal{P}_E$ from $\mathcal{M}$.
3. Scan $\mathcal{P}$ and find the subset $\mathcal{P}_L$, each member of which covers the largest number of members of $\mathcal{M}$ not yet covered by $\mathcal{P}$.
4. Scan $\mathcal{P}_L$ and find the first member P_i with the fewest literals. Remove P_i from $\mathcal{P}$, and remove the terms covered by P_i from $\mathcal{M}$. Add P_i to $\mathcal{P}_S$.
5. Repeat steps 2 through 4 until $\mathcal{M}$ is empty; $\mathcal{P}_S$ is then the final cover.

It is obvious that *GREEDYCOV* always generates a complete cover that defines a functionally correct SOP (POS) expression. It favors a P_i that covers many cells and has few literals; thus, it tends to produce small and frequently minimal covers. Heuristic techniques like *GREEDYCOV* are considered to be **greedy** because they operate on the principle: "Make the choice that has the largest immediate benefit." Hence, the P_i chosen for inclusion in the current solution $\mathcal{P}_S$ is one that reduces the number of uncovered members of $\mathcal{M}$ by the maximum amount. This locally optimal decision may not lead to a globally optimal solution of the covering problem. However, it often leads to a solution that is quite acceptable for practical purposes.

Example 5.8

Heuristic selection of a prime-implicate cover via *GREEDYCOV*

Let us apply the *GREEDYCOV* heuristic to the POS covering problem of Examples 5.5 and 5.6, where we needed to select a cover for 13 0-cells (maxterms) from the set $\mathcal{P} = \{P_A, P_B, \ldots, P_P\}$ of 16 prime implicates. There is no essential P_i in this instance, so $\mathcal{P}_S$ remains the empty set $\varnothing$ after step 2. From the cover table of Figure 5.38a, we can see that there are two prime implicates, P_G and P_J, that cover four maxterms, and none that cover more than four. Thus, we have $\mathcal{P}_L = \{P_G, P_J\}$ as the initial set from which to choose a term of the solution.

Among choices of equal cost, the first one scanned is chosen.

To choose between $P_G = \overline{a} + b + \overline{e}$ and $P_J = b + d + \overline{e}$, which have the same number of literals, we need a **tie-breaking rule**. The simple tie-breaking rule implicit in *GREEDYCOV* is to select the first choice encountered among several equal-cost choices, as determined by the order in which the members of $\mathcal{P}_L$ are scanned. We assume that the scanning is in the normal alphabetical order of the subscripts of the prime terms. Thus, P_A is selected before P_B, P_B is selected before P_C, and so on. Because no other significance is attached to the subscripts, this is tantamount to making an arbitrary choice among the available members of $\mathcal{P}_L$. In the present instance, P_G is selected before P_J, so we begin our solution with $\mathcal{P}_S = \{P_G\}$, and delete the four maxterms M_{17}, M_{19}, M_{21}, and M_{23} covered by $\mathcal{P}_G$.

Now there are four three-literal prime implicants P_A, P_L, P_N, and P_O that cover three of the remaining maxterms; we select the first of these, P_A, to add to $\mathcal{P}_S$. At this point, six maxterms—M_6, M_9, M_{11}, M_{14}, M_{15}, and M_{31}—remain to be covered, as indicated in the reduced table of Figure 5.38b. P_L is the first prime implicate that covers three of the latter—M_{14}, M_{15}, and M_{31}—so we add it to the partial solution, making $\mathcal{P}_S = \{P_A, P_G, P_L\}$. Examining M_6, M_9, and M_{11}, we find only two prime implicates that cover two of them: P_B and P_N. Each has three literals, so P_B, which covers M_9 and M_{11}, is added to the solution. The final uncovered maxterm M_6 is covered by both $P_E = a + b + \overline{c} + e$ and $P_K = \overline{c} + \overline{d} + e$. P_K has fewer literals and so is added to $\mathcal{P}_S$ to yield the final result:

$$\mathcal{P}_S = \{P_A, P_B, P_G, P_K, P_L\}$$

Prime implicates	00001	00100	00101	00110	01001	01011	01110	01111	10001	10011	10101	10111	11111	
P_A = 00-0-	×	×	×											Choice 2
P_B = 010--					×	×								
P_C = 111--													×	
P_D = -1-10							×							
P_E = 001-0		×		×										
P_F = -00-1	×								×	×				
P_G = 10--1									×	×	×	×		Choice 1
P_H = 0-00-	×				×									
P_I = -000-	×								×					
P_J = -0-01	×		×						×		×			
P_K = --110				×			×							
P_L = -111-							×	×					×	
P_M = 1-11-												×	×	
P_N = 0-0-1	×				×	×								
P_O = 01-1-						×	×	×						
P_P = 1-1-1											×	×	×	

(a)

Prime implicates	00110	01001	01011	01110	01111	11111	
P_B = 010--		×	×				Choice 4
P_C = 111--						×	
P_D = -1-10				×			
P_E = 001-0	×						
P_F = -00-1							
P_H = 0-00-		×					
P_I = -000-							
P_J = -0-01							
P_K = --110	×			×			Choice 5
P_L = -111-				×	×	×	Choice 3
P_M = 1-11-						×	
P_N = 0-0-1		×	×				
P_O = 01-1-			×	×	×		
P_P = 1-1-1						×	

(b)

FIGURE 5.38 (a) A prime implicant cover table; (b) a reduced table.

This *happens* to be an optimal solution, in fact the one found earlier (Figure 5.18c) by the K-map method. Note that the tabular method (Example 5.6) gave us a slightly different but also optimal solution, with P_N replacing P_B.

Heuristics are Nonoptimal . . .

The fact that a greedy heuristic does not always yield a minimal cover is illustrated by applying *GREEDYCOV* to the cyclic K-map of Figure 5.11a, which is repeated in Figure 5.39a, with its eight prime implicants labeled in a specific order. Each prime implicant is inessential and covers exactly two 1-cells (minterms). *GREEDYCOV* therefore chooses among them using the same tie-breaking rule: Among equal choices, select the one with the lowest subscript.

We begin by setting $\mathcal{P}_S = \{P_A\}$. This leaves six uncovered 1-cells; five prime implicants P_B, P_C, P_D, P_G, and P_H cover two of them. We add the first of these, P_B, to $\mathcal{P}_S$. The remaining uncovered 1-cells are m_0, m_2, m_7, and m_{10}. Two of these are covered by P_C and P_E, so P_C, which covers m_2 and m_{10}, is selected, making $\mathcal{P}_S = \{P_A, P_B, P_C\}$. The final two 1-cells m_0 and m_7 are not adjacent, so two prime implicants are needed to cover them, yielding the solution $\mathcal{P}_S = \{P_A, P_B, P_C, P_D, P_E\}$. This is nonminimal, as the four-element cover shown in Figure 5.11b demonstrates. *GREEDYCOV* made a poor choice of prime implicants because of its tie-breaking rule. A different ordering of the subscripts of the prime implicants might allow *GREEDYCOV* to find a minimal cover in this case, but there is no way to determine the best order *a priori*. Hence, we conclude that a heuristic method like *GREEDYCOV* will, in general, produce a nonminimal cover.

. . . But Often Work Quite Well

Despite its nonoptimality, *GREEDYCOV* generally works well in practice, for several reasons:

1. It involves far less computation than any exact minimization algorithm for the covering problem, such as Petrick's method or the branching method.
2. It provides a solution that is always complete and often near-minimal, if not indeed minimal.
3. We can sometimes estimate how close the greedy solution is to a minimal one, and thus determine whether exact minimization is worthwhile.

Heuristics that allow solution quality to be gauged are especially useful.

To illustrate the last point, consider again the POS covering problem posed by Figure 5.38a. There are 13 maxterms to cover, and the largest P_i can cover only four maxterms. Hence, we will need at least $\lceil 13/4 \rceil = 4$ prime implicates in any solution. Four is therefore a lower bound on the number of prime terms required. Because *GREEDYCOV* yielded a five-term solution, we conclude that this solution has at most one more term than any minimal cover, and so can reasonably be called near-minimal.

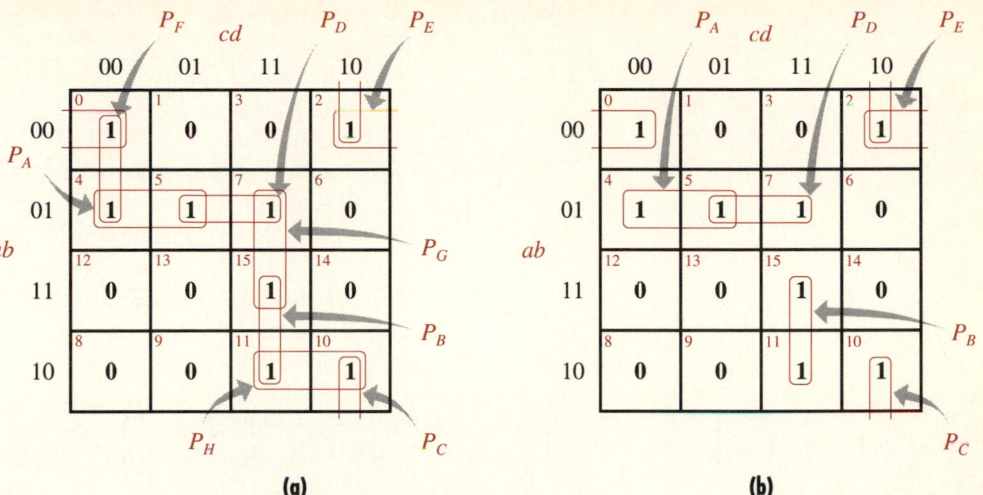

FIGURE 5.39 (a) The prime implicants of z_2; (b) a nonminimal SOP cover produced by the *GREEDYCOV* heuristic.

Multilevel Design

Practical designs for some logic functions must contain more than two levels.

Although heuristic minimization methods can provide minimal or at least near-minimal designs for many problems, two-level circuits are sometimes impractical, either because they require too many gates, even after minimization, or because they fail to meet the fan-in and fan-out constraints of the implementation technology. We will often see examples of combinational circuits that have more than two levels, but contain fewer gates than any minimal two-level design for the same function, and also meet lower fan-in and fan-out limits.

No practical and exact minimization techniques are known for general multilevel circuits. Consequently, heuristic techniques are normally used to tackle the multilevel minimization problem, with the aid of CAD software such as the MIS (Multilevel Interactive Synthesis) program [Brayton et al., 1987].

EXCLUSIVE-OR Circuits

We can see the cost advantages of multilevel circuits compared to two-level designs in extreme form in the case of the n-bit EXCLUSIVE-OR or parity function $z_{XOR}(x_1, x_2, \ldots, x_n)$. This function has precisely 2^{n-1} prime implicants, each of which is essential and corresponds to a minterm of z_{XOR} that has an odd number of true literals. Hence, a two-level SOP implementation of, say, the NAND-NAND type requires 2^{n-1} n-input NAND gates and a single 2^{n-1}-input NAND as the output gate. If we could use EXCLUSIVE-OR gates of unlimited fan-in to implement z_{XOR}, a single n-input EXCLUSIVE-

OR gate would suffice. We can also easily decompose a large EXCLUSIVE-OR gate into a tree-like structure of smaller EXCLUSIVE-OR gates. We can then map these gates into a functionally equivalent all-NAND circuit.

EXCLUSIVE-OR has only very costly two-level designs, but has low-cost multilevel ones.

Suppose we want to realize $z_{XOR}(x_1, x_2, \ldots, x_8)$ with $n = 8$. An SOP implementation has the 129-gate NAND-NAND structure shown in Figure 5.40a. This two-level design is impractical due to its huge number of gates and their excessive fan-in, not to mention the huge fan-out required of the primary input lines. A multilevel realization of the same function composed of seven two-input EXCLUSIVE-OR gates appears in Figure 5.40b. Finally, Figure 5.40c shows an all-NAND design obtained by mapping each two-input EXCLUSIVE-OR gate into a two-level NAND circuit N_2. Due to the presence of input inverters, N_2 should be treated as having three rather than two levels; we cannot "hide" N_2's input inverters when its input signals are not primary inputs of the overall circuit. This circuit has a total of $7 \times 5 = 35$ gates, and the circuit's maximum fan-in and fan-out is only two. However, the price paid for this relatively inexpensive design is a large number (nine) of logic levels. (A PLA implementation of a multilevel EXCLUSIVE-OR circuit is described in Example 4.8 of Section 4.7.)

Next we briefly illustrate some methods of multilevel design that are implemented in logic synthesis programs like MIS. Typically, these techniques are applied repeatedly in heuristic ways to different parts of a circuit, or to a Boolean expression representing it, in an attempt to reduce hardware costs, propagation delays, or some combination of the two.

Factoring

Factoring out common subexpressions can reduce gate and connection costs.

Factoring and the inverse operation (multiplying out or expansion) are central operations in most forms of multilevel logic design. The basic idea is to take a Boolean expression E in, say, SOP form and factor out a subexpression E^* that appears (perhaps implicitly) in several places in E. We replace several copies of E^* with a single one, thereby reducing the circuit's hardware cost but increasing its propagation delay. The inverse, multiplying-out process tends to "flatten" a circuit, reducing delay but increasing the number of gates.

Consider the Boolean expression

$$z_1(x_1, x_2, x_3, x_4, x_5) = x_1x_3 + x_1x_4 + x_1x_5 + x_2x_3 + x_2x_4 + x_2x_5 + x_1x_2 \tag{5.44}$$

which corresponds to the eight-gate SOP realization shown in Figure 5.41a. Because the literal x_1 appears in four product terms, we can factor it out, a step justified by the distributive law $ab + ac = a(b + c)$. We can also factor x_2 out of the three remaining product terms of (5.44) to obtain

$$z_1 = x_1(x_2 + x_3 + x_4 + x_5) + x_2(x_3 + x_4 + x_5) \tag{5.45}$$

This sum-of-products-of-sums form corresponds to the three-level, five-gate circuit of Figure 5.41b, which has lower cost but greater delay than the two-level design.

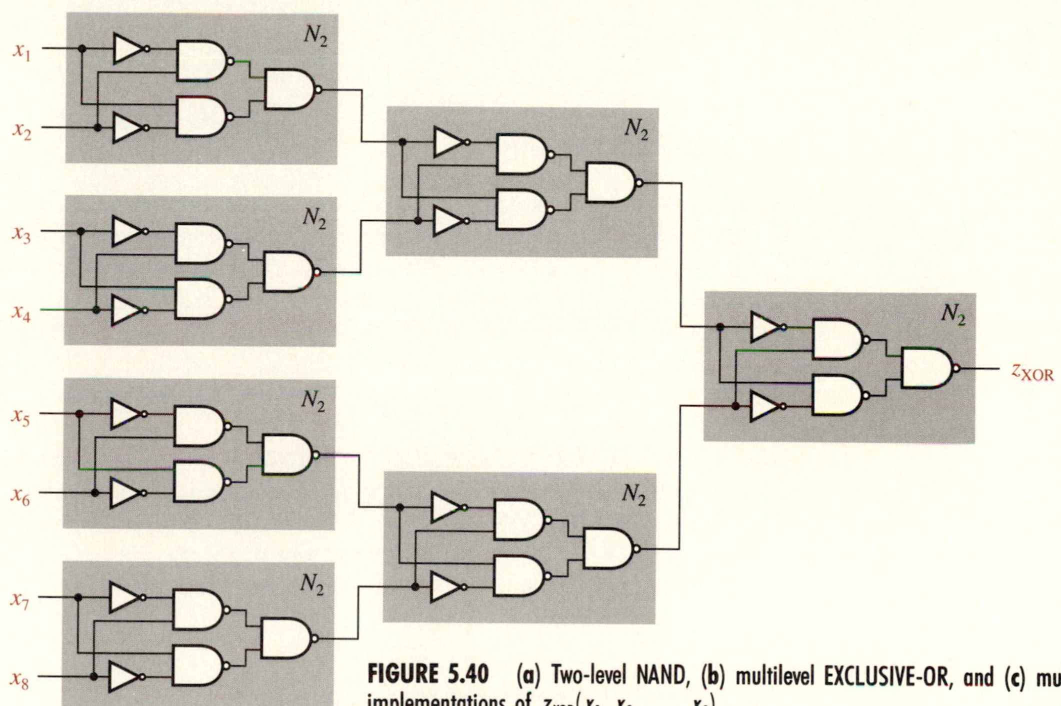

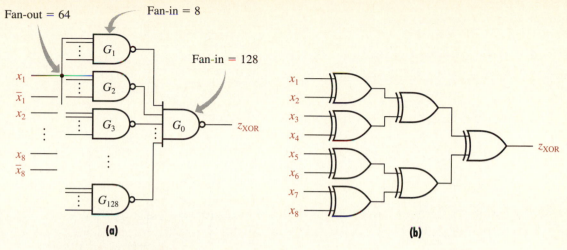

FIGURE 5.40 (a) Two-level NAND, (b) multilevel EXCLUSIVE-OR, and (c) multilevel NAND implementations of $z_{XOR}(x_1, x_2, \ldots, x_8)$.

On inspecting (5.45), we see that it contains two almost identical subexpressions $(x_3 + x_4 + x_5)$ and $(x_2 + x_3 + x_4 + x_5)$. We can reduce the larger of these to the smaller by partially expanding (5.45) thus:

$$z_1 = x_1(x_3 + x_4 + x_5) + x_1x_2 + x_2(x_3 + x_4 + x_5) \tag{5.46}$$

Now we factor out the common subexpression $(x_3 + x_4 + x_5)$, yielding

$$z_1 = (x_1 + x_2)(x_3 + x_4 + x_5) + x_1x_2 \tag{5.47}$$

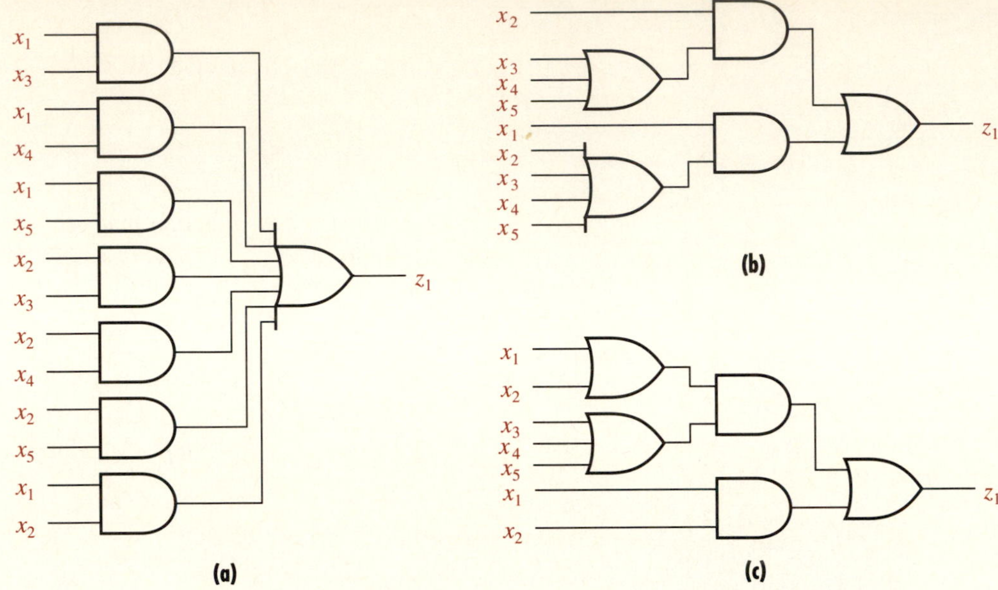

FIGURE 5.41 Implementations of $z_1(x_1, x_2, x_3, x_4, x_5)$: (a) a two-level SOP form; (b), (c) multilevel factored forms.

This expression defines another three-level, five-gate circuit (Figure 5.41c) that has slightly lower connection cost than the preceding one.

A Boolean expression can be factored in a huge number of ways.

The factoring of a Boolean expression is complicated by the fact that the same expression may be factored in many different ways—far more than a similar numerical expression. If we factor the "divisor" D from the Boolean expression E, we can write

$$E = D \cdot Q + R$$

where Q and R denote "quotient" and "remainder" expressions, respectively. For example, the three expressions for z_1 we obtained in Equations (5.45), (5.46), and (5.47) have the form $D \cdot Q + R$, with $D = x_3 + x_4 + x_5$ but with different expressions for Q and R. In the case of (5.45), we have $Q = x_2$ and $R = x_1(x_2 + x_3 + x_4 + x_5)$, whereas (5.47) has $Q = x_1 + x_2$ and $R = x_1 x_2$.

Knowledge-Based Transformations

At this point, we might ask whether we have found the simplest possible realization of z_1. We have explored only a tiny part of the "design space" of WF circuits that implement z, so we cannot really answer this question. Boolean expressions can often be transformed in complex ways based on prior knowledge of equivalent expressions or known "good" realizations of certain expressions. This knowledge resides in a designer's experience and may be built into CAD programs to a limited extent. Applying such knowledge is, perhaps, the most heuristic of methods, but is often necessary to obtain the best designs.

Returning to the preceding example, we might observe that the original SOP expression

$$z_1 = x_1x_3 + x_1x_4 + x_1x_5 + x_2x_3 + x_2x_4 + x_2x_5 + x_1x_2$$

contains the subexpression $x_1x_2 + x_2x_3 + x_1x_3$, which is recognizable (known) as the carry or majority function that occurs in full-adder designs. We also know—the proof is trivial—that it is a self-dual function, permitting us to write

$$x_1x_2 + x_2x_3 + x_1x_3 = (x_1 + x_2)(x_2 + x_3)(x_1 + x_3) \tag{5.48}$$

Now, comparing the right-hand side of (5.48) to our last factored expression (5.47) for z_1, we see that both contain the factor $(x_1 + x_2)$. This suggests that we make the following series of transformations of (5.47):

$$\begin{aligned} z_1 &= (x_1 + x_2)(x_3 + x_4 + x_5) + x_1x_2 \\ &= (x_1 + x_2)(x_3 + x_4 + x_5) + x_1x_2 + x_2x_3 + x_1x_3 \\ &= (x_1 + x_2)(x_3 + x_4 + x_5) + (x_1 + x_2)(x_2 + x_3)(x_1 + x_3) \\ &= (x_1 + x_2)(x_3 + x_4 + x_5) + (x_1 + x_2)(x_3 + x_1x_2) \\ &= (x_1 + x_2)(x_3 + x_4 + x_5 + x_1x_2) \end{aligned} \tag{5.49}$$

The final expression, which we have derived in roundabout fashion using a fair amount of our knowledge of Boolean algebra, yields the three-level circuit given in Figure 5.42. This has only four gates and is the best design we have so far. Note that we have managed to transform the original two-level sum-of-products expression into a three-level product-of-sums-of-products expression. Note too that several of the intermediate steps in deriving (5.49) actually increase the circuit cost, even though the final cost is reduced.

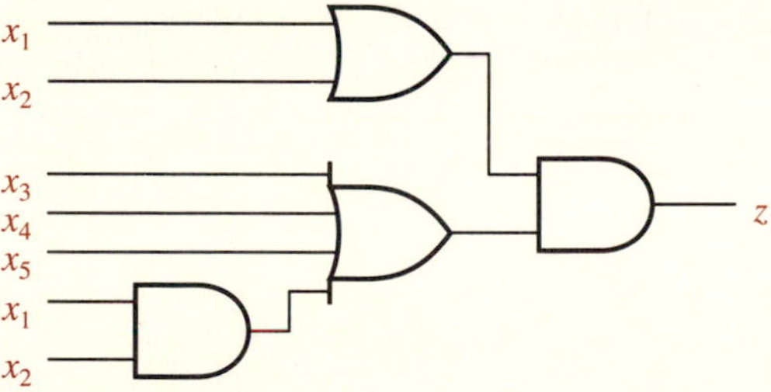

FIGURE 5.42 Another multilevel implementation of z_1.

Local Transformations

Some computer-aided logic design tools incorporate knowledge in the form of transformation rules of the above kind. Such rules may be expressed as

$$E_1 \rightarrow E_2 \tag{5.50}$$

where E_1 and E_2 are expressions or circuits that specify the same Boolean functions. Thus, (5.50) is basically a law of Boolean algebra, possibly not an obvious one. The synthesis program is used to scan a circuit for occurrences of the subcircuit C_1, which it then replaces with C_2, perhaps with guidance from a human designer. The consequences of a sequence of such **local transformations** cannot be predicted, but by trying many of them, a large number of functionally equivalent circuits can be generated; that is, a large portion of the design space can be explored fairly rapidly. If the gate or connection cost of the circuits are evaluated, one that has the lowest cost can be selected as the final design.

Good designs may often be generated by a sequence of small local transformations.

Figure 5.43 shows a small library of circuit transformations employed in an experimental CAD system developed at IBM [Darringer, 1981]. Observe that these local transformations apply to NAND subcircuits that involve up to five inputs, two outputs, and three levels of logic. All are easily verified to transform a NAND circuit without altering the output function(s) produced. The circuit E_2 on the right of $E_1 \rightarrow E_2$ acts as a template that is compared to similar subcircuits of the current circuit C. When a match is found, the matching subcircuit E_2 of C is replaced by E_1. For example, the transformation designated NAND3 in Figure 5.43 has the form

$$\overline{a}\overline{b} + \overline{c} \rightarrow \overline{(a + b)c}$$

and so is a type of factoring rule.

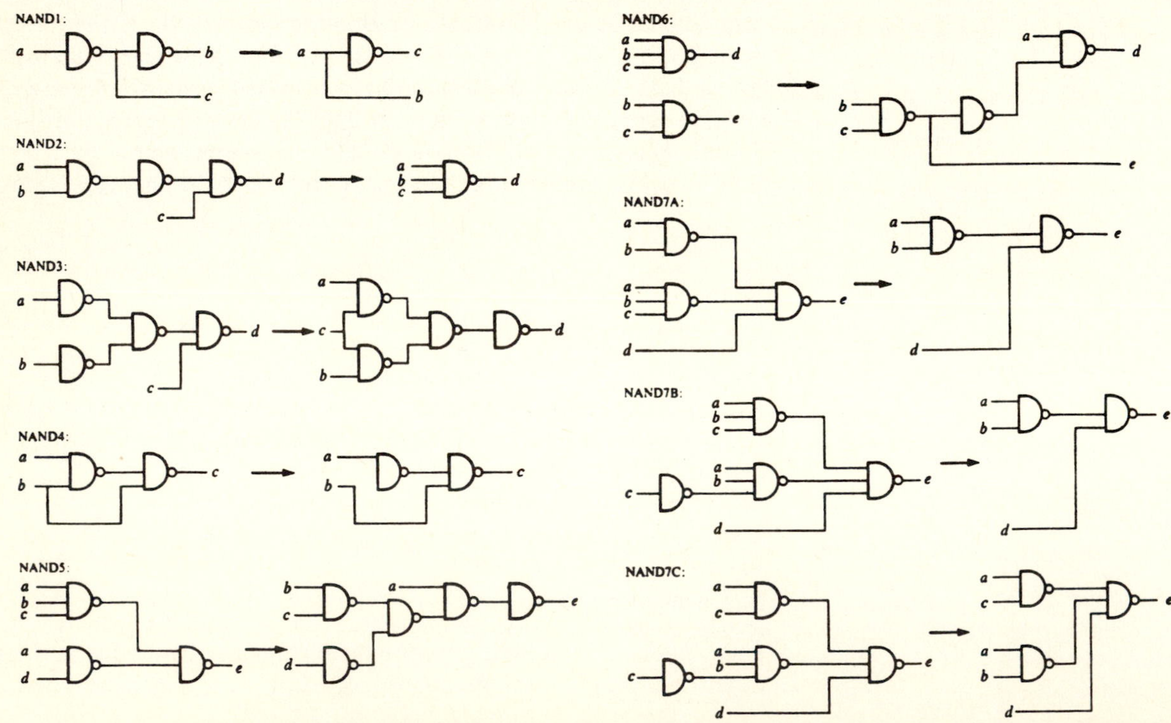

FIGURE 5.43 A CAD library of local transformation rules for NAND circuits. [Darringer, 1981]

Corresponding parts of E_1 and E_2 may be ignored in the transformation $E_1 \rightarrow E_2$, as long as functional equivalence is maintained. Consider the first transformation NAND1 in Figure 5.43. If the c output of E_1 and E_2, and the gate (inverter) feeding only c in E_2 are ignored, NAND1 reduces to saying that a wire may be replaced by two inverters in series; that is, $\bar{\bar{a}} \rightarrow a$.

Perhaps the most useful heuristic design transformation is to decompose a large complex design into small parts of manageable size. Each component part is chosen to have a well-defined function and to be suited to the logic design technique (two-level minimization, knowledge-based heuristic design, and so on) that we want to employ. In the remainder of this chapter, we examine a number of circuits that have well-defined and useful functions and are among the most frequently encountered components of digital systems.

Useful Circuits

This section addresses fundamental combinational design problems that occur often in practice: how to transfer data between parts of a system and how to add and subtract binary numbers, a basic form of data processing. We will see examples of both two-level and multilevel designs, and discuss the impact of fan-in and fan-out constraints.

5.8 Data Transfer Logic

First, we study the basic but vital task of controlling the transfer of binary information between several points in a circuit. This is a moving or copying operation in which no changes are made to the data's logic values; that is, no data processing takes place.

Shared Data Paths

Data transfer paths are frequently shared to reduce hardware cost.

Often, several data sources have a common destination, or one source must send data to several destinations. Because a particular destination can take signals from only one source at a time, the source-to-destination data transfer paths must be shared to some degree. Shared data paths are also employed primarily to reduce the number of lines needed for communication purposes, thereby reducing overall hardware costs. In either case, special circuits, the design of which we examine next, are needed to control the sharing of the data transfer paths in question.

Figure 5.44 shows some basic methods of sharing the access paths used in data transfer operations. We assume for the moment that we are only interested in data transfers in one direction, from one or more data sources or transmitters to one or more data destinations or receivers.

The simplest case is illustrated by Figure 5.44a, in which a single source must "broadcast" its data to a set of $m \geq 1$ destinations simultaneously.

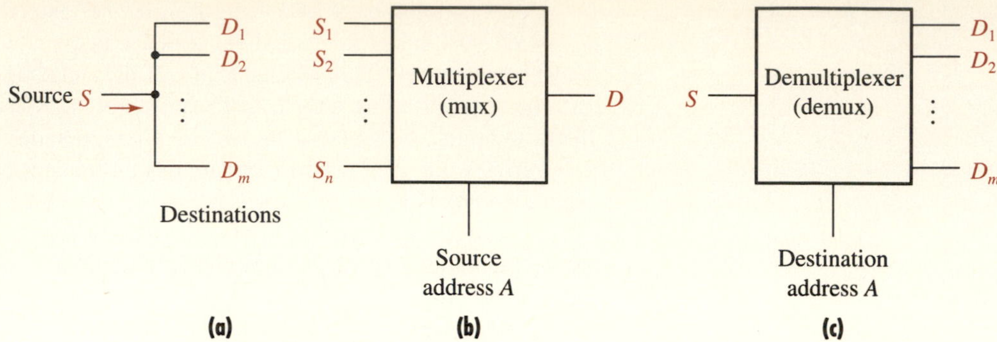

FIGURE 5.44 Data transfer connections: (a) from one source to many simultaneous destinations; (b) from one of many sources to one destination; (c) to one of many destinations from one source.

The logic circuit needed is a **wiring network**, and is implemented simply by fanning out the source's output line S (or k output lines, in the case of parallel data transfers) to each of the destinations $D_1, D_2, \ldots, D_m$. If necessary, buffers or drivers can be introduced into the wiring network to meet fan-out constraints.

Multiplexers and Demultiplexers

A multiplexer connects several sources to the same destination.

It is not normally possible to reverse the direction of data flow in the simple wiring network of Figure 5.44a in order to connect several sources to a single destination, because that would allow conflicting signals (simultaneous 0 and 1) from two sources to appear on the network. Instead, a logic circuit called a **multiplexer** or **data selector** is used, the general form of which is indicated in Figure 5.44b. A multiplexer M, often called a **mux** for short, allows one and only one of n sources to be logically connected to a common destination D at any time. The source S_i used at any particular time is determined by signals placed on M's **address** or **select inputs** A, which appear at the bottom of the figure. These A signals are interpreted as a binary integer i and serve to activate the circuits that select the desired source S_i, whose numerical address is i. By changing the A signals from i to j, the input source for a data transfer through the multiplexer can be changed from S_i to S_j.

A demultiplexer connects one source to a single selected destination.

The inverse of a multiplexer is a **demultiplexer**, abbreviated **demux**, which, as shown in Figure 5.44c, serves to connect a common source S to a single selected destination D_i. Again, a set of address lines specify which destination is to be selected at any time.

Figure 5.45 presents conceptual implementations of a multiplexer and a demultiplexer by means of multiposition mechanical switches. The central switching arm or armature can be rotated to any of n "on" positions to select a particular source or destination. Figure 5.45a shows the multiplexer's source S_2 connected to the common destination D. Early telephone switching systems employed electromechanical switches of this type.

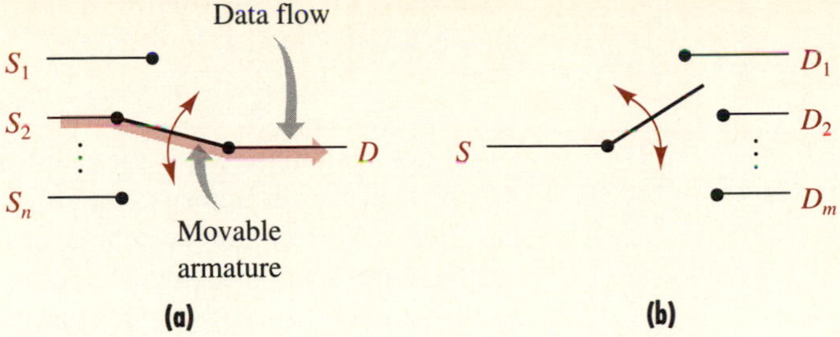

FIGURE 5.45 (a) A multiplexer and (b) a demultiplexer modeled as mechanical switches.

Multiplexer Design

We begin with an example of a simple multiplexer M that is required to connect one of two 1-bit sources x_0 and x_1 to a common destination z. Only one address bit a is needed to identify the selected source: $a = 0$ selects x_0 to connect to z, and $a = 1$ selects x_1. The desired circuit has three input lines and one output line. The output function $z(a, x_0, x_1)$ realized by M is easily specified on a three-variable K-map (Figure 5.46a). A minimum SOP cover is marked on the K-map and consists of the two essential prime

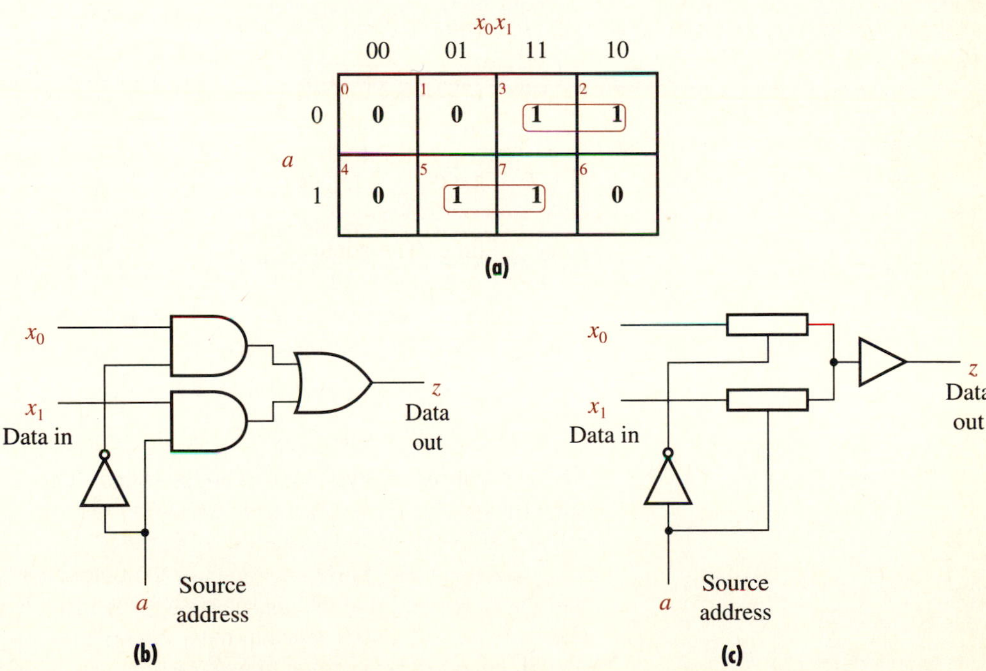

FIGURE 5.46 A small two-way multiplexer: (a) its K-map; (b) a gate implementation; (c) a switch implementation.

implicants. (There is a third, absolutely inessential prime implicant.) The minimal SOP equation for z is

$$z = \overline{a}x_0 + ax_1 \tag{5.51}$$

and a minimal AND-OR realization of M appears in Figure 5.46b.

A switch-level implementation of the same multiplexer appears in Figure 5.46c. This current-switching circuit links each source to the common destination via an on-off switch. At any time, one switch is turned off and so applies Z to z; the other switch is turned on and applies its x_i signal to z. This multiplexer design is common in MOS VLSI circuits for which the switches are either transmission gates or specially configured "pass" transistors [Weste and Eshragian, 1985].

n-Way Multiplexers

Multiplexers have simple gate- and switch-level designs.

Multiplexers can easily be designed with any number of data inputs (sources) n. Because k address bits allow 2^k distinct addresses to be specified, it is usual to make $n = 2^k$, so that the maximum number of sources can be accommodated by a given k. Thus, we could extend the preceding multiplexer example to accommodate four sources instead of two by changing (5.51) to

$$z = \overline{a}_1\overline{a}_0x_0 + \overline{a}_1a_0x_1 + a_1\overline{a}_0x_2 + a_1a_0x_3$$

We can see from the foregoing equations that a 2^k-input multiplexer has the defining minimal SOP equation

$$z = \sum_{\text{All } i} \underbrace{\dot{a}_{k-1}\dot{a}_{k-2}\ldots\dot{a}_1\dot{a}_0}_{\text{Binary address } i} x_i \tag{5.52}$$

An optimal two-level design for any n can be derived directly from a Boolean equation like (5.52).

Suppose, for example, that $k = 4$ and $n = 16$. The defining minimal SOP equation for this multiplexer is

$$\begin{aligned} z = {} & \overline{a}_3\overline{a}_2\overline{a}_1\overline{a}_0x_0 + \overline{a}_3\overline{a}_2\overline{a}_1a_0x_1 + \overline{a}_3\overline{a}_2a_1\overline{a}_0x_2 + \overline{a}_3\overline{a}_2a_1a_0x_3 \\ & + \overline{a}_3a_2\overline{a}_1\overline{a}_0x_4 + \overline{a}_3a_2\overline{a}_1a_0x_5 + \overline{a}_3a_2a_1\overline{a}_0x_6 + \overline{a}_3a_2a_1a_0x_7 \\ & + a_3\overline{a}_2\overline{a}_1\overline{a}_0x_8 + a_3\overline{a}_2\overline{a}_1a_0x_9 + a_3\overline{a}_2a_1\overline{a}_0x_{10} + a_3\overline{a}_2a_1a_0x_{11} \\ & + a_3a_2\overline{a}_1\overline{a}_0x_{12} + a_3a_2\overline{a}_1a_0x_{13} + a_3a_2a_1\overline{a}_0x_{14} + a_3a_2a_1a_0x_{15} \end{aligned}$$

which contains 16 five-literal product terms. This equation (with different signal names) is implemented in AND-OR form in the commercial 74X150 IC, whose logic circuit appears in Figure 5.47.

Note the use of buffers to increase the output drive of those input signals with large fan-out, and the unusually large fan-in (16) of the output OR gate. For $n > 16$, the fan-in and fan-out requirements of two-level multiplexers tend to be excessive, and multilevel designs are more practical. Such designs

Thus, for $n = 4$, a minimal two-level SOP circuit that realizes $c_{out} = c_3$ contains a total of $2^5 = 32$ gates, more than are in the entire 4-bit ripple-carry design of Figure 5.55. Moreover, the fan-in (31) of the output OR gate is very large. For $n = 16$, a figure of practical interest, the minimum gate count of an SOP realization of $c_{out} = c_{15}$ jumps to an astronomical $2^{17} = 131,072$, with correspondingly huge fan-in and fan-out figures. Thus, to realize practical fast adders, we *must* employ multilevel logic circuits and settle for less than the theoretical maximum speed of operation.

Carry Lookahead

Carry signals can be defined efficiently in terms of generate and propagate functions.

Fast addition with reasonable amounts of hardware requires a method of generating carry signals that is intermediate in complexity between the very fast but costly two-level approach and the slow but inexpensive ripple-carry technique. The most widely used scheme of this sort, called **carry lookahead**, employs special carry circuits to generate each c_i rapidly. The carry circuits do not take their inputs directly from the x_j and y_j primary inputs; instead, they use as their inputs two sets of auxiliary functions $g_i = x_i y_i$ and $p_i = x_i + y_i$, defined for $i = 0, 1, \ldots, n-1$ and called the **generate** and **propagate functions**, respectively. These functions are so called because $g_i = 1$ whenever a carry ($c_i = 1$) is *generated* by the ith operand-pair x_i, y_i, independent of the value of c_{i-1}; $p_i = 1$ whenever an input carry ($c_{i-1} = 1$) is *propagated* forward as $c_i = 1$. These auxiliary functions allow the full-adder equations

$$\begin{aligned} s_i &= x_i \oplus y_i \oplus c_{i-1} \\ c_i &= x_i y_i + x_i c_{i-1} + y_i c_{i-1} \end{aligned}$$

to be rewritten in the form

$$\begin{aligned} s_i &= g_i \oplus p_i \oplus c_{i-1} \\ c_i &= g_i + p_i c_{i-1} \end{aligned} \tag{5.57}$$

as can be immediately verified.

We can derive recursively from (5.57) an SOP expression for each c_i as a function of g_j and p_j, which is significantly simpler than the corresponding SOP expression for c_i in terms of x_j and y_j. The first such expression is

$$c_0 = g_0 + p_0 c_{in} \tag{5.58}$$

With $i = 1$, (5.57) implies

$$c_1 = g_1 + p_1 c_0 \tag{5.59}$$

Substituting the expression for c_0 from (5.58) into (5.59) yields

$$\begin{aligned} c_1 &= g_1 + p_1(g_0 + p_0 c_{in}) \\ &= g_1 + p_1 g_0 + p_1 p_0 c_{in} \end{aligned} \tag{5.60}$$

Comparing the SOP expressions (5.56) and (5.60) for c_1, we see that the latter is much simpler. Repeating the foregoing process for $i = 2$ leads to

$$\begin{aligned} c_2 &= g_2 + p_2c_1 \\ &= g_2 + p_2(g_1 + p_1g_0 + p_1p_0c_{\text{in}}) \\ &= g_2 + p_2g_1 + p_2p_1g_0 + p_2p_1p_0c_{\text{in}} \end{aligned}$$

and so on. In general, we can convert c_i to the desired SOP form

$$\begin{aligned} c_i = g_i + p_ig_{i-1} + p_ip_{i-1}g_{i-2} + p_ip_{i-1}p_{i-2}g_{i-3} + \cdots \\ \cdots + p_ip_{i-1}p_{i-2}\cdots p_1g_0 + p_ip_{i-1}p_{i-2}\cdots p_1p_0c_{\text{in}} \end{aligned} \qquad (5.61)$$

by starting with (5.57), substituting the SOP expression for c_{i-1}, and multiplying out the result to SOP form. The resulting expressions are obviously minimal in the literals g_i, p_i, and c_{in}. A two-level logic circuit that implements $c_0, c_1, \ldots, c_{n-1}$ in this form is called a **carry-lookahead generator**.

Carry-Lookahead Adders

A carry-lookahead adder generates all carries simultaneously.

A **carry-lookahead adder** is a fast circuit that implements n-bit binary addition using the foregoing carry-lookahead technique. It consists of three principal subcircuits:

1. A set of $2n$ two-input AND and OR gates (or functionally equivalent circuits) that implement the n generate and propagate functions $g_i = x_iy_i$ and $p_i = x_i + y_i$, respectively
2. A two-level carry-lookahead generator that implements the n carry functions $c_0, c_1, \ldots, c_{n-1}$ defined by (5.61)
3. A summation circuit that consists of n three-input EXCLUSIVE-OR gates (or equivalent) that implement the n sum functions $s_i = g_i \oplus p_i \oplus c_{i-1}$

Before we examine the general characteristics of this type of fast adder, we present a specific example.

Example 5.10 **Designing a 4-bit carry-lookahead adder**

Consider the design of a carry-lookahead adder from AND, OR, and EXCLUSIVE-OR gates for the case $n = 4$. Four AND and four OR gates are needed to implement the four pg functions:

$$\begin{aligned} g_0 &= x_0y_0 & p_0 &= x_0 + y_0 \\ g_1 &= x_1y_1 & p_1 &= x_1 + y_1 \\ g_2 &= x_2y_2 & p_2 &= x_2 + y_2 \\ g_3 &= x_3y_3 & p_3 &= x_3 + y_3 \end{aligned}$$

From (5.61), the four carry functions needed to implement carry lookahead have the following minimal SOP form:

$$c_0 = g_0 + p_0 c_{in}$$
$$c_1 = g_1 + p_1 g_0 + p_1 p_0 c_{in}$$
$$c_2 = g_2 + p_2 g_1 + p_2 p_1 g_0 + p_2 p_1 p_0 c_{in}$$
$$c_3 = g_3 + p_3 g_2 + p_3 p_2 g_1 + p_3 p_2 p_1 g_0 + p_3 p_2 p_1 p_0 c_{in}$$

Finally, the four sum functions are

$$s_0 = p_0 \oplus g_0 \oplus c_{in}$$
$$s_1 = p_1 \oplus g_1 \oplus c_0$$
$$s_2 = p_2 \oplus g_2 \oplus c_1$$
$$s_3 = p_3 \oplus g_3 \oplus c_2$$

Fast adders need much more hardware than slow ones.

Realizing all these equations in the most direct fashion gives the logic circuit shown in Figure 5.56. The total number of gates is 26; the number of logic levels or depth of the circuit is four. The comparable ripple-carry adder of Figure 5.55 contains only 20 gates but has depth eight. This illustrates

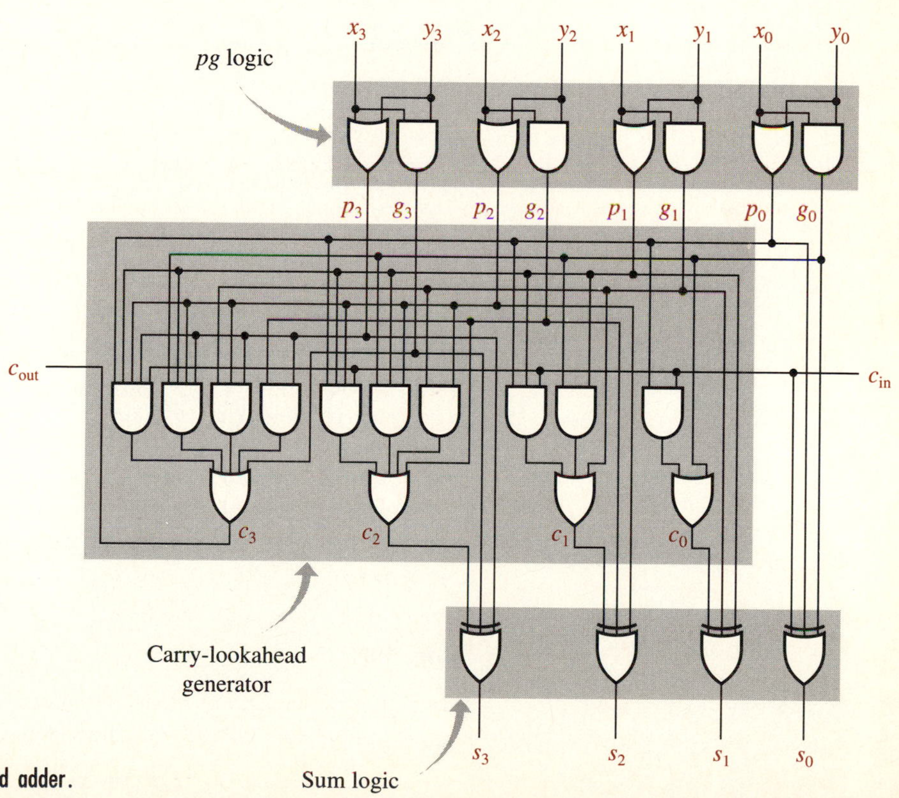

FIGURE 5.56 A 4-bit carry-lookahead adder.

a typical trade-off between hardware complexity and speed of operation—higher speed requires more hardware—of a type that pervades all aspects of digital design. The circuit of Figure 5.56 is essentially the same as that used in the commercial "fast-adder" ICs designated 74X83 and 74X283.

Adder Complexity

It is apparent from (5.61) that the carry-lookahead equation for c_i contains $i + 2$ product terms, and so a two-level realization requires $i + 2$ gates that have a maximum fan-in of $i + 2$. The total number of gates needed for an n-bit carry-lookahead generator is therefore

$$\sum_{i=0}^{n-1}(i + 2) = n^2/2 + 3n/2 \tag{5.62}$$

An additional $3n$ gates are necessary to generate the p, g, and s functions, leading to a total gate count of $n^2/2 + 9n/2$ for an n-bit carry-lookahead adder; extra gates may also be needed to meet fan-out constraints.

TABLE 5.1 Comparison between n-bit carry-lookahead and ripple-carry adders.

Parameter	*Carry-lookahead adder*	*Ripple-carry adder*
Number of gates	$n^2/2 + 9n/2$	$5n$
Number of levels (depth)	4	$2n$
Maximum fan-in	$n + 1$	3
Maximum fan-out	$\lfloor (n + 1)^2/4 \rfloor + 1$	4

Table 5.1 summarizes the characteristics of the particular ripple-carry and carry-lookahead designs considered above. Although the number of gates in the ripple-carry adder grows linearly, or in proportion to n, the number of gates in the carry-lookahead adder grows as the square of n, a significantly higher growth rate. Moreover, the maximum fan-in and fan-out of the ripple-carry design are constant, but those of the carry-lookahead adder grow with n. The total number of gates and the maximum fan-in and fan-out figures ultimately limit the size of the largest practical carry-lookahead adder to $n = 8$ or so.

Subtracters

Subtracters and adders use very similar circuits.

Subtraction of (unsigned) binary numbers was discussed in Section 2.5 and is quite similar to addition. The basic 1-bit (full) subtraction step is

$$b_i d_i = x_i - y_i - b_{i-1} \tag{5.63}$$

where b_i denotes a borrow bit. Equation (5.63) can be realized by the Boolean equations

$$\begin{aligned} d_i &= x_i \oplus y_i \oplus b_{i-1} \\ b_i &= \overline{x}_i y_i + \overline{x}_i b_{i-1} + y_i b_{i-1} \end{aligned} \tag{5.64}$$

from which the circuit of a 1-bit (full) subtracter is easily derived.

There is little difference between the full adder and full subtracter equations; each leads to logic circuits with similar characteristics. To convert the ripple-carry adder of Figure 5.55 to a "ripple-borrow" subtracter, it is only necessary to complement the x_i signals. The basic borrow equation can also be converted into the "borrow-lookahead" form

$$b_i = g_i + p_i b_{i-1}$$

where $g_i = \overline{x}_i y_i$ and $p_i = \overline{x}_i + y_i$. A borrow-lookahead subtracter that is similar to the design of Fig. 5.56 can readily be constructed.

Adder-Subtracter Units

A fixed-point arithmetic-logic unit (ALU) of the kind found in every computer's central processor performs the addition and subtraction operations $X+Y$ and $X-Y$, as well as bitwise logical operations like $\overline{X}$ and $X \oplus Y$, which we will not consider here. Because subtraction can lead to negative results, the ALU must deal with signed numbers. As explained in Section 2.6, it is inconvenient to use the signed-magnitude (SM) code for both signed addition and subtraction; it is necessary to compare the magnitudes $|X|$ and $|Y|$ of X and Y in order to determine whether to subtract $|X|$ from $|Y|$ or $|Y|$ from $|X|$. Hence, SM code requires the use of a magnitude comparator, a separate adder and subtracter, and "glue" logic circuits to tie together these components.

2C code leads to efficient adder-subtracter designs.

The combined addition and subtraction of two's-complement (2C) numbers is much simpler to implement than SM addition-subtraction (see Section 2.6). The negation of a 2C number Y_{2C} to form $-Y_{2C}$ is defined by the equation

$$-Y_{2C} = \overline{y}_{n-1}\overline{y}_{n-2}\ldots\overline{y}_1\overline{y}_0 \;+\; 00\ldots01 \qquad (\text{mod } 2^n)$$

from which it follows that

$$\begin{aligned} X_{2C} - Y_{2C} = {} & x_{n-1}x_{n-2}\ldots x_1x_0 \;+\; \overline{y}_{n-1}\overline{y}_{n-2}\ldots\overline{y}_1\overline{y}_0 \\ & + \; 00\ldots01 \qquad (\text{mod } 2^n) \end{aligned} \tag{5.65}$$

Thus, the 2C subtraction $X-Y$ can be implemented by an adder with a set of n inverters in its Y input lines. The addition of one required by (5.65) can be achieved by applying 1 to the adder's input carry line c_{in}. Figure 5.57 shows a slight modification of the resulting circuit, which accommodates addition and subtraction. The 4-bit ripple-carry design of Figure 5.55 forms the basic adder subcircuit. The inverters needed for subtraction according to

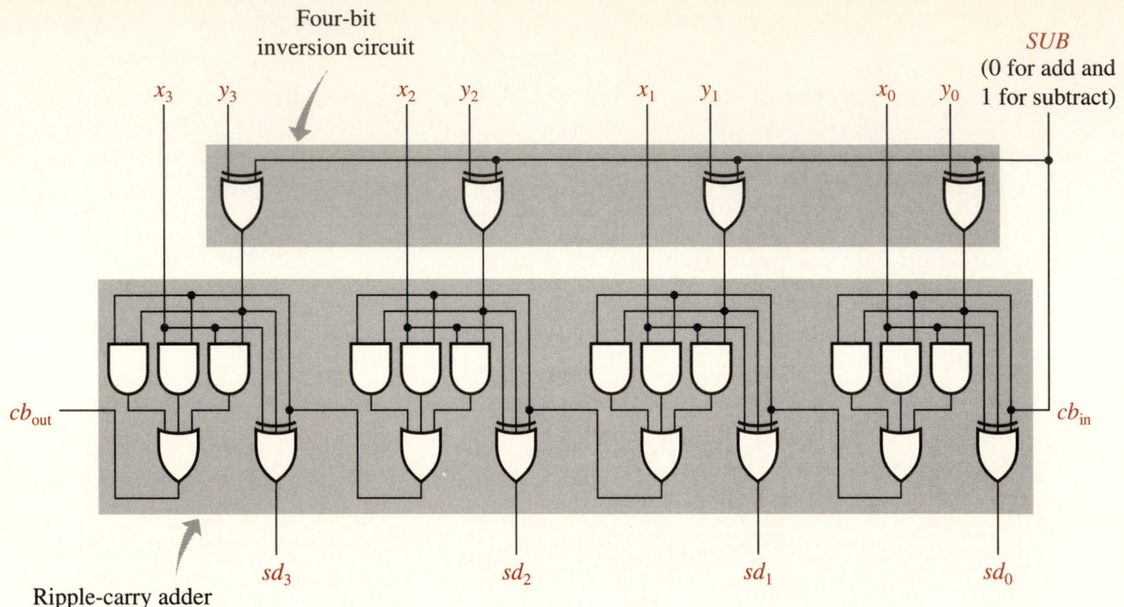

FIGURE 5.57 A 4-bit adder-subtracter for two's-complement numbers.

(5.65) are realized by n two-input EXCLUSIVE-OR gates, with one input of each gate connected to a common control line SUB. When $SUB = 1$, the output of the ith EXCLUSIVE-OR is $y_i \oplus 1 = \bar{y}_i$. Because $SUB = c_{\text{in}} = 1$, 2C subtraction is performed according to (5.65). When SUB changes to 0, the output of the same EXCLUSIVE-OR becomes $y_i \oplus 0 = y_i$, and c_{in} becomes 0, causing 2C addition to take place. Hence, this circuit can be switched between the arithmetic operations $X + Y$ and $X - Y$, simply by changing SUB.

Functional Decomposition

Reusing old designs lowers costs and reduces errors.

The foregoing adder and subtracter examples illustrate some general approaches to the design of very complex circuits. Foremost among these is the decomposition or breaking of a large circuit into several simpler ones, for which efficient designs—for example, optimal two-level implementations—are known or can readily be obtained. Thus, an n-bit ripple-carry adder is composed of n identical copies of a full adder, which is very easy to design optimally.

The 2C subtracter design (Figure 5.57) decomposes the subtraction process into two main subcircuits: an n-bit controllable inversion circuit and an n-bit adder. In designing the subtracter around an adder, we use a previously developed design. This is an example of **design reuse**, a technique that pervades the digital design field. Reuse of old designs is facilitated by the maintenance of libraries or catalogs of tried-and-proven circuits, a task that CAD tools can readily support. Such reuse not only reduces the time to produce a new design, it also greatly reduces the likelihood of design errors.

The carry-lookahead adder also employs a decomposition that breaks the overall circuit into three parts: the pg logic, the carry-lookahead generator, and the sum logic, as shown in Figure 5.56. Each of the three subcircuits is implemented by an optimal one- or two-level design. We can regard the pg and carry-lookahead logic as together implementing c_i from x_i and y_i by a single circuit, defined by the following expression that corresponds directly to (5.61):

$$\begin{aligned} c_i = {} & x_i y_i + (x_i + y_i)x_{i-1}y_{i-1} + (x_i + y_i)(x_{i-1} + y_{i-1})x_{i-2}y_{i-2} + \cdots \\ & + (x_i + y_i)(x_{i-1} + y_{i-1})(x_{i-2} + y_{i-2}) \cdots (x_1 + y_1)x_0y_0 \\ & + (x_i + y_i)(x_{i-1} + y_{i-1})(x_{i-2} + y_{i-2}) \cdots (x_1 + y_1)(x_0 + y_0)c_{\text{in}} \end{aligned} \tag{5.66}$$

Equation (5.66) is a sum-of-products-of-sums expression, which, as we have demonstrated, leads to an efficient three-level design combining high speed with relatively low hardware cost. It also represents a certain factoring of a two-level expression for c_i into a more desirable, multilevel form.

CHAPTER 5 SUMMARY

Two-Level Design

5.1 ■ Minimal sum-of-products (product-of-sums) expressions contain the fewest product (sum) terms, and they minimize the number of literals as a secondary goal.

■ We can use the properties of Boolean algebra to simplify SOP or POS expressions by removing unnecessary literals or entire product or sum terms. The key procedure is to combine terms that are **logically adjacent.** Logically adjacent terms are identical except that one variable appears complemented in one term and uncomplemented in the other.

■ The primary goal of the minimization process is to obtain a logic circuit that has at most two levels (depth two) and contains the smallest possible number of gates. Minimizing the number of gates reduces component cost. The secondary minimization goal is to reduce the number of literals used, which reduces connection cost.

■ A **prime implicant** is a product term that has no redundant literals; a **prime implicate** is a sum term with no redundant literals.

■ Solving the minimization problem requires finding the smallest set of prime implicants (implicates) that cover the minterms (maxterms) of the function being minimized.

5.2 ■ The **Karnaugh-map** (or **K-map**) **method** is a systematic way of obtaining prime terms and using them to construct minimal expressions and circuits. A K-map is a modified truth table intended to allow minimal SOP and POS expressions to be obtained by visual inspection.

■ A product term is represented on a K-map by a group of 2^k adjacent 1-cells, called a **product group.** A prime implicant is a product group that is not contained in any larger product group.

■ An **essential prime implicant** covers a minterm not covered by any other minterm. An **absolutely inessential prime implicant** covers minterms that are all covered by essential prime implicants.

- Minimizing a Boolean function consists of several steps: first, identify all prime implicants using a K-map; second, include all essential prime implicants in the cover; third, discard all absolutely inessential prime implicants; and fourth, select a minimal set of the remaining (inessential) prime implicants that covers all of the minterms of the function not already covered by essential prime implicants.

5.3 ■ **Petrick's method**—a systematic, nonexhaustive way of finding minimum prime implicant covers in less obvious cases—represents all covering conditions by a POS proposition.

- The POS covering proposition is converted to an SOP expression. A product term of this SOP expression that has the fewest literals represents a cover that contains a minimum number of prime terms.

5.4 ■ Beyond five or six variables, K-maps become unwieldy and impractical.

- Incompletely specified Boolean functions are defined with some unrestricted or unspecified output values, called **don't-cares**. The designer does not care if a don't-care output is 0 or 1 when the corresponding input combination is present.
- Because don't-cares can assume any value, the designer can choose 0 or 1 for each don't-care X separately to reduce the size of the minimal cover.

Computer-Aided Design

5.5 ■ CAD programs are essential for solving very large design problems.

- The **tabular** or **Quine-McClusky method** is a two-level minimization technique that can be used for minimizing functions too complex for K-maps. CAD programs like McBOOLE also use this method. It uses binary notation for terms, and tables to obtain minimum covers.

5.6 ■ An implementation of a set of Boolean functions—that is, a multiple-output function—may use nonminimal implementations for some functions while still achieving a minimal total solution. These solutions exploit the sharing of implicants among functions. The tabular approach can be extended to minimize multiple-output functions.

5.7 ■ For large two-level minimization problems, we can often find a quick solution using **heuristics**. These methods underlie most CAD programs used in digital design. Heuristic methods try to find acceptable, but not necessarily optimal, solutions.

- Good designs can sometimes be generated by applying a sequence of simple local transformations.

Useful Circuits

5.8 ■ Data transfer logic controls the basic, vital task of transferring binary information between several points in a circuit. No changes are made to the data's logic values during the transfer.

- To reduce hardware cost, data transfer paths are often shared. A **multiplexer** connects several sources to the same destination; a **demultiplexer** connects one source to one of several possible destinations. **Tri-state buses** support bidirectional communication among many devices.

5.9 ■ One of the most basic data-processing operations is adding two fixed-point binary numbers. A circuit that implements this operation is called an ***n*-bit adder.**

- The **full adder** adds three input bits to produce sum and carry output bits and is a basic component of arithmetic circuits.
- There are many possible implementations for an n-bit adder, illustrating the tradeoff between speed and cost. A **ripple-carry adder** is a slow, cheap adder formed by n full adders in series. A **carry-lookahead adder** has extra logic to generate all carries simultaneously, and so is faster and more expensive.
- A simple modification enables an n-bit adder to perform subtraction as well as addition, if the two's-complement (2C) code is used.

Further Readings

Classical two-level minimization algorithms receive comprehensive coverage in S. Muroga's text, *Logic Design and Switching Theory* (Wiley-Interscience, New York, 1979). More recent developments, especially in the area of heuristic and multilevel design methods, are covered in the collection of reprinted articles edited by A. R. Newton, *Logic Synthesis for Integrated Circuit Design* (IEEE Press, New York, 1987). Integrated circuit manufacturers' data books, such as the *TTL Logic Data Book* (Texas Instruments Inc., Dallas, Tex., 1988) covering the 74X integrated circuit family are the standard sources on the logic design and characteristics of multiplexers, decoders, adders, and other useful circuit types. Similar in style and scope are catalogs of standard-cell libraries produced by vendors of VLSI cell libraries and design services—for example, *Low-Power LP900C CMOS Standard-Cell Data Book* (AT&T Microelectronics, Allentown, Pa., 1991).

Chapter 5 Problems

Two-Level Design

Two-level circuits; canonical, SOP, and POS forms; minimization; prime implicants and implicates; essential terms; Karnaugh maps; K-map minimization; Petrick's method; incompletely specified functions; multiple-output functions.

5.1 Demonstrate that all but two of the nondegenerate two-variable Boolean functions can be realized by a single AND, NAND, OR, or NOR, assuming that both input variables are available in true or complemented form. Identify the two functions that do not have such a one-level realization.

5.2 In our examples of two-level logic circuits, we have ignored circuit structures of the form AND-NAND, OR-NOR, NAND-NOR, and NOR-NAND. Give the reasons for this omission.

5.3 Give the logic equations for a full adder in **(a)** the canonical minterm and maxterm forms and **(b)** the corresponding decimal representations.

5.4 Determine algebraically which of the following five product functions are implicants, prime implicants, or essential prime implicants of $z_0(a, b, c, d) = \sum(0, 1, 2, 4, 5, 6, 7, 8, 9, 10, 15)$: $A = \overline{a}\,\overline{d}$; $B = \overline{a}c$; $C = bcd$; $D = \overline{a}b\overline{c}$; $E = a\overline{b}cd$.

5.5 For the function $z_0(a, b, c, d) = \sum(0, 1, 2, 4, 5, 6, 7, 8, 9, 10, 15)$, establish algebraically which of the following four sum functions are implicates, prime implicates, or essential prime implicates: $Q = a+b+\overline{c}$; $R = \overline{a}+\overline{b}+d$; $S = a+\overline{b}+\overline{c}+\overline{d}$; $T = \overline{a} + b + c + d$.

5.6 In 1959, the following comment and question came from a member of the audience following a presentation of his method by S. R. Petrick. "For almost all functions it is possible to obtain the minimal sum (simplest normal form) without ever obtaining all prime implicants, as it is only necessary to obtain the essential prime implicants. Where does your method have any advantage over a method which obtains only the essential prime implicants?" Discuss briefly the validity of the audience member's comment, and answer his question.

5.7 Convert each of the following expressions algebraically to minimal SOP form: **(a)** $z_1(a, b, c) = \bar{a}\bar{b}\bar{c} + a\bar{b}\bar{c} + a\bar{b}c + ab\bar{c}$; **(b)** $z_2(a, b, c, d) = \bar{a}\bar{c}d + bcd + bc\bar{d} + \bar{a}\bar{b}\bar{c}\bar{d} + ab\bar{c}\bar{d} + a\bar{b}\bar{c}\bar{d}$; **(c)** $z_3(a, b, c, d) = \bar{a}\bar{c}\bar{d} + \bar{a}\bar{c}d + \bar{a}bc + bc\bar{d} + a\bar{b}\bar{c}$.

5.8 Convert each of the following expressions algebraically to minimal POS form: **(a)** $f_1(x_1, x_2, x_3) = (x_2 + x_3)(\bar{x}_1 + x_2)(x_1 + \bar{x}_3)(x_1 + x_2 + x_3)$; **(b)** $f_2(v, w, x, y) = (v + x + y)(v + x + \bar{y})(\bar{w} + \bar{x} + \bar{y})(v + \bar{x} + y)(v + w + \bar{x} + \bar{y})(\bar{v} + w + x + \bar{y})(\bar{v} + w + \bar{x} + \bar{y})$; **(c)** $F_3(A, B, C, D) = ((A + B + C)(\bar{B} + C + \bar{D}) + (A + B + C)(\bar{C} + \bar{D}))(A + B + \bar{C})$.

5.9 Use algebraic manipulation to transform the following Boolean expression into minimal SOP and POS forms:

$$z(x_1, x_2, x_3, x_4) = \bar{x}_1\bar{x}_3\bar{x}_4 + \bar{x}_1x_3x_4 + \bar{x}_1x_3\bar{x}_4 + x_3x_4 + x_1\bar{x}_2\bar{x}_3\bar{x}_4$$

Cite by name the main laws of Boolean algebra used in your transformation process.

5.10 Verify that (5.15) is a minimal SOP expression for $z(a, b, c, d, e)$. Determine whether it is the only such expression.

5.11 Algebraically transform the following expressions to their complements; that is, change each z_i to $\bar{z}_i$, and express $\bar{z}_i$ in minimal (preferably) or near-minimal SOP form: **(a)** $z_1(a, b, c, d) = (a + \bar{b}c)(a + d)$; **(b)** $z_2(a, b, c, d) = ab + ac + ad + bc + bd + cd$; **(c)** $z_3(a, b, \ldots, h) = (((a + bc) + de) + ef) + gh$.

5.12 Repeat Problem 5.11, this time expressing each z_i in minimal or near-minimal POS form.

5.13 A Boolean function $z(x_1, x_2, \ldots, x_n)$ is called **positive** or **positive unate** if there is some SOP expression for z in which, for every i, $0 \leq i \leq n$, the true (positive) literal x_i appears, but its complement $\bar{x}_i$ does not. For example, $z(x_1, x_2, x_3, x_4) = x_1x_2 + x_2x_3x_4$ is positive. Similarly, z is called **negative** or **negative unate** if there is some SOP expression in which, for every i, only the complemented literal $\bar{x}_i$ appears. Prove that every prime implicant of a positive or negative function is essential, which implies that the function has a unique minimal SOP form.

Unate functions have unique minimal SOP and POS forms.

5.14 The function $z(x_1, x_2, \ldots, x_n)$ is said to be **unate** if there is an SOP expression for z in which, for every i, either $\bar{x}_i$ or x_i appears, but not both. For example, $f(a, b, c, d) = abc + \bar{b}\bar{d} + a\bar{d}$ is unate, as are the positive and negative functions defined in the preceding problem. Use the result of that problem to prove that every unate Boolean function has a unique minimal SOP form. Furthermore, prove that every such function has a unique POS form. Thus, unate functions are extremely easy to minimize.

5.15 Consider the K-map in Figure 5.58, in which there are 10 marked 1-cell groups $A, B, \ldots, J$. Write down a product term that represents each of these 10 groups.

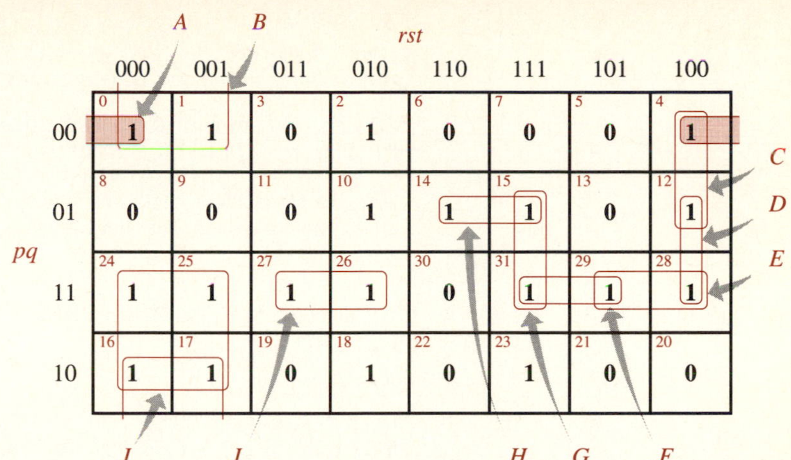

FIGURE 5.58 The K-map of a five-variable Boolean function $f(p, q, r, s, t)$.

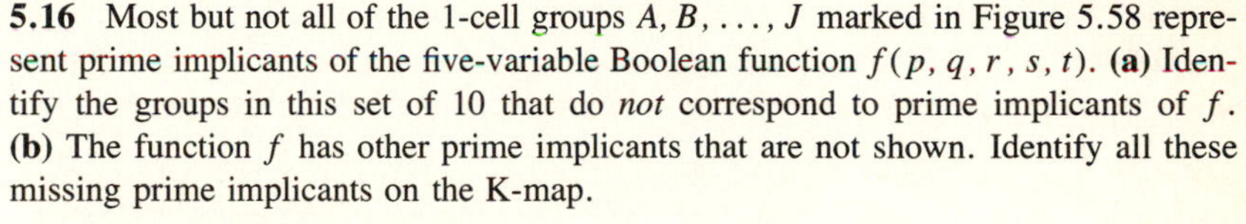

5.16 Most but not all of the 1-cell groups $A, B, \ldots, J$ marked in Figure 5.58 represent prime implicants of the five-variable Boolean function $f(p, q, r, s, t)$. **(a)** Identify the groups in this set of 10 that do *not* correspond to prime implicants of f. **(b)** The function f has other prime implicants that are not shown. Identify all these missing prime implicants on the K-map.

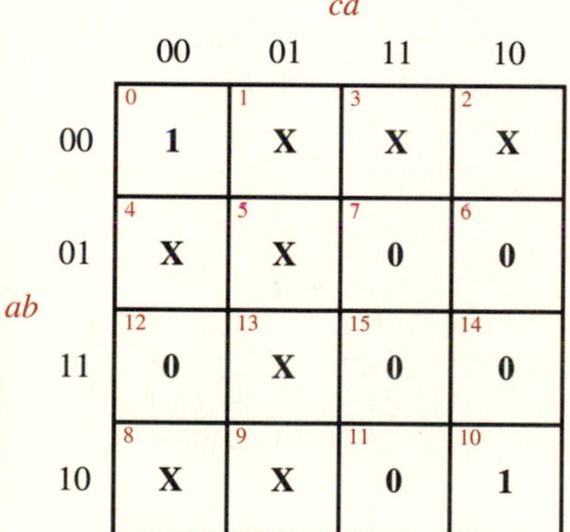

FIGURE 5.59 The K-map of an incompletely specified four-variable Boolean function.

5.17 Figure 5.59 shows the K-map of an incompletely specified four-variable function. Assign values to the don't-cares to construct four completely specified functions z_1, z_2, z_3, and z_4 with the following properties: **(a)** z_1 is a function of exactly two variables; **(b)** z_2 has exactly five minterms; **(c)** z_3 has exactly four prime implicants; **(d)** z_4 has an absolutely inessential prime implicant.

5.18 Assign values to the don't-cares in the K-map of Figure 5.59 to construct three completely specified functions f_1, f_2, and f_3 with the following properties: **(a)** f_1 has exactly eight maxterms; **(b)** f_2 has exactly six prime implicates; **(c)** f_3 has the same number of prime implicants and prime implicates.

5.19 Use the K-map method to find minimal SOP and POS expressions for the three-variable function $f(x_1, x_2, x_3) = \sum(0, 1, 2, 5, 6)$. Using AND and OR gates, construct a two-level circuit for each expression.

5.20 Use K-maps to find minimal SOP expressions for the three functions defined in Problem 5.7.

5.21 Using four-variable K-maps, find minimal SOP expressions for each of the following functions: **(a)** $z_1 = \sum(0, 2, 3, 4, 5, 6, 8, 11, 12, 13)$; **(b)** $z_2 = \sum(1, 3, 4, 6, 7, 9, 10, 12, 13)$; **(c)** $z_3 = \sum(0, 1, 2, 3, 5, 7, 8, 10)$.

5.22 Use K-maps to find minimal POS expressions for the three functions defined in Problem 5.8.

5.23 Using K-maps of appropriate size, find minimal POS expressions for each of the following functions: **(a)** $f = \sum(0, 3, 5, 6)$; **(b)** $g = \prod(0, 2, 5, 7, 8, 10, 13, 15)$; **(c)** $h = \prod(0, 1, 3, 4, 9, 10, 13, 14, 16, 19, 20, 23, 25, 26, 29, 30)$.

5.24 An alarm circuit contains four switches S_1, S_2, S_3, and S_4. The alarm A is sounded if and only if at least two of the switches are closed. Design a minimum-cost two-level circuit composed entirely of NAND gates that implements the alarm circuit.

5.25 Use a K-map to determine *all* minimal POS expressions for the function $z_1(a, b, c, d) = \prod(0, 2, 4, 5, 7, 10, 11, 13, 15)$.

5.26 Via the K-map approach, find a minimal two-level realization of the four-variable function defined by the following Boolean expression:

$$z_2(a, b, c, d) = \bar{a}\bar{b}(\bar{c}d + \bar{c}\bar{d}) + \bar{a}b\bar{c}d + ab(c + d) + ac\bar{d}$$

Consider both SOP and POS realizations. Using AND and OR gates, draw a logic diagram of your design.

5.27 Use a K-map to determine all prime implicants of the five-variable function $z(a, b, c, d, e)$ defined by Equations (5.13), (5.14), and (5.15). List in product form all the prime implicants and identify those that are essential.

5.28 Find a minimal SOP expression and also a minimal POS expression for the five-variable function $z_3(a, b, c, d, e) = \sum(0, 1, 2, 5, 8, 9, 17, 25)$.

5.29 Use Petrick's method to find all minimum-sized SOP covers for Figure 5.58, where all cover terms must be taken from the 10-member set $A, B, \ldots, J$ that appears in the figure. Consider only the 15 1-cells that are covered by at least one of $A, B, \ldots, J$.

5.30 Consider the five-variable function f defined in Figure 5.58 along with some of its prime implicants. **(a)** Indicate all of f's prime implicants on the K-map. **(b)** Identify and remove all the prime implicants that are either essential or absolutely inessential. Use Petrick's method to construct a minimum-cost SOP cover from the remaining prime implicants. **(c)** What are the gate and literal costs of a minimal SOP cover for f?

5.31 A prime implicant covering problem yields the following covering proposition:

$$\mathbf{P} = (\mathbf{A} + \mathbf{F})(\mathbf{A} + \mathbf{B} + \mathbf{D})(\mathbf{A} + \mathbf{B} + \mathbf{C} + \mathbf{E} + \mathbf{F})(\mathbf{C} + \mathbf{D})$$
$$(\mathbf{C} + \mathbf{F})(\mathbf{E})(\mathbf{E} + \mathbf{F})(\mathbf{A} + \mathbf{D}) \quad (5.67)$$

Suppose that prime implicants A, B, and C have literal cost three, and prime implicants D, E and F have literal cost two. Evaluate (5.67) to find a minimum-cost cover.

5.32 Use four-variable K-maps to obtain minimal SOP expressions for the following incompletely specified functions: **(a)** $F_1(a, b, c, d) = \sum(1, 2, 4, 15) + \Delta(5, 6, 8, 10, 11, 13, 14)$; **(b)** $F_2(a, b, c, d) = \prod(1, 4, 5, 6, 12, 13) \cdot \Delta(2, 8, 10, 11, 15)$.

5.33 Repeat Problem 5.32, this time finding minimal POS expressions for the given four-variable functions.

5.34 Use the K-map method to find a minimum-cost two-level circuit for each of the following functions: **(a)** $Z_a(A, B, C, D) = \sum(0, 6, 7, 8, 9, 14) + \Delta(1, 5, 11, 15)$; **(b)** $Z_b(A, B, C, D, E) = \sum(0, 1, 4, 11, 27, 30) + \Delta(2, 9, 10, 13, 14, 15, 16, 17, 18, 20, 26)$.

5.35 Consider the incompletely specified five-variable Boolean function defined by the K-map of Figure 5.60. Obtain a minimal SOP expression for this function.

ab \ cde	000	001	011	010	110	111	101	100
00	0 **1**	1 **0**	3 **0**	2 **1**	6 **1**	7 **0**	5 **0**	4 **1**
01	8 **1**	9 **0**	11 **0**	10 **X**	14 **1**	15 **1**	13 **0**	12 **1**
11	24 **0**	25 **X**	27 **1**	26 **0**	30 **0**	31 **1**	29 **X**	28 **X**
10	16 **X**	17 **0**	19 **0**	18 **0**	22 **0**	23 **0**	21 **0**	20 **1**

FIGURE 5.60 The K-map of an incompletely specified five-variable Boolean function.

5.36 A six-variable K-map is shown in Figure 5.61. It is formed by appending to the five-variable K-map a reflected copy (shown colored) of the five-variable map. Enter the following function into the six-variable K-map:

$$z_4(a, b, c, d, e, f) = \sum(4, 8, 9, 10, 11, 14, 26, 29, 30, 34, 36, 38, 42, 43, 46, 47, 48, 49, 50, 52, 54, 57, 59, 61)$$

Mark all prime implicants of z_4 on the K-map and identify those that are essential.

abc \ def	000	001	011	010	110	111	101	100
000	0	1	3	2	6	7	5	4
001	8	9	11	10	14	15	13	12
011	24	25	27	26	30	31	29	28
010	16	17	19	18	22	23	21	20
110	48	49	51	50	54	55	53	52
111	56	57	59	58	62	63	61	60
101	40	41	43	42	46	47	45	44
100	32	33	35	34	38	39	37	36

FIGURE 5.61 A K-map for six-variable Boolean functions.

5.37 Using the K-map method, design a minimal SOP realization of a 2-bit adder for unsigned integers. The inputs to the adder are two 2-bit numbers $X = x_1x_0$ and $Y = y_1y_0$, and an input carry bit c_{in}. The adder's outputs are the two-bit sum $Z = z_1, z_0$ and an output carry bit c_{out}. Your answer should include a set of K-maps with all (multiple-output) prime implicants identified, and a set of minimal SOP expressions for the adder.

5.38 Design a nearly minimal five-input, four-output AND-OR circuit to convert a single digit from the two-out-of-five code to the 4-bit BCD code, both of which are specified in Figure 5.62. Estimate how far your solution is from a truly minimal one, indicating precisely what you mean by a minimal solution.

Digit	BCD	2-out-of-5
0	0000	01100
1	0001	10001
2	0010	10010
3	0011	00011
4	0100	10100
5	0101	00101
6	0110	00110
7	0111	11000
8	1000	01001
9	1001	01010

FIGURE 5.62 The BCD and two-out-of-five decimal codes.

5.39 Using K-maps, design a two-level all-NAND circuit that realizes the following three four-variable functions and has the minimum cost:

$$z_1(a, b, c, d) = \sum(0, 1, 2, 7, 8, 9)$$

$$z_2(a, b, c, d) = \sum(0, 2, 6, 7, 8, 9, 10, 13, 15)$$

$$z_3(a, b, c, d) = \sum(0, 2, 6, 7, 8, 10)$$

5.40 Prove that the two-level circuit with five inputs and three outputs shown in Figure 5.37 has lower cost than any SOP or POS realization of the same set of functions.

Computer-Aided Design

CAD programs; tabular (Quine-McCluskey) minimization method; multiple-output functions; iterative consensus method; heuristic minimization; multilevel design.

5.41 Figure 5.63 shows the output generated by McBOOLE when asked to find a minimal SOP design for the 2-bit multiplication function that computes the numerical product $Z = z_3z_2z_1z_0$ of two 2-bit unsigned numbers $X = x_1x_0$ and $Y = y_1y_0$. Verify that McBOOLE's solution is optimal and construct a two-level all-NAND circuit that implements this solution.

```
/*
011x    0010
x1x1    0001
1111    1000
1110    0110
1x01    0010
101x    0100
10x1    0010
/*
```

FIGURE 5.63 A 2-bit multiplier design computed by the McBOOLE minimization program.

5.42 Figure 5.64 gives the solution to the five-variable SOP minimization problem posed in Example 5.4 (Section 5.4), which was produced by ESPRESSO. **(a)** Verify that this solution is functionally correct. **(b)** Verify also that although the number of prime implicants is the minimum possible, ESPRESSO's heuristics failed to minimize the literal cost. How far is ESPRESSO's solution from the exact minimum?

```
# espresso -Dexact -s ex5.4
# UC Berkeley, Espresso Version #2.3, Release date 01/31/88
# PLA is ex5.4 with 5 inputs and 1 outputs
# ON-set cost is c=18(18) in=90 out=18 tot=108
# OFF-set cost is c=0(0) in=0 out=0 tot=0
# DC-set cost is c=0(0) in=0 out=0 tot=0
# exact Time was 0.16 sec, cost is c=8(8) in=26 out=8 tot=34
.i 5
.o 1
.p 8
111-- 1
000-- 1
11-1- 1
-111- 1
00-0- 1
-000- 1
0-101 1
1-001 1
.e
```

FIGURE 5.64 Solution to Example 5.4 computed by the ESPRESSO minimization program.

5.43 Suppose you have a computer program DUMBOOLE that computes a minimal SOP expression for a given Boolean function z, which is entered as a list of minterms. It never occurred to DUMBOOLE's programmer to allow the entry of maxterms or to write code to process POS expressions. Can you still use DUMBOOLE to compute a minimal POS expression for z? Explain how you can or why you cannot.

5.44 Consider the 6×9 prime implicant covering table T given in Figure 5.65. **(a)** Identify all dominance relations among the rows and columns of T by making six lists of the form "Row i dominates rows . . . " and nine lists of the form "Column i dominates columns . . . " **(b)** Using only the dominance information you have identified above, check off all rows that can be eliminated from the prime implicant table. Use Petrick's method to find an optimal cover of the unchecked columns by the unchecked rows.

Minterms

Prime implicants	a	b	c	d	e	f	g	h	i
A		×		×			×		
B	×		×		×		×		
C					×	×		×	
D		×	×			×	×		×
E	×	×	×	×					
F	×	×		×	×		×		×

FIGURE 5.65 A small prime implicant covering table.

5.45 Use the tabular covering approach along with the branching method to solve the minimal covering problem defined in Example 5.3.

5.46 Using the tabular method, find a minimal POS expression for the incompletely specified five-variable function defined by Figure 5.60.

5.47 Use the tabular method to find a minimum SOP expression for the following Boolean function:

$$z(a, b, c, d, e) = \sum(0, 2, 5, 6, 7, 8, 10, 12, 13, 20, 21, 24, 26) + \Delta(1, 14, 15, 16, 17, 19, 22, 28)$$

Check your answer using a K-map.

5.48 Using the tabular method in place of the K-map approach, carry out the six-variable SOP minimization introduced in Problem 5.36. The function to be minimized is

$$z_4(a, b, c, d, e, f) = \sum(4, 8, 9, 10, 11, 14, 26, 29, 30, 34, 36, 38, 42, 43, 46, 47, 48, 49, 50, 52, 54, 57, 59, 61) \qquad (5.68)$$

5.49 Use the tabular method to compute a minimal POS expression for the five-variable function defined in Problem 5.47.

5.50 A 13×19 covering table produced during SOP minimization of a five-variable Boolean function appears in Figure 5.66. Find a minimum-cost SOP expression for the function in question from this table.

Prime implicants	Minterms: 00000	00010	00110	01001	01010	01011	01100	01110	01111	10000	10001	10010	10011	10110	11000	11011	11100	11110	11111
P_A = 0--10		×	×		×			×											
P_B = -0-10		×	×									×		×					
P_C = -111-								×	×									×	×
P_D = --110			×					×						×				×	
P_E = -1100							×										×		
P_F = 11-00															×		×		
P_G = -00-0	×	×								×		×							
P_H = 1-000										×					×				
P_I = 1-011													×			×			
P_J = 010-1				×		×													
P_K = -1-11						×			×							×			×
P_L = 01-1-					×	×		×	×										
P_M = 100--										×	×	×	×						

FIGURE 5.66 A large prime implicant covering table.

5.51 Apply the tabular approach to obtain a minimal POS (not SOP) expression for the six-variable function defined by Equation (5.68). (*Hint:* This function has 40 maxterms and 25 prime implicates. Derivation of the latter by hand is extremely tedious. Consider using a computer to assist you in this problem.)

5.52 Use the tabular method to design a minimal four-input, five-output AND-OR circuit to convert a single digit from the 4-bit BCD code to the two-out-of-five code, both of which are specified in Figure 5.62. Give a PLA implementation of your design.

5.53 Consider the three-output four-input minimization problem defined in Problem 5.39. Solve this problem by the tabular method and give the solution in the form of a PLA.

5.54 A two-level circuit C with four inputs and four outputs is to be designed to convert a decimal digit from standard BCD (or 8421) code to excess-3 code. The circuit may be composed of all NANDs or all NORs (not a mixture). Use the tabular method to obtain the lowest-cost two-level logic circuit for C. Give both a minimal Boolean (POS or SOP) expression and a logic circuit for C.

Consensus theorem

5.55 Let the variable x appear in the product expression for a product term $P = xP^*$, and let $\bar{x}$ appear in the product term $Q = \bar{x}Q^*$. The **consensus** $P \not{c}\, Q$ of P and Q is defined as P^*Q^* and is written as a product term with no duplicated literals. For example, $a\bar{b}\bar{c} \not{c}\, b\bar{c}d = a\bar{c}d$. An SOP expression for z is said to be a **complete sum** if it is the OR sum of every prime implicant of z. The following useful consensus theorem is due to W. V. Quine: An SOP expression $E = \sum P_i$ is a complete sum for z if and only if (a) no term P_i of E implies or is covered by any other term P_j of E; (b) the consensus of any two product terms of E either does not exist or is contained in some P_i in E. Prove the "only if" part of this theorem—that is, if E is a complete sum, it must have properties (a) and (b).

5.56 The consensus theorem stated in the last problem is the basis of the following **iterative consensus method** to compute all prime implicants of z. Unlike the tabular method, which starts with all the minterms, the iterative consensus method can start with any SOP expression for z. Thus, it is more efficient than the tabular method when z is provided in reduced SOP form.

Procedure

Procedure *ITCONS* for computing prime implicants

1. Form a list $\mathcal{L}$ of all product terms that appear in the given SOP expression E for z. Scan $\mathcal{P}$ from top to bottom and compare every term P_i with every term P_j below it in $\mathcal{L}$.
2. If P_i implies P_j—that is, if P_i is covered by P_j—then delete P_i from $\mathcal{L}$.
3. Determine whether P_i and P_j have a consensus $P_i \not{c}\, P_j$. If so, compare $P_i \not{c}\, P_j$ to all members of P. If $P_i \not{c}\, P_j$ does not imply any term that appears in $\mathcal{L}$, then add $P_i \not{c}\, P_j$ to the bottom of the list.
4. Continue until every term has been compared to every term below it. The final contents of $\mathcal{L}$ are the prime implicants of z.

Use the iterative consensus procedure to find the prime implicants of

$$z(a, b, c, d) = \bar{a}\bar{c}\bar{d} + b\bar{c}d + \bar{a}bd + abd + ac + a\bar{b}\bar{c}\bar{d}$$

5.57 Use the heuristic procedure *GREEDYCOV* to solve the covering problem defined by Figure 5.65. Assume that rows A and B have (literal) cost 2, C and D have cost 4, and E and F have cost 6. Determine how far your solution is from the optimum.

5.58 Use *GREEDYCOV* to solve the covering problem specified in Figure 5.66. Estimate how good your solution is.

5.59 A fire alarm system is to be connected to eight alarm switches $x_1{:}x_8$. If exactly one of these switches is activated, a siren S_1 should go off. If two or more switches are activated simultaneously, S_1 and a second siren S_2 should go off. The logic C for this system is to be implemented in a field-programmable gate array (FPGA) whose only cell types are AND, OR, NAND, and NOR gates, all with a maximum fan-in of two and a fan-out drive capability of 10. We want to implement C using as few gates as possible; speed of operation is of minor concern. **(a)** Roughly estimate the gate cost of implementing C using two-level logic. **(b)** Using heuristic factoring or other appropriate methods, obtain a low-cost multilevel realization of C.

5.60 Use the laws of Boolean algebra to prove that the NAND transformations NAND5 and NAND7C in Figure 5.43 preserve circuit function unchanged.

5.61 Apply a sequence of local transformations taken from Figure 5.43 to the circuit of Figure 5.67a to transform it into the more economical design of Figure 5.67b. Identify clearly the transformation used at each step, which may be any transformation NANDi $E_1 \rightarrow E_2$ or its inverse $(\text{NAND}i)^{-1}$ $E_2 \rightarrow E_1$ that appears in Figure 5.43. Draw the transformed circuit generated in each step.

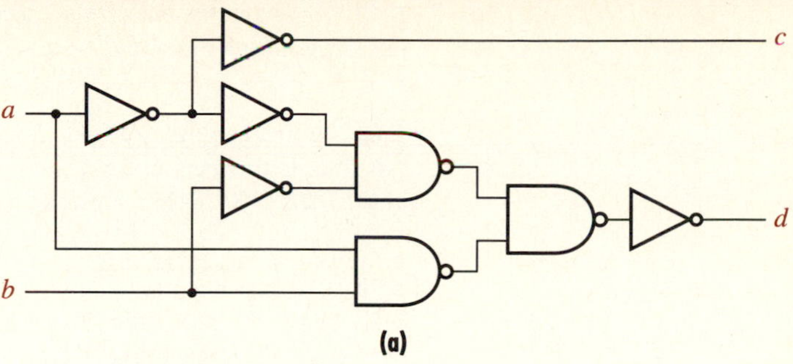

(a)

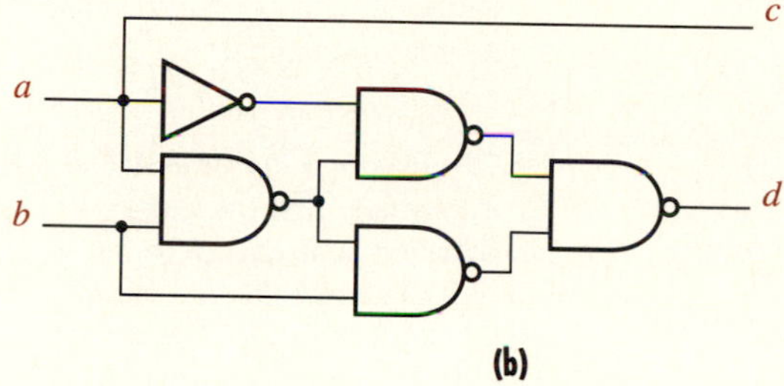

(b)

FIGURE 5.67 A NAND circuit (**a**) before and (**b**) after transformation.

Useful Circuits

Data transfer logic; multiplexers; buses; demultiplexers; decoders; ripple-carry adders; carry-lookahead adders; subtracters.

5.62 Carry out the logic design of a three-way, 1-bit multiplexer that uses only NOR gates (including inverters).

5.63 Consider again the design of a three-way, 1-bit multiplexer. This time, implement it using **(a)** a ROM and **(b)** a PLA of appropriate size. Comment on the relative efficiency of these two designs.

5.64 Carry out the logic design of a CMOS-style three-way, 2-bit multiplexer that uses transmission gates and inverters only.

5.65 Redesign the four-way, 2-bit multiplexer of Figure 5.48 to reduce the number of input data buses (sources) from four to three and to make all input lines fully buffered. Make as few changes to the original circuit as possible.

5.66 Design a nine-way, 1-bit multiplexer that uses only NAND gates (including inverters).

5.67 Design a four-way, 3-bit demultiplexer that uses only NOR gates.

5.68 Design a one-out-of-nine decoder that uses only NAND gates. Show how it can be used as a nine-way, 1-bit demultiplexer.

5.69 A time-multiplexed line L is to connect six possible 1-bit sources to three possible destinations using a multiplexer M and a demultiplexer DM. Using only NAND and NOR gates, construct a logic circuit to implement this circuit.

5.70 Consider the preceding problem of connecting six sources to three destinations via a single time-shared line L. Suppose we now want to be able to connect a selected source S, either to a single destination D_a or simultaneously to two destinations D_a and D_b. Carry out the logic design of the required circuit, again restricting the components used to NANDs and NORs. Give logic equations for the three output data lines in minimal SOP form.

5.71 Redesign the time-shared circuit of Figure 5.51 to replace L with a tri-state bus. Assume that, as before, communication will be between one source and one destination at a time. What are the advantages and disadvantages of the tri-state bus in this application?

5.72 A row or output (OR plane) column line of a PLA forms a kind of tri-state bus. Explain why this is so and why these lines never assume the indeterminate value U.

5.73 Prove that the carry function $c_{n-1}(x_{n-1}, x_{n-2}, \ldots, x_0, y_{n-1}, y_{n-2}, \ldots, y_0, c_{\text{in}})$ is unate, as defined in Problem 5.14. Use this fact to show that c_{n-1} has exactly $p(n) = 2^{n+1} - 1$ prime implicants. (*Hint:* Obtain a recursive numerical formula for $p(n)$ along the lines of the recursive Boolean formula for c_{n-1}.)

5.74 Design a 4-bit ripple-carry adder that is composed of NORs, EXCLUSIVE-ORs, and inverters only and has fully buffered inputs.

5.75 Draw an extension to the logic diagram of Figure 5.56 that makes it into a 5-bit carry-lookahead adder.

5.76 Redesign the 4-bit carry-lookahead adder of Figure 5.56 entirely from NAND and NOR gates, without making the circuit's overall depth more than six.

5.77 Give the carry-lookahead equations for c_0, c_1, c_2, c_3 in minimal POS form. Prove that, in general, these POS expressions have no significant cost advantage over the corresponding SOP expressions.

5.78 Consider an n-bit carry-lookahead adder designed in the style of Figure 5.56. Identify the internal signal(s) with the largest fan-out and show that this fan-out value is $\lfloor (n + 1)^2/4 \rfloor + 1$, as asserted in Table 5.1.

5.79 Suppose that an n-bit adder is to be implemented by a gate array containing 500 NAND gates whose maximum fan-in and fan-out figures are 16. Estimate, showing all your calculations, the largest value of n for an n-bit adder implemented with the gate array, using each of the following design styles: carry-lookahead and ripple-carry.

5.80 **(a)** Using any gate types, draw the logic circuit for a 4-bit ripple-borrow subtracter for unsigned numbers. **(b)** Devise an easy way to combine a ripple-carry adder and a ripple-borrow subtracter into a single circuit that computes either $X+Y$ or $X-Y$, as specified by a single control signal *SUB*.

5.81 Design a 4-bit borrow-lookahead subtracter, with the adder of Figure 5.56 serving as a model. Give a circuit diagram and sets of equations that define all the propagate, generate, borrow, and difference functions that your circuit calculates.

5.82 Redesign the adder-subtracter circuit of Figure 5.57 so that when $SUB = 1$, it computes either $X-Y$ or $Y-X$, as determined by a second control signal *SWITCH*.

SEQUENTIAL LOGIC

Delays and Latches

6.1 Signal Storage
6.2 Propagation Delays
6.3 Latches
6.4 Timing and State Behavior

Clocks and Flip-Flops

6.5 Basic Concepts
6.6 Flip-Flop Design
6.7 Flip-Flop Behavior

Asynchronous Circuits

6.8 Hazards
6.9 The Huffman Model
6.10 Some Design Issues

Overview

The ability to store information distinguishes sequential logic circuits from combinational circuits and is a crucial property of most digital systems. In this chapter, we study the basic types of memory elements—latches and flip-flops—used in sequential circuits. We briefly discuss asynchronous sequential circuit theory, which provides a formal theory of designing such memory elements. This discussion also motivates discussion of synchronous sequential circuits, considered in Chapter 7, which employ timing signals called *clocks* to increase their reliability and simplify the design process.

6

Delays and Latches

To begin, we examine the sources of delays in a digital circuit and show that they are fundamental to describing the circuit's sequential behavior. Latches, the simplest types of logic circuits that serve as memory elements, are defined and their properties are examined. The concept of state behavior is also introduced.

6.1 Signal Storage*

Information storage can be implemented by many types of physical devices, including mechanical springs and electrical capacitors. We begin with a brief presentation of a technology-independent, switch-level model of the binary storage function performed by such devices.

Physical Memory

Many storage devices have two energy states: charged and discharged.

A binary storage device requires two stable states to represent 0 and 1, as well as a provision for storing (writing) a new state and for retrieving (reading) the old state. Such memories often store information in the form of two energy states, allowing storage and retrieval to be accomplished by a transfer of energy. Figure 6.1 indicates how a mechanical spring can be used in this way. The spring is held fixed at its lower end (Figure 6.1a). The upper end is movable, so we can transfer energy to the spring by applying a weight to this end. With no weight applied, the spring stores zero energy, which we equate with logical 0. If we apply a weight as shown in Figure 6.1b, the spring is compressed, and we can consider it to store logical 1. (Obviously, the roles of 0 and 1 here are interchangeable.) The compressed spring

*This section may be omitted without loss of continuity.

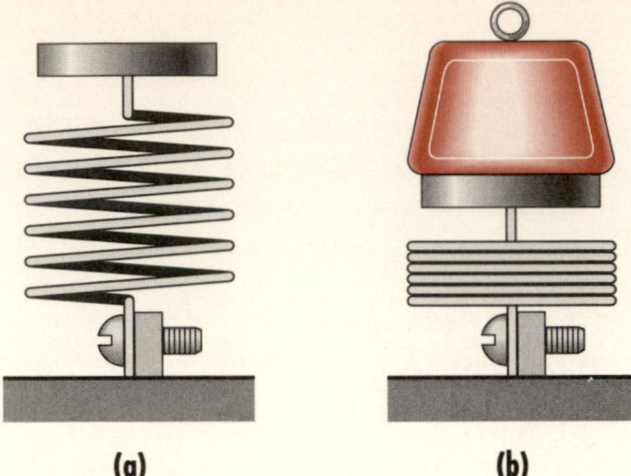

FIGURE 6.1 A spring as a binary storage device: (a) relaxed (storing 0); (b) compressed (storing 1).

stores a charge of energy that exerts an upward force on the weight. This energy is released (discharged) when the weight is removed and may be transferred to an external device connected to the spring's upper end.

Electrical phenomena exhibit similar binary storage effects. For example, suppose a person's hair is electrically charged by rubbing it with a plastic comb under very dry conditions. The rubbing converts the relaxed (uncharged or 0) state depicted in Figure 6.2a to a charged one in which individual strands of hair stand on end (Figure 6.2b). In the latter state, the hair "stores" a set of electrically charged (positive or negative) particles that are not present in the uncharged hair. An electrical capacitor acts in much the same way, storing a packet of electrical particles in the 1 state, and storing none in the 0 state. Wires provide an easy means to transfer charge packets to and from the capacitor. Such capacitors are the basis of several types of fast memory devices found in digital systems.

FIGURE 6.2 Human hair as a binary storage device: (a) relaxed (storing 0); (b) electrically charged (storing 1).

Wells

A well is an abstract model of a binary storage cell.

The foregoing phenomena can be abstracted to obtain a technology-independent model of a primitive storage device. We define a **well** W as a storage element that has two configurations or states—namely, storing 0 (denoted y_0) and storing 1 (denoted y_1) [Hayes, 1987]. The well has two terminals and is represented by the symbol shown in Figure 6.3a. Like other logic devices, we can input or output 0s and 1s via the well's terminals to control its behavior.

For simplicity, we assume that one terminal of a well is permanently held at the 0 level, so that all information storage and retrieval functions can be associated with the other, variable terminal z. This constraint is less severe than it might seem. It corresponds to making one end of a spring immovable, as in Figure 6.1. The capacitors that serve as storage cells in MOS ICs are transistor-like devices that usually have one terminal permanently connected to a power supply terminal, and so satisfy this constraint.

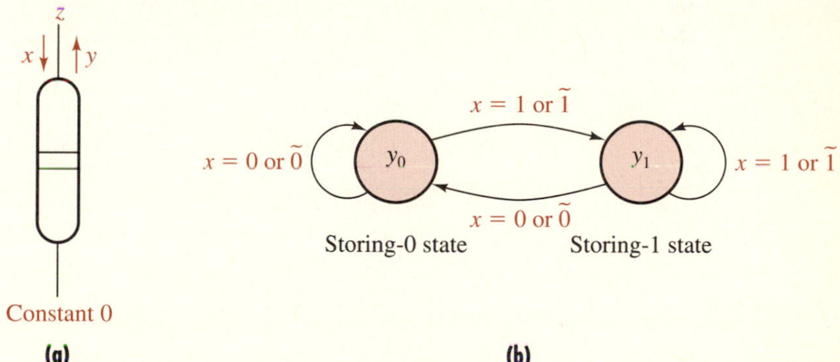

FIGURE 6.3 A well: **(a)** its graphic symbol; **(b)** a state diagram of well behavior.

Well Behavior

As in Section 4.3, we place a tilde (˜) over 0 or 1 to indicate a weak or attenuated version of a logic signal. If we apply a weak or strong external signal x to its variable terminal z, a well W assumes the state that corresponds to the 0 or 1 part of x, according to the following rules:

1. Input $x = 0$ or $x = \tilde{0}$ makes the well's state y_0 (storing 0).
2. Input $x = 1$ or $x = \tilde{1}$ makes the well's state y_1 (storing 1).

An x signal with the same value as W's current state has no effect on W. Hence, $x = 0$ or $\tilde{0}$ does not affect state y_0, and $x = 1$ or $\tilde{1}$ does not affect state y_1.

Figure 6.3b represents the states y_0 and y_1 of W using circles; arrows indicate all the possible transitions between the states, as specified above. The values of x that cause a particular transition are marked on the corresponding

State diagrams depict sequential behavior graphically.

arrow. For example, the right-pointing arrow marked "$x = 1$ or $\tilde{1}$" indicates that these values of x cause the well's state to change from y_0 to y_1. Figures of this type, termed **state diagrams**, provide a graphic picture of the dynamic or time-dependent behavior of a well or other sequential device. Unless external signals change it, W can maintain its current state indefinitely. Thus, a well is a two-state storage device with an infinitely long memory.

If the external input signal x is removed from the well (or changed to the high-impedance state Z by appropriate switch settings), W becomes a signal source that can drive other devices via an internally-generated y signal whose 0 or 1 value is determined by the well's current state. A well in state y_0 makes y assume the 0 level; state y_1 makes y assume the 1 level.

Because any external signal value except the high-impedance state Z is assumed to alter W's state, the y output signal from a well is assigned the new strength size "very weak," and its possible values will be denoted by $\tilde{\tilde{0}}$ and $\tilde{\tilde{1}}$. When the well is in the storing-0 state y_0, we obtain $y = \tilde{\tilde{0}}$; the storing-1 state y_1 produces $y = \tilde{\tilde{1}}$. Thus, the y signal serves to inform external circuits connected to z of the logic value of the quantity stored in W. For example, in the case of a mechanical spring, $y = \tilde{\tilde{1}}$ if the spring is in the compressed state, and $y = \tilde{\tilde{0}}$ if the spring is in the relaxed state.

Storage Delay

An important aspect of the real storage devices modeled by well W is that the time required to change the well's state decreases with the strength of the external (driving) signal x applied to W. This concept is illustrated by the fact that a spring is compressed quickly by a strong external force, but it is compressed more slowly by a weak external force. We assume that a strong 0 or 1 signal can change the state of a well instantly—that is, with no delay. We also assume that a weak applied signal ($\tilde{0}$ or $\tilde{1}$) takes some nonzero time Δ_i seconds to change the well's state. This delay determines the time at which the state y of W, and the effective output signal $z = \#(x, y)$, change value. (Recall from Section 4.3 that # is the connection function of wired logic, so that z assumes the logic value specified by the stronger of x and y.)

A delay element has a finite memory.

We conclude that when a weak signal derived from an attenuator A is applied to a well W via a common z line, there is a nonzero delay Δ_i before the level of z changes. Hence, the attenuator-well circuit configuration of Figure 6.4a introduces delays into the transmission of signals from x to z, as indicated graphically in Figure 6.4c. Note that the duration of the delay before z is affected by a change in x and may also depend on the direction of the signal changes. Hence, a 0-to-1 change can be transmitted faster than a 1-to-0 change (the case illustrated in the figure), or vice versa. The usual electrical implementation of an attenuator-well circuit is shown in Figure 6.4b. In the next section, we will introduce a simpler model (a delay element) that captures the timing behavior of these circuits.

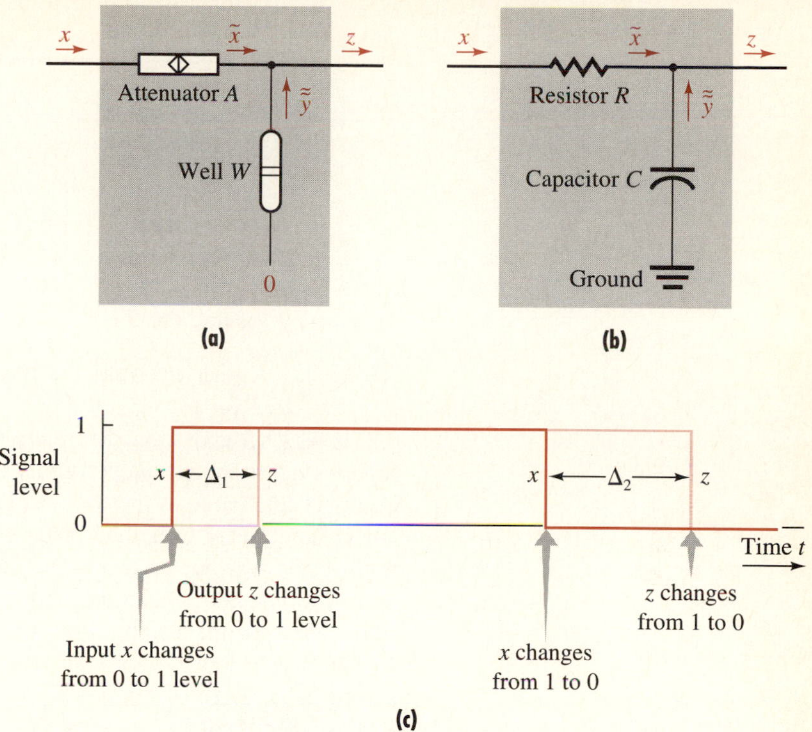

FIGURE 6.4 (a) An attenuator-well circuit; (b) its equivalent electrical circuit; (c) 0-to-1 and 1-to-0 signal transitions.

In order to use a well W as a practical storage element of infinite memory span, the path from x to W must be controlled, so that W can be isolated from external state-changing signals. The next example shows a simple but practical way of doing this, which is the basis for the large-scale random-access memories (RAMs) found in computers and other digital systems.

Example 6.1

Understanding a dynamic random-access memory (DRAM)

Perhaps the most important information-storage component found in digital systems is the **dynamic random-access memory** or **DRAM** (pronounced *dee-ram*). A DRAM is an integrated circuit designed to store very large amounts of binary information in a form that allows it to be rapidly stored and retrieved. Each bit of information is stored in the form of an electrical charge in a tiny capacitor-based memory cell. A DRAM is constructed from a large number of such 1-bit cells—as many as 2^{24} or more—which are laid out in a two-dimensional array on a single IC.

The basic DRAM cell consists of a single MOS transistor switch T that controls a capacitor C, which performs the information-storage function. The cell can be modeled abstractly by the switch-attenuator-well circuit of Figure

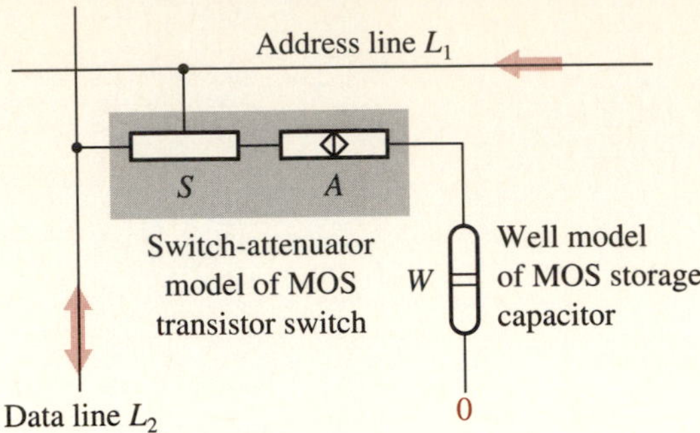

FIGURE 6.5 A switch-level model of a dynamic random-access memory (DRAM) cell.

6.5, in which S and A together represent T, and W represents C. The switch S captures the switching property of T; the attenuator A represents the nonzero electrical resistance in the data path of a real transistor. A and W define an input/output delay Δ as described in Figure 6.4, which is termed the **access time** of the cell. Here, Δ is a measure of the time required to transfer data to or from the cell, operations called **writing** and **reading**, respectively. The cell is accessed for reading or writing purposes via two lines, an address line L_1 and a data line L_2, as indicated in the figure. These lines are typically shared by many cells in much the same way as the address and data lines of a ROM (Section 4.7).

DRAMs need regular refreshing to restore leaking data.

The operation of a cell within a DRAM IC is illustrated by Figure 6.6. The cell is selected for reading or writing by a control circuit that activates (sets to 1) the address line L_1. This turns the switch S on, thus creating a path from the data line L_2 into the well. In a read operation (Figure 6.6a), the y signal generated by the well is passed through the attenuator and transferred (in possibly weaker form) to a buffer called a **sense amplifier**, where it is amplified and made available on the external data bus line B. Note the use of tri-state logic (Section 4.3) to control B. In the case of a write operation (Figure 6.6b), the cell is selected as before, but this time the internal data line L_2 is driven by a (strong) external signal x, obtained via an input buffer from B. This signal determines the next state of W in the manner illustrated by Figure 6.3.

Although it is not apparent in Figure 6.6, certain physical effects in real DRAM cells, especially leakage of a stored charge, cause the y_1 state to change spontaneously over time (a fraction of a second) into the state y_0. To prevent the stored information from being lost in this manner, the cells periodically (every few milliseconds) have their contents read out and then written back in amplified form by special control circuits, a process called **refreshing**. The need for automatic refreshing in DRAMs is the origin of the name *dynamic*. RAMs that do not require refreshing are called **static RAMs (SRAMs)**.

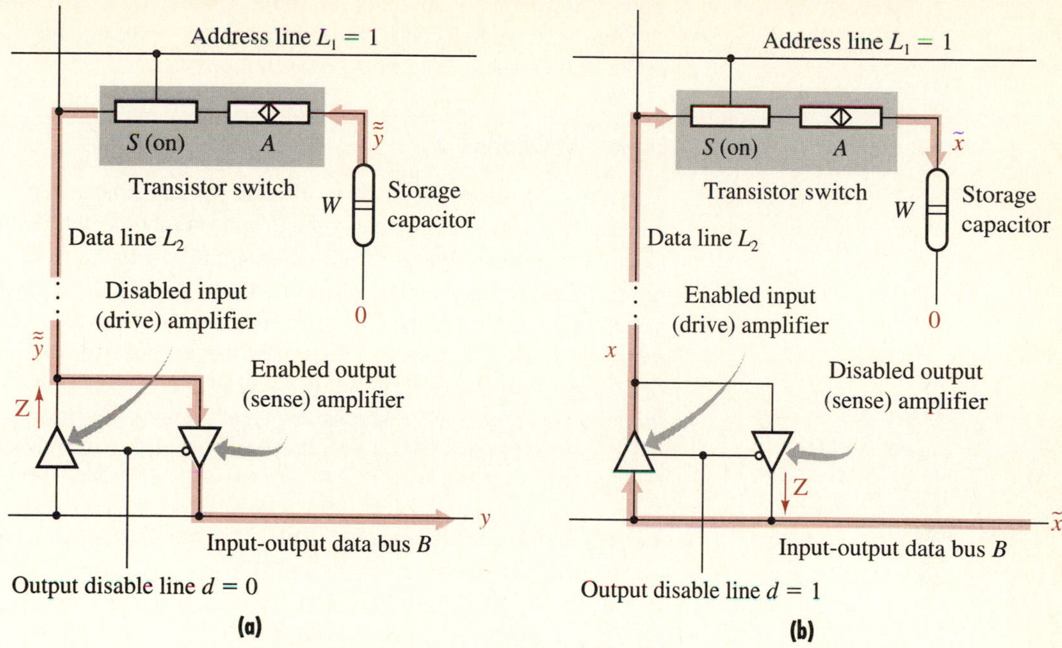

FIGURE 6.6 Operation of a DRAM: (a) reading from a cell; (b) writing to the cell.

The properties of signal attenuation and storage are inherent in all physical devices to some degree. So, although ideal lines and switches do not attenuate or store the signals they handle, the corresponding physical devices—for example, wires and transistors—do. The extent to which these effects need to be taken into account varies with the application. For most purposes in this book, a transistor can be modeled by an ideal switch. However, as indicated by the DRAM cell model of Figure 6.5, it is sometimes desirable to model explicitly the attenuation associated with a transistor's data path—in this case, to capture the cell's nonzero access time. Other aspects of transistor circuit behavior may impose further constraints on how such circuits are modeled at the switch level. In general, the more accurate or detailed the behavior that must be captured, the more complex the corresponding switch-level models must be.

6.2 Propagation Delays

The physical media such as electric currents that carry logic signals through a digital circuit travel at a finite rate. Consequently, logic signals are delayed in their passage from one part of a circuit to another. Such delays are ignored (assumed to be zero) in the design of combinational circuits. Nonzero signal delays are the source of the time-dependent behavior that distinguishes sequential from combinational logic, and so occupy a fundamental position in the analysis and design of sequential circuits.

Signal Waveforms

The physical quantities such as electric voltage that carry logic signals are inherently analog rather than digital and cannot switch instantaneously between the 0 and 1 levels. An analog signal changing from one signal level to the other passes through a continuum of intermediate levels, and does so at a finite rate determined by various physical characteristics of the circuit technology used. Figure 6.7 illustrates these issues for a signal that consists of a single 0-to-1 transition followed by a 1-to-0 transition, which is referred to as **pulse**. Figure 6.7a shows the "real" analog waveform of an electrical pulse, while Figure 6.7b shows its idealized binary approximation.

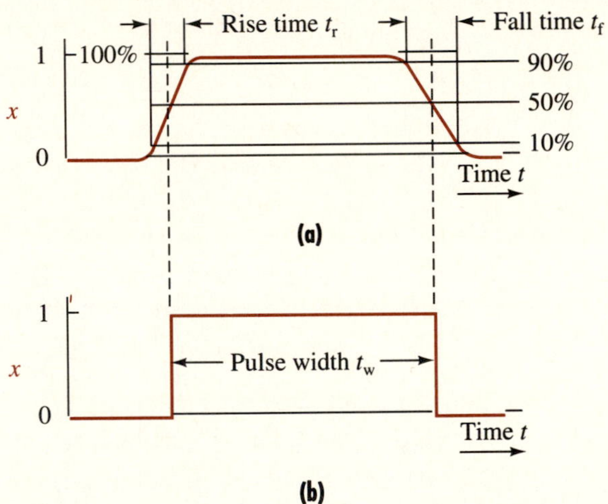

FIGURE 6.7 (a) The waveform of an analog pulse; (b) its digital approximation.

It is standard practice to measure time delays from certain reference points on the rising (0-to-1) and falling (1-to-0) edges of an analog waveform, which are indicated in Figure 6.7. The width t_w of a pulse is measured from the points on the rising and falling edges that lie halfway between the minimum and maximum signal levels. The duration of a 0-to-1 change, referred to as the **rise time** t_r, is measured from the time when the rising signal reaches 10 percent of the maximum analog value to the time it achieves 90 percent of the maximum. Similarly, the **fall time** t_f is the time a signal changing from 1 to 0 takes to drop from 90 percent to 10 percent of the maximum analog signal value. We normally assume that the signals of interest are purely binary, as in Figure 6.7b, and that $t_r = t_f = 0$.

Delay Types

As binary signals travel through a logic circuit, they are subject to delays from various sources. The most important of these are the **gate delays** that physical gates impose on the signals that pass through them. In the case of off-the-shelf ICs, gate delays are specified in manufacturers' data books and are roughly the same for all gates implemented using the same physical technology. For example, all gates in the 74AS (advanced Schottky) family of bipolar ICs have propagation delays of around 2 ns. The basic gate delay parameter is the propagation delay or switching time t_{pd}, which we defined in Section 3.5 as the time elapsing between a change in an input signal and the appearance of a resulting change in the gate's output signal.

There are also **line delays** associated with wires, conductors within ICs, and other interconnection media. Line delays tend to be much smaller than gate delays, however, and so can either be assumed to be zero, or else can be lumped into the propagation delay of an associated gate. We will ignore line delays in our analysis, and assume that all the propagation delays of interest in logic circuits can be represented by gate delays.

Delays are (mainly) due to gates and to interconnecting lines.

As mentioned in Section 3.5, physical delays can be modeled by logic devices called *delay elements*. Figure 6.8a shows the standard graphic symbol[1] used in logic diagrams to represent the most basic delay element. The behavior of this device is described by the time-dependent Boolean equation

$$z(t) = x(t - t_{pd}) \tag{6.1}$$

which is illustrated in Figure 6.8b. Its function is to delay the transfer of a signal change from x to z by a fixed amount t_{pd}, so that a pulse is shifted in time as it passes through the delay element. Later, we will see other delay elements that delay 0-to-1 and 1-to-0 signal changes by different amounts, and therefore alter the width of the transmitted pulse. The curved arrows in Figure 6.8b indicate the cause-and-effect relationship between x and z.

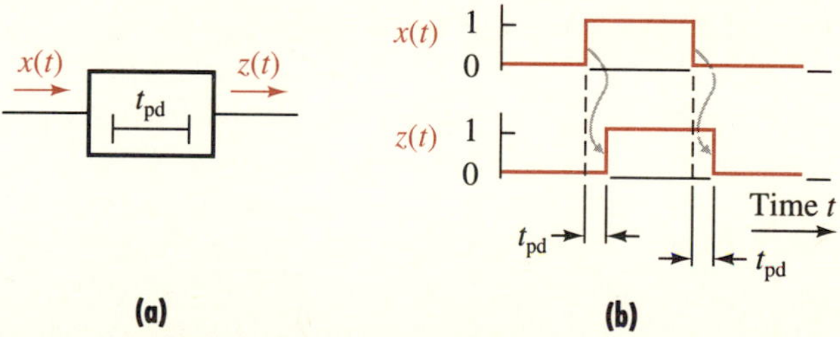

FIGURE 6.8 A basic delay element: (a) its graphic symbol; (b) its behavior.

[1]The symbols we use for delay elements and other sequential components such as latches and flip-flops closely follow the IEC 617-12 standard for logic symbols [Kampel, 1985] and compatible national standards such as the IEEE Standard 91–1984 [IEEE, 1984].

Gate Delays

We can easily incorporate delay elements of the foregoing kind into our models of logic gates. Figure 6.9 shows two representations of a physical NAND gate that has a propagation delay $t_{pd} = 10$ ns. In Figure 6.9a, a 10-ns delay element DE is inserted in the output line of an ideal (zero-delay) NAND gate. Clearly, this circuit's behavior is defined by the equations

$$y(t) = \overline{x_1(t)x_2(t)x_3(t)}$$
$$z(t) = y(t - t_{pd}) = \overline{x_1(t - t_{pd})x_2(t - t_{pd})x_3(t - t_{pd})}$$

The gate delay t_{pd} lumps together physical delays from various sources.

The delay element DE is an abstract "lumped" representation of the many sources of delay distributed throughout the physical devices that actually implement the NAND gate. Thus, the internal line y in Figure 6.9a does not correspond to any physical line. It is convenient to represent the delay DE implicitly, as shown in Figure 6.9b, where we write the NAND's propagation delay adjacent to the NAND symbol and omit the delay element. Figures 6.9a and 6.9b are therefore two functionally equivalent logical models of the same physical gate.

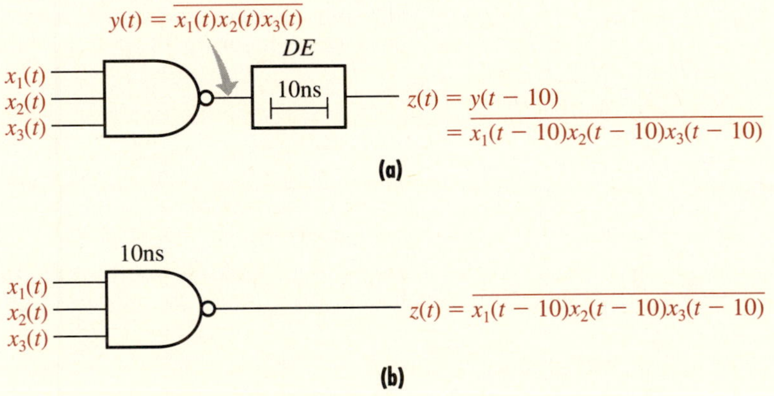

$y(t) = \overline{x_1(t)x_2(t)x_3(t)}$

DE

$x_1(t)$ $x_2(t)$ $x_3(t)$ 10ns $z(t) = y(t - 10)$ $= \overline{x_1(t - 10)x_2(t - 10)x_3(t - 10)}$

(a)

10ns

$x_1(t)$ $x_2(t)$ $x_3(t)$ $z(t) = \overline{x_1(t - 10)x_2(t - 10)x_3(t - 10)}$

(b)

FIGURE 6.9 Two models of a NAND gate that has a 10-ns propagation delay: **(a)** a model with an explicit delay element; **(b)** a model without an explicit delay element.

In a multigate circuit, the propagation delays of the individual gates combine to determine the delays experienced by signals as they pass through the circuit. Consider the circuit C of Figure 6.10a, which consists of two 10-ns gates, an AND gate G_1 and a NAND gate G_2. Figure 6.10b shows representative waveforms of the circuit's behavior, again with arrows showing cause-and-effect relationships. A signal change that occurs at the primary input line x_2 affects the primary output z after a delay of 10 or 20 ns, depending on the input/output paths over which the signal propagates. Consider, for example, the effect of x_2 changing from 0 to 1 at time $t = t_1$ with $x_1 = 1$,

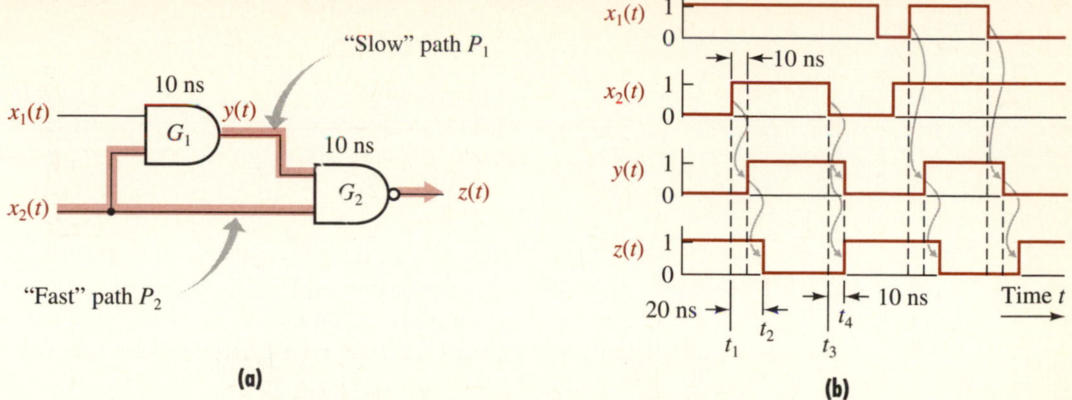

FIGURE 6.10 Propagation delays in a two-gate circuit.

$y = 0$, and $z = 1$. Because $y = 0$, this change does not directly affect the output z of G_2, which remains at 1 for the time being. G_1 is directly affected, however, so that after 10 ns, its output y changes from 0 to 1. Now two 1s are being applied to G_2, so that 10 ns later, at $t_2 = t_1 + 20$, z changes from 1 to 0. At this point, $x_1 = x_2 = y = 1$ and $z = 0$. At a later time t_3, x_2 is changed back to 0. This causes y and z to change to 0 and 1, respectively, at $t_4 = t_3 + 10$, as indicated.

It can be seen from this example that z responds faster to a 1-to-0 change in x_2 than to a 0-to-1 change in the same line. This difference can be attributed to the fact that the 1-to-0 change propagates through the "slow" path P_1, which contains two gate delays, whereas the 0-to-1 change in x_3 propagates through the "fast" path P_2, which contains only one gate delay. Figure 6.10b also shows that the corresponding changes in x_1 always affect the output z after a fixed 20-ns delay, because there is only one path from x_1 to z.

Rise and Fall Delays

Rising and falling signal transitions may be delayed differently.

The example of Figure 6.10 demonstrates that individual gate delays interact in subtle ways. It also shows that the two edges of a pulse can be delayed by different amounts when travelling between the same two points. In fact, if we look closely at any delay source with input x and output z, we may expect the kind of timing behavior depicted in Figure 6.11a. Here, the 0-to-1 transition (positive edge) of the signal x is shifted by the amount t_{PLH}, called the **rise delay**, while the 1-to-0 transition (negative edge) of x is shifted by t_{PHL}, called the **fall delay**. (In these standard subscripts, P denotes propagation, HL denotes high-to-low, and LH denotes low-to-high.)

t_{PLH} denotes a rise (low-to-high) delay; and t_{PHL} denotes a fall (high-to-low) delay.

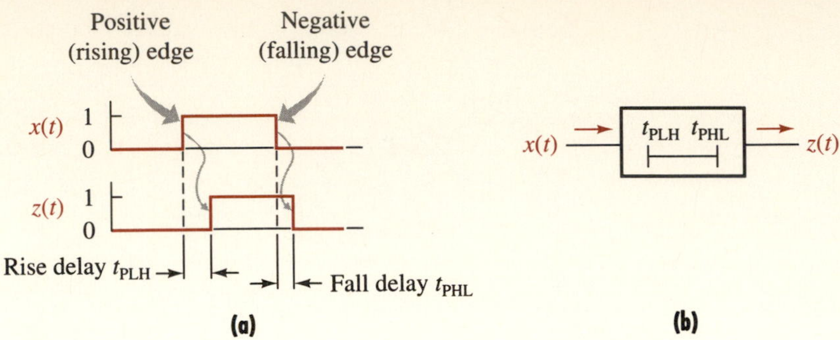

FIGURE 6.11 Rise-fall delays: (a) a timing diagram; (b) a delay element.

The timing behavior of Figure 6.11a is modeled by the logic element of Figure 6.11b, which is known as a **rise-fall delay element**. When

$$t_{PLH} = t_{PHL} = t_{pd} \tag{6.2}$$

the rise-fall delay element reduces to the basic delay element of Figure 6.8. Equation (6.2) generally holds for gates that are implemented in CMOS technology, but in TTL technology, rise and fall delays differ appreciably and are usually specified separately by manufacturers. If $t_{PLH} \neq t_{PHL}$ and we want to represent a gate's delay by a single number t_{pd}, we may use the approximation

$$t_{pd} = (t_{PLH} + t_{PHL})/2 \tag{6.3}$$

However, such approximations can lead to large errors when used for timing analysis.

Uncertain Delays

Delay times are subject to considerable uncertainty.

The durations of gate and line delays are difficult to control in the manufacturing process. Gate delays vary substantially, even among gates of the same type within the same circuit family. For example, the rise and fall delays of a particular logic gate in an IC can deviate by 50 percent or more from the average delay figures published for gates of its type. Even harder to predict are line delays, because they depend on a line's physical characteristics, such as its length, shape, and electrical capacitance.

Because of uncertainty about the values of gate delays, IC manufacturers specify their average values, referred to as **typical** or **nominal values**. They also usually provide maximum values, which are useful for determining propagation delays under worst-case conditions; minimum delay values may also be given. We indicate typical, maximum, and minimum values here by the superscripts *typ, max,* and *min,* respectively. Table 6.1 shows the published delay parameters in nanoseconds of the gates that appear in various

TABLE 6.1 Published propagation delays (in nanoseconds) for various IC implementations of a three-input NAND gate.

		t_{pd}			t_{PLH}			t_{PHL}		
Part no.	**IC family**	**Min**	**Typ**	**Max**	**Min**	**Typ**	**Max**	**Min**	**Typ**	**Max**
74ALS10	Advanced low-power Schottky TTL				2		11	2		10
74AS10	Advanced TTL				1.0		4.5	1.0		4.5
74HC10	High-speed CMOS		10	19						
74F10	F-type TTL				1.6	3.3	5.0	1.0	2.8	4.3
74LS10	Low-power Schottky TTL					9	15		10	15
74S10	Schottky TTL					3.0	4.5		3.0	5.0

versions of the 74X10 triple three-input NAND IC [Texas Instruments, 1986, 1987, 1988a, 1988b]. We can see that, in several of the TTL subfamilies, signal rise and fall delays differ slightly, and so are specified individually. In the 74HC CMOS family, however, the rise and fall times are essentially the same, so only t_{pd} values are given.

Pulse Propagation

The min-max ambiguity in the propagation delay of gate G, to which an input signal x is applied, introduces uncertainty about the time when changes in x affect the output signal of G. This uncertainty increases with each successive gate through which the changes propagate. It is of particular interest to know the effect of delays on a pulse, which is a signal that comprises two closely spaced transitions. A 1-pulse consists of a 0-to-1 transition followed by a 1-to-0 transition; a 0-pulse consists of a 1-to-0 transition followed by a 0-to-1 transition. The delays experienced by a pulse as it passes through a circuit with uncertain gate delays can be determined by systematically combining the minimum and maximum propagation delays of each gate encountered, as illustrated by the following example.

Example 6.2

Calculating pulse propagation delays through a circuit

Figure 6.12a shows a circuit composed of two type-74F10 NAND gates whose propagation delay parameters are defined in Table 6.1. We want to determine the times of the signal changes that result from the application of a 0-pulse of width $t_2 - t_1 = 9$ ns to input line $x_2(t)$, with $x_1(t) = x_3(t) = 1$.

The change of $x_2(t)$ from 1 to 0 at time t_1 causes the output $y(t)$ of the first gate G_1 to change from 0 to 1 at some time between $t_1 + t_{PLH}^{min} = t_1 +$ 1.6 ns and $t_1 + t_{PLH}^{max} = t_1 + 5.0$ ns. Thus, there is a transition region of width $t_{PLH}^{max} - t_{PLH}^{min} = 3.4$ ns, shown shaded in the figure, during which the status of $y(t)$ is uncertain. All we can say is that at some time in this transition region, $y(t)$ rises from 0 to 1. Similarly, $y(t)$ returns from 1 to 0 in response to the change of x_2 from 0 to 1 at some time between $t_2 + t_{PHL}^{min} = t_2 +$

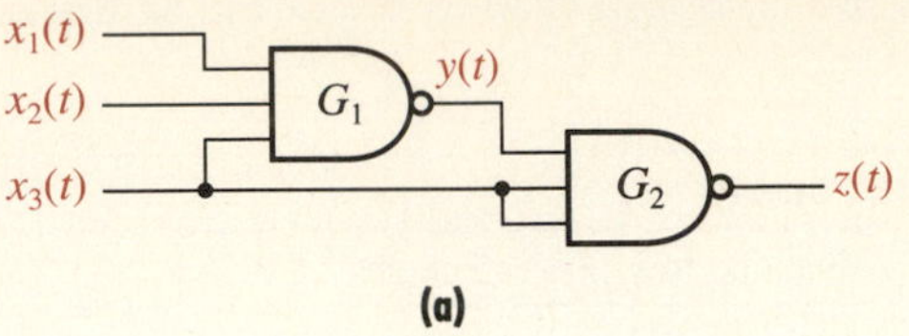

(a)

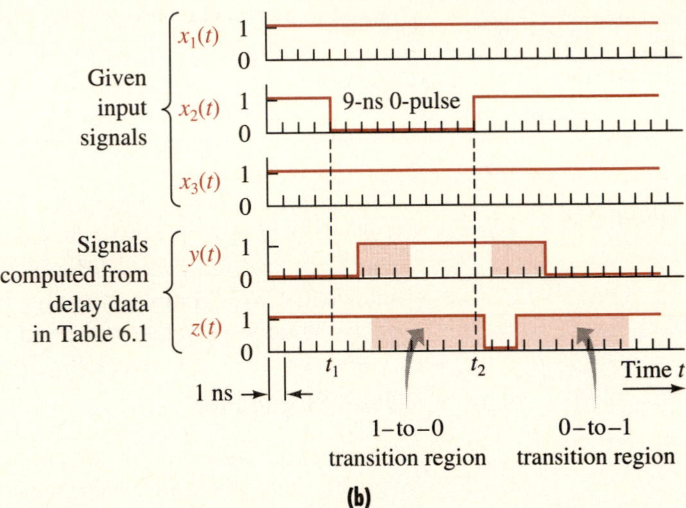

(b)

FIGURE 6.12 Pulse propagation through type-74F10 NAND gates.

1.0 ns and $t_2 + t_{\rm PHL}^{\rm max} = t_2 + 4.3$ ns, so the transition region for the falling $y(t)$ signal is of width $t_{\rm PHL}^{\rm max} - t_{\rm PHL}^{\rm min} = 3.3$ ns. On passing through the second gate G_2, the width of the uncertainty region in which the output $z(t)$ changes in response to the changes of $x_2(t)$ approximately doubles. For example, $z(t)$ falls from 1 to 0 as early as $1.6 + 1.0 = 2.6$ ns, or as late as $5.0 + 4.3 = 9.3$ ns after t_1. The "typical" delay through the two-gate circuit for both 0-to-1 and 1-to-0 changes is $t_{\rm PLH}^{\rm typ} + t_{\rm PHL}^{\rm typ} = 3.3 + 2.8 = 6.1$ ns.

Inertial Delay

We see from Figure 6.10 that even constant delays can alter signal waveforms by changing pulse widths. For example, the first 1-pulse applied to x_2 in Figure 6.10b produces a 0-pulse on z that is narrower by one gate delay (10 ns). This effect is more pronounced when uncertain delays are assumed. Thus, the 0-pulse of width $t_2 - t_1 = 9$ ns present on x_2 in Figure 6.12b results in a 0-pulse on z whose width can vary between 2.3 and 15.7 ns. If the width of the 0-pulse on x_2 is reduced to 4 ns, the shaded transition regions of y merge, allowing the possibility that $y(t)$ remains unchanged.

In other words, the gate to which the narrow pulse is applied, in this case G_1, effectively fails to respond to it.

Inertial effects tend to block narrow pulses.

If a pulse applied to a physical gate is very narrow—typically of width t_w comparable to, or less than, the gate's t_{pd}—the pulse may not pass through the gate. This is due to the gate's natural **inertia** or resistance to change, which implies that an input pulse must have a certain minimum energy, and therefore some minimum width, in order to elicit a response from the gate. Hence, very narrow pulses tend to be filtered out automatically by physical gates. Unfortunately, the general uncertainty about actual delay values and the technology dependence of the inertial characteristics of gates make it hard to say precisely for what values of t_w pulses will or will not be transmitted through a gate. We can, however, make the following general statements concerning pulse propagation:

Rule of thumb for pulse propagation through a gate

1. Pulses of width $t_w \leq t_{pd}$ are likely to be blocked by a gate of delay t_{pd}.
2. Pulses of width $t_w \geq 3t_{pd}$ will almost certainly be passed by a gate of delay t_{pd}.

Thus, if a signal $x(t)$ changes value at time t, it should not change value again until time $t + 3t_{pd}$ or later to guarantee that the changes in question form a pulse of sufficient width not to be blocked or excessively compressed by a gate.

Delay Simulation

Detailed analysis of the timing behavior of large logic circuits is complex and is usually done with the aid of CAD simulation programs that model delays in the various ways discussed above. The simplest of these timing simulators assign the same fixed propagation delay t_{pd} to all gates. Because we can make $t_{pd} = 1$ time unit using some appropriate time scale, these are known as **unit-delay simulators**.

Advanced simulators allow each gate to have a different propagation delay, distinguish rise and fall delays, or allow minimum and maximum delay values to be specified. The more detailed delay models require large amounts of computation but detect more potential timing problems, such as excessive propagation delays or inadequate pulse widths. In our analysis, we assume that all gates have a fixed (unit) delay t_{pd}, unless otherwise stated.

6.3 Latches

We turn next to the role of propagation delays in endowing gate-level logic circuits with memory. We will see that feedback is essential to providing a practical and stable memory element that is capable of storing information indefinitely.

Feedback

A (pure) delay element *DE* of size t_{pd} of the kind depicted in Figure 6.8 is a primitive storage device, the memory span of which is restricted to the finite (and very short) time period t_{pd}. The output $z(t)$ of *DE* at time t is completely determined by the input signal $x(t - t_{pd})$ present at $t - t_{pd}$. Subsequent values of x affect future values of z, but input signals that occur before $t - t_{pd}$ cannot influence z at any time after the present time t; they are, in effect, "forgotten" by *DE*.

Any logic circuit in which signal flow is unidirectional, a so-called **feed-forward circuit**, has a finite memory span bounded by the maximum value of the combined propagation delays along any path from a primary output to a primary input. For instance, the maximum input/output propagation delay through the circuit of Figure 6.12, and therefore its memory span, is $t_{PLH}^{max} + t_{PHL}^{max} = 5.0 + 4.3 = 9.3$ ns. Note that the maximum propagation delay is *not* $2t_{PLH}^{max} = 10.0$ ns in this case, because 0-to-1 transitions cannot appear concurrently at the outputs of both gates, due to the NAND gate's inversion property.

Feedback is needed for unrestricted memory.

In order to construct a circuit with unbounded memory span from unidirectional logic elements, it is necessary to create a closed signal path or a **feedback loop**.[2] We implicitly excluded feedback from our definition of a well-formed combinational circuit (Section 4.1) because, as we have seen, it can lead to inconsistent behavior. Feedback is a basic property of sequential circuits, but it must be used with care.

Stability

Some forms of feedback cause instability.

Figure 6.13 illustrates the problem caused by feedback in a purely combinational logic circuit—that is, one with gate delays of zero. There is a feedback loop from the output to the input of the NAND gate, implying that the Boolean equation

$$z(t) = \overline{x(t) \cdot z(t)} \tag{6.4}$$

must be satisfied. The appearance of the variable z on both the right and left sides of (6.4) is a typical indication of feedback. Equation (6.4) is satisfied by $x(t) = 0$ and $z(t) = 1$, because $\overline{0 \cdot 1} = \overline{0} = 1$. Consequently, the signal configuration depicted in Figure 6.13a is consistent and stable. On the other hand, if $x = 1$, as in Figure 6.13b, we obtain a logically inconsistent situation, where if $z(t) = 1$, then by (6.4), $z(t) = \overline{1 \cdot 1} = \overline{1} = 0$; similarly, if $z(t) = 0$, Equation (6.4) requires $z(t) = \overline{1 \cdot 0} = \overline{0} = 1$. Thus, we have a contradiction, because $z(t)$ cannot be 1 and 0 *at the same time*. The situation depicted in Figure 6.13b must therefore be viewed as logically impossible.

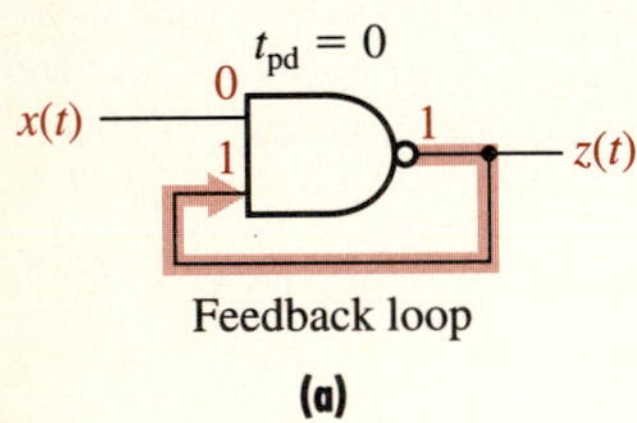

(a)

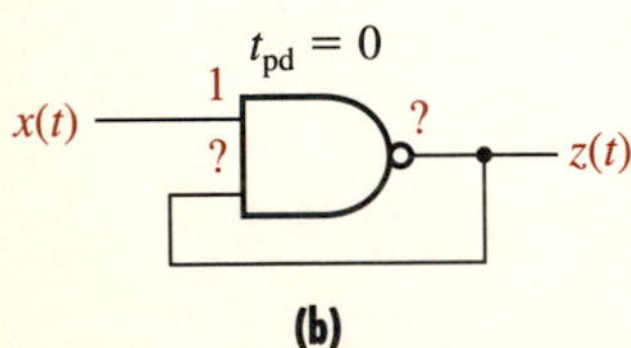

(b)

FIGURE 6.13 A NAND gate with both feedback and zero delay: (a) consistent signals; (b) inconsistent signals.

[2]The well model of a storage cell examined in Section 6.1 possesses infinite memory span in the above sense, without feedback. However, it is a *bidirectional* device, for which each terminal can serve as an input or an output.

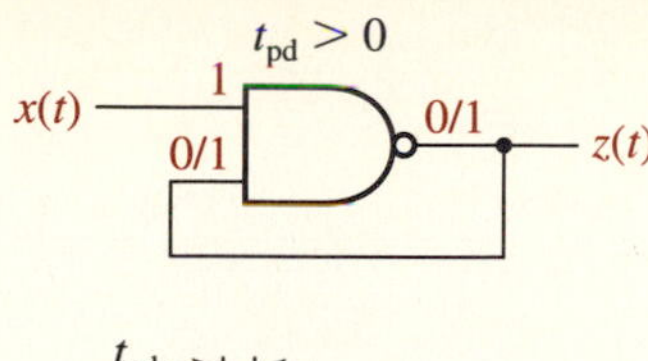

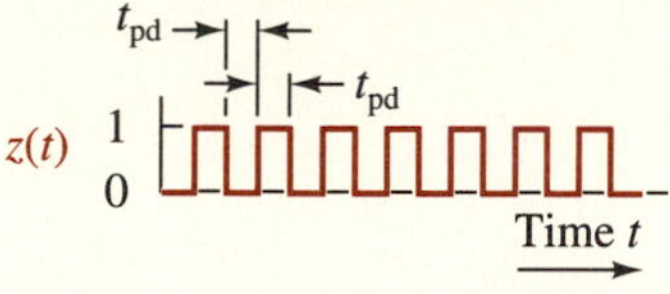

FIGURE 6.14 A NAND gate with feedback and nonzero delay that exhibits unstable behavior (oscillation).

The inconsistency present in this example disappears if the NAND gate has a nonzero propagation delay t_{pd}, which also makes it a better model of the behavior of a physical gate. Now Equation (6.4) changes to

$$z(t) = \overline{x(t - t_{pd}) \cdot z(t - t_{pd})} \tag{6.5}$$

The output signal $z(t)$ is no longer a function of its present value. Instead, it depends on the past value $z(t - t_{pd})$, which can differ from $z(t)$. Specifically, when $x(t - t_{pd}) = 1$, we can satisfy (6.5) with $z(t) = \overline{z(t - t_{pd})}$. Hence, if z changes from 1 to 0 at some time t, this change will cause z to change from 0 to 1 at time $t + t_{pd}$. According to Equation (6.5), this second change will change z from 1 to 0 at $t + 2t_{pd}$, and so on. Hence, the value of z must change every t_{pd} time units, as depicted in Figure 6.14. This type of regular and spontaneous changing, called **oscillation**, is an extreme form of unstable behavior. Unlike the situation depicted in Figure 6.13b, it is not logically inconsistent. In fact, oscillatory circuits with behavior like that of Figure 6.14 play a small but useful role in generating the clock signals that control synchronous sequential circuits.

Spontaneous oscillation of the above kind involves narrow pulses of width t_{pd} that tend to be filtered out by the gates through which they pass, as noted in the preceding section. Consequently, such oscillation usually dies out quickly. (Clock generators designed to oscillate continuously require more elaborate circuits.)

Metastability

A metastable signal is temporarily stuck between 0 and 1.

Instead of oscillating, the signal $z(t)$ in Figure 6.14, or any output signal that is subject to opposing input signals, may linger for some time Δ_M at a physical value U that is intermediate between logical 0 and 1. This condition, in which a signal is pulled equally toward 0 and 1 and, in consequence, remains stationary at some in-between value, is called **metastability**. If we consider a tossed coin to have two stable states—heads and tails—when it falls to the ground, the metastable state corresponds to the coin landing on its edge and remaining upright. This is obviously a very unusual but not impossible event, a characteristic shared by metastable logic signals.

As illustrated in Figure 6.15, the effect of metastability on a logic signal is a rise delay (as in the figure) or a fall delay that is unusually long. Small fluctuations (noise) that are present in all signals tend to push the signal away from U toward 0 or 1, so that most of the time Δ_M is negligible. Metastability can affect a changing signal $z(t)$ in any logic circuit, however, and is responsible for some rare and hard-to-find failures. Note that some logic signals change state billions of times a second, providing ample opportunity for occasional metastability.

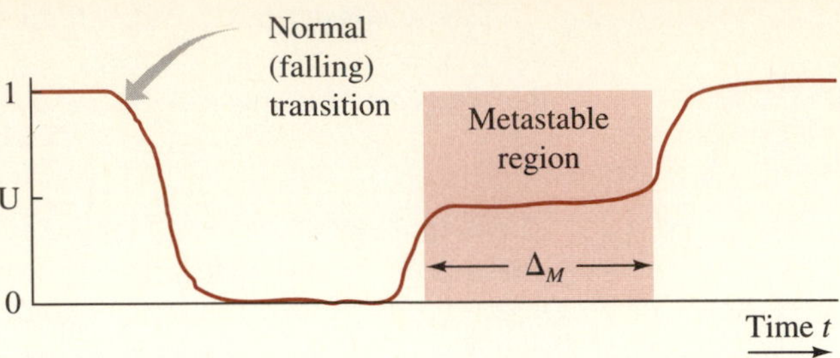

FIGURE 6.15 A signal waveform that exhibits metastable behavior.

Bistable Circuits

To design useful sequential devices, we need a memory circuit that has unrestricted memory span, and so is capable of storing the value of a binary quantity Q indefinitely. This circuit must have two stable internal configurations or **states**, defined as follows:

State $Q = 0$: the memory circuit stores the value 0

State $Q = 1$: the memory circuit stores the value 1

The variable Q is called the **state variable** or simply the stored data. The above states are required to be **stable** in the sense that once the memory is placed in a particular state ($Q = 0$ or 1) and its inputs are held at appropriate fixed values, Q remains at its current value indefinitely. A change in Q requires a specific change to be made to the input signals. While changing from one stable state to the other, the memory may pass through one or more unstable or transient states.

A bistable circuit has two stable states.

A logic circuit that has precisely two stable states of the above kind is called a **bistable circuit**. (The term *flip-flop* is also widely used for a general bistable circuit. Following modern design practice, however, we reserve *flip-flop* for a particular class of bistable circuits with special timing characteristics, introduced in Section 6.5.)

t_{pd} $y(t) = 0$ t_{pd}

$(t - t_{pd}) = 1$

State $Q = 0$

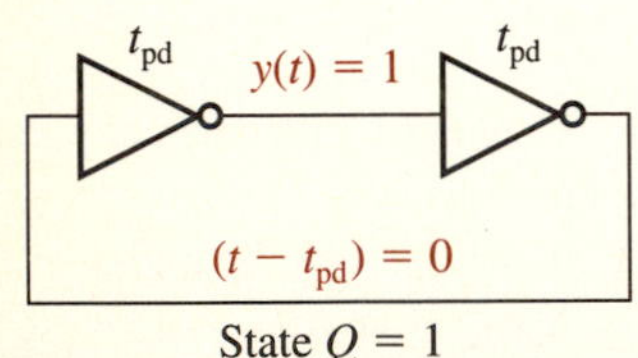

FIGURE 6.16 An ideal bistable logic circuit.

Figure 6.16 shows a conceptual model of a bistable logic circuit that consists of two inverters connected to form a simple closed loop. If the output of one of the inverters is taken to be $y(t)$, then the output of the other is $\bar{y}(t - t_{pd})$. Thus, we can associate state $Q = 0$ with $y(t) = 0$, and state $Q = 1$ with $y(t) = 1$, as indicated in the figure. Unlike the situation in Figure 6.14, the state signals are stable, because the signals fed back from the output of each inverter to the input of the other tend to hold the inverters' output signals at their current values.

To make a practical bistable memory circuit, we need an input/output mechanism for changing the state from $Q = 0$ to $Q = 1$, and vice versa. In the case of Figure 6.16, we must modify the two inverters by introducing one or more input lines that can switch (flip) the gate output signals. We also need to add at least one output line to observe the state variable Q.

SR Latches

A direct way of converting the bistable circuit of Figure 6.16 into a practical memory element is depicted in Figure 6.17a, where the inverters have been replaced by two two-input NOR gates. One input of each NOR forms part of the original feedback loop. The other is an independent primary input used to switch the device's state. The new primary inputs are called the **set** (S) and **reset** (R) **lines**, and the entire circuit is called a **set-reset latch** or **SR latch**.[3] The internal state variable y is connected to a primary output, denoted Q. Because the complementary signal $\overline{y}$ is automatically generated inside the latch, it is also connected to a second primary output line, denoted $\overline{Q}$. Thus, the state variable is provided in both true and complemented form, which facilitates the computation of (combinational) logic functions that involve Q as an input variable.

An SR latch is a bistable circuit composed of two NOR gates.

The SR latch of Figure 6.17a is one of the simplest memory circuits. It is shown redrawn in Figure 6.17b in a way that emphasizes the SR latch's symmetry. In this version, the NOR gates are considered to be "cross-coupled," with each NOR feeding back its output to the other NOR gate. Figure 6.17c shows the graphic symbol widely used to represent an SR latch in logic diagrams, while hiding the latch's internal details. The minor variant of this symbol, preferred by the IEC 617-12 standard, is given in Figure 6.17d.

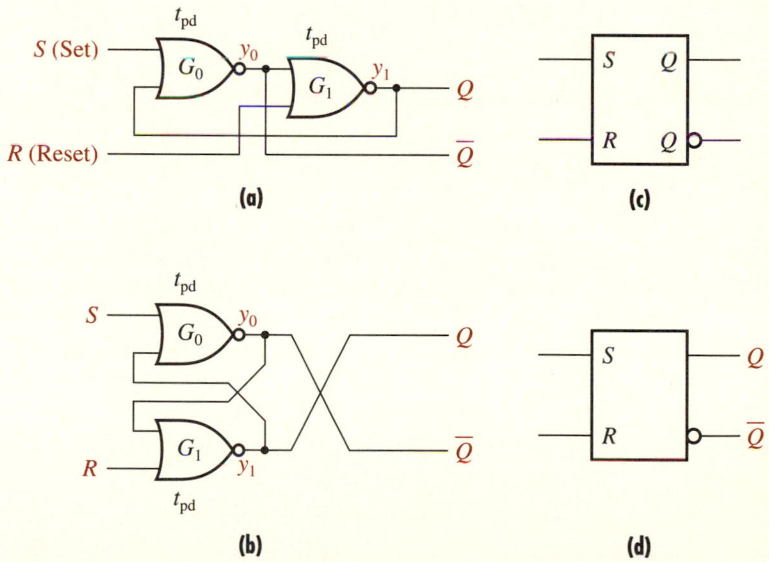

FIGURE 6.17 An SR (set-reset) latch: (**a**) its logic circuit; (**b**) its logic circuit redrawn in cross-coupled form; (**c**) its graphic symbol; (**d**) an alternative (IEC 617-12) symbol.

[3]Some authors prefer the name *reset-set* or *RS latch* for this circuit. It is also widely referred to as a *flip-flop* rather than a latch.

Note that in each symbol, the function that appears on the output side of the lower output line is $\overline{Q}$. The output function that precedes the inversion symbol—that is, "inside" the latch—is taken to be Q in each case, so the two latch symbols are consistent.

The bistable behavior of the SR latch is easily related to that of the inverter circuit of Figure 6.16. When S and R are both 0, the latch's two NOR gates effectively become inverters, and the circuits of Figures 6.16 and 6.17 are functionally the same. The SR latch is then stable, with either $Q = 0$ or $Q = 1$. The input combination $(S, R) = (0, 0)$ is called **quiescent**, because it leaves the latch indefinitely in either of its stable states. To avoid inconsistencies of the type depicted in Figure 6.13, we must take the gate delays into account. We therefore always assume that both NOR gates have propagation delay $t_{\text{pd}} > 0$.

SR Latch Operation

Suppose that at time t, we have $Q = R = S = 0$, and S is changed from 0 to 1. The output $y_0 = \overline{Q}$ of gate G_0 changes from 1 to 0, as required by the sequential NOR equation

$$y_0(t + t_{\text{pd}}) = \overline{S(t) + y_1(t)} = \overline{1 + 0} = 0 \tag{6.6}$$

Note that this change in y_0 occurs after t_{pd} seconds, due to the propagation delay of G_0. The new value of y_0, in turn, affects the output y_1 of the second gate G_1 after another t_{pd} seconds. Signal y_1 is initially 0, and changes to 1 as follows:

$$y_1(t + 2t_{\text{pd}}) = \overline{R(t + t_{\text{pd}}) + y_0(t + t_{\text{pd}})} = \overline{0 + 0} = 1$$

At this point, we have $y_0 = \overline{Q} = 0$ and $y_1 = Q = 1$, so the SR latch has changed state. The new state is stable, with $(S, R) = (1, 0)$, because the NOR gates have consistent input/output signals. No further change occurs if S now changes back to 0, restoring the quiescent input combination $(S, R) = (0, 0)$. From the symmetry of S and R with respect to Q and $\overline{Q}$, it is apparent that changing R from 0 to 1, with $S = 0$, changes the state of the latch from $Q = 1$ to $Q = 0$.

The behavior of the SR latch described above may be summarized as follows:

1. It has two stable states defined by $Q = 0$, which is called the **reset state**, and $Q = 1$, which is called the **set state**.
2. The input combination $(S, R) = (0, 0)$ is quiescent in that it leaves the latch in either of its stable states indefinitely.
3. The input combination $(S, R) = (1, 0)$ sets the latch by changing its state from $Q = 0$ to $Q = 1$. If the initial state is already $Q = 1$, setting the latch has no effect.

Applying $(S, R) = (1, 0)$ to an SR latch stores 1; $(S, R) = (0, 1)$ stores 0; $(S, R) = (0, 0)$ has no effect.

4. The input combination $(S, R) = (0, 1)$ resets the latch by changing its state from $Q = 1$ to $Q = 0$. If the initial state is already $Q = 0$, resetting the latch has no effect.
5. A complete change of state takes approximately $2t_{pd}$ time units, where t_{pd} is the average gate delay.

It is important to observe that the SR latch's outputs Q and $\overline{Q}$ are complementary only after the latch's state has stabilized. For example, during a set operation that commences at time t, y_0 changes from 1 to 0 at $t + t_{pd}$, as defined by Equation (6.6), while y_1 is still 0. During the next t_{pd} seconds, until y_1 has time to respond to the input changes, the latch's state, as defined by (y_0, y_1), is (0,0). This is an unstable state that appears only temporarily during state transitions. We use y_i to reflect all the possible states, stable and unstable, and Q to denote the stable states only, as is standard practice. Figure 6.18 describes all the foregoing (normal) behavior modes of an SR latch by means of a timing diagram.

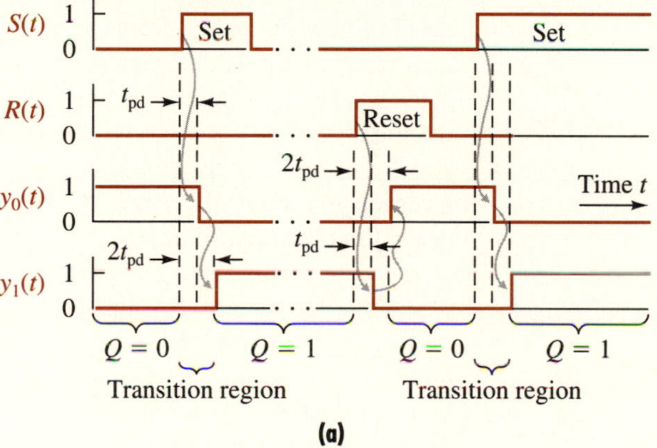

(a)

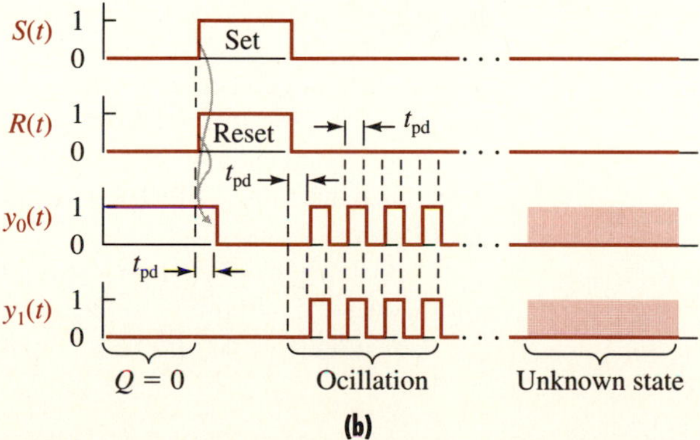

(b)

FIGURE 6.18 Timing diagrams for an SR latch: (a) normal operation; (b) the effect of the "forbidden" input combination $(S, R) = (1, 1)$.

Anomalous Behavior

Applying $(S, R) = (1, 1)$ to an SR latch produces unpredictable behavior.

A natural question that arises at this point is, what happens if we apply the input combination $(S, R) = (1, 1)$ to an SR latch? As might be expected from such an attempt to set and reset the latch simultaneously, the resulting state is unpredictable. Making $S = R = 1$ obviously forces the outputs of both NOR gates to 0, so the latch enters the state $(y_0, y_1) = (0, 0)$, where it will remain as long as both primary inputs are held at 1. If the quiescent input pattern $(S, R) = (0, 0)$ is applied, the outputs of both NORs become 1. These values feed back to the inputs and cause both NORs to produce 0. Hence, if the propagation delays of the two NOR gates were exactly the same, the SR latch would oscillate spontaneously between the two states $(y_0, y_1) = (0, 0)$ and $(y_0, y_1) = (1, 1)$, which is obviously unacceptable behavior in a memory circuit.

The preceding analysis assumes that the gates of the SR latch have only pure delays of exactly equal duration. In practice, the inertia of the gates, as well as minor differences in their propagation delays, tend to damp out the oscillations and make the latch settle into one of its two stable states. We do not know the exact delays of the latch's gates, so we cannot predict the final state of the latch when it eventually stabilizes. Hence, application of $(S, R) = (1, 1)$ followed by $(S, R) = (0, 0)$ can cause the latch to oscillate for some time before entering an unknown stable state (Figure 6.18b). Such a condition, in which the next state of a sequential circuit depends on the precise gate delays present, is called a **race**, because various internal signals "race" against one another to determine the circuit's final state. Race conditions are clearly undesirable in logic circuits.

Thus, we see that an SR latch has unpredictable behavior under certain operating conditions. As a result, other latch designs (described below) are more often used in practice. When SR latches are used, steps must be taken to avoid applying the input combination $(S, R) = (1, 1)$ to them, a constraint that complicates the design process.

D Latches

We turn next to another latch circuit, called a **delay circuit** or **D latch**,[4] which does not have the restriction on its input patterns found in the SR case. A common implementation of a D latch appears in Figure 6.19. This design is essentially the same as that used in the 74LS75, an SSI IC containing four such latches.

A D latch has a data input D and an enable input C.

A D latch has two primary inputs: a data input D and a control input C, usually referred to as the **enable line**. Like the SR latch, a D latch has two complementary primary outputs, labeled Q and $\overline{Q}$, which indicate the

[4]The *D* in the latch's name is also interpreted as *data*, leading to the name *data latch*. Other names used for a D latch, which exists in many different implementations, are *level-sensitive latch*, *polarity-hold latch*, and *sample-and-hold latch*.

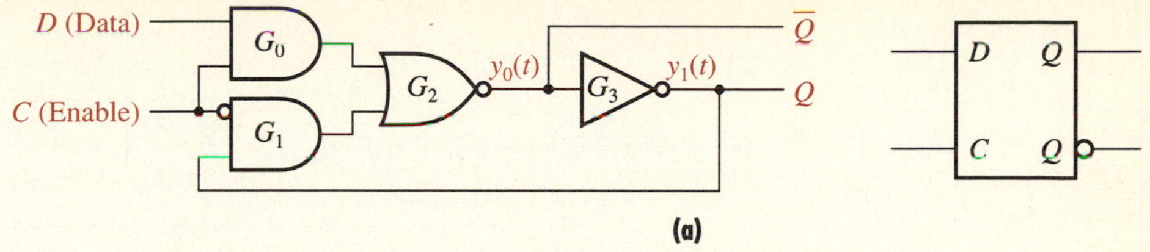

(a)

FIGURE 6.19 A delay (D) latch: (a) its logic circuit; (b) its graphic symbol.

circuit's stable states. The D input line specifies the next "data" value to be stored in the D latch. The latch contains a feedback loop, which can be thought of as capturing and storing—that is, **latching**—the data signal D. The role of C is to open and close this feedback loop to allow new data to be entered or retained.

The D latch displays an important characteristic of most memory circuits (and, indeed, of most sequential circuits); its input signals are of two types: data and control. In general, a **data input** explicitly specifies the data value that is to be stored in the memory circuit; a **control input** determines when the data are stored in the latch. In the case of the SR latch, the role of data input is split between the S and R lines, and there is no separate control input as such.

D Latch Operation

When enabled, a D latch is transparent to D input signals.

The operation of a D latch is illustrated by Figure 6.20. For it to respond to the D input, the latch must be **enabled**; that is, its C input must be in the active or 1 state. In this configuration (Figure 6.20a), the latch is said to be **transparent** to the D signal, in that D "flows" through the circuit unchanged except for inversion by some gates, while at the same time the feedback loop is logically opened at the NOR gate. $\overline{Q}$ takes the value of $\overline{D}$ after a couple of gate delays; Q takes the value of D one gate delay later. The outputs of the D latch can therefore be defined by the equations

$$\overline{Q}(t) = \overline{D}(t - t_1)$$
$$Q(t) = D(t - t_2)$$

where t_1 and t_2 denote propagation delays within the D latch. Thus, the state Q is simply a delayed version of the D input signal; hence the name *delay latch*.

As long as $C = 1$, the D latch responds immediately to changes that occur on its D input line. This transparency property makes the D latch well suited to data capture in high-speed circuits. We will see later that transparency also has undesirable aspects, and is specifically eliminated in the design of memory elements that are designated "flip-flops" rather than latches.

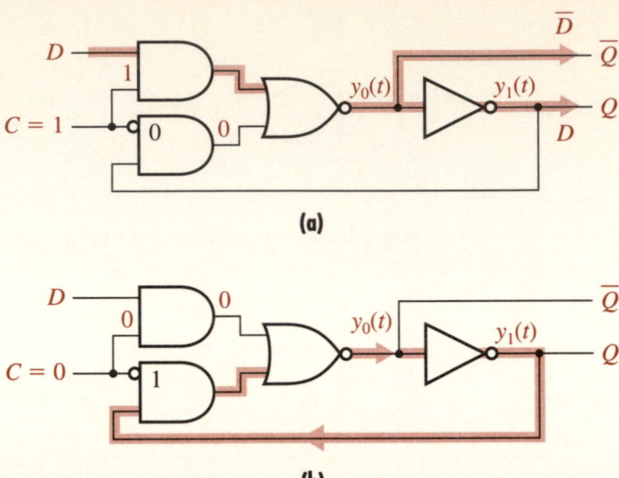

FIGURE 6.20 Operation of a D latch: (**a**) the enabled (transparent) condition; (**b**) the disabled (quiescent) condition.

The effect of $C = 0$, shown in Figure 6.20b, is to disconnect the D input from the latch and to close the feedback loop. In fact, the feedback loop now becomes equivalent to the inverter circuit of Figure 6.16, and so holds Q at its current value until the latch is again enabled. Thus, $C = 0$ is, in effect, the quiescent condition, for which the D latch stores its current state indefinitely. Note that when C goes from 1 to 0, the latch retains as Q the value present in the feedback loop immediately before the loop is closed. Figure 6.21 shows a timing diagram that summarizes the behavior of the D latch, assuming that each of its four gates has the same propagation delay t_{pd}. Observe that $\overline{Q}(t) = \overline{D}(t - 2t_{\text{pd}})$ and $Q(t) = D(t - 3t_{\text{pd}})$. In Section 6.10, we reexamine the timing behavior of this D latch in more detail and see that it requires modification to eliminate certain timing problems (hazards).

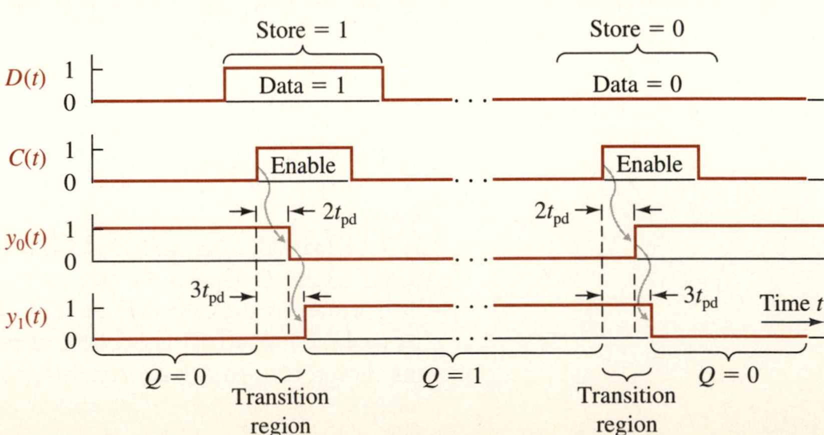

FIGURE 6.21 A timing diagram for normal operation of a D latch.

Example 6.3

Constructing a D latch from an SR latch

To illustrate the relationship between the SR and D latches, we show how an SR latch is modified by the addition of a few gates to form a bistable element that is another type of D latch. First, a two-input AND gate is inserted into each of the S and R inputs of the SR latch. One input of each AND gate is used to form an external enable input C. The result, shown inside the box in Figure 6.22, is called a **gated SR latch** or **clocked SR latch**. Making $C = 0$ clearly disables the original SR latch by making $S = R = 0$. When $C = 1$, the other two inputs of the AND gates, denoted by S^* and R^*, behave like S and R, respectively. The circuit is now enabled and functions as an SR latch of the kind described earlier.

To complete the modification, we add an inverter to the inputs of the gated SR latch, as shown in Figure 6.22. The new D primary input is connected to the S^* input of the gated latch and also to the inverter; the inverter's output is connected to the gated latch's R^* input. Thus, $D = 1$ makes

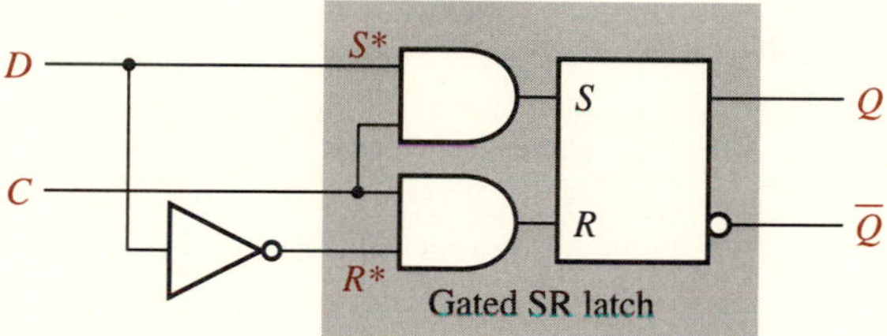

FIGURE 6.22 A D latch constructed from an SR latch.

$(S^*, R^*) = (1, 0)$, which sets the gated latch when it is enabled. Similarly, $D = 0$ makes $(S^*, R^*) = (0, 1)$, which resets the enabled latch. It is easy to see that the entire circuit has the same behavior as the D latch of Figure 6.19, except for its signal propagation delays. Notice that the troublesome input combination $(S, R) = (1, 1)$ is avoided, because the added inverter forces S^* and R^*, and therefore S and R, to have opposite values whenever the latch is enabled.

6.4 Timing and State Behavior

We next examine how the timing constraints and the functional behavior of latches can be expressed in concise forms, called *state tables* and *state diagrams,* which are suitable for practical analysis and design. Later, we will use behavioral representations of this kind with all types of sequential circuits.

Timing Parameters

The propagation delays of a latch or, indeed, of any sequential circuit can be determined by analyzing the influence of each of the circuit's internal gate delays, as we did in Section 6.2. To reduce the need for a tedious analysis of this sort, circuit manufacturers provide propagation delay parameters (t_{pd}, t_{PLH}, and t_{PHL}) linking representative pairs of input/output signals of the overall circuit. We define these parameters the same way we defined the delay parameters for gates. Thus, the t_{PLH} propagation delay from input x_i to output z_j is the time required for a signal change in x_i, with all other inputs held at appropriate constant values, to cause a low-to-high (0-to-1) transition to appear on z_j. The corresponding t_{PHL} parameter is the time required for a change in x_i to cause a high-to-low (1-to-0) transition on z_j. For example, in the case of the SR latch, our analysis (Figure 6.18) shows that t_{PHL} is one gate delay for S and $\overline{Q}$, and that t_{PLH} is two gate delays for R and $\overline{Q}$.

Example 6.4

Reading timing parameters from manufacturers' data sheets

Table 6.2 shows the complete published propagation delay parameters for the 74LS75 D latch [Texas Instruments, 1988b], which has the logic circuit of Figure 6.19. The figure includes four sets of t_{PLH} and t_{PHL} values, corresponding to the four possible pairings of the two input signals D and C with the two output signals Q and $\overline{Q}$. For instance, the last line of this table specifies that the delay t_{PHL} required for input C to cause output $\overline{Q}$ to go from 1 to 0 has a typical value of 7 ns and a maximum value of 15 ns. The signal change in $\overline{Q}$ that occurs here requires $\overline{Q}$ to be 1 initially, while D is held at 1 and C is changed from 0 to 1. These conditions are found in the transition region at the bottom left of the timing diagram in Figure 6.19, where y_0 represents $\overline{Q}$. Our previous analysis has revealed that the 1-to-0 propagation delay for this case is $2t_{pd}$, which corresponds to

$$t_{PHL}^{typ} = 7 \text{ ns}$$

in the 74LS75 specification sheet. We also obtain

$$t_{PLH}^{typ} = 2t_{pd} \text{ ns}$$

for the same signal pair C,y_0, whereas the corresponding specification for the 74LS75 is 16 ns. Thus, our analysis of the generic D latch based on the logic diagram of Figure 6.19 and the assumption that each gate has the same (unit) propagation delay t_{pd} produces only a crude approximation to the timing behavior of a 74LS75-type latch. More data on the 74LS75's internal delays would be needed to construct an accurate timing model for it.

The propagation-delay times provided by manufacturers are not always easy to use. For example, 16 different propagation delay values are given

TABLE 6.2 Published timing data for the 74LS75 D latch [Texas Instruments, 1988b].

recommended operating conditions

	SN54LS75 SN54LS77			SN74LS75			UNIT
	MIN	NOM	MAX	MIN	NOM	MAX	
Supply voltage, V_{CC}	4.5	5	5.5	4.75	5	5.25	V
High-level output current, I_{OH}			−400			−400	μA
Low-level output current, I_{OL}			4			8	mA
Width of enabling pulse, t_w	20			20			ns
Setup time, t_{su}	20			20			ns
Hold time, t_h	5			5			ns
Operating free-air temperature, T_A	−55		125	0		70	°C

switching characteristics, V_{CC} = 5 V, T_A = 25°C

PARAMETER¶	FROM (INPUT)	TO (OUTPUT)	TEST CONDITIONS	'LS75			'LS77			UNIT
				MIN	TYP	MAX	MIN	TYP	MAX	
t_{PLH}	D	Q	C_L = 15 pF, R_L = 2 kΩ, See Figure 1		15	27		11	19	ns
t_{PHL}					9	17		9	17	
t_{PLH}	D	$\bar{Q}$			12	20				ns
t_{PHL}					7	15				
t_{PLH}	C	Q			15	27		10	18	ns
t_{PHL}					14	25		10	18	
t_{PLH}	C	$\bar{Q}$			16	30				ns
t_{PHL}					7	15				

¶ t_{PLH} = propagation delay time, low-to-high-level output
t_{PLH} = propagation delay time, high-to-low-level output

for the 74LS75 latch in Table 6.2. As we show next, a smaller set of timing figures (the setup, hold, and enable pulse-width times, which also appear in Table 6.2) define the timing of memory elements adequately for most design purposes.

Setup, Hold, and Enable Times

Proper operation of a latch requires that certain rules be observed concerning the times at which its input signals change value. Recall that, due to inertial delays, if an input signal x is deactivated too soon after it is activated, or vice versa—in other words, if the width of an x-pulse is too small—the circuit may not respond to x and may fail to change state, as required. Excessively long input pulses also lead to undesirable behavior, because they permit multiple state changes to occur in a circuit when only one change is desired. In general, signal changes that occur on different input lines to a latch should be spaced far enough apart that desired changes of state always occur reliably, in spite of some variability in the latch's internal propagation delays.

To address timing concerns of the foregoing sort, minimum values of three standard delay parameters that relate a data signal D to its enabling control signal C are specified for all latches and flip-flops. The parameters in question are the setup, hold, and enable times, and are defined in Figure 6.23 for a D latch like that of Figure 6.19. The key point of reference is the time at which the latch begins to store or latch the current data signal; we refer to this as the **latching point**. In Figure 6.23, the latching point is the 1-to-0 (falling) transition of the enable signal C. Recall that this transition closes

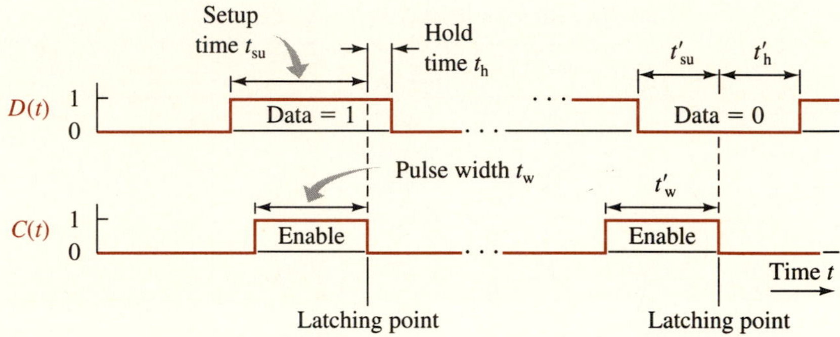

FIGURE 6.23 Definition of the setup time t_{su}, hold time t_h, and enable time t_w for a D latch.

the latch's feedback loop, capturing whatever data value the loop currently contains. Note also that C is usually connected to a periodic and precisely timed clock signal, whose rising and falling edges make excellent reference points.

Correct operation requires meeting minimum setup, hold, and enable times.

The **setup time** t_{su} is the time that the data signal D is held at a new value immediately before the latching point. It is measured from the latching point to the last transition of D that precedes the latching point. The setup time must not be less than a specified minimum t_{su}^{min} to ensure complete entry into the latch of the new D value, allowing for worst-case propagation delays. The **hold time** t_h is an additional period after the latching point, during which D is maintained at its new value. The hold time should not be less than a specified value t_h^{min} to ensure that all changes that result from deactivation of the latch stabilize before D is altered, again allowing for worst-case propagation delays. The third parameter is the **enable (pulse-width) time** t_w, which is the time during which C is at its active or enabling value. This enable time must be at least t_w^{min} to ensure that the latch's inertial delay is overcome, allowing it to execute a complete state change.

Violation of any of the minimum setup, hold, and pulse-width times causes indeterminate behavior in a latch. For example, an enable pulse that is too narrow ($t_w < t_w^{min}$) can result in the latch's failure to make a required state change. Insufficient setup or hold times can cause incorrect state changes by not allowing all the effects of the current data signal to propagate through the circuit.

Setup, hold, and enable times cannot be determined with much accuracy from a logic circuit like the one in Figure 6.20. Consequently, they are usually calculated from a more detailed (electronic) circuit model, or else are measured experimentally. In either case, they are normally included in the specifications of the latch used by the logic designer. For example, in Table 6.2 we see that the setup, hold, and enable parameters have the following minimum allowed values in the case of the 74LS75 D latch:

$$t_{su}^{min} = 20 \text{ ns} \qquad t_h^{min} = 5 \text{ ns} \qquad t_w^{min} = 20 \text{ ns}$$

These parameters are defined for most types of memories, including both latches and flip-flops.

States

So far we have used timing or waveform diagrams to represent the detailed behavior of the few sequential circuits we have encountered. Because the number of possible input/output waveforms is unlimited, it is desirable to have a concise, finite way of representing a sequential circuit's full range of behavior. Such a representation is provided by state tables and diagrams, which (1) identify all the key signal combinations or states that occur in the circuit, and (2) display all possible transitions between these states. What is defined as a state depends on the level of abstraction at which a circuit is analyzed. For example, sometimes we will be interested only in a circuit's stable states, and we exclude the unstable or transient states from consideration.

In general, the **state** of a circuit at time t is defined as the current logic values of some set of signals of interest. We distinguish two main cases:

The (internal) state is the set of stored data in a circuit.

1. The **internal state** $Y(t) = (y_0(t), y_1(t), \ldots, y_{p-1}(t))$ is the combination of the p internal signal values that represent all the information (at an appropriate level of abstraction) currently stored in the circuit. (Later we will drop the commas and outer parentheses, and write $Y(t)$ more concisely as $y_0(t)y_1(t)\ldots y_{p-1}(t)$.) The internal state is usually referred to simply as the *state* of the circuit.
2. The **total state** $X(t), Y(t) = (x_0(t), x_1(t), \ldots, x_{n-1}(t), y_0(t), y_1(t), \ldots, y_{p-1}(t))$ combines the internal state with the current value of all n externally applied or primary input signals.

The internal state represents what the circuit remembers from its behavior prior to time t. It is apparent that the total state completely determines the next action to be taken by the circuit, so that both X and Y can be regarded

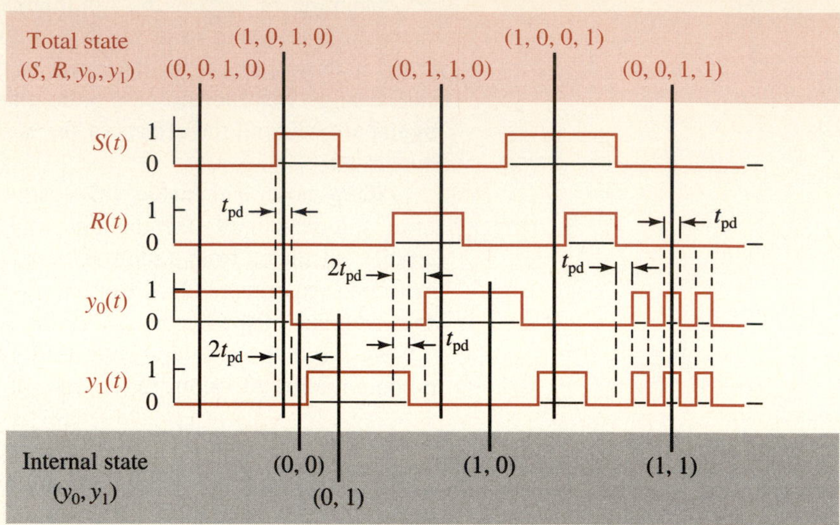

FIGURE 6.24 Identification of all internal states and some total states in an SR latch.

as inputs to the circuit. To distinguish the primary input variables X from the internal state variables Y, the latter are called **secondary input variables** or **state variables**.

Each combination of primary and secondary variable values determines a possible state. Consider, for example, the behavior of the basic SR latch, which is defined by timing diagrams in Figure 6.18; the same information is redrawn in Figure 6.24. The four possible combinations of the values of $(y_0(t), y_1(t))$ are identified as the latch's (internal) states; the total states correspond to the 16 possible combinations of $(S(t), R(t), y_0(t), y_1(t))$, not all of which appear in the figure.

State Tables and Diagrams

State tables define a memory's next-state functions.

A **state table** has a column for every combination $X(t)$ of primary input variables, and a row for every combination $Y(t)$ of secondary input variables—that is, for every internal state. If there are n primary input variables and p secondary input variables, a state table has 2^n rows and 2^p columns, giving a total of $2^n \times 2^p = 2^{n+p}$ entries in the table, one for every total state. Figure 6.25a shows a four-row-by-four-column state table for an SR latch, with the total state $(S(t), R(t), y_0(t), y_1(t))$. The entry in column $X(t)$ and row $Y(t)$ of a state table includes the next internal state $Y(t + \tau)$ that appears in response to the total state $X(t), Y(t)$, where τ is a suitable but unspecified time unit, such as an elementary propagation delay. In Figure 6.25a, for instance, τ is chosen to be the gate delay t_{pd}. State tables are also used to specify primary output signals $Z(t)$ in addition to next states, as we will see later. For the

		Current input $(S(t), R(t))$				
		(0, 0)	(0, 1)	(1, 0)	(1, 1)	
	(0, 0)	(1, 1)	(1, 0)	(0, 1)	(0, 0)	
Present state	(0, 1)	(0, 1)	(0, 0)	(0, 1)	(0, 0)	Next state
$(y_0(t), y_1(t))$	(1, 0)	(1, 0)	(1, 0)	(0, 0)	(0, 0)	$(y_0(t + t_{pd}), y_1(t + t_{pd}))$
	(1, 1)	(0, 0)	(0, 0)	(0, 0)	(0, 0)	

(a)

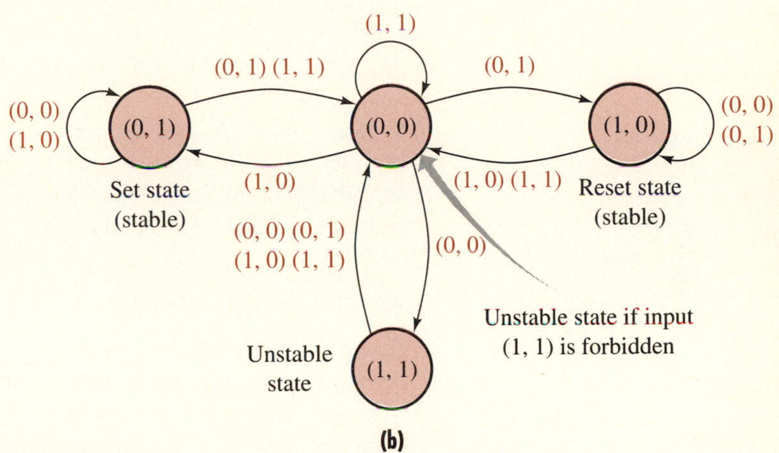

(b)

FIGURE 6.25 **(a) A detailed state table and (b) a state diagram for an SR latch, showing stable and unstable states and the response to $(S, R) = (1, 1)$.**

latches under consideration here, however, $Z(t)$ coincides with the internal state $Y(t)$, and so is not shown explicitly in the state table.

An entry $Y(t + \tau)$ in column $X(t)$ and row $Y(t)$ of a state table can be thought of as representing a **state transition**, symbolized by

$$Y(t) \xrightarrow{X(t)} Y(t + \tau) \tag{6.7}$$

and read as follows: If (primary) input $X(t)$ is applied to the circuit with (internal) state $Y(t)$ present, the next state will be $Y(t + \tau)$. The symbolic format of (6.7) leads to another, nontabular representation of state behavior. A **state diagram** is a figure that has circles (nodes) that represent all internal states, and arrows between nodes that represent all possible state transitions (Figure 6.25b). Each arrow of a state diagram is labeled with the value(s) of the primary inputs that produce the corresponding transition. (Mathematically speaking, a state diagram is a type of labeled, directed graph [Harary, 1969].) Thus, (6.7) corresponds to an arrow labeled $X(t)$ from node $Y(t)$ to node $Y(t + \tau)$ in the state diagram.

State tables and diagrams convey the same behavioral information.

State tables and diagrams defined at the same level of abstraction contain exactly the same information about a sequential circuit's behavior. Hence, given a state table, one can easily construct from it the corresponding state diagram, and vice versa. Using either representation, the sequence of states visited in response to any sequence of k consecutive primary input values is easily determined by identifying, in turn, the k corresponding entries in the state table or, equivalently, the k corresponding transitions in the state diagram.

State diagrams allow repetitive behavior (loops) to be easily visualized.

An advantage of a state diagram is that it shows certain behavioral features in a graphical form that humans can easily recognize. For example, repetitive behavior can be directly related to sequences of state transitions that form closed loops in a state diagram. A self-loop—that is, an arrow that starts and ends on the same node—represents a quiescent total state. For example, in the state diagram for the SR latch, we have five self-loops: $(0,1) \xrightarrow{(0,0)} (0,1)$, $(0,1) \xrightarrow{(1,0)} (0,1)$, $(0,0) \xrightarrow{(1,1)} (0,0)$, $(1,0) \xrightarrow{(0,0)} (1,0)$, and $(1,0) \xrightarrow{(0,1)} (1,0)$. State (1, 1) has no self-loops and so is inherently transient, meaning that if it is entered at time t, it must be exited immediately at time $t + \tau$, independent of the value of $X = (S, R)$. State tables have the advantage of conciseness and are more suitable than state diagrams for manipulation by CAD programs.

Example 6.5

Analyzing the state behavior of an SR latch

Figure 6.25 shows the complete state behavior of an SR latch that has no restrictions on the allowable input combinations. Considering the state diagram of Figure 6.25b, the leftmost state, labeled $(y_0, y_1) = (0, 1)$, denotes the set condition in which the latch is considered to be storing a 1. As indicated by the self-loop arrow from the (0,1) state back to itself, the latch's internal state remains stable at (0,1) as long as either of the input combinations $(S, R) = (0, 0)$ or $(S, R) = (1, 0)$ is applied to the circuit. Application of $(S, R) = (0, 1)$ or (1, 1) causes the state to change from (0,1) to (0,0) after some implicit time delay. Continued application of $(S, R) = (0, 1)$ causes the state to become (1,0), the reset condition. However, continued application of $(S, R) = (1, 1)$ to the latch when in state (0,0) causes the latch to lock into state (0,0), as shown by the self-loop marked (1,1). On the other hand, continued application of $(S, R) = (0, 0)$ to the SR latch, starting in state (0,0), causes it to oscillate back and forth indefinitely between the states (0,0) and (1,1), as implied by the arrows labeled (0,0) that link these two states. Note how the left-right symmetry of Figure 6.25b reflects the symmetry between the set and reset conditions. Note too that if the input $(S, R) = (1, 1)$ is forbidden, state (1,1) is never entered and state (0,0) reduces to an unstable state that is present briefly during any transition between the set and reset states.

Usually, we are not interested in a latch's unstable states.

Figure 6.26 shows a simplified—that is, more abstract—state table and state diagram for the SR latch. Here, the primary input combination $(S, R) = (1, 1)$ has been omitted, because it is not allowed when designing with SR latches. As noted above, this design constraint also eliminates the state (1,1). We are usually interested only in a latch's stable states; hence, we also omit the normal but unstable (0,0) state. With only the set and reset states remain-

		Current input $(S(t), R(t))$			
		(0, 0)	(0, 1)	(1, 0)	
Present state	0	0	0	1	Next state
$Q(t)$	1	1	0	1	$Q(t + 2t_{pd})$

(a)

(0, 1)

(0, 0)
(1, 0)

1

0

(0, 0)
(0, 1)

Set

(1, 0)

Reset

(b)

FIGURE 6.26 (a) A simplified state table and (b) a state diagram for the SR latch, showing only stable states and permissible input combinations.

ing, we now see that the single variable denoted by Q suffices to represent the latch's internal state; $Q = 1$ denotes the set state, and $Q = 0$ denotes the reset state. Note also that in this state description, the time unit for a single transition between two states has become $2t_{pd}$ rather than t_{pd}.

Latch Applications

SR and D latches are among the simplest and least expensive types of memory elements used in logic circuits. They are used as data storage elements in circuits in which there is no direct feedback from the outputs of any latch through an external circuit (as opposed to the internal feedback loop within a latch) to the latch's inputs. This requirement is met by the temporary storage needs of many data processing and input/output circuits. Figure 6.27 shows how a computer input device (a keyboard) and an output device (a display

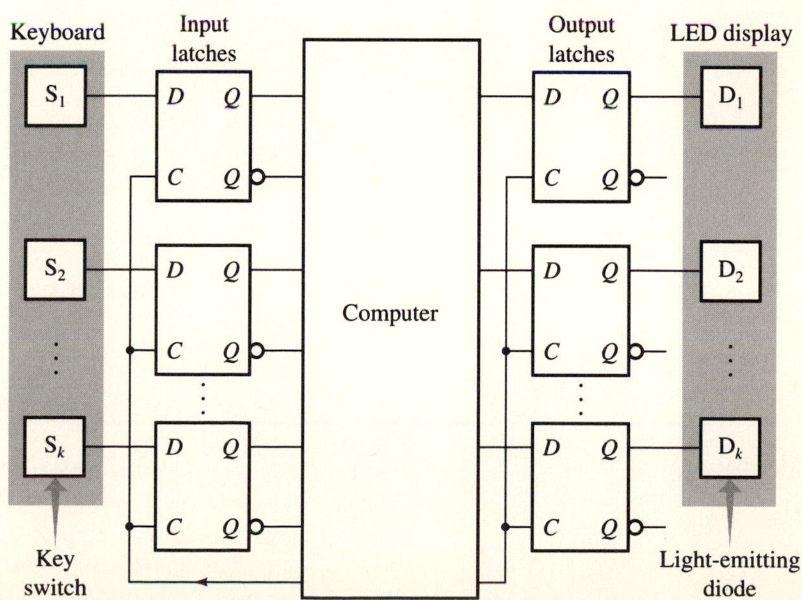

FIGURE 6.27 Typical use of latches in a computer's input/output circuits.

panel of light-emitting diodes, known as LEDs) employ latches. Each column of latches L has a common control line and serves as a temporary storage area for a data word en route from the input device to the computer, or from the computer to the output device. A data word is latched into L, where it remains until a new data word arrives to replace it.

Latches are suitable for circuits that have no complex feedback.

Latches are unsuitable as memory devices in circuits for which input signal changes can easily violate their setup, hold, or enable time constraints. Latches are also unsuitable for use in sequential circuits that have the feedback structure depicted in Figure 6.28, which allows the latch's outputs to feed back via some combinational logic to its inputs. This feedback structure, which is considered in detail later, is typical of control logic. When a latch in such a circuit is enabled, its output signals (the state values Q and $\overline{Q}$) can "race" around the feedback loop back to the D input, as indicated in the figure. Unfavorable combinations of propagation delays can then cause multiple state changes to occur, leading to an indeterminate new latch state. This race problem is a direct consequence of latch transparency, which allows an unlimited number of data changes to pass through an enabled latch. To restrict the memory elements to a single state change per enabling step, we need nontransparent memories such as flip-flops, which we consider next. Unlike a latch, a flip-flop can maintain its old state constant at its output terminals while a new state is being entered (written) into it via its input terminals. The inability to be read from and written into concurrently is a fundamental limitation of a latch.

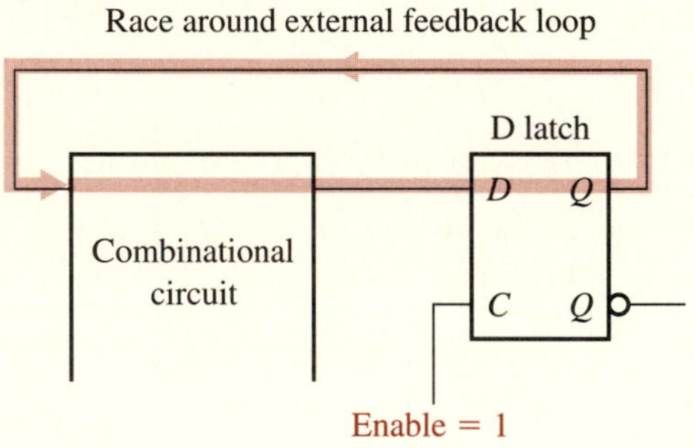

Race around external feedback loop

D latch

D Q

Combinational circuit

C Q

Enable = 1

FIGURE 6.28 A race condition due to latch transparency in a circuit that contains global feedback.

Clocks and Flip-Flops

Sequential logic circuits that have global feedback require a bistable memory element with more elaborate timing logic than that of a latch. This need is met by the *flip-flop,* which employs a clock signal to control precisely the times at which data are transferred to or from memory. Sections 6.5, 6.6, and 6.7 present the basic flip-flop types and analyze their behavior.

6.5 Basic Concepts

We pointed out in Section 6.3 that while a latch is enabled, changes to its input data signals flow in an unrestricted manner to the latch's data output lines. This transparency property is inherent in latches such as a D latch or a gated SR latch. Combined with the unavoidable uncertainty surrounding the propagation delays experienced by signals before and after they enter the latch, transparency leads to difficulties like the race condition of Figure 6.28, which restrict the range of applications to which latches are suited.

Consider a bistable memory circuit ME with data input(s) X, data output Q, and control (clock) input C. The number of data and control lines present, as well as the way the circuit is controlled, vary with the memory type, as we will see. When ME is suitably enabled by C, a state transition of the form

$$Q(t) \xrightarrow{X(t)} Q(t+\tau) \tag{6.8}$$

between two stable states is required to take place. We consider ME a **reliable** component in a particular circuit if (6.8) is always performed correctly, provided all ME's setup, hold, and enable time constraints are met. In other words, if $Q(t)$ and $X(t)$ are set to specific values, and ME is operated in the proper way, the next state $Q(t + \tau)$ is always the same. As in the latch case discussed in the preceding sections, the minimum setup, hold, and enable times are chosen to allow for the worst-case variations in the memory circuit's internal propagation delays. A latch operates reliably only in circuits that do not contain a feedback path from the latch's outputs to its inputs. More complicated memory elements (flip-flops) are needed to operate reliably in more general sequential circuits.

Reliable operation may require better timing control than that offered by a latch.

Flip-Flops

For the state transition (6.8) to take place reliably, it is essential that the new state value $Q_{\text{new}} = Q(t + \tau)$ not interfere with the old value $Q_{\text{old}} = Q(t)$ in the course of the state change operation. Similarly, multiple changes that affect $Q(t)$ during a state transition should not cause an incorrect final state $Q^*(t + \tau) \neq Q(t + \tau)$. The memory circuit must have $Q(t)$ available unchanged as Q_{old} while it computes Q_{new}, which is some logic function of Q_{old} and $X(t)$ that eventually replaces Q_{old}.

A **flip-flop** is defined as a bistable memory circuit that employs a special control signal C (occasionally, several such signals) to specify the times at which the memory responds to changes on its input data signals, and the times at which the memory changes its output data signals, so that reliable, nontransparent operation is achieved. Because of C's role as a timekeeping or synchronizing signal for the flip-flop, it is called a **clock signal**. The enable signal C of a D latch (Figure 6.19) or a gated SR latch (Figure 6.22) is a simple type of clock signal, and is generally referred to by that name in the literature.

A flip-flop is a bistable memory circuit with precise timing control.

Clocking Methods

Many ways of designing and controlling flip-flops have evolved over the years; they differ in their logic design details—especially how they use their enable or clock signal C [Langdon, 1974]. The role of C in a latch is illustrated by Figure 6.29a. During the period $t_1 : t_2$ when the clock is enabled $(C = 1)$, any changes made to a data signal may enter the latch immediately. After some propagation delay, these changes affect the latch's data output Q (and also $\overline{Q}$) during the period $t_3 : t_4$. Thus, ignoring the brief and somewhat uncertain transition periods when the data and clock signals are actually changing value, the latch responds to all input changes that occur when C is at the 1 level. It is unaffected by input changes that occur when C is at the inactive 0 level. For these reasons, latches are said to be **level-sensitive** or **level-triggered**.

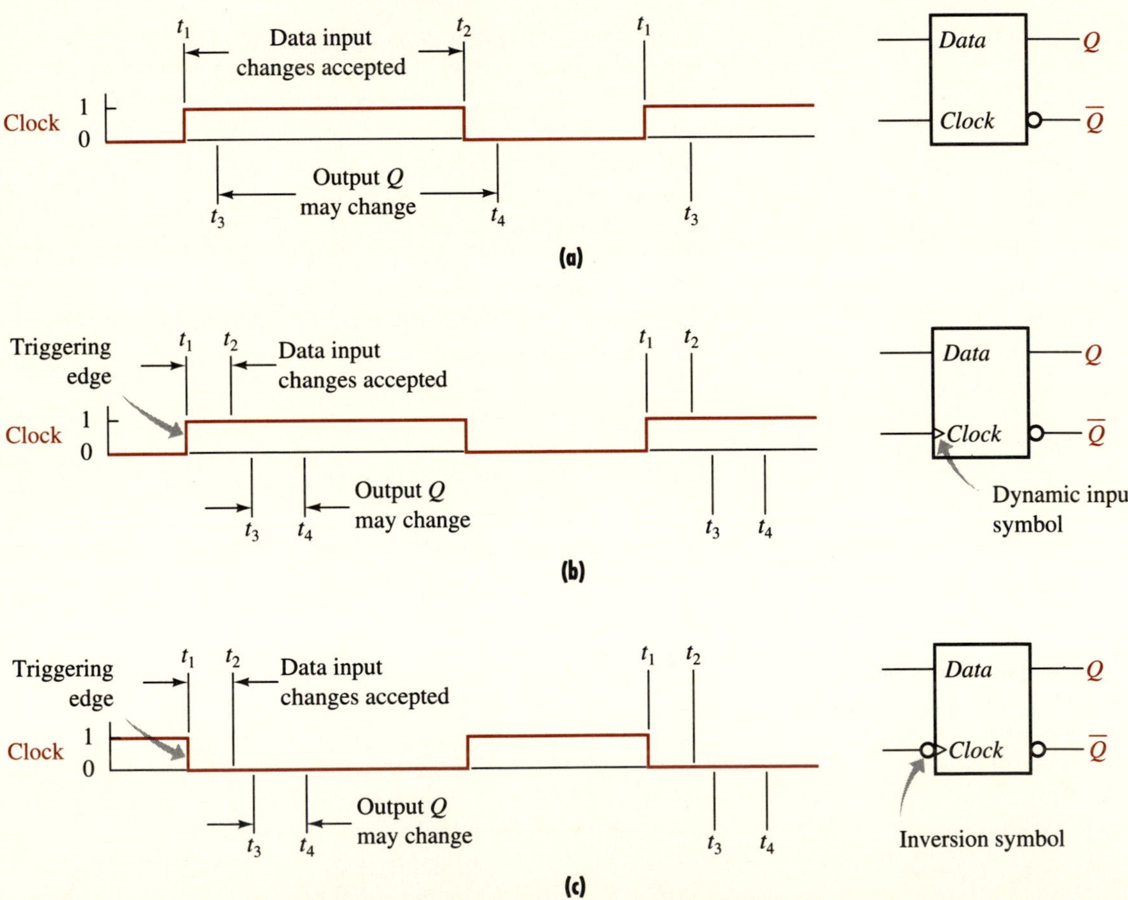

FIGURE 6.29 Basic clocking methods and symbols: **(a)** a level-sensitive latch; **(b)** a positive edge-triggered flip-flop; **(c)** a negative edge-triggered flip-flop.

To obtain flip-flop behavior, we must ensure that the period $t_1 \colon t_2$ (when input data changes are accepted) and the period $t_3 \colon t_4$ (when the output data changes) do not overlap. One way a flip-flop can meet this requirement is by accepting input changes when $C = 1$, and changing its output when $C = 0$. This "pulse" mode of operation was used in some early designs for bistables, and will be illustrated later. The clocking method most commonly used in modern flip-flop design is **edge triggering**, in which a transition or edge of the clock signal C causes the actions required in $t_1 \colon t_2$ and $t_3 \colon t_4$ to take place, as shown in Figures 6.29b and 6.29c.

Flip-flops use the clock edge to time input and output changes (edge triggering).

Figure 6.29b defines the behavior of a **positive edge-triggered flip-flop**, which uses the positive or rising edge of the clock to initiate the entire state transition process. The data input signals present when the clock goes from 0 to 1 are captured and trigger a sequence of internal events, culminating with the data outputs Q and $\overline{Q}$ assuming new values. Note that, as in the latch case, the data outputs are held constant at all times *except* $t_3 \colon t_4$. Unlike the latch, however, the flip-flop's data outputs are constant while it responds to changes on its input data lines. The internal propagation delays of the edge-triggered flip-flop ensure that the $t_1 \colon t_2$ and $t_3 \colon t_4$ periods do not overlap, so that the old state $Q(t)$ can be read while a new state $Q(t + \tau)$ is being written. A **negative edge-triggered flip-flop** behaves just like the positive kind, except that the negative or falling edge of the clock initiates the state transition process (Figure 6.29c).

Flip-flops have the same box symbols as latches, with small but crucial modifications to specify their clocking modes. The wedge-shaped **dynamic input symbol** $>$ is placed at the end of the clock line in the flip-flop symbol to indicate positive edge triggering. It implies that application of a positive edge (0-to-1 signal change) to C produces a transient 1-signal inside the flip-flop that enables a state transition. In other words, for a brief period following the arrival of a triggering clock edge, the flip-flop is enabled; the duration of this enabled period P depends on the flip-flop's internal structure and delays. C has no effect on an edge-triggered flip-flop at times other than P. Negative edge triggering is indicated by combining the dynamic input symbol with the inversion symbol, as shown in Figure 6.29c.

The symbol $>$ on a clock input denotes edge triggering.

Noise Sensitivity

Edge-triggered flip-flops are much less sensitive than latches to spurious and short-lived 0- and 1-pulses that are a normal feature of real digital circuits and are collectively known as **noise**. Figure 6.30 illustrates a major source of noise signals called **glitches**. Glitches, which we consider further in Section 6.8, are caused by unfavorable combinations of signal switching times and uncertain propagation delays.

Edge triggering reduces sensitivity to noisy input signals.

Suppose that the signals x_1 and x_2 are intended to be complementary, so that they change values simultaneously at the indicated times t_1, t_2, t_3, t_4. Because of variations in propagation delays, these signals do not always switch at *exactly* the same time, so there may be brief periods when x_1 and

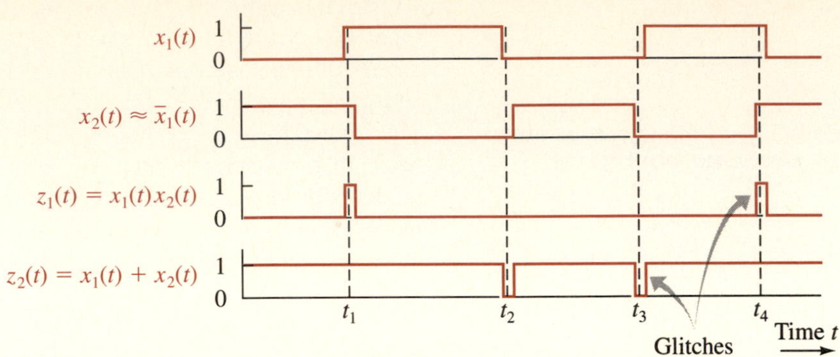

FIGURE 6.30 Noise signals (glitches) on lines z_1 and z_2.

x_2 have the same value. For instance, at t_1, x_2 switches from 1 to 0 a little after x_2 switches from 0 to 1. Thus, x_1 and x_2 are both 1 for a short time. An AND gate with inputs x_1 and x_2 and output z_1 may react to this situation by producing a brief glitch in the form of a 1-pulse, as shown in the figure. A similar glitch is produced by the AND gate at t_4, but not at t_2 or t_3. An OR gate with the same inputs and the output z_2 can produce spurious 0-pulses at the latter two times, however, because it has 0 applied briefly to both its inputs. In general, any gate can generate a glitch in response to certain closely spaced input signal changes.

Noise signals are inherently transient. However, they produce permanent errors in a digital circuit if they are captured by a memory circuit. Because edge-triggered flip-flops are sensitive to input changes for only a brief period of time, they are immune to noise for most of the clock cycle, and prevent any glitches that appear at their inputs from being transferred to the flip-flop outputs. Latches, on the other hand, accept noise signals for the entire active period of the clock and, because of their transparency, permit glitches to pass to their outputs unchecked. These differences are seen clearly in Figure 6.29, in which the flip-flops have a much narrower time window $t_1{:}t_2$ than the latch, during which they respond to changes on their input data lines. Furthermore, any noise-producing output signal changes occur in an edge-triggered flip-flop after t_2, when the noise cannot affect the flip-flop's behavior. Hence, edge triggering confers a high degree of noise immunity on logic circuits that use this clocking philosophy.

D Flip-Flops

A **delay flip-flop** or **D flip-flop** is the counterpart of a D latch, in which a nontransparent clocking mode is used. An edge-triggered D flip-flop can be constructed from a pair of D latches connected as in Figure 6.31a. This design is principally used in CMOS circuits such as the 74HC74. (TTL logic circuits favor an alternative D flip-flop design, presented in Section 6.6.) This positive edge-triggered flip-flop employs two D latches and an inverter.

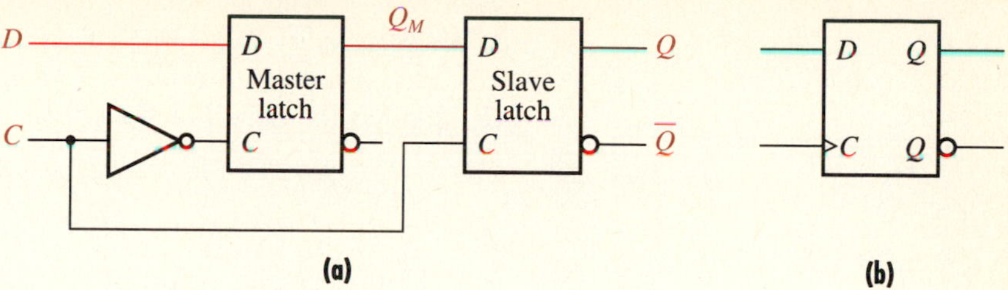

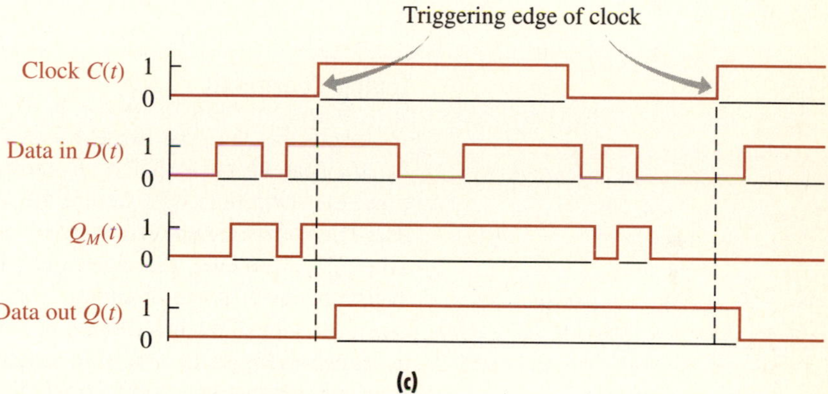

FIGURE 6.31 A positive edge-triggered D flip-flop: (a) its logic circuit; (b) its graphic symbol; (c) its timing diagram.

A master-slave configuration of D latches forms an edge-triggered D flip-flop.

The external clock signal C is connected through the inverter to the clock input of the first D latch, known as the "master" latch, while C is connected directly to the second "slave" latch. The flip-flop's input data signal is applied to the master's data input line; the flip-flop's output data signals Q and $\overline{Q}$ are taken from the corresponding outputs of the slave. As might be expected, this particular configuration is known as a **master-slave circuit**. Note that the $\overline{Q}$ output of the master latch is not needed.

The D flip-flop of Figure 6.31 operates as follows: When the external clock signal C is 1, the master latch is disabled, so its output line Q_M remains stable. Because Q_M is also the data input to the slave latch, the latter's data outputs remain stable, even though the slave is enabled by $C = 1$. When C changes to 0, the slave latch is disabled, so its data outputs remain unchanged. However, the master is now enabled, and it begins to respond transparently to data signals applied to D. The propagation delays through the clock inverter and the master's internal circuits ensure that the slave is fully disabled before any changes pass through the master to the slave's input data line. Arbitrary changes may occur on D with no effect on Q and $\overline{Q}$, thanks to the complementary clock signals applied to the latches.

Finally, when C changes from 0 to 1, the master captures the data on D and becomes disabled. At the same time, the slave becomes enabled. The value of D then appears on Q_M and is transferred through the slave to Q and, in complementary form, to $\overline{Q}$. Consequently, the output data signal $Q(t)$ of a D latch is a delayed version of an earlier input signal $D(t - t_1)$, so the D flip-flop has the same functional behavior

$$\begin{aligned} Q(t) &= D(t - t_1) \\ \overline{Q}(t) &= \overline{D}(t - t_2) \end{aligned} \tag{6.9}$$

as a D latch, despite major differences in their clocking methods.

Timing Constraints

Like latches, flip-flops must satisfy setup, hold, and enable time constraints.

To ensure that the slave does not "see" multiple input data changes through the transparent master latch, D must be held constant for some minimum time, the setup time $t_{\mathrm{su}}^{\mathrm{min}}$, before the triggering clock edge arrives. The input data D must also remain constant for some minimum time afterward, the hold time $t_{\mathrm{h}}^{\mathrm{min}}$. Moreover, the clock must be held at 1 for a minimum time $t_{\mathrm{w}}^{\mathrm{min}}$. Thus, the same types of timing parameters used to specify the operational characteristics of latches (Figure 6.23) also apply to edge-triggered flip-flops, as shown in Figure 6.32 for the case of a positive-edge triggered D flip-flop. The two positive edge-triggered D flip-flops that appear in the 74LS74A IC have the following timing parameters [Texas Instruments, 1988b]:

$$t_{\mathrm{su}}^{\mathrm{min}} = 20 \text{ ns} \qquad t_h^{\mathrm{min}} = 5 \text{ ns} \qquad t_{\mathrm{w}}^{\mathrm{min}} = 25 \text{ ns}$$

The setup and hold times of a flip-flop define a window around the triggering edge during which the input data must be held steady. If, for example, the setup time is insufficient, as illustrated on the right side of Figure 6.32, the flip-flop may see $D = 0$ or $D = 1$, depending on its internal propagation delays, and the next state of Q will be uncertain.

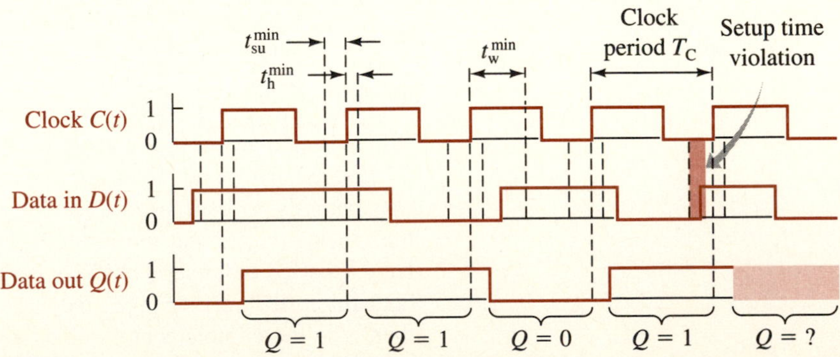

FIGURE 6.32 Operation of a positive edge-triggered D flip-flop, with a periodic clock signal.

Synchronous Operation

Figure 6.32 illustrates a typical sequence of state changes that occur in a flip-flop. The controlling clock signal is **periodic**, meaning that it goes to 1 and then to 0 in a regular pattern called a **clock cycle** that repeats every T_C seconds, where T_C is the **clock period**. The durations of the 0 and 1 periods within a clock cycle are determined by the flip-flop's own timing parameters, as well as by the propagation delays of any external logic circuits connected to the flip-flop. In particular, we must have $C = 1$ for at least t_w seconds, where $t_w \geq t_h$, and we must have $C = 0$ for at least t_{su} seconds. The roles of $C = 0$ and $C = 1$ are reversed in the case of a negative edge-triggered flip-flop.

Synchronous logic circuits are timed by periodic clock signals.

If all the flip-flops of a logic circuit are properly clocked by the same periodic clock signal C, the flip-flops are said to be operating **synchronously**, and the circuit is referred to as a **synchronous sequential circuit**. As we see from Figure 6.32, synchronous operation causes a flip-flop to have a single well-defined stable state, indicated by $Q = 0$ or $Q = 1$, in every clock cycle. This state is present at all times, except for a brief transition period that follows the triggering edge of the clock. Thus, instead of worrying about the flip-flop's state at every instance of time, it is sufficient to view the state once per clock cycle. Hence, a continuous (analog) sequence of state values can be replaced by a discrete sequence of discrete state values defined at discrete times $t = 1, 2, 3, \ldots, i, \ldots,$ where $t = i$ refers to the ith clock cycle. Furthermore, the clock period can be made a constant T_C that is independent of a circuit's internal propagation delays. This simple model of timing behavior is a key reason most sequential circuits are designed to operate synchronously.

Initialization

Finally, it might be asked, what determines the state $Q = 0$ or 1 of a flip-flop or latch when operation begins? Merely switching on the power to the circuit does not guarantee that a flip-flop will assume a known initial state. An individual flip-flop may "power up" to a fixed state, but two different flip-flops of the same type may always power up to different states. It is desirable, therefore, to be able to initialize every flip-flop to a known starting state after the circuit has been powered up.

Flip-flops are usually designed with one or two extra control inputs intended only for state initialization. A control signal that forces the flip-flop into the state $Q = 0$ is referred to as a **clear input**; one that forces the flip-flop into the state $Q = 1$ is called a **preset input**. These control signals operate independent of the clock signal, the effect of which they override. They do not depend on the clock in any way, so preset and clear inputs are said to be **asynchronous inputs**, whereas the clock-controlled D input is

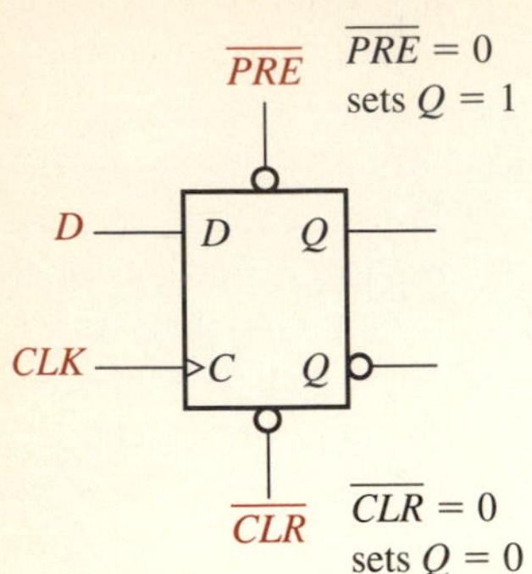

FIGURE 6.33 The symbol for a positive edge-triggered D flip-flop with preset and clear.

a **synchronous input**. The basic flip-flop circuits presented earlier are easily modified to incorporate preset or clear lines; examples are given in Section 6.6.

Figure 6.33 shows a graphic symbol for the D flip-flop used in the 74LS74A, which has both preset and clear inputs. In this particular case, preset and clear are active when set to 0 and inactive when set to 1; hence, the use of an inversion symbol in these lines. Thus, to initialize the flip-flop to $Q = 0$, a 0-pulse is applied to the clear line $\overline{CLR}$, and the preset line $\overline{PRE}$ is held at 1. To obtain an initial $Q = 1$, a 0-pulse is applied to $\overline{PRE}$, and $\overline{CLR}$ is held at 1. The minimum duration of these initialization pulses, which is typically about t_w^{min}, is specified in logic data books. During subsequent synchronous operation, $\overline{CLR}$ and $\overline{PRE}$ must be held static at the 1 (inactive) level, and play no further role in flip-flop operation.

JK Flip-Flops

Practical set-reset flip-flops are of the edge-triggered JK type.

We could, in principle, construct an **SR flip-flop**, which has the same state behavior as an SR latch but with edge-triggered clocking. Such a flip-flop is never used in practice, however, because of its anomalous behavior when $(S, R) = (1, 1)$, inherited from the SR latch. Practical set-reset flip-flops are designed in a form called the **JK flip-flop**,[5] which has two data inputs, denoted J and K, that serve as the set and reset lines, respectively. It also has the usual Q and $\overline{Q}$ output lines, as well as a clock input line C.

Figure 6.34 shows the symbol for a positive edge-triggered JK flip-flop, along with a (stable) state diagram and a timing diagram illustrating its behavior. As with all edge-triggered flip-flops, the next state is determined by the values of the data inputs that are set up when the triggering clock edge appears. These values are captured from J and K and then used in a nontransparent way to change the flip-flop's Q and $\overline{Q}$ output signals.

The state behavior of the JK flip-flop is similar to that of an SR flip-flop in most respects, with the J and K inputs corresponding to S and R, respectively. Thus, $(J, K) = (0, 0)$ is a quiescent input condition that causes no change of state. The input combinations $(J, K) = (1, 0)$ and $(J, K) = (0, 1)$ cause the flip-flop to enter the set ($Q = 1$) and reset ($Q = 0$) states, respectively. The JK design deviates from SR behavior in its response to the input combination $(J, K) = (1, 1)$. Recall that $(S, R) = (1, 1)$ is forbidden in SR latches and flip-flops, because it produces an indeterminate next state. The (1,1) input combination causes the JK flip-flop to change to a determinate

[5]The origin of the name JK is obscure. Although various people with the initials J.K. have been credited with inventing the JK flip-flop, the JK name appears to have originated at Hughes Aircraft Company in the 1950s as an arbitrary choice of letters assigned to the data inputs of a "Hughes type" of flip-flop [EDN, 1968].

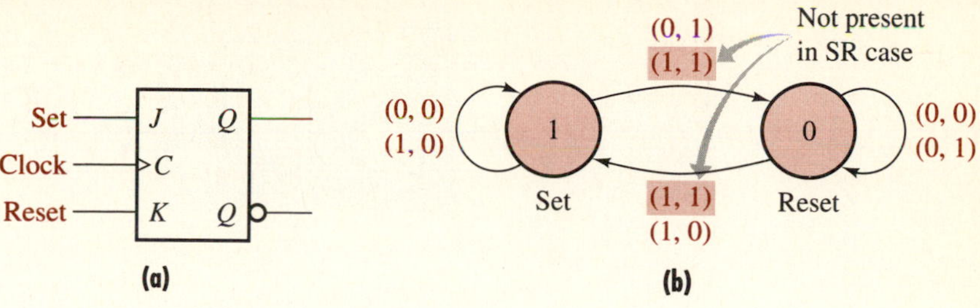

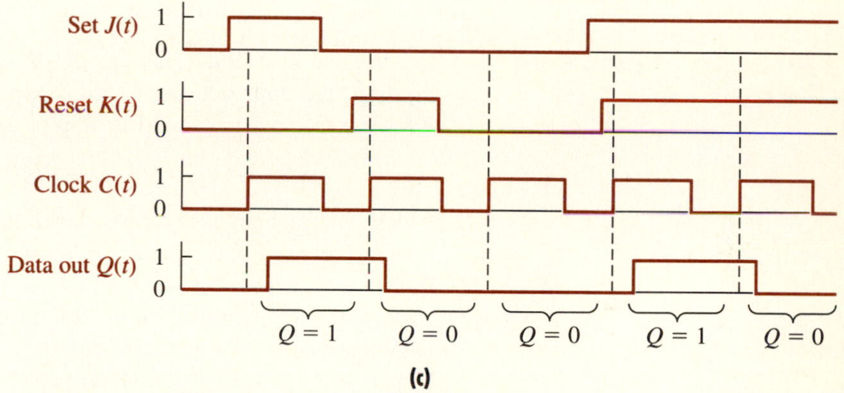

FIGURE 6.34 A positive edge-triggered JK flip-flop: (a) its graphic symbol; (b) its state diagram; (c) its timing diagram.

next state, and there are no restrictions on its use. If $Q = 0$, application of $(J, K) = (1, 1)$ has the same effect as $(J, K) = (1, 0)$ and sets the state to $Q = 1$. Similarly, if $Q = 1$, then $(J, K) = (1, 1)$, like $(J, K) = (0, 1)$, resets the state to $Q = 0$.

JK Flip-Flop Design

We now briefly discuss the design of JK flip-flops; this topic is pursued further in Section 6.6. Figure 6.35a shows how a hypothetical SR flip-flop can be converted into its JK equivalent. The two added AND gates permit the current state values Q and $\overline{Q}$ to change an externally applied $(J, K) = (1, 1)$ pattern into (1,0) if the SR flip-flop is reset, or into (0,1) if the SR flip-flop is set. Thus, the (1,1) signal cannot reach the S and R inputs, so the anomalous behavior inherent in the SR design is avoided. For the remaining three combinations of J and K, we obtain $(J, K) = (S, R)$, and the behaviors of the JK and SR flip-flops coincide. Note that if (J, K) is held at (1,1), the flip-flop changes state every clock cycle, which is sometimes called **toggling**. If the dynamic input symbol on the C is dropped from Figure 6.35a, the resulting circuit becomes a **JK latch**.

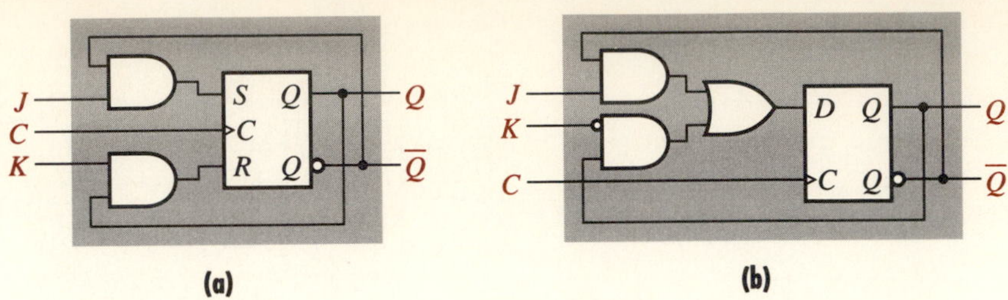

FIGURE 6.35 A positive edge-triggered JK flip-flop based on (a) an SR flip-flop and (b) a D flip-flop.

A D flip-flop is easily converted to JK form (and vice versa).

We can also obtain the JK flip-flop of Figure 6.35b by modifying a D flip-flop, the approach used for JK flip-flops in the 74HC series of CMOS ICs. The D input to the D flip-flop is defined by the Boolean equation

$$D = J\overline{Q} + \overline{K}Q \tag{6.10}$$

which is implemented by a two-level combinational circuit, as shown in the figure. The validity of (6.10) can be seen as follows: When $(J, K) = (0, 0)$, Equation (6.10) reduces to $D = Q$, so the next state will be the same as the current state, as required for the JK flip-flop's quiescent condition. The input combinations $(J, K) = (1, 0)$ and $(J, K) = (0, 1)$ result in $D = \overline{Q} + Q = 1$ and $D = 0 + 0 = 0$, as required to change Q to 1 and 0, respectively. Finally, when $(J, K) = (1, 1)$, Equation (6.10) becomes $D = \overline{Q}$, so that the flip-flop changes state.

6.6 Flip-Flop Design*

Some additional examples of D and JK flip-flop designs are presented here. Two new clocking methods for flip-flops, *pulse triggering* and *multiphase clocking*, are also described.

Master-Slave JK Flip-Flops

The double-latch master-slave construction of the D flip-flop in Figure 6.31a can also be applied directly to JK flip-flop design. The resulting circuit (Figure 6.36a) has a relatively simple structure and was used in early computer designs. It employs two SR latches, the first of which (the master) is modified in the manner of Figure 6.35a to make it respond to the (1,1) input data combination by toggling in the JK manner. Because the master never produces (1,1) on its output lines, an SR rather than a JK latch suffices as the slave. As in the master-slave D flip-flop, the slave latch here holds the main

*This section may be omitted without loss of continuity.

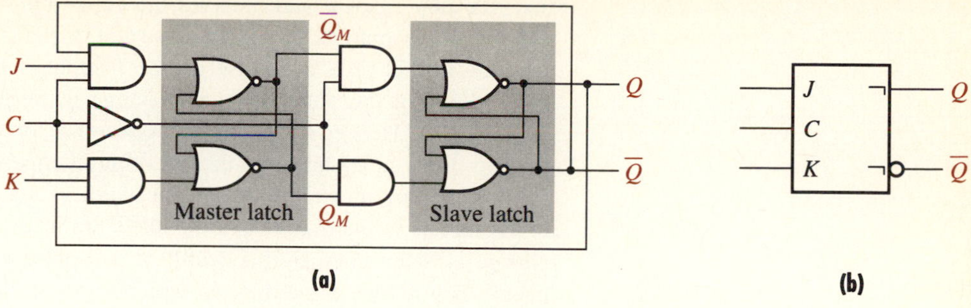

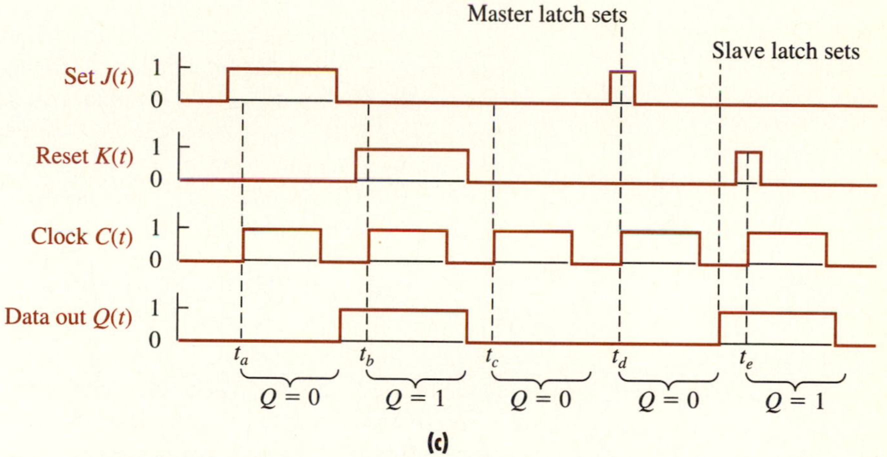

FIGURE 6.36 A master-slave JK flip-flop: (a) its logic circuit; (b) its graphic symbol; (c) its timing diagram.

data outputs of the flip-flop constant while the master latch changes state. After its data inputs have been disabled, the master transfers its state to the slave. However, unlike the master-slave D flip-flop, this design is *not* edge-triggered. It employs a different clocking mode known as *pulse triggering*, described below.

Pulse Triggering

A master-slave JK flip-flop composed of SR latches is pulse-triggered.

The key characteristic of **pulse triggering** is that the events that determine a flip-flop's new state can span the entire width of a clock pulse; they are not clustered around a single clock edge, as in edge-triggered circuits. To see this, consider the behavior of the flip-flop depicted in Figure 6.36c. At time t_d, the external clock C changes from 0 to 1, while the flip-flop is in the reset state $Q = 0$. This enables the master latch and, because $(J, K) = (1, 0)$ at this point, the master immediately changes to the set state. If, as indicated in Figure 6.36c, (J, K) subsequently becomes $(0, 0)$, the master latch remains set. When C eventually returns to 0, the master becomes disabled. The slave is then enabled and receives the (1,0) input from the master, causing the flip-flop's outputs to change to $(Q, \overline{Q}) = (1, 0)$. The data setting the master—in this case, a 1-pulse on the J input—is accepted when the clock goes to 1, but the outputs do not change in response until after the clock returns to 0. Hence, between the appearance of input data that determine the flip-flop's next state and the resulting change on the flip-flop's outputs, there is a delay roughly equal to the width of a 1-pulse of the clock. If the clock signals to the master and the slave are inverted, the duration of a state change is determined by the width of a 0-pulse of the clock. This process clearly is not edge triggering, in which the time taken to change the overall state is unaffected by the duration of the clock pulses, or by changes in J or K while $C = 1$.

In general, the master latch of Figure 6.36 can respond to input changes any time $C = 1$, but the output does not change until C becomes 0. Although this is also true of the primary output changes in the master-slave D flip-flop (Figure 6.31), the latter's master latch captures the data on its D input only during a narrow window around the triggering edge. Because the JK flip-flop has both active and quiescent input combinations, it can capture an active data condition that occurs long before the negative edge of the clock appears, and have quiescent input data values when the time comes for the outputs to change, as illustrated for times t_d and t_e in Figure 6.36c. Consequently, there is no fixed delay between the output changes of a pulse-triggered flip-flop and the input data changes that trigger them. In contrast, the master-slave D flip-flop has no quiescent data values, so this situation cannot arise. To ensure that the master-slave JK flip-flop changes state reliably, it is necessary to hold the data inputs constant for the entire duration of the active clock pulse, which implies that $t_{su} \geq t_W$. This reliable mode of operation is illustrated by the state changes initiated at times t_a, t_b, and t_c in Figure 6.36c. Although slightly more complex in structure, edge-triggered JK designs are preferable to pulse-triggered designs because of the former's more precise timing behavior.

Pulse triggering has been superseded by edge triggering.

The pulse-triggered mode of clocking is indicated by a special IEC 617-12 symbol ¬, called the **postponed output symbol**, which is attached to the Q and $\overline{Q}$ outputs of the flip-flop symbol, as in Figure 6.36b. The dynamic input symbol >, which indicates edge triggering, must not be used in this case. The reader is cautioned that this master-slave JK flip-flop is sometimes incorrectly referred to as edge-triggered in the literature, and the postponed output symbol is often omitted, especially in pre-1985 publications.

Summary of Clocking Modes

We have seen three distinct ways of clocking memory elements: *level triggering,* which applies to latches only; *edge triggering,* which is the normal triggering mode for flip-flops; and *pulse triggering,* which also applies to flip-flops but is rarely used nowadays. A fourth flip-flop clocking mode, called *data lockout,* is also recognized by the IEC 617-12 standard and is described in Problem 6.26.

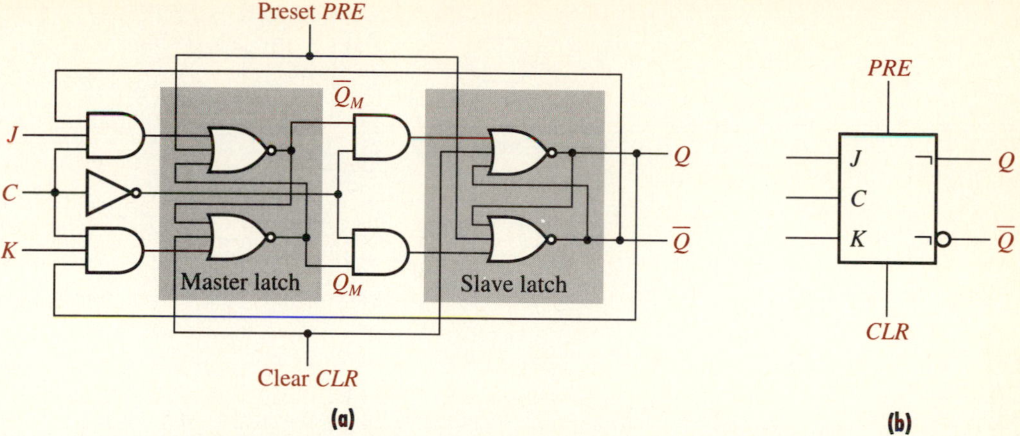

FIGURE 6.37 A master-slave JK flip-flop with preset and clear: (a) its logic circuit; (b) its graphic symbol.

Asynchronous preset and clear inputs are easily added to the master-slave JK design, as shown in Figure 6.37. These signals are designed for use in the quiescent state with $J = K = 0$, but operate independently of the value of the clock signal C. The clear line CLR is connected to the NOR gates that feed the true (Q) output of both the master and the slave latches. Making $CLR = 1$ forces Q to 0 in each latch. Subsequently, depending on the value of C, both $\overline{Q}$ outputs go to 1. CLR must be maintained at the 1 level until the latches have time to respond, which is approximately the minimum clock pulse width t_w^{min}. CLR must be reset to 0 for normal synchronous operation of the flip-flop. The preset input PRE is connected to the two latches in a similar way to force their $\overline{Q}$ outputs to 1.

Another JK Flip-Flop

We conclude our discussion of JK flip-flop structure by examining the circuit used in most recent TTL implementations. This elegant edge-triggered design does not employ master-slave latches. Instead, the output data values are held in a single latch L, and input circuits with carefully designed propagation delays are used to achieve edge triggering.

Example 6.6

Negative edge-triggered JK flip-flop

The specific JK flip-flop presented here is used in the 74LS107A and other 74X-series TTL ICs [Texas Instruments, 1988b]. It is classified as a "negative edge-triggered JK flip-flop with clear," and has the logic circuit and graphic symbol that appear in Figure 6.38. It contains a single NOR-type SR latch L, which is the flip-flop's data storage circuit, along with gating circuits that ensure JK rather than SR behavior and that make the flip-flop edge-triggered.

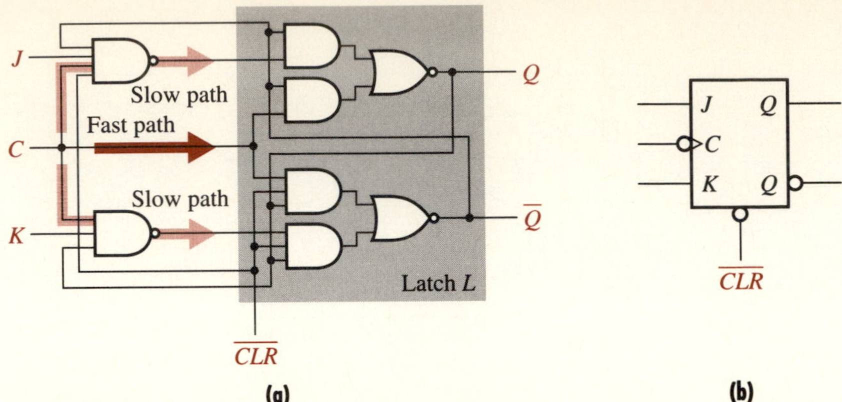

FIGURE 6.38 The negative edge-triggered JK flip-flop with clear used in the 74LS107A: (a) its logic circuit; (b) its graphic symbol.

A key feature of this type of flip-flop is the way differing internal gate delays are employed to achieve edge triggering. The clock signal C has a direct "fast" path to the latch L, which appears in the center of Figure 6.38, and "slow" paths to L via the two NAND gates. These NANDs are designed to have significantly longer propagation delays (by a factor of four or so) than the other gates in the flip-flop. When C changes from 1 to 0, a 0 clock signal is applied immediately to L via the fast path, causing the latch to respond to the currently set up J and K data signals. After sufficient time has elapsed for L to change state, the effect of the 1-to-0 change in C reaches L via a slow path and blocks any further state changes.

Figure 6.39 illustrates the behavior of the JK flip-flop in detail; the asynchronous clear logic has been removed to simplify the diagram. In this analysis, we assume that the two NAND gates G_1 and G_2 have propagation delay 4τ, and that all the other gates have delay τ. (The exact values of the propagation delays are unimportant, and minor deviations from these nominal delay values will not adversely affect flip-flop operation.) The reset state of the flip-flop with the quiescent data inputs $(J, K) = (0, 0)$ is depicted in Figure 6.39a. When the clock signal C is held at 0 or 1, the outputs of NAND gates G_1 and G_2 are both 1, and the signals in the feedback loop of the latch L can pass through AND gates G_3 and G_6. Thus, as long as C is stable, the latch L (and therefore the entire JK flip-flop) remains stable. This, of course, is required behavior for any edge-triggered flip-flop: L should change only when C makes a transition in the designated direction.

Now suppose that J changes from 0 to 1, followed by a similar change from 0 to 1 by C at time $t = 0$ (Figure 6.39b). The colored lines in the figure change value as indicated at the sequence of encircled times $t = \tau, 4\tau, 5\tau$ determined by the various gate propagation delays. The lower input of NOR gate G_7 changes from 0 to 1 at $t = \tau$ due to the effect of the clock-signal change propagated through G_4. At $t = 5\tau$, the upper input of G_7 changes

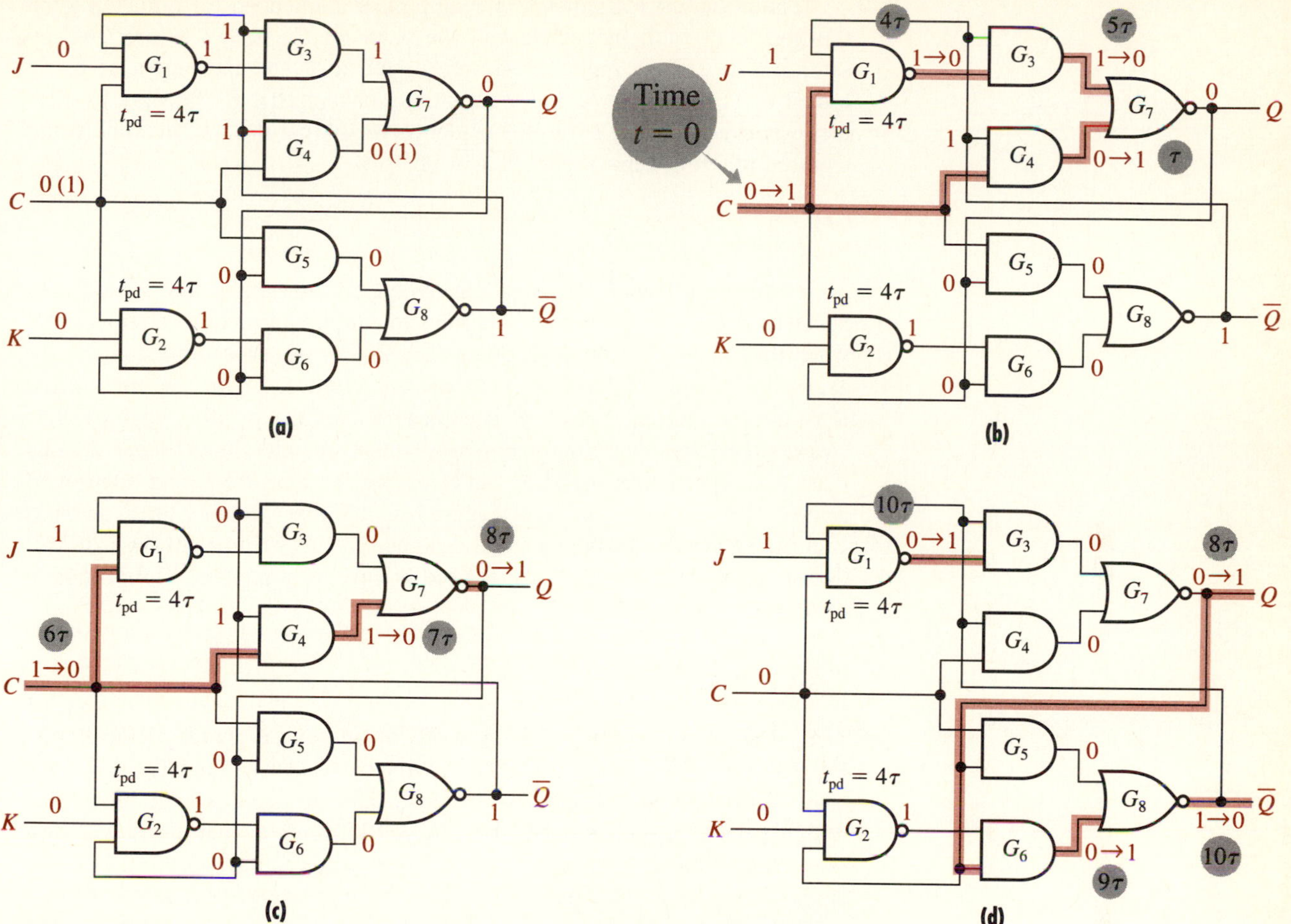

FIGURE 6.39 JK flip-flop behavior: (a) its reset state; (b), (c), and (d) steps in a set operation.

from 1 to 0 due to the effect of the clock signal propagated through gates G_1 and G_3 via a slow path. The result is that the output of G_7 remains at 0, so the latch does not change state as long as J and C are held at 1.

Next, let C change from 1 to 0 at $t = 6\tau$, as in Figure 6.39c, so that a triggering clock edge is applied to the JK flip-flop. The sequence of signal changes depicted in the figure occurs, affecting the colored lines at the indicated times and resulting in the flip-flop's Q output becoming 1 at $t = 8\tau$. Figure 6.39d shows the remaining sequence of actions as the new value of Q is fed back to the other half of L. At $t = 10\tau$, $\overline{Q}$ changes from 1 to 0 and the flip-flop enters the set state. At about the same time, the output of NAND gate G_1 becomes 1 in response to C and the new set state is locked in. The flip-flop is stable under the indicated input combination, and remains stable if J returns to 0.

The clear line $\overline{CLR}$ shown in Figure 6.38 is intended for initializing the flip-flop to $Q = 0$, independent of the value of C. If $\overline{CLR}$ is activated by setting it to 0, both inputs of G_8 are forced to 0 and the output of G_1 is forced to 1, making $\overline{Q} = 1$ and eventually causing the JK flip-flop to enter the reset state. We leave it to the reader to verify that this circuit exhibits the required toggling behavior when $J = K = 1$.

A different way of designing edge-triggered flip-flops is to insert ungated latches into the clock or input data lines to capture and retain certain signals when they change in the appropriate (positive or negative) direction. If, for example, a signal C is applied to the set input S of an SR latch while $R = Q = 0$, the latch sets (Q becomes 1) after the positive edge of C is received, and remains set for any subsequent changes in S. Other circuits must be used to reset the latch via R before the next triggering edge of C arrives. The result is an edge-triggered circuit that contains three or more latches. SR or JK flip-flops can be designed in this manner [Mowle, 1976], but the multiple-latch technique is used mainly to implement D flip-flops in certain IC technologies.

Another D Flip-Flop

The master-slave design of Figure 6.32 for an edge-triggered D flip-flop is less suited to TTL than to CMOS, in which we can take advantage of the nearly ideal switching properties of CMOS transmission gates. An alternative design approach, outlined here, is used in TTL implementations of a D flip-flop—for instance, in the 74LS74A IC. The circuit in question (Figure 6.40) employs the same clocking method, positive edge triggering, as the master-

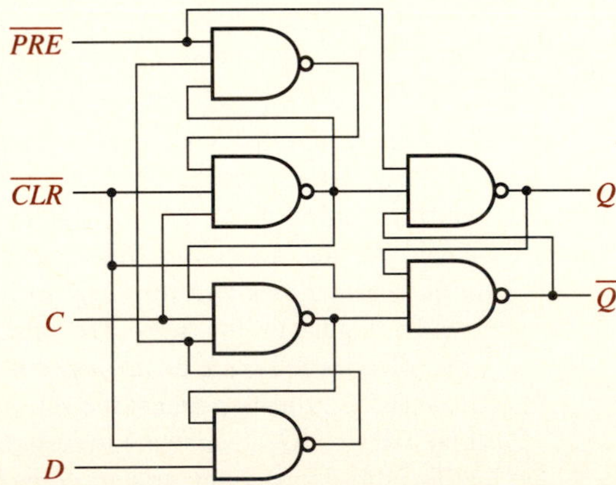

FIGURE 6.40 The logic circuit of a positive edge-triggered D flip-flop with preset and clear.

slave D latch of Figure 6.31a, but has asynchronous preset and clear inputs for initialization. Its graphic symbol is given in Figure 6.31b and its overall timing behavior is as described by Figure 6.32.

This D flip-flop is used in TTL circuits.

The D flip-flop consists of three latches, each of which is composed of two cross-coupled NAND gates, and is referred to as an $\overline{\mathrm{SR}}$ **latch**. $\overline{\mathrm{SR}}$ latches are dual versions of the NOR latches we studied in Section 6.3, in which the roles of 0 and 1 are reversed. For example, the quiescent input combination of the $\overline{\mathrm{SR}}$ latch is (1,1) rather than (0,0). A 0 rather than a 1 is applied to the set input line $\overline{S}$ to set the latch; a 0 on $\overline{R}$ resets it. The main $\overline{\mathrm{SR}}$ latch used to store the state $Q/\overline{Q}$ of the flip-flop comprises the two rightmost gates of Figure 6.40. The other two latches record the various combinations of the C and D signals that occur in the operation of the flip-flop. The detailed behavior may be derived in the manner of Figure 6.39, assuming every gate of the flip-flop has the same propagation delay (Problem 6.33).

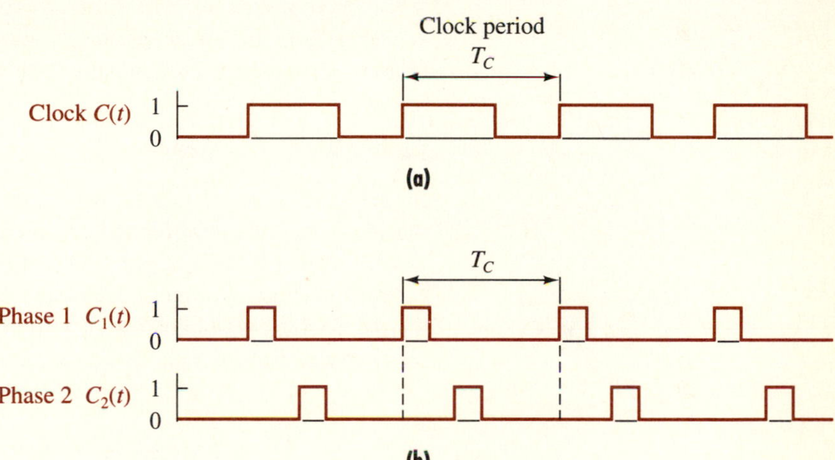

FIGURE 6.41 Clock signals: (a) single-phase; (b) two-phase.

Multiphase Clocks

So far, we have assumed that circuit timing is controlled by a single clock signal C that has a periodic waveform, as shown in Figure 6.41a. As we have seen, flip-flops controlled by C must be carefully designed to respond differently to various parts of the clock cycle. In most cases, the flip-flop is composed of several latches and gates, and is designed to change state in response to one edge of C. An alternative clocking method is to replace C with two or more separate signals $C_1, C_2, \ldots, C_p$, called **clock phases**. The phases have the same period, and are usually arranged so that when any phase is 1, all other phases are 0; that is, the clock phases are **nonoverlapping**.

Figure 6.41b illustrates the most common multiphase scheme, which employs two nonoverlapping clock phases and is therefore termed **two-phase clocking**. The clocking schemes we considered earlier can now all be called **single-phase clocking**. The phases of a multiphase clock are usually derived from a common circuit that generates a periodic signal P of the desired frequency. Each phase C_{i+1} is a copy of P that is slightly delayed, relative to the preceding phase C_i.

Multiphase clocking simplifies flip-flop design.

Two-phase clocking simplifies flip-flop design at the expense of more complicated clock generation and distribution circuitry. In particular, it is possible to use a master-slave approach in which the master latch is enabled by the first phase C_1 and the slave latch is enabled by the second phase C_2. The nonoverlapping property of the clock phases means that the latches are always enabled at differing times, so that nontransparency is assured. The circuits required in single-phase flip-flops for edge detection, hazard and race avoidance, and the like can be dispensed with, resulting in simpler bistable circuits that require less area on an IC chip. Consequently, in the design of VLSI circuits such as those found in computer systems, two-phase clocking is very common. Single-phase clocking is generally used for designs that employ off-the-shelf components such as 74X-series ICs.

Two-Phase D Flip-Flops

Figure 6.42 shows a flip-flop composed of two D latches connected in a master-slave configuration and controlled by a two-phase clock. Its structure is very similar to that of the edge-triggered, master-slave flip-flop of Figure 6.31. The main difference is that in the Figure 6.42 design, both latches can be disabled simultaneously (when $C_1 = C_2 = 0$), whereas in the earlier single-phase design, one of the two latches is always enabled.

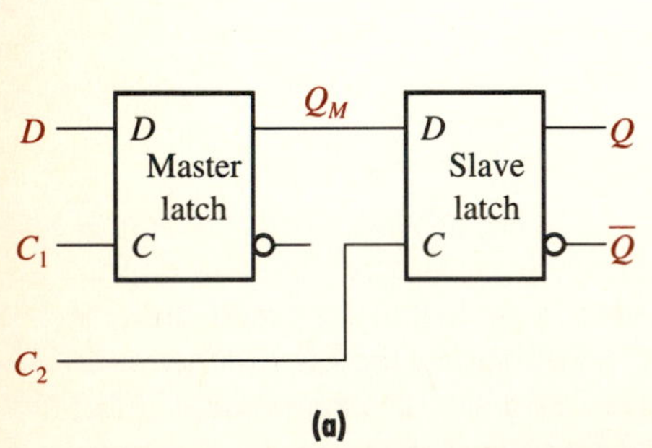

(a)

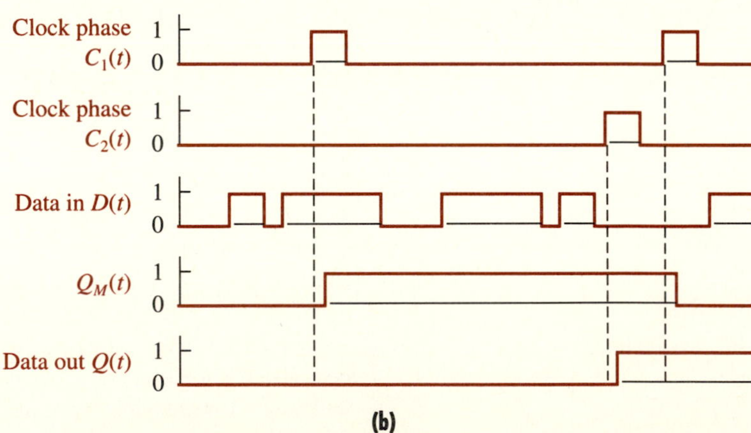

(b)

FIGURE 6.42 A master-slave D flip-flop with two-phase clocking: (a) its logic circuit; (b) its timing diagram.

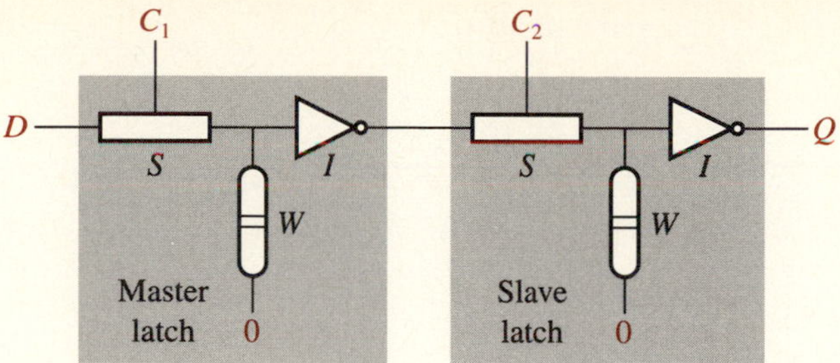

FIGURE 6.43 A switch-level model of a dynamic MOS D flip-flop with two-phase clocking.

Example 6.7

Two-phase dynamic flip-flop

The simplicity achievable with two-phase design is illustrated by Figure 6.43, which shows a switch-level model of a D flip-flop often used in MOS VLSI designs [Weste and Eshraghian, 1985]. Each "latch" here consists of an on-off switch S controlled by a clock phase C_i, a well W that can store 0 or 1, and an inverting buffer I. The switch S serves as the input control or enable device. $C_i = 1$ turns on the switch and allows the data signal D to flow into W. $C_i = 0$ turns off the switch, trapping the data in W. The inverter I amplifies the data signals as they pass through the latch. The use of differing phases to clock the two latches ensures nontransparency.

The switch S in this D flip-flop can be implemented by a switching transistor or, in the CMOS case, by a transmission gate. The well W typically models the electrical capacitance associated with the input circuits of I. Note the similarity between this circuit and the DRAM cell of Figure 6.5. Because the stored signal in W tends to leak away with time, the circuit of Figure 6.43 is called a **dynamic flip-flop** and, like the DRAM cell, requires special measures (refreshing) to ensure that the stored data are not lost.

6.7 Flip-Flop Behavior

We conclude our discussion of flip-flops by examining their state behavior. This behavior is illustrated by some simple but useful sequential circuits composed entirely of flip-flops.

State Behavior

As we saw in Section 6.5, the main features of flip-flop behavior can be described by means of state tables and state diagrams. These equivalent representations specify the transitions between (stable) states that occur in response to the various possible input data combinations. Each transition is triggered by a clock signal at a time determined by the clock's waveform and the

		Current input $D(t)$		
		0	1	
Present state	0	0	1	Next state
$Q(t)$	1	0	1	$Q(t+\tau)$

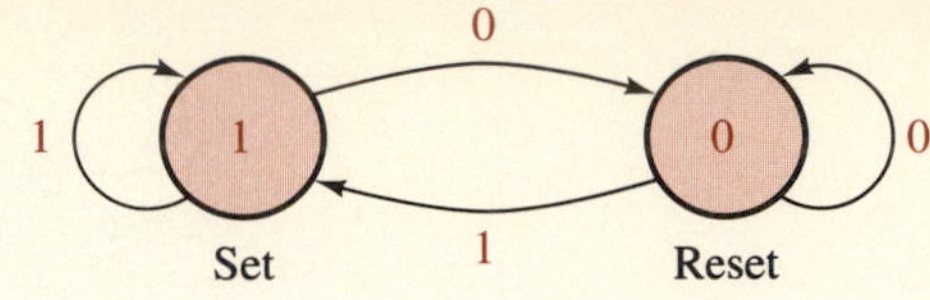

(a)

		Current input $(J(t), K(t))$				
		(0, 0)	(0, 1)	(1, 0)	(1, 1)	
Present state	0	0	0	1	1	Next state
$Q(t)$	1	1	0	1	0	$Q(t+\tau)$

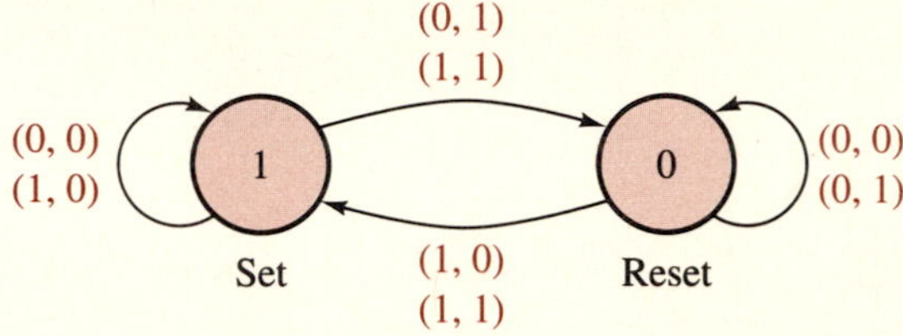

(b)

FIGURE 6.44 State tables and diagrams for **(a)** a D flip-flop and **(b)** a JK flip-flop.

triggering mode of the flip-flop. Unless otherwise specified, we assume that only single-phase clocks and positive or negative edge triggering are used. We also assume that the clock signals are designed so that the minimum setup time, hold time, and pulse-width constraints are met to avoid unpredictable behavior. Chapter 7 adds the further requirement of a centralized and periodic clock signal.

State tables and diagrams are convenient ways to specify behavior.

Figure 6.44 gives state tables and diagrams for D and JK flip-flops. Each clearly exhibits the bistable behavior characteristic of all latches and flip-flops. The basic state behavior of the JK flip-flop is identical to that of the SR latch shown in Figure 6.18, with J and K replacing S and R, respectively. In addition, the forbidden input data combination (1,1) is now allowed and causes the JK flip-flop to change state or toggle. The state behavior of the D flip-flop is that of a clocked delay element, in that the next output (state) signal $Q(t + \tau)$ is the same as the current input signal $D(t)$; $Q(t + \tau)$ is therefore a delayed copy of $D(t)$.

Characteristic Equations

We can also describe the state behavior of a flip-flop by sequential Boolean equations, which are equations that involve Boolean variables that are functions of time t. For the moment, we denote the present state by $Q(t)$ and the next state by $Q(t + \tau)$, where τ is the period of the clock signal used to operate the flip-flop in synchronous mode. Thus, the D flip-flop's behavior is concisely described by

$$Q(t + \tau) = D(t) \tag{6.11}$$

The JK flip-flop can be described by the slightly more complex equation

$$Q(t + \tau) = \overline{J}(t)\overline{K}(t)Q(t) + J(t)\overline{K}(t) + J(t)K(t)\overline{Q}(t) \tag{6.12}$$

The first term $\overline{J}(t)\overline{K}(t)Q(t)$ in (6.12) indicates that $Q(t + \tau) = Q(t)$ when $J = K = 0$; this is the quiescent condition. The second term specifies the basic set condition $JK = 10$, making $Q(t+\tau) = 1$ independent of its previous value. The third term in (6.12) describes the toggle condition $Q(t+\tau) = \overline{Q}(t)$ that occurs whenever $J = K = 1$. Finally, if $JK = 01$, Equation (6.12) implies that $Q(t + \tau) = 0$. The JK equation (6.12) obviously simplifies to the following minimal SOP form

$$Q(t + \tau) = J(t)\overline{Q}(t) + \overline{K}(t)Q(t) \tag{6.13}$$

Characteristic equations also describe state behavior.

An equation like (6.11) and (6.13) that defines the next state as a function of the current state and the current data inputs is referred to as the **characteristic equation** of the flip-flop. The foregoing characteristic equations can also be rewritten in the complementary forms

$$\overline{Q}(t + \tau) = \overline{D}(t)$$

for the D flip-flop and

$$\begin{aligned}\overline{Q}(t + \tau) &= \overline{J(t)\overline{Q}(t) + \overline{K}(t)Q(t)} \\ &= (\overline{J}(t) + Q(t))(K(t) + \overline{Q}(t))\end{aligned}$$

for the JK case.

Registers

A register is a set of flip-flops with common control logic.

A set of n identical flip-flops connected to store an n-bit word constitute an n-bit **register**. Figure 6.45 shows how a 4-bit register is constructed from D or JK flip-flops. Each flip-flop stores an independent data bit Q_i, but all control lines (clock, preset, and clear) are common. Data are loaded into the register by placing a new word on the input data lines, then applying an appropriate clock signal to the common control line C. Positive edge triggering is used in the D-type register of Figure 6.45a; the JK-type register of Figure 6.45b employs negative edge triggering. Each of these registers has 16 states, which correspond to all possible combinations of the four state variables $Q_3Q_2Q_1Q_0$. Application of a 0-pulse of appropriate duration to the common clear line $\overline{CLR}$ initializes the D register to the state $Q_3Q_2Q_1Q_0 = 0000$. A 0-pulse on the preset line $\overline{PRE}$ of the JK register makes $Q_3Q_2Q_1Q_0 = 1111$.

Because all input and output data bits of the registers of Figure 6.45 are transferred simultaneously or in parallel, these registers are said to be of the **parallel type**. They contrast with **shift registers**, which also serve as memory elements for n-bit words but employ **serial access**, meaning that data are transferred to and from the shift register one bit at a time or serially.

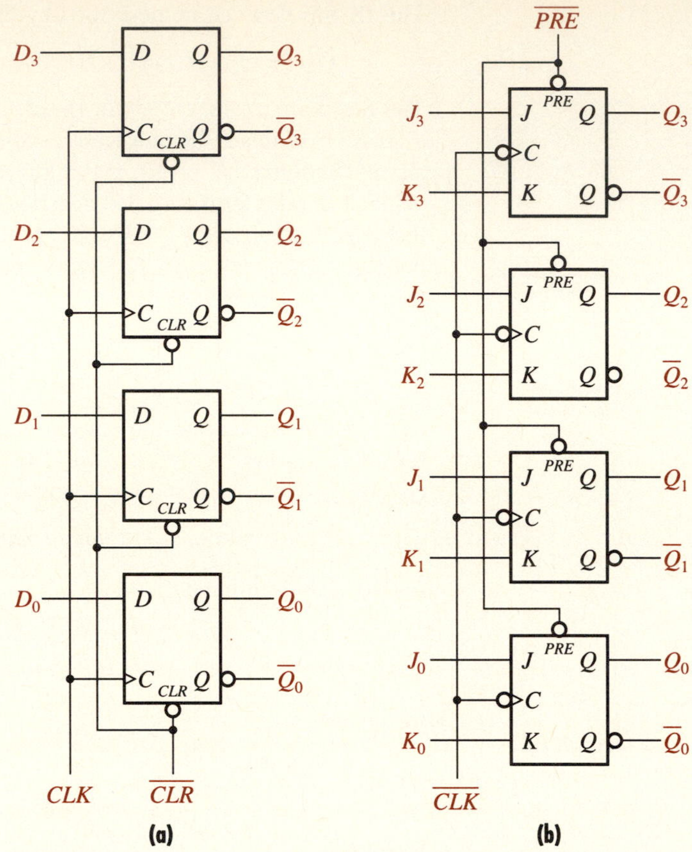

FIGURE 6.45 Examples of 4-bit (parallel) registers composed of (a) D flip-flops and (b) JK flip-flops.

Example 6.8

A 4-bit D-type shift register

Figure 6.46 shows a 4-bit shift register *SHR* that is constructed from D flip-flops. *SHR* has an external input data line x and an output data line z. The output Q_i of each flip-flop, except the rightmost one, is connected to the input D_i of the flip-flop on its right; x is applied to the leftmost flip-flop. Consequently, when the shift register is clocked, its state changes from $Q_3Q_2Q_1Q_0$ to $xQ_3Q_2Q_1$. Thus, each time *SHR* is clocked, its stored data are shifted one bit position to the right. The data bit on x is shifted into the leftmost flip-flop; the data bit Q_0 in the rightmost flip-flop is shifted out of *SHR*. As in all sequential circuits, correct operation of the shift register requires that the flip-flops' setup time, hold time, and pulse-width requirements be satisfied.

Suppose that *SHR* is initialized to the state $Q_3Q_2Q_1Q_0 = 0000$ by applying a 0-pulse to its clear line $\overline{CLR}$. Let x be held at 1, and let four 1-pulses

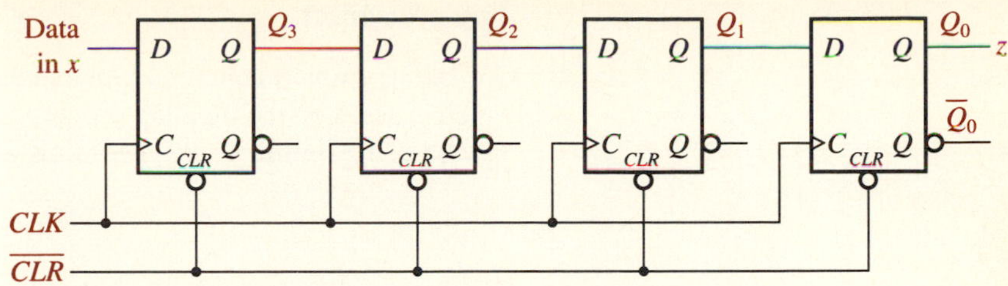

FIGURE 6.46 A 4-bit shift register *SHR* that is composed of D flip-flops.

of adequate width and spacing be applied to C. Then the following sequence of state transitions takes place:

$$0000 \rightarrow 1000 \rightarrow 1100 \rightarrow 1110 \rightarrow 1111 \qquad (6.14)$$

Hence, an arbitrary 4-bit data word $X = x_3x_2x_1x_0$ can be loaded serially into *SHR* in four successive clock periods. If the clock line is then deactivated, X is stored indefinitely as the current internal state $Q_3Q_2Q_1Q_0$. Four additional clock pulses are needed to retrieve X from *SHR* via the z line. The state behavior of *SHR* or, more generally, of any n-bit shift register, may be described by the following characteristic equations:

$$Q_{n-1}(t + \tau) = x(t)$$
$$Q_i(t + \tau) = Q_{i+1}(t) \qquad \text{for } i = n-2, n-3, \ldots, 0$$

Shift registers need flip-flops rather than latches.

A shift register provides a nice illustration of the differences between latches and flip-flops. Suppose that the four D flip-flops in *SHR* (Figure 6.46) are replaced with four D latches. Again, let the initial state be 0000 and let x be held at 1. Now suppose that the clock signal is activated by setting C to 1. Because all the latches become transparent, the circuit will pass through the entire sequence of state transitions given by Equation (6.14) if C is held at 1 long enough. To ensure that the x signal is loaded into the leftmost D latch only, a 1-pulse of fairly precise duration is required on C. In fact, this pulse on C must cause every latch to execute precisely one state change, which may be difficult in view of the unavoidable variations in propagation delays between different latches. This constraint does not apply to a shift register based on (edge-triggered) flip-flops: a single clock pulse of any duration beyond t_w^{min} allows each flip-flop to change state just once. Consequently, latches are unsuitable for use in shift registers.

Another useful class of all-flip-flop circuits are **counters**, which count a sequence of 1-pulses applied to an input line. As we will see, the input line on which the pulses are counted is generally the clock line C rather than the usual "data" lines. Counters can be designed to count up or down and to use various number codes.

Ripple Counters

One of the simplest counters, illustrated by Figure 6.47, is the **ripple counter,** which is based on the toggling property of JK flip-flops. It is intended to count upward using ordinary unsigned binary integers, according to the rule

$$N \rightarrow N + 1 \qquad \text{(modulo } 2^n\text{)} \tag{6.15}$$

where n is the number of flip-flops in the counter. The internal state $Q_{n-1}Q_{n-2}\ldots Q_0$ of the circuit represents the count N being computed, and the data Q_i stored in the ith flip-flop FF_i constitute the bit of weight 2^i in N.

A ripple counter is a cascade of flip-flops, each toggling its successor.

The ripple counter operates as follows. Each of the n flip-flops FF_i has a constant 1 (easily obtained from the circuit's power supply) applied to its J and K lines. Hence, applying a triggering positive edge to FF_i's C line causes it to toggle. The sequence of N data pulses on x to be counted are applied to the C input of the rightmost or least significant flip-flop FF_0. Consequently, each 1-pulse that arrives on x causes the outputs Q_0 and $\overline{Q}_0$ of this flip-flop to change value. The bit Q_0 can, by itself, be interpreted as N (modulo 2); it becomes 1 in response to every odd-numbered pulse on x and 0 in response to every even-numbered pulse. Now the output $\overline{Q}_0$ of FF_0 is applied to the C input of the flip-flop FF_1 to its left. Consequently, when $\overline{Q}_0$ changes from 0 to 1, which it does in response to every second (even-numbered) x-pulse, flip-flop FF_1 toggles. Thus, the output Q_1 of FF_1 toggles in response to every fourth x-pulse. Consequently, Q_1Q_0 represents N (modulo 4). In the same way, the third flip-flop FF_2 toggles after the arrival of every eighth x-pulse, and so on.

When all flip-flops of the counter are set to 1, corresponding to a count of $N = 2^n - 1$, the arrival of the next pulse triggers a sequence of n consecutive reset operations that "ripple" through all the flip-flops from right to left, eventually changing N to zero. This rippling of signals on the C lines, which resembles carry propagation through a ripple-carry adder (Section 5.9), means that the time taken by the ripple counter to complete the state transition (6.15) increases with the number of bits n in the counter, as we can see from Figure 6.47b.

A ripple counter, like all modulo-m counters, has a state diagram in which the transitions among the m states form a single closed loop or cycle (Figure 6.47d). Each transition arrow labeled 1 indicates that a 1-pulse on the x input line causes a state transition in which the count N, represented by the state variables, is incremented by one. Note that the 1-pulses must satisfy the setup, hold, and clock pulse-width requirements of the flip-flops used. However, despite being applied to the C (clock) inputs of the flip-flops, the x-pulses counted are *not* clock signals and can vary in width and frequency. The self-loops on the counter's states are all labeled 0, indicating that $x = 0$ (the absence of a 1-pulse to be counted) has no effect on the counter's state. Finally, note that the counter has the initial state $N = 0000$, which can

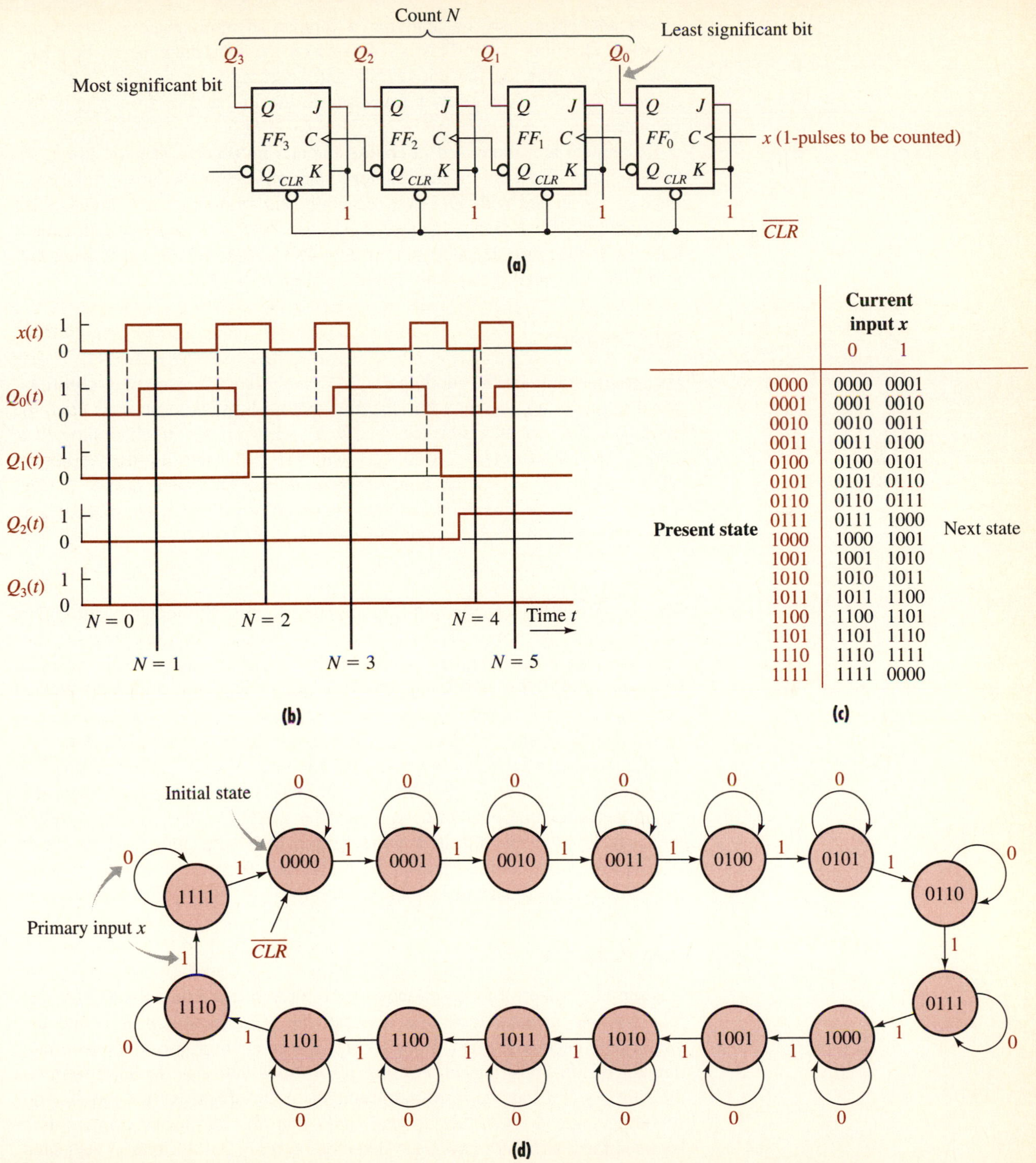

Present state	Next state, $x = 0$	Next state, $x = 1$
0000	0000	0001
0001	0001	0010
0010	0010	0011
0011	0011	0100
0100	0100	0101
0101	0101	0110
0110	0110	0111
0111	0111	1000
1000	1000	1001
1001	1001	1010
1010	1010	1011
1011	1011	1100
1100	1100	1101
1101	1101	1110
1110	1110	1111
1111	1111	0000

FIGURE 6.47 A 4-bit ripple counter: **(a)** its logic circuit; **(b)** its timing diagram; **(c)** its state table; **(d)** its state diagram.

be set asynchronously by applying a 0-pulse of appropriate width to the common clear line $\overline{CLR}$. This is indicated on the state diagram by the arrow labeled $\overline{CLR}$ entering the initial state and "coming from nowhere." This is intended to denote concisely a transition to the initial state from *any* of the 16 states.

Counters are often used in control circuits in which a specific line z_i is to be set to 1 if and only if the counter's state represents the binary number i. Each z_i associated with an n-bit modulo-2^n binary counter can therefore be generated by connecting a 1-out-of-2^n decoder to the counter's n data output lines. A 1-out-of-16 decoder such as the 74X154 (see Figure 5.54) is needed to decode the state of the 4-bit counter of Figure 6.47.

Ring Counters

A ring counter rotates a single 1 through all its flip-flops.

For small values of the modulus m, a counter known as a **ring counter**, which requires no decoding but contains m flip-flops, is sometimes used. A modulo-m ring counter is basically an m-bit shift register in which the count state N is of the form 00...010...0, with a 1 in the Nth flip-flop and 0s in all the remaining $m - 1$ flip-flops. The count N is incremented to $N + 1$ by a 1-bit shift of the counter's contents. A modulo-6 counter, for example, is required to pass through the state sequence

$$000001 \rightarrow 000010 \rightarrow 000100 \rightarrow 001000 \rightarrow 010000 \rightarrow 100000 \rightarrow 000001$$

The outputs of the leftmost flip-flop are connected to the inputs of the rightmost flip-flop to form a closed path or ring around which the stored 1 can circulate; hence, the counter's name. In effect, a ring counter stores a single 1, which is shifted each time the count is incremented. It can be regarded as counting with a 1-out-of-n number code instead of the usual binary code.

Figure 6.48 shows a modulo-6 ring counter that is composed of six JK flip-flops. The counter can be set to the initial state 000001 by activating the (asynchronous) $\overline{RESET}$ input line, which presets the rightmost flip-flop and clears the remaining flip-flops. Note that the x input is directly connected to the clock input of each flip-flop so that, unlike the ripple counter, all flip-flops here change state simultaneously. Note also the cyclic structure of the state diagram, which is characteristic of counters.

Other Counters

Modulo-m counters encode the count information (the internal state) in various ways to achieve trade-offs between the number of flip-flops and the amount of decoding logic needed. The ripple counter uses the minimum number of flip-flops $\lceil \log_2 m \rceil$ but requires a full 1-out-of-m decoder, whereas the ring counter has the maximum number of flip-flops (m) but requires no decoding logic. Many other counter types that aim at various compromises between these extremes have been invented. A design of this type is presented in the following example.

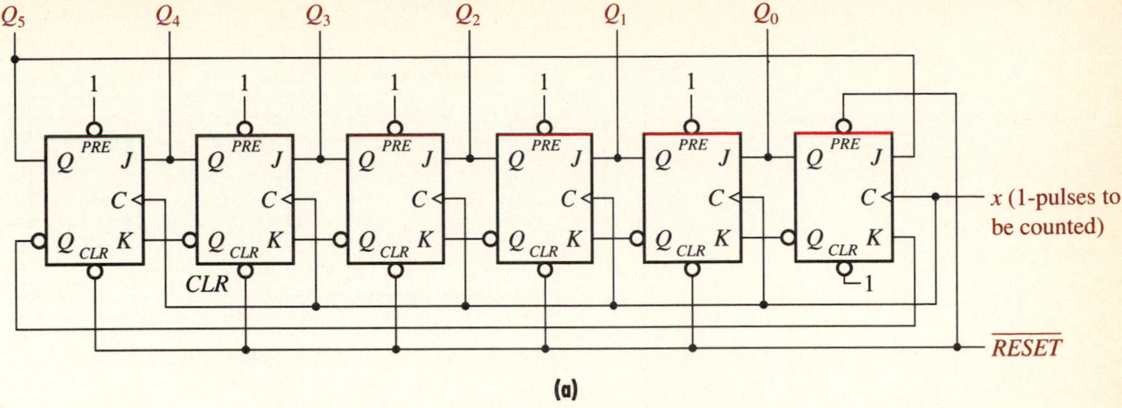

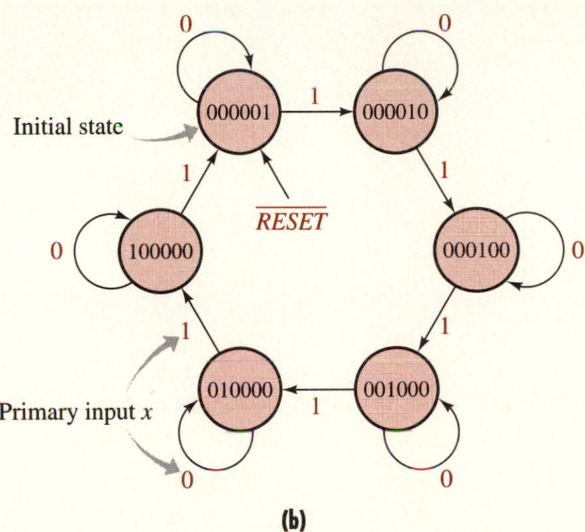

FIGURE 6.48 A modulo-6 ring counter: (a) its logic circuit; (b) its state diagram.

Example 6.9

A 4-bit (modulo-8) Johnson counter

A useful counter based on a shift register is a **Johnson counter**,[6] a 4-bit version of which is shown in Figure 6.49a. It is similar to a ring counter in that the output Q of the leftmost flip-flop is fed back to the rightmost one. In this case, however, the $\overline{Q}$ output, rather than Q output, is entered into the rightmost flip-flop. A modulo-8 Johnson counter can be formed from the 4-bit D-type shift register of Figure 6.46, with the clock line serving as

[6]Other names for this well-known circuit are a *twisted-ring counter,* a *switch-tail counter,* and a *Möbius counter,* all of which allude colorfully to the counter's feedback structure.

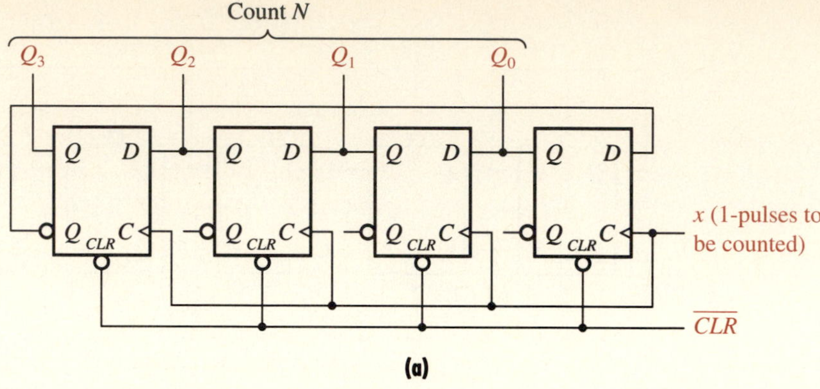

(a)

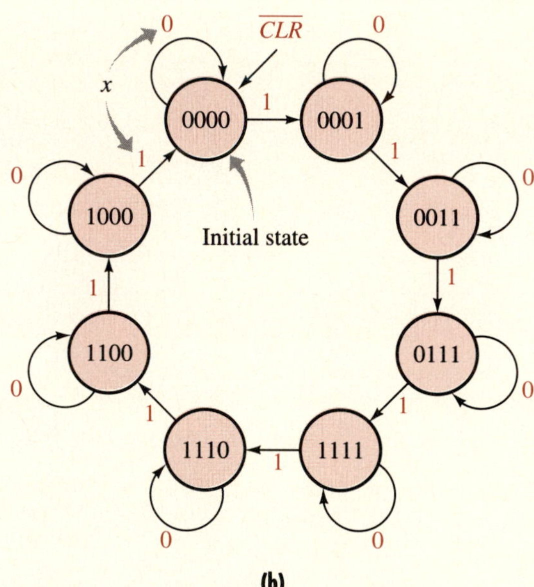

(b)

FIGURE 6.49 A 4-bit Johnson counter: (a) its logic circuit; (b) its state diagram.

the x line, the pulses of which are to be counted. It is easy to see that after the counter is initialized to $Q_3Q_2Q_1Q_0 = 0000$, it will go through the following sequence of states N in response to 1-pulses on x:

$$0000 \rightarrow 0001 \rightarrow 0011 \rightarrow 0111 \rightarrow 1111 \rightarrow 1110 \rightarrow 1100 \rightarrow 1000 \rightarrow 0000 \tag{6.16}$$

Because N returns to 0000 after eight 1-pulses have arrived, half of the 16 possible combinations of $Q_3Q_2Q_1Q_0$ are unused, implying that the circuit has eight effective states instead of 16. If we interpret (6.16) as a number sequence, we can view the 4-bit Johnson counter as counting modulo 8 in a nonstandard binary code. This code is easy to decode, however (see Problem 6.40), and so is more useful than it appears at first sight.

Like the ring counter, all flip-flops of a Johnson counter are triggered simultaneously, so the state transition time for the counter is the same as that of a single flip-flop. Hence, the Johnson counter operates faster than a ripple counter. An n-bit Johnson counter has the modulus $m = 2n$, so it uses only half the number of flip-flops of a comparable modulo-m ring counter.

Asynchronous Circuits*

Sequential circuits are usually divided into two broad categories, called *synchronous* and *asynchronous*. In this section, we examine asynchronous circuits and their design and reliability problems. We will see that these problems limit the role of asynchronous methods to a few special circuit types such as memory circuits, and motivate the almost universal use of synchronous sequential circuits.

6.8 Hazards

We observed in Section 6.2 that the time a logic signal takes to travel through a circuit depends on propagation delays that are subject to considerable uncertainty. The propagation delay of a gate is a function of its family type, and different gates of the same type can have significantly different delays. A signal change may propagate along several different paths at once, so that its effects are received at different destinations at different times. It is difficult or impossible to ensure that several signals change simultaneously, or to maintain simultaneity if it is achieved. The logic designer must therefore assume that the times at which signals change are known only within certain bounds.

Static Hazards

Uncertain delays in logic circuits have the ability to introduce temporary or transient signal pulses where none are supposed to exist. The circuit conditions that allow such signals to appear are called **hazards**, and the spurious pulses or noise signals that result are referred to as **glitches**. Glitch pulses are generally narrow in width—a gate delay or less—and so are often filtered out by the inertial property of physical gates. Note, too, that the normal functional properties of a gate can also block glitches. For example, a 0 signal on one input of an AND or NAND gate blocks a glitch that appears on any

*This part (Sections 6.8, 6.9, and 6.10) may be skipped without loss of continuity.

other input; a 1 on an input of an OR or NOR gate blocks the propagation of a glitch via any other input.

Suppose that the input signals $x_1(t)$ and $x_2(t)$ of AND gate G are required to change simultaneously from 1 to 0 and 0 to 1, respectively, so that the output $z(t)$ remains at 0. In practice, delay uncertainties may cause the input signals to change value anywhere within the indicated transition regions. If $x_1(t)$ changes first, as illustrated by Figure 6.50b, then $z(t)$ remains at 0, as required. If, on the other hand, $x_2(t)$ changes first, then for a brief period the 1-values of $x_1(t)$ and $x_2(t)$ overlap, allowing $x_1(t)x_2(t)$ to become 1. If

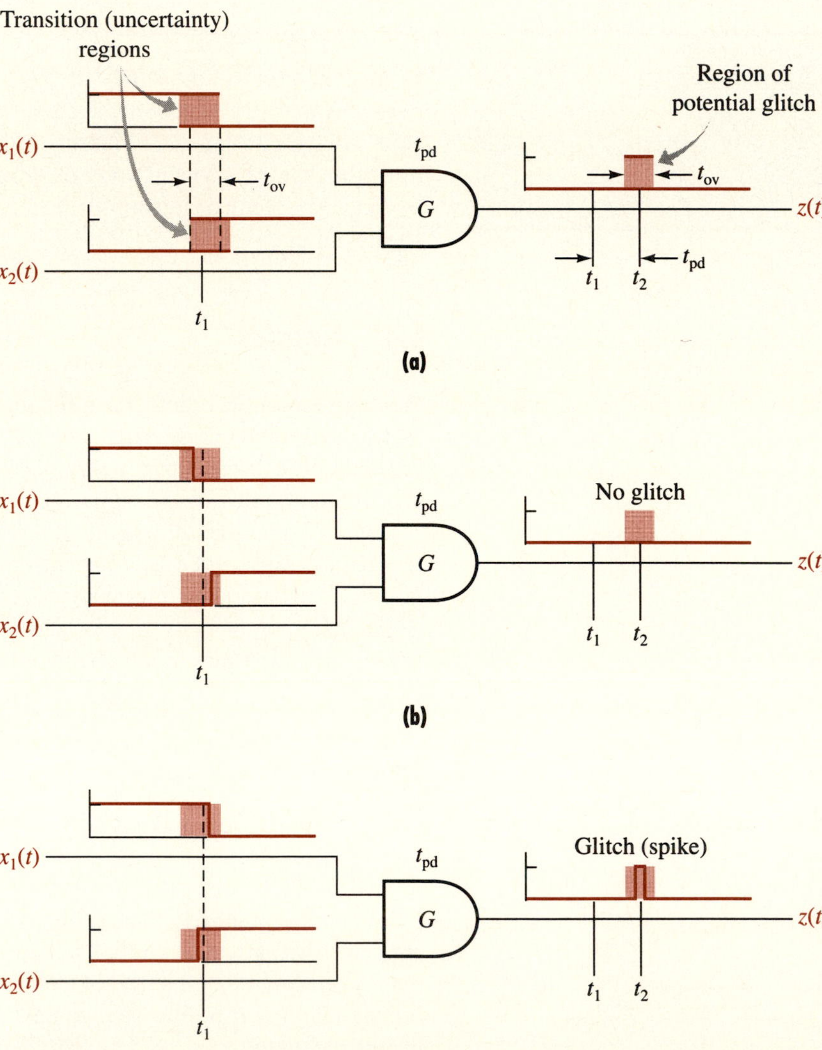

FIGURE 6.50 (a) A static hazard condition due to multiple input changes in an AND gate; (b) normal output; (c) erroneous output that contains a glitch.

Static hazards can produce glitches (spurious pulses) in static logic signals.

the overlap period t_{ov} is sufficiently long relative to G's inertial delay, the glitch depicted in Figure 6.50c is produced. The hazard in Figure 6.50 is called a **static 0-hazard**, because it affects a signal $z(t)$ whose correct value is unchanging or static—a static 0, in this case. A similar **static 1-hazard** exists if G in Figure 6.50 is an OR gate, in which case the glitch, if any, is a 0-pulse that appears in an otherwise static 1-signal.

A common cause of hazards is the existence of two or more signal paths of unequal delay between the same two points a and b of a logic circuit. If the paths contain different numbers of gates, and all gates have approximately the same propagation delay, the signals propagating through them are likely to arrive out of step at b. Some examples of this situation appear in Figure 6.51. The primary input $x_2(t)$ changes from 1 to 0; the other primary inputs $x_1(t)$, $x_3(t)$, and $x_4(t)$ are held at 1, 1, and 0, respectively, as shown by the small, inset timing diagrams. The figure also shows a possible set of circuit responses to the change in $x_2(t)$. There are two paths from x_2 to the output of OR gate G_5, one containing three gates (G_1, G_3, and G_5) and the other containing two gates (G_4 and G_5). The presence of the inverter G_1 in only one of the paths, as well as any other delay variations present, allow the output signals of AND gates G_3 and G_4 to change at slightly different times. This results in a glitch in the output of G_5 depicted in the figure. This particular signal should be a constant 1, so we have another example of a static 1-hazard.

Dynamic Hazards

Figure 6.51 also demonstrates that spurious pulses can occur in signals that are in transition from 0 to 1 or from 1 to 0. We say that these glitches are caused by **dynamic hazards**. The particular dynamic hazard shown in the circuit of Figure 6.51 produces a glitch in the output signal of G_5 shortly after the output of G_6 changes from 0 to 1. The primary output gate G_7 responds to G_6 by changing its output z from 0 to 1, but the glitch from G_5 causes z to return temporarily to 0. Hence, instead of going through a single 0-to-1

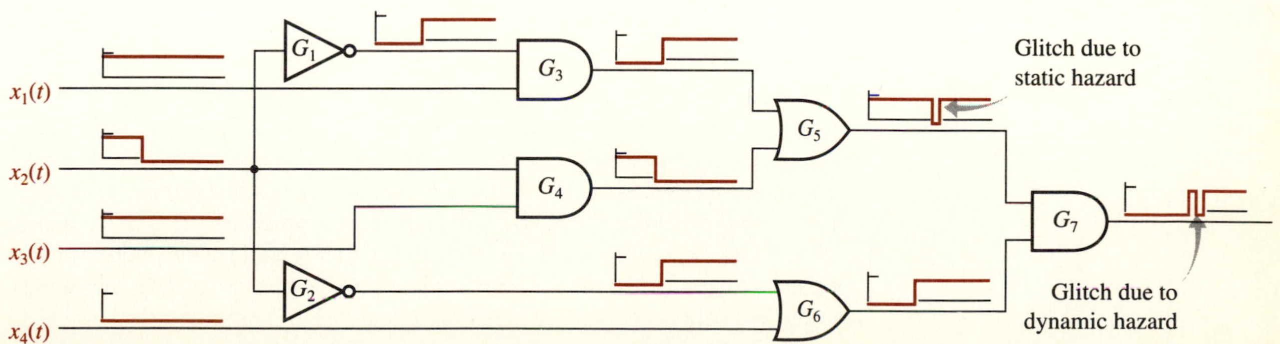

FIGURE 6.51 A combinational circuit that exhibits both static and dynamic hazards.

Dynamic hazards can produce glitches (multiple transitions) in changing signals.

transition, z performs a multiple 0-to-1-to-0-to-1 transition before settling down to its final 1 value. Note that the inertial delay of G_7 may be such that it ignores the incoming 0-pulse from G_5, in which case no glitch appears at the circuit's primary output.

We stress that the existence of a hazard in a logic circuit does not mean that glitches will occur in any specific physical realization of the circuit or, if they do, that they will cause erroneous primary output signals. It all depends on the actual delays in the circuit and the precise times at which signals change. Nevertheless, it is necessary to deal with hazards in logic design to ensure that the glitches, if any, that they produce cause no harm. As the preceding examples demonstrate, subtle hazards can exist in quite simple circuits.

Hazard Control

The explicit identification and, where feasible, the elimination of hazards and the glitches they create is very difficult in the case of general logic circuits [Unger, 1969]. This is due to both the inherent uncertainty in the propagation delays that need to be considered and the huge number of possible combinations of these delays. It is therefore feasible to perform explicit hazard analysis for only fairly small circuits that operate under restrictive conditions.

One way to eliminate at least some hazard-produced glitches is to avoid multiple (nearly) simultaneous input changes of the type depicted in Figure 6.50. This is not always feasible, and does not help cases like that of Figure 6.51, where a single input change can produce glitches. Another way to counter glitches is to insert delays, usually in the form of extra gates, to slow down specific glitch-causing signals. For example, if an extra delay of appropriate size is inserted in the output line of gate G_4 in Figure 6.51, its 1-to-0 transition can be slowed down enough to eliminate the glitch in the output signal of G_5 and, hence, the glitch in the output signal of G_7. This ad hoc solution is unsatisfactory, however, because it will not help if the signal transitions on the outputs of G_3 and G_4 are switched, so that G_3's output signal should be slowed down relative to G_4's.

Another solution for the particular circuit of Figure 6.51 is to introduce a new AND gate G_8, the inputs of which are the primary inputs x_1 and x_3, and the output of which is connected as a new input to G_5. This changes the function realized by G_5 from $y = x_1\overline{x}_2 + x_2x_3$ to $x_1\overline{x}_2 + x_2x_3 + x_1x_3$, which is equivalent to y, because the new term x_1x_3 generated by G_8 is an absolutely inessential prime implicant of y, as is easily verified. However, for the input patterns depicted in Figure 6.51, we have $x_1x_3 = 1$, which means that G_8's output will hold output y of G_5 at 1, independent of the changes that occur on the other two inputs of G_5. Consequently, no glitches of the indicated type will occur in the circuit with G_8 present.

Summary

As the foregoing examples show, hazards occur in many forms in logic circuits. Some hazards, referred to as **logic hazards**, can be eliminated by special logic design methods, such as the inclusion of redundant gates or extra delays. These techniques (which are explored further in the end-of-chapter problem set) are limited in scope; for example, one method applies to two-level combinational circuits with the single-input change restriction. Other, so-called **function hazards**, such as those associated with multiple input changes, are found in *all* realizations of a combinational or sequential function.

Hazards are best controlled by clock signals (synchronous design).

Glitches are temporary phenomena that occur while logic signals are in transition or unstable; therefore, they tend to disappear quickly as all the relevant signals stabilize. Consequently, the most practical approach to the hazard problem in general logic circuits is to permit hazards to be present, but to design circuits so that all key actions, such as the setting of a new state, occur after any hazard-induced glitches have dissipated. This is a basic characteristic of synchronous sequential circuits, which employ periodic clock signals to indicate when the signals of interest have stabilized.

6.9 The Huffman Model

Next we examine the differences between asynchronous and synchronous circuit structures and modes of operation. A standard model for sequential circuit analysis is introduced and illustrated by some simple asynchronous examples.

Synchronous vs. Asynchronous Circuits

A sequential logic circuit can be composed of gates, latches, and/or flip-flops interconnected in possibly complex configurations that usually include some feedback. The circuit is considered to be **asynchronous** if it does not employ a periodic clock signal C to synchronize its internal changes of state. State changes therefore occur in direct response to signal changes on primary (data) input lines, and different memory elements can change state at different times. The Johnson counter of Figure 6.49 is an asynchronous circuit, according to this definition; the x signal applied to the clock inputs of all its flip-flops is not assumed to be periodic and is a data signal rather than a control signal. The ripple counter of Figure 6.47 is more obviously asynchronous, because no common signal is applied to the flip-flop clock inputs. Hence, the triggering of the flip-flops is not synchronized in this case.

In general, an asynchronous circuit does not need the precise timing control supported by flip-flops. It may therefore contain latches rather than flip-flops. In many cases, an asynchronous circuit simply relies on the

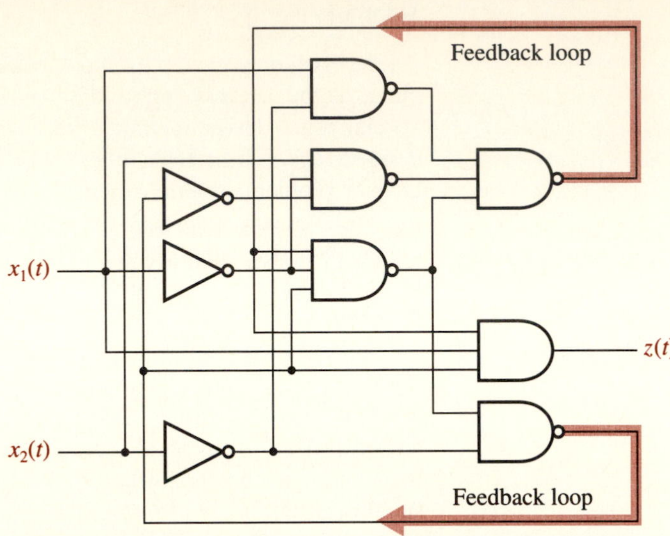

FIGURE 6.52 An example of a small asynchronous sequential circuit; nonzero propagation delays are implicit.

propagation delays of its component gates and connections, combined with the circuit's feedback structure, to implement its memory functions. An example of such a circuit that contains no explicit memory elements appears in Figure 6.52.

Asynchronous circuits do not employ clock signals; synchronous circuits do.

We define a **synchronous circuit** as one that employs an independent, periodic clock signal C to synchronize all internal changes of state. Such a circuit has flip-flops or, in some instances, gated latches, for its memory elements. The memory of an asynchronous circuit, on the other hand, can reside in the propagation delays associated with gates and lines. A circuit designed to be synchronous becomes asynchronous if the signals applied to the clock inputs of the memory elements are not treated as clock signals. Consequently, all sequential logic circuits can be analyzed or operated as asynchronous circuits. We conclude, therefore, that the distinction between synchronous and asynchronous—there is a surprisingly large and contentious literature on this issue [Langdon, 1974]—involves both a circuit's structure and its mode of operation. Synchronous circuits must contain clocked memory elements and be controlled by independent timing signals; no such constraints are placed on asynchronous circuits.

We can think of an asynchronous circuit as transparent to all input changes, in the same sense as a latch that has been enabled. Consequently, input signal changes can enter the circuit unhindered, interact in complex ways, and produce large numbers of glitches. In a synchronous sequential circuit, on the other hand, flip-flops—especially of the edge-triggered variety—are relatively insensitive to and tend to filter out such glitches on their data lines.

Consequently, asynchronous circuits tend to be much more error-prone than synchronous ones. Because of the design simplification and higher operating reliability that results from clock control, the majority of digital circuits are designed for synchronous operation.

Applications

Asynchronous design methods have only a small role in modern logic design.

Despite the predominance of synchronous circuits, there are several areas of logic design in which asynchronous behavior is of interest:

1. *Analysis and synthesis of memory circuits:* The signals inside a latch, flip-flop, or similar small sequential circuit interact asynchronously. At this low level of abstraction, a clock is seen as just another input signal; its particular timing function is apparent only at higher levels. In general, logic circuits that are viewed "below" the level of the clock signal appear to function asynchronously.
2. *Circuits that have unpredictable input signal timing:* A circuit is inherently asynchronous if the times at which its input signals change value cannot be placed within certain bounds. An example is a counter of randomly occurring 1-pulses, for which it is impossible to predict the times at which the data signals of interest occur. Consequently, such data cannot be controlled by an independent and periodic clock signal.
3. *Circuits that have multiple independent clocks:* It is impractical to distribute a single clock signal over large distances (say, a meter or more) due to the variability of propagation delays along different signal paths. Circuits separated by such distances—for example, the central processing unit and an input/output controller of a large computer—have separate internal clocks and therefore operate asynchronously with respect to each other. Communication between such circuits requires asynchronous design methods.

The latch, flip-flop, and counter structures examined earlier in this chapter can now be seen as examples of asynchronous circuits. The fairly complex interactions of their internal signals under various delay assumptions is typical of asynchronous behavior.

Speed Considerations

Because of the wide range of propagation delays in logic devices and their interconnections, different sequential subcircuits of a digital system such as a computer often operate at different inherent speeds. Furthermore, one circuit may process some signal combinations faster than other combinations, so the speed of a particular circuit is not necessarily constant. Clock signals eliminate all such speed differences by, in effect, constraining all system components to operate at the speed of the slowest component. By allowing each

Asynchronous circuits are potentially faster than synchronous circuits.

component to operate independently at its maximum speed, which requires asynchronous design techniques, faster overall operation may be achievable. Thus, in principle, asynchronous circuits can operate faster than functionally equivalent synchronous circuits.

To replace the synchronization function of the clock, large asynchronous circuits need special subcircuits to generate completion signals that indicate when a particular asynchronous component has completed its current operation and has its results available for use by other parts of the system. Asynchronous logic circuits of this type are said to **self-timed** or **speed-independent circuits.** The provision of reliable completion signals is quite difficult; it requires special control circuits, which are themselves a source of significant cost and signal delay. Thus, the potential speed advantages of eliminating the clock are, in practice, outweighed by the cost and complexity of the asynchronous control circuits that replace it.

Delay Considerations

When analyzing the behavior of an asynchronous circuit like that of Figure 6.52, we must account for the times at which individual signals change value due to both external effects on the primary input signals and internal propagation delays in the circuit's lines and gates. Here we are faced with a bewildering array of possibilities. Each gate of the circuit has its own propagation delay, which constitutes a small memory element. These delays can all differ in value from one another and, to compound the problem, their exact values cannot be known with much accuracy. Hence, a detailed timing model such as the gate-delay model used in our previous analysis of latches and flip-flops can become exceedingly complex. For example, if a distinct propagation delay Δ_i is associated with each of the nine gates that appear in Figure 6.52, the number of possible internal states to be dealt with could be as high as $2^9 = 512$.

A consequence of the uncertainty about the precise magnitude of Δ_i is that when two or more signals are required to change value at approximately the same time, we cannot always be sure of the order in which the signals will change. For example, a pair of signals required to make the transition $00 \rightarrow 11$ will do so in the sequence

$$00 \rightarrow 01 \rightarrow 11 \tag{6.17}$$

if the right variable changes first. A slight difference in propagation delays may cause the left variable to change first, resulting in the transition sequence

$$00 \rightarrow 10 \rightarrow 11 \tag{6.18}$$

Care must be taken to ensure that the two different intermediate values 01 and 10 of the changing variables do not lead to different final outcomes.

Model Definition

To deal with the above-mentioned issues when analyzing the state behavior of an asynchronous circuit, certain simplifying assumptions are made. The basic model for this purpose, known as the **Huffman model** after its inventor, David A. Huffman, is summarized below [Huffman, 1954]:

Assumptions of the Huffman model of asynchronous circuits

1. *Feedback delays:* A delay element, the output of which defines a secondary input variable, is placed in each feedback line; all gates and other lines are assumed to have zero propagation delay.
2. *Delay uncertainty:* In a state change that involves two or more secondary variables, the variables can change value in arbitrary order.
3. *Fundamental mode:* A primary input signal can change value between the 0 and 1 levels only after the circuit has reached a stable state.
4. *Single input changes:* Primary input signals are assumed to change value one at a time.

Feedback Delays

Feedback delays reduce the number of states in circuit models.

The Huffman model greatly reduces the complexity of a circuit's sequential behavior by restricting all propagation delays to the feedback lines, the number of which is much less than the number of gates. It may also be called the **feedback-delay model** to contrast it with the gate-delay model introduced earlier. The nine-gate circuit of Figure 6.52, for example, has two feedback loops, so only two delay elements and four internal states are implied by its Huffman model (Figure 6.53). The gates themselves are assumed to have zero propagation delay, allowing them to be analyzed as ideal combinational devices. The Huffman model also applies in simplified form to synchronous sequential circuits, in which flip-flops replace delay elements in the feedback paths.

All the gate and line delays of the underlying physical circuit are represented by lumped delay elements that are placed in the feedback paths, as in Figure 6.53, for example. This not only reduces the number of distinct propagation delays (and therefore the number of states) to manageable levels, it also enables the effects of the feedback signals to be precisely analyzed. The output y_i of one of the delay elements Δ_i constitutes a state variable of the circuit. The size of Δ_i is not specified exactly, but is assumed to be of the order of the total input/output propagation delay in the forward direction through the combinational part of the underlying physical circuit. For example, in the circuit of Figure 6.53, Δ_1's magnitude is roughly equal to the propagation delay through the three levels of gates between the primary inputs x_1, x_2 and the input y_1^+ of Δ_1. As we will see later, the feedback delay assumption is most realistic when latches or flip-flops are placed in the feedback lines.

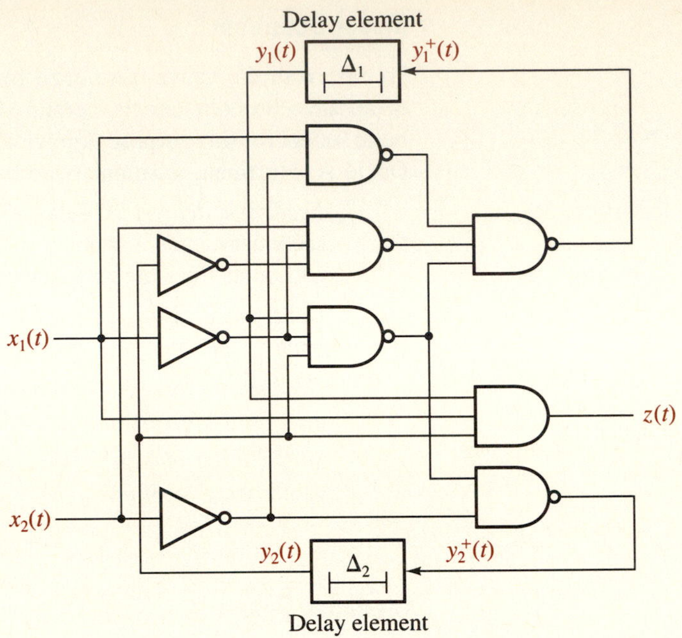

FIGURE 6.53 The Huffman model of the asynchronous circuit of Figure 6.52.

Inexact feedback delays permit races; critical races must be avoided.

The state variables are allowed to change value in any order, thereby accounting for the inherent inexactness of Δ_i. When two or more state variables change value together, a so-called **race condition** exists. If the final values of these variables depend on the order in which each y_i changes—that is, on the "winner" of the race—the race is said to be **critical**; otherwise, it is **noncritical**. Asynchronous circuits must be designed to avoid critical races, a requirement that greatly complicates the design process.

Input Changes

Except in special cases such as clock generators, logic circuits are required to settle into a stable state in response to every input signal condition. However, a change to some input signal x_j can trigger a sequence of changes among the state variables, which cause the circuit to pass through several transient or unstable states before settling into a final stable state. If x_j is allowed to change a second time before the final stable state is reached, it is possible that the circuit will never stabilize. Moreover, analysis of the effects of applying multiple input changes to an unstable circuit is extremely complicated. Thus, the so-called "fundamental" mode assumption made in the Huffman model, which says that primary input signals should be changed only after the circuit has reached a stable state, is perfectly reasonable.

The single-input-change assumption permits only one primary input variable to change at a time. Hence, if we want to change some input pair x_1x_2

from 00 to 11, this change is required to take place in two steps, as in (6.17) and (6.18). This constraint is most easily met by small circuits such as latches and flip-flops that have only two or three primary input lines. The setup time and hold time constraints which require, for example, that data input D must not change when enable signal C changes are examples of single-input-change restrictions in the normal operation of a memory circuit. In more general circuits, such restrictions are harder to justify, because several independent signals can easily change sufficiently close to one another to be considered simultaneous. Consequently, the single-input-change restriction of the Huffman model is often unrealistic.

Pulse-Mode Circuits

Fundamental-mode circuits respond to level changes; pulse-mode circuits respond only to pulses.

The input signals implied by the fundamental mode of operation are the logic levels 0 and 1, and a single change from 0 and 1 or vice versa suffices to change the circuit's state. A class of sequential circuits that are sometimes contrasted with fundamental-mode circuits are **pulse-mode circuits**, which require two consecutive transitions between 0 and 1—that is, a 0-pulse or a 1-pulse—to alter the circuit's state. A pulse-mode circuit is designed to respond to pulses of certain duration; the constant signals between the pulses are "null" or "spacer" signals, which do not affect the circuit's behavior.

Logic circuits in which most or all input/output signals are pulses are mainly of historical interest [Caldwell, 1956]. A synchronous circuit may be considered a special type of pulse-mode circuit, because its clock input line, but not its other inputs, is controlled by periodic pulses. Synchronous circuits have also been referred to as "clocked fundamental-mode" circuits [Langdon, 1974], so the pulse-mode/fundamental-mode distinction is, at best, imprecise and will not be used further.

6.10 Some Design Issues

Next we examine further the problems encountered in analyzing and synthesizing small asynchronous circuits. These problems provide motivation for the almost universal use of clocks and synchronous techniques in sequential logic design.

Circuit Analysis

To analyze the behavior of a general asynchronous circuit, we first insert a delay element Δ_i in each feedback path, as in Figure 6.53. The exact size of Δ_i is not important, because allowance is made for race conditions, which effectively permit the value of Δ_i to vary over a modest range. The output signal of Δ_i is a state or secondary input variable y_i. The input signal to Δ_i, which we denote y_i^+, is the next value of the state variable and is a Boolean function of the primary inputs $x_1, x_2, \ldots, x_n$ and the secondary inputs or present state $y_1, y_2, \ldots, y_p$. In other words, y_i^+ is a function of the total state of the circuit. The next-state function $y_i^+(x_1, x_2, \ldots, x_n, y_1, y_2, \ldots, y_p)$ can be represented in any standard format for combinational functions such as a Boolean equation or a truth table. Each primary output function z_k is also an ordinary combinational function. By evaluating y_i^+ for each total state, we can determine the internal state transitions of the circuit, leading to a state table or diagram.

To illustrate the foregoing concepts, we reexamine the behavior of the basic ungated SR latch, which was discussed in Section 6.3 using the gate-delay model. The Huffman model for the circuit appears in Figure 6.54a. Here a single delay in the circuit's feedback path replaces the two gate propagation delays assumed earlier (Figure 6.17), so this is *not* the same circuit model as before. The next state of the circuit is defined by the following Boolean equation:

$$\begin{aligned} y^+ &= \overline{R + \overline{S + y}} \\ &= \overline{R}(S + y) \end{aligned} \tag{6.19}$$

which expresses y^+ as the function of the primary inputs S and R and the secondary input y, defined by the two zero-delay NOR gates. Note that

$$y(t) = y^+(t - \Delta)$$

Equation (6.19) yields the table shown in Figure 6.54b, which can be interpreted either as a K-map for y^+, or as a partial state table for the latch, showing the next state as a function of the present state and present input signals. The K-map-style layout of the table implies that a single-input change moves the circuit's operating point from a column to a logically adjacent one, a minor advantage in analyzing asynchronous circuit behavior.

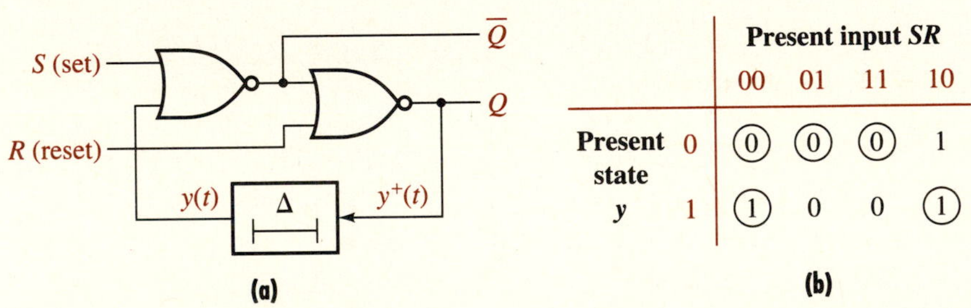

FIGURE 6.54 The Huffman model of an SR latch: (**a**) its logic circuit; (**b**) its state table (flow table), with stable states circled.

Stable and Unstable States

Stable (circled) states do not change with constant inputs; unstable states do.

We can easily see that if a particular next-state entry is the same as the present state—that is, if $y^+ = y = S_1$ for some input combination I_1—the circuit's state is **stable** under I_1. That is, as long as the input is held constant at I_1, the state remains at S_1 and all signals of the circuit are static. If, on the other hand, $y^+ \neq y$, the state will change from $S_1 = y$ to $S_2 = y^+$ after a brief delay and S_2 will become the new y. If S_2 is stable under I_1, no further changes occur; otherwise, unstable state S_2 changes to S_3 according to (6.19), and so on. Stable states are circled in the state table.

In Figure 6.54b, for example, $y = 0$ is stable under the input combination $SR = 11$, as indicated by the circled 0 entry in the third column of the first row. If $y = 0$ and SR changes from 11 to 10 (a single-input change), y becomes 1. The circuit's total state or operating point moves from the third to the fourth column of the first row, and then to the second row, as follows:

$$\begin{array}{ccc} & 11 & 10 \\ \rightarrow & \textcircled{0} \rightarrow & 1 \\ & & \downarrow \\ & & \textcircled{1} \end{array} \tag{6.20}$$

Because the final y^+ is the same as the original y, this state is stable and no further changes of state occur as long as SR is held at 10. The state transition sequence depicted in (6.20) will be written more conveniently as follows:

$$\xrightarrow{11} \textcircled{0} \xrightarrow{10} 1 \xrightarrow{10} \textcircled{1} \tag{6.21}$$

A flow table includes unstable states.

The label on top of each arrow in (6.21) is the primary input combination that causes the indicated state transition. A state table like that of Figure 6.54b, which distinguishes unstable and stable states, is often referred to as a **flow table**.

As discussed in Section 6.3 for the gate-delay model, problems arise if the input combination $SR = 11$ is applied to an SR latch. These problems are also reflected in the Huffman model, but in a somewhat different form. Suppose again that the latch's operating point is $SR = 11$ and $y = 0$, a stable condition. Let both inputs of the latch be changed to 0 at about the same time. If R changes first, the input condition $SR = 10$ appears, which takes the latch to the state $y = 1$, as specified by (6.21), where it remains when S becomes 0.

$$\xrightarrow{11} \textcircled{0} \xrightarrow{10} 1 \xrightarrow{10} \textcircled{1} \xrightarrow{00} \textcircled{1} \tag{6.22}$$

However, if S changes first, SR goes from 11 to 01 and eventually to 00, which leaves y unchanged.

$$\xrightarrow{11} \textcircled{0} \xrightarrow{01} \textcircled{0} \xrightarrow{00} \textcircled{0} \tag{6.23}$$

Thus, the next state that results from an input transition between 11 and the quiescent input 00 is uncertain, confirming the need to avoid the $SR = 11$ combination when using SR latches. Note also that when $SR = 11$, the Q and $\overline{Q}$ output signals are both 0, another undesirable aspect of this input combination.

Next we examine a slightly larger circuit that exhibits more complex state behavior.

Example 6.10

Analyzing an asynchronous circuit

Consider again the logic circuit that appears in Figure 6.53, which has two feedback paths that define the state variables y_1y_2 and allow up to four internal states. The next state $y_1^+y_2^+$ of the circuit is defined by the Boolean equations

$$\begin{aligned} y_1^+ &= \bar{x}_1y_1y_2 + \bar{x}_1x_2\bar{y}_2 + x_1\bar{x}_2 \\ y_2^+ &= \bar{x}_1y_1y_2 + x_2 \end{aligned} \tag{6.24}$$

and the output function is given by $z = x_1y_1y_2$. Again, K-maps provide a useful way of representing these functions, (Figures 6.55a and 6.55b). The rows denote the present state variables, and the columns the primary inputs. We can then merge the two K-maps to obtain the state table of Figure 6.55c, in which each entry has the form $y_1^+y_2^+, z$, where $y_1^+y_2^+$ represents the circuit's next state, and stable states are circled.

Example of a noncritical race

Certain primary input changes result in both state variables changing value. In such cases, it must be asked if a critical race exists. Consider, for instance, what happens when the circuit is in the initial stable state $y_1y_2 = 10$, with the input combination $x_1x_2 = 10$. Let x_1x_2 change to 11. Figure 6.55c requires the state to change from 10 to 01. Suppose that y_1 changes

(a) y_1^+

y_1y_2 \ x_1x_2	00	01	11	10
00	[0] 0	[1] 1	[3] 0	[2] 1
01	[4] 0	[5] 0	[7] 0	[6] 1
11	[12] 1	[13] 1	[15] 0	[14] 1
10	[8] 0	[9] 1	[11] 0	[10] 1

(b) y_2^+

y_1y_2 \ x_1x_2	00	01	11	10
00	[0] 0	[1] 1	[3] 1	[2] 0
01	[4] 0	[5] 1	[7] 1	[6] 0
11	[12] 1	[13] 1	[15] 1	[14] 0
10	[8] 0	[9] 1	[11] 1	[10] 0

(c)

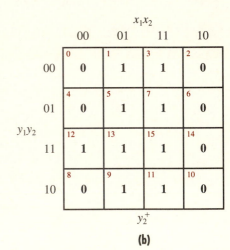

Present state y_1y_2	Present input x_1x_2: 00	01	11	10
00	(00), 0	11, 0	01, 0	10, 0
01	00, 0	(01), 0	(01), 0	10, 0
11	(11), 0	(11), 0	01, 1	10, 1
10	00, 0	11, 0	01, 0	(10), 0

FIGURE 6.55 (a), (b) Secondary output functions y_1^+, y_2^+, and (c) a state table for the circuit of Figure 6.54.

before y_2, making $y_1y_2 = 00$. Then, with x_1x_2 held at the new value 11, the state must become 01, which is stable under $x_1x_2 = 11$. Hence, we have the complete state transition sequence

$$\xrightarrow{10} \textcircled{10} \xrightarrow{11} 00 \xrightarrow{11} 01 \xrightarrow{11} \textcircled{01} \tag{6.25}$$

in which we reach the correct end state 01 after briefly passing through the state 00.

Suppose now that y_2 changes before y_1, with the same initial conditions. The state goes from 10 to 11, which is unstable under $x_1x_2 = 11$, and leads to the next state 01, as required by the state table.

$$\xrightarrow{10} \textcircled{10} \xrightarrow{11} 11 \xrightarrow{11} 01 \xrightarrow{11} \textcircled{01} \tag{6.26}$$

Thus, (6.25) and (6.26) represent two state transition sequences with the same final state, and so define a noncritical race condition.

Example of a critical race

The circuit under consideration does contain a critical race, however, and so will not operate correctly if certain unfavorable combinations of delays are present in the physical implementation. Let $y_1y_2 = 00$ and $x_1x_2 = 00$, a stable combination, and let x_1x_2 change to 01. This requires y_1y_2 to become 11. If y_1y_2 first changes to 10, it next changes to the desired state 11, where it remains.

$$\xrightarrow{00} \textcircled{00} \xrightarrow{01} 10 \xrightarrow{01} 11 \xrightarrow{01} \textcircled{11} \tag{6.27}$$

If, however, y_1y_2 changes to 01, it locks into this incorrect state because $y_1y_2 = 01$ is stable under $x_1x_2 = 01$.

$$\xrightarrow{00} \textcircled{00} \xrightarrow{01} \textcircled{01} \tag{6.28}$$

Clearly, (6.27) and (6.28) imply the existence of a critical race condition.

Another undesirable aspect of this circuit is the occurrence of glitch-like signals on the output line z for certain state transition sequences. For example, consider again the behavior specified by (6.26). If we add to each arrow label $X(t)$ the corresponding output value $z(t)$ thus, $X(t)/z(t)$, we obtain the following:

$$\xrightarrow{10/0} \textcircled{10} \xrightarrow{11/0} 11 \xrightarrow{11/1} 01 \xrightarrow{11/0} \textcircled{01}$$

We see that during this particular transition from stable state 10 to stable state 01, z goes from 0 to 1 and back to 0 again, resulting in a transient 1-pulse on z.

The Design Process

The analysis illustrated by the preceding examples can be reversed to obtain a general design method, originally formulated by Huffman in the 1950s, for synthesizing small asynchronous sequential circuits. This **Huffman design**

method is complex and of limited practicality. We will therefore content ourselves with a brief outline and an example. (A complete treatment of asynchronous design can be found in [Unger, 1969].)

The starting point of the Huffman design process is an informal description of the desired behavior. This description is first converted into a state table (flow table), with the internal states of the circuit represented by abstract symbols such as $A, B, C, \ldots$. The result is a precise and formal behavioral specification of the target circuit. Secondary variables $y_0, y_1, y_2, \ldots$ are then assigned to represent the states in binary form; this assignment must be designed to avoid critical races. The combinational next-state functions $y_0^+, y_1^+, y_2^+, \ldots$ and the primary output functions $z_0, z_1, z_2, \ldots$ are determined from the state table and are represented by equations, or in a tabular format such as a K-map. The final step is to implement these functions by suitable combinational circuits such as two-level designs. Additional steps may be taken to simplify the initial state table and to avoid hazard conditions.

Example 6.11

Designing a hazard-free D latch

To illustrate the Huffman method of asynchronous circuit design, we consider a small but useful example: a D latch L of the type introduced in Section 6.3. We derive the latch from basic principles and examine its timing behavior for hazards. The input/output signals of L are specified in Figure 6.56a. Recall that it has two inputs: a data signal D and a control signal C, which is referred to as the enable or clock input; we assume there is just one output Q. No special timing properties are attached to the C signal here. Both it and D are treated as independent input signals that can change at arbitrary times (but not simultaneously).

The desired behavior of L may be summarized informally as follows: When $C = 0$, the latch stays in its current state and the output Q does not change. When C changes to 1, L proceeds to store the current value (0 or 1) of D and the output Q becomes D.

The first step is to determine an abstract but complete set of possible internal states for L. These should represent all the possible combinations of

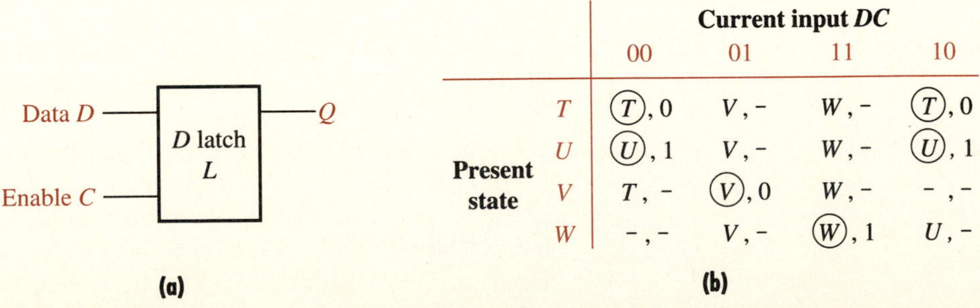

		Current input DC			
		00	01	11	10
Present state	T	(T), 0	V, -	W, -	(T), 0
	U	(U), 1	V, -	W, -	(U), 1
	V	T, -	(V), 0	W, -	-, -
	W	-, -	V, -	(W), 1	U, -

(b)

FIGURE 6.56 (a) Input/output signals and (b) a state table for the D latch L.

State identification is an ad hoc, intuitive process.

things that L should "remember," such as the stored data item, active/inactive conditions, and the like. In general, derivation of the states is an intuitive process that depends on the problem at hand. In the present case, we might tentatively identify four states as follows:

State T: Disabled and storing 0

State U: Disabled and storing 1

State V: Enabled and storing 0

State W: Enabled and storing 1

(Because the names C and D are already assigned to the inputs of L, we use the symbols T, U, V, W in place of A, B, C, D for the states of L.)

The state transitions and output values associated with the above four states can now be derived from the word specification of L's behavior. For example, when the present state is T and the input combination is $DC = 00$, the latch remains in state T and outputs $Q = 0$, thus:

$$\xrightarrow{00} T \xrightarrow{00/0} \textcircled{T}$$

This defines the leftmost entry $\textcircled{T}$, 0 in the top row of the state table of Figure 6.56b. When C becomes 1, thereby enabling the latch, the state changes to V, as specified by the second entry $V, -$ in the top row of the state table. We can treat V as a transient state, leaving the corresponding output signal Q as an unspecified or don't-care value to be assigned later. This is indicated by the dash in $V, -$. As in combinational logic design, such don't-care values can be used to reduce circuit size. Still with present state T, another possible input change is D becoming 1 while C remains at 0. This does not affect the latch and results in the entry $\textcircled{T}$, 0 at the right end of the top row in the state table. The remaining entry in the top row corresponds to $DC = 11$; note that it can follow $DC = 01$ or $DC = 10$ but not $DC = 00$ without violating the single-input-change assumption. $DC = 11$ causes a transition to state W, resulting in the entry $W, -$. Continuing in this way, the rest of the state table of Figure 6.56 is easily filled in. The state V is stable only when $DC = 01$. Consequently, the rightmost entry in row 3 of the state table, which requires DC to change from 01 to 10, a double-input change, is left completely unspecified.

States with compatible next states and output signals can be merged.

We can see from Figure 6.56b that the states T and V are functionally similar and can be merged into a single state. This is because T and V have the same next-state entry in every column, and also have the same output values wherever both are specified. When an item is specified as α in one row and is unspecified (–) in the other row, we can change the unspecified entry to α without violating the state table specifications. For the same reason, states U and W can be merged. States T and U cannot be so merged because they have different output values in both the leftmost and rightmost columns. The result of the foregoing mergers is the two-state table that appears in Figure 6.57a.

		Current input DC			
		00	01	11	10
Present state	T	(T), 0	(T), 0	U, –	(T), 0
	U	(U), 1	T, –	(U), 1	(U), 1

(a)

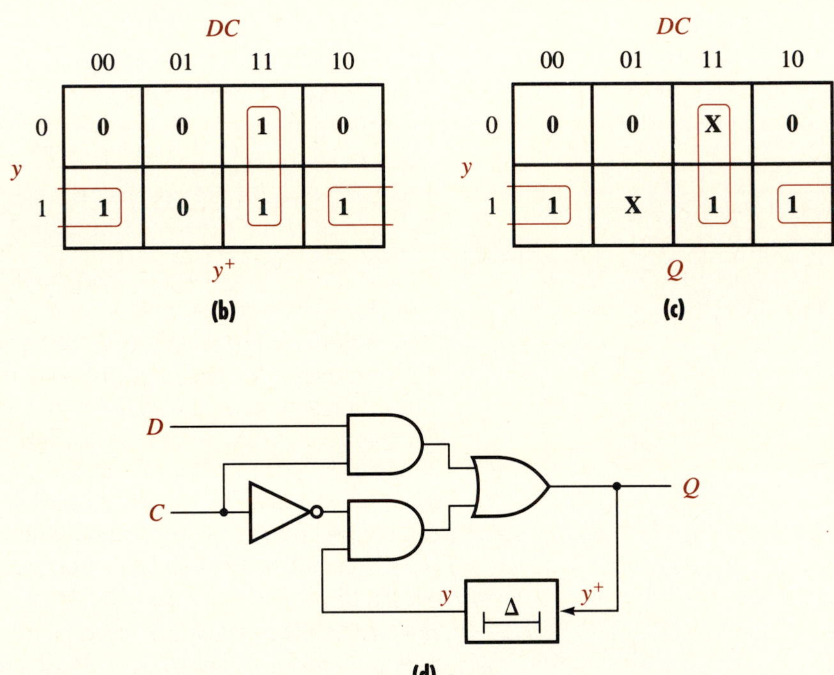

FIGURE 6.57 (a) A reduced state table for the D latch; (b) a K-map for y^+; (c) a K-map for Q; (d) a preliminary logic circuit (contains a hazard).

Next we must assign secondary variables to represent the states. With just two states, a single variable y suffices and there is no possibility of a race. We make the arbitrary state assignment $y = 0$ for T and $y = 1$ for U. Replacing T and U with 0 and 1, respectively, yields the tables (K-maps) for the next state function y^+ and the output function Q that appear in Figures 6.57b and 6.57c. It is a trivial problem to obtain the minimum-cost SOP realization for y^+:

$$y^+ = DC + y\overline{C}$$

The output Q can be conveniently expressed as $Q = y^+$. The resulting latch circuit implemented by AND, OR, and NOT gates appears in Figure 6.57d. This is essentially the same as the D-latch design considered earlier (Figure 6.19).

It is not hard to see that the combinational part of Figure 6.57d contains a static hazard. Suppose that $(D, C) = (1, 1)$, so the latch is enabled and storing 1. The input combination applied to the OR gate is (1,0), making $Q = 1$. Now let the enable signal C change to 0; this causes the OR gate's input to become (0,1). If, due to differences in signal propagation delays, the upper signal to the OR gate changes from 1 to 0 before its lower input changes from 0 to 1, the OR gate briefly sees the input combination (0,0), and may therefore produce a glitch in the form of a 0-pulse on Q. In fact, this spurious 0-pulse can become trapped in the latch's feedback loop, causing an incorrect transition to the reset state $Q = 0$. A spurious 0-pulse can also appear on Q when C changes from 0 to 1 with $D = 1$, but in this case Q returns to the correct state.

Hazards can sometimes be avoided by using redundant prime implicants.

The static 1-hazard in this particular D latch is neatly eliminated by adding an extra AND gate to the latch to generate the third, redundant prime implicant Dy of y^+ (and Q). This prime implicant Dy and the corresponding AND gate are colored in Figure 6.58. Now when the above glitch-inducing input condition occurs (C changes from 1 to 0), the fact that D and y are both 1 ensures that the output Dy of the new AND is 1. This 1-signal holds the output of the OR gate at a steady 1 while its other two input lines change in response to the changes in C. Thus, we obtain a hazard-free design that deliberately includes a nonminimal AND-OR circuit. An alternative way to eliminate the hazard is to add a special delay circuit, called a "clock skew" circuit, to the latch, which ensures that the $\overline{C}$ signal rises before C falls, and $\overline{C}$ falls before C rises. Practical circuits like the 74LS75 employ one of the above methods to eliminate hazards from this type of latch [McCluskey, 1986], although that is not always apparent from manufacturers' data books.

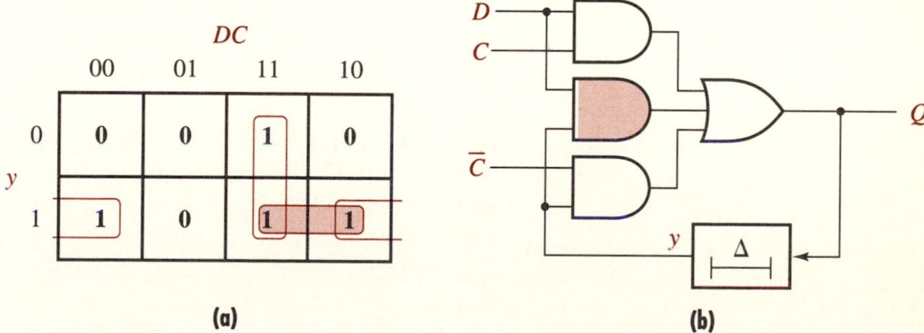

FIGURE 6.58 (a) A modified K-map for y^+; (b) the corresponding hazard-free D latch.

The general procedure used to design the preceding D latch is summarized below. As we will see in Chapter 7, similar procedures are used to design small synchronous sequential circuits. However, the clock signals of the latter eliminate the need to consider races and hazards in the design process. This greatly simplifies such steps as state assignment, which is concerned with selection of the binary patterns used to encode the circuit's states. In the synchronous case, any encoding that assigns a distinct pattern to each state will do. Example 6.11's technique of adding extra gates to eliminate hazards is readily extended to any circuit whose combinational part has only two levels (Problem 6.42). Few types of hazards can be handled in this way, however.

Procedure

General procedure for asynchronous sequential circuit synthesis

1. *State table construction:* Construct a state (flow) table that defines the desired behavior in precise symbolic terms. The internal states of the circuit may be represented by the symbols $A, B, C, \ldots$.
2. *State reduction:* Reduce the size of the state table by identifying and merging states that are functionally compatible with one another.
3. *State assignment:* Assign secondary variables $y_0, y_1, y_2, \ldots$ to represent the states of the reduced state table in binary form. This assignment must avoid critical races.
4. *Next-state function specification:* Derive the specifications of the next-state functions $y_0^+, y_1^+, y_2^+, \ldots$ and the primary output functions $z_0, z_1, z_2, \ldots$.
5. *Combinational circuit design:* Obtain efficient combinational logic circuits that implement y_i^+ and z_j. Modify the design as necessary to avoid hazards.

An Edge Counter

To illustrate how races influence the state assignment process, we turn to an asynchronous design problem that involves several state variables. A small counter *CTR* is to be designed that counts (modulo 4) the number of positive and negative edges or, equivalently, the number of level changes occurring in an input signal x. This should be contrasted with the counters discussed in Section 6.7, which count 1-pulses, each of which contains one positive and one negative edge. Our present circuit has one input line x and two output lines z_0 and z_1, as indicated by Figure 6.59a. The output combination $z_1 z_0$ denotes, in ordinary binary code, the number of edges of x encountered to date.

Clearly, *CTR* should contain four states, symbolically denoted S_0, S_1, S_2, S_3, where S_i corresponds to an edge count of i. A state table for *CTR* appears in Figure 6.59b. It assumes that the circuit is initialized to the state

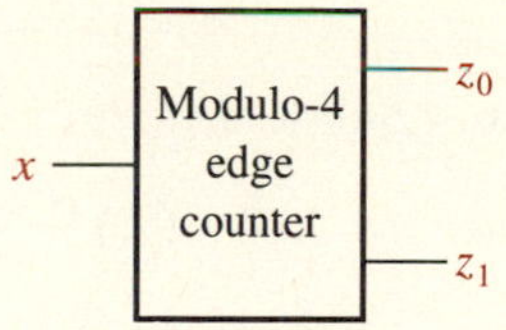

		Present input x 0	1
	S_0	(S_0), 00	S_1, –
Present state	S_1	S_2, –	(S_1), 01
y_0y_1	S_2	(S_2), 10	S_3, –
	S_3	S_0, –	(S_3), 11

(a) x → Modulo-4 edge counter → z_0, z_1

(b)

FIGURE 6.59 A modulo-4 edge counter: (a) its input/output lines; (b) its state table.

S_0 when $x = 0$. Thus, the circuit is stable, with next state S_0 and output combination $z_1z_0 = 00$ when the present state is S_0 and $x = 0$. This yields the first entry in the top row and the left column of the state table. The next state S_0 is circled to indicate a stable condition. When x changes from 0 to 1, marking the arrival of the first edge to be counted, the next state becomes S_1 and the output z_1z_0 changes. Eventually, the state stabilizes at S_1 and the output stabilizes at 01, denoting a count of one. The corresponding transition sequence

$$\xrightarrow{0/00} (S_0) \xrightarrow{1/-} S_1 \xrightarrow{1/01} (S_1)$$

gives rise to the top two entries in the right column of the state table. It involves passing through an unstable condition when the output signals are changing; these output values are left unspecified for the moment. The next incoming edge occurs when x changes from 1 to 0, giving rise to the transition sequence

$$\xrightarrow{1/01} (S_1) \xrightarrow{0/-} S_2 \xrightarrow{0/10} (S_2)$$

and so on. It is not hard to see that the number of rows in the state table cannot be reduced below four, so *CTR* has four states and requires two state variables y_0 and y_1.

State Assignment

Arbitrary state assignments are likely to produce races.

The next step is to assign the four possible patterns of y_1y_0 to the four states of *CTR*. The "natural" state assignment follows:

$$\begin{aligned} y_1y_0 &= 00 \text{ represents } S_0 \\ y_1y_0 &= 01 \text{ represents } S_1 \\ y_1y_0 &= 10 \text{ represents } S_2 \\ y_1y_0 &= 11 \text{ represents } S_3 \end{aligned} \qquad (6.29)$$

This, however, leads to critical races. Consider, for example, the following transitions:

$$\xrightarrow{1/11} \textcircled{S_3} \xrightarrow{0/-} S_0 \xrightarrow{0/00} \textcircled{S_0}$$

Critical races are avoided by careful state assignment.

With the state assignment of (6.29) they become

$$\xrightarrow{1/11} \textcircled{11} \xrightarrow{0/-} 00 \xrightarrow{0/00} \textcircled{00} \tag{6.30}$$

Both state variables are required to change from 1 to 0. The Huffman model therefore allows 01 and 10 as intermediate states, depending on which state variable changes first. If 10, representing S_2, is the intermediate state (implying that the second state variable wins the race), *CTR* enters state S_2 with $x = 0$. From Figure 6.60b we see that this is a stable condition, so that instead of (6.30), we obtain the erroneous sequence

$$\xrightarrow{1/11} \textcircled{11} \xrightarrow{0/-} 10 \xrightarrow{0/10} \textcircled{10} \tag{6.31}$$

meaning that the count changes from three to two instead of from three to zero.

Critical races like the above can be eliminated by the following state assignment:

$$\begin{aligned} y_1y_0 &= 00 \text{ represents } S_0 \\ y_1y_0 &= 01 \text{ represents } S_1 \\ y_1y_0 &= 11 \text{ represents } S_2 \\ y_1y_0 &= 10 \text{ represents } S_3 \end{aligned} \tag{6.32}$$

which switches the patterns assigned to S_2 and S_3. With this new assignment only one state variable changes as the count is incremented from i to $i + 1$ (modulo 4); consequently, there can be no critical races of any type.

Combinational Design

Using (6.32), we can now proceed to define the combinational functions that specify the state-variable transitions that correspond to the state table (Figure 6.59b). The result appears in Figure 6.60a in a format known variously as a **transition table** or an **excitation table**. This table combines the information in the state table with that of the state assignment (6.32). Thus, the first row of the transition table shows the response of *CTR* to $x = 0$ when the present state is $y_1y_0 = 00$ or S_0: the next state is again S_0 and the output is $z_1z_0 = 00$. To find what happens when x changes to 1, we go to the fifth row, which shows the next state as $y_1y_0 = 01$—that is, S_1—with don't-cares as the output signals. Hence, the transition table specifies the combinational functions required to construct *CTR*.

It remains to devise a logic circuit to implement the transition table. This would be a simple task if we were not concerned about hazards. Figure

Present input	Present state		Next state		Present output	
x	y_1	y_0	y_1^+	y_0^+	z_1	z_0
0	0	0	0	0	0	0
0	0	1	1	1	X	X
0	1	0	0	0	X	X
0	1	1	1	1	1	0
1	0	0	0	1	X	X
1	0	1	0	1	0	1
1	1	0	1	0	1	1
1	1	1	1	0	X	X

(a)

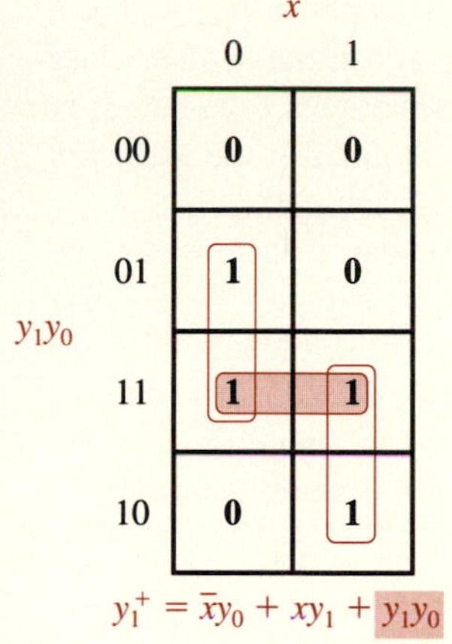

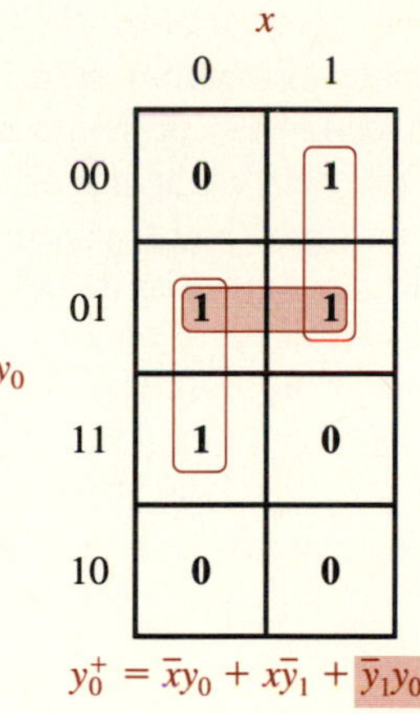

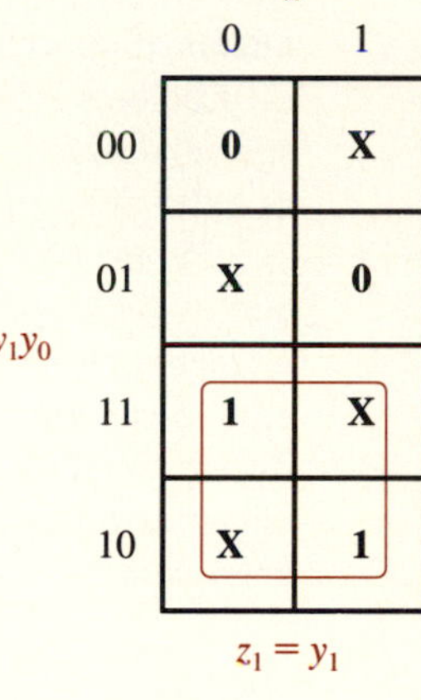

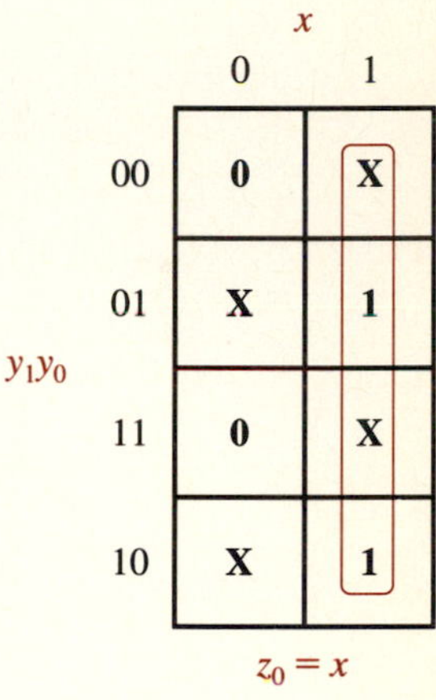

(b)

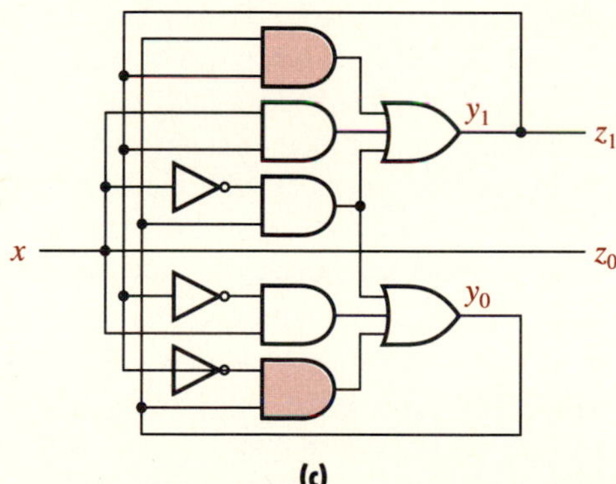

(c)

FIGURE 6.60 A modulo-4 edge counter: **(a)** its transition table; **(b)** its next-state and output functions; **(c)** its logic circuit (incomplete).

6.60b shows K-maps for all the required functions, from which we can easily obtain two-level designs via the method of Example 6.11. As in that example, the colored terms and gates are included to prevent certain types of hazards. Unfortunately, this circuit also contains another class of hazards, the so-called

essential hazards, which are inherent in the counter's state behavior. These hazards are explored further in the problem set (Problem 6.56), and can be eliminated only by carefully controlling the circuit's propagation delays.

Summary

We can deduce from the preceding examples that the design of reliable asynchronous circuits is greatly complicated by the need to handle hazards and critical races, both of which result from uncertainty surrounding the circuit's internal delays. Races depend on the order in which the state variables change value during state transitions. They can be avoided by careful assignment of state-variable combinations to the states to be implemented. This may necessitate using more than n state variables (or feedback loops) to implement a circuit of up to 2^n variables (Problem 6.55). Hazards constitute another class of problems whose incidence can be reduced and sometimes eliminated by special design techniques, such as the use of redundant gates or the inclusion of extra delays in certain signal paths, especially feedback paths.

Figure 6.61a shows the Huffman model of a general asynchronous circuit, which combines zero-delay combinational logic with a set of delay elements in a very general fashion. The race and hazard problems of asynchronous logic can be attributed to the fact that these delay elements can take different values $\Delta_1, \Delta_2, \ldots, \Delta_p$ that allow signals to propagate through the circuit at different speeds, so they can interact in unpredictable and undesirable ways.

Synchronous Circuits

The foregoing design problems would largely disappear if all the feedback delays were of *exactly* the same size Δ, as suggested in Figure 6.60b. This would, in effect, synchronize all signal changes, because the outputs of the identical delay elements would then be in lock-step with one another. Such a synchronization method is impractical due to the impossibility of manufacturing physical devices with the necessary identical delays.

Synchronous circuits use a clock signal to control races and hazards.

If, as in Figure 6.61c, the feedback delay elements in an asynchronous circuit model are replaced by clocked D flip-flops—recall that the *D* means *delay* here—we obtain an extremely good approximation to feedback delay elements of a fixed size Δ. This follows from the fact that the common clock signal forces the outputs of all the flip-flops to change at essentially the same time. In effect, the clock period defines the delay Δ, and this one delay is shared by all the flip-flops. The use of clocked flip-flops in the feedback paths has the further advantage that when disabled, they do not respond to (narrow) input pulses and so mimic the glitch-filtering properties of inertial delays. Thus, the synchronous circuit design of Figure 6.61c, which is quite practical, avoids the race and hazard problems that bedevil asynchronous design. Consequently, synchronous circuits are used wherever possible in practice, so we will focus on synchronous logic design in the remaining chapters.

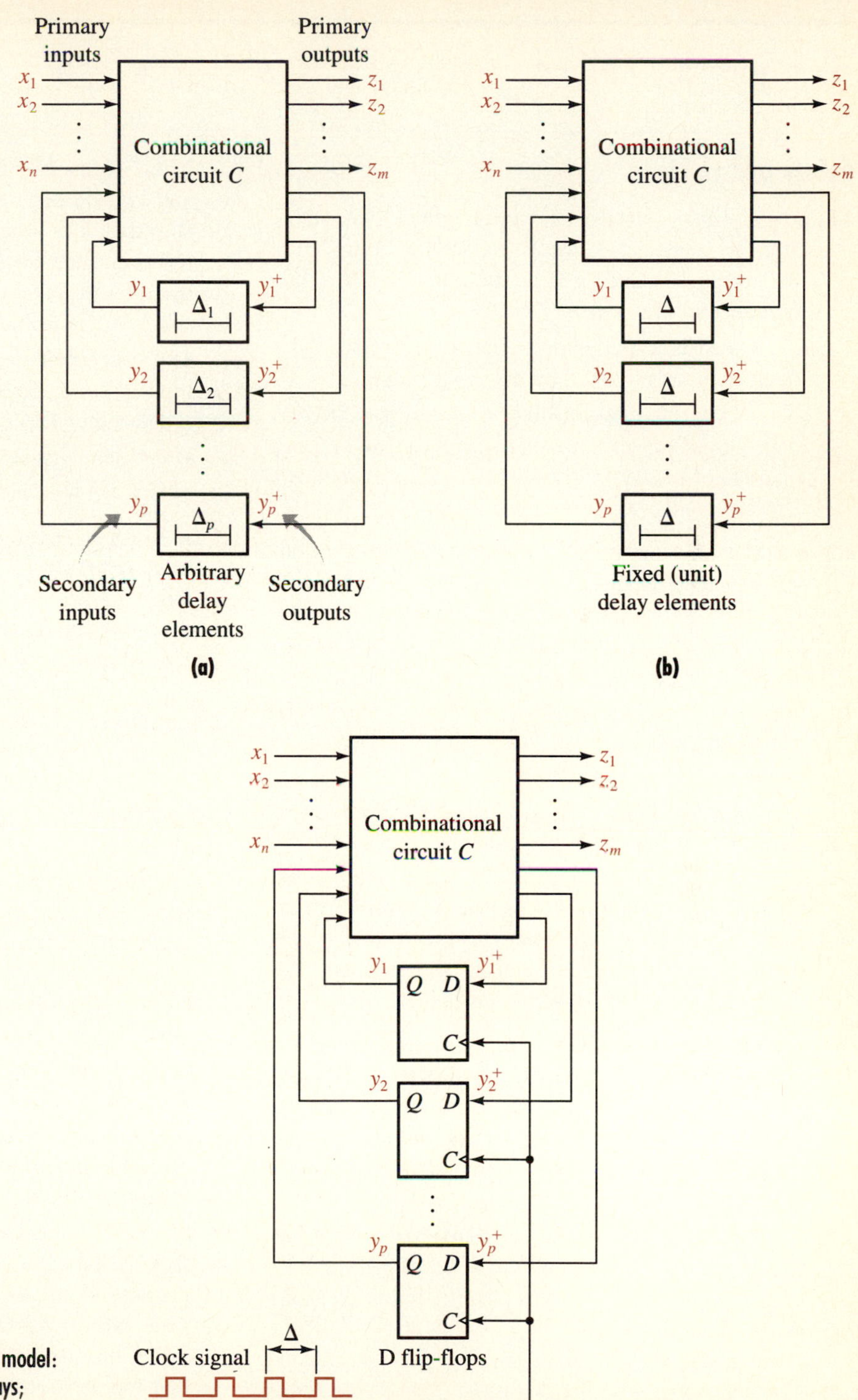

FIGURE 6.61 Variants of the Huffman model: (a) basic asynchronous with arbitrary delays; (b) synchronous with unit delays; (c) synchronous with D (delay) flip-flops.

CHAPTER 6 SUMMARY

Delays and Latches

6.1 ▪ A binary storage device requires two stable states to represent 1 and 0, as well as a provision for storing (writing) a new state and for retrieving (reading) an old state.

▪ A **well** is a technology-independent model of a memory element based on energy storage.

▪ A **dynamic random-access memory** or **DRAM** is an integrated circuit designed to store very large amounts of binary information in a form that allows it to be rapidly stored and retrieved.

6.2 ▪ The physical quantities that carry logic signals are inherently analog rather than digital, and cannot switch instantaneously between the 0 and 1 levels.

▪ Among the types of delays a signal experiences as it travels are **gate delay**, which physical gates impose on the signals that pass through them, and **line delay**, which is associated with interconnection media.

▪ Signal propagation delays are modeled in logic circuits by means of delay elements.

▪ Actual gate and line delays are difficult to estimate. Manufacturers often provide **typical**, **minimum**, and **maximum delays** for each device they produce. Designers must ensure that a circuit behaves correctly with **worst-case delays**.

6.3 ▪ Propagation delays endow gate-level logic circuits with memory. **Feedback** is essential to the process of providing a practical and stable memory element that is capable of storing information indefinitely. However, some forms of feedback cause **instability**.

▪ A **metastable signal** is temporarily stuck between 0 and 1. Metastability can occur occasionally in any logic signal and results in unusually long propagation delays.

▪ A **bistable circuit** has two stable states. An SR latch is a bistable circuit composed of two NOR gates. It has two input signals S (set) and R (reset) and two output signals Q and $\overline{Q}$, which define the stored data or state of the latch. The SR latch is set, reset, and left unchanged by the SR input combinations 10, 01, and 00, respectively. The combination $SR = 11$ is forbidden.

▪ A **D latch** does not have the restriction on its input patterns found in an SR latch. A D latch has a data input D and an enable input C.

▪ A latch, such as the D latch, that includes an enable signal is called a **gated latch**.

6.4 ▪ Proper operation of a latch requires that certain rules be observed concerning the times at which its input signals change value. The **setup time** is the time that the data signal is held at a new value immediately before the latching point. The **hold time** is an additional period after the latching point, during which the data signal is maintained at its new value. The **enable** or **pulse-width time** is the time during which the control signal is at its active or enabling value.

▪ In general, the **state** of a circuit at a particular time is defined as the current logic values of some set of signals of interest. The **internal state** is the combination of the internal signal values that represent all the information currently stored in the circuit. The **total state** combines the internal state with the current value of all externally applied or primary input signals.

▪ The operation of a sequential circuit is described as a sequence of **state transitions** controlled by inputs. For each total state, the circuit produces a set of output values.

▪ **State tables** define a circuit's next-state functions; they have columns for every combination of primary input variables, and rows for every internal state. **State diagrams** have circles (nodes) that represent all internal states, and arrows between nodes that represent all possible state transitions.

Clocks and Flip-Flops

6.5 ■ Sequential logic circuits that have global feedback require a bistable memory element with more elaborate timing logic than that of a latch. This need is met by the **flip-flop**, which employs a clock signal to control precisely the times at which memory changes state and produces new outputs.

■ Gated latches use **level triggering** to control the timing of state changes. Level triggering is **transparent** in that the state may continue changing as long as the enable signal is asserted.

■ In **edge triggering**, the most common clocking method in modern flip-flop design, a transition or edge of the clock signal causes the required actions to take place.

■ Edge triggering reduces sensitivity to **noise,** which consists of spurious, short-lived pulses, by restricting state changes to the small time window around the latching transition.

■ A **delay** or **D flip-flop** is the counterpart of a D latch, in which edge triggering is used. **JK flip-flops** are practical forms of set-reset flip-flops, which have two data inputs that serve as the set and reset lines. The JK flip-flop toggles (changes state) if both the set and reset inputs are asserted.

6.6 ■ There are several other clocking methods for flip-flops; for example, pulse triggering and multiphase clocking. Pulse triggering is rarely used.

6.7 ■ The state behavior of a flip-flop can be described by a state table, a state diagram, or a **characteristic equation**. A characteristic equation is a Boolean equation that gives the flip-flop's new state as a function of its current state and current inputs.

■ A set of identical flip-flops with common control logic is called a **register**. A **shift register** is distinguished from an ordinary (parallel) register by its use of serial or bit-by-bit data transfer.

■ Minor modifications to a shift register can produce a variety of useful counting circuits, such as ripple, ring, and Johnson counters.

Asynchronous Circuits

6.8 ■ **Hazards** can produce incorrect values, or glitches, in logic signals. **Static hazards** cause temporary changes in signals that should not change. **Dynamic hazards** produce spurious pulses in changing signals.

■ Some, but not all, hazards can be removed with design techniques that include redundant gates. Hazards are best controlled by the clock signals used in synchronous design.

6.9 ■ A circuit is considered **asynchronous** if it does not employ a periodic clock signal to synchronize its internal changes of state. A **synchronous circuit** employs an independent, periodic clock signal that synchronizes all internal state changes.

■ Asynchronous techniques are needed to design memory devices, circuits with unpredictable input timing, and circuits with several clocks.

■ Asynchronous circuits are potentially faster than synchronous circuits, but are very difficult to analyze and design.

■ The basic model for analyzing asynchronous circuits is the **Huffman model**, which restricts the location of circuit delays and the times when primary inputs can change.

6.10 ■ In analyzing asynchronous circuit behavior, we distinguish **stable states**, which do not change with constant primary inputs, from **unstable states**, which do.

■ A **race** occurs when two or more state variables change in response to a single input change. The race is **critical** if the final stable state depends on the order in which the state variables change.

■ Asynchronous circuits must be designed to avoid critical races and various types of hazards.

Further Readings

Latches and flip-flops receive a thorough treatment with proper modern notation in E. J. McCluskey's text, *Logic Design Principles* (Prentice-Hall, Englewood Cliffs, N.J., 1986). There is a useful discussion of the subtleties of triggering methods and their symbolism in the standards document *IEEE Standard Graphic Symbols for Logic Functions* (Institute of Electrical and Electronics Engineers, New York, 2nd ed., 1987). A lengthy and clear chapter on asynchronous circuit theory appears in A. D. Friedman and P. R. Menon, *Theory and Design of Switching Circuits* (Computer Science Press, Woodland Hills, Calif., 1975). The historical development of the Huffman model and alternative models, along with the controversies they have provoked, are discussed and analyzed in G. C. Langdon, Jr., *Logic Design: A Review of Theory and Practice* (Academic Press, New York, 1974).

Chapter 6 Problems

Delays and Latches

Storage concepts; propagation delays; delay calculations; feedback and stability; state behavior; SR and $\overline{S}\overline{R}$ latches; D latches; other latches.

6.1 Define the terms *attenuator* and *well*. Identify a mechanical and an electrical quantity that is modeled by each of these concepts.

6.2 A DRAM has the **destructive readout property**, meaning that the process of reading a memory cell may destroy its contents. Explain why this is so at the switch level and outline a method for handling this problem.

6.3 Using the propagation delay data given in Table 6.1, redraw the timing diagrams of Figure 6.12 for **(a)** 74S10 NANDs and **(b)** 74AS10 NANDs.

6.4 Consider the three-input, two-output NAND circuit of Figure 6.62. Let the gates have the following timing parameters:

$$t_{\rm PLH}^{\rm min} = 2 \text{ ns} \qquad t_{\rm PLH}^{\rm max} = 5 \text{ ns} \qquad t_{\rm PHL}^{\rm min} = 1.5 \text{ ns} \qquad t_{\rm PHL}^{\rm max} = 4.5 \text{ ns}$$

Calculate the same four parameters for the entire circuit, considering each of the six input/output pairs x_i, z_j, where $i = 0, 1, 2$ and $j = 0, 1$.

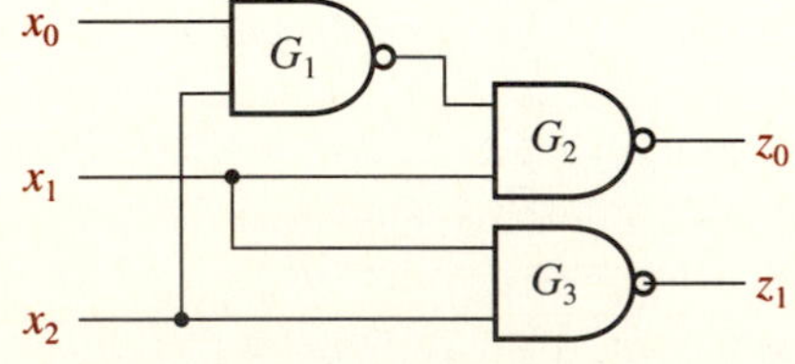

FIGURE 6.62 A small NAND circuit.

A circuit's longest path does not necessarily define its longest delay.

6.5 Suppose that all gates have a fixed propagation delay τ. The depth or number of levels d of a combinational logic circuit usually implies the circuit's maximum input/

output propagation delay t_{pd}^{max} is $d\tau$. This is not always the case, however. Analyze the five-level circuit of Figure 6.63 and show that $t_{pd}^{max} = (d-1)\tau = 4\tau$. This means that no sequence of signal changes ever passes entirely through any of the five-gate paths in the circuit; such a path is therefore referred to as a **false path**. False paths are often overlooked, leading to pessimistic assessments of a circuit's worst-case delay.

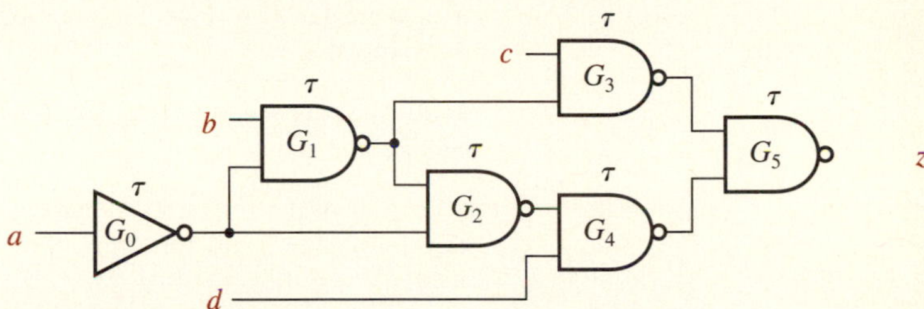

FIGURE 6.63 A five-level circuit that has a maximum propagation delay of less than 5τ.

6.6 Consider a one-input, one-output circuit C formed by chaining together (connecting in series or cascading) k inverters. Each inverter is characterized by the following delay parameters: $t_{PLH}^{min} = 2$ ns, $t_{PLH}^{max} = 7$ ns, $t_{PHL}^{min} = 3$ ns, $t_{PHL}^{max} = 10$ ns. **(a)** Let a 1-pulse of width 50 ns be applied to the primary input of C. Determine the minimum and maximum size of the resulting pulse at the output of C for $k = 1$ and 2. **(b)** Repeat part (a) for a 0-pulse of width 50 ns. **(c)** If, as in part (a), a 1-pulse of width 50 ns is applied to C, what is the maximum chain length k such that the output pulse from C can be guaranteed to have width 20 ns or greater?

Inertial delays and other complex delays can often be simulated.

6.7 An **inertial delay element** *ID* is a delay element that responds only to input changes that persist for some minimum time t_{id}. In other words, input pulses of width less than t_{id} are ignored—that is, filtered out—by *ID*. Pulses of width t_{id} or more are transmitted through *ID*, with a pure delay of t_{id}. Some CAD systems allow a designer to assign an inertial delay value to a gate, but most lack this feature. In the latter case, we can construct a "workaround" model for an inertial delay from a pure delay and some zero-delay gates, as shown in Figure 6.64. Prove that this circuit does indeed simulate a well-behaved inertial delay element by analyzing its response to 0- and 1-pulses of various widths that are applied to line x.

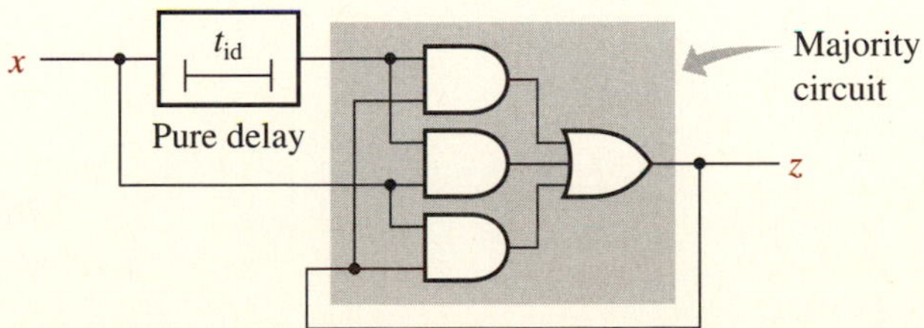

FIGURE 6.64 A circuit that simulates an inertial delay.

6.8 Discuss briefly the significance of feedback in sequential circuit design. Demonstrate by an example (preferably) or a suitable argument that it is possible for a combinational function Z to be realized by a circuit that (1) contains feedback, and (2) is composed of zero-delay gates, none of which can be removed without altering Z.

6.9 Show that if a narrow pulse of width w is "trapped" in the inverter loop of Figure 6.16, it can circulate indefinitely, causing the latch to oscillate. Show also that if w is greater than some minimum pulse width w^{min}, no oscillation occurs. What is the value of w^{min}?

6.10 What is metastability, and why can it slow down the operation of a digital circuit?

6.11 Fill in the responses y_0 and y_1 of the SR latch (Figure 6.17) to the S and R waveforms that appear in Figure 6.65. Discuss any abnormal conditions that occur.

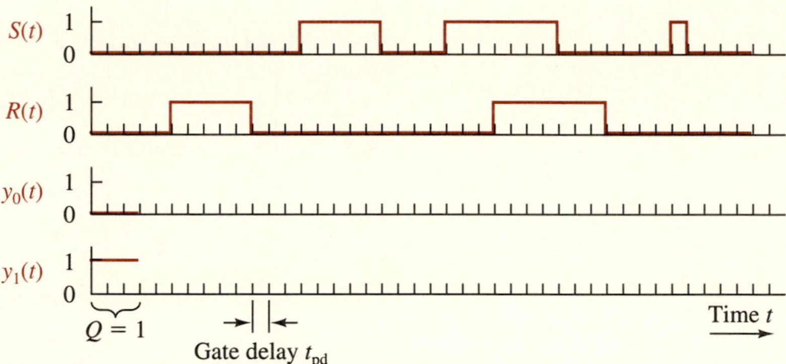

FIGURE 6.65 A timing diagram for Problem 6.11.

An $\overline{SR}$ latch is a NAND version of the basic set-reset latch.

6.12 Figure 6.66 shows a useful variant of the set-reset latch composed of two cross-coupled NAND gates and known as the $\overline{\mathbf{SR}}$ **latch**. Draw a timing diagram for this latch along the lines of Figure 6.18, showing both normal set and reset operations as in Figure 6.18a, and oscillatory behavior as in Figure 6.18b.

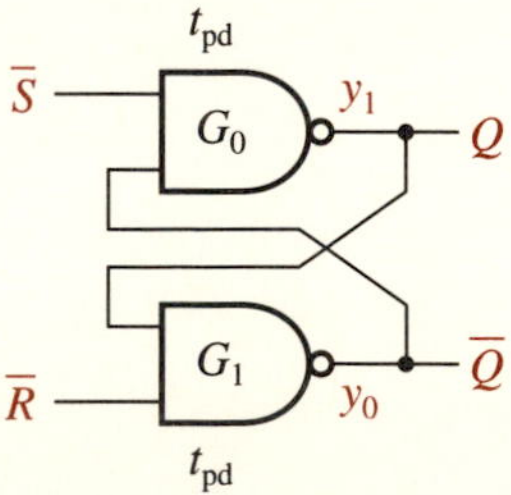

FIGURE 6.66 A logic circuit of an $\overline{SR}$ latch.

6.13 Fill in the responses of the D latch (Figure 6.19) to the D and C waveforms that appear in Figure 6.67. Discuss any abnormal conditions that can occur.

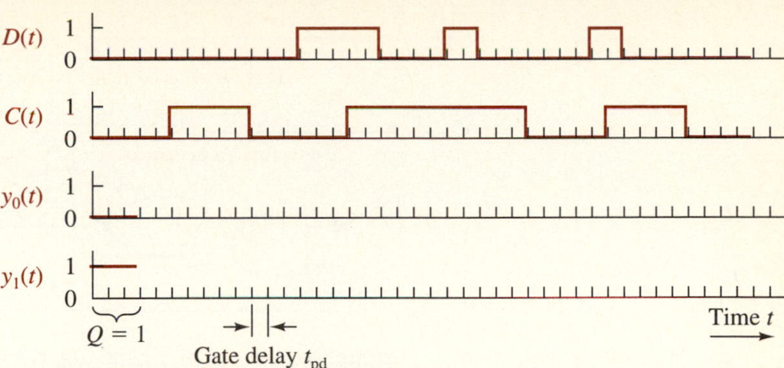

FIGURE 6.67 A timing diagram for Problem 6.13.

6.14 **(a)** Compute all values of t_{PHL} and t_{PLH} for the basic SR latch that appears in Figure 6.17, assuming all gates are characterized by $t_{pd} = 1$. **(b)** Compute all values of t_{PHL} and t_{PLH} for the gated SR latch of Figure 6.22 under the same assumptions.

6.15 Define what is meant by the minimum setup and hold times of a latch.

6.16 Construct a detailed state table and state diagram in the style of Figure 6.25 for the $\overline{\text{SR}}$ latch of Figure 6.66. Include all stable and unstable states and all input combinations.

6.17 Construct a state table and state diagram in the style of Figure 6.26 for the D latch of Figure 6.19.

CMOS latches and flip-flops use transmission gates extensively.

6.18 Figure 6.68 gives the logic diagram for the D latch that appears in the 74HC75 IC. This design employs transmission gates, which function as ideal switches and are a unique feature of CMOS technology (see Section 4.4 for details). Redraw the timing chart of Figure 6.21 for this D latch, assuming all the transmission gates and buffers in Figure 6.68 have the same delay t_{pd}.

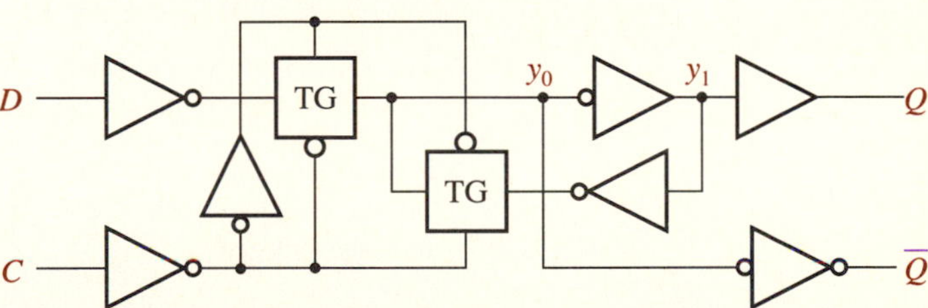

FIGURE 6.68 The logic circuit of the 74HC75 CMOS D latch.

6.19 Figure 6.69 shows the logic circuit of a gated set-reset latch taken from a certain CMOS standard cell library. The names of the latch's three input signals x_1, x_2, x_3 have been deliberately disguised. Analyze the circuit's behavior to determine the function of and the standard symbol for each input.

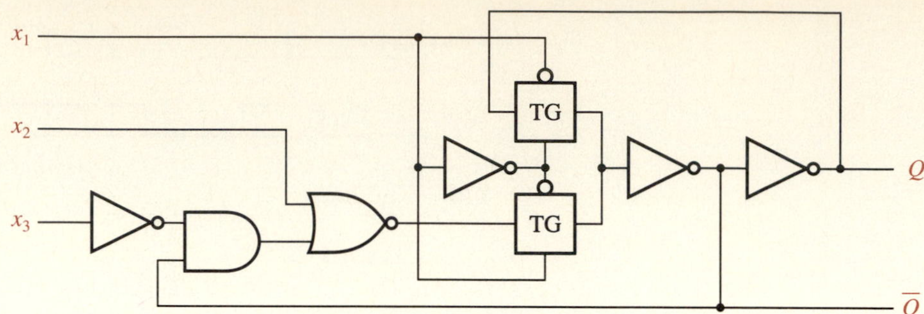

FIGURE 6.69 A mystery CMOS latch for Problem 6.19.

6.20 A circuit called the **JH latch** has been proposed. It is formed from the SR latch by replacing with a NAND gate the NOR gate with input S. The S and R inputs are then renamed J and H, respectively. Obtain a detailed state diagram for this circuit and comment on the circuit's usefulness as a memory element.

Clocks and Flip-Flops

Clocking modes; edge triggering; D and JK flip-flops; other flip-flops; shift registers; ripple, ring, and Johnson counters.

6.21 The period $t_1{:}t_2$, during which data inputs of the pulse-triggered flip-flop of Figure 6.29b can change, spans the entire active period of the clock ($C = 1$). Explain why the period $t_3{:}t_4$, during which output lines of the pulse-triggered flip-flop change, does not span the entire inactive (0) period of the clock C.

6.22 Fill in the output response lines in Figure 6.70 for D-type bistable circuits with each of the following triggering mechanisms: negative level triggering, positive pulse triggering, negative edge triggering.

6.23 Using a positive edge-triggered JK flip-flop and one or more additional gates, show how to implement a negative edge-triggered D flip-flop.

6.24 Show how to implement a positive edge-triggered D flip-flop using two SR latches and some additional gates.

6.25 Explain clearly why the master-slave D flip-flop of Figure 6.31 is considered to be edge-triggered, whereas the master-slave JK configuration shown in Figure 6.36 is not edge-triggered.

Data lockout is yet another flip-flop triggering mode.

6.26 A fourth triggering method for bistable elements, called **data lockout**, combines features of both pulse triggering and edge triggering. As in the pulse-triggering method, input data signals are enabled by the (positive) edge of the clock C. However, internal circuits automatically disable or "lock out" the data inputs a short time later, as in the edge-triggering case, without further reference to C. The output of the data-lockout flip-flop changes after C returns to 0, as in the pulse-triggered case. Data-lockout bistable circuits are denoted by the standard flip-flop symbols, with a dynamic signal indicator on the clock input *and* postponed signal symbols on the data outputs. **(a)** Fill in the response to Figure 6.70 for both positive and negative data-lockout D-type flip-flops. **(b)** Write a brief note comparing and contrasting the data-lockout flip-flop with the basic pulse-triggered and edge-triggered flip-flop types.

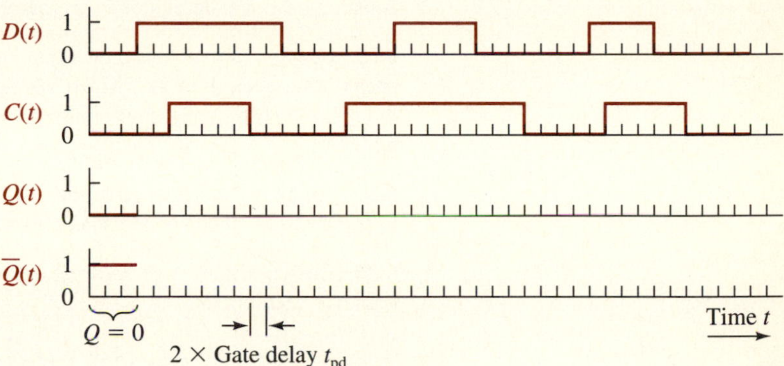

FIGURE 6.70 A timing diagram for Problem 6.22.

6.27 Suppose that the two D latches in the master-slave circuit of Figure 6.31 are changed to positive edge-triggered D flip-flops. (To indicate this requires adding a dynamic symbol to the C input of each latch.) Describe the triggering mode of the resulting circuit and comment on the circuit's usefulness.

6.28 Construct a set of characteristic equations and a state diagram for **(a)** an SR flip-flop and **(b)** a D flip-flop.

6.29 Many flip-flops have asynchronous preset or clear inputs, or both. Analyze the JK flip-flop of Figure 6.38 to determine the behavior of its clear function. **(a)** What happens if the clock is triggered while the clear line is active ($\overline{CLR} = 0$)? **(b)** Show how to modify the flip-flop's circuit to remove the clear function and replace it with preset. **(c)** Estimate the minimum width that a clear pulse must have relative to the circuit's gate delays.

Here are two more flip-flop types: the toggle and set-dominant flip-flops.

6.30 A well-known flip-flop type, called a **toggle** (or **T**) **flip-flop**, is defined in Figure 6.71. Like a D flip-flop, it has a single data input T. Applying a constant $T = 1$ to this flip-flop causes it to change state once in every clock cycle; $T = 0$ is the quiescent condition. **(a)** Catalogs of 74X-series ICs do not include T flip-flops because they can easily be obtained from JK flip-flops. Explain why this is so. **(b)** Describe how to modify a T flip-flop to make it a JK flip-flop.

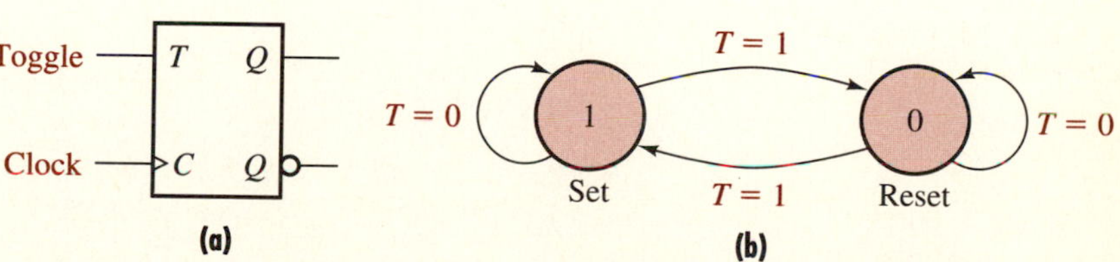

FIGURE 6.71 A toggle (T) flip-flop: **(a)** its graphic symbol; **(b)** its state diagram.

6.31 Another variant of the set-reset flip-flop that avoids the problems posed by simultaneously setting both data inputs S and R to 1 is the **set-dominant flip-flop**, which behaves like an SR or JK flip-flop, except that $(S, R) = (1, 1)$ always sends a set-dominant flip-flop to the set state $Q = 1$. Show how to modify a positive edge-triggered SR flip-flop to make it set-dominant. Construct a set of characteristic equations and a state diagram for your design.

6.32 Suppose that all gates, including the two NANDs, have approximately the same propagation delay τ in the 74LS107A JK flip-flop design of Figure 6.38. Reexamine the set operation depicted in Figure 6.39 and show the flip-flop will not now behave reliably.

6.33 Analyze the behavior of the 74LS74A D flip-flop of Figure 6.40 in the style of Figure 6.39, showing every step in a complete reset-to-set state transition.

6.34 Design a four-stage shift register composed of JK flip-flops. Give a state diagram for your circuit.

6.35 We observed that a shift register such as that of Figure 6.46 will not work properly if its edge-triggered flip-flops are replaced by level-triggered latches. Will it work reliably if the flip-flops are pulse-triggered?

6.36 A modulo-10 or **decimal counter** is to be designed using n JK flip-flops. What should n be for each of the following three counter types: ripple, ring, and Johnson?

6.37 What is the largest value of n such that an n-bit Johnson counter and an n-bit ripple counter have exactly the same modulus?

6.38 Redesign the 4-bit ripple counter of Figure 6.47 so that it counts down rather than up—that is, so that each x-pulse causes the count N to change as follows:

$$N \rightarrow N - 1 \qquad (\text{modulo } 2^4)$$

6.39 Design a 5-bit Johnson counter in the style of Figure 6.49. Again use D flip-flops as your only components. How many states has the counter? Construct a state diagram for your design.

6.40 An attractive feature of the modulo-m Johnson counter is that the count N can easily be decoded into m output signals $z_0, z_1, \ldots, z_{m-1}$ such that $z_i = 1$ if and only if $N = i$. In fact, each output z_i can be generated by a single NAND gate. Using NAND gates, design the decoding logic needed for the modulo-8 Johnson counter (Figure 6.49).

Asynchronous Circuits

Hazards and glitches; Huffman model; asynchronous circuit analysis; asynchronous circuit design; races; essential hazards.

6.41 Define each of the following terms: glitch, static 0-hazard, dynamic hazard, critical race, noncritical race.

6.42 Consider a two-level AND-OR realization N of a logic function $z(X)$ and assume that only one input variable can change at a time. Let A and B denote two input combinations applied to N that differ in exactly one variable x_i. Prove the following two results, which give necessary and sufficient conditions for N to contain static hazards. **(a)** Suppose that $z(A) = z(B) = 0$. N contains a static 0-hazard for the input transition $A \rightarrow B$ if and only if N contains an AND gate G that has x_i and $\overline{x}_i$ applied to it and is such that A and B both apply 1 to all other inputs of G. **(b)** Suppose that $z(A) = z(B) = 1$. N contains a static 1-hazard for the input transition $A \rightarrow B$ if and only if N contains no AND gate whose output is 1 for both A and B.

Static hazards due to single input changes can be eliminated from two-level circuits.

6.43 Prove that if a sum-of-products expression E for $z(X)$ consists of all prime implicants of $z(X)$, then a two-level circuit that implements E contains no static hazards with respect to single input changes. (This result does not hold for multiple input changes, however.)

6.44 Consider a two-level NAND-NAND circuit that implements the Boolean expression

$$z(a, b, c, d) = \overline{a}\overline{c}\overline{d} + \overline{a}\overline{b}c + \overline{a}bd + bcd$$

Using a timing diagram, identify a single-variable input transition that defines a static hazard. Show how to modify the circuit to eliminate all such hazards.

6.45 Identify a specific static hazard associated with the enable signal C in the D latch of Figure 6.19. Can this hazard cause the latch to enter an incorrect stable state? Explain how the propagation delays of the latch's gates can be assigned to minimize the likelihood of a glitch being produced by the hazard.

6.46 The D latch design of Figure 6.22, which is based on an SR latch, is frequently used in practice. Analyze it carefully to determine whether or not it is free from the hazard problem associated with the D latch of Figure 6.19.

6.47 What is meant by the fundamental mode of operating a sequential circuit? Explain by means of a small example what undesirable behavior can result from nonfundamental-mode operation.

6.48 Show that every asynchronous sequential circuit that is well formed under the gate-delay model has a Huffman model in the style of Figure 6.61a.

6.49 Identify the key differences between our two timing models for the ungated SR latch: the gate-delay model of Figure 6.17 and the feedback-delay (Huffman) model of Figure 6.54. Show how these differences predict distinctly different behavior patterns under certain circumstances.

Switch debouncing is an important application of SR latches.

6.50 Mechanical on-off switches such as the keys of a keyboard "bounce" imperceptibly when operated, so that instead of changing state once when pressed, a key opens and closes several times before settling at its final state. Consecutive bounces are separated by several milliseconds, and so are very slow, relative to electronic circuit speeds. The bounce problem is solved by connecting each key to a **debouncing circuit** of the kind depicted in Figure 6.72, which consists of an SR latch and a pair of pull-down attenuators. **(a)** Explain briefly how this circuit performs its debouncing function. **(b)** Construct a state table for the debouncing circuit that shows all stable and unstable states.

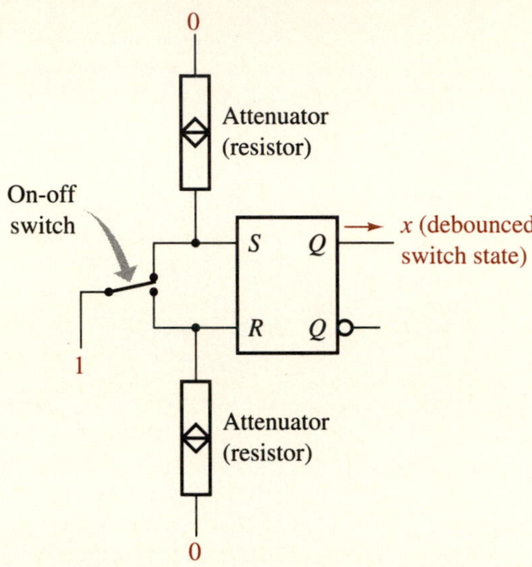

FIGURE 6.72 A debouncing circuit for a mechanical switch.

6.51 Construct a reduced state (flow) table for the Huffman model of the positive edge-triggered D flip-flop whose behavior is illustrated by Figure 6.31. Also, give a state assignment for this state table that is free of critical races. (*Hint:* Four states are sufficient.)

6.52 Using the Huffman model, derive a transition table and a state (flow) table for the circuit of Figure 6.73.

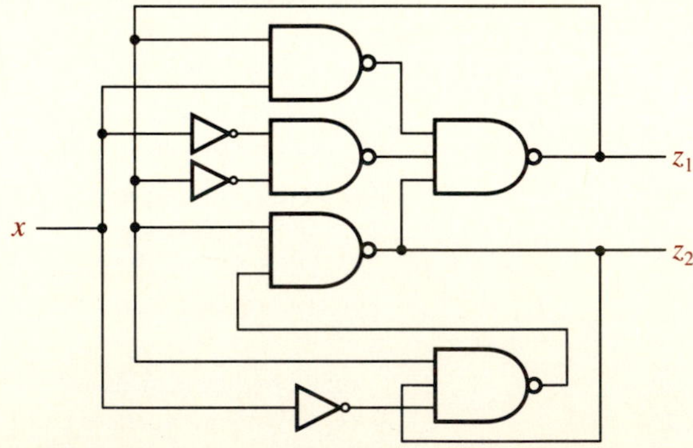

FIGURE 6.73 An example of an asynchronous circuit.

6.53 An asynchronous circuit that employs two-level AND-OR logic has the following next-state and output equations:

$$y_1^+ = y_1y_2 + y_1x_1 + \bar{x}_1\bar{x}_2\bar{y}_1$$
$$y_2^+ = x_1x_2 + \bar{x}_2y_1\bar{y}_2 + \bar{x}_1\bar{x}_2\bar{y}_1\bar{y}_2$$
$$z = y_1y_2 + x_2\bar{y}_1$$

Construct a logic circuit and a state (flow) table for this circuit.

6.54 A variant of the toggle (T) flip-flop of Figure 6.71 has the following specifications: As before, $T = 0$ is the quiescent state and $T = 1$ causes the state to change. However, when $T = 1$, the state should now change on both the rising *and* the falling edges of the clock signal C. Carry out the logic design of the modified flip-flop, giving a reduced flow table and a complete all-NAND implementation.

6.55 Prove that no reassignment of the two state variables y_1 and y_2 to the four states in the state table of Figure 6.55c can yield a circuit free of critical races. This leads to the gloomy conclusion that some n-state tables need more than $\lceil \log_2 n \rceil$ state variables for a race-free state assignment.

Essential hazards are design independent and hard to control.

6.56 A state table is said to contain an **essential hazard** if there is a stable state S for which three consecutive changes (for example, 0-to-1-to-0-to-1) in some input variable x_i take the circuit from S to a different stable state than the first change in x (0-to-1 in the example) would by itself. These hazards are inherent, and can be addressed either by modifying the state behavior, or by selectively inserting delays in feedback loops [Friedman and Menon, 1975]. Find an example of an essential hazard in the state table of Figure 6.59b for the modulo-4 edge counter.

6.57 **(a)** Using the definition given in Problem 6.56, construct a simple example of a state table that has three states and that contains an essential hazard. **(b)** Show that the D latch of Figure 6.58b contains no essential hazards.

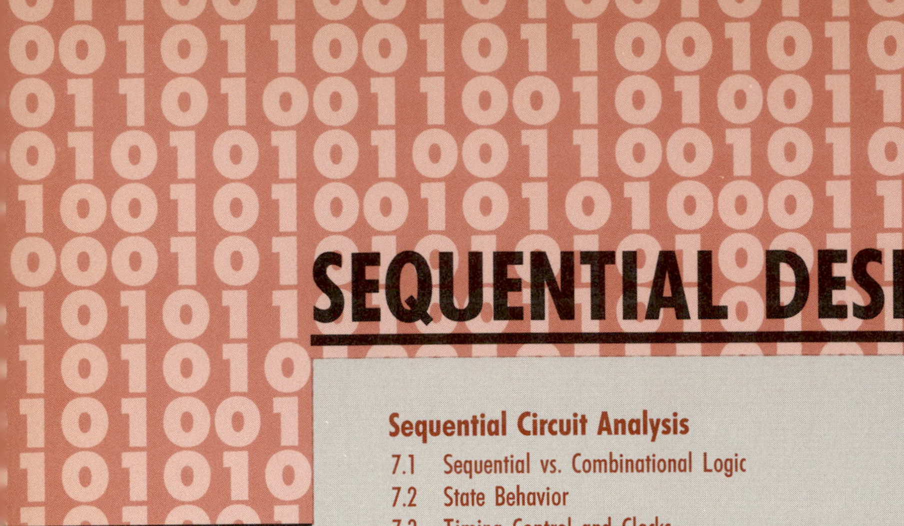

SEQUENTIAL DESIGN

Sequential Circuit Analysis

7.1 Sequential vs. Combinational Logic
7.2 State Behavior
7.3 Timing Control and Clocks

Sequential Circuit Synthesis

7.4 The Design Process
7.5 The Classical Method
7.6 State Minimization

Design and Testing

7.7 Faults and Tests
7.8 Test Generation
7.9 Design for Testability

Overview

Having examined gates and flip-flops—the basic building blocks of sequential circuits—in depth, we consider next the analysis and design of synchronous or clocked sequential circuits. We focus on small circuits, often referred to as finite-state machines, the state behavior of which can be completely defined at the gate level. This chapter presents, with numerous examples, a general methodology for the design of such circuits. It also discusses the important problem of testing logic circuits for physical faults.

Sequential Circuit Analysis

We start with a discussion of the general structure and behavior of synchronous sequential circuits, and contrast them with combinational circuits. This section then presents the concepts needed to analyze the state behavior of a sequential circuit, and examines in detail synchronization by means of clock signals.

7.1 Sequential vs. Combinational Logic

Sequential logic circuits differ from combinational circuits in that their output signals depend upon past as well as present input signals. This implies that a sequential circuit has memory, and so can store information derived from past inputs. The stored information constitutes the circuit's internal state. The circuits of interest here have clocked flip-flops as memories and use a periodic clock signal to synchronize the times at which all flip-flops change their individual states, and therefore the state of the entire circuit. Logic circuits of this kind are referred to as *synchronous circuits* and are contrasted with *asynchronous circuits,* which do not use clocks.

Sequential Circuit Structure

The building blocks of synchronous sequential logic circuits are gates and flip-flops. We will use all the usual gate types (AND, OR, NOT, NAND, NOR, EXCLUSIVE-OR, and EXLUSIVE-NOR), introduced in Chapter 3, as well as the edge-triggered D and JK flip-flop types, covered in Chapter 6. The gates and flip-flops may be connected in many ways to form a sequential circuit, with the flip-flops collectively constituting the memory M and the gates forming one or more combinational subcircuits $C_1, C_2, \ldots, C_q$. Although we exclude feedback from well-formed combinational circuits, we allow feedback between C_i and M in the sequential case.

Synchronous sequential circuits have memories that are composed of clocked flip-flops.

Figure 7.1 shows some representative well-formed sequential circuit types, indicating the permissible connections between C_i and M. In Figure 7.1a, no global feedback is present, limiting the behavior of this circuit to cases in which the next state Y^+ is independent of the present state Y. The next state Y^+ is a function of the primary inputs X only, and is determined by the **input logic** C_1. The primary outputs Z depend on Y only, as deter-

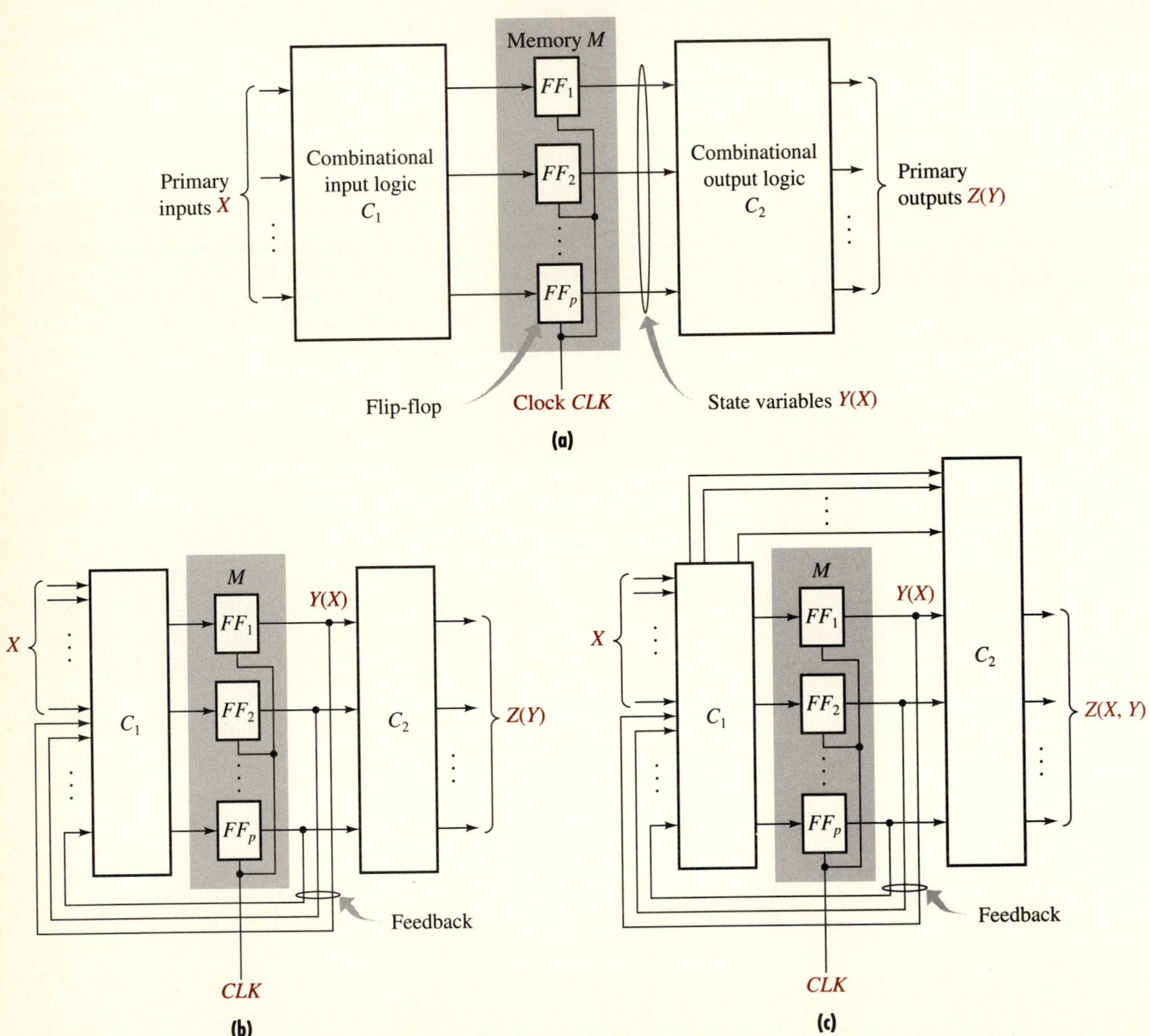

FIGURE 7.1 Synchronous sequential circuit structures: (a) no global feedback present; (b) primary outputs independent of primary inputs (the Moore model); (c) a general case (the Mealy model).

mined by the **output logic** C_2. There is no feedback in this circuit, apart from that within the flip-flops themselves. Sequential circuits of this type are quite limited in their behavior and can, in most cases, use gated latches instead of flip-flops.

In Moore-type circuits, the primary output Z is a function of the internal state Y only.

A more general circuit structure is depicted in Figure 7.1b, in which there is global feedback from the memory M to the input subcircuit C_1, which permits the next state to be a function of both X and Y. Again, however, Z depends on Y only. The most general case is illustrated by Figure 7.1c. Here Z, as well as Y^+, are functions of both X and Y. The sequential circuit model of Figure 7.1c is sometimes called the **Mealy model**. It is contrasted with the more restricted **Moore model** of Figure 7.1b, in which the output logic C_2 receives its inputs from M only. These circuit types are named after G. H. Mealy and Edward F. Moore, who first studied their behavior at AT&T Bell Laboratories in the 1950s [Mealy, 1955; Moore, 1956].

If we combine C_1 and C_2 of Figure 7.1c and the connections between them into a single combinational circuit C, we get the Huffman model discussed in Section 6.9, so the Mealy and Huffman models of synchronous sequential circuits are essentially the same. The structures of Figures 7.1c and 7.2 cover all the main synchronous sequential circuit types and are the only ones we consider. By restricting the input variables on which their combinational parts depend, each can be reduced to the special cases of Figures 7.1a and 7.1b. For example, by making the output Z of C_2 or C independent of X, we obtain a Moore-type circuit.

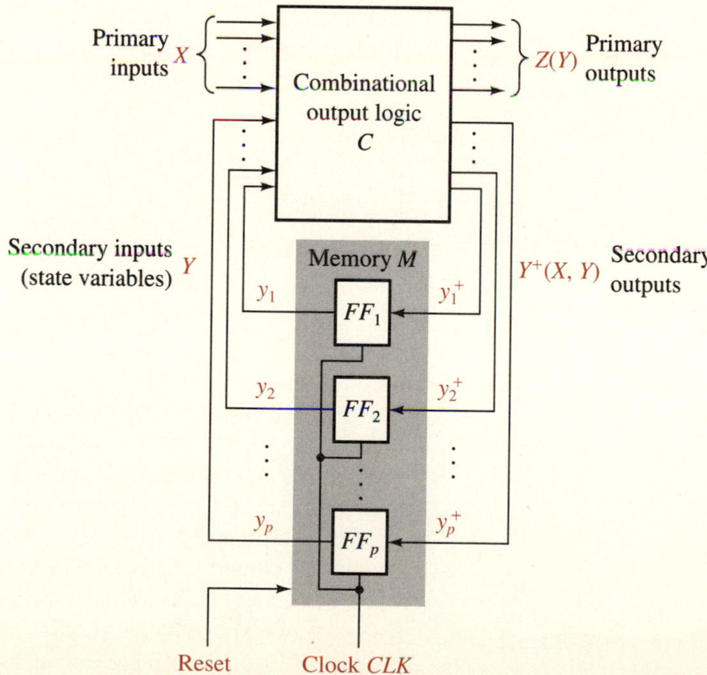

FIGURE 7.2 Another view of general synchronous sequential circuit structure (the Huffman model).

Main Signal Types

Figure 7.2 identifies the key types of signals that appear in a (synchronous) sequential circuit. The n externally controllable input signals $X = x_1, x_2, \ldots, x_n$ applied to the combinational logic C are referred to as the circuit's *primary input variables*. There are m externally observable output signals $Z = z_1, z_2, \ldots, z_m$, which are derived from C and are called the *primary output variables*. The p flip-flops that constitute the memory M supply p input signals $Y = y_1, y_2, \ldots, y_p$ to C and are called the *secondary input variables* or, more often, the (internal) *state variables*. The state Y represents all the information memorized by the circuit from its previous activities.

The combinational logic C generates a set of secondary output signals that determine the next state $Y^+ = y_1^+, y_2^+, \ldots, y_p^+$ of the sequential circuit. The form of these secondary output signals depends on the flip-flop type used. For D flip-flops, the secondary outputs directly define Y^+ as the D input signals of the flip-flops. In the JK case, Y^+ is indirectly defined via p pairs of J and K signals produced by C. We explicitly indicate the functional dependencies among the X, Y, Y^+, and Z signals by writing Y^+ and Z as $Y^+(X, Y)$ and $Z(X, Y)$, respectively.

At least $\lceil \log_2 P \rceil$ flip-flops are needed by a P-state circuit.

With p flip-flops, up to 2^p distinct states can be represented, each of which is assigned its own pattern of state-variable bits. Consequently, to implement a P-state sequential circuit, p must satisfy the inequality

$$p \geq \lceil \log_2 P \rceil \tag{7.1}$$

The smallest p that satisfies (7.1) is often chosen to minimize memory size and hardware cost. However, as we will see in Chapter 8, design and maintenance costs can sometimes be reduced by using as many as P flip-flops to store the P states.

Timing Signals

The key control signal in every synchronous sequential circuit is the clock signal *CLK*. As Figure 7.3 shows, *CLK* has a regular waveform that repeats once every T seconds, where T is the clock period or cycle. During each clock period, *CLK* exhibits a positive edge and a negative edge, one of which triggers a state change in the circuit controlled by *CLK*. Obviously, the triggering edge is determined by the type of flip-flop used. It is convenient to designate the clock cycle as beginning at one triggering edge of *CLK* and ending at the next; in Figure 7.3, a clock cycle begins and ends at the negative (falling) edge of *CLK*. During each cycle, the clock supplies a 1-pulse of duration T_1 and a 0-pulse of duration $T_0 = T - T_1$. The ratio T_1/T, which is typically around 0.5, is known as the clock's **duty cycle**.

A clock signal triggers a single change of state in every clock cycle.

When operated in normal synchronous mode, a sequential circuit experiences at most one change of state every clock cycle. The clock, therefore, has the effect of quantizing the analog time parameter t into a sequence of

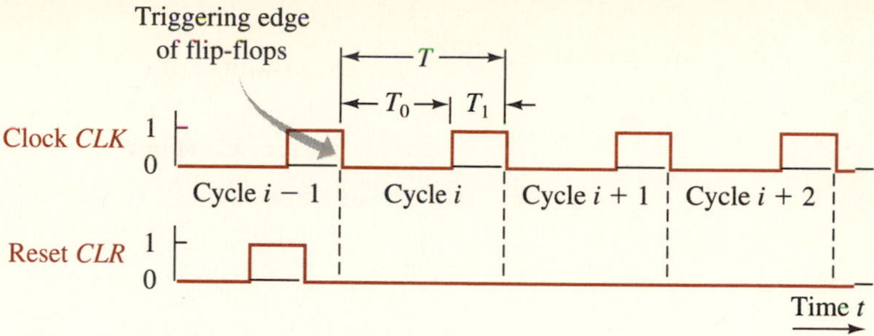

FIGURE 7.3 Typical waveforms of a clock signal and an asynchronous reset signal.

discrete time steps, each corresponding to a clock cycle. Hence, it is usual in this context to measure time in terms of sequences of clock cycles, denoted by the integers $i = \ldots, -1, 0, 1, 2, \ldots$. If we treat some some generic clock cycle i as the "present" time, as we often will, then we can refer to the cycle $i + 1$ as the "next" time period.

Recall that flip-flops can be equipped with asynchronous reset inputs to simplify state initialization. A clear signal *CLR* forces a flip-flop to the reset state $Q = 0$; a preset signal *PRE* makes $Q = 1$. In sequential circuits, the corresponding reset lines of all the flip-flops are connected to form a single common *CLR* or, less frequently, a common *PRE* line. Figure 7.3 illustrates the use of a *CLR* signal. It is activated once at the beginning of the circuit's operation, typically just after the power is switched on. A pulse of appropriate width and direction is applied to *CLR*, causing all the flip-flops to reset, independent of the state of *CLK* and the circuit's other input signals. In Figure 7.3, *CLR* happens to be activated in the middle of clock cycle $i - 1$, and initializes the circuit to some state S_0. When the next triggering clock edge arrives in cycle i, normal synchronous operation begins, with S_0 serving as the present state.

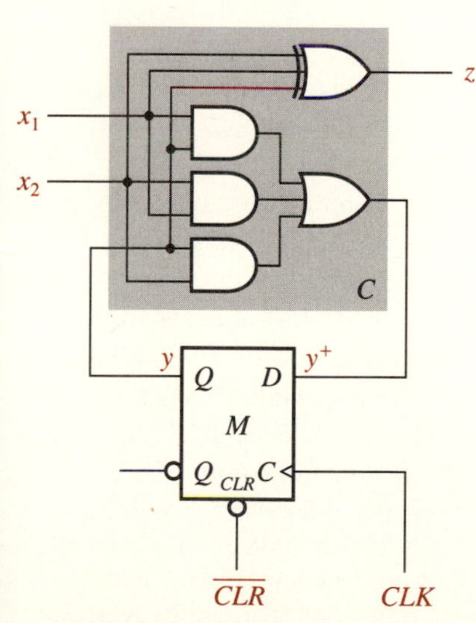

FIGURE 7.4 An example of a small sequential circuit, the serial adder *SADD*.

Sequential Adders

We illustrate the foregoing concepts using the small sequential circuit *SADD* that appears in Figure 7.4. This circuit has two primary inputs x_1 and x_2 and one primary output z. Comparing it with Figure 7.2, we see that *SADD* has the same overall structure, with a primary output function $z(x_1, x_2, y)$ and a secondary output function $y^+(x_1, x_2, y)$. *SADD* contains just one D-type flip-flop, implying that it has at most two internal states. The flip-flop's data output y is the circuit's sole state variable. Comparing Figure 7.4 with Figure 7.1c, we see that *SADD* also conforms to the general Mealy model. Its output circuit C_2 consists of the single EXCLUSIVE-OR gate that generates z, and its input circuit C_1 consists of the remaining four gates of C, which generate the secondary output signal $y^+ = D$.

Writing down the Boolean equations for C directly from the logic diagram, we get

$$\begin{aligned} z &= x_1 \oplus x_2 \oplus y \\ y^+ &= x_1x_2 + x_1y + x_2y \end{aligned} \tag{7.2}$$

The reader should recognize these as the equations for a full adder, where z is the sum function s_i, y is the carry-in function c_{i-1}, and y^+ is the carry-out function c_i. Consequently, based on our prior knowledge of addition logic, we can identify the combinational part C of *SADD* as a full adder. It is then apparent that the task of the flip-flop of Figure 7.4 is to store the carry bit produced in the previous clock cycle. The stored carry is added to the primary input signals x_1 and x_2 applied to *SADD* in the next clock cycle to produce a new sum bit and a new carry bit. The sum is made available to the outside via z; the new carry bit is stored as y in the flip-flop.

We conclude that *SADD* can add two arbitrarily long unsigned binary numbers that are applied serially to the two primary input lines. One pair of input bits and any previously stored carry bit are added in each clock cycle, and one result (sum) bit and one carry bit are produced. *SADD* is therefore called a **serial adder** and is the simplest but slowest implementation of binary addition.

The simplest adder is a sequential circuit—a serial adder.

Combinational vs. Sequential Adders

Some basic differences between combinational and sequential logic can be seen by comparing the serial adder of Figure 7.4 to the various "parallel" adders studied in Section 5.9. Figure 7.5a repeats the 4-bit ripple-carry adder *RC* of Figure 5.55, and shows it performing the binary addition 1001 + 0101 = 1110 or, in decimal terms, 9 + 5 = 14. *RC* consists of four copies of a full adder C and performs the addition in one combinational step, which we can equate with a single clock cycle.

A serial adder adds two n-bit numbers in n clock cycles; a combinational adder does it in one.

The serial adder *SADD*, on the other hand, contains just one copy of C and requires four consecutive clock cycles $i = 1, 2, 3, 4$ to perform the same 4-bit addition. As shown in Figure 7.5b, the operands must be applied to *SADD* bit by bit, with the least significant bits first. Just after the start of cycle 1, the two least significant bits of the input operands of *SADD*, which are both 1 in the figure, are applied to the full adder C, along with an initial carry bit $y = 0$. The resulting sum bit 0 appears immediately on z, and the new carry bit 1 is applied to the input of the D flip-flop, into which it is loaded at the start of cycle 2 when the flip-flop is triggered by the positive edge of the clock *CLK*. In cycle 2, the next two operand bits $x_1x_2 = 00$ are applied to C, along with the carry $y = 1$ produced in cycle 1. This addition results in $z = 1$ and a new carry bit of 0, which is stored for use in cycle 3, and so on. The precise times at which the various input signals change must satisfy all the timing constraints (setup, hold, and pulse-width times) of the D flip-flop.

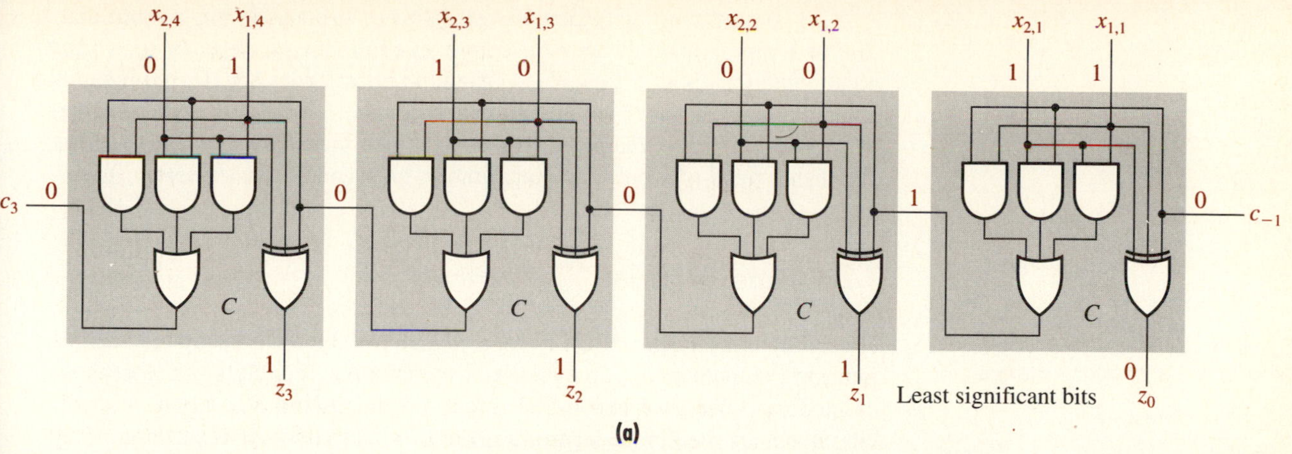

(a)

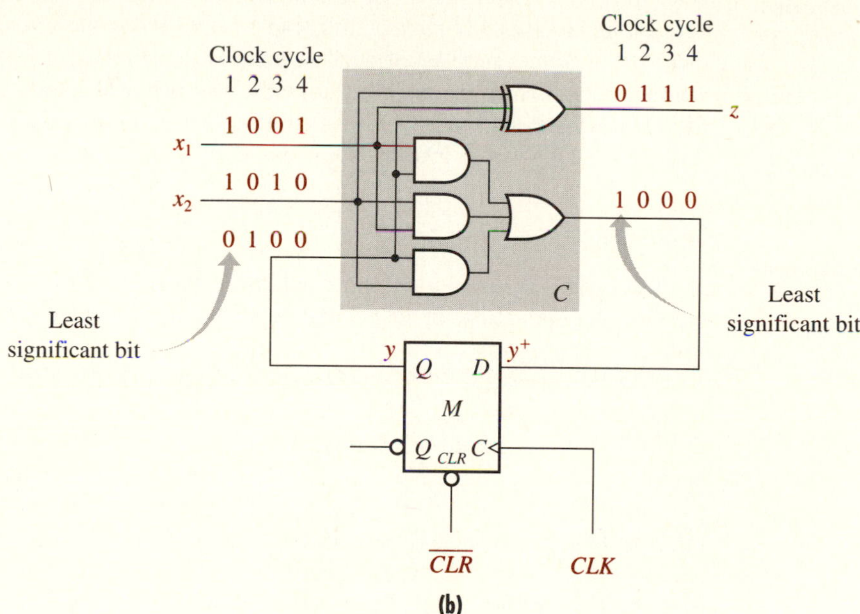

(b)

FIGURE 7.5 Four-bit binary addition as performed by (**a**) a combinational (ripple-carry) adder and (**b**) the sequential (serial) adder *SADD.*

For proper operation of the serial adder, the initial value of the state variable y should be 0 before the first (least significant) bits of x_1 and x_2 are applied (in clock cycle 1 in the case of Figure 7.5b). This requires that the D flip-flop be cleared (in clock cycle 0) before each new n-bit addition begins. Initialization is easily arranged by activating the flip-flop's asynchronous clear line $\overline{CLR}$. *SADD* has an inversion built into its clear line, so a single 0-pulse (whose minimum width approximates that of a clock pulse) is applied to $\overline{CLR}$ in order to initialize y to 0. During subsequent operation of the adder, $\overline{CLR}$ must be held constant at its inactive (1) level.

On generalizing this example from 4-bit to n-bit addition, we see that, to a first approximation, an n-bit combinational adder such as *RC* performs binary addition about n times faster than the serial adder *SADD*, but requires about n times as much combinational hardware.[1] Thus, by switching between combinational and sequential designs like these, we can often achieve significant trade-offs between computation time and hardware cost.

7.2 State Behavior

It is not possible to list explicitly all a sequential circuit's input/output sequences.

The behavior of a sequential circuit is defined by the relationship between its input and output signals. In the case of combinational circuits—at least those of moderate size—we can list all input signal possibilities, along with the output signals they produce, in the form of a truth table. The corresponding behavioral description of a sequential circuit would require us to list all possible input/output *sequences* that can be applied to the circuit. Because there is an infinite number of such sequences and the sequences can be of arbitrary length, such an explicit truth table-like listing is totally impractical, even for the simplest circuits. For example, in the serial adder case (Figure 7.4), this listing would take the form of Figure 7.6.

Input sequence 1	Input sequence 2	Ouput sequence
···000000	···000000	···000000
···000000	···000001	···000001
···000001	···000000	···000001
···000001	···000001	···000010
···000010	···000000	···000010
···000010	···000001	···000011
⋮	⋮	⋮

FIGURE 7.6 The behavior of the serial adder described in tabular form.

State Tables

To deal with this infinity of input/output sequences, we observe that the number of combinations of the primary input and internal state values is finite. With n input variables X and p state variables Y, the number of different X, Y combinations is 2^{n+p}, which can be dealt with completely, provided n and p are not too large. If we list each X, Y primary input/present-state combination, along with the corresponding Y^+, Z next-state/primary output combination, in a truth table-like format, we obtain a state table, examples of which we encountered in Chapter 6. By threading through k entries of the state table, we can find the output response sequence and the next-state sequence that result from the application of any input sequence of length k to the circuit in question. Thus, a state table contains all the circuit's input/output sequences in an implicit and finite form.

Figure 7.7a shows a state table that specifies the behavior of the serial adder *SADD* (Figure 7.4) in the conventional format. Each row represents a distinct state combination Y, and each column represents a distinct primary input combination X. An entry of the form Y^+, Z at the intersection of row

[1]This comparison is inexact due to the fact that the add time of the ripple-carry increases with n due to its longer propagation delays. Increasing n effectively stretches *RC*'s "cycle" time. Faster combinational designs like the carry-lookahead adder aim to hold this time constant, but at the expense of more hardware. Note that carry signals can travel much faster between the stages of *RC* than they can through the carry flip-flop of *SADD*.

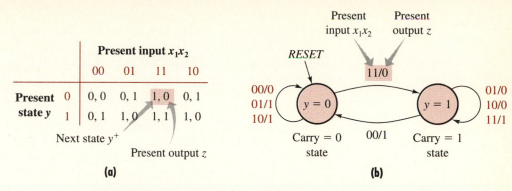

		Present input x_1x_2			
		00	01	11	10
Present state y	0	0, 0	0, 1	1, 0	0, 1
	1	0, 1	1, 0	1, 1	1, 0

Next state y^+ Present output z

(a)

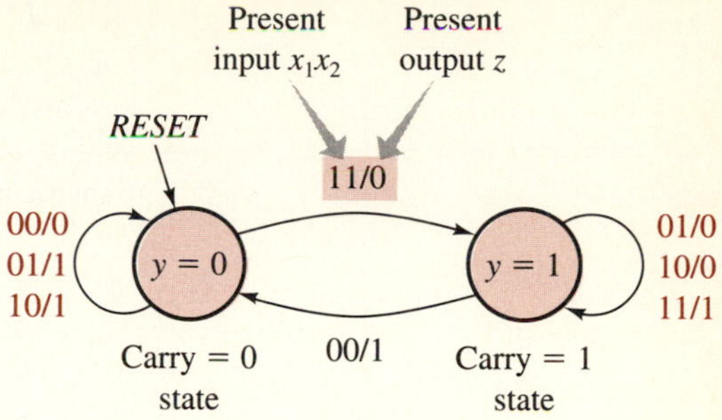

(b)

FIGURE 7.7 Behavioral descriptions of the serial adder: **(a)** a state table; **(b)** a state diagram.

Y and column X gives the next state Y^+ and the present output Z that result from applying X to the circuit with Y as its present state. All the information needed to compute this particular state table is contained in Equations (7.2). For example, consider the boxed entry in Figure 7.7a. The present state is $Y = y = 0$ and the present input is $X = x_1x_2 = 11$. From Equations (7.2) for *SADD*'s combinational logic C, we obtain

$$\begin{aligned} z &= x_1 \oplus x_2 \oplus y \\ &= 1 \oplus 1 \oplus 0 = 0 \\ y^+ &= x_1x_2 + x_1y + x_2y \\ &= 1 \cdot 1 + 1 \cdot 0 + 1 \cdot 0 = 1 \end{aligned}$$

yielding the next-state/primary output values $y^+, z = 1, 0$ entered in the table.

State Diagrams

State tables and diagrams describe implicitly all a sequential circuit's input/output sequences.

A state diagram, examples of which we also saw in Chapter 6, conveys, in graphical form, the same information as a state table. In this case, an internal state Y is represented by a node or a circle, and arrows between the nodes define state transitions. Each state-transition arrow is labeled with the primary input and output signals associated with the transition; we write this label in the form X/Z, where a slash serves to separate the input from the output signals. We also use the equivalent in-text format

$$Y \xrightarrow{X/Z} Y^+$$

Figure 7.7b shows a complete state diagram for the serial adder. The specific state transition discussed above corresponds to the arrow with the boxed label, and may also be written as

$$0 \xrightarrow{11/0} 1$$

To avoid cluttering the state diagram with arrows that have the same source and destination states, such arrows may be replaced by a single arrow with several labels. For instance, the single self-loop arrow in Figure 7.7b from state $y = 0$ back to itself has three separate labels that denote the three state transitions

$$0 \xrightarrow{00/0} 0 \qquad 0 \xrightarrow{01/1} 0 \qquad 0 \xrightarrow{10/1} 0$$

The arrow marked *RESET* indicates that the state it enters is the circuit's initial or reset state. Because it is an asynchronous signal and may be applied in any state, *RESET* is drawn as coming from nowhere in particular.

Figure 7.8 demonstrates how the state diagram can be used to determine the response of a sequential circuit to a particular input sequence, in this case, the sequence

Cycle:	**...**	**1**	**2**	**3**	**4**
x_1:	...	1	0	0	1
x_2:	...	1	0	1	0

of length 4 that is applied to the serial adder *SADD* in Figure 7.5b. We begin at the specified initial state $y = 0$, which is established via the asynchronous clear line $\overline{CLR}$. If we begin with a different initial state, we will, in general, obtain a different response. The first input combination $x_1x_2 = 11$ applied to *SADD* in cycle 1 causes a transition to the state $y = 1$, as indicated by the heavy arrow marked "Cycle 1" in Figure 7.8. The label on this transition is 11/0, telling us that $z = 0$. The input in cycle 2 returns the circuit to the state $y = 0$, in which it remains for the next two cycles. Listing the four occurrences of z encountered as we trace through the adder's state diagram gives us the desired output response—namely, 0111.

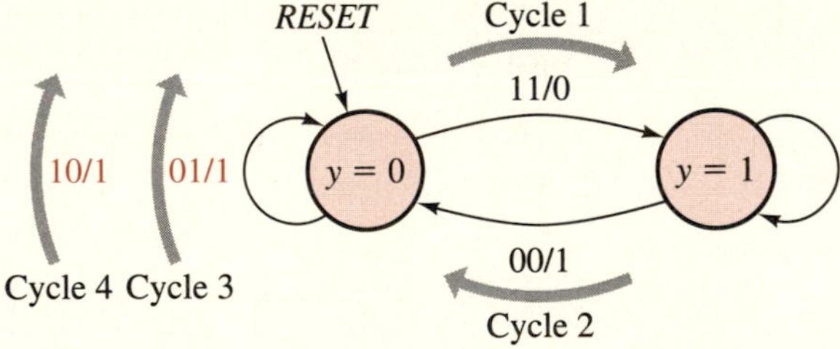

FIGURE 7.8 Determining the response of the serial adder to a specific input sequence of length 4.

Timing Considerations

Edge-triggered flip-flops have several properties that simplify the design of sequential circuits. They react to new input data values that appear on the secondary outputs of the combinational logic C only in a narrow time window around the triggering clock edge at the start of each clock cycle. At other times, changes in these signals do not affect the memory logic M. As we will discuss in detail later, the clock's period and pulse width must be designed so that the propagation delays through C and M meet certain timing constraints. Once these easily determined constraints are satisfied, the problem of races, or multiple state changes in the same clock cycle, is eliminated from the design of synchronous circuits. Edge-triggered flip-flops are also relatively insensitive to noise on their primary input or internal lines due to hazards in C, power-supply fluctuations, or other causes. (The race and hazard problems that arise in asynchronous circuits, which lack the protection afforded by clock signals, are examined in Sections 6.8, 6.9, and 6.10.)

A timing diagram for the serial adder *SADD* appears in Figure 7.9, which shows the signal waveforms that correspond to the state transitions marked on Figure 7.8. Observe how the various input/output signals of C appear as pulses, because they can assume different values within each clock period, the duration of which is T. The key value of an input data signal x_j in a clock period i is the value present during a brief active window around the triggering edge of the clock *CLK*. The width of this window is determined by the setup and hold times of the flip-flops.

In general, a primary output signal z_h is a function of x_j, so a change in x_j can immediately affect z_h. For example, the narrow 1-pulse in x_1, which is an externally generated glitch in the middle of cycle 3 of Figure 7.9, produces a corresponding glitch in the form of a 0-pulse in z. The 1-pulse does not affect the state y, because it has disappeared by the time the flip-flop is triggered at the start of cycle 4. To produce a single well-defined output value or state change in any clock period, each data input should be

Input signals should be held stable long enough to produce proper output and state changes.

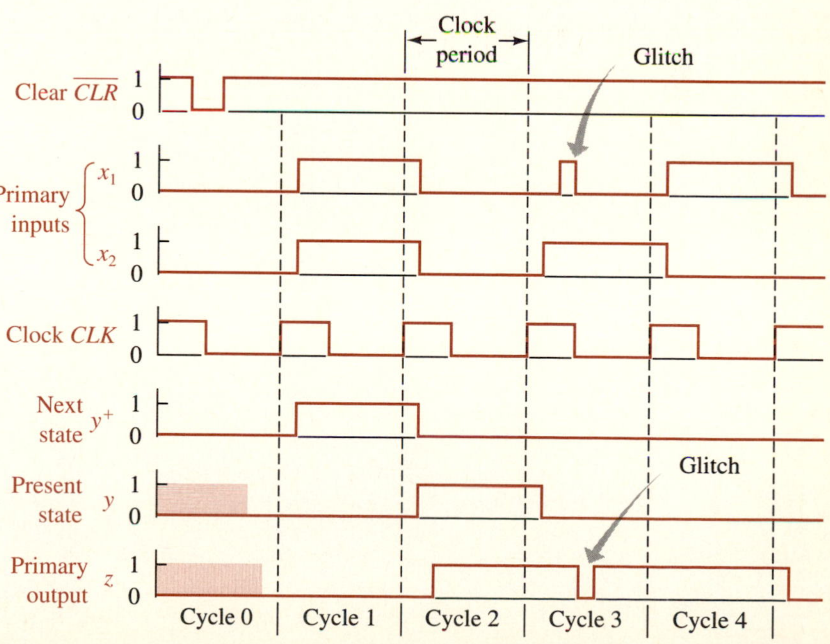

FIGURE 7.9 A timing diagram for the serial adder *SADD*, corresponding to Figure 7.8.

held constant for most of the clock period—that is, until it must be changed to set up the next clock cycle. Hence, for proper operation of the circuit, all x_j signals should be well-defined 0- or 1-pulses, the widths of which are one or more full clock periods. Then the Z signals will also be 0- or 1-pulses of width approximately kT, for some integer k.

When the primary and secondary input variables are changing value, the circuit's other signals can change temporarily in unpredictable ways. Such transition periods are brief, however; in a properly designed circuit, all signals quickly settle to their new final values. These new stable values appear around the middle of clock cycle i and are taken to be the signals' values "at time i."

Moore Circuits

We defined Moore-type sequential circuits as those with a primary output $Z(Y)$ that depends only on the current state Y. To illustrate the differences between Moore and Mealy circuits, we will construct a Moore realization of a serial adder. The state behavior of the Mealy version presented earlier is repeated in Figure 7.10a. There are two internal states S_0 and S_1, which

		Present input x_1x_2			
		00	01	11	10
Present state	S_0	S_0, 0	S_0, 1	S_1, 0	S_0, 1
	S_1	S_0, 1	S_1, 0	S_1, 1	S_1, 0

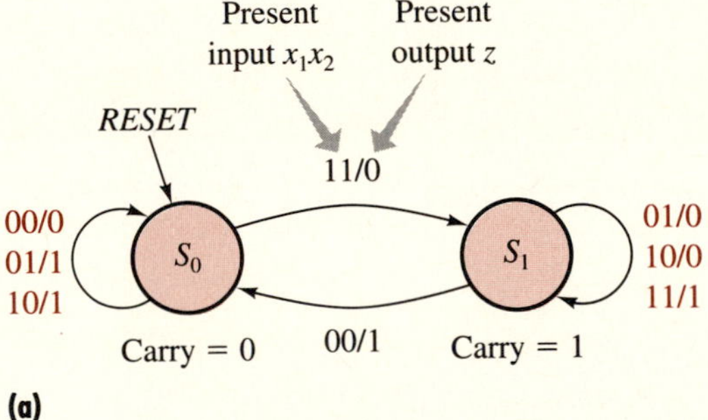

(a)

		Present input x_1x_2				Present output z
		00	01	11	10	
Present state	S_{00}	S_{00}	S_{01}	S_{10}	S_{01}	0
	S_{01}	S_{00}	S_{01}	S_{10}	S_{01}	1
	S_{10}	S_{01}	S_{10}	S_{11}	S_{10}	0
	S_{11}	S_{01}	S_{10}	S_{11}	S_{10}	1

(b)

FIGURE 7.10 State behavior of a serial adder: (a) the Mealy version; (b) the Moore version.

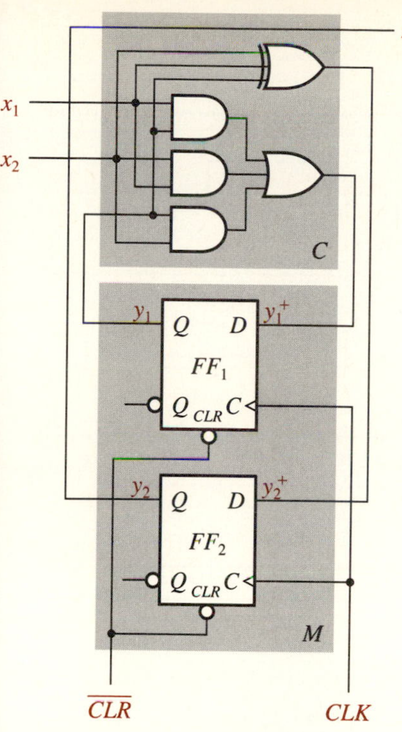

FIGURE 7.11 An implementation *SADD2* of a Moore version of the serial adder.

correspond to stored carries of 0 and 1, respectively. Observe that each state can produce both possible values 0 and 1 of the present output signal z. To meet the Moore constraints, we split the two original states in two, and associate a fixed value of z with each split state, as shown in Figure 7.10b. Thus, S_0 is split into the states S_{00} and S_{01}, for which the outputs are $z = 0$ and $z = 1$, respectively. In the same way, S_1 has been split into the states S_{10} and S_{11}, with outputs $z = 0$ and $z = 1$, respectively. It is no longer necessary to indicate state transitions in the form X/Z—although this may still be done—because Z is fixed by the present state. Instead, it is usual to attach the value of Z once to each state, as Figure 7.10b shows. All entries in every row of the state table have a common output value, shown on the right of the table. Each node in the state diagram now has the corresponding output value z written inside the node.

The state transitions for the Moore design are easily derived from the Mealy case. Each split state has the same outgoing or next-state transitions as the original unsplit state. Incoming transitions, originally marked $X/0$, go to the split part of the next state with $z = 0$; those marked $X/1$ go to the split part with $z = 1$. For example, consider the transition

$$S_0 \xrightarrow{11/0} S_1 \tag{7.3}$$

in Figure 7.10a. This gives rise to the two arrows marked 11 that emerge from S_{00} and S_{01}, the two states into which S_0 has been split. Because (7.3) specifies that the output is $z = 0$ and next state is S_1, the 11 arrows from S_{00} and S_{01} go to S_{10}, the state with $z = 0$ derived from S_1.

Moore-type circuits tend to have more states than do Mealy circuits.

The Moore state behavior requires two state variables y_1 and y_2—that is, double the number of states in the Mealy case. Suppose state S_{ij} is assigned the bit pattern $y_1y_2 = ij$. Then y_1 denotes the stored carry bit and y_2 denotes the output value z. A Moore-style adder with the foregoing state assignment is implemented by the circuit *SADD2* that appears in Figure 7.11. The reader can easily verify this fact by comparing *SADD2* with its Mealy equivalent *SADD* (Figure 7.5b). The combinational part C of the two designs is exactly the same, a consequence of the state assignment we used for *SADD2*, which has y_1 storing the carry signal and y_2 storing the sum signal. Note, however, that in *SADD2*, $z = y_2$ rather than $y_2^+ = x_1 \oplus x_2 \oplus y_2$, so z is delayed by one clock cycle, compared to the corresponding signal in *SADD*. Looking at it another way, all signals are generated by *SADD2* as in the Mealy design, except that z passes through an extra flip-flop *FF*2 in *SADD2*.

A timing diagram for *SADD2* with the same input signals as Figure 7.9 appears in Figure 7.12. Notice that the change in z from 0 to 1, which occurs in cycle 2 in *SADD*, is delayed until cycle 3 in *SADD2*. Note also that the glitch-like 1-pulse from x_1 in cycle 3 has no effect on z in *SADD2*. In general, the fact that $Z(Y)$ does not depend on the primary input X in a Moore circuit means that changes in X that do not change the state cannot affect Z. Hence, some of the protection against glitches that the clock provides for Y is extended to Z. As this example shows, Moore circuits have the slight disadvantage that their output responses are delayed by up to a clock period

Moore circuits are less sensitive to input glitches.

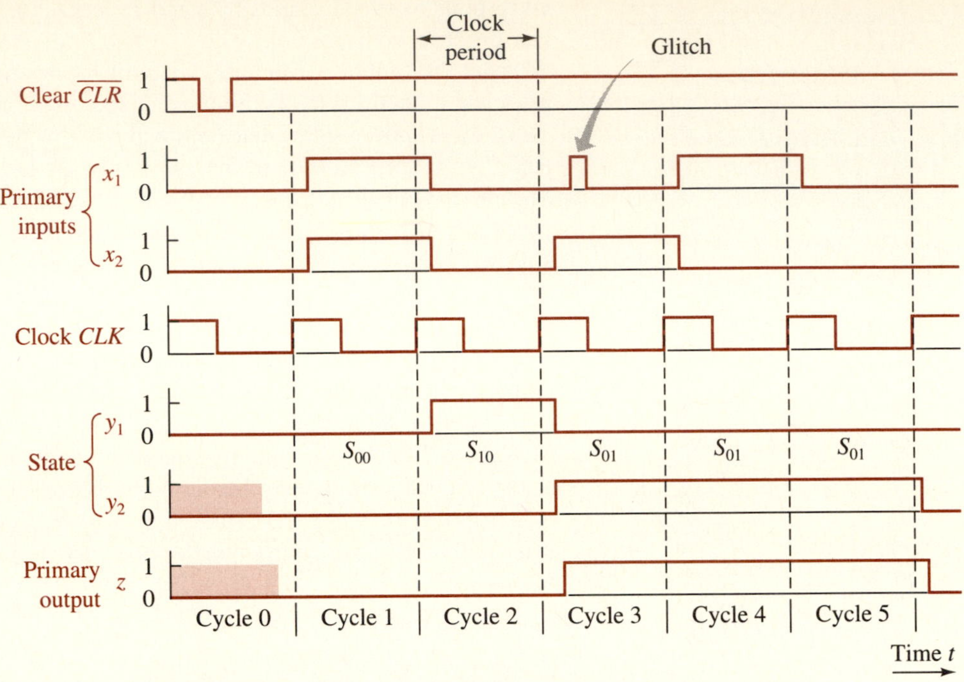

FIGURE 7.12 A timing diagram for the Moore-type serial adder *SADD2*.

due to the fact that for an input signal to change Z, it must first change Y. Mealy circuits, on the other hand, can respond to changes in X immediately.

Example 7.1

Analyzing a circuit that contains D flip-flops

Suppose we want to determine the input/output behavior of a given logic circuit—a typical "reverse-engineering" problem. The circuit in question is a single-input, single-output circuit called *DC*, the logic diagram for which appears in Figure 7.13a. This diagram is redrawn in Figure 7.13b, following the Huffman style of Figure 7.2, to clarify its feedback structure. Comparing *DC* to the Mealy model (Figure 7.1c), we see that $C = C_1$ and there is no separate output logic, apart from the line feeding z. *DC* contains two EXCLUSIVE-OR gates and three D flip-flops, so the circuit has at most eight internal states. We can compute the response of *DC* to any input sequence directly from the logic diagram, but it is usually easier to use a state table or diagram, the derivation of which we consider next.

We begin by writing the Boolean equations defining *DC*'s primary and secondary outputs:

$$\begin{aligned} z &= x \oplus y_1 \oplus y_3 \\ y_1^+ &= y_2 \\ y_2^+ &= y_3 \\ y_3^+ &= x \oplus y_1 \oplus y_3 = z \end{aligned} \tag{7.4}$$

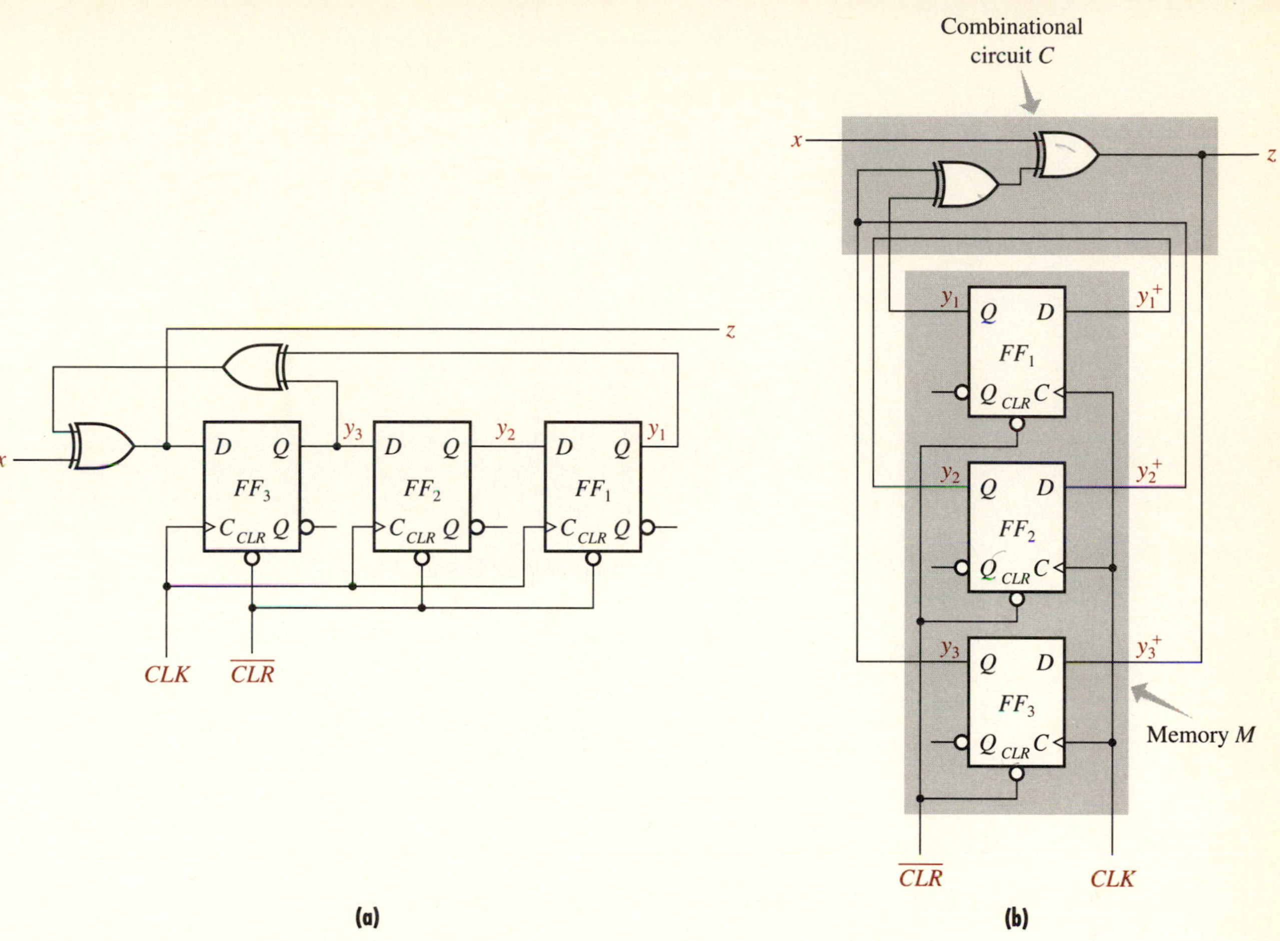

FIGURE 7.13 (a) The sequential circuit *DC* of Example 7.1; (b) the circuit redrawn in the style of Figure 7.2.

We can represent *C*'s function conveniently in the form of a transition table.

From these equations—or directly from the logic diagram—we can derive a truth table for y_1^+, y_2^+, y_3^+, and z as functions of x, y_1, y_2, and y_3, as shown in Figure 7.14. This table is also referred to as a **state transition table**, because each row defines a state transition of the form

$$y_1 y_2 y_3 \xrightarrow{x/z} y_1^+ y_2^+ y_3^+$$

For example, setting $x, y_1 y_2 y_3 = 0, 001$ in (7.4) yields $y_1^+ y_2^+ y_3^+, z = 011, 1$, as specified by the second row of the transition table. A state table for *DC* is obtained by replacing the bit patterns for the eight states by abstract symbols such as $S_0 = 000, S_1 = 001, \ldots, S_7 = 111$, and reformatting the transition table, as in Figure 7.15a. The equivalent state diagram appears in Figure 7.15b. The state S_0 that corresponds to $y_1 y_2 y_3 = 000$ serves as the initial state obtained by activating the clear input line $\overline{CLR}$.

Primary input	Present state			Next state			Primary output
x	y_1	y_2	y_3	y_1^+	y_2^+	y_3^+	z
0	0	0	0	0	0	0	0
0	0	0	1	0	1	1	1
0	0	1	0	1	0	0	0
0	0	1	1	1	1	1	1
0	1	0	0	0	0	1	1
0	1	0	1	0	1	0	0
0	1	1	0	1	0	1	1
0	1	1	1	1	1	0	0
1	0	0	0	0	0	1	1
1	0	0	1	0	1	0	0
1	0	1	0	1	0	1	1
1	0	1	1	1	1	0	0
1	1	0	0	0	0	0	0
1	1	0	1	0	1	1	1
1	1	1	0	1	0	0	0
1	1	1	1	1	1	1	1

FIGURE 7.14 The transition table for Example 7.1.

The input/output behavior—that is, the response of *DC* to any input sequence—is readily obtained from either the state table or the state diagram. For instance, if we apply the three-cycle input sequence $x(i), x(i+1), x(i+2) = 1, 0, 0$ to *DC* with initial state S_0, the resulting output sequence is $z(i), z(i+1), z(i+2) = 1, 1, 1$; the corresponding next-state sequence is S_1, S_3, S_7. If we apply the same input sequence to *DC* starting in state S_1, the response obtained is $z(i), z(i+1), z(i+2) = 0, 0, 1$, and the sequence of states visited is S_2, S_4, S_1.

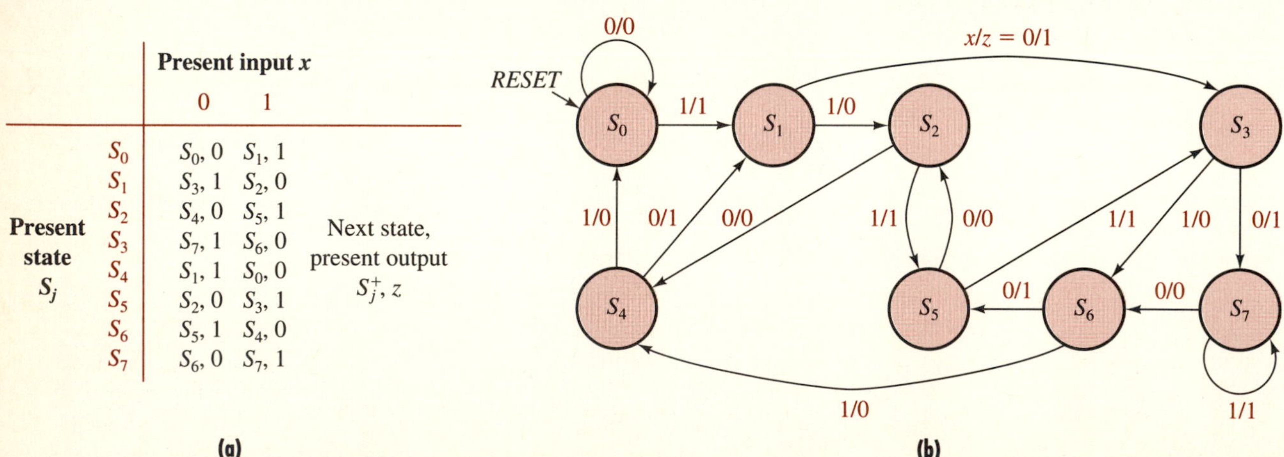

Present state S_j	Present input x = 0	1
S_0	S_0, 0	S_1, 1
S_1	S_3, 1	S_2, 0
S_2	S_4, 0	S_5, 1
S_3	S_7, 1	S_6, 0
S_4	S_1, 1	S_0, 0
S_5	S_2, 0	S_3, 1
S_6	S_5, 1	S_4, 0
S_7	S_6, 0	S_7, 1

Next state, present output S_j^+, z

(a)

(b)

FIGURE 7.15 (a) A state table and (b) a state diagram for Example 7.1.

We can know a circuit's behavior without knowing what the circuit is used for.

Although the foregoing state analysis provides us with complete knowledge of a sequential circuit's input/output behavior, it cannot tell us the circuit's *purpose*. In fact, *DC* is a useful circuit designed to decode a serial data stream that has been encoded using a "cyclic" single-error-detecting code [Johnson, 1989]. It can also be interpreted as a special type of binary division circuit. A suitably encoded input data stream P applied to the circuit, starting in state S_0, will produce an output stream W, which is a decoded version of P. It will also leave *DC* in the same final state S_0. An error that changes one bit of P will take *DC* to some nonzero final state, which therefore serves as an error indicator. (The properties of *DC* are explored further in the problem set at the end of this chapter.)

Flip-Flop Characteristics

The sequential circuit examples considered so far in this chapter all use D flip-flops, which have the nice property that the next value of the state variable y_j stored in the jth D flip-flop is the secondary output signal y_j^+. Moreover, it is only necessary to directly apply y_j^+ to the flip-flop's D input line in order to ensure that the correct next state is obtained. This follows from the D flip-flop's characteristic equation, which we can now write in the form

$$Q(i + 1) = D(i) \tag{7.5}$$

instead of the form $Q(t + \tau) = D(t)$ used in Chapter 6. Equation (7.5) implies that $y_j(i + 1) = Q_j(i + 1) = D_j(i) = y_j^+(i)$. Omitting the time parameters entirely, we write the equation that defines the next-state value y_j^+ as a function of the flip-flop's data input signal in the following compact form:

Characteristic equation of a D flip-flop that stores y_i

$$y_j^+ = D_j \tag{7.6}$$

This is basically a simplified version of the D flip-flop's characteristic equation (7.5). Observe that this flip-flop behaves like a delay element that delays $y_j^+(i)$ by one clock cycle to form $y_j(i + 1)$.

When JK flip-flops are used, two secondary output signals $J_j(i)$ and $K_j(i)$ must be generated by C in place of $D_j(i)$. These signals are applied to the jth flip-flop's data inputs to produce the desired next state $y_j(i + 1)$. The characteristic equation of the JK flip-flop

$$Q(i + 1) = J(i)\overline{Q}(i) + \overline{K}(i)Q(i) \tag{7.7}$$

defines the next-state value $y_j^+(i)$ in terms of the flip-flop's J and K data signals as follows: $y_j(i + 1) = Q_j(i + 1) = J_j(i)\overline{y}_j(i) + \overline{K}_j(i)y_j(i) = y_j^+(i)$. Again omitting the time parameters, we can rewrite (7.7) in the following concise form:

Characteristic equation of a JK flip-flop storing y_i

$$y_j^+ = J_j\overline{y}_j + \overline{K}_j y_j \tag{7.8}$$

Despite their more complicated characteristic equations, JK flip-flops sometimes produce simpler sequential circuits than D flip-flops, as we will see later.

When the sequential circuit's memory consists of p JK flip-flops, $2p$ secondary output functions $J_1, K_1, J_2, K_2, \ldots, J_p, K_p$ are produced by the combinational logic C. To analyze the state behavior of this type of circuit, we first determine the J_j and K_j signals for each combination of the primary input X and the present state Y. We then use (7.8) to obtain the value of y_j^+ for all j, thereby identifying the circuit's next state Y^+.

Example 7.2

Analyzing a circuit that contains JK flip-flops

Consider the sequential circuit in Figure 7.16, which has a single primary input x and a single primary output z, and contains two JK flip-flops. Its combinational part C generates four secondary output signals J_1, K_1, J_2, and K_2, which are applied to the flip-flops. We can write a set of logic equations for C directly from the logic diagram:

$$\begin{aligned} z &= \bar{x} + \bar{y}_2 \\ J_1 &= \bar{x} + \bar{y}_1 + \bar{y}_2 \\ K_1 &= \bar{x} + \bar{y}_2 \\ J_2 &= 1 \\ K_2 &= x + y_2 \end{aligned} \tag{7.9}$$

To obtain the next-state equations that specify y_j^+, we substitute J_j and K_j from (7.9) into the JK characteristic equation (7.8), yielding

$$\begin{aligned} y_1^+ &= (\bar{x} + \bar{y}_1 + \bar{y}_2)\bar{y}_1 + \overline{(\bar{x} + \bar{y}_2)}y_1 \\ y_2^+ &= \bar{y}_2 + \overline{(x + y_2)}y_2 \end{aligned}$$

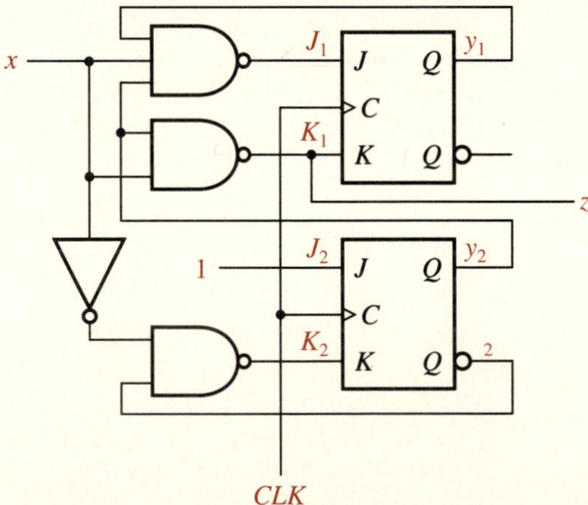

FIGURE 7.16 A logic diagram of a sequential circuit that contains JK flip-flops.

On simplifying these equations, we obtain

$$\begin{aligned} y_1^+ &= \bar{y}_1 + xy_2 \\ y_2^+ &= \bar{y}_2 \end{aligned} \tag{7.10}$$

We can now easily construct a transition table for the circuit; see Figure 7.17a. A symbolic state table with the state assignment $A = 00$, $B = 01$, $C = 10$, and $D = 11$ appears in Figure 7.17b; the corresponding state diagram is in Figure 7.17c.

Primary input x	Present state y_1	y_2	Next state y_1^+	y_2^+	Primary output z
0	0	0	1	1	1
0	0	1	1	0	1
0	1	0	0	1	1
0	1	1	0	0	1
1	0	0	1	1	1
1	0	1	1	0	0
1	1	0	0	1	1
1	1	1	1	0	0

(a)

Present state	Present input x = 0	1
A	D, 1	D, 1
B	C, 1	C, 0
C	B, 1	B, 1
D	A, 1	C, 0

(b)

(c)

FIGURE 7.17 A sequential circuit that contains JK flip-flops: **(a)** its transition table; **(b)** its state table; **(c)** its state diagram.

Equivalent States

The state diagram in Figure 7.17c suggests that a strong symmetry exists between the state pairs A, D and B, C. In fact, as we will prove later, any input sequence S_{in} applied to the circuit, starting in state A, produces exactly the same output sequence S_{out} that appears when S_{in} is applied to the same circuit, starting in state C. For example,

$$A \xrightarrow{0/1} D \xrightarrow{0/1} A \xrightarrow{1/1} D$$

$$C \xrightarrow{0/1} B \xrightarrow{0/1} C \xrightarrow{1/1} B$$

Equivalent states respond in the same way to every input sequence.

Consequently, by observing only the primary output signal z, we cannot tell if the starting state is A or C; these states are therefore said to be **equivalent states**, and we write $A \equiv C$. (Of course, we can distinguish A and C if we observe *internal* signals such as the state variables, because y_1 is 0 for A and 1 for C.) Similarly, we have $B \equiv D$.

It follows that for purposes of determining external input/output behavior, we can replace the state table and diagram of Figure 7.17 with the equivalent

reduced state table and diagram shown in Figure 7.18, in which state E replaces A and C, and state F replaces B and D. We obviously now have more concise representations of the circuit's behavior than those of Figure 7.16.

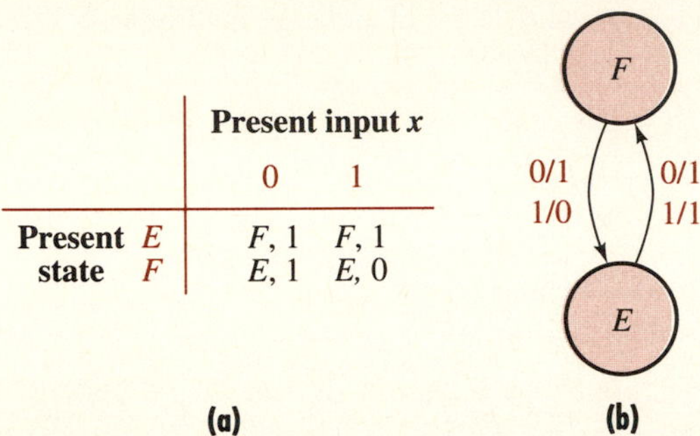

		Present input x	
		0	1
Present state	E	F, 1	F, 1
	F	E, 1	E, 0

(a) (b)

FIGURE 7.18 (a) A reduced state table and (b) a reduced state diagram equivalent to those of Figure 7.17.

Summary

The following procedure *SCANAL* summarizes the main steps we used to analyze a sequential circuit—that is, determine the circuit's behavior, given its logic diagram. The applicability of this approach is limited by circuit size. It becomes impractical for circuits with more than half a dozen or so primary and secondary variables, mainly due to the excessive size of the various tables employed, especially the transition table. For larger circuits, we need symbolic representations of groups of signals and states, and we rely on program-like descriptions of sequential circuits, based on hardware description languages or flowcharts, a topic to which we return in Chapter 8.

Procedure

Procedure *SCANAL* for analyzing a given sequential circuit

1. *Combinational function identification:* Determine the primary and the secondary output functions generated by the combinational circuit C. Represent these by Boolean equations or tables.
2. *State transition identification:* From the characteristics of the flip-flops and the secondary output functions applied to them, construct a transition table that specifies in binary form all state transitions of the circuit and all primary output values.
3. *Formal behavioral specification:* Assign symbolic names to all internal states and construct a state table or diagram from the transition table.

7.3 Timing Control and Clocks

This section considers the finer timing details of sequential circuit behavior. We show how to determine the period and waveform of the clock signal to ensure that no timing problems occur.

Why Clocks?

Clock signals determine precisely when states change, and minimize sensitivity to glitches.

The times at which a sequential circuit's primary and secondary output signals change value depend on when its primary input signals (including the clock) change, as well as on the circuit's internal propagation delays. Obviously, for reliable operation the input signals must change at times that conform to the desired state behavior. In general, an input that changes while the circuit is in the wrong state will cause an error, with possibly serious consequences. In addition, allowance must be made for signal noise and uncertainty in the circuit's internal delays. In synchronous circuits, as demonstrated next, a properly designed clock system minimizes timing problems of this kind.

The main propagation delays through the combinational part C of a general (Mealy-type) sequential circuit SC are depicted symbolically in Figure 7.19. Signals that affect the primary outputs $Z(X, Y)$ experience delays, collectively denoted t_{c1}; those that affect the secondary outputs $Y^+(X, Y)$ experience delays denoted t_{c2}. In general, a change to X or Y causes the signals that compose $Z(X, Y)$ and $Y^+(X, Y)$ to change. Individual signals may change at different times, and $Z(X, Y)$ and $Y^+(X, Y)$ may pass through several transient values before settling at their desired new values. The temporary values on individual lines take the form of short pulses, referred to as *glitches* when they are a potential source of errors. The flip-flops that form the memory part M of SC do not respond to $Y^+(X, Y)$ until triggered by the

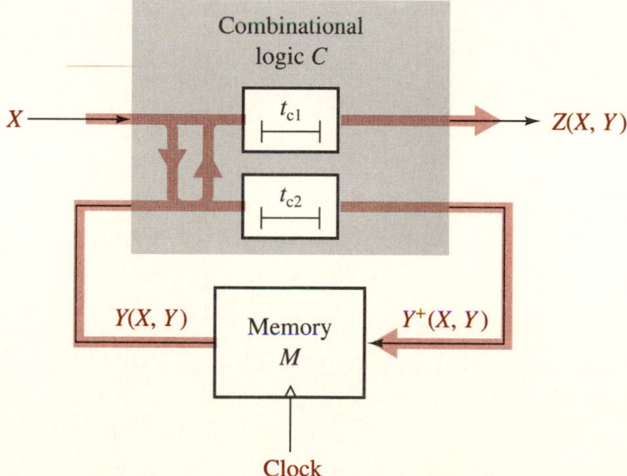

FIGURE 7.19 Overall propagation delays in a sequential circuit SC.

clock, so glitches that appear at the inputs of M before the triggering time are ignored. *The essential role of the clock is to confine flip-flop triggering to a brief time slot after the* $Y^+(X, Y)$ *signals have settled at their new values, allowing for noise and for the worst-case values of the delays that contribute to* t_{c2}*, as well as for the worst-case propagation delays associated with the flip-flops themselves.* As long as these requirements are met, we do not have to worry about transient signal values or the specific propagation delays affecting individual signals. This is the reason synchronous circuits are so much easier to design and are inherently more reliable than asynchronous circuits.

Primary Output Control

It can be seen from Figure 7.19 that, unlike the secondary outputs $Y^+(X, Y)$, the primary outputs $Z(X, Y)$ of SC are not controlled by a clock. Consequently, signal changes that affect the output logic can produce glitches on $Z(X, Y)$ that can be transmitted to external devices. Glitches produced within combinational circuits are discussed in Chapter 6 (see Sections 6.8, 6.9, and 6.10). Sequential circuits contain some additional sources of glitches. Suppose, for example, that the outputs of two flip-flops are connected to a two-input AND gate that produces the output signal z. Glitches can be produced by unequal propagation delays through the flip-flops when their outputs change value. Even if the flip-flops' internal delays are well matched, the flip-flops may change state at slightly different times due to **clock skew**, the arrival of the same clock edge at slightly different times (skewed) at different flip-flops. Clock skew is caused by unequal delays in the clock transmission paths to the various flip-flops.

Glitches can appear on primary output lines due to uncertain delays in C.

How do we deal with glitches on the primary output lines Z of SC? If Z is connected to other sequential circuits controlled by the same clock as SC, then the clocking of the other circuits takes care of the transient behavior of Z, just as SC must accommodate transients in its own input signals X. Consider, for example, the serial adder *SADD* of Figure 7.4, the output logic of which consists of the EXCLUSIVE-OR gate that generates the sum function $z = x_1 \oplus x_2 \oplus y$. It is not hard to see that this EXCLUSIVE-OR is prone to generating glitches. For example, if any two of its inputs change in opposite directions, the output z should remain unchanged. However, a small difference in the times at which the signals switch will result in a transient parity change that can lead to a glitch on z. Typically, an adder of this type is a component of a larger system (a serial arithmetic unit) that includes shift registers to store both input and output sequences applied to the adder. A small serial system of this sort is depicted in Figure 7.20, which attaches to the adder three 4-bit shift registers based on the design of Figure 6.46. All the circuits involved, the shift registers and the adder, are connected to a common clock signal CLK. The triggering edge of CLK is positioned to ensure that any glitches from the adder's input and output combinational logic have dissipated before the flip-flops are clocked.

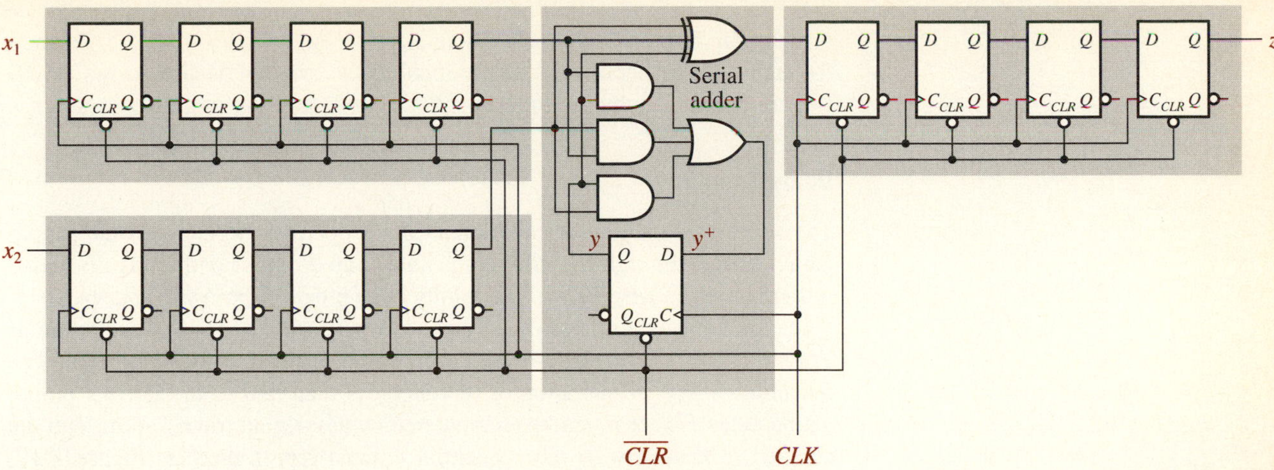

FIGURE 7.20 The use of shift registers as input/output circuits for the serial adder *SADD*.

Glitches can be blocked by placing latches or flip-flops in primary output lines.

If Z is a final output of a digital system, it can be connected to latches, as shown in Figure 7.21. The latches are controlled by the system clock and are not enabled until any transients in the output logic have disappeared. Flip-flops may be used in place of latches, but latches with level triggering that is compatible with the edge triggering of M suffice, assuming there is no global feedback from the latches. In some instances, Z may be used directly to drive output devices such as light-emitting diode (LED) displays that function satisfactorily, despite occasional glitches on their inputs.

We observed in Section 7.2 that Moore-type sequential circuits are less prone to output glitches than the Mealy type. Often, the primary outputs of a Moore circuit are derived directly from M, so the circuit has a natural structure that is similar to Figure 7.21; in such cases, the flip-flops also serve as output latches. Our Moore version of the serial adder *SADD2* (Figure

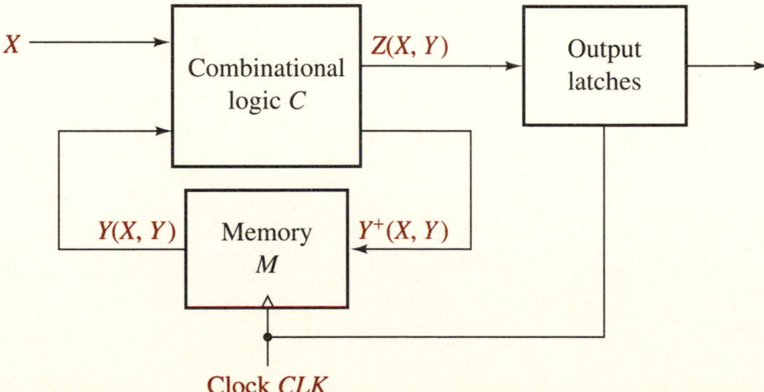

FIGURE 7.21 The use of latches to block glitches on the primary outputs.

7.11) is a case in point. If we unwind the circuit's feedback paths, we see that the flip-flop *FF2* is directly connected to the primary output z. In fact, the entire circuit happens to be identical to the original Mealy design *SADD* (Figure 7.4), with the addition of *FF2* in the role of an output latch for z.

Delay Sources

The speed at which a combinational circuit operates depends on the propagation delays through the circuit's primary input/output paths. As discussed in Section 6.2, these delays are hardly ever precisely known, and vary significantly between parts of the circuit. The best we can do is place the propagation delays of interest between guaranteed upper and lower bounds. The maximum speed of operation of the circuit is then tied to the longest propagation delay t_{pd}^{max} that can be encountered by any signal traveling through the circuit. In dealing with the sequential circuits represented by Figure 7.19, which consist of a combinational part C and a clocked memory part M, we must consider the propagation delays through C, as well as the timing parameters and triggering modes of the flip-flops that compose M. Together, these determine the minimum length T^{min} of a clock cycle, and therefore the maximum speed at which the sequential circuit can operate reliably.

Figure 7.22 illustrates the important timing events in the operation of a general sequential circuit that employs positive edge-triggered flip-flops. The data inputs to all the flip-flops in M must be held steady for a setup

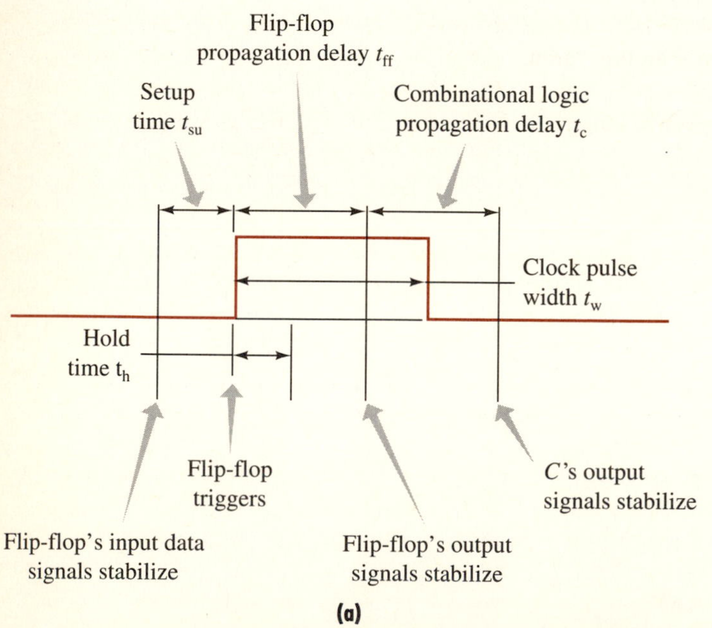

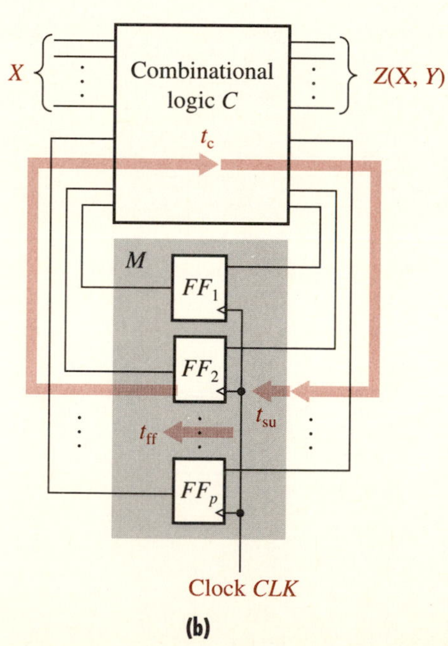

FIGURE 7.22 (a) Events and delays related to a clock pulse; (b) delay sources.

time t_{su} before the triggering edge of the clock signal CLK is applied to each flip-flop's clock line. This setup time allows the flip-flop to read in its current input data values (D, or J and K, depending on the flip-flop type) in preparation for the next state change. Once triggered, the flip-flop proceeds to change its output data signals Q and $\overline{Q}$, a process that takes an amount of time we call the **flip-flop propagation delay** and denote by t_{ff}. This is not a standard timing parameter in logic data books, but may be directly derived from the figures provided for the propagation delays t_{PLH} and t_{PHL} between CLK and Q and between CLK and $\overline{Q}$. Note that $t_{ff} > t_h$, the flip-flop's hold time, and that, in practice, we frequently find $t_h = 0$. Another overlapping requirement is that the width of the clock's 1-pulse should be at least t_w^{min}, the minimum clock-pulse width specified for the flip-flops.

The new output state signals Y of the flip-flops are applied to the input side of the combinational circuit C and generate new secondary output signals Y^+ that return to the input side of the flip-flops after a delay t_c, which we term the **combinational propagation delay**. If Z is connected to output circuits that are to be clocked by CLK, as in Figure 7.21, then we replace t_c with $\max\{t_{c1}, t_{c2}\}$, where t_{c1} and t_{c2} are the delays defined in Figure 7.19.

Path Delay Constraints

The delay in a circuit's feedback path is the sum of t_{ff}, t_c, and t_{su}.

A data signal that appears at the output of a flip-flop, passes around the circuit's main feedback loop, and returns to the input of the same or another flip-flop experiences a total propagation delay which is the sum of t_{su}, t_{ff} and t_c (Figure 7.22b). Obviously, the period T of the clock signal CLK must be no shorter than this "around-the-loop" propagation delay under the worst-case conditions, when all the delay parameters reach their maximum (max) values. Hence, we obtain the following **longest path constraint** that relates the clock period T to the propagation delays of C and M:

The longest path constraint on the clock period T

$$T > t_{ff}^{max} + t_c^{max} + t_{su}^{max} \tag{7.11}$$

If (7.11) is violated by a clock that is too fast, it is possible for a flip-flop to be triggered before all the changes that result from the previous triggering reach the flip-flops and become properly "set up" at the flip-flop's inputs, resulting in erroneous operation. If T is a fixed design goal, C must be designed with a propagation delay t_c^{max} that satisfies (7.11). Conversely, if the circuit is given, we can use (7.11) to estimate its minimum clock period for safe operation or, equivalently, its maximum safe clock frequency.

Timing problems also arise if logic signals travel too fast around the main feedback loop or, equivalently, if the clock is too slow, so that new input data signals arrive at a flip-flop that is still in a "hold" period, responding to the previous set of input changes. To avoid this possibility, the following **shortest path constraint** must also be met:

The shortest path constraint on the clock period T

$$t_h^{min} < t_{ff}^{min} + t_c^{min} \tag{7.12}$$

This constraint is usually satisfied without any explicit effort on the logic designer's part, so that the longest path constraint (7.11) normally determines the clock period T.

Clock Signal Design

In practice, the clock period T is usually computed from the longest path formula

$$T = t_{\text{ff}}^{\max} + t_{\text{c}}^{\max} + t_{\text{su}}^{\max} + E \tag{7.13}$$

derived from (7.11), with an added tolerance (error) term E to allow for some inaccuracy in the assumed delay parameters for C and M. E also allows for the possibility of clock skew, in which the same clock signal triggers several flip-flops at slightly different times (Problem 7.27). The parameters $t_{\text{su}}^{\max}$, $t_{\text{h}}^{\min}$, and $t_{\text{ff}}^{\max}$ in (7.13) depend on the flip-flop types used, and are derived as follows: The setup time must be at least equal to the minimum setup time $t_{\text{su}}^{\min}(FF_i)$ of the slowest flip-flop FF_i; therefore, we require

$$t_{\text{su}}^{\max} = \max\{t_{\text{su}}^{\min}(FF_1), t_{\text{su}}^{\min}(FF_2), \ldots, t_{\text{su}}^{\min}(FF_p)\}$$

Similarly, t_{ff} must at least equal the maximum delay through any flip-flop.

$$t_{\text{ff}}^{\max} = \max\{t_{\text{ff}}^{\max}(FF_1), t_{\text{ff}}^{\max}(FF_2), \ldots, t_{\text{ff}}^{\max}(FF_p)\}$$

The $t_{\text{su}}^{\min}(FF_i)$ and $t_{\text{ff}}^{\max}(FF_i)$ figures are obtained directly from the flip-flop specifications. Once the flip-flop types are fixed, only the propagation delay through the combinational logic C determines the clock period T. Consequently, considerable effort is often devoted to reducing $t_{\text{c}}^{\max}$ in order to achieve a desired value of T.

Example 7.3

Specifying a clock signal

Suppose you are designing a sequential circuit *SC* that will use TTL ICs from the 74LS low-power Schottky series. In particular, the memory part M of *SC* is to be constructed using the 74LS107A, an IC that contains two independent negative edge-triggered JK flip-flops. (The internal logic of these flip-flops was presented in Example 6.6 of Section 6.6.) The manufacturer's timing data for the 74LS107A are reproduced in Figure 7.23 [Texas Instruments, 1988]. From the "recommended operating conditions," we obtain the following minimum setup, hold, and enable times:

$$t_{\text{su}}^{\min} = 20 \text{ ns} \qquad t_{\text{h}}^{\min} = 0 \text{ ns} \qquad t_{\text{w}}^{\min} = 20 \text{ ns}$$

The enable time $t_{\text{w}}^{\min}$ is the minimum width of the 1 (high) level of a clock pulse whose falling edge triggers the flip-flop. The flip-flop delay t_{ff} is deduced from the propagation delay figures that appear under "switching characteristics" in Figure 7.23. The worst-case delay given for $t_{\text{PLH}}^{\max}$ and $t_{\text{PHL}}^{\max}$,

recommended operating conditions

			SN54LS107A			SN74LS107A			UNIT
			MIN	NOM	MAX	MIN	NOM	MAX	
V_{CC}	Supply voltage		4.5	5	5.5	4.75	5	5.25	V
V_{IH}	High-level input voltage		2			2			V
V_{IL}	Low-level input voltage				0.7			0.8	V
I_{OH}	High-level output current				– 0.4			– 0.4	mA
I_{OL}	Low-level output current				4			8	mA
f_{clock}	Clock frequency		0		30	0		30	MHz
t_w	Pulse duration	CLK high	20			20			ns
		$\overline{CLR}$ low	25			25			
t_{su}	Setup time before CLK↓	data high or low	20			20			ns
		$\overline{CLR}$ inactive	25			25			
t_h	Hold time-data after CLK↓		0			0			ns
T_A	Operating free-air temperature		– 55		125	0		70	°C

switching characteristics, V_{CC} = 5 V, T_A = 25°C (see note 3)

PARAMETER	FROM (INPUT)	TO (OUTPUT)	TEST CONDITIONS	MIN	TYP	MAX	UNIT
f_{max}			R_L = 2 kΩ, C_L = 15 pF	30	45		MHz
t_{PLH}	$\overline{CLR}$ or CLK	Q or $\overline{Q}$			15	20	ns
t_{PHL}					15	20	ns

NOTE 3: Load circuits and voltage waveforms are shown in Section 1.

FIGURE 7.23 Published timing data for the 74LS107A flip-flop. [Texas Instruments, 1988b]

measured from the clock input *CLK* to either output Q or $\overline{Q}$, is 20 ns, so we take $t_{\text{ff}}^{\max} = 20$ ns. Hence, the longest path formula (7.10) requires the clock period T to satisfy the equation

$$T = 20 + t_{\text{c}}^{\max} + 20 + E \text{ ns} \tag{7.14}$$

Because $t_{\text{h}}^{\min} = 0$ and the minimum values of both t_{ff} and t_{c} will certainly be nonzero, the shortest path constraint (7.12) is automatically satisfied.

It remains to determine $t_{\text{c}}^{\max}$, the maximum propagation delay through the combinational circuit C, as well as the error margin E. Suppose that C has depth 3 and is to be implemented from 74LS00 ICs, each of which contains four two-input NANDs. The propagation delays for these NANDs are as follows:

$$t_{\text{PLH}}^{\text{typ}} = 9 \text{ ns} \qquad t_{\text{PLH}}^{\max} = 15 \text{ ns} \qquad t_{\text{PHL}}^{\text{typ}} = 10 \text{ ns} \qquad t_{\text{PHL}}^{\max} = 15 \text{ ns}$$

A typical error margin E for the clock period is 20 percent.

Assuming, conservatively, that the longest delay in C occurs along a path through three NANDs, each of delay 15 ns, we can take $t_c^{max} = 45$. Hence, from (7.14) we conclude that

$$T = 20 + 45 + 20 + E = 85 + E \text{ ns}$$

A reasonable figure for E is about 20 percent of $t_{ff}^{max} + t_c^{max} + t_{su}^{max}$, or $85 \times 0.2 = 17$ ns in the present case. If we take $E = 15$ ns, we obtain the nice round figure of $T = 100$ ns $= 0.1\ \mu$s for the clock period, which corresponds to a clock frequency of $1/0.1 = 10$ MHz.

A suitable waveform of period $T = 100$ ns for the clock signal is given in Figure 7.24. The clock is 1 (active or high) for 25 ns to meet the 74LS107A's enable time constraint $t_w^{min} = 20$ ns, and also to allow 5 ns for error tolerance. The clock is 0 for the remaining 75 ns of the clock cycle, which includes another 10-ns error margin.

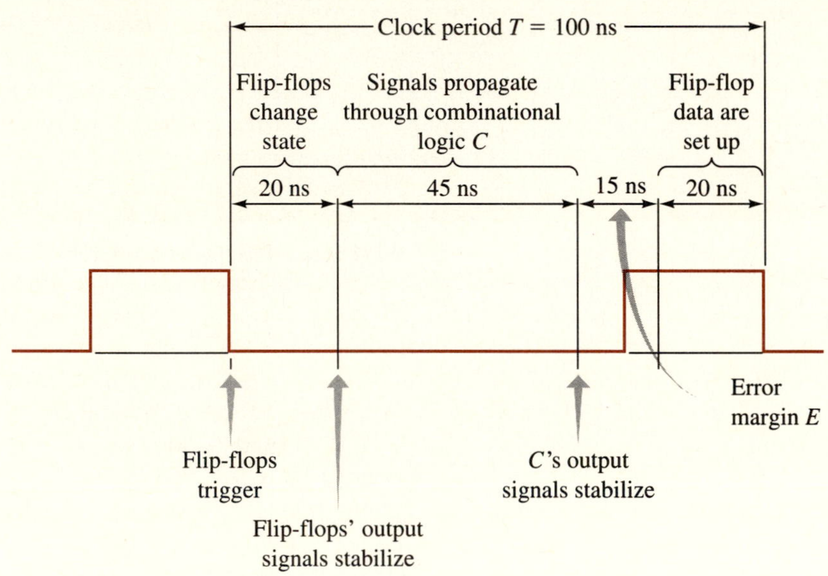

FIGURE 7.24 The clock waveform for the example circuit.

The design of a clock signal that allows a circuit to operate at its maximum reliable rate is a difficult task. The designer is often aided by a class of simulation programs called **timing analyzers**, which can determine all the (worst-case) propagation delays of the circuit under various operating conditions [Hitchcock, 1988].

From here on we assume that the circuits under consideration have a properly designed clock system and acceptable signal quality on their primary output lines. We will measure a clock cycle from the triggering edge of the clock signal to the next triggering edge of the same kind. Then, without worrying about timing details, we can assume that internal state changes occur at the beginning of a clock cycle, and that the corresponding primary output changes appear toward the middle of the clock cycle.

Sequential Circuit Synthesis

In this section, we examine the design process for sequential circuits, the state behavior of which can be fully defined at the gate level. We present a systematic technique for implementing a sequential circuit, starting with the construction of a state table or diagram and culminating in a complete logic circuit. Some practical cost issues, such as minimizing the number of flip-flops and gates used, are also addressed.

7.4 The Design Process

The main goal of sequential circuit design is to obtain a logic circuit that has a desired state behavior.

Like all synthesis problems, the design of a (synchronous) sequential circuit involves deriving a structure that has a specified behavior. Meeting the behavioral specification—that is, ensuring functional correctness—is always the primary objective of the design process. As in combinational circuit design, minimizing the total cost of the components used and maximizing the circuit's operating speed are important additional objectives. Because of the greater difficulty of designing sequential circuits, it is frequently desirable to reduce the complexity of the design process itself at the expense, say, of using extra components. Another key design goal, which is the subject of Section 7.9, is good testability, implying that the circuit can be tested quickly and efficiently for faults. Yet another common goal is flexibility or expandability, so that later changes to the behavior specifications can be accommodated with relatively little redesign.

The General Approach

The main steps required to synthesize a sequential circuit can be obtained by reversing the order of the steps in the analysis procedure *SCANAL*, presented in Section 7.2.

1. *State behavior specification:* The end point of sequential circuit analysis—namely, construction of a state table or diagram—is the starting point in the design process. This should describe precisely the input/output behavior required of the target circuit. This step identifies the internal states needed and the transitions that can occur between the states.

2. *State assignment:* Once we have a complete state table, we can assign p binary state variables $Y = y_1, y_2, \ldots, y_p$ to represent all P symbolic states, where $p \geq \lceil \log_2 P \rceil$. The choice of bit patterns of Y that correspond to the symbolic states is called the *state assignment*.
3. *Combinational function specification:* A transition table, which specifies the required state behavior in binary form, is constructed from the state assignment and state table or diagram. This table directly defines the primary output functions. We also need to define the input functions that must be applied to the flip-flops to produce the required next-state values on their outputs. These functions are, of course, the secondary output signals implemented by the combinational logic C.
4. *Combinational circuit design:* The remaining task is to design C, a standard combinational design problem. Often we want to use the fastest possible circuit to implement C, which implies a two-level design, to which the various combinational design methods covered in Chapter 5 are applicable.

Creating an accurate and complete state table or diagram is often the hardest design step.

Most of the steps in sequential logic design can be implemented by exact algorithms, and therefore can be automated—at least in principle. The exception is the initial step of deriving a formal state table from an informal behavioral specification, because the latter, by definition, cannot be formalized and therefore cannot be incorporated into an exact algorithm. Often the state table or state diagram has to be derived laboriously from a vague and incomplete word specification of the desired behavior. As the state behavior is being specified, the designer is often forced to make assumptions about the operation of the target circuit, which are not included in the original problem specification. Hence, the designer's ingenuity is most strongly displayed in this first design step.

Two Circuit Types

Sequential circuits are divided into datapath and control circuits, depending on their function.

Just as we divided sequential circuit structure into several structural categories (Figure 7.1), we can also identify a few broad behavioral categories. The small sequential circuits of interest here are sometimes designed to implement a well-defined arithmetic or logic function. Examples of such sequential **data-processing** or **datapath circuits** encountered earlier are the serial adder (Figure 7.4) and the cyclic code decoder (Figure 7.13). For these circuits, a formal functional description is usually relatively short and easy to construct. **Control circuits**, which produce the sequences of control signals for larger digital systems such as a multifunction datapath unit, form another important class of applications for small sequential circuits. They often have rather complex specifications that include many input sequences, sometimes in the form of (micro) instructions; consequently, control circuits can be quite hard to design.

To introduce all the main issues in sequential circuit design, we examine in considerable detail the design of a small control unit.

Problem Specification

The traffic lights that control the intersection of two streets employ a digital system *TC*, the task of which is to switch the lights on and off in a sequence that responds to the flow of vehicles and pedestrians. We consider a simplified example, in which a minor north-south (NS) street intersects an equally minor east-west (EW) street (Figure 7.25). The only inputs to *TC* are signals from a set of *WALK* buttons that are pushed by pedestrians who want to cross the street. When no pedestrians are present, the lights allow the NS traffic to flow for one minute; that is, they are green in the NS direction and red in the EW direction. Then the lights change automatically to allow the EW traffic to flow for one minute. Thus, the green lights alternate between the NS and EW directions at one-minute intervals. If, however, a pedestrian pushes a *WALK* button, all the lights turn red for one minute, allowing pedestrians to cross either street. To keep the problem small, we will deliberately ignore the yellow or amber warning lights that are a vital part of real traffic-control systems.

Problem specifications are sometimes quite informal.

A fairly informal verbal description of the foregoing kind is a typical starting point in the sequential circuit design process. We first must make some assumptions to make the system specifications more precise and complete. The pedestrian buttons will be represented collectively by a single binary variable x. The signal $x = 1$ indicates to *TC* that some pedestrian button has been pressed, and $x = 0$ indicates that no such button has been pressed. *TC* will be assumed to have two primary output signals: z_1 to make the lights green ($z_1 = 1$) or red ($z_1 = 0$) in the EW direction, and z_2 to make

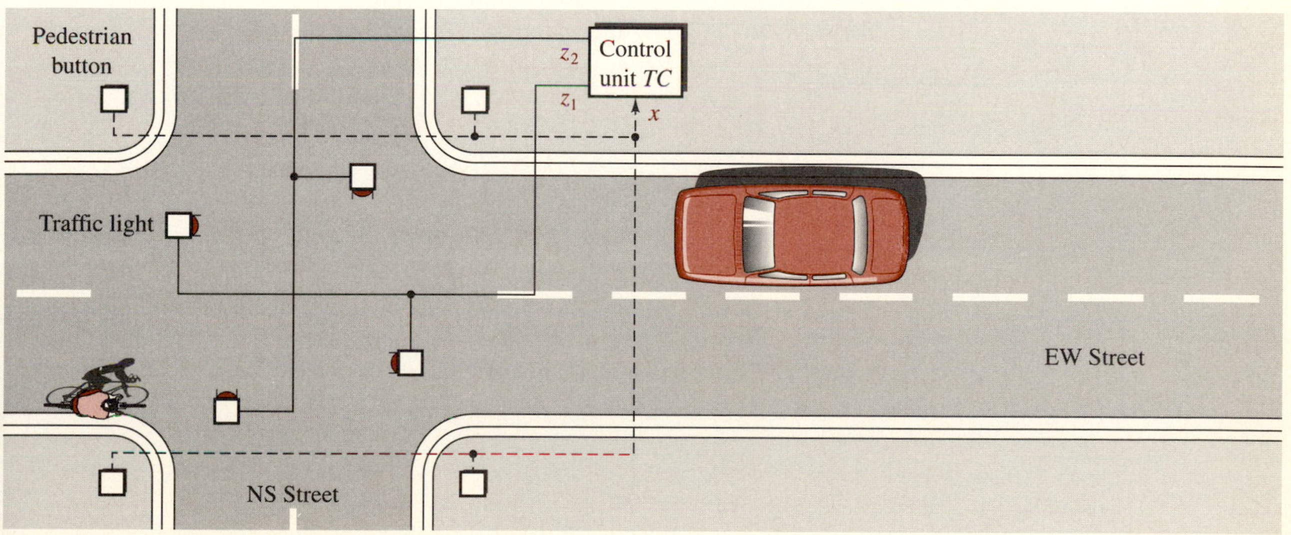

FIGURE 7.25 The layout of the traffic-light control problem.

the lights green ($z_2 = 1$) or red ($z_2 = 0$) in the NS direction. Hence, the all-red condition, when pedestrians can cross, is signalled by $z_1 = z_2 = 0$. Because the minimum green or red period is one minute, we choose one minute as the period for the clock that synchronizes *TC*.

Defining the State Behavior

Next, we pose a key question: What are the internal states of *TC*? The traffic lights have three obvious configurations that define three possible states S_0, S_1, and S_2 of *TC*:

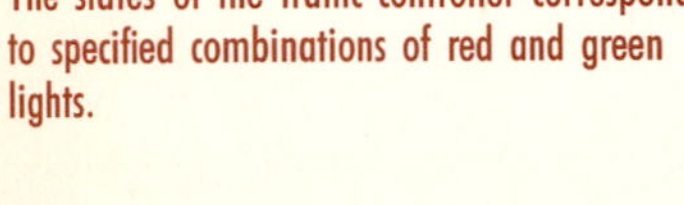

The states of the traffic controller correspond to specified combinations of red and green lights.

1. State S_0: Green in the NS direction and red in the EW direction
2. State S_1: Green in the EW direction and red in the NS direction
3. State S_2: Red in both the EW and NS directions

Because these states determine the output signals z_1 and z_2 completely, a Moore-style state diagram seems most natural for *TC* (Figure 7.26).

Now consider the possible transitions among the states. When $x = 0$—that is, when no pedestrians are waiting to cross—*TC* should alternate between the states S_0 and S_1, as shown by the arrows labeled $x = 0$ going back and forth between the S_0 and S_1 nodes in Figure 7.26. When x becomes 1, the state should change to S_2, where it should remain as long as $x = 1$. This is indicated in Figure 7.26 by all the 1-arrows going to S_2. When x eventually returns to 0, normal green-red operation should resume. We make the arbitrary assumption that the circuit goes from S_2 to S_0 under these conditions. The precise times at which the outputs z_1 and z_2, and therefore the lights, change are determined by the clocking details. A pedestrian who pushes x will have to wait a variable amount of time up to one minute before he or she can cross the street in the next all-red condition.

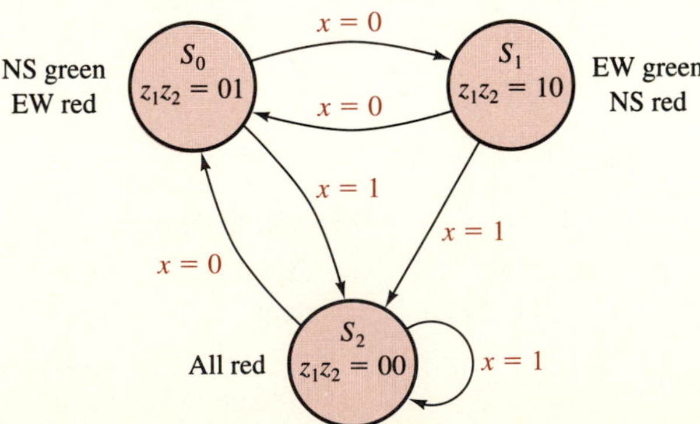

FIGURE 7.26 A preliminary state diagram for the traffic-light controller.

Improving the State Behavior

Often, we find unexpected state behavior that the problem specifications overlook.

Although it appears to meet the specifications for the traffic-control unit, a little thought shows several flaws in the state behavior just proposed. A large number of pedestrians who repeatedly push the *WALK* buttons, or a single malicious pedestrian, could easily block all vehicular traffic indefinitely. The self-loop associated with the all-red S_2 state is the culprit for this, and so is very undesirable. A second flaw—which is, perhaps, less serious but is nevertheless likely to enrage some drivers—is that the return transition from S_2 to normal green-red operation is always to the same state S_0. This state will therefore tend to be visited more often than S_1, unfairly penalizing the EW traffic, which can move only when *TC* is in state S_1. In fact, it is possible for the states to alternate between states S_0 and S_2, effectively locking out S_1.

To rectify these problems, we will restrict the all-red condition to one clock period at a time by forcing a return to S_0 or S_1 from the all-red state, independent of the value of x. Furthermore, to ensure fair treatment of the traffic in each direction, we will require a transition from S_0 to the all-red state, to be followed immediately by a transition to S_1, while a transition from S_1 to the all-red state is to be followed immediately by a transition to S_0. These objectives can be met by splitting the original S_2 into two symmetrical all-red states S_2 and S_3. The new S_2 can only follow S_0, and is followed itself by S_1; S_3 can only follow S_1, and is followed in turn by S_0. We will take the original state diagram, with this modification, as our final specification of *TC*'s behavior (Figure 7.27).

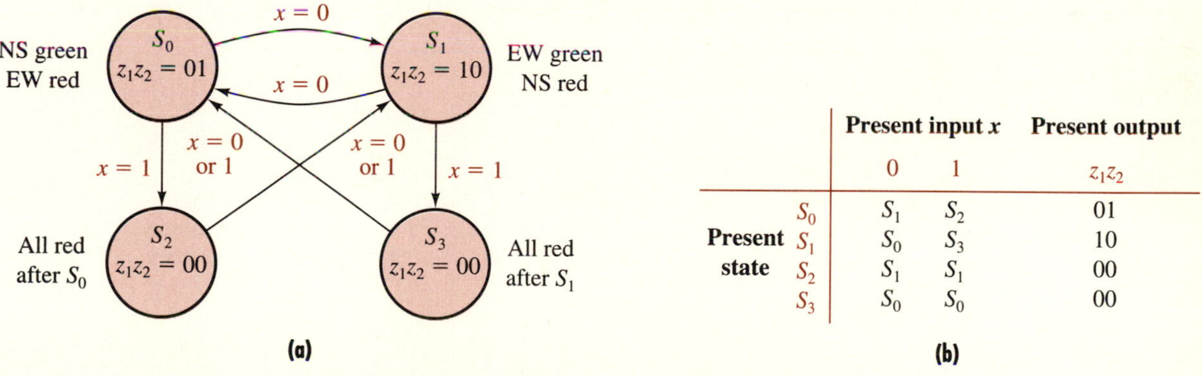

		Present input x 0	Present input x 1	Present output z_1z_2
Present state	S_0	S_1	S_2	01
	S_1	S_0	S_3	10
	S_2	S_1	S_1	00
	S_3	S_0	S_0	00

FIGURE 7.27 (a) The revised state diagram and (b) the state table for the traffic-light controller *TC*.

State Assignment

We are now ready to assign binary patterns to the states of *TC*, which are represented symbolically in Figure 7.27. We need two state variables y_1 and y_2 to represent the four states S_0, S_1, S_2, and S_3. Two flip-flops suffice to store the state variables, and we choose positive edge-triggered D flip-flops for this purpose. JK flip-flops can be used, if preferred; the flip-flop choice has little impact on design difficulty or circuit cost.

It is not obvious how best to assign the bit patterns of y_1y_2 to the four states. We select the following simple and essentially arbitrary assignment: The bit pattern that represents the binary number i is assigned to the state S_i. Hence,

$$y_1y_2 = 00 \text{ represents } S_0$$
$$y_1y_2 = 01 \text{ represents } S_1$$
$$y_1y_2 = 10 \text{ represents } S_2$$
$$y_1y_2 = 11 \text{ represents } S_3$$

Mapping these bit patterns into the original state table (Figure 7.27b) yields the binary version of the state table shown in Figure 7.28a. Observe that the present states are values of $y_1y_2 = y_1(i)y_2(i)$, while the next states are values of $y_1^+y_2^+ = y_1^+(i)y_2^+(i)$, which become $y_1(i+1)y_2(i+1)$ in the next clock cycle.

		Present input x		Present output
		0	1	z_1z_2
Present state y_1y_2	00	01	10	01
	01	00	11	10
	10	01	01	00
	11	00	00	00

(a)

Primary input	Present state		Next state		Primary outputs	
x	y_1	y_2	y_1^+	y_2^+	z_1	z_2
0	0	0	0	1	0	1
0	0	1	0	0	1	0
0	1	0	0	1	0	0
0	1	1	0	0	0	0
1	0	0	1	0	0	1
1	0	1	1	1	1	0
1	1	0	0	1	0	0
1	1	1	0	0	0	0

(b)

Primary input	Present state		Next state		Secondary outputs		Primary outputs	
x	y_1	y_2	y_1^+	y_2^+	D_1	D_2	z_1	z_2
0	0	0	0	1	0	1	0	1
0	0	1	0	0	0	0	1	0
0	1	0	0	1	0	1	0	0
0	1	1	0	0	0	0	0	0
1	0	0	1	0	1	0	0	1
1	0	1	1	1	1	1	1	0
1	1	0	0	1	0	1	0	0
1	1	1	0	0	0	0	0	0

(c)

FIGURE 7.28 (a) A binary version of the state table for *TC*; (b) its transition table; (c) its excitation table.

Combinational Function Definition

C's functions depend on the state behavior, the state assignment, and the flip-flop types.

The next step is to determine the primary and secondary output functions to be realized by the combinational part C of TC. The combinational functions to be implemented here have the general form $f_i(x, y_1, y_2)$. If we rearrange the binary state table of Figure 7.28a as a truth table—this is called a *transition table*—with inputs x, y_1, y_2, we obtain complete definitions of the next-state functions y_1^+, y_2^+ and primary output functions z_1, z_2, as shown in Figure 7.28b.

We have selected D-type flip-flops, so the next-state functions y_1^+, y_2^+ immediately define the inputs D_1, D_2 to the flip-flops, a consequence of the D flip-flop's characteristic equation $y_i^+ = D_i$. The D_1 and D_2 signals are, of course, the secondary outputs of C, which serve to "excite"—that is, activate—the flip-flop inputs. Adding the definitions of D_1 and D_2 to the transition table, we obtain a complete functional definition of C in the form known as an **excitation table** (Figure 7.28c).

Combinational Logic Design

It remains to implement TC's combinational logic C, whose functions are now fully specified by the excitation table. We can obtain a minimum-cost, two-level implementation of C via the K-map method, because we have only three input variables to deal with. Figure 7.29 shows K-maps of the four

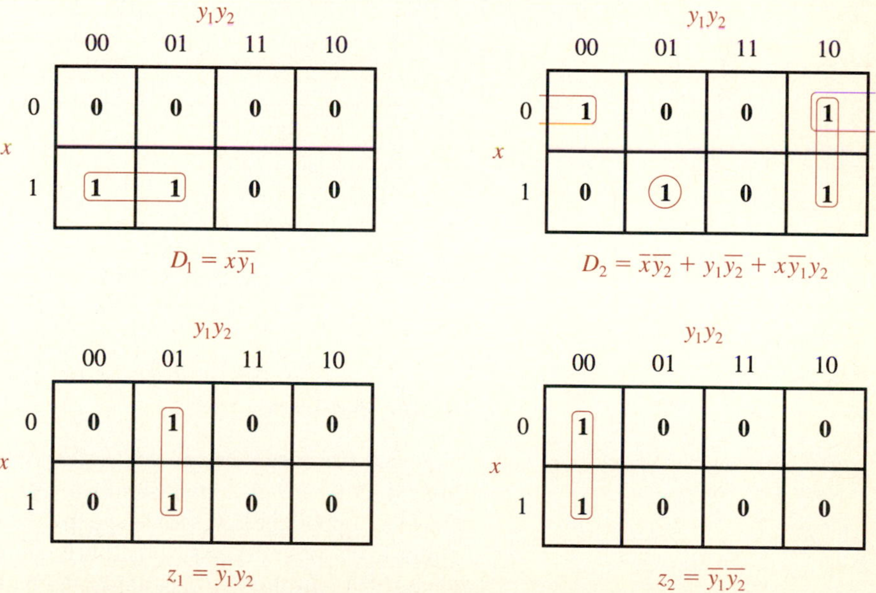

y_1y_2
00 01 11 10
x
0: 0 0 0 0
1: 1 1 0 0
$D_1 = x\overline{y_1}$
y_1y_2
00 01 11 10
x
0: 1 0 0 1
1: 0 1 0 1
$D_2 = \overline{x}\overline{y_2} + y_1\overline{y_2} + x\overline{y_1}y_2$
y_1y_2
00 01 11 10
x
0: 0 1 0 0
1: 0 1 0 0
$z_1 = \overline{y_1}y_2$
y_1y_2
00 01 11 10
x
0: 1 0 0 0
1: 1 0 0 0
$z_2 = \overline{y_1}\overline{y_2}$

FIGURE 7.29 K-maps that define the combinational logic C for TC.

Standard combinational methods can be used to design C.

output functions of C taken directly from the excitation table. It is a trivial matter to identify a minimal set of product terms to cover each function, leading to the following unique set of minimal SOP equations for C:

$$D_1 = x\overline{y}_1$$
$$D_2 = \overline{x}\overline{y}_2 + y_1\overline{y}_2 + x\overline{y}_1y_2$$
$$z_1 = \overline{y}_1y_2$$
$$z_2 = \overline{y}_1\overline{y}_2$$

A straightforward circuit implementation of these equations is given in Figure 7.30. The output logic circuit C_2 of TC has no connections to the primary input x, as expected in Moore-type sequential circuits.

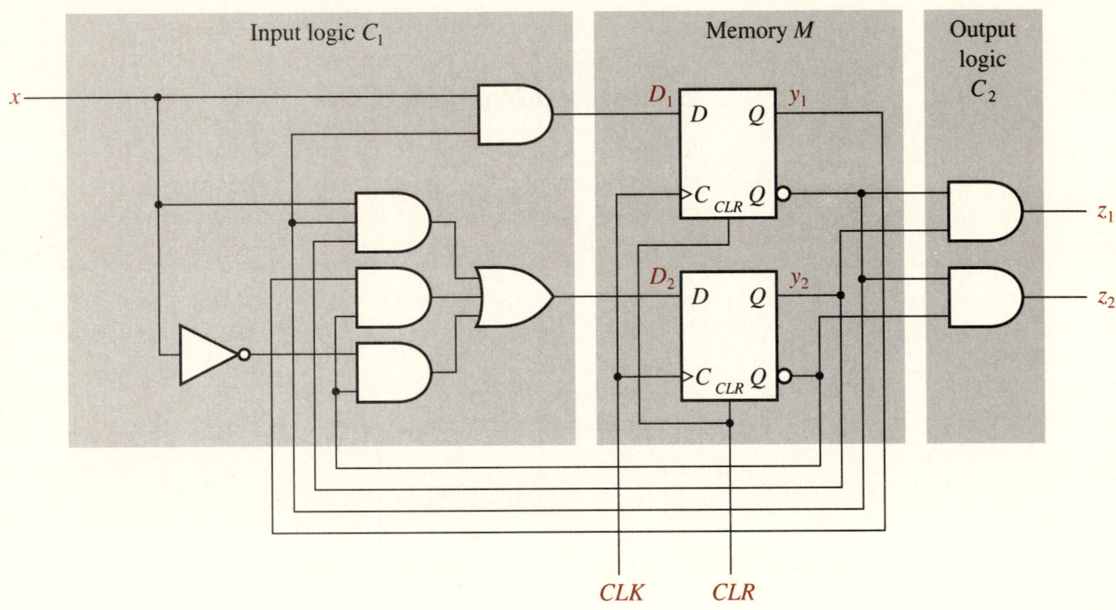

FIGURE 7.30 The logic circuit for the traffic-light control unit *TC*.

Design Verification

To check that a new design behaves as required, it is highly desirable to simulate it by means of a logic simulation program. Logic simulators are among the most important CAD tools and are essential if we are to eliminate design errors from logic designs, even fairly small ones. A computer simulation of the traffic-control unit TC is illustrated by Figure 7.31. The simulator we use here is DesignWorks; the host is a Macintosh computer.

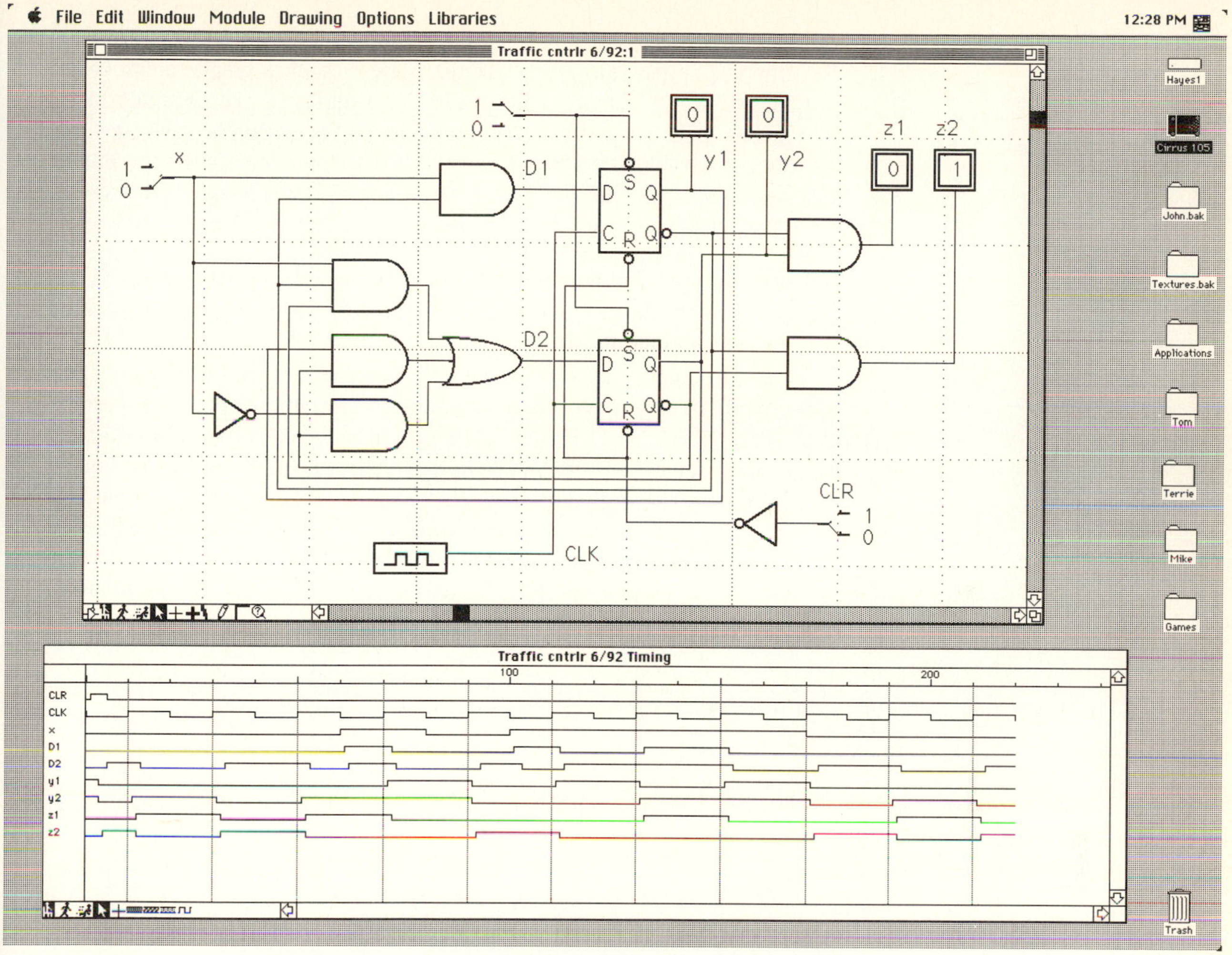

FIGURE 7.31 A computer screen image during simulation of *TC*.

The circuit schematic in the upper window is constructed by taking the necessary components (gates, D flip-flops, and so on) from a predefined component library via the DesignWorks Libraries menu. A component is selected from this menu by standard "pick and place" techniques using a mouse and a cursor. The circuit connections (wires) are made by dragging lines from point to point using the mouse. We have "instrumented" the circuit by placing switches on the primary input lines, and probes (the small double boxes that display 0 or 1) on lines of interest, such as primary and secondary input/output lines.

Simulation is essential for verifying the correctness of all but the smallest sequential circuits.

Simulation to verify the circuit's behavior requires supplying input signal values to exercise the circuit and observing and checking the response values of selected signals. The inputs to this circuit are obtained from a clock and a set of on-off switches, both of which are built-in DesignWorks components. The designer turns a switch on and off by placing the cursor on it and clicking the computer's mouse key. (This technique does not allow precise control over switching times, a characteristic of simpler simulators like DesignWorks.) In response to a changed primary input signal, the simulator calculates the corresponding value of all selected signals, assuming a fixed, user-specifiable propagation delay within each component. DesignWorks displays the resulting signal changes in the timing diagram that forms the lower window of Figure 7.31; current signal values are also displayed in the probe boxes of the upper (circuit) window. Like most logic simulators, DesignWorks is not able to directly generate state tables or diagrams; the state behavior must be inferred from timing diagrams.

This particular simulation exercise was designed to take *TC* through all its states, while turning x on and off at various representative times. Inspection of the timing diagram of Figure 7.31 shows that the state transitions and the values of z agree with the behavior specifications for *TC* given in Figure 7.27. For example, in the first few clock cycles, we see the primary outputs z_1 and z_2 switching back and forth between the EW and NS green conditions, while $x = 0$. At time 60—the time scale used is somewhat arbitrary—when $y_1y_2 = 01$, which denotes state S_1, and $z_1z_2 = 10$, meaning EW is green and NS red, we first switch x to 1. The next positive edge of *CLK* arrives at time 70, causing the state to change to $y_1y_2 = 11 = S_3$, thereby creating the all-red condition $z_1z_2 = 00$, as required. We have thus verified that *TC*'s state transition

$$S_1 \xrightarrow{x\,=\,1} S_3$$

is properly realized by our design. All the other state transitions are similarly checked.

7.5 The Classical Method

We examine next a systematic design technique for small synchronous sequential circuits, which has as its objective the use of the minimum number of flip-flops; minimizing the number of gates is a secondary design objective. This technique, the origins of which go back to the work of Huffman and others in the 1950s [Moore, 1964], was implicitly followed in the traffic-control problem we studied in Section 7.4.

An Overview

The classical synthesis procedure *SCSYN* for sequential circuits is summarized below. Its starting point is a formal specification of the state behavior desired of the target circuit. The specification often takes the form of a state table, so this approach to sequential circuit synthesis is sometimes called the **state-table method**. The goal is a logic circuit SC with the general (Huffman) structure depicted in Figure 7.2. It consists of a combinational logic circuit C and a memory M, which is composed of edge-triggered flip-flops that are connected to a common clock signal. The binary pattern $Y(i) = y_1(i)y_2(i)\cdots y_p(i)$ stored in M during any clock cycle i constitutes the (internal) state of SC. The next value $y_j(i+1) = y_j^+(i)$ of each y_j is a function of the primary input variables $X(i)$ and the present state $Y(i)$. The circuit's primary output signals $Z(i)$ and the secondary outputs, which determine the next-state functions, are computed by C, which therefore represents the data-processing part of SC. $Z(i)$ is, in general, a function of $X(i)$ and $Y(i)$. In the special case of Moore circuits, $Z(i)$ is a function of $Y(i)$ only.

Procedure

Procedure *SCSYN* for designing small sequential circuits

1. *State table specification:* Construct a state table (or diagram) that precisely defines the behavior desired of SC. The circuit's internal states may be represented symbolically as $A, B, C, \ldots$.
2. *State minimization:* Reduce the state table to an equivalent one that contains the minimum number of states P.
3. *State assignment:* Select $p = \lceil \log_2 P \rceil$ flip-flops for M and assign P patterns of p secondary variables $Y = y_1, y_2, \ldots, y_p$ to represent the reduced set of internal states.
4. *Excitation table specification:* Find the primary output functions $Z(X, Y)$ and the secondary (excitation) functions to be applied to M, which depend on the type of flip-flop used. These functions define the behavior of the combinational logic C.
5. *Combinational circuit design:* Construct minimal two-level logic circuits that implement C.

Sequential circuit design is probably best described by examples, and a facility for it is best acquired by practice. In the remainder of this section, we examine a series of design examples of gradually increasing difficulty that illustrate *SCSYN* and highlight its main steps.

State-Table Specification

The first step of *SCSYN* is to convert the usual informal specification of the target behavior into a state table or, equivalently, a state diagram. As noted earlier, this step is not formalizable by its very nature, and may require considerable ingenuity on the designer's part. The trick is to determine the essential information generated in the current cycle i that must be stored in M for future use. For example, in the case of a serial adder, carry information must be stored, a fact that follows from an understanding of the addition

Specifying state behavior efficiently requires careful analysis of the problem.

operation. The traffic-light controller, like many control units, is somewhat easier because the required states correspond to specified configurations of the controlled devices.

Even with a good understanding of the desired functional behavior, there is no guarantee that the state table derived will be the best possible. Consider, for example, the Mealy-type state table of Figure 7.32a, which contains three states. This is a valid state table for a serial adder, so it must be reducible to the two-state table for the adder derived earlier. In the three-state table, the states B and C are equivalent,[2] allowing them to be merged into a single state. This follows from the fact that for each combination $X(i)$ and $Y(i)$, the output values $Z(i)$ of B and C are the same; in addition, the next-state values $Y(i+1)$ of B and C are either the same, or one is B and the other is C. Hence, any input sequence applied with initial state B will produce exactly the same response that it would if C were the initial state. The two-state table of Figure 7.32b results from discarding the C row and replacing all remaining C entries with B.

We conclude that state tables are not unique, and that the table we first construct may, like Figure 7.32a, have more states than necessary. This, however, is much less important than ensuring that the state description is functionally correct. A careful analysis of the problem under consideration will often lead to a minimum or near-minimum number of states.

		Present input x_1x_2			
		00	01	11	10
Present state	A	$A, 0$	$A, 1$	$B, 0$	$A, 1$
	B	$A, 1$	$C, 0$	$C, 1$	$B, 0$
	C	$A, 1$	$B, 0$	$C, 1$	$C, 0$

(a)

		Present input x_1x_2			
		00	01	11	10
Present state	A	$A, 0$	$A, 1$	$B, 0$	$A, 1$
	B	$A, 1$	$B, 0$	$B, 1$	$B, 0$

(b)

FIGURE 7.32 (a) Nonminimal and (b) minimal state tables for a serial adder.

Example 7.4

Designing a sequential two's-complementer *N2C*

We begin with an easy case: the design of a tiny sequential circuit *N2C* that negates two's-complement (2C) numbers. *N2C* implements a sequential procedure called *SCANEG2c*, which appears in the problem set of Chapter 2 and is repeated below. This is a serial algorithm that converts a number X_{2C} bit by bit into $-X_{2C}$ in proper 2C code. Although slow, *SCANEG2c* has the advantage of being applicable to numbers of arbitrary length n.

[2] A general definition of equivalence and a method of identifying the equivalent states in a state table are presented in Section 7.6.

Procedure

Procedure *SCANEG2c* to negate a 2C number

1. Input the 2C number $X_{2C} = x_{n-1}x_{n-2} \cdots x_1x_0$. Set a control flag f to an initial value of 0.
2. Scan X_{2C} from right to left—that is, from $i = 0$ to $n-1$. If $f \neq 0$, replace x_i with $\overline{x}_i$; otherwise, if $x_i = 1$, set f to 1. When the scan is complete, X_{2C} has been changed to its two's-complement.

The scanning process obviously translates to entering X_{2C} serially, least significant bit first, into the target circuit *N2C* while at the same time generating a serial output stream of bits, forming $-X_{2C}$. Thus, *N2C* will have a single data input line x and a single data output line z. The *SCANEG2c* procedure gives a strong hint of the nature of *N2C*'s internal states. There should be a state S_0 to indicate that a 1 has not yet been encountered among the input bits from X_{2C}. There should be a second state S_1 to indicate that at least one 1 has been encountered in X_{2C}. We conclude that we need two internal states S_0 and S_1, which we will represent by the state variable f, corresponding to the flag mentioned in *SCANEG2c*. S_0 is clearly the reset state.

The necessary state transitions can be deduced directly from the negation procedure. As long as we are in state S_0, the input $x = 0$ will leave the circuit in S_0 and produce the output $z = 0$. When x first becomes 1, a transition is made to S_1, and z becomes 1. Hence, the two possible state transitions from S_0 are as follows:

$$S_0 \xrightarrow{0/0} S_0 \qquad S_0 \xrightarrow{1/1} S_1$$

A trap state can be exited only via an asynchronous reset.

Once in S_1, *N2C* remains there indefinitely—that is, until the circuit is asynchronously reset. (States like S_1 that may be entered but never exited are called **trap states** or, when they represent abnormal conditions, **dead** or **hang-up states**.) As implied by *SCANEG2c*, for each input value x_i, *N2C* now outputs its complement $\overline{x}_i$. Therefore, the possible transitions from S_1 are as follows:

$$S_1 \xrightarrow{0/1} S_1 \qquad S_1 \xrightarrow{1/0} S_1$$

The foregoing properties of *N2C* are combined in the state table of Figure 7.33; the corresponding state diagram is also given. It's usually a good idea to check the state behavior with some sample input sequences. Clearly,

		Present input x	
		0	1
Present state	S_0	S_0, 0	S_1, 1
	S_1	S_1, 1	S_1, 0

(a)

RESET
Trap state
0/0
S_0
1/1
S_1
1/0
0/1
$f = 0$
$f = 1$

(b)

FIGURE 7.33 (a) A state table and (b) a state diagram for the two's-complementer *N2C*.

$N1_{2C} = 00000000$, denoting zero is converted into $-N1_{2C} = 00000000$ in $n = 8$ clock cycles, during which $N2C$ never leaves its initial state S_0. $N2_{2C} = 01110110 = 55_{10}$ is converted into $-N2_{2C} = 10001010$ and causes $N2C$ to spend one cycle in S_0 before it goes to S_1, where it remains.

The last few design steps for this example are rather trivial. Figure 7.34a shows an excitation table for $N2C$, which assumes that the state variable f is stored in a D flip-flop. The excitation table comes directly from the state table, with $f = 0$ and $f = 1$ denoting S_0 and S_1, respectively. The secondary output—that is, the flip-flop's input (excitation) function—is defined as usual by $D = f^+$. It remains to implement the two combinational functions $D(x, f)$ and $z(x, f)$. We see immediately from the excitation table that $D = x + f$ and $z = x \oplus f$, implying the realization of Figure 7.34b.

Primary input	Present state	Next state	Secondary output	Primary output
x	f	f^+	D	z
0	0	0	0	0
0	1	1	1	1
1	0	1	1	1
1	1	1	1	0

(a)

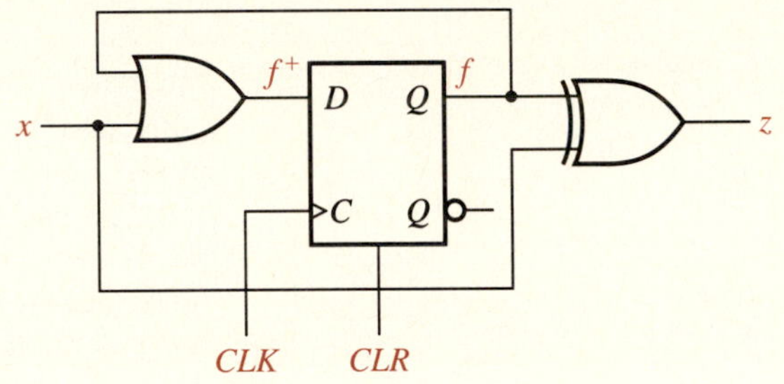

(b)

FIGURE 7.34 (a) An excitation table and (b) a logic diagram for *N2C*.

Recognizers

Recognizers are sequential circuits designed to detect specific input sequences.

We consider next the design of a special class of control circuits known as **recognizers**. They are intended to output a specific signal, say, $z = 1$, only when an input sequence received by the circuit belongs to a well-defined set—the set of "recognized" sequences. A "combination" lock that opens in response to a particular sequence of input symbols is a familiar—but in the logic design context, misnamed—example of a recognizer.

Example 7.5

Designing a recognizer circuit *R3* for a 3-bit sequence

A recognizer circuit *R3* has a single input signal x and a single output z, along with the usual control inputs, the clock signal *CLK*, and an asynchronous initialization signal *RESET*. The behavior required of *R3* is to set z to 1 after it receives at least three 1-signals on its x input line. Thus, *R3* distinguishes input sequences that contain three or more 1s from those that do not.

Several points need to be clarified before we can specify the state behavior of *R3*. As in all synchronous sequential circuits, each clock cycle is assumed to present a new value of x. We will assume that the 1-values of x to be counted are not necessarily in consecutive cycles; they may be separated by any number of cycles in which $x = 0$. We also need to know when to start counting. We will assume that the 1s are counted from the last time *R3* was reset via an asynchronous reset. Once the output z is set to 1, it remains at 1 until the next asynchronous reset occurs. Finally, we will assume that D flip-flops with preset and clear inputs are to be used in this design, along with any of the standard gate types.

R3's state behavior is readily derived from the foregoing specifications. To identify the internal states, we pose the question: What information does *R3* need to remember? The answer is clearly the number of 1s received on line x so far. There are four cases:

1. State S_0: No 1s have been received yet
2. State S_1: A single 1 has been received
3. State S_2: Two 1s have been received
4. State S_3: Three or more 1s have been received

Hence, S_i is the state entered after i 1s have been received by *R3*. The reset state of *R3* is S_0, and each new 1-signal recognized on x changes the state from S_i to S_{i+1} until state S_3 is reached. Three 1s take the circuit to S_3; subsequent 1s and 0s cause *R3* to stay in S_3 indefinitely, producing $z = 1$. (S_3 is another example of a trap state.) The input $x = 0$ causes *R3* to remain in its current state and output $z = 0$, except in the case of S_3, where $z = 1$. Figure 7.35 presents a state table and a state diagram for *R3*.

Observe that the state behavior of *R3* is similar to that of a modulo-4 counter, except that $x = 1$ returns the counter from S_3 to S_0. In contrast, *R3* must be reset to S_0 asynchronously. It is not hard to see that four is the minimum number of states needed by *R3*, so the state representations of Figure 7.35 are fully reduced. It should also be noted that, because each row of the state table contains the same output value, the circuit's behavior is of the Moore type. We could therefore represent the behavior in the slightly different format of Figure 7.10b. However, we will leave it in the more general Mealy format for illustrative purposes.

Some circuits are, like *R3*, inherently of the Moore type.

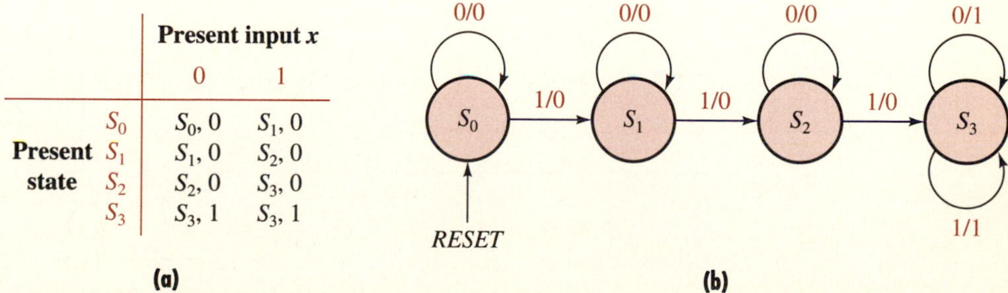

		Present input x	
		0	1
Present state	S_0	S_0, 0	S_1, 0
	S_1	S_1, 0	S_2, 0
	S_2	S_2, 0	S_3, 0
	S_3	S_3, 1	S_3, 1

FIGURE 7.35 (a) A state table and (b) a state diagram for the recognizer circuit *R3*.

Obviously, two D flip-flops *FF1* and *FF2* suffice to store the state y_1y_2 of *R3*. As in the traffic-controller case, which also had four states, we make the following rather arbitrary state assignment:

$$\begin{aligned} S_0 &= 00 \\ S_1 &= 01 \\ S_2 &= 10 \\ S_3 &= 11 \end{aligned} \tag{7.15}$$

Together, the state table and assignment (7.15) yield the transition table of Figure 7.36a. Because D flip-flops are being used again, each next-state signal has the form $y_i^+ = D_i$, and the transition table immediately defines

Primary input x	Present state y_1	y_2	Next state y_1^+	y_2^+	Primary output z
0	0	0	0	0	0
0	0	1	0	1	0
0	1	0	1	0	0
0	1	1	1	1	1
1	0	0	0	1	0
1	0	1	1	0	0
1	1	0	1	1	0
1	1	1	1	1	1

(a)

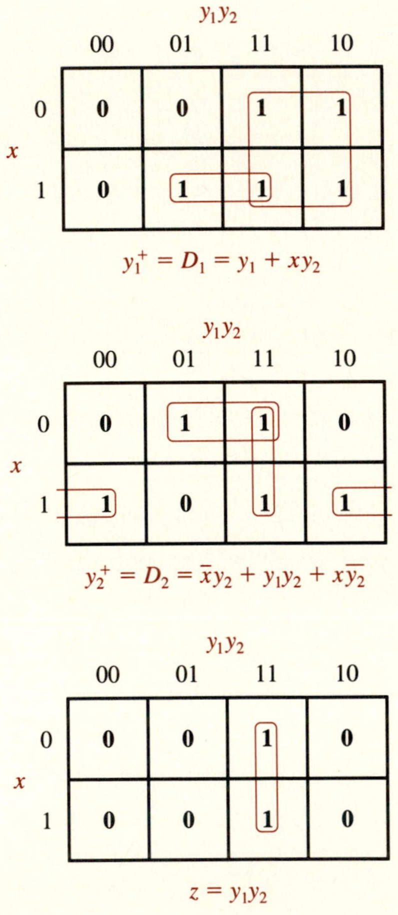

x \ y_1y_2	00	01	11	10
0	0	0	1	1
1	0	1	1	1

$y_1^+ = D_1 = y_1 + xy_2$

x \ y_1y_2	00	01	11	10
0	0	1	1	0
1	1	0	1	1

$y_2^+ = D_2 = \bar{x}y_2 + y_1y_2 + x\overline{y_2}$

x \ y_1y_2	00	01	11	10
0	0	0	1	0
1	0	0	1	0

$z = y_1y_2$

(b)

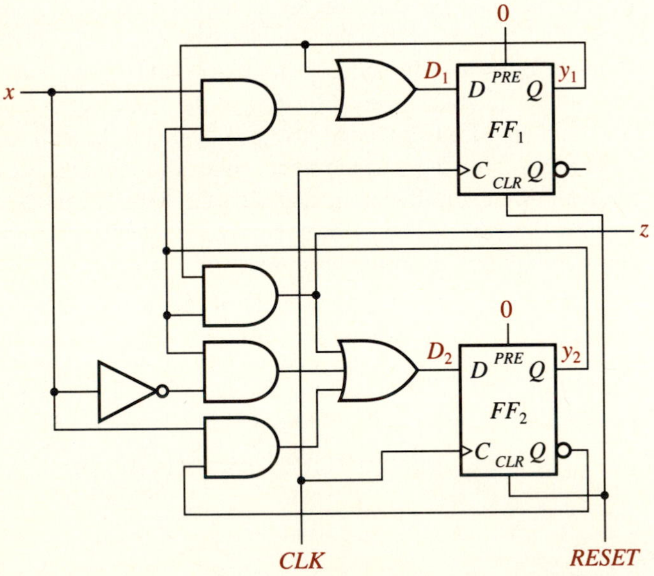

(c)

FIGURE 7.36 Design steps for *R3*: **(a)** a transition table; **(b)** K-maps for *C*; **(c)** a logic circuit.

the excitation functions for the two flip-flops; it also defines the primary output function z. We can obtain a minimal two-level realization of the combinational part C of *R3* by employing K-maps, as shown in Figure 7.36b. Optimal sum-of-products or product-of-sums solutions are easily found by inspection. The SOP solution shown has the advantage of sharing a prime implicant between D_2 and z. Observe how the complemented ($\overline{Q}$) outputs of the flip-flops are used to supply each $\overline{y}_i$ needed by C. Observe also that we have obtained a Moore-type sequential circuit, because the output function $z = y_1 y_2$ is independent of x.

State Assignment

The second major step of *SCSYN* is to make a state assignment for the P states in the reduced state table, using the minimum number of flip-flops—namely, $p = \lceil \log_2 P \rceil$—of the chosen type. From the state assignment and the state transition information in the state table, we can specify all the functions to be generated by C. If D flip-flops are used, C must generate p functions of the form $D_1, D_2, \ldots, D_p$. JK flip-flops, on the other hand, require $2p$ functions of the form $J_1, K_1, J_2, K_2, \ldots, J_p, K_p$. Combining these with the m primary output functions $Z = z_1, z_2, \ldots, z_m$, we conclude that we need C to generate either $p + m$ or $2p + m$ functions, depending on the flip-flop type.

Example 7.6

Another version of the recognizer *R3*

Consider again the recognizer circuit *R3*, the state behavior of which is defined in Figure 7.35a. We obtained the design of Figure 7.36c by assigning binary patterns to the two state variables y_1 and y_2, according to (7.15). It is apparent that the state assignment affects the definition of the functions generated by C, and so influences the complexity of the circuits that realize C. To illustrate this, Figure 7.37 gives another design for *R3*, with the following different state assignment:

$$\begin{aligned} S_0 &= 10 \\ S_1 &= 00 \\ S_2 &= 11 \\ S_3 &= 01 \end{aligned} \tag{7.16}$$

As can be verified from the design details given in Figure 7.37, two fewer AND gates are needed to generate the secondary outputs D_1 and D_2. An additional AND gate is needed for the primary output z, for a net saving of one two-input AND gate, compared to the design of Figure 7.36. Note that

Primary input x	Present state y_1	y_2	Next state y_1^+	y_2^+	Primary output z
0	0	0	0	0	0
0	0	1	0	1	1
0	1	0	1	0	0
0	1	1	1	1	0
1	0	0	1	1	0
1	0	1	0	1	1
1	1	0	0	0	0
1	1	1	0	1	0

(a)

x \ y_1y_2	00	01	11	10
0	0	0	1	1
1	1	0	0	0

$y_1^+ = D_1 = \bar{x}y_1 + x\bar{y}_1\bar{y}_2$

x \ y_1y_2	00	01	11	10
0	0	1	1	0
1	1	1	1	0

$y_2^+ = D_2 = y_2 + x\bar{y}_1\bar{y}_2$

x \ y_1y_2	00	01	11	10
0	0	1	0	0
1	0	1	0	0

$z = \bar{y}_1 y_2$

(b)

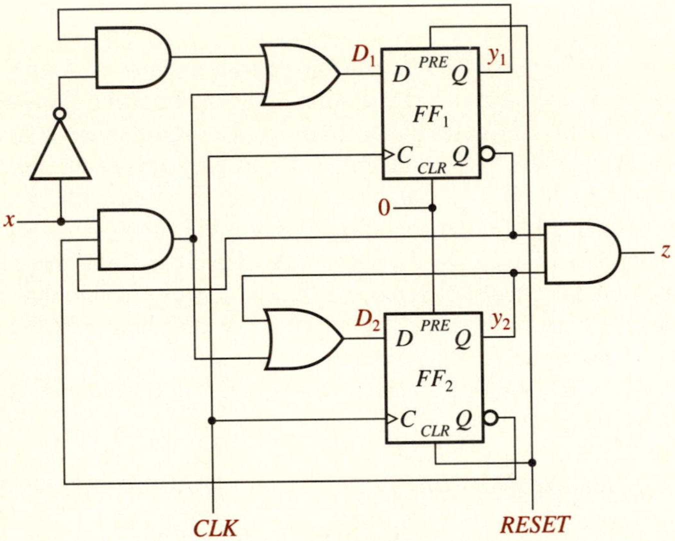

(c)

FIGURE 7.37 The design for *R3*, using a second state assignment: (**a**) its transition table; (**b**) its K-maps; (**c**) its logic circuit.

y_1^+ and y_2^+ are able to share the AND gate that realizes the multiple-output prime implicant $x\bar{y}_1\bar{y}_2$. Also note that the reset connections must be changed as indicated to clear the flip-flops, so that the initial state becomes $y_1y_2 = 10$ rather than 00.

Assignment Choices

There are many ways in which states can be assigned to binary patterns. There are $\binom{2^P}{P} = \frac{2^P!}{P!(2^P-P)!}$ ways to choose distinct sets of P binary patterns from the 2^P patterns defined by the state variables $y_0y_1 \ldots y_{p-1}$, where $k! = k \times (k-1) \times (k-2) \times \cdots \times 2 \times 1$. These pattern sets can be permuted in $P!$ possible ways, leading to $\binom{2^p}{P}P! = \frac{2^p!}{(2^p-P)!}$ assignments. For example, for $P = 9$ and $p = 4$, we obtain

$$\binom{2^4}{9}9! = \frac{16!}{(16-9)!}$$
$$= 16 \times 15 \times 14 \times \cdots \times 9 \times 8$$
$$= 4{,}151{,}347{,}200$$

Essentially equivalent assignments are obtained by permuting or complementing the bits in any positions within $y_0y_1 \ldots y_{p-1}$. There are $p!$ ways to permute the p state variables, and 2^p ways to select subsets of state variables to complement. Hence, the total number ν_{sa} of nonequivalent state assignments with P states and p state variables is given by the following formula [Harrison, 1965]:

$$\nu_{sa} = \frac{\binom{2^p}{P}P!}{p!2^p} = \frac{(2^p-1)!}{(2^p-P)!p!} \tag{7.17}$$

Assuming $p = \lceil \log_2 P \rceil$, we find that the ν_{sa} grows very rapidly with P. For example, with $P = 9$, we require $p = 4$, and (7.17) gives $\nu_{sa} =$ 10,810,800.

State assignments affect circuit cost in unpredictable ways.

No practical way is known to determine the state assignments that lead to the lowest-cost implementations, although some interesting theoretical work has addressed this problem [Hartmanis and Stearns, 1966]. A few rules of thumb for making state assignments are known, but their value is dubious (Problem 7.37). Recently, CAD programs have been written to compute efficient state assignments heuristically [Devadas et al., 1988]. In this chapter, we usually settle for the simple expedient of assigning bit-pattern i to state S_i, as in the preceding examples. Chapter 8 introduces a useful state assignment technique, the "one-hot" method, which greatly simplifies the design of the combinational part of a sequential circuit at the expense of a far from minimal number of flip-flops.

Excitation Tables

As we have seen, the behavior of the combinational part C of a small sequential circuit is often described by an excitation table, which specifies the input signals that must be applied to the flip-flops in order to produce the desired next states. These signals are the secondary outputs of C. The excitation table can be obtained from a transition table like Figure 7.37a by adding a set of secondary output columns. Knowing the transition from y_j to y_j^+ required in any row, we can determine the input signals D_j, J_j, K_j, etc., to the corresponding flip-flop to produce this transition. The choice of flip-flop signals and their values depends on the flip-flop type, an issue that we now examine.

JK flip-flops have many useful don't-cares among their inputs.

In general, for any state variable y_j in any row of the transition table, we find one of four possible combinations of the present value y_j and the desired next value y_j^+. Suppose $y_j = y_j^+ = 1$; then the jth flip-flop FF_j should remain unchanged in the set state. We therefore need $D_j = 1$ if a D flip-flop is used, and either $J_j = K_j = 0$ (the quiescent input) or $J_j = 1$ and $K_j = 0$ (the set input) if a JK flip-flop is used. Thus, in the JK case, we have $J_j = \mathrm{X}$ and $K_j = 0$, because we don't care what the value of J_j is as long as K_j is 0. If $y_j = 1$ and $y_j^+ = 0$, so that FF_j should be reset, we need $D_j = 0$ if a D flip-flop is used. The corresponding JK inputs are $J_j = 0$ and $K_j = 1$ (the reset input) or $J_j = K_j = 1$ (the toggle input); in other words, $J_j = \mathrm{X}$ and $K_j = 1$. The other two combinations of y_j and y_j^+ have similar requirements.

Figure 7.38 gives the complete excitation tables for the D and JK flip-flops; these are the tabular equivalents of the flip-flops' characteristic equations. For good measure, we include the excitation table required by the SR flip-flop. This differs from the JK case only in the fact that the input combination 11 is disallowed. This table may be referred to when extending the transition table of a sequential circuit into an excitation table. Once the excitation table of a new design is complete, the behavior of its combinational logic C has been defined, and we can proceed to find a suitable implementation, using the techniques of Chapter 5.

Desired transition		Flip-flop inputs				
Q	$\rightarrow Q^+$	D	J	K	S	R
0	0	0	0	X	0	X
0	1	1	1	X	1	0
1	0	0	X	1	0	1
1	1	1	X	0	X	0

FIGURE 7.38 Excitation tables for the basic flip-flop types.

It might be thought that JK flip-flops are less desirable than D flip-flops because twice as many excitation functions must be generated by C. However, as Figure 7.38 demonstrates, many of the input signals to JK flip-flops assume the don't-care value X, which can often be used to advantage to simplify C.

Example 7.7

Designing a serial adder with a JK flip-flop

Consider the design of a serial adder, which was presented earlier in a version that employed a D flip-flop as its memory element (Figure 7.4). This time we will use a JK flip-flop. A state table for the adder can be found in Figure 7.32. As before, we use the state assignment $A = y = 0$ and $B = \overline{y} = 1$, so that y represents a stored carry bit. The resulting excitation table appears in Figure 7.39a. The flip-flop inputs J and K have been derived from Figure 7.38, using as many don't-cares as possible.

A set of K-maps for C, derived from the adder's excitation table, appears in Figure 7.39b. They lead immediately to the minimal sum-of-products forms

$$J = x_1 x_2$$
$$K = \overline{x}_1 \overline{x}_2$$

which define C's secondary outputs. The primary output z is the same sum function as before. The resulting logic circuit appears in Figure 7.39c and has somewhat simpler next-state logic than the first design with the D flip-flop.

Primary input		Present state	Next state	Secondary outputs		Primary output
x_1	x_2	y	y^+	J	K	z
0	0	0	0	0	X	0
0	0	1	0	X	1	1
0	1	0	0	0	X	1
0	1	1	1	X	0	0
1	0	0	0	0	X	1
1	0	1	1	X	0	0
1	1	0	1	1	X	0
1	1	1	1	X	0	1

(a)

y \ x_1x_2	00	01	11	10
0	0	0	1	0
1	X	X	X	X

$J = x_1x_2$

y \ x_1x_2	00	01	11	10
0	X	X	X	X
1	1	0	0	0

$K = \overline{x_1}\,\overline{x_2}$

y \ x_1x_2	00	01	11	10
0	0	1	0	1
1	1	0	1	0

$z = x_1 \oplus x_2 \oplus y$

(b)

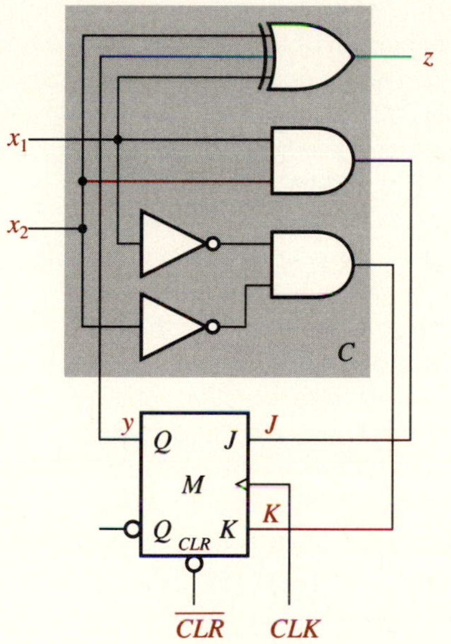

(c)

FIGURE 7.39 Design of a serial adder using a JK flip-flop: **(a)** its excitation table; **(b)** K-maps for C; **(c)** its logic circuit.

More on State Behavior

We turn again to the task of obtaining an initial state table for a new design, the behavior of which is specified informally. We can only give some broad guidelines for tackling this sometimes difficult problem because, as observed already, it is inherently unformalizable.

A general approach is to try to identify by classes the main sequences of actions taken by the target sequential circuit *SC* and the input sequences associated with them. For example, a serial adder has one class of input sequences that produce an output carry of 1, and another class that produce an output carry of 0. In the case of recognizer circuits like *R3*, the input sequences also fall into two main classes: those that lead to $z = 1$—that is, recognition—and those that do not. Next we can subdivide each input sequence class into 1-cycle steps of the form $X_a(1)X_b(2)\ldots X_k(h)$, where $X_j(i)$ denotes some primary input combination applied to *SC* in clock cycle i. We can now introduce a sequence of symbolic states $S_a(1)S_b(2)\ldots S_k(h)$ to represent *SC*'s response to $X_a(1)X_b(2)\ldots X_k(h)$ thus:

$$S_{\text{init}} \xrightarrow{X_a} S_a \xrightarrow{X_b} S_b \xrightarrow{X_c} \cdots \xrightarrow{X_k} S_k \tag{7.18}$$

It's useful to identify representative classes of input sequences.

The states introduced in the above way can be new states, or they may be previously identified states that fit into the sequence in question. It is desirable to reuse old states as much as possible to minimize the proliferation of new states. The success of this approach is largely dependent on keeping the number and length of the state-transition sequences like (7.18) reasonably small. It is generally not a problem if we inadvertently introduce some redundant states—that is, states equivalent to previously defined ones—because the equivalent states can be identified systematically later in a state-reduction process (step 2 of *SCSYN*).

As we stated already, a good way to identify representative behavior sequences like (7.18) is to work out a few simple cases in detail—that is, to simulate the desired behavior of *SC* by hand. This is particularly true in the case of datapath circuits, in which the types of data and the operations to be performed are specified fairly precisely. We can examine each operation in turn and apply it to representative input data sequences. Of special interest are the circuit's treatment of boundary cases, such as the largest and smallest allowable input numbers in the case of arithmetic circuits.

We conclude with a couple of examples that further illustrate these issues, as well as the complete classical design process for sequential circuits. The first example is another type of recognizer circuit, which is required to recognize and distinguish two specific 4-bit input sequences that occur any time after a reset operation.

Example 7.8

Designing a two-sequence recognizer *R4*

The problem here is to design a one-input, two-output sequential circuit *R4* that recognizes and distinguishes the occurrence of two specific 4-bit sequences in its input stream. The quiescent output pattern is $z_1z_2 = 00$. If the sequence $X_1 = 0101$ is applied to the input line x, the output should become 01 when the last bit of X_1 appears, and the circuit should return to its reset state A. If a second sequence $X_2 = 1110$ is applied to x, the output should become 10 when the last bit of X_2 appears, and again the circuit should return to state A. Note that X_1 and X_2 may be preceded by any patterns that do not include either of them. Figure 7.40 shows *R4* with some representative input/output signals that indicate the desired behavior.

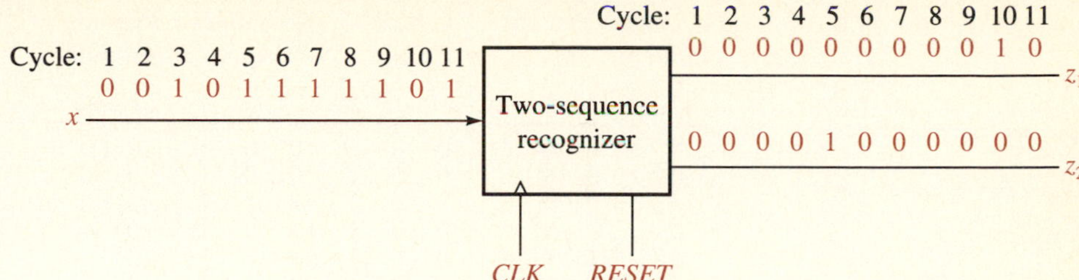

FIGURE 7.40 The two-sequence recognizer circuit *R4*.

It is apparent from this description that *R4* must have the following state transitions, among others, in order to recognize the two specified sequences:

$$\text{To recognize } X_1: \quad A \xrightarrow{0/00} B \xrightarrow{1/00} C \xrightarrow{0/00} D \xrightarrow{1/01} A$$

$$\text{To recognize } X_2: \quad A \xrightarrow{1/00} E \xrightarrow{1/00} F \xrightarrow{1/00} G \xrightarrow{0/10} A$$

Figure 7.41 shows these eight transitions placed in a seven-state diagram, the other six state transitions of which were derived as follows: If $x = 0$ in state B, the next state should be B, because $x = 0$ may indicate the beginning of a new copy of X_1. Thus, B has a self-loop labeled 0/00. For the same reason, there are transitions to B labeled 0/00 from states D, E, and F. It remains to specify the behavior in states C and G when $x = 1$. Because $x = 1$ in state B takes the circuit to C, the application of $x = 1$ in state C could represent the second bit in a copy of X_2. Consequently, the appropriate next state is F, the second state in the sequence of states required to recognize X_2. State G has a self-loop marked 1/00, because each additional 1 received can still form the last of the three 1-bits in a copy of X_2. Any subsequent 0 will result in recognition of X_2 and a return to the reset state A.

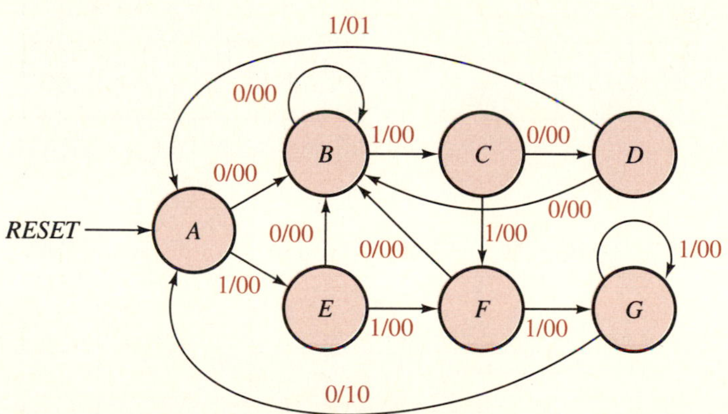

FIGURE 7.41 The state diagram for *R4*.

Next, we implement *R4*, using D flip-flops and NAND gates. We choose the obvious state assigment $A = 000$, $B = 001$, etc., with three state variables $y_1y_2y_3$. The excitation table for *R4* in Figure 7.42a comes directly from the state diagram in the usual manner. The primary and secondary (excitation) functions are transferred to K-maps in Figure 7.42b, which show

Primary input	Present state			Next state			Secondary outputs			Primary outputs	
x	y_1	y_2	y_3	y_1^+	y_2^+	y_3^+	D_1	D_2	D_3	z_1	z_2
0	0	0	0	0	0	1	0	0	1	0	0
0	0	0	1	0	0	1	0	0	1	0	0
0	0	1	0	0	1	1	0	1	1	0	0
0	0	1	1	0	0	1	0	0	1	0	0
0	1	0	0	0	0	1	0	0	1	0	0
0	1	0	1	0	0	1	0	0	1	0	0
0	1	1	0	0	0	0	0	0	0	1	0
0	1	1	1	X	X	X	X	X	X	X	X
1	0	0	0	1	0	0	1	0	0	0	0
1	0	0	1	0	1	0	0	1	0	0	0
1	0	1	0	1	0	1	1	0	1	0	0
1	0	1	1	0	0	0	0	0	0	0	1
1	1	0	0	1	0	1	1	0	1	0	0
1	1	0	1	1	1	0	1	1	0	0	0
1	1	1	0	1	1	0	1	1	0	0	0
1	1	1	1	X	X	X	X	X	X	X	X

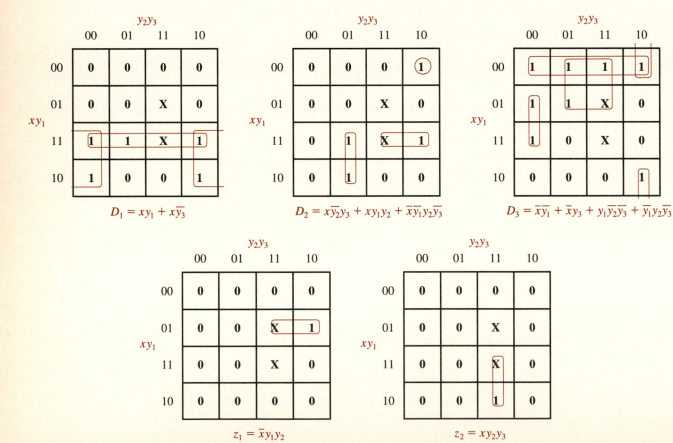

(a)

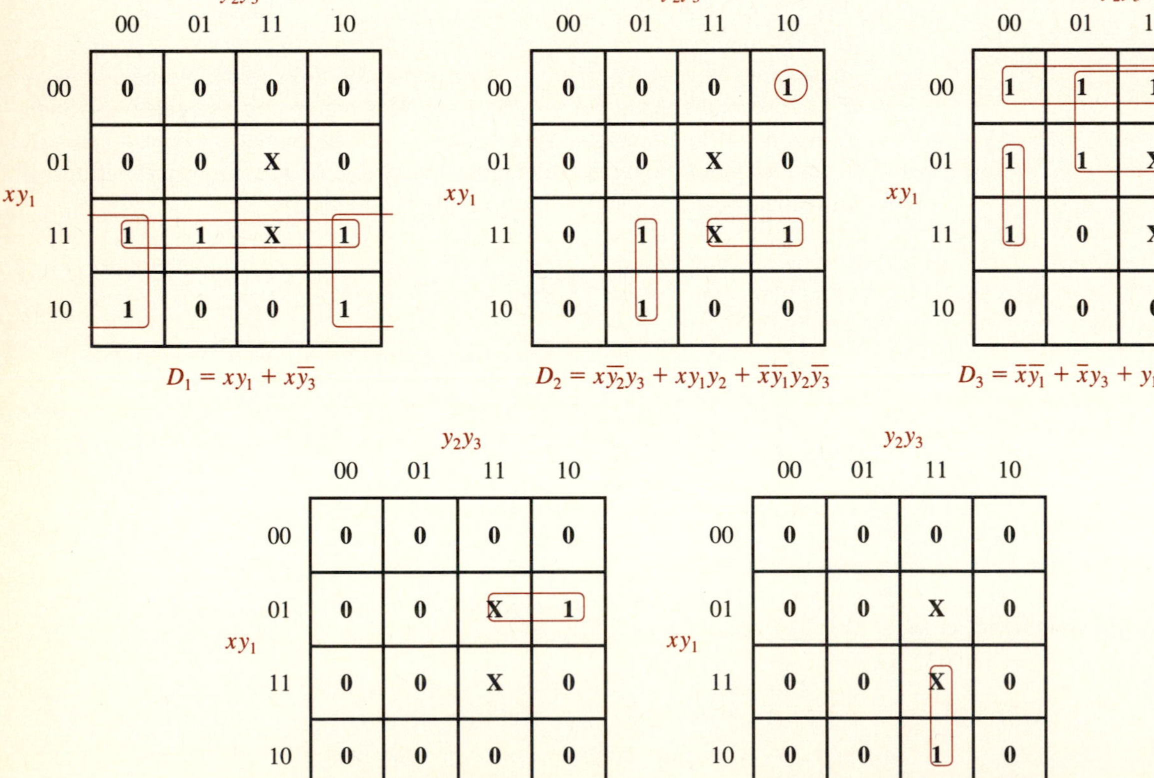

$D_1 = xy_1 + x\overline{y_3}$
$D_2 = x\overline{y_2}y_3 + xy_1y_2 + \overline{x}\overline{y_1}y_2\overline{y_3}$
$D_3 = \overline{x}\overline{y_1} + \overline{x}y_3 + y_1\overline{y_2}\overline{y_3} + \overline{y_1}y_2\overline{y_3}$
$z_1 = \overline{x}y_1y_2$
$z_2 = xy_2y_3$

FIGURE 7.42 (a) An excitation table and (b) K-maps for *R4*. (b)

a set of minimal SOP covers. The corresponding logic circuit, which uses two-level NAND-NAND logic, appears in Figure 7.43. (Recall that each gate in a two-level AND-OR circuit can be replaced by a NAND gate, without altering the output functions.) ■

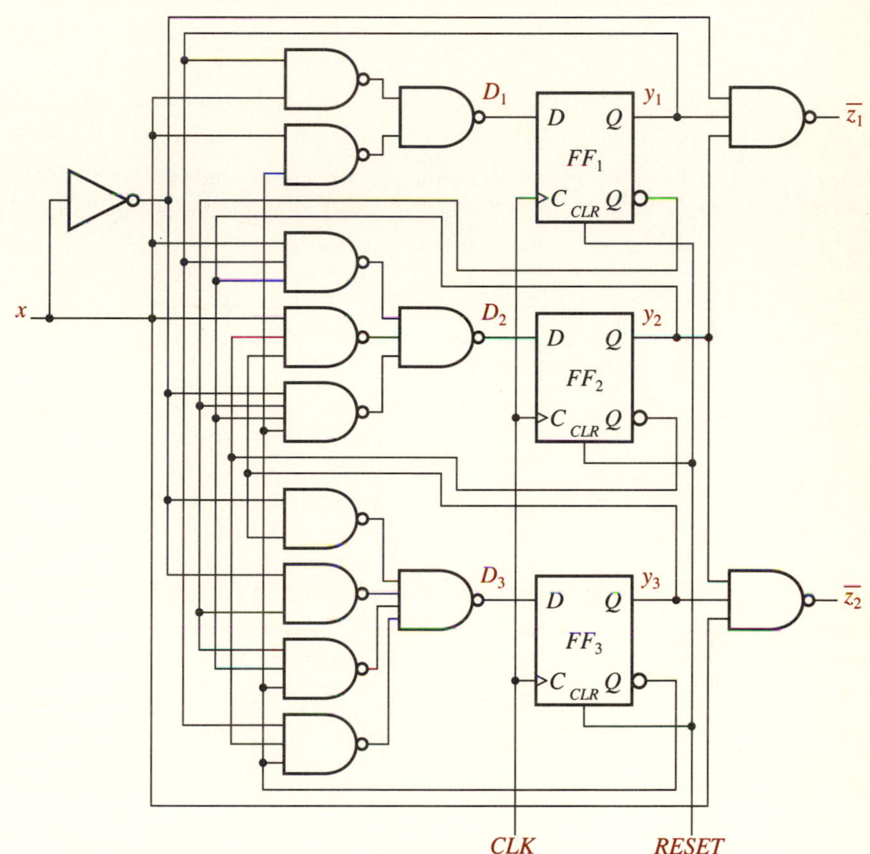

FIGURE 7.43 The logic circuit for *R4* corresponding to Figure 7.42.

The next example concerns a sequential circuit that performs a specific arithmetic operation: multiplication of an unsigned number by a constant.

Example 7.9

Designing a serial multiply-by-3 circuit *M3*

The problem is to design a sequential circuit *M3* that has two input signals x and ON and a single output signal z, and that multiplies a number by the constant 3. An unsigned binary number n of arbitrary length is entered serially via x, with the least significant bit first. The output that appears

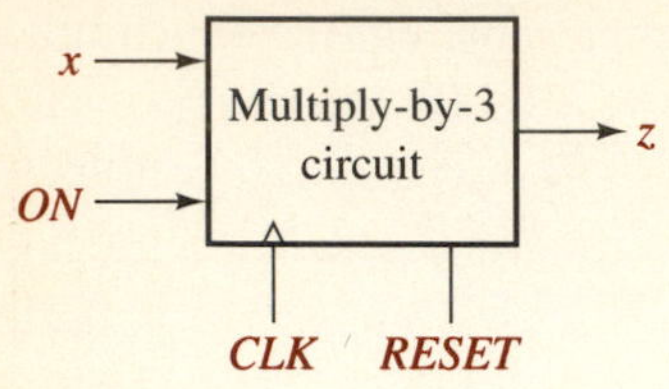

FIGURE 7.44 The multiply-by-3 circuit *M3*.

serially on z is the binary representation of $n \times 3$. *ON* is a control signal that is set to 1 to indicate that the current x is the first (least significant) bit of x; it is reset to 0 immediately after the last (most significant) bit of x has been entered. Thus, *ON* marks the start and end of the bit sequence that represents n. If we assume the usual *CLK* and *RESET* control signals are present, *M3* has the external connections shown in Figure 7.44.

An example of the input/output behavior desired of *M3* is given below:

Cycle:	**...**	**0**	**1**	**2**	**3**	**4**	**5**	**6**
ON:	...	0	1	1	1	1	0	0
x:	...	X	1	0	0	1	0	0
z:	...	0	1	1	0	1	1	0

This behavior corresponds to $9_{10} \times 3_{10} = 27_{10}$. Note that the least significant bits of x and z are on the left in cycle 1, rather than in their usual place on the right.

Our experience with serial adders suggests that *M3* should compute an output bit—in this case, a product bit z—and some carry information CY in each clock cycle. Specifically, in cycle i, *M3* should compute the product

$$p(i) = x(i) \times 3 + CY(i)$$

It should output the least significant bit of $p(i)$ as $z(i)$, and store the remaining bit or bits of $p(i)$ as the carry-in value $CY(i + 1)$, to be used in cycle $i + 1$ when computing $p(i + 1)$ and $z(i + 1)$.

As in the serial adder, CY is stored as the circuit's internal state. However, in this case, it is not so obvious what the possible values of the carry signals are. We can determine them by experimenting with various input sequences. For instance, if we list the values of a carry-in parameter *CY* associated with the above example ($n = 9$), we get

Cycle:	**...**	**0**	**1**	**2**	**3**	**4**	**5**	**6**
ON:	...	0	1	1	1	1	0	0
x:	...	X	1	0	0	1	0	0
z:	...	0	1	1	0	1	1	0
CY:	...	0	0	1	0	0	1	0

indicating that we need to be able to store carries of zero and one, as in the serial adder. However, there may be larger values of *CY* to record. We can identify them by considering the largest possible input number—namely, the number consisting entirely of 1s.

Cycle:	**...**	**0**	**1**	**2**	**3**	**4**	**5**	**6**
ON:	...	0	1	1	1	1	1	1
x:	...	X	1	1	1	1	1	1
z:	...	0	1	0	1	0	1	1
CY:	...	0	0	1	10	10	10	10

In this case, carry-in values of two ($CY = 10_2$) occur. We now see that two is the largest value of CY, because if we multiply an incoming one by three and add a carry of two, the result p is five, which is 101_2. The rightmost bit of 101_2 is sent out as $z = 1$, while the left two bits of 101_2 are stored as the new carry $CY = 10_2$—that is, two. Thus, we cannot increase CY beyond two.

At this point, we can identify all the states of *M3*: three states A, B, and C, corresponding to carry values of 0, 1, and 2, respectively, and a fourth state D, denoting the idle or reset condition, when multiplication is not being performed. State D may be entered synchronously when $ON = 0$, or asynchronously when $RESET = 1$; we will associate the inactive output $z = 0$ with this state.

The state transitions and z values that occur during multiplication can be identified by considering all combinations of $x \times 3 + CY$, where CY is the carry-in value associated with the current state and $ON = 1$. Many of these transitions appear in the sample multiplications listed above. These considerations lead to the state table of Figure 7.45. Note that when ON changes from 1 to 0 during multiplication, the state should not return to D until any stored carry information has been transferred to z; otherwise, the last few bits of the product may be incorrect. For example, if ON becomes 0 when the state is C, *M3* executes the terminal transition sequence

$$C \xrightarrow{\text{X0/0}} B \xrightarrow{\text{X0/1}} D \xrightarrow{\text{X0/0}} D \xrightarrow{\text{X0/0}} D \cdots$$

and settles into state D.

		Present input x ON				
		00	01	11	10	Comment
	A	$D, 0$	$A, 0$	$B, 1$	$D, 0$	Carry $CY = 0$
Present	B	$D, 1$	$A, 1$	$C, 0$	$D, 1$	Carry $CY = 1$
state	C	$B, 0$	$B, 0$	$C, 1$	$B, 0$	Carry $CY = 2$
	D	$D, 0$	$A, 0$	$B, 1$	$D, 0$	Reset state

FIGURE 7.45 A state table for *M3*.

Proceeding to the next step, we make the state assignment $A = 00$, $B = 01$, $C = 10$, $D = 11$, using the state variable pair y_1y_2. This assignment is not quite arbitrary, because y_1y_2 directly indicates the value of CY, a possibly useful feature for debugging. Now we can combine the state assignment with the information in the state table to obtain an excitation table (Figure 7.46) for the chosen flip-flop types, in this case, JK flip-flops. For the reasons discussed earlier, we include all possible don't-cares in the flip-flop input signals.

An optimal or near-optimal two-level realization C of these incompletely specified functions can be obtained, using the methods of Chapter 5. Exact multiple-output function minimization is surprisingly hard to do by hand in

Primary input		Present state		Next state		Secondary outputs				Primary output
x	ON	y_1	y_2	y_1^+	y_2^+	J_1	K_1	J_2	K_2	z
0	0	0	0	1	1	1	X	1	X	0
0	0	0	1	1	1	1	X	X	0	1
0	0	1	0	0	1	X	1	1	X	0
0	0	1	1	1	1	X	0	X	0	0
0	1	0	0	0	0	0	X	0	X	0
0	1	0	1	0	0	0	X	X	1	1
0	1	1	0	0	1	X	1	1	X	0
0	1	1	1	0	0	X	1	X	1	0
1	0	0	0	1	1	1	X	1	X	0
1	0	0	1	1	1	1	X	X	0	1
1	0	1	0	0	1	X	1	1	X	0
1	0	1	1	1	1	X	0	X	0	0
1	1	0	0	0	1	0	X	1	X	1
1	1	0	1	1	0	1	X	X	1	0
1	1	1	0	1	0	X	0	0	X	1
1	1	1	1	0	1	X	1	X	0	1

FIGURE 7.46 An excitation table for *M3*.

this case—try it and see! Figure 7.47a shows minimal SOP and POS covers for C that were computed in less than one second by the Espresso minimization program. The minimal covers are presented in a self-explanatory truth table format that is easily mapped into a PLA. The input file is essentially the excitation table of Figure 7.46, with a dash (–) denoting a don't-care value. The SOP solution is the cheaper of the two, requiring eight prime terms, compared to the nine prime terms of the minimal POS cover. Figure 7.47b displays Espresso's SOP solution on a set of K-maps for C. Most of the product terms are multiple-output prime implicants that are shared by two or three functions.

An optimal SOP implementation of *M3* that uses NAND gates for C appears in Figure 7.48a in the format needed for simulation via the DesignWorks program. Figure 7.48b shows a timing diagram generated by DesignWorks that simulates the 9×3 illustration discussed above. We have annotated this diagram to show the input/output and state sequences in symbolic form. As can be seen, the operation in question is verified by this simulation run. Observe too that DesignWorks flags glitches that may occur in the output line z in cycles 1 and 4 when the main inputs x and ON are changing value.

Finite-State vs. Infinite-State Machines

Our examples indicate that deriving finite-state tables and diagrams from informal specifications can be difficult. In fact, in some easily stated cases, it is impossible, as we now demonstrate. Suppose that we require a recognizer circuit *RN* that will recognize all input sequences that consist of n 0s followed

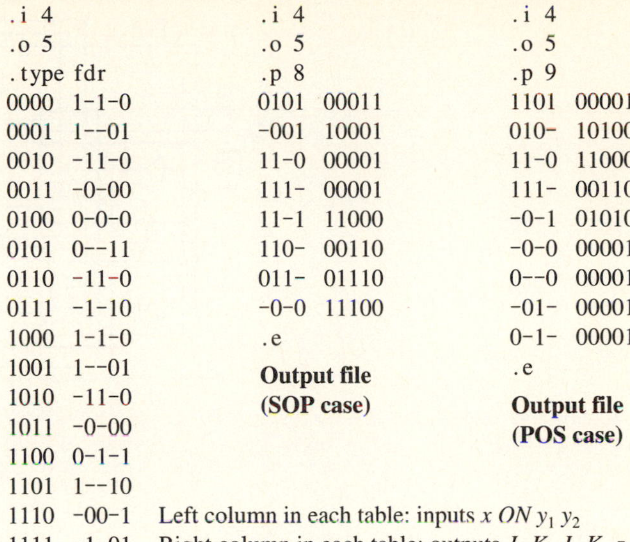

Input file

```
.i 4
.o 5
.type fdr
0000 1-1-0
0001 1--01
0010 -11-0
0011 -0-00
0100 0-0-0
0101 0--11
0110 -11-0
0111 -1-10
1000 1-1-0
1001 1--01
1010 -11-0
1011 -0-00
1100 0-1-1
1101 1--10
1110 -00-1
1111 -1-01
```

Output file (SOP case)

```
.i 4
.o 5
.p 8
0101 00011
-001 10001
11-0 00001
111- 00001
11-1 11000
110- 00110
011- 01110
-0-0 11100
.e
```

Output file (POS case)

```
.i 4
.o 5
.p 9
1101 00001
010- 10100
11-0 11000
111- 00110
-0-1 01010
-0-0 00001
0--0 00001
-01- 00001
0-1- 00001
.e
```

Left column in each table: inputs $x\ ON\ y_1\ y_2$

Right column in each table: outputs $J_1\ K_1\ J_2\ K_2\ z$

(a)

FIGURE 7.47 (a) Two-level minimization of *M3*'s combinational logic by Espresso; (b) K-maps that show the minimal SOP solution.

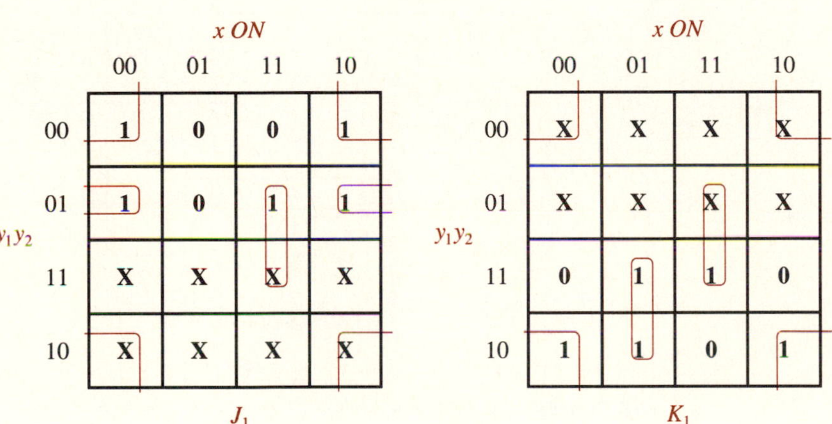

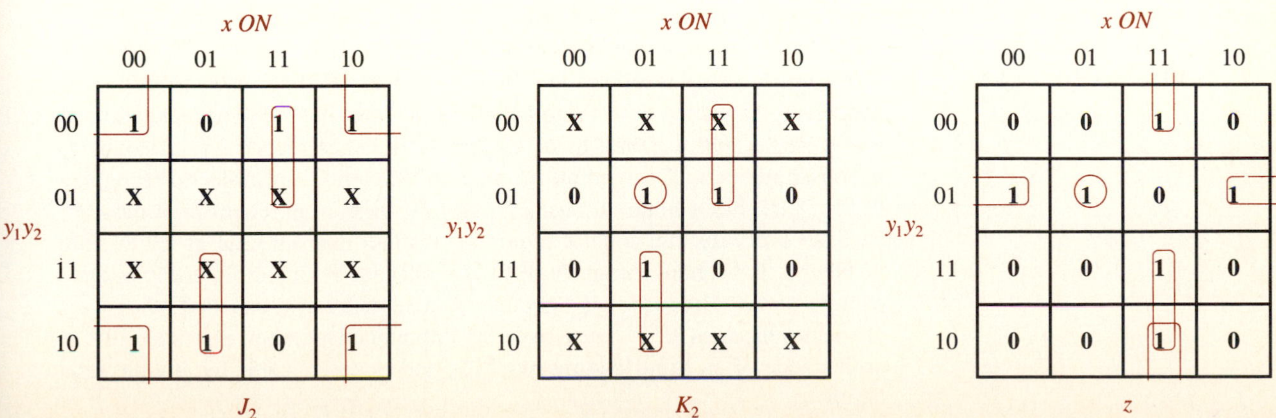

(b)

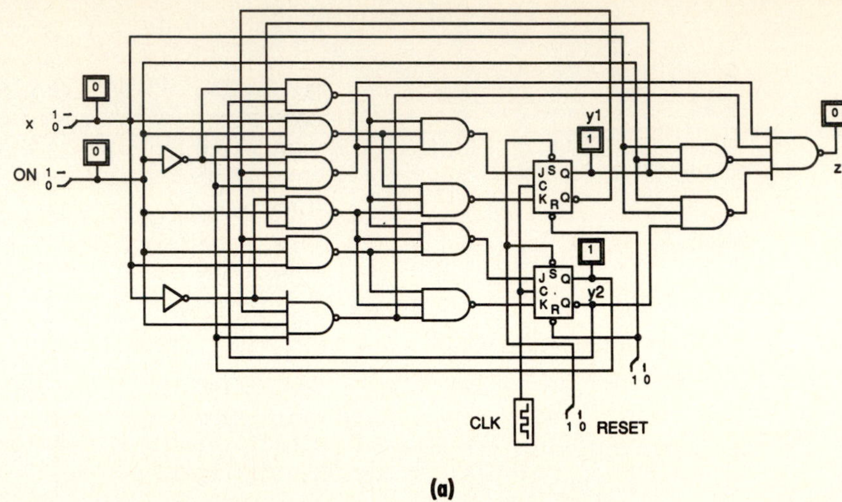

(a)

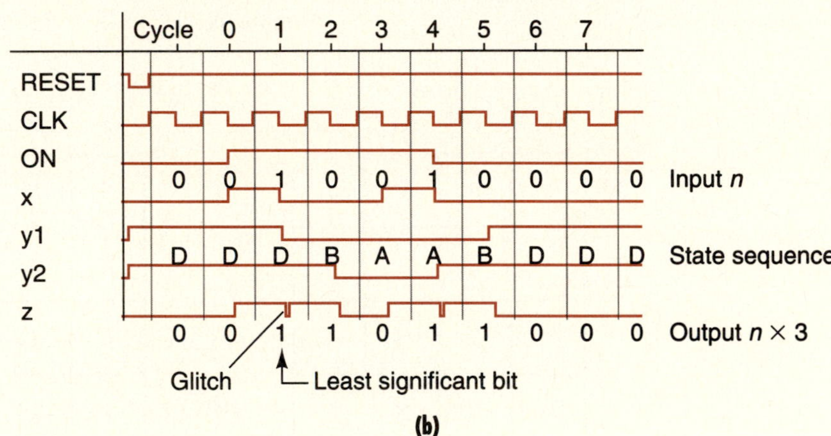

(b)

FIGURE 7.48 (a) An implementation of *M3* for use with DesignWorks; (b) simulation of the multiplication 9 × 3.

A finite-state machine can add but not multiply two numbers of arbitrary length.

by n 1s, where n is any integer greater than or equal to 1. Thus, the set of recognized sequences has the form 01, 0011, 000111, and so on. Let us denote the sequence of n 0s followed by n 1s by 0^n1^n. Any realization of *RN* must be able to count to n, so that after it has received (and counted) n consecutive 0s, it can count up to n subsequent 1s in order to recognize 0^n1^n. If *RN* has a finite number of states P, then it can count to at most P, because every distinct count n requires a distinct internal state of *RN* to store it. Hence, *RN* cannot recognize 0^n1^n if n exceeds P; it simply does not have enough states. Because n is unbounded, a sequential circuit that recognizes all occurrences of 0^n1^n must have an unbounded number of states; this is an instance of an **infinite-state machine** and is unrealizable by any physical circuit.

Another example of an infinite-state machine is a serial multiplier to compute the product of two arbitrary numbers. In the multiplier of Example 7.9, one of the numbers is fixed at 3, and a finite-state realization is readily obtained. However, with two arbitrarily long operands, it is easy to see that the stored carry information also becomes arbitrarily large, and eventually exceeds the capacity of any finite memory. The problem here is not with the potentially infinite length of the operands, because a two-state serial adder can add two numbers of arbitrary length. Rather, it is with the maximum amount of information that must be stored from one basic step to the next, which is strictly limited in the case of finite-state machines. The finiteness of state machines is of little concern to the logic designer, however, because a finite upper bound can easily be placed on the size of all operands of practical interest.

7.6 State Minimization*

We consider here the problem of reducing a state table or diagram to an equivalent one that contains the fewest states. We present a straightforward algorithm to solve this problem.

Equivalent States

Equivalent states cannot be distinguished from each other by observing input/output sequences.

Earlier we saw several examples in which the same state behavior was represented by different numbers of states. This is due to the existence of equivalent states, such as the states B and C in the adder state table of Figure 7.32a. Two states S_1 and S_2 are defined to be **equivalent** or **indistinguishable**, denoted $S_1 \equiv S_2$, if, on applying any input sequence $SEQ_a = X_1X_2\ldots X_k$ of any length k to the circuit starting in initial state S_1, we get exactly the same response sequence $Z_1Z_2\ldots Z_k$ that we would get if S_2 were the initial state. Thus, there is no way we can determine whether the initial state was S_1 or S_2 by observing the circuit's primary outputs; this is the sense in which equivalent states are indistinguishable. Moreover, if $X_1X_2\ldots X_k$ is applied to the equivalent initial states S_1 and S_2, then the final states S_1^k and S_2^k reached in each case must also be equivalent. If not, we could find another input sequence SEQ_b that would distinguish S_1^k from S_2^k. Appending SEQ_b to SEQ_a forms a composite input sequence SEQ_aSEQ_b that would distinguish S_1 from S_2, which contradicts their equivalence.

Hence, if $S_1 \equiv S_2$, the application of *any* input sequence $X_1X_2\ldots X_k$ to S_1 and S_2 results in the transition sequence

$$S_1 \xrightarrow{X_1/Z_1^1} S_1^1 \xrightarrow{X_2/Z_1^2} S_1^2 \xrightarrow{X_3/Z_1^3} \cdots \longrightarrow S_1^{k-1} \xrightarrow{X_k/Z_1^k} S_1^k$$

$$S_2 \xrightarrow{X_1/Z_2^1} S_2^1 \xrightarrow{X_2/Z_2^2} S_2^2 \xrightarrow{X_3/Z_2^3} \cdots \longrightarrow S_2^{k-1} \xrightarrow{X_k/Z_2^k} S_2^k$$

*This section may be skipped without loss of continuity.

such that $Z_1^i = Z_2^i$ and $S_1^i \equiv S_2^i$ for all i, where $1 \leq i \leq k$. These considerations lead directly to the following result, which is the basis of a method for identifying all equivalent states of a sequential circuit.

Theorem 17

Two states S_1 and S_2 of a sequential circuit are equivalent if and only if for every possible input combination X: (a) the corresponding primary output signals $Z_1 = Z(X, S_1)$ and $Z_2 = Z(X, S_2)$ are the same—that is, $Z_1 = Z_2$; and (b) the corresponding next states $S_1^+(X, S_1)$ and $S_2^+(X, S_2)$ are equivalent—that is, $S_1^+ \equiv S_2^+$.

An Illustration

To introduce our procedure of finding equivalent states, consider the small state table of Figure 7.17b, which is repeated in Figure 7.49a. There are $\binom{4}{2} = 6$ distinct state pairs: (A, B), (A, C), (A, D), (B, C), (B, D), and (C, D). We want to examine each of these pairs to see if they satisfy the criteria stated in the theorem. To do this systematically, we construct the six-box table

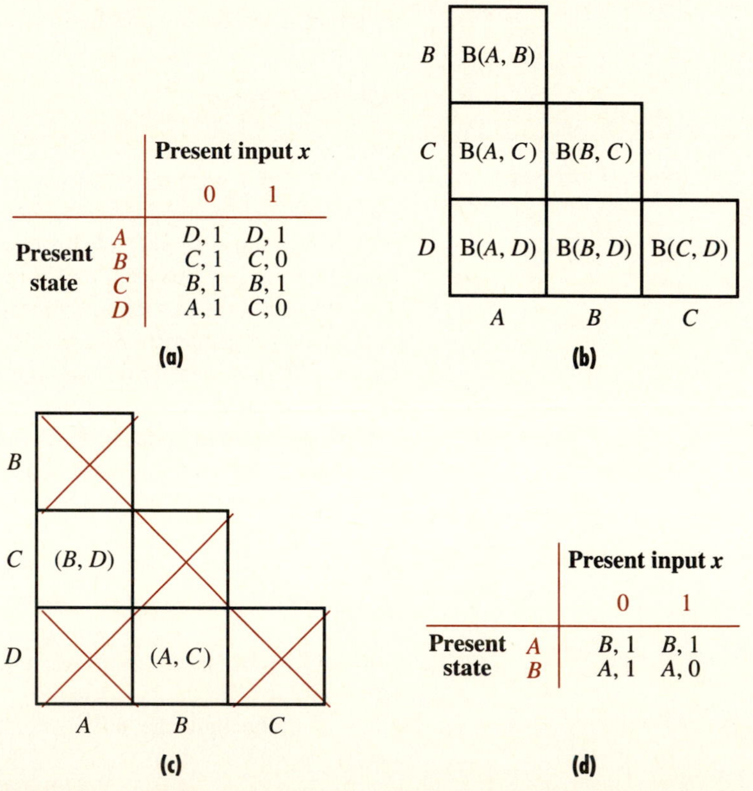

(a)

Present state	Present input x = 0	1
A	D, 1	D, 1
B	C, 1	C, 0
C	B, 1	B, 1
D	A, 1	C, 0

(d)

Present state	Present input x = 0	1
A	B, 1	B, 1
B	A, 1	A, 0

FIGURE 7.49 **(a)** A state table with four states; **(b)** its implication chart; **(c)** the implication chart after one pass through the state table; **(d)** a minimal two-state table.

shown in Figure 7.49b, which is called an **implication chart**. It is drawn in a staircase fashion to have one box $B(S_i, S_j)$ for each pair of (present) states S_i and S_j, with S_i and S_j serving as horizontal and vertical coordinates of $B(S_i, S_j)$.

Some input sequence must eventually take two nonequivalent states to states with different output values.

We now scan the state table from top to bottom, examining each pair of present states in turn, and seeing if the corresponding primary output values differ. The first state pair encountered is (A, B). With the input combination $x = 1$, the corresponding output response pair is (1,0), implying immediately that $A \not\equiv B$. In other words, if a sequential circuit *SC* with this state behavior is set to initial state A or B, but we do not know which, if we apply $x = 1$ to *SC* we will be able to identify the initial state directly from the output signal: A will produce output $z = 1$, whereas B will produce output $z = 0$. Hence, we cross out the box $B(A, B)$, as shown in Figure 7.49c, to indicate that $A \not\equiv B$ and that these states need not be considered further.

The state pair (A, C) considered next produces the same response: $z = 1$ when $x = 0$ and when $x = 1$. For each value of x, the next-state pair is (B, D). We cannot say at this point whether A and C are equivalent; it depends on whether B and D are equivalent. We therefore postpone a decision on A and C, and enter their next-state pair (B, D) into box $B(A, C)$ of the implication chart. Continuing in this way, we obtain Figure 7.49c after one complete pass through the state table. We have shown that four state pairs—those whose boxes are crossed out—are nonequivalent. The two pairs with the noncrossed-out boxes—namely, $B(A, C)$ and $B(B, D)$—remain to be resolved.

Now $B(A, C)$ contains only (B, D), while $B(B, D)$ contains only (A, C). Neither of these next-state pairs implies a contradictory output value for any value of x; otherwise, one of the boxes $B(A, C)$ or $B(B, D)$ would have been crossed out in our first pass through the state table. Hence, if we apply the same input sequence to two copies of *SC*, we can never reach two states that produce different output values, so we are forced to conclude that $A \equiv C$ and $B \equiv D$. We may now delete rows C and D from the original state table, and replace all occurrences of C and D with A and B, respectively. The result is the two-state table shown in Figure 7.49d. This is referred to as the **minimal state table** that is equivalent to the original one.

		Present input		
		...	x_h	...
Present state	S_i	...	S_i^+, Z_i	...
	S_j	...	S_j^+, Z_j	...

(a)

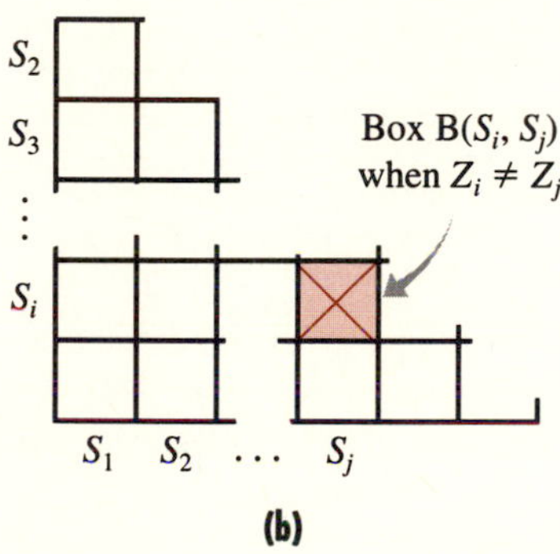

(b)

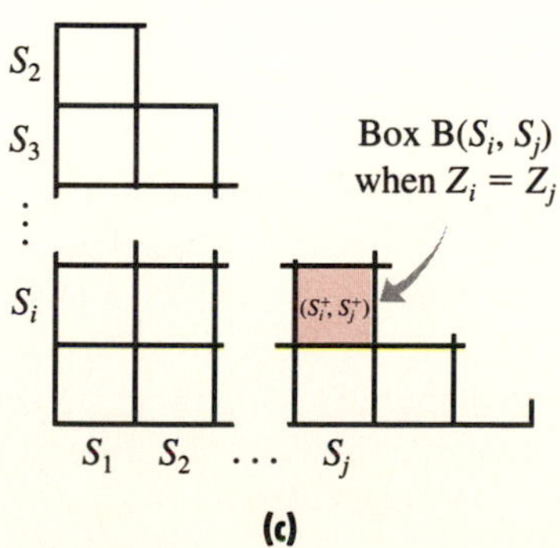

(c)

FIGURE 7.50 (a) The general state table *ST*; (b),(c) the implication chart for *ST*.

Finding Equivalent States

Based on the foregoing analysis, we next describe a formal procedure, called *EQSTATE*, for identifying all pairs of equivalent states in a given state table. It involves checking the possible present outputs and next states for every pair of present states in the table. As in the example, an implication chart is used to record actual or potentially distinguishable states (Figure 7.50). Several passes may have to be made through the implication chart to resolve all cases.

Procedure

Procedure *EQSTATE* to find all equivalent pairs of states

1. Given a $P \times M$ state table *ST*, construct an implication chart that contains a box $B(S_i, S_j)$ for each pair of distinct states (S_i, S_j) in *ST*.

2. For every input combination X_h and every present-state pair (S_i, S_j) in *ST*, check the primary output responses Z_i and Z_j and the next states S_i^+ and S_j^+ that correspond to S_i and S_j, respectively. If $Z_i \neq Z_j$, cross out box $B(S_i, S_j)$ in the implication chart to indicate that $S_i \not\equiv S_j$. If $Z_i = Z_j$, $(S_i^+, S_j^+) \neq (S_i, S_j)$, and $S_i^+ \neq S_j^+$, enter (S_i^+, S_j^+) into box $B(S_i, S_j)$. Otherwise, leave $B(S_i, S_j)$ empty.

3. Examine each nonempty box $B(S_i, S_j)$ that is not yet crossed out. If it contains an entry (S_u, S_v) such that $B(S_u, S_v)$ has been crossed out, cross out $B(S_i, S_j)$.

4. Repeat step 3 until no more occurrences of $B(S_i, S_j)$ can be crossed out. Two states are then equivalent if and only if the corresponding box in the implication chart is not crossed out.

Suppose that the state behavior is represented by a state table *ST* with P rows (states) and M columns (input combinations), as in Figure 7.50a. The implication chart contains a box $B(S_i, S_j)$ for each pair of distinct states S_i and S_j in *ST*; there are $\binom{P}{2}$ such pairs. For each of the M input combinations X_h, we check the primary output response pair Z_i and Z_j and next-state pair S_i^+ and S_j^+ that correspond to S_i and S_j, respectively, which are specified by *ST*. If $Z_i \neq Z_j$, we can immediately conclude that $S_i \not\equiv S_j$ and cross out box $B(S_i, S_j)$ in the implication chart (Figure 7.50b). If $Z_i = Z_j$, we enter the next-state pair (S_i^+, S_j^+) into $B(S_i, S_j)$, as shown in Figure 7.50c, but only if $(S_i^+, S_j^+) \neq (S_i, S_j)$ and $S_i^+ \neq S_j^+$. In other words, we record only the next-state pairs that may eventually lead to nonequivalent states.

Once all state pairs and primary input combinations have been considered (step 2 of *EQSTATE*), each $B(S_i, S_j)$ has either been crossed out, indicating that $S_1 \not\equiv S_2$, or it contains zero or more next-state pairs. In the latter case, the equivalence of S_i and S_j depends on the equivalence of the next-state pairs in $B(S_i, S_j)$, as asserted by theorem 17. If $B(S_i, S_j)$ contains no next-state pairs and is not crossed out, we must have either $(S_i^+, S_j^+) = (S_i, S_j)$ or $S_i^+ = S_j^+$ in all columns of the state table, implying that $S_i \equiv S_j$. Hence, an empty box at this stage indicates a pair of equivalent states.

Next, we scan the implication chart again (step 3), examining only those nonempty boxes that have not yet been crossed out. If any such box $B(S_i, S_j)$

contains a state pair S_u, S_v such that the corresponding box $B(S_u, S_v)$ has been crossed out, $B(S_i, S_j)$ itself is now crossed out, because we have shown that we can reach a pair of nonequivalent states S_u and S_v from S_i and S_j, so the latter cannot be equivalent.

At most $P - 1$ passes through a P-state implication chart identify all equivalent state pairs.

After scanning the complete implication table, we repeat this process again, because the new information we have about nonequivalent states may enable us to identify more nonequivalent states. This scanning continues until no new information can be gleaned from the implication chart; that is, no new boxes can be crossed out. At that point, all crossed-out boxes correspond to nonequivalent states; all boxes that are not crossed out correspond to equivalent states. Because we continue to scan the implication chart only as long as we can cross out at least one box, *EQSTATE* must eventually terminate after at most $\binom{P}{2}$ repetitions of step 3. (In fact, it can be shown that at most $P - 1$ repetitions of step 3 suffice.)

Equivalence Classes

It is possible for more than two states to be mutually equivalent. A set of $r \geq 1$ states $\mathscr{E} = \{S_a, S_b, \ldots, S_r\}$ is called an **equivalence class** of states if all pairs of states in $\mathscr{E}$ are equivalent to each other, and no state not in $\mathscr{E}$ is equivalent to any state in $\mathscr{E}$. The P states of a sequential circuit can always be grouped into P or fewer disjoint equivalence classes. For example, the states A, B, C, and D from Figure 7.49 fall into two equivalence classes, $\mathscr{E}_1 = \{A, C\}$ and $\mathscr{E}_2 = \{B, D\}$. If a state table or diagram *ST* contains no equivalent states—that is, if all its equivalence classes consist of a single state—*ST* is said to be in **minimal form**. It is usually desirable to reduce a state representation to minimal form before implementing it by a logic circuit.

Procedure *EQSTATE* computes all pairs of equivalent states in a given state table. If we know all such pairs, we can easily find the corresponding equivalence classes, using the following property of the state-equivalence relation:

$$\text{If} \quad S_1 \equiv S_2 \quad \text{and} \quad S_2 \equiv S_3 \quad \text{then} \quad S_1 \equiv S_3 \tag{7.19}$$

A set of pair-wise equivalent states form an equivalence class that can be replaced by one state.

The property defined by (7.19) is called **transitivity**, and the operator $\equiv$ is said to be a **transitive relation**. We can identify all equivalence classes from the pairs of equivalent states computed by *EQSTATE* in the following simple way:

> Scan all the states in sequence and assign each state S_i to an existing equivalence class $\mathscr{E}_j$ if $S_i \equiv S_k$ for some state S_k in $\mathscr{E}_j$; otherwise, assign S_i to a new equivalence class.

If S_i is assigned to equivalence class $\mathscr{E}_j$ based on $S_i \equiv S_k$, the transitive property of $\equiv$ guarantees that S_i is equivalent to every other member of $\mathscr{E}_j$.

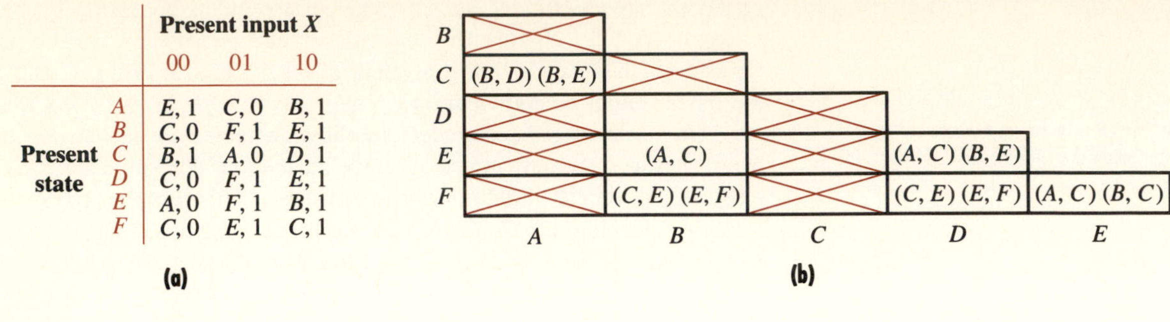

		Present input X		
		00	01	10
Present state	A	E, 1	C, 0	B, 1
	B	C, 0	F, 1	E, 1
	C	B, 1	A, 0	D, 1
	D	C, 0	F, 1	E, 1
	E	A, 0	F, 1	B, 1
	F	C, 0	E, 1	C, 1

(a)

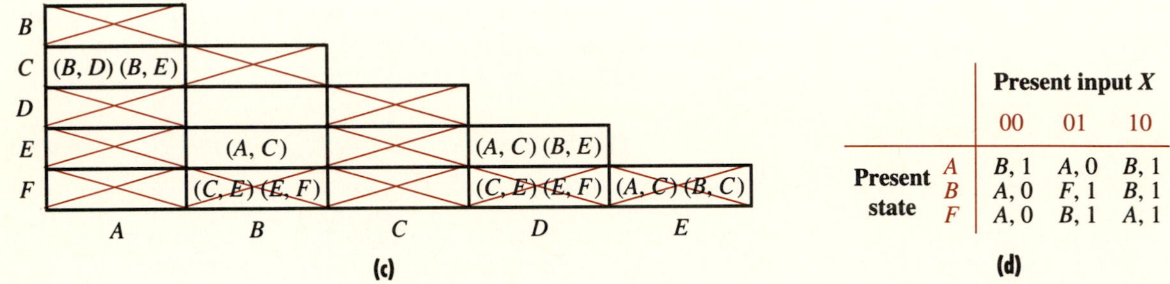

		Present input X		
		00	01	10
Present state	A	B, 1	A, 0	B, 1
	B	A, 0	F, 1	B, 1
	F	A, 0	B, 1	A, 1

(d)

FIGURE 7.51 The state minimization process for Example 7.10: (a) the state table; (b) the implication chart after step 2; (c) the implication chart after step 3; (d) a minimal state table.

Example 7.10

Converting a state table to minimal equivalent form

The problem is to convert the 6×3 state table of Figure 7.51a to a minimal equivalent form. (The fact that the number of input combinations is not a power of two has no significance here—it merely indicates that certain combinations are not used.) This **state minimization problem** requires determining all equivalence classes in the given table, which can then be replaced by a minimum-row table that contains one state per equivalence class.

We will apply *EQSTATE* to the given state table. The implication chart that results from steps 1 and 2 appears in Figure 7.51b. The first box B(A, B) is crossed out because there are different output responses in the first two rows of the state table for input combination 00 (and also for input 01). The first and third rows, corresponding to states A and C, contain no such present-output conflicts, so we consider the next-state pairs in each of the three columns of rows A and C of the state table. Columns 10 and 00 yield next-state pairs (B, D) and (B, E), respectively, which are then entered into box B(A, C) as shown. No entry is made for the next-state pair (A, C) from column 01, because it is the same as the present-state pair and therefore can provide no information about the equivalence or nonequivalence of A and C. Box B(B, D) remains empty after completion of step 1, because the next states reachable from the present-state pair (B, D) are (C, C), (F, F), and

(E, E) for inputs 00, 01, and 10, respectively. We conclude right away that $B \equiv D$.

We now proceed to step 3 and scan the entries in the implication chart of Figure 7.51b that are neither crossed out nor blank. The first of these, B(A, C), contains (B, D) and (B, E), neither of whose boxes B(B, D) and B(B, E) is crossed out; therefore, B(A, C) cannot yet be crossed out. The same holds true for B(B, E), which contains (A, C). In the case of B(B, F), however, we have an entry (C, E), the box B(C, E) of which has been crossed out. Hence, we now cross out B(B, F), because $C \not\equiv E$ implies $B \not\equiv F$. For similar reasons, we can also cross out B(D, F) and B(E, F), yielding Figure 7.51c. A further pass through step 3 of *EQSTATE* results in no change to the four boxes B(A, C), B(B, D), B(B, E), and B(D, E) that have not been crossed out, and procedure *EQSTATE* terminates. This means that $A \equiv C$, $B \equiv D$, $B \equiv E$, and $D \equiv E$. From transitivity, we conclude that $B \equiv D \equiv E$.

Hence, the machine under consideration here has three equivalence classes:

$$\mathcal{E}_1 = \{A, C\}$$
$$\mathcal{E}_2 = \{B, D, E\}$$
$$\mathcal{E}_3 = \{F\}$$

These classes lead to the three-state table of Figure 7.51d, for which we have chosen the old state names A, B, and F to represent the corresponding equivalence classes; new names such as the names assigned to the classes themselves may be used, if preferred. The reduced table is obtained from the original one (Figure 7.51a) simply by deleting all rows except A, B, and F, and replacing each occurrence of C, D, and E in the remaining next-state entries with the representative of its equivalence class. We have reduced the number of states from six to three; therefore, two rather than three flip-flops are needed to realize the given state behavior.

As in the case of combinational circuits, the initial specification of a sequential circuit is often incomplete in that some primary output and/or next-state entries are don't-cares. One reason for the don't-cares is the fact that the number of states P may not be a power of two; for example, the recognizer machine *R4* of Figure 7.43 has $P = 7$ specified states, in addition to a don't-care state, which is indicated by Xs in the transition table. The general problem of minimizing the number of states of an incompletely specified state table is much harder than the completely specified case just considered. A complete discussion of this problem can be found in [Friedman and Menon, 1975].

Design and Testing*

Digital circuits, like all physical devices, are subject to failure and must be tested periodically to determine if they are working properly. Testing adds significantly to the cost of a system over its lifetime. In some critical applications in which failures cannot be tolerated for very long, testing and related maintenance activities are the main cost source. If testing is insufficient in scope or frequency, excessive numbers of system "crashes" will occur, at a cost that can range from a minor inconvenience to a major disaster.[3]

In the next few sections, we address the problem of testing for faults that occur during or after the manufacture of a digital system. We will study the basic techniques for modeling faults and deriving tests at the gate level, along with some design methods to enhance a circuit's testability.

7.7 Faults and Tests

This section introduces the fundamental concepts of testing and describes how to model faults in combinational logic circuits.

The Testing Process

Digital circuits can fail to operate properly, either because they have been designed incorrectly or because they develop physical faults during manufacture or use. Design errors are addressed by improving the quality of the design process—for example, by ensuring that designers are properly trained and by making effective use of such CAD tools as logic simulators and timing verifiers. The incidence of physical faults can be reduced but never entirely eliminated by the use of reliable components, by good quality control in the manufacturing process, and by ensuring proper operating conditions in the field. The unavoidable failures are addressed by the use of test procedures, which determine the presence and, if desired, the location of the faults.

Testing is done by applying input patterns to the unit under test and observing the response.

The major features of a typical testing process are depicted in Figure 7.52. The circuit being tested, often called the **unit under test** or **UUT**, is subject to faults F that alter its behavior. A sequence of input combinations T, called **test patterns** or **test vectors**, is applied to the UUT, and the resulting output signals, the sequence R of **test responses**, is observed. R is then compared to the expected fault-free response R'. If $R = R'$, the UUT passes the test and is assumed to be fault-free. If, on the other hand, $R \neq R'$, the

*Sections 7.7, 7.8, 7.9, and 7.10 may be omitted as a group without loss of continuity.

[3]For example, an article in the *Los Angeles Times* of June 18, 1980, reported that the undetected failure of a 46-cent circuit in a communications system belonging to the U.S. Strategic Air Command (SAC) caused a false alarm, signalling a Soviet missile attack. Installations with over 1,000 intercontinental ballistic missiles were placed on alert; SAC bomber crews ran to their planes and started the engines, but did not take off.

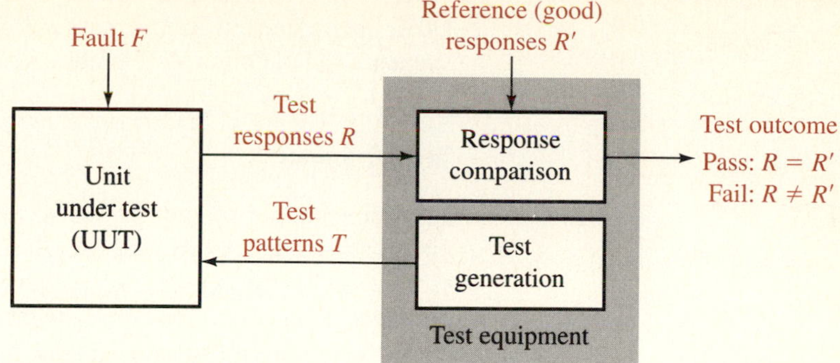

FIGURE 7.52 Testing a circuit for physical faults.

test fails and the UUT is known to be faulty. If T is appropriately designed, R may contain enough information to locate or isolate F. CAD tools, in the form of **test-generation programs**, are widely used to compute T, R, and R'. The test data may be precomputed and stored in a suitable piece of **automatic test equipment** or **ATE** that applies the tests and evaluates the results. Computer-controlled ATE is used, for example, to do high-volume testing of ICs or printed-circuit boards (PCBs) as they are manufactured. Some logic circuits are designed to test themselves and are said to employ **built-in self-testing**, referred to as **BIST**.

Fault Sources

Physical faults have numerous sources. Examples of faults that occur during manufacture include missing or defective parts and improper assembly. ICs are subject to processing defects from such sources as dust particles that contaminate a clean-room manufacturing facility. Interconnections between parts, such as solder joints and plug-in connectors, are also prone to failure. After manufacture, a circuit can fail due to various wear-and-tear phenomena. These include mechanical breakage such as dropping a piece of equipment on the floor. Less obvious are electrical faults caused by insulation breakdown, contact burning, and the like. These produce **short circuits**, which connect normally disconnected signal lines, and **open circuits**, which disconnect signal lines that are normally connected. Because of their extremely small internal components and wires, ICs are subject to many unusual failure modes. For example, if a tiny leak develops in an IC package, water vapor can enter from the air, causing rapid corrosion and circuit failure.

Common fault sources are improper manufacture or handling, overheating, and power supply fluctuations.

Finally, a major source of faults is the environment in which a circuit is operated. Electronic circuits generate appreciable amounts of heat in normal operation and will fail if there is inadequate heat dissipation due to improper installation or failure of a cooling device such as a fan. Fluctuations in the voltage produced by a circuit's power supply are another major source of failures, often of an intermittent variety that is especially hard to detect. Other environmental factors that can lead to failure are mechanical vibration, sudden acceleration or deceleration, or the like.

Testing Stages

To deal with the foregoing faults, testing is performed repeatedly over a circuit's lifetime. An IC chip, for instance, is first tested immediately after fabrication, when it is part of a semiconductor wafer that contains perhaps hundreds of other copies of the same chip. It is tested again after it is separated from the wafer and installed in its plastic or ceramic package. After it is incorporated into a PCB, the board is tested, and so on. It is worth noting that the cost of testing tends to increase by a factor of about 10 at each stage. Thus, if it costs $1 to find a faulty chip at the wafer stage, it can be expected to cost $10 and $100, respectively, to find the same fault at the IC package and PCB stage. After the system housing the chip is installed in its final location in the field, it and the rest of the system are tested regularly as a part of routine maintenance.

The later a fault is detected, the higher is the cost of testing.

As discussed in Chapter 1, digital systems can be analyzed at several complexity levels that are determined by the components and signals regarded as primitive. Each of these levels introduces a different set of testing problems and techniques. Testing carried out at the electrical circuit or transistor level is concerned with measuring voltage waveforms, electrical propagation delays, and the like; it is referred to as **parametric testing**. At the logic level, we are primarily concerned with testing a circuit's logical behavior, for which the test patterns and responses are viewed as sequences of 0s and 1s; this type of testing is referred to as **logic testing**. We are concerned here with logic testing only.

Fault Models

For analysis at the logic level, we need logical fault models.

Logic testing recognizes gates, latches, and flip-flops as the primitive components, and 0s and 1s as the primitive signals. In order to perform logic testing at this level of complexity, we require **logical fault models** that allow faulty behavior to be defined entirely in terms of the recognized logic components and signals. It will be observed that many of the physical faults cited above are expressed in electrical terms that cannot be directly mapped to the logic level. For example, if two wires carrying the independent logic signals a and b are joined by a short circuit, the resulting signal on the common a–b wire may be undefined—for instance, if $a = 1$ and $b = 0$. Similarly, if line a is broken by an open circuit, the signal on the floating input end at the break is undefined. Faults like these can be modeled at the switch level, however.

The most widely used logical fault model is the single stuck-line (SSL) model.

A fault model known as the **single stuck-line** or **SSL model**, introduced in the 1950s, has long been the standard for gate-level testing. The SSL model defines two types of faults for each line x in a logic circuit: the stuck-at-0 fault and the stuck-at-1 fault, which cause x to be permanently stuck at the logic value 0 or 1, respectively. Only one line is assumed to be faulty at a time (the single-fault assumption), and all gates and other components are assumed to operate correctly. Figure 7.53 shows the effect of the two possible SSL faults on a line x_5, which is an input to a three-input AND gate G in some circuit. The fault x_5 stuck-at-0, which we abbreviate to $x_5/0$, causes the output of G to change from $x_6 = x_1x_2x_3$ to 0; the fault x_5 stuck-at-1, abbreviated to $x_5/1$, changes x_6 from $x_1x_2x_3$ to $1x_2x_3 = x_2x_3$. Note that the other lines that are connected directly to x_5, which share a common fan-out point x_1, are assumed to be unaffected by faults on x_5.

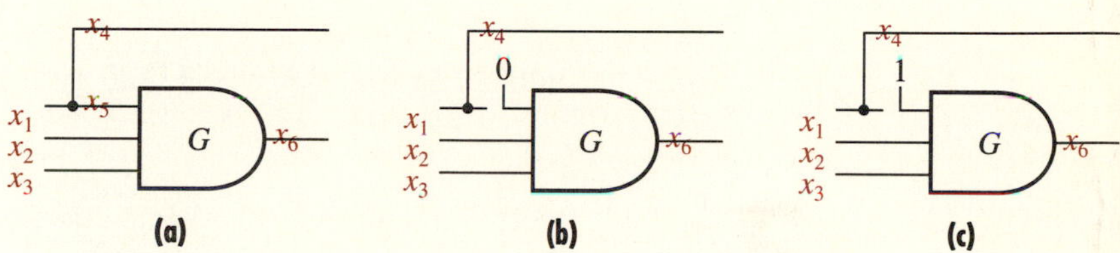

FIGURE 7.53 (a) A fault-free circuit; (b) fault x_5 stuck-at-0, denoted $x_5/0$; (c) fault x_5 stuck-at-1, denoted $x_5/1$.

Testing for Stuck-Line Faults

In order for a test pattern T to detect, say, $x_5/0$, it must apply a 1 to x_5 under normal conditions. In addition, it must apply signals to the other lines in the circuit that cause a set of changes in response to the fault that will cause an incorrect signal value—that is, an **error**—to appear on an observable primary output line. For example, it is easy to see that, to detect $x_5/0$, T must apply 1 to x_1, x_2, and x_3 in Figure 7.53a. That will make $x_6 = 1 \cdot 1 \cdot 1 = 1$ when no fault is present, and $x_6 = 0 \cdot 1 \cdot 1 = 0$ when the fault $x_5/0$ is present. Thus, the fault causes an error on the output of G when the test T is applied. In effect, an error signal has been propagated from the fault site x_5 through gate G to line x_6. If x_6 is a primary output that can be accessed by a tester, the fault is detected at this point. Otherwise, the error must be propagated in the same fashion through more gates until a suitable output line is reached.

The SSL model has the advantage of being easy to use and of involving a relatively small number of faults. In a circuit that contains r distinct lines, there are $2r$ possible SSL faults, and the number of tests needed grows slowly with r. Consider, for example, the problem of computing test patterns to detect all SSL faults in a three-input NAND gate, assuming its output line is directly observable. There are four lines (three input and one output), so there are eight possible SSL faults.

A fault table lists faults and the tests that detect (cover) them.

Figure 7.54 compares the faults in the NAND gate with the $2^3 = 8$ input combinations that are possible tests for them. Each row of this **fault**

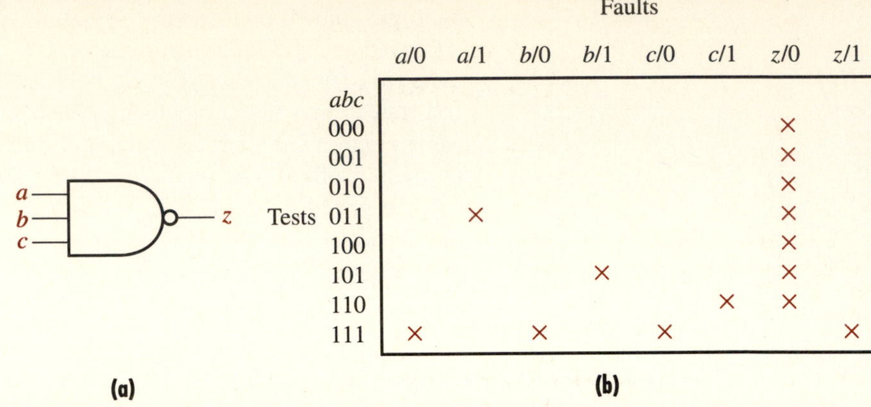

Tests abc	a/0	a/1	b/0	b/1	c/0	c/1	z/0	z/1
000							×	
001							×	
010							×	
011		×					×	
100							×	
101				×			×	
110						×	×	
111	×		×		×			×

FIGURE 7.54 (a) A three-input NAND gate; (b) a fault table for all its SSL faults.

table represents a distinct input combination—that is, a potential test pattern T_i—and each column represents a distinct fault F_j. An × is placed at the intersection of row T_i and column F_j if and only if T_i detects the presence of F_j. If test T_i detects fault F_j, T_i is said to **cover** F_j. Thus, the input pattern $abc = 000$ detects only the SSL fault z stuck-at-0, resulting in an × in the column headed $z/0$ of row 000. The input $abc = 111$, on the other hand, covers all three input stuck-at-0 faults, as well as the output stuck-at-1; this results in the four occurrences of × in the bottom row of the fault table. Note the similarity between the fault tables and the prime implicant (cover) tables presented in Section 5.5.

Test Set Minimization

Exhaustive testing is practical for small circuits only.

In test generation, we are usually interested in finding a small (preferably minimal) set of tests that cover all SSL faults in a given circuit. It is obvious that all 2^n distinct input patterns form a complete set of tests for an n-input combinational circuit. The use of all 2^n input patterns as tests for a combinational circuit is called **exhaustive testing**. This testing method detects far more than the SSL faults. In fact, *any* fault, SSL or non-SSL, that alters the circuit's truth table is detected by exhaustive testing. Because 2^n (the number of test patterns) grows exponentially, exhaustive testing is limited in practice to circuits in which $n \leq 20$ or so.

In most cases, only a small fraction of the possible input patterns are needed as tests. Consider again the NAND fault table of Figure 7.54. The column $a/1$ contains a single ×, which occurs in row 011. Clearly, 011 is the only possible test for $a/1$. Unique tests of this sort are said to be **essential tests**. (Compare it with the definition of essential prime implicants.) There are altogether four essential test patterns here, forming the test set $\mathcal{T}_3 = \{011, 101, 110, 111\}$. The members of $\mathcal{T}_3$ cover all eight faults in the

NAND gate, and therefore form a complete, minimal test set for the gate. In fact, $\mathcal{T}_3$ is the only test set with this property. This result is generalized in the following easily proven theorem.

Theorem 18

Every n-input AND, NAND, OR, and NOR gate G has a unique minimal set $\mathcal{T}_G$ of $n+1$ tests that detect all SSL faults in G. For the AND and NAND cases, these tests are

All SSL faults in an n-input (N)AND or (N)OR gate can be detected with $n + 1$ test patterns.

$$\begin{matrix} 011 \ldots 1 \\ 101 \ldots 1 \\ \ldots \\ 111 \ldots 0 \\ 111 \ldots 1 \end{matrix}$$

For the OR and NOR cases, the tests are

$$\begin{matrix} 100 \ldots 0 \\ 010 \ldots 0 \\ \ldots \\ 000 \ldots 1 \\ 000 \ldots 0 \end{matrix}$$

Hence, of the 2^n possible input patterns, an n-input gate of the (N)AND/(N)OR types needs only $n + 1$ patterns to detect all its SSL faults. EXCLUSIVE-(N)OR gates have very different test requirements, however; see Problem 7.50.

Advantages of the SSL Fault Model

It should be apparent from the foregoing discussion that only a small fraction of the physical faults that occur in real circuits can be modeled exactly by SSL faults. Nevertheless, practical experience over many years suggests that a set of tests $\mathcal{T}$ that covers all SSL faults does a good job of covering most types of physical faults as well. In fact, in industry it is common to require tests for logic circuits to guarantee detection of 99 percent of all possible SSL faults.

The following intuitive argument provides some theoretical support for such requirements. In most logic circuits, the average fan-in per gate $\phi_{\text{in}}^{\text{typ}}$ is quite small, typically between 2 and 3. If $n = \phi_{\text{in}} = 1$, then $n+1 = 2^n = 2$, so by the last theorem, every one-input gate is exhaustively tested by $\mathcal{T}$. If $n = \phi_{\text{in}} = 2$, then $n+1 = 2^n - 1 = 3$, and $\mathcal{T}$ applies at least 75 percent of the test patterns needed for exhaustive testing of every two-input gate. In fact, many two-input gates will have the fourth pattern applied to them as an indirect consequence of applying tests to other gates in the circuit. Similarly, $\mathcal{T}$ applies at least 50 percent of the possible test patterns to each three-input gate. Hence, $\mathcal{T}$ comes close to exhaustively testing all the individual gates in a typical circuit.

Circuit Testing

Fault tables like Figure 7.54b constitute one possible tool for deriving test sets for combinational circuits. An n-input, r-line circuit requires a fault table with dimensions of $2^n \times 2r$; therefore, its construction is tedious and its size is excessive for all but the smallest circuits.

Two faults are equivalent (indistinguishable) if they are detected by exactly the same tests.

A small ($n = 4$ and $r = 10$) circuit C_1 composed of four gates is shown with its SSL fault table in Figure 7.55. The primary inputs are a, b, c, and d; the primary output is z. A number of interesting testing issues are illustrated by this table. Several columns have exactly the same $\times$ pattern, implying that the corresponding faults pass and fail precisely the same tests. Two such faults x/α and y/β are therefore called **indistinguishable** or **equivalent faults**, denoted $x/\alpha \equiv y/\beta$, because they cannot be distinguished by the testing process. From the table we see that, for example, $a/0 \equiv d/1$, because each has an $\times$ in the same three rows. Like the states of a sequential machine,

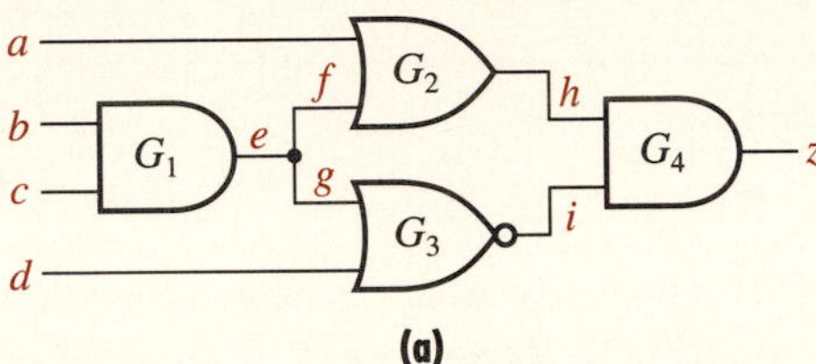

(a)

Faults

Tests abcd	a/0	a/1	b/0	b/1	c/0	c/1	d/0	d/1	e/0	e/1	f/0	f/1	g/0	g/1	h/0	h/1	i/0	i/1	z/0	z/1
0000		×										×				×				×
0001																				×
0010		×										×				×				×
0011																				×
0100		×										×				×				×
0101																				×
0110													×					×		×
0111																		×		×
1000	×							×		×				×	×		×		×	
1001							×											×		×
1010	×			×				×		×				×	×		×		×	
1011							×											×		×
1100	×					×		×		×				×	×		×		×	
1101							×											×		×
1110			×		×				×				×					×		×
1111																		×		×

(b)

FIGURE 7.55 (a) A small logic circuit C_1; (b) a fault table for all its SSL faults.

faults can be grouped into equivalence classes known as **equivalent fault classes**. From Figure 7.55b we get the following 10 fault classes for C_1:

$$\begin{aligned}
&\mathcal{F}_1 = \{a/0, d/1, e/1, g/1, h/0, i/0, z/0\} \\
&\mathcal{F}_2 = \{a/1, f/1, h/1\} \quad \mathcal{F}_3 = \{b/0, c/0, e/0\} \\
&\mathcal{F}_4 = \{b/1\} \quad \mathcal{F}_5 = \{c/1\} \quad \mathcal{F}_6 = \{d/0\} \\
&\mathcal{F}_7 = \{f/0\} \quad \mathcal{F}_8 = \{g/0\} \quad \mathcal{F}_9 = \{i/1\} \quad \mathcal{F}_{10} = \{z/1\}
\end{aligned} \tag{7.20}$$

For test selection purposes, it suffices to retain one representative of each fault class; hence, we could delete half the columns from the fault table while retaining all essential information.

Undetectable Faults

The fault $f/0$ in Figure 7.55 has an unusual property: its column in the fault table contains no occurrences of $\times$. This means that no test exists for this particular fault, which is therefore said to be an **undetectable fault**. The undetectability of $f/0$ can be explained as follows: Suppose $f = 1$ in the fault-free circuit; every test for $f/0$ must set f to 1. This implies that $e = g = 1$. The application of $g = 1$ to the NOR gate G_3 makes $i = 0$, which, in turn, makes the primary output $z = 0$. A fault that forces f to 0 does not affect z, so its presence cannot be detected at z.

Undetectable faults occur in circuits that contain redundant logic.

A circuit that, like C_1, contains one or more undetectable SSL faults is called a **redundant circuit**, because the line that has the undetectable fault, and possibly some of the circuits connected to that line, can be removed without affecting the circuit's normal behavior. For example, line f can be removed from C_1 in Figure 7.55a. This reduces the OR gate G_2 to a single-input gate that realizes the identity function $h = a$; hence, we can remove G_2 entirely and feed its a input signal directly into the output AND gate G_4. Therefore, line f and G_2 are truly redundant. Circuits with such redundancies are generally undesirable, both because they contain unnecessary components that add to hardware cost and because they make testing more difficult.[4] For example, most test-generation algorithms waste a great deal of time attempting in vain to find a test for an undetectable fault such as $f/0$.

Test Set Minimization

If a fault table is available, we can go about deriving a complete set of tests systematically. Figure 7.56 shows a reduced fault table for the circuit C_1 of Figure 7.55a, obtained by removing all but the first member of each fault class listed in (7.20); the undetectable fault class $\mathcal{F}_7 = \{f/0\}$ has also been deleted. Three of the remaining faults, $b/0$, $b/1$, and $c/1$, are in columns

[4]Redundant gates can be deliberately inserted in a circuit to eliminate hazards, as discussed in Section 6.10. However, this method must be weighed against the accompanying introduction of undetectable SSL faults.

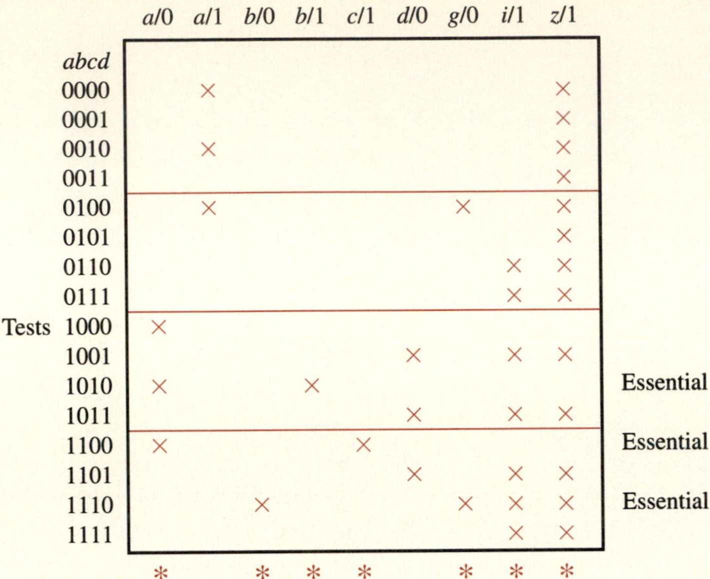

Tests abcd	a/0	a/1	b/0	b/1	c/1	d/0	g/0	i/1	z/1	
0000		×							×	
0001									×	
0010		×							×	
0011									×	
0100		×					×		×	
0101									×	
0110								×	×	
0111								×	×	
1000	×									
1001						×		×	×	
1010	×			×						Essential
1011						×		×	×	
1100	×				×					Essential
1101						×		×	×	
1110			×				×	×	×	Essential
1111								×	×	
	*		*	*	*		*	*	*	

FIGURE 7.56 A reduced fault table for C_1.

with a single ×. Therefore, the rows that contain those occurrences of ×—namely 1010, 1100 and 1110—define essential tests that must appear in any complete test set for C_1. Now we can mark as covered all columns with an × in any of these three rows; this is done using asterisks in Figure 7.56. Two columns $a/1$ and $d/0$ remain to be covered. Each can be covered by any of three tests: 0000, 0010, or 0100 in the case of $a/1$, and 1001, 1011, or 1110 in the case of $d/0$. Arbitrarily taking the first test in each case, we obtain the following five-member test set for all SSL faults in C_1:

$$\mathcal{T}_{C_1} = \{1010, 1100, 1110, 0000, 1001\}$$

It is obvious that $\mathcal{T}_{C_1}$ contains the minimum number of tests.

The general problem of selecting a minimal set of tests to cover a given set of faults is the same as that of selecting a minimal set of prime implicants to cover a given set of minterms. Solution methods for the latter, such as Petrick's method or the covering heuristic *GREEDYCOV* (Chapter 5), can also be applied directly to the test-set minimization problem.

7.8 Test Generation

The approaches to test generation discussed so far are not feasible for large circuits, especially sequential ones, because of the amount of computation they require. Next we examine some practical techniques to generate tests for SSL faults in relatively large combinational and sequential circuits.

In practice, a set of tests $\mathcal{T}_C$ to detect all SSL faults in a combinational circuit C is usually obtained by direct analysis of the circuit structure. Obviously, $\mathcal{T}_C$ must apply the test set $\mathcal{T}_G$ specified by theorem 18 to every gate G in C. We assume that a tester can only access those input and ouput lines of C designated here as primary. These include the circuit's usual primary inputs and outputs. They may also include normally internal lines that can be accessed by the test equipment during testing—for example, by special probes. In the typical case, G's input and output lines will not be primary inputs or outputs of C. Consequently, to apply a test pattern from $\mathcal{T}_C$ to the input lines of G, it is necessary to construct a test pattern T for C, which, when applied to C's primary inputs, controls the input signals of G in the required way; this is called the **controllability problem**. At the same time, the test pattern must propagate the response of G, including any error signals, through C to a primary output, where it can be observed and checked; this is the **observability problem**. The controllability and observability of a particular gate varies in complex ways with its location in the circuit, and is a major reason why generating a test set for a large circuit is difficult.

Path Sensitization

Many practical test-generation methods for combinational circuits take the following approach. An SSL fault F is selected for which a test pattern T is required. Then a path P is traced through the circuit that leads from the site of F to an (observable) primary output. For each gate G along P, gate input signal values are computed that enable G to propagate an error signal from its input to its output, thus bringing the error one step closer to the output. Once P and the signals on its gates that enable an error due to F to propagate along P have been determined, P is referred to as a **sensitized path**. At this point, it is necessary to work backward through the circuit to determine the values of the yet unspecified signals needed to produce all the sensitizing signals to be applied to the gates along P; this process is called **justification**. Eventually, the required test pattern T is obtained when the path sensitization and justification steps assign suitable values to all primary input lines.

Test generation is often based on systematic generation of sensitized paths for error propagation.

It is possible that contradictory requirements will be encountered during justification. For example, a 0 might be required by some input line of a NAND gate whose output line was set to 0 while constructing P; this contradicts NAND behavior. To resolve such conflicts, it is necessary to **backtrack**—that is, to return to some previous decision in which a choice was made—and make an alternative decision. For example, it may be possible to adjust the signals feeding the NAND gate so that all are 1, or it may be possible to reroute P to avoid the problematic NAND gate entirely. In a large logic circuit that contains tens of thousands of gates, the number of choices encountered while generating a test is huge. Furthermore, the path-sensitization process must be repeated many times to obtain a test set

If error propagation is blocked, we must backtrack and start again.

that adequately covers all the faults of interest. As a result, test-generation can require enormous amounts of computation, which must be performed by computer. **Automatic test pattern generation** or **ATPG** programs therefore form another important class of CAD tools used by logic designers.

An Illustration

To illustrate test generation by path sensitization, we consider again the circuit C_1 of Figure 7.55a. Because it contains 10 lines, there are 20 possible SSL faults to consider, assuming we do not know the fault classes. Let us find a test for the fault $b/0$. As depicted in Figure 7.57a, this requires applying the pattern $bc = 11$ to G_1, because the all-1 pattern is the unique test for any stuck-at-0 fault on an input of an AND gate. A 1 is then produced on line e, which will change to a 0 if the target fault $b/0$ is present. Because e fans out to two lines f and g, we have a choice of two paths to reach the primary output z. We arbitrarily select the upper path through OR gate

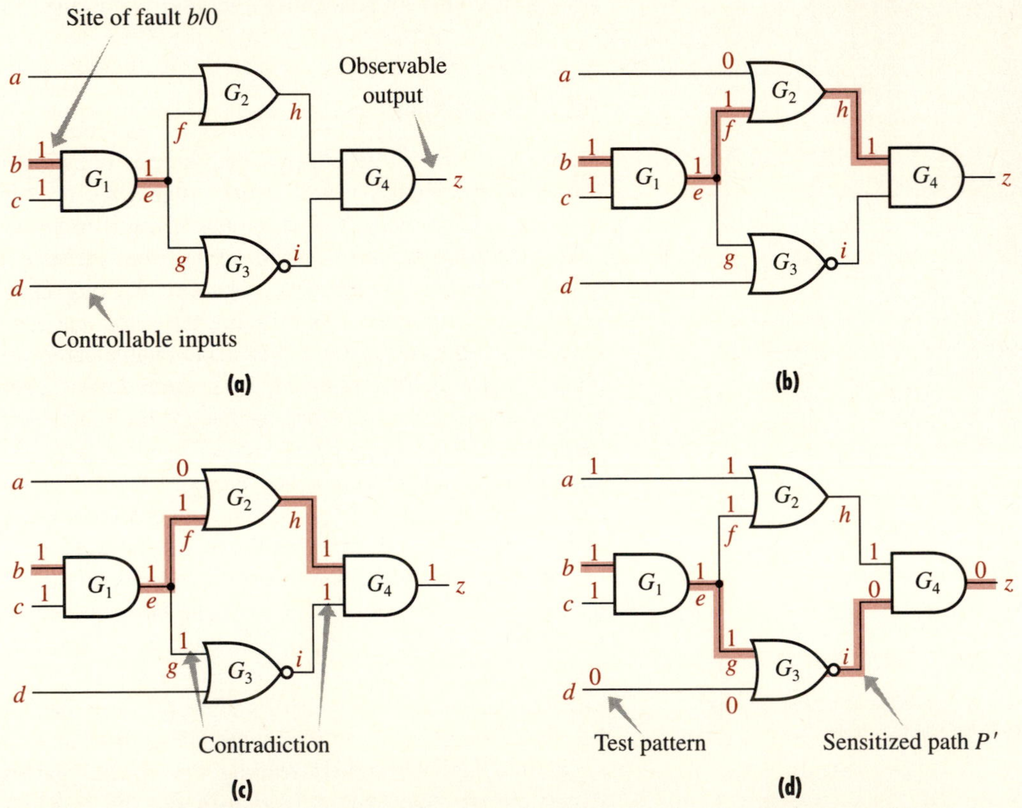

FIGURE 7.57 Finding a test for $b/0$: **(a)** sensitization of G_1; **(b)** the path extension through G_2; **(c)** the path extension through G_4 that leads to contradiction; **(d)** the successful path sensitization through G_3.

G_2. Because we are applying a 1 to G_2's f input, we must apply a 0 signal to G_2's other input a that will propagate a change in f through the gate to h. Setting $a = 1$ will not do so, because it forces h to 1, independent of the value of f. Setting $a = 0$, on the other hand, makes h sensitive to the value of f, so that a change in f from 1 to 0 changes h from 1 to 0. We now have the situation shown in Figure 7.57b, in which the signals assigned to the heavy lines form part of a potential sensitized path from b to z. It remains to extend the partially sensitized path through the second AND gate G_4 to z. This requires putting a 1 on input i of G_4; a 0 on this line would force z to 0, independent of h.

It now appears that we have the desired sensitized path $P = befhz$. However, our assignment of signals to the gates along P leads to a contradiction at the off-path NOR gate G_3, which an ATPG program would detect during justification. The condition $e = 1$ requires $g = 1$, which forces G_3's output i to 0. However, to sensitize P, we were required to make $i = 1$. In fact, we had no choice in the signals assigned to gates G_1, G_2, and G_4. This means that it is impossible to sensitize a path through these gates for the fault $b/0$. We must therefore backtrack and try an alternative path—namely, the path $P' = begiz$, which goes through G_3 instead of G_2. Again, the signals that must be assigned to all gates happen to be unique, but this time no contradiction is created. The resulting test $T(a, b, c, d) = 1110$ is shown in Figure 7.57d. It applies 11100 to path *begiz* when no fault is present. When the fault $b/0$ is present, these signals change to 00011, producing the desired error signal on z.

It is apparent that if any signal changes on a sensitized path due to a fault, the resulting error will propagate to the primary output at the end of the path. If a test T applies a 0 to some line x along a sensitized path it creates, the fault $x/1$ is detected by T; similarly, the stuck-at-0 fault is detected on any line to which T applies a 1. Consequently, if T creates a sensitized path that contains k lines, we can easily identify k different faults detected by T. In the case of Figure 7.57d, for instance, the test $T = 1110$ detects, in addition to the target fault $b/0$, the faults $e/0$, $g/0$, $i/1$, and $z/0$, all of which lie on the path $P' = begiz$ sensitized by T. Thus, we detect five SSL faults for the price of one. The relative ease with which a large set of faults can be detected by a single path-sensitization effort is a major advantage of this approach.

D Notation

Path sensitization of the foregoing kind involves a number of difficulties, including the fact that it may be necessary to sensitize several paths simultaneously to detect certain difficult faults (Problem 7.58). A theoretically complete path-sensitizing method, called the **D algorithm**, was developed by John Paul Roth of IBM in the 1960s [Roth, 1966], and is the basis of many test-generation programs used in industry since then.

D denotes a signal value that is 1 in the fault-free circuit and 0 in the faulty circuit.

A useful innovation of Roth's, which also gives the D algorithm its name, is a special logic value D (for *discrepancy*) that is 1 in the good (fault-free) circuit, and 0 when some fixed fault of interest is present. $\overline{\text{D}}$ denotes a signal that is 0 in the good circuit and 1 in the faulty circuit. Figure 7.58 shows the D notation used to describe the situation of Figure 7.57d. Note how the sensitized path corresponds to a sequence of Ds and $\overline{\text{D}}$s that propagates from the fault site to the primary output. The D algorithm (and other test-generation algorithms) are often expressed in terms of the five signal values 0, 1, D, $\overline{\text{D}}$, and X, where X denotes an unspecified or don't-care signal.

The discrepancy or error values D/$\overline{\text{D}}$ can be operated on by any Boolean function f, such as the gate functions AND, OR, NOT, and so on, by separately applying f to the faulty and fault-free values that D/$\overline{\text{D}}$ represent, which are always 0 or 1. For example, the function of the NOR gate G_3 in Figure 7.58 can be written as

$$\overline{\text{D} + 0} = \overline{\text{D}} \tag{7.21}$$

because in the absence of a fault G_3 computes $\overline{1 + 0} = 0$; if the fault $b/0$ is present, G_3 computes $\overline{0 + 0} = 1$. The D notation in (7.21) allows these two cases to be merged into one concise equation. If both D and $\overline{\text{D}}$ are applied to the NOR gate, the result is $\overline{\text{D} + \overline{\text{D}}} = 0$, which represents $\overline{1 + 0} = 0$ and $\overline{0 + 1} = 0$ in the fault-free and faulty cases, respectively. This is an example of **error blocking** or **masking**, when one or more error signals reach the inputs of a gate, but cannot propagate through the gate. Note that, unlike don't-cares, all occurrences of D in a circuit must take the same value at the same time.

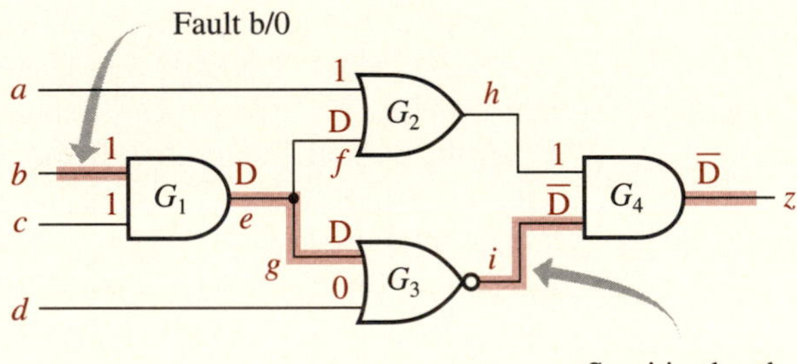

FIGURE 7.58 Roth's D notation applied to C_1 in Figure 7.57d.

D Algorithm

The D algorithm tries to propagate D/$\overline{\text{D}}$ signals from the fault site to a primary output.

We now present a test-generation procedure *DALG*, based on a slightly simplified version of the D algorithm. Its goal is to find a test for an SSL fault F in a combinational circuit. Test generation begins at the fault site, with the assignment of a D or $\overline{\text{D}}$, as appropriate, to the faulty line. This D/$\overline{\text{D}}$ signal is then systematically driven forward through the circuit until it reaches a primary output. At any point in this process, the set of D/$\overline{\text{D}}$ signals closest to the primary outputs—that is, separated from a primary output by the fewest levels of gates—constitutes the so-called **D frontier**. This frontier is pushed forward until some D/$\overline{\text{D}}$ signal reaches a primary output line, thus completing the forward propagation phase. Each gate is processed by assigning D/$\overline{\text{D}}$ to its output and appropriate 0 and 1 signals to its unassigned inputs. These sensitizing input signals are all 1s in the AND and NAND case, and all 0s in the OR and NOR case.

Next, all signals implied or forced by the current gate assignment are identified; this is step 3 (implication) of *DALG*. For example, if a 1 is assigned to the output of a NOT gate that feeds a gate on the sensitized path, the input of the NOT gate must be set to 0, as implied by its output signal. If all D/$\overline{\text{D}}$ signals are blocked from further propagation, the D algorithm backtracks to a previously untried gate, effectively selecting a different error-propagation path. Occasionally, all possibilities may be exhausted without driving the D frontier to the primary outputs—that is, without constructing a complete sensitized path. In that case, we conclude that the fault under consideration is undetectable.

Procedure

Procedure *DALG* to generate a test for a given fault

1. *Initialization:* A combinational circuit C and a fault F are given for which a test is desired. Mark all distinct signals in C with an implicit X, which denotes "unassigned." If the fault is stuck-at-1 (0), assign $\overline{\text{D}}$ (D) to the fault site.
2. *Propagation:* If there is a D or $\overline{\text{D}}$ at a primary output, go to step 4. Otherwise, select an untried gate from the D frontier, sensitize a path through it, and go to step 3. If at the D frontier there is no D or $\overline{\text{D}}$ that can be propagated, backtrack to a different path. If no untried path exists, exit in failure.
3. *Implication:* Assign all 0, 1, D, and $\overline{\text{D}}$ signal values that are implied (forced) by the current signal assignments. Return to the previous step (2 or 4).
4. *Justification:* If there are no unjustified gates, exit in success. Otherwise, select an unjustifed gate from the J frontier, assign an untried set of justifying input signals to it, and go to step 3. If a contradiction is encountered, backtrack. If no untried justification pattern exists, exit in failure.

DALG either finds a test, or determines that none exists.

Once a sensitized path has been found, step 4 (justification) of *DALG* is executed. The idea here is to work backward through the circuit to assign input values to all "unjustified" gates, defined as those gates for which output signals are specified, but for some of which inputs are not specified. In this case, the set of unjustified gates closest to the primary outputs constitutes the so-called **J frontier**. A typical gate can be justified by several different assignments. For example, a NAND gate with a 1 on the output is justified by any input assignment that applies at least one 0 to the NAND. On the other hand, the output 0 is justified only by the all-1 input assignment. After justifying a gate by suitable signal assignments, *DALG* determines the implications, if any, of those assignments. If *DALG* encounters a contradiction, it backtracks as long as untried justifying input patterns remain. If all possibilities are exhausted, *DALG* again terminates in failure. In most cases, a circuit is eventually completely justified, at which point the signals on the primary inputs define a test for the given fault. Some of the input values may be X, in which case an arbitrary assignment of 0s and 1s can be made to the Xs.

The similarity between the propagation and justification phases of *DALG* is worth noting. Propagation drives the D frontier forward to the primary outputs, whereas justification drives the J frontier backward to the primary inputs. In each phase, logical contradictions can be encountered due to poor decisions, in which case the D algorithm backtracks to an alternative (as yet untried) decision. In large logic circuits, this backtracking can become very time-consuming.

Example 7.11

Generating tests for a combinational circuit via *DALG*

Suppose *DALG* is required to find a test pattern for the SSL fault that consists of the output line of G_3 stuck-at-1—that is, $c/1$—in the all-NAND circuit C_2 of Figure 7.59a. For clarity, we omit the Xs assigned to all signal values in step 1. A $\overline{\text{D}}$ is assigned to the faulty line c and propagated through G_5. This requires a sensitizing 1 on input line b and produces a D on the output e of G_5. Now we consider the implications of these assignments. The fault-free 0 placed on the output of the NOT gate G_3 requires the input signal x_4 to be 1. The D on the output of G_5 implies a $\overline{\text{D}}$ on the output of the inverter G_7. We cannot determine both the faulty and fault-free values of any other signals, so step 3 ends with the situation depicted in Figure 7.59b. Both G_5 and G_7 have a D/$\overline{\text{D}}$ signal on their outputs and each is separated by only one gate from the primary outputs, so these signals define the current D frontier. We arbitrarily choose one error signal, the $\overline{\text{D}}$ on the output of G_7, to propagate further. (If this leads to a contradiction, we will backtrack to the other gate.) We now propagate through G_{10}, which yields a 1 on input line f and a D on z_2. This time, implication yields no further signal values. Because z_2 is a primary output, the propagation part has been successfully completed. On returning to step 2, we immediately exit to the justification step 4.

Examination of Figure 7.59c reveals that three gates, G_2, G_4, and G_6, need to be justified. Note that G_1's output value is not specified, so it is not considered yet; it may remain unspecified. Of the three gates that need justification, G_6 is closest to a primary output and so forms the J frontier all by itself. To justify its 1 output, at least one of its two unassigned inputs d or x_5 must have a 0 applied to it. We apply a 0 to the first line d; note that this is another decision that might have to be undone later. Returning to the implication step 2, we find that G_4's inputs must both be 1 to make $d = 0$.

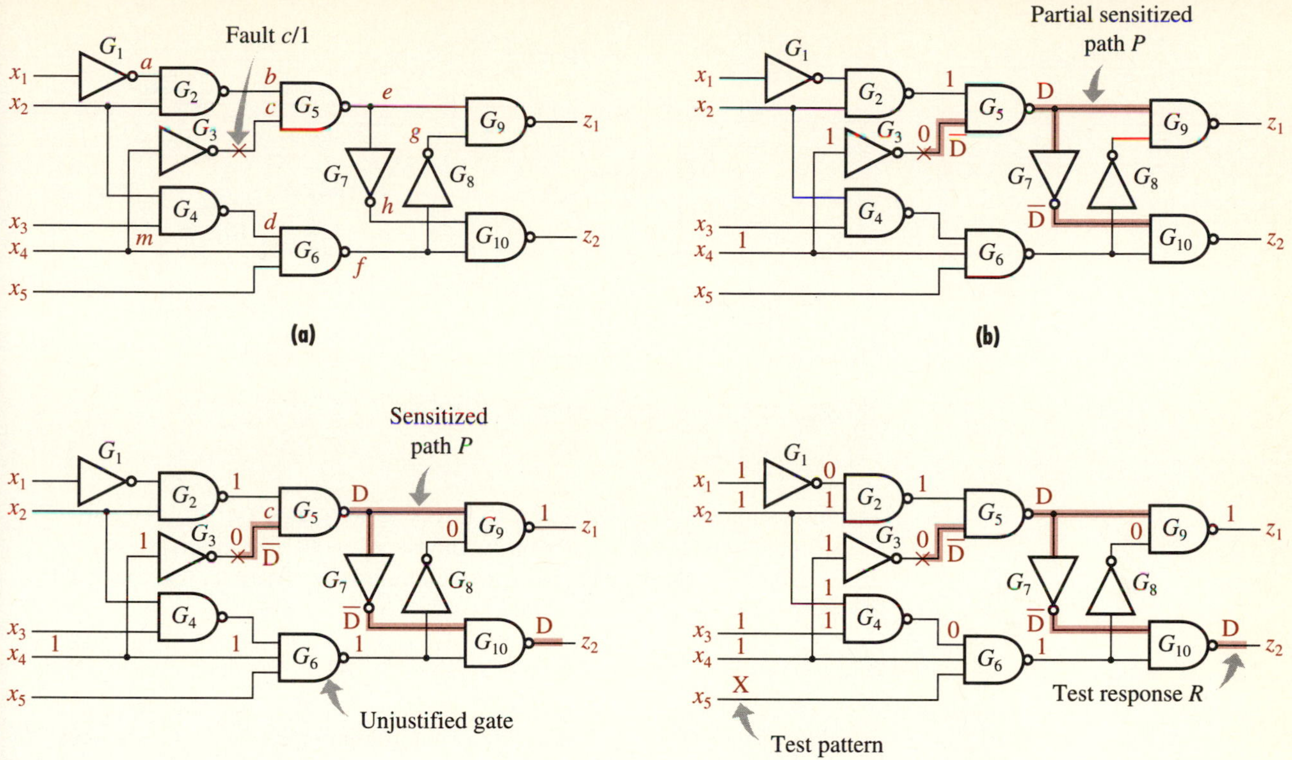

FIGURE 7.59 Derivation of a test for $c/1$ in C_2: **(a)** the initial condition; **(b)** after propagation through G_5; **(c)** after final propagation; **(d)** after justification.

We make this assignment so that G_4 no longer needs justification. We also consider G_6 to be justified with the input combination 01X. Retaining x_5 as a don't-care signal X simply increases our later options. At this point, G_2 has output 1 and inputs X1. The NAND function implies that the input a = X must be set to 0. This implies $x_1 = 1$, and gives us the final configuration that appears in Figure 7.59d. The primary input pattern $x_1x_2x_3x_4x_5$ = 1111X defines a test for $c/1$. Either of the test patterns it implies, 11110 or 11111, will serve as a test for the fault in question.

Figure 7.60 summarizes the test-generation process for a second fault in the same circuit, this time the stuck-at-0 fault $m/0$. We start by applying 1 to primary input x_4, which creates a D at the fault site. To propagate this error signal through G_6 requires $x_5 = 1$ and $d = 1$, which produces a $\overline{\text{D}}$ on line f. This, in turn, implies a D on line g. At the same time, $x_4 = 1$ forces c to 0 and e to 1, creating the condition needed to propagate the D on line g through G_9 and causing it to appear as $\overline{\text{D}}$ on the observable ouput z_1. We cannot propagate the sensitized path through z_2, because the circuit

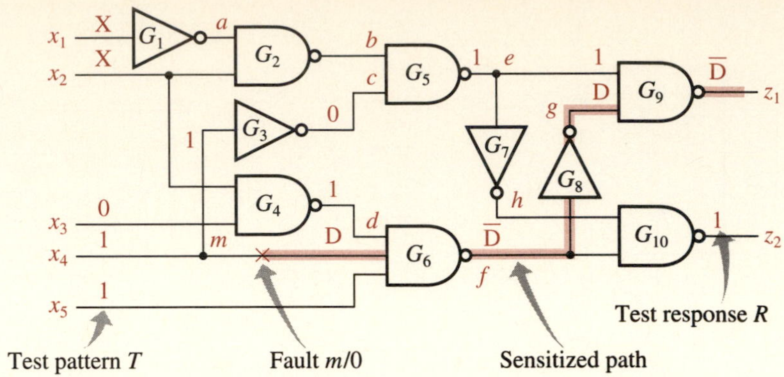

FIGURE 7.60 Derivation of a test for $m/0$ in C_2.

conditions force it to 1. It remains to justify G_1, G_2, and G_4. At least one of G_4's inputs must be 0; we arbitrarily choose $x_3 = 0$. The inputs x_1 and x_2 to G_1 and G_2 are don't-cares, because their values do not affect the sensitized path.

Practical implementations of the D algorithm employ heuristic shortcuts to reduce the amount of needed computation, especially backtracking. Alternative algorithms have been developed with this objective in mind. The most successful of these is the PODEM technique—the name stands for Path-Oriented DEcision Making—which was devised by Prabhakar Goel, also of IBM, around 1980 [Goel, 1981]. PODEM shares some features with the D algorithm, but has been found to be faster for some types of circuits [Abramovici, Breuer, and Friedman, 1990].

Iterative Array Model

The preceding test-generation method can be extended in a fairly direct manner to synchronous sequential circuits. Now a test for a particular fault F is a sequence $TS = T_1, T_2, \ldots, T_k$ of $k \geq 1$ input patterns rather than a single pattern. TS is applied to the sequential UUT SC in k consecutive clock cycles, and typically causes an error response to appear in the last cycle of TS, when F is present. Testing is synchronized by the UUT's clock, which is therefore assumed to function correctly. Failure of the clock circuitry will usually cause an easily detected catastrophic failure, in which the circuit fails to respond properly to any input.

The general (Huffman) model of a synchronous sequential circuit SC appears once more in Figure 7.61a. It has a set of feedback paths that contain the flip-flops that form the memory part M of the circuit. Suppose we

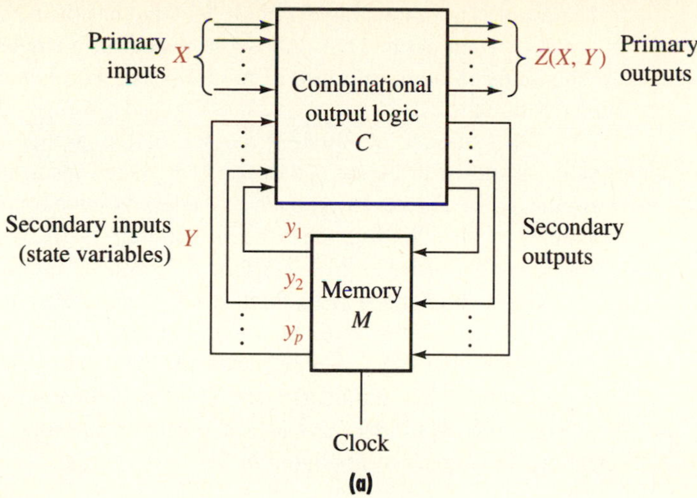

(a)

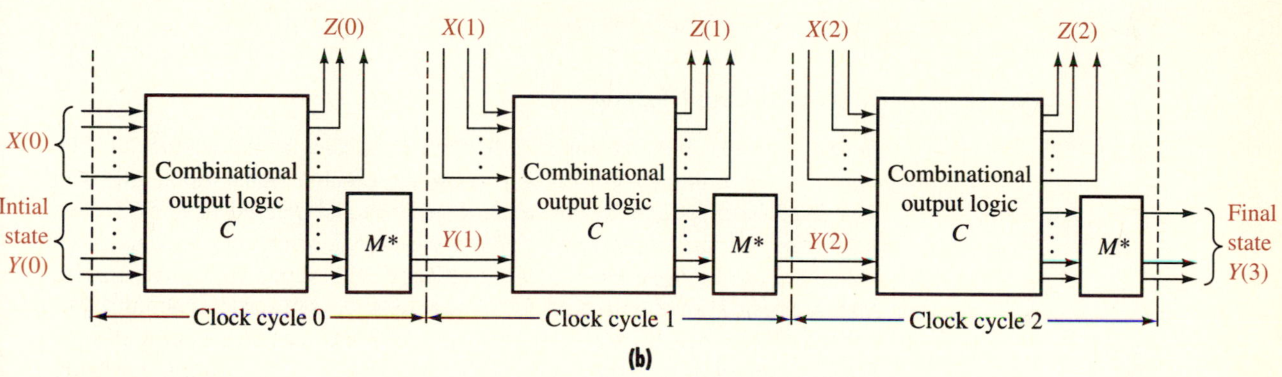

(b)

FIGURE 7.61 (a) The general sequential circuit SC; (b) the iterative combinational model SC^* of SC.

conceptually break all the feedback paths at the left (output) side of M, so that the state variables Y become primary outputs of SC. Moreover, suppose we replace M by a feedback-free combinational circuit M^*, the outputs of which are zero-delay versions of the sequential Boolean functions generated by the flip-flops. For example, if M is composed of D flip-flops, replace each flip-flop with a combinational circuit that implements the D flip-flop's characteristic equation

$$y(i + 1) = D(i) \tag{7.22}$$

(This D signal should not be confused with the error value D.) A noninverting buffer, or even a plain wire, suffices to implement (7.22). If M contains JK flip-flops, we replace the flip-flop with a combinational implementation of the JK equation, such as

$$y(i + 1) = J(i)\bar{y}(i) + \bar{K}(i)y(i)$$

The result is a combinational circuit SC^* that has the same functional behavior as the original circuit, but only with respect to a single clock period. M^* is called a **pseudo-combinational model** of M; SC^* is a pseudo-combinational model of SC.

If we cascade or iterate k copies of the pseudo-combinational model SC^*, as shown in Figure 7.61b for $k = 3$, the result represents the original circuit SC of Figure 7.61a over k successive clock cycles. Some structural features of the flip-flops have been lost, but as we will see, this loss does not adversely affect test generation. SC^* can be interpreted as the result of applying a **time-space transformation** to SC that replaces one copy of the sequential circuit that must be viewed at k time steps (clock cycles) with k copies of a circuit that can be viewed in one step, just like a combinational logic circuit. This multistage model of a sequential circuit, known as an **iterative array model**, provides a useful framework for constructing a k-pattern test sequence for faults in SC.

For test generation, a sequential circuit is modeled as an iterative combinational circuit.

Sequential Circuit Testing

Suppose a test is required for an SSL fault F in a sequential circuit SC. The fault is inserted into the corresponding pseudo-combinational model SC^* and replicated in each stage of the iterative array model derived from SC^*. Test generation begins with the first clock cycle $i = 0$, which requires an initial value $Y(0)$ to be specified for the secondary state variables Y. We assume that SC^* can always be forced to a known initial state that serves as $Y(0)$. Once $Y(0)$ is specified, we are faced with the problem of determining a value for $X(0)$ that will produce a D or $\overline{\text{D}}$ error signal as $Z(0)$. This is basically a combinational test-generation problem, and can be solved using the D algorithm or any other suitable method.

It is possible that a D or $\overline{\text{D}}$ cannot be driven to the primary output Z, but can be driven to a secondary output as part of $Y(1)$. In other words, the fault may be unobservable at Z, but can be recorded in the circuit's internal state Y. A sequential test-generation algorithm, therefore, attempts first to drive an error to Z; if it is unsuccesful, it attempts to drive an error to Y. In the latter case, the entire combinational test-generation process is repeated, with a new initial state $Y(1)$. Again an effort is made to produce an error on Z or, failing that, on Y. With each new test pattern, the test-generation process advances one stage through the iterative array model, which corresponds to advancing time by one clock cycle in the underlying sequential circuit. Eventually, either an error signal is produced on Z or the test-generation process terminates in failure. Such failure occurs when all feasible input sequences have been tried or, as is more often the case in practice, when some reasonable limit on computing time has been exceeded.

Example 7.12 **Generating a test for a serial adder**

To illustrate sequential test generation, consider again the serial adder depicted in Figure 7.4. Two stages of an iterative array model of this circuit are shown in Figure 7.62. The D flip-flop that stores the carry has been replaced by a buffer that realizes a zero-delay version of the flip-flop's function—that is, the identity function $y(i+1) = D(i)$. Note that although neither reset nor the clock circuits appear in this model, their presence is assumed for generating, and subsequently applying, the test sequences. The initial state is taken to be $y(0) = 0$ here, and can be produced by activating the asynchronous $\overline{CLR}$ input of the original circuit.

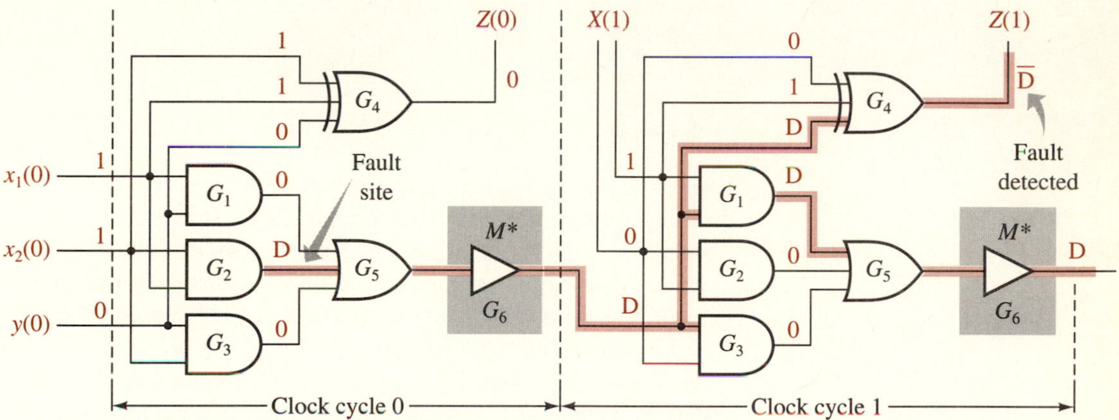

FIGURE 7.62 Generating a test for the fault $G_2/0$ in the serial adder.

Let $G_2/0$, denoting the output of AND gate G_2 stuck-at-0, be the target fault. There is no direct path from G_2 to the primary output z; hence, it is impossible to drive an error signal from G_2 to z in one clock cycle. The best we can do is drive a D from G_2 to the output of M^*, where, as shown in Figure 7.62, it becomes the next state $y(1)$. This error signal is now available as an input in the next clock cycle, and can easily be driven to z via the EXCLUSIVE-OR gate G_4, leading to the two-pattern test $X(0), X(1) = 11, 10$. Figure 7.62 shows by colored lines the sensitized paths generated by a typical test-generation algorithm for this problem. Note that any values applied to its off-path inputs sensitize an EXCLUSIVE-OR gate; therefore, any combination of values can be used for x_1 and x_2 in cycle 1 to extend the sensitized path through G_4. Whether D or $\overline{\text{D}}$ appears on z depends on the parity of the sensitizing signals applied to G_4.

Suppose that the initial state $Y(0)$ is 1 rather than 0 in the above problem. In order to force the outputs of G_1 and G_3 to the sensitizing value 0 required for error propagation through the OR gate G_5, we must have $x_1 = x_2 = 0$ in cycle 0. This, however, forces the output of G_2 to 0 so that the target fault $G_2/0$ cannot be activated. Consequently, no error signal can be generated in the circuit under these circumstances. This does *not* mean that the fault is undetectable. With $x_1x_2y = 001$, the adder's next state becomes 0 after one clock cycle. With $y = 0$, we can obtain a two-cycle test, as described above. Hence, the fault $G_2/0$ is detectable when the initial state of the adder is $y(0) = 1$, but a test sequence of length three, such as $X(0), X(1), X(2) = 00, 11, 10$, is necessary. This three-cycle test sequence will also serve as a test for $G_2/0$ if the initial state is 0; hence, it will work if the initial state is unknown—that is, $y(0) = \text{X}$. It follows that we do not need to know the initial state in order to derive a test for a fault in a sequential circuit. However, unknown initial states generally require longer test sequences that are often very hard to compute.

Summary

Tests can be generated for a sequential circuit by modifying a combinational test-generation method so that it can be applied repeatedly to an iterative array model of the sequential circuit. The basic goal of driving a D/$\overline{\text{D}}$ signal to a primary output is modified by the addition of the secondary goal of driving a D/$\overline{\text{D}}$ to a secondary output, if the first goal cannot be met. As Example 7.12 demonstrates, it may also be necessary to begin the test-generation process with a sequence of iterations that drive the circuit from an initial, possibly unknown, state to a one in which error signals that lead to fault detection can be produced. Because of the huge choice of initial states, next states, input/output signals, and internal signals, the problem of generating a complete set of tests for SSL faults is far more difficult to solve for sequential circuits than for combinational circuits. The computation time rises so rapidly that heuristic shortcuts that, say, try only a few possible combinations of initial states or next-state values are invariably used in practice, even for relatively small circuits.

7.9 Design for Testability

So far, we have only discussed how we might test circuits that are designed without built-in features to support testing. Most early digital systems were of this sort; the problem of testing was an afterthought. To obtain a satisfactory set of tests for most circuits at reasonable cost, we must make ease of testing a design goal from the outset, a topic to which we now turn.

Testability Goals

High fault coverage is a major design-for-testability goal.

Testability can be succinctly defined as a design characteristic that allows the operational status of a unit to be determined in a timely manner. Quantifying testability has proven to be a very difficult problem. A key measure of testability is high **fault coverage**, defined as the percentage of faults of interest, such as all SSL faults, that are detected by a particular test procedure. For combinational circuits, 100 percent SSL fault coverage is a realistic design goal; a somewhat lower goal, such as 95 percent fault coverage, may be set for sequential circuits. The fault coverage of a given set of tests $\mathcal{T}$ can be determined by computer-aided simulation of the UUT, with the faults of interest inserted one after another, a process called **fault simulation**. CAD programs that attempt to compute testability in terms of the controllability and observability of individual logic signals have not proven very helpful in the design process [Abramovici, Breuer, and Friedman, 1990].

Logic designers have various rules of thumb for testability improvement.

Besides maximizing fault coverage, it is desirable to minimize the hardware overhead for testing, which can be measured in terms of the extra circuits, extra input/output pins, and the like that are added to the circuit to support testing. Other useful testability goals are minimizing the amount of test data, minimizing the test-generation effort, and maximizing the speed at which tests can be applied. All these testability criteria are related to one another and to the structure of the UUT in complex, poorly understood ways.

Ad Hoc Methods

It is generally agreed that features to enhance testability should be introduced into the system during the design process; this is called **design for testability** or **DFT**. Experienced logic designers have developed numerous guidelines for ensuring that a circuit is reasonably testable; these ad hoc DFT rules are summarized in Figure 7.63. Perhaps the most basic requirement is to be

1. Provide at least one control input line and supporting logic to initialize the circuit to a known starting state.
2. Insert test points to increase controllability and/or observability of selected places (bottlenecks) in the circuit.
3. Partition large hard-to-test subcircuits like counters and shift registers into smaller, more easily tested ones.
4. Avoid redundancy that introduces undetectable faults.
5. Allow feedback paths to be opened and closed during testing.
6. Allow internal clock circuits to be disabled during testing.

FIGURE 7.63 Some ad hoc design rules for enhancing testability.

able to reset a sequential circuit to a known state before and during testing. This is most easily accomplished by a *RESET* control line that connects to the asynchronous clear or preset lines of the UUT's flip-flops and takes the circuit to a fixed initial state, typically 00...0. Figure 7.64b demonstrates how the circuit of Figure 7.64a is provided with reset logic to produce the all-0 initial state. This circuit has the unusual feature of an internal signal x that controls the flip-flops' *CLR* inputs. To override x for testing purposes, an AND gate G is inserted into x and connected to a new control input, called $\overline{RESET}$. In normal operation, $\overline{RESET} = 1$ and does not affect circuit behavior; changing $\overline{RESET}$ to 0 clears the flip-flops, independent of the value of x.

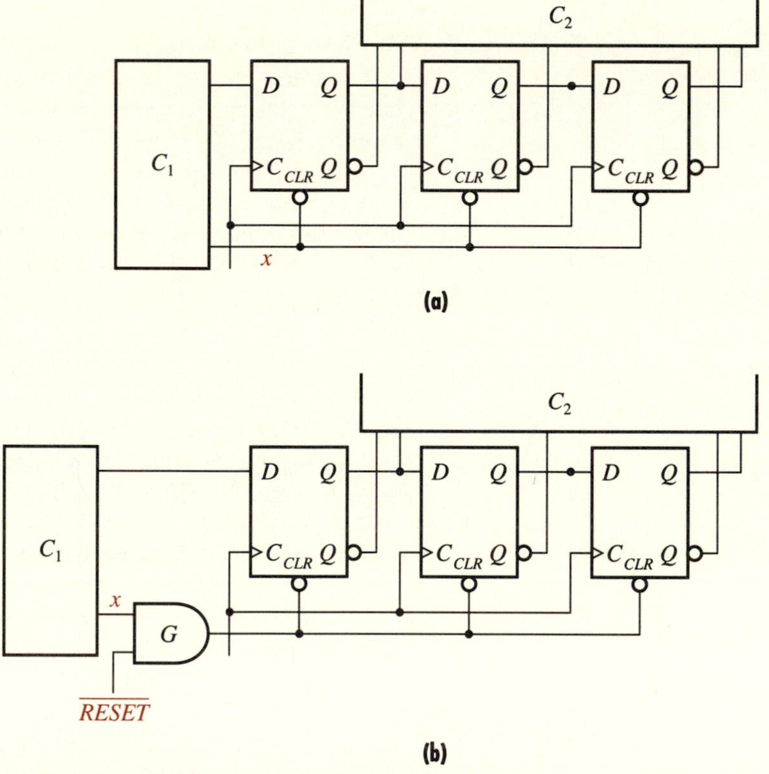

FIGURE 7.64 A sequential circuit (a) before and (b) after adding reset logic.

In Figure 7.64, the added logic for initialization can force a particular signal x to 0, independent of its normal value. This increases the controllability of x for testing purposes. We can generalize this idea to that of making an arbitrary line x controllable with respect to either 0 or 1. Suppose we introduce two external control signals CP_0 and CP_1 (CP denotes *control*

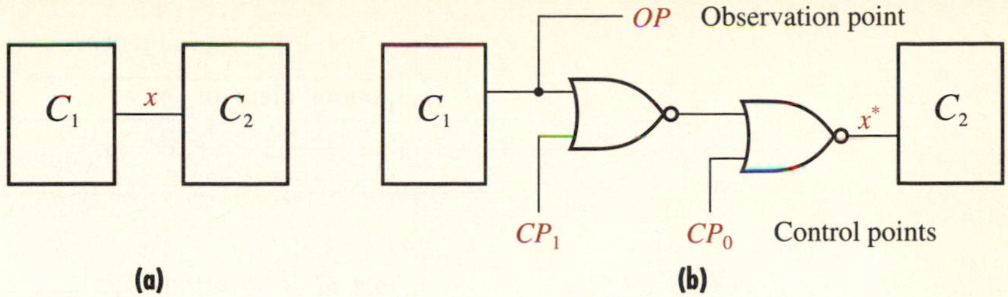

FIGURE 7.65 (a) Before and (b) after adding logic to make a line x fully controllable and observable.

point) to force a line x to 0 and 1, respectively. Then we must insert in x a logic circuit with inputs x, CP_0, and CP_1 that produces an output x^*, which satisfies the equation

$$x^* = \overline{CP_0}(x + CP_1) \tag{7.23}$$

With $CP_0CP_1 = 00$, we obtain $x^* = x$ for normal operation; $CP_0CP_1 = 10$ and $CP_0CP_1 = 01$ produce $x^* = 0$ and $x^* = 1$, respectively, for use during testing. An implementation of (7.23) by two NOR gates appears in Figure 7.65b. We can make x fully observable, as shown in the figure, simply by connecting it to a new primary output OP (for *observation point*). Testability-enhancing logic of the kind shown in Figure 7.65b is referred to as a **test point**.

Figure 7.66 lists representative places in a logic circuit that are suitable locations for test points. In general, they involve signals that are hard to control or observe, or signals with values that are particularly useful to the ATE (automatic test equipment) during test application. The number of such signals may be very large (hundreds or thousands), and each requires access to special input/output lines. In practice, the number of extra input/output connections that can be added to an IC or a PCB for testing purposes is less than 10, so that use of test points of the kind shown in Figure 7.65b is severely limited by the availability of the input/output pins. It is also affected to some extent by the gate cost of the test points, as well as the

1. Flip-flops that control a circuit's major states
2. "Buried" components that lie far from primary inputs or primary outputs
3. Lines of high fan-in and fan-out, such as internal buses
4. Redundant lines associated with undetectable faults
5. Lines flagged as having low controllability or observability by testability measurement programs

FIGURE 7.66 Some good locations for test points.

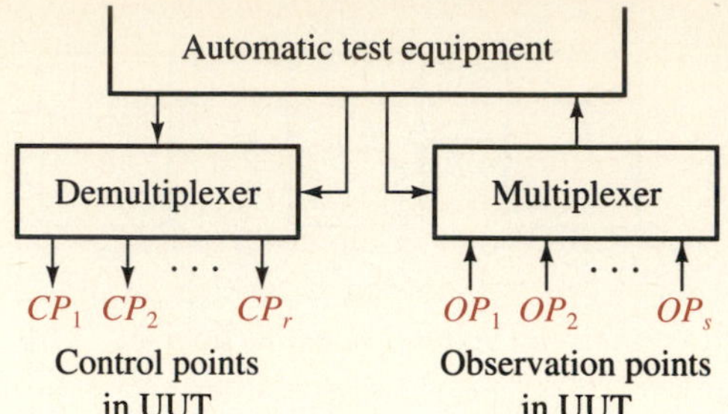

FIGURE 7.67 Input/output pin reduction using multiplexers and demultiplexers.

extra propagation delays they produce. The pin requirements can be reduced by means of multiplexers and demultiplexers, as depicted in Figure 7.67. Here a group of s observation points is channeled through a multiplexer into a single line observed by the ATE; a demultiplexer channels control signals from a single output line of the ATE to r control points in the UUT. This approach allows only one signal to be controlled and one to be observed at any time, however, which seriously restricts the method's usefulness.

Testing a Counter

Certain types of circuits are inherently difficult to test unless they are (re)designed with testability in mind. An example is the long (16-bit) counter shown in Figure 7.68a. To test the most significant output bit z_{15} for the stuck-at-0 fault $z_{15}/0$, a 1 must be applied to z_{15} via a sequence of test

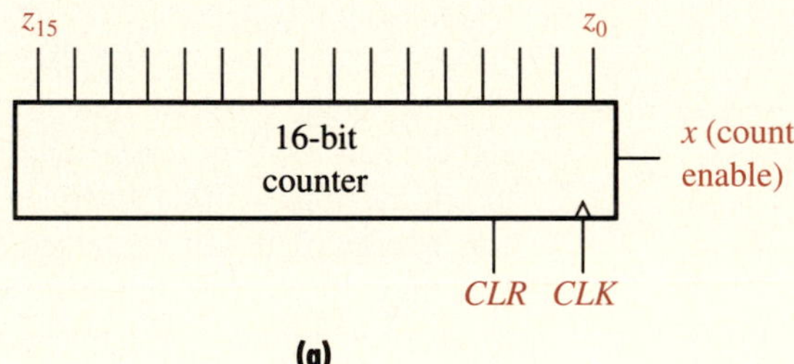

(a)

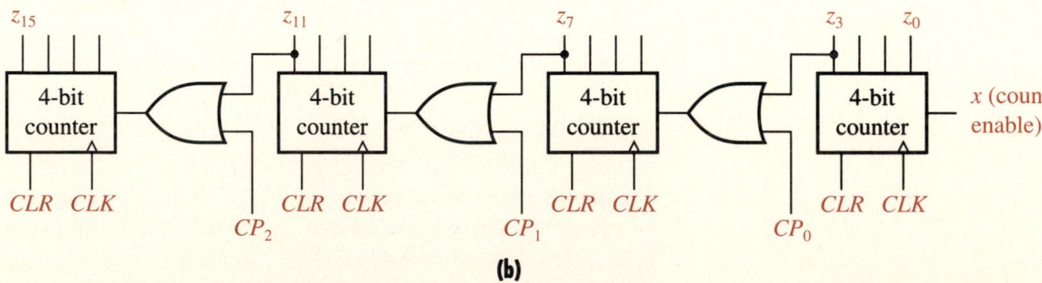

(b)

FIGURE 7.68 (a) Before and (b) after partitioning a counter to improve its testability.

patterns entered through the circuit's data input x, which is the "count enable" line used to increment the counter. After resetting the counter to the state $00\ldots0$, it is neccessary to apply $2^{15} = 32{,}768$ 1s to x in order to produce the first state $10\ldots0$ that sets z_{15} to 1. If the counter is embedded within a larger circuit, the problem of testing z_{15} and many other lines in the counter becomes even more difficult.

Temporarily partitioning a big circuit into small parts can simplify testing.

The foregoing problem is solved by partitioning the big counter into several smaller ones, as depicted in Figure 7.68b. The small counters—in this case, four 4-bit counters—are separated by test points. The circuit operates as a 16-bit counter when each $CP_i = 0$. When $CP_i = 1$, the corresponding 4-bit counter is incremented independent of the value of x. By applying only $2^3 = 8$ 1s to CP_2, we can make $z_{15} = 1$ in the partitioned counter.

Scan Design

The ad hoc DFT techniques we have been discussing suffer from lack of generality and a tendency to require too many input/output pins. What is needed, therefore, is a more systematic approach to DFT that is easily incorporated in the overall design process, enhances the controllability and observability of key signals, and requires few extra input/output pins. The only systematic technique that meets all these criteria is known as **scan design**. It was developed around 1970 [Williams and Angell, 1972] and is best suited to relatively small state machines of the kind addressed earlier in this chapter. Such machines are characterized by moderate numbers of flip-flops (no more than a few hundred), but may have thousands of gates. They are typically encountered in the control circuits of computers, an application for which scan design is widely used.

Scan design makes a circuit's memory totally controllable and observable for testing.

The essence of scan design is to make the memory M of a sequential circuit completely controllable and observable with respect to the external ATE. Any desired state can be written serially into M (a process called **scan in**), while the contents of M can be read out serially (a process called **scan out**). Scanning bypasses the circuit's normal combinational logic and primary input/output lines. It is done serially to minimize the number of input/output pins and supporting logic required for test purposes. This implies that M must support serial access during testing, but not during normal operation. Consequently, M is designed with modified flip-flops that can be reconfigured into a kind of shift register, referred to as a **scan register** or **scan chain**. Figure 7.69 illustrates the overall structure of a sequential circuit that employs the scan approach. During the normal mode of operation, which is set by disabling a special control signal *TEST*, the circuit's structure conforms to the Huffman model and the *SCAN* lines are not used. In the test mode specified by $TEST = 1$, the flip-flops are reorganized into a scan chain, as indicated by heavy lines in Figure 7.69b. This scan chain is accessed by two extra input/output pins, called *SCAN IN* and *SCAN OUT*, which serve as the tester's

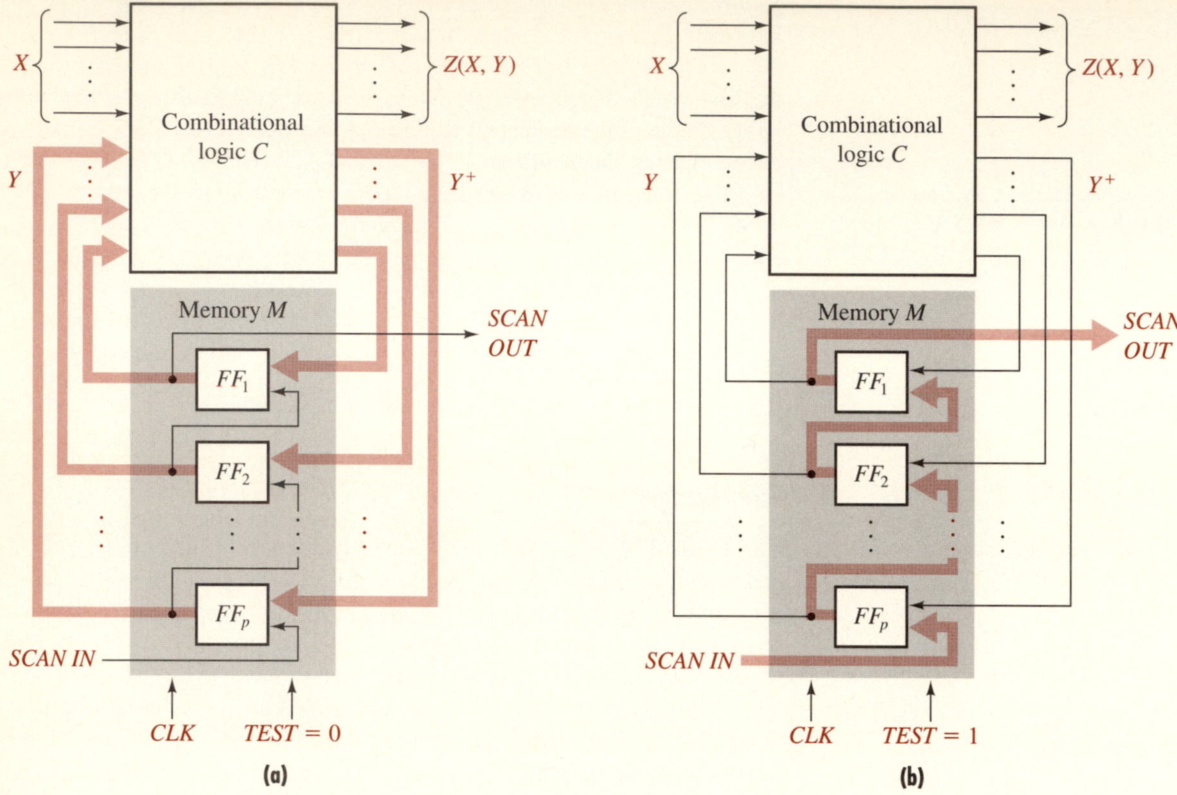

FIGURE 7.69 A sequential circuit that employs scan design: (**a**) its normal mode of operation; (**b**) its scan mode for testing.

control and observation points, respectively, for M. The tester is assumed to have direct access to the primary input/output lines X and Z in the usual fashion.

Testing via Scan Logic

Testing circuits that use scan design can be slow.

We see from Figure 7.69b that the scan registers provide the tester with direct, albeit slow, access to the secondary inputs and outputs Y, Y^+ of the UUT. Consequently, the tester "sees" the UUT as the combinational circuit C, with inputs X, Y and outputs Z, Y^+. The main testing task, therefore, is to apply a suitably complete sequence of test patterns to C and verify the resulting responses. Each of the test patterns T_i should be applied to the inputs X of the UUT in the presence of some initial state Y_i that was entered earlier via the *SCAN IN* line, and should produce a next state Y_{i+1} and output response Z_i. In other words, it should test for the behavior $Y_i \xrightarrow{T_i/Z_i} Y_{i+1}$. A state transition of this form is tested in the following way.

Procedure

Procedure *SCANTEST* to test a state transition

1. Set the test mode via $TEST = 1$. Enter the initial state Y_i serially into the scan chain. Apply T_i to X.
2. Reset to the normal mode via $TEST = 0$. Clock the circuit to execute a single state transition that causes the new state and output signals Y_{i+1}^* and Z_i^* to be generated.
3. Again set the test mode via $TEST = 1$. Read the new output value Z_i^*. Read the new state Y_{i+1}^* serially from the scan chain.
4. Compare the observed signals Z_i^* and Y_{i+1}^* with the expected signals Z_i and Y_{i+1}, respectively, and record any error.

Clearly, any sequential test for the UUT can be applied, state transition by state transition, in the above manner. The required test-pattern sequence can be derived readily by an efficient test-generation method. There is some overhead in extra circuitry due to the more complex flip-flops needed to construct the scan registers; this typically adds 5 percent to 15 percent to the total hardware cost. The overhead in extra input/output pins is very small, however, and can be reduced to as little as one (the *TEST* line) by multiplexing the *SCAN IN* and *OUT* lines with the regular input/output lines X and Z. The main disadvantage of scan design is the slow speed of the testing process. The time to test one state transition increases with the length of the scan chain, so the method is limited to circuits with scan chains of moderate length (no more than a few hundred flip-flops). A variety of techniques for designing the flip-flops used in scan chains have been developed. A straightforward approach is demonstrated in the next example.

Example 7.13

Inserting a scan chain into a sequential circuit

Consider the recognizer circuit *R4* that we designed earlier (Example 7.8). It contains three D flip-flops and its logic circuit appears in Figure 7.43 of Section 7.5. To make it scannable, we can replace each flip-flop with the circuit of Figure 7.70, which consists of a D flip-flop with a two-way, 1-bit multiplexer, allowing the flip-flop's input data to come from two sources: the original input data signal D_i and a scan signal S_i that, with the data output

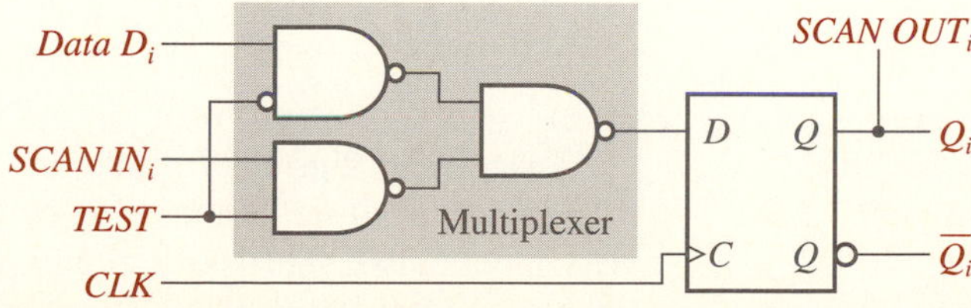

FIGURE 7.70 A flip-flop modified for use in a scan chain.

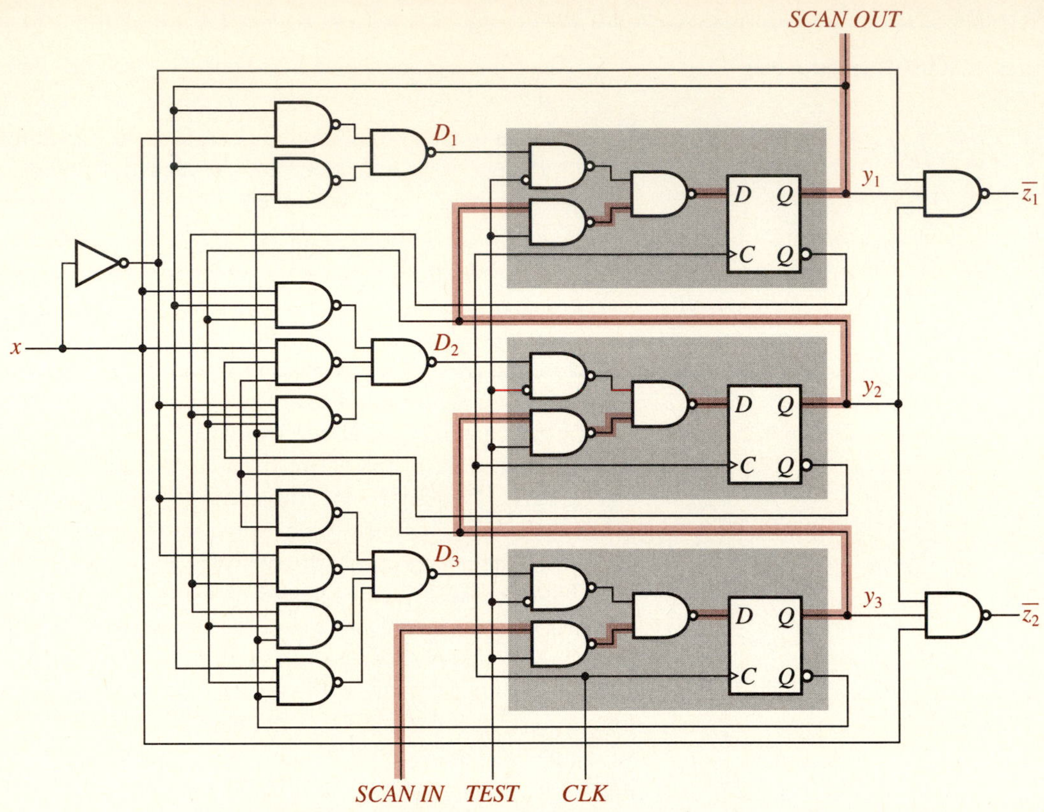

FIGURE 7.71 Circuit *R4* (Figure 7.43) modified to incorporate scan design.

Y_i, connects the flip-flop to a scan chain. Figure 7.71 shows three copies of this scan flip-flop inserted into the logic circuit for *R4*. When the control line *TEST* = 1, the three modified flip-flops are linked into a scan chain via their *SCAN IN* and *SCAN OUT* connections, as indicated by the colored lines in the figure. Notice that the reset line has been removed, because the circuit can be set to an arbitrary state, including 000, via a scan-in operation. The scan flip-flop of Figure 7.70 can be made faster and simpler in various ways: for example, by merging the multiplexer with the internal logic of the flip-flop or by replacing the edge-triggered flip-flop with a pair of latches controlled by a two-phase clock [Abramovici, Breuer, and Friedman, 1990].

Boundary scan is a standard method of adding scan logic to an IC or PCB.

Scan design can, in principle, be applied to any signal in a logic circuit. It is particularly useful in the case of "boundary" signals, which are the primary input/output signals of a circuit package such as an IC or a PCB. These signals are often difficult or impossible for a tester to probe directly. However, if the boundary signals are incorporated into a scan chain, a technique known as **boundary scan**, they can be shifted serially to a distant, more accessible test point. The key idea is to attach a special boundary-scan cell to each boundary signal x; this cell contains a latch to store x and associated logic controlling *SCAN IN* and *SCAN OUT* signals. The *SCAN IN/OUT* signals are linked to form a scan chain or shift register, making the scanned signals and the internal circuits to which they are connected completely controllable and observable for testing purposes (Figure 7.72).

A standard implementation of boundary scan, the IEEE P1149 *Standard Test Access Port and Boundary-Scan Architecture*, has been adopted by many circuit manufacturers [Maunder & Tulloss, 1990]. In addition to specifying the design of boundary-scan cells, the P1149 standard describes how to incorporate scannable instruction and data registers into a boundary-scan framework to allow test patterns for internal circuits to be entered, and their responses extracted, via the boundary-scan cells.

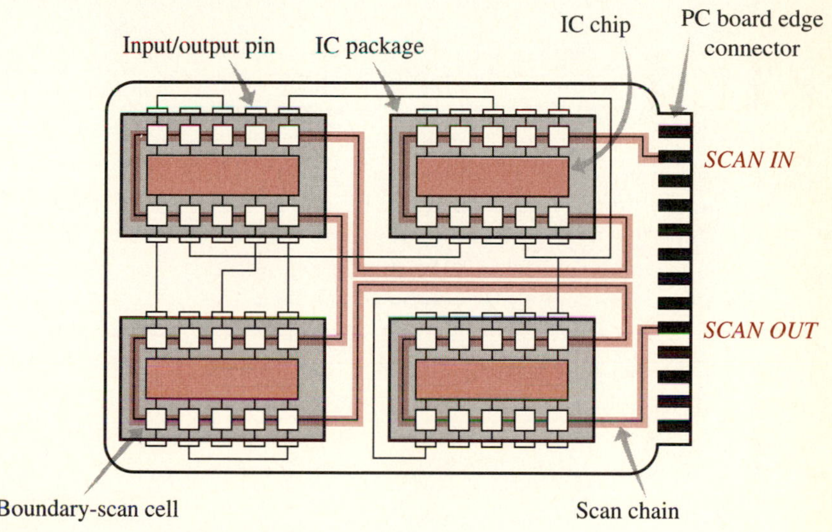

FIGURE 7.72 A printed-circuit board that incorporates boundary scan.

CHAPTER 7 SUMMARY

Sequential Circuit Analysis

7.1 ■ Sequential logic circuits differ from combinational circuits in that their output signals depend on past as well as present input signals, implying that they have memory.

■ **Synchronous circuits** have clocked flip-flops as memories and use a periodic clock signal to synchronize the times at which all flip-flops change their individual states, and therefore the state of the entire circuit. **Asynchronous circuits** do not use clocks.

■ The main signal types that appear in a synchronous sequential circuit are **primary input variables**, which are the externally controllable input signals, and the **primary output variables**, which are the externally observable output signals. The **secondary input variables**, also called the **state variables**, are the outputs of the flip-flops that constitute the circuit's memory. The **secondary output signals** are generated by the combinational logic and determine the next state of the sequential circuit.

■ In **Moore-type** sequential circuits, the primary outputs are a function of the internal state only. The more general **Mealy-type** circuits have ouputs that are functions of both the internal state and the primary input signals.

7.2 ■ Sequential circuit behavior is concisely defined using either a **state table** or a **state diagram**. Each row of a state table gives the next state and primary outputs as a function of the current state and primary inputs, that is, each row represents a single state transition. A state diagram uses nodes to represent states and arrows to represent state transitions.

■ States may be represented by symbolic names such as A, B, C, ... A **state assignment** associates a unique binary pattern of state variables with each state. A state table in which all states are denoted by their binary patterns is called a **transition table**.

■ Because of their simple characteristic equation, D flip-flops are easier to use when designing and analyzing sequential circuits. However, JK flip-flops may result in simpler circuits despite having twice as many inputs.

7.3 ■ Clock signals determine precisely when states change, and minimize sensitivity to hazards, race conditions, and noise. Triggering on the clock edge confines flip-flop state changes to a brief time slot after the flip-flop's input signals have settled at their new values, allowing for worst-case delays.

■ Due to **clock skew** (the arrival of the same clock edge at slightly different times at different flip-flops) and other causes, glitches can appear on primary output lines. The clock can be used to trigger output latches or flip-flops to remove such glitches.

■ The speed at which a sequential circuit operates depends on the propagation delays through the circuit's combinational logic and flip-flops, as well as the flip-flops' setup and hold times. These factors determine the period and waveform of the clock signals.

Sequential Circuit Synthesis

7.4 ■ The main goal of sequential circuit design is to obtain a logic circuit that has a desired state behavior.

■ The key steps in sequential circuit synthesis are deriving a formal description of state behavior, assigning bit patterns to represent the states, and designing the combinational logic that implements the next-state and output functions.

■ The combinational logic depends on the state behavior, the state assignment, and the flip-flop types. We can represent the combinational logic completely by an **excitation table**, which is a state (transition) table that also lists the secondary outputs required for each state transition.

- Simulation is essential for verifying the correctness of all but the smallest sequential circuits.

7.5
- The **classical method** for the design of sequential circuits employs the minimum number of flip-flops and attempts to minimize the combinational logic.
- The classical method consists of the following steps: state table specification; minimization of the state table by merging equivalent states; assignment of a specific bit pattern to represent each state; development of the excitation table; and the design of minimal combinational logic to produce the desired primary and secondary outputs.
- Developing a state table that correctly describes the desired behavior is the most difficult step.

7.6
- **Equivalent** or **indistinguishable states** cannot be distinguished from one another by observing input/output sequences.
- To determine equivalent states systematically, we examine all pairs of states with the aid of an implication chart. State pairs that eventually lead to two states with distinguishable output values are nonequivalent; otherwise, they are equivalent.
- A set of states $\mathscr{E}$ is called an **equivalence class** of states if all pairs of states in $\mathscr{E}$ are equivalent to each other, and a state not in $\mathscr{E}$ is not equivalent to any state in $\mathscr{E}$. An equivalence class can be replaced by a single representative state.

Design and Testing

7.7
- Like all physical devices, digital circuits are subject to failure and must be tested periodically to determine if they work properly. Testing is done by applying input patterns, called **test patterns**, to the **unit under test (UUT)**, and observing the resulting output signals, called the **test responses**.
- Common fault sources are improper manufacture or handling, overheating, and power supply fluctuations. The later a fault is detected, the higher is the cost of testing and repair.
- The **single stuck-line (SSL) model** is the standard fault model for gate-level testing. It defines two types of faults for each line in a logic circuit: the stuck-at-0 fault and the stuck-at-1 fault, which cause the line to be permanently stuck at the logic value 0 or 1, respectively.
- A **fault table** lists all faults occurring in a circuit and all the tests that detect them.
- The use of all input patterns as tests for a combinational circuit is called **exhaustive testing**; this method is practical for small circuits only. It is more efficient to generate a set of test patterns that will **cover**, or detect, all SSL faults.

7.8
- For large circuits, generating test patterns via fault tables is impractical. Many methods use a systematic search for signal paths that will propagate inputs to and outputs from the site of SSL faults. These paths are **sensitized**, that is, conditioned by other inputs, to correctly propagate the signals.
- The **D algorithm** is a popular test-generation algorithm that makes use of the special value D, which is 1 in a fault-free circuit and 0 in a faulty circuit.

7.9
- It is usually more efficient to design circuits to be testable than to develop tests for circuits designed without considering testing. **Design-for-testability** (DFT) procedures should be used whenever possible, especially in large circuits.
- High **fault coverage**, which is defined as the percentage of faults of interest that are detected by a particular test procedure, is a major DFT goal.
- Designers have developed ad hoc design rules for improving testability. These typically attempt to increase the controllability or observability of selected parts of a circuit.
- Scan design is a systematic DFT method that is suitable for sequential circuits of moderate size. A circuit's flip-flops are designed so that they can be linked to form a shift register during a special test mode of operation, allowing test inputs and outputs to be shifted serially into and out of the circuit.

Further Readings

The classical analysis and synthesis methods for small sequential circuits are covered in almost all textbooks on switching theory and logic design. Good treatments are found in the following books: D. J. Comer, *Digital Logic and State Machine Design*, 2nd ed. (Saunders, Philadelphia, 1990); D. L. Dietmeyer, *Logical Design of Digital Systems*, 3rd ed. (Allyn & Bacon, Boston, 1988); and J. F. Wakerly, *Digital Design Principles and Practices* (Prentice-Hall, Englewood Cliffs, N.J., 1990). Anthony K. Stevens's *Introduction to Component Testing* (Addison-Wesley, Reading, Mass., 1986) is a concise introduction to testing issues written from a practical perspective. *Digital System Testing and Testable Design* (Computer Science Press, Potomac, Md., 1990), by M. Abramovici, M. A. Breuer, and A. D. Friedman, presents a comprehensive discussion of all aspects of test generation and design for testability.

Chapter 7 Problems

Sequential Circuit Analysis

Sequential circuit structure; Mealy and Moore circuits; state tables and diagrams; timing diagrams; transition and excitation tables; deriving state behavior from logic circuits; clock design.

7.1 Which of the three sequential circuit structures depicted in Figure 7.1 has **(a)** the circuit in Figure 7.73, and **(b)** the circuit in Figure 7.74?

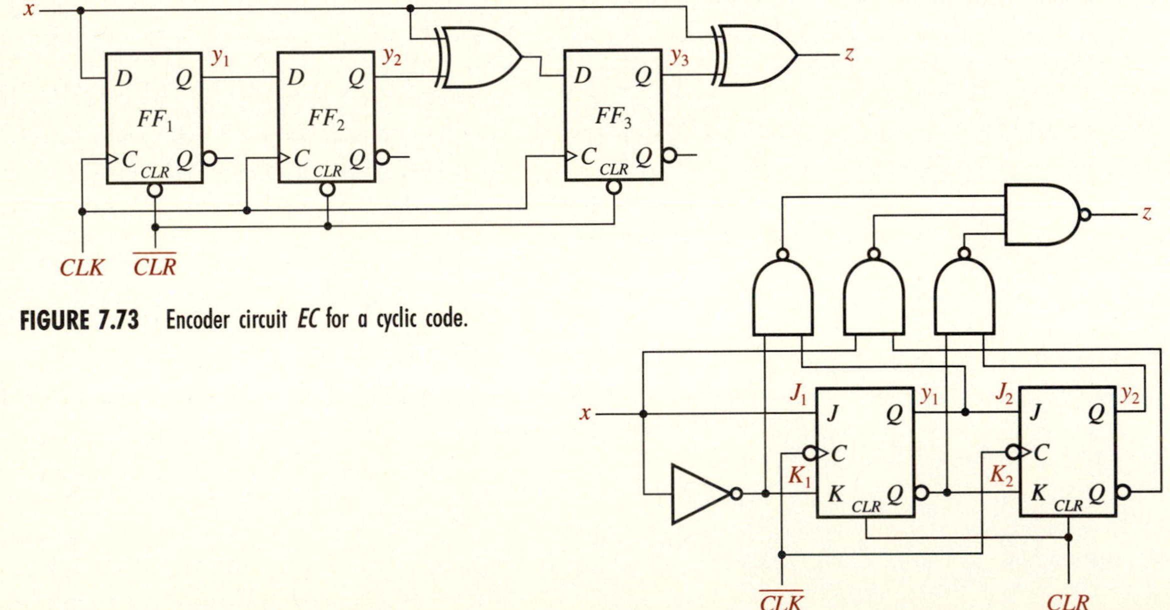

FIGURE 7.73 Encoder circuit *EC* for a cyclic code.

FIGURE 7.74 A sequential circuit with JK flip-flops.

7.2 Give an example of a simple computational task that cannot be performed by any sequential circuit with the structure of Figure 7.1a—that is, by any circuit without global feedback.

7.3 What are the main structural and behavioral differences between Moore and Mealy machines? Illustrate your answer with some simple examples.

7.4 A sequential circuit *SC1* has six primary inputs, four primary outputs, and 10 secondary inputs derived from six D-type flip-flops. How many secondary outputs does it have? How many rows and columns are in *SC1*'s state table?

7.5 A sequential circuit *SC2* has 18 states and is implemented with a minimum number of JK flip-flops. How many secondary inputs and outputs does *SC2* have?

7.6 **(a)** Draw a state diagram that corresponds to the state table of Figure 7.51a. **(b)** Draw a state table that corresponds to the state diagram of Figure 7.41.

7.7 Construct a Moore-type state table and state diagram for the sequential two's-complementer *N2C* defined in Example 7.4.

7.8 The state behavior for the recognizer circuit *R3* (Figure 7.35) is to be modified by replacing the transition $S_2 \xrightarrow{1/0} S_3$ with $S_2 \xrightarrow{1/1} S_3$. This means that z becomes 1 immediately on receipt of the third $x = 1$ signal to be counted, instead of waiting until the next state S_3 is entered. Analyze this revised behavior from the viewpoint of primary output signal timing and identify some potential problems and their solutions.

7.9 Construct a timing diagram in the style of Figure 7.9 for the Mealy version of the serial adder as it performs the addition $3 + 6 = 9$.

7.10 Draw a timing diagram in the style of Figure 7.12 for the Moore version of the serial adder as it performs the addition $3 + 6 = 9$.

7.11 Construct a timing diagram in the style of Figure 7.9 for the two's-complementer *N2C* (Example 7.4), which shows it negating the number $10100_{2C} = -12_{10}$.

7.12 Explain the differences among a truth table, a state table, a transition table, and an excitation table.

7.13 Write down the characteristic equations for JK and D flip-flops in terms of the complemented output signal $\overline{Q}$.

7.14 A toggle or T flip-flop has a single data input T such that $T = 1$ causes the flip-flop to change state or toggle, while $T = 0$ is the quiescent state. Obtain the characteristic equation for a T flip-flop.

7.15 **(a)** Construct a transition table for the Moore-type serial adder, the state table for which appears in Figure 7.10, assuming the following state assignment:

$$S_{00} = 00$$
$$S_{01} = 11$$
$$S_{10} = 01$$
$$S_{11} = 10$$

and using the two state variables $y_1 y_2$. **(b)** For this adder, construct an excitation table that uses D flip-flops.

7.16 Repeat Problem 7.15(b) for the Moore-type serial adder, assuming this time that JK flip-flops are to be used.

7.17 Analyze the sequential circuit of Figure 7.74 by obtaining **(a)** logic equations for its primary and secondary outputs; **(b)** equations for its next-state functions $\{y_i^+\}$; **(c)** a transition table; and **(d)** a symbolic state table.

The circuits *EN* (Figure 7.73) and *DC* (Figure 7.13) are useful for encoding and decoding data streams.

7.18 Repeat Problem 7.17 for the circuit of Figure 7.73, which is the encoder that corresponds to the decoder of Example 7.1 (Figure 7.13).

7.19 Repeat Problem 7.17 for the circuit of Figure 7.75. Explain why, in practice, D and JK flip-flops are rarely combined in the same circuit in this manner.

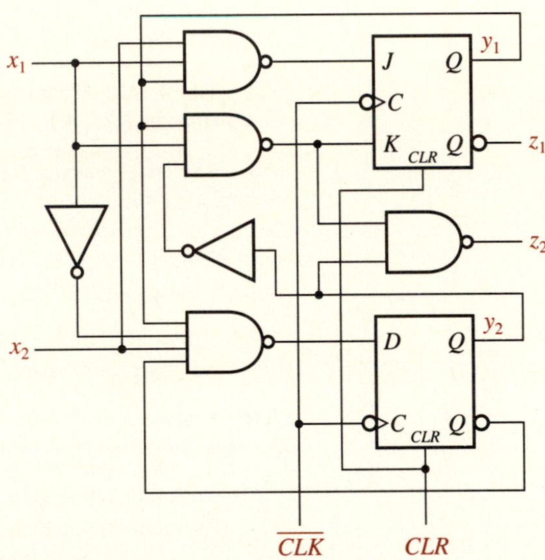

FIGURE 7.75 A sequential circuit that mixes D and JK flip-flops (not a recommended practice).

7.20 In coding theory, a serial data stream $S_{\text{in}} = d_k, \ldots, d_2, d_1, d_0$ is sometimes interpreted as a polynomial of the form

$$P(v) = d_k \cdot v^k \oplus \ldots \oplus d_2 \cdot v^2 \oplus d_1 \cdot v^1 \oplus d_0 \cdot v^0 \qquad (7.24)$$

where $\cdot$ and $\oplus$ denote the logical operations AND and EXCLUSIVE-OR, respectively. (Here v serves merely as a placeholder and is not to be confused with a logic signal.) A cyclic code generator transforms S_{in}, represented by $P(V)$, into an encoded form S_{out}, represented by $W(V) = P(V) \times G(V)$, by multiplying (7.24) by another polynomial $G(V)$, where AND and EXCLUSIVE-OR play the role of the multiply and add operations, respectively, on individual bits. The circuit *EC* of Figure 7.73 performs this type of multiplication with $G(V) = v^3 \oplus v \oplus 1$. If $S_{\text{in}} = 1010101$ is applied to *EC*, determine the resulting output sequence S_{out}, assuming that *EC* is initialized to the 000 state. Also determine the corresponding sequence of states through which *EC* passes.

7.21 As discussed in Problem 7.20, cyclic codes employ code words generated by a special polynomial multiplication of the form $W(V) = P(V) \times G(V)$. Decoding is accomplished by a corresponding special division of the encoded bit stream $W(V)$ by the original generator polynomial $G(V)$. The circuit *DC* of Figure 7.13 performs such a division with $G(V) = v^3 \oplus v \oplus 1$. As expected, the remainder $R(V)$ from this division should be 0. A nonzero remainder that corresponds to a final state different from 000 in *DC*'s memory indicates that an error was introduced into the incoming bit stream S_{in}. Any error that affects only a single bit of S_{in} in this way makes the remainder nonzero, so that cyclic codes form a class of single-error-detecting codes. Clearly, the circuits *EC* of Figure 7.73 and *DC* of Figure 7.13 form an encoder-decoder pair for a particular cyclic code. **(a)** If S_{in} = 0001011 is applied to *DC* with initial state 000, determine the resulting output sequence S_{out} and the corresponding state sequence. **(b)** Repeat part (a), assuming that a single-bit error changes S_{in} to 0001001.

7.22 Consider the sequential circuits *EC* of Figure 7.73 and *DC* of Figure 7.13, which are discussed in Problems 7.20 and 7.21. Suppose that the output of *EC* is connected to the input of *DC* to form a composite six-flip-flop circuit *EDC* with input x^* and output z^*. What function is implemented by z^*? How many nonequivalent states has *EDC*?

7.23 No (asynchronous) clear or preset inputs are shown in the sequential circuit of Figure 7.16. Nevertheless, if the circuit's initial state is unknown, it can be placed in either of its equivalent states *E* or *F* by applying appropriate input sequences to x and observing the results on z. Explain how to do this.

7.24 Suppose that the circuit of Figure 7.16 is implemented by 74LS-series ICs, the timing characteristics of which are described in Example 7.3. Assume that all the gates have the 74LS00's timing parameters. Determine the minimum clock period necessary for reliable operation. Draw the clock waveform annotated in the style of Figure 7.24.

7.25 Repeat Problem 7.24 for the circuit of Figure 7.74.

7.26 Repeat Problem 7.24 for the circuit of Figure 7.43. Assume that the D flip-flop is implemented by a suitable 74LS-series IC with the following parameters: t_{su}^{min} = 15 ns; t_{h}^{min} = 10 ns; and t_{ff}^{max} = 30 ns.

Clock design should account for uncertain delays due to clock skew.

7.27 Consider a sequential circuit with positive edge-triggered flip-flops, in which the clock signal *CLK* is subject to skew in the amount of $\pm\Delta t_{LH}$ on the positive edge, and $\pm\Delta t_{HL}$ on the negative edge. This means that the positive edge of *CLK*, if due at time t_1, can arrive as early as $t_1 - \Delta t_{LH}$ or as late as $t_1 + \Delta t_{LH}$, due to skew; the negative edge of *CLK* can be similarly skewed. **(a)** Modify the longest and shortest path constraints, (7.11) and (7.12), respectively, on the clock period T to include the parameters Δt_{LH} and Δt_{HL}. **(b)** What allowance should be made for Δt_{LH} and Δt_{HL} in determining the width of the clock's 1-pulse?

Sequential Circuit Synthesis

Classical design method; state table/diagram construction; state assignment; logic implementation with D and JK flip-flops; state minimization.

7.28 Suppose we want to add the normal yellow warning lights to the traffic lights of Example 7.1. Redesign the state diagram (Figure 7.27) for the traffic-light control unit *TC* to activate yellow lights at the appropriate times. For simplicity, assume that the red and yellow periods are all of equal duration (one minute). Use as few states as you can and state all the assumptions you make.

7.29 Using a JK flip-flop in place of the D flip-flop, redesign the two's-complementer *N2C* (Example 7.4). Give an excitation table and logic circuit for your design.

7.30 Consider the state diagram that appears in Figure 7.76. Construct an excitation table from it using the state assignment $A = 00$, $B = 01$, $C = 10$, $D = 11$. Using any standard gate types and D flip-flops, implement the excitation table and draw the resulting logic diagram.

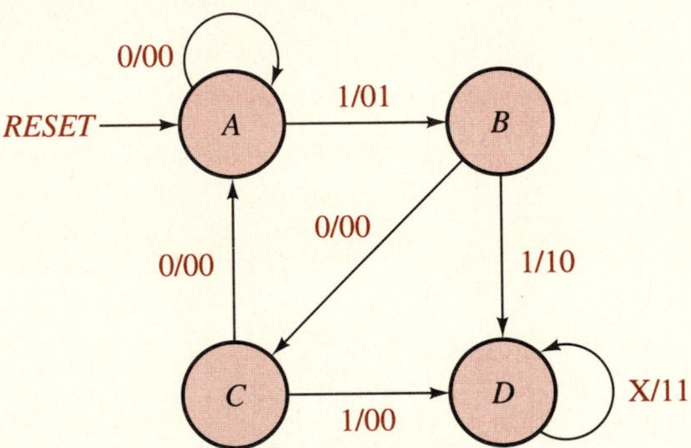

FIGURE 7.76 A four-state diagram.

7.31 Repeat Problem 7.30, this time using the state assignment $A = 10$, $B = 00$, $C = 01$, $D = 11$, and JK flip-flops.

7.32 Using JK flip-flops and NOR gates only, design a serial subtracter analogous to the serial adder discussed in the text. The inputs to the circuit are two unsigned binary numbers n_1 and n_2; the output is the difference $n_1 - n_2$. Give a state table, an excitation table, and a logic circuit for your design.

7.33 You are to design a serial adder *A4* that computes the sum of four separate input streams rather than the usual two. The adder has four primary inputs x_1, x_2, x_3, x_4 and a single primary output z. Using D flip-flops and any standard gates, construct a state table, an excitation table, and a logic circuit diagram for *A4*.

7.34 A single-input, single-output sequential circuit is to be designed that recognizes only the two input sequences 0110 and 11111 applied to its input lines immediately after an asynchronous reset. Use D flip-flops and NAND gates only. Construct a state table, an excitation table, and the final logic circuit diagram.

7.35 Repeat Problem 7.34, assuming now that the input sequences 0110 and 11111 should be recognized any time they occur in the input stream.

7.36 Using JK flip-flops in place of D flip-flops, apply the classical design method to the traffic-control unit *TC* described in the text (Section 7.4). Which flip-flop type is more appropriate for this problem and why?

7.37 The following rules of thumb are sometimes proposed for state assignment, with a view to reducing the number of gates in the combinational part of a sequential circuit:

Here are two general rules for making state assignments.

Rule 1: Give logically adjacent assignments to two states S_1 and S_2 that have the same next state S_3 for some common input combination X'.

Rule 2: Give logically adjacent assignments to two states S_1 and S_2 that are both next states of some state S_0.

Explain the rationale behind these rules. Briefly analyze the advantages and disadvantages of these rules, illustrating your answer with example circuits from this chapter.

7.38 Apply the two state assignment rules given in Problem 7.37 to the design of the recognizer circuit *R4* of Example 7.8. Obtain a logic circuit composed of NAND gates and D flip-flops in the style of Figure 7.43 and compare the hardware complexity of the two designs.

7.39 Design a single-input, single-output synchronous sequential circuit *D5* that treats its input sequence as an unsigned binary number n and produces an output sequence that is the input sequence divided by 5. Using D flip-flops and NOR gates only, design *D5*. Include in your answer a state diagram, a state assignment, a transition table, and the final logic circuit diagram. Keep your circuit as simple as you can.

7.40 A useful circuit referred to as a modulo-8 up-down counter *C8* is to be designed. It has the input/output lines shown in Figure 7.77. The 1-signals that appear on the count in line *CNT* are counted in normal binary code (modulo 8), with the resulting count appearing on the three output lines $Z = z_2z_1z_0$. A control signal *DOWN* determines the direction of the counting. When $DOWN = 0$, the count Z is incremented in each clock cycle where $CNT = 1$; Z is decremented instead if $DOWN = 1$. Using the classical design method, the objectives of which are to minimize the number of gates and flip-flops, design *C8*. Include a state table, an excitation table, and a logic diagram in your answer.

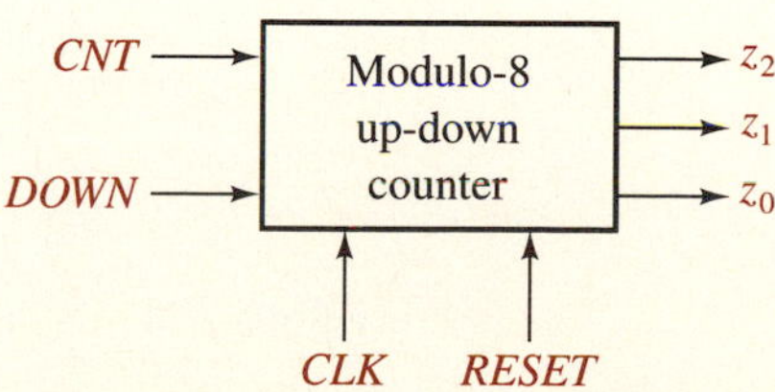

FIGURE 7.77 The modulo-8 up-down counter *C8*.

7.41 Design a sequential circuit *MATCH* with two inputs a and b and a single output z that detects certain matching runs of two or more 1s that appear simultaneously on a and b. Specifically, suppose $k \geq 2$ and that for clock cycles t and $t + k$ we have $a = b = 0$. Suppose also that for each intermediate cycle $t+i$, where $1 \leq i \leq k-1$, we have $a = b = 1$. Then z should be 0 for cycles t through $t + k$, and become 1 in the cycle $t + k + 1$. Thus, the appearance of $z = 1$ indicates that both a and b had 1s in some k preceding cycles. The design should use JK flip-flops and NAND gates exclusively. Obtain a state table, an excitation table, and a logic diagram for *MATCH*.

7.42 A sequential circuit *MON* is to be designed that monitors the condition of a patient in a hospital bed. The input to *MON* is a binary number n that ranges in value from 1 to 7 and indicates the patient's condition. The expected value of n is 3, but values of 2 and 4 are not considered abnormal. A new value of n is sent to *MON* automatically every five seconds. If n goes below 2 or above 4 on two or more occasions, the machine should activate an alarm at a nurse's station. The nurse responds by administering medication to the patient and resetting the monitor. Using JK flip-flops and NOR gates only, carry out the logic design of *MON*. Obtain a state table, a state assignment, and a complete logic diagram.

7.43 Consider the following three tasks and determine which ones cannot be performed by a finite-state machine. Provide brief but clear supporting arguments. **(a)** Recognize (accept) all even-length binary input sequences whose second half is the reverse of its first half; for example, 11, 001100, and 1011000110001101. Sequences of this kind are called **palindromes**. **(b)** Recognize all binary input sequences that contain exactly the same number of 0s and 1s. **(c)** Recognize all binary input sequences that contain an even number of 1s and an odd number of 0s.

7.44 Define what is meant by equivalent states and equivalence classes of states. Suppose *all* states of a given state machine M are found to be mutually equivalent. What is special about M?

7.45 Find a minimum-state table that can replace the seven-state table given in Figure 7.78a. Assume that A is the initial (reset) state.

Present state	Present input x: 0	1
A	B, 0	A, 1
B	C, 1	D, 0
C	E, 1	C, 1
D	A, 1	D, 0
E	E, 1	C, 1
F	F, 0	G, 1
G	G, 1	F, 0

(a)

Present state	Present input X: 00	01	10	11
A	E, 1	C, 0	B, 1	E, 1
B	C, 0	F, 1	E, 1	B, 0
C	B, 1	A, 0	D, 1	F, 1
D	G, 0	F, 1	E, 1	B, 0
E	C, 0	F, 1	D, 1	E, 0
F	C, 1	F, 1	D, 0	H, 0
G	D, 1	A, 0	B, 1	F, 1
H	B, 1	C, 0	E, 1	F, 1

(b)

FIGURE 7.78 Two examples of state tables.

7.46 Determine all the equivalence classes of states for the state table of Figure 7.78b.

7.47 Prove that the state diagram for *R4* that appears in Figure 7.41 has the minimum number of states; that is, the diagram is in minimal form.

Design and Testing

Fault types; SSL faults; equivalent faults; fault tables; path sensitization; D algorithm; combinational and sequential circuit testing; design for testability; scan design.

7.48 **(a)** List five common physical failure modes for ICs. **(b)** Why are physical faults modeled by logical faults such as the single stuck-line (SSL) model?

7.49 Consider the task of testing a single NAND gate G with four inputs x_1, x_2, x_3, x_4 and output z. **(a)** Give a minimum set of tests to detect all SSL faults that affect G. How many of these tests can be deleted if we want to detect only the stuck-at-1 faults that affect G? **(b)** Repeat part (a), now assuming that G is a four-input NOR gate.

7.50 Prove that all SSL faults in an n-input EXCLUSIVE-OR or EXCLUSIVE-NOR gate can be detected by two or three test patterns, depending upon whether n is even or odd. Are the needed tests unique, as in the (N)AND/(N)OR case?

7.51 Identify all the equivalent fault classes of a three-input NAND from the fault table of Figure 7.54. Generalize this result to determine the number of such classes for n-input gates of the following types: NAND, NOR, and EXCLUSIVE-NOR.

7.52 We showed that the fault table of Figure 7.55b contains 20 SSL faults and 10 equivalent fault classes. It is found in practice that for most combinational circuits, the number of fault classes is approximately half the total number of possible SSL faults. Give a brief intuitive explanation of why this is so.

7.53 Suppose that all four gates in the circuit C_1 of Figure 7.55a are changed to NAND gates. Construct a fault table with respect to all SSL faults for the resulting circuit and identify its equivalent fault classes.

7.54 Apply Petrick's method (Section 5.3) and the heuristic covering procedure *GREEDYCOV* (Section 5.7) to the fault table of Figure 7.55b in order to determine a set of test patterns to detect all SSL faults in C_1. Comment on the effort required by the two methods and the size of the test sets obtained.

7.55 **(a)** Consider the five-member set of signal values $\mathcal{V}_5 = \{0, 1, \mathrm{D}, \overline{\mathrm{D}}, \mathrm{X}\}$ used in test generation. Define suitable truth tables for the AND, OR, and NOT operations ab, $a + b$, and $\overline{a}$, where a and b take values from $\mathcal{V}_5$. Prove that $\mathcal{V}_5$ does not form a Boolean algebra by demonstrating two of Huntington's postulates (Table 3.2) that are violated by $\mathcal{V}_5$ with the foregoing operations. **(b)** If we delete D, $\overline{\mathrm{D}}$, and X from $\mathcal{V}_5$, we obviously obtain the two-valued Boolean algebra $\mathcal{B}_2$ as a subalgebra of $\mathcal{V}_5$. Is $\mathcal{B}_2$ the only Boolean algebra that is a subalgebra of $\mathcal{V}_5$? Prove your answer.

7.56 Use an informal path-sensitizing method to obtain tests for the following SSL faults in the circuit C_1 of Figure 7.57: $i/0$, $i/1$, and $a/0$. Mark the sensitized paths in a separate logic diagram for each fault.

7.57 Use the D algorithm to obtain tests for the following SSL faults in the circuit C_2 of Figure 7.59: $b/1$, $f/1$, and $x_3/0$. List the actions taken in all steps executed.

Some faults can be detected only by sensitizing several paths simultaneously.

7.58 Use path sensitization to derive a test for the fault $i/0$ in the circuit of Figure 7.79. Show that neither path from line i to the primary output z can be individually sensitized. Show, however, that *both* of these paths can be sensitized simultaneously, leading to a test based on **multiple-path sensitization**. Comment briefly on the general significance of multiple-path sensitization for test-generation methods like the D algorithm.

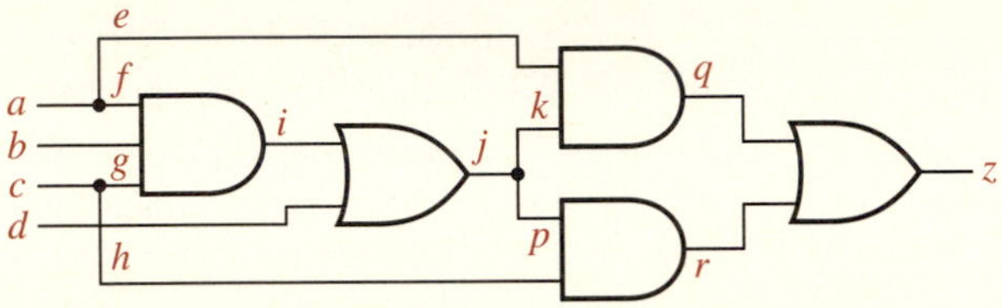

FIGURE 7.79 A circuit that requires multiple-path sensitization.

7.59 **(a)** Use the D algorithm to obtain tests for the faults $i/0$ and $z/1$ in the circuit of Figure 7.79. **(b)** Describe how you would use the D algorithm to determine if this circuit is redundant.

7.60 Construct iterative combinational array models for the following sequential circuits: **(a)** the serial adder (with JK flip-flop) that appears in Figure 7.39; **(b)** the two's-complementer *N2C* of Figure 7.34. In each case, show the full array circuit spanning two consecutive clock cycles in the style of Figure 7.62.

7.61 Using the iterative array model and the D algorithm, derive a test for the fault $D_1/1$ in the recognizer circuit *R3* of Figure 7.36. Assume that when testing begins the initial state is $y_1y_2 = 00$.

7.62 Repeat Problem 7.61 for the fault $y_2/0$ in *R3* (Figure 7.36). Again assume that the initial state is $y_1y_2 = 00$.

7.63 A test sequence for a sequential circuit is said to be a **strong test** for a fault F if it detects F for every possible initial state of the circuit. Describe how to extend the D algorithm to obtain strong tests of this sort. Derive a strong test for $D_1/1$ in the circuit *R3* (Figure 7.36).

A homing sequence can initialize any circuit from an unknown starting state.

7.64 An input sequence S_{hom} to a sequential circuit is called a **homing sequence** if the final state after application of S_{hom} can always be determined from the response to S_{hom}. Thus, S_{hom} can take the circuit from an unknown initial state to an identifiable final state. It is common to place a homing sequence at the beginning of a test sequence for a sequential circuit. It can be shown that every n-state sequential machine that is **strongly connected**, meaning every state is reachable from every other state by some sequence of input transitions, has a homing sequence of maximum length $n(n-1)/2$. **(a)** Construct an algorithm to compute S_{hom} from a strongly connected state table or diagram. **(b)** Use your algorithm to obtain S_{hom} from the state diagram for *R3* in Figure 7.36.

7.65 A homing sequence (see the preceding problem) that takes a sequential circuit to a fixed final state, independent of the initial state, is called a **synchronizing sequence**. Construct a simple state table that has no synchronizing sequence, but has a homing sequence. Devise a simple general technique for modifying any given state table to add a synchronizing sequence.

7.66 Explain the reasoning behind each of the ad hoc DFT rules given in Figure 7.63 and illustrate each rule with a simple example.

7.67 Circuits composed entirely of EXCLUSIVE-OR gates are widely used in the design of circuits for processing error codes. They are also especially easy to test. Demonstrate this by examining the testability of the class of n-input, single-output circuits constructed from two-input EXCLUSIVE-ORs. How controllable and observable are these circuits? How many tests do they require for SSL faults?

Triple modular redundancy (TMR) is a design method to achieve fault tolerance.

7.68 Identify a testability problem associated with each of the following circuits and describe a DFT approach to improve its testability. **(a)** A 32-bit shift register with a single serial input x and a single serial output z. **(b)** A circuit *TMR* contains three identical copies of a combinational circuit C. The output z of each copy of C is connected to a three-input "voter" circuit V, whose output $z^* = 1$ if and only if at least two occurrences of z are 1. Thus, $z^* = 0$ whenever at least two occurrences of z are 0. *TMR* is said to employ **triple modular redundancy** and serves as a fault-tolerant implementation of C. Under fault-free conditions, *TMR* produces the output $z^* = z$. If any one of the three Cs is faulty, *TMR* produces the same output signal as its two nonfaulty copies of C; the good copies "outvote" the bad copy.

7.69 Define what is meant by scan design, illustrating your answer with a small example. Identify a general type of logic circuit for which scan design is particularly well suited, and another circuit type for which it is *not* suited, explaining your reasoning in each case.

7.70 Modify the multiply-by-3 circuit *M3* of Figure 7.48a to incorporate scan design by redesigning the logic around *M3*'s two flip-flops.

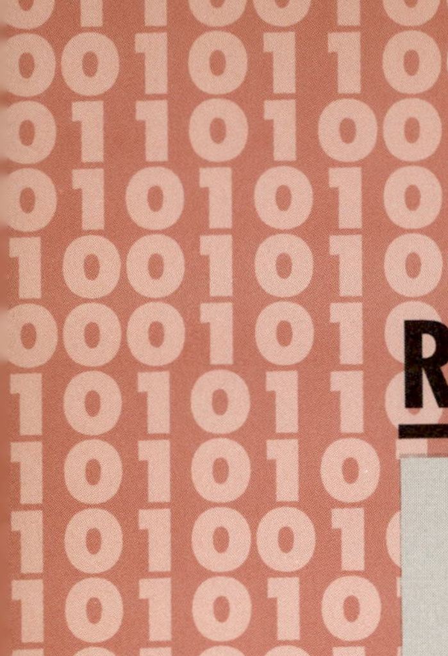

REGISTER-LEVEL DESIGN

The Register Level

8.1 General Characteristics
8.2 Combinational Components
8.3 Sequential Components

Datapath and Control Design

8.4 Datapath Units
8.5 Control Units
8.6 Programmable Controllers

ASM Design Methodology

8.7 Algorithmic State Machines
8.8 Integrating Data and Control

Overview

We move now to logic design that uses as building blocks higher-level components such as multiplexers, decoders, and registers. The target systems—for example, a multifunction arithmetic-logic unit—are fairly complex and are implemented using fixed or programmable logic elements. They can also be partitioned into separate control and data-processing parts, which, as we will see, involve quite different design problems.

8

The Register Level

We begin by examining the characteristics of the register level of design, including the main combinational and sequential component types employed. The methods used to describe designs at this higher level and the design process itself are introduced.

8.1 General Characteristics

Gate-level circuits such as multiplexers and registers become primitives at the register level.

A digital system can be viewed at various levels of abstraction or complexity; for example, the gate or logic level, which serves as the main focus of this book. In Chapter 3, we also examined the switch level, which is below or less abstract than the gate level. In this chapter, we move to a level higher than the gate level, where we treat as primitives small and medium-sized gate-level circuits of the kind examined in the preceding chapters. A central component at this level is an n-bit storage device or register; for this reason, the level in question is known as the **register** or **register-transfer level.**

Example: The 74X181 ALU

The 74X181 ALU has 63 gates and a truth table with 16,384 rows.

As an illustration of the need for higher-level descriptions, consider the 74X181, a standard medium-scale integrated circuit described as an arithmetic-logic unit/function generator; for brevity, we refer to it simply as the 74X181 ALU [Texas Instruments, 1988b]. The 74X181 is also found in standard cell libraries for VLSI design. Figure 8.1 shows the gate-level logic circuit for the 74X181, which contains a formidable collection of 63 gates of various types. The functions performed by this circuit are not obvious from the logic circuit alone. For example, we can write the following logic equation, which

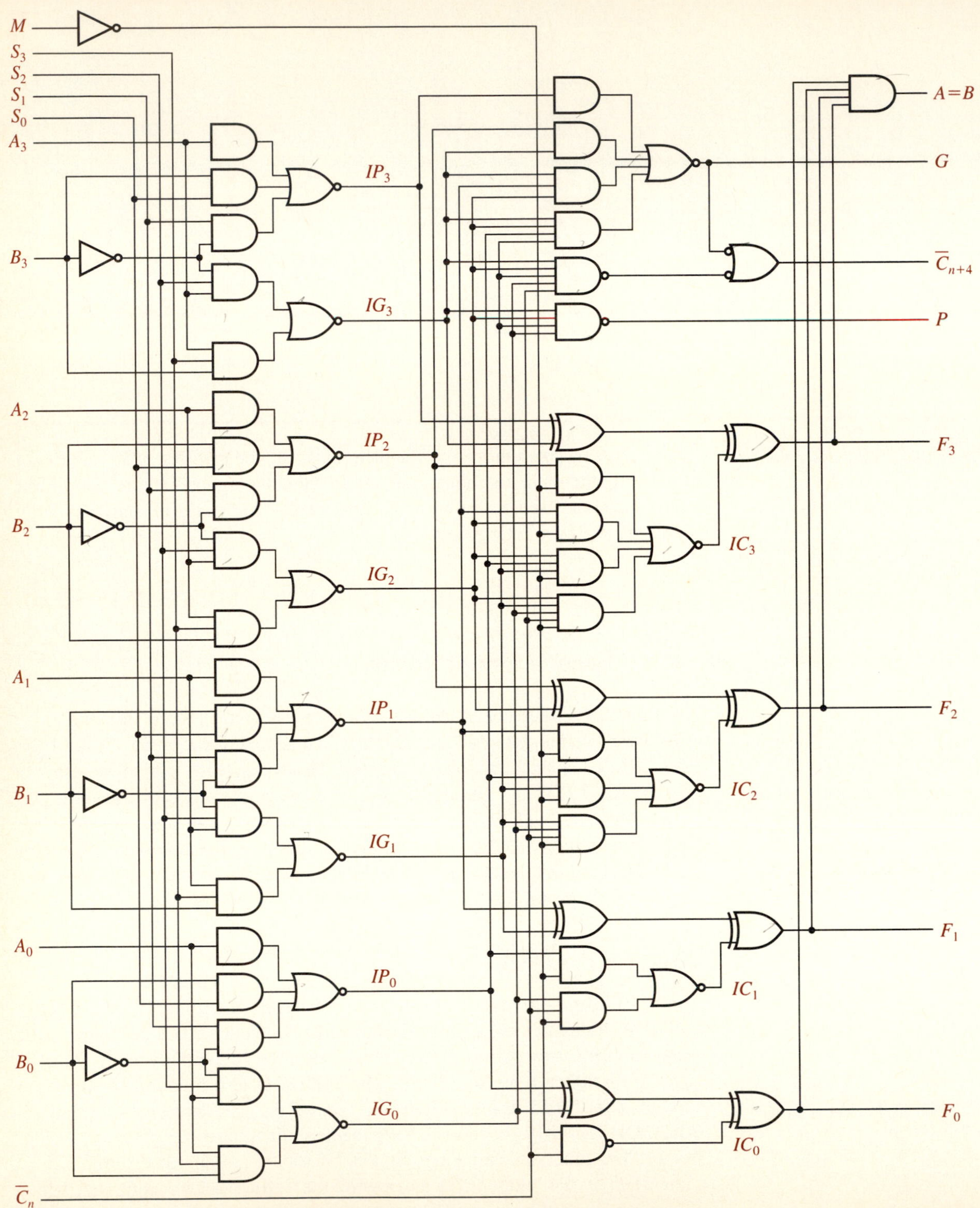

FIGURE 8.1 The gate-level logic circuit of the 74X181 arithmetic-logic unit/function generator.

defines one of the 74X181's output signals, directly from the logic diagram:

$$\begin{aligned} F_0 &= \overline{(\overline{M}\,\overline{C}_n)} \oplus (IP_0 \oplus IG_0) \\ &= (M + C_n) \oplus \overline{(A_0 + B_0S_0 + \overline{B}_0S_1)} \oplus \overline{(A_0\overline{B}_0S_2 + A_0B_0S_3)} \quad (8.1) \end{aligned}$$

This eight-variable function is complex, however, and sheds little light on the circuit's behavior. The entire circuit can be described by eight equations like (8.1), some of which depend on all 14 input variables. A complete truth table for the 74X181 would contain $2^{14} = 16{,}384$ rows!

We can better understand the 74X181's structure and behavior by identifying high-level structures within the circuit, as well as the arithmetic and logic operations it was designed to perform. The 74X181 implements various arithmetic operations (mainly two's-complement addition and subtraction) on two 4-bit numbers $A = A_3A_2A_1A_0$ and $B = B_3B_2B_1B_0$, producing the function $F = F_3F_2F_1F_0$. A few of the arithmetic operations performed by the 74X181 are

$$\begin{aligned} F &= A + B + C_n \\ F &= A + \overline{B} \\ F &= A - 1 \end{aligned} \quad (8.2)$$

The expressions that appear in (8.2) involve typical register-level operators (plus and minus) and multibit operands (4-bit numbers), in contrast to the single-bit logical operators and operands of (8.1).

We can also specify logical operations and operands at the register level. Consider, for example, the 4-bit EXCLUSIVE-OR operation

$$F_i = A_i \oplus B_i \qquad \text{for } i = 0, 1, 2, 3 \quad (8.3)$$

Here A, B, and F are treated as logical "words" rather than binary numbers. We can rewrite (8.3) more succinctly as the **word operation**

$$F = A \oplus B$$

in which $\oplus$ is interpreted as a word or bitwise version of the EXCLUSIVE-OR operation.

The 74X181's input/output lines and logic can be divided into data (datapath) and control parts.

In general, the 74X181 can perform any one of 32 different arithmetic or logical operations. The bit pattern applied to the "mode" line M and the four "select" lines $S = S_3S_2S_1S_0$ determine which operation is to be performed at any time. In particular, $M = 1$ selects a logical operation; $M = 0$ selects an arithmetic operation. Thus, the input/output lines of the 74X181 are divided into two groups: data lines like A, B, and F, and control lines like S and M. There are several other input/output lines that we treat as control lines, but for which the data/control distinction is not so clear-cut: the input and output carry signals $\overline{C}_n$ and $\overline{C}_{n+4}$, and the generate and propagate signals G and P, respectively, associated with binary addition using the carry-lookahead technique (Section 5.9). The output line named $A = B$ is used in conjunction with the carry lines to allow the 74X181 to be employed as a comparator that can compare the numerical magnitudes of the input words A and B.

High-Level Structure

Figure 8.2 shows a possible register-level logic diagram of the 74X181. The number of components has been reduced by more than an order of magnitude (from 63 to 5), so this diagram is far simpler than the gate-level logic diagram (Figure 8.1). We have grouped various gate-level subcircuits into primitive components, based on the functions they perform. The right half of Figure 8.1 yields a multifunction "datapath"—that is, data-processing—component *DP*, which basically performs 4-bit lookahead carry generation and summation; compare this figure with Figure 5.56.

The left half of Figure 8.1 implements two 4-bit logic functions $IP(A, B, S_0, S_1)$ and $IG(A, B, S_2, S_3)$ that are determined by the control variable S. The circuits serve as three-way multiplexers to allow various combinations of the input data operands A, B, and $\overline{B}$ (and also 4-bit representations of zero and one) to be applied to the inputs of *DP*. They also specify the propagate/generate logic (Figure 5.56) for carry-lookahead addition. Thus, if $S = 1001$ and $M = 0$, we can deduce from Figure 8.1 that $IP_i = \overline{A_i + B_i}$ and $IG_i = \overline{A_i B_i}$, the complements of the propagate and generate functions, respectively, and the two's-complement sum $A + B + C_n$ appears on the F output bus of *DP*. If M is now changed to 1, the internal carry signals produced by *DP* are forced to 1, and the IP and IG signals are combined by the EXCLUSIVE-OR gates of *DP* to produce the logical function $F = \overline{A \oplus B}$.

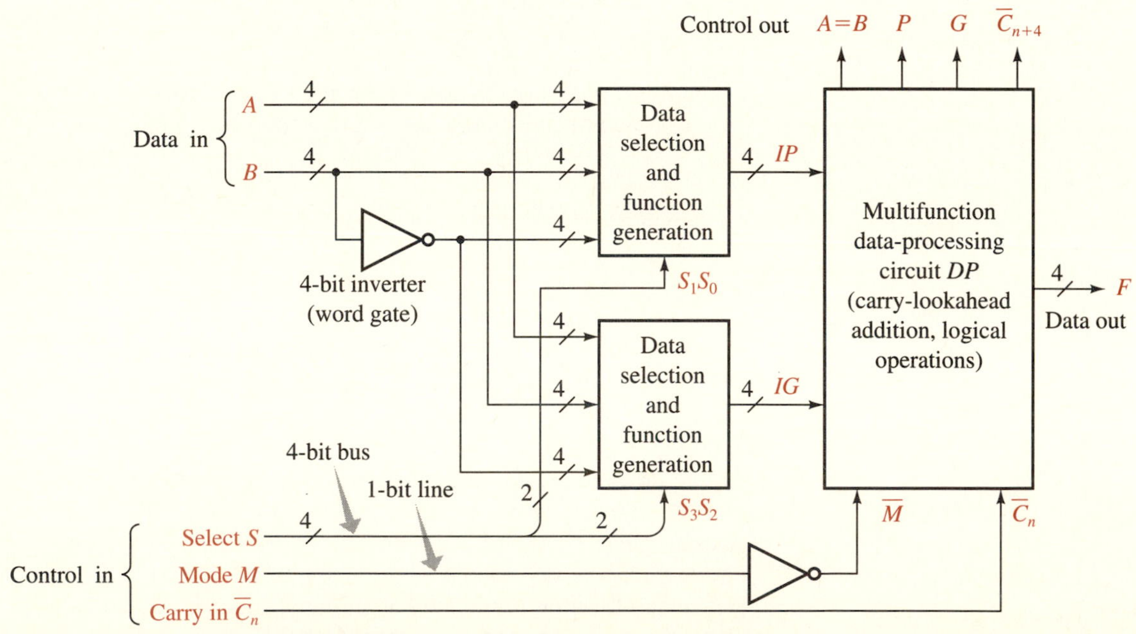

FIGURE 8.2 A register-level model of the 74X181 ALU.

An n-bit bus is denoted by a single line, optionally labeled slash-n.

Along with the grouping of gates into register-level components, associated signal lines have been grouped in Figure 8.2 into multibit buses, such as the 4-bit data buses A, B, and F and the control bus S. Each bus may be drawn as a single line in register-level circuit diagrams. The number of single-bit lines n in a bus can be specified by a short slash and an adjacent n thus: $\xrightarrow{n}$. As in gate-level diagrams, the joining of all lines that compose a bus (a wire junction or point of fan-out) is indicated by a black dot.

When several lines merge to form a bus, or when individual lines diverge from a bus, the lines involved are connected by a short curved or angled line. This is illustrated in Figure 8.2 for the case of the 4-bit control bus $S = S_3S_2S_1S_0$, in which only half the lines are used to control each of the two data selection circuits. The upper data selection circuit, for instance, is controlled by S_1S_0, as indicated in the figure. It is very important to label clearly the individual lines or subbuses that break off from a bus; otherwise, the signals involved may not be identifiable. When the direction of signal flow in a bus is unclear, we insert a direction-indicating arrowhead.

The Top Level

The highest level at which we can view any component is as a single primitive element, represented by a block or a **black box**—black suggests the unspecified or unknown nature of the box's contents—as in Figure 8.3a. The 74X181 is often treated as a single component in this manner. The fact that the diagrams of Figures 8.2 and 8.3a are both considered *register-level*

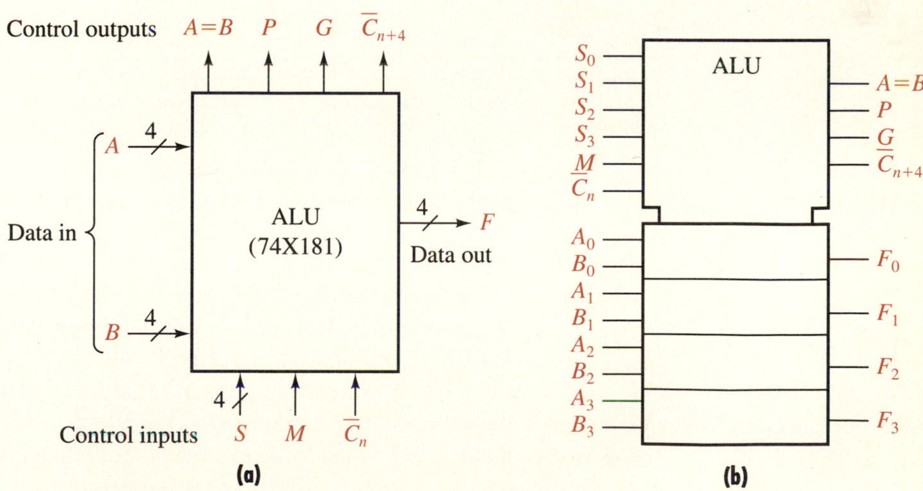

FIGURE 8.3 (a) The box symbol and (b) the IEC 617 (IEEE 91) standard symbol for the 74X181 ALU.

models of the same physical circuit indicates that the register level is less precisely defined than the gate level. There are no "canonical" circuit structures, and there is less adherence to graphics standards for drawing register-level logic diagrams. Such standards exist, nevertheless, notably in the international IEC 617 standard and its U.S. counterpart, the ANSI/IEEE 91 standard for logic symbols, which were introduced in Chapter 3. Another relevant standard is the ANSI/IEEE 991 standard for drawing logic circuit diagrams, which covers such subtle issues as bus representation [IEEE, 1987]. We will not attempt to adhere rigorously to these standards—they require attention to details that are beyond the scope of this book—but we will continue to maintain broad consistency with them.

Figure 8.3b shows the standard IEC 617 graphic symbol for the 74X181 ALU. It separates the device's control and datapath functions. The upper box that has the narrowed bottom constitutes a **control block**, which determines the function performed by all the boxes beneath it; the latter form the data-processing part of the circuit. In the 74X181 case, the data-processing part is drawn as four individual boxes, each of which performs the same operation as a 1-bit portion, or a **bit slice**, of the 74X181's 4-bit operands. The IEC 617 symbol retains the individual identities of all input/output lines, whereas our simpler box symbol of Figure 8.3a coalesces some of the data and control lines into multibit buses.

Descriptive Methods

We have seen that logic diagrams can be adapted to handle high-level descriptions of digital circuits—at least those of moderate complexity. The resulting diagrams are primarily structural models and do not completely describe the circuit's behavior. As noted above, both binary truth tables and logic equations are at a level of abstraction too low to describe the functional behavior of the 74X181 in a useful way. They may, however, be adapted for this purpose, as illustrated by Figure 8.4, which reproduces the tabular description of the 74X181 found in IC data books [Texas Instruments, 1988b]. This table gives, for all 32 combinations of the input variables $S_3S_2S_1S_0$ and M, either an arithmetic or logic equation that defines the function performed by DP. When $S_3S_2S_1S_0M = 01101$ (01101 corresponds to LHHLH in the figure), the EXCLUSIVE-OR operation $F = A \oplus B$ is realized by DP. Changing M from 1 to 0 while leaving the S variables unchanged causes DP to implement the subtraction operation $F = A - B - \overline{C}_n$. This ad hoc combination of truth tables, logic equations, and arithmetic equations provides a fairly clear and concise picture of the 74X181's operation. It is also typical of the mixed approaches that are commonly used to describe structure and behavior at the register level.

Register-level circuits are often described by an ad hoc mixture of tables, equations, and the like.

SELECTION				ACTIVE-HIGH DATA		
				M = H	M = L; ARITHMETIC OPERATIONS	
S3	S2	S1	S0	LOGIC FUNCTIONS	$\overline{C}_n$ = H (no carry)	$\overline{C}_n$ = L (with carry)
L	L	L	L	F = $\overline{A}$	F = A	F = A PLUS 1
L	L	L	H	F = $\overline{A + B}$	F = A + B	F = (A + B) PLUS 1
L	L	H	L	F = $\overline{A}$B	F = A + $\overline{B}$	F = (A + $\overline{B}$) PLUS 1
L	L	H	H	F = 0	F = MINUS 1 (2's COMPL)	F = ZERO
L	H	L	L	F = $\overline{AB}$	F = A PLUS A$\overline{B}$	F = A PLUS A$\overline{B}$ PLUS 1
L	H	L	H	F = $\overline{B}$	F = (A + B) PLUS A$\overline{B}$	F = (A + B) PLUS A$\overline{B}$ PLUS 1
L	H	H	L	F = A ⊕ B	F = A MINUS B MINUS 1	F = A MINUS B
L	H	H	H	F = A$\overline{B}$	F = A$\overline{B}$ MINUS 1	F = A$\overline{B}$
H	L	L	L	F = $\overline{A}$ + B	F = A PLUS AB	F = A PLUS AB PLUS 1
H	L	L	H	F = $\overline{A \oplus B}$	F = A PLUS B	F = A PLUS B PLUS 1
H	L	H	L	F = B	F = (A + $\overline{B}$) PLUS AB	F = (A + $\overline{B}$) PLUS AB PLUS 1
H	L	H	H	F = AB	F = AB MINUS 1	F = AB
H	H	L	L	F = 1	F = A PLUS A†	F = A PLUS A PLUS 1
H	H	L	H	F = A + $\overline{B}$	F = (A + B) PLUS A	F = (A + B) PLUS A PLUS 1
H	H	H	L	F = A + B	F = (A + $\overline{B}$) PLUS A	F = (A + $\overline{B}$) PLUS A PLUS 1
H	H	H	H	F = A	F = A MINUS 1	F = A

† Each bit is shifted to the next more significant position.

FIGURE 8.4 Functional description of the 74X181 ALU [Texas Instruments, 1988b].

Hardware Description Languages

HDLs provide precise specifications well suited to CAD use.

A disadvantage of logic diagrams like Figures 8.1, 8.2, and 8.3, and tabular descriptions like Figure 8.4, is that they cannot be easily manipulated by CAD programs. To address this deficiency and to provide precise, unambiguous models of logic circuits at the register and other levels, various artificial languages that resemble computer programming languages have been developed to describe circuits in a logical, technology-independent fashion. These languages, known as **hardware description languages** or **HDLs**, allow a complex piece of hardware to be described precisely (but not necessarily concisely) in a format resembling a computer program. Like a program, an HDL description can be manipulated easily by CAD tools such as behavior simulators and logic synthesizers. The latter, for example, can (semi-) automatically translate a high-level design like Figure 8.2 into a lower-level one like Figure 8.1, following a procedure analogous to the compilation of a program from a high-level programming language into a low-level (executable) machine language.

Numerous HDLs have been devised over the years—for example, the ISP[1] language (later renamed ISPS) developed at Carnegie-Mellon University in the 1960s [Siewiorek, Bell, and Newell, 1982], and the Verilog language,

[1]ISP stands for *instruction-set processor,* reflecting ISP's major use in describing various computer processors and their instruction sets.

a widely used HDL due to Cadence Design Systems, Inc. [Thomas and Moorby, 1991]. In the 1980s, the U.S. Department of Defense (DOD) sponsored a new language, called VHDL, based on the Ada programming language. VHDL, which stands for the VHSIC[2] Hardware Description Language, was designed as an all-encompassing language with a rigidly specified structure. Its definition is enshrined in the IEEE Standard 1076, which was published in 1987 [IEEE, 1987]. VHDL descriptions are now widely used for design documentation and to supplement or replace circuit diagrams (schematics) in the design process itself.

Example 8.1

Describing an arithmetic-logic circuit in VHDL

The VHDL language contains most of the constructs of Pascal, Ada, and other programming languages. However, many of the details, especially those intended for specifying hardware features, are unique to VHDL and make it one of the larger and more complex artificial languages of its kind. To illustrate its usage, Figure 8.5 presents a partial description of the 74X181 ALU in VHDL. A full description of the 74X181's behavior alone takes more than five pages of VHDL text [Coelho, 1989]. This length hints at VHDL's complexity and is a reason why we do not attempt to present VHDL in its complete form here.

VHDL descriptions tend to be very long.

A piece of digital hardware at any complexity level, from a primitive gate to an entire system, is called a **design entity**, and is specified by a composite VHDL statement of the form

entity *Name* **is** . . . **end;**

Figure 8.5 begins with an **entity** statement that identifies the hardware item of interest as *74X181_ALU*, a user-specified name, and lists the ALU's primary input/output signals and their types. The input/output signal list is introduced by the keyword **port** (a term often used to refer to a group of associated input/output lines in a digital system). The signal direction, **in** or **out**, is indicated in the **port** statement, as well as the signal type (in this case **bit**, meaning that all the named signals assume the binary values 0 and 1 only). Alternative signal types might include an unknown value X, or we might want to treat the signals as integers, in which case the keyword **bit** would be replaced in the **port** statement by the keyword **integer**.

The structure and behavior of the target circuit is described within a compound VHDL statement of the form

architecture *Name* **is begin** . . . **end;**

which takes up the remainder of Figure 8.5. This introductory example describes the 74X181's functional behavior only, but we could also specify the circuit's internal structure at essentially any desired level of abstraction at or

[2]VHSIC refers to *Very High-Speed Integrated Circuits*, the name of an earlier DOD-sponsored research program.

```
entity 74X181_ALU is
   port (A3, A2, A1, A0, B3, B2, B1, B0, S3, S2, S1, S0, M, C̄n: in bit;
         A=B, P, G, C̄n+4, F3, F2, F1, F0: out bit);
end 74X181_ALU;

architecture Behavior of 74X181_ALU is
begin
  -- Describe any internal structure of circuit here (omitted)
  -- Description of functional behavior follows
  process (A3, A2, A1, A0, B3, B2, B1, B0, S3, S2, S1, S0, M, Cn);
  type word is array (3 downto 0) of bit;
  variable A, B, F: word;
  variable S: integer:= 0;
  Bits_to_word(A3, A2, A1, A0, A);  -- Define A as a binary word
  Bits_to_word(B3, B2, B1, B0, B);  -- Define B as a binary word
  Bits_to_word(S3, S2, S1, S0, S);  -- Define S as a binary word
  if (M = '1') then   -- Select a logic operation
      case S is
      when 0 => F <= not A after 2 ns;
      when 1 => F <= not (A or B) after 2 ns;
        .
        .
        .
      when 15 => F <= A after 2 ns;
      end case; end if;
  if (M = '0') then   -- Select an arithmetic operation
      case S is
      when 0 => if (Cn='1') then F <= A + 1 after 2 ns;
          else F <= A after 2 ns; end if;
      when 1 => if (Cn='1') then F <= (A or B) + 1 after 2 ns;
          else F <= A and B after 2 ns; end if;
        .
        .
        .
      when 15 => if (Cn='1') then F <= A after 2 ns;
          else F <= A - 1 after 2 ns; end if;
      end case; end if;
      -- Compute other output signals A=B,P,G,C̄n+4
      end process;
end Behavior;
```

FIGURE 8.5 A portion of a VHDL description of the 74X181 ALU.

above the gate level. Timing information such as component delays and signal rise and fall times can also be included in an **architecture** specification.

A basic VHDL construct is a **signal assignment** such as

$$F <= \textbf{not } A; \tag{8.4}$$

VHDL's signal assignment operator <= involves no storage function (unlike :=).

which states that the signal F—in this case a 4-bit word or "bit-vector" in VHDL parlance—on the left of the VHDL signal assignment operator

<=, which represents a left arrow, is assigned the current value of the expression on the right-hand side of (8.4). This expression defines the 4-bit word $\overline{A}_3\overline{A}_2\overline{A}_1\overline{A}_0$ in the obvious fashion, using VHDL's logical complement operator **not**. To specify a propagation delay of, say, 2 ns in the operation defined by (8.4), the keyword **after** may be used, thus:

$$F <= \textbf{not } A \textbf{ after } 2 \text{ ns}; \tag{8.5}$$

VHDL's **case** and **if** statements describe conditional operations.

We can use the familiar **if-then-else** statement to add additional logical conditions to a signal assignment; for instance,

if M ='0' **then** F <= **not** A **after** 2 ns;
else F <= A **or** (**not** B) **after** 2.8 ns; **end if**;

Here M serves as a control signal, the two values 0 and 1 of which determine the logical operation to be used to calculate the value of the data signal F. When a large number of control conditions exist, a sequence of **if** statements can be replaced by a more concise **case** statement:

case C **is**
when C_1 => Z <= $f_1(X)$;
when C_2 => Z <= $f_2(X)$;
⋮
when C_n => Z <= $f_n(X)$;
end case;

This means that if the multibit control variable C has the current value C_i, then the multibit Z signal is assigned the corresponding value $F_i(X)$. Several examples of **case** statements in this VHDL style appear in Figure 8.5.

Other standard language features illustrated by Figure 8.5 are the use of predefined functions (*Bits_to_word* and *Bits_to_integer*), and the association of types like **bit** and **integer** with signals and other variables. The **process** structure that forms the bulk of the **architecture** statement encloses a long list of signal assignments that define the 74X181's input/output behavior as a combinational logic circuit with propagation delays. All the signal assignments within the **process** statement may be executed concurrently when using this VHDL description to simulate the 74X181's behavior. ■

HDLs may replace or supplement schematic diagrams in the design process.

Hardware design languages like Verilog and VHDL are an integral part of most CAD tools used in register-level design. They can replace circuit diagrams, flowcharts, and other graphic techniques as the primary documentation medium for circuit structure and behavior. As noted earlier, they are especially well suited to the design verification process because, with appropriate supporting software, an HDL description can be directly executed to simulate the circuit it describes. HDLs also provide a convenient way to describe the input/output data used by logic synthesis tools. This book uses a VHDL-derived language to describe the behavior of some components and circuits at the register level.

Register-Level Structure

Register-level circuits process and store words that denote numbers, instructions, and the like.

As Example 8.1 suggests, register-level design is oriented toward the processing of words as primitive signals or units of information, in contrast to the gate level, which treats individual bits as the primitive signals. Such words represent either numbers or nonnumerical data such as instructions or control words. The word length in bits varies, but is often a multiple of four, with 16 and 32 bits being representative figures. Four bits is a convenient unit for data representation because it efficiently accommodates a single decimal (BCD) digit; alphanumeric character codes (Section 2.10) are designed to use 8 or 16 bits. Medium-scale integrated (MSI) circuits such as the 74X181 are also often designed to process 4-bit words, because this data size requires moderate numbers of gates and primary input/output connections—63 and 24, respectively, in the 74X181 case.

Another key aspect of the register level is a "data-flow" representation that views a circuit as processing data words that flow between computational devices and storage devices. The main word-storage devices are the registers after which this level of abstraction is named. The computational devices are typically multifunction combinational circuits like the 74X181 ALU, which operates on 4-bit data words that are either numbers or logical words. These components, and ultimately the entire circuit, require control signals to select the functions to be performed at any time, as well as logic circuits to generate the control signals at the proper times and in the proper sequence. This leads to a dichotomy between data and control signals, which is a fundamental characteristic of the register level.

Data and Control

Many circuits consist of separate datapath and control units.

Most circuits may be divided, both structurally and behaviorally, into a data-processing part *DP* and a control part *CU* at the register level; see Figure 8.6. Typically, *DP*, which is often referred to as the circuit's **datapath**, is designed to perform a specified set of operations on n-bit data words that represent numbers, logical items, or other operand types. It also may provide temporary storage for the input/output data, as well as any intermediate results that it generates. The control part *CU*, called the **control unit** or **controller**,

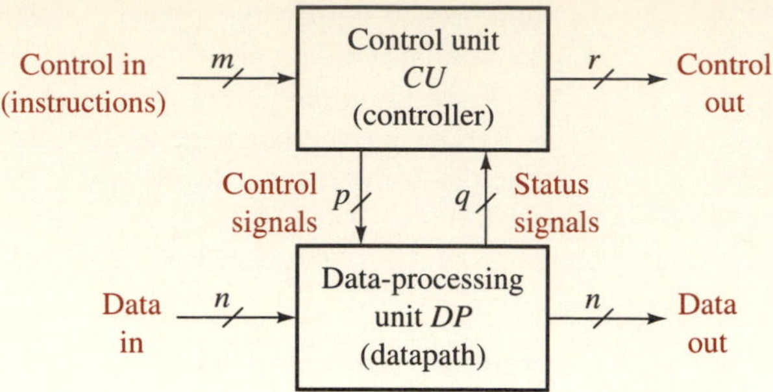

FIGURE 8.6 Data and control units of a digital system.

receives external signals (instructions) that specify the overall function to be performed. In response, *CU* generates a sequence of control signals that it applies to *DP* to implement the detailed steps of the current instruction. The datapath may, in turn, provide feedback to *CU* to indicate the status of the current operation ("proceeding normally," "abnormal condition has occurred," and so on).

Register-transfer statements are the basic HDL constructs at the register level.

Using an HDL, we can specify the behavior of a general register-level circuit in the form of a set of **register-transfer statements**. The HDL constucts are so called because they typically indicate how data words that are stored in input registers are transferred through and transformed by combinational processing circuits to create results that are stored in output registers. For example, Figure 8.7 depicts a small processor, in which the datapath's behavior (and, to a limited extent, its structure) corresponds to the following three register-transfer statements, which follow the VHDL style:

$$\begin{aligned} &\textbf{if } ADD = '1' \textbf{ then } C := A + B \\ &\quad \textbf{else if } SUB = '1' \textbf{ then } C := A - B; \\ &\textbf{if } \text{overflow\_occurs} \textbf{ then } OVF <= '1'; \end{aligned} \tag{8.6}$$

The three registers and their contents (the relevant data words) are represented by the variables A, B, and C. The symbol $:=$ (which is suggestive of the left arrow symbol $\leftarrow$ used in some HDLs) denotes the **register assignment operator**. The signals on the right-hand side of $:=$, which represent input data processed by the combinational ALU, are transferred to and stored in the output register C on the left of $:=$.

The register assignment operation denoted by $:=$ is normally carried out during a single clock cycle, which implies that the output data are formed and permanently stored in a register toward the end of the cycle. This is in contrast to the **signal assignment operation**, denoted by $<=$ in VHDL but most often by $=$, which involves no storage function. An output operand of $<=$, such as the overflow *OVF* signal in (8.6), retains the indicated value only for

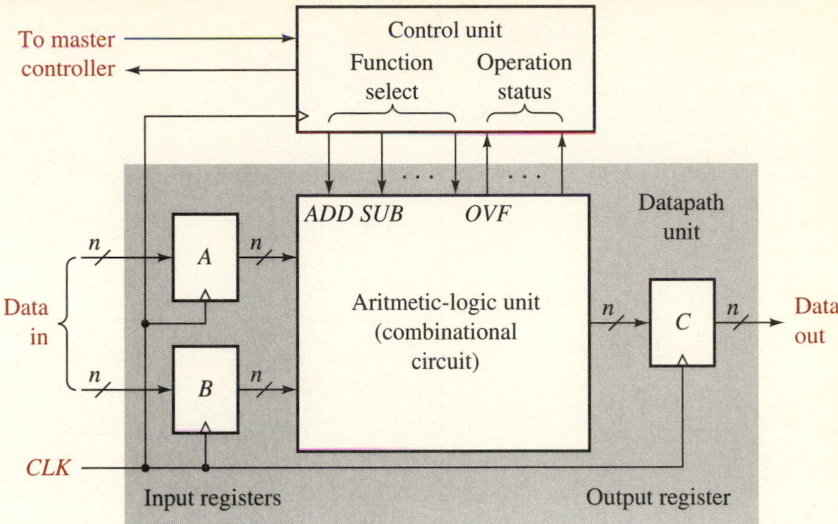

FIGURE 8.7 The structure of a small processor.

the duration of the current clock cycle. If, as is often the case, *OVF* is to be stored in a register (the *OVF* flip-flop), the signal assignment *OVF* <= '1' should be replaced in (8.6) by the register assignment *OVF* := '1'. Observe that *OVF* is a typical example of a status signal that is sent from a datapath unit to its control unit.

Flowcharts

A flowchart is a useful graphical tool for specifying behavior at the register level.

Another common means of representing register-level behavior is a diagram called a **flowchart**, which is a type of directed graph with operations indicated by boxes (nodes) and the sequence of the operations indicated by arrows. This is similar to the flowchart of a computer program, but it is interpreted somewhat differently, an issue to which we will return in Section 8.7. Figure 8.8 shows a flowchart equivalent to the HDL description of (8.6). Actions are specified by rectangular boxes, while decisions (corresponding to **if-then-else** statements in the HDL description) are specified by diamonds. A diamond has two exits, one of which is followed when the indicated condition is satisfied and the other when it is not. The entries in the boxes are brief statements, often written in the same shorthand notation used in the HDL.

A control unit *CU* tends to be more complex than the datapath it controls, primarily due to the fact that the control signals generated by *CU* are often single bits or short words that have little or no structure. *CU* can therefore be designed as a gate-level sequential circuit or a finite-state machine. In its simpler forms, *CU* is designed to generate a well-defined set of states

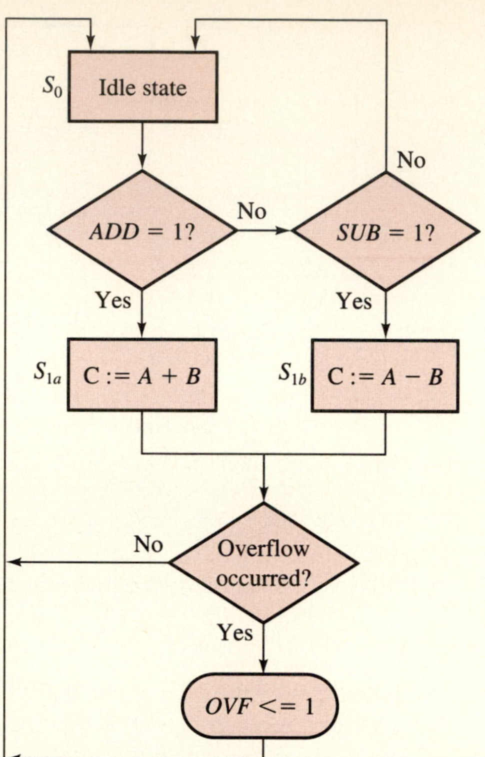

FIGURE 8.8 A flowchart of the small processor.

$S_1, S_2, \ldots, S_k$, which are such that S_i specifies a step in a sequence of actions to be performed by *DP*. In the design of Figure 8.7, for instance, S_i selects a particular operation to be performed by the multifunction ALU. The flowchart of Figure 8.8 defines three states: an idle state S_0 and two states S_{1a} and S_{1b}, representing the states in which addition and subtraction, respectively, are performed. The circuit is always in one of these states for the duration of each clock cycle. Observe that states are associated with rectangular boxes that have square corners. The rectangular box with rounded corners indicates a secondary, conditional action, in this case, activation of the overflow signal *OVF* in state S_{1a} or S_{1b}, if overflow has occurred.

Finally, we note that the data/control structure of a digital system is hierarchical. Within *DP* or *CU*, except in the simplest cases, we can find data and control units that are related to each other, as in Figure 8.6. This implies that the concepts of data and control signals are relative rather than absolute. A signal set viewed as a control word at one level in the design hierarchy may appear as one or several sets of data words at a different design level.

Register-Level Components

Register-level components correspond to MSI circuits.

The principal component types that characterize the register level are listed in Table 8.1. Most will be recognized as circuits with an internal design examined in earlier chapters; now they are of interest as primitive components in the design process. Their complexity varies from tens to hundreds of gates. When implemented as single ICs, these circuits fall into the traditional medium-scale integrated (MSI) circuit range. They are also implemented as standard cells within more complex ICs. Register-level components are designed to process data words of various sizes. All bits of a data word are processed simultaneously in essentially the same way. Often, the components contain some control logic, which, with any external control signals, acts in unison on the data words being processed.

We divide register-level component types into two groups: combinational and sequential. The combinational components implement logical and numerical operations, as well as route data through a system. The sequential components perform word storage and simple sequential operations such as shifting and counting. In the next two sections, we examine in detail the behavior and use of the components listed in Table 8.1.

TABLE 8.1 Basic register-level component types: (**a**) combinational; (**b**) sequential.

Component type	*Functions*
Word gate	Logical (Boolean) operations: AND, OR, NOT, NAND, EXCLUSIVE-OR, etc.
Multiplexer	Data selection and routing; Boolean function generation
Decoder/encoder	Data format conversion
Adder	Binary addition and subtraction
Comparator	Data comparison
ALU	Arithmetic and logical operations

(a)

Component type	*Functions*
(Parallel) register	Data storage
Shift register	Data storage and format conversion
Counter	Control state generation; arithmetic operations; timing

(b)

8.2 Combinational Components

We turn now to the combinational components recognized as primitive design elements at the register level. First, we consider the way in which these and all register-level components are connected.

Buses

A bus is a primitive word-transmitting connector.

A **bus** is a set of $n \geq 1$ logic lines used to interconnect two or more components. The parameter n is called the **size** or **width** of the bus. The term *bus* is often reserved for a major signal highway that is shared by many different components of a circuit. In the latter case, the bus is often itself considered a component, with its own internal data and control structure. The simplest bus consists of n lines that transmit n bits of information simultaneously, or in parallel, between a source and a destination. The usual graphic symbolism for such a bus is shown in Figure 8.9a; the shorthand version, which consists of a single line marked by slash-n, is preferred in register-level circuits. This line carries a single register-level signal—namely, an n-bit word. Note that the names $X = x_1{:}x_n = x_1x_2\ldots x_n$ are assigned both to the bus lines themselves and to the signals they carry. Tri-state logic provides an efficient way to implement bidirectional buses, as discussed in Section 5.8.

Figure 8.9b shows how bus lines may be regrouped to form new buses or subbuses. Thus, the n incoming lines $x_1x_2\ldots x_n$ are grouped to form a single n-bit bus X. They are subsequently ungrouped into three buses of one, two, and $n-3$ lines each. To avoid ambiguity, the lines of the new buses

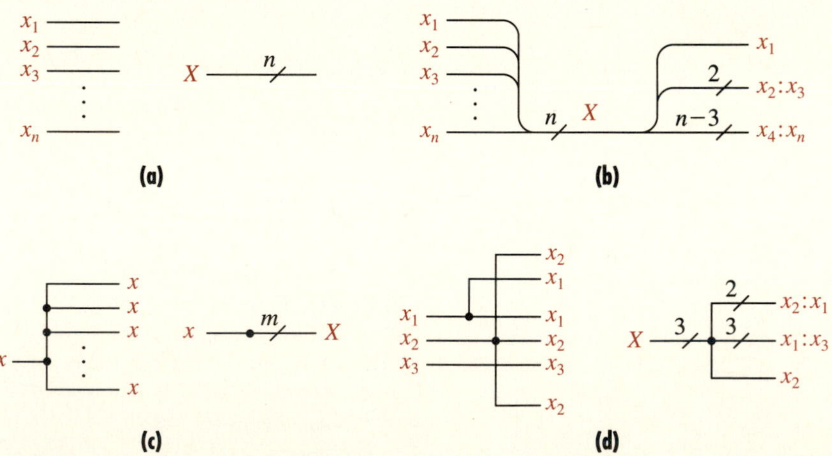

FIGURE 8.9 Bus symbolism: (a) a basic n-line bus; (b) grouping and ungrouping of lines; (c) simple fan-out; (d) complex fan-out.

and their order should be clearly indicated by appropriate signal names or labels. A bus notation for a single line that fans out to n destinations appears in Figure 8.9c. A multiline bus can fan out in complex ways, as suggested by the example in Figure 8.9d; again, careful labeling is needed to avoid ambiguity.

Word Gates

An n-bit word gate is a group of n copies of a logic gate.

Just as a bus groups n independent logic lines into a single connecting line, a **word gate** groups n independent gates into a single logic element. The m inputs and the output of the basic gate type all become n-line buses, so that the word gate can be seen as performing the gate operation bitwise on bus signals. We represent this in a natural way by the usual gate symbol, with all the input/output lines replaced by n-line buses. For example, Figure 8.10a defines a three-input, n-bit NOR word gate; Figure 8.10b shows its graphic symbol.

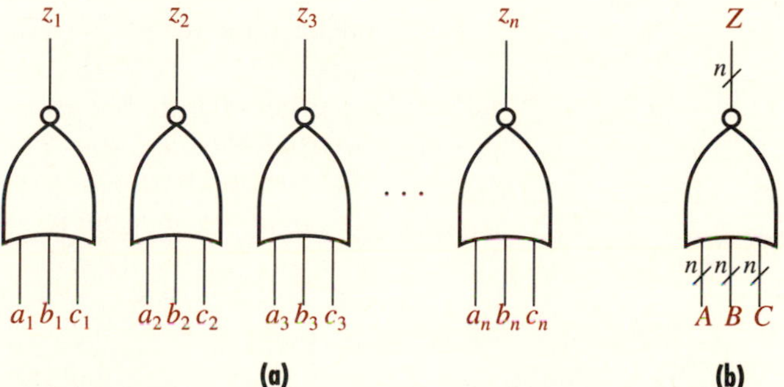

FIGURE 8.10 A three-input, n-bit NOR word gate: (a) its gate-level circuit; (b) its register-level symbol.

Example 8.2

Describing an n-bit 2C adder-subtracter at the register level

The use of word gates and multibit buses to provide a very concise description at the register level is illustrated by Figure 8.11. This figure shows an adder-subtracter designed to add and subtract two n-bit two's-complement numbers; compare the gate-level design for $n = 4$ that appears in Figure 5.57 (Section 5.9). Gates G_1 and G_2 are n-bit word buffers, each of which is composed of n ordinary buffers, one per input line. (These buffers reduce the fan-out requirements of the adder-subtracter's input signals to one, thus minimizing the load imposed on those signals.) G_4 is a 1-bit word buffer—that is, an ordinary buffer.

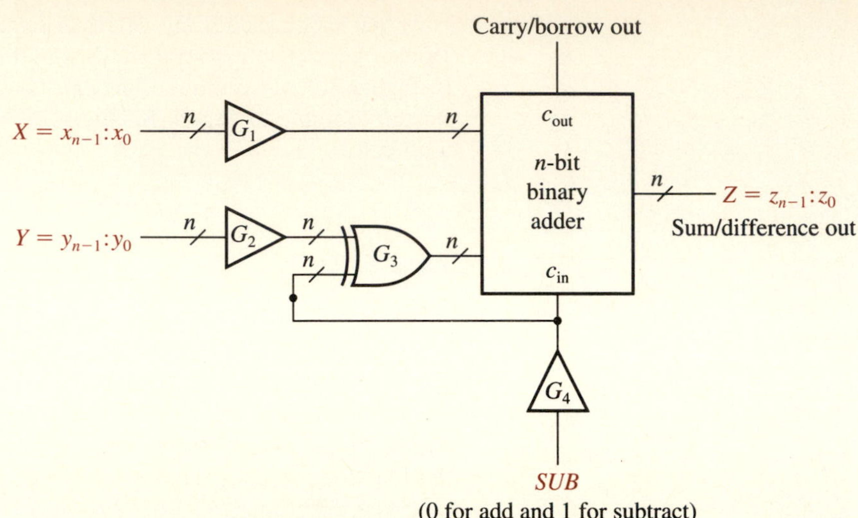

FIGURE 8.11 A register-level logic diagram for an n-bit two's-complement adder-subtracter.

The input control line *SUB* determines which of the two operations, addition or subtraction, is to be executed. G_3 is a two-input, n-bit EXCLUSIVE-OR word gate composed of n two-input EXCLUSIVE-OR gates. One of its inputs is the buffered n-bit Y signal. The other is obtained by fanning out the buffered *SUB* signal to every EXCLUSIVE-OR gate. Consequently, the output function of G_3 can be expressed as $Y \oplus SUB$, where $Y \oplus 0 = Y = y_{n-1}y_{n-2} \ldots y_0$ and $Y \oplus 1 = \overline{Y} = \overline{y}_{n-1}\overline{y}_{n-2} \ldots \overline{y}_0$. The overall arithmetic behavior of the adder-subtracter is described succinctly by the VHDL statement:

if SUB = '0' **then** $c_{out}, Z <= X + Y$ **else** $c_{out}, Z <= X - Y$;

Multiplexers

A multiplexer routes a data word from a selected source to a common destination.

As we saw in Section 5.8, the chief function of a multiplexer M (also called a *data selector*) is to control the transfer of data words from one of several sources to a common destination. A multiplexer normally has $n = 2^k$ input data buses connecting it to its sources $S_0, S_1, \ldots, S_{n-1}$, and a single output bus connecting it to the destination D. A k-bit input address or "select" bus A determines the source to connect through M to D; when A carries the binary number i, S_i is the selected source. The width m of the data buses is an independent parameter. A multiplexer with n input data buses, each of width m, is called an n-way, m-bit multiplexer.

Figure 8.12 shows two symbols used for a four-way, 2-bit multiplexer; this is a slightly simplified version of the 74X153 IC [Texas Instruments, 1988b]. Note that $S_i = S_{i,0}S_{i,1}$, $D = D_0D_1$, and $A = A_1A_0$. The multiplexer's function is described concisely by the following VHDL **case** statement:

$$
\begin{array}{l}
\textbf{case } A \textbf{ is} \\
\textbf{when } 0 => D <= S_0; \\
\textbf{when } 1 => D <= S_1; \\
\textbf{when } 2 => D <= S_2; \\
\textbf{when } 3 => D <= S_3; \\
\textbf{end case};
\end{array} \tag{8.7}
$$

where A is interpreted as an integer. Note that a gate-level truth table for this multiplexer contains $2^{10} = 1{,}024$ rows, and is much less expressive than the HDL description (8.7).

Multiplexers are available as standard cells or ICs in sizes ranging from $n = 2$ to $n = 16$ and $m = 1$, 2, or 4. As we will see later (Section 8.4), multiplexers are easily expanded to handle more, or wider, data buses.

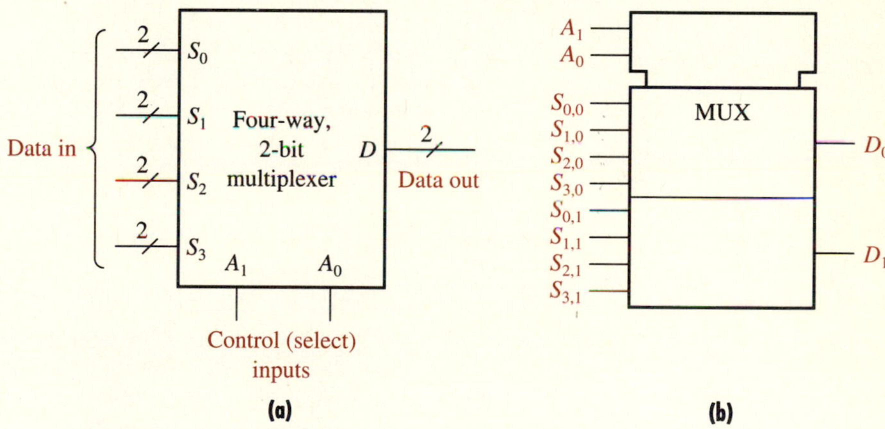

FIGURE 8.12 (a) The box symbol and (b) the IEC 617 (IEEE 91) symbol for a four-way, 2-bit multiplexer.

Multiplexers as Function Generators

A multiplexer is sometimes used as a general-purpose combinational component.

A multiplexer has a secondary use as a general-purpose component to implement combinational logic functions. The basic idea, illustrated by Figure 8.13b, is to apply the input variables $X = x_1, x_2, \ldots, x_k$ of a desired function $z(x_1, x_2, \ldots, x_k)$ to the control inputs $A_0, A_1, \ldots, A_{k-1}$ of an n-way

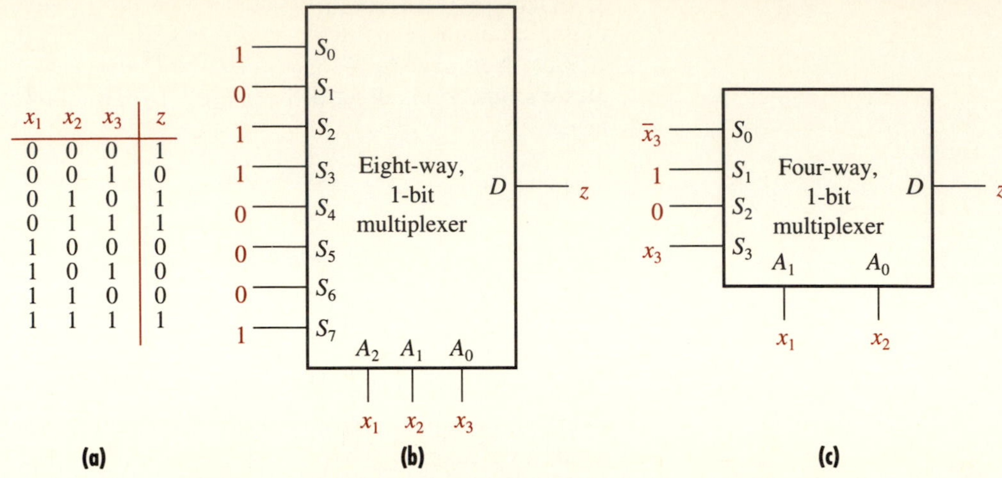

FIGURE 8.13 Use of multiplexers to implement the three-variable function z: (a) the truth table for z; (b) its first implementation; (c) its second implementation.

multiplexer with $n = 2^k$. The constants 0 and 1 are then applied to the data (source) inputs $S_0, S_1, \ldots, S_{2^n-1}$ in such a way that $S_i = 1$ if and only if X is the binary representation of i, for $0 \leq i \leq 2^n - 1$ and $z(x) = 1$. Hence, the 1s on the data lines represent the minterms of $z(X)$; the 0s on the data lines represent $z(X)$'s maxterms. The eight-way, 1-bit multiplexer of Figure 8.13b is easily seen to generate the three-variable function

$$z = \overline{x}_1\overline{x}_3 + x_2x_3$$

which has the truth table of Figure 8.13a, and the four minterms $m_0 = \overline{x}_1\overline{x}_2\overline{x}_3$, $m_2 = \overline{x}_1x_2\overline{x}_3$, $m_3 = \overline{x}_1x_2x_3$, and $m_7 = x_1x_2x_3$.

If one input variable x_k is available in both true and complemented form, or if an inverter is used for that variable, we can realize a k-variable function that uses an $(n/2)$-way multiplexer, where, as before, $n = 2^k$. In this case, we apply x_k or $\overline{x}_k$ to the multiplexer's data inputs in addition to 0 and 1; the other $k - 1$ input variables $X^* = x_1, x_2, \ldots, x_{k-1}$ are applied to the select lines as before. The technique is illustrated by Figure 8.13c, in which $x_k = x_3$, and a four-way multiplexer suffices to generate the same three-variable function z produced by the eight-way multiplexer of Figure 8.13b.

To see how this works, observe that for the output signal $z(X) = z(X^*, x_k)$, both $z(X^*, 0)$ and $z(X^*, 1)$ must assume the value applied to the same input data line S_i, where i is the number specified by X^*. To realize an arbitrary k-variable function, we must be able to obtain any of the four possible combinations of 0s and 1s that happen to be assigned by z to the output value-pair $z(X^*, 0), z(X^*, 1)$, for any particular X^*. We can do this by placing one of the four signals 0, 1, x_k, or $\overline{x}_k$ on data input S_i, according to the following rules:

Rules for assigning signals to data input S_i of a multiplexer to generate $Z(X^*, x_k)$.

$z(X^*, 0)$	$z(X^*, 1)$	S_i
0	0	0
0	1	x_k
1	0	$\overline{x}_k$
1	1	1

(8.8)

Consider, for example, the topmost input S_0 to the four-way multiplexer of Figure 8.13c. This is selected by $X^* = x_1x_2 = 00$. The definition of z requires $z(0, 0, 0) = 1$ and $z(0, 0, 1) = 0$; hence, by the foregoing rules with $x_k = x_3$, we must apply $\overline{x}_3$ to S_0, as has been done.

Example 8.3

Implementing a full adder by means of a multiplexer

Another example of how a multiplexer can be used to realize a combinational function appears in Figure 8.14, where the four-way, 2-bit multiplexer of Figure 8.12 implements the full adder equations

$$\begin{aligned} s_i &= x_i \oplus y_i \oplus c_{i-1} \\ c_i &= x_iy_i + x_ic_{i-1} + y_ic_{i-1} \end{aligned} \tag{8.9}$$

and so forms a full adder. One line of each 2-bit data bus is used for the sum function s_i; the other line of each bus is used for the output carry function c_i.

x_i	y_i	c_{i-1}	s_i		c_i	
0	0	0	0	c_{i-1}	0	0
0	0	1	1		0	
0	1	0	1	$\overline{c}_{i-1}$	0	c_{i-1}
0	1	1	0		1	
1	0	0	1	$\overline{c}_{i-1}$	0	c_{i-1}
1	0	1	0		1	
1	1	0	0	c_{i-1}	1	1
1	1	1	1		1	

(a)

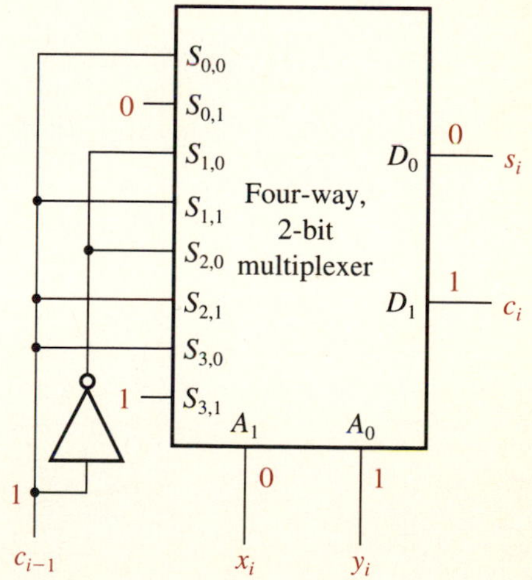

$S_{0,0}$ 0 $S_{0,1}$ $S_{1,0}$ $S_{1,1}$ $S_{2,0}$ $S_{2,1}$ $S_{3,0}$ 1 $S_{3,1}$ Four-way, 2-bit multiplexer D_0 0 s_i D_1 1 c_i A_1 A_0 0 1 1 c_{i-1} x_i y_i

(b)

FIGURE 8.14 A multiplexer implementation of a full adder.

The input signals to be applied to the S buses are determined in Figure 8.14a, according to rules (8.8) above. Figure 8.14b shows the response to the input combination $x_i y_i c_{i-1} = 011$. The control signals $x_i y_i = 01_2 = 1_{10}$ connect input bus $S_1 = S_{1,0}S_{1,1}$ to the output bus $D = D_0 D_1$, resulting in $s_i = \bar{c}_{i-1} = 0$ and $c_i = c_{i-1} = 1$, as required by (8.9).

Decoders

A decoder activates a unique output line for every input pattern.

Decoders, which also serve as demultiplexers, were introduced in Section 5.8. A 1-out-of-n, or k-line-to-n-line, decoder has k input lines $X = x_{k-1}, x_{k-2}, \ldots, x_0$ and $n = 2^k$ output lines $Z = z_0, z_1, \ldots, z_{2^k-1}$. Its behavior is defined by the property that when X denotes the binary number i, the corresponding output line z_i with index (subscript) i is 1 and all other output lines are 0. There is also usually a control line EN, called the *enable line*, which must be active ($EN = 1$) for the decoder to operate; when $EN = 0$, all output lines are 0. Figure 8.15a gives a symbol for a 1-out-of-16 decoder and shows how the input combination $X = 1010_2 = 10_{10}$ makes

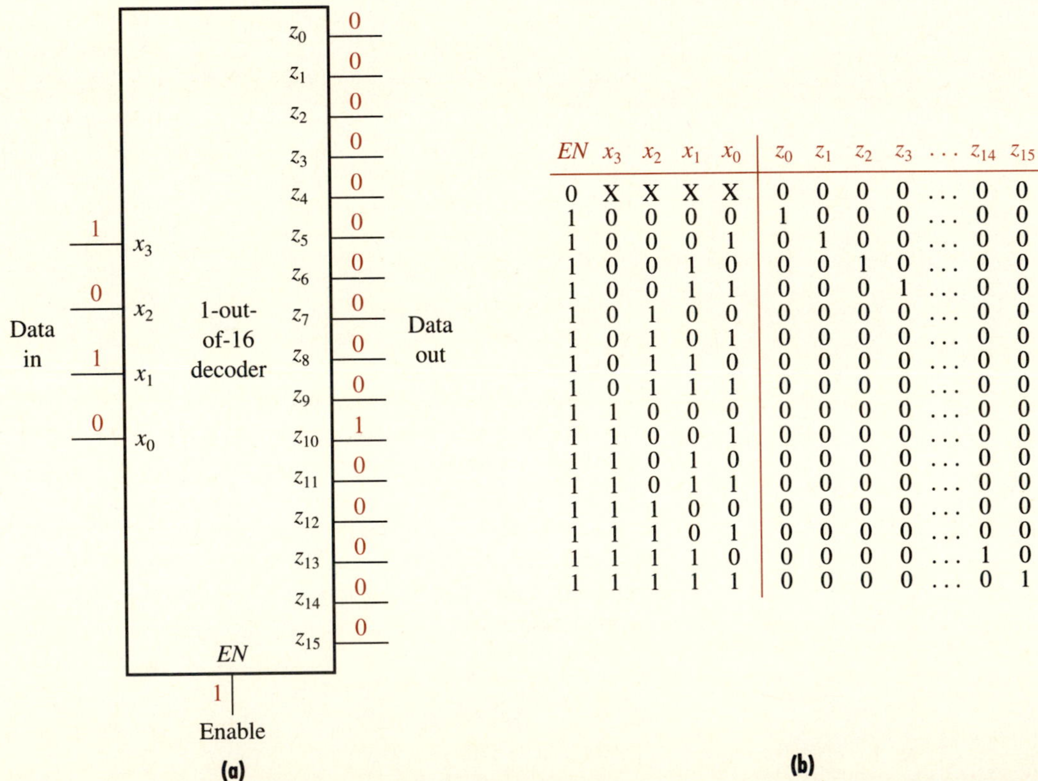

EN	x_3	x_2	x_1	x_0	z_0	z_1	z_2	z_3	...	z_{14}	z_{15}
0	X	X	X	X	0	0	0	0	...	0	0
1	0	0	0	0	1	0	0	0	...	0	0
1	0	0	0	1	0	1	0	0	...	0	0
1	0	0	1	0	0	0	1	0	...	0	0
1	0	0	1	1	0	0	0	1	...	0	0
1	0	1	0	0	0	0	0	0	...	0	0
1	0	1	0	1	0	0	0	0	...	0	0
1	0	1	1	0	0	0	0	0	...	0	0
1	0	1	1	1	0	0	0	0	...	0	0
1	1	0	0	0	0	0	0	0	...	0	0
1	1	0	0	1	0	0	0	0	...	0	0
1	1	0	1	0	0	0	0	0	...	0	0
1	1	0	1	1	0	0	0	0	...	0	0
1	1	1	0	0	0	0	0	0	...	0	0
1	1	1	0	1	0	0	0	0	...	0	0
1	1	1	1	0	0	0	0	0	...	1	0
1	1	1	1	1	0	0	0	0	...	0	1

FIGURE 8.15 A 1-out-of-16 or 4-to-16 decoder: **(a)** its graphic symbol; **(b)** its truth table.

$z_{10} = 1$. The rather simple internal structure of a decoder of this type, the 74X154, can be found in Figure 5.54.

Our symbol for the decoder appears in Figure 8.15a; the IEC 617 (IEEE 91) symbol is essentially the same. The decoder's function is described by the truth table of Figure 8.15b. It is also described somewhat more concisely as follows:

if $EN = 0$ **then** $z_0z_1 \ldots z_{2^k-1} <= 00\ldots0;$
else case X **is**
  when $0 => z_0z_1 \ldots z_{2^k-1} <= 10\ldots0;$
  when $1 => z_0z_1 \ldots z_{2^k-1} <= 01\ldots0;$
  $\vdots$
  when $2^k - 1 => z_0z_1 \ldots z_{2^k-1} <= 00\ldots1;$
  end case

Encoders

An n-input **binary encoder** is a logic circuit that, given an n-bit input word X that contains one active signal x_i, generates an output word Z, which is a binary representation of i, the index of the active input signal. Thus, an encoder is the inverse of a decoder, and typically has $n = 2^k$ input lines for X and k output lines for Z. A four-input encoder, for instance, has $n = 4$ and $k = 2$, and maps the input combinations 1000, 0100, 0010, and 0001 onto the output combinations 00, 01, 10, and 11, respectively. To identify the situation in which no input signal is active—that is, X is all 0s—an encoder has an output control signal *GS*, called the "group signal." There is also an "enable in" control signal *EI* which, with *GS*, facilitates expansion to accommodate more inputs and outputs.

An encoder outputs a pattern that indicates the active input line of highest priority.

An encoder *ENC* is intended to identify a single active input signal. However, there is nothing to prevent several of its X input lines from being active at the same time, because they may be driven from independent external sources. To deal with this situation, most encoders are designed as **priority encoders,** which have the property that when several inputs are active at the same time, the output number i that appears on Z is the index of the active input line x_i with the highest "priority." Priority is based on a simple ordering of the inputs: x_i has higher priority than x_j if $i > j$. Hence, for an n-input encoder, x_{n-1} has the highest priority, and x_0 the lowest. For the $n = 4$ case above, with $X = x_0x_1x_2x_3$ and $EI = 1$, we have the following behavior, where X denotes a don't-care value: all eight input combinations of the form X = XXX1 produce Z = 11; the four combinations XX10 produce Z = 10; X100—that is, 0100 and 1100—produce Z = 01; only 1000 produces Z = 00.

Figure 8.16 defines a 1-out-of-8 or 8-line–to–3-line priority encoder, which is similar to the 74X148. Input x_7 has the highest priority, so that

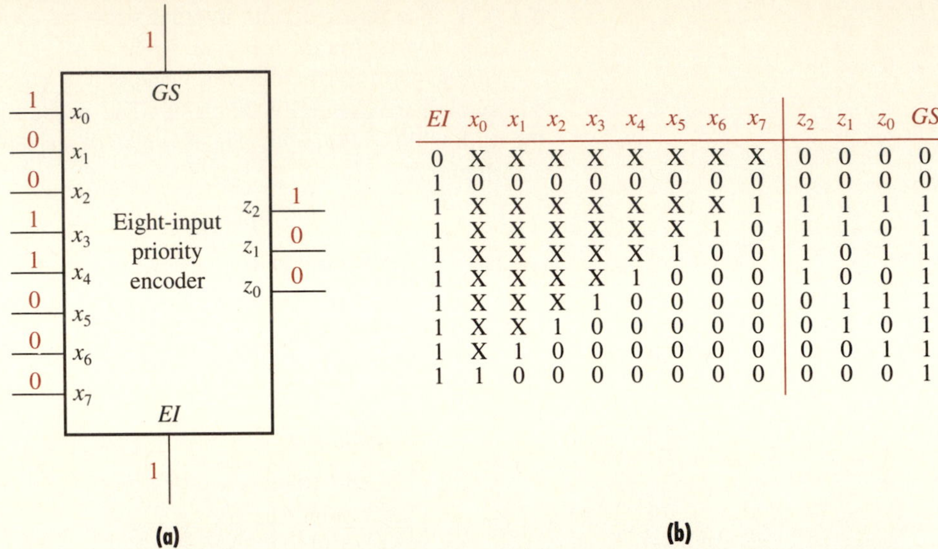

EI	x_0	x_1	x_2	x_3	x_4	x_5	x_6	x_7	z_2	z_1	z_0	GS
0	X	X	X	X	X	X	X	X	0	0	0	0
1	0	0	0	0	0	0	0	0	0	0	0	0
1	X	X	X	X	X	X	X	1	1	1	1	1
1	X	X	X	X	X	X	1	0	1	1	0	1
1	X	X	X	X	X	1	0	0	1	0	1	1
1	X	X	X	X	1	0	0	0	1	0	0	1
1	X	X	X	1	0	0	0	0	0	1	1	1
1	X	X	1	0	0	0	0	0	0	1	0	1
1	X	1	0	0	0	0	0	0	0	0	1	1
1	1	0	0	0	0	0	0	0	0	0	0	1

(a) (b)

FIGURE 8.16 An eight-input priority encoder: (a) its circuit symbol; (b) its truth table.

with $EI = 1$, any input of the form X = XXXXXXX1 produces the output $Z = 111$, X = XXXXXX10 produces $Z = 110$, and so on. Thus, when $x_4 = 1$ and $x_i = 0$ for all $i \geq 5$, as shown in Figure 8.16b, the encoder produces $Z = 100_2 = 4_{10}$. The encoder of Figure 8.16 also has an output control signal GS to indicate the presence of an active X signal. When EI = 0 or X = 00000000, $GS = 0$; otherwise, $GS = 1$.

The behavior of a general 2^k-input priority encoder can be defined by a truth table, as in Figure 8.16b; it can also be defined as follows:

if $EI = 0$ **then** $z_{k-1}z_{k-2}\ldots z_1z_0GS$ <= 00...000;
  else case $X = x_0x_1x_2\ldots x_{2^k-3}x_{2^k-2}x_{2^k-1}$ **is**
  when 000...000 => $z_{k-1}z_{k-2}\ldots z_2z_1z_0GS$ <= 00...0000;
  when 100...000 => $z_{k-1}z_{k-2}\ldots z_2z_1z_0GS$ <= 00...0001;
  when X10...000 => $z_{k-1}z_{k-2}\ldots z_2z_1z_0GS$ <= 00...0011;
  ...
  when XX...X10 => $z_{k-1}z_{k-2}\ldots z_2z_1z_0GS$ <= 11...1101;
  when XX...XX1 => $z_{k-1}z_{k-2}\ldots z_2z_1z_0GS$ <= 11...1111;
end case

Arithmetic Components

Adders, multipliers, comparators, and ALUs are examples of combinational arithmetic components.

Because of the importance of numerical computations, it should come as no surprise that logic circuits to perform various arithmetic operations are standard components of design libraries (Figure 8.17). Binary adders (Figure 8.17a) have been encountered several times already. They are designed to add two unsigned n-bit binary numbers X and Y, along with an input carry

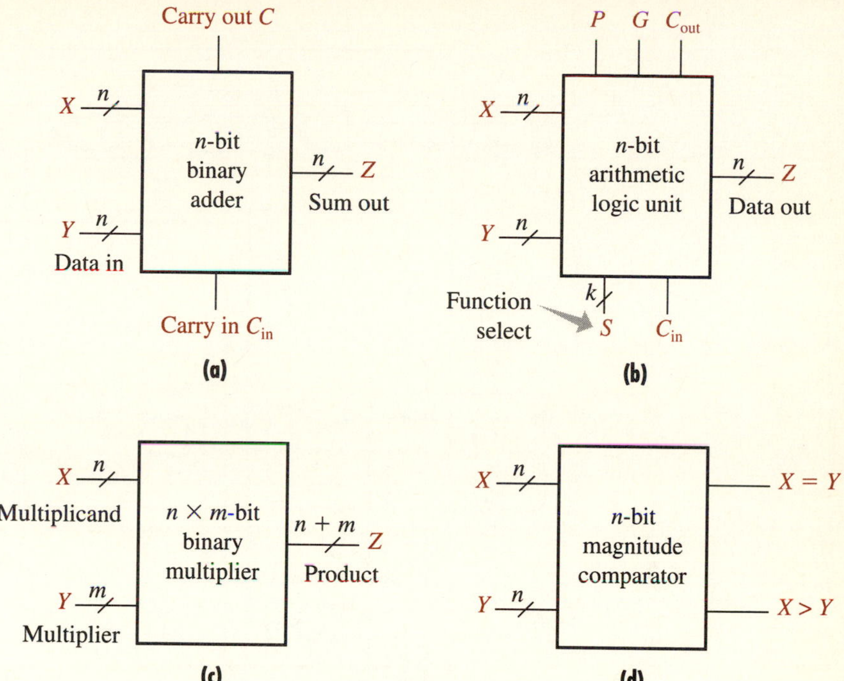

FIGURE 8.17 Some arithmetic components: (a) an adder; (b) an ALU; (c) a multiplier; (d) a magnitude comparator.

bit C_{in}, to produce an $n + 1$ bit result C_{out}, Z representing the arithmetic sum $X + Y + C_{in}$. They can also handle signed numbers directly if the two's-complement code is used. The simple scheme depicted in Figure 8.11 suffices to perform both addition and subtraction of 2C numbers. With the addition of circuits to perform logical (word gate) operations, an adder-subtracter is easily extended to a multifunction combinational arithmetic-logic unit (Figure 8.17b). The 74X181 ALU examined in Section 8.1 is a typical example, performing 2C addition and subtraction and a selection of logical operations on 4-bit data words. Another arithmetic function that can be implemented by a combinational circuit, albeit a complex one, is an n-by-m multiplier (Figure 8.17c).

The last component shown in Figure 8.17 is a comparator designed to compare the numerical magnitudes of two unsigned binary numbers. This version has two outputs denoted $X{=}Y$ and $X{>}Y$, which we will also call z_1 and z_2, respectively. Now $z_1 = 1$ indicates that the input data words X and Y are identical, and $z_2 = 1$ indicates that X is bigger than Y. It can be inferred from $z_1z_2 = 00$ that X is smaller than Y; $z_1z_2 = 11$ is a don't-care condition. Figure 8.18 shows the logic circuit for a 4-bit comparator in the 74X series, the 74X85. Observe that this magnitude comparator has three output signals, any one of which can be inferred from the other two. Comparators are used to provide status inputs to control circuits concerning the result of arithmetic operations.

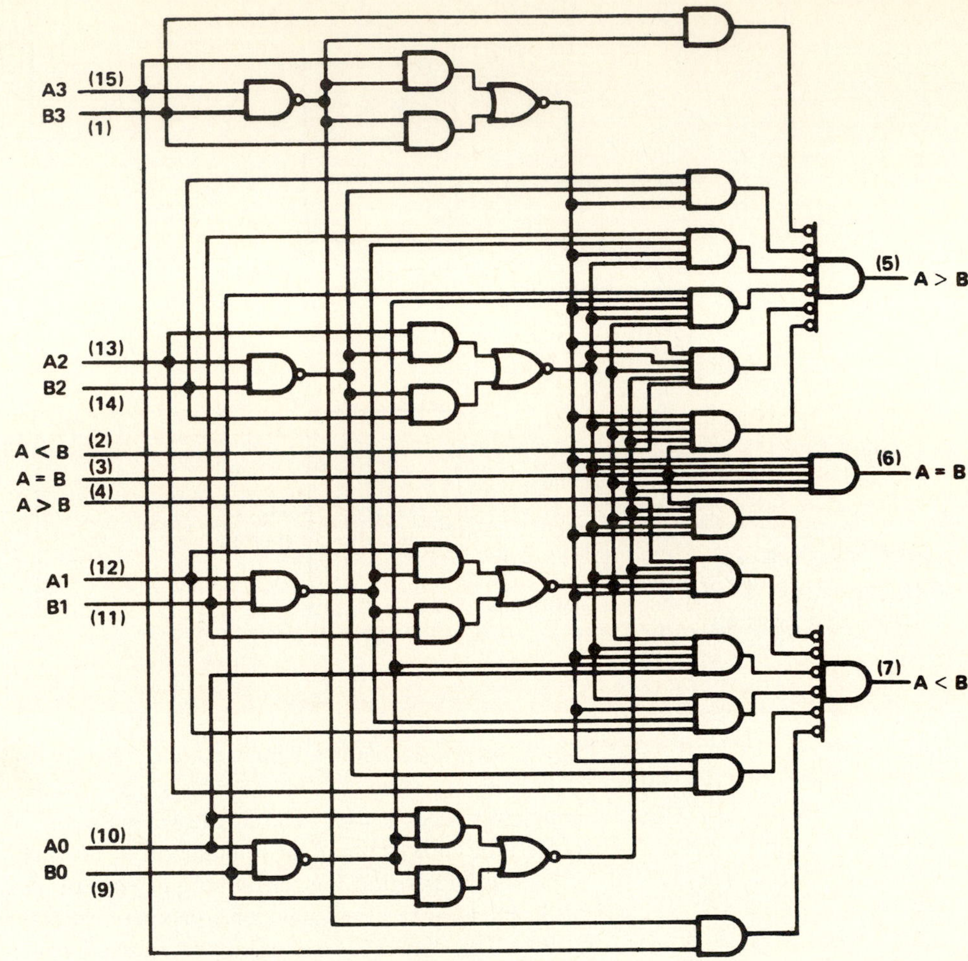

FIGURE 8.18 A gate-level logic circuit for the 74X85 4-bit comparator [Texas Instruments, 1988b].

74X-Series Components

Table 8.2 lists a few representative register-level combinational components from the 74X series [Texas Instruments, 1988b], several of which we have seen already. These circuits, whose part numbers and gate-level logic circuits have been standardized, are widely available, both as medium-scale integrated (MSI) circuits implemented in various IC technologies and as standard cells in CAD libraries. The many SSI ICs that contain sets of logic gates of a particular type and size (Table 4.3 in Section 4.5) can be seen as implementing word gates.

TABLE 8.2 A sampling of register-level combinational components in the 74X series.

IC part number	*Logic functions provided by the IC*	*Logic circuit*
74X83	Four-bit binary adder with carry-lookahead	Figure 5.56
74X85	Four-bit magnitude comparator	Figure 8.18
74X145	BCD-to-decimal decoder/driver	
74X148	Eight-input (eight-line to three-line) priority encoder	
74X150	Sixteen-way, 1-bit multiplexer	Figure 5.47
74X153	Four-way, 2-bit (dual four-line to one-line) multiplexer	Figure 8.68
74X154	Four-line to 16-line decoder/demultiplexer	Figure 5.54
74X156	Dual two-line to four-line decoder/demultiplexer	
74X181	Four-bit arithmetic logic unit/function generator	Figure 8.1
74X182	Four-bit carry-lookahead generator	Figure 8.35
74X284	Four-bit by 4-bit parallel binary multiplier	
74X682	Eight-bit magnitude comparator	

Programmable Logic Devices

The various programmable components (ROMs, PLAs, and PALs) introduced in Section 4.7 can serve as general-purpose logic function generators at the register level. They are especially suited to the implementation of nonstandard functions such as decoders or encoders for unusual alphanumeric codes. They are also useful for complex arithmetic operations such as multiplication and division. PLDs are limited mainly by the number of input/output lines they can accommodate, as well as the available number of crosspoints.

PLDs such as FPGAs form an important class of flexible register-level components.

Field-programmable gate arrays (FPGAs) form another, increasingly important class of programmable components. Several widely used FPGAs are built around flexible combinational or sequential modules or cells that can be customized to form standard or nonstandard components of moderate complexity. An example, illustrated in Figure 8.19, is the combinational cell that serves as a building block for logic circuits in a commercial FPGA series, the Actel ACT-1 [Actel, 1991]. This cell has eight independent primary inputs and one primary output, and contains three two-way, 1-bit multiplexers and a two-input OR gate. The cell approximates a single four-way multiplexer configured as a general-purpose logic function generator. It is customized to perform a particular logic function by programming the signals applied to its input lines. An ACT-series FPGA IC contains hundreds of such cells, with all cell functions and intercell connections individually programmable.

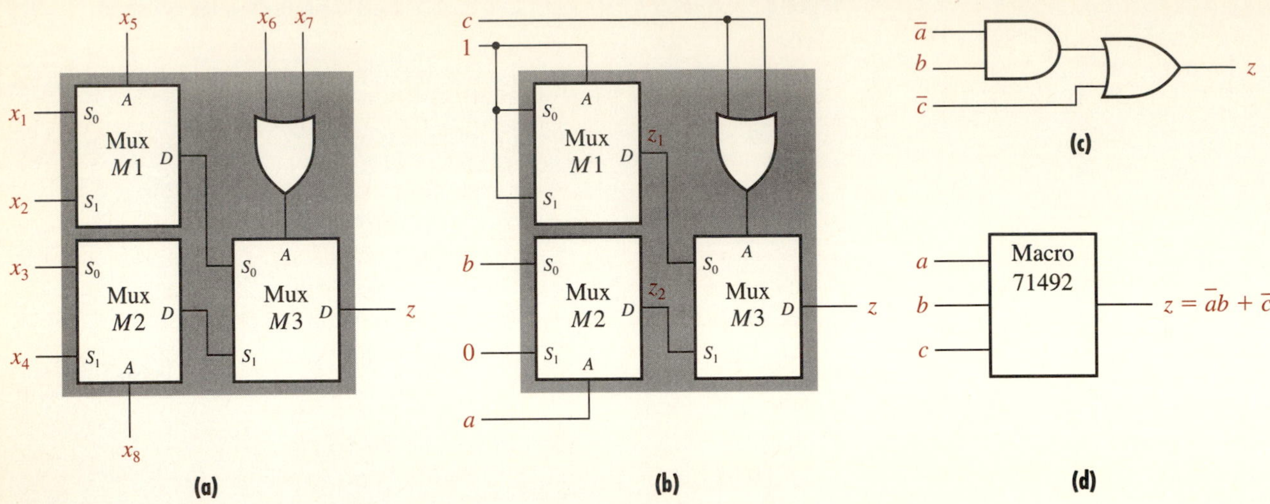

FIGURE 8.19 (a) The ACT-1 FPGA cell; (b) programmed implementation of the function $z = \bar{a}b + \bar{c}$; (c) a gate-level representation; (d) a register-level representation.

Example 8.4

Realizing a three-variable function using an FPGA cell

The ACT-1 FPGA cell can be made to realize any Boolean functions of up to three variables, and many useful functions of up to eight variables. Figure 8.19b shows its use to realize the three-variable function $z = \bar{a}b + \bar{c}$. The first multiplexer *M1* generates $z_1 = 1$; *M2* produces $z_2 = \bar{a}b + a0 = \bar{a}b$. The primary output z is computed by *M3* as

$$z = \bar{c}z_1 + cz_2 = \bar{c}1 + c\bar{a}b = \bar{c} + \bar{a}b$$

which is the desired function.

Programmed cells like that of Figure 8.19 are typically treated as standard components, often referred to as **macros,** and are placed in a CAD library for the designer's use. Often the designer only needs to see such macros in their gate-level or primitive register-level representations during the design process.

8.3 Sequential Components

We consider next the main sequential logic elements of the register level, which are, not surprisingly, registers. A few other types of sequential components, notably counters, are also available.

Parallel Registers

A register is a one-word memory formed from a set of flip-flops with shared control lines.

A register, examples of which were introduced in Section 6.7, is a storage element for n-bit words. It consists of n flip-flops or, less frequently, latches with a common set of control lines, including a clock line and a reset line. The register's key design characteristics, such as its triggering mode, setup and hold times, and the like, are inherited from its constituent flip-flops. Figure 8.20 shows a basic type of register, the 74X273, which consists of eight edge-triggered D-type flip-flops. All the input/output connections of each flip-flop in this register are directly accessible primary input/output lines; hence, the register is considered to have parallel input/output.

It is useful to equip a parallel register R with an input control line LD, denoting load, that can be activated ($LD = 1$) to transfer an external data word X into R. When this line is inactive ($LD = 0$), the contents of R remain unchanged. Figure 8.21a shows how to do this using an 8-bit JK register. The 8-bit input word $X = x_0 : x_7$ is fed to the register's J input; $\overline{X}$ is fed to the K input. Two 8-bit AND word gates are inserted into the J and K lines so

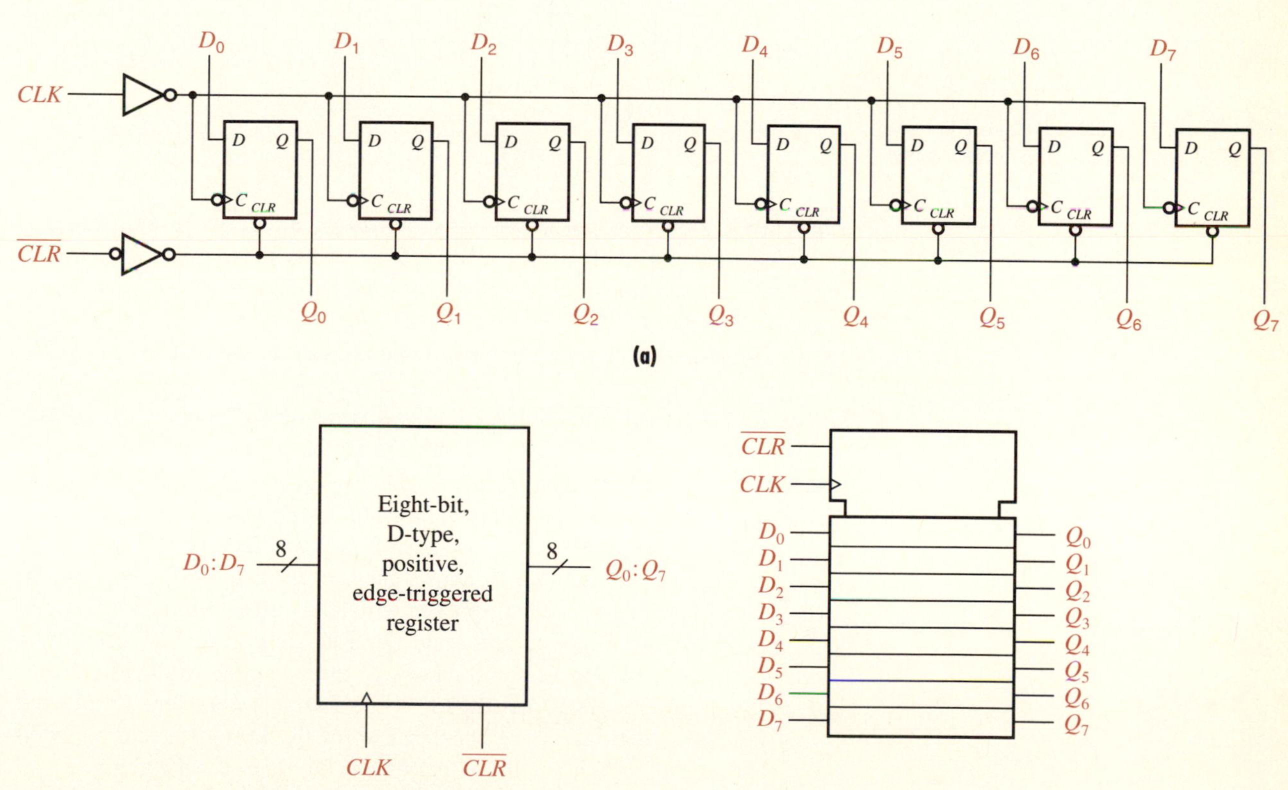

FIGURE 8.20 An 8-bit D-type register, the 74X273: (**a**) its gate-level circuit; (**b**) its box symbol; (**c**) its IEC 617 (IEEE 91) symbol.

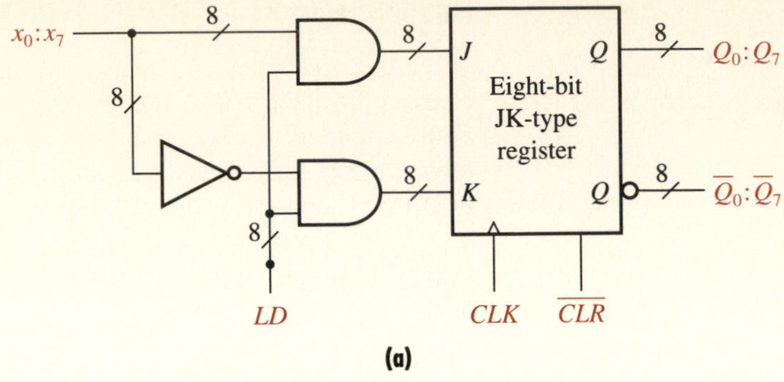

(a)

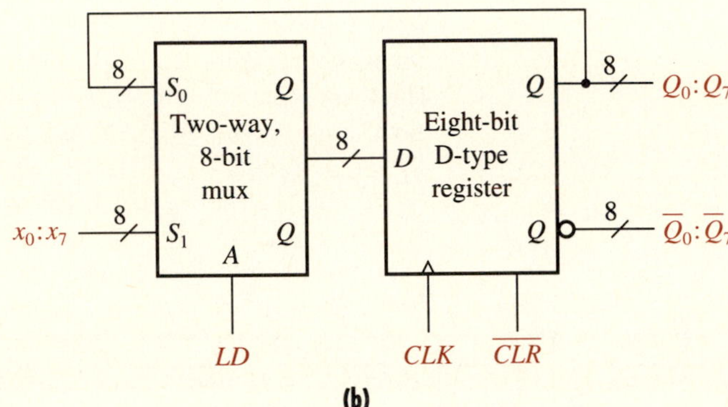

(b)

FIGURE 8.21 Parallel load logic for (a) a JK-type register; (b) a D-type register.

that when $LD = 0$, J and K are both forced to 0 in every flip-flop, causing its state to remain unchanged. When $LD = 1$ and the register is properly triggered by the clock signal CLK, the ith JK flip-flop is set by $x_i = 1$ and reset by $x_i = 0$.

A JK register can capture a data word X^* that is applied to it in some clock cycle i, with $LD = 1$. If LD is reset to 0 in cycle $i + 1$, the register retains its copy of X^* indefinitely and is unaffected by later changes made to the X signals. When $LD = 0$, the JK register sees only its quiescent data input, the all-0 word. Because a D flip-flop has no quiescent input combination—it always responds to the input data D whenever it is clocked—a different technique is needed to control a D register like that of Figure 8.20 via a load line LD. Figure 8.21b shows how to do this. We insert into the register's D lines a two-way multiplexer, which has its data inputs connected to X and to the flip-flop's current state Q. The multiplexer's select line is controlled by LD. Setting $LD = 0$ causes Q to be applied to D, effectively leaving the register's state unchanged, while $LD = 1$ transfers X into the D register. Thus, with $LD = 0$, the contents of the D register are reloaded every clock cycle, which has the effect of leaving them unchanged.

Because the D register, or any type of register, retains its data unchanged when the clock is inactive, it might be thought that using the LD signal to turn the register's clock input on and off would be an easy way to implement the parallel load function. This **clock-line gating** technique requires inserting an AND gate (and, possibly, additional logic) into the clock line, with controlling inputs derived from *LD* and the external clock source *CLK*. Clock-line gating invites trouble and, in general, should be avoided in synchronous circuit design. The reason is that the addition of gates and control signals to clock lines introduces large propagation delays that tend to skew the clock signals as they fan out through the circuit. In other words, clock-line gating causes the various copies of the clock's triggering edges to arrive at different destinations at different times. Because correct sequential operation depends on fairly precise synchronization of the triggering times of all the circuit's memory elements, such clock skew can upset this synchronization and lead to improper operation.

Care is needed when loading data into registers; clock gating should be avoided.

Example 8.5

Designing a small register file

We can combine a set of m parallel n-bit registers to form a memory for m words, each of which may be separately addressed. Such a memory is called an $m \times n$ **register file**, and serves as a temporary data store or **scratchpad memory** of the kind found in a computer's CPU. We want to design a register file to store four 32-bit words so that we can simultaneously read a word from one register A_1 while we are writing into another A_2 (which may be the same as A_1). Inputs A_1 and A_2 constitute separate ports of access to the memory, which is therefore sometimes called **dual-ported memory.**

Figure 8.22 shows an appropriate register-level design for this purpose. It is composed of four 32-bit D-type registers, each of which consists of 32 D flip-flops with a common clock line and a common clear line. A four-way, 32-bit demultiplexer controlled by the write address A_2 determines which register is written into, and a four-way, 32-bit multiplexer controlled by the read address A_1 determines which register is read from.

Shift Registers

An n-bit shift register is accessed serially in n clock cycles.

The entire contents of a parallel register can be read or written in a single clock cycle. This contrasts with a **shift register**, which is characterized by its use of serial data transfers. It takes n clock periods to read or alter the entire contents of an n-bit shift register via serial shifting. An example of a simple shift register appears in Figure 6.46 (Section 6.7).

A more complicated shift register, the 74X166 8-bit parallel-load shift register, is shown in Figure 8.23. It has two ways to enter data, which are specified by the control line called $SHIFT/\overline{LOAD}$. When $SHIFT/\overline{LOAD} = 1$, data must be entered one bit at a time via the line *SERIAL INPUT*. When $SHIFT/\overline{LOAD} = 0$, all eight flip-flops are loaded simultaneously from the eight input data lines labeled $A, B, \ldots, H$. All data must be read out via the single output data line Q_H. Data are shifted continuously through the

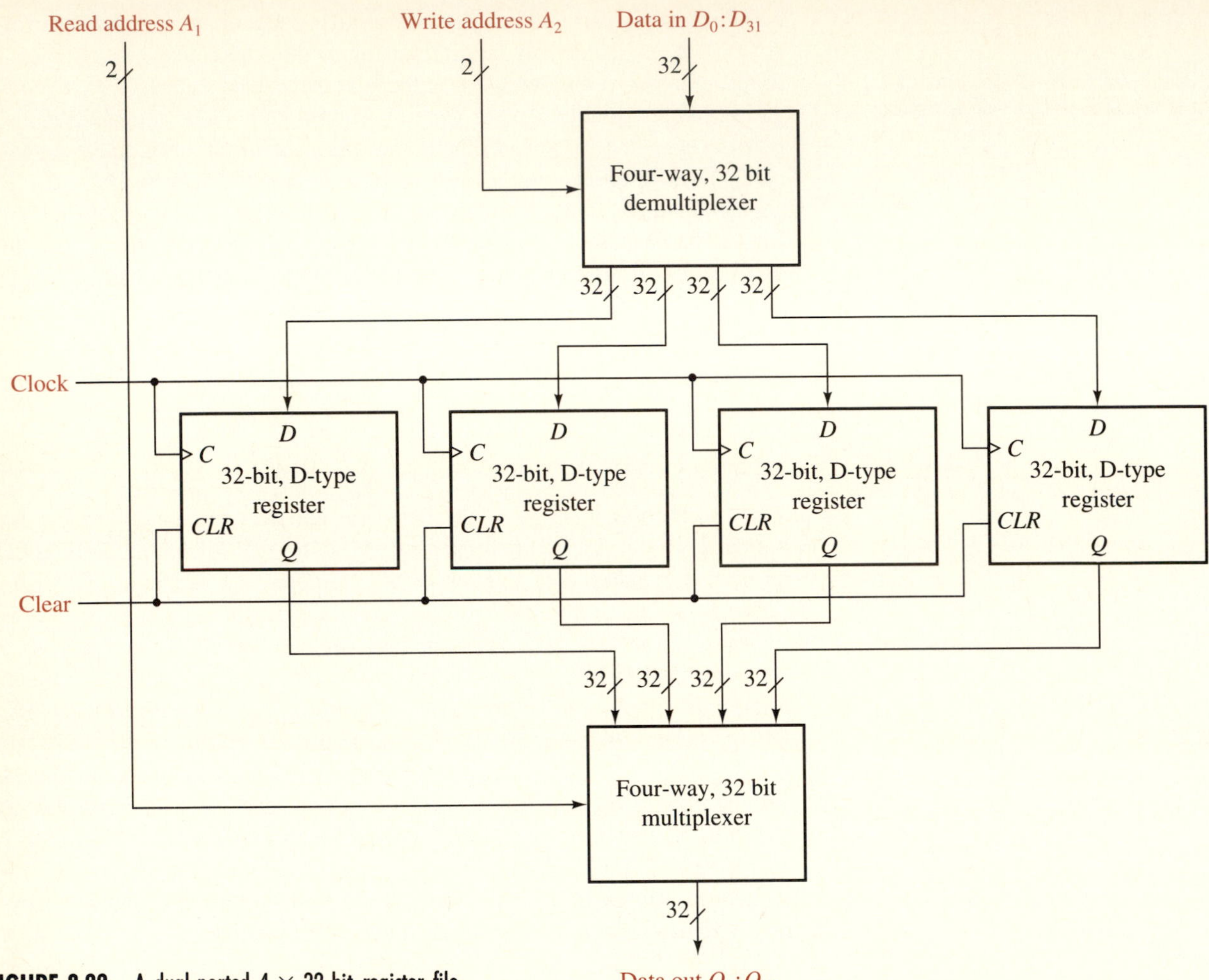

FIGURE 8.22 A dual-ported 4 × 32-bit register file.

shift register, one bit per clock cycle. The final control input line, *CLOCK INHIBIT*, is intended to allow the shifting to be stopped without turning off the clock signal *CLOCK*, a technique that is occasionally useful when designing asynchronous circuits. Setting *CLOCK INHIBIT* to 1 effectively blocks the *CLOCK* signal, causing the shift register to "freeze" in its current state. As noted above, such clock-line gating should be avoided in synchronous circuit design.

The 74X166 can be represented in circuit diagrams by a single box with four control lines, but this does not convey the full meaning of the control functions. In Figure 8.23b, the shift register's basic shift-register function is represented by a box with the usual control inputs *CLK* and $\overline{CLR}$ and, in addition, two "mode" control signals: *SH* and *LD*, which denote shift and (parallel) load, respectively. A NOR gate is added to show the relationship between *CLOCK* and *CLOCK INHIBIT*, and the two modes of operation

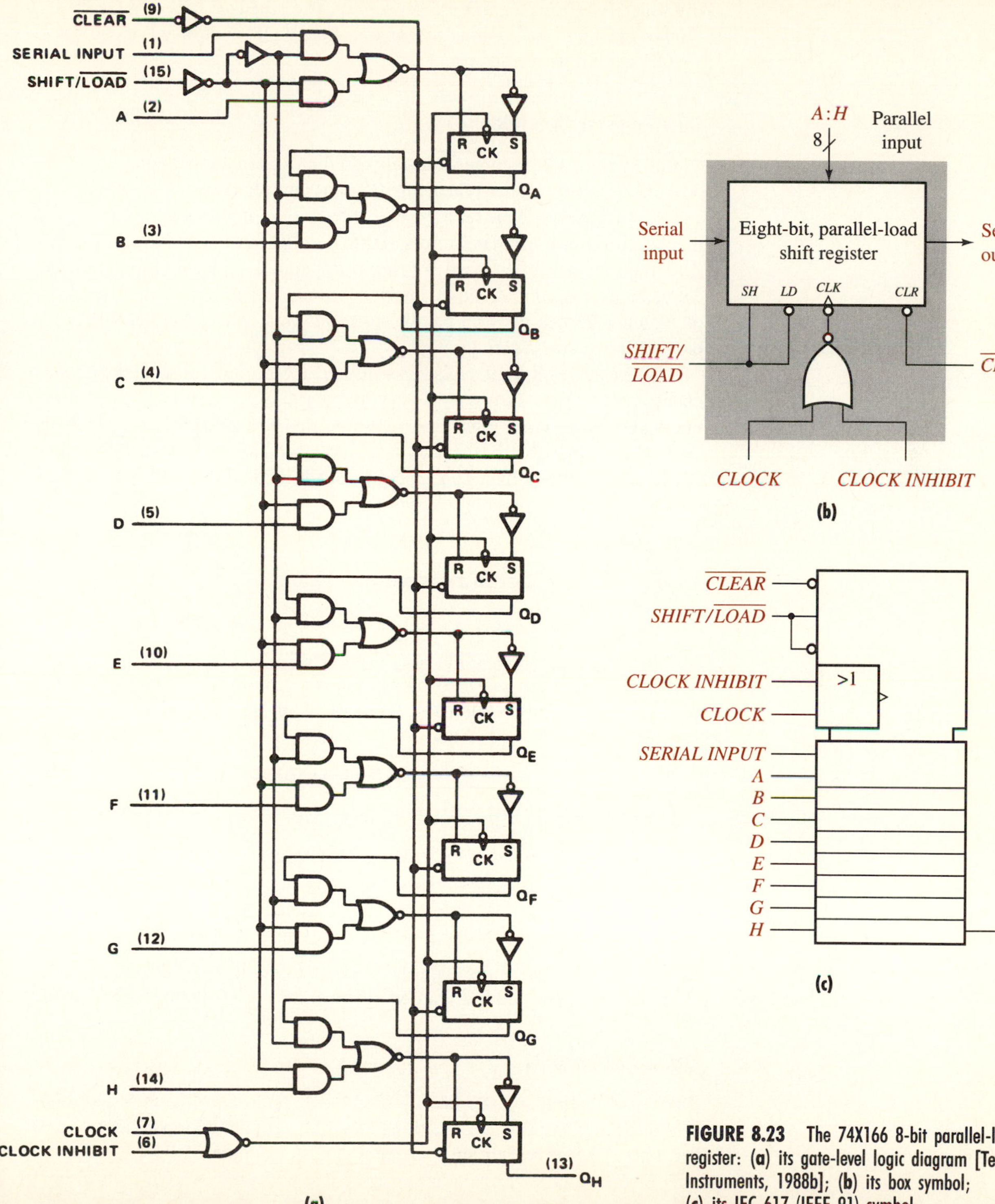

FIGURE 8.23 The 74X166 8-bit parallel-load shift register: (a) its gate-level logic diagram [Texas Instruments, 1988b]; (b) its box symbol; (c) its IEC 617 (IEEE 91) symbol.

controlled by *SHIFT/$\overline{LOAD}$* are separated. The standard IEC 617 (IEEE 91) symbol of Figure 8.23c similarly separates the mode control and incorporates a NOR gate in its control part.

Shift Register Operation

Digital systems like computers usually do their internal processing in parallel or word by word, whereas their input/output devices (keyboards, telephones, and so on) process data serially, or bit by bit. A major function of a shift register is to transform data from serial to parallel form and vice versa. Thus, shift registers are often used for serial-to-parallel or parallel-to-serial data conversion in the input/output interface circuits of a computer. The circuit of Figure 8.23 can perform parallel-to-serial data conversion because it has parallel inputs and a serial output.

When an n-bit shift register is shifted to the left, its content or state changes from $Y = y_{n-1}y_{n-2}\ldots y_2y_1y_0$ to $y_{n-2}y_{n-3}\ldots y_1y_0x$, where x is the signal on its serial input line. Observe that the leftmost y_{n-1} bit is discarded, and a new bit x is shifted into the rightmost position; x is the signal on the serial data-in line. If x is set to 0, the **logical left shift** operation is defined by

$$y_{n-2}y_{n-3}\ldots y_1y_0 0 := y_{n-1}y_{n-2}\ldots y_2y_1y_0 \tag{8.10}$$

The corresponding **logical right shift** operation is

$$0y_{n-1}\ldots y_3y_2y_1 := y_{n-1}y_{n-2}\ldots y_2y_1y_0 \tag{8.11}$$

In this case, the rightmost bit y_0 is discarded.

Shift operations implement multiplication and division by two in modular arithmetic.

If Y denotes an unsigned binary integer, Equation (8.10) can be interpreted as multiplication (mod 2^n), while (8.11) is interpreted as division (mod 2^n). For example, suppose that $n = 8$ and

$$Y = 11001011_2 = 203_{10}$$

A left shift yields

$$Y = 10010110_2 = 150_{10} = 203_{10} \times 2 \quad (\text{mod } 2^8)$$

while a right shift produces

$$Y = 01100101_2 = 101_{10} = 203_{10} \div 2 \quad (\text{mod } 2^8)$$

In multiplication, the most significant bit is discarded; in division, the least significant (fraction) bit is discarded—hence, the need for modular arithmetic.

Other Components

Two more flexible register-based components appear in Figure 8.24. Registers that provide many combinations of serial and parallel input/output access are known as **universal shift registers.** The example of Figure 8.24a has four distinct modes of operation, as specified by its two *MODE* control lines.

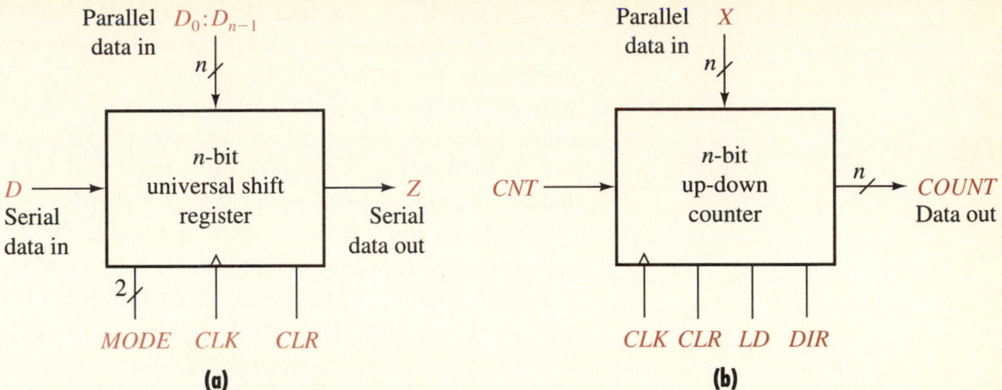

FIGURE 8.24 Two multifunction sequential components: (a) a universal shift register; (b) an up-down counter.

It has a parallel input mode, specified by $MODE = 11$, which causes all flip-flops to be loaded in one clock cycle from the parallel data input bus $D_0 : D_{n-1}$. It also has both left ($MODE = 01$) and right ($MODE = 10$) shifting modes with serial input/output. If the circuit is operated as a right shift register, for example, each time it is clocked, the serial input signal D is transferred to the leftmost flip-flop FF_0, the contents of each FF_i are transferred to FF_{i+1}, and the new contents of the rightmost flip-flop FF_{n-1} are transferred to the serial data output Z. The old contents of FF_{n-1} are, of course, discarded. Changing the shift mode from left to right reverses the shift direction and connects D and Z to the rightmost and leftmost flip-flops, respectively. A fourth mode of operation, defined by $MODE = 00$, causes the universal shift register to do nothing. A VHDL description of the shift register's functions follows:

```
variable q : bit_vector(0 to n − 1);       -- Define internal state
  if (CLR = '1') then for i in range 0 to n − 1 loop q(i) := 0 end loop
    else case MODE is
      when 1 => begin
          for i in range 0 to n − 2 loop q(i) := q(i + 1) end loop
          q(n − 1) := D;  Z <= q(0) end
      when 2 => begin
          for i in range 1 to n − 1 loop q(i) := q(i − 1) end loop
          q(0) := D;  Z <= q(n − 1) end
      when 3 => begin
          for i in range 0 to n − 1 loop q(i) := D(i);  end loop
          Z <= q(n − 1) end
      end case
```

The second component in Figure 8.24 is a counter designed to count up (increment) or down (decrement), as specified by the input control line *DIR* (direction). (The internal design of basic, up-only counters was examined in Section 6.7.) This **up-down counter** contains an n-bit register that stores the count Z and combinational logic to control and execute the increment and decrement functions. In response to a suitable 1-signal on the *CNT* line, the count Z is changed to $Z + 1$ if $DIR = 0$, or $Z - 1$ if $DIR = 1$. Additional control inputs allow the counter to be reset to the zero state ($CLR = 1$), or to have the input data X entered as the next state ($LD = 1$). The counting is done modulo m, where $m = 2^n$ in the case of a **binary counter**, and $m = 10$ in the case of a **decade counter**. Up-down counters may also be designed in which the modulus is a controllable input variable.

Standard ICs

Table 8.3 lists some typical register-level sequential components from the 74X series of integrated circuits [Texas Instruments, 1988b]. It contains many types of registers and counters, as well as a few components that are useful building blocks for general-purpose processors.

TABLE 8.3 A sampling of register-level sequential components in the 74X series.

IC part number	*Logic functions provided by the IC*	*Logic circuit*
74X91	Eight-bit all-serial shift register	
74X93	Four-bit binary (modulo-16) counter (ripple type)	
74X162	Four-bit decade (modulo-10) counter (synchronous type)	
74X166	Eight-bit parallel-load shift register	Figure 8.23
74X169	Programmable 4-bit up-down binary counter (synchronous type)	
74X172	8×2-bit register file with tri-state outputs	
74X273	Eight-bit D-type parallel register	Figure 8.20
74X323	Eight-bit universal shift register	
74X681	Four-bit accumulator (arithmetic-logic unit with registers)	

PLDs are flexible design components at the register level.

Some programmable logic devices are useful as general-purpose building blocks for sequential circuits, because flip-flops have been added to their input/output lines. Typically, these flip-flops are placed in some of the output lines of the OR plane, and the flip-flop output signals are connected to primary output pins; they are also fed back as secondary inputs to the AND plane. Hence, such a PLD can, by itself, implement a small state machine. Figure 8.25 shows an example of a commercial PAL chip, the PAL16R4 [Breuniger

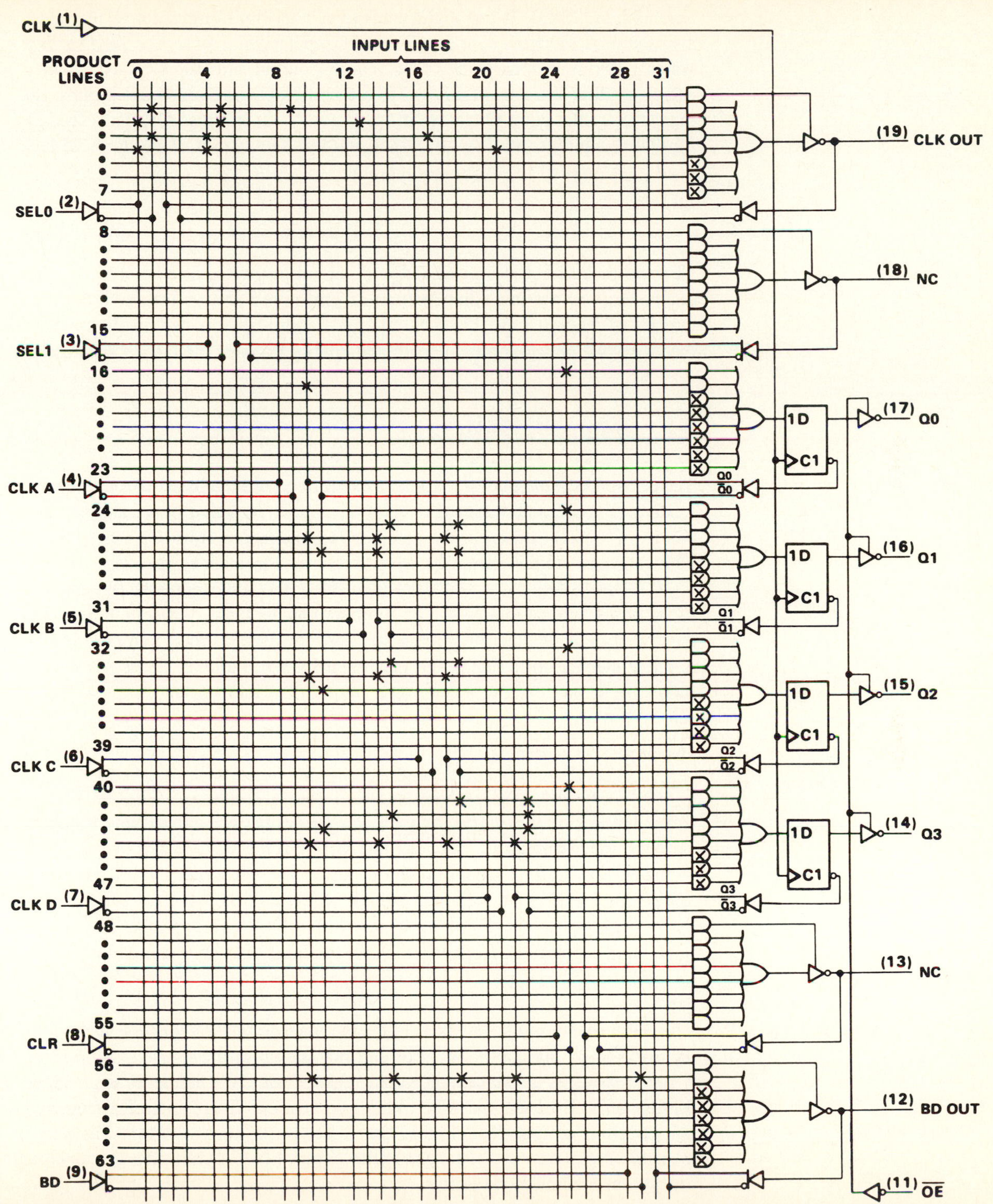

FIGURE 8.25 The PAL16R4 implementation of a counting circuit [Breuniger and Schiele, 1988].

and Schiele, 1988]. It has eight input lines and eight output lines; the latter also serve as secondary inputs. Thus, 16 input signals enter the AND array, which has dimensions of 32×64. The 64 output lines of the AND plane, called product lines in Figure 8.25, are combined in eight fixed groups, as indicated by the eight OR gates in the figure. Four of these OR gates feed D flip-flops, which form a small output register. The circuit of Figure 8.25 has been programmed to implement a counter.

Datapath and Control Design

Next we discuss the design of a digital circuit at the register level, using the components of the preceding section as primitives. We will assume that the circuits of interest have the data-control dichotomy depicted in Figure 8.6; therefore, we consider the datapath and control parts of the circuit separately.

8.4 Datapath Units

The data-processing part of a digital system frequently consists of a number of relatively large combinational circuits $C_1, C_2, \ldots C_p$ that move and process words. The data are typically entered into and taken from C_i via (shared) buses and are stored internally, when necessary, in a set of registers R_1, R_2, $\ldots R_q$. Only a few data word sizes are needed; they determine the dimensions of the buses and the registers. The combinational circuits, buses, and registers have well-defined functional roles that are indicated by their names, such as arithmetic-logic unit (ALU), left ALU multiplexer, memory address decoder, input/output bus, instruction register, output data register, and the like.

Structure

Figure 8.26 shows two representative datapath structures. That of Figure 8.26a is one of the simplest, with a single-function or multiple-function combinational circuit C, such as an ALU circuit, and a pair of registers A and B for storing input/output operands. In one clock cycle, the ALU computes a result of the form $f_i(A, B)$, which can then be stored back into the A register at the beginning of the next clock cycle. Thus, the circuit implements a class of register transfer operations that have the general form

$$A := f_i(A, B) \qquad (8.12)$$

where the selection of the function f_i, the loading of the registers, and so on are directed by an external control unit not shown in the figure. The register A serves both as an input and output register for C, whereas B is an input register only. Because A can be used to accumulate the results of a sequence of operations of type (8.12), it is referred to as an **accumulator**. The name *accumulator* is also applied to an entire circuit like Figure 8.26a that is built around an accumulator register.

An accumulator circuit computes results and collects them in an accumulator register.

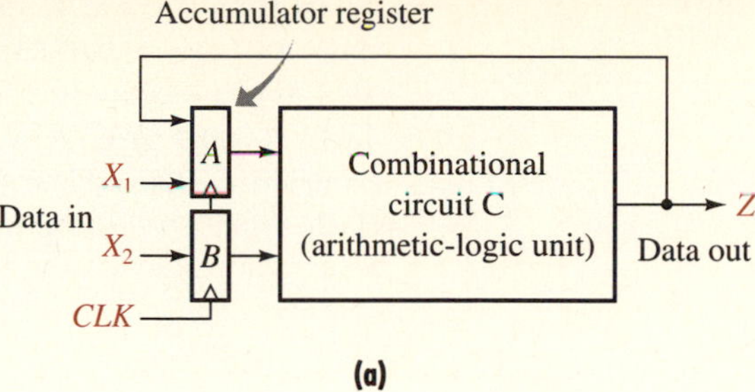

(a)

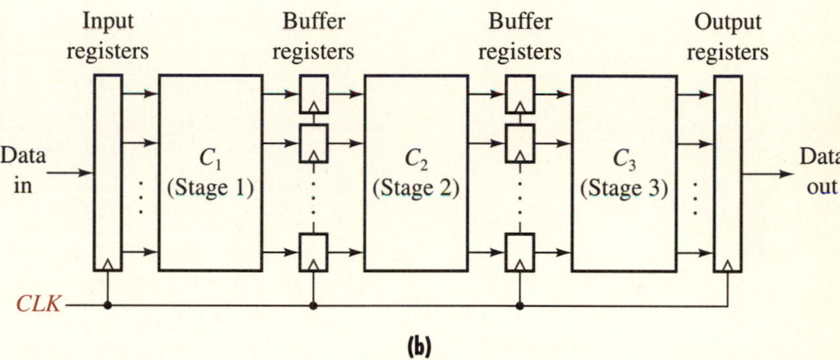

(b)

FIGURE 8.26 Some datapath configurations: (a) an accumulator; (b) a three-stage pipeline.

A pipeline circuit computes in stages, the operation of which can be overlapped.

The circuit of Figure 8.26b is an example of a datapath structure called a **pipeline processor**. It consists of several combinational circuits, called the **stages** of the pipeline, through which a set of data words S can flow from left to right. At each stage, some processing is performed on S, and the final results are available only after the data have flowed through all stages—three in this case. The pipeline stages are separated by buffer registers. A pipeline resembles a manufacturing assembly line, in which each stage performs a part of the manufacturing process; a pass through the entire line is necessary for complete processing. The stages of the pipeline may all contain different sets of data at the same time, with the data sets insulated from one another by the buffer registers. Consequently, pipelined datapaths achieve a certain degree of parallel processing. They are typically used in high-speed computers to implement complex arithmetic operations like the addition and subtraction of floating-point numbers.

As is typical of register-level descriptions, Figure 8.26 omits small, and sometimes obvious, design details, which, however, must be filled in to actually build any of the circuits in question. For example, the interconnection paths shown in Figure 8.26a imply that the accumulator register A can be loaded either from input bus X_1 or from the output bus Z of C. If A is a normal parallel register with a single input data bus X, a multiplexer with inputs from X_1 and Z must be inserted in X. A somewhat more detailed version of this circuit appears in Figure 8.27, with various missing details, such as the multiplexer for A, added. Such obviously necessary signals as the *CLR* and *FUNCTION SELECT* lines omitted from the earlier diagram are shown here. It should be stressed that all registers are controlled by a common clock signal so that the circuit operates in strictly synchronous fashion.

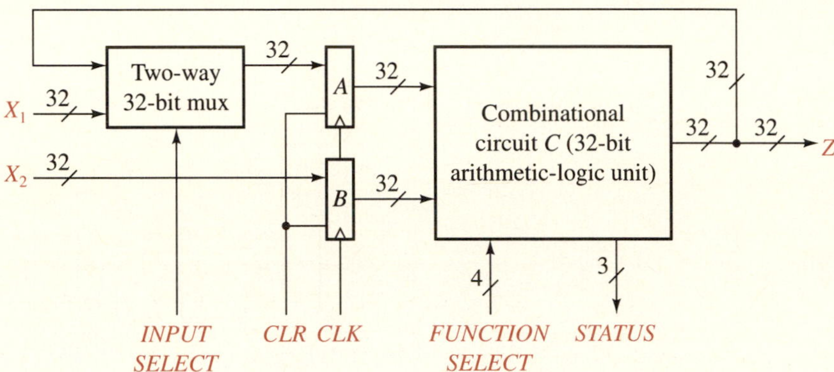

FIGURE 8.27 A more detailed logic diagram for the accumulator-based datapath of Figure 8.26a.

Scaling

A common feature of all register-level components is that they act on data words of specified sizes. The word size n is a variable comparable to the fan-in (number of input lines) of a gate, and can vary over a moderate range of values for each component type. Some register-level components are characterized by several independent size parameters. For example, the size of a multiplexer is defined by the number m of its input data buses and the width n of each data bus. It is impractical to provide different versions of each component for every possible value of m, n, and the like, so register-level components are designed to be easily **scaled**—that is, expanded or reduced—to accommodate different size parameters. We therefore turn next to the scaling techniques needed for the most common component types; this is often a central part of the datapath design process.

Component expansion is an important aspect of register-level design.

In general, reducing a scaling parameter such as the number of data inputs or outputs of a component presents little difficulty. Unwanted input lines are connected to constant signal values that render them inactive; unwanted output lines are left disconnected. Expansion is carried out in a number of ways that depend on the component type. As we will see, this expansion is greatly facilitated by a component's control lines.

Bit Slicing

Consider a generic component *COMP* with input/output data buses of various sizes, as shown in Figure 8.28a. Suppose that all the data words being processed increase uniformly in size by some factor m as the component is scaled up. Then m copies of *COMP* may often be combined as shown in Figure 8.28b by merging the corresponding data buses so that each becomes m times wider, and sharing the control lines K. The resulting circuit is said to be **bit-sliced**, with each copy of the component forming a **slice** of the total circuit. Another view of this expansion is that a fixed section (slice) of every data word is assigned to a particular copy of *COMP*, and all copies of *COMP* work in tandem on the expanded data words. A bit-sliced circuit constructed from m copies of *COMP* is essentially similar to *COMP,* with all buses, registers, and data-processing circuits expanded by a factor of m.

Bit slicing exploits independence among the bits of the data words being processed.

A simple instance of bit slicing is the juxtaposition of m copies of an ordinary logic gate to form an m-bit word gate (Figure 8.10). Each ordinary gate forms a slice and processes one bit of the relevant m-bit words. The connection of a set of flip-flops to form a register, as in Figure 8.20, or a set of small registers to form a large register, also constitutes bit slicing. In this case, the data storage parts are juxtaposed and the control signals are joined. The key property that makes this form of bit slicing possible is the absence of data connections, and therefore of functional interaction, between the circuit slices. Each slice handles its own data, independent of the others, as in the illustration of Figure 8.28b, in which there are no data connections between the slices.

It is sometimes necessary to connect adjacent slices via one or two data signals so that the functions of the component slices extend properly across

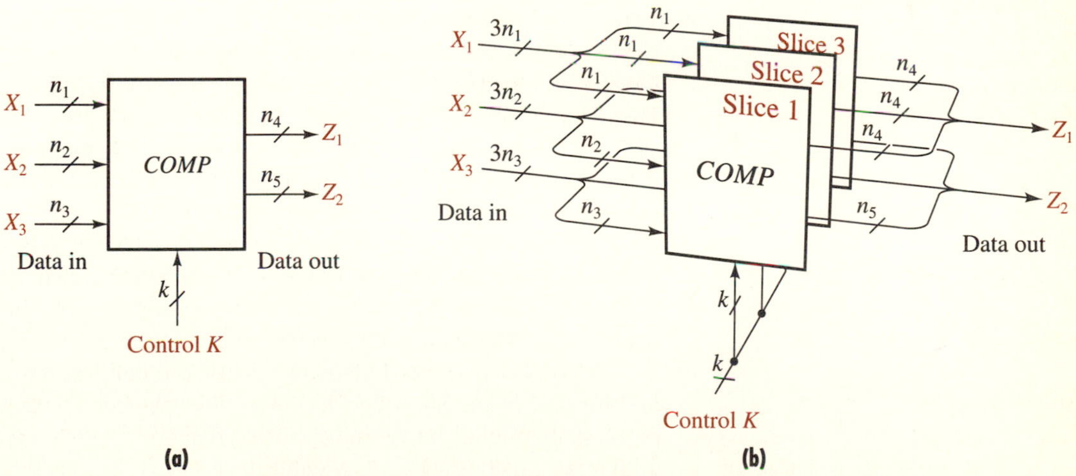

FIGURE 8.28 Component expansion by bit slicing: (a) a single component; (b) an expanded version that employs three slices.

the entire bit-sliced circuit. An m-bit ripple-carry adder composed of m full adder cells is considered to be bit-sliced, but each cell must be connected to its neighbors via lines for carry propagation. Similarly, a shift register is expanded in bit-sliced fashion by connecting serial data outputs to serial data inputs, as shown in Figure 8.29. This example uses m n-bit shift register slices to implement a single mn-bit shift register. The interslice connections are clearly necessary to allow right-shifting across the entire circuit.

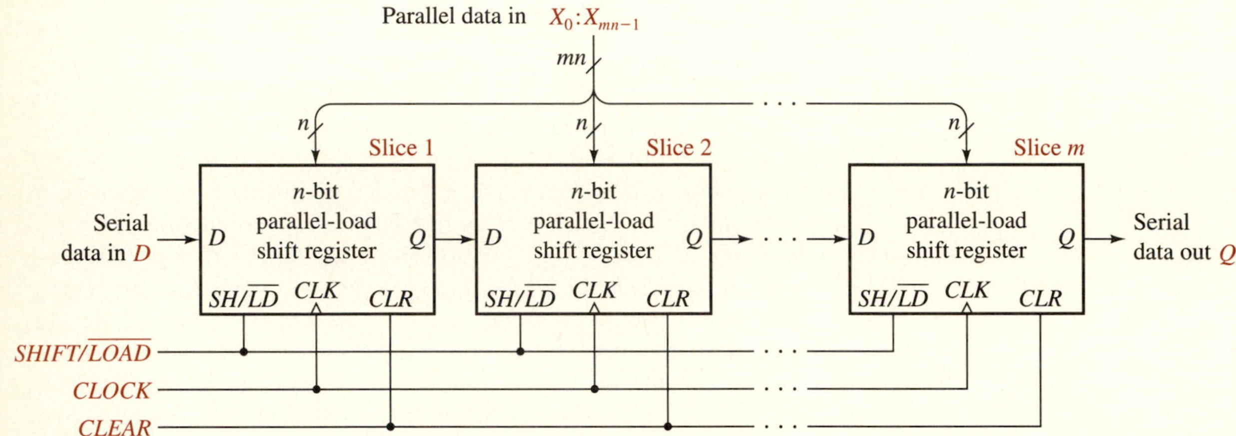

FIGURE 8.29 Shift register expansion.

Example 8.6

Designing a bit-sliced accumulator-shifter circuit

Relatively complex datapath circuits can be designed using bit slicing. Consider again the accumulator circuit of Figure 8.26a. Suppose we want a 4-bit version of this circuit, in which the A register is changed to a right-shift register and the only function of the combinational logic C is binary addition. The required accumulator-shifter circuit can be constructed from four copies of the 1-bit slice of Figure 8.30a, with the interslice connections shown in Figure 8.30b. Each slice contains a 1-bit portion of both the A and B registers—that is, two D flip-flops, a 1-bit (full) adder, and a multiplexer. The interslice data connections are carry lines linking the full adders and shift lines linking the A flip-flops. The original control signals are connected to all the slices in the same way. Except for their common control lines, there is no explicit link between the multiplexers of the individual slices.

The multiplexer has 1-bit data inputs from both the external $X1$ bus and the output Z of the full adder. Its output line connects to the D input of the A flip-flop. To implement the right-shifting function, we add to the multiplexer a third input called *SHIFT IN*, which is taken from the neighboring slice on the left. The output Q of each A flip-flop must then be connected to the *SHIFT IN* line of the neighboring slice on the right. The resulting three-way multiplexer

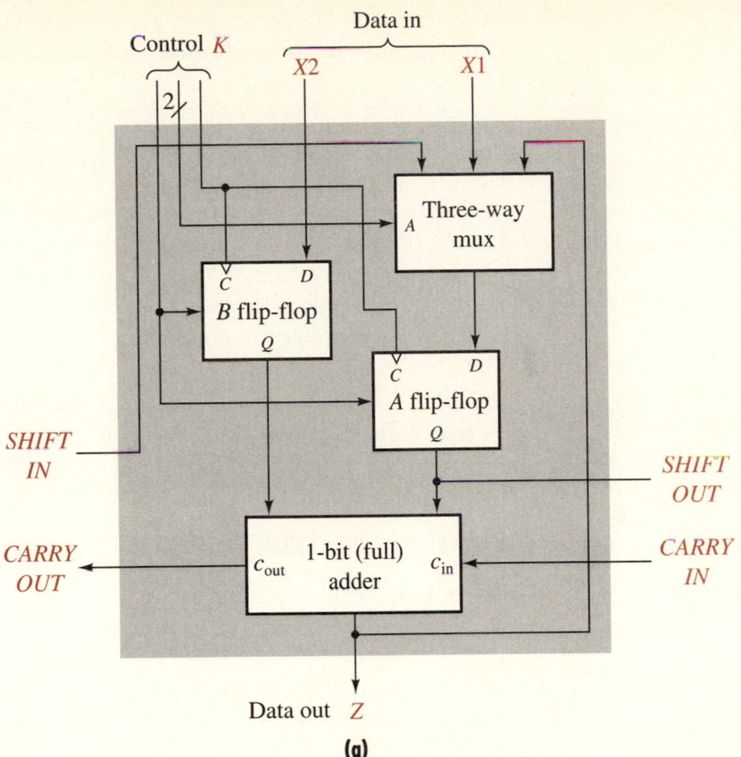

(a)

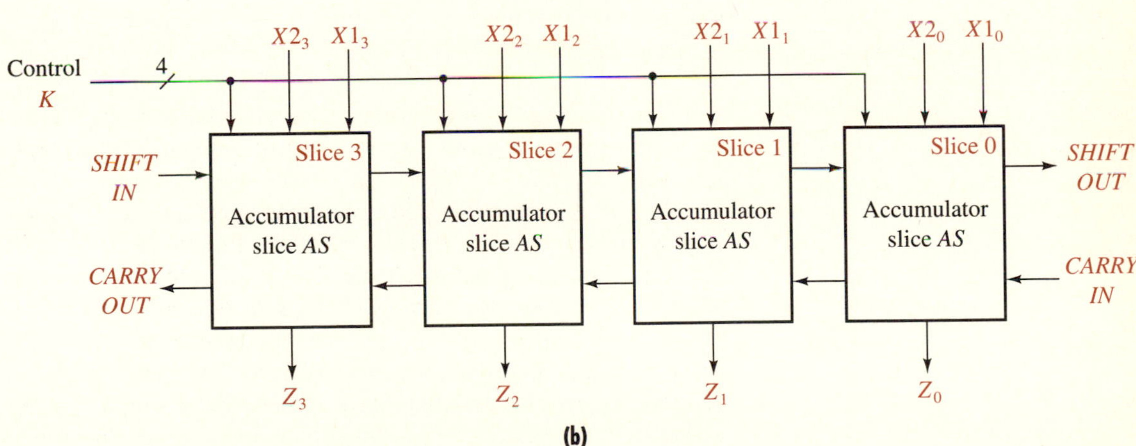

(b)

FIGURE 8.30 An accumulator-shifter circuit: (**a**) a 1-bit slice; (**b**) a 4-bit circuit.

can be set via two select control lines in K to perform three different data-routing functions: load $X1$ into A, load Z into A, or load *SHIFT IN* into A. As can be seen from Figure 8.30b, the *SHIFT IN* and *SHIFT OUT* lines in adjacent slices are joined, as are adjacent *CARRY IN* and *CARRY OUT* lines.

Observe that the feedback loop evident in Figure 8.30a is not seen in Figure 8.30b because it has been made internal to the slices.

Some component size parameters do not scale in the simple, uniform fashion of bit slicing. For example, if the number of input lines of a decoder is increased by one from n to $n + 1$, the number of output lines doubles from 2^n to 2^{n+1}. If the number of input data buses to a multiplexer increases, typically by doubling, the number of output buses remains constant at one. Thus, we need additional expansion methods, which we consider next for some key component types, including multiplexers, decoders, encoders, and various arithmetic circuits.

Multiplexer Expansion

Multiplexer expansion uses convergent trees of small multiplexers.

To accommodate a larger number n' of data sources, several smaller n-way multiplexers are connected in a converging structure of the kind shown in Figure 8.31. The n' sources are connected to the data inputs of $\lceil n'/n \rceil$ copies of the n-bit multiplexer. The multiplexers select $\lceil n'/n \rceil$ of the sources, which they proceed to connect to their D buses. These $\lceil n'/n \rceil$ D buses are, in turn, connected to the inputs of $\lceil n'/n^2 \rceil$ additional copies of the n-bit multiplexer, thereby reducing the number of selected sources by a further factor of n. This selection process continues until the number of selected sources is reduced to one. The output bus of the final multiplexer is then connected to the destination. The resulting circuit has a "tree" structure that may be contrasted with the linear (all-in-line) structure of bit-sliced circuits. Observe that as the multiplexer tree grows in size, so does its propagation delay, but at a relatively slow rate.

An example of such a tree circuit that forms a 12-way, 2-bit multiplexer appears in Figure 8.31. Here the building block is the four-way, 2-bit multiplexer of Figure 8.12, so we have $n' = 12$, $n = 4$, and $m = 2$. Now $\lceil 12/4 \rceil = 3$; hence, three copies of the four-way multiplexer are needed to select three of the 12 possible sources. A final four-way multiplexer, which has an unused input data bus, selects one of the latter three sources. The selection process is controlled by four select lines $A = A_3A_2A_1A_0$, arranged as shown in the figure. The two low-order bits A_1A_0 are connected to all three multiplexers on the primary input side. The high-order bits A_3A_2 connect to the single multiplexer on the output side. As in the four-way multiplexer, setting A to the binary number i selects the source S_i.

Suppose, for instance, that S_9 is to be selected. As shown in Figure 8.31, the binary number $1001_2 = 9_{10}$ must be applied to the four external address lines $A = A_3A_2A_1A_0$. The three input multiplexers, reading from top to bottom, select the data sources S_1, S_5, and S_9, respectively, in

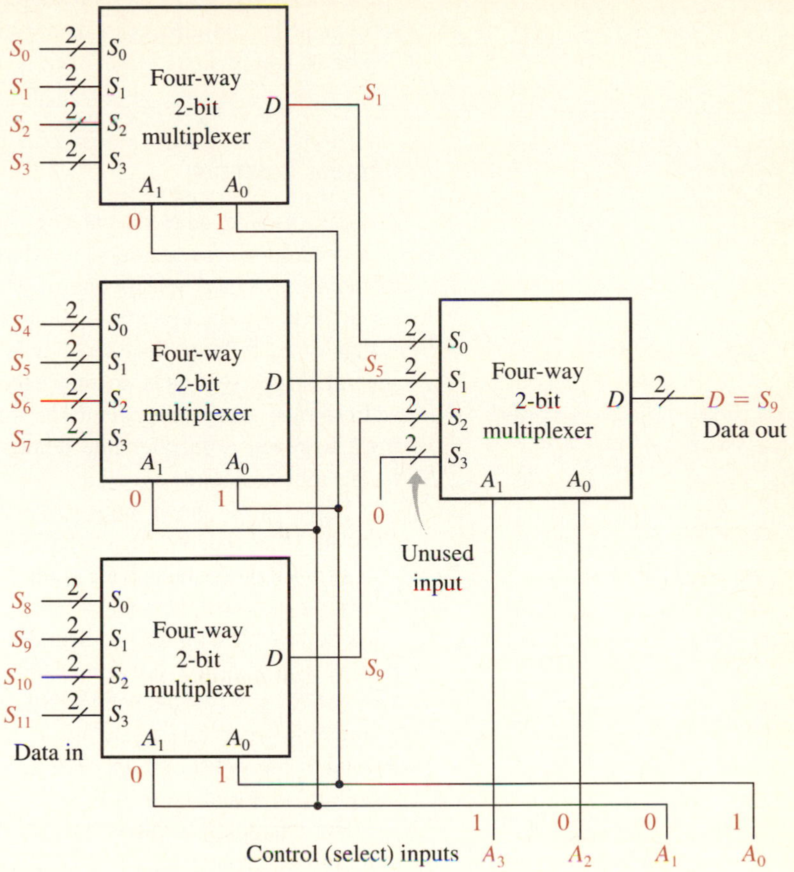

FIGURE 8.31 A 12-way multiplexer tree constructed from four-way multiplexers.

response to $A_1A_0 = 01$. Hence, the three data sources S_1, S_5, and S_9 are connected to the top three input data buses of the output multiplexer. Because the latter is controlled by the two high-order address bits $A_3A_2 = 10$, its third input from the top—that is, S_9—is selected and has its 2-bit data word transferred to the final destination D.

The expansion process illustrated by Figure 8.31 is unaffected by the width m of the data bus, which happens to be two. When necessary, the bus width is easily increased by bit slicing. For example, to expand m from 2 to 4, two copies C_1 and C_2 of the entire circuit of Figure 8.31 are required. The 2-bit data buses labeled S_i of C_1 and C_2 are grouped to form a 4-bit bus, which then becomes the new S_i. The output data buses of C_1 and C_2 are similarly grouped. The input address bus $A = A_3A_2A_1A_0$ is connected to the corresponding control inputs of both circuits; in other words, A fans out to both C_1 and C_2. Hence, C_1 and C_2 operate in tandem, each transferring half the data from the common source to the common destination. In general, to expand the multiplexer's data bus width from m to m' requires $\lceil m'/m \rceil$ copies of the smaller multiplexer.

Decoder Expansion

Decoder expansion uses divergent trees of small decoders.

A 1-out-of-n decoder selects one out of n choices. It is a key component of a random-access memory, in which a k-bit memory address word serves as the input X and selects one of $2^k = n$ memory locations to be accessed for a read or write operation. Such **address decoders** often have very large values of n. A large decoder may be built from small decoders arranged in several stages or levels so that each stage decodes some of the incoming address bits. Again the circuit has a tree-like structure. Unlike a multiplexer tree, however, a decoder tree diverges rather than converges as one moves from input to output.

Figure 8.32 shows how to construct a 1-out-of-256 decoder using 17 1-out-of-16 decoders of the type in Figure 8.15. The first stage, which consists of a single decoder DEC_0, decodes the four high-order bits $x_7{:}x_4$ of the incoming 8-bit data word $X = x_7{:}x_0$; the second stage, which consists of the 16 decoders $DEC_1{:}DEC_{16}$, decodes the low-order bits $x_3{:}x_0$. The 16 outputs $y_0{:}y_{15}$ of DEC_0 are connected to the enable inputs of the 16 decoders $DEC_1{:}DEC_{16}$. The data input bus of each of these decoders is connected to the low-order bits $x_3{:}x_0$ of X. The enable input EN of DEC_0 is used to enable the entire circuit. It is easily verified that $X = i$ when $EN = 1$ makes $z_i = 1$ and $z_j = 0$ for all $j \neq i$.

For example, suppose that $X = 00110001_2 = 49_{10}$. If $EN = 0$, the outputs $Y = y_0{:}y_{16}$ of DEC_0 are all 0. Hence, all the second-stage decoders are also disabled, making the primary output signals $Z = z_0{:}z_{255}$ all 0. Setting $EN = 1$ causes DEC_0 to decode $x_7{:}x_4 = 0011$ and make $y_3 = 1$, thereby activating decoder DEC_4 with outputs $z_{48}{:}z_{63}$. DEC_4 now decodes its input data $x_3{:}x_0 = 0001$, which causes it to activate its second output line—namely, z_{49}, as required. Note that an attempt to build this decoder in a single stage along the lines of Figure 5.54 would violate all reasonable fan-in and fan-out constraints. The two-stage design of Figure 8.31 has high fan-out (16) from the data inputs $x_3{:}x_0$; many circuit technologies require buffers in these lines to increase their fan-out capability.

The decoder expansion technique illustrated by Figure 8.32 is readily generalized to handle decoding tasks of arbitrary size. By adding a third stage that consists of 256 1-out-of-16 decoders, a 1-out-of-4096 decoder is obtained, and so on. A decoding circuit composed of 1-out-of-n decoders has a diverging tree structure in which the number of decoders that appear in stage i is n times the number in the preceding stage $i - 1$. As in the multiplexer case, the propagation delay of the decoder tree grows slowly with the number of inputs. It is worth noting that the enable control signals play an essential part in the expansion of decoders.

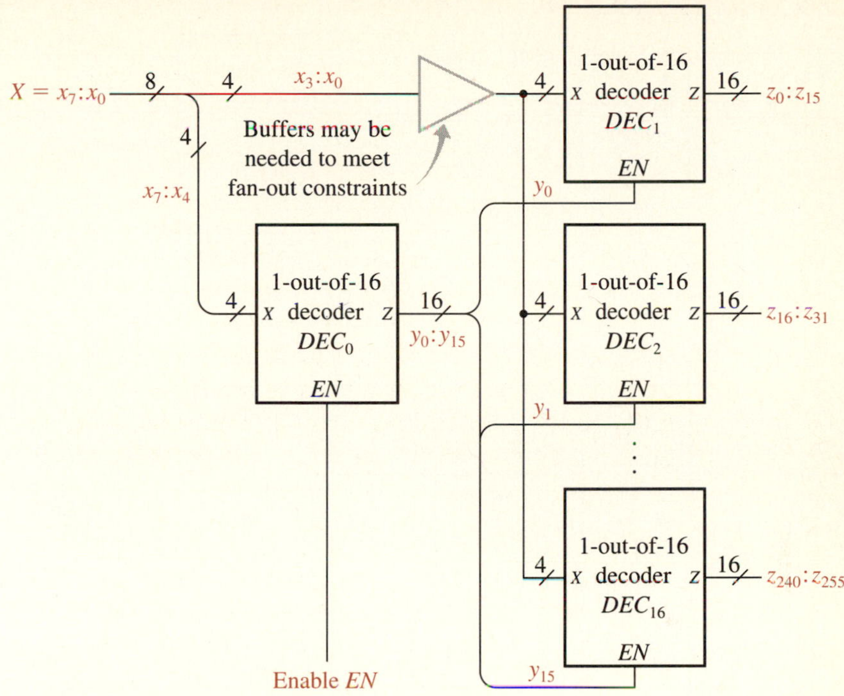

FIGURE 8.32 A 1-out-of-256 decoder tree constructed from 1-out-of-16 decoders.

Encoder Expansion

Encoder expansion is difficult; it requires several different component types.

The priority requirement makes encoder expansion more difficult than that of other register-level components. Several output control signals may be provided to support expansion, and other components such as word gates or multiplexers may be needed in the expansion process. The 74X148 eight-input priority encoder, for instance, has the general structure of Figure 8.16 and two output control signals: a group signal *GS*, which indicates the presence of an active input data signal x_i, and a second signal, called *EO* for "enable out," which is defined by the logic equation

$$EO = EI \cdot \overline{GS} \tag{8.13}$$

In other words, *EO* is active whenever none of the X_i inputs are active and the encoder is enabled.

Figure 8.33 shows one way to expand the number of inputs of a priority encoder from eight to 64. Nine copies of an eight-input priority encoder similar to but slightly simpler than the 74X148 are used, each of which has both *GS* and *EO* output signals. Three eight-way, 1-bit multiplexers are also needed. The priority encoders' *EO* output signals are not used explicitly in this expansion method; see Problem 8.38 for an alternative expansion method that employs *EO*. Eight of the encoders $ENC_0{:}ENC_7$ serve

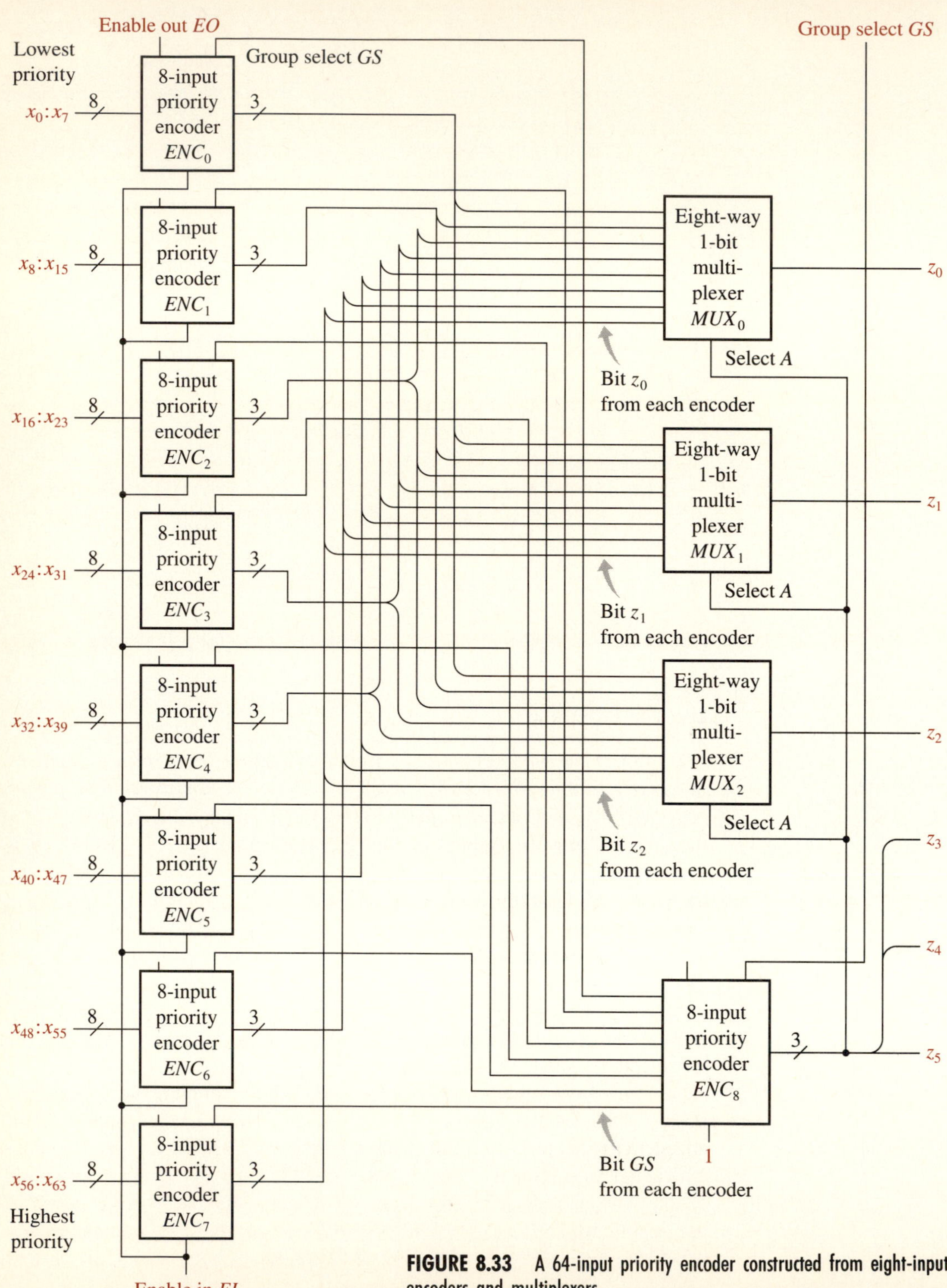

FIGURE 8.33 A 64-input priority encoder constructed from eight-input encoders and multiplexers.

to encode the 64 input signals $x_0{:}x_{63}$ in groups of eight. These encoders are all simultaneously enabled by $EI = 1$. The GS output of each of ENC_i, $0 \le i \le 7$, is connected to input x_i of the last encoder ENC_8. The data output of ENC_8 is therefore the index i in binary form of the input encoder ENC_i, with the highest priority active input x_j. This number i provides the three high-order bits z_5, z_4, z_3 of the final 6-bit encoded output Z. To obtain the remaining three bits of Z, we need to determine the highest priority active input j of ENC_i. This information is available in the required form at the Z outputs of ENC_i. The three multiplexers in Figure 8.33 serve to route the three bits of j from ENC_i to the three low-order primary output bits z_2, z_1, z_0, as indicated. Note that i is available at the output of ENC_8 and is connected to the select inputs of the multiplexers.

To illustrate the operation of the 64-bit encoder, suppose that x_{51} alone is 1, and EI is set to 1 to enable the circuit. The output signal GS of ENC_6 is active and is applied to data input 6 of ENC_8, resulting in the primary output signals $i = z_5z_4z_3 = 110_2 = 6_{10}$. Now i is applied to the select inputs of the multiplexers, which proceed to route 011 from the output data of ENC_6 to $z_2z_1z_0$. The result is $z_5z_4z_3z_2z_1z_0 = 110011_2 = 51_{10}$, as desired. This result is unaffected by other occurrences of an active x_j, provided $j < 51$.

Adder Expansion

Adders and adder-based ALUs like the 74X181 (Figures 8.1, 8.2, and 8.3) are most easily expanded using bit slicing with ripple-carry propagation, as illustrated by Figure 8.30. A set of m n-bit components can be combined to accommodate mn-bit data simply by connecting the carry output C_{out} of the ith stage to the carry input C_{in} of the next, $(i + 1)$st, stage.

Fast arithmetic circuits use expansion techniques with carry lookahead.

For faster operation, arithmetic slices employ carry lookahead internally in place of ripple-carry propagation. The same carry-lookahead technique used in a datapath slice, such as the 74X83 4-bit adder (Section 5.9), can also be used between slices, if each slice generates the appropriate propagate and generate output signals. The 74X181, for example, which is a 4-bit ALU/function-generator slice, has this feature, producing the signal-pair P and G to support carry lookahead. An auxiliary component, called a **carry-lookahead generator**, is needed to compute the carry input signal for each slice, except for the low-order slice, which directly uses the external carry-in signal. Figure 8.34 demonstrates how four 74X181s can be expanded into a 16-bit ALU with carry lookahead across all four stages. A standard carry-lookahead generator IC, the 74X182, is available; it uses the propagate and generate signals from all four 74X181s to compute the carry-in signals for the three higher order slices. It also generates P and G signals that can be used to expand the carry-lookahead process over even larger data word sizes.

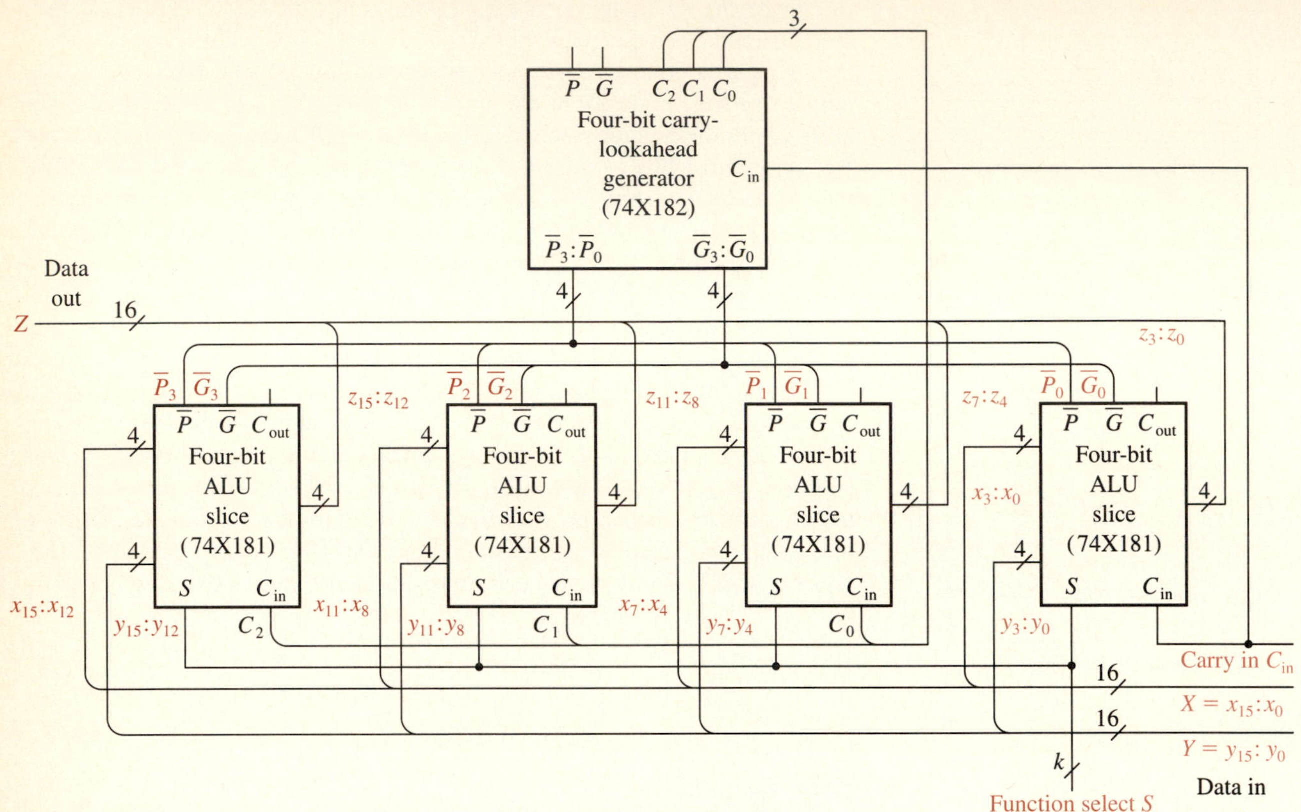

FIGURE 8.34 Design of a 16-bit ALU from 4-bit ALU slices, using carry lookahead between slices.

The internal logic of the 74X182 carry-lookahead generator is shown in Figure 8.35. The 74X182 has essentially the same structure and function as the carry-lookahead subcircuit of the 74X83-style adder that appears in Figure 5.56. In the present case, a two-level AND-NOR circuit is used to generate the carry signals, and the input signals are complemented. As described in Section 5.9, the carry signal C_i produced in stage or slice i is defined by the recurrence relation

$$C_i = G_i + P_i C_{i-1} \tag{8.14}$$

where P_i and G_i are the propagate and generate signals, respectively, produced by the current slice. The P_i signal is defined to be 1 whenever a 1 is propagated through slice i from its C_{in} input; $G_i = 1$ whenever a carry is generated within the slice itself.

The above definitions of P_i and G_i are independent of the slice's width, so our earlier analysis of carry lookahead, which was implicitly for 1-bit

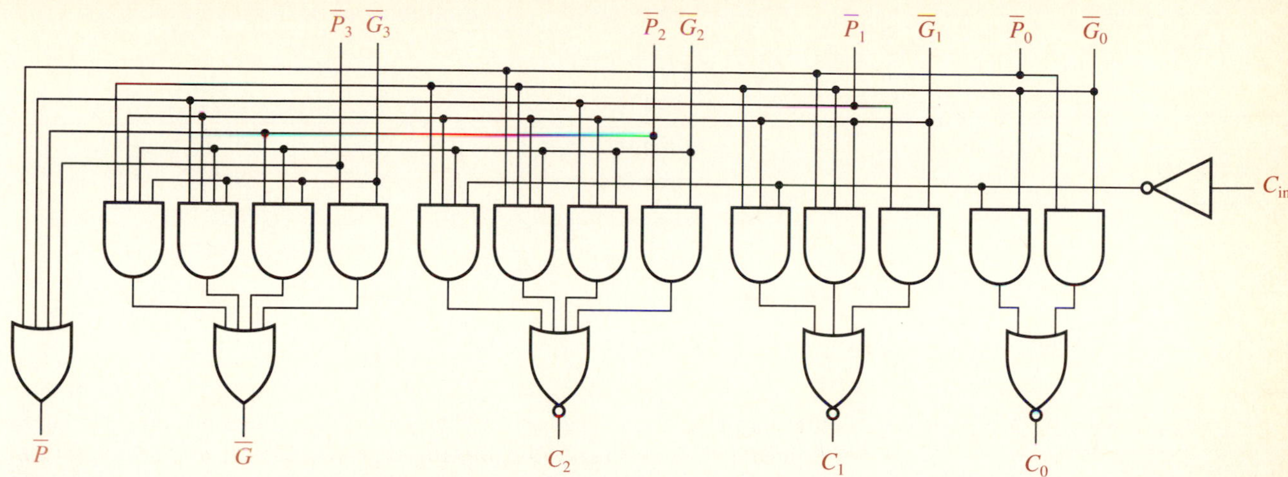

FIGURE 8.35 The logic circuit of the 74X182 carry-lookahead generator.

slices, can now be seen to apply directly to 4-bit or n-bit slices. We can transform (8.14) by some basic laws of Boolean algebra as follows:

$$\begin{aligned} C_i &= (G_i + P_i)(G_i + C_{i-1}) \\ &= \overline{\overline{G}_i\overline{P}_i + \overline{G}_i\overline{C}_{i-1}} \end{aligned} \tag{8.15}$$

Setting $i = 0$ in (8.15) defines the AND-NOR realization of C_0 in Figure 8.35. The other two carry functions C_1 and C_2 are obtained by repeated use of (8.15), as is easily checked.

To illustrate the applications of the various register-level components discussed so far, we examine next the design of a multifunction datapath circuit C of the type that lies at the heart of a computer's CPU.

Example 8.7

Designing a multifunction datapath unit

The function of the datapath of a CPU is to execute a set of instructions, the number of which ranges from a dozen or so to a couple of hundred, depending on the type of computer. Such a datapath DP can be built around a combinational ALU for two's-complement numbers of appropriate size, like the 16-bit 74X181-based ALU of Figure 8.34. In simple cases, an adder suffices in place of a full-fledged ALU, and will be used in this example. To simplify the problem further, we assume that DP must execute only six instructions, each of which operates on one or two incoming data words X and Y and produces a result Z. The particular instruction to be executed at any time is designated by a separate 3-bit control word or (micro) instruction. This instruction is decoded to determine the appropriate data and control signals to apply to the adder.

The desired behavior of these instructions is specified as follows:

case *Instruction* **is**
  when 000 => $Z := X1 + Y1;$ -- ADD X1,Y1
  when 001 => $Z := X1 - Y1;$ -- SUB X1,Y1
  when 010 => $Z := Y1 - X1;$ -- SUB Y1,X1
  when 011 => $Z := X1 + 1;$ -- INC X1
  when 100 => $Z := X1 - 1;$ -- DEC X1
  when 101 => $Z := X1;$ -- LOAD X1
end case

All these instructions can be realized by various forms of 16-bit addition and subtraction. The basic addition and subtraction operations are implemented in the standard two's-complement fashion, described by the following equation:

$$X1 \pm Y1 = X1 + (Y1 \oplus K) + SUB \tag{8.16}$$

where, for addition, $SUB = 0$ and K is the 16-bit all-0 word; for subtraction, $SUB = 1$ and K is the 16-bit all-1 word. $Y1 \oplus K$ denotes a 16-bit EXCLUSIVE-OR operation that allows $Y1$ to be replaced by its ones' complement $\overline{Y1}$ during subtraction. Observe that K can be obtained by fanning out the SUB signal from one to 16 lines.

The increment (INC) and decrement (DEC) operations are special cases of add and subtract, with the second input operand $Y1$ of Equation (8.16) set to the 16-bit two's-complement version of ± 1—that is, to $\pm 1_{2C}$. Hence, the increment instruction INC X1 is implemented by the register-transfer operation

$$Z := X1 + 1_{2C}$$

The load instruction LOAD X1, which merely transfers $X1$ to the output register Z, can be realized by $Z := X1 + Y1$, with $Y1 = 0_{2C}$—that is, by

$$Z := X1 + 0_{2C}$$

An adder can implement several useful functions such as subtract, load, increment, and decrement.

Thus, the adder should be able to take its input data from various sources, including the variables $X1$ and $Y1$ stored in the input registers, and the constant words 0_{2C} and 1_{2C}. Note also that there are two subtraction instructions that differ in the order of $X1$ and $Y1$. To handle all these possibilities, multiplexers are placed at the data inputs of the adder-subtracter to allow its input operands to be selected from any of several source-pairs.

A complete design for DP appears in Figure 8.36. Two four-way multiplexers allow one of the input operand-pairs $(X1, Y1)$, $(Y1, X1)$, $(X1, 0_{2C})$, and $(X1, 1_{2C})$ to be selected, depending on the current instruction type. A 1-out-of-8 decoder activates a control signal s_i that uniquely identifies the current instruction. This signal s_i then determines the appropriate operand-pair to use and the signal to be applied to the function select line SUB. A

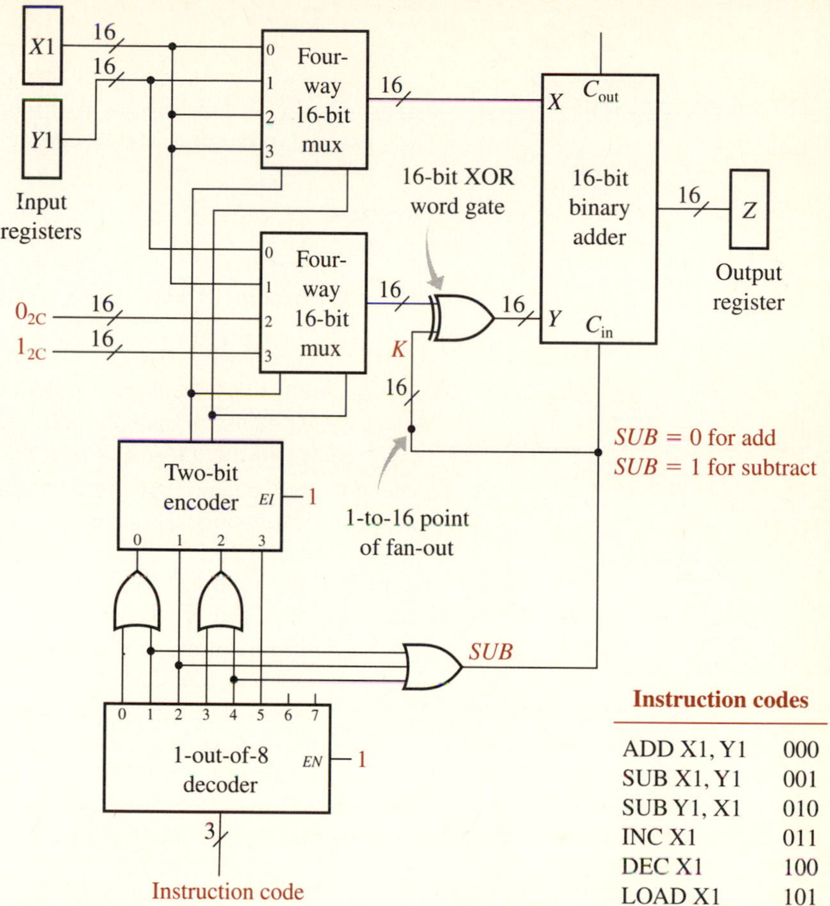

FIGURE 8.36 A 16-bit datapath unit to execute a small instruction set.

2-bit encoder and a few gates are needed for this purpose, as shown in the figure.

Consider, for instance, what happens when the instruction word is 010, denoting SUB Y1,X1. This activates output line 2 of the decoder, which sets $SUB = 1$ and, at the same time, activates input line 1 of the encoder. The encoder responds by placing 01 to the address inputs of the two multiplexers, which then apply $Y1$ and $X1$ to the upper and lower data inputs, respectively, of the adder. The other subtraction instruction, encoded as 001, also sets $SUB = 1$, but it applies 00 instead of 01 to the multiplexers, thus reversing the order of $X1$ and $Y1$. Note that if the data word size or the number of instructions is increased, the circuit components in Figure 8.36 will have to be expanded, but the overall circuit structure can remain much the same as before.

8.5 Control Units

Control units are sequential circuits that generate and select control signals.

A control unit is a digital circuit intended to generate and interpret the logic signals that control the operation of a multifunction data-processing unit such as a datapath processor. The design of such control units is our next topic.

Structure

A control unit *CU* is viewed here as a synchronous sequential circuit with the specialized input/output signals depicted in Figure 8.37a. It directs the operation of a datapath unit *DP* by activating a set of output (control) signals that are sent from *CU* to *DP*. The number of such signals can range from one or two to several hundred in the case of a complex data processor. The main task of *CU* is to activate its output control signals in the proper order at the proper times. Thus, the control unit is responsible for both operation selection and operation timing or sequencing. The controlled unit may also send back information on the status of its operations to *CU*. In addition, *CU* may be linked by a few control lines to a higher level controller; a "master" controller *MC* such as, say, a human operator who can start or stop *CU* or make decisions that are beyond *CU*'s scope. An example of such a decision is what to do in the event of an error condition such as arithmetic overflow or a hardware failure in *DP*.

The control unit *CU* has a well-defined and fairly small set of internal states and so is often referred to as a *finite-state machine* or simply a *state*

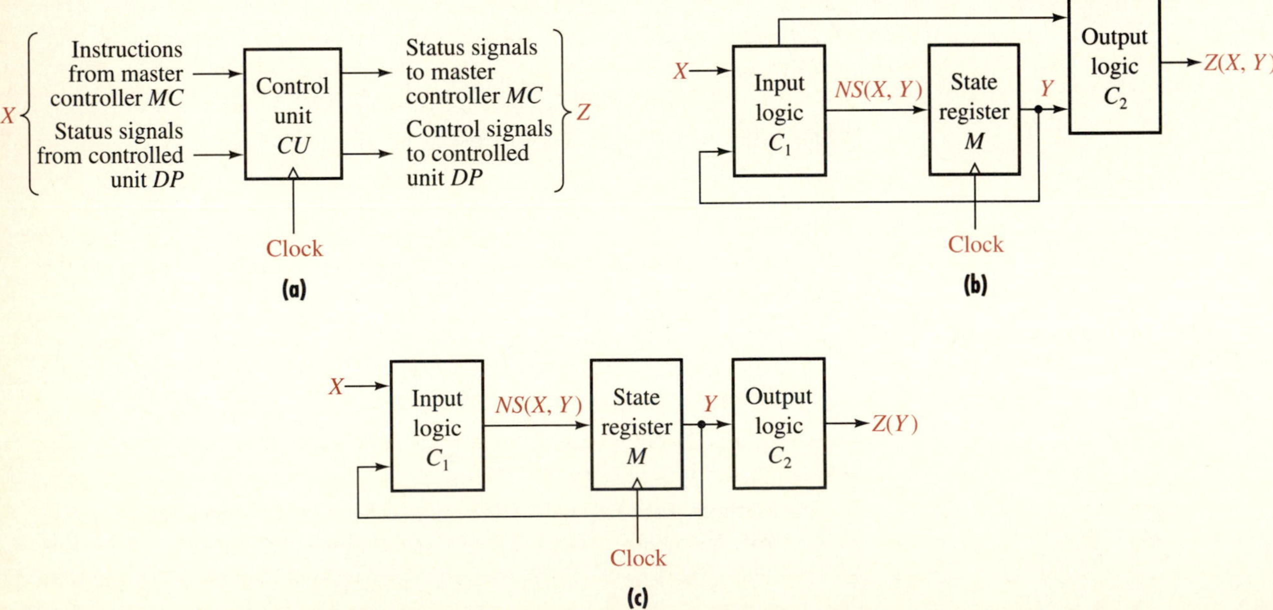

FIGURE 8.37 Control unit structure: (a) input/output signals; (b) a Mealy machine; (c) a Moore machine.

machine. Its internal structure conforms to the general (Huffman) model for a synchronous sequential circuit discussed earlier, with a combinational part C and a memory part M; see Figure 8.37b. The memory part M of CU is a set of flip-flops referred to as the **state register**. It is controlled by a clock signal, which, as usual, is the central timing reference. The combinational logic can be subdivided into the input (next-state) logic C_1, which produces the excitation functions $NS(X, Y)$ applied to M, and the output logic C_2, which produces the outgoing control signals $Z(X, Y)$. Another feature of control units is that they tend to cycle repeatedly through most of their states, leaving this cycle only when exceptional conditions occur, such as normal or abnormal termination by the controlled (datapath) unit. Consequently, control units are sometimes called **sequencers**.

The output signals of the general control unit of Figure 8.37b are functions of the primary inputs X as well as the state variables Y; hence, they can change value whenever the primary input signals change. The slightly simpler structure depicted in Figure 8.37c is also commonly used for control unit design. Here the output signals $Z(Y)$ are solely a function of the state variables Y and are independent of the primary inputs X. In this case, C_2 is effectively a type of decoder for the state register, and the times at which the Z signals change value are controlled by the system clock.

The state machines of Figures 8.37b and 8.37c are known as Mealy machines and Moore machines, respectively. Despite its more restricted structure, a Moore machine can be used to implement any sequential circuit behavior exhibited by a Mealy machine, but generally requires more states to do so, as we will see later. An advantage of a Moore machine is that the primary output signals Z are shielded by the clock from noise induced by changes in the primary input signals. As discussed in Section 7.3, additional clocked latches or registers may be inserted in the primary output lines of either a Mealy or a Moore machine to remove glitches due to hazards in the output logic.

Control Unit Design

Several approaches to the logic design of control units can be identified; next we consider three of the more important:

Here are three general methods to design control units.

1. The classical (finite) state machine procedure *SCSYN*, developed in Chapter 7, which minimizes the number of flip-flops in the state register
2. The "one-hot" method, which is based on a simplified state assignment with one flip-flop per state
3. Methods that employ programmable logic circuits and organize control information systematically into (micro) programs

As we will see in this section, these methods involve various trade-offs between circuit size and design difficulty. The one-hot and programmable logic approaches fix aspects of the control unit's structure in order to simplify the design process. CAD tools are available to automate some or all of the design process, once the necessary control algorithms have been specified.

State Machines

The classical method for designing sequential circuits can be used directly or in various modified forms to implement the logic of a control unit CU. This so-called **state-machine approach** is characterized by the fact that the main subcircuits of CU—the input logic C_1, the output logic C_2, and the state register M—are all specifically tailored to the (state) behavior of CU. As we saw earlier, the design process can be difficult and time-consuming. However, it has the advantage that the number of logic gates and flip-flops can be kept close to the minimum, thus reducing circuit size and cost. For example, if CU requires precisely P internal states, we can implement it with the absolute minimum number of flip-flops—namely, $p = \lceil \log_2 P \rceil$.

Example: A LIFO Stack

A LIFO stack is a memory that is accessed from one end via push and pop.

To illustrate the state-machine approach applied to control unit design, we consider the design of a controller for a special type of memory known as a stack. A **last-in, first-out** or **LIFO stack** consists of a sequence of n registers organized so that data can only be written into or read from a fixed end of the sequence, known as the **top** of the stack. When a new item is written into the stack, the previously stored items are—at least conceptually—pushed deeper into the stack to accommodate the new word. When the stack is read, the data word currently stored at the top of the stack is completely removed or "popped," allowing the remaining words to move one position closer to the top of the stack. An obvious analogy may be made if we think of a stack of trays in a cafeteria, which when removed by incoming customers exhibit the same kind of last-in, first-out access procedure. The write and read operations used to access a stack memory are called **push** and **pop**, respectively.

The particular stack of interest here stores a maximum of $n = 4$ words and has the structure outlined in Figure 8.38a. The datapath part DP is implemented as a four-word shift register and can be constructed from four standard w-bit left-right shift registers, where w is the data word size. The stack control unit CU responds to external push and pop command signals by sending appropriate shift signals to DP. We assume that a push requires a right shift, during which the contents of the data-in bus are loaded into the top of the stack. Similarly, a pop initiates a left shift, during which the contents of the top of the stack are transferred to the data-out bus. If the *POP* line is activated when no data are stored—that is, when the stack is empty—the control line *ERROR* should be activated; an attempt to push a new word into the stack when it is already storing four words should also activate the *ERROR* line.

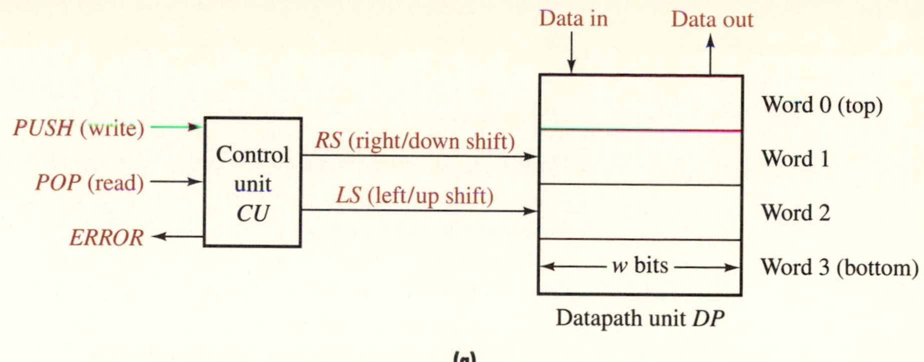

(a)

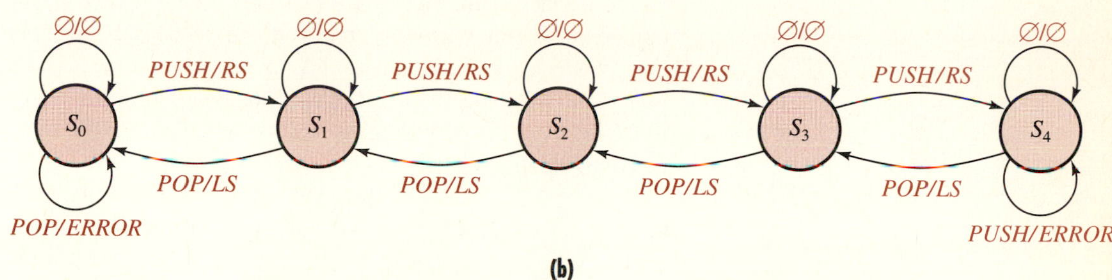

(b)

FIGURE 8.38 A four-word LIFO stack memory: **(a)** its system structure; **(b)** its state diagram.

State Description

A formal state representation of the stack's behavior can be deduced from the foregoing description. Figure 8.38b shows a state diagram in which, as before, each state S_i has a set of outgoing transition arrows with labels of the form A/B, with A denoting a set of input signals (Boolean functions) and B denoting the corresponding output signals produced by the combined presence of S_i and A. The diagram format has been slightly modified, however, by defining A and B as lists of symbolic logic signals, so that A/B appears as

$$a_1, a_2, \ldots, a_r / b_1, b_2, \ldots, b_s \tag{8.17}$$

We simplify the description of state behavior by listing only active signals.

The state transition with label (8.17) takes place if and only if all the occurrences of a_i in the input list A are currently active. The output list B specifies all those signals that are activated when the corresponding state transition occurs.

For example, the label $PUSH/RS$ that appears on several transition arrows in Figure 8.38b states that when only the $PUSH$ signal is active ($PUSH = 1$), the corresponding state transition occurs and only the output signal RS is activated ($RS = 1$). Note that $PUSH/RS$ is an abbreviated representation of the more complete label

$$PUSH = 1, POP = 0/RS = 1, LS = 0, ERROR = 0$$

indicating the relevant values of all five of the control unit's input/output signals. In many control unit designs, only one or two input/output signals are active at a time, so the A/B notation is quite compact. The label $-/B$, in which the input list consists of the don't-care symbol $-$, indicates a state transition that takes place every clock cycle, independent of the primary input values. We use the empty-set symbol $\varnothing$ to specify an empty list. Hence, $A/\varnothing$ indicates that all output signals are inactive. The label $\varnothing/\varnothing$ that appears in Figure 8.38b means that the corresponding transition activates no output signals and occurs only when all inputs are inactive.

The stack controller has five states $S_0{:}S_4$, where S_i indicates that the number of words currently stored in the stack is i. In most cases, activating $PUSH$ in state S_i causes RS to be set to 1; activating POP sets LS to 1. The error indicator $ERROR$ is activated in S_0 if $POP = 1$ and in S_4 if $PUSH = 1$. For simplicity, we assume that $PUSH$ and POP are never activated simultaneously.

Because CU has five states, it can be implemented with a minimum of three flip-flops. The design steps required in the classical procedure $SCSYN$ (Section 7.5) are to assign the states to five of the eight possible combinations of the three state variables, determine the primary and secondary output functions that define the combinational part C of CU, and design (minimal) two-level logic circuits that realize C. A typical design of this sort that contains three D flip-flops appears in Figure 8.39. It is a straightforward process to verify that this circuit has the desired state behavior with the state vector $y_2y_1y_0$ and the state assignment $S_0 = 000, S_1 = 001, S_2 = 010, S_3 = 011, S_4 = 100$.

One-Hot State Assignment

A major complication in sequential circuit design with the minimum number of flip-flops is that there is usually no direct relation between the state assignment and the functions assigned to the various states. The state vectors in M appear as random patterns of 0s and 1s, and adversely affect, in ways that are hard to determine, both the size of the combinational logic C and the difficulty of obtaining C.

The one-hot assignment has one flip-flop per state, which simplifies design.

An alternative approach is to use a fixed but nonminimal state assignment that simplifies the design process. An example of such a method, and one that is well suited to control unit design, is the colorfully named **one-hot method**, which uses one flip-flop per state—that is, P flip-flops instead of $\lceil \log_2 P \rceil$ for a P-state machine. For each state S_i, the corresponding state variable y_i is set to 1, or is "hot," while all other flip-flops are set to 0.

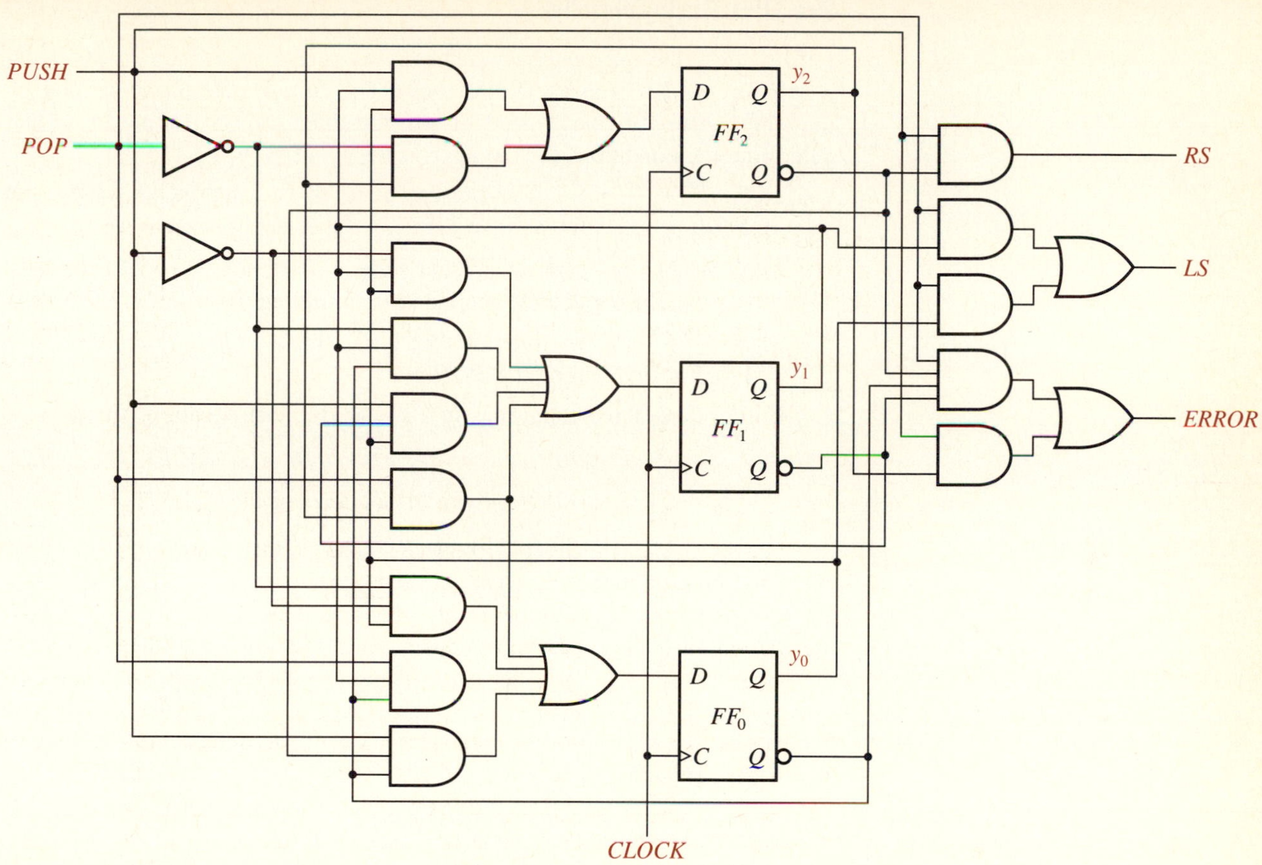

FIGURE 8.39 A typical state-machine implementation of the stack memory controller that uses a minimum number of flip-flops.

Thus, for the five states of the stack control unit discussed above, we have the following one-hot assignment using the state vector $y_0y_1y_2y_3y_4$:

$$
\begin{aligned}
S_0 &= 10000 \\
S_1 &= 01000 \\
S_2 &= 00100 \\
S_3 &= 00010 \\
S_4 &= 00001
\end{aligned}
\tag{8.18}
$$

The one-hot state assignment has the obvious disadvantage of possessing more flip-flops than are necessary. Its main advantage is that the next-state and output equations that define the combinational part C of the circuit are relatively simple and can be quickly deduced directly from a state table or diagram.

The LIFO Stack Controller

Let us apply the one-hot assignment (8.18) to the state diagram of Figure 8.38b for the LIFO stack control unit. The initial state S_0 is defined by the single state variable $y_0 = 1$. There are three arrows leading to S_0 that determine the conditions for the next value y_0^+ of y_0 to be 1:

1. $S_0 \xrightarrow{POP/ERROR} S_0$: The present state is S_0 (that is, $y_0 = 1$) and $POP = 1$.
2. $S_0 \xrightarrow{\varnothing/\varnothing} S_0$: Again the present state is S_0 (that is, $y_0 = 1$) and both inputs *PUSH* and *POP* are inactive, implying that $\overline{PUSH} \cdot \overline{POP} = 1$.
3. $S_1 \xrightarrow{POP/LS} S_0$: The present state is S_1 (that is, $y_1 = 1$) and the active input is *POP*, implying $POP = 1$.

Combining these three conditions in a single Boolean equation, we get

$$
\begin{aligned}
y_0^+ &= y_0 \cdot POP + y_0 \cdot \overline{PUSH} \cdot \overline{POP} + y_1 \cdot POP \\
&= y_0 \cdot POP + y_0 \cdot \overline{PUSH} + y_1 \cdot POP
\end{aligned}
$$

For the state S_1, there are also three situations that require $y_1^+ = 1$: first, when the present state is S_0 (that is, $y_0 = 1$) and $PUSH = 1$; second, when $y_1 = 1$, and both *PUSH* and *POP* are 0; and, finally, when $y_2 = 1$ and *POP* is 1. Combining these possibilities, we obtain the next-state function for y_1:

$$y_1^+ = y_0 \cdot PUSH + y_1 \cdot \overline{PUSH} \cdot \overline{POP} + y_2 \cdot POP$$

By considering each state separately in this way, we obtain a full set of five next-state equations:

$$
\begin{aligned}
y_0^+ &= y_0 \cdot POP + y_0 \cdot \overline{PUSH} + y_1 \cdot POP \\
y_1^+ &= y_0 \cdot PUSH + y_1 \cdot \overline{PUSH} \cdot \overline{POP} + y_2 \cdot POP \\
y_2^+ &= y_1 \cdot PUSH + y_2 \cdot \overline{PUSH} \cdot \overline{POP} + y_3 \cdot POP \\
y_3^+ &= y_2 \cdot PUSH + y_3 \cdot \overline{PUSH} \cdot \overline{POP} + y_4 \cdot POP \\
y_4^+ &= y_3 \cdot PUSH + y_4 \cdot PUSH + y_4 \cdot \overline{POP}
\end{aligned} \tag{8.19}
$$

These are converted into flip-flop excitation equations in the usual way, depending on the flip-flop type. D flip-flops are the most convenient to use, for if D_i is the data input of the ith flip-flop, the excitation equations follow directly from (8.19) via the relation

$$y_i^+ = D_i \tag{8.20}$$

One-hot excitation and output equations can be taken directly from the state diagram.

To determine the output equations for the stack control unit *CU*, we examine the output-signal labels in the state diagram (Figure 8.38b). We find that *RS* is set to 1 in every state except S_4 whenever $PUSH = 1$. Thanks to the one-hot assignment, this means that

$$
\begin{aligned}
RS &= y_0 \cdot PUSH + y_1 \cdot PUSH + y_2 \cdot PUSH + y_3 \cdot PUSH \\
&= (y_0 + y_1 + y_2 + y_3) \cdot PUSH \\
&= \bar{y}_4 \cdot PUSH
\end{aligned} \tag{8.21}
$$

Similarly, the output control signal LS is set to 1 by $POP = 1$ in all states except S_0. Hence,

$$LS = (y_1 + y_2 + y_3 + y_4) \cdot POP$$
$$= \bar{y}_0 \cdot POP \tag{8.22}$$

Finally, the error signal is defined by

$$ERROR = y_0 \cdot POP + y_4 \cdot PUSH \tag{8.23}$$

A logic circuit that implements CU according to the one-hot assignment and Equations (8.19), (8.20), (8.21), (8.22), and (8.23) appears in Figure 8.40. Note that although the number of flip-flops has increased from three to

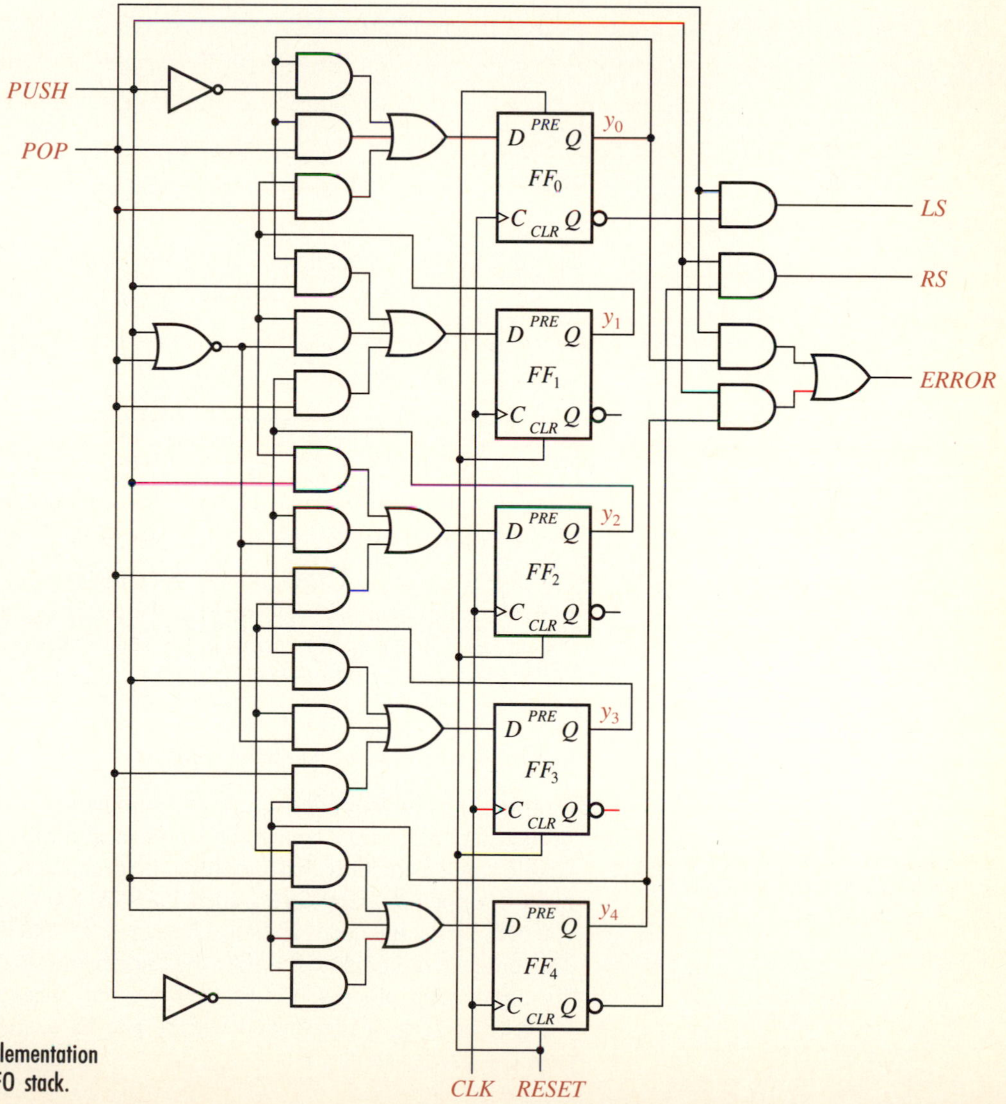

FIGURE 8.40 One-hot implementation of the control unit for the LIFO stack.

five, or by two thirds, compared to the design of Figure 8.39, the number of gates has increased by only about a third. Note also the unusual reset connections to initialize the circuit to state S_0.

One-Hot Equations

In the general case, the combinational logic for a one-hot design may be obtained as follows: Let $S_{i,1}, S_{i,2} \ldots, S_{i,P_i}$ denote the set of all distinct states that have S_i as a possible next state. For each such state $S_{i,j}$, let $I_{(i,j),1}, I_{(i,j),2} \ldots, I_{(i,j),Q(j)}$ be all input combinations that have $S_{i,j}$ as the initial state and S_i as the final state. Then the next-state equation for y_i, the state variable that defines S_i in the one-hot assignment, is

General one-hot next-state equation.

$$\begin{aligned} y_i^+ = {} & y_{i,1} \cdot (I_{(i,1),1} + I_{(i,1),2} + \cdots + I_{(i,1),Q(1)}) \\ & + y_{i,2} \cdot (I_{(i,2),1} + I_{(i,2),2} + \cdots + I_{(i,2),Q(2)}) \\ & + y_{i,3} \cdot (I_{(i,3),1} + I_{(i,3),2} + \cdots + I_{(i,3),Q(3)}) \\ & \ldots \\ & + y_{i,P_i} \cdot (I_{(i,P_i),1} + I_{(i,P_i),2} + \cdots + I_{(i,P_i),Q(P_i)}) \end{aligned} \tag{8.24}$$

Let $I_{(i,j),1}/z_j, I_{(i,j),2}/z_j \ldots, I_{(i,j),U(j)}/z_j$ be all outgoing transition labels from state S_i that require setting $z_j = 1$. Then, summing over all states $S_{i,1}, S_{i,2} \ldots, S_{i,V_i}$, with z_j activated at least once in an outgoing transition, we obtain

General one-hot output equation.

$$\begin{aligned} z_j = {} & y_{i,1} \cdot (I_{(i,1),1} + I_{(i,1),2} + \cdots + I_{(i,1),U(1)}) \\ & + y_{i,2} \cdot (I_{(i,2),1} + I_{(i,2),2} + \cdots + I_{(i,2),U(2)}) \\ & + y_{i,3} \cdot (I_{(i,3),1} + I_{(i,3),2} + \cdots + I_{(i,3),U(3)}) \\ & \ldots \\ & + y_{i,V_i} \cdot (I_{(i,V_i),1} + I_{(i,V_i),2} + \cdots + I_{(i,V_i),U(V_i)}) \end{aligned} \tag{8.25}$$

The foregoing logic equations can be simplified or converted to standard forms, such as minimal SOP or POS forms, if desired. Observe that if $I_{i,j}$ appears as a Boolean expression in the state description, the expression may be inserted directly into Equations (8.24) and (8.25).

Example 8.8

One-hot design for a counter-based multiplier

Consider the problem of computing the product $P = X \times Y$ of two unsigned, n-bit binary numbers X and Y. This type of multiplication has a conceptually simple implementation that employs three counters *CX*, *CY*, and *CP*. The input operands X and Y are loaded into *CX* and *CY*, respectively; the third counter *CP* is intended to hold the result P and is initially set to zero. The product is computed simply by incrementing *CP* a total of P times. To control this process, *CX* is decremented once per clock cycle. When *CX* reaches zero, it is reloaded with X and its decrementing resumes. Each time *CX* reaches zero, *CY* is also decremented. Clearly, *CY* reaches zero after

Multiplication by counting is slow but flexible and inexpensive.

CX has been decremented a total of $X \times Y$ times. If we increment *CP* at the same time we decrement *CX*, then, when *CY* reaches zero, *CP* must contain the desired result *P*. This method of multiplication is obviously slow, but it requires relatively little datapath hardware and can easily accommodate unusual number lengths or codes.

The multiplier has a datapath unit *DP* that consists primarily of the two n-bit down counters *CX* and *CY* and a $2n$-bit up counter *CP*. Because a copy of X is needed at various points in the calculation, we will also provide an additional n-bit register *RX* to store X. The multiplier requires a controller *CU* that generates control signals to increment or decrement the counters, to load *CX* and *CY*, and to reset *CP*. The control unit *CU*, in turn, requires two status signals from *DP* that indicate when either *CX* or *CY* reaches zero. We shall also assume that there is a master controller *MC* that supplies the operands X and Y, initiates multiplication by activating a control signal *BEGIN*, and requires a termination signal *END* from *CU* when the multiplication is complete. Figure 8.41 gives the general structure of *DP*, *CU*, and their interconnections.

From the foregoing word description of the counter-based multiplier, we can deduce that *CU* has the following states:

1. S_0: An idle state when no operations are being performed
2. S_1: An initialization state triggered by *BEGIN*= 1 and causing the input operand X to be loaded into *RX*; in this state, *CP* can be reset to zero

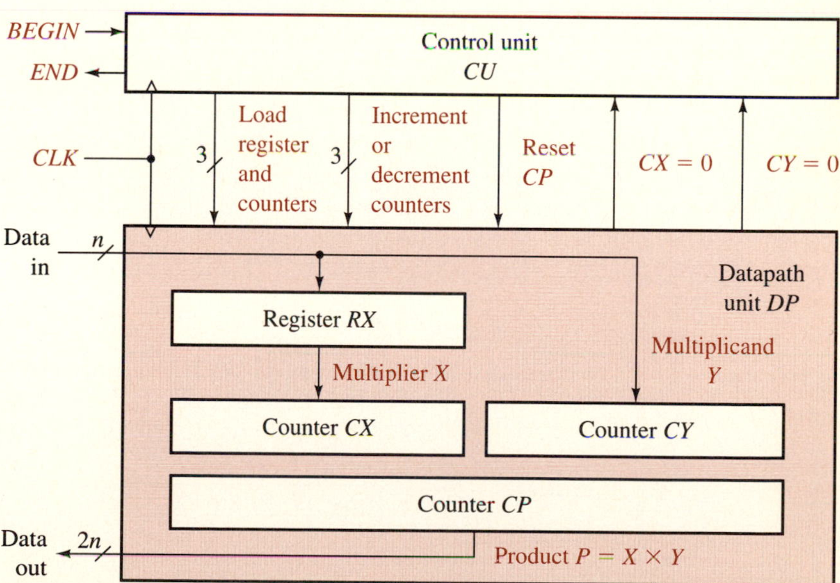

FIGURE 8.41 An overview of the structure of a counter-based multiplier.

3. S_2: A second initialization state that causes the input operand Y to be loaded into CY; also, the contents of RX are transferred to CX, thus completing initialization

4. S_3: A counting state during which CX is decremented and CP is incremented

5. S_4: The state entered after CX reaches zero: CY is decremented and X is reloaded into CX from RX

A state diagram for *CU* can readily be constructed; see Figure 8.42. The three input signals named *LOAD* are used to parallel-load the specified device; *RESET_CP* resets the product counter *CP*; *DEC_CX* and *DEC_CY* decrement *CX* and *CY*, respectively; and *INC_CP* increments *CP*. In state S_3, for instance, the output signals *DEC_CX* and *INC_CP* are activated whenever the status signal named *CX_ZERO* is 0 (inactive), indicating that the contents of counter *CX* are nonzero; the control state remains at S_3. When *CX_ZERO* changes to 1, the state of *CU* changes to S_4 and the control signals *DEC_CX* and *INC_CP* are changed to 1, causing *CX* to be decremented and *CP* to be incremented. Recall that a state transition arrow with the empty input list denoted by ∅ merely indicates that the next state is independent of the control signals other than, of course, the clock.

For the five states of the multiplier discussed above, we have the usual one-hot assignment $S_0 = 10000, S_1 = 01000, \ldots, S_4 = 00001$, using the state vector $y_0y_1y_2y_3y_4$. The corresponding five next-state equations implied by (8.24) are as follows:

$$
\begin{aligned}
y_0^+ &= y_0 \cdot \overline{BEGIN} + y_2 \cdot (CX\_ZERO + CY\_ZERO) \\
&\quad + y_4 \cdot CY\_ZERO \\
y_1^+ &= y_0 \cdot BEGIN \\
y_2^+ &= y_1 \\
y_3^+ &= y_2 \cdot (\overline{CX\_ZERO} \cdot \overline{CY\_ZERO}) + y_3 \cdot \overline{CX\_ZERO} \\
&\quad + y_4 \cdot \overline{CY\_ZERO} \\
y_4^+ &= y_3 \cdot CX\_ZERO
\end{aligned}
\tag{8.26}
$$

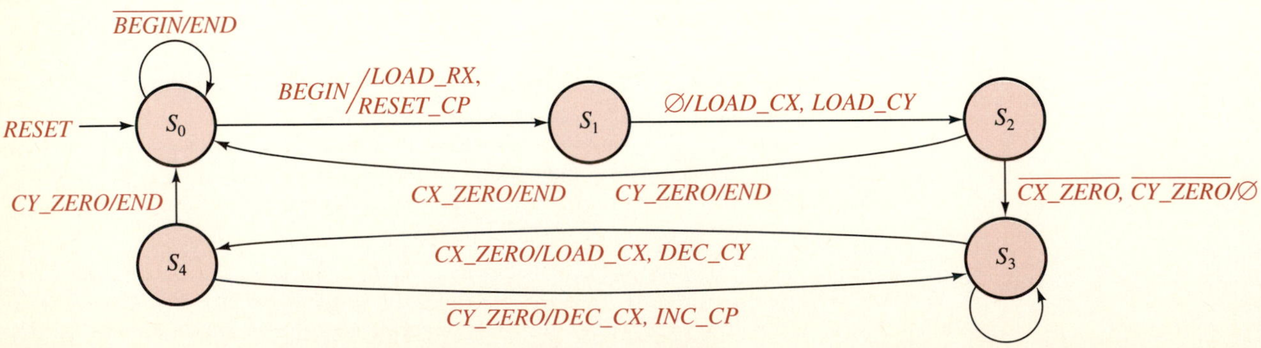

FIGURE 8.42 A state diagram for the control unit of the counter-based multiplier.

From (8.25), the output equations are

$$
\begin{aligned}
LOAD\_RX &= y_0 \cdot BEGIN \\
LOAD\_CX &= y_1 + y_3 \cdot CX\_ZERO \\
LOAD\_CY &= y_1 \\
DEC\_CX &= y_3 \cdot \overline{CX\_ZERO} + y_4 \cdot \overline{CY\_ZERO} \\
DEC\_CY &= y_3 \cdot CX\_ZERO \\
INC\_CP &= y_3 \cdot \overline{CX\_ZERO} + y_4 \cdot \overline{CY\_ZERO} \\
RESET\_CP &= y_0 \cdot BEGIN \\
END &= y_0 \cdot \overline{BEGIN} + y_2 \cdot (CX\_ZERO + CY\_ZERO) \\
&\quad + y_4 \cdot CY\_ZERO
\end{aligned}
\tag{8.27}
$$

A logic circuit that implements CU according to the one-hot assignment and Equations (8.26) and (8.27) appears in Figure 8.43. This is the logic diagram used as input to the simulation program (LogicWorks) that checks this design. Figure 8.43 shows the circuit in state $y_0y_1y_2y_3y_4 = 00010 = S_3$, which occurs in the middle of a typical multiplication process, with both CX and CY nonzero. The only active output signals are DEC_CX and INC_CP, which are both 1.

Moore Machines

Moore machines are especially suitable for control unit design.

In Section 7.2, we distinguished between two major types of sequential circuits or state machines: Mealy and Moore machines. The more general Mealy model allows a circuit's output signals Z to depend on both the primary inputs X and the secondary or state inputs Y, as indicated by the notation $Z(X, Y)$. This implies that the Z signals change in response to changes in X, as indicated in Figure 8.44a. This timing diagram shows a clock cycle that represents the state transition fragment

$$\longrightarrow Y' \xrightarrow{X'/Z'} Y'' \xrightarrow{X''/Z''} \tag{8.28}$$

where positive edge triggering by the clock is assumed. The time at which the new output value $Z'' = Z(X'', Y'')$ appears depends on the time at which X changes relative to Y. If the change in X is unduly delayed relative to that of Y due to external conditions or spurious internal delays, undesired glitch signals such as $Z^* = Z(X', Y'')$ (in color in the figure) can appear as shown. Moreover, any noise or glitches that occur in X can be passed directly to Z.

The fact that Z fluctuates with X in the above manner complicates the design of a control unit CU because CU's output signals are inputs to a datapath unit DP and must satisfy the latter's timing constraints, such as the setup and hold times of its flip-flops. It can be useful to make Z only depend

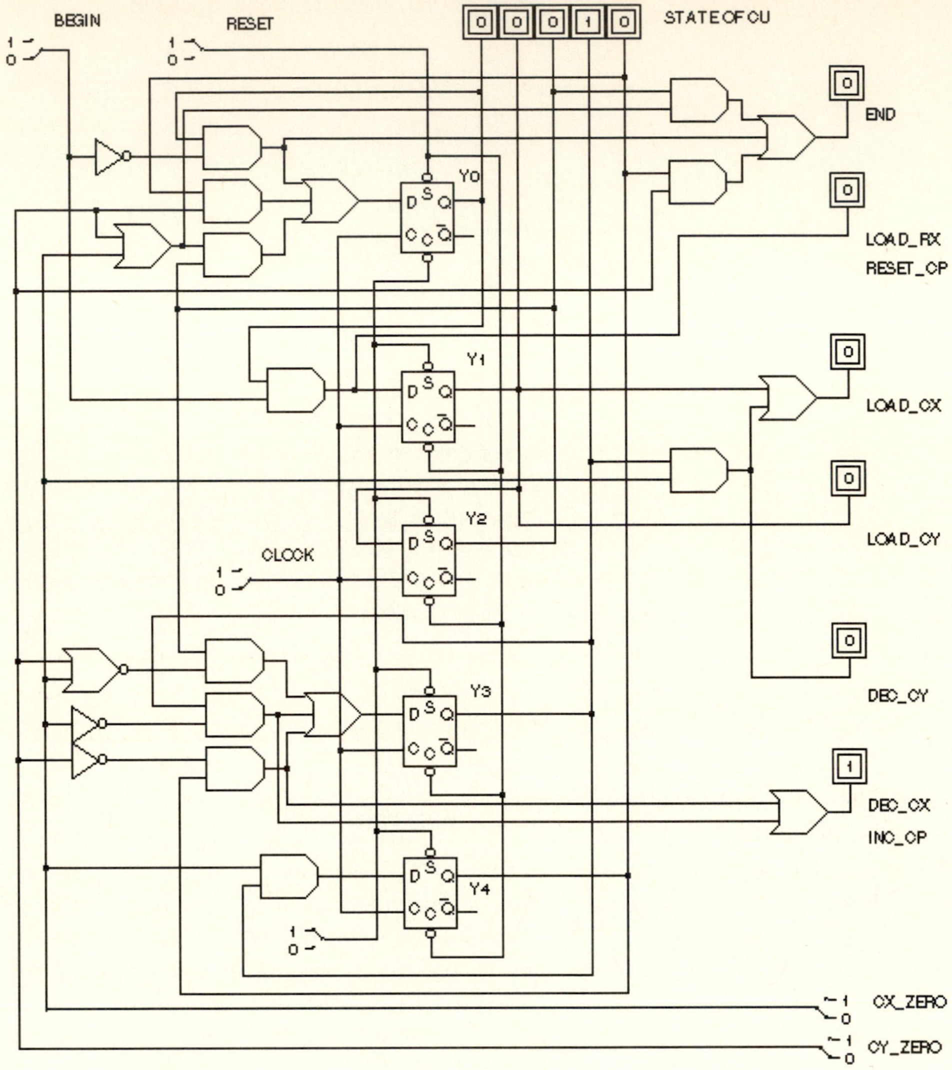

FIGURE 8.43 Simulated one-hot implementation of the control unit for the counter-based multiplier.

on Y, so that changes in X no longer directly affect Z. In this case, the control unit is a Moore state machine rather than a Mealy machine. Figure 8.44b illustrates the timing behavior of a Moore machine that implements the state behavior given in Equation (8.28). Note the "clean" single change in Z and the fact that the clock, via the change in the state Y, determines the time at which Z changes.

The more precise output timing control of the Moore model is not without its price, however. The value of Z is fixed by that of Y, so that more states

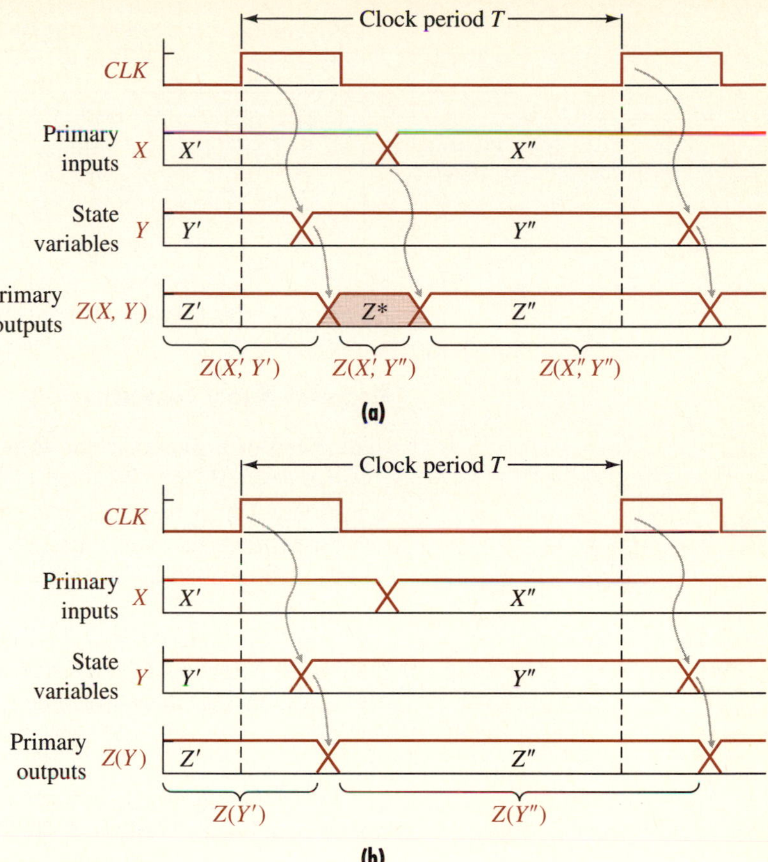

FIGURE 8.44 Timing behavior of (a) a Mealy machine and (b) a Moore machine.

may be needed to represent a given range of behavior. We can omit the output signal information from (8.28), replacing it with

$$\longrightarrow Y' \xrightarrow{X'} Y'' \xrightarrow{X''} \tag{8.29}$$

because Z' and Z'' are the only possible outputs generated by this Moore machine in states Y' and Y'', respectively. Thus, a Moore version of a state table or state diagram omits the primary output entries and replaces them with a list showing the fixed output value that corresponds to every state.

A small example of a Moore machine appears in Figure 8.45a, which has one primary input, two primary outputs, and four internal states (this is the traffic-light control unit discussed in Section 7.4). Here, for instance, when the machine is in state S_0, the primary output is $Z = z_1z_2 = 10$. Despite their differences in response timing and circuit structure, any sequential machine behavior can be abstractly described by either a Mealy or a Moore model. We demonstrate this concept via procedures to convert state tables or diagrams from either model to the other.

		Present input x 0	1	Present output z_1z_2
Present state	S_0	S_1	S_2	01
	S_1	S_0	S_3	10
	S_2	S_1	S_1	00
	S_3	S_0	S_0	00

(a)

		Present input x 0	1
Present state	S_0	S_1, 10	S_2, 00
	S_1	S_0, 01	S_3, 00
	S_2	S_1, 10	S_1, 10
	S_3	S_0, 01	S_0, 01

(b)

FIGURE 8.45 The traffic-light controller: (a) its Moore state table; (b) its Mealy state table.

Moore-to-Mealy Conversion

Conversion of a state machine from the Moore format to the Mealy format is quite easy: It only involves moving output labels from states to state transitions. The Mealy machine has the same set of states and the same number of next-state transitions as the given Moore machine. Let the input X_j take it from S_i to S_k; the Moore machine therefore has the state transition

$$S_i \xrightarrow{X_j} S_k$$

If Z_k is the output value associated with state S_k in the above state transition, the corresponding Mealy machine is assigned the transition

$$S_i \xrightarrow{X_j/Z_k} S_k$$

Conversion from Moore to Mealy format is easy; only output labels must be moved.

This Moore-to-Mealy conversion process is illustrated by Figure 8.45b, which is a Mealy version of the Moore state table of Figure 8.45a. For example, the second entry in the first row of the Moore table specifies the state transition $S_0 \xrightarrow{1} S_2$; the output associated with S_2 is 00. From this, we obtain the Mealy-type transition $S_0 \xrightarrow{1/00} S_2$, which is entered into the corresponding position in the Mealy table. The other entries of the Mealy table are obtained similarly.

The two machines have the same state behavior in the sense that, if the input sequence $x = 0, 1, 1, 0$ is applied to each machine, starting in state S_0, each machine goes through the same state sequence S_1, S_3, S_1, S_0 and generates the same output sequence $z_1z_2 = 10, 00, 10, 01$. The foregoing conversion process equates the present Mealy output with the output of the next Moore state, so that the output of the Moore machine is delayed relative to the corresponding Mealy machine output.

Mealy-to-Moore Conversion

Mealy-to-Moore conversion may introduce many new states.

Conversion from Mealy format to Moore format is a bit more difficult. For each state S_k of the Mealy machine, we introduce as many states as there are transitions of the form $S_i \xrightarrow{X_j/Z_{kh}} S_k$, leading to S_k with different output values Z_{kh}. Thus, if Z_{kh} assumes the p distinct values $Z_{k1}, Z_{k2}, \ldots, Z_{kp}$, we replace S_k with p new states $S_{k1}, S_{k2}, \ldots, S_{kp}$. To each such state S_{kh}, we assign the fixed output Z_{kh} in the Moore model. We also supply each S_{kh} with an input transition of the form $S_i \xrightarrow{X_j} S_{kh}$ (that is, all the original transitions into the state S_k are repeated for $S_{k1}, S_{k2}, \ldots, S_{kp}$).

This conversion process is applied to the Mealy state diagram of the LIFO stack control unit (Figure 8.38), which is repeated in Figure 8.46a, and produces the Moore state diagram of Figure 8.46b. Near each Moore state, we write a list of the output signals that are activated when the machine is in that state. As before, ∅ indicates that no output signals at all are active. Notice the large increase in the number of states due to the fact that each original Mealy state has been split into three new states. For example, state S_0 has three outgoing arrows with different output labels, so we split it into three states $S_{0,0}$, $S_{0,1}$, and $S_{0,2}$, which have the associated active output lists ∅, *LS*, and *ERROR*, respectively. State minimization techniques (Section 7.6) can reduce the number of states in some cases.

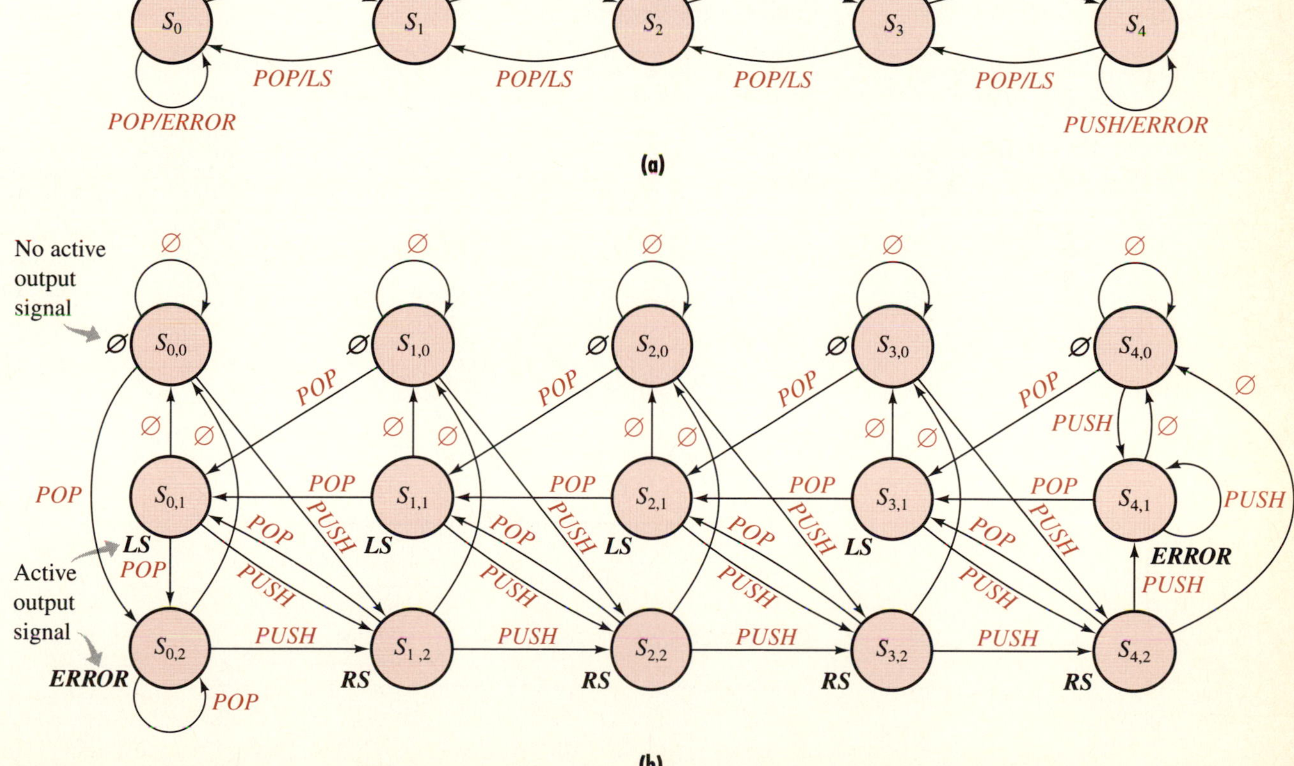

FIGURE 8.46 State diagrams for the LIFO stack control unit: (a) the Mealy version; (b) the Moore version.

8.6 Programmable Controllers

Programmable logic devices, either alone or in combination with other logic circuits, are especially suited to the design of control units. We now briefly discuss the design of such programmable controllers and contrast them with the hardwired designs we have considered so far.

Control units are often designed with the control signals they are required to generate stored explicitly in a PLD such as a ROM or a PLA. These signals are read out in the proper order when the control unit is in operation. By changing the stored control information in such a **programmable control unit**—that is, by writing or programming a new control algorithm (program) into the control unit—the same underlying hardware can be used for many different applications. Control units of this kind are often termed **microprogrammable units**, especially when they implement the control part of a digital system, like a computer's CPU that is itself programmable at a higher level of abstraction. In contrast, a nonprogrammable or **hardwired** control unit of the kind discussed in the preceding sections has its output sequences buried throughout its structure, and so cannot be adapted for a new control application without a thorough redesign.

Control Unit Structure

Programmable control units store control information as instruction sequences.

The structure common to most (micro) programmable control units is outlined in Figure 8.47. This circuit *PCU* is a Moore-type sequential machine with a state register M from which the current output (control) signals Z are derived, usually with little or no output logic C_2; compare it with Figure 8.37. The input logic C_1 contains a component called the **control memory** *CM*, also known as the **control store**, which is a programmable device—a ROM, for example—that contains a program-like description of the desired behavior of *PCU*. This description is a set of words, referred to as **micro-**

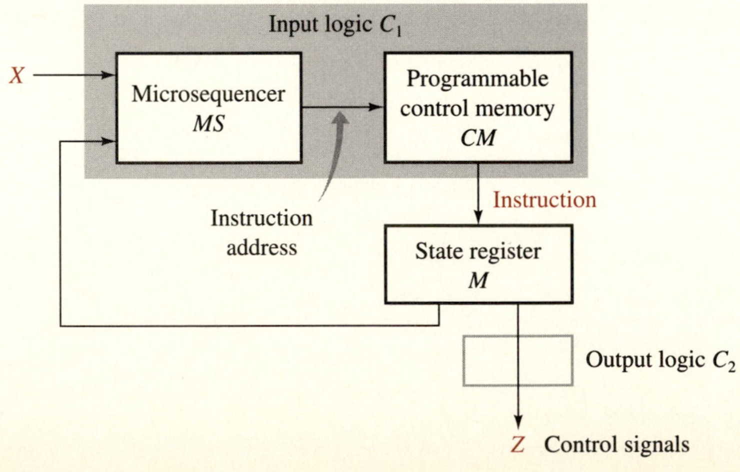

FIGURE 8.47 The general structure of a (micro) programmable control unit.

instructions. By systematically transferring instruction words from *CM* to *M*, one per clock cycle, a sequence of control signals is applied by *M* to the output lines *Z*.

Every instruction stored in *CM* has a unique address that must be applied to *CM* in order to transfer the correct instruction into *M*. A component called a **microsequencer** *MS* selects the next address to be applied to *CM*. This address is determined both by the current instruction stored in *M*—that is, by the control unit's current state—and by the external input *X*. Because of its role in storing the current microinstruction, the state register is usually called the **microinstruction register** in this context.

Read-only memories are natural devices for realizing the control memory *CM*. Each instruction of the target machine is assigned an address in the ROM, and the corresponding next-state and output data are programmed into the ROM—that is, stored in word form in its OR plane. However, combinational PLAs and PALs may also be used for this purpose, especially if the number of states is small or the size of *CM* is to be kept to a minimum. These devices can be viewed as read-only memories with restricted address spaces; in other words, n input address lines but fewer than 2^n stored words.

The size of the *CM* increases rapidly with the number of input/output lines and possible next states. Each extra address line, for example, doubles the number of next-state rows in *CM*. If PLAs, PALs, or other programmable devices have built-in registers, as in the case of the PAL16R4 (Figure 8.25), it may be possible to implement the entire control unit of Figure 8.47 around a single PLD.

Simple Design Method

Storing the complete state table is an easy but impractical way to design programmable controllers.

A conceptually simple but generally impractical way to construct a programmable control unit *PCU* would be to store its (Moore) state table in *CM*, as suggested by Figure 8.48. (This is analogous to storing the truth table of a combinational circuit in a ROM.) *CM* and *M* must then be designed to treat the contents of a row of the state table as an "instruction." Each such instruction is addressed and read into *M* under the control of the current

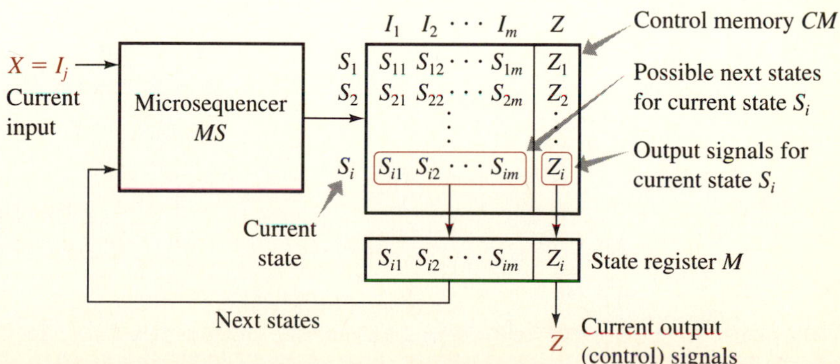

FIGURE 8.48 Embedding of a Moore state table in a programmable control unit.

state S_i. The part of M that stores Z_j supplies the output control signals; the part that stores the set of possible next states is fed back to CM via the microsequencer MS, which, in conjunction with the current input combination $X = I_j$, selects the particular next state needed to address CM in the following clock cycle.

Figure 8.49 shows the above technique applied to the design of the traffic-light controller of Section 7.4. This control unit has four states: a single primary input x and two primary outputs z_1, z_2. Each state is encoded in two bits in the standard way, with $S_0 = 00$, $S_1 = 01$, $S_2 = 10$, and $S_3 = 11$. These binary codes serve as the instruction addresses. For each current state, there are only two possible next states. Hence, to store all possible next states as well as the two current output bits requires 6 bits. The control memory of this control unit, therefore, stores a total of four 6-bit words.

A two-way, 2-bit multiplexer MUX provides a convenient way to implement the microsequencer MS. The 2-bit (data) word channeled through MS to CM is the address of the next instruction. To read out the row that corresponds to state S_i, the 2-bit binary number i must be applied by MS to CM as the current address. Application of the address i to CM causes the corresponding 6-bit word to be read into the state register M. Two bits of M then supply the output signals to the device being controlled.

In Figure 8.49b, the current state shown is $S_0 = 00$. Its next states are $S_1 = 01$ when $x = 0$ and $S_1 = 10$ when $x = 1$; the corresponding output combination is $z_1z_2 = 01$. Hence, the current instruction word that appears in the instruction register M is 011001, which is the encoded form of the first row of the truth table. If $x = 0$, MUX selects its lower input bus, marked 0, and transfers its contents, 01, to CM's address bus so that the next state visited will be $S_1 = 01$, as required to implement the state table's first entry $S_0 \xrightarrow{0} S_1$.

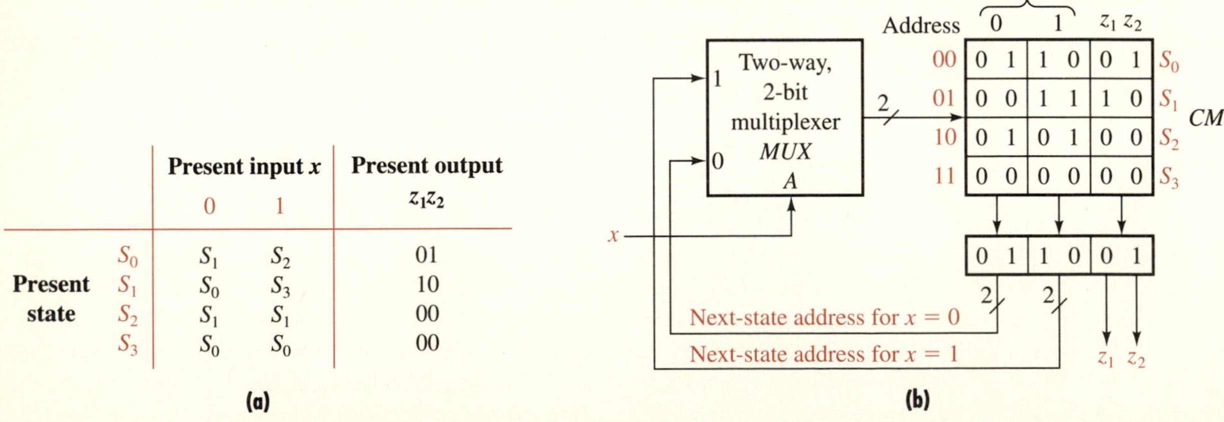

		Present input x 0	Present input x 1	Present output z_1z_2
Present state	S_0	S_1	S_2	01
	S_1	S_0	S_3	10
	S_2	S_1	S_1	00
	S_3	S_0	S_0	00

FIGURE 8.49 (a) A Moore state table and (b) a programmable implementation of the traffic-light controller state machine.

Practical Issues

Control unit behavior can be corrected or changed just by changing the contents of *CM*.

The key advantage of a programmable control unit is the fact that the same hardware can be used to implement *any* state table of appropriate size, merely by changing the contents of *CM*. We can embed any table with at most four states, one input bit, and two output bits in the circuit of Figure 8.49b by reprogramming *CM*. The controller's hardware remains fixed; we only need to change the "software" or "firmware," as *CM*'s contents are often called. This flexibility is also a great advantage when it comes to debugging a new design. A typical small logic design error, like an incorrect next state or an erroneous output signal, can be corrected quickly by making a small change to the contents of *CM*.

It is not feasible to store a complete state table in a control memory for all but the smallest designs because of the table's large size, nor is it necessary to do so. Control units generally have only a small number of possible next states because, as noted earlier, only a few input signals are active at any time. Thus, many of the possible input combinations never arise, which means that only a small number of different next states are used. Hence, we need only include the addresses of these few active next states in the instruction format, along with the output signals Z, as the following example shows.

Example 8.9

Programmable controller for a LIFO stack memory

Consider again the small LIFO stack memory introduced in Section 8.5. It is designed to store up to four words, which are accessed in LIFO sequence via push and pop operations. Previously, we saw two different hardwired designs of its control unit (Figures 8.39 and 8.40). We now want to design a programmable controller for the same stack memory.

We start the design process with the Moore state diagram of Figure 8.46b, which has 15 states, two primary inputs X, and one primary output Z. A version of the stack controller in the style of Figure 8.49 requires 15 rows, each of which is identified by a 4-bit address. There are four input signal combinations, suggesting that we need to store in each row the addresses of four possible next states—that is, a total of 16 address bits. The three output bits bring the total width of *CM* to 19 bits. In practice, not all of this next-state storage space is actually needed, because the number of distinct next states that appear in any row of the state table may be less, often far less, than the number of possible input combinations. From Figure 8.46b, we see that no state of this particular control unit has more than three distinct next states; hence, it suffices to store only these next states in *CM*, thus reducing its width to 15 bits. This approach is taken in the stack controller design of Figure 8.50.

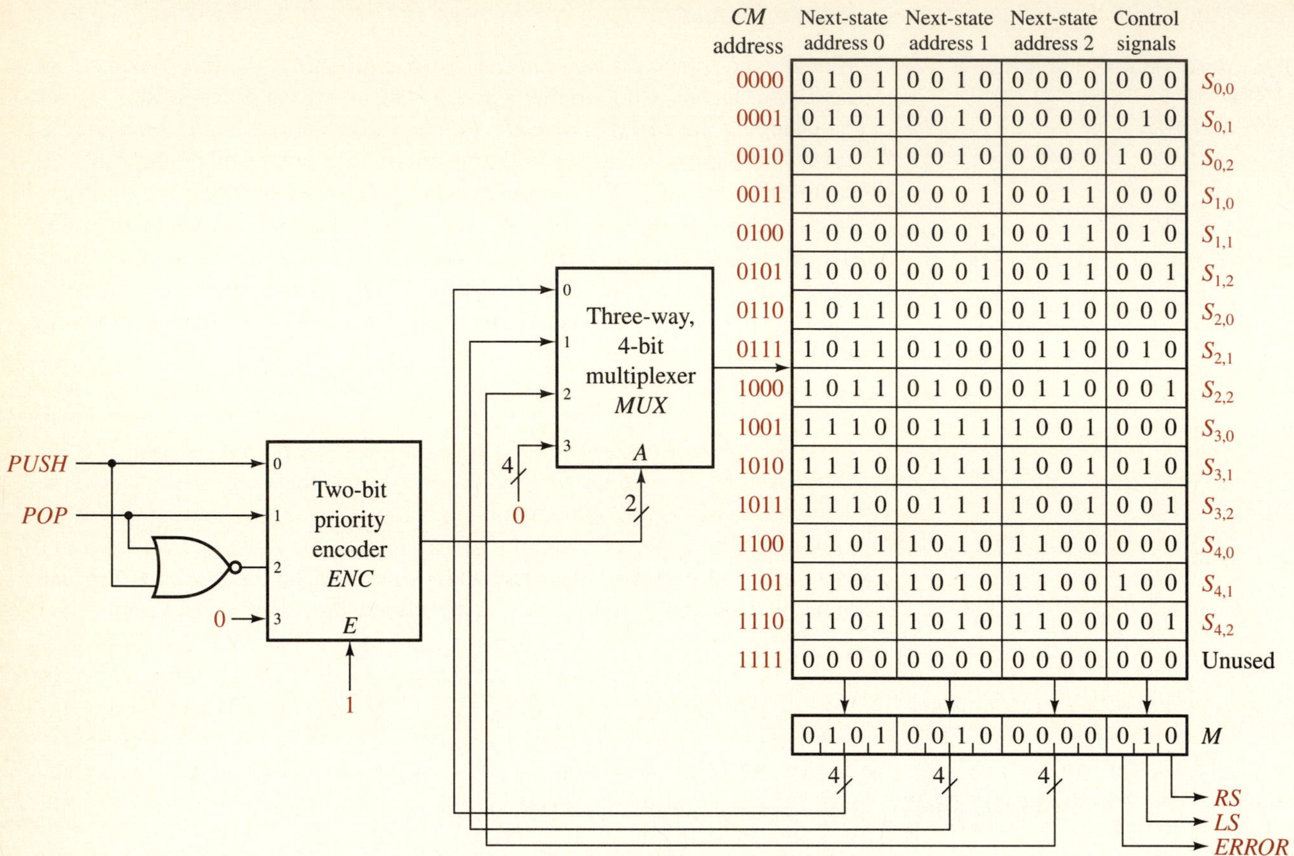

CM address	Next-state address 0	Next-state address 1	Next-state address 2	Control signals	
0000	0 1 0 1	0 0 1 0	0 0 0 0	0 0 0	$S_{0,0}$
0001	0 1 0 1	0 0 1 0	0 0 0 0	0 1 0	$S_{0,1}$
0010	0 1 0 1	0 0 1 0	0 0 0 0	1 0 0	$S_{0,2}$
0011	1 0 0 0	0 0 0 1	0 0 1 1	0 0 0	$S_{1,0}$
0100	1 0 0 0	0 0 0 1	0 0 1 1	0 1 0	$S_{1,1}$
0101	1 0 0 0	0 0 0 1	0 0 1 1	0 0 1	$S_{1,2}$
0110	1 0 1 1	0 1 0 0	0 1 1 0	0 0 0	$S_{2,0}$
0111	1 0 1 1	0 1 0 0	0 1 1 0	0 1 0	$S_{2,1}$
1000	1 0 1 1	0 1 0 0	0 1 1 0	0 0 1	$S_{2,2}$
1001	1 1 1 0	0 1 1 1	1 0 0 1	0 0 0	$S_{3,0}$
1010	1 1 1 0	0 1 1 1	1 0 0 1	0 1 0	$S_{3,1}$
1011	1 1 1 0	0 1 1 1	1 0 0 1	0 0 1	$S_{3,2}$
1100	1 1 0 1	1 0 1 0	1 1 0 0	0 0 0	$S_{4,0}$
1101	1 1 0 1	1 0 1 0	1 1 0 0	1 0 0	$S_{4,1}$
1110	1 1 0 1	1 0 1 0	1 1 0 0	0 0 1	$S_{4,2}$
1111	0 0 0 0	0 0 0 0	0 0 0 0	0 0 0	Unused
M	0 1 0 1	0 0 1 0	0 0 0 0	0 1 0	

FIGURE 8.50 A programmable implementation of a control unit for the stack memory.

As in Figure 8.49, we use a multiplexer *MUX*, this time one with 4-bit data buses, to select from *M* the address of the next state/row to be used. However, we need additional logic to control the select inputs *A* of *MUX* in response to the primary input signals *X*. An encoder *ENC*, the inputs of which are obtained from *X* and the outputs of which go to *MUX*'s *A* lines, is added to the microsequencer for this purpose. For each expected active input combination, *ENC* must be supplied with an input 1-signal that generates the appropriate bit pattern (00, 01, or 10) that causes *MUX* to select the correct address from *M*. Thus, the *PUSH* and *POP* signals are individual inputs to *ENC* and are encoded into addresses 00 and 01, respectively. To create an input to *ENC* that is active when both *PUSH* and *POP* are inactive, we insert a NOR gate that applies the input signal $\overline{PUSH + POP}$ to input 2 of *ENC*, causing the latter to generate the address 10.

Suppose, for example, that the current data in *M* are

$$0101 \quad 0010 \quad 0000 \quad 010 \tag{8.30}$$

corresponding to state $S_{0,1}$, which is assigned to row 0001 of CM. Now let $POP = 1$ and $PUSH = 0$. This causes ENC to produce $A = 01$, in turn causing the multiplexer to select next-state address 1—that is, 0010—from (8.30). The contents of the row of CM with this address correspond to state $S_{0,2}$ and will be transferred to M in the next cycle of the control unit. The foregoing sequence of actions correctly implements the state transition

$$S_{0,1} \xrightarrow{POP} S_{0,2}$$

and produces the outputs $LS = 1$ in $S_{0,1}$ and $ERROR = 1$ in $S_{0,2}$, as required by the stack controller's state behavior. This correspondence is the basis of most general approaches used to derive the contents of CM from a state description.

When large and complex controllers are designed, the state behavior is often specified in the form of a **control program** written in an HDL such as VHDL, Verilog, or other programming languages. The control program is compiled—that is, translated automatically—into binary form for storage in the control memory CM of a general-purpose programmable control unit. Compilers for this application attempt to minimize the size of CM because this normally has the greatest effect on the hardware cost of the control unit.

Hardwired vs. Programmable Control

Hardwired control units are smaller and faster; programmable control units are easier to design and maintain.

Comparing a programmable control unit like that of Figure 8.50 with an equivalent minimum-state hardwired one (Figure 8.39), we see that the latter uses significantly less hardware, as measured by the number of its gates. The hardwired design can be expected to be faster, because the multilevel input logic C_1, composed of MS and CM, can be replaced by a fast two-level implementation. A one-hot design (Figure 8.40) uses more than the minimum amount of hardware in order to simplify the design process; this approach is practical for only very small control units, however.

The programmable control unit has a definite edge in terms of ease of design and maintenance. Its explicit stored representation of next states and output control sequences makes such a machine easier to understand and therefore easier to design. Furthermore, changes to the design, either to correct design errors or to introduce new features into an old design, are easy to accommodate. Similar small changes to a hardwired control unit require a more complicated and costly redesign and remanufacturing of the hardware.

ASM Design Methodology

This section discusses the design of a complete digital circuit that comprises datapath and control parts. For this purpose, we introduce a systematic and easily understood descriptive technique, the ASM chart, and a comprehensive design approach based on the use of these charts.

8.7 Algorithmic State Machines

As we saw in the introduction to this chapter, over the years several ways have evolved for describing the structure and behavior of a (sequential) register-level circuit. State tables or diagrams, hardware description languages (HDLs), and flowcharts are used for mainly behavioral descriptions; logic (block) diagrams or schematics and HDLs are preferred for describing structure. In principle, any of these methods alone could be adapted to most design tasks. However, each method has features that make it suitable for some design tasks but not others.

HDLs have the advantages of precision, powerful multilevel control structures, and a computer-friendly form suitable for use with CAD tools. On the other hand, HDL descriptions tend to be verbose and difficult for humans to read or write. State tables are also precise, easily manipulated by computers, and composed of just a few simple, easily understood constructs. Their simplicity makes them suitable for small designs, but they tend to become unwieldy when applied to large circuits. Graphical description methods like state diagrams and flowcharts have more "eye appeal" and so are convenient for human designers, but they are difficult to manipulate by computer.

Register-level behavior can be described by various combinations of HDLs and diagrams.

At the present time, a typical digital design environment provides support for several of these methods. For example, the larger CAD systems allow a design to be described either graphically or via a formal language such as VHDL. CAD programs that automatically translate between these two formats also exist.

Basic Concepts

We present a formal descriptive method that was developed originally in the 1960s at Hewlett-Packard Company [Clare, 1973]. Although intended for manual design, this technique has the flavor of a CAD system and the added merit of great clarity and simplicity. It aims to provide a precise and systematic way of designing "algorithmic" state machines, and so is known as the **algorithmic state machine** or **ASM** method. ASMs are synchronous finite-state machines of the usual type and can be composed of several linked datapath and control units.

What mainly distinguishes the ASM method is its use of a specialized diagram, called an **ASM chart**, as the starting point in the design process. An ASM chart, which is closely related to a state diagram or a very concise HDL description, is a graphical description of the desired behavior. It contains within its box-like symbols a condensed description of system actions and decisions in terms of control and status signals. The end product is a complete design composed of datapath and control units, which can be implemented directly by fixed or programmable (ROM-based) logic circuits.

ASM Chart Elements

An ASM chart is a diagram composed of state, decision, and conditional output boxes.

An ASM chart is composed of three element types: state boxes, decision boxes, and conditional output boxes, which have the graphic symbols depicted in Figure 8.51. A **state box** is a rectangle that specifies a set of actions to be taken during a single clock cycle—for example, loading a particular set of registers or activating a primary output line. The state box corresponds to a node (circle) in a state diagram, or to the value stored in the state register of a sequential circuit; it thus represents a single control state of the machine. It is labeled with a symbolic name and contains a concise description of the actions taken by the machine in that particular state. In their simplest form, these actions are just a list of control signals to be activated (asserted) in the relevant state. A state box has one entry point and one exit point, indicated by arrows.

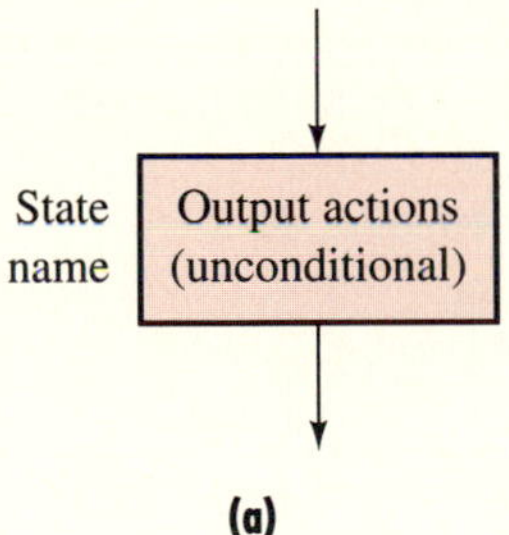

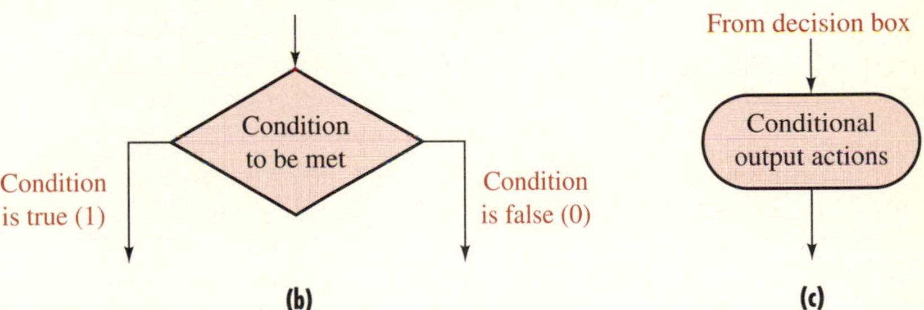

FIGURE 8.51 Elements of an ASM chart: **(a)** a state box; **(b)** a decision box; **(c)** a conditional output box.

A **decision box** is a diamond with a single entry and two exits. It contains the name of a logical (Boolean) variable that serves as a condition to be tested—for example, a status signal generated by the datapath part of the system. The value of this signal, which is either 1 (true) or 0 (false), determines which of the decision box's two exits is to be followed in the current clock cycle. A decision box that contains the variable a may be equated to the following typical **if** statement in an HDL:

$$\textbf{if } a = '1' \textbf{ then } \text{Take exit 1 } \textbf{else } \text{Take exit 0;}$$

Finally, there is a **conditional output box**, which is an oval-shaped box—a rectangle with rounded corners. Its single entry point must be connected to an exit from a decision box that tests some condition—say, $a = 1$. The conditional output box specifies actions to be taken only if the condition $a = 1$ holds in the current clock cycle.

ASM Chart Structure

An ASM chart is a flowchart composed of state, decision, and conditional output boxes, linked by arrows. It is intended to describe completely and unambiguously, in terms of register-level actions and decisions, the behavior required of a sequential machine composed of datapath and control units. The chart defines the state behavior of the target machine's control unit in an explicit manner, similar to a state diagram. The datapath unit's behavior may be inferred from that of the control unit, so the datapath too is implicitly defined by the ASM chart.

State and decision boxes alone suffice to specify the behavior of Moore-type control units. Because output control signals depend unconditionally on the current state, all necessary actions can be entered into the state boxes. This point is illustrated by Figure 8.52, which gives the ASM chart for a very simple case, a JK flip-flop; a state diagram and an HDL description of the flip-flop are also given for comparison. This ASM chart has two state boxes that represent the flip-flop's set and reset states. A state box's entry specifies whether or not the flip-flop's main output variable Q is activated. In the set state box, the entry Q means that Q is active or set to 1 in that state; this could be indicated more explicitly by writing $Q <= 1$. The reset state box contains the empty-set symbol $\varnothing$, indicating that Q is inactive—that is, $Q <= 0$. Because the flip-flop's complementary output $\overline{Q}$ is dependent on Q, it is not considered a separate output signal and so does not appear in the ASM chart. We could, however, equally well make $\overline{Q}$ the primary output variable. In that case, because $\overline{Q}$ is active when $\overline{Q} = 1$ (implying $Q = 0$), we would insert $\overline{Q}$ in the reset state box and $\varnothing$ in the set state box.

The changes of state of the JK flip-flop are conditional on the values of the input variables J and K, which are therefore associated with a pair of decision boxes in Figure 8.52. When the current state is the reset state, a decision based on the value of J determines the next state. If $J = 1$, the left exit, marked '1', of the J decision box is followed, leading to the set state; otherwise, the right exit, marked '0', which leads back to the reset state, is followed. Similarly, the K decision box determines the next state when the current state is the set state.

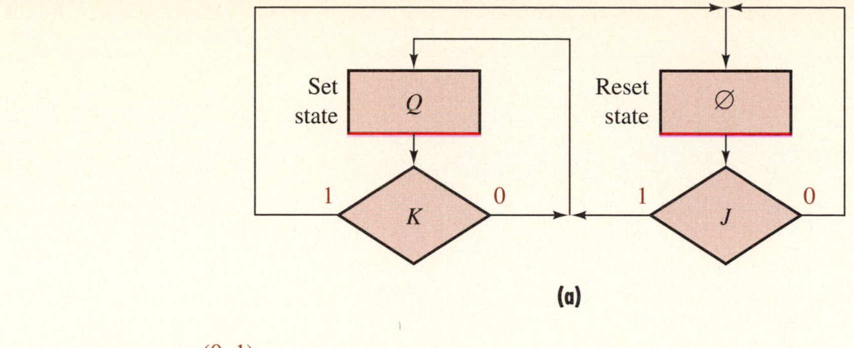

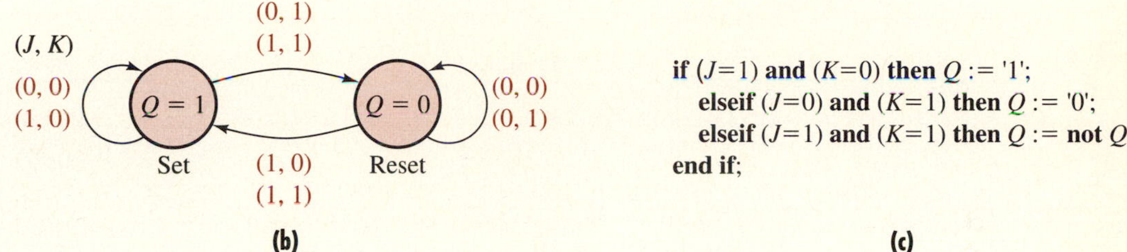

FIGURE 8.52 Three descriptions of a JK flip-flop: (**a**) an ASM chart; (**b**) a Moore state diagram; (**c**) an HDL description.

Conditional Outputs

ASM charts of Mealy machines use conditional output boxes; those of Moore machines do not.

The signals specified within a state box are unconditional; that is, they are always activated whenever the corresponding state is the current state. Conditional output boxes are needed in the ASM charts of Mealy control units because these machines specify output actions that occur in certain states only when certain input variables take particular values. A conditional output box is placed in a path that leads from a state box S_i through one or more decision boxes. Its contents are of the same type that appear in state boxes—that is, a list of output signals to be activated. The actions specified in the conditional output box CO_j take place only when all the conditions leading from S_i to CO_j hold.

The foregoing concepts are illustrated by Figure 8.53a. This ASM chart describes the behavior of the LIFO stack controller (Section 8.5) whose Mealy state diagram appears in Figure 8.38b. We have added state transitions in the form of three new self-loops to the original state diagram to indicate what happens under the error condition in which the *PUSH* and *POP* inputs are activated together. Observe that each conditional output box receives its input from a decision box. A conditional output box that contains *ERROR*, for example, means that the output signal of that name is activated (set to 1) when all the conditions leading to the box are satisfied. In the case of state

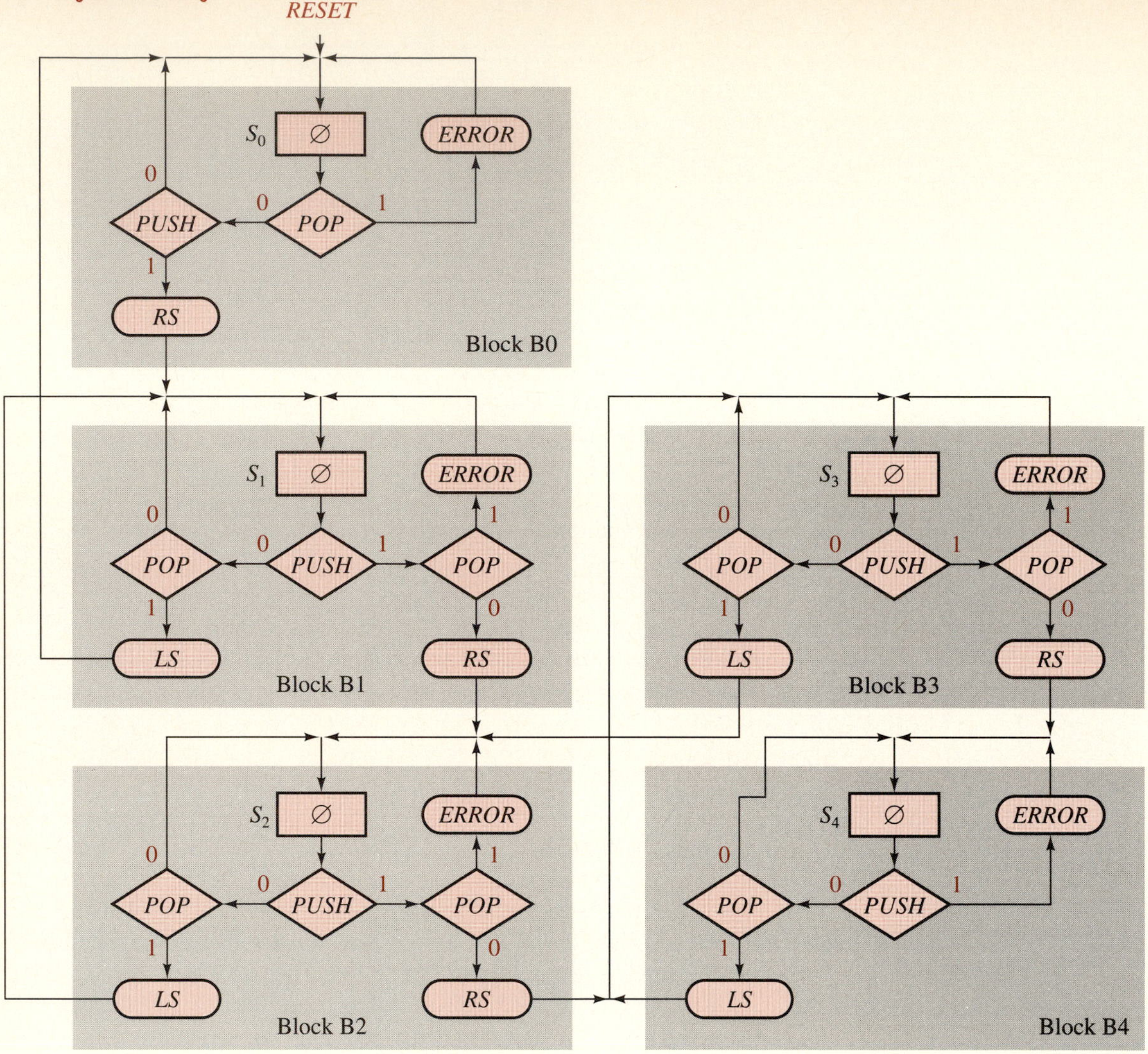

(a)

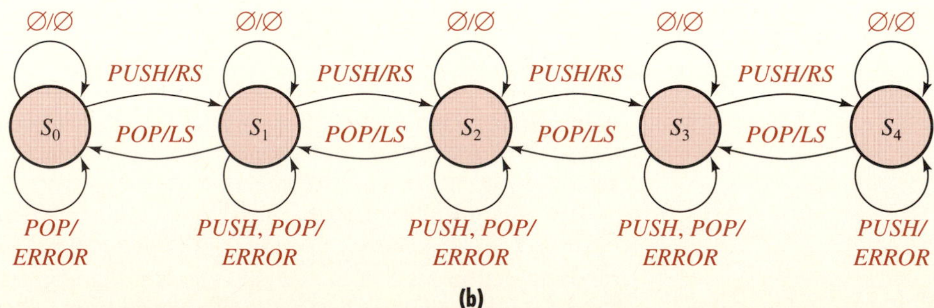

(b)

FIGURE 8.53 A LIFO stack controller: (**a**) its ASM chart; (**b**) its Mealy state diagram.

S_0, there is a path from the S_0 state box through a *POP* decision box to the *ERROR* conditional output box, which is enabled only when $POP = 1$. The path from S_1 to the adjacent *ERROR* box passes through two decision boxes and is enabled only when $PUSH = 1$ and $POP = 1$.

Interpreting ASM Charts

In a state diagram, a node S_i and the labeled arrows that enter and leave it define all the actions associated with state S_i. In an ASM chart, the same information is contained in a portion of the chart referred to as a **block**, in which S_i is the sole state box. The block has a single entry point that leads directly to the S_i box; it may contain many decision boxes and conditional output boxes and have multiple exits. The five blocks that occur in the ASM chart of the LIFO stack control unit are marked in Figure 8.53. For example, block B0 contains the S_0 state box, *PUSH* and *POP* decision boxes, and a pair of conditional output boxes. These particular *PUSH–POP* decisions and the corresponding conditional output actions are relevant only when the controller is in state S_0.

A block of an ASM chart defines all actions that occur in one clock cycle.

Just as a state diagram is considered to have an operating point, the current state, that moves from node to node every clock cycle, an ASM chart has a similar operating point: the current block. All the actions and decisions that take place in a single clock cycle are specified within a single block. It is important to note that an ASM chart does *not* define the sequence or times of occurrence of the actions specified within a block. In particular, the sequence in which decisions are drawn or paths are traced in the ASM chart has no relation to the sequence in which corresponding events occur in a circuit implementation. In this respect, ASM charts differ fundamentally from the flowcharts used to describe general algorithms or computer programs.

It should be apparent that a given state behavior can be described by many different ASM charts, depending on the order in which decision variables are tested, and the like. Figure 8.54 gives a valid ASM chart that is structurally different from but functionally the same as block B1 in Figure 8.53.

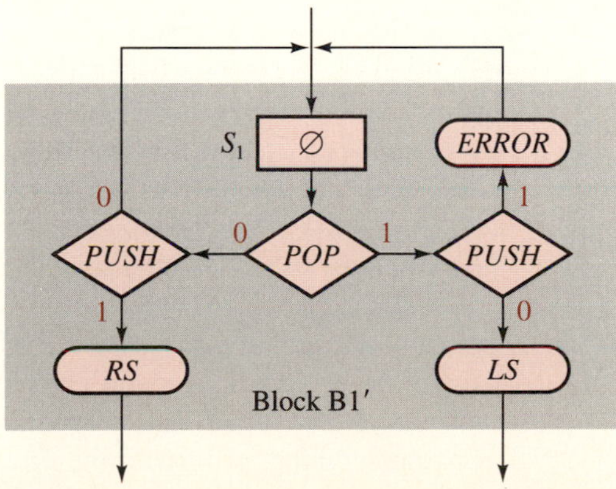

FIGURE 8.54 An alternative ASM chart for block B1 of Figure 8.53.

As noted already, ASM charts are closely related to other methods of behavior specification. This relationship is demonstrated by Figure 8.55, which gives typical correspondences among an ASM chart, a state diagram, and a formal language (VHDL) description. The machine in question is assumed to have the primary (control) inputs $a_0, a_1, \ldots, a_p$ and the primary (status) outputs $b_0, b_1, \ldots, b_p, c$. It is assumed that c is always activated when state S is entered; activation of the b_i signals is conditional on the values of the a_i signals.

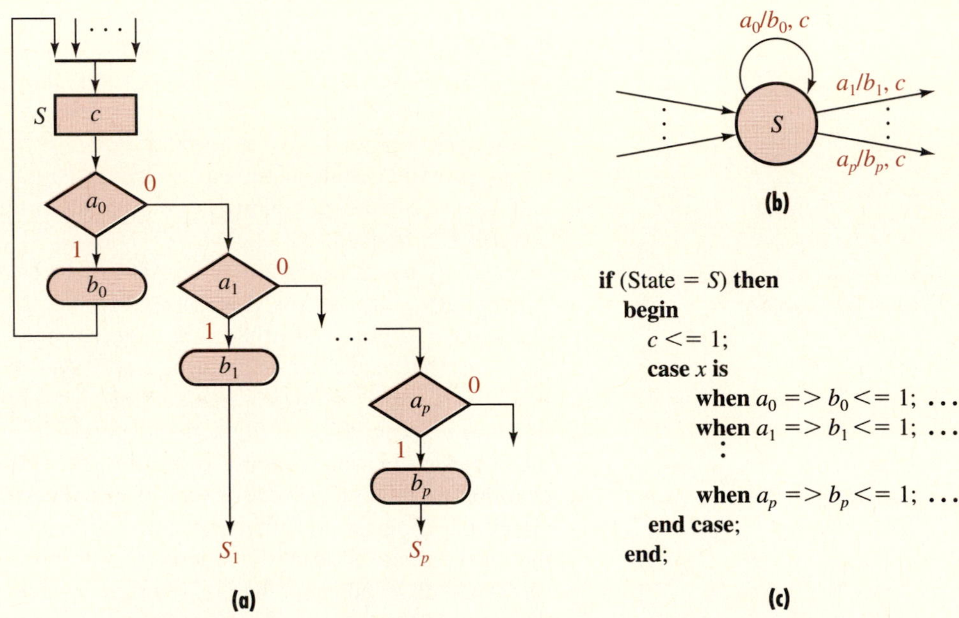

FIGURE 8.55 Representations of state behavior that use **(a)** an ASM chart; **(b)** a state diagram; and **(c)** VHDL.

ASM charts drawn with inconsistent specifications are invalid. A precise definition of an invalid chart cannot be given, but a representative example appears in Figure 8.56. Obviously, block 2 is invalid because it contains two states. Less obvious is the fact that block 0 is also invalid due to a logical rather than a structural contradiction. If $a_1 = a_2 = 0$, both paths leading to the conditional output boxes are enabled. Because one of the latter specifies that output b_1 be set to 1 and the other specifies that $\bar{b}_1$ be set to 1, we have a contradictory situation. Moreover, when $a_1 = a_2 = 0$, the next state is not unique; the enabled paths lead to both S_3 and S_7. This type of problem can usually be avoided by separating a set of decisions in the manner depicted in Figure 8.55a. Again note that the order of the decision boxes in the ASM chart has nothing to do with the order in which the decisions are made; all decisions in a single block are viewed as simultaneous at the register level.

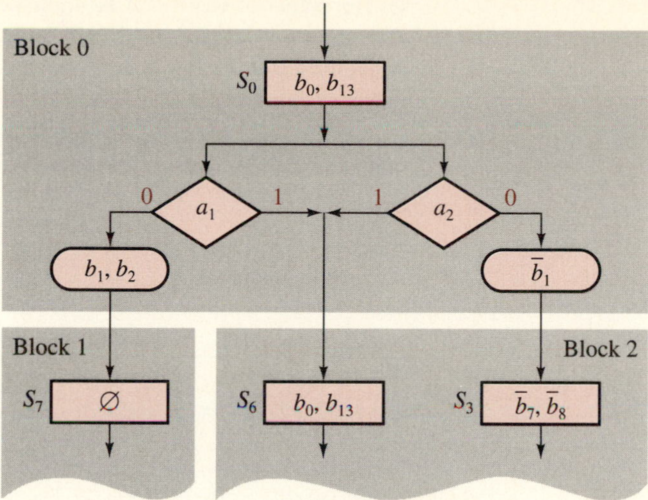

FIGURE 8.56 An example of an invalid ASM chart.

8.8 Integrating Data and Control

We now examine the interaction between the datapath and the control parts of a digital system. We will discover how to use ASM charts to derive a design for a complete system.

CU-DP Timing

Register-transfer operations occur in the clock cycle after the enabling control signal is activated.

A control unit *CU* and the datapath unit *DP* it controls are typically linked by many signals. Those going from *CU* to *DP* are control signals or commands of the type "load register *R*," "reset counter *K*," or "add registers *A* and *B* to *C*," all of which initiate register-transfer operations within the datapath unit. The signals sent by *DP* to *CU* are status signals such as "counter *K* contains zero" or "overflow *OVF* is active." Status signals serve as primary inputs to *CU* that it can use for conditional tests, such as "Is counter $K \neq 0$?" All the memory elements in *CU* and *DP* are triggered by a common system clock.

Consider the following typical HDL statements that describe a register-transfer operation followed by a conditional test on one of the affected registers:

$$\begin{aligned} &A := A + B; \\ &\textbf{if } (A = 0) \textbf{ then} \ldots \textbf{else} \ldots; \end{aligned} \tag{8.31}$$

Let the add operation $A := A + B$ be enabled by a control signal *ADD_A*, and let the status of register A be indicated by a status signal $A = ZERO$, which DP sets to 1 when $A = 0$, and to 0 otherwise (Figure 8.57). Suppose CU activates ADD_A when it is in some state S_1 that occurs in clock cycle i. The datapath does not respond until the next clock cycle $i + 1$, because it must receive the triggering edge of the system clock—the leading or positive clock edge in Figure 8.57c—to perform any register-transfer operation. Hence, CU

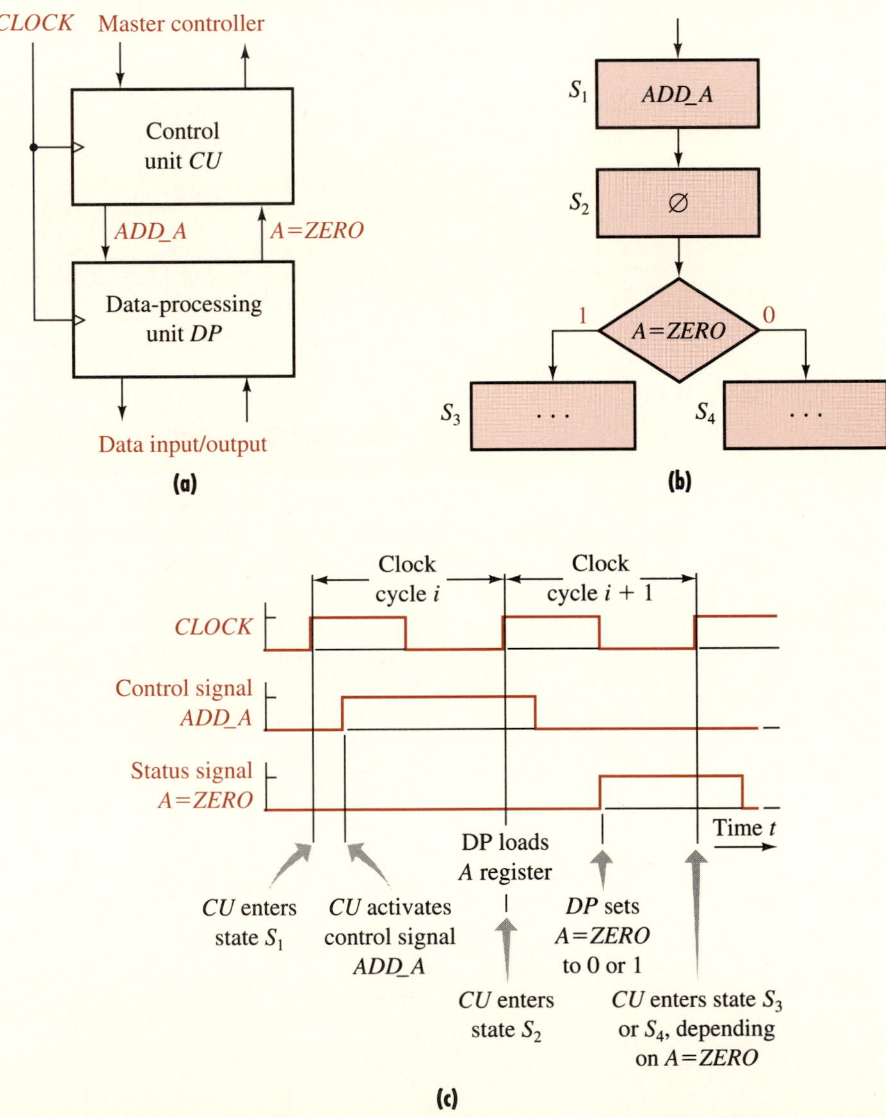

FIGURE 8.57 Control and status signal interaction: **(a)** a circuit diagram; **(b)** an ASM chart; **(c)** a timing diagram.

cannot test the status signal $A = ZERO$ until cycle $i + 1$, when it is in a second state S_2. The test in question determines the next state, S_3 or S_4, of the control unit.

We conclude that, in general, a new value of a control signal established by CU in cycle i does not change the state of DP until the start of the following clock cycle $i + 1$. The resulting new state of DP also cannot be tested by CU until cycle $i + 1$.

Asynchronous Signals

Some input signals applied to a synchronous circuit are inherently asynchronous; that is, they change at times unrelated to the synchronizing clock signal *CLOCK* of the receiving circuit. The most common examples are the asynchronous (pre)set and clear inputs to registers, counters, and similar sequential components. If such signals are transmitted between *CU* and *DP*, they act immediately on arrival without waiting for a triggering clock edge. Many externally generated signals such as a master controller's *BEGIN* control signal, input data signals from keyboards, and the like are also untimed and must be synchronized with *CLOCK*. If they are not so synchronized, they are likely to violate the setup or hold times of a receiving device, leading to unpredictable behavior. A frequent problem caused by violating the setup time of a latch or flip-flop is metastability (discussed in Section 6.3). The output Q of a metastable device assumes an intermediate value between 0 and 1, where it can remain for an unpredictable time t_{meta} before going to a stable 0 or 1 state.

Synchronizer flip-flops are used to capture untimed input signals.

To deal with these problems, an asynchronous input signal x bound for a destination z is first entered into a fast D flip-flop $FF1$—one with short setup and hold times—which is controlled by the clock *CLOCK* of the receiving unit, as depicted in Figure 8.58a. This flip-flop, which is called a **synchronizer**, receives x and produces an output signal x' in the next

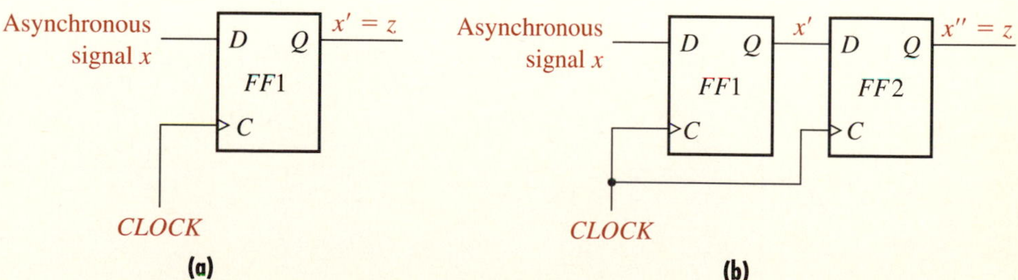

FIGURE 8.58 Synchronizer circuits: (**a**) a basic one-flip-flop design; (**b**) a more reliable two-flip-flop design.

clock cycle that usually satisfies the circuit's timing constraints. Of course, if the synchronizer flip-flop's own setup and hold times are violated, x' may become metastable for some time t_{meta}. A more reliable synchronizer design, shown in Figure 8.58b, feeds x' to a second D flip-flop $FF2$, the output of which, one cycle later, will be valid (synchronized with the clock), as long as $t_{meta} \leq T - t_{su}$, where T is the clock period and t_{su} is the setup time of $FF2$.

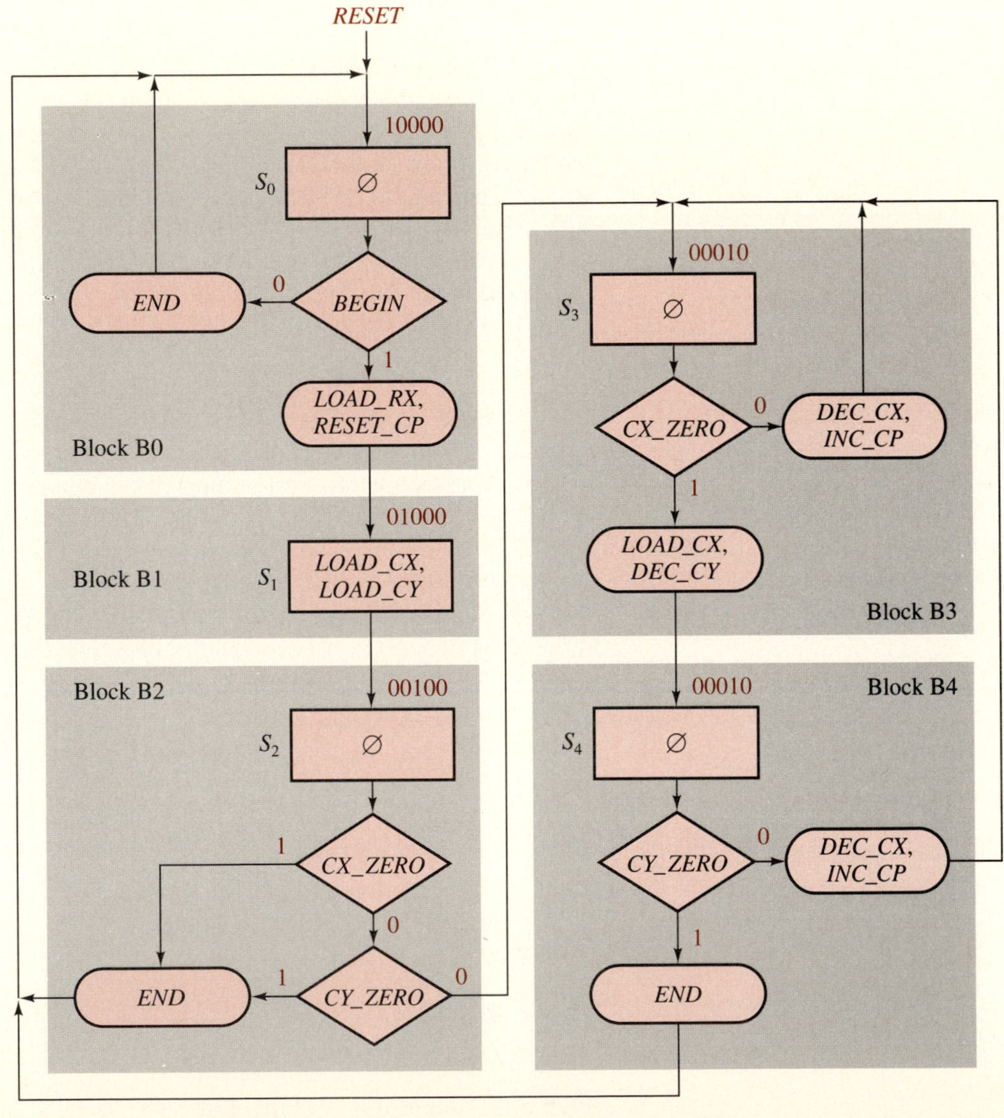

FIGURE 8.59 An ASM chart for the counter-based multiplier.

Datapath Design

We now develop a complete datapath unit design, starting from an ASM chart, which defines the target behavior in precise, algorithmic form. This chart is usually constructed directly from the initial, informal design specifications that indicate the main components to be used and the general method to be followed. We can also derive the ASM chart indirectly from some other formal description such as a state diagram or a VHDL description, if available, by exploiting the correspondence shown in Figure 8.55.

The datapath components and operations are implicit in the ASM chart.

As an example, consider again the counter-based multiplier of Example 8.8 (Section 8.5). The main datapath components of this system are three counters *CX*, *CY*, and *CP*, as well as a register *RX* (Figure 8.41). Figure 8.59 gives an ASM chart that defines the multiplier's behavior; this chart was derived directly from the state diagram obtained earlier (Figure 8.42), and uses the same names for the states and the input/output signals. The latter consist mainly of status and control signals that pass between the control unit *CU* and the datapath unit *DP*. There are also *BEGIN* and *END* signals that connect *CU* to an external (master) controller. All the foregoing signals are defined in Table 8.4. In addition, two data buses *DATA_IN* and *DATA_OUT* link *DP* to the outside world.

TABLE 8.4 Input/output signals for the control unit of the counter-based multiplier.

Type	*Signal*	*Condition or operation specified*
Input	*BEGIN*	Initiate multiplication
	CX_ZERO	Contents of counter *CX* are 0
	CY_ZERO	Contents of counter *CY* are 0
Output	*DEC_CX*	Decrement counter *CX*
	DEC_CY	Decrement counter *CY*
	END	Multiplication is complete
	INC_CP	Increment counter *CP*
	LOAD_CX	Load counter *CX* from register *RX*
	LOAD_CY	Load counter *CY* from input data bus
	LOAD_RX	Load register *RX* from input data bus
	RESET_CP	Reset counter *CP* to zero

Component Selection

The next task is to identify the major register transfer components of *DP* and the ICs available to implement them. We will assume that we have available a small library of components based on the 74X series. For concreteness, we will also assume a data word size of 8 bits; scaling to other word sizes is easy, as it is for most datapath designs.

First, we examine all the register-transfer operations mentioned or implied in the ASM chart (Figure 8.59): load register *RX*, load counter *CX*, load counter *CY*, decrement counter *CX*, decrement counter *CY*, and increment counter *CP*. From these, we conclude that the following major datapath components are needed:

1. An 8-bit parallel register for *RX*
2. Eight-bit binary counters with parallel load and down-counting capabilities for *CX* and *CY*
3. A 16-bit binary counter for *CP*; in this case normal up-counting suffices, and parallel loading is not required

Component Modification

Library components often must be adapted to fit a particular design.

In general, not all the available components match exactly the requirements of a particular design; they require scaling (word-size adjustment), fan-out buffering, and other small modifications. Let us assume that on checking our parts catalog, we find that the only suitable ICs for implementing the above components are the 8-bit D-type register and the 8-bit binary up-only counter depicted in Figure 8.60. The D register, which resembles the 74X273 (Figure 8.20), has parallel inputs, but it has no load control of the kind needed by *RX*. To allow the register to be loaded from the input data bus *DATA_IN* in some clock cycles and to remain unchanged in others, we will employ the scheme of Figure 8.21b, which places a two-way multiplexer in the input path of the D register. One data input of the multiplexer is connected to *DATA_IN*, the other to the outputs of the D register. We hold the data contents of the

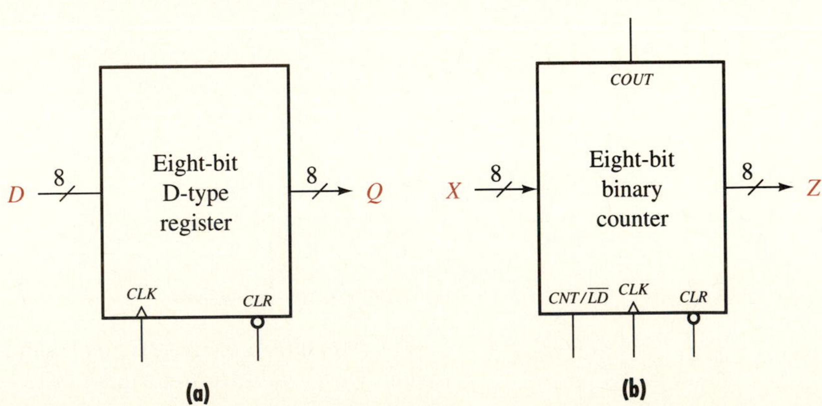

FIGURE 8.60 ICs used in the counter-based multiplier: (a) an 8-bit D-type register; (b) an 8-bit binary counter.

register unchanged by reloading its old contents every clock cycle via the multiplexer. Hence, we require a two-way, 8-bit multiplexer to implement *RX*; the 74X157 is a suitable IC for this purpose. The resulting design for *RX* appears in Figure 8.61.

We face a similar problem (and some new ones) in using the counter IC of Figure 8.60b in *DP*. This IC has, in addition to the usual clock and clear inputs, a control input called $CNT/\overline{LD}$, meaning "count on 1 and load on 0." If set to 1, $CNT/\overline{LD}$ causes the counter to increment its output *Z* when next

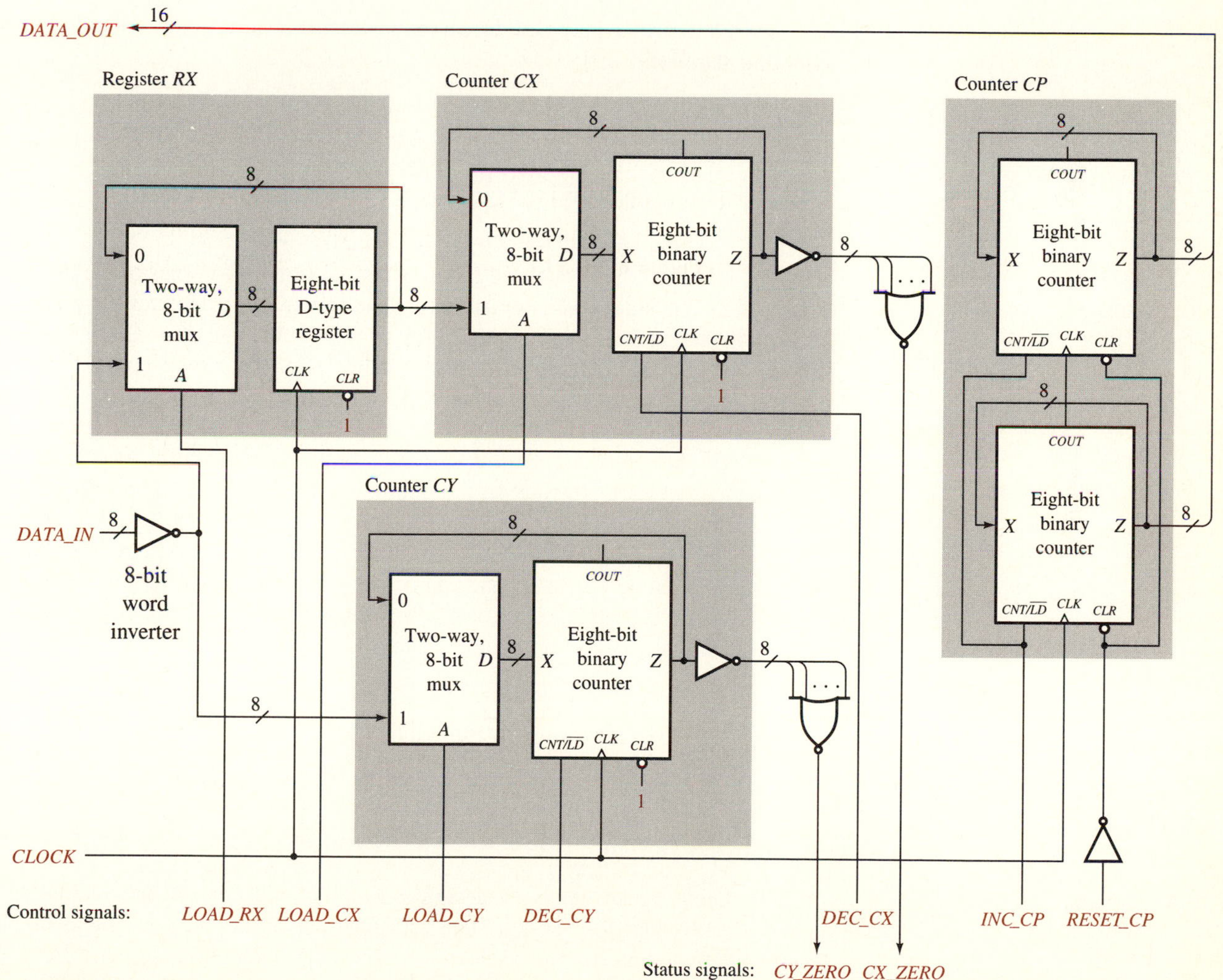

FIGURE 8.61 A logic diagram for the datapath unit *DP* of the counter-based multiplier.

triggered by the positive edge of the clock signal *CLK*. When $CNT/\overline{LD} = 0$, the counter instead loads the current data on X, which becomes the new count Z. If we use an external control signal *CEN* to enable counting, that signal must be connected to $CNT/\overline{LD}$. When $CEN = 0$ or is inactive, the counter will execute a parallel load in each subsequent clock cycle. To prevent this from changing Z, we feed back Z to the data inputs of the counter. In the case of *CX* and *CY*, a two-way, 8-bit multiplexer is needed, as shown in Figure 8.61, because these counters, unlike *CP*, must also be loadable from external sources.

To make the 8-bit counter count down instead of up, we simply complement its output signals Z; this is done by the two 8-bit word inverters appearing in *CX* and *CY* in Figure 8.61. We must also initialize the counter with the complement $\overline{Z}_0$ of the desired inital count Z; a word inverter is placed in *DATA_IN* for this purpose. Observe how the upward-going, modulo-2^8 counting sequence

$$11111110, 11111111, 00000000, 00000001, 00000010, 00000011, \ldots$$

becomes the downward-going sequence

$$00000001, 00000000, 11111111, 11111110, 11111101, 11111100, \ldots$$

when all the counter's output signals are complemented.

The 16-bit product counter *CP* is constructed from two copies of the 8-bit counter IC. The latter has an output carry line *COUT* that generates a 1-pulse whenever the count is incremented from $Z = 11111111$ to $Z = 00000000$. (The 8-bit counter, of course, counts modulo 2^8.) Thus, we connect the *COUT* output line of the low-order 8-bit counter to the *CLK* input of the high-order 8-bit counter to obtain the desired 16-bit counter.

A few other gates are needed to complete the datapath unit. Two eight-input NOR gates are attached to the output lines of *CX* and *CY* to serve as zero detectors that generate the status signals *CX_ZERO* and *CY_ZERO*. We now see that *DP* requires about a dozen, mostly MSI-level, ICs: one register, four counters, three multiplexers, and a few gate-level ICs for "glue" logic to complement signals and perform other minor functions. This design can also be easily mapped into a single custom IC or a PLD.

Control Unit Design

At this point, we are ready to design the multiplier's control unit *CU*. We noted already that if a one-hot state assignment is used, we can obtain the logic equations that define the next-state and output functions directly from a state diagram or a state table. Reflecting the close correspondence between

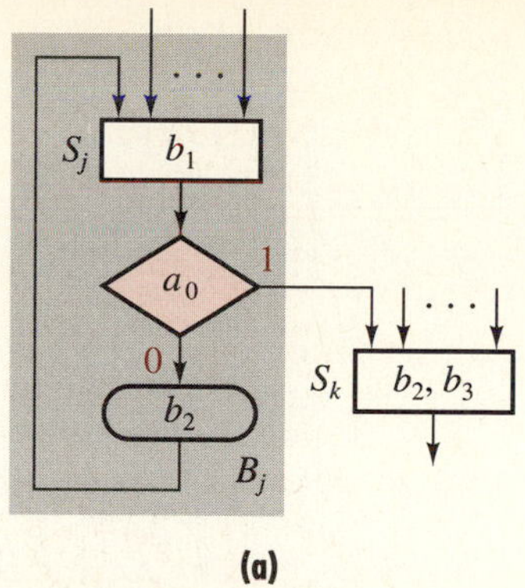

(a)

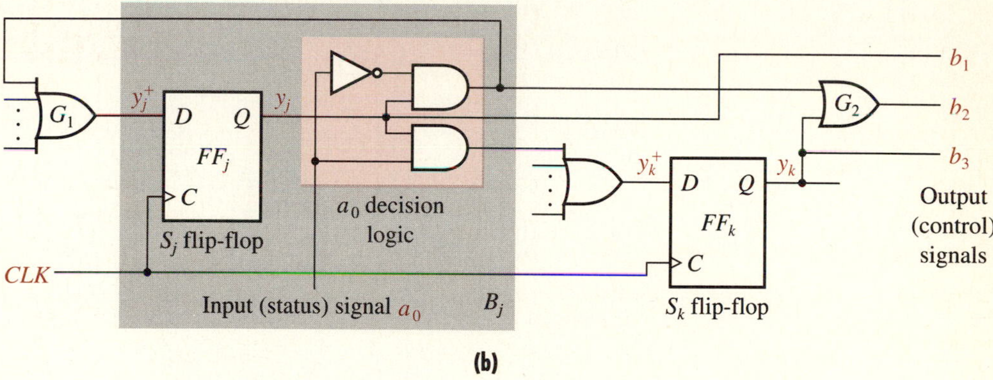

(b)

FIGURE 8.62 (a) A portion of an ASM chart; (b) the corresponding one-hot circuit implementation.

an ASM chart and a state diagram, we can also map an ASM chart directly into the combinational logic needed for a one-hot state implementation, as we now see.

The relevant chart-to-circuit correspondence is illustrated by Figure 8.62. Each state box corresponds to a D flip-flop, thanks to the one-hot correspondence between state S_j and a single state variable y_j: the circuit is in state S_j if and only if $y_j = 1$. A decision box is implemented by a pair of AND gates and an inverter, as shown by the colored region of Figure 8.62b. The relevant decision variable a_0 "steers" the state variable y_j—that is, the "hot" one—through the AND gates to one of two possible places: to y_j^+ (and also b_2) if $a_0 = 0$, or to y_k^+ if $a_0 = 1$. Each next-state signal y_i^+ and output signal b_i is derived from the outputs of the flip-flops or the decision logic. If a y_i^+ or b_i signal has several sources, the source signals are connected to an OR gate, the output of which is the required control signal. For instance, output signal b_2 of Figure 8.62b is activated by both $y_k = 1$ and $\overline{a}_0 y_j = 1$; these signals are combined by gate G_2 to produce b_2.

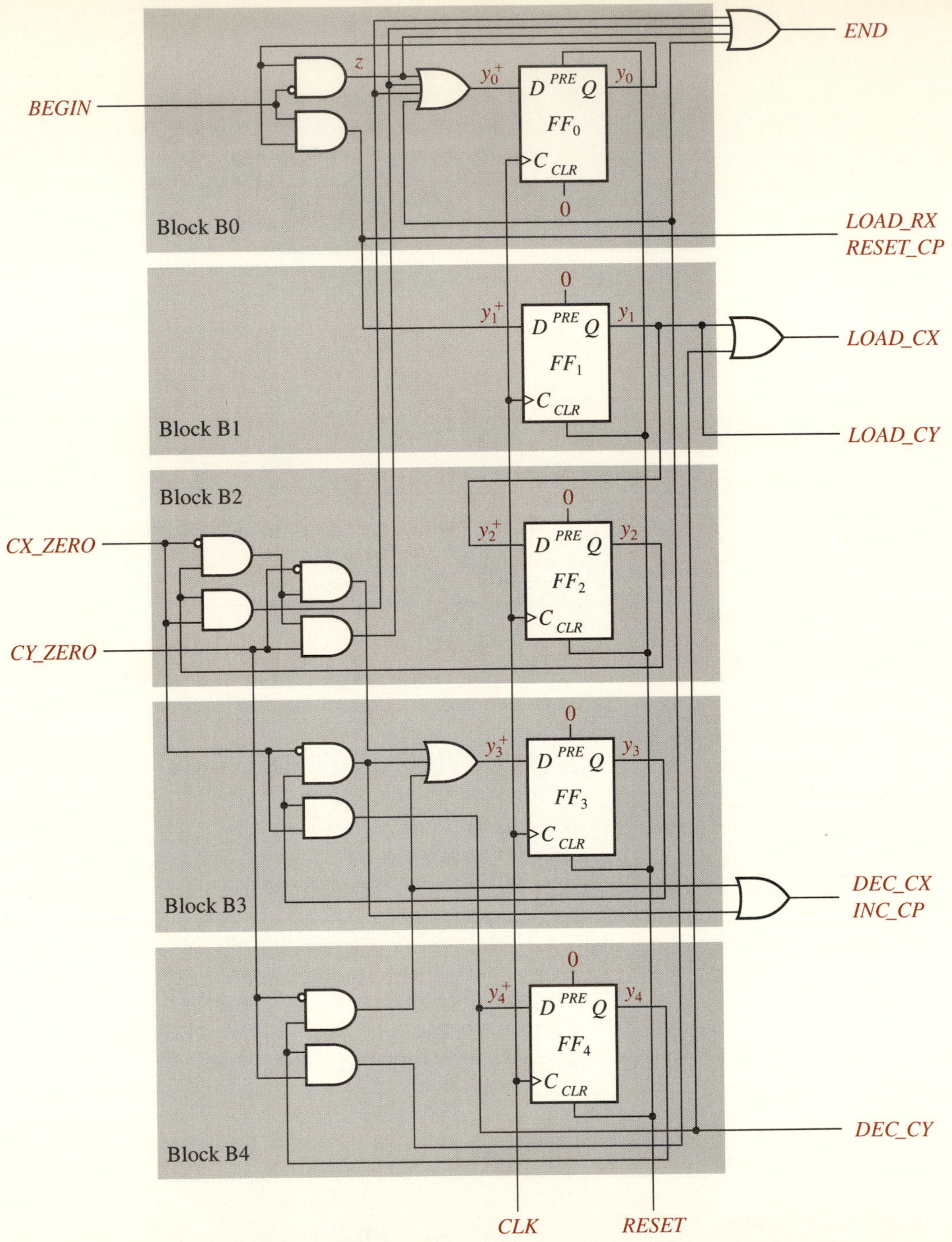

FIGURE 8.63 One-hot implementation of the control unit for the counter-based multiplier, derived from its ASM chart.

A one-hot control unit can be constructed directly from the ASM chart.

Figure 8.63 shows the control unit obtained by applying the correspondence of Figure 8.62 to the ASM chart of the counter-based multiplier (Figure 8.59). There are five D-type flip-flops that correspond to the multiplier's five states $S_0 = 10000$, $S_1 = 01000$, $S_2 = 00100$, $S_3 = 00010$, and $S_4 = 00001$. The five decision boxes give rise to five AND-gate pairs. All are wired together according to the layout of the ASM chart, with OR gates introduced as needed to combine next-state or output signals. The subcircuits of the control unit that correspond to the various blocks of the ASM chart are marked in Figure 8.63. For example, the decision box labeled *BEGIN* in block B0 produces the AND-gate pair at the top left of Figure 8.63. The upper AND gate's output z is activated if $y_0 = 1$; that is, the circuit is in state S_0, and $BEGIN = 0$. This signal is one of several that activates the *END* line, and so z is connected to an OR gate on the top right that generates *END*. The condition $z = 1$ should also lead to the next state S_0; therefore, z is one of several signals feeding the OR gate that generates y_0^+. As might be expected, this control unit is broadly similar in structure and complexity to the one-hot design derived earlier (Figure 8.43) from the multiplier's state table.

As a further illustration of the ASM approach, we will design a sequential multiplier for binary numbers based on repeated shifting and adding, as in long multiplication by hand. This multiplication method has been widely used in the design of arithmetic circuits for electronic calculators and computers. It requires small logic circuits and can complete an $(n \times n)$-bit multiplication in n steps, in marked contrast with the foregoing counting scheme, which, in the worst case, requires around $2^n \times 2^n = 2^{2n}$ steps to produce the same result.

Example 8.10

Designing a shift-and-add multiplier

The problem here is to design a multiplication circuit for n-bit unsigned binary numbers, in which the multiplicand Y is multiplied by one digit of the multiplier X at a time and the resulting partial products are combined to obtain the final product $P = X \times Y$. The precise algorithm required is best developed by analyzing some small numerical examples. Consider the task of multiplying $1011_2 = 11_{10}$ by $1101_2 = 13_{10}$ to produce $10001111_2 = 143_{10}$. Figure 8.64a shows a typical hand calculation in which the result, 0 or Y, of each 1-bit multiplication of Y by x_i is shifted left i places to account for the weight 2^i of x_i. (Recall that an i-bit shift of a binary number N to the left or right corresponds to multiplication or division, respectively, of N by 2^i.)

The shift-and-add method multiplies two n-bit numbers in n steps, compared to up to 2^{2n} steps for the multiply-by-counting method.

A direct hardware implementation of this technique in the n-bit case would require simultaneous storage of n shifted partial products, followed by a single summation of n $2n$-bit operands. The same result can be achieved

1011	$Y = y_3y_2y_1y_0$
1101	$X = x_3x_2x_1x_0$
1011	2^0x_0Y
0000	2^1x_1Y
1011	2^2x_2Y
1011	2^3x_3Y
10001111	$P = 2^0x_0Y + 2^1x_1Y + 2^2x_2Y + 2^3x_3Y$

(a)

1011	$Y = y_3y_2y_1y_0$
1101	$X = x_3x_2x_1x_0$
0000	$A = 0$
1011	2^4x_0Y
01011	$A = A + 2^4x_0Y$ (add)
01011	$A = 2^{-1}A$ (shift)
0000	2^4x_1Y
001011	$A = A + 2^4x_1Y$ (add)
001011	$A = 2^{-1}A$ (shift)
1011	2^4x_2Y
0110111	$A = A + 2^4x_2Y$ (add)
0110111	$A = 2^{-1}A$ (shift)
1011	2^4x_3Y
10001111	$A = A + 2^4x_3Y$ (add)
10001111	$A = 2^{-1}A = P$ (shift)

(b)

FIGURE 8.64 An example of shift-and-add multiplication (**a**) with deferred summation; (**b**) with accumulation of sums.

with much less hardware by doing the addition on the fly, as in Figure 8.64b. Here each partial product is added to the sum of the preceding partial products and so need not be stored for more than one step. To simplify the implementation further, the accumulated sum is right-shifted while the partial product is held fixed. As we will see, this allows an ordinary two-operand adder of n-bit numbers to be used.

We therefore take the scheme of Figure 8.64b as the basis for our multiplier circuit, whose structure is outlined in Figure 8.65. It will contain the following storage elements: two n-bit registers Q and M to store the input operands X and Y, respectively, and an accumulator register A to store the accumulated partial products. As the numerical example illustrates, the accumulator should be able to store the $2n$-bit product P. To facilitate the 1-bit shifting operations, A will be made a right-shift register. An n-bit adder is needed to add the contents of M and the leftmost n bits of A. Thus, the adder's inputs will be connected to M and A, and its outputs to A. The carry bit generated by each n-bit add operation must be retained in A because after shifting, it becomes the most significant bit of the A operand used in the next addition step. Thus, A needs to store a total of $2n + 1$ bits. We index the operand bits and register flip-flops as integers in the following way:

$$M = M_{n-1}M_{n-2}\ldots M_1M_0 \quad \text{which stores the multiplicand } Y$$
$$Q = Q_{n-1}Q_{n-2}\ldots Q_1Q_0 \quad \text{which stores the multiplier } X$$
$$A = A_{2n}A_{2n-1}\ldots A_nA_{n-1}A_{n-2}\ldots A_1A_0 \quad \text{which stores the product } P$$

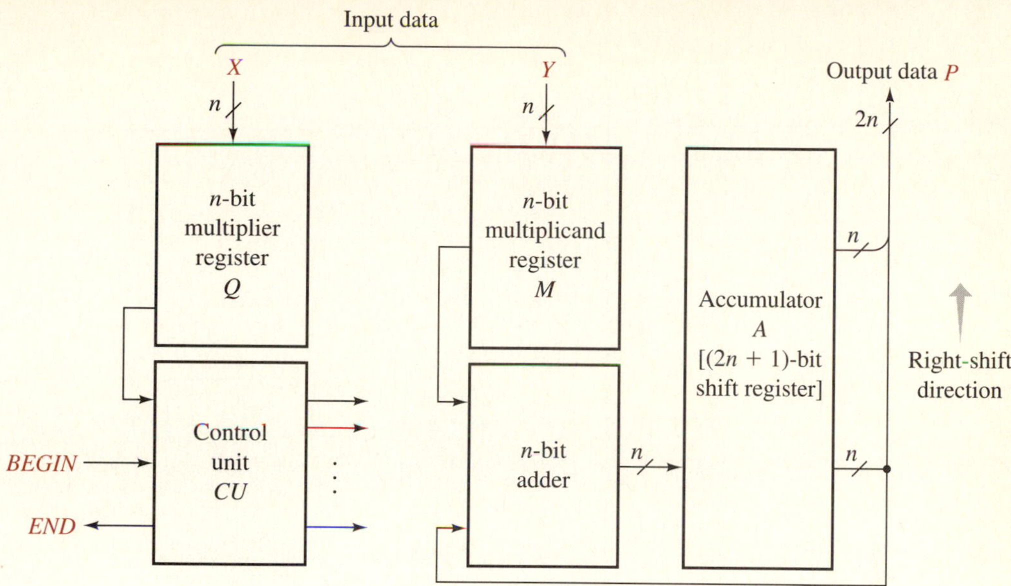

FIGURE 8.65 A high-level view of the shift-and-add multiplier.

The n-bit subregister $A_{2n-1}A_{2n-2}\ldots A_{n+1}A_n$ of A provides one input operand for each add step. The resulting $(n+1)$-bit sum is placed in $A_{2n}A_{2n-1}A_{2n-2}\ldots A_{n+1}A_n$, so that A_{2n} stores the output carry bit.

We see that multiplication is accomplished in n steps $i = 0, 1, 2, \ldots, n-1$, each involving two register-transfer operations: the addition of $Q_i \cdot M$ to $A_{2n-1}A_{2n-2}\ldots A_{n+1}A_n$, followed by a right shift of the entire contents of A. If $Q_i = 0$, then $Q_i \cdot M$ is zero, so no actual addition needs to be performed in this case. We could therefore define the multiplier's control unit *CU* to have $2n$ states or so. However, *CU* can be reduced to just three states by making the iteration counting a part of the datapath unit *DP*, which suggests placing a counter *COUNT* in *DP* for this purpose.

An iteration counter is useful when implementing repetitive behavior.

At this point, we can construct an ASM chart for the multiplier; see Figure 8.66. For simplicity, we have omitted the input/output operations needed to transfer data to and from the circuit. There are three blocks B0, B1, and B2. The first block B0 initializes the circuit after the *BEGIN* signal triggers the multiplication process; the second block B1 implements the conditional add step; and the third block B2 implements the shift step. We also increment the iteration counter *COUNT* from i to $i+1$ in B1 and test it for termination in B2. The current value of Q_i is tested in B1 to determine whether M should be added to A. The control signals identified in this ASM chart are listed in Table 8.5.

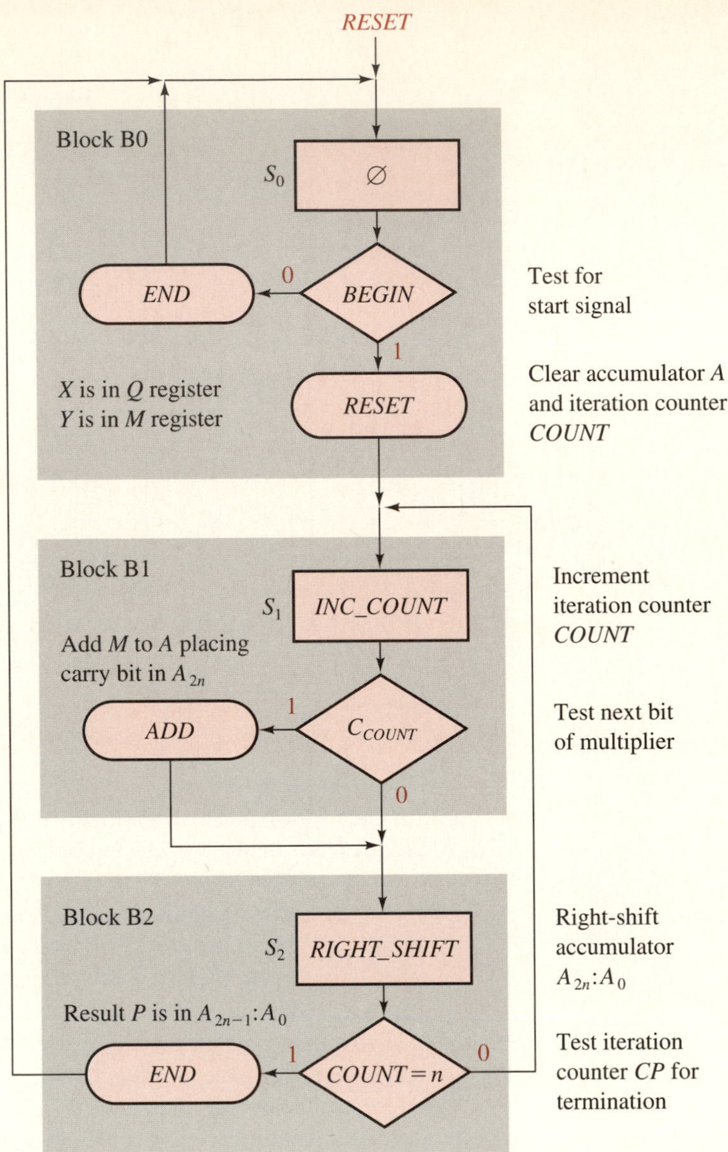

FIGURE 8.66 An ASM chart for the shift-and-add multiplier.

A straightforward implementation of the multiplier datapath for $n = 8$ is given in Figure 8.67, which uses a full range of 74X-style ICs. The 8-bit adder is composed of two 4-bit adders like the 74X83. The accumulator A consists of two 8-bit universal shift registers: *SR0* for storing A_7:A_0 and *SR1*

TABLE 8.5 Input/output signals for the control unit of the shift-and-add multiplier.

Type	*Control signal*	*Condition or operation specified*
Input	*BEGIN*	Initiate multiplication
	COUNT=*n*	Iteration count *COUNT* is *n*
	Q_{COUNT}	Active bit of multiplier *X* is 1
Output	*ADD*	Add multiplicand register *M* to accumulator *A*
	CARRY_CONTROL	Control carry input to accumulator
	END	Multiplication is completed
	HOLD	Maintain the accumulator's current state
	INC_COUNT	Increment iteration counter *COUNT*
	LOAD_RX	Load register *RX* from input data bus
	RESET	Clear *A* and *COUNT*
	RIGHT_SHIFT	Right-shift *A*

for storing A_{15}:A_8; an additional D flip-flop stores A_{16}. The D flip-flop sends its output "rightward" to the serial data input *SI* of *SR1*, which, in turn, sends its least significant bit to the *SI* input of *SR0*. The two shift registers have four operating modes controlled by a pair of *MODE* control lines. The four possible patterns 00, 01, 10, and 11 of *MODE* specify, respectively, hold (do nothing), right shift, left shift, and load. Control signals for each of these operations, except the unused left-shift mode, are generated by the control unit *CU* and are sent to *A*'s *MODE* inputs via a 2-bit encoder. The specified actions do not occur until the appropriate triggering edge of *CLOCK* arrives. Observe that the addition is implemented by loading (*MODE* = 11) into *SR1* the sum that is continuously available at the adder outputs. Because *SR0* should be unaffected by this addition, its output data are fed back to its inputs, so that A_7:A_0 is not changed by a load operation. The D flip-flop receives the adder's carry-out signal during add operations, but it receives 0 during shift operations; a two-way, 1-bit multiplexer with the select signal *CARRY_CONTROL* controls this process.

The operation $A := A + M$ is implemented by loading the adder's sum output into *A*.

A standard 4-bit counter is used to generate the iteration count *COUNT*. Its main task is to signal the multiplier's control unit *CU* when the count reaches $n = 8$. This occurs when the counter's most significant output bit changes from 0 to 1; hence, this output signal directly implements the control signal *COUNT* = *n*. The counter is also used in conjunction with an eight-way, 1-bit multiplexer to select the active bit Q_{COUNT} of the *Q* register and send to it *CU*. A complete gate-level design for *CU* also appears in Figure 8.67. It contains only three states, so its design is easy and is left as an exercise.

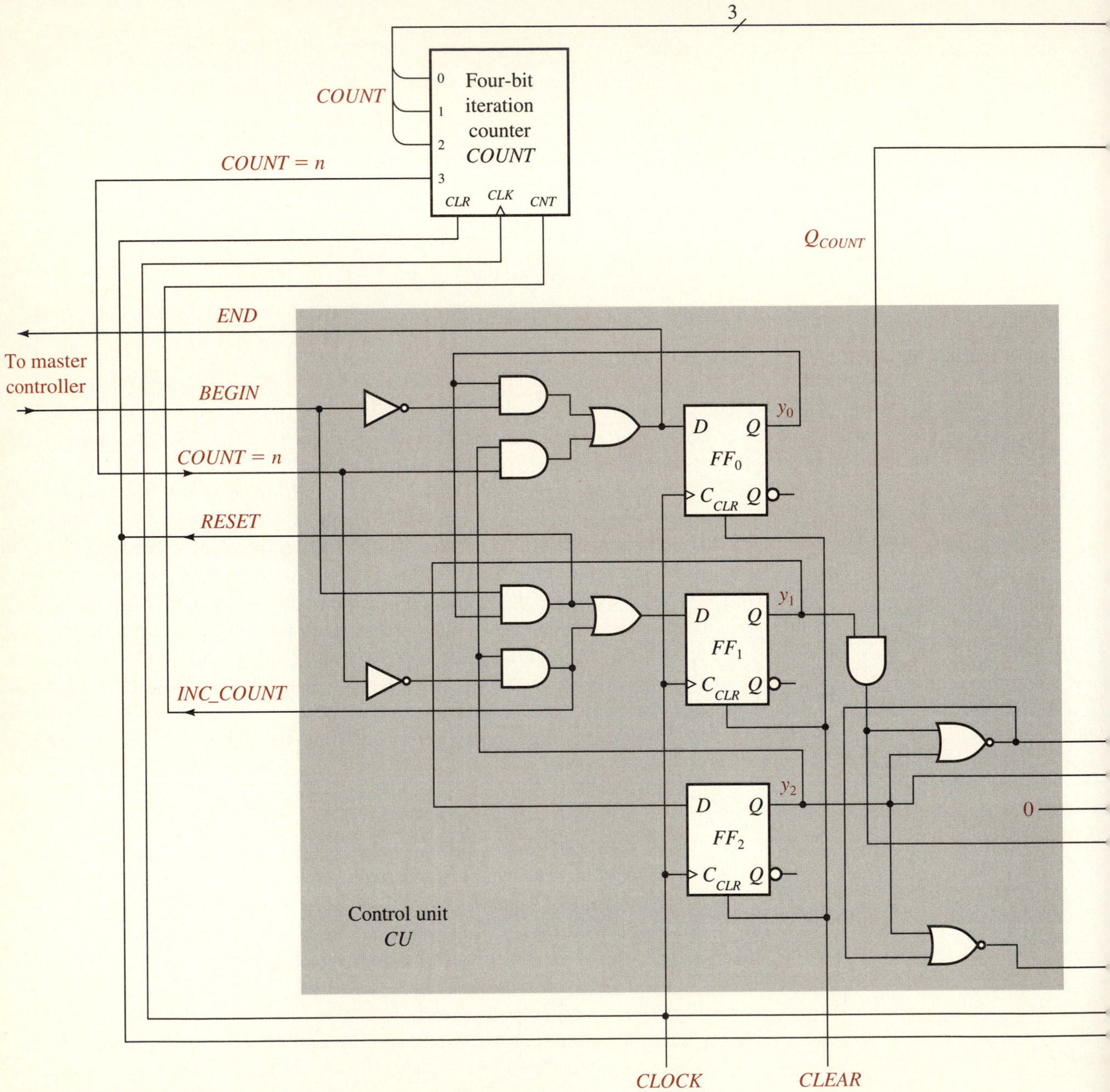

FIGURE 8.67 The complete design for the 8-bit shift-and-add multiplier.

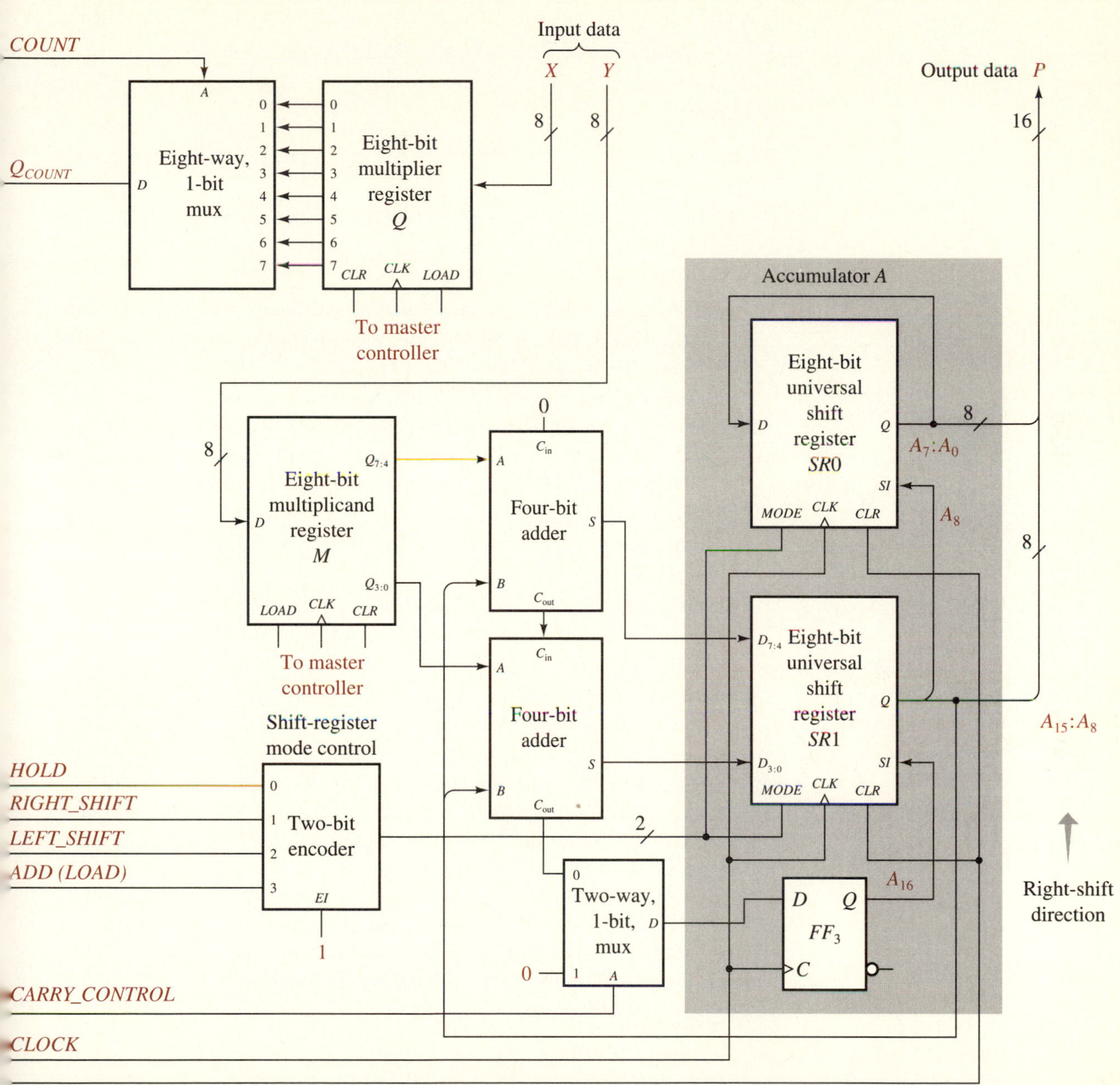
COUNT
Input data
X
Y
Output data P
Eight-way, 1-bit mux
Eight-bit multiplier register Q
CLR
CLK
LOAD
To master controller
Q COUNT
16
Accumulator A
Eight-bit universal shift register SR0
MODE
SI
A_7:A_0
A_8
Eight-bit multiplicand register M
$Q_{7:4}$
$Q_{3:0}$
To master controller
Four-bit adder
Cin
Cout
Eight-bit universal shift register SR1
$D_{7:4}$
$D_{3:0}$
A_{15}:A_8
Shift-register mode control
HOLD
RIGHT_SHIFT
LEFT_SHIFT
ADD (LOAD)
Two-bit encoder
EI
Two-way, 1-bit, mux
A_{16}
FF_3
Right-shift direction
CARRY_CONTROL
CLOCK
RESET

Circuits like that of Figure 8.67, which contains the equivalent of several hundred gates, are sufficiently complex that an initial design is likely to contain logic or timing errors. Hence, it is very desirable to simulate it using a suitable logic simulation tool to identify and correct such errors.

Summary

We can summarize the steps of the ASM design process, illustrated by the foregoing examples, as follows:

1. Devise a suitable algorithm *A* that defines the register-level behavior of the target circuit. The basic operations of *A* should correspond closely to the functions of the available logic components.
2. Outline the circuit's overall structure and identify its major control and status signals. Construct an ASM chart that defines *A* in terms of these signals.
3. Determine the necessary datapath functions from the ASM chart and identify the components and interconnections that these functions entail. Implement the datapath unit using the most appropriate available components, modifying them as needed.
4. Decide on the control unit style to be used—for example, the one-hot method with D-type flip-flops. Design the control unit, guided by the ASM chart.
5. Integrate the datapath and control unit designs, adding glue logic where necessary to match signals at the *CU-DP* interface, to meet timing or fan-out constraints, and the like.
6. Simulate the design and revise it as needed until correct operation, along with satisfactory cost and performance, are achieved.

Computer-aided simulation is a key step in ensuring the correctness of a complex new design.

Once a prototype has been designed and successfully tested, significant improvements often become apparent. For example, the multiplexer-register combination used in Example 8.10 to extract Q_{COUNT} from the multiplier operand *X* could be replaced by an 8-bit shift register, the contents *X* of which are accessed by shifting them out bit by bit. Instead of introducing a separate shift register for this purpose, we could store *X* in the *SR0* part of *A*, making *SR0* play the role of *Q*. Notice that the accumulated partial product expands to fill *SR0* at the same rate *X* would be shifted out. This technique, which is discussed further in the problem set, is generally used in the design of shift-and-add multipliers.

CHAPTER 8 SUMMARY

The Register Level

8.1 ■ The **register** or **register-transfer level** of a circuit is the level immediately higher than the gate level. A central component of the register level is the n-bit **register**, a storage device composed of n flip-flops.

■ Register-level circuits are often described by an ad hoc mixture of block diagrams, tables, equations, and descriptive text.

■ Also useful for describing register-level circuits are artificial languages known as **hardware description languages** or **HDLs**, which resemble computer programming languages and describe circuits in a logical, technology-independent fashion. HDLs may replace or supplement schematic diagrams in the design process.

■ Register-level design treats words as primitive signals or units of information, in contrast to the gate level, which treats individual bits as the primitive signals. These words represent either numbers or nonnumerical data such as instructions or control words. The word length in bits varies, but is often a multiple of four, with 16 and 32 bits being representative figures.

■ Register-level design is also characterized by a data-flow representation that views a circuit as processing data words that flow between computational devices and storage devices. The main word-storage devices are the registers after which this level of abstraction is named.

■ Most circuits can be divided, both structurally and behaviorally, into a data-processing part and a control part at the register level. The data-processing part, called the **datapath**, is designed to perform a specified set of operations on data words and provide temporary storage for the input/output data and intermediate results. The control part, called the **control unit** or **controller**, generates at the proper time and in the proper sequence the main control signals required by the circuit.

■ Using an HDL, we can specify the behavior of a general register-level circuit in the form of a set of **register-transfer statements**. These statements indicate how data words stored in input registers are transferred through and transformed by combinational processing circuits to create results that are stored in output registers.

8.2 ■ A key primitive design element at the register level is the **bus**, a set of logic lines used to interconnect two or more components.

■ Another component is the **word gate**, which groups independent logic gates into a single logic element.

■ A **multiplexer** routes a data word from a selected source to a common destination. It is also sometimes used as a general-purpose combinational component. **Decoders**, which also serve as demultiplexers, activate a unique output line for each input pattern. **Encoders** output a pattern that indicates the active input line of highest priority.

■ Arithmetic components at the register level include n-bit adders, n-bit arithmetic-logic units (ALUs), $n \times m$-bit multipliers, and n-bit magnitude comparators.

■ The various programmable components such as PLAs, PALs, and ROMs introduced in Chapter 4 can serve as general-purpose logic function generators at the register level.

8.3 ■ Among sequential logic elements of the register level, **parallel registers** are storage elements for n-bit words. A parallel register consists of flip-flops or latches with a common set of control lines.

■ Unlike a parallel register, a **shift register** is characterized by its use of serial data transfers. It takes n clock periods to read or alter the entire contents of an n-bit shift register via serial shifting. Universal shift registers can be accessed in parallel or serially from left to right or from right to left.

■ Another flexible component at the register level is an **up-down counter**, which counts up (increment) or down (decrement) as specified by an input control line.

■ Many PLDs also include flip-flops and can be used to implement special-purpose register-level sequential components.

Datapath and Control Design

8.4 ■ The data-processing part of a digital system frequently consists of a number of relatively large combinational circuits that move and process words.

■ Register-level components are designed to be easily **scaled** (expanded or reduced) to accommodate different word sizes. In **bit slicing** each circuit is made up of a number of identical components, or bit slices, with each component responsible for producing one bit of a result. Bit slicing exploits the independence among the bits of the data words being processed.

■ **Multiplexer expansion** uses convergent trees of small multiplexers to accommodate a larger number of data sources. **Decoder expansion** uses divergent trees of small decoders; **encoder expansion**, while possible, requires several different component types and is therefore more difficult.

8.5 ■ **Control units** are sequential circuits that generate, select, and interpret control signals.

■ Control units can be designed by following one of several general design methods. The classical **state-machine approach** minimizes the number of flip-flops. The **one-hot method** is based on a state assignment with one flip-flop per state, which simplifies design.

8.6 ■ PLDs are especially suited to the design of control units. These **programmable control units** are contrasted with **hardwired control units**. By changing the stored control information in a programmable control unit—by writing or programming a new control algorithm into it—the same hardware can be used for many different applications.

■ Most programmable control units are composed of a **control memory**, which stores a program-like description of the unit's desired behavior in the form of a list of **microinstructions**. The **microsequencer** selects the next address to be applied to the control memory. This address is determined both by the current instructions stored in the state register and by external inputs.

ASM Design Methodology

8.7 ■ A systematic design approach known as the **algorithmic state machine** or **ASM method** addresses finite-state machines that are composed of several linked datapath and control units.

■ The **ASM chart**, the starting point in the design process, is closely related to a state diagram or a very concise HDL description. It is a diagram composed of state, decision, and conditional output boxes.

■ A **state box** is a rectangle that specifies a set of actions to be taken during a single clock cycle. A **decision box**, a diamond with a single entry and two exits, contains the name of a logical variable that serves as a condition to be tested. A **conditional output box** is an oval-shaped box that specifies actions to be taken only if a preceding condition holds.

■ A **block** of an ASM chart defines all actions that occur in one clock cycle.

8.8 ■ A control unit and the datapath unit it controls are typically linked by many signals. Those signals going from the control unit to the datapath unit are control signals or commands. The signals sent by the datapath unit to the control unit are status signals, which serve as primary inputs to the control unit.

- The steps of the ASM design process can be summarized as follows: (1) Devise a suitable algorithm that defines the register-level behavior of the target circuit. (2) Outline the circuit's overall structure and identify its major control and status signals. Construct an ASM chart that defines the algorithm in terms of these signals. (3) Determine the necessary datapath functions from the ASM chart. Implement the datapath unit, using the most appropriate components, modifying them as needed. (4) Decide on the control unit style and design it. (5) Integrate the datapath and control unit designs, adding glue logic where necessary. (6) Simulate the design and revise it as needed.

Further Readings

Integrated circuit manufacturers' data books and manuals are among the more useful sources of information on register-level components and their usage. An old but excellent example is *The TTL Applications Handbook* (Fairchild Semiconductor, Mountain View, Calif., 1973), edited by P. Alfke and I. Larsen. Robert B. Seidensticker's *The Well-Tempered Digital Design* (Addison-Wesley, Reading, Mass., 1986) presents a concise collection of "proverbs" on good design practice for more advanced logic designers. An overview of VHDL, with many 74X-series components as examples, can be found in D. R. Coelho, *The VHDL Handbook* (Kluwer Academic Publishers, Boston, 1989). The influential and still readable work that introduced the ASM methodology is *Designing Logic Systems Using State Machines* (McGraw-Hill, New York, 1973), by C. R. Clare.

Chapter 8 Problems

The Register Level

General characteristics; symbols; HDL usage; combinational components: multiplexers, comparators, decoders, encoders; sequential components: parallel registers, shift registers, counters.

8.1 Define the gate and register levels of abstraction and list the main similarities and differences between them.

8.2 Give one application, *not* listed in Table 8.1, of each of the following register-level components: a multiplexer, a decoder, a shift register.

8.3 Analyze the gate-level model of the 74X181 ALU that appears in Figure 8.1 to verify that when $S = 0110$, the 4-bit EXCLUSIVE-OR ($M = 1$) and two's-complement subtraction ($M = 0$) operations are performed, as required by Figure 8.4. For each of these operations, determine the functions realized by the *IP*, *IG*, and *IC* signals. Also determine F_0, F_1, F_2, and F_3 as functions of the primary inputs of *DP*: *IP*, *IG*, *M*, and C_n.

8.4 Draw suitable box symbols for each of the following combinational components: **(a)** a 32-way, 4-bit multiplexer; **(b)** a 1-out-of-16 decoder; **(c)** a 16-input priority encoder; **(d)** a 12-bit binary adder-subtracter. Include all normal data and control buses and specify their widths.

8.5 Draw a concise register-level logic diagram for a circuit C_1 that is described in words as follows: C_1 contains an 8-bit adder that computes the sum $S = A + B$. Both S and $\overline{S}$ are primary output buses of C_1. In addition, the high-order half S_{high} of S is sent to one of four destinations Z_0, Z_1, Z_2, Z_3, as specified by a 2-bit destination address A. The low-order half S_{low} of S is used to activate a set of 16 light switches $L_0 : L_{15}$ so that $L_i = 1$ if and only if S_{low} denotes the number i.

8.6 Figure 8.68 shows the logic diagram of the 74X153 multiplexer. Design suitable register-level primitive symbols for this circuit in the styles of Figures 8.3a and 8.3b.

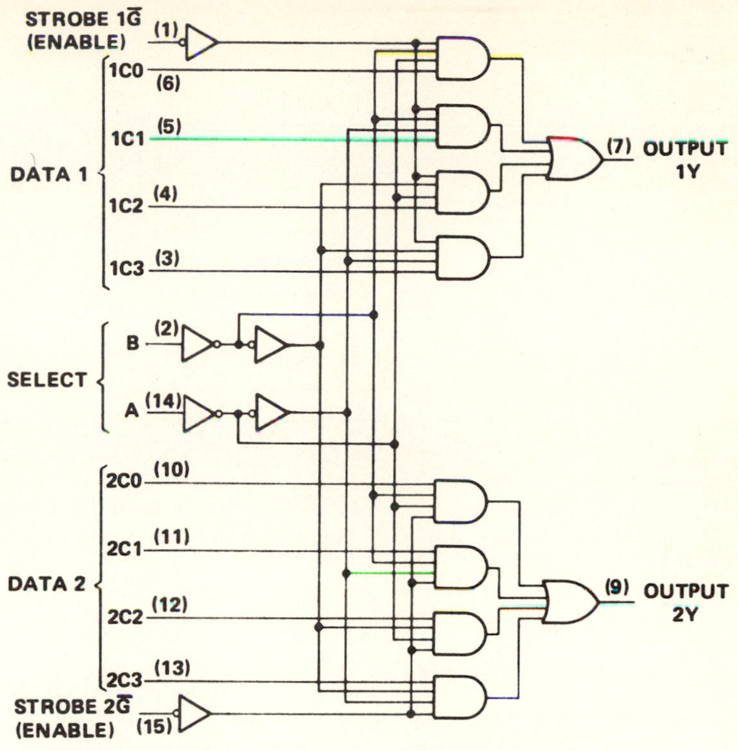

FIGURE 8.68 A gate-level logic circuit for the 74X153 multiplexer [Texas Instruments, 1988b].

8.7 Figure 8.18 gives the gate-level representation of the 74X85 magnitude comparator. This circuit compares the magnitudes of two 4-bit unsigned numbers A and B, yielding three outputs ($A < B$, $A = B$, and $A > B$) that indicate the relative magnitudes of the two numbers. **(a)** Draw a suitable box symbol for the entire circuit in the style of Figure 8.3a. **(b)** Attempt to construct a circuit diagram that is intermediate between that of Figure 8.18 and the single-box representation, using Figure 8.2 as a model. Comment on the difficulties encountered in partitioning the 74X85 along the lines of Figure 8.2.

8.8 Discuss the advantages and disadvantages of using a hardware description language to describe the structure of a digital system, contrasting it with the use of logic diagrams (schematics) for the same purpose. Consider extremely large designs that require hundreds of pages of HDL text to describe them fully.

8.9 A half-adder is a simple component that adds two bits x and y, producing two output bits $sum = x \oplus y$ and $carry = xy$. Construct a VHDL description of this circuit along the lines of Figure 8.5, including an **entity** part that defines the half-adder's name and input/output signals, and an **architecture** part that defines its functional behavior. Assume that each output signal experiences a propagation delay of 2 ns.

8.10 Consider the 74X153 multiplexer of Figure 8.68. Construct a VHDL description of this device's input/output interface and behavior in the style of Figure 8.5.

8.11 Repeat Problem 8.10 for the 74X85 comparator (Figure 8.18). Note that VHDL permits the relational symbols $<$, $=$, and $>$ to be applied to integers with their usual meanings.

8.12 **(a)** Use an eight-way, 1-bit multiplexer (Figure 8.13b) to implement the three-variable EXCLUSIVE-OR function $a \oplus b \oplus c$. **(b)** Suppose that the four-variable EXCLUSIVE-OR function $a \oplus b \oplus c \oplus d$ is to be implemented using multiplexers. Show how this can be done using as few eight-way, 1-bit multiplexers as possible.

8.13 Using **(a)** an eight-way multiplexer and **(b)** a four-way multiplexer, show how to implement a 1-bit full subtracter.

8.14 Using a multiplexer of suitable size, implement a 2-bit adder along the lines of Example 8.3.

8.15 A very large multiplexer *LUMMUX* is available; it has 1,024 input data buses, each 8 bits wide. We want to use one copy of *LUMMUX* to implement an adder for k-bit binary numbers, as in Example 8.3. What is the largest value of k for which this is possible?

8.16 Suppose we want to design a comparator *COMP* that compares two 2-bit unsigned numbers A and B and produces an output signal z_1 that indicates $A = B$, and a second signal z_2 that indicates $A > B$. Show with the aid of a diagram how to implement *COMP*, using a single m-way, n-bit multiplexer of appropriate size.

8.17 Taking all data and control signals into account, construct truth tables for **(a)** a 1-out-of-10 decoder, and **(b)** a four-input priority encoder.

8.18 Describe how to implement the eight-input priority encoder defined in Figure 8.16 by means of a multiplexer of suitable size. Compare the gate cost of this design to that of a conventional two-level NAND-gate implementation of the encoder by estimating the number of gates present in the multiplexer.

8.19 Show how to use the ACT-1 FPGA cell of Figure 8.19 to realize the EXCLUSIVE-OR function $z = x_1 \oplus x_2 \oplus x_3$.

8.20 **(a)** Prove conclusively that the ACT-1 FPGA cell of Figure 8.19 can realize any Boolean function of up to three variables. **(b)** This FPGA cell can realize some but by no means all functions of up to eight variables. Find some function of k variables, $3 < k \leq 8$, that this cell cannot realize. (*Hint:* Consider the standard gate functions.)

8.21 Consider again the ACT-1 FPGA cell shown in Figure 8.19. Describe how to use one or more copies of this cell to implement a basic clocked flip-flop, clearly identifying the flip-flop type you choose.

8.22 Draw suitable box symbols for each of the following sequential components: **(a)** an 8-bit parallel D-type register with true outputs only; **(b)** a 4-bit shift register with serial input x, serial output s, and parallel output Z; **(c)** a modulo-20 up-down counter with parallel load. All the components should have clock and clear inputs, and the widths of all data buses should be indicated.

8.23 Draw a logic circuit for a 6-bit parallel register R composed of JK flip-flops. Assume that R has separate J and K input buses, as well as Q and $\overline{Q}$ output buses. R also has the usual clock, preset, and clear control lines. Give a suitable box symbol for R.

8.24 Design a 4-bit shift register *SHR* composed of either D or JK flip-flops. *SHR* has a serial input x and a pair of serial outputs z and its complement $\overline{z}$. The control inputs are a clock signal *CLK* and a clear signal $\overline{CLR}$. Give a suitable box symbol for *SHR*.

8.25 In most applications that require set-reset flip-flops, the JK type is preferred. Suggest a reason why the shift register of Figure 8.23a uses SR flip-flops rather than JK flip-flops. Will the circuit work properly if the edge-triggered SR flip-flops are replaced by level-triggered SR latches?

8.26 Give a logic diagram for a $4n$-bit universal shift register constructed from four copies of the component in Figure 8.24a. Show all connected and merged lines and buses and assign an appropriate name to every logic signal.

An arithmetic shift multiplies or divides a 2C number by two.

8.27 A left-shift register can be used to multiply a signed two's-complement number by two. Shifting that handles signs correctly is called **arithmetic shifting**, in contrast to the **logical shifting** considered earlier. For example, if $Y = 11001011_{2C} = -00110101_2 = -53_{10}$ is subjected to an arithmetic left shift, the result should represent $-53_{10} \times 2 = -106_{10} \pmod{2^8}$ in proper 2C form. Similarly, an arithmetic right shift of Y should perform signed division by two. Show how to modify a standard universal shift register to support arithmetic shifting of both positive and negative 2C integers. (*Hint:* Use the sign of the stored number to determine the new bit to be entered into the shift register.)

8.28 Design a simple modulo-16 (up) counter from a 4-bit register, a 4-bit adder, and other standard gate-level and register-level components. Using an external data word X, include a parallel-load circuit to initialize the counter.

8.29 Show how to modify a modulo-16 up counter *COUNT* to form a modulo-16 up-down counter, using as little extra logic as possible. All the modifications are to be external to *COUNT*, and you may employ any standard gate-level or register-level components.

8.30 From the short parts lists given in Tables 8.2 and 8.3, estimate the number of 74X-series ICs needed to implement the 4 × 32 register file of Figure 8.22.

Datapath and Control Design

Datapath units; expansion of: multiplexers, decoders, encoders, comparators, arithmetic circuits; carry-lookahead; control units; state machines; one-hot realizations; Moore vs. Mealy machines; programmable controllers.

8.31 Define each of the following terms: accumulator; bit slicing; control memory; datapath unit; state machine; Moore machine.

8.32 Consider the eight-way, 1-bit multiplexer that appears in Figure 8.13b. Using as few copies of this circuit as you can and no other types of components, design a 16-way, 1-bit multiplexer *MUX16*. Describe how you would use *MUX16* to design a 16-way, 4-bit multiplexer.

8.33 Show how to use the 74X153 multiplexer (Figure 8.68) to implement **(a)** a two-way, 2-bit multiplexer, and **(b)** an eight-way, 4-bit multiplexer.

A barrel shifter is a combinational circuit for data shifting.

8.34 A **barrel shifter** is a useful combinational circuit designed to shift a data word left or right by a specified number of bits. It is thus a high-speed replacement for a shift register. Design a barrel shifter *BS* for eight-digit (32-bit) hexadecimal numbers. The data signals of *BS* are a 32-bit input word *X* and a 32-bit output word *Z*, which is the shifted version of *X*. The control inputs to *BS* are as follows: an enable bit *ENABLE*; a 3-bit word *SIZE*, which is the number of digit positions to be shifted; a bit *DIR* that defines the direction of the shift; and a bit *FILL* that defines the data to be introduced into the positions vacated by the shift (trailing or leading 0s and 1s). Using generic register-level components, give a complete block diagram for your design and describe briefly how your circuit works.

8.35 How many 1-out-of-2 decoders are needed to construct **(a)** a 1-out-of-16 decoder, and **(b)** a 1-out-of-2^n decoder? Give the logic diagram for a 1-out-of-16 decoder constructed in this way.

8.36 A 1-out-of-16 decoder *DEC16* is required, but the only decoders available are of the 1-out-of-4 type *DEC4*, each implemented as a small-scale IC. Two alternative designs for *DEC16* are proposed. The first design uses five copies of *DEC4* and no additional logic; the second design uses two copies of *DEC4* and several two-input, 16-bit AND word gates. **(a)** Draw a logic diagram for each of these designs for *DEC16*. **(b)** The cost of *DEC4* is \$1.00. Integrated circuits are available that contain four two-input AND gates and cost 25 cents each. The mounting and wiring costs are 25 cents per IC and 5 cents per input or output connected. (Assume that the costs for connecting power and ground are included in the mounting cost of 25 cents per chip figure.) Determine the costs of each design for *DEC16*. Which one is more economical?

8.37 Eliminate as many components as you can from the 64-input priority encoder circuit of Figure 8.33 to produce a 32-input priority encoder.

When expanding encoders, we can trade hardware for speed.

8.38 The encoder scaling technique depicted in Figure 8.33 is called **parallel expansion**. Figure 8.69 shows an alternative technique called **series expansion**, which is used here to build a 16-input priority encoder from two eight-input priority encoders and some additional gates. Extend this serial expansion technique to produce a 64-input priority encoder composed of eight-input encoders and a few additional gates. Compare the two 64-input encoder designs in terms of speed and component cost.

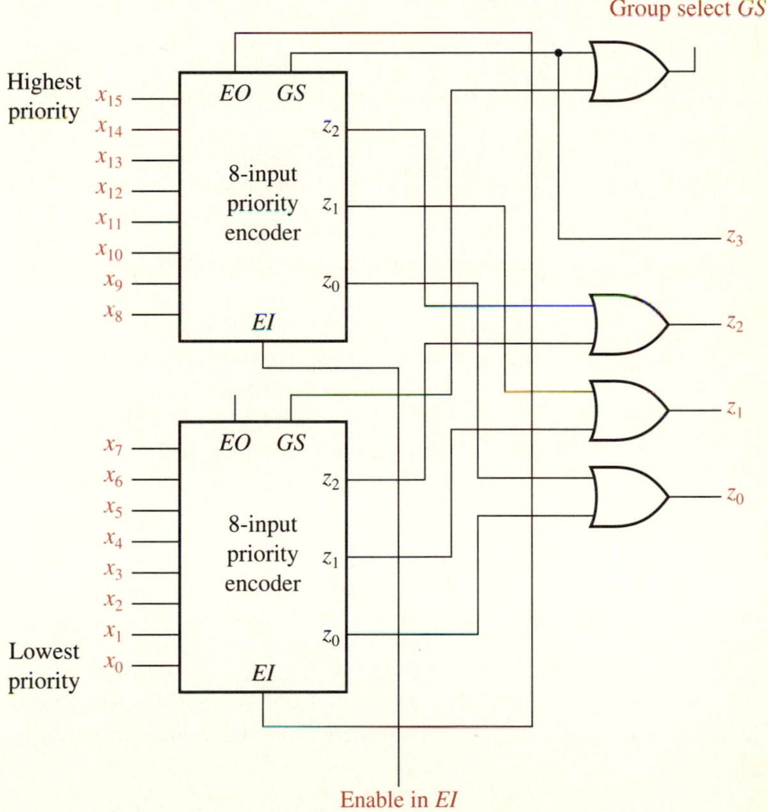

FIGURE 8.69 The serial technique for encoder expansion.

8.39 This problem concerns the design of circuits that compare two 16-bit unsigned numbers A and B, using multiple copies of the 4-bit magnitude comparator *COMP4* depicted in Figure 8.70. The complete logic circuit of a commercial version of *COMP4*, the 74X85 IC, appears in Figure 8.18. **(a)** Show how to cascade four copies of *COMP4* to perform the 16-bit comparison. Note that the input signals labeled $A<B$, $A=B$, and $A>B$ are intended to facilitate this type of expansion. Explain briefly how your circuit works. **(b)** Show also how to arrange five copies of *COMP4* in a tree structure to perform the desired 16-bit comparison. Note that no additional logic elements are needed, and that the $A<B$, $A=B$, and $A>B$ inputs are not used in this case.

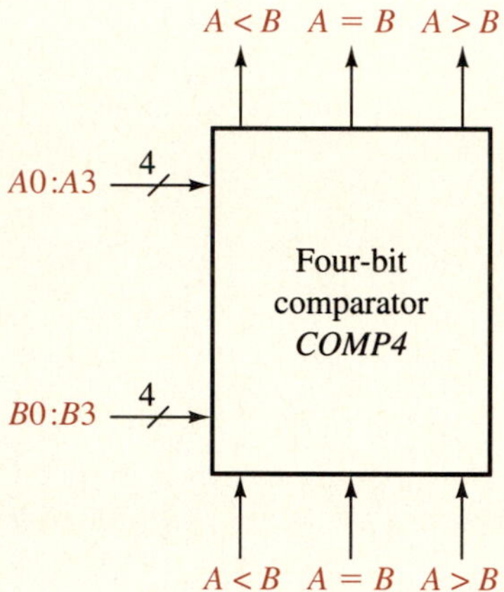

FIGURE 8.70 A 4-bit magnitude comparator.

8.40 Consider the design of a 16-bit ALU using four copies of the 74X181. It can be designed with ripple carry propagation or, using the 74X182, with carry lookahead. Assume that every internal gate in the components used has a propagation delay of 1 ns. For each 16-bit ALU, identify and describe (in words) the worst-case (critical) delay path through the circuit and calculate the critical delay in nanoseconds.

8.41 An ALU such as the 16-bit bit-sliced design of Figure 8.34 usually has several status output bits, called **flags**, that indicate special or unusual features of current results on the data output bus Z. Design circuits for the ALU of Figure 8.34 to generate the following signals: **(a)** a flag S to indicate the numerical sign of Z; **(b)** a flag *OVF* to indicate the occurrence of overflow after performing a two's-complement arithmetic operation; and **(c)** a flag P that indicates the parity of Z. Assume that no connections can be made to any purely internal signals of the ALU slices.

8.42 Redesign the 4×32 register file of Figure 8.22 so that it contains eight 32-bit registers.

8.43 Modify the 16-bit datapath unit of Example 8.7 (Figure 8.36) to accommodate the following changes: **(a)** an increase in data word size from 16 to 32 bits; **(b)** two new instructions: ADD Z, X1 and SUB Z, X1, which perform the operations $Z := Z + X1$ and $Z := Z - X1$, respectively. Assume primitive components such as multiplexers, decoders, registers, and the like of the appropriate sizes are available. Assign codes to all instructions and give a complete block diagram of the new design.

8.44 Suppose that a generic parallel-load shift register *SHR* like that in Figure 8.24a is available as a design component. Assume that *SHR* can perform a right- but not a left-shift operation. Modify *SHR* by adding as little extra logic as possible to form a new expandable component *SHLR* that has both left- and right-shifting capability.

8.45 An up-down counter such as that of Figure 8.24b typically has two additional control lines *EI* (enable in) and *EO* (enable out) to simplify expansion. Determine the functions of these signals and show how they can be used to link two n-bit up-down counters to form a $2n$-bit counter. Give a general expression for and explanation of the logic function implemented by *EO*.

8.46 Construct a state diagram for the control unit of the shift-and-add multiplier that appears in Figure 8.67. Verify it by comparison with the ASM chart of Figure 8.66.

8.47 Using the minimum number of D flip-flops, implement the control unit for the counter-based multiplier of Example 8.8. Compare the cost of your circuit in terms of hardware complexity and design effort with the one-hot design presented in Figure 8.43.

8.48 Construct a one-hot implementation of the seven-state recognizer circuit *R4* described in Example 7.8 (Section 7.5). Use D-flip-flops and NAND gates. Compare the gate count of your design to that of Figure 7.43.

8.49 Redesign the stack control unit of Figure 8.38 to accommodate six instead of four words in the memory and to respond appropriately to the simultaneous activation of *PUSH* and *POP*. Give a state diagram for the redesigned control unit, and a logic diagram for a one-hot implementation.

8.50 The multiplier design technique using counters presented in Example 8.8 can be readily adapted to integer division. Carry out the design of the control unit for such a division circuit that divides Y by X to compute a quotient Q and a remainder R, defined by

$$Y = X \times Q + R \qquad \text{where } 0 \le R < X$$

Describe your approach in words and outline the structure of the datapath unit. Give a complete state diagram for the controller.

8.51 Discuss the differences between Mealy and Moore machines in terms of their structure and their timing of state transitions and primary output responses. Describe a typical design situation in which a Mealy machine is clearly preferable, and a situation in which the reverse is true.

8.52 Construct a Moore-type state diagram for the counter-based multiplier control unit of Example 8.8. Identify the functions of all its states and relate them to those of the five states defined in the original Mealy state diagram (Figure 8.42). Why are the extra states needed?

8.53 Compare and contrast in depth the three designs, two hardwired and one programmable, given in the text for the control unit of the LIFO stack memory; see Figures 8.39, 8.40, and 8.50. In your answer consider hardware cost, speed of operation, design time, and maintenance costs.

8.54 Analyze the sequential PAL design of Figure 8.25 and obtain a state diagram for the counting circuit it implements.

8.55 Construct a programmable design for the seven-state recognizer circuit *R4* described in Example 7.8 (Section 7.5). Using the PAL16R4 PAL defined in Figure 8.25, outline its implementation.

8.56 Design a programmable machine like that of Figure 8.50 which makes into a left-right shift register the part of the state register *SR* that stores the three control signals $Z = z_1z_2z_3$. Hence, the Z signals are either unchanged, shifted one bit to the left, or shifted one bit to the right in each clock cycle.

8.57 After completing the design of a programmable control unit for the LIFO stack controller of Figure 8.50, it is decided that we must define a new state called S_{dead} that is entered after both *PUSH* and *POP* are activated simultaneously. The control unit will remain in this state indefinitely until the unit is reset. Redesign the control unit program (Figure 8.50) to implement this change. Retain intact as much of the original program as possible.

8.58 A central function of the stack controller (Figure 8.50) is to count and store implicitly in its state the number of available data storage locations in the memory unit M. This has the drawback of causing the number of control states to grow rapidly with the size of M. The problem can be eliminated by adding a separate counter *COUNT* to the datapath unit that counts the number of unused locations. Then, by testing the contents of *COUNT*, the control unit can determine how to respond to a push or pop request. **(a)** Redesign the stack memory in this manner, giving new versions of its structure and state diagram in the style of Figure 8.38. **(b)** Construct either a hardwired or a programmable implementation of the new stack control unit.

ASM Design Methodology

ASM charts: elements, structure, interpretation; correspondence with state diagrams; ASM design problems: datapath and control unit design, one-hot implementation.

8.59 **(a)** Describe the basic functions of each of the three major elements of an ASM chart: the state box, the decision box, and the conditional output box. **(b)** What general feature distinguishes the ASM chart of a Mealy machine from that of a Moore machine?

8.60 Explain why the ASM chart fragments that appear in Figure 8.71 are invalid. Describe a possible fix for each problem you identify.

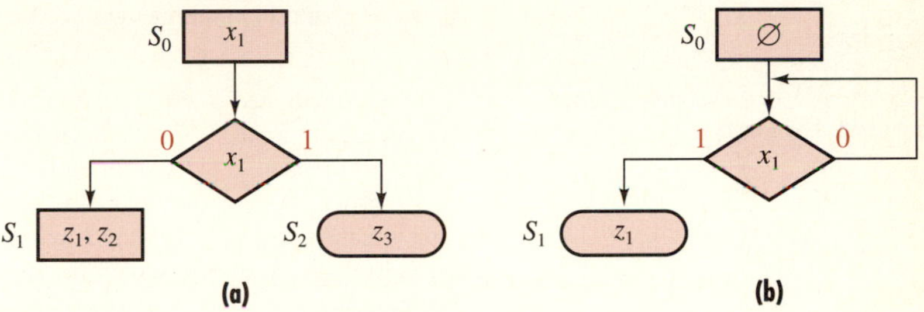

FIGURE 8.71 A pair of invalid ASM charts.

8.61 Draw an ASM chart that describes the four-state traffic controller, the state table for which appears in Figure 8.45.

8.62 Construct an ASM chart that describes a 4-bit shift register *SR* composed of four D flip-flops, like that of Figure 6.46.

8.63 Derive an ASM chart for the serial, multiply-by-3 circuit examined in Example 7.9 (Section 7.5). Indicate all its blocks.

8.64 Derive a state diagram in standard form from the ASM chart of the shift-and-add multiplier (Figure 8.66).

8.65 Use the one-hot technique to implement the control unit part of the LIFO stack controller described by the ASM chart of Figure 8.53a.

8.66 Construct a one-hot control unit for the shift-and-add multiplier of Example 8.10, using its ASM chart (Figure 8.66) as the starting point.

8.67 The design of a counter-based multiplier is described in detail in the text. This approach can be readily adapted to unsigned, n-bit integer division. Construct an ASM chart for such a division circuit that divides Y by X to compute a quotient Q and a remainder R, defined by

$$Y = X \times Q + R \qquad \text{where } 0 \leq R < X$$

List and state the function of all the control signals used. Give a high-level block diagram of the circuit in the style of Figure 8.65.

8.68 The following modifications are to be made to the shift-and-add multiplier of Example 8.10: **(a)** the number size is to be increased from 8 to 16 bits, and **(b)** the number formats are to be changed from unsigned to signed-magnitude integers. Thus, the circuit should now compute a 32-bit result $P = X \times Y$, which has a sign in the leftmost bit position. Describe in detail all the changes that must be made to the ASM chart (Figure 8.66) and to the logic diagram (Figure 8.67).

Here is a lower cost version of the shift-and-add multiplication technique.

8.69 As mentioned in the text, the amount of hardware needed for the shift-and-add multiplier of Figure 8.67 can be reduced by merging the A and Q registers. The multiplier operand X is loaded into the *SR0* part of A, which is initially empty. Then each right-shift operation moves the active value $X_i = Q_{COUNT}$ into the A_0 position, from which it can be sent to the control unit. The partial products being accumulated expand to fill *SR0* at the same rate as X is shifted out of *SR0*. Thus, at the end of the multiplication, X has disappeared and the product P occupies the accumulator A as before. Carry out the design of this modified multiplier, following the style of the original design. Your answer should include an ASM chart and a complete logic circuit for the data-processing unit.

8.70 A coin-operated vending machine dispenses 25-cent stamps and accepts nickels (5 cents), dimes (10 cents), and quarters (25 cents). You may insert exact change, or an excess amount and receive change. It is assumed you will not insert an excess amount if you have correct change; that is, you might insert three dimes to pay for the 25-cent stamp, but not a dime followed by a quarter. After inserting the coins, you press a button S to obtain the stamp. If you have inserted enough money, you will receive the stamp and any change due. If you have not, your money will be returned.

Use the ASM approach to design a synchronous sequential *VM* that controls the dispensing of the stamp and the change. The coins must be inserted one at a time into a single slot. A coin detector issues one of three signals, N, D, or Q for nickels, dimes, and quarters, respectively, to *VM*. The outputs of *VM* are Z_S to deliver the stamp, Z_N for delivering a nickel in change, and Z_R to return all deposited coins. The circuit should reset automatically following each S pulse. Your design should be clear and easily understandable; hardware minimization is not a concern.

8.71 Analyze Problem 8.70 and suggest at least three low-cost improvements to the specifications that would produce a more flexible and user-friendly vending machine. One of these requirements should be that the postal authorities be able to reprogram in 5-cent increments the amount of the dispensed stamp to account for increases in postage rates. Construct an ASM chart that describes all your changes.

Here is a technique for extending shift-and-add multiplication to 2C numbers.

8.72 Figure 8.72 gives a concise, HDL description of a multiplication method *MULT2c* for two's-complement numbers, attributed to James E. Robertson of the University of Illinois [Robertson, 1955]. It is similar to the shift-and-add scheme for multiplication of Example 8.10, with the modifications described in Problem 8.69. The main differences in the 2C case are the special treatment of the sign bit, for which a "flag" flip-flop F is used, and the fact that, if X is negative, a final subtraction (referred to as the correction step) is performed. The description of Robertson's method in Figure 8.72 assumes that the word size is $n = 8$ and that all numbers are 2C fractions. After initialization, $A = 000000000X$ and $M = Y$; at the end of the multiplication procedure, $A = P$. The shifting is such that, at any time, $A(15)$ contains the currently active bit of the multiplier X. **(a)** Explain in general terms or by means of an example why the correction step is needed. **(b)** Construct an ASM chart for *MULT2c*. **(c)** Design from generic 74X-style logic circuit parts a register-level circuit that implements your ASM chart.

```
procedure MULT2c;
{Declare registers: A(0:15), M(0:7), COUNT(0:2), F}
  begin
  M := Y; A(8:15) := X;                    {Input the operands X and Y}
  A(0:7) := 0; COUNT := 7; F := 0;         {Initialize registers}
  for i := 1 to COUNT do
    begin
    A(0:7) := A(0:7) + M(0:7) × A(15);     {Add partial product to A}
         F := [M(0) and A(15)] or F;       {Set flag F}
    A(0) := F, A(1:15) := A(0:14)          {Right-shift accumulator A}
    end;
    A(0:7) := A(0:7) − M(0:7) × A(15),     {"Correction" step for negative X}
    A(15) := 0;                            {Clear least significant bit of A}
  end;
{Product P is now in accumulator A}
end.
```

FIGURE 8.72 An HDL description of a two's-complement multiplication procedure (Robertson's method).

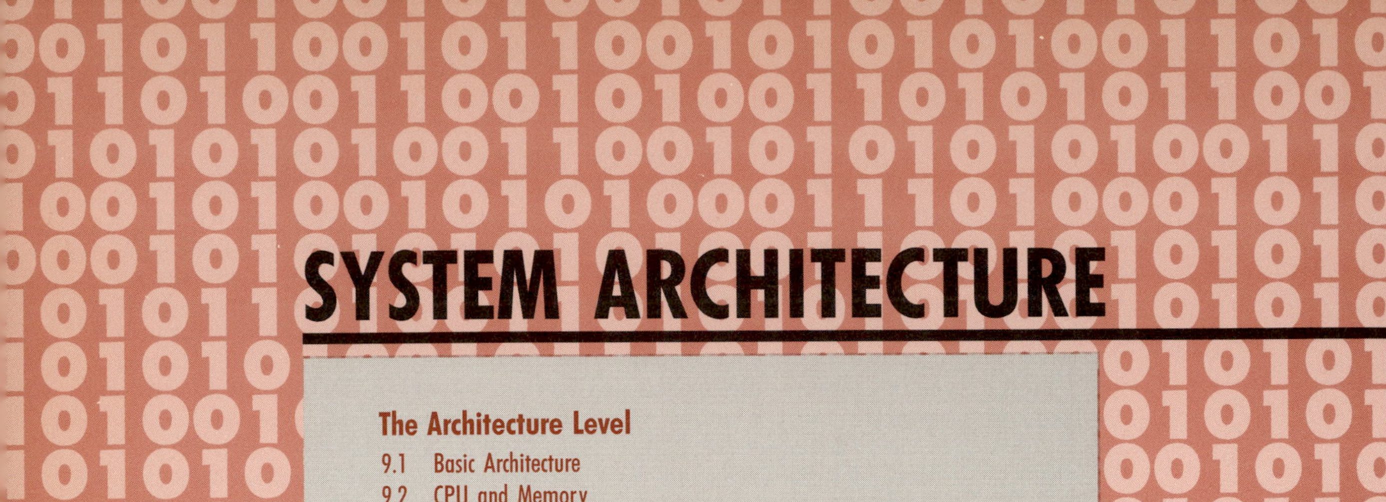

SYSTEM ARCHITECTURE

The Architecture Level

9.1 Basic Architecture
9.2 CPU and Memory
9.3 Input/Output Operations

The Central Processing Unit

9.4 CPU Operation
9.5 Instruction Sets and Programming

Computer-Based Systems

9.6 Microcontrollers
9.7 Input/Output Interfacing

Overview

The goal of many design projects is a complete, self-contained digital system centered around a programmable computer. This final chapter examines the organization and programming of a general-purpose computer. To this end, we move to the highest level of the design hierarchy, known variously as the architecture, system, or processor level. We also discuss the use of single-chip microcomputers as built-in controllers for digital systems.

9

The Architecture Level

This section discusses the general nature of the architecture of a complex digital system, focusing on basic computer organization. It introduces the principal component types (the central processor, main memory, and input/output circuits) recognized at that level, and discusses their design and application.

9.1 Basic Architecture

The register level described in Chapter 8 is characterized by the use of components like registers and arithmetic-logic circuits that store and process individual data words. At the architecture level, we deal with memories and processing units that store and process long sequences or blocks of word-organized data. We are also concerned with the connections between the system and the outside world. These connections require the presence of input/output circuits to handle communication and any associated data processing, such as converting data between analog and digital forms, changing the word size used, and the like.

As in the case of the register level, we can distinguish the system's data and control functions. System-level information takes the form of large blocks of data and control words. The control words constitute instructions; a block of instructions that perform a specific task is called a **program**. The fact that it can be programmed to carry out essentially any computational task gives a computer enormous flexibility and generality. Advances in VLSI hardware technology have made it possible to place all the electronic parts of a computer in a single low-cost IC, or a few such ICs, to create a **microcomputer**. This, in turn, makes it feasible to employ computers as specialized controllers in an essentially unlimited number of applications. Indeed, the single-chip microcomputer has become a standard device for controlling systems, ranging from microwave ovens to the automatic pilot of an aircraft.

Programmability plus low-cost VLSI implementation explain the explosion in microcomputer applications.

CPU and Memory

The CPU is a computer's key component because it executes the programs.

Figure 9.1 shows the overall architecture of a typical digital computer viewed from the high-level perspective of a user or manager of the system. The part responsible for executing instructions is the **central processing unit (CPU)**, which contains datapath (data-processing) logic in the form of multifunction arithmetic-logic circuits, as well as sets of registers to store instructions and data while they undergo execution. It also contains the control logic needed to interpret the CPU's instructions and decide the order in which the instructions of a program are to be executed. A typical CPU is designed to handle from 20 to 200 or so different types of instructions, but only executes one instruction at a time. A CPU that is implemented on a single IC chip is referred to as a **microprocessor**.

Programs are instruction sequences that are stored in main memory.

A computer program is usually far too large to be stored all at once in the CPU, so a computer has a separate storage unit called the **main memory**, the purpose of which is to hold a program's instructions and associated data while they are being executed. Main memory consists of electronic circuits designed to allow storage and retrieval of programs and data, one item (word) at a time. The unit of storage is a fixed-length word containing one or more bytes. Any randomly chosen memory word can be accessed with equal ease; hence, the main memory is also referred to as a **random-access memory** or **RAM**. If the main memory is designed to store only predetermined instructions and data, then, as discussed in Section 4.7, it can take the form of a read-only memory or ROM, which is also randomly accessible.

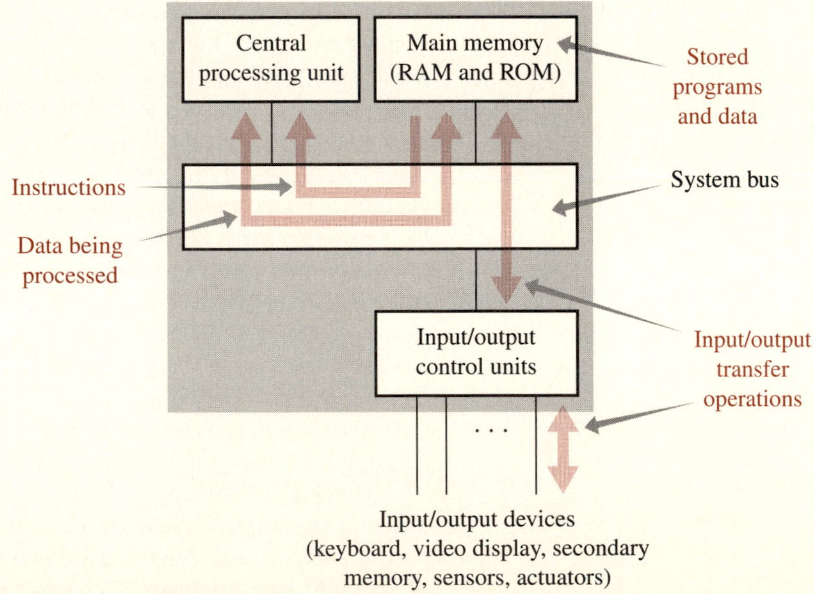

FIGURE 9.1 Overall architecture of a digital computer.

The CPU fetches instructions from main memory and executes them.

It follows from the preceding discussion that the basic mode of operation of a computer requires transferring instructions (only copies are actually transferred) one at a time from the main memory to the CPU, where they are executed. The data on which the instructions act are also stored in main memory. Hence, normal computer operation involves a continual flow of instructions and data between the CPU and its memory, as suggested by arrows in Figure 9.1. Instructions flow into the CPU from main memory. Data words that represent the input operands of instructions also flow into the CPU from main memory; data words that represent the results of instruction execution flow from the CPU to main memory. Typically, millions of instructions can be processed in this way during every second of a computer's operation. Indeed, the parameter known as millions of instructions (executed) per second, or **MIPS**, is a popular measure of a computer's performance. To support the rapid transfer of information within the computer, the CPU and main memory have a high-speed communication link, of which the usual form is a **system bus** that consists of a hundred or more wires and their supporting interface circuits.

The Input/Output Subsystem

I/O devices, including secondary memories, link a computer with the outside world.

The last major architecture-level components shown in Figure 9.1, the **input/output** or **I/O control units**, serve to interface the computer to external devices; the latter are referred to collectively as **I/O devices** or **peripherals**. Among the more familiar I/O devices in personal computers are a keyboard, which is an input device, and a printer, which is an output device. A video display unit is primarily an output device, but in conjunction with a mouse or a similar pointing instrument, it can also be made to serve as an input device.

Most computers require more storage space than can be provided in main memory. Consequently, they employ secondary memory units, which form part of the computer's I/O equipment; main memory is the system's "primary" memory. Examples of secondary memory devices are hard or flexible (floppy) disk drives and magnetic tape units. These memories employ storage media (magnetic disks and tapes) that have much higher storage capacity and lower cost per stored bit than main memory. However, secondary memories also transfer data at much slower rates than CPU/main-memory transfers. The purpose of the I/O control circuits and their supporting programs is to manage efficiently the transfer of information between main and secondary memory, as well as between main memory and other I/O devices.

Embedded Computers

The basic computer architecture depicted in Figure 9.1 can be modified in many ways to create special-purpose or application-specific digital systems. If the set of programs or **software** needed to implement a particular task—for example, control of a car's braking system—is fixed and embedded per-

Many devices are controlled by embedded computers called microcontrollers.

manently in a microcomputer's main memory, then we obtain a sophisticated logic circuit dedicated to the task in question. Embedded microcomputer-based devices of this kind are called **microcontrollers** or **microprocessor-based systems**.

Microcontrollers have very low cost when manufactured in high volume. They also have a huge and growing range of applications and frequently employ specialized I/O devices to allow them to interact with their environment. These I/O devices include sensors to measure physical parameters such as temperature, position, and acceleration, and electromechanical actuators such as electrical motors. Despite their diverse applications and I/O circuit types, all microprocessor-based systems are firmly grounded in common computer architectural concepts. Knowledge of these concepts is therefore of fundamental importance to designers and users of digital systems.

A Little History

The basic computer organization of Figure 9.1 has its origins in the work of the Englishman Charles Babbage (1792–1871), who attempted to build a general-purpose digital computer, the Analytical Engine, in the mid-nineteenth century. The design of the Analytical Engine included a programmable central processor and a large memory. It was never completed, however, mainly due to the difficulty of building so complex a system from purely mechanical parts like gears and levers. The late nineteenth and early twentieth centuries saw the successful introduction of less ambitious electromechanical calculators—forerunners of today's pocket calculators—notably the punched-card tabulator invented by the American Herman Hollerith (1860–1929).

The first electronic computers appeared in the 1940s.

Around 1940, a number of prototype digital computers that used various electromechanical and electronic technologies were successfully built in the United States and Europe [Randell, 1982]. Perhaps the most influential of these machines was the ENIAC (Electronic Numerical Integrator And Calculator) constructed at the University of Pennsylvania in the period 1943 to 1946. The ENIAC spawned a series of electronic computers that employed vacuum tubes as their basic switching devices; these constitute the so-called "first generation" of computers.

Modern Computers

Computers are traditionally classified as mainframes, minicomputers, or microcomputers.

The invention of the transistor at AT&T Bell Laboratories in 1947 soon gave rise to a second generation of computers that were significantly smaller, faster, and more reliable than their vacuum-tube counterparts. However, all early computers were physically large and costly machines that could easily fill a large room. They were housed in computer centers so they could be shared by many users and operated by specialists. Such multimillion-dollar machines and their more recent successors are referred to as **mainframe computers.**

Several major innovations appeared in the mid-1960s. IBM Corporation introduced a series of mainframe computers known as the System/360, which clearly distinguished the architecture of a system from its implementation. The **architecture** of this and subsequent computer series was defined as a standard specification that covered the information formats, instruction types, and the high-level structure and behavior of all members of the series. Particular computer models in the series could differ greatly in their lower-level structure and in their circuit technology; the latter are regarded as implementation details.

Another innovation of the 1960s pioneered by Digital Equipment Corporation was the **minicomputer**, a small (desk-sized) computer with limited hardware and software facilities, compared to a mainframe. Minicomputers were of sufficiently low cost (tens or hundreds of thousands of dollars) to allow them to be used as shared office computers, or for industrial process control. Around this time, integrated transistor circuits began to displace the earlier, discrete transistor circuits, further reducing computer size and cost (Figure 9.2).

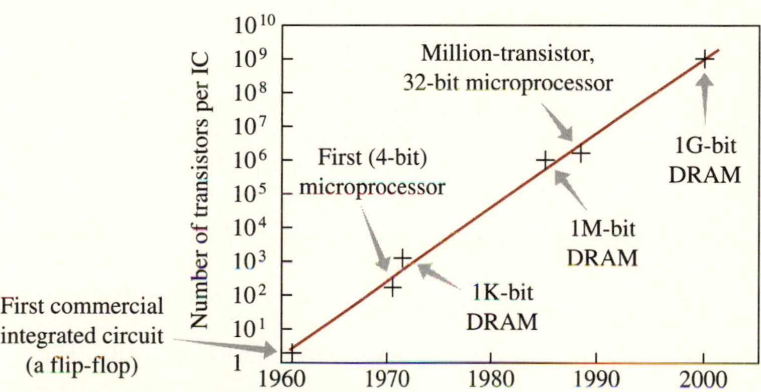

FIGURE 9.2 The evolution of IC technology.

Microprocessors and Microcomputers

A major milestone occurred in the 1970s with Intel Corporation's introduction of microprocessors (CPU ICs) and main-memory ICs of relatively high complexity (Table 9.1). The concept of a microcomputer, an entire computer—CPU, main memory, and I/O control circuits—constructed from a few low-cost ICs, also emerged. Integrated circuit technology evolved rapidly, making it possible to incorporate into a single VLSI chip more and more of the features of minicomputers and mainframe computers. For example, from 1971 to 1991 the storage capacity of a commercial dynamic random-access memory (DRAM) IC increased by a factor more than a thousand, from 2^{10} or 1K bits

The maximum number of transistors per IC has doubled every year or so since 1971.

TABLE 9.1 Milestones in the development of computer technology.

Date	*Key technology*	*Representative systems*
1850s	Mechanical	Charles Babbage's Difference and Analytical Engines
1900s	Electromechanical	Telephone switching systems, desk calculators
1930s	Electromechanical	First operational general-purpose digital computers
1940s	Electronic tubes	First-generation electronic computers
1950s	Discrete transistors	Second-generation computers
1960s	Small-scale ICs	Third-generation computers, minicomputers
1970s	Large-scale ICs	Microprocessors, microcomputers, microcontrollers, high-capacity memory ICs
1980s	Very large-scale ICs	Personal computers, massive computer networks

per chip to 2^{22} or 4M bits per chip.[1] Over the same period, microprocessors went from calculator-like devices capable of processing information only 4 bits at a time to 32-bit microprocessors with performance levels exceeding the most powerful CPUs of earlier generations. As the manufacturing cost per IC changes little once high production volumes are reached, these developments have meant a steady improvement in the performance/cost ratio of computer components.

By 1980, single-chip microprocessors and microcomputers were being mass-produced in several types. The simpler devices had word sizes of 4 to 8 bits or so and cost only a few dollars each. They were being used for single or "dedicated" applications such as the control of toys, domestic appliances, and scientific instruments; this is the area in which the term *microcontroller* is often used. The more complex microprocessors that cost a few tens or hundreds of dollars began to appear in low-cost, general-purpose computers, a development that produced the **personal computer**, exemplified by the PC family IBM introduced in 1981. The IBM PC series is based on Intel's 80X86 family of 32-bit microprocessors. It is also characterized by a system control program or **operating system** known as DOS (Disk Operating System), which defines the software interface a computer presents to its users.

By 1990, most computers could be placed in a few families defined by certain processor and operating system standards, such as compatibility with the IBM PC or with Unix, a widely used operating system. By then, too, the

[1]When referring to memory capacity, the scaling prefixes K (kilo) and M (mega) have the special meanings of $2^{10} = 1{,}024$ and $2^{20} = 1{,}048{,}576$, respectively. K and M represent the powers of two that are closest to a thousand and a million. In this context, G (giga) denotes 2^{30} or approximately a billion.

old distinctions among mainframe, minicomputer, and microcomputer had lost most of their significance. "Large" general-purpose computer systems now usually take the form of networks of microcomputers, the collective computing power of which can greatly exceed that of any single-CPU system.

Microcontrollers

Microprocessor technology has led to low-cost personal computers and microcontrollers.

A **microcontroller** is a "computer on a chip" that is programmed to perform a dedicated set of tasks. This programmability, coupled with relatively low design and manufacturing cost, is the key to the flexibility of microcontrollers and their enormous range of applications. Customizing a microcontroller requires the writing of control programs that define a particular application. These programs are written in standard programming languages such as the C language, using a variety of computer design aids like language translators (compilers) and prototype development systems. Although the design of a microcontroller is mainly a software task, it has some hardware aspects that typically involve interfacing the microcontroller with application-specific devices such as keyboards, video displays, communication networks, and the like.

Microcontrollers are customized by software design; PLDs are customized by hardware design.

The technique of programming off-the-shelf microcontrollers for specific applications contrasts with the design of application-specific integrated circuits (ASICs), in which hardware is custom-designed for an application. ASICs such as PLAs (Section 4.7) are "programmed" in a very different way to customize the underlying hardware. The ASIC approach permits the hardware cost and operating speed of a digital system to be optimized for a particular application, but generally the associated design and production costs are higher.

9.2 CPU and Memory

This section presents an overview of the structure and operation of two key components of a computer system at the architectural level: the CPU and main memory. We also examine the design of random-access memories (RAMs).

CPU Functions

The primary task of the CPU is to fetch instructions from main memory M and execute them. The CPU contains all the instruction-processing circuits needed by its particular instruction set. It is clearly divided into data and control units, as shown in Figure 9.3. The CPU's control part is designed to decode instructions and determine their order of execution. This unit is known as the **instruction unit** or the **I-unit**; it is also referred to as the **program-control unit**. The other main part of the CPU, known as the **execution unit** or **E-unit**, contains the datapath logic needed to execute the instructions, as

The CPU has a control part (the I-unit) and a datapath part (the E-unit).

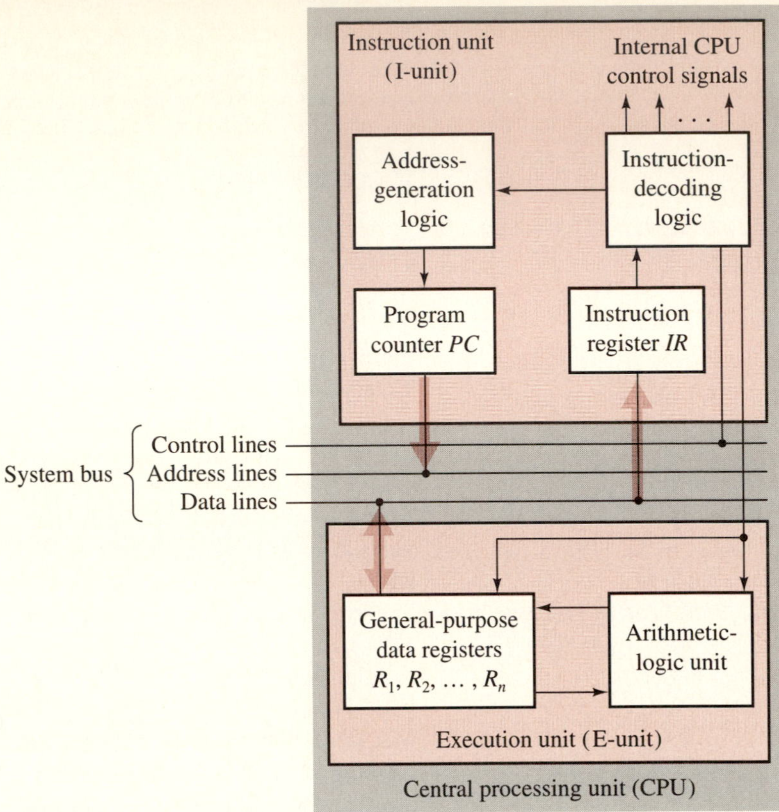

FIGURE 9.3 Internal organization of a CPU.

well as registers to store input and output operands temporarily. While such operands could be stored exclusively in the main memory M, time is saved by providing registers for their storage in the CPU.

Another function of the CPU is to control the system bus, which is its main highway for communicating with M, I/O devices, and the outside world. The system bus is organized into several subbuses for transferring various types of information. It typically contains around a hundred individual lines, some of which are bidirectional, tri-state lines.

The Instruction Cycle

We can most easily describe the internal operation of a CPU at the register level. The CPU is a complex, synchronous sequential circuit whose clock period is the computer's fundamental time unit. During one clock cycle, the CPU is able to perform a register-transfer operation such as fetching the instruction word I stored at main memory address A_1 and transferring it over the system bus to the CPU's instruction register IR. This operation may be described symbolically as follows:

$$IR := M(A_1); \quad \text{-- Fetch instruction} \tag{9.1}$$

Here we are using the VHDL-like hardware description language we introduced in Chapter 8; recall that := is the register assignment operation.

In words, the register-transfer statement (9.1) says, "load the contents $M(A_1)$ of memory location A_1, which is assumed to contain an instruction I, into the CPU's instruction register IR." Once in IR, the instruction is decoded to determine the next actions to be taken by the CPU. This might be to perform an arithmetic or logical operation on data stored in M or in the CPU registers. For example, the "add-register" instruction requires the register-transfer operation

$$R_1 := R_1 + R_2; \quad \text{-- Execute instruction} \tag{9.2}$$

which states that the contents of CPU registers R_1 and R_2 should be treated as numbers and added together; the resulting sum should be placed in R_1. If required by I, this result can be transferred to M, a task that requires a "store-to-memory" operation of the form

$$M(A_2) := R_1; \quad \text{-- Store result} \tag{9.3}$$

An instruction cycle involves fetching an instruction, executing it, and storing the results.

The sequence of steps (9.1), (9.2), and (9.3)—that is, fetch the instruction, execute it, and store its results—constitutes an **instruction cycle**. In normal operation, a CPU goes through a continuous sequence of instruction cycles, the length and timing details of which vary with the instruction type. Some instructions require no transfer of results to M. Other instructions require the execution step to be preceded by one or more operations of the form

$$R_i := M(A_j); \quad \text{-- Fetch data} \tag{9.4}$$

designed to load data (input operands) from M into a CPU register for use by the current instruction.

Programming Languages

We will often describe instructions by means of register-transfer statements like (9.1) through (9.4). Some programming languages employ formats of this general kind for their instructions; most, however, have their own distinctive formats.

Programs can be written in a symbolic format called assembly language.

Computers are sometimes programmed in a low-level language called **assembly language**, which represents the instruction as a word with symbolic components (fields) that define the operation to be performed and the locations assigned to input operands and results. Unlike high-level programming languages (FORTRAN, Pascal, and so on), assembly languages are machine-specific; that is, each CPU family tends to have its own unique version of assembly language. For example, in the Motorola 68000 microprocessor, a particular add instruction for binary numbers is expressed in assembly language as follows:[2]

$$\text{ADD} \quad \text{D2,D1} \tag{9.5}$$

with ADD denoting the desired E-unit operation and implicitly specifying the use of 16-bit two's-complement operands. *D1* and *D2* are the names of two of the general-purpose registers in the 68000's E-unit. (Here *D* stands for data register.) The action of the add instruction can be described by the register-transfer statement

$$D1 := D1 + D2;$$

The part of an instruction that identifies the operation to be performed is called the **opcode** (operation code); the parts that identify operands are usually referred to as **addresses**. Hence, in (9.5), ADD is the opcode, while D1 and D2 are the operand addresses. Every instruction contains an opcode, but the number of its operand addresses varies with the instruction type. Consequently, instructions often come in several different sizes.

Programs must be translated into machine language prior to execution.

As it is actually stored and executed within a 68000-based computer, the add instruction is just a 16-bit sequence of 0s and 1s; specifically,

$$1101001001000010 \tag{9.6}$$

which contains binary-coded versions of the opcode and the operand addresses. The binary instruction format represented by (9.6) is referred to as **machine language**. An assembly-language program must be converted to a machine-language program for execution by the host computer. Roughly speaking, an assembly-language program is a symbolic and, for humans, a more readable version of the machine-language program that is actually executed.

The task of translating symbolic instructions to machine-language form is carried out automatically by various types of language-translation programs known as **compilers** (which have a high-level programming language as input) and **assemblers** (which have an assembly language as input). Early computers were programmed directly in assembly language (or even in machine language, in a few instances), but almost all programs today are written in higher-level languages like C, Pascal, and Ada, the latter being a close relative of VHDL.

[2]The order of the operands in instructions like this one is not standardized. Some assembly languages represent $D1 := D1 + D2$ by ADD D1,D2 instead of ADD D2,D1.

The Program Counter

Before it can be executed, a machine-language version of a program must be placed somewhere in main memory M. The order in which the instructions within the program are stored in M corresponds roughly to the order in which they are to be executed. The CPU can therefore obtain instructions from the program in the proper order by fetching, one at a time, the contents of the set of consecutive main-memory locations $A, A+1, A+2, \ldots$, each containing an instruction (or part of an instruction) to be executed.

The program counter *PC* automatically maintains the address of the next instruction word.

To keep track of these consecutively stored instructions, all the CPU needs is a counter of appropriate size; this key device is known as the **program counter** *PC*. The program counter addresses, or points to, the next instruction (or a portion thereof) required by the CPU from the currently active program. The memory control circuits automatically perform a memory read or fetch operation to bring the indicated instruction into the CPU. Thus, the operation of loading the instruction register *IR* with the opcode of a new instruction is implemented by the opcode-fetch operation

$$IR := M(PC);$$

Additional memory fetch operations may be required to obtain operand data. After each such fetch, the CPU's I-unit simply increments *PC* to determine the next instruction address it needs.

It is occasionally desirable to switch execution from one part of a program to another, nonconsecutive part. To make this possible, special "branch instructions" are included in the CPU's instruction set. The simplest of these performs the operation

$$PC := ADR; \tag{9.7}$$

which loads a new, nonconsecutive address *ADR* into *PC*. In the 680X0 series, for instance, there is a "jump instruction" of precisely this type, which is written as

JMP ADR

in assembly language, and as

0100111011111000 xxxxxxxxxxxxxxxx

in machine language, where xxxxxxxxxxxxxxxx denotes a 16-bit binary version of the symbolic address *ADR*. Note that the machine-language jump instruction is twice as long (32 bits) as the 680X0 add instruction (9.6) we encountered earlier.

Instruction branching can also be made dependent on certain conditions occurring inside or outside the CPU; these conditions give programs the flexibility to respond to changes or unexpected conditions encountered in the course of a computation. An example of such an exceptional condition is arithmetic overflow, which might require a branch to an overflow-handling error program.

Instruction Types

A typical CPU is designed to execute a fairly large set of instruction types. These can be grouped into the following major classes:

The three main instruction classes are data-transfer, data-processing, and program-control instructions.

1. **Data-transfer instructions** move data unchanged from one place to another. Typical of this class are memory-to-CPU **load instructions** of the form $R_i := M(A)$ and CPU-to-memory **store instructions** of the form $M(A) := R_i$. Input/output instructions are also of this class, with an input/output device serving as the data source or destination.
2. **Data-processing instructions** specify various arithmetic and logical (nonnumerical) operations on data, and their execution uses the E-unit's arithmetic-logic circuits.
3. **Program-control instructions** modify the execution order of the instructions within a program. Most such instructions take the form of conditional or unconditional branch instructions like (9.7).

The number of instruction types varies widely among the various processor families. In theory, one or two basic types suffice for all operations, just as the NAND or NOR gate types suffice to implement all logic functions. Complex operations can be programmed by sequences of simpler ones. For example, multiplication can be performed by a sequence of additions, as we saw in Chapter 8. In practice, CPU instruction sets include representative types from each of the three main classes: data transfer, data processing, and program control. The instruction types are selected to achieve a good balance between usefulness in programming applications, and implementation complexity as it affects CPU cost and performance. Thus, almost every CPU implements an add instruction for binary numbers; few implement the much more complex and rarely used square-rooting operation. Thus, when square roots must be computed, multi-instruction programs are invoked for this purpose.

Instruction formats vary among computer families.

Instructions take many forms, depending on the allowed formats of their data operands and the manner in which these operands are addressed. Numerical data can employ the various fixed-point (signed and unsigned integer) and floating-point formats discussed in Chapter 2. The Motorola 68000 microprocessor, for example, allows fixed-point numbers to be 8, 16, or 32 bits long. Instructions with the (symbolic) opcode ADD treat their operands as two's-complement binary numbers; the ABCD (Add Binary Coded Decimal) opcode, on the other hand, specifies BCD data. Later microprocessors in the same family allow even longer fixed-point numbers, as well as numbers that employ the IEEE 754 floating-point format (Section 2.12).

Main Memory

Main memory is built from random-access memory ICs of various types.

M is a random-access memory that contains a large array of individually addressable storage locations. The term *random-access memory* implies that each word-storage location of M can be accessed directly via its address A, in contrast with the slower, serial-access method employed in secondary memories like magnetic disks. The internal structure of a RAM is illustrated in Figure 9.4. There are n address bits, each of whose 2^n patterns is the address A of a specific memory location. A 1-out-of-2^n decoder, the address decoder, selects the storage location $M(A)$ specified by address A. If each such location can store an m-bit data word, M is said to be a $(2^n \times m)$-bit RAM and has a total capacity of $m2^n$ bits.

A RAM performs two basic operations: read and write. A **read operation**, also referred to as a **fetch** or **load operation**, transfers a previously stored data word from the currently addressed storage location $M(A)$ to the external data bus and from there to an external device such as the CPU. A **write operation**, also referred to as a **store operation**, causes a new data word to be transferred from the external data bus into $M(A)$. Because read and write operations are not performed simultaneously, a single bidirectional bus suffices to transfer data words to or from the RAM.

The development of low-cost RAM ICs of very high capacity has had a profound effect on the nature and cost of computers and computer-based digital systems. Random-access memory chips are now available in a wide variety of sizes and speeds. **Dynamic RAM** chips, known as **DRAMs**, employ a single-transistor storage cell, for which the theory of operation is discussed in Section 6.1 (Example 6.1). **Static RAMs** (**SRAMs**) employ multitransistor cells and avoid one of the chief disadvantages of DRAMs: the decay of stored data that requires continual automatic refreshing of a DRAM. Static RAMs are also faster than DRAMs, but their more complex storage cells make SRAMs more expensive.

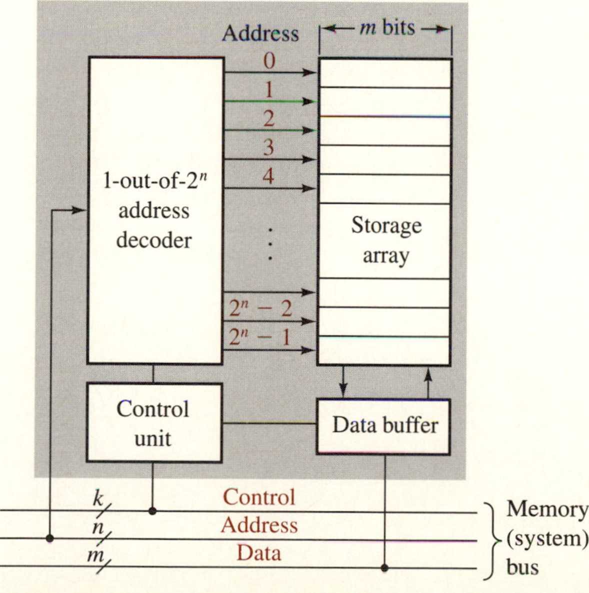

FIGURE 9.4 Structure of a $(2^n \times m)$-bit random-access memory (RAM).

A **read-only memory** or **ROM**, which we encountered earlier (Section 4.7), has the same structure and addressing mechanism as a RAM, except that there is no provision for writing new data into a ROM. The ROM is written (programmed) once during manufacture; its stored data are essentially permanent thereafter. ROMs are often used to build the part of main memory that stores a system's most frequently used control software.

Memory Behavior

Some details of the timing behavior of a static RAM (and a ROM) are depicted in Figure 9.5. DRAM behavior is more complicated; for example, reading a DRAM erases the data in the cells being read (destructive readout), so the data are restored automatically via an internal write operation, with the data buffer depicted in Figure 9.4 serving as a temporary holding register for the data in question. In general, a control signal named *CS* (chip select) is used to enable a read or write operation. The operation type, read or write, is

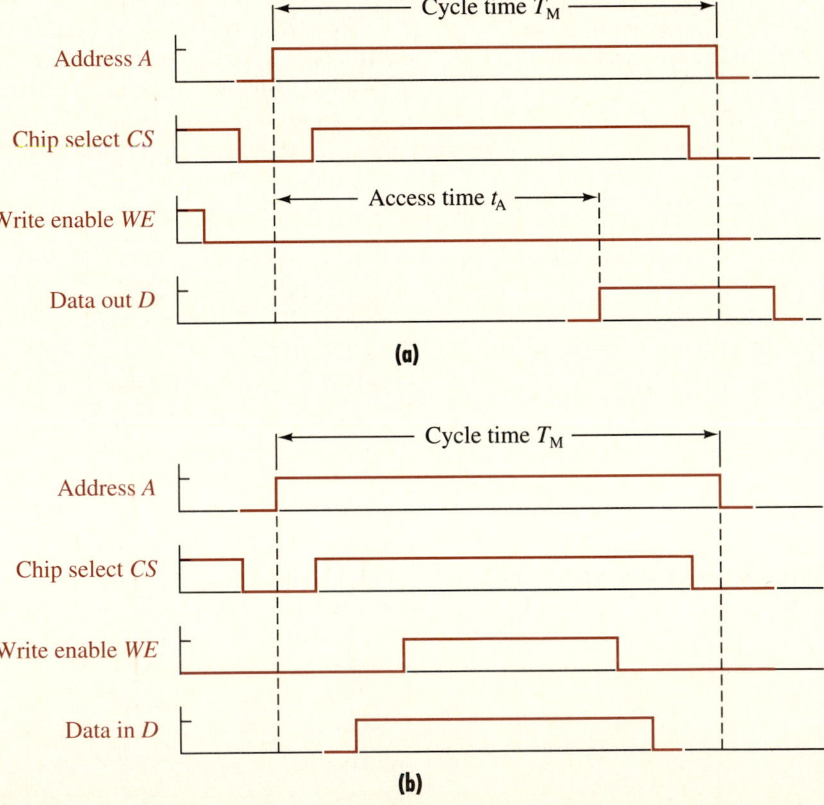

FIGURE 9.5 Typical timing diagrams for a RAM: **(a)** a read operation; **(b)** a write operation.

specified by a second control signal WE (write enable), which is 1 for write and 0 for read. (Commercial RAM ICs often employ the opposite convention, and so have a write enable signal denoted by $\overline{WE}$.) Obviously, a write-enable signal is not needed by a ROM.

To perform a read operation, the write-enable line WE is first disabled, the target address A is placed on the address bus, and CS is activated. After some time, the desired data word $M(A)$ is retrieved and placed on the data bus D, making it available to the requesting external device. The time between the application of the address to M and the appearance of $M(A)$ at the output bus D of M, making due allowance for all signals to stabilize, is termed the memory's (read) **access time** t_A and is a fundamental indicator of the memory's speed. A write operation proceeds in a similar fashion (see Figure 9.5b). In this case, an external device supplies both the address word A and the input data word D to be stored.

Slow memories require extra CPU cycles (wait states) in each main memory cycle.

The signals depicted in Figure 9.5 must meet various setup, hold, and pulse-width constraints, which are similar to those of latches and flip-flops and can be found in memory IC manufacturers' data sheets. Memory ICs must be operated with a minimum clock period known as the **memory cycle time** T_M. This period may or may not correspond to the clock period T_C of the CPU. If, as is often the case, $T_M > T_C$—that is, if the memory is slower than the CPU—it is necessary to slow down the CPU with extra, idle clock cycles, referred to as **wait states**, when memory read or write operations are being performed.

Memory Design

A common problem facing the designer of a digital system is to use off-the-shelf RAM or ROM ICs of a particular capacity 2^n and word size m to implement a memory with higher capacity 2^N or with a larger word size M, where n, m, N, and M are usually powers of two. To facilitate this type of memory expansion, RAM ICs are manufactured with a variety of powers-of-two word sizes, from $m = 1$ upward. Main memories are designed to operate with data-word sizes of $M = 8$, 16, or more.

Memory word size is easily increased by bit slicing.

It is quite easy to increase RAM word size via the bit-slicing technique discussed in Section 8.4. To obtain a word size of $M = km$, we need k copies of a RAM with word size m. As shown in Figure 9.6, the k RAMs are connected in bit-sliced fashion, with each component RAM IC contributing an m-bit slice of each data word. The control and address lines are connected in common to all the RAM ICs as indicated. In this example, a tri-state data bus D is assumed; D enters the high-impedance state Z whenever $CS = 0$. When $CS = 1$, $D =$ Z unless an additional control signal OE (output enable) is set to 1. For read operations, both CS and OE are set to 1, which allows the RAM to drive the bidirectional data bus D. Write operations

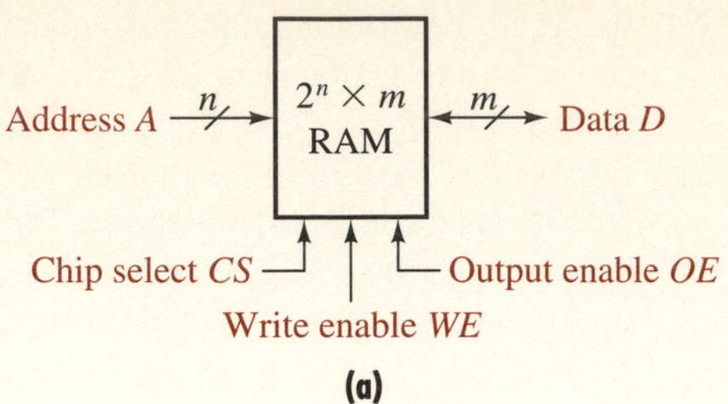

(a)

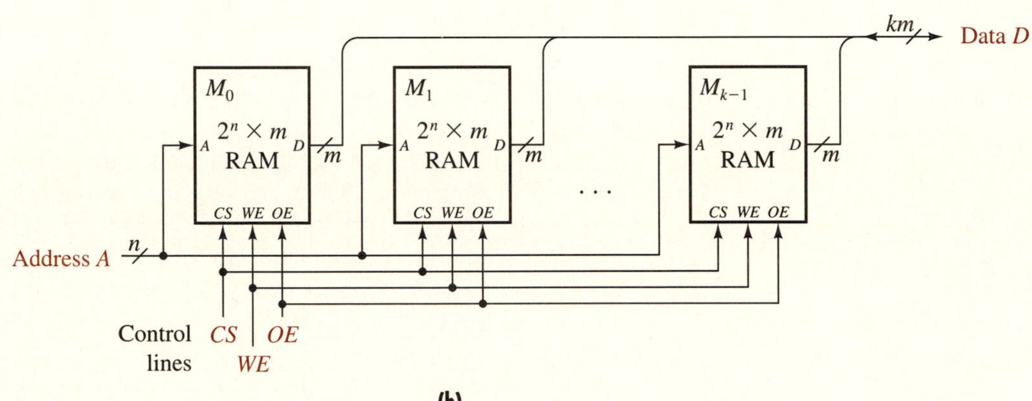

(b)

FIGURE 9.6 (a) A RAM IC with a bidirectional data bus; (b) increasing RAM word size from *m* to *km*.

require $CS = 1$ and $OE = 0$, which allows D to be driven by an external data source. Increasing RAM capacity requires more address lines; each such line permits the storage capacity to be doubled. If d new address lines are introduced, a 1-out-of-2^d decoder may be needed to augment the RAM's internal address decoder, as we see in the next example.

Example 9.1

Designing a 4K-byte RAM using 1K-byte RAM ICs

Suppose we want to construct a (4K × 8)-bit RAM with (1K × 8)-bit versions of the RAM IC of Figure 9.6a as the basic components. Four copies M_0:M_3 of the 1K-byte RAM are obviously needed, along with an auxiliary one-out-of-four address decoder (see Figure 9.7). Now $2^{12} = 4K$, so the target design has a 12-bit address word $A = A_{11}$:A_0. We assume here that the four RAM ICs have full 10-bit address buses. (This is not usually the case with RAMs in the 1M-bit range and beyond; see Problem 9.11.) These address buses are connected to the 10 low-order bits A_9:A_0 of the incoming

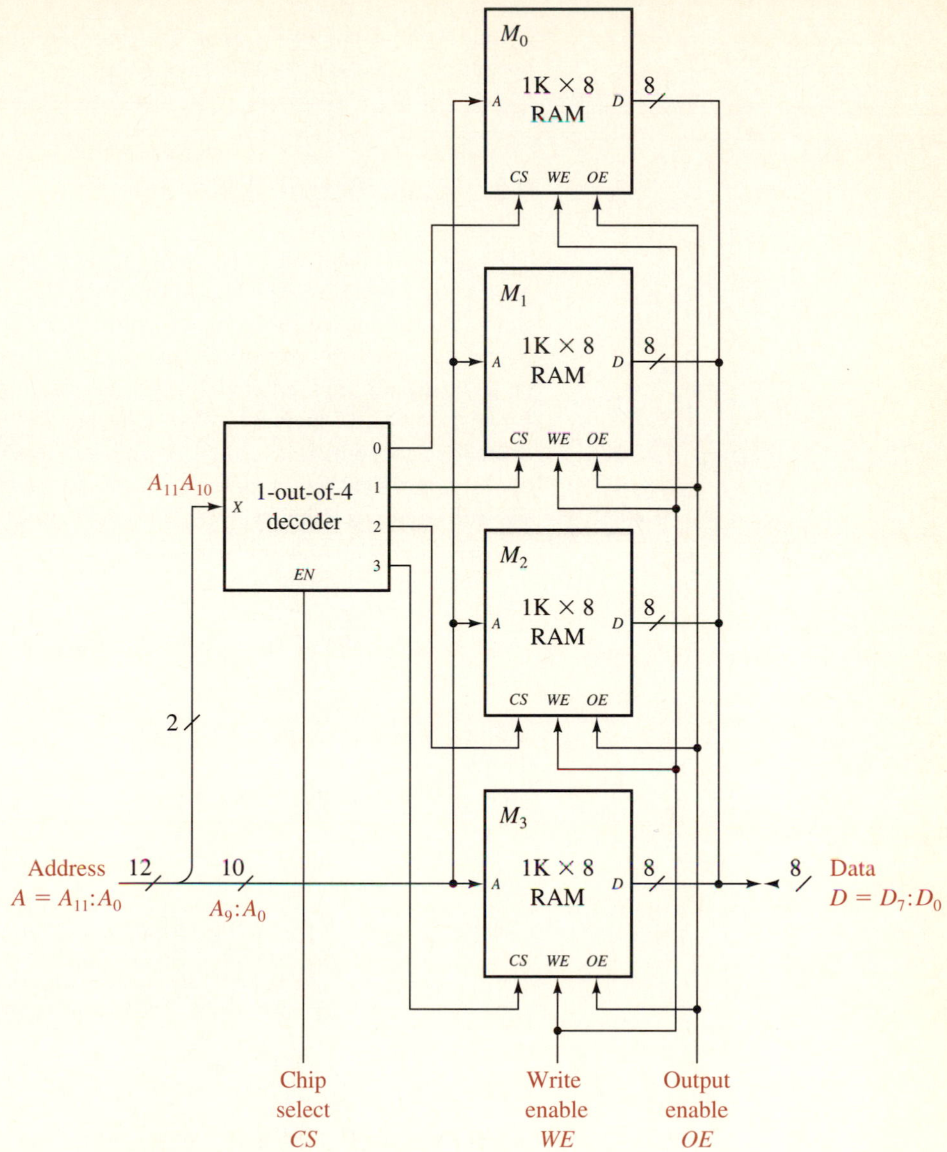

FIGURE 9.7 Increasing RAM capacity by a factor of four.

address bus A. The remaining two address bits $A_{11}A_{10}$ are decoded by the one-out-of-four decoder. Each decoded subaddress $A_{11}A_{10}$ activates the chip-select control line CS of one of the four component RAMs. Thus, all addresses of the form $00A_9A_8 \cdots A_0$ select M_0, $01A_9A_8 \cdots A_0$ selects M_1, and so on. The use of tri-state logic allows the data lines of each RAM IC to be tied together as indicated. Because only one RAM IC has 1 applied to its *CS* input, only one of the four ICs can drive the data *D* bus during a read operation, or respond to a write operation.

9.3 Input/Output Operations

The CPU and main memory of a computer are central to its information-processing role. However, the information they process is derived as input from sources outside the computer proper, while the processed information is sent as output to external destinations. The term **input/output operation** or **I/O operation** is applied loosely to information transfers of this kind between the CPU–memory subsystem and the outside world. Input/output operations are sometimes called "peripheral" operations, but this term should not be interpreted as diminishing the importance of the I/O part of a computer. Indeed, in many types of microcomputers, I/O operations constitute the dominant system activity.

Some hint of the wide variety of I/O equipment to which computers are linked is shown in Table 9.2. Familiar from their use in personal computers are keyboards (input devices), video displays (output devices), and secondary

TABLE 9.2 Representative types of input/output devices used in digital systems.

Class	*Device types*
Input	Manual switches, optical switches, keypads, keyboards Sensors of temperature, pressure, acceleration, position, chemical concentration, etc. (analog sensors require analog-to-digital converters)
Output	LCD and LED display panels, CRT displays Printers Electromechanical actuators, relay switches, electric (stepping) motors Analog devices: electric (DC and AC) motors, speakers, power supplies, etc. (these require digital-to-analog converters)
Input and output	Computer terminals Secondary memories: magnetic disks and tapes, optical disks Communication devices: standard buses, modems, fax machines, computer networks

memory units such as magnetic disk memories, which act as both input and output devices. Less familiar, perhaps, are the many types of I/O devices attached to the specialized computers that serve as microcontrollers. Some I/O devices are analog rather than digital and require the use of special hardware or software for analog-to-digital or digital-to-analog conversion, which we discuss later.

Example 9.2 **The input/output devices in a microcontroller-based thermostat**

A thermostat that controls a domestic heating and cooling system is a fairly typical microcontroller application (Figure 9.8). This device's purpose is to maintain the temperature of a house or a room within comfortable limits, which may vary with the time of day, the day of the week, and the season. The main input to the thermostat is a temperature sensor such as a bimetallic strip that bends with temperature increase or decrease to turn an electrical switch on and off. The main outputs are a few relay switches that turn on and off a furnace, an air conditioner, a fan, or the like. A small LCD display panel outputs the current temperature, the time of day, and other pertinent information.

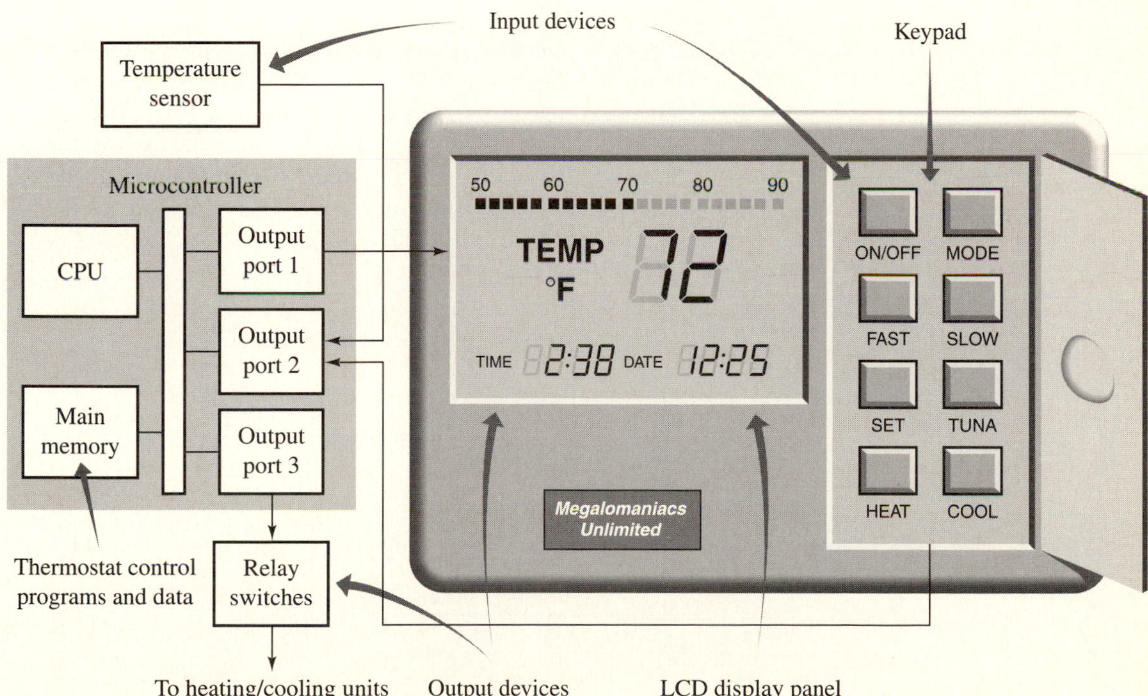

FIGURE 9.8 Input/output devices of a microcontroller-based thermostat.

To allow the temperature schedule to be tailored to individual desires, a keypad input device and a simple programming "language" are also provided. A user can then program the thermostat via the keypad; that is, the user can enter into the microcontroller's main memory a series of instructions specifying the temperature control functions to be performed. These "user" programs share the microcontroller's main memory with such "system" programs as a time-of-day clock, a calendar, and a program to read the temperature sensor, and programs to perform certain time and temperature calculations. The CPU spends its time executing these I/O programs to update the clock and calendar and to read the temperature sensor and compare it to its current schedule, and, from time to time, to turn a heating or cooling unit on or off.

I/O Operations

I/O devices link a computer to its environment via I/O ports.

Communication between a CPU and an I/O device D can be implemented along the same lines as communication between the CPU and M. To link D to the system bus, we require an **I/O port** of the general form illustrated by Figure 9.9. This is basically an addressable data register—in effect, a tiny RAM. It is assigned a unique address, so it can be a source or a destination of data-transfer instructions executed by the CPU. An I/O port that is dedicated to transferring data in only one direction may be called an **input port** or an **output port**, as appropriate. In addition to its connection to the system bus, an I/O port is connected to one or more I/O devices. Three such ports, two for output data and one for input data, are used by the microcontroller of Figure 9.8.

A load or read instruction that addresses an I/O port constitutes an input instruction. A store or write instruction to an I/O port is an output instruction. In an input operation, an input device transfers a sequence of data words to

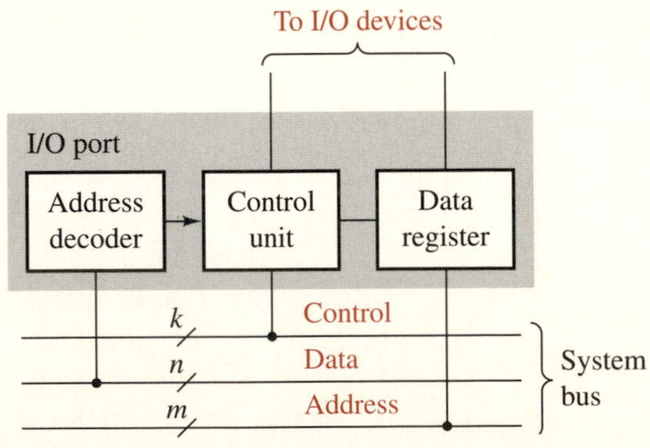

FIGURE 9.9 Basic structure of an input/output port.

an input port's data register. These data words are then transferred to the computer's main memory M by executing a sequence of input instructions, so that each data word is placed in M before the next data word arrives at the I/O port. An output operation reverses the data-transfer direction and requires execution of a sequence of output instructions that address the relevant port.

For I/O operations like these to function properly, the actions of the I/O device and the CPU must be synchronized because, in general, I/O devices proceed independent of and at a slower pace than the CPU. An electromechanical device such as a printer, for example, has a clock period measured in milliseconds, while a human's "clock" period is even longer. This speed disparity makes it possible for a single CPU to control many I/O devices concurrently. CPU and I/O device synchronization is accomplished by an exchange of "handshaking" control signals, as described later.

A Basic Input Device: The Switch

Many input devices contain a mechanical or electromechanical on-off switch, also referred to as a key or a button, that produces a single binary signal x. Such switches are an integral part of keypads, keyboards, and most electromechanical I/O devices. To obtain a binary voltage-like logic signal from a switch requires the circuit shown in Figure 9.10a; the theory of switch-level circuits of this kind is covered in Section 4.3. The pull-up attenuator A, which is typically implemented by an electrical resistor or an appropriately configured transistor, serves as the source of a weak 1 when the switch

When used as input devices, on-off switches require a pull-up attenuator.

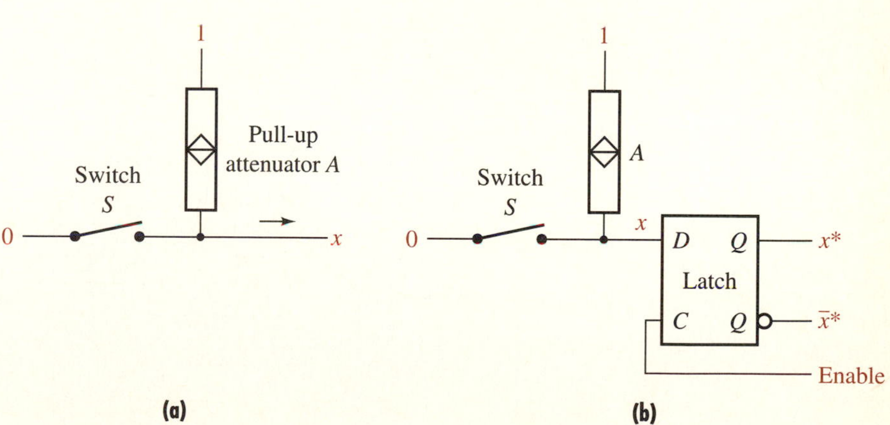

FIGURE 9.10 (a) An on-off switch as an input device; (b) a latched switch.

S is turned off, at which point the attenuator makes $x = 1$. Turning on S connects x to a strong 0 source and makes $x = 0$. Note that if A is omitted, x goes to the high-impedance state Z when the switch is opened.

The signal x produced by the switch can be connected directly to a latch or flip-flop within an I/O port, as shown in Figure 9.10b. The latch's enable signal serves as a timing signal, sometimes called a **strobe**, which is activated to capture the switch state x at a specific point of time. This enable signal may be controlled by a local I/O controller or, more remotely, by the CPU. In either case, it serves to synchronize the switch's activity with the CPU's internal clock.

A Basic Output Device: The Seven-Segment Display

Output display devices often build images from binary light-operated cells.

A basic binary output device is depicted in Figure 9.11a. This is a single cell or segment of a seven-segment **liquid crystal display** or **LCD**, which has two states, light and dark, that serve to indicate $z = 0$ and $z = 1$. An LCD cell consists of a special type of liquid sandwiched between translucent conducting plates; its light-transmitting properties change when an electric field applied to the liquid crystal changes. From a logical viewpoint, the LCD resembles a well (Section 6.1) and is so indicated in Figure 9.11a. LCDs are familiar from their use as display devices in digital watches. Similar displays can be constructed from light-emitting diodes (LEDs) and other optoelectronic devices.

The seven-segment display depicted (in simplified form) with its interfacing circuitry in Figure 9.11b is intended to output a decimal digit by means of seven LCD or LED cells that can be selectively made light or dark. To display a particular digit (the digit 3 is shown in the figure), the digit is

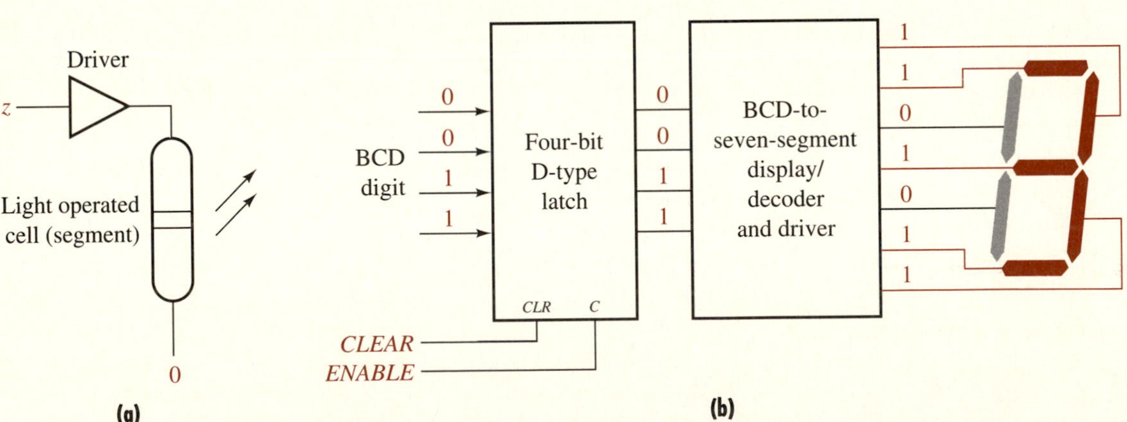

FIGURE 9.11 (a) A model of an LCD cell (segment); (b) a seven-segment display interface.

supplied in BCD form to a suitable decoding circuit that is then used to drive the seven-segment display. The entire circuit may be attached to or built into the I/O port of a microcomputer system.

Other I/O Devices

Input devices often contain large arrays of switches.

More sophisticated I/O devices are built from arrays of switches (keys) and optoelectronic cells. A **keypad** consists of a two-dimensional array or matrix of switches, typically organized as in Figure 9.12. This example has 16 wires forming eight rows and eight columns, with an on-off switch placed at the intersection of each row and column. Pressing switch $S_{i,j}$ makes a connection between row i and column j; releasing the switch breaks the connection. Here the eight rows are connected to an output port P_1 of the host computer, while the eight columns are connected to an input port P_2. This arrangement makes it possible for the CPU to execute an I/O program that "reads" the keyboard to determine which switch, if any, is pressed at any time. To determine the status of $S_{i,j}$, for instance, the CPU can write a word of the form 1..101..1, which contains a single 0 in position i to output port P_1. This causes a 0 to be transmitted to the corresponding jth bit position of input port P_2, if and only if $S_{i,j}$ is pressed. Hence, by reading the data at P_2 and then determining if bit $j = 0$, the CPU can resolve the status of the switch in question. By writing a 0 to each row in turn and reading the result—that is, by "scanning" the keys—the status of the entire keypad can readily be determined. Figure 9.12 shows the result of scanning row 3 with the switch $S_{3,2}$ pressed. (Some other design issues associated with keyboard usage are explored in Problems 9.14, 9.15, and 9.16.)

The foregoing scanning process requires execution of a program—a keypad input routine—that systematically addresses every switch in the keypad. Such an I/O program forms part of the operating system of a microcomputer that is equipped with a keypad. It is executed sufficiently often, perhaps many times each second, to ensure that any user action affecting a key is detected. Each scan might involve a few hundred instruction cycles, which, however, takes less than a millisecond to execute. Thus, the management of a slow input device like a keypad consumes only a tiny fraction of the host computer's time. This time can be further reduced by adding an I/O controller to the keyboard in the form of a small state machine that determines switch states automatically and makes that information available directly to the CPU (Problem 9.16).

Output devices often employ large arrays of LCD cells.

Large **display panels** can similarly be constructed from two-dimensional arrays of LCDs or LEDs. A small array of this type, such as a 5×7 dot-matrix display, can represent a single alphanumeric character and, with 35 instead of seven cells, can provide a more complex but more flexible replacement for the seven-segment display of Figure 9.11b. A much larger array of, say,

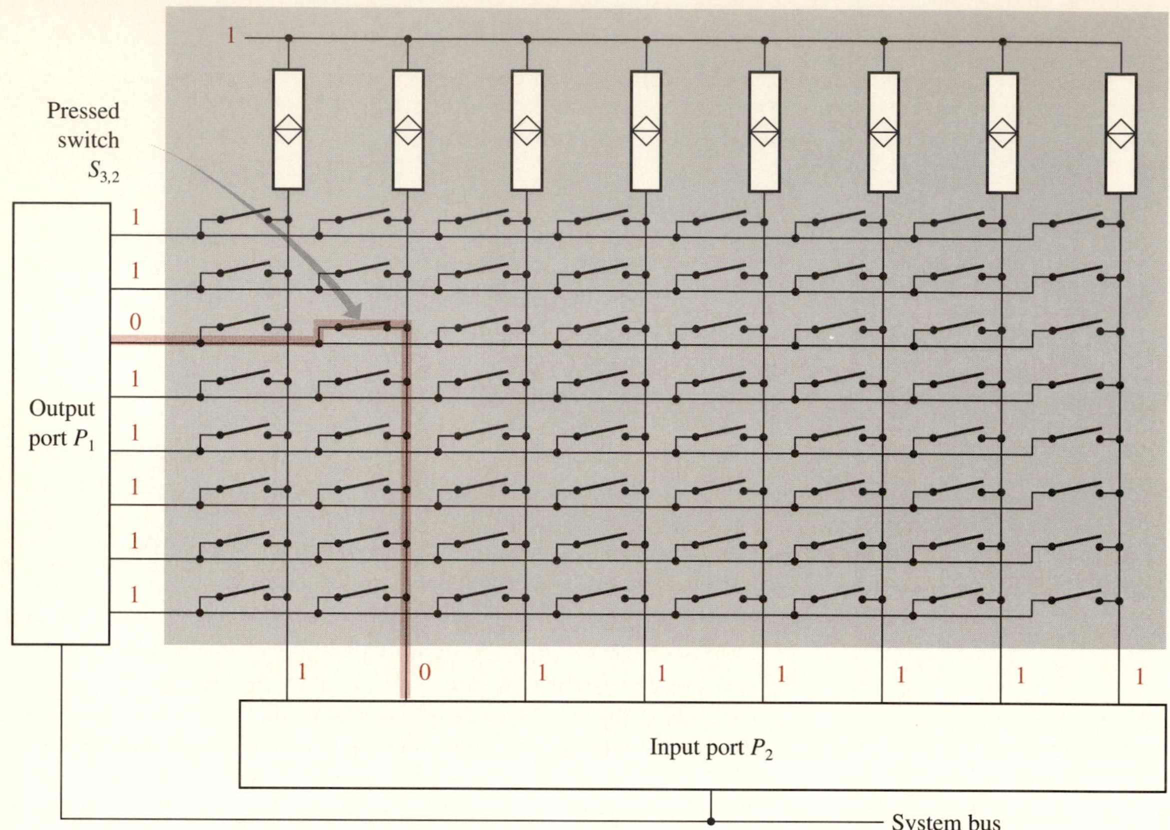

FIGURE 9.12 A 64-switch keypad and its interface circuit (simplified).

1000 × 1000 dot-like optoelectronic cells can serve as a video display screen for a portable computer, replacing the traditional cathode ray tube display. The scanning process outlined above for reading a keypad can be adapted to the task of writing data to a display panel. Because of its size, however, scanning such a display panel is often too time consuming for the CPU, so this task is often delegated to a special-purpose I/O controller (a video display unit).

Secondary Memory

Bigger computer systems like personal computers need large amounts of memory to store all their programs and data. The IC technology used in RAMs is too expensive for bulk storage of this kind. It is also **volatile**, meaning that it loses its stored information whenever the power supply is cut off. Several alternative technologies provide very large amounts of compact, nonvolatile storage at relatively low cost. These are used to construct secondary memories that form part of the computer's I/O system. As is usually the case with computer hardware, lower cost means lower speed, so secondary memories tend to have much longer access times than IC (RAM) memories.

Magnetic-surface recording is the memory technology commonly used for secondary storage devices. Data words are stored serially in concentric circular tracks along the surface of a disk coated with a magnetic material such as ferric oxide. Tiny regions (cells) in the magnetic material can be magnetized in two directions, corresponding to the logic values 0 and 1. The disk is rotated at high speed, and a device called a read-write head is placed near the surface of the spinning disk so that it can read from or write into any cell as it passes underneath (Figure 9.13). The read-write head can be moved back and forth to access any track. A control unit is needed to manage the movements of the disk and its read-write head and to convert data between the magnetic form used for storage and the electrical form needed to communicate with the host computer.

Magnetic disk memories are cheap and nonvolatile, but are slower than RAMs.

There are several key points of difference between IC RAM memories and magnetic disk memories. Magnetic media retain their stored information when the power is switched off, and so this memory technology is nonvolatile. The time required to access data in a magnetic disk memory depends on the relative positions of the cells to be accessed and the read-write head. First, the read-write head must be positioned over the correct track, the so-called "seek" time t_S, which is typically around 10 ms. Once the correct

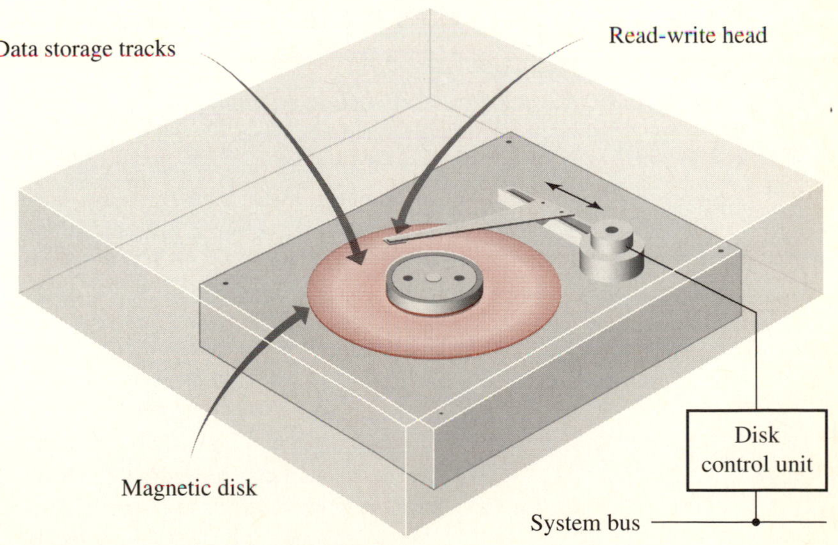

FIGURE 9.13 A magnetic disk memory.

track is reached, a delay called the disk latency t_L results until the first cell to be accessed rotates into the proper position under the read-write head. If the speed of the disk is r rotations per second, the average latency is the time for half a rotation—that is, $t_L = 1/(2r)$. To read out or write in m consecutive bits takes $m/(rN)$ seconds, where N is the bit capacity of a single track. Hence, the average total time to access m bits is given by

$$t_A = t_S + 1/(2r) + m/(rN) \tag{9.8}$$

This mode of access is called **serial access** in contrast with the random-access method used in IC memories.

Because of the mechanical motion involved, disk access times are several orders of magnitude longer than RAM access times. On the other hand, the cost per megabyte (8×2^{20} bits) of storage is one or two orders of magnitude greater for the RAM. Representative 1990 figures for the access time t_A are 100 ns for a DRAM and 20 ms for a hard disk memory. The corresponding cost-per-megabyte figures are around \$100 for the DRAM and \$5 for the hard disk. Typically, data stored in secondary memories are accessed in large blocks (pages) of P words at a time, so that all P words share the same seek and latency times. Because of the long access times required, programs stored in secondary memory cannot be accessed or executed directly by the CPU. They must first be copied into the system's main (primary) memory M. This process can be handled automatically by memory management routines within the computer's operating system.

The Central Processing Unit

This section examines the internal components of a CPU in more detail and demonstrates their operation. The CPU implements the fetching and execution of instructions, which are the main information-processing activities of a computer. Instruction types are discussed, along with some basic programming concepts at the assembly-language level.

9.4 CPU Operation

As we saw in the preceding section, a CPU contains an I-unit (instruction unit) responsible for the sequencing, fetching, and decoding of instructions, which are stored pending their execution in the computer's main memory M. The I-unit issues control signals to the E-unit (execution unit), which contains the datapath logic and registers needed to execute the various instruction types. The CPU also exercises varying degrees of control over input/output operations via control signals and instructions. The CPU spends most of its time fetching instructions from M and executing them in its E-unit.

Example 9.3 **Tracing through the instruction cycles for add and store instructions**

Figure 9.14 gives a snapshot of a CPU and its main memory in the course of executing a program, a fragment of which is shown in symbolic (assembly-language) form in M. We assume for this example that the memory word size is 16 bits, which can be represented by four hexadecimal digits. The address length also happens to be 16 bits, so the maximum capacity of M is 2^{16} = 64K words.

We begin by describing the instruction cycle for a representative data-processing instruction

$$\text{ADD} \quad \text{R2,R1} \tag{9.9}$$

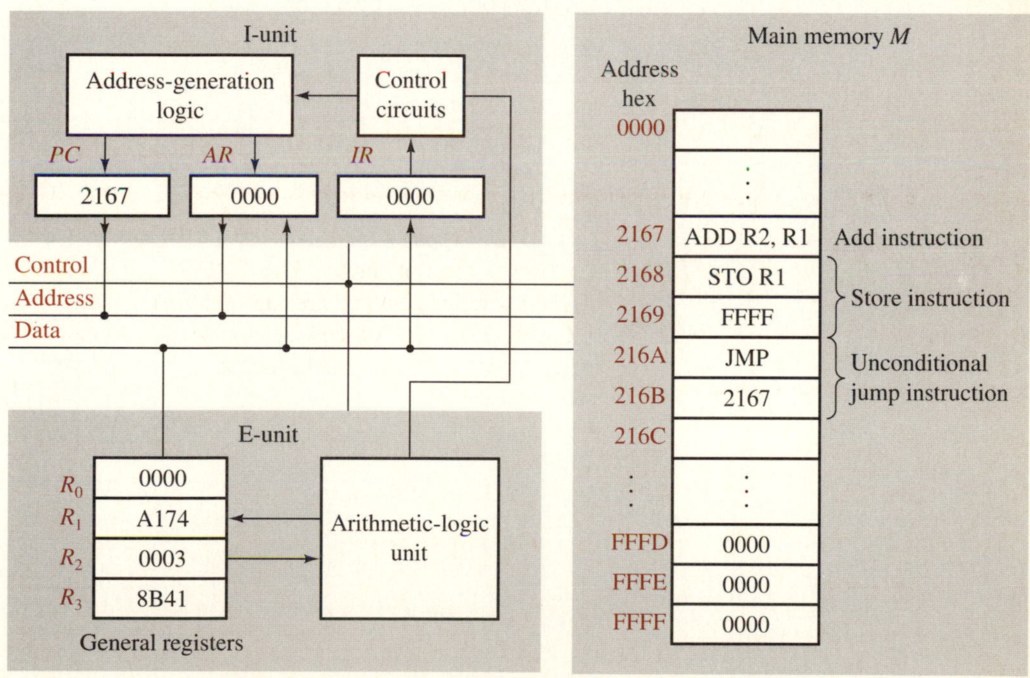

FIGURE 9.14 A snapshot of CPU operation.

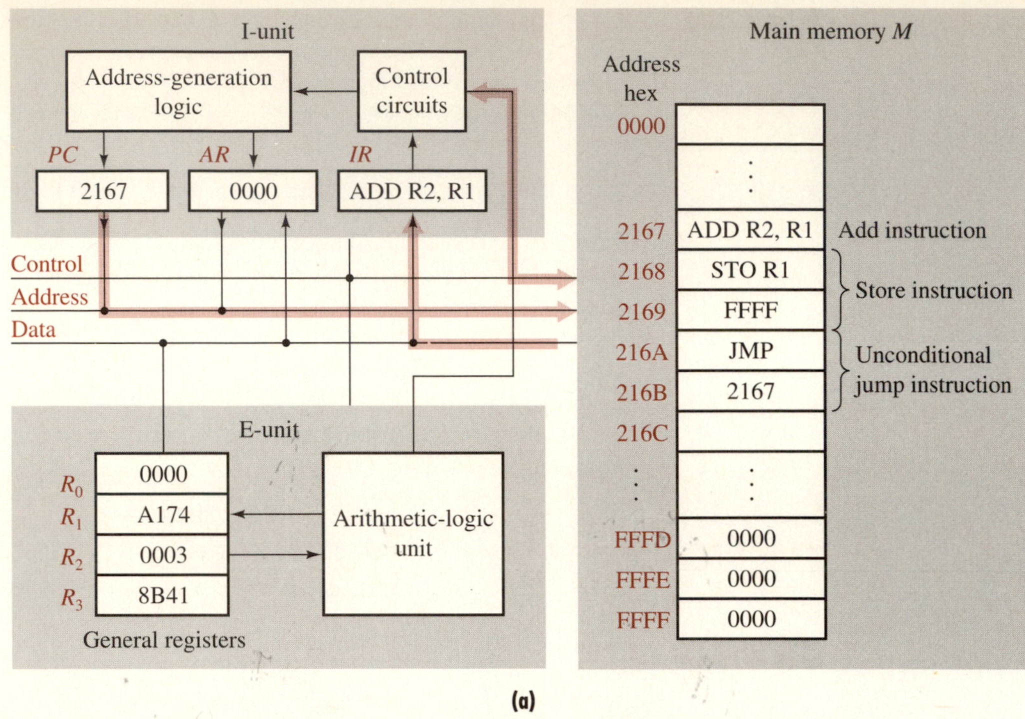

(a)

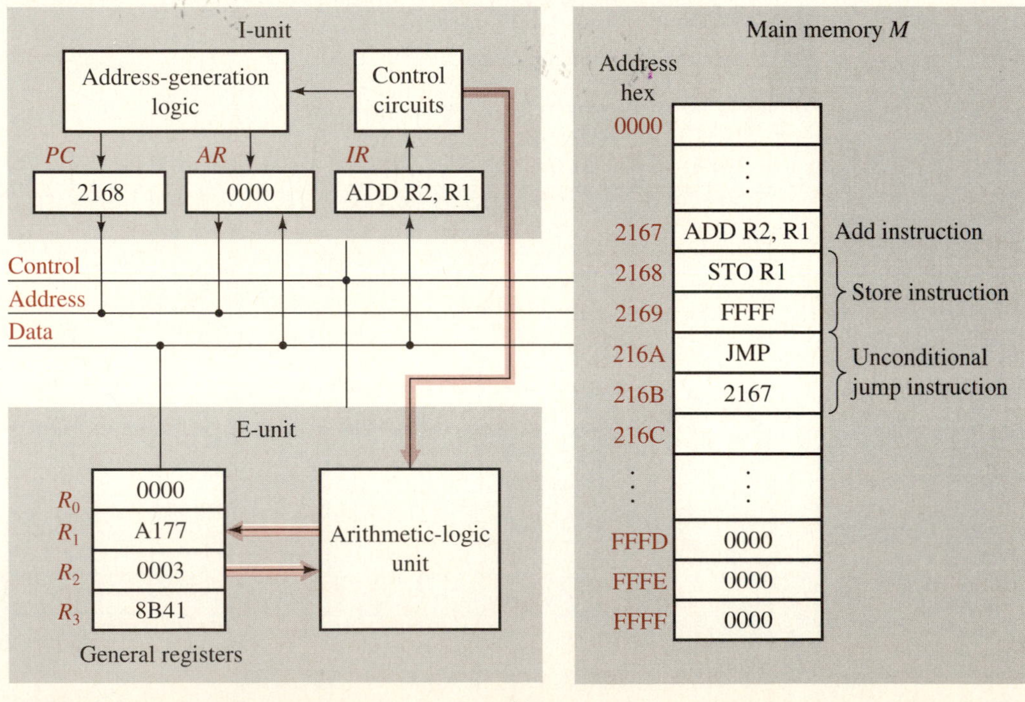

(b)

FIGURE 9.15 (a) Fetching and (b) executing a register-to-register add instruction.

which performs the register-to-register add operation $R_1 := R_1 + R_2$. The state of the CPU just before the instruction cycle for (9.9) begins is shown in Figure 9.14. The process of fetching and executing the add instruction is illustrated in Figure 9.15. The instruction is assumed to be a 16-bit word that is stored in the 16-bit main memory location with address 2167_{16}, so the initial contents of the program counter *PC* should be 2167_{16}.

Instruction words are fetched from consecutive main memory locations pointed to by *PC*.

First, the CPU transfers this address from *PC* to the address part of the system bus and orders *M* (via the appropriate control signals) to perform a read operation. The memory accesses the specified location and places its contents on the data part of the system bus, from which the CPU transfers it to the instruction register *IR* (Figure 9.15a), thus completing the operation $IR := M(2167_{16})$. The I-unit proceeds to decode the contents of *IR*, and to increment *PC* to point to the next consecutive instruction word in memory, which is at location 2168_{16}. The decoded *IR* contents cause the I-unit to send one or more control signals to the E-unit, triggering the desired addition process. These signals create the necessary logical connections between the general registers R_1, R_2 and the ALU. They also instruct the ALU to perform (signed binary) addition. The state of the CPU at the end of the instruction cycle is shown in Figure 9.15b. Note that register R_1 now contains $A174_{16} + 0003_{16} = A177_{16}$.

The main steps of the next instruction cycle are depicted in a similar manner in Figure 9.16. The relevant instruction in this case is

$$\text{STO} \quad \text{R1,FFFF} \tag{9.10}$$

which performs the store operation $M(FFFF_{16}) := R_1$, which transfers the results of the previous instruction from register R_1 in the CPU to the specified location in *M*. Because the memory address is one word (16 bits) long, the store instruction is two words long; hence, two memory read operations are required to fetch it. In the first read operation (Figure 9.16a), the word indicating the opcode (store) and the name of the source register is fetched from memory location $PC = 2168_{16}$ and placed in *IR*, as before. This step is usually referred to as the **opcode fetch**, and the entire 16-bit instruction word is treated as the opcode.

Some instructions require two or more read operations to fetch them from main memory.

Next, the I-unit decodes the opcode in *IR*, from which it learns that the instruction has a second part that contains an operand address needed for its execution; *PC* is also routinely incremented to 2169_{16}. The I-unit initiates a second read cycle, which, as shown in Figure 9.16b, brings to the CPU the word $FFFF_{16}$ stored in $M(2169)_{16}$. This word, which constitutes the destination address of the data to be stored, is placed in a register termed the (memory) **address register** *AR*. The CPU is now ready to execute the store operation proper, which consists of a memory write cycle, with register R_1 serving as the data source and *AR* serving as the address source. The contents of R_1 and *AR* are therefore placed on the data and address lines of the system bus, as indicated in Figure 9.16c, and the CPU orders *M* to perform a write operation. At the end of this step, the contents of memory location $M(FFFF_{16})$ have changed from 0000 to $A177_{16}$.

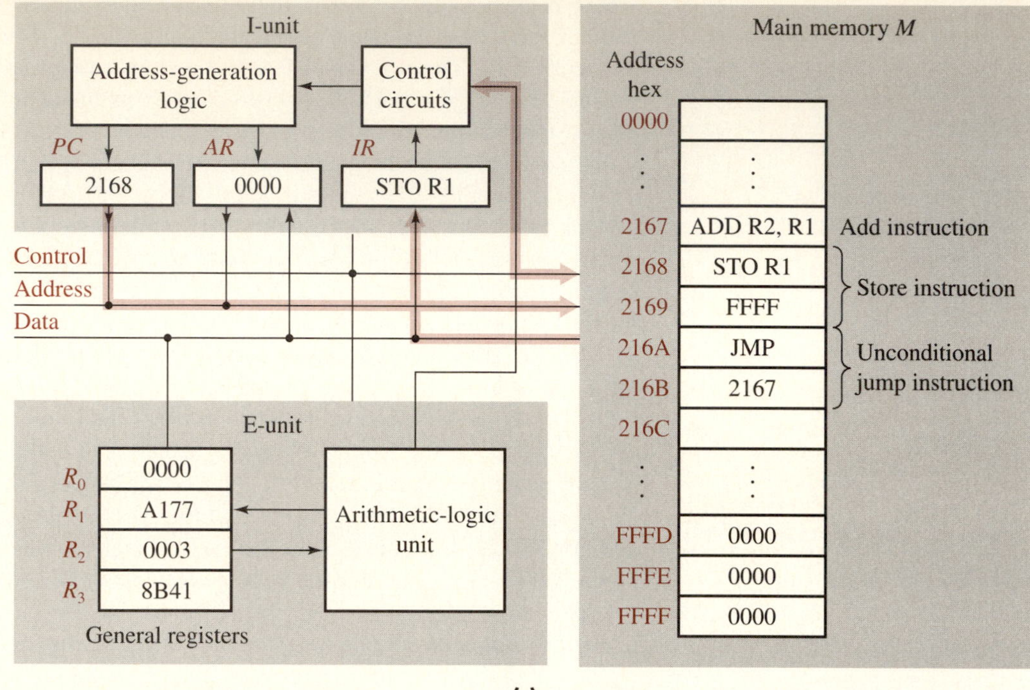

(a)

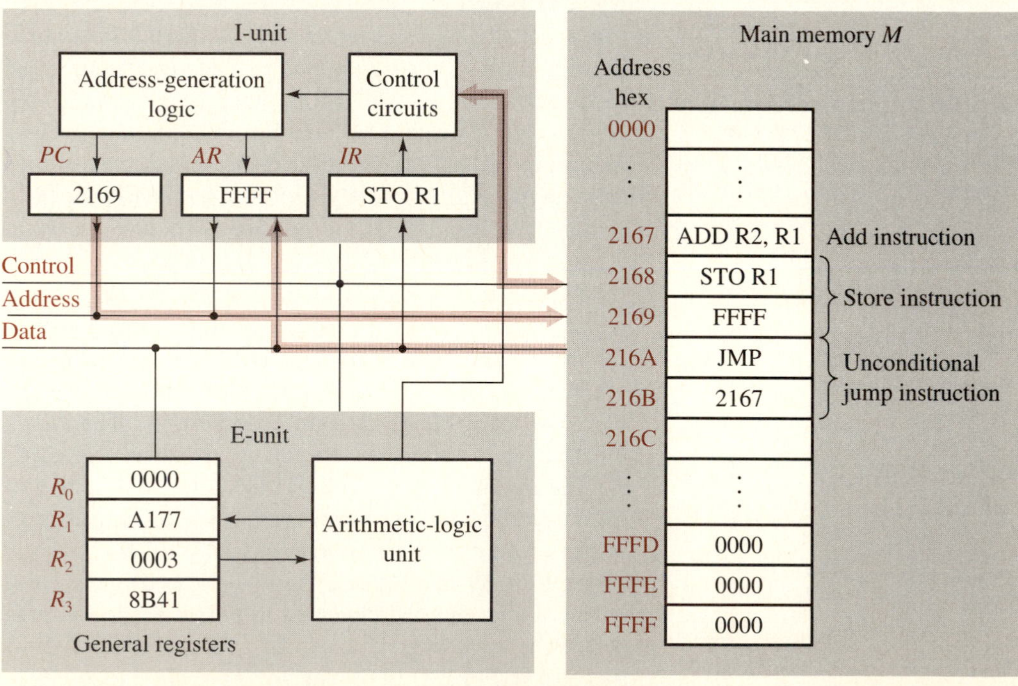

(b)

FIGURE 9.16 (a), (b) Fetching and (c) executing the store instruction.

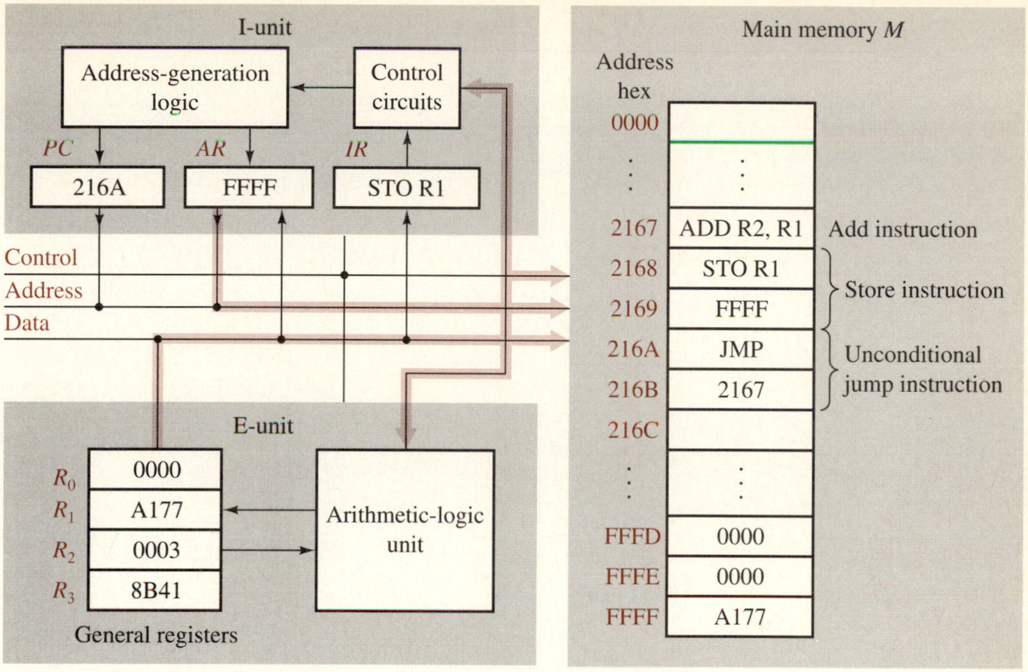

I-unit
Address-generation logic
Control circuits
PC 216A
AR FFFF
IR STO R1
Control
Address
Data
E-unit
R_0 0000
R_1 A177
R_2 0003
R_3 8B41
General registers
Arithmetic-logic unit
Main memory *M*
Address hex
0000
2167 ADD R2, R1 Add instruction
2168 STO R1
2169 FFFF Store instruction
216A JMP
216B 2167 Unconditional jump instruction
216C
FFFD 0000
FFFE 0000
FFFF A177

(c)

FIGURE 9.16 (Continued)

The General Instruction Cycle

The behavior of the CPU when executing instruction cycles is summarized by the flowchart in Figure 9.17. Each state S_i corresponds to one or more clock cycles of the CPU, depending on the access time of the main memory M. To initialize the CPU, it is reset asynchronously to a state in which all its registers are cleared. This means that PC $= 0$, so $M(0)$ becomes the location of the first instruction that is executed after a reset. The CPU then cycles continuously through instructions as directed by *PC*, and each cycle varies with the type of instruction encountered.

The first step of every instruction cycle is to bring the word *PC* points to into the I-unit, where it is placed in the instruction register *IR* and decoded. The first or opcode word of an instruction always contains its opcode, which explicitly or implicitly specifies the main operation to be performed and the number, type, and general location (CPU registers or main memory) of any operands used. The opcode word may also contain additional information such as register names and special ways of constructing addresses, which we consider later. If the instruction requires a data operand that is stored in memory, the next step is to fetch that operand or, more likely, its address, and load it into a CPU register such as *AR*. Once all the data needed by the

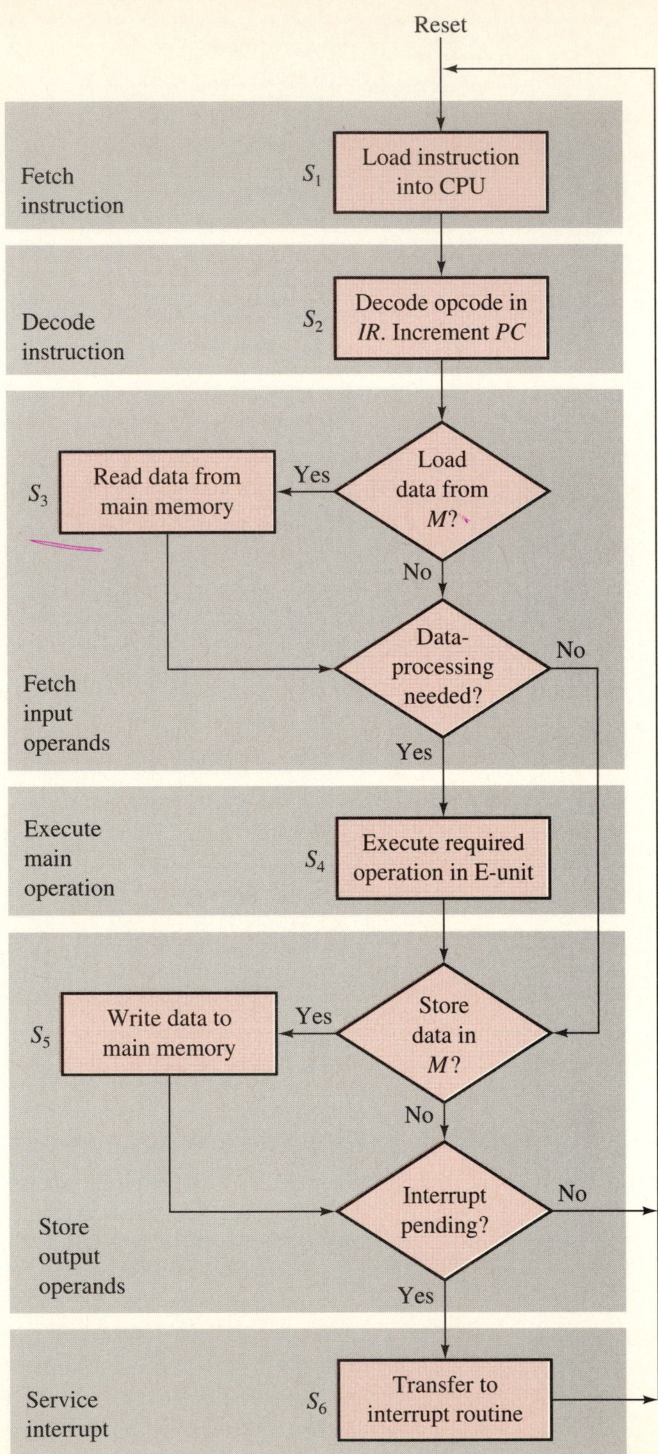

FIGURE 9.17 An overview of the CPU's behavior in an instruction cycle.

current instruction are in the CPU, data-processing operations like addition can be performed. The final step of the instruction cycle may be a store operation to transfer data to M.

Consider the register-to-memory add instruction $M(\text{ADR1}) := R_1 + M(\text{ADR1})$, which has two of its operands in main memory. It has the assembly-language format

ADD R1,ADR1

The instruction cycle for this add instruction requires the following steps, assuming a system configuration like that of Figures 9.14, 9.15, and 9.16:

1. Fetch the opcode word of the instruction and decode it.
2. Fetch the operand address word ADR1.
3. Fetch the data word $M(\text{ADR1})$ and place it in a suitable CPU register R.
4. Perform the addition $R := R_1 + R$.
5. Store the contents of R in $M(\text{ADR1})$.

Observe that four of these five steps involve instruction or data transfers; only step 4 is doing useful work. The register-to-register add instruction discussed above (Figure 9.15) requires only steps 1 and 4; the store instruction (Figure 9.16) requires all steps except step 4.

The CPU continuously executes instruction cycles, in which the details vary with the instruction type.

We conclude that instruction lengths and execution times vary with the types of operations they perform and the locations (addresses) of their operands. The shortest and fastest instructions have all their operands stored in CPU registers rather than in main memory. In the small CPU considered above, there are only four registers $R_0{:}R_3$ in the E-unit. Hence, any of these registers can be specified by a 2-bit address, making it possible to include several register addresses and an opcode within a 16-bit opcode word. However, it requires $k + 1$ words to form an instruction that refers to k memory locations. Such an instruction will require more time to fetch and execute than the equivalent instruction, the operands of which are in CPU registers.

The Branch Instruction Cycle

Branch instructions load new data into *PC*.

We can also see from Figure 9.16 how a branch instruction is executed. There is an unconditional jump instruction

JMP 2167

stored in memory locations $216A_{16}$ and $216B_{16}$. Execution of this instruction begins when $PC = 216A_{16}$. The I-unit, as usual, fetches the first (opcode) word from $M(216A_{16})$ and decodes it. It then fetches the second half of the instruction (the address field $M(216B_{16}) = 2167_{16}$) and completes execution of the instruction by loading the branch address 2167_{16} into PC. This means that the next instruction will be the ADD R2,R1 instruction residing at address 2167_{16}.

It is apparent that the program fragment given in Figure 9.16 forms an infinite loop, in which the add, store, and jump instructions are executed forever, or until an external event like a reset occurs.

Input/Output Instructions

I/O instructions and interrupts enable the CPU to interact with I/O devices.

A CPU can interact with its environment via input and output instructions, which are executed in essentially the same manner as memory load and store instructions, respectively. Input instructions access data made available at I/O ports. Once read in, the data in question can be processed by other instructions to determine their significance. This is a fairly slow process and is inadequate in common situations. For example, a secondary memory device like a magnetic disk unit transfers data at a fixed and relatively fast rate, so that data are lost if the computer cannot react quickly to a request from a disk memory for the CPU to send or receive data or control information. Even more urgent is a signal that a power failure is about to occur, which can be generated by a voltage sensor attached to the computer's power supply. Such a signal provides a brief warning for the computer either to save current data in a nonvolatile memory or activate a backup power supply.

To allow a CPU to respond to service requests of the above sort, it is provided with direct input signals from external devices. Activation of these signals temporarily interrupts the CPU so that it can provide some service required by the external device in question; this is referred to as an **interrupt facility**. As Figure 9.17 shows, the CPU checks for the presence of interrupts at the end of every instruction cycle. If an interrupt is pending—that is, if there is an active interrupt request awaiting a response from the CPU—a transfer is made to a program P_{new} designed to handle that particular interrupt type. This usually means saving essential information about the state S_{old} of the interrupted program P_{old} so that the CPU can later be restored to S_{old} and execution of the interrupted program can resume.

E-Unit Structure

We now examine the structure of an execution unit in more detail. The internal structure of a hypothetical E-unit with many typical features appears in the register-level block diagram of Figure 9.18. Its main components are registers for temporary storage of data, a combinational arithmetic-logic circuit C, and various buses, multiplexers, and other minor circuits (not all of which are shown) needed to transfer data inside the CPU. Data are processed in words of some maximum length n, which is typically a power of two in the range four (for small microcontrollers) to 64 (for powerful scientific computers).

An E-unit can be designed around an accumulator or a set of general-purpose registers.

Some E-units contain only one or two registers, which are referred to as **accumulators**. An accumulator such as AC in Figure 9.18 acts as the main source and/or destination of operands during instruction execution. Other E-units contain a set of registers or a **register file** $R_0, R_1, \ldots, R_{2^m-1}$, which are

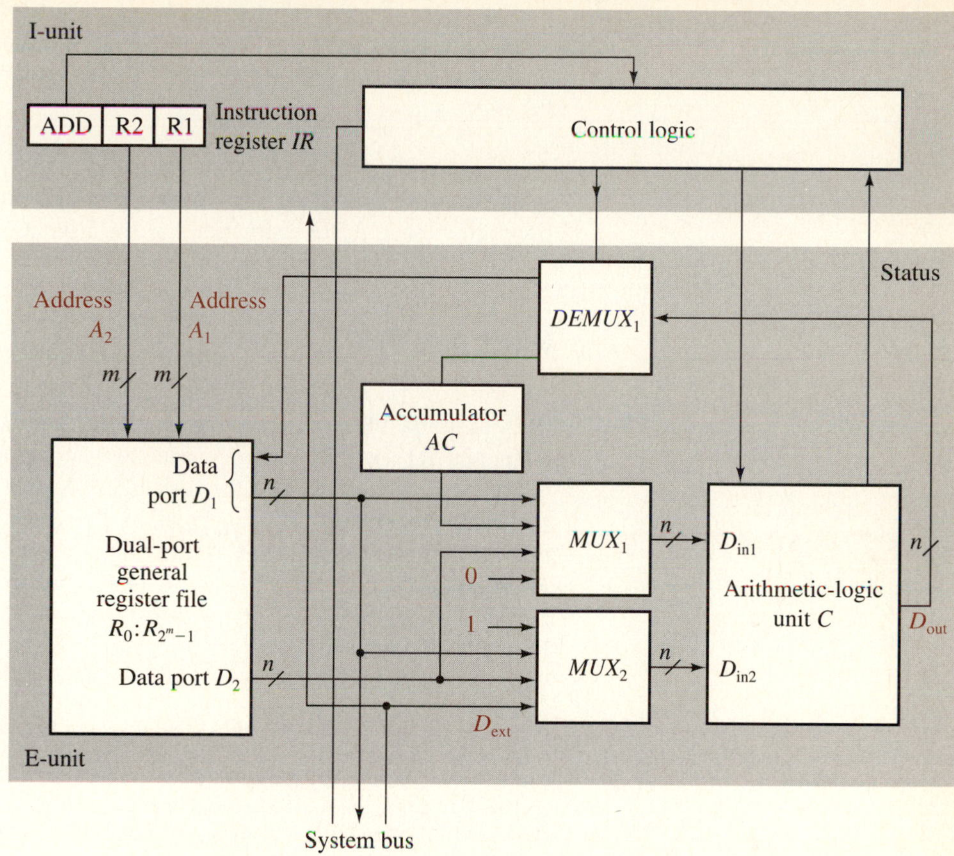

FIGURE 9.18 Internal structure of an E-unit.

organized like a $2^m \times n$ RAM. Each storage cell of the register file is a full-fledged flip-flop, unlike a DRAM cell (a switched capacitor) or an SRAM cell (a latch). The addresses needed to access the register file are supplied by an instruction such as

ADD R2,R1

shown in Figure 9.18, which specifies the operation $R_1 := R_1 + R_2$. Because this operation has two addresses—namely, R_1 and R_2—execution of the add instruction is faster when two separate read or write operations can be carried out simultaneously.

A dual-port register file allows two registers to be accessed simultaneously.

A register file or memory with the ability to access two independent registers at the same time is referred to as **dual-port**. In the example under consideration (Figure 9.18), we have a dual-port general register file that supports two simultaneous read or write operations. It has a pair of address buses A_1 and A_2, and a pair of data buses D_1 and D_2. The D_2 bus is attached to an output port, making this particular register file a read-only memory with respect to A_2. On the other hand, the D_1 port handles both reads and writes so that the register file is a RAM-like read-write memory with respect to A_1. It can be seen that this organization permits register-transfer operations of the form

$$R(A_1) := f(R(A_1), R(A_2)) \tag{9.11}$$

to be executed in a single CPU clock cycle. Like most "two-address" instructions, the above ADD instruction has the format of (9.11).

E-Unit Data Flow

We can now trace the flow of data through the E-unit of Figure 9.18 during the execution of a data-processing instruction such as ADD R2,R1. The addresses R1 and R2, suitably encoded into binary form—say, 0001 and 0010—are transferred from the instruction register *IR* to the register file in the first phase of the execution cycle, and read commands are issued for both addresses. At the same time, control signals are applied to the multiplexers MUX_1 and MUX_2, causing them to select their input data from the D_1 and D_2 outputs, respectively, of the register file. Also, control signals derived from the ADD opcode in *IR* are applied to the multifunction ALU circuit C; these signals configure C to perform the required type of addition:

$$D_{\text{out}} := D_{\text{in1}} + D_{\text{in2}}$$

Finally, the demultiplexer $DEMUX_1$ is set to transfer data from D_{out} back to the D_1 input port of the register file. Now when the input operands—that is, the contents of registers R_1 and R_2—emerge from the register file, they are channeled through MUX_1 and MUX_2 to C, where they are added, and the sum is transferred through $DEMUX_1$ into R_1. Note that reading and writing R_1 in a single clock cycle is perfectly feasible, provided the register is implemented using flip-flops and the flip-flops' timing constraints are met.

A few E-unit configurations can realize many instruction types.

The E-unit structure of Figure 9.18 is quite flexible and is intended to accommodate many instruction and operand types. The multiplexers allow the operands of the ALU C to have various sources and destinations. The upper data input D_{in1} of C can receive (variable) data from either of the register file's data ports D_1 or D_2, from the accumulator *AC*, or in the form of the all-0 constant, denoted 0 in the figure. The lower data input D_{in2} of C is connected to D_1, D_2, the all-1 constant, and to the data part of the system bus.

The E-unit's connection to the system bus allows external data D_{ext}, either from main memory M or from an I/O device, to be loaded into the E-unit. For example, execution of the load-accumulator instruction

LD ALPHA,AC

requires the CPU to issue a read command to M that causes the data word at location ALPHA to be returned to the E-unit as D_{ext}. The data word is then passed unchanged through MUX_2, C, and $DEMUX_1$ into AC. Here C only performs a simple data-transfer function

$$D_{\text{out}} := D_{\text{in2}} \tag{9.12}$$

This can be implemented by configuring C to perform addition again, but with MUX_1 applying the all-0 word to D_{in1}, because

$$D_{\text{out}} := 0 + D_{\text{in2}}$$

produces precisely the same result as (9.12).

I-Unit Design

An I-unit manages a large number of control signals in the E-unit and the system bus.

Next, we examine the structure of some typical instruction units. The I-unit is responsible for controlling the entire instruction cycle depicted in Figure 9.17. The specific actions taken within each such cycle depend on the nature of the instruction being executed, which is revealed to the I-unit when it decodes the opcode part of the instruction. A particular instruction type such as ADD, LOAD, or STORE may have many minor variants that depend on the types of operands (unsigned binary, signed binary, decimal), their length (full word, half word, single bit, etc.), and their location (register file, accumulator, main memory, or input/output port). The inputs to the I-unit are the opcode of the current instruction and certain control signals from the E-unit and the system bus.

The E-unit sends to the I-unit a few bits of control information, referred to variously as **status bits**, **flags**, or a **condition code**, which indicate the outcome of the E-unit's most recent actions. The I-unit has a **status register** *SR* to store this information. A typical instance of a status bit is an overflow flag *OVF*, which is set to 1 if, during the addition or subtraction of signed numbers, arithmetic overflow occurs; *OVF* is set to 0 if no overflow occurs. Another common status bit is a zero flag Z, which is set to 1 whenever C produces the all-0 result; if C produces a nonzero result, it sets Z to 0. The contents of *SR* are tested by conditional branch instructions, from which they can determine a few key facts about the results produced by preceding instructions that affect *SR*. The outputs of the I-unit are many control signals (tens or even hundreds may be used, depending on CPU complexity) that control C, the register file, the various (de)multiplexers, and the system bus.

Hardwired I-Units

Hardwired I-units are designed around special-purpose state machines.

An I-unit is constructed in one of two ways, which are illustrated in simplified form by Figure 9.19. In the first case (Figure 9.19a), the I-unit is a fixed sequential circuit or state machine that implements all the actions required to control the CPU. This is termed a **hardwired I-unit** and is designed according to the techniques discussed in earlier chapters, usually with an eye to

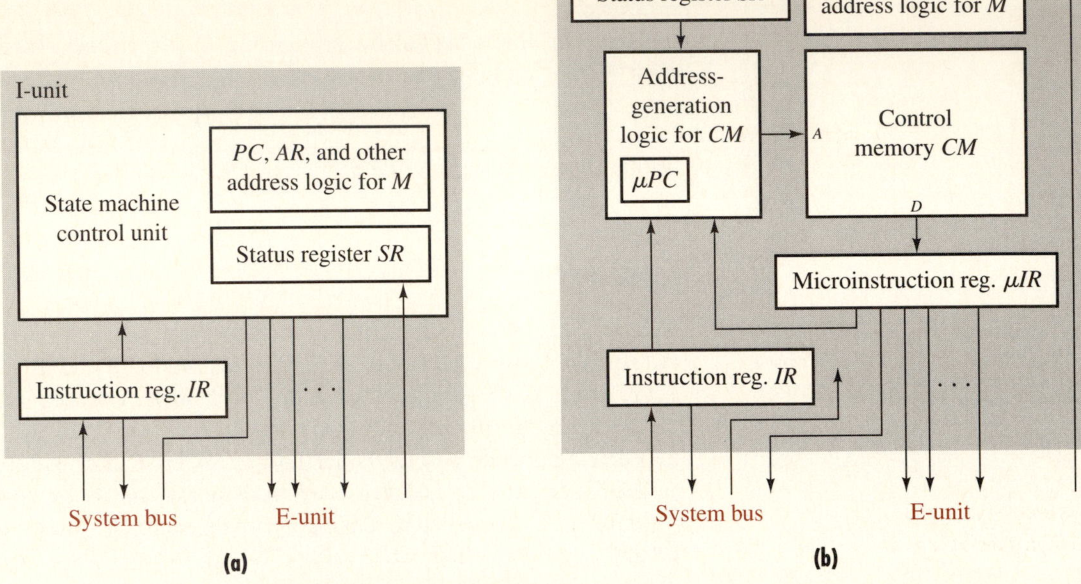

FIGURE 9.19 Internal structure of an I-unit: (**a**) hardwired; (**b**) microprogrammed.

minimizing the amount of hardware used and ensuring that the CPU can operate at or near the maximum speed allowed by its particular implementation technology.

Because of the large number of control signals and the often intricate sequences in which they must be issued, a hardwired I-unit constitutes the most complex part of a computer. As a result, it is costly to design and prone to design errors. Many a computer manufacturer has begun delivery of a new type of CPU to its customers, only to have a tiny error show up in the field, resulting in an extremely expensive recall-and-replacement process.

Microprogrammable I-Units

Microprogrammed I-units are designed around a programmable control memory.

An alternative way of designing an I-unit is depicted in Figure 9.19b. It is an example of a programmable control unit of the kind discussed in Section 8.6. This I-unit is said to be **microprogrammable** to distinguish its level of programming from that associated with the CPU as a whole. A microprogrammable I-unit resembles a computer within a computer; it contains a special memory, the **control memory** *CM*, from which it fetches and executes control information, in much the same manner as the CPU fetches and executes instructions from the main memory *M*.

A microprogrammable I-unit has its output control signals formatted into words called **microinstructions**, which are stored in a ROM (usually), RAM, PLA, or similar structure forming *CM*. The microinstruction word is often quite wide (a hundred or more bits is not uncommon), reflecting the large number of control signals found in a CPU. Microinstruction formats vary considerably, but a typical one is

$$next\_adr\ c_0\ c_1\ c_2\ c_3\ c_4\ c_5\ \ldots\ c_{q-2}\ c_{q-1}\ c_q \tag{9.13}$$

Here each c_i represents the desired value, active or inactive, of some control signal, and *next_adr* is the address in *CM* of the next microinstruction to be executed. When the microinstruction is read from *CM*, it is placed in the **microinstruction register** μIR, as shown in Figure 9.19b. The c_i values can then be transferred directly from μIR to the corresponding control lines. The *next_adr* field can be transferred from μIR to the I-unit's address-generation logic, where it determines the location of the next microinstruction to be used.

A group of related microinstructions that performs a specific task is referred to as a **microprogram**. The contents of *CM* are thus a set of microprograms, most containing just one or two microinstructions, which perform such functions as fetching the next instruction, executing an add instruction, performing a main memory read operation, and responding to an interrupt request. A few microprograms—for instance, those that control floating-point arithmetic operations—can be quite long. Typically, there is one microprogram per instruction type, and the microprogram is seen as interpreting that instruction. Reflecting their intermediate position between the hardware they control and the instruction-level software they interpret, microprograms are often referred to as **firmware**.

Borrowing ideas from the programming level, a microprogrammable I-unit normally contains a **microprogram counter** μPC in its address-generation logic for *CM*. This address counter is incremented automatically to point to the next consecutive microinstruction of the microprogram that is currently being executed. The address field *next_adr* within the microinstruction format of (9.13) can then be reserved for use by microinstructions that require a branch to be made to a nonconsecutive address within *CM*.

Microprogrammed vs. Hardwired Control

Microprogrammable I-units are flexible and simplify design; hardwired I-units are more compact and faster.

As with all programmable logic circuits (Section 8.6), the key advantage of a microprogrammable I-unit is the ease with which new designs can be created and modified. This is a consequence of the fact that a fixed hardware structure is employed (that of Figure 9.19b), while the computer-specific design data are programmed in a software-like format into *CM*. The design of the I-unit becomes a type of low-level programming called, naturally enough,

microprogramming. Once the microprograms have been written, it is only necessary to load them into *CM* to tailor the I-unit's hardware to a particular CPU design.

Design changes to correct errors or introduce new features are readily handled by replacing old microprograms with new ones. Similar changes to a hardwired CPU require redesign and remanufacture of the I-unit's hardware, a slow and costly process. On the other hand, a hardwired I-unit can be made smaller and somewhat faster than a comparable microprogrammed one. In a VLSI implementation, the microprogrammable I-unit generally occupies more chip area than the hardwired version. However, the microprogrammable unit's geometric structure is very regular and easy to lay out using CAD tools. The "random" interconnection structure of a typical hardwired controller complicates chip layout and usually requires a lot of time-consuming manual design to minimize its area.

9.5 Instruction Sets and Programming

We next consider the makeup of a CPU's instruction set, as well as the address and data types used. The instruction set of a widely used CPU family, the Motorola 680X0 series, serves as a running example. We also introduce some basic programming concepts.

Instruction Set Size

From a user's perspective, a CPU's behavior is defined by its instruction set, a term that covers the different types of instructions (add, subtract, load, store, jump, and so on) and data formats (32-bit two's-complement numbers, 16-bit BCD numbers, 8-bit nonnumerical or "logical" words, 32-bit floating-point numbers, etc.) that are directly processed by the CPU, as well as the ways in which instructions and data interact. To the question, "What instruction and data types are essential?" the answer, as we saw earlier, is "very few," because a complex instruction can be programmed from a sequence of simple ones.

As hardware costs have dropped with the evolution of VLSI, instructions that perform multiplication and other complex operations have become quite common. As a result, some computer families, termed **complex instruction set computers** or **CISCs**, have large instruction sets with several hundred instruction types. There are also **reduced instruction set computers** or **RISCs** that try to restrict instruction sets to 20 or so simple but fast types and stress

RISCs stress the efficient use of small instruction sets to achieve high performance.

other means of improving performance such as large register sets and efficient compilation of programs into machine language. The RISC-CISC distinction is not clear-cut, however, and is sometimes used as more of a marketing device than a technical classification.

The Bare Essentials

An instruction set should contain some data-transfer, data-processing, and program-control instructions.

Every instruction set should include a few carefully chosen representatives of the three main classes mentioned earlier: data transfer, data processing, and program control. Table 9.3 shows a small set of this kind; the opcodes given are representative of those used in assembly-language programs. Opcodes and other features of assembly language vary from one computer family to another. The language used for a particular machine is normally that defined by the principal manufacturer—for example, Intel Corporation in the case of the 80X86 microprocessor family used in PC-series computers.

In the data-transfer category, the fundamental instructions are LD (load), which transfers data from M to the CPU, and ST (store), which transfers data from CPU to M. Also common are instructions that transfer information from one register to another. In some computer families, the data-transfer instructions are collectively identified by the MOV (move) opcode.

The basic data-processing instructions found in all computers are ADD and SUB (subtract). They are present in versions that operate on various types of fixed-point binary numbers, both signed and unsigned. In most instances, the signed number code employed is two's-complement. Besides

TABLE 9.3 A very reduced set of instruction types.

Type	*Opcode*	*Description*
Data transfer	LD	Load: Transfer word from M to CPU
	ST	Store: Transfer word from CPU to M
	MOV	Move: General data-transfer instruction
Data processing	ADD	Add: Compute sum of two (signed) binary numbers
	SUB	Subtract: Compute difference of two binary numbers
	AND	And: Compute bitwise logical product of two words
	NOT	Not: Compute bitwise logical complement of word
	XOR	Exclusive-or: Compute bitwise exclusive-or function of two words
	SH	Shift: Shift word in specified manner
Program control	JMP*cond*	Conditional jump: Test flag and branch if condition is satisfied
	CALL	Call: Save program state and branch
	RET	Return: Restore program state and branch

these numerical or arithmetic instructions, some logical instructions are also found; the AND, NOT, and XOR instructions listed in Table 9.3 are typical. In principle, any set of logical operations that are functionally complete such as {AND, NOT} or {NAND} will produce a similarly complete set of logical instructions. NAND is an inconvenient operation in programming, for the same reasons that the NAND operator | is awkward to use in gate-level design, while EXCLUSIVE-OR, although not adding to the completeness of AND and NOT, is quite useful in itself. Another handy data-processing instruction is SH (shift), which shifts the contents of a word left or right in some specified manner; this is an instruction that has both numerical and nonnumerical interpretations.

Program-control instructions give a processor the ability to make decisions.

Finally, a program-control instruction of the form

JMP*cond* ADR

meaning "test a status flag *cond* and, depending on the test's outcome, branch to the instruction at memory address ADR," is essential to support conditional decisions within computer programs. Such decisions appear in statements of the if–*cond*–then form in programs written in high-level programming languages. Another useful pair of program-control instructions are CALL and RET (return), which support the transfer of control between one program and another as occurs, for instance, when a procedure or subroutine is invoked in a program, or an interrupt request is processed. CALL saves essential state information before branching to a new program; RET restores the previously saved state and transfers control back to the first program.

The 680X0 Microprocessor Family

For illustrative purposes, we will use the instruction set of a series of microprocessors and microcontrollers, which trace their origins to the Motorola 68000 single-chip CPU introduced in 1979. This important series includes the 68020, 68030, and 68040 microprocessors (the latter appeared in 1990) and various microcontrollers intended for applications that require embedded microprocessors. These devices all share a common core instruction set, although the later members have built-in features, such as floating-point instructions, which were not present in earlier members of the series. We shall refer to all these microprocessors collectively as the "680X0" family.

The 680X0 has a general-register organization without accumulators.

The 680X0 is considered to follow the CISC design style and has a microprogrammed CPU. Excluding floating-point instructions, it employs around a hundred different opcode types, of which about half fall into the data-processing category and about a third are program-control instructions. Various operand sizes are permitted, with 16 bits treated as the nominal "word" length. In this chapter, we use only a small part of the 680X0's instruction set, which includes the most common and useful instructions.

Example 9.4

A 680X0 program to test and replace a word in memory

To illustrate the 680X0 series' basic instruction types, consider the problem of testing a word stored in main memory and replacing the word with its logical (ones') complement if it is nonzero. We will solve this problem by a combination of data-transfer, data-processing, and program-control instructions.

Many different kinds of data transfers are specified by the 680X0's MOVE opcode. For instance, the load operation D1 := M(ADR), where D1 denotes one of the eight general-purpose "data" registers named D0:D7 in the CPU, is specified by the load instruction

```
MOVE    ADR,D1
```

The corresponding store instruction is

```
MOVE    D1,ADR
```

which specifies M(ADR) := D1. We will use these instructions to transfer the data of interest between M(ADR) and D1. We will operate on the contents of D1 using a couple of data-processing instructions: ADD and EOR. (680X0 assembly language uses EOR rather than the more common abbreviation XOR for exclusive-or.) As we will see, these instructions have many uses beyond the obvious ones indicated by their opcodes.

The basic purpose of the ADD instruction is to compute the sum of two binary numbers. For example,

```
ADD    D0,D1
```

performs the register-transfer operation D1 := D1 + D0, implicitly treating the contents of registers D0 and D1 as 16-bit two's-complement words. As in the case of most data-processing instructions, the sum produced by ADD is automatically used by the CPU to set or reset the CPU's zero flag Z (and several other flags). If ADD results in a zero sum, Z is set to 1; otherwise, Z is reset to 0. We will use this fact to check whether the contents of D1 are zero. The instruction

```
ADD    #0,D1                    (9.14)
```

implements the operation D1 := D1 + 0. Here the # prefix indicates that #0 denotes an actual numerical value—namely, zero—and not a symbolic address. Operands whose actual values are included in instructions in this way are referred to as **immediate operands**. Thus, (9.14) adds zero to the contents of register D1 and so does not change D1. However, this instruction causes the value of Z to be recomputed, so that the new value of Z indicates whether or not the final value of D1 is zero.

An immediate operand has its value stored in the instruction itself.

The value to which Z is set by ADD (or by any other data-processing instruction) can be checked by various 680X0 conditional branch instructions, such as

```
BEQ    CONT
```

where BEQ denotes "branch if Z is equal to 1," and the entire instruction states "branch to CONT if Z is equal to 1." Another possibility is

```
BNE    CONT
```

where BNE denotes "branch if Z is not equal to 1." CONT is an arbitrary symbolic label the programmer assigns to the next instruction to be executed if the branch condition is satisfied. If this condition is not satisfied, the instruction placed immediately after the branch instruction is executed.

Finally, we will use the data-processing exclusive-or instruction

```
EOR    #$FFFF,D1                                   (9.15)
```

to form the ones' complement of the contents of D1. Here the immediate operand, again indicated by the # prefix, consists of 16 1s, corresponding to the hexadecimal word $FFFF_{16}$. The $ prefix is a Motorola convention for denoting a hexadecimal number in assembly language. The exclusive-or instruction (9.15) complements every bit of the word stored in D1, because

$$\overline{D1} = D1 \oplus 1111111111111111_2$$

Comments increase the readability of assembly-language programs.

Putting all the above ideas together leads to the following program fragment in 680X0 assembly language that complements the memory word M(ADR), if it is nonzero:

```
        MOVE    ADR,D1       ; Load M(ADR) into CPU register D1
        ADD     #0,D1        ; Dummy add instruction to change
                             ;   flag Z
        BEQ     CONT         ; Branch to CONT if Z = 1
                             ;   (implies D1 = 0)
        EOR     #$FFFF,D1    ; Complement the contents of D1
        MOVE    D1,ADR       ; Store D1 in M(ADR)
CONT    ...                  ; Next instruction
                                                        (9.16)
```

The part of each statement on the right preceded by a semicolon is a comment for the reader; it is ignored when the assembly language is translated into machine code. The final statement has the item CONT at the left, which serves as the label or address of the statement. Such labels are optional and are used primarily as symbolic target addresses for branch instructions—in this case, the BEQ instruction.

The 680X0 instruction set provides several more natural ways to test and complement an operand than the roundabout ADD–EOR construction used above. It also allows most instructions to specify operands that are located in M so that their transfer to the CPU for processing is implicit rather than explicit. There is an instruction TST (test) that compares a specified operand to zero and sets the Z flag, based on the test outcome. There is also a NOT instruction that forms the logical complement of its operand. These instruction types enable us to replace (9.16) with the following equivalent but shorter program fragment:

```
        TST    ADR     ; Set Z = 1 if M(ADR) is zero;
                       ; ,set Z = 0 otherwise
        BEQ    CONT    ; Branch to CONT if Z = 1
                       ;   (implies M(ADR) = 0)
        NOT    ADR     ; Complement M(ADR)
CONT    ...            ; Next instruction
```

(9.17)

Machine-Language Code

An assembly-language instruction generally consists of several parts: a label, an opcode followed by a list of operands, and a comment:

Label Opcode $\text{Operand}_1, \ldots, \text{Operand}_k$ Comment

The opcode and operands of each statement are translated by an assembler program into executable machine code. A symbolic label (which is optional) is translated into the physical address in M where the instruction is to be stored. Where no such label is given, the assembler assigns the instruction to the next available memory location. In this way, every instruction, whether or not a programmer has given it a label, is mapped into a specific address in M. As noted earlier, comments are ignored in the assembly process; they are merely for the convenience of a human reader.

An assembly listing shows how assembly-language code is translated into machine language.

The output of the translation of an assembly-language program, called an **assembly listing**, is illustrated in Figure 9.20a for the program segment (9.16). Here the machine-language program encoded in hexadecimal format appears on the left, while the corresponding assembly-language program is on the right. Figure 9.20b shows how the binary machine-language program is stored in M.

Each 680X0 assembly-language opcode is translated into a 16-bit word, represented by four hexadecimal digits in machine language. The binary name i of a data register Di is only 3 bits long, so it is easily included in the opcode word. As can be seen from Figure 9.20a, the load instruction MOV ADR,D1 is translated to the two-word (32-bit) machine-language instruction denoted 3238 03E8. The opcode

$$3238_{16} = 0011\underline{001}000111000_2 \tag{9.18}$$

is decoded by the I-unit as "move a word of data from memory to CPU register D1." The three underlined bits 001 in (9.18) tell the I-unit that D1 is the destination register. The second word of the MOVE instruction, $03E8_{16}$, is the address in main memory of the data to be loaded into D1. This corresponds to the symbolic address ADR in the assembly-language program. A register-to-register instruction such as MOVE D2,D5 or ADD D0,D1 translates into a one-word (16-bit) machine-language instruction. A memory-to-memory instruction such as MOVE ADR1,ADR2, which performs the operation $M(\text{ADR2}) := M(\text{ADR1})$, translates into a three-word (48-bit) machine-language instruction because it contains two 16-bit memory addresses. Hence, the length of an instruction varies with the number and types of its operands.

Address Assignment

The correspondence between the symbolic address ADR and the physical address $03E8_{16}$ in M is specified explicitly by the programmer, using the instruction

$$\text{ADR} \quad \text{EQU} \quad 1000 \qquad (9.19)$$

where EQU means "is equivalent to" and 1000 is the decimal memory address, which the assembler changes to its hexadecimal equivalent 03E8. Note how all subsequent occurrences of ADR are translated into 03E8 in Figure 9.20a. The EQU instruction (9.19) is not itself translated into machine code, because it is an instruction for the assembly process rather than the target processor. EQU is called an **assembly-language directive**, to distinguish it from an executable instruction like MOVE or ADD. Another example of an assembly-language directive in Figure 9.20a is ORG (origin), which directs the assembler to assign the program to consecutive memory locations, beginning with the address given in the ORG statement—in this case, 0000. The directives ORG 0100 or ORG $0064 (again, the $ prefix denotes a hexadecimal operand) would make $0100_{10} = 0064_{16}$ the program's starting address.

Directives are assembly-language instructions that control the assembly process.

The 680X0, like the majority of CPUs, addresses main memory M by byte; that is, the smallest addressable memory item is an 8-bit byte. A machine-language program is therefore stored as a linear sequence of bytes, as depicted in Figure 9.20b. Because MOVE ADR,D1 is the first executable instruction and consists of 4 bytes, it is assigned to the memory locations with addresses 0000_{16}:0003_{16}. The next instruction ADD #$0000,D1 also happens to be four bytes long and is assigned to addresses 0004_{16}:0007_{16}, and so on. By the time the programmer-specified symbolic address CONT is reached, 20 bytes of machine code have been generated, so CONT is automatically made equivalent to the physical address $20_{10} = 0014_{16}$ by the assembler.

We can see from Figure 9.20b how the CPU's program counter PC must keep track of instructions by byte addresses. The opcode in the instruction register IR tells the I-unit the length of the current instruction, so it knows how many operand bytes, if any, to fetch, and by how much to increment PC to point to the next consecutive instruction. The actual number of bytes fetched simultaneously by the CPU depends on system implementation details, especially the width of the system bus, and may vary among the members of a given processor series. The 68040 CPU, for instance, is designed with a 32-bit data bus that links the CPU to main memory, making it possible to

Machine code				Assembly code	
0000				ORG	0000
03E8			ADR	EQU	1000
0000	3238	03E8		MOVE	ADR, D1
0004	0641	0000		ADD	#$0000, D1
0008	6700	0006		BEQ	CONT
000C	0A41	FFFF		EOR	#$FFFF, D1
0010	31C1	03E8		MOVE	D1, ADR
0014	. . .		CONT	. . .	
Memory address	Opcode	Operand address(es)	Memory address (label)	Opcode and operand address(es)	

(a)

Address	Contents of *M*	Hex contents	
0000000000000000	00110010	32	MOVE ADR, D1
0000000000000001	00111000	38	
0000000000000002	00000011	03	
0000000000000003	11101000	E8	
0000000000000004	00000110	06	ADD #$0000, D1
0000000000000005	01000001	41	
0000000000000006	00000000	00	
0000000000000007	00000000	00	
0000000000000008	01100111	67	BEQ CONT
0000000000000009	00000000	00	
000000000000000A	00000000	00	
000000000000000B	00000110	06	
000000000000000C	00001010	0A	EOR #$FFFF, D1
000000000000000D	01000001	41	
000000000000000E	11111111	FF	
000000000000000F	11111111	FF	
0000000000000010	00110001	31	MOVE D1, ADR
0000000000000011	11000001	C1	
0000000000000012	00000011	03	
0000000000000013	11101000	E8	
0000000000000014			
	1 byte		

FIGURE 9.20 A program segment for a 680X0 CPU: (**a**) its assembly listing; (**b**) the executable program stored in *M*.

fetch an entire 16-bit or 32-bit instruction in one clock cycle. These implementation details do not concern the programmer, but directly affect the computer's overall speed of operation.

Data Formats

The 680X0 series, like all computers, supports a number of different data types. The main such types are called **integer** and **real** and correspond to fixed-point and floating-point numbers, respectively. Integers may be 8, 16, or 32 bits long, while real numbers may be 32, 64, or 96 bits long (Figure 9.21). The integer formats are used for signed and unsigned binary numbers; in the former case, two's-complement code is employed. The real formats conform to the IEEE 754 floating-point standard discussed in Section 2.12. Because decimal instructions are relatively uncommon, only two 8-bit decimal types are provided: **packed** (as in BCD code) and **unpacked** (as in ASCII code). There is also a data type called **bit-field** that is designed to handle nonnumerical data items with lengths ranging from 1 to 32 bits. Observe that all the data lengths, except that of the variable-length bit-field type, are exact multiples of a byte so that they can be stored efficiently in main memory.

Operand data types are defined, sometimes implicitly, by the opcode.

Only certain data types may be used with each instruction type; these are specified in the programmer's reference manual or similar publication for the particular CPU model under consideration. When several data types are possible, the applicable type is specified by either a special field in assembly-language statements or a modifier appended to the opcode; there is a distinct opcode word for each data type at the machine-language level. In some microprocessor families, two versions of the same instruction that differ only in their data types have very different assembly-language opcodes. In the 680X0 series, data types are defined by appending to a common opcode the one-letter type designations given in Figure 9.21, as in the following five variants of the MOVE instruction:

```
MOVE.B   ADR,D1      ; Load byte at M(ADR)
                     ;   to CPU register D1
MOVE.W   ADR,D1      ; Load word (2 bytes) at M(ADR)
                     ;   to register D1
MOVE.L   D0,D1       ; Move 4 bytes from register D0
                     ;   to register D1
MOVE.B   ADR1,ADR2   ; Move byte at M(ADR1)
                     ;   to M(ADR2)
MOVE.L   ADR2,ADR1   ; Move 4 bytes at M(ADR2)
                     ;   to M(ADR1)
```

(9.20)

Assuming that ADR1 = 1000_{10} = $03E8_{16}$ and ADR2 = 2000_{10} = $07D0_{16}$, these five assembly-language instructions become the following in hexadecimal-coded machine language:

```
1238  03E8
3238  03E8
2200
11F8  03E8  07D0
21F8  07D0  03E8
```

Note that all the opcode (leftmost) words are different in these examples, and that MOVE.W has the same interpretation as the plain MOVE opcode used earlier. When a memory address *ADR* refers to more than 1 byte, *ADR* is taken to be the location of the first byte only. A k-byte operand referred

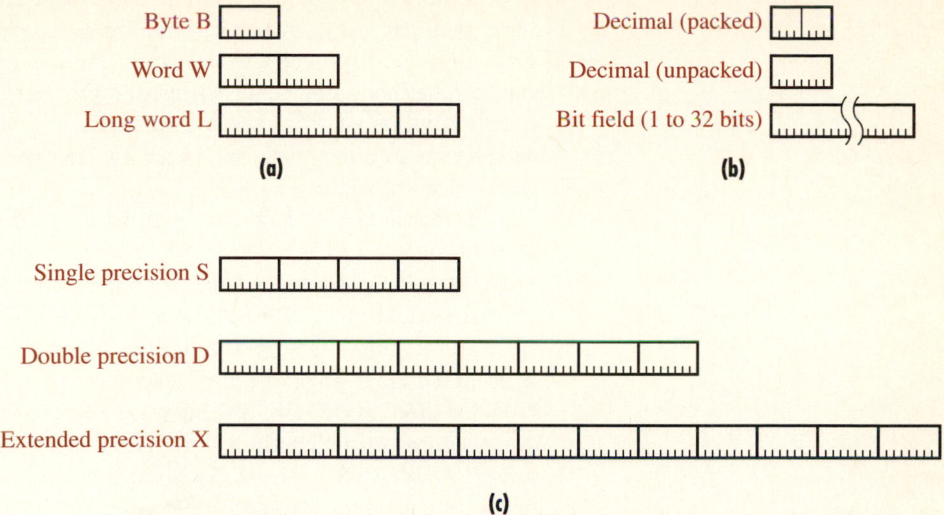

FIGURE 9.21 680X0 data formats: (**a**) integer (fixed-point); (**b**) decimal and bit-field; (**c**) real (floating-point).

to by *ADR* is assumed to be stored at the k consecutive addresses ADR, $ADR + 1$, $ADR + 2, \ldots, ADR + k - 1$, with the most significant byte stored in $M(ADR)$.

Basic Addressing Modes

Immediate, direct, and indirect addressing allow increasingly more flexibility in specifying data values.

The manner in which an operand's current value is specified in an instruction is termed an **addressing mode**, of which many kinds exist. So far we have primarily used the **direct addressing mode**, in which the location of the desired operand, such as a memory word ADR or a register R3, is explicitly specified in the instruction. The operand value V needed by the instruction forms the contents of the named memory location or register. We have also encountered **immediate addressing**, in which the operand value V is included in the instruction itself.

In the 680X0 program example of Figure 9.20, both operands ADR and D1 in the load instruction

```
MOVE    ADR,D1
```

use direct addressing. In the exclusive-or instruction

```
EOR    #$FFFF,D1                (9.21)
```

however, the first operand employs immediate addressing, as indicated by the # prefix; in other words, the first operand has the value $V = \text{FFFF}_{16}$. If the # was deleted from (9.21), an assembler would assume that the first operand is the word $M(\text{FFFF}_{16})$ stored at the address FFFF_{16} in main memory.

Comparing immediate and direct addressing for the same operand symbol in an instruction I_j, we see that the immediate mode allows the operand to have only one fixed value every time I_j is executed. With direct addressing, on the other hand, the value assigned to the operand can be changed between each execution of I_j. Thus, direct addressing allows the same instruction to refer to different data values at different times, thereby increasing the instruction's flexibility.

Further flexibility is gained by **indirect addressing**, in which an operand symbol such as ADR1 identifies a storage location in the CPU or M that contains not the operand value V, but the (direct) address of the location that stores that value. When ADR1 points to a CPU register, that register is called an address register. The desired operand value is then M(ADR1) and is accessed via a single memory read or write operation. When ADR1 points to a location in M, two memory accesses are needed to obtain the data value addressed indirectly by ADR1: the first to obtain the contents—say, ADR2 = M(ADR1)—of the location specified by ADR1, and the second to access the location M(ADR2) that contains the needed operand value V.

Indirect addressing is indicated in the 680X0 series by enclosing the operand in question in parentheses, thus:

MOVE (OPND),D1 (9.22)

The 680X0 contains a set of eight address registers A0:A7 explicitly designed to support indirect addressing. To use one of these address registers, the (OPND) parameter in (9.22) must be of the form (Ai), where $0 \leq i \leq 7$. When OPND is a main memory address, execution of (9.22) requires two consecutive fetch operations, which we can express in our register-transfer notation as follows;

$$\text{AR} := M(\text{OPND}); \quad \text{D1} := M(\text{AR})$$

Here AR denotes a memory address register in the CPU employed by the I-unit as a temporary holding area for memory addresses. We could also represent (9.22) more compactly by

$$\text{D1} := M(M(\text{OPND}))$$

Relative and Indexed Addressing

Addressing relative to the program counter PC facilitates program relocation.

It is often useful to make a memory address ADR relative to the contents of some register such as the program counter PC, an addressing mode known as **relative addressing**. This is particularly true for branch address calculation. In the 680X0 series, for example, most branch instructions use relative addressing. Consider the statement

BEQ CONT

which appears in the program of Figure 9.20 and means "branch conditionally (only if flag Z = 1) to the statement labeled CONT." The physical branch address EA required by the machine code, the so-called **effective address**

EA, is computed by adding a number *rel*(CONT) derived from CONT to the current contents of the program counter *PC*; in other words,

$$\text{Effective branch address} = \text{Address in instruction} + PC \qquad (9.23)$$

In this case, EA = *rel*(CONT)+ *PC*. As can be seen from the assembly listing, after fetching and decoding the 2-byte BEQ instruction, we will have $PC = 0010_{16}$, which is 2 bytes beyond the address 0008_{16} of the BEQ instruction. The physical address assigned to CONT is 0014_{16}—that is, 6 bytes beyond the current value of *PC*. The assembler therefore inserts *rel*(CONT) = 0006_{16} into the operand part of the machine-language version of the BEQ instruction. Hence, during execution of BEQ CONT, the I-unit computes the effective branch address as 0006 + *PC*.

The advantage of calculating addresses relative to the program counter is that it makes a program executable anywhere it is placed in main memory, as long as its instructions remain in the same positions relative to one another. This would not be so if the fixed, nonrelative address 0014_{16} were assigned to CONT; in that case, the instruction labeled by CONT would always have to be stored at the same physical address 0014_{16}. Thus, addressing relative to the *PC* makes a program relocatable, a feature that an operating system can use to manage efficiently large numbers of programs that share main memory.

Indexed addressing is useful for accessing an array.

Closely related to relative addressing is the notion of **indexed addressing**, in which an address ADR obtained from an instruction is systematically modified by a quantity I, called an **index**, to yield a sequence of k effective addresses of the form ADR, ADR + 1, . . . , ADR + k − 1. In this case, the effective address is given by the formula

$$\text{Effective address} = \text{Address in instruction} + \text{Index value} \qquad (9.24)$$

The index value I is typically held in a register (called an **index register**) and is incremented or decremented on each new reference to ADR. Indexing provides an efficient way for a programmer to refer to or perform a common operation on all elements of an array of data stored in memory. Observe the similarities in the effective address calculations of (9.23) and (9.24).

We illustrate the indexed addressing mode with an assembly-language instruction

```
MOV    AL,DISP(SI)                                  (9.25)
```

for the Intel 80X86 microprocessor series. As indicated by its MOV (move) opcode, this is a data-transfer instruction. AL denotes a general-purpose (accumulator) register in the CPU, DISP is an immediate operand, and SI (source index) denotes an index register that is also in the CPU. In the 80X86 convention, which is the opposite of that of the 680X0, the first address AL specifies the data destination; the second, composite address DISP(SI) specifies the data source. The destination AL is addressed directly, whereas indexed addressing is used for the source. The latter has the effective address given by

$$\text{Effective address} = \text{DISP} + \text{SI}$$

in accordance with (9.24). Hence, the move instruction (9.25) performs the following register-transfer operation:[3]

$$\text{AL} := M(\text{DISP} + \text{SI})$$

Figure 9.22 summarizes the five addressing modes discussed above. Observe the different ways in which the immediate direct, indirect, and indexed modes all refer to the same operand value V. These addressing modes are often modified or combined in many ways; such combinations lead to 18 distinct addressing modes in the case of the 68040 microprocessor, for example [Tabak, 1991].

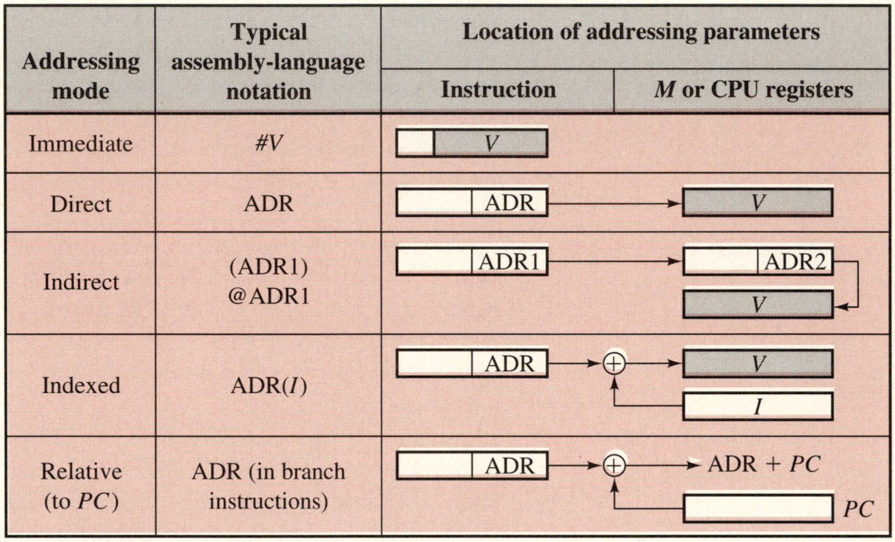

Addressing mode	Typical assembly-language notation	Location of addressing parameters	
		Instruction	*M* or CPU registers
Immediate	#*V*	*V*	
Direct	ADR	ADR	*V*
Indirect	(ADR1) @ADR1	ADR1	ADR2 *V*
Indexed	ADR(*I*)	ADR	*V* *I*
Relative (to *PC*)	ADR (in branch instructions)	ADR	ADR + *PC* *PC*

FIGURE 9.22 A summary of five common addressing modes.

Example 9.5

A 680X0 program to sum an array of numbers

Suppose we want to compute the sum of a sequence of n binary integers, where n can be quite large. Such a sequence is conveniently stored as an array of n consecutively addressed items in M. Our approach will be to fetch each integer of the array in turn and add it to an accumulating sum stored in a CPU register. To this end, we assign one of the 680X0's eight

[3]We are ignoring the 80X86's segment registers, which also figure implicitly in the effective address calculation, and resemble index registers managed by the operating system. The instruction (9.25) is normally interpreted as AL := M(DISP + SI + 16 × DS), where DS is a data segment register.

available address registers—namely, A0—to store the address of the array whose symbolic name, corresponding to its start address, will be *ARRAY*. Assuming each integer is one 680X0 word in length—that is, 2 bytes long—we will store *ARRAY* in $2n$ consecutive locations of M and use the default word (W) data type when manipulating array elements.

The key data-processing step is addition of an array element to an accumulator register; we assign the general-purpose data register D1 to be our accumulator. We want, therefore, to execute a binary add instruction of the general form

```
ADD    EA,D1
```

a total of n times, where EA is the effective address of an array element pointed to by A0. The opcode ADD, which could also be written as ADD.W, assumes its operands are 2 bytes long. Thus, if we use A0 as an indirect address register for the data array, we will need to increment its contents through the following address sequence

$$ARRAY, ARRAY + 2, \ldots, ARRAY + 2(n - 1)$$

Figure 9.23a gives the assembly listing of a program to compute the array sum, using the somewhat restricted set of instruction types that we have examined so far. Several EQU directives assign numerical values to the symbolic parameters *ARRAY* (the array's starting address), *SIZE* (the array's size n in words), and *SUM* (the program's final result). The first executable instruction is a MOVE instruction to place *ARRAY* in the address register A0. Then the *SIZE* parameter is placed in the general-purpose data register D0, which serves as a counter to keep track of the number of add steps executed. It will be decremented after each addition so that the condition D0 = 0 will indicate termination. To initialize the accumulator D1, we use the MOVE #0, D1 instruction to transfer the all-0 pattern to D1.

Repetitive computation, like stepping through a data array, can be implemented by a program loop.

At this point, we enter a program loop that starts with the ADD instruction labeled LOOP and ends with BNE. The instructions in the loop will be executed repeatedly until the desired sum has been computed. The first of these

```
LOOP    ADD    (A0),D1
```

uses indirect addressing to access the array word whose address is stored in A0; it then executes the operation

$$\text{D1} := M(\text{A0}) + \text{D1}$$

The CPU next increments the array address explicitly by a second ADD instruction, which, as shown in Figure 9.23a, adds two to the contents of A0. Next, the contents of D0 are reduced by one via the subtract instruction SUB. Like most arithmetic instructions, SUB affects the CPU's Z flag, setting it to 1 when the result it places in D0 is 0. Thus, we can test this flag to determine when D0 becomes 0. The BNE (branch if Z is not equal to 1) instruction is used for this purpose; it branches back to the instruction labeled LOOP as

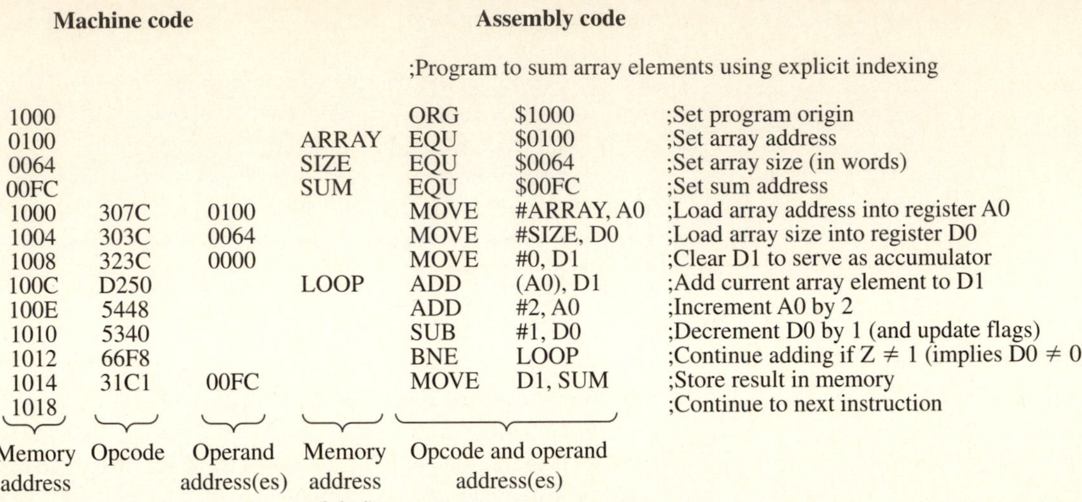
Machine code
Assembly code
;Program to sum array elements using explicit indexing
1000 ORG $1000 ;Set program origin
0100 ARRAY EQU $0100 ;Set array address
0064 SIZE EQU $0064 ;Set array size (in words)
00FC SUM EQU $00FC ;Set sum address
1000 307C 0100 MOVE #ARRAY, A0 ;Load array address into register A0
1004 303C 0064 MOVE #SIZE, D0 ;Load array size into register D0
1008 323C 0000 MOVE #0, D1 ;Clear D1 to serve as accumulator
100C D250 LOOP ADD (A0), D1 ;Add current array element to D1
100E 5448 ADD #2, A0 ;Increment A0 by 2
1010 5340 SUB #1, D0 ;Decrement D0 by 1 (and update flags)
1012 66F8 BNE LOOP ;Continue adding if Z ≠ 1 (implies D0 ≠ 0)
1014 31C1 00FC MOVE D1, SUM ;Store result in memory
1018 ;Continue to next instruction
Memory address
Opcode
Operand address(es)
Memory address (label)
Opcode and operand address(es)

(a)

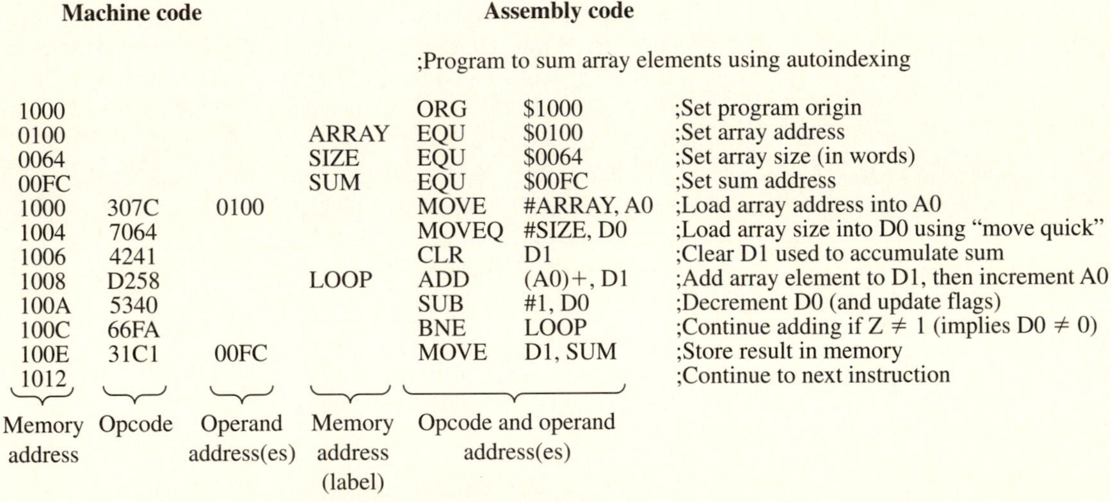
Machine code
Assembly code
;Program to sum array elements using autoindexing
1000 ORG $1000 ;Set program origin
0100 ARRAY EQU $0100 ;Set array address
0064 SIZE EQU $0064 ;Set array size (in words)
00FC SUM EQU $00FC ;Set sum address
1000 307C 0100 MOVE #ARRAY, A0 ;Load array address into A0
1004 7064 MOVEQ #SIZE, D0 ;Load array size into D0 using "move quick"
1006 4241 CLR D1 ;Clear D1 used to accumulate sum
1008 D258 LOOP ADD (A0)+, D1 ;Add array element to D1, then increment A0
100A 5340 SUB #1, D0 ;Decrement D0 (and update flags)
100C 66FA BNE LOOP ;Continue adding if Z ≠ 1 (implies D0 ≠ 0)
100E 31C1 00FC MOVE D1, SUM ;Store result in memory
1012 ;Continue to next instruction
Memory address
Opcode
Operand address(es)
Memory address (label)
Opcode and operand address(es)

(b)

FIGURE 9.23 A 680X0 program to sum elements of an array: (**a**) an initial version with explicit address indexing; (**b**) a streamlined version with autoindexing.

long as $Z \neq 1$. When SUB reduces D0 to 0, Z becomes 1 and the branch condition tested by BNE is no longer satisfied. The final MOVE instruction is then executed, storing the contents of D1 in the memory location SUM.

The manner in which the assembly-language instructions are translated into machine language can be seen in Figure 9.23a. In most cases, the translation is straightforward, although 680X0 opcodes are often encoded in a complicated fashion. Notice how the three items defined by the EQU directives appear as bytes 3 and 4 in the machine instructions that use them as addresses. The translation of branch addresses is less obvious, due to the use of relative addressing. Figure 9.23a shows that the instruction BNE LOOP is translated into the machine instruction 66F8, implying that $F8_{16}$ is the relative branch address that points to the statement LOOP. To see that this is correct, note that all 680X0 integer operations, including address calculations, are made with two's-complement arithmetic. Now $F8_{16} = 11111000_{2C} = -8_{10}$. Hence, the I-unit subtracts eight from PC to move the point of execution after BNE from the address 1018_{16} back to the address 1010_{16} = LOOP.

Autoindexing automatically increments or decrements a designated address register.

Because address indexing of the type that occurs in this example is so useful, some computers, including the 680X0, provide a feature called **autoindexing** that automatically increments or decrements the contents of an address register at the start or end of an instruction. We can replace the instruction pair

```
ADD    (A0),D1
ADD    #2,A0
```

with the single instruction

```
ADD  (A0)+,D1                    (9.26)
```

in which the plus in (A0)+ means that the contents of A0 are incremented as the final step of the instruction execution phase; this is called **postincrementing**. The amount by which A0 is incremented is the length in bytes of the operand being addressed, which is 2 in this case. The 680X0 also supports a **predecrementing** mode, denoted by −(A0), which automatically decrements the address register at the beginning of instruction execution. Because they add significantly to the complexity of the I-unit, the number of 680X0 instructions that permit autoindexing is quite restricted.

A revised version of the array summation program is listed in Figure 9.23b, in which the above postincrementing scheme is used. Several new types of instructions are also introduced to replace MOVE instructions with simpler or shorter data-transfer instructions, of which the 680X0 has numerous kinds. For example, the 4-byte MOVE #0,D1 instruction that clears D1 is replaced by a 2-byte CLR (clear) instruction designed expressly for this purpose. A 2-byte MOVEQ (move quick) instruction designed to move short (1-byte), immediate operands to registers replaces another 4-byte MOVE instruction. The result of these changes is a shorter (18 bytes versus 24 bytes of machine code) and faster program than that of Figure 9.23a.

The 680X0 Programming Model

A programmer sees the computer as a set of registers, memory location, and I/O ports on which the instruction set can act.

A programmer (or a compiler/assembler) views a computer from a particular architectural viewpoint known as a **programming model**. This level of abstraction recognizes the system as containing a certain set of CPU registers, main memory locations, and I/O locations that can be used by programs for storing information. It also recognizes the operation types, data formats, and addressing modes provided by the instruction set.

This view of the 680X0 architecture is presented in Figure 9.24; with minor variations, it applies to all 680X0-series CPUs. There are two main sets of registers: eight data registers D0:D7 and eight address registers A0:A7. These are general-purpose "user" registers that may be deployed by the programmer in any manner supported by the instruction set. (Address register A7 has a special use that we will discuss shortly.) As we saw in the preceding

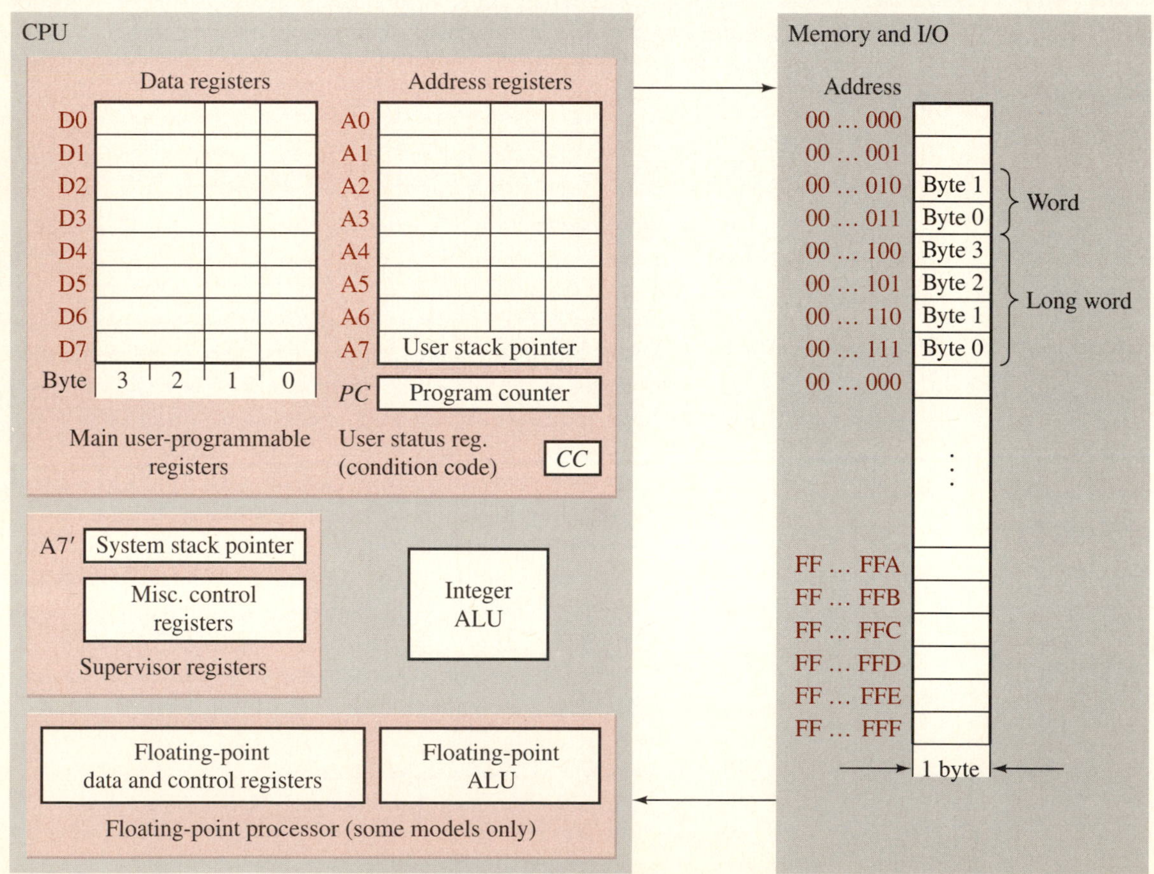

FIGURE 9.24 A programmer's view of Motorola 680X0 system hardware.

examples, a data register can serve as a temporary data store, a counter, and so on. The address registers support various addressing modes and can be used as address registers for indirect addressing, as index registers, and the like. Two specialized address registers called stack pointers facilitate the transfer of control between programs; they are discussed below. The 680X0 also has some "supervisor" registers that are reserved for use by the operating system. The 680X0's main instruction types, excluding those associated with floating-point operations, are summarized in Table 9.4.

TABLE 9.4 A summary of the 680X0's instruction types (excluding floating-point instructions).

Instruction type	*Opcode*	*Description*
Data transfer	EXG	Exchange (swap) contents of two registers
	MOVE	Move (copy) data unchanged from source to destination
	SWAP	Swap left and right halves of register
Data processing	ABCD	Add decimal (BCD) numbers with carry (extend) flag
	ADD	Add binary (two's-complement) numbers
	AND	Bitwise logical AND
	AS*x*	Arithmetic shift left (x = L) or right (x = R) with sign extension
	CLR	Clear operand by resetting all bits to 0
	DIV	Divide binary numbers
	EOR	Bitwise logical EXCLUSIVE OR
	EXT	Extend the sign bit of subword to fill register
	LS*x*	Logical shift left (x = L) or right (x = R)
	MUL	Multiply binary numbers
	NBCD	Negate decimal number (subtract with carry from zero)
	NEG	Negate binary number (subtract from zero)
	NOT	Bitwise logical complement
	OR	Bitwise logical OR
	RO*x*	Rotate left (x = L) or right (x = R)
	SBCD	Subtract decimal (BCD) numbers
	SUB	Subtract binary (two's-complement) numbers
Program control	B*cc*	Branch relative to *PC* if specified condition *cc* is true
	BRA	Branch unconditionally, relative to *PC*
	BSR	Call (branch to) subroutine at address relative to *PC*; save *PC* state (return address) in stack
	CMP	Compare two operand values and set flags based on result
	DB*cc*	Loop instruction: Test condition *cc* and perform no operation if condition is true; otherwise, decrement specified register and branch to specified address
	JMP	Branch unconditionally to specified address
	JSR	Call (jump to) subroutine at specified address; save *PC* state (return address) in stack
	NOP	No operation, but instruction execution continues
	RTS	Return from subroutine
	S*cc*	Set operand to 1s (0s) if condition *cc* is true (false)
	TST	Test an operand by comparing it to zero and setting flags

In the 680X0 series, data registers and main memory can be accessed by byte, word, or long word.

The 680X0's data registers are designed to store various data types; each can be treated as 1, 2, or 4 bytes. When used with an instruction such as MOVE.B, for example, which moves 1-byte operands, only the low order byte part (byte 0 in Figure 9.24) of a CPU register is used. Instructions like MOVE or MOVE.W that specify word operands utilize register bytes 0 and 1, while MOVE.L, which specifies long-word operands, uses all 4 bytes. In similar fashion, a memory address *ADR* for a word operand finds byte 1 in *M*(*ADR*) and byte 0 in *M*(*ADR*+1). If the operand is a long word, byte 3 is in *M*(*ADR*) and the remaining three bytes are stored at the next three addresses, as shown in Figure 9.24.

680X0-series CPUs have an integer ALU that executes all data-processing instructions except those involving floating-point numbers. To handle the latter, several alternatives are possible. Early 680X0 processors such as the 68000 CPU require floating-point operations to be implemented in software—that is, by encoding them in algorithms that use integer instructions only. Later CPUs such as the 68020 can use an auxiliary processor chip (the 68881 in the 68020 case) called a **coprocessor**, which contains a set of eight floating-point registers and processing logic to support a comprehensive set of floating-point instructions [Motorola, 1987]. Beginning with the 68040, the floating-point logic is fully integrated into the main CPU chip, which, as a consequence, contains more than a million transistors. The 68040's floating-point instructions include add, subtract, multiply, and divide in the data-processing category. There are also instructions to move and test floating-point numbers.

The memory and I/O subsystems often share the available addresses (memory-mapped I/O).

Main memory is regarded as a long array of storage locations, each byte of which is individually addressable. The maximum number of addresses allowed varies with the CPU model but can range up to 2^{32} bytes or 4 gigabytes. Input/output ports share the same set of addresses as main memory; that is, a particular address pattern is assigned to either a memory location or an I/O port. Such a shared addressing scheme for I/O devices is called **memory-mapped** I/O. Consequently, the 680X0 series has no I/O instructions per se. Instead, any instruction such as MOVE or ADD that can reference main memory can also be used as an I/O instruction. In contrast, microprocessors such as the 80X86 series have instructions that refer only to I/O devices, permitting separate memory and I/O address spaces; this is called **I/O-mapped** I/O.

Stacks

As Figure 9.24 indicates, the 680X0 has a couple of special address registers called *stack pointers*. They support a software data structure in main memory known as a stack. A **stack** is a sequence of storage locations, the contents of which can be accessed by writing to or reading from one end of the sequence, referred to as the **top** of the stack. (There is an obvious analogy with a stack of trays in a cafeteria, which has the property that trays can only be added or removed from the top of the stack.) A stack is typically created to store the state information associated with the transfer of control between two programs, as in interrupt or subroutine (procedure) processing. Most instruction sets contain subroutine call and return instructions that automatically perform push and pop operations whenever needed.

A stack is accessed from its top via push (write) and pop (read) instructions.

A stack is accessed via two basic operations: PUSH, which writes a new data item into the first unused location at the top of the stack, and POP, which reads the contents of the first occupied location at the top of the stack, a location that is then treated as empty. When control is to be transferred from program P_1 to program P_2, the information about P_1 that needs to be saved is pushed into a stack. The saved information can be subsequently retrieved via POP operations when execution of P_1 is resumed. Because of the last-in, first-out (LIFO) order in which items are stored in a stack, multiple program transfers can be nested easily. To access a stack, a special address register, the **stack pointer**, is needed to keep track of the top of the stack. This function is assigned to the 680X0's user address register A7 (although the other address registers may also be utilized as stack pointers), and to a supervisor stack pointer denoted A7′.

In addition to call and return instructions, some computer families have explicit PUSH and POP opcodes in their instruction sets. These instructions read or write to a user-definable stack and automatically update a stack pointer. The 680X0 does not have PUSH or POP instructions per se; however, its autoindexing addressing modes make it easy to turn MOVEs into stack instructions, as we now demonstrate.

Example 9.6 **Implementing stack operations by 680X0 instructions**

Suppose we want to push a 2-byte word from data register D4 to the top of a user-defined stack in a 680X0-series computer, where A3 has been designated as the stack pointer (see Figure 9.25a). Assume that the stack's top has the lowest address—that is, the stack region grows toward the low-address end of the 680X0's main memory. Then the instruction

```
MOVE   D4,-(A3)     ; Push instruction
```

performs the desired push operation. First, the pointer register is predecremented by the operand length, in this case two, so that A3 points to the word address $A106_{16} - 2 = A104_{16}$ immediately above the top of the stack. The contents of D4 (viewed as a word register) are now moved to the memory word-location addressed by A3; this address defines the new top of the stack (Figure 9.25b).

Now suppose we want to pop a word from the top of the stack into data register D7, where again A3 has been assigned as the stack pointer. As Figure 9.26 shows, the instruction

```
MOVE   (A3)+,D7     ; Pop instruction
```

carries out the required operation. First, it copies the word addressed by A3 into D7. Because of the postincrement mode of autoindexing, A3 is then incremented by two so that it points to the new top word in the stack.

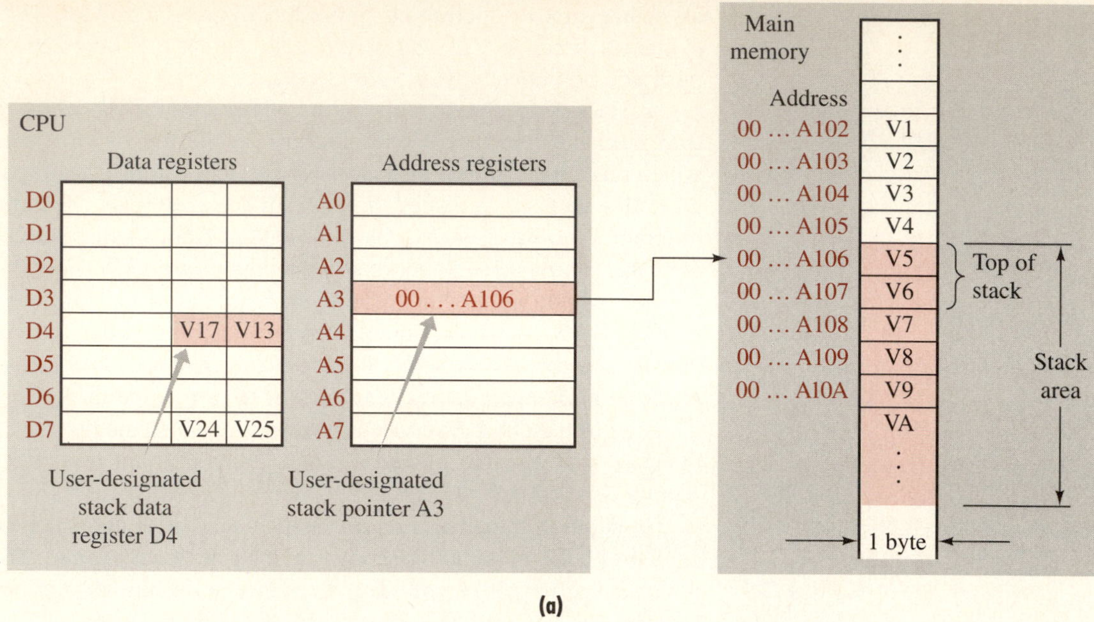

(a)

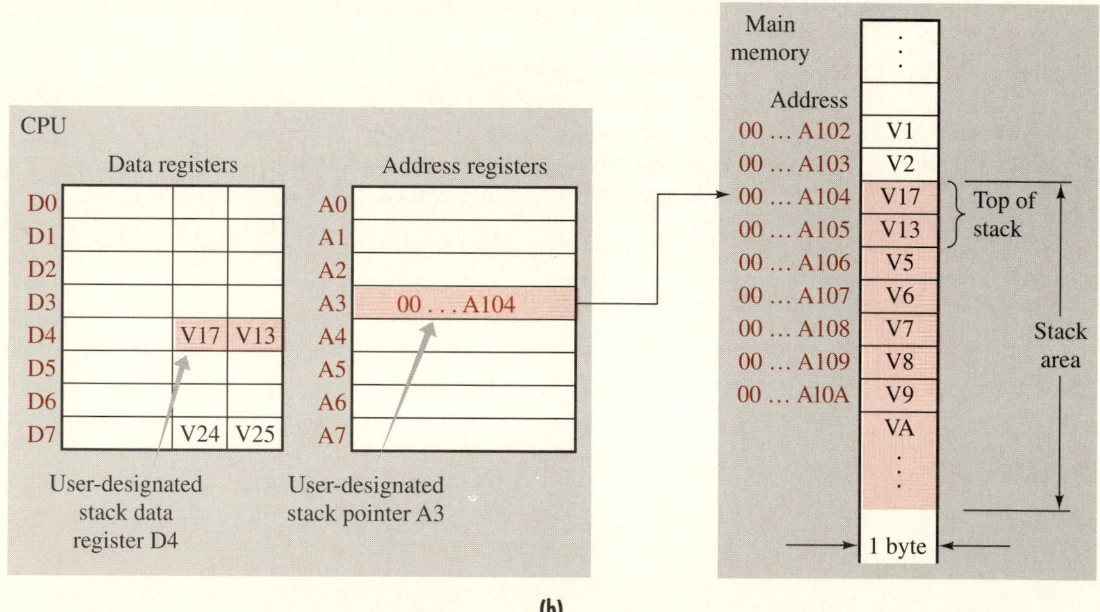

(b)

FIGURE 9.25 Stack control in the 680X0, using the push instruction MOVE D4, −(A3): (a) before and (b) after execution.

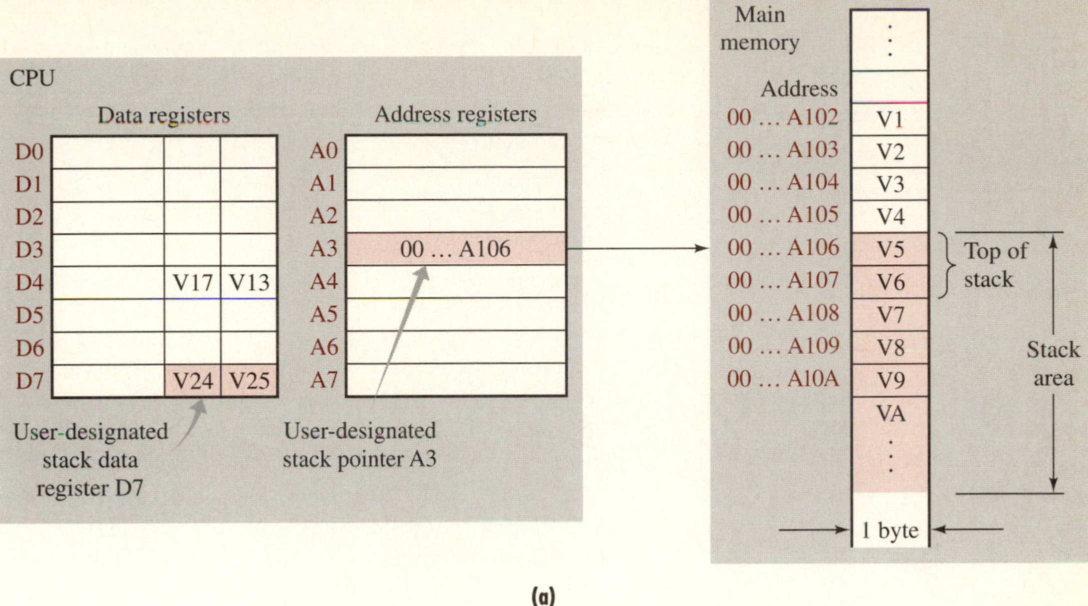

(a)

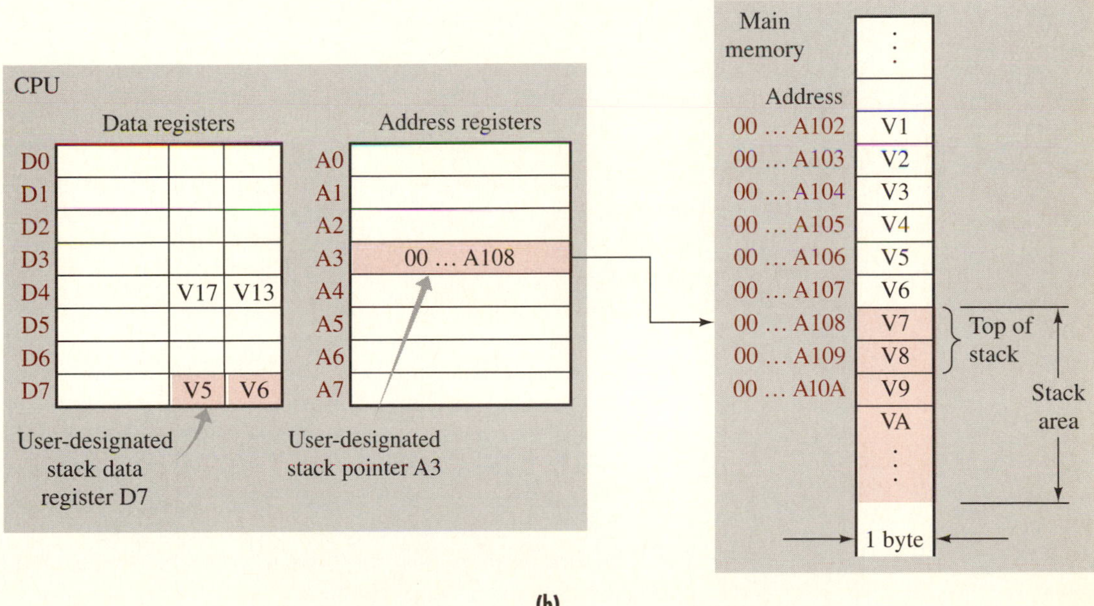

(b)

FIGURE 9.26 Stack control in the 680X0, using the pop instruction MOVE (A3)+,D7: (a) before and (b) after execution.

Computer-Based Systems

This last section discusses the design of special-purpose systems in which computers in the form of single-chip microcontrollers are the key components. Some issues involved in interfacing a microcontroller to input/output devices are also examined.

9.6 Microcontrollers

Microcontrollers are computers that are embedded, often invisibly, into many types of specialized devices.

Microcomputers have an enormous range of applications in which they serve as controllers for other systems. These applications range from commonplace and fairly simple items like telephone sets and washing machines to complex industrial applications such as the control of robots on an assembly line. We refer to specialized computers of this sort as **microcontrollers**. They are also called **embedded computers** to distinguish them from more general, stand-alone computers like workstations or personal computers. The fact that microcontrollers are embedded in the system they control often makes them invisible to the users of that system. The driver of a car is scarcely aware that half a dozen or so microprocessors are secreted throughout the car, busily executing programs that manage such critical functions as fuel injection, ignition timing, gear shifting, braking, and the dashboard instrument panel. The laser printer attached to a PC may contain a CPU that is comparable in power to the PC's but is dedicated to the tedious and computationally complex task of composing pages of text and figures to be printed.

Microcontroller Design

A microcontroller is a one-chip IC with programmable ROM and I/O circuits.

The overall organization of a microcontroller differs little from those of the general-purpose computers we have been discussing. Many microcontrollers are based on microprocessors with 4- or 8-bit architectures and modest performance levels. Increasingly, however, the same architectures used for large general-purpose processors like the 680X0 and 80X86 are being adapted to produce dedicated microcontrollers. This makes microcontrollers compatible with the software used in general-purpose computers, along with giving them the potential for achieving very high performance. Fast computation might seem unnecessary to control a device like a television set, but it becomes necessary if apparently simple operations like changing channels are to be controlled in sophisticated ways, such as by voice command. Computer processing of human speech to identify its meaning is a difficult problem that requires a huge amount of specialized data processing.

Microcontrollers must interface with a wide range of input/output devices that link them to the equipment being controlled; as a result, they must be designed to satisfy a wide range of interfacing needs. In some cases, the application employs analog I/O signals, so analog-to-digital or digital-to-analog conversion circuits must be provided. All-digital I/O devices also vary in their signal formats and transmission speeds. Consider, for instance, the word size w used to transfer digital data to or from a microcontroller. A secondary memory device like a magnetic disk unit produces a high-speed stream of serial data for which $w = 1$. A typewriter-style printer, on the other hand, is designed to receive a relatively slow stream of alphanumeric data, which is transmitted in parallel fashion, with $w = 8$.

Microcontrollers are also called upon to measure time, sometimes with high accuracy, so that they can control **real-time devices**, which are devices that send or receive data subject to time deadlines. For example, because a magnetic disk spins at a fixed rate, it also transmits data to or receives data from a host microcontroller at a fixed rate. If the microcontroller fails to respond in time to the disk unit's needs, data will be lost. Although many I/O devices are slow relative to the internal operating speed of the microcontroller, often the microcontroller controls a large number of I/O devices simultaneously. Consequently, the microcontroller must keep close track of its processing time and subdivide it appropriately among the devices competing for its services, a process called **time multiplexing** or **time sharing**.

To meet these diverse requirements, a typical microcontroller contains a core CPU, a main memory, and a number of multipurpose I/O control circuits. The main memory usually consists of a large ROM part in which the microcontroller's special-purpose application programs are stored, and a small RAM part for temporary data storage. The I/O circuits can include some that are designed for serial digital data and others designed for parallel data, analog-digital conversion logic, and time measurement circuits. They can also be expected to have flexible interrupt mechanisms for communication with the CPU.

Input/Output Ports

An I/O port's characteristics such as the data transfer direction can be reconfigured online.

The characteristics of a microcontroller's I/O ports, such as the signal directions, word length, and data transmission speeds they handle, are designed to be alterable by instructions issued from the CPU. Such I/O control circuits are said to be **programmable**. Figure 9.27 shows how a typical I/O port can be programmed to act as either an input or an output port. The port's current configuration is specified by a control register *CR*, which is itself a programmable output register. The port is configured by writing an appropriate control word to *CR*. Control signals from *CR* then establish the required logical connections to the port's main data register *DR*. Tri-state logic is normally employed to facilitate bidirectional data transmission.

Figure 9.27a shows the I/O port configured as an input port. The current contents of *CR* specify that the data direction is *in*—that is, from the I/O device to the CPU—and the port's tri-state logic is set accordingly. The I/O device can send a data word into the port's data register *DR* in response to an input or load instruction that is addressed to *DR*. The data word can then be transferred from *DR* to the system bus, from which it goes to main memory

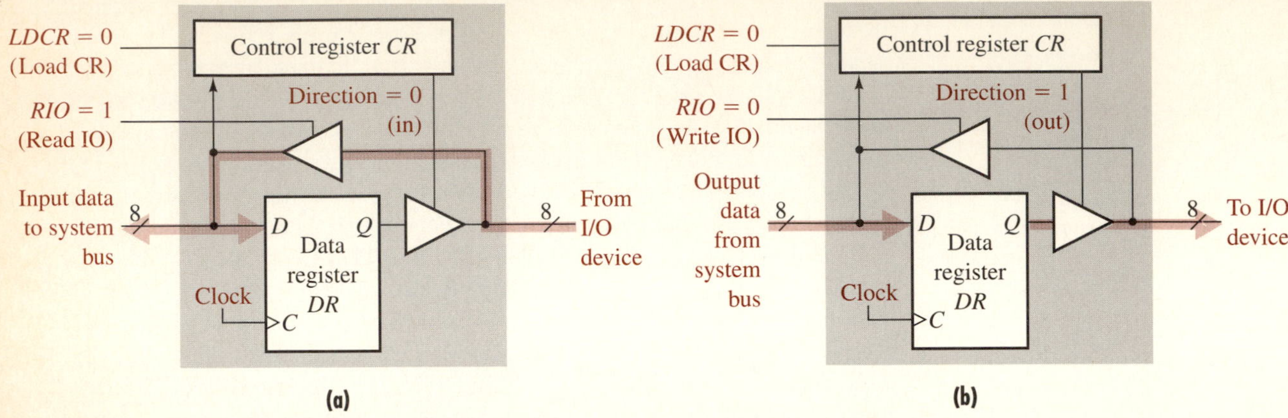

FIGURE 9.27 A programmable I/O port configured as (**a**) an input port and (**b**) an output port.

or, in some cases, directly to the CPU. As the CPU and I/O devices operate largely asynchronously, *DR* and some associated control signals serve as a buffer to synchronize the internal and external activities of the microcontroller. By loading a new control word into *CR*, the port is reconfigured as an output port, as illustrated by Figure 9.27b. Now an output or store instruction addressed to *DR* causes it to transfer a data word to *DR*, and from there to an output device. Note that a programmable I/O port's configuration can be changed repeatedly during system operation so that at some times it is an input port and at other times an output port.

The Motorola 68705

To illustrate the structure of a representative single-chip microcontroller, we examine the Motorola 68705 (Figure 9.28). This microcontroller is based on the 6800 microprocessor series, a precursor of the 680X0 that is characterized by an 8-bit data word size and an accumulator-based CPU that contains a very small set of registers. The main components of the 68705 are the CPU, an on-chip main memory *M* with a capacity slightly under 4K bytes, a set of four programmable I/O ports, and a programmable timing unit. The CPU's E-unit comprises an 8-bit integer ALU, a single 8-bit data register (the accumulator *AC*), and a 5-bit status register *SR*. The I-unit contains two main address registers: a program counter *PC* and a stack pointer *SP*, which are analogous to the registers of the same names in the 680X0. The length of these address registers is only 12 bits, however, implying a maximum capacity of $2^{12} = 4K$ bytes for main memory and I/O ports. (As in the 680X0, memory-mapped I/O is used.) There is also an 8-bit index register IX to support an indexed addressing mode.

The 68705's read-only memory is electrically programmable off-line by the user.

Most of the space in the 68705's main memory *M* is devoted to a programmable read-only memory (PROM) intended to store the programs that

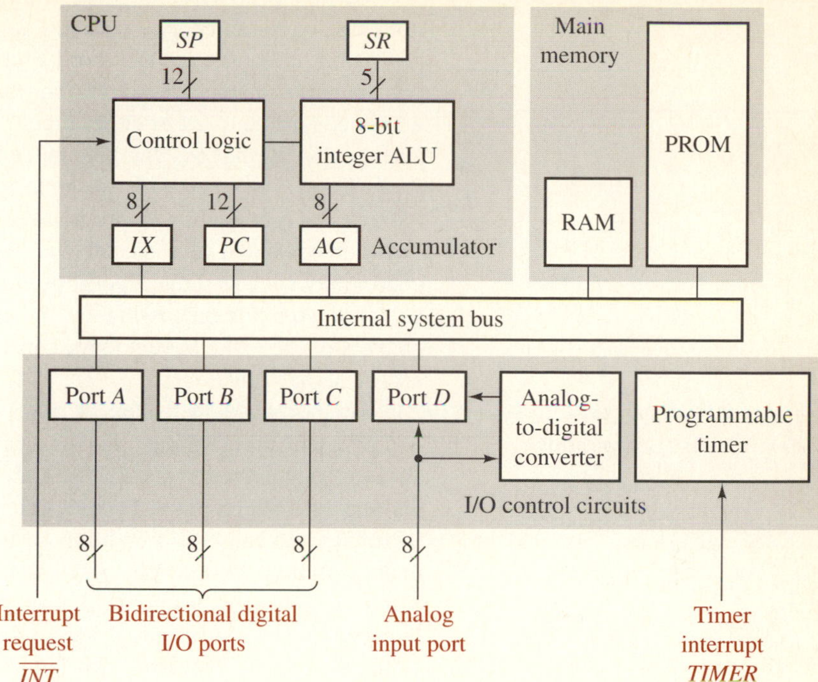

FIGURE 9.28 Organization of the Motorola 68705 microcontroller.

tailor the 68705 to its application. The PROM is electrically programmable, meaning that its contents can be changed off-line by means of an electrical programming device. In normal, on-line operation, it behaves like any other nonwriteable memory device. Because of its user programmability, a PROM facilitates the development process for new microcontroller designs. Equivalent mask-programmable ROMs tend to be preferred in production models because of their somewhat lower cost and higher reliability. There is also a 112-byte portion of RAM space in M. This is used for temporary data storage and compensates to some extent for the lack of CPU registers in the 68705. In general, the CPU can access main memory locations faster when they are on the same IC chip as the CPU than when a chip-to-chip transfer must be made.

The 68705's I/O Circuits

The 68705 has four programmable I/O ports and an analog-to-digital converter.

The 68705, like all microcontrollers, has a fairly large set of I/O ports and supporting control logic. Data can be transferred to or from the external lines of each port by executing move instructions; this is the basic method by which the microcontroller communicates with I/O devices. There are four I/O ports named A, B, C, and D in the 68705, which are intended for 8-bit (parallel) word transfers. Each has an addressable data register DR that is directly connected to eight of the 40 pins in the standard package that houses the 68705. These pins, in turn, can be wired to I/O devices.

All the 68705's I/O ports are programmable on-line, meaning their functions can be modified by executing appropriate instructions. This is accomplished by moving a control word to a control register associated with each port to define the port's function. For example, ports A, B, and C can be programmed on-line to act as either input or output ports; that is, the direction of each port's external bus can be changed by software, as illustrated by Figure 9.27. The 68705's port D is an input-only port that can accept analog or digital data, as specified by its programmable control register. Each analog input line can be routed to a shared analog-to-digital converter that quantizes a continuous voltage signal into 256 digital levels, which it then encodes into an 8-bit word for processing by the microcontroller. The analog-to-digital conversion step is relatively slow, consuming 30 clock cycles to digitize each analog signal value.

A microcontroller can measure time accurately by means of a programmable timer.

Another standard feature of microcontrollers exemplified by the 68705 is a **programmable timer**. This is an internal I/O device that is built around an 8-bit down-counter *TDR*, the timer data register, which counts pulses derived from the on-chip CPU clock signal. *TDR* is assigned an address within the 68705's memory-I/O address space, and so can be written into or read under program control. When activated, *TDR* counts down at a precise rate determined by the clock period T_C, a typical value for which is a microsecond. If a program writes an initial value t_1 into *TDR* and subsequently reads the value t_2 from *TDR* (before the latter reaches zero), then the elapsed time between the read and write operations is given by $(t_2 - t_1)T_C$. Because the contents of *TDR* can range from zero to 255, driving the counter directly from the system clock with $T_C = 1\mu s$ permits a maximum time-measurement period of 255 μs, with a resolution of 1 μs. To allow the measurement of longer time periods, a built-in scaling mechanism may be used; see Problem 9.44. Alternatively, software routines can be written to simulate a larger *TDR*. The timer can be programmed so that when it reaches zero, it interrupts the CPU. This enables *TDR* to force the CPU to perform some action after a certain maximum amount of time has elapsed. In this warning role, *TDR* is referred to as a **watchdog timer**.

Example 9.7

An instrument to monitor pulse rate

A representative application of a low-cost, single-chip microcontroller like the 68705 is illustrated by Figure 9.29. This is a portable pulse meter intended to monitor a person's pulse rate while exercising. It does so by processing an analog signal derived from a sensitive pressure sensor that can be clipped to a finger. The sensor's output is an analog voltage, the hills and valleys of which follow the fluctuations of the pulse. This analog signal is sampled at regular intervals and converted to digital form. By counting the hills over a known time period, an average pulse rate r in beats per minute is easily calculated. This quantity r is then sent to the LCD display panel, as depicted in the figure. A single 8-bit output port suffices to control a multidigit display

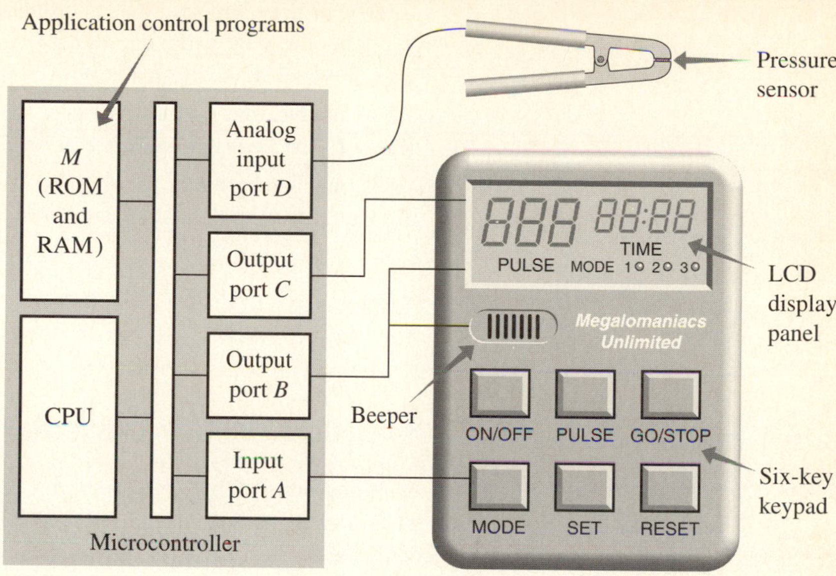

FIGURE 9.29 Application of a single-chip microcontroller to a pulse meter.

of this sort. Half the port can be used to output a 4-bit decimal digit in the manner depicted in Figure 9.11b; the lines of the other half-port select the particular seven-segment device to display the digit. By transferring the digits serially to each of the seven-segment elements in turn—that is, by multiplexing the output port—a multidigit image can be displayed.

A small keypad allows the human user to control various functions of the pulse meter. Besides simply displaying the current value of the pulse rate r, various special modes of operation can be easily implemented in software. For example, the minimum or maximum pulse rates over a user-specified time period can be entered; to this end, another four digits are added to the LCD display panel to show minutes and seconds. Yet another function that is easily added is a beeper that can be used to issue an audible warning when the measured pulse rate exceeds a level previously entered by the user. Indeed, the number of possible functions of this kind that can be added is limited only by the designer's imagination, thanks to the presence of the microcontroller.

Programs stored in the microcontroller's ROM determine the system's range of functions.

The major portion of the pulse meter's design resides in the software stored in its ROM. This consists of a set of fairly simple programs that perform the following actions, typically in round-robin fashion:

1. Input and check the keypad signals from port A. If a new key has been pressed, branch to a program that implements the actions invoked by that key, such as "turn the meter on or off," "initiate or terminate the pulse-rate measurement," or "change the unit's mode of operation."
2. Input the signal from the pressure sensor from port D, convert it to digital form, and store it.

3. Perform any internal calculations required by the current operating mode. For example, compute a new value of the pulse rate r by averaging a set of measured values over some time period.
4. Update the output data and send them to the LCD display panel or to the beeper via output ports B and C.

Provision may be made for the software to respond to interrupts such as a time-out signal generated by an internal timing circuit.

9.7 Input/Output Interfacing

A microcontroller-based system requires some application-specific hardware and software to link the microcontroller to the I/O devices it controls. This hardware-software link is the system's **I/O interface**; the task of designing it is called **interfacing**. This section attempts to give the reader a flavor of some basic interface design issues.

The I/O Interface

Interfacing is the design of the hardware and software that links a microcontroller to its I/O devices.

The computer side of the I/O interface contains a set of I/O ports and their associated circuits. These ports are defined by their size, signal direction, data-transfer rates, programmability, and certain electrical properties. The I/O device side of the interface can be similarly characterized, but its parameters may not be directly compatible with those of the microcontroller. Where incompatibility exists, electrical or logical "glue" circuits can be designed into the interface to eliminate the problem. Alternatively, programs to perform the same tasks can be added to the system's application software. In cases in which, as in Figure 9.29, there is essentially perfect hardware compatibility between the microcontroller and its I/O devices, all the design effort goes into the system software. Often, however, a combination of hardware and software yields the best solution to an interfacing problem, as we will see shortly in the case of analog-to-digital conversion.

Common I/O functions are data conversion, signal selection, and signal synchronization.

The I/O functions common to most microcontroller applications are data conversion, signal selection, and signal synchronization. Data conversion refers to matching the electrical or logical characteristics of the communicating circuits—for example, converting signals from analog to digital form or vice versa. Selection logic is needed to address individual words or signals and to permit sharing of the limited number of I/O lines and other resources of an I/O port. Finally, discrepancies in the operating speeds and timing requirements of the microcontroller and the I/O devices may require special synchronization hardware. The extent to which a logic designer must provide special interface circuits and software for the foregoing tasks depends on the amount by which the characteristics of the I/O devices deviate from those of the host computer.

To illustrate some of the design trade-offs in interfacing a microcontroller to the outside world, consider the task of converting analog signals into digital form for processing by computer. Many physical quantities are analog or continuous in nature, such as pressure (an input to the pulse meter discussed above), temperature, position, and so on. Usually, such analog signals are first mapped into an analog electrical voltage, the fluctuations of which follow those of the original signal; this is done by a **sensor** or **transducer**. The analog voltage V must then be converted into an n-bit binary integer, the value of which best approximates V, a process called **quantization**. The integer word size n typically lies between eight and 16; larger word sizes are hard to achieve due to physical limitations on the accuracy possible in voltage measurements.

Analog-to-Digital Conversion: The Direct Method

The "direct" method of analog-to-digital conversion is fast but expensive.

First, we describe an all-hardware technique, the so-called **direct method**, for analog-to-digital conversion (Figure 9.30). This conceptually straightforward technique relies on a device called a **voltage comparator**[4] (Figure 9.30a), which compares two voltages V_a and V_b and outputs a binary signal z to indicate their relative values: $z = 1$ if $V_a > V_b$, and $z = 0$ if $V_a \leq V_b$. Thus, the voltage comparator activates its output when the first voltage V_a (indicated by a plus sign on the standard triangle symbol for a voltage comparator) exceeds a threshold level set by the second voltage V_b (indicated by a minus sign). To build a k-bit analog-to-digital converter, we use 2^k analog comparators $C_0, C_1, \ldots, C_{2^k-1}$, whose threshold voltages $V_0, V_1, \ldots, V_{2^k-1}$, where $V_i < V_{i+1}$ correspond to the 2^k quantization levels that we want to encode. The outputs of the 2^k analog comparators are fed to a 2^k-input priority encoder *ENC* that generates the desired k-bit result.

Figure 9.30b illustrates the direct analog-to-digital method for $k = 3$. The voltage comparator C_0 with the lowest threshold voltage V_0 is connected to the lowest priority input of the encoder, C_1 with the next lowest threshold voltage V_1 is connected to the encoder input of next lowest priority, and so on. Hence, the voltage comparator C_i whose index i is the largest for which the comparator's output is 1, determines the output of *ENC*; the latter is the k-bit binary representation of i. In the figure, C_i produces output 1 for $0 \leq i \leq 4$; hence, the input analog voltage V lies between V_4 and V_5. The

[4]This should not be confused with a digital magnitude comparator, a logic element that compares the magnitudes of two binary numbers (Section 8.2).

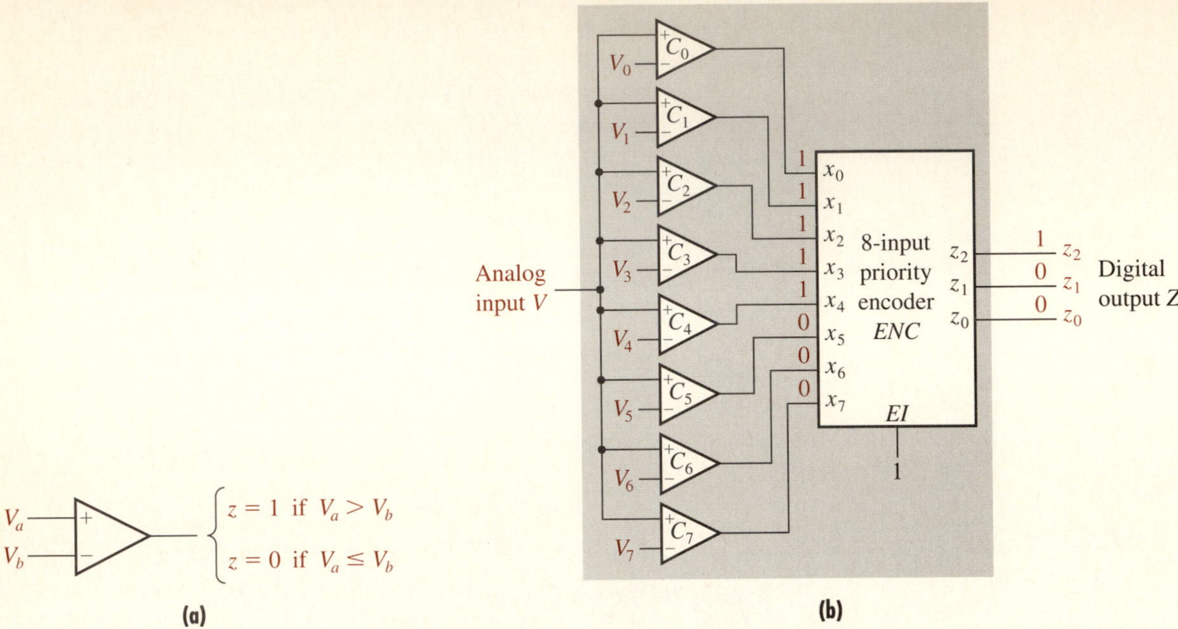

FIGURE 9.30 (**a**) An analog voltage comparator; (**b**) a 3-bit analog-to-digital converter that uses the direct method.

digital output produced by *ENC* is therefore $Z = 100_2 = 4_{10}$. This direct analog-to-digital converter is fast but relatively expensive due to the large number of comparators it employs. Hardware costs limit the direct conversion method to $n \leq 5$ or so.

Analog-to-Digital Conversion: The Ramp Method

The "ramp" analog-to-digital conversion method is slow but requires little hardware.

A slower but less expensive alternative to the direct method is depicted in Figure 9.31. The basic idea of this **ramp method** for analog-to-digital conversion is to use a single voltage comparator to compare the input voltage V to 2^n different reference (threshold) voltages in 2^n steps, rather than all at once, as in the previous design. It exploits the fact that hardware n-bit digital-to-analog converters are relatively simple, fast, and inexpensive, even for large n. The components needed for the second analog-to-digital converter are an n-bit digital-to-analog converter and a voltage comparator C, which are connected to an output and an input port, respectively, as shown in Figure 9.31a for $n = 8$.

A program-generated number VREF corresponding to a threshold voltage V_i is sent to the output port. From VREF, the digital-to-analog converter produces the analog value V_i, which is then compared to V by the voltage

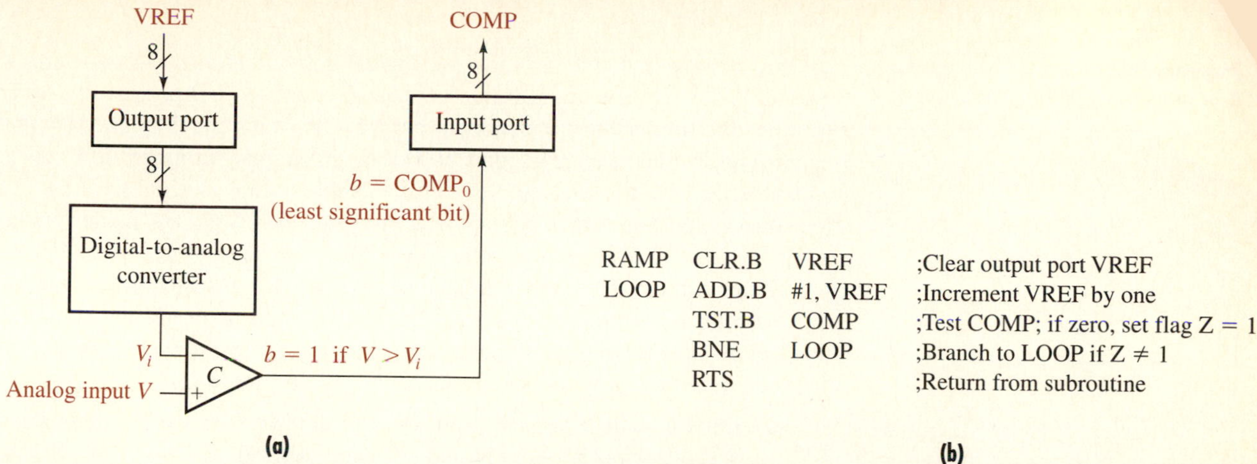

VREF
8
Output port
8
Digital-to-analog converter
V_i
Analog input V
C
$b = 1$ if $V > V_i$
$b = \text{COMP}_0$
(least significant bit)
Input port
8
COMP
(a)
RAMP CLR.B VREF ;Clear output port VREF
LOOP ADD.B #1, VREF ;Increment VREF by one
TST.B COMP ;Test COMP; if zero, set flag Z = 1
BNE LOOP ;Branch to LOOP if Z ≠ 1
RTS ;Return from subroutine
(b)

FIGURE 9.31 (a) Hardware and (b) software for an analog-to-digital converter that uses the ramp method.

comparator. The latter's binary output b is sent to the host microcontroller via the input port, where its value can be checked. VREF is incremented from 0 in steps of one, producing a ramp-like voltage waveform from which the conversion method takes its name. This process continues until b changes value from 1 to 0, at which point the current value of VREF is taken as the digital approximation to V.

A short subroutine called RAMP that controls the hardware of the ramp analog-to-digital method appears in Figure 9.31b. It is written in 680X0 assembly language, mainly using instructions we have encountered already, and should be self-explanatory. The instruction TST (test) reads a byte of the form 0000000b from input port COMP, which it compares to 00000000; TST uses the 680X0's zero flag Z to indicate the result of this comparison. The conditional branch instruction BNE ("branch if Z is not equal to zero") then tests Z to determine when the analog-to-digital conversion is complete. The final instruction RTS ("return from subroutine") causes control of the CPU to be returned to the program that called the RAMP subroutine. The RAMP subroutine's three-instruction loop is executed on average 2^7 times to digitize V.

The direct and ramp analog-to-digital converters provide a nice illustration of the hardware/software trade-offs possible in the design of microcomputer-based systems. The software approach is the slower of the two, but many applications have very modest interfacing and data-processing needs that can easily be met by software. Moreover, programmed control has the advantage of being easily changed to introduce better control methods or to handle new applications. For example, the conversion program of Figure 9.31 determines the analog voltage V by comparing it exhaustively to all possible (digital) values. The average comparison time can be significantly reduced by employing a more sophisticated algorithm to search the digital values for one that approximates V (Problem 9.49).

I/O Programming

res all I/O operations ectly by CPU programs.

As the analog-to-digital example of Figure 9.31 illustrates, I/O operations primarily involve the transmission of data to and from I/O ports, with some simple decision making performed by the CPU and the I/O device. The control style illustrated by Figure 9.31b is called **programmed I/O** and is characterized by the fact that all key decisions in the execution of an I/O operation, such as when to start and end the operation, are made by the CPU.

Programmed I/O requires the CPU to execute I/O instructions at regular, predetermined intervals, if for no other reason than to check the status of the I/O devices it controls. Typically this involves the following sequence of actions, which must be built into the CPU's main program:

1. Input a status word *SW* from the I/O device; *SW* is often assigned a separate input port. This step is called **polling**.
2. Examine *SW* to determine the I/O device's status, such as "inactive and not requiring service," "ready to send input data to the microcontroller," "ready to receive output data from the microcontroller," and "requires special attention because an exceptional condition has occurred."
3. If no service is required by the I/O device, the CPU continues with other tasks. If service is required, the CPU selects and executes the appropriate program to provide the desired service. For example, it reads a word of data from the device and stores it in main memory.

Input/output programs are often quite short and involve the transfer of only one or two words across the system's I/O interface; the program of Figure 9.31 is typical in this regard. However, if many I/O devices need regular service—for example, a few fast devices or many slow ones—they can take up an unacceptably large part of the CPU's time. In particular, the CPU can waste a great deal of time checking the status of I/O devices that do not require its services.

Interrupts

Interrupts allow the CPU to respond quickly to a service request from an I/O device.

A faster alternative to status polling is provided by interrupts, which enable I/O devices to attract the CPU's attention at any time. One or more interrupt request lines, like the 68705's $\overline{\text{INT}}$ line (Figure 9.28), go from the I/O device to the CPU to initiate an interrupt. Interrupt request lines are checked by the CPU during every instruction cycle. The CPU's I-unit contains logic to respond to an interrupt request and to save automatically the current contents of *PC* in a stack region of *M*. A program P_{new} that handles the interrupt must be in memory with a starting address that is known to the I-unit. The I-unit then loads this address into PC to initiate execution of P_{new}. After P_{new} has completed its task, control should be returned to the interrupted program P_{old} by executing a return-from-subroutine instruction like the RTS instruction in Figure 9.31b.

Some CPUs and microcontrollers have more powerful and flexible interrupt facilities than the one just outlined. For example, the saving of CPU data and address registers used by P_{old} can be done automatically by the CPU in response to the interrupt request signal so that the first instruction of P_{new} can respond to the interrupting device. The CPU pushes the relevant information into a main-memory stack from which it is subsequently popped by a RETURN instruction executed by P_{new}. Some CPUs are designed to handle large numbers of I/O devices, all of which are able to generate interrupt requests. This requires many interrupt-handling routines, as well as special hardware and software to identify interrupt sources and to enable the CPU to select for service one of many competing requests.

Computer-Aided Design

Typical steps in the design of a microcontroller-based system.

Several specialized CAD tools are available to assist in the design of a new microcontroller-based or, indeed, any microprocessor-based system. The most useful of these is a computer system that can support the design and execution of the software needed by the target processor and can also assist in the development and debugging of the hardware for the I/O interface. Such a CAD tool is generally known as a **microprocessor development system** or **MDS**. (We may also substitute the word *microcontroller* for *microprocessor* in MDS.) A typical MDS appears in Figure 9.32. It consists of a development computer C in the workstation or personal computer class, equipped with software for editing, translating (compilers and assemblers), executing, and testing microcontroller software.

An MDS also has facilities to create and test prototype hardware designs. A cable linking the target, microcontroller-based design to the MDS's CPU and main memory allows the latter to override the microcontroller's on-chip CPU and (P)ROM, making it possible to drive the microcontroller directly

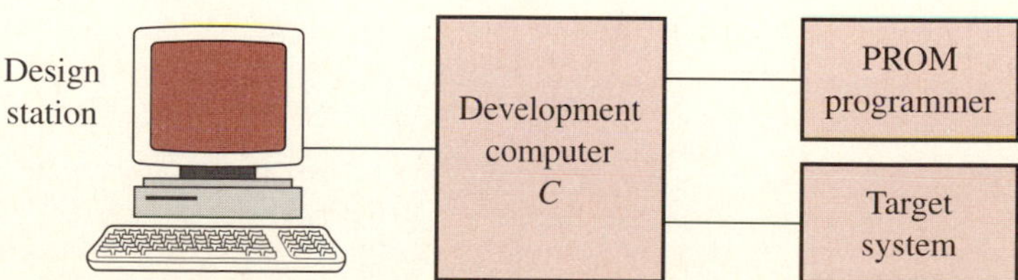

FIGURE 9.32 The structure of a microprocessor development system (MDS).

from the MDS. The designer can then use the MDS to alter and monitor the performance of the current design, a process called **in-circuit emulation**. Another useful adjunct to an MDS is a PROM programming unit, which allows the ROM memories of microcontrollers like the 68705 to be easily reprogrammed in response to design changes. Some MDSs are tailored to a particular computer family, but others—the so-called universal development systems—have hardware and software to support designs that use many different microcontroller types.

The design of a microcontroller-based system normally proceeds along the following lines:

1. *Planning:* The specifications of the application are determined. These specifications include the functions to be implemented, the performance and cost constraints, compatibility requirements, the CAD support available, and the production schedule.
2. *Architectural design:* The overall structure and behavior of the target microcontroller are defined and its hardware and software requirements are estimated. A candidate microcontroller family that best meets the design constraints is selected.
3. *Detailed design:* The interface hardware is defined. The algorithms underlying the application software are developed and programmed via an MDS.
4. *Prototype construction and testing:* A working prototype is built and tested against the system specifications, again using an MDS. The prototype is modified (debugged) until a satisfactory design is obtained. The final design is then documented.

Although the effort devoted to these design steps is highly dependent on the application, a major portion of the design cost is often incurred in the final debugging phase. Hence, design practices that minimize errors can significantly reduce costs as well as time to market.

CHAPTER 9 SUMMARY

The Architecture Level

9.1 ■ At the architecture level, memories and processing units store and process long sequences or blocks of word-organized data.

■ A block of instructions that performs a specific task—a **program**—can be organized to carry out essentially any computational job. The component of the computer that is responsible for executing instructions is the **central processing unit** or **CPU**.

■ **Main memory** holds the program and associated data while they are being processed. Main memory is constructed primarily from **random-access memory (RAM)** ICs, in which any location can be read from or written into in a single step.

■ The **input/output (I/O) control units** interface the computer to external devices such as printers, monitors, and keyboards, which are called **I/O devices** or **peripherals**.

■ **Computer architecture** is defined as a standard specification that covers the information formats, instruction types, and the high-level structure and behavior of all members of a computer series.

■ A **microcontroller** is a "computer on a chip" that is programmed to perform a dedicated set of tasks. This programmability and its relatively low cost are the key to the microcontroller's flexibility.

9.2 ■ The CPU consists of a control part (the **instruction unit** or **I-unit**) and a datapath part (the **execution unit** or **E-unit**).

■ The CPU executes a program by completing an **instruction cycle** for each instruction. The instruction cycle consists of the following steps: fetching the opcode from memory; reading the operands needed; executing the instruction; and writing the results to memory.

■ Programs can be written in a machine-specific, symbolic format called **assembly language**. The part of an instruction that identifies the operation to be performed is the **opcode**; the parts that identify operands are referred to as **addresses**.

■ Before they can be executed, programs must be translated into **machine language.** This is done automatically by language-translation programs called **assemblers** and, in the case of high-level programming languages, **compilers**.

■ The three main types of instructions found in every computer are **data-transfer**, which move data from place to place, **data-processing**, which perform arithmetic and logical operations on data, and **program-control instructions**, which alter the order of instruction execution.

■ Main memory performs two basic operations: a **read operation,** which transfers a stored data word from storage to an external bus, and a **write operation**, which causes a new data word to be transferred from the external bus into storage.

9.3 ■ Information transfers between the CPU-memory subsystem and the outside world are called **input/output (I/O) operations**. I/O devices link a computer to its environment via **I/O ports**, which are basically addressable data registers.

■ Example input devices include switches and keyboards. Example output devices include seven-segment displays, monitors, and printers. Some devices, such as disk drives, can be used for both input and output.

The Central Processing Unit

9.4 ■ An E-unit contains registers for temporary storage of data, an arithmetic-logic circuit for processing data, and various buses, multiplexers, and other minor circuits.

■ When an E-unit has only one or two registers, they are called **accumulators**; they act as the main source or destination of operands during instruction execution. Other E-units contain a set of registers, called a **register file**.

- The I-unit is responsible for controlling the instruction cycle. The key component of the I-unit is the **program counter**, which keeps track of the address of the currently executing instruction.
- An I-unit can be **hardwired** (designed around a special-purpose state machine) or **microprogrammable** (designed around a programmable control memory).

9.5 ■ Writing programs for a microprocessor requires an understanding of its **programming model**, which includes the CPU's instruction and data types, the addressing modes used to specify the location of data, and the memory and I/O addresses available.

- I/O devices can be assigned addresses in the processor's normal memory address space. Such **memory-mapped** devices are addressed with the usual instructions for accessing memory. **I/O-mapped** devices use a special I/O address space and are accessed with special instructions. All CPUs support memory-mapped I/O, but some do not support I/O-mapped I/O.
- Some computers have been designed with large instruction sets—sometimes several hundred instruction types. These are **complex instruction set computers** or **CISCs**. There are also **reduced instruction set computers** or **RISCs**, which restrict instruction sets to 20 or so simple but fast types.

Computer-Based Systems

9.6 ■ Because microcontrollers are embedded, often invisibly, into many types of specialized devices, they are often called **embedded computers**. Programs stored in the microcontroller's memory determine the controlled system's range of functions.

- The inclusion of various I/O and timing facilities on microcontroller chips makes them a very cost-effective solution for many control applications.

9.7 ■ The link between a computer and its I/O devices is called the **I/O interface**; the task of designing it is called **interfacing**. Common I/O interface functions are data conversion, signal selection, and signal synchronization.

- **Programmed I/O** requires all I/O operations to be controlled directly by CPU programs. **Interrupts** enable I/O devices to signal the CPU directly when they need service.
- A **microprocessor development system (MDS)** is a CAD tool for the design and debugging of new microcontroller applications.

Further Readings

There is extensive literature describing how computers work, such as *Computer Architecture and Organization, 2nd ed.* (McGraw-Hill, New York, 1988), by J. P. Hayes. Many texts and manufacturers' data books deal with the architecture and programming of specific microprocessor and microcontroller series. The Motorola 680X0 is covered in T. Mimar's *Programming and Designing with the 68000 Family* (Prentice-Hall, Englewood Cliffs, N.J., 1991); the Intel 80X86 is the subject of *Microcomputer Systems: The 8086/8088 Family* (Prentice-Hall, Englewood Cliffs, N.J., 1985), by Y. Liu and G. A. Gibson. *Design with Microcontrollers*, by J. B. Peatman (McGraw-Hill, New York, 1988) is a comprehensive text on the use of microcontrollers. The 68705 microcontroller and its applications are covered in S. J. Cahill, *The Single-Chip Microcomputer* (Prentice-Hall, Englewood Cliffs, N.J., 1987).

Chapter 9 Problems

The Architecture Level

Computer organization; CPU structure; instruction types; main memory (RAM) design; I/O devices: switches, keyboards, output displays, secondary memory.

9.1 Define and briefly explain the following common acronyms: CPU, DRAM, IR, LCD, MIPS, ROM.

9.2 List the main changes that have occurred in CPU technology since the first electronic computers appeared in the 1940s. Why has the cost/performance ratio of computers declined by many orders of magnitude over this period of time?

9.3 **(a)** State the role played by the program counter *PC* in the operation of a computer. **(b)** A few computers have been designed in which the *PC* has been removed from the CPU and placed instead in the main memory unit. Determine one key advantage and one disadvantage of such a "remote" *PC*.

9.4 Besides a program counter *PC* and an instruction register *IR*, a certain simple computer has only one memory-address register and one data register in its CPU. What roles should be assigned to these two registers to ensure that a small instruction set like that of Table 9.3 can be executed? Illustrate your answer for one representative instruction type.

9.5 Place each of the following instructions in one of our three main instruction-type categories: data transfer, data processing, or program control. **(a)** Subtract the contents of register R_2 from register R_1 and place the result in R_1; **(b)** subtract 0 from the contents of R_1 and place the result in memory location $M(A)$; **(c)** interchange the contents of registers R_1 and R_2; **(d)** skip the next instruction if R_7 contains the all-0 pattern; **(e)** input the keyboard data at I/O port P_3; **(f)** reset the keyboard data at I/O port P_3 to 0.

9.6 Specify by means of register-transfer statements the main actions occurring in the instruction cycle associated with fetching and executing each of the following instructions. Assume the architecture of Figure 9.3 with a memory word size of 16 bits. All instructions are of length $16(k + 1)$ bits, where k is the number of memory addresses specified in the instruction. **(a)** ADD R3,ADR5, which adds the contents of register R_3 to memory location M(ADR5) and places the result in R_3; **(b)** ADD ADR5,R3, which adds the contents of memory location M(ADR5) to register R_3 and places the result in M(ADR5); **(c)** ADD ADR4,ADR5, which adds the contents of memory location M(ADR5) to memory location M(ADR4) and places the result in M(ADR4).

9.7 Consider a RAM IC of the type depicted in Figure 9.6a with $n = 16$ and $m = 4$. Using this IC as the basic component, show how to construct RAMs with capacities of **(a)** 2^{18} 4-bit words; **(b)** 1M bytes; and **(c)** 256K 32-bit words.

9.8 Suppose the bidirectional tri-state data bus in the RAM IC of Figure 9.6a is replaced by two unidirectional m-bit buses, DATA IN and DATA OUT. Redesign the memory system of Figure 9.7 to accommodate this change.

9.9 The main memory M of a computer is to be designed with a capacity of 16M bytes. Three different DRAM ICs of the type depicted in Figure 9.6a are available as components of M. The three types are Type A, a (4M × 1)-bit DRAM costing \$14 per IC; Type B, a (1M × 2)-bit DRAM costing \$6; and Type C, a (256K × 4)-bit DRAM costing \$3. Address decoding logic, including ICs and wiring, is estimated to cost \$$10d$, where d is the number of address bits to be decoded. Determine which type of DRAM IC to use in order to minimize the cost of M. Draw a logic diagram for the resulting design.

9.10 Consider the three RAM IC types A, B, and C defined in Problem 9.9. We want to build from one of these IC types a memory with the largest possible storage capacity that costs, at most, \$60. Determine the RAM type to use and the structure of the memory if **(a)** the memory word size can be arbitrarily large, and **(b)** the memory word size must be exactly 4 bits.

RAM addresses are often split into row and column addresses that are transmitted separately.

9.11 Figure 9.33 shows how a certain commercial (4M × 1)-bit DRAM IC is packaged. As with most high-capacity DRAM ICs, the 22-bit memory address is divided into two parts: an 11-bit row address and an 11-bit column address. The row address is first sent to the DRAM via pins A0:A10 and is latched (strobed) onto the chip by the row-address strobe signal $\overline{\text{RAS}}$. The column address is then sent to the DRAM via A0:A10 and is latched by $\overline{\text{CAS}}$. **(a)** Why do you think the address is split in two in this way? **(b)** Explain how the $\overline{\text{RAS}}$ and $\overline{\text{CAS}}$ control signals eliminate the need for a separate chip select signal.

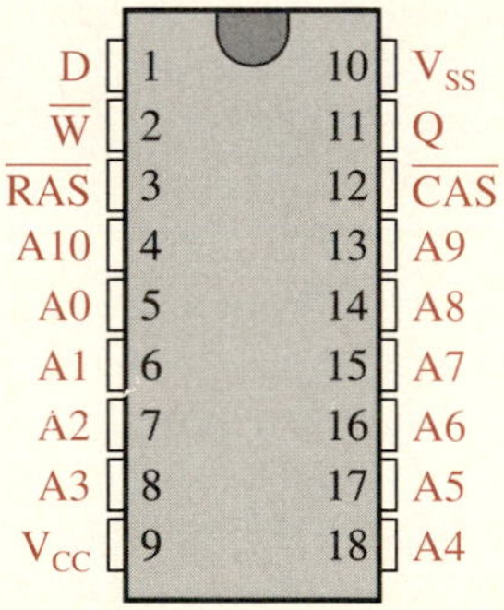

Signal	Pin	Pin	Signal
D	1	10	V_{SS}
$\overline{\text{W}}$	2	11	Q
$\overline{\text{RAS}}$	3	12	$\overline{\text{CAS}}$
A10	4	13	A9
A0	5	14	A8
A1	6	15	A7
A2	7	16	A6
A3	8	17	A5
V_{CC}	9	18	A4

Pins	Function
A0:A10	Address inputs
$\overline{\text{CAS}}$	Column-address strobe
D	Data input
Q	Data output
$\overline{\text{RAS}}$	Row-address strobe
V_{CC}	Power (5 V)
V_{SS}	Ground
$\overline{\text{W}}$	Write enable

FIGURE 9.33 Package and pin assignment used for a 4M-bit DRAM IC.

9.12 Draw a logic diagram that shows how to build a (16M × 2)-bit DRAM, using the 4M-bit DRAM of Figure 9.33 as the basic component. Explain briefly how your design works; in particular, describe its row/column strobe logic.

9.13 RAM chips are sometimes designed with two chip select lines CS_1 and CS_2, both of which must be enabled—that is, $CS_1 \cdot CS_2 = 1$—to initiate a read or write operation. This makes it possible to arrange a memory as a two-dimensional array of ICs, in which each IC has a row address enabled by CS_1 and a column address enabled by CS_2. Explain the advantages of this technique for the design of a 64M × 8 RAM that is to be constructed from 1M × 8 RAM ICs with two CS lines.

Mechanical switches suffer from contact bounce, which requires debouncing hardware or software.

9.14 An inherent problem with mechanical on-off switches is **contact bounce**, which results from the contacts opening and closing rapidly several times as they are being pressed together or pulled apart. This bouncing effect, which is imperceptible to a human user, causes the switch's logical state to change several times over a short period (10 ms or so) before settling at its final value. When switches are used as input devices to a computer, a hardware or software technique to eliminate the effects of contact bounce must be provided. Show how to attach a latch or similar logic device to an on-off transfer switch (Figure 1.6b) in order to filter out the effects of contact bounce; this is termed a **debouncing circuit**.

9.15 Problems arise when two or more keys in certain positions are pressed simultaneously in the switch array of Figure 9.12. Identify one such problem and suggest a possible remedy.

9.16 Apply the ASM methodology of Section 8.7 to the design of a control unit K that implements the scanning routine for reading the 8 × 8 keypad of Figure 9.12. The input to K is a periodic signal x that causes K to scan the keypad and identify any pressed key, assuming one key, at most, is pressed at a time. The control unit then outputs a 16-bit word W that identifies the pressed key or indicates that no key is pressed. This word W is sent to an input port, where it can be read by the host CPU. Give an ASM chart for your design and a block diagram for K. State how you would modify K to handle the situation in which several keys are pressed simultaneously.

9.17 Several seven-segment LEDs are typically combined to form a multidigit output device. To minimize the wiring involved, the seven segments of each digit can be driven from a common 7-bit bus B and the necessary segment signals to each digit D_i are multiplexed over B. Each digit D_i is supplied with a separate select line s_i that allows it to be addressed individually. At any time, only one digit is activated, but if the multiplexing is sufficiently fast, a flicker-free image is obtained from the entire display. Design the register-level hardware needed to multiplex eight seven-segment LEDs and connect them to an output port of a computer that supplies a binary version of the number to be displayed.

9.18 RAMs are considered more reliable than magnetic disk memories of comparable capacity. Explain why this is so, taking into account such factors as power-supply failure, overheating, vibration, and the like.

9.19 A certain magnetic disk memory has an average seek time of 12 ms and its disks rotate at 3,600 revolutions per minute. Each disk contains a total of 200 tracks, each with a capacity of 1M bits. **(a)** What is the worst-case time required to read a block of 1K bits from the memory? **(b)** What is the average time required to read a 1K-bit block? **(c)** How big should each block be if the average access time per block is to be reduced by 20 percent?

The Central Processing Unit

CPU operation; instruction cycle; E- and I-unit design; microprogramming; instruction sets; the 680X0 series; assembly-language programming; stacks.

9.20 Define and briefly explain each of the following terms: CISC, E-unit, interrupt, I-unit, microprogramming, RISC.

9.21 Consider the CPU and main-memory configuration depicted in Figure 9.14. Describe in the manner of Figures 9.15 and 9.16 the main steps involved in fetching and executing the unconditional branch instruction stored at memory address $216A_{16}$.

9.22 Suppose the store instruction at memory address 2168_{16} in Figure 9.14 is replaced by a load instruction that specifies the same register and memory address for its operand. Following the style of Figure 9.16, show the main steps involved in fetching and executing this load instruction.

9.23 What is the smallest number of clock cycles m required by any instruction cycle described by the flowchart of Figure 9.17? Identify a useful instruction that uses the minimum number of clock cycles, and specify the path it follows through the flowchart.

9.24 Using generic register-level components, carry out the logic design of the dual-port register file for the E-unit depicted in Figure 9.18. Assume that there are eight registers in the file and that each is 16 bits wide. As indicated, data can be read from both ports D_1 and D_2, but only port D_1 is used to write into the file.

9.25 Define three types of status or condition code flags found in a typical CPU. For each of these flags, describe a programming task in which testing the state of that flag is a key operation.

9.26 State the function of each of the following components of a microprogrammable I-unit: the control memory, the microprogram counter, and the microinstruction register.

9.27 List the advantages and disadvantages of a microprogrammable I-unit versus a hardwired one, taking into account each of the following factors: ease of design, CPU cost, CPU speed, and ease of testing.

In principle, every computation can be done with only one or two carefully chosen instruction types.

9.28 **(a)** Describe in general terms how integer multiplication and division can be programmed for a CPU whose only arithmetic instructions are add and subtract. **(b)** Consider the strange instruction GLOP a,b which implements the operation AC := $1/(a - b)$, where AC (accumulator) denotes a general-purpose CPU register. Using GLOP only, it is possible to program all the four basic arithmetic operations $a + b$, $a - b$, $a \times b$, and a/b. For example, the fact that $a - b = 1/(1/(a - b) - 0)$ implies that the following two-instruction program implements subtraction:

```
GLOP    a,b
GLOP    AC,0
```

Devise a similar program for addition that uses GLOP only and does not contain a minus sign among its operands.

9.29 Suppose the number of instruction types in the set of Table 9.3 is to be reduced from 12 to six. Which six instructions would you retain, and why?

9.30 The 680X0 instruction set contains the unconditional branch instruction BRA (for "branch always"), which uses the same addressing mode as the conditional branch instructions BEQ and BNE discussed in the text. It also contains a second unconditional branch instruction JMP, which has an immediate operand that serves as the branch address. Thus, "branch to the statement labeled LOOP" can be implemented in an assembly-language program by either BRA LOOP or JMP LOOP. Under what circumstances would you use BRA rather than JMP, and vice versa?

9.31 The 680X0 has a special branch instruction JSR SUB, meaning "jump to subroutine at address SUB." JSR pushes the program counter contents onto a stack using stack pointer register SP and then causes a jump to the instruction at memory location SUB. The operation of JSR is described in 680X0 programming manuals in the following fashion:

$$-(SP) := PC; \quad PC := SUB$$

(a) Show how to simulate JSR using two MOVE instructions, assuming that the latter can have SP and PC as operands (which is not actually the case). **(b)** The subroutine or procedure starting at SUB should have a return from subroutine instruction RTS (with no operands) as its final instruction. Its function is to restore to PC the address saved there earlier by JSR; this instruction should also update SP. Using one or more of the 680X0's MOVE instructions, show how to simulate RTS, again assuming that MOVE can have SP and PC as operands.

9.32 The 680X0 stack shown in Figure 9.25 grows toward the low-address end of M. A stack can easily be arranged to grow in the opposite direction—that is, toward the high-address end of M. Specify the push and pop instructions needed for the latter case.

9.33 Modify the program segment (9.16) to compute the two's-complement rather than the ones'-complement of $M(ADR)$.

9.34 Consider the three-line assembly language program (9.17), which complements a memory word. Assembly of this 680X0 program maps the BEQ CONT statement into four machine-code bytes, the hexadecimal representation of which has the form 6700 $a_1a_2a_3a_4$. What specific values are assigned to the hexadecimal digits $a_1a_2a_3a_4$ if the physical address of the BEQ instruction is A14C?

9.35 Modify the array summation program of Figure 9.23a to compute the sum of 100 32-bit (long) words. Assume the ARRAY start address is $01FF_{16}$ and that the SUM address is unchanged.

9.36 Modify the array summation program of Figure 9.23b to compute the sum of 200 8-bit words. Assume the ARRAY start address is $01FF_{16}$ and that the SUM address is unchanged.

Bit slicing is useful for programming operations that employ very long operands.

9.37 Suppose we want to add two very long n-byte integers, where n is, say, 10 or more. No available integer data type will allow a single add instruction to be used. Instead, we must chop the numbers into pieces on which an available add instruction can act. We then splice the results together to form an n-byte sum. In doing so, we are simply implementing in software a form of the bit-slicing technique discussed in Section 8.4. To support this, the 680X0 (and most other computers) have "add with carry" instructions that form the sum of two numbers plus a carry bit produced by a preceding add instruction and saved in a carry flag flip-flop C. The 680X0 has such an instruction called ADDX (add extended). ADDX may be used with the byte, word, and long-word data formats, but only with the following two addressing modes:

```
ADDX    Di,Dj
ADDX    -(Ai),-(Aj)
```

Using ADDX, design an assembly-language program that treats the entire data array *ARRAY* of Figure 9.23 as a single number *NUM* of length *SIZE* words and performs the increment operation $NUM := NUM + 1$.

9.38 Consider the following very simplified timing model for a 680X0-style processor. Each one-word memory read or write operation takes one time unit. The decoding of each instruction and any execution steps that take place entirely within the CPU also takes one time unit. (We are ignoring floating-point and other slow instructions.) Thus, an instruction like CLR D1 (clear register) takes two time units: one to fetch the instruction and one to decode and execute it. The load instruction MOVE D1,SUM, on the other hand, takes four time units: two to fetch the instruction (which is two words long), one to decode it, and one to load a word from memory into D1. With this timing model, calculate the total execution time for the program of Figure 9.20 when **(a)** $M(\text{ADR}) = 0$, and **(b)** $M(\text{ADR}) \neq 0$.

9.39 Consider again the timing model for 680X0 instructions given in Problem 9.38. **(a)** Using this model, determine the number of time units required to fetch and execute each individual executable instruction of the array-summation program in Figure 9.23a. **(b)** Determine the total execution time for the entire program.

9.40 Repeat Problem 9.39 for the array-summation program of Figure 9.23b.

Computer-Based Systems

Microcontrollers; I/O circuits; I/O interfacing; I/O programming; interrupts; direct memory access (DMA).

9.41 Briefly explain each of the following terms: in-circuit emulation, interrupt-driven I/O, MDS, microcontroller, programmed I/O.

9.42 A microcontroller can often replace electromechanical operations with cheaper and more reliable electronic implementations. It can also enable new functions that cannot be implemented by purely mechanical means. Choose one of the following formerly nonelectronic applications and list some ways in which its functionality has been significantly enhanced by embedded microcontrollers: **(a)** a typewriter; **(b)** a cash register; **(c)** a thermostat for an apartment or house.

9.43 The data register *DR* at a programmable I/O port is sometimes designed so that the CPU can write into it or read from it, independent of the port's current direction (input or output). Suggest a reason why this ability is useful. Redesign the logic of the I/O port of Figure 9.27 to allow *DR* to be accessed by the CPU in this way.

9.44 The 68705 microcontroller contains a built-in, programmable "prescaler" that allows the period of the input (count enable) signal to the timer data register TDR to be set to $2^k T_C$, where $0 \leq k \leq 7$, k is a programmable scale factor, and T_C is the CPU clock period. Using this prescaler only and assuming $T_C = 1$ μs, what is the maximum time period that can be measured by TDR? What is the corresponding resolution?

9.45 Outline a program that will increase the effective length of the 68705 microcontroller's timer data register TDR from 8 to 16 bits. If the extended TDR is decremented every microsecond, what is the maximum time period P that can be measured? If P is required to be at least one week, what is the minimum required length L of TDR (in bits)?

9.46 We require an I/O program P that transfers a block of *SIZE* bytes from an input device D via an I/O port to main memory M, where *SIZE* is a programmable variable. The region in M where the input data are to be stored has the (symbolic) start address *BUFFER*. Assume that new data are made available by D as fast as they can be read into memory by P. **(a)** Code P, using a reasonable approximation to 680X0 assembly language. **(b)** Suppose it takes one time unit to transfer a byte of information between the CPU and an I/O port, or between the CPU and M. Opcode decoding and any execution steps that stay within the CPU also take one time unit. With this simple timing model, estimate the total execution time of your program.

9.47 Consider the contact bounce phenomenon described in Problem 9.14 for the case of a keyboard input device. Assume that the keyboard is configured as in Figure 9.12 and is read by a scanning program executed by the CPU. Sketch an assembly-language program to perform debouncing—that is, to minimize the impact of contact bounce—entirely in software.

9.48 Frequently, a microcontroller is required to take some predetermined action when a "ready" signal F produced by an I/O device changes value. Use the 680X0's test (TST) instruction and other instructions we have encountered to write a short assembly-language program to monitor the status of F and begin execution of a program named *SERV* when F changes from 0 to 1. Suggest an alternative and more efficient way to handle this task.

The successive approximation method of analog-to-digital conversion is both fast and cheap.

9.49 The analog-to-digital converter of Figure 9.31 can be speeded up considerably without changing the interface hardware by replacing the ramp program with one that implements a method called **successive approximation**. The idea is to determine the binary value of V bit by bit, using a "binary search" approach. This may be compared to the way a human searches for a particular word W in a dictionary by first matching the first letter of W, then matching the second letter, and so on. A dictionary search that corresponds to the ramp method is to scan every entry from A to Z, which is clearly much slower. The successive approximation method for analog-to-digital conversion is described concisely in Figure 9.34. Using a reasonable approximation to 680X0 assembly language with plenty of comments, implement this algorithm. How much faster is your program, on the average, than the ramp program of Figure 9.31?

procedure *Successive_approximation*;
begin
$VREF_{n-1}VREF_{n-2} \ldots VREF_0 := 00 \ldots 0$;
for $i := n-1$ **downto** 0 **do**
  begin
  $VREF_i := 1$;
  if $VREF > v$ **then** $VREF_i := 0$;
  $i := i-1$;1
  end;
{$VREF$ = binary approximation to v}
end;

FIGURE 9.34 Analog-to-digital conversion by the successive-approximation method.

9.50 Sketch in assembly language a typical interrupt-service routine that performs a one-word I/O data transfer in response to an interrupt request. Assume that the data are placed in a memory region (an I/O buffer area) called *BUF*, the size and position of which are fixed. What are the advantages and disadvantages of replacing *BUF* with a stack?

BIBLIOGRAPHY

Abramovici, M., M. A. Breuer, and A. D. Friedman. *Digital System Testing and Testable Design.* Potomac, Md.: Computer Science Press, 1990.

Actel Corporation. *ACT Family Field Programmable Gate Array Databook.* Sunnyvale, Calif.: March 1991.

Alfke, P., and I. Larsen, eds. *The TTL Applications Handbook.* Mountain View, Calif.: Fairchild Semiconductor, 1973.

Alford, R. C. *Programmable Logic Designer's Guide.* Indianapolis: Howard W. Sams, 1989.

AT&T Microelectronics. *LP900C Standard-Cell Data Book.* Allentown, Pa., 1991.

Boole, G. *An Investigation of the Laws of Thought.* London: Macmillan, 1854. Reprint. Dover Books, New York, 1954.

Brayton, R. K. et al. *Logic Minimization Algorithms for VLSI Synthesis.* Boston: Kluwer Academic Publishers, 1984.

Brayton, R. K. and R. Rudell. "MIS: Multiple-Level Interactive Logic Optimization System," *IEEE Transactions on Computer-Aided Design,* Vol. CAD-6, pp. 1062-1081, Nov. 1987.

Breeding, K. J. *Digital Design Fundamentals,* 2d ed. Englewood Cliffs, N.J.: Prentice-Hall, 1992.

Breuniger, R. K. and L. E. Schiele. *Designing with Texas Instruments Field-Programmable Logic.* Texas Instruments, Dallas, Tex., 1988.

Brown, F. M. *Boolean Reasoning.* Boston: Kluwer Academic Publishers, 1990.

Burks, A. W. and J. B. Wright. "Theory of Logical Nets," *Proceedings of the IRE,* Vol. 41, pp. 1357-1365, Oct. 1953. Reprint. E. F. Moore, ed., pp. 193-212, 1964.

Cahill, S.J. *The Single-Chip Microcomputer.* Englewood Cliffs, N.J.: Prentice-Hall, 1987.

Caldwell, S. H. *Switching Circuits and Logical Design.* New York: John Wiley & Sons, 1958.

Clare, C. R. *Designing Logic Systems Using State Machines.* New York: McGraw-Hill, 1973.

Coelho, D. R. *The VHDL Handbook.* Boston: Kluwer Academic Publishers, 1989.

Comer, D. J. *Digital Logic and State Machine Design,* 2d. ed. Philadelphia: Saunders, 1990.

Coraor, L. D. *The Student Edition of MICRO-LOGIC II.* Reading, Mass.: Addison-Wesley, 1989.

Cutler, R. B. and S. Muroga. "Derivation of Mimimal Sums for Completely Specified Functions," *IEEE Transactions on Computers,* Vol. C-36, pp. 277-292, March 1987.

Dagenais, M. R., V. K. Agarwal, and N. C. Rumin. "McBOOLE: A New Procedure for Exact Logic Minimization," *IEEE Transactions on Computer-Aided Design,* Vol. CAD-5, pp. 229-238, Jan. 1986. Reprint. A. R. Newton, ed., pp. 75-83, 1987.

Darringer, J. A. et al. "Logic Synthesis Through Local Transformations," *IBM Journal of Research & Development,* Vol. 25, pp. 272-280, July 1981. Reprint. A. R. Newton, ed., 1987, pp. 14-22.

Devadas, S. et al. "MUSTANG: State Assignment of Finite-State Machines Targeting Multi-Level Logic Implementations," *IEEE Transactions on Computer-Aided Design,* Vol. 7, pp. 1290-1300, Dec. 1988.

Dewhurst, C. *LogicWorks User's Guide.* New Westminster, B.C.: Capilano Computing Systems, 1991.

Dietmeyer, D. L. *Logical Design of Digital Systems,* 3d ed. Boston: Allyn and Bacon, 1988.

Eccles, J. C. *The Understanding of the Brain,* 2d ed. New York: McGraw-Hill, 1977.

Fitch, E. C. and J. B. Surjaatmadja. *Introduction to Fluid Logic.* Washington, D.C.: Hemisphere Publishing Corp., 1978.

Fredericksen, T. M. *Intuitive Digital Computer Basics.* New York: McGraw-Hill, 1988.

Friedman, A. D. and P. R. Menon. *Theory and Design of Switching Circuits.* Woodland Hills, Calif.: Computer Science Press, 1975.

Goel, P. "An Implicit Enumeration Algorithm to Generate Tests for Combinational Logic Circuits," *IEEE Transactions on Computers,* Vol. C-30, pp. 215-222, March 1981.

Hamming, R. W. *Coding and Information Theory,* 2d ed. Englewood Cliffs, N.J.: Prentice-Hall, 1986.

Harary, F. *Graph Theory.* Reading, Mass.: Addison-Wesley, 1969.

Harrison, M. A. *Introduction to Switching and Automata Theory.* New York: McGraw-Hill, 1965

Hartmanis, J. and R. E. Stearns. *Algebraic Structure Theory of Sequential Machines.* Englewood Cliffs, N.J.: Prentice-Hall, 1966.

Hayes, J. P. "An Introduction to Switch-Level Modeling," *IEEE Design and Test,* Vol. 4, No. 4, pp. 18-25, Aug. 1987.

Hayes, J. P. *Computer Architecture and Organization,* 2d ed. New York: McGraw-Hill, 1988.

Heath, F. G. "Origins of the Binary Code," *Scientific American,* Vol. 227, No. 2, pp. 76-83, Aug. 1972.

Hitchcock, Sr., R. B. "Timing Verification and the Timing Analysis Program," *Proceedings of the 19th Design Automation Conference,* pp. 594-604, 1988.

Huffman, D. A. "The Synthesis of Sequential Switching Circuits," *Journal of the Franklin Institute,* Vol. 257, pp. 275-303, 1954. Reprint. E. F. Moore, ed., pp. 3-62, 1964.

Huntington, E. V. "Sets of Independent Postulates for the Algebra of Logic," *Transactions of the American Mathematical Society,* Vol. 5, pp. 288-309, 1904.

Huntington, E. V. "New Sets of Independent Postulates for the Algebra of Logic, with Special Reference to Whitehead and Russell's *Principia Mathematica,*" *Transactions of the American Mathematical Society,* Vol. 35, pp. 274-304, 1933.

IEEE (Institute of Electrical and Electronics Engineers) Inc. *Standard Graphic Symbols for Logic Functions: ANSI/IEEE Std 91.* New York, 1984. Reprint. IEEE, 1987.

IEEE Inc. *Standard for Binary Floating-Point Arithmetic. IEEE Std 754.* New York, 1985.

IEEE Inc. *Electrical and Electronics Graphic Symbols and Reference Designations,* 2d ed. New York, 1987.

IEEE Inc. *VHDL Reference Manual. IEEE Std 1076.* New York, 1988.

Johnson, B. W. *Design and Analysis of Fault Tolerant Digital Systems.* Reading, Mass.: Addison-Wesley, 1989.

Kampel, I. *A Practical Introduction to the New Logic Symbols,* 2d ed. London: Butterworths, 1986.

Karnaugh, M. "The Map Method for Synthesis of Combinational Logic Circuits, *Transactions of the AIEE, Part 1*, Vol. 72, pp. 593-599, 1953.

Knuth, D. E. *The Art of Computer Programming,* Vol. 2. Seminumerical Algorithms, Reading, Mass.: Addison-Wesley, 1969.

Langdon, Jr., G. C. *Logic Design: A Review of Theory and Practice.* New York: Academic Press, 1974.

Lindley, P. L. "JK Mystery Solved," *EDN,* p. 30, Aug. 1968.

Liu, Y-C. and G. A. Gibson. *Microcomputer Systems: The 8086/8088 Family,* 2d ed. Englewood Cliffs, N.J.: Prentice-Hall, 1986.

Maunder, C. M. and R. E. Tulloss. *The Test Access Port and Boundary-Scan Architecture.* Los Alamitos, Calif.: IEEE Computer Society Press, 1991.

McCluskey, E. J. "Minimization of Boolean Functions," *Bell System Technical Journal,* Vol. 35, pp. 1417-1444, 1956.

McCluskey, E. J. *Logic Design Principles.* Englewood Cliffs, N.J.: Prentice-Hall, 1986.

Mealy, G. H. "A Method for Synthesizing Sequential Circuits," *Bell System Technical Journal,* Vol. 34, pp. 1045-1079, 1955.

Mimar, T. *Programming and Designing with the 68000 Family.* Englewood Cliffs, N.J.: Prentice-Hall, 1991.

Moore, E. F. "Gedanken Experiments on Sequential Machines," *Automata Studies,* Annals of Mathematics Studies, No. 34. Princeton, N.J.: Princeton Univ. Press, pp. 129-153, 1956.

Moore, E. F., ed. *Sequential Machines: Selected Papers.* Reading, Mass.: Addison-Wesley, 1964.

Motorola Inc. *M68000 32-Bit Microprocessor Programmer's Reference Manual,* 4th ed. Englewood Cliffs, N.J.: Prentice-Hall, 1984.

Motorola Inc. *MC68881/MC68882 Floating-Point Coprocessor User's Manual.* Englewood Cliffs, N.J.: Prentice-Hall, 1987.

Mowle, F. J. *A Systematic Approach to Digital Logic Design.* Reading, Mass.: Addison-Wesley, 1976.

Muroga, S. *Logic Design and Switching Theory.* New York: Wiley-Interscience, 1979.

Newton, A. R., ed. *Logic Synthesis for Integrated Circuit Design.* New York: IEEE Press, 1987.

Peatman, J. B. *Design with Microcontrollers.* New York: McGraw-Hill, 1988.

Petrick, S. R. "On the Minimization of Boolean Functions," *Proceedings of the International Conference on Information Processing.* Paris: Unesco, pp. 422-423, 1960.

Quine, W. V. "A Way to Simplify Truth Functions," *American Mathematical Monthly,* Vol. 62, pp. 627-631, 1955.

Randell, B., ed. *The Origins of Digital Computers: Selected Papers,* 3d ed. Berlin: Springer-Verlag, 1982.

Robertson, J. E. "Two's Complement Multiplication in Binary Parallel Computers," *IRE Transactions on Electronic Computers,* Vol. EC-4, pp. 118-119, Sept. 1955.

Roth, J. P. "Diagnosis of Automata Failures: A Calculus and a Method," *IBM Journal of Research & Development,* Vol. 10, pp. 278-291, July 1966.

Sapiro, S. and R. J. Smith. *Handbook of Design Automation.* Englewood Cliffs, N.J. Prentice-Hall, 1986.

Savage, J. E., S. Magidson and A. M. Stein. *The Mystical Machine: Issues and Ideas in Computing.* Reading, Mass.: Addison-Wesley, 1986.

Scott, N. R. *Computer Number Systems and Arithmetic.* Englewood Cliffs, N.J.: Prentice-Hall, 1985.

Seidensticker, R. B. *The Well-Tempered Digital Design.* Reading, Mass.: Addison-Wesley, 1986.

Shannon, C. E. "A Symbolic Analysis of Relay and Switching Circuits," *Transactions of the AIEE,* Vol. 57, pp. 713-723, 1938.

Siewiorek, D. P., C. G. Bell and A. Newell. *Computer Structures: Readings and Examples.* New York: McGraw-Hill, 1982.

Simon, H. A. *The Sciences of the Artificial,* 2d ed. Cambridge, Mass.: MIT Press, 1981.

Stevens, A. K. *Introduction to Component Testing.* Reading, Mass.: Addison-Wesley, 1986.

Tabak, D. *Advanced Microprocessors.* New York: McGraw-Hill, 1991.

Texas Instruments Inc. *ALS/AS Logic Data Book.* Dallas, Tex., 1986.

Texas Instruments Inc. *F Logic Data Book.* Dallas, Tex., 1987.

Texas Instruments Inc. *High-Speed CMOS Logic Data Book.* Dallas, Tex., 1988a.

Texas Instruments Inc. *TTL Logic Data Book.* Dallas, Tex., 1988b.

Thomas, D. E. and P. R. Moorby. *The Verilog Hardware Description Language.* Boston: Kluwer Academic Publishers, 1991.

Unger, S. H. *Asynchronous Sequential Switching Circuits.* New York: Wiley-Interscience, 1969.

Wakerly, J. F. *Digital Design Principles and Practices.* Englewood Cliffs, N. J.: Prentice-Hall, 1990.

Waser, S. and M. J. Flynn. *Introduction to Arithmetic for Digital Systems Designers.* New York: Holt, Rinehart and Winston, 1982

Weste, N. and K. Eshragian. *Principles of CMOS VLSI Design.* Reading, Mass.: Addison-Wesley, 1985.

Williams, M. J. Y. and J. B. Angell. "Enhancing Testability of Large-Scale Circuits via Test Points and Additional Logic," *IEEE Transactions on Computers,* Vol. C-22, pp. 46-60, Jan. 1973.

INDEX

74AS ICs, 399
74F10 gates, 403-404
74HC ICs, 240, 403
74HC153 multiplexer, 360
74LS ICs, 240
74LS74A latch, 432, 440-441
74LS75 latch, 412, 416-417, 471
74LS107A flip-flop, 437-438, 516-518
74X IC series, 238-239, 624-625, 634-636
74X83 adder, 372, 647-648
74X85 comparator, 623-624
74X148 encoder, 645-647
74X150 multiplexer, 211-213, 356-357, 362
74X154 decoder/demultiplexer, 365-366
74X157 multiplexer, 687
74X166 shift register, 629-631
74X181 ALU, 599-609, 647
74X182 carry-lookahead generator, 647-648
74X273 register, 627-628, 686
74X283 adder, 372
680X0 microprocessor family, 756-759
  address assignments for, 760-761
  addressing modes in, 763-769
  data formats for, 761-763
  machine-language code for, 759-760
  programming model for, 770-772
  stacks with, 773-775

ABEL hardware description language, 255
Absolutely inessential prime terms, 299
Absorption laws, 172-174
Abstracted subsystems, 17
Access time of memory, 33, 396, 729, 740
Accumulator-shifter circuit, 640-642
Accumulators, 636-637, 748
Accuracy
  in number conversion, 64-65
  of digital systems, 4-5
  of numbers, 56-57, 107-109
ACT-series FPGAs, 625-626
ADD instruction, 755
Adder-subtracter circuits, 84-85, 373-374, 615-616
Adders
  in adder-subtracter units, 373-374
  for BCD numbers, 98
  for binary numbers, 11-12
  with carry lookahead, 369-372
  combinational, 25, 366-375, 496-498
  complementary circuits for, 140-141
  complexity of, 372
  fan-in and fan-out in, 157, 372
  fast, 368-369, 372
  functional decomposition in, 374-375
  gates arrays for, 244-246
  with JK flip-flops, 538-539
  logic gates in, 153-154
  multiplexers for, 619-620
  n-bit, 366
  PLAs for, 252-254, 260
  ripple-carry, 71, 367-368
  sequential, 26, 495-498
  serial, 26, 495-498, 538
  SSI ICs for, 241-243
  tests for, 575-576
  types of, 26
Addition
  of arrays of numbers, 766-769
  of BCD numbers, 98-100
  of floating-point numbers, 111-113
  in modular arithmetic, 75-77
  OR operation as, 145
  of signed-magnitude numbers, 79
  of two's-complement numbers, 83-85
  of unsigned binary numbers, 67-71
ADDn procedure, 69-71
Address decoders, 225, 644-645
Address registers, 743, 770
Addresses
  assignments for, 760-761
  computation of, 76-77
  in instructions, 724, 745
  with multiplexers, 354
  in tables, 74
Addressing modes, 763-769
ADDsm procedure, 79-80
Adjacent terms in minimization, 282-283, 290, 319
  after statement in VHDL, 608
Algebraic minimization, 282-284
Algebras, 162. *See also* Boolean algebra
  function, 184-185
  set, 168-171
  switching, 166-167
Algorithmic state machines (ASMs), 674-681. *See also* ASM charts
Algorithms, 27, 65-66
Alphanumeric data, 102
Amplification. *See* Buffers
Analog systems, 1-6
Analog-to-digital conversion, 4-5
  direct method, 783-784
  ramp method, 784-785
  successive approximation method, 798
Analysis of systems, 16
Analytical Engine, 718
AND circuits and gates, 132, 144-145
  behavior of, 134
  in complete gate sets, 213-215
  SSL faults in, 561
AND instruction, 756
AND operator, 179
AND-OR circuits, 211-213, 216-217
AND-OR-INVERT circuits, 274
Antifuses in PLDs, 248
Application-specific integrated circuits (ASICs), 247, 285, 721
Arabic numerals, 51
Architecture
  design level for, 18-20
  of systems, 715-721
Architecture specifications in VHDL, 606-608
Arithmetic components, 622-624
Arithmetic-logic circuits, VHDL language for, 606-609
Arithmetic-logic unit/function generator, 599-605
Arithmetic-logic units (ALUs), 373
Arithmetic shifting, 705

Arrays, addressing of, 765-769
ASCII (American Standard Code for Information Interchange) code, 102-103
ASICs (application-specific integrated circuits), 247, 285, 721
ASM charts, 675-677
  conditional outputs with, 677-679
  for datapath design, 685
  interpretation of, 679-681
  for shift-and-add multipliers, 693-694
ASMs (algorithmic state machines), 674-681. *See also* ASM charts
Assemblers, 724
Assembly language, 723-724
Assembly-language directives, 760
Assembly listings, 759
Assignments in VHDL, 607, 610. *See also* State assignments
Associative laws, 172-173
Asynchronous circuits
  applications for, 459
  circuit analysis for, 463-464
  delay considerations in, 460-462
  design issues in, 463-476
  edge counters, 472-473
  hazards in, 453-457
  Huffman model for, 457-463
  speed considerations in, 459-460
  stable and unstable states in, 464-467
  vs. synchronous, 457-459
Asynchronous inputs for flip-flops, 431, 437
Asynchronous input signals, 683-684
ATE (automatic test equipment), 557, 579
ATPG (automatic test pattern generation) programs, 566
Attenuator-well circuits, 394-396
Attenuators, 227-230, 735
Audio recordings, 3-4
Autoindexing, 769
Automatic test equipment (ATE), 557, 579
Automatic test pattern generation (ATPG) programs, 566
Axioms for Boolean algebra, 162-164

Babbage, Charles, 718
Backtracking, 565, 570, 572
Bacon, Francis, 88-89
Bandwidth, 33
Bar codes, 92-93
Barrel shifters, 706
Base-2^k numbers, 94-96
Bases
  of floating-point numbers, 108
  of number systems, 53, 94-96
BCD. *See* Binary-coded decimal (BCD) numbers
Behavior of systems, 14-16. *See also* State behavior
BEQ instruction, 757
Biasing of floating-point numbers, 109-110
Bidirectional devices, 406
Bidirectional terminals, 130
Biformed alphabets, 88-89
Binary-coded decimal (BCD) numbers, 96-97, 101
  addition of, 98-100
  vs. binary, 97
  subtraction of, 99-100
Binary codes
  base-2^k numbers, 94-96
  binary-coded decimal numbers, 96-101
  for characters, 101-103
  error-detecting, 104-107
  for floating-point numbers, 107-113
  Gray codes, 91-92
  for optical encoders, 89-91
  for secret writing, 88-89
  universal product code, 92-93
Binary counters, 634
Binary encoders, 621
Binary point, 55-56
Binary systems and numbers, 6-8, 66
  adders and addition in, 11-12, 67-71
  advantages of, 8
  and base-2^k numbers, 94-96
  vs. BCD, 97
  bits in, 6-7
  converting to decimal, 57-61
  converting decimal to, 61-66
  for decision making, 10
  division of, 73
  for logical models, 11
  multiplication of, 73-74
  for reliability, 10
  signed, 77-87
  subtraction of, 71-73
  switches, 9
  in tabular method, 319
  unsigned, 67-71
  word size for, 55
BINDECi procedure, 59
BINDECm procedure, 60-61
Bipolar transistors, 232
Biquinary code, 101
BIST (built-in self-testing), 557
Bistable circuits, 408
Bit-field data, 762
Bit slices, 604, 639-642, 729
Bits (binary digits), 6-7
  check, 105-106
  hidden, 110
  sign, 82
  significant, 55-57
  status, 751
  in VHDL, 606, 608
Black boxes, 17, 603
Block diagrams, 13-14, 16
Blocks
  in ASM charts, 679
  control, in graphic symbols, 604
BNE instruction, 758, 785
Boole, George, 160
Boolean algebra
  axioms for, 162-163
  circuits in, 181-184
  expressions in, 164, 177-184
  function algebra, 184-185
  functions in, 177-178
  gate-level interpretation of, 165-166
  Huntington's postulates for, 163-164
  minimization in. *See* Minimal forms and minimization
  vs. number algebra, 164
  origins of, 160-161
  set algebra, 168-171
  switch-level interpretation of, 164-165
  switching algebra, 166-167
  theorems and proofs for, 171-177
  time-dependent, 399
Boolean rings, 194
Borrow digits in subtraction, 71, 373
Borrow-lookahead subtracters, 373
Bottom-up design, 24
Boundary scan, 584-585
Branch addresses, 77
Branching (branch-and-bound) method, 324-326
Branching instructions, 725, 747-748, 756
Breadboards, 41
Breakage, faults from, 557
Bridge circuit, 188
Buffering of input signals, 361-362
Buffers, 138
  CMOS, 234
  complementary circuits for, 138-139
  for demultiplexers, 361-362
  for DRAMs, 396
  for multiplexers, 356
  for pipelines, 637
  single-input gates, as, 143
  tri-state, 263
Bugs in design, 30
Built-in self-testing (BIST), 557
Burglar alarm circuit, 134-136
Buses
  for interconnection, 614-615
  system, 717
  tri-state logic for, 362-364

CAD. *See* Computer-aided design (CAD)
Calculators, 718
CALL instruction, 756
Canonical forms, 202, 279-280
  disjunctive, 204-205
  product-of-maxterms, 206
  sum-of-minterms, 204-205
Capacitors
  in DRAMs, 395-396
  as storage devices, 392-393
Carries
  in adders, 26, 67, 71, 367-372
  with BCD numbers, 99

Carry lookahead
adders with, 369-372
generators for, 647
case statements in HDL, 255, 608, 617
Cells
in FPGAs, 266-267
in Karnaugh maps, 289-295
Central processing units (CPUs), 29, 716, 740
execution units in, 748-751
functions of, 721-723
instruction cycles with, 741-748
instruction units in, 751-754
Character codes, 101-103
Character conversions, ROM for, 259
Characteristic equations for flip-flops, 444-445, 507-509
Characteristic of floating-point numbers, 110
Check bits, 105-106
Circuits. *See also* Systems
in Boolean algebra, 181-184
for burglar alarm, 134-136
combinational. *See* Combinational circuits and logic
complementary, 136-141
depth of, 160
sequential. *See* Sequential circuits and logic
series-parallel, 136
switches for, 125-127
switching, 131-136
CISCs (complex instruction set computers), 754-755
Classes
equivalence, of states, 553-555
equivalent fault, 563
Classical method for sequential circuit synthesis, 528-529
excitation tables in, 537-539
for recognizers, 532-536
state assignments in, 535-537
state behavior in, 539-546
state-table specification in, 529-532
Clear input for flip-flops, 431, 437
Clock arithmetic, 76-77
Clock cycles, 431
Clock-line gating technique, 629
Clock periods, 431
Clock phases, 441
Clock skew, 471, 512
Clocked SR latches, 415
Clocks and clock signals, 424-425
as delay sources, 514-515
design of, 516-519
digital and analog, 2-3
duty cycles of, 494-495
for JK flip-flops, 436-437, 441-443
methods for, 426-427
and path delay constraints, 515-516
for primary output control, 512-514
propagation delays in, 429
purpose of, 511-512
for sequential circuits, 27, 511-519
and synchronous operation, 431
Closed sets, 75
Closure property, 162-163
CMOS (complementary MOS) technology, 137. *See also* Complementary circuits.
compatibility of, 240
D flip-flops in, 428-430
logic in, 234
for multiplexers, 359-360
and *n*MOS logic, 236-237
propagation delays in, 403
rise and fall times in, 402-403
transmission gates in, 235-236, 483
Code words, 104, 319
Codes
converting, ROMs for, 259
numbers as. *See* Binary codes
for secret writing, 88-89
Column dominance in tabular method, 323-324
Combinational circuits and logic, 24-25
for adders and subtracters, 25, 366-375, 496-498
in asynchronous circuits, 474-476
complete gate sets for, 213-219
for data transfer logic, 353-366
heuristic methods for, 342-353
and IC implementation, 238-246
Karnaugh maps for, 289-300
logic gates for, 152-154
minimal forms in, 279-289
minterms in, 202-203
MOS circuits, 232-237
for multiple-output circuits, 330-342
non-well-formed, 201-202
product-of-maxterms form, 206-208
product-of-sums form, 211
programmable, 246-256
for programmable logic devices, 260-266
propagation delays in, 515
for read-only memories, 257-260
vs. sequential, 491-498
sum-of-minterms form, 203-208
sum-of-products form, 209-211
tabular method, 316-330
tests for, 570-572
for traffic light problem, 525-526
two-level forms, 208-209
well-formed, 199-201
wired logic, 219-231
Combinational components, register-level arithmetic, 622-624
buses, 614-615
decoders, 620-621
encoders, 621-622
multiplexers, 616-620
programmable logic devices, 625-626
word gates, 615-616
Combinational functions, 177-185, 199
Comments in assembly language, 758
Commutative laws, 164
Commutativity, 162-164
COMP instruction, 785
Compact disk systems, 3-4
Comparators
magnitude, 184, 623-624
voltage, 783-784
Compatibility
of component families, 20
in design, 34
of IC families, 150, 240
Compilers, 724
Complementary circuits
for buffering, 138-139
vs. current-switching, 136-137
for EXCLUSIVE-OR gate, 141
for inverters, 138
for NAND gate, 139-141
Complementation, 131, 133
Complementers, 84. *See also* Inverters.
Complements of sets, 169
Complete gate sets, 213-219
Complete sum expressions, 385
Complex instruction set computers (CISCs), 754-755
Complexity management in hierarchical design, 22-23
Components, 13
compatible, 20
defective, 557
failures in, 556-557
in hierarchical design, 22-23
register-level, 613, 685-688
reliable, 425
as systems, 17
Computer-aided design (CAD), 16-17
hardware description languages for, 605, 674
heuristic methods in, 342-353
in logic minimization, 317-318
for multiple-output circuits, 330-342
for PLDs, 255-256
schematic capture, 38-39
simulation, 40-41, 316-317, 405
for state assignment, 537
for synthesis, 42
for tabular method, 316-330
tools for, 36-37
Computers. *See* Digital systems; Systems
Condition codes, 751
Conditional instructions, 756
Conditional outputs with ASMs, 676-679
Congruent modulo *m*, 118
Connection function, 222
Connections, 13
faults in, 557
minimization of, 338-342
parallel, 132

propagation delays in, 399
series, 131
for switches, 130
Consensus theorem, 385
Consistent postulates, 164
Contact bounce, 487-488, 793
Continuous data, 1-6
Contradiction, proof by, 175
Control blocks, 604
Control circuits, 450, 520
Control instructions, 29
Control memory (control store), 668-671, 752
Control points for tests, 578-579
Control programs, 673
Control units, 609-611
design of, 653-654, 688-698
for I/O, 717
and Moore machines, 663-666
one-hot method for, 656-664
programmable, 668-673
state descriptions for, 655-656
structure of, 652-653
timing for, 681-684
Control variables of switches, 127
Controllability problem, 565, 581
Controllers. *See* Control units.
Conversion
analog to digital, 4-5, 783-785, 798
of binary and base-2^k numbers, 95-96
of binary and decimal numbers, 57-66
of codes, ROMs for, 259
of Moore and Mealy machines, 666-667
data shift registers for, 632
Coprocessors, 109, 772
Correctness
as design goal, 30
simulation programs for, 316
Costs, 30-32
of gates, 154, 285
with hierarchical design, 22
and IC area, 243
and minimal forms, 285
with multiple-output circuits, 330-332
and propagation delay, 32
of testing, 556, 558
Counter-based multipliers
control units for, 688-691
one-hot design for, 660-664
Counters
binary, 634
decade, 634
decimal, 486
edge, 472-473
Johnson, 451-453
ring, 450
ripple, 448-450
testing of, 580-581
up-down, 634
Cover tables, 321-324
Covering
of faults, 560
heuristic methods for, 343-345
Petrick's method, 300-302
CPUs. *See* Central processing units (CPUs)
Crashes, 556
Critical paths, 368
Critical races
in asynchronous circuits, 467
in edge counters, 474
in Huffman model, 462
Cross-coupled gates, 409
Crosspoint switches in PLDs, 248
Crosstalk, 104
CUPL hardware description language, 255
Current-switching circuits vs. complementary, 136-137
Custom design projects, 24
Cycles
instruction, 722-723, 741-748
on Karnaugh maps, 303

D algorithm, 567-572
D flip-flops, 428-430, 440-441
excitation table for, 538
in registers, 446-447
in stack controller, 658
state behavior of, 504-507
two-phase, 442-443
D frontier, 569-570
D latches, 412-413
hazard-free, 468-472
operation of, 413-414
from SR latches, 415
D-type shift registers, 451-452
DALG procedure, 568-572
Data
formats for, 761-763
at register level, 609-611
Data books for ICs, 239
Data-flow representations, 609
Data input with D latches, 413
Data latches. *See* D latches
Data lockout triggering method, 484
Data-processing circuits, 520
Data-processing instructions, 29, 726
Data registers, 770
Data selectors. *See* Multiplexers (muxes)
Data transfer logic
instructions for, 29, 726
multiplexers and demultiplexers for, 354-366
shared data paths, 353-354
Data variables of switches, 127
Datapath circuits, 520
Datapath units, 602, 609-610, 685
adder expansion in, 647-651
bit slicing in, 639-642
components for, 685-688
decoder expansion in, 644-645
encoder expansion in, 645-647
multifunction, 649-651
multiplexer expansion in, 642-644
scaling of, 638
structure of, 636-638
timing for, 681-684
De Morgan's laws, 172-174
Dead states, 531
Debouncing circuits, 487-488, 793
Debugging, 30
Decade counters, 634
DECBINf procedure, 63-66
DECBINi procedure, 62
Decimal counters, 486
Decimal notation, 208
Decimal numbers, 7, 51-52
BCD conversion, 97
binary conversion, 57-66
Decimal point, 52
Decision boxes for ASMs, 675-676
Decision making, 10, 756
Decoders, 364, 730
address, 225, 644-645
design of, 364-366, 620-621
expansion of, 620-621, 644-645
Decoding of serial data, 507, 590
Decomposition, functional, 374-375
Defective parts, 557
Degenerate functions, 185
Delay circuits. *See* D latches
Delay elements, 159-160, 399
Delay parameters for latches, 417-419
Delays, 397
in asynchronous circuits, 460-462
and clocks, 429, 511-512, 514-515
in design, 32-33
in flip-flops, 436, 515-517
hazards from, 453-457
in Huffman model, 461-462
inertial, 404-405, 481
in latches, 416-417
in logic gates, 158-159, 399-402
propagation, 158
in ripple-carry adders, 368
rise and fall, 401-403
in ROMs, 258
in signal storage, 394-397
and signal waveforms, 398
simulation of, 405
in synchronous circuits, 476
timing charts for, 316
types of, 399
uncertain, 402-403
Demultiplexers (demuxes), 354
design of, 364-366
Depletion-mode MOS transistors, 237
Depth of circuits, 160
Descriptions of systems, 11, 15, 39-41
Descriptive methods for registers, 604-605
Design, 16
of asynchronous circuits, 463-476
of clock signals, 516-519
compatibility in, 34

computer-aided. *See* Computer-aided design (CAD)
of control units, 653-654, 688-698
cost goals in, 30-32
of datapath units, 685-688
of demultiplexers, 360
of flip-flops, 434-443
of ICs, 241-243
levels in, 17-24
of logic gates, 154-155
of memory, 729-732
of microcontrollers, 776-777
of multiplexers, 355-356
of multipliers, 340-342, 660-664, 691-698
performance goals in, 32-33
power consumption goals in, 34
of programmable controllers, 669-673
of recognizers, 532-536
reliability goals in, 34-35
of sequential circuits, 463-476, 519-528
for testability (DFT), 35, 576-585
trade-offs in, 35-36
verification of, 183-184, 316, 526-528
Design errors, 23, 30, 556
Design reuse, 375
DesignWorks simulator, 316, 526-528
Destructive readout property, 480
Difference digits in subtraction, 71
Digital systems, 1. *See also* Systems
vs. analog, 1-6
binary. *See* Binary systems and numbers
clocks, 2-3
computers, 28-29
latches for, 423-424
quantization in, 5-6
structure and behavior of, 13-17
Digitization, 5-6
Digits, 2, 51
Diminished radix-complement (DRC) codes, 85-86
Direct addressing mode, 763
Direct method of analog-to-digital conversion, 783-784
Direct-polarity indication, 149
Directives, assembly-language, 760
Disable control lines, 224
Disabled decoders, 365
Disabled outputs, 224
Discrete data. *See* Digital systems
Discretionary wiring in PLDs, 248
Disks
compact, 3-4
magnetic, 738-740
Display panels, 737
Distributive law, 164
Distributivity, 162-164, 170-171
Division
by shifting, 632
of two's-complement numbers, 84-85
of unsigned binary numbers, 73
Documentation of systems, 40
Don't care values, 308-309
in JK flip-flops, 537-538
minimization with, 309-313
Double-precision number systems, 107
DRAM (dynamic random-access memory), 28-29, 716-717, 726-728
capacity of, 719-720
destructive readout property of, 480
operation of, 395-397
DRC (diminished radix-complement) codes, 85-86
Driven terminals, 222
Driving, strength in, 227
Dual expressions, 180
Dual-port register files, 749
Dual-ported memory, 629
Dual theorems, 181
Duality, principle of, 180-181
Duals of Boolean functions, 197
Duodecimal number system, 116
Duty cycle, 494-495
Dynamic flip-flops, 443
Dynamic hazards, 455-456
Dynamic input symbol (>), 427, 436
Dynamic random-access memory (DRAM), 28-29, 716-717, 726-728
capacity of, 719-720
destructive readout property of, 480
operation of, 395-397

E-units (execution units), 721, 740, 748-751
E3 (excess-3) code, 100-101
EBCDIC (Extended Binary-Coded Decimal Interchange Code), 102-103
Edge counter, 472-473
Edge-triggered flip-flops, 427-430, 432-440, 501
Editors in computer-aided design, 36, 39
Effective address, 764-765
8421 decimal code, 97
Electric circuits, switches in, 125-126
Electromagnetic interference (EMI), 104
Electromechanical relays, 128
Embedded computers. *See* Microcontrollers
Empty set, 169
Emulation, in-circuit, 788
Enable signals, 224
Enable times of latches, 418-419
Encoders
binary, 621-622
optical, 89-91
expansion of, 645-647, 707
Engineering costs, 30
Enhancement-mode MOS transistors, 237
ENIAC computer, 718
entity statement in VHDL, 606
Environment factors, faults from, 557-558
EOR (exclusive OR) instruction, 758
EQSTATE procedure, 551-552, 554-555
EQU directive, 760, 769
Equivalence classes, 553-555
Equivalence operator, 271
Equivalent Boolean expressions, 177-179
Equivalent faults and fault classes, 562-563
Equivalent states, 509-510, 549-555
Error blocking, 568
Error correction, 107
Error-detecting codes, 101, 104
circuits for, 106-107, 218
parity, 104-106
Errors, 559. *See also* Faults
in analog estimates, 2
in design, 23, 30, 556
and design verification, 183-184, 316
from noise, 10
from overflow, 70-71
simulation programs for, 316
U value from, 221, 223
ESPRESSO program, 317, 546-547
Esential hazards, 475-476, 489
Essential prime implicants, 296-299, 322
Essential tests, 560
Even parity, 105, 148
Excess-3 (E3) code, 100-101
Excitation tables, 474, 524-525, 537-539
EXCLUSIVE-OR circuits and gates, 20, 147-148
CMOS, 236
complementary circuits for, 141
multilevel, 347-348
sum-of-minterms form of, 205-206
Execution of instructions, 741-745
Execution units (E-units), 721, 740, 748-751
Exhaustive testing, 560
Exponentiation in binary system, 58
Exponents of floating-point numbers, 108-110
Expressions
Boolean, 164, 177-184
logic, 11, 134
Extended Binary-Coded Decimal Inter-change Code (EBCDIC), 102-103

Factoring in multilevel design, 348-350
Factory-programmed PLDs, 247
FADD procedure, 112
Fall delay, 401-403
Fall time, 398, 401-402
False paths, 481
Family selection in ICs, 239-240
Fan-in and fan-out
in adders, 157, 372
in logic gates, 155-157
in minimal forms, 286
in multiplexers, 356, 361
with two-level forms, 208-209
Fast adders, 368-369, 372
Fault coverage in testability goals, 577
Fault simulation, 577
Fault tables, 559-560
Faults, 556
indistinguishable, 562-563

models of, 558-559
sources of, 557-558
stuck, 221, 223, 559-560
tests for, 560-561
undetectable, 563
Feedback, 406
in Huffman model, 461-462
with latches, 424
in non-WF circuits, 201-202
in sequential circuits, 27
in synchronous circuits, 476
Feedback loops, 406
Feedforward circuits, 202, 406
Fetch cycles, 723
Fetching instructions, 727, 741-745
Field-programmable gate arrays (FPGAs), 266-267, 625-626
Field-programmable PLDs, 247-248, 261-263
Finite-state machines, 27, 546-549, 652-653
Firmware, 753
Fixed-point numbers, 57
Fixed word size, 55
Flags, 106, 708, 751
Flip-flops, 424-425
behavior of, 443-453
vs. bistable circuits, 408
characteristics of, 444-445, 507-509
clocks with, 512
D. *See* D flip-flops
delays with, 436, 515-517
design of, 434-443
initialization of, 431-432, 437
JK. *See* JK flip-flops
noise sensitivity of, 427-428, 501
set-dominant, 486
state behavior of, 443-444, 504-509
as synchronizers, 683-684
synchronous operation of, 431
timing constraints on, 430
toggle, 485
Floating-point numbers, 57
addition of, 111-113
IEEE 754 standard, 110-113
for multiple precision, 107
pipelines for, 637
representation of, 108-111
Floating-point processors, 109, 772
Floating-point registers, 772
Floating state, 220
Flow tables, 465
Flowcharts, 65-66
ASM charts, 676-677
in register-level design, 611-612
Formal descriptions of systems, 15
Formal proofs, 174-177
Formal specifications, 65-66, 510
Formats for data, 761-763
Fractions, 54
conversions with, 60-64
in decimal system, 52
radix point in, 56
word size for, 55
Frameworks for computer-aided design, 37
Full-adder equations, 367
Full adders, 26, 71, 367
design of, 153-157
gates array as, 244-246
multiplexer as, 619-620
NAND, 154-155
operation of, 67-69, 367
PLA as, 252-254, 260
SSI ICs for, 241-243
Full custom design projects, 24
Full subtraction, 71, 372-373
Fully parenthesized WF (FPWF) Boolean expressions, 179
Function algebra, Boolean, 184-185
Function generators, multiplexers as, 617-620
Function hazards, 457
Function tables, 196
Functional correctness
as design goal, 30
simulation programs for, 316
Functional decomposition, 374-375
Functionally complete gate sets, 213
Functions, 15, 177-178
Fundamental mode assumption in Huffman model, 462
Fuse maps, 256
Fuses in PLDs, 248

Gate arrays, 244-246
Gate delays, 399-402
Gate-level interpretation of Boolean algebra, 165-166
Gate-level testing, 558-559
Gated SR latches, 415
Gates. *See* Logic gates
General-purpose digital computers, 28-29
Generate function for carry lookahead adders, 369
Glitches
and clocks, 476, 511-512
and edge-triggered flip-flops, 501
from hazards, 453-457
in output signals, 513-514
sources of, 427-428
Goel, Prabhakar, 572
Graphics boards, cost example with, 31
Gray codes, 91-92, 120, 291
Gray, Frank, 91
Greedy techniques, 344
GREEDYCOV procedure, 343-346, 564
Ground in digital circuits, 150-151, 238
Guard bands in universal product code, 92

Half adders, 26, 67
Hang-up states, 531
Hardware, 28
Hardware description languages (HDLs), 40, 605-609
for CAD, 605, 674
for PLDs, 255
Hardwired systems, 28
control units, 668, 673
instruction units, 751-754
Hazard-free D latches, 468-472
Hazards
control of, 456-457
dynamic, 455-456
essential, 475-476, 489
and redundant circuits, 563*n*
static, 453-455
Heuristic methods, 42, 342
for covering, 343-345
factoring, 347-348
knowledge-based transformations, 350-351
local transformations, 351-353
for multilevel design, 347-350
nonoptimal solutions with, 346
Hexadecimal number system, 54
in assembly language, 758
and binary numbers, 94-96
Hidden bits in floating-point numbers, 110
Hierarchical design, 17-19
advantages of, 22-24
levels in, 18-22
High-impedance state Z, 220-221, 223
Hold times
of flip-flops, 430
of latches, 418-419
Hollerith, Herman, 718
Homing sequences, 596
Huffman design method, 457-464, 467-468, 474
Huntington's postulates, 163-164
Hybrid systems, 4
Hydraulic circuits, switches in, 126-127

I-units (instruction units), 721, 740, 751-754, 786-787 ICs. *See* Integrated circuits (ICs)
Ideal switches, 129-130
Idempotence theorem, 172-173, 176
Identity function, 138, 143
IDENTITY gate, 190, 219
IEEE 754 standard floating-point number format, 110-111
if-then-else statements in VHDL, 608, 611, 621
Immediate addressing mode, 763
Immediate operands, 757
Implementation of functions, 164
Implicants, 287. *See also* Prime implicants
Implicates, 288-289. *See also* Prime implicates
Implication charts in state minimization, 551
Implication in DALG procedure, 569
In-circuit emulation, 788
Incompletely specified functions, 308-313
Independent postulates, 164
Index registers, 765

Indexed addressing mode, 764-769
Indirect addressing mode, 764
Indistinguishable faults, 562-563
Indistinguishable states, 549-555
Inertial propagation delays, 404-405, 481
Inessential prime terms, 299
Infinite numbers, floating-point representation of, 111
Infinite-state machines, 546-549
Informal proofs, 172-174
Information signals, 130
Initialization
- of CPUs, 745
- in DALG procedure, 569
- of flip-flops, 431-432, 437

Input buffers for demultiplexers, 361-362
Input logic, 492-493
Input ports, 734
Input signals. *See* State behavior
Input terminals, 130
Input timing, asynchronous circuits for, 459
Input variables
- of latches, 420
- in sequential circuit analysis, 494

Input/output (I/O) operations and subsystems, 29, 717, 732-733
- control units, 717
- devices for, 717
- instructions for, 748
- interfacing, 782-788
- memory-mapped, 772
- for microcontrollers, 777-780
- polling, 786
- ports, 734
- secondary memory, 738-740
- seven-segment displays for, 736-737
- switches for, 735-736
- in thermostats, 733-734

Insignificant digits, 55
Instruction register, 745
Instruction-set processors, 605n
Instruction sets, 754-756
Instruction units (I-units), 721, 740, 751-754, 786-787
Instructions and instruction cycles, 29, 722-723, 726
- addresses in, 724, 745
- branch, 725, 747-748, 756
- execution of, 741-745
- general, 745-747
- input/output, 748

Integers, 54, 761
- conversions with, 58-62
- radix points with, 56
- in VHDL, 606, 608

Integrated circuits (ICs), 21-22
- design using, 241-243
- family selection in, 239-240
- gate arrays, 244-246
- small-scale integration, 21, 238-243
- very large-scale integration, 243

Interconnections. *See* Connections
Interfacing in systems, 29, 782-788
Interference, 104
Internal lines, 153-154
Internal state of latches, 419
Interrupts, 748, 786-787
Intersection of sets, 168
Inverse axioms, 163
Inversion, 131, 133
Inverted inputs, gates with, 146-147
Inverters, 133, 138, 151, 222-224
Involution law, 172-173
I/O. *See* Input/output (I/O) operations and subsystems
I/O-mapped I/O, 772
Irredundant expressions, 287
ISP language, 605
ITCONS procedure, 386
Iteration counters, 693
Iterative array test model, 572-574
Iterative consensus method, 386

J frontier, 569-570
JK flip-flops, 432-433
- characteristics of, 445, 507-509
- clocking modes for, 436-437
- design of, 433-434
- don't care states in, 537-538
- master-slave, 434-435
- multiphase clocks for, 441-443
- pulse triggering of, 435-436
- in registers, 445-446

JK latches, 434
JMP (jump) instructions, 725, 747-748, 756
Johnson counters, 451-453
Justification process, 565, 569-570

K memory capacity, 720n
Karnaugh maps, 289
- cell grouping in, 290-291
- cycles on, 303
- essential prime implicants in, 296-299
- with five variables, 304-308
- with four variables, 291-292
- large groups in, 295
- logical and physical adjacency in, 290
- minimization with, 292-300
- for multiple-output functions, 314-315
- one-cell groups in, 293
- prime implicants in, 293-296, 299-300
- two-cell groups in, 293-295

Keyboards, 717
- debouncing circuits for, 487-488
- latches for, 423

Keypads, 737
Knowledge-based transformations, 350-351

Labels in programs, 758-759
Large-scale integration (LSI), 243
Last-in, first out stacks. *See* LIFO stacks
Latches, 405
- applications for, 423-424
- as bistable circuits, 408
- D, 412-415, 468-472
- and feedback, 406
- for glitch suppression, 513-514
- level-sensitive, 426
- and metastability, 407-408
- noise sensitivity of, 427-428
- setup, hold, and enable times of, 417-419
- and shift registers, 447
- SR, 409-412, 415, 422-423
- and stability, 406-407
- state tables for, 420-423
- states of, 419-420
- timing and state behavior of, 415-424

Latching, 413
Latching point, 418
Latency performance factor, 32
LCDs (liquid crystal displays), 736-737
LD (load) instruction, 755
Leading zeroes, 55
Leakage in DRAMs, 396
Least significant bits, 56
LEDs (light-emitting diodes), 423-424, 736-737
Left shift operation, 632
Level-sensitive latches, 426
Levels
- of circuits, 160
- in design, 17-24

Libraries
- of local transformations, 352
- standard-cell, 244

LIFO stacks, 654-655
- ASM charts for, 675-677, 679-681
- one-hot method for, 658-660
- programmable controller for, 671-673
- state description of, 655-656

Light-control circuit, 12
Light-emitting diodes (LEDs), 423-424, 736-737
Line delays, 399, 402
Liquid crystal displays (LCDs), 736-737
Literals and costs, 285, 320
Load instructions, 726-727
Local transformations, 351-353
Logic circuits. *See* Boolean algebra
Logic design, 11
Logic design level, 19-21
Logic expressions, 134
Logic functions, 178
Logic gates, 20, 132, 142-143
- AND and OR, 144-145
- attenuator-based, 229-230
- and Boolean expressions, 181-183
- with combinational circuits, 152-154
- complete sets of, 213-219
- and costs, 285

delay with, 158-159, 399-402
design goals in, 154-155
EXCLUSIVE-OR and EXCLUSIVE-NOR, 147-148
fan-in and fan-out with, 155-157
in full adders, 153-154
NAND and NOR, 145-147
open-circuit, 230-231
polarity indication for, 149-150
positive and negative logic for, 148
reduction of, with multiple-output circuits, 331-332
restoring, 190
single-input, 143-144
speed of, 160
truth tables for, 143
voltage considerations in, 150-151
Logic hazards, 457
Logic minimization. *See* Minimal forms and minimization.
Logic minimization programs, 317-318
Logic partitioning, 241-242
Logic simulation programs, 316-317, 526-528
Logic testing, 558
Logical behavior of circuits, 132
Logical descriptions vs. physical, 11
Logical fault models, 558
Logical models, 11
Logical shifting, 632
Logically adjacent terms, 282-283, 290
Logically complete gate sets, 213
LogicWorks simulation program, 38
Longest path constraints, 515
Loops, programming, 767-769
Low-power Schottky TTL ICs, 239
Lumped delay elements, 400, 461
M memory capacity, 720n
Machine language, 724, 759-761
Macros, 626
Magnetic-surface recording, 739
Magnetic tape units, 717
Magnitude comparators, 373
Main memory, 716-717, 726-728
Mainframe computers, 718
Mantissa, 108-110
Mapping ROM, 259
Mappings, 15, 241-242. *See also* Karnaugh maps
Mask-programmed PLDs, 247
Masking of errors, 568
Master latches, 429
Master-slave circuits, 429, 434-435
Maxterms, 206-208
McBOOLE program, 317-318
MDSs (microprocessor development systems), 787-788
Mealy model and machines, 493, 503, 653, 666-667
Member of set, 168
Memory, 28-29, 716-717, 726-728. *See also* Dynamic random-access memory (DRAM), flip-flops, latches
asynchronous circuits for, 459
behavior of, 728-729
capacity of, 719-720
for controllers, 668-671
design of, 729-732
scratchpad, 629
secondary, 738-740
for signal storage, 391-392
in synchronous sequential circuits, 491
Memory access time, 392
Memory addresses, computing, 76-77
Memory cycle time, 729
Memory-mapped I/O, 772
Metal-oxide semiconductor (MOS) circuits. *See* MOS circuits
Metastability, 407-408
Metatheorems, 180
Microcomputers, 715
Microcontrollers, 29, 717-718, 720-721
design of, 776-777
input/output ports for, 777-780
Motorola 68705, 778-782
for pulse rate monitor, 780-782
for thermostats, 733-734
Microinstruction register, 669, 753
Microinstructions, 668-669, 753
Microprocessor-based systems, 718
Microprocessor development systems (MDSs), 787-788
Microprocessors, 29, 31-32, 716, 719-720
Microprogram counter, 753
Microprogramming, 754
of control units, 668
of instruction units, 752-754
Microprograms, 753
Microsequencers, 669
Millions of instructions per second (MIPS), 33, 717
Minicomputers, 719
Minimal covers, 296, 301-302, 319
Minimal forms and minimization, 280-282
algebraic minimization, 282-284
alternative formulations in, 285-286
and canonical forms, 279-280
cost implications of, 285
with don't care values, 309-313
fan-in/fan-out constraints in, 286
indistinguishable states in, 549-555
Karnaugh maps for, 292-300
for multiple-output functions, 315
of output connections, 338-342
prime implicants in, 287-288
prime implicates in, 288-289
programs for, 317-318
redundancy in, 286-287
in testing, 560-561, 563-564
Minimal SOP (POS) expressions and circuits, 285
Minimal state tables, 551, 553-555
Minimization theorem, 282-283
Minterms, 202-203
MIPS (millions of instructions per second), 33, 717
Missing parts, 557
Mixed logic circuits, 149
Mixed numbers, 54
binary to decimal conversion, 60-61
decimal to binary conversion, 62-64
Mixed systems, 4
Mobius counters, 451-453
Modular arithmetic, 74-77
Modulo-2 adders and circuits, 11-12, 83
Modulo-10 circuits, 98
Modulo-*m* arithmetic, 76
Modulo-*m* counters, 450
Modulus, 76
Moore circuits and machines, 493, 653, 663-666
converting, with Mealy, 666-667
latches in, 513-514
state behavior, 502-507, 671
Morse, Samuel F. B., 120
Morse code, 120
MOS circuits, 232
CMOS, 234
*n*MOS, 236-237
strength in, 227
for switches, 232-233
transmission gates, 235-236
MOS transistors, 128, 395-396
Most significant bits, 56
Motorola 68040 microprocessor, 22-23, 756
Motorola 680X0 microprocessor family. *See* 680X0 microprocessor family.
Motorola 68705 microcontroller, 778-782
MOV (move) instruction, 755
Multifunction datapath units, 649-651
Multilevel design
EXCLUSIVE-OR circuits, 347-348
factoring in, 348-350
heuristic methods for, 347-350
MULTIMIN procedure, 339-343
Multiphase clocking, 434, 441-443
Multiple independent clocks, 459
Multiple-output circuits
minimization of, 338-342
prime implicants for, 333-337
tabular method for, 330-342
Multiple-output functions, 314-315
Multiple-path sensitization, 596
Multiple-precision number systems, 107-109
Multiplexers (muxes), 354, 616
in adder-subtracter circuits, 84
design of, 355-356
expansion of, 642-644
for full adders, 619-620
as function generators, 617-620
input buffers for, 361-362
in time-multiplexed circuits, 362-363
for transmission gates, 359-360
in tri-state circuits, 224-226, 362-364
Multiplication

AND operation as, 144
for exponentiation, 58
by shifting, 632
of two's-complement numbers, 84-85
of unsigned binary numbers, 73-74
Multiplication tables, 74, 259-260
Multipliers, 73-74
control units for, 688-691
counter-based, 660-664
design of, 340-342
ROM-based, 74, 259-260
shift-and-add, 691-698
Multiply-by-3 circuit, 543-546

NAND circuits and gates, 145-147
adder using, 154
CMOS, 234
complementary circuits for, 139-141
delays in, 158-159, 400-401, 403-404
in SSI ICs, 238 SSL
faults in, 560-561
stability of, 406-407
symbols for, 217-218
NAND-NAND circuits, 217
NEG2c procedure, 80-83
Negation, 78, 530-532
Negative edge-triggered flip-flops, 427, 437-440
Negative logic, 148
Negative numbers, 82. *See also* Signed numbers
Negative switches, 131
Negative unate functions, 378
Netlists, 39
Network editors in computer-aided design, 39
Networks. *See* Circuits, Switches; Systems
Next-state function, 464
Nines'-complement (9C) code, 86
*n*MOS logic, 229, 236-237
Noise. *See also* Hazards
and clocks, 512
errors from, 10
with flip-flops, 427-428, 501
and metastability, 407
in TTL standard, 151
Nominal delay values, 402-403
Non-well-formed (non-WF) circuits, 201-202
Noncode words, 104
Noncritical races, 462, 466-467
Noninverting gates, 151
Nonoptimal solutions. *See* Heuristic methods
Nonoverlapping clock phases, 441
Nonrecurrent engineering (NRE) costs, 30
Nonrestoring gates, 190
Nonvolatile memory, 739
NOR circuits and gates, 145-147
CMOS, 237
symbols for, 217-218
NOR-NOR circuits, 217
Normalized form of floating-point numbers, 109
Not a number (NaN) category, 111

NOT circuits and gates, 133
behavior of, 134
complementary circuits for, 138
as single-input gates, 143
NOT instruction, 756, 758
NOT operator, 179
Number systems
base and word size in, 53-57
binary to decimal conversion, 57-61
decimal to binary conversion, 61-66
positional numbers, 51-53
Numerals, 51

Observability problem, 565, 581
Observation points for tests, 579
Octal number system, 94-96
Odd parity, 105, 147
Odometers, 5
On-off switches, 9
1-cells in Karnaugh maps, 290, 293
One-hot method in control-unit design, 653, 656-657
for counter-based multipliers, 660-664
for LIFO stacks, 658-660
for multiplier control unit, 688-691
One-out-of-*m* decoders, 364-365
One's complement (1C) code, 80-82
Opcode fetch, 743
Opcodes, 724, 745, 755
Open-circuit (OC) logic, 220, 230-231
Open circuits, 220, 557
Open-collector circuits, 221
Operating systems, 720
Operator symbols for functions, 162
Operators, precedence rules for, 179
Optical encoders, 89-91
OR circuits and gates, 132-133, 144-145
behavior of, 134
wired, 226
OR operator, 179
OR-AND circuits, 211
ORG directive, 760
Oscillation, 202, 407
Output control, clocks for, 512-514
Output logic, 493
Output ports, 734
Output signals. *See* State behavior
Output terminals, 130
Overflow
with adder-subtracter circuits, 84
in binary addition, 70-71
with floating-point numbers, 111
and modular arithmetic, 74-77
with signed-magnitude numbers, 79
with signed numbers, 86-87
OVFDET procedure, 86-87

Packed data, 762
Packed decimal format, 103
PAL (programmable array logic) elements, 266
PAL16R4 PLD, 634-636

PALASM hardware description language, 255
Palindromes, 594
Parallel adders, 496
Parallel circuits, 132
Parallel registers, 627-629
Parallel-to-serial data conversion, 632
Parametric testing, 558
Paramount prime implicants, 333-334
Parentheses in well-formed expressions, 179
Parity and parity functions, 104-106, 147-148
circuits for, 106-107
PLAs for, 263-266
Parity error flags, 106
Partitioning
logic, 241-242
in testing, 581
Pass transistors, 233*n*
Path delay constraints, 515-516
Path sensitization, 565-567, 596
Patient monitor, control logic for, 305-308
Performance goals in design, 32-33
Periodic clock signals, 431
Peripherals, 717
Personal computers, 720
Petrick's method, 300-304
Physical adjacency in Karnaugh maps, 290
Physical descriptions vs. logical, 11
Physical design level, 19, 21-22
Physical faults, 556
Physical switches, 127-129
Pipeline processors, 637
PLAs (programmable logic arrays), 249, 260, 285, 721
example of, 250-252
for full adders, 252-254
for parity function, 263-266
reprogramming, 254-255
PLDs. *See* Programmable logic devices (PLDs)
PLS153 field-programmable PLA, 261-263
*p*MOS logic, 229, 236
PODEM technique, 572
Polarity control in PLAs, 263
Polarity-hold latches. *See* D latches
Polarity indication for logic gates, 149-150
Polarity symbol, 149
Polling, 786
Pop operation with stacks, 654, 773
Ports, 734, 777-780
POS (product-of-sums) expressions, 211
algebraic minimization of, 282-285
minimal, 280-281
Positional numbers, 51-53
Positive edge-triggered flip-flops, 427, 430, 432-435
Positive logic, 148
Positive switches, 131
Positive unate functions, 378
Postincrementing, 769
Postponed output symbol, 436
Postulates for Boolean algebra, 162-164

Power consumption goals in design, 34
Power supplies, faults from, 557
Precedence rules, 179
Precision
in number conversion, 64-65
of digital systems, 4-5
of numbers, 56-57, 107-109
Predecrementing, 769
Preset input for flip-flops, 431, 437
Primary output control, clocks for, 512-514
Prime implicant covers, 296
Petrick's method for, 300-304
in tabular method, 322
Prime implicant groups, 293
Prime implicant tables, 321-322
Prime implicants, 287-288
essential, 296-299, 322
in hazard-free D latch design, 471
identification of, 293-295, 320-321, 334-335
incompletely specified functions, 309-313
on Karnaugh maps, 293-295
for multiple-output circuits, 333-337
selection of, 299-300, 321-322, 335-337
in tabular method, 320-322
Prime implicates, 288-289
Prime terms, 287
Principle of duality, 180-181
Printed-circuit boards, 22, 238
Printers, 717
Priority encoders, 621-622, 645-647
Procedures, 27, 65-66
Process structure in VHDL, 608
Processor design level, 18-20
Product expressions, 144
Product groups on Karnaugh maps, 293
Product-of-maxterms form, 206-208
Product-of-sums (POS) form, 211
algebraic minimization of, 282-285
minimal, 280-281
Program-control instructions, 29, 726
Program-control units. *See* Instruction units (I-units)
Program counter, 725
Programmable array logic elements (PALs), 266
Programmable controllers
design of, 669-673
vs. hardwired, 673
structure of, 668-669
Programmable I/O circuits, 777
Programmable logic arrays (PLAs), 249, 260, 285, 721
example of, 250-252
for full adders, 252-254
for parity function, 263-266
reprogramming, 254-255
Programmable logic circuits, 653
Programmable logic devices (PLDs), 74, 231, 246, 625-626
computer-aided design of, 255-256
crosspoint switches in, 248
FPGAs, 266-267
PALs, 266
PLAs. *See* Programmable logic arrays (PLAs)
programming process for, 247-248
ROMs. *See* read-only memories (ROMs), 257-260
Programmable ROMs (PROMs), 258, 778-779
Programmable systems, 27-28
Programmable timers, 780
Programmed I/O, 786
Programming languages, 723-724
Programming models, 770-772
Programming units for PLDs, 247
Programs, 27-28, 715
PROMs (programmable ROMs), 258, 778-779
Proofs in Boolean algebra
formal, 174-177
informal, 172-174
Propagate function for carry lookahead adders, 369
Propagation delays. *See* Delays
Propagation in DALG procedure, 569
Propositional calculus, 161, 168
Propositional calculus, 161, 168
Pseudo-combinational model, 574
Pseudocode, 65-66
Pull-up/pull-down circuits and devices, 136-137
attenuators as, 228-229
for input switches, 735
Pulse-mode circuits, 463
Pulse rate monitor, 780-782
Pulse triggering of JK flip-flops, 435-436
Pulse-width times of latches, 418-419
Pulses, 398-399
propogation of, 403-405
rise and fall times of, 398, 401-403
Push operation with stacks, 654, 773

Quality control, 556
Quantization, 5-6, 783
Quiescent input combinations, 410
Quine-McCluskey method. *See* Tabular method
Quotients in division, 73

r-ary number systems, 53
Race conditions
in asynchronous circuits, 462, 466-467, 472-474
with latches, 412, 424
Radiation, errors from, 104
Radix-complement (RC) codes, 85-86
Radix of number system, 53
Radix point, 54-56
RAM (random-access memory), 28-29, 716-717, 726-728
capacity of, 719-720
destructive readout property of, 480
operation of, 395-397
Ramp method of analog-to-digital conversion, 784-785
Read-only memories (ROMs), 28, 74, 249, 257-258, 728
applications of, 259-260
for control memory, 669
limitations of, 260
in microcontrollers, 778-779
operation of, 258-259
programmable, 258, 778-779
for table lookups, 74
Read operation, 396-397, 727-729
Read-write head, 739
Read-write memory, 28-29, 716-717, 726-728
capacity of, 719-720
destructive readout property of, 480
operation of, 395-397
Real data type, 761
Real-time devices, 777
Recognizers, 532-536, 540-543
Reduced cover tables, 322
Reduced instruction set computers (RISCs), 754-755
Reduced state tables, 510
Redundancy in minimal forms, 286-287
Redundant circuits, 563
Reflected Gray codes, 91-92
Refreshing
of DRAMs, 396
of dynamic flip-flops, 443
Register assignment operator in VHDL, 610
Register files, 629, 748-749
Register-level models, 603-604
Register-transfer level, 599
Register-transfer statements, 610, 723
Register level, 599-604
algorithmic state machines, 674-681
arithmetic components for, 622-624
bit slicing, 639-642
combinational components of, 614-626
components for, 613, 685-688
control units for, 652-673, 688-698
data and control in, 609-611
datapath units for, 602, 609-610, 636-651, 685-688
descriptive methods for, 604-605
hardware description languages for, 605-609
ICs for, 634-636
integrating data and control, 681-698
sequential components of, 626-636
Registers,
parallel, 445, 627-629
shift, 445-447, 629-634
Relative addressing mode, 764-769
Relays, 128, 161
Reliability
in binary systems, 10
of components, 425
as design goal, 34-35
Remainders with division, 73
Repetition, loops for, 767-769

Reprogramming of programmable arrays, 254-255
Research and development (R&D) costs, 30
Reset-set latches. *See* SR latches
Response time. *See also* Delays
in combinational circuits, 25, 152
in sequential circuits, 25
RET (return) instruction, 756
Reuse of designs, 375
Reverse-engineering problem, 504-507
Right shift operation, 632
Ring counters, 450
Ring-sum operator, 148
Ring, Boolean, 194
Ripple-borrow subtracters, 72-73, 373
Ripple-carry adders, 71, 367-368
Ripple counters, 448-450
RISCs (reduced instruction set computers), 754-755
Rise delay, 401-403
Rise-fall delay elements, 402
Rise time, 398, 401-402
ROM. *See* Read-only memories (ROMs)
Roman numerals, 52-53
Roth, John Paul, 567-568
ROUND procedure, 64-65, 117
Rounding, 64-65, 117
Row covers in tabular method, 322
Row dominance in tabular method, 323
RS latches. *See* SR latches
RTS (return from subroutine) instruction, 785

Sample-and-hold latches. *See* D latches
Scale factors, 60
Scan design for testability, 581-585
Scan registers (scan chains), 581-585
SCANAL procedure, 510, 519-520
SCANEG2c procedure, 119, 530-532
Scanning of keypad, 737
SCANTEST procedure, 583
Schematic capture, 38-39
Schematic diagrams, 13-14
Schottky TTL ICs, 239
Scientific notation, 108
Scratchpad memory, 629
SCSYN procedure, 528-529, 653, 656
Sea of gates, 244-246
Secondary input variables, 420, 494
Secondary memory, 738-740
Secret writing, codes for, 88-89
Seek time of secondary memory, 739
Segment registers, 766*n*
Select inputs of multiplexers, 354
Self-complementing decimal codes, 122
Self-dual expressions, 180
Self-evident theorems, 171
Self-loops, 422, 448
Self-timed circuits, 460
Sense amplifiers, 396
Sensitized paths, 565-567, 596
Sensors, 783
Sequencers. *See* Control units for registers
Sequences
homing, 596
input/output, 498
synchronizing, 597
Sequential circuits and logic, 25-26
for adders, 26, 495-498
classical design method for, 528-529
delays in, 397-405
design issues, 463-476, 519-528
hazards in, 453-457
Huffman model for, 457-463
main signal types in, 494
signal storage, 391-397
state behavior, 498-510, 522-523
state minimization, 549-555
structure of, 27, 491-493
timing signals in, 415-424, 494-495, 511-519
Serial access, 740
Serial adders, 495-498
with JK flip-flops, 538-539
state diagram for, 499-500
state table for, 498-499
tests for, 575-576
timing diagrams for, 501-504
Serial-to-parallel data conversion, 632
Series connections, 131
Series-parallel circuits, 136
Set algebras, 168-171
Set-dominant flip-flops, 486
Set-reset latches. *See* SR latches
Sets, 75, 168-169
Setup times
of flip-flops, 430
of latches, 418-419
Seven-segment displays, 736-737
Sexagesimal number system, 116
SH (shift) instruction, 756
Shannon, Claude E., 161
Shared data paths, 353-354
Shared terms in multiple-output functions, 314
Shift-and-add multiplication circuits, 691-698
Shift registers, 445-447, 629-634
Shifters, barrel, 706
Shifting, arithmetic and logical, 705
Short circuits, 220, 557
Shortest path constraint, 515
Sign bits in two's-complement code, 82
Sign conventions, 78
Signal storage
delays in, 394-397
physical memory for, 391-392
wells for, 393-394
Signals, 153-154
and delays, 398
strength of, 226-230, 394-397
in VHDL, 607, 610
Signed-magnitude (SM) code, 77-80, 84-85, 373
Signed numbers, 57, 373
comparison of, 87
ones' complement code, 80-82, 119
overflow with, 86-87
radix-complement codes, 85-86
signed-magnitude code, 77-80
two's-complement code, 80-85
Significand, 110
Significant digits, 55-57
Sign of mantissa, 108
Simulators and simulation programs, 40-41, 316-317, 405
in computer-aided design, 36
for design verification, 526-528
hardware description languages for, 605
Single-board computers, 22
Single-input-change assumption in Huffman model, 462-463
Single-input logic gates, 143-144
Single-phase clocking, 442
Single-precision numbers, 107
Single stuck-line (SSL) fault model, 558-561
Sink terminals, 222
Slash-*n* notation, 603
Slave latches, 429
Slicing, bit, 639-642
Sloppy minimization, 286
SM (signed-magnitude) code, 77-80, 84-85, 373
Small-scale integration (SSI), 21, 238-239, 241-243
Software, 28, 717-718
SOP (sum-of-products) expressions, 209-211, 280-284
Source terminals, 222
Specification methods, 65-66
Speed
in asynchronous circuits, 459-460
as design goal, 32-33
of logic gates, 160
Speed-independent circuits, 460
Springs as storage device, 391-392
SR flip-flops, 433
SR latches, 409
anomalous behavior in, 412
D latches from, 415
Huffman model of, 464
operation of, 410-411
state behavior of, 422-423
SR latches, 441, 482
SRAMS (static RAMs), 396, 727
SSL (single stuck-line) fault model, 558-561
ST (store) instruction, 755
Stability
in asynchronous circuits, 464-467
in bistable circuits, 406-408
Stack pointer registers, 771-773
Stacks, 654-655, 772-775
Stages, pipeline, 637
Standard-cell design, 244
Standard parts in hierarchical design, 23
Standard TTL ICs, 239

State assignments
in asynchronous circuits, 473-474
in synchronous circuits, 535-537
rules for, 593
in SCANAL procedure, 520
State behavior, 498-510
State boxes for ASMs, 675-676
State diagrams, 415, 444, 499-500
State machines, 27, 652-654
State minimization problem, 549-555
State registers, 653
State tables, 415, 444, 498-499
State transition tables, 505-506
State transitions, 421, 510
State variables, 408, 494
States, 27, 408
dead, 531
equivalent, 509-510, 549-555
stable and unstable, 464-467
Static hazards, 453-455
Static RAMs, 396, 727
Status registers, 751
Status signals, 681-683, 751
Storage of signals, 391-397. *See also* Memory
Store instructions, 726-727
Strength values, 226-227
Strobes, 736
Strongly-connected sequential machines, 596
Structure of circuits and systems, 13-16
vs. behavior, 16
of combinational circuits, 199-213
of sequential circuits, 27, 491-493
Stuck faults, 221, 223, 558-561
SUB (subtract) instruction, 755
SUBn procedure, 72
Subroutines, 756
Subscripts for number bases, 54
Subsets, 168-171
Subsystems, 17
Subtracters, 372-374
Subtraction
of BCD numbers, 99-100
of two's-complement numbers, 83-85
of unsigned binary numbers, 71-73
Successive approximation method, 798
Sum modulo 2, 11, 76, 148
Sum-of-minterms form, 203-208
Sum-of-products (SOP) form, 209-211, 280-284
Supervisor registers, 771
Support costs, 30
Switch level,
basic circuits, 131-136
complementary circuits, 136-141
connection function, 222
logic values, 219-223
Switch-level interpretation of Boolean algebra, 164-165
Switch-tail counters, 451-453
Switches, 9
crosspoint, 248
debouncing circuits for, 487-488, 793
in electric circuits, 125-126
in hydraulic circuits, 126-127
ideal, 129-130
physical, 127-129
positive and negative, 131
Switching functions, 178
Synchronization of input/output, 782
Synchronizer flip flops, 683-684
Synchronizing sequences, 597
Synchronous circuits
vs. asynchronous, 457-459
delays in, 476
iterative array model for, 572-574
Synthesis of systems, 16, 42
Synthesizers in computer-aided design, 36, 605
System bus, 717
System design level, 18-20
Systems
architecture of, 715-721
CAD for, 787-788
combinational vs. sequential, 24-29
components as, 17
computers, architecture of, 28-29, 715-740
computer-based, 776-788
CPU operation in, 721-723, 740-754
data formats for, 761-763
input/output, 717, 732-740, 782-788
instruction sets for, 754-775
memory for, 716-717, 726-732
microcontrollers in, 717-718, 776-782
programmable, 27-28
programming languages for, 723-724
structure and behavior of, 13-17
synthesis of, 16, 42

T flip-flops, 485
Table lookup, 74, 259
TABMIN procedure, 326-330
Tabular method
branching in, 324-326
code words in, 319
column dominance, 323-324
cover tables in, 321-324
essential prime implicants in, 322
general approach, 319
for multiple-output circuits, 330-342
prime implicants in, 320-322
row dominance, 323
TABMIN procedure, 326-330
Tags, 334-336
Tally number system, 52
Technology independence, 20, 129
Technology mapping, 241-242
Ten's complement (10C) code, 86
Terminals of switches, 130
Ternary number system, 45, 54
Test-generation programs, 557
Test patterns (test vectors), 556
Test points, 579-580
Test responses, 556
Testability
ad hoc methods in, 577-580
designing for, 35, 576-585
goals in, 577
scan design , 581-585
Testing
of circuits, 562-563
of counters, 580-581
minimization in, 560-561, 563-564
of sequential circuits, 556-564, 574-576
for single stuck lines, 558-561
Tests
D algorithm, 567-572
generation of, 564-575
iterative array model for, 572-574
path sensitization in, 565-567
stages in, 558
Theorems
basic, 171-177
consensus, 385
dual, 181
minimization, 282-283
Thermostats, microcontroller-based, 733-734
Tie-breaking rules with heuristics, 344
Time delays. *See* Delays
Time-dependent Boolean equations, 399
Time-multiplexed circuits, 362-363, 777
Time sharing, 777
Time-space transformations, 574
Timing analyzers, 518
Timing diagrams and control, 5-6, 158. *See also* Clocks and clock signals
Timing parameters for latches, 416-417
TMR (triple modular redundancy) design method, 597
Toggle flip-flops, 485
Toggling, 434, 448
Top-down design, 24
Top of stack, 654, 772
Total state of latches, 419
Trade-offs in design, 35-36
Traffic light problem
combinational logic design for, 525-526
design process for, 520-521
design verification of, 526-528
state assignment in, 523-524
state behavior in, 522-523
Trailing zeroes, 55
Transceivers, tri-state, 363-364
Transducers, 783
Transformations
local, of logic circuits, 351-353
time-space, 574
Transistor-transistor logic (TTL), 239-240
delays in, 402-403
voltage standard, 150-151
Transistors, 128, 229, 232-233, 237, 395-396

Transition tables, 474, 505-506, 525
Transitivity property, 553
Transmission gates
 in CMOS circuits, 235-236, 483
 in multiplexers, 359-360
Transparent latches, 413
Trap states, 531
Tri-state logic, 221, 224-226
 for buses, 362-364
 for memory design, 731-732
 for PLAs, 263
Triple modular redundancy (TMR) design method, 597
Truncation in converting numbers, 64-65
Truth tables, 15, 132
 for design verification, 183
 and Karnaugh maps, 289-290
TST (test) instruction, 758, 785
TTL circuits and ICs, 239-240
 delays in, 402-403
Twisted-ring counters, 451-453
Two-cell groups in Karnaugh maps, 293-295
Two-level design
 Karnaugh maps for, 289-300
 minimal forms in, 279-289
 Petrick's method for, 300-304
Two-level forms, 208-209
Two-level NAND and NOR circuits, 216-217
2-out-of-5 code, 101
Two-phase clocking, 442
Two's-complement (2C) code, 80-81, 373
 addition and subtraction with, 83-86
 negation circuit for, 530-532
 vs. ones'-complement, 81-82
 properties of, 82-83
Typical delay values, 402-403

U logic value, 219-223
Unate functions, 378
Unbiased rounding, 117
Uncertain delays, 402-403
Uncertain state, 219-221, 223
Underflow, 123
Undetectable faults, 563
Union of sets, 168-169
Unipolar transistors, 232
Uniqueness theorems, 173-176
Unit-delay simulators, 405
Unit under test (UUT), 556
Universal gates, 139
Universal product code (UPC), 53, 92-93
Universal set, 169
Universal shift registers, 632-633
Unknown state, 219-221, 223
Unpacked data, 762
Unpacked decimal format, 103
Unsigned binary numbers
 addition of, 67-71
 division of, 73
 multiplication of, 73-74
 subtraction of, 71-73
Unstable states, 464-467
Up-down counters, 634
UPC (universal product code), 53, 92-93
User programs, 734
User registers, 770
UUT (unit under test), 556

Value-strength parameters, 226-227
Venn, John, 169
Venn diagrams, 169-171, 196
Verification of design, 183-184, 316, 536-538
Verilog hardware description language, 255, 605
Very large-scale integration (VLSI), 21, 243
VHDL hardware description language, 255
 for arithmetic-logic circuit, 606-609
 for shift registers, 633
Video display units, 717, 738
Vigesimal number system, 116
Volatile memory, 738
Voltage comparators, 783-784

Wait states, 729
Watchdog timers, 780
Waveforms, 5-6, 398
Weak signals, 227-228, 394-397
Weight in positional number systems, 52
Well-behaved circuits, 200
Well-formed (WF) Boolean expressions, 179
Well-formed (WF) combinational circuits, 199-201
Wells for signal storage, 393-394
Width
 of buses, 614
 of pulses, 405
Wired logic, 219, 231
 and attenuators, 227-230
 connection function for, 222
 open-circuit, 230-231
 and signal strength, 226-227
 tri-state, 224-226
Wiring networks, 354
Word gates, 615-616
Word size, 54-55, 729-730
Words in register-level design, 609
Workstations for computer-aided design, 37
Wraparound in modular arithmetic, 75,77
Write operations, 396-397, 727-728
Writing, secret, 88-89

X-cells, 308-309
XOR circuits. *See* EXCLUSIVE-OR circuits and gates
XOR instruction, 756

Z logic value, 219-226
0-cells in Karnaugh maps, 290
Zero properties, 162-163, 173
Zero uniqueness, proof of, 174-175
Zeroes
 floating-point, 109-110
 in ones'-complement, 81-82
 signed-magnitude, 79
 two's-complement, 82
 and word size, 55
Zone fields in numbers, 103